THE AH RECEPTOR IN BIOLOGY AND TOXICOLOGY

THE AH RECEPTOR IN BIOLOGY AND TOXICOLOGY

Edited by

RAIMO POHJANVIRTA

A JOHN WILEY & SONS, INC., PUBLICATION

Copyright © 2012 by John Wiley & Sons, Inc. All rights reserved

Published by John Wiley & Sons, Inc., Hoboken, New Jersey
Published simultaneously in Canada

No part of this publication may be reproduced, stored in a retrieval system, or transmitted in any form or by any means, electronic, mechanical, photocopying, recording, scanning, or otherwise, except as permitted under Section 107 or 108 of the 1976 United States Copyright Act, without either the prior written permission of the Publisher, or authorization through payment of the appropriate per-copy fee to the Copyright Clearance Center, Inc., 222 Rosewood Drive, Danvers, MA 01923, (978) 750-8400, fax (978) 750-4470, or on the web at www.copyright.com. Requests to the Publisher for permission should be addressed to the Permissions Department, John Wiley & Sons, Inc., 111 River Street, Hoboken, NJ 07030, (201) 748-6011, fax (201) 748-6008, or online at http://www.wiley.com/go/permission.

Limit of Liability/Disclaimer of Warranty: While the publisher and author have used their best efforts in preparing this book, they make no representations or warranties with respect to the accuracy or completeness of the contents of this book and specifically disclaim any implied warranties of merchantability or fitness for a particular purpose. No warranty may be created or extended by sales representatives or written sales materials. The advice and strategies contained herein may not be suitable for your situation. You should consult with a professional where appropriate. Neither the publisher nor author shall be liable for any loss of profit or any other commercial damages, including but not limited to special, incidental, consequential, or other damages.

For general information on our other products and services or for technical support, please contact our Customer Care Department within the United States at (800) 762-2974, outside the United States at (317) 572-3993 or fax (317) 572-4002.

Wiley also publishes its books in a variety of electronic formats. Some content that appears in print may not be available in electronic formats. For more information about Wiley products, visit our web site at www.wiley.com.

Library of Congress Cataloging-in-Publication Data:

The AH receptor in biology and toxicology / edited by Raimo Pohjanvirta.
   p. ; cm.
  Includes bibliographical references and index.
  ISBN 978-0-470-60182-2 (cloth)
1. Polycyclic aromatic hydrocarbons–Toxicology. 2. Polycyclic aromatic hydrocarbons–Metabolism. 3. Dioxins–Toxicology. 4. Dioxins–Metabolism. 5. Nuclear receptors (Biochemistry) 6. Transcription factors I. Pohjanvirta, Raimo.
   [DNLM: 1. Dioxins–toxicity. 2. Receptors, Aryl Hydrocarbon–agonists. 3. Environmental Pollutants–toxicity. 4. Receptors, Aryl Hydrocarbon–physiology. 5. Risk Assessment. 6. Signal Transduction. WA 240]
  RA1242.P73A3 2011
  572.8'845–dc23
                                      2011019938

Printed in the United States of America

10 9 8 7 6 5 4 3 2 1

CONTENTS

PREFACE ix

CONTRIBUTORS xi

PART I   HISTORICAL BACKGROUND

1 History of Research on the AHR 3
   Thomas A. Gasiewicz and Ellen C. Henry

PART II   AHR AS A LIGAND-ACTIVATED TRANSCRIPTION FACTOR

2 Overview of AHR Functional Domains and the Classical AHR Signaling Pathway: Induction of Drug Metabolizing Enzymes 35
   Qiang Ma

3 Role of Chaperone Proteins in AHR Function 47
   Iain A. Murray and Gary H. Perdew

4 AHR Ligands: Promiscuity in Binding and Diversity in Response 63
   Danica DeGroot, Guochun He, Domenico Fraccalvieri, Laura Bonati, Allesandro Pandini, and Michael S. Denison

5 Dioxin Response Elements and Regulation of Gene Transcription 81
   Hollie Swanson

6 The AHR/ARNT Dimer and Transcriptional Coactivators 93
   Oliver Hankinson

7 Regulation of AHR Activity by the AHR Repressor (AHRR) 101
   Yoshiaki Fujii-Kuriyama and Kaname Kawajiri

8	**Influence of HIF-1α and Nrf2 Signaling on AHR-Mediated Gene Expression, Toxicity, and Biological Functions** *Thomas Haarmann-Stemmann and Josef Abel*	109
9	**Functional Interactions of AHR with Other Receptors** *Sara Brunnberg, Elin Swedenborg, and Jan-Åke Gustafsson*	127
10	**The E3 Ubiquitin Ligase Activity of Transcription Factor AHR Permits Nongenomic Regulation of Biological Pathways** *Fumiaki Ohtake and Shigeaki Kato*	143
11	**Epigenetic Mechanisms in AHR Function** *Chia-I Ko and Alvaro Puga*	157

PART III AHR AS A MEDIATOR OF XENOBIOTIC TOXICITIES: DIOXINS AS A KEY EXAMPLE

12	**Role of the AHR and its Structure in TCDD Toxicity** *Raimo Pohjanvirta, Merja Korkalainen, Ivy D. Moffat, Paul C. Boutros, and Allan B. Okey*	181
13	**Nongenomic Route of Action of TCDD: Identity, Characteristics, and Toxicological Significance** *Fumio Matsumura*	197
14	**Interspecies Heterogeneity in the Hepatic Transcriptomic Response to AHR Activation by Dioxin** *Paul C. Boutros*	217
15	**Dioxin-activated AHR: Toxic Responses and the Induction of Oxidative Stress** *Sidney J. Stohs and Ezdihar A. Hassoun*	229
16	**Dioxin Activated AHR and Cancer in Laboratory Animals** *Dieter Schrenk and Martin Chopra*	245
17	**Teratogenic Impact of Dioxin Activated AHR in Laboratory Animals** *Barbara D. Abbott*	257
18	**The Developmental Toxicity of Dioxin to the Developing Male Reproductive System in the Rat: Relevance of the AHR for Risk Assessment** *David R. Bell*	267
19	**TCDD, AHR, and Immune Regulation** *Nancy I. Kerkvliet*	277
20	**Effects of Dioxins on Teeth and Bone: The Role of AHR** *Matti Viluksela, Hanna M. Miettinen, and Merja Korkalainen*	285
21	**Impacts of Dioxin-Activated AHR Signaling in Fish and Birds** *Michael T. Simonich and Robert L. Tanguay*	299
22	**Adverse Health Outcomes Caused By Dioxin-Activated AHR in Humans** *Sally S. White, Suzanne E. Fenton, and Linda S. Birnbaum*	307

23	The Toxic Equivalency Principle and its Application in Dioxin Risk Assessment *Jouko Tuomisto*	317
24	AHR-active Compounds in the Human Diet *Stephen Safe, Gayathri Chadalapaka, and Indira Jutooru*	331
25	Modulation of AHR Function by Heavy Metals and Disease States *Anwar Anwar-Mohammed and Ayman O.S. El-Kadi*	343
26	Transgenic Mice with a Constitutively Active AHR: A Model for Human Exposure to Dioxin and Other AHR Ligands *Patrik Andersson, Sara Brunnberg, Carolina Wejheden, Lorenz Poellinger, and Annika Hanberg*	373

PART IV AHR AS A PHYSIOLOGICAL REGULATOR

27	Structural and Functional Diversification of AHRs During Metazoan Evolution *Mark E. Hahn and Sibel I. Karchner*	389
28	Invertebrate AHR Homologs: Ancestral Functions in Sensory Systems *Jo Anne Powell-Coffman and Hongtao Qin*	405
29	Role of AHR in the Development of the Liver and Blood Vessels *Sahoko Ichihara*	413
30	Involvement of the AHR in Cardiac Function and Regulation of Blood Pressure *Jason A. Scott and Mary K. Walker*	423
31	Involvement of the AHR in Development and Functioning of the Female and Male Reproductive Systems *Bethany N. Karman, Isabel Hernández-Ochoa, Ayelet Ziv-Gal, and Jodi A. Flaws*	437
32	The AHR in the Control of Cell Cycle and Apoptosis *Cornelia Dietrich*	467
33	The AHR Regulates Cell Adhesion and Migration by Interacting with Oncogene and Growth Factor-Dependent Signaling *Angel Carlos Roman, Jose M. Carvajal-Gonzalez, Sonia Mulero-Navarro, Aurea Gomez-Duran, Eva M. Rico-Leo, Jaime M. Merino, and Pedro M. Fernandez-Salguero*	485
34	The Physiological Role of AHR in the Mouse Immune System *Charlotte Esser*	499
35	AHR and the Circadian Clock *Shelley A. Tischkau*	511

INDEX   523

PREFACE

The Ah receptor (AHR) is a fascinating and intriguing protein. Initially discovered in the 1970s as the mediator of a primarily adaptive response, induction of xenobiotic metabolizing enzymes, to the environmental toxicants dioxins and polycyclic aromatic hydrocarbons, it soon proved to be also intimately involved in their toxicity. Mechanistic studies showed a close functional resemblance of the AHR to nuclear receptors, and it was therefore included in that protein superfamily. However, molecular cloning of the AHR in the early 1990s revealed that instead of having the typical nuclear receptor structure, the AHR shares key domains with the bHLH/PAS protein family, the first members of which had been discovered only a few years earlier. It proved to be phylogenetically old with orthologs of mammalian AHR existing in insects and worms. Importantly, these ancient orthologs do not bind dioxin but have key roles in morphogenesis and neuronal differentiation. The generation of AHR-deficient mice then gave a major impetus to expand AHR studies into its physiological functions also in mammals. In that field, the first findings associated it with liver development. The reason for an abnormally small liver in *Ahr*-null mice was found to reside in a circulatory failure arising from a persisting fetal structure, *ductus venosus*, which prompted studies on the role of the AHR in the regulation of blood vessel formation and blood pressure. Another phenotype of *Ahr*-null mice is their poor breeding performance, which has guided research into the endocrine deviations occurring in these animals. Both *in vivo* and *in vitro* studies have demonstrated notable alterations in immune responses to experimental challenges in AHR knockout mice or immune cells originating from them. In particular, together with the opposite model, sustained activation of AHR by TCDD, these studies have detected a novel physiological regulatory role for the AHR in relation to autoimmunity. In recent years, AHR research has continued to expand and ramify to new areas including bone and tooth development, tumor growth, and regulation of such cellular phenomena as adhesion, migration, proliferation, and apoptosis. Furthermore, emerging topics include involvement of the AHR in, for example, epigenetic effects and regulation of circadian rhythms, as well as development and search of selective AHR modulators.

In light of these remarkable advances made in our understanding of the functions of the AHR and also of its action mechanisms, it is surprising that hitherto there has been no book available in the market that would have provided a comprehensive review of the various facets of this protein. It is the objective of this book to bridge the existing gap. This book arose as an international teamwork of renowned experts on AHR biology or toxicology, each writing on his/her own main subspeciality. Although the book will cover all the major established aspects of AHR function and give a detailed account of its research history, the focus in it is on the most recent findings with even a lot of data included from as yet unpublished studies. Thereby, this book will serve as an excellent and essential source of information on the AHR for scholars and students alike in biochemistry, physiology, toxicology, and related disciplines.

I want to thank all the distinguished scientists who participated in the generation of this book for their contributions. I am also obliged to Jonathan Rose for originally suggesting me this project and for his valuable advice and technical tips throughout the process.

RAIMO POHJANVIRTA

Helsinki
September 1, 2011

CONTRIBUTORS

Barbara D. Abbott, Developmental Biology Branch, Toxicity Assessment Division, National Health and Environmental Effects Research Laboratory, Office of Research and Development, U.S. Environmental Protection Agency, Research Triangle Park, NC, USA

Josef Abel, IUF-Leibniz Research Institute for Environmental Medicine, Düsseldorf, Germany

Patrik Andersson, Global Safety Assessment, AstraZeneca R&D, Mölndal, Sweden

Anwar Anwar-Mohamed, Faculty of Pharmacy and Pharmaceutical Sciences, 3126 Dentistry/Pharmacy Centre, University of Alberta, Edmonton, Alberta, Canada

David R. Bell, European Chemicals Agency, Helsinki, Finland

Linda S. Birnbaum, National Cancer Institute, National Institutes of Health, Research Triangle Park, NC, USA

Laura Bonati, Dipartimento di Scienze dell'Ambiente e del Territorio, Università degli Studi di Milano-Bicocca, Milano, Italy

Paul C. Boutros, Informatics and Biocomputing Platform, Ontario Institute for Cancer Research, MaRS Centre, Toronto, Ontario, Canada

Sara Brunnberg, Department of Biosciences and Nutrition, Karolinska Institutet, Novum, Huddinge, Sweden

Jose M. Carvajal-Gonzalez, Departamento de Bioquímica y Biología Molecular, Facultad de Ciencias, Universidad de Extremadura, Badajoz, Spain

Gayathri Chadalapaka, Department of Veterinary Physiology & Pharmacology, Texas A&M University, College Station, TX, USA

Martin Chopra, Food Chemistry and Toxicology, University of Kaiserslautern, Kaiserslautern, Germany

Danica DeGroot, Department of Environmental Toxicology, University of California, Davis, CA, USA

Michael S. Denison, Department of Environmental Toxicology, University of California, Davis, CA, USA

Cornelia Dietrich, Institute of Toxicology, University Medical Center of the Johannes Gutenberg-University, Mainz, Germany

Ayman O. S. El-Kadi, Faculty of Pharmacy and Pharmaceutical Sciences, 3126 Dentistry/Pharmacy Centre, University of Alberta, Edmonton, Alberta, Canada

Charlotte Esser, Division of Molecular Immunology, Leibniz Research Institute for Environmental Medicine, Heinrich-Heine University of Düsseldorf, Düsseldorf, Germany

Suzanne E. Fenton, National Toxicology Program, National Institute of Environmental Health Sciences, National Institutes of Health, Research Triangle Park, NC, USA

Pedro M. Fernandez-Salguero, Departamento de Bioquímica y Biología Molecular, Facultad de Ciencias, Universidad de Extremadura, Badajoz, Spain

Jodi A. Flaws, Department of Veterinary Biosciences, College of Veterinary Medicine, University of Illinois at Urbana-Champaign, Urbana, IL, USA

Domenico Fraccalvieri, Dipartimento di Scienze dell' Ambiente e del Territorio, Università degli Studi di Milano-Bicocca, Milano, Italy

Yoshiaki Fujii-Kuriyama, Tokyo Medical and Dental University, Medical Research Institute, Tokyo, Japan; University of Tokyo, Institute of Molecular and Cellular Biosciences, Tokyo, Japan

Thomas A. Gasiewicz, Department of Environmental Medicine, University of Rochester Medical Center, Rochester, NY, USA

Aurea Gomez-Duran, Departamento de Bioquímica y Biología Molecular, Facultad de Ciencias, Universidad de Extremadura, Badajoz, Spain

Jan-Åke Gustafsson, Department of Biosciences and Nutrition, Karolinska Institutet, Novum, Huddinge, Sweden; Center for Nuclear Receptors and Cell Signaling, Department of Biology and Biochemistry, University of Houston, Houston, TX, USA

Thomas Haarmann-Stemmann, IUF-Leibniz Research Institute for Environmental Medicine, Düsseldorf, Germany

Mark E. Hahn, Biology Department, Woods Hole Oceanographic Institution, Woods Hole, MA, USA

Annika Hanberg, Institute of Environmental Medicine, Karolinska Institutet, Stockholm, Sweden

Oliver Hankinson, Molecular Toxicology Interdepartmental Program, Department of Pathology and Laboratory Medicine, Center for the Health Sciences, University of California at Los Angeles, Los Angeles, CA, USA

Ezdihar A. Hassoun, College of Pharmacy, University of Toledo, Toledo, OH, USA

Guochun He, Department of Environmental Toxicology, University of California, Davis, CA, USA

Ellen C. Henry, Department of Environmental Medicine, University of Rochester Medical Center, Rochester, NY, USA

Isabel Hernández-Ochoa, Department of Veterinary Biosciences, College of Veterinary Medicine, University of Illinois at Urbana-Champaign, Urbana, IL, USA

Sahoko Ichihara, Department of Human Functional Genomics, Life Science Research Center, Mie University, Tsu, Mie, Japan

Indira Jutooru, Department of Veterinary Physiology & Pharmacology, Texas A&M University, College Station, TX, USA

Sibel I. Karchner, Biology Department, Woods Hole Oceanographic Institution, Woods Hole, MA, USA

Bethany N. Karman, Department of Veterinary Biosciences, College of Veterinary Medicine, University of Illinois at Urbana-Champaign, Urbana, IL, USA

Shigeaki Kato, Institute of Molecular and Cellular Biosciences, University of Tokyo, Tokyo, Japan; ERATO, Japan Science and Technology Agency, Kawaguchisi, Saitama, Japan

Kaname Kawajiri, Research Institute for Clinical Oncology and Hospital, Saitama Cancer Center, Saitama, Japan

Nancy I. Kerkvliet, Department of Environmental and Molecular Toxicology, Oregon State University, Corvallis, OR, USA

Chia-I Ko, Department of Environmental Health and Center for Environmental Genetics, University of Cincinnati College of Medicine, Cincinnati, OH, USA

Merja Korkalainen, Department of Environmental Health, National Institute for Health and Welfare, Kuopio, Finland

Qiang Ma, Receptor Biology Laboratory, Toxicology and Molecular Biology Branch, Health Effects Laboratory Division, National Institute for Occupational Safety and Health, Centers for Disease Control and Prevention, Morgantown, WV, USA

Fumio Matsumura, Department of Environmental Toxicology, University of California Davis, Davis, CA, USA

Jaime M. Merino, Departamento de Bioquímica y Biología Molecular, Facultad de Ciencias, Universidad de Extremadura, Badajoz, Spain

Hanna M. Miettinen, Department of Environmental Health, National Institute of Health and Welfare, Kuopio, Finland

Ivy D. Moffat, Department of Pharmacology and Toxicology, University of Toronto, Toronto, Canada

Sonia Mulero-Navarro, Departamento de Bioquímica y Biología Molecular, Facultad de Ciencias, Universidad de Extremadura, Badajoz, Spain

Iain A. Murray, Center for Molecular Toxicology and Carcinogenesis, Department of Veterinary and Biomedical Sciences, The Pennsylvania State University, University Park, PA, USA

Fumiaki Ohtake, Institute of Molecular and Cellular Biosciences, University of Tokyo, Tokyo, Japan; ERATO, Japan Science and Technology Agency, Kawaguchisi, Saitama, Japan

Allan B. Okey, Department of Pharmacology and Toxicology, University of Toronto, Toronto, Canada

Allesandro Pandini, Division of Mathematical Biology, National Institute for Medical Research, London, UK

Gary H. Perdew, Center for Molecular Toxicology and Carcinogenesis, Department of Veterinary and Biomedical Sciences, The Pennsylvania State University, University Park, PA, USA

Lorenz Poellinger, Department of Cell and Molecular Biology, Karolinska Institutet, Stockholm, Sweden

Raimo Pohjanvirta, Department of Food Hygiene and Environmental Health, Faculty of Veterinary Medicine, University of Helsinki, Helsinki, Finland

Jo Anne Powell-Coffman, Department of Genetics, Development, and Cell Biology, Iowa State University, Ames, IO, USA

Alvaro Puga, Department of Environmental Health and Center for Environmental Genetics, University of Cincinnati College of Medicine, Cincinnati, OH, USA

Hongtao Qin, Cold Spring Harbor Laboratory, Cold Spring Harbor, NY, USA

Eva M. Rico-Leo, Departamento de Bioquímica y Biología Molecular, Facultad de Ciencias, Universidad de Extremadura, Badajoz, Spain

Angel Carlos Roman, Departamento de Bioquímica y Biología Molecular, Facultad de Ciencias, Universidad de Extremadura, Badajoz, Spain

Stephen Safe, Center for Environmental and Genetic Medicine, Institute for Biosciences and Technology, Texas A&M Health Science Center, Houston, TX, USA; and Department of Veterinary Physiology & Pharmacology, Texas A&M University, College Station, TX, USA

Dieter Schrenk, Food Chemistry and Toxicology, University of Kaiserslautern, Kaiserslautern, Germany

Jason A. Scott, Department of Pharmaceutical Sciences, University of New Mexico Health Sciences Center, Albuquerque, NM, USA

Michael T. Simonich, Department of Environmental and Molecular Toxicology, Oregon State University, Corvallis, OR, USA

Sidney J. Stohs, Creighton University Health Sciences Center, Omaha, NE, USA

Hollie Swanson, University of Kentucky, Lexington, KY, USA

Elin Swedenborg, Department of Biosciences and Nutrition, Karolinska Institutet, Novum, Huddinge, Sweden

Robert L. Tanguay, Department of Environmental and Molecular Toxicology and Environmental Health Sciences Center, Oregon State University, Corvallis, OR, USA

Shelley A. Tischkau, Department of Pharmacology, Southern Illinois University School of Medicine, Springfield, IL, USA

Jouko Tuomisto, Department of Environmental Health, THL (National Institute for Health and Welfare, formerly National Public Health Institute) and University of Eastern Finland (formerly the University of Kuopio), Kuopio, Finland

Matti Viluksela, Department of Environmental Health, National Institute of Health and Welfare, Kuopio, Finland

Mary K. Walker, Department of Pharmaceutical Sciences, University of New Mexico Health Sciences Center, Albuquerque, NM, USA

Carolina Wejheden, Institute of Environmental Medicine, Karolinska Institutet, Stockholm, Sweden

Sally S. White, National Toxicology Program, National Institute of Environmental Health Sciences, National Institutes of Health, Research Triangle Park, NC, USA

Ayelet Ziv-Gal, Department of Veterinary Biosciences, College of Veterinary Medicine, University of Illinois at Urbana-Champaign, Urbana, IL, USA

PART I

HISTORICAL BACKGROUND

1

HISTORY OF RESEARCH ON THE AHR

THOMAS A. GASIEWICZ AND ELLEN C. HENRY

1.1 INTRODUCTION

Since Alan Poland's laboratory at the University of Rochester published the first report on the identification of the Ah receptor (AHR) in 1976 [1], there have been nearly 6000 additional publications (as determined by a PubMed search) detailing the characteristics of this protein, its function and regulation, and consequences of its activation in a variety of biological systems. As a graduate student at Rochester at the time of Alan's discovery, I can still recall his dedication, excitement, and enthusiasm for what were particularly novel, but at the time somewhat controversial, findings. Little did we know that such a "cottage industry" of research activity (as Alan described it years later) would sprout from these initial studies. Over the years, many individuals contributed to the existing body of work on the AHR, and each one of these contributions has added some piece to our understanding of this transcription factor. Nevertheless, there have been major milestones along the way that have significantly transformed the field and that are likely to define the direction of future investigations. Many of these milestones have come about through a logical progression and focus in a particular research stream. Some were more serendipitous findings. Still other advances have been enabled by some major development in other areas of research that initially had little to do with the AHR. Finally, in some cases the importance of seemingly minor discoveries has yet to be appreciated.

The purpose of this chapter is to give an overview of the emergence and evolution of milestones in research on the AHR (Fig. 1.1), what these tell us about AHR biology in cells and tissues, and how this biology may relate to human disease. The involvement of the AHR in the toxicity of many xenobiotics is certainly one important part of this field. Much knowledge has been and will be gleaned about the possible normal function of this protein from our knowledge of the toxicity caused by these xenobiotics. However, we primarily touch on this vast toxicology literature only inasmuch as has contributed to our understanding of a biological function of this protein. Readers are referred to several excellent reviews that have discussed toxicity mediated by the AHR in both humans and a variety of animal models [2–6]. It is also worthwhile to point out that an earlier publication by Allan Okey [7] was very useful in providing references and reminders about some of the historical events in this area of research. It is hoped that this chapter complements and extends that provided by Dr. Okey.

1.2 A HYPOTHESIS IS CONCEIVED, AND GENETICALLY DEFINED GROWTH AND EXPANSION OF THE HYPOTHESIS LEADS TO THE IDENTIFICATION OF THE Ah LOCUS

In the early 1950s while working as a graduate student in the lab of James and Elizabeth Miller at the University of Wisconsin, a young Allan Conney became intrigued by the findings of H. L. Richardson and colleagues at the University of Oregon Medical School that treatment of rats with the carcinogen 3-methylcholanthrene (3-MC) inhibited the hepatocarcinogenic activity of 3′-methyl-4-dimethylaminoazobenzene [8]. From this grew a hypothesis that treatment with 3-MC or other polycyclic aromatic hydrocarbons (PAHs) would alter the metabolism of the carcinogenic aminoazo dyes and thus protect from subsequent carcinogenic activity. Allan went on to find that indeed a single injection of 3-MC

The AH Receptor in Biology and Toxicology, First Edition. Edited by Raimo Pohjanvirta.
© 2012 John Wiley & Sons, Inc. Published 2012 by John Wiley & Sons, Inc.

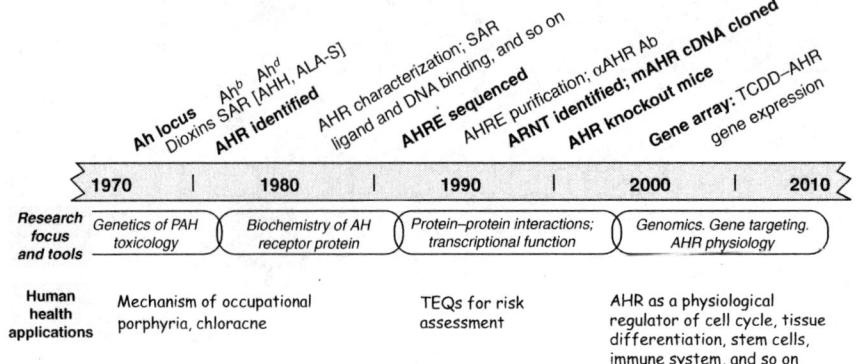

FIGURE 1.1 Timeline of major milestones in AHR research. Owing to limitations of space, only a subset of the most significant advances is shown, along with the changing research tools and some of the human health considerations that drove the research approaches.

could induce the activity of not only aminoazo dye N-demethylase and azo-link reductase [9], but benzo[a]pyrene (BP) hydroxylase as well [10]. Further work by Dr. Conney while at the National Institutes of Health (NIH) in Bethesda showed that there was a selective stimulation of the metabolism of some drugs and xenobiotics but not others [11], suggesting that the induction of these enzymes was under separate regulatory control. This was the first bit of evidence consistent with the notion that there may be a selective signaling pathway that mediates this enzyme induction.

A major step in the pathway leading eventually to the identification of the AHR came when it was recognized by Dan Nebert at the NIH in the late 1960s that aryl hydrocarbon hydroxylase (AHH; previously called BP hydroxylase) was inducible by a number of PAHs in several inbred mouse strains but not others [12, 13]. This was consistent with an earlier observation that some inbred strains differed in sensitivity to skin ulcerations in response to 7,12-dimethylbenzanthracene [14]. Nebert's group further found that the inheritance of this inducibility among mouse strains segregated primarily as an autosomal dominant trait [15, 16]. Based on this responsiveness to aromatic hydrocarbons, a new genetic locus, the Ah locus, was defined. Later it was found that the inheritance of this trait in mice is actually more complex when crosses in other strains were examined [17]. Nevertheless, it was clear that in some mouse strains 3-MC and other PAHs induced AHH activity, while in other strains no such induction was observed; the responsive Ah^b allele was typified by C57BL/6 mice, whereas the "nonresponsive" Ah^d allele was typified by DBA/2 mice. Additional work by Dr. Nebert and a number of other groups demonstrated that besides AHH inducibility, the Ah locus also regulates carcinogenic, mutagenic, teratogenic, and toxic responses to PAHs that are closely related to the ability of these compounds to be metabolized by the induced enzymes [17, 18].

1.3 A FOCUS ON PUBLIC HEALTH LEADS TO THE IDENTIFICATION OF THE Ah RECEPTOR

In the early 1970s, Alan Poland was a physician interested in the mechanisms by which workers in factories producing 2,4,5-trichlorophenol (2,4,5-T) succumbed to industrially acquired acne (chloracne) and porphyria cutanea tarda [19]. While 2,3,7,8-tetrachlorodibenzo-p-dioxin (TCDD) was known to be the active acnegenic contaminant in 2,4,5-T [20, 21], the cause of the porphyria was still unclear at the time. To test whether TCDD was also responsible for the porphyria, Dr. Poland examined the effects on ALA synthetase, the rate-limiting step in the heme biosynthetic pathway, using a chick embryo model. TCDD was extremely potent in inducing ALA synthetase in this model [22]. Since it was found that many compounds that induced ALA synthetase also induced microsomal mixed-function oxidase activity, Dr. Poland's lab did additional work to show that TCDD was also a potent inducer of AHH activity [23]. Notably, they observed that the structure–activity relationship (SAR) for the ability of different dioxin congeners to induce AHH was the same as that for ALA synthetase, strongly indicating that these responses were mediated by the same initial step or binding site [22, 23]. Importantly, these were the first SAR studies of these compounds, predating the identification of the AHR, detailing a structural basis for their toxicity. Since that time, these structural requirements have been further refined and are still widely used to show that a particular response is mediated by the AHR. This principle also served as a basis for the many screening assays developed later to assess the presence of dioxin-like chemicals in the environment, as well as for risk assessment of these chemicals.

Several important additional steps were taken when Dr. Poland's lab demonstrated that TCDD is about 30,000 times more potent than 3-MC at inducing AHH in rat liver [24], and then, teaming up with Dr. Nebert's group, showed that TCDD could overcome the "nonresponsiveness" to 3-MC in a number of mouse strains [25]. These studies clearly indicated that the "nonresponsive" mice possess the regulatory and structural genes for the expression of AHH activity. However, the question remained about exactly what was the defect in these mice that made them "nonresponsive" to 3-MC but responsive to TCDD. The additional finding that responsive and "nonresponsive" mouse strains differed in their sensitivity to TCDD as well [26] was clearly consistent with the hypothesis that the product of the *Ah* locus is a binding site for both 3-MC and TCDD and initiates events leading to the expression of the gene encoding AHH. Furthermore, these data indicated that the mutation in the *Ah* locus in "nonresponsive" mice is responsible for the production of a "receptor" with reduced affinity for 3-MC, TCDD, and other inducing chemicals [27].

Synthesis of radiolabeled TCDD by Andrew Kende at the University of Rochester was the important technological advancement that ultimately led to the identification of the AHR protein by Dr. Poland's lab [1]. This landmark study utilizing detailed cell fractionation studies, SAR for both binding and enzyme induction, basic protein biochemistry, and taking advantage of the genetic differences between mouse strains identified a high-affinity and saturable TCDD binding protein that exhibited the same SAR for binding other dioxins and PAHs as their ability to induce AHH activity. This investigation still stands (at least in our opinion) as the single most important contribution to this field.

1.4 THE 1980s: GROWTH AND MATURATION OF THE PARADIGM

Dr. Poland had postulated that the binding of TCDD or other PAH ligands initiates a series of events whereby the AHR–ligand complex translocates to the nucleus and, in a manner analogous to that described for steroid hormone receptors, alters transcription and translation of "induction-specific" RNA [1]. In the years subsequent to the initial identification of the AHR and the hypothesis regarding its mechanism of action, there was a flurry of activity from Dr. Poland's lab and many others to further characterize the properties of this protein, examine its presence in a variety of tissues and different species, and detail the events leading to its ability to modulate gene expression. Work by Bill Greenlee while a graduate student in Dr. Poland's lab [28], and by Allan Okey while on sabbatical in the Nebert lab [29], demonstrated ligand-dependent nuclear translocation of the AHR. This was later confirmed by Rick Pollenz, working in Dr. Poland's lab (then at the University of Wisconsin), who used immunofluorescent approaches and recently developed antibodies against the AHR [30, 31]. Studies in a number of other labs, including that of Mike Denison (while at Cornell University and then with Dr. Okey), Tom Gasiewicz at Rochester, Jan-Åke Gustafsson at the Karolinska Institute, Allan Okey in Toronto, and Lorenz Poellinger (originally with Dr. Gustafsson), identified and confirmed properties of the AHR in a variety of tissues and species. These investigations noted similarities and differences in both physical–chemical characteristics and relative binding affinities for TCDD in a variety of species including humans [32–42]. A striking and consistent finding from these studies was that the apparent characteristics of the AHR complex changed following ligand binding and translocation into the nucleus [36, 43]. These data suggested that other proteins were associated with the AHR under physiological conditions and that these associations may be necessary for AHR function as a gene regulatory protein.

From these studies, an overall model of action for the AHR was emerging that, as Dr. Poland had predicted, was similar to that for the steroid hormone receptors. But it was still unclear whether the AHR directly bound to DNA or indirectly modified gene expression, and, if the former, whether binding to DNA was assisted by another protein or proteins. Utilizing a series of mutant cell lines and examining differences in their ability to induce AHH in response to TCDD, Oliver Hankinson and his group at the University of California at Los Angeles and others collaborating with Dr. Hankinson obtained results implying that AHH induction requires formation of the TCDD–AHR complex as well as the interaction of this complex with some other component(s) of the cell nucleus [44–46]. The same conclusions were reached by Jim Whitlock and his group at Stanford examining differences in the ability of mutant cells to metabolize benzo[*a*]pyrene [47, 48]. Although it became recognized that the TCDD–receptor complex is a DNA binding protein [36, 49], the mechanisms that designated specificity were still not known.

Through a series of extensive and detailed investigations of the mechanisms whereby the cytochrome P_1-450 gene (*Cyp1a1*) (the product responsible for AHH activity) was induced by TCDD, the Whitlock lab identified the specific nucleotide sequence to which the nuclear AHR complex binds within the 5′-flanking region of the gene [50–52]. The discovery immediately had broader implications for other genes and species, including humans. One important aspect was that the dioxin-responsive element (DRE), as it was called, to which the AHR bound, could activate transcription from a heterologous promoter and even within a heterologous species, that is, not just mice, but rats as well as humans. The discovery was confirmed by the Fujii-Kuriyama laboratory at Tohoku University who named the enhancer element the "xenobiotic-responsive element" (XRE) in recognition of the fact that other xenobiotics, in addition to TCDD, could activate transcription through this AHR-dependent pathway [53]. The identification of the core recognition sequence (5′-TNGCGTG-3′), more recently termed AHR-responsive element (AHRE), was arguably the second most important landmark in the history of AHR research. Many groups have subsequently identified AHREs in the upstream regions of numerous genes in addition to *Cyp1a1* [54–57] and refined our understanding of variations that likely account, at least in part, for noted differences in the regulation of a variety of AHR-responsive genes within a given species as well as between species. For example, differences in the nucleotides flanking the core AHRE have been shown to affect both binding of the AHR complex as well as transcription [58, 59]. While data showing that the AHR complexes from different species are able to bind the mouse AHRE suggested some conservation of this interaction [60], more recent data suggest that the set of target genes regulated by the AHR may not be well conserved across species [61]. In addition, several investigators discovered an AHRE-II enhancer element (CATG[N6]C[T/A]TG) that is conserved across human, mouse, and rat orthologs and that appears to regulate some sets of genes for ion channels and transporters [62, 63]. Furthermore, as first demonstrated by the work in Steve Safe's lab at Texas A&M, it has become clear that inhibitory effects on certain genes can be produced by the positioning of the AHREs relative to other transcription factor binding sites [64–67]. Similarly, the presence of negative regulatory elements within AHR-responsive genes has been shown to inhibit AHR-dependent transcription without affecting AHRE binding [68, 69].

Shortly after the discovery of the AHRE, it was observed that the AHRE binding form of AHR has different physicochemical properties from the cytosolic form [70–72], indicating that there was an additional, as yet unidentified, component that helped to designate specificity of binding. This was further consistent with Dr. Hankinson's data, indicating that certain mutants of a mouse hepatoma cell line have a functionally defective ligand–AHR complex that fails to accumulate in the nucleus [44, 46]. In 1991, Hankinson's group reported the cloning of a human gene capable of restoring AHR nuclear uptake and function in these cells and named it the AHR nuclear translocator, or ARNT [73]. Shortly thereafter, they recognized that ARNT heterodimerizes with the AHR to generate the active AHRE binding transcription factor [74]. Later, it was determined that in most cells ARNT resides predominantly, if not exclusively, in the nucleus and does not appear to be directly involved in translocating the AHR [31]. The discovery of ARNT was an exciting and significant breakthrough for a number of reasons. First, it solidified a mechanism whereby AHR could directly act as a transcription factor and selectively regulate gene expression. The AHR–ARNT complex became the first example of a heterodimer within the basic helix–loop–helix–Per/Arnt/Sim (bHLH–PAS) family of proteins [75]. Furthermore, this provided one of the first insights that the AHR signaling pathway could interact with other intracellular signaling pathways making the biology of the AHR much more complex—and indeed more intriguing.

Several years after ARNT was cloned and the amino acid sequence defined, it was recognized that the hypoxia-inducible factor (HIF) protein was composed of two subunits, HIF1α and HIF1β, and that in fact HIF1β was identical to ARNT [76]. There has been much subsequent work on the interactions between the hypoxia and AHR signaling pathways [56, 77–79].

About the same time that the AHRE was discovered and defined, Dr. Poland's lab, then at the University of Wisconsin, was working vigorously to purify and sequence the AHR. The development of the radiolabeled photoaffinity ligand 2-azido-3-[^{125}I]iodo-7,8-dibromodibenzo-*p*-dioxin by Dr. Kende provided the essential tool leading to this purification and the preparation of the first AHR-specific antibodies [80–83]. Several years later, but a year after ARNT was identified, two labs, Dr. Bradfield's in Wisconsin and Dr. Fujii-Kuriyama's, reported the cloning of the receptor's cDNA [84, 85].

1.5 DETAILS, DETAILS

The cloning and sequencing of both the AHR and ARNT were very significant developments that enabled rapid progress on several fronts. First of all, it became clear that the AHR was not similar in sequence and structure to the steroid hormone receptors as many had believed. In fact, AHR and ARNT contained domains with high sequence homology to two recently identified proteins in *Drosophila*, PER and SIM [86–89]. Since that time, the bHLH–PAS family of transcription factors has expanded rapidly to include a number of other proteins such as HIF, CLOCK, and MOP that are involved in physiological responses to cellular and tissue environmental signals including circadian rhythms, oxygen tension, and oxidation–reduction status [75, 90]. At about the same time that the AHR and ARNT were being purified, cloned, and sequenced, additional data were accumulating from many investigators, including Chris Bradfield, Gary Perdew, and Lorenz Poellinger, to indicate that several other proteins — hsp90, p23, and XAP2 (also called ARA9 or AIP) — interact with the AHR to regulate its intracellular stability, ligand binding ability, and cytoplasmic–nuclear trafficking [91–101]. Mapping of specific domains within the AHR and ARNT that were coupling sites for these proteins provided confirmation of the interactions and a further understanding of the mechanisms by which ligand binding regulated AHR function. The mapping of these domains also provided a greater mechanistic understanding of the specificity for ligand binding, AHRE recognition, and transactivation of gene expression [102–110].

Second, the cloning and sequencing of the AHR have expanded our understanding of the evolution of this protein and its possible function in a variety of species. Early studies demonstrated some similarity in TCDD binding ability of the AHR contained in different lab animals [37, 111]. However, more detailed physicochemical analyses demonstrated clear differences in molecular mass of AHR proteins from a variety of species and among rodent strains [112–114]. Subsequent cloning of the AHR genes from a number of animals, mainly by Mark Hahn's lab at the Woods Hole Oceanographic Institution [115–121], revealed some interesting features: (1) Mammals appear to have a single AHR gene, while other vertebrates such as fish and birds may have up to five. Additional investigation of these divergent forms may provide interesting clues to functions of the AHR. (2) The N-terminal region of the AHR, containing the DNA binding domain, nuclear localization sequence, and dimerization domain within the bHLH–PAS region, is, for the most part, highly conserved. The C-terminal region, containing the transactivation domain, is more variable. Although yet to be determined, these variations may eventually help explain the differences in the spectra of AHR-responsive genes among species, as well as the differences in sensitivity of these species to the toxic effects of various xenobiotics (see Section 1.6). (3) Invertebrates possess single AHRs that do not recognize the typical xenobiotics that bind and activate vertebrate AHRs. Based on these findings, Dr. Hahn proposed that the ability of the AHR to bind to dioxins, PAHs, and other xenobiotics, and thus also the ability to regulate xenobiotic metabolizing enzymes, occurred through gene duplications and gene loss during the evolution of vertebrates [118]. (4) Differences in AHR binding affinities and sensitivities to AHH induction among mouse strains were verified and found attributable to a mutation of alanine to valine within the ligand binding domain (LBD) [115]. (5) The latter finding was of particular interest because cloning of a human receptor (hAHR) [122] indicated a valine at position 381 and a proline at position 480 that are also present in the low-affinity "nonresponsive" mAHR^d allele [115]. Furthermore, the relative TCDD binding affinity of a cloned hAHR is about an order of magnitude lower than that of most lab animals, in particular the mAHR^b allele. This knowledge may prove to be important in estimating the relative risk to humans of toxic AHR ligands such as TCDD (see below). On the other hand, the development of these approaches will also facilitate the assessment of human AHR polymorphisms to determine whether some humans may possess high-affinity receptors or other AHR mutations that predispose (or protect) these individuals to (from) certain diseases (see below).

1.6 LET US NOT FORGET THE TOXICOLOGY

1.6.1 New Questions

Although much research during the 1970s, 1980s, and early 1990s focused on the properties of the AHR and the mechanisms whereby this protein modulates gene expression, the toxicology of xenobiotic AHR ligands continued to be a focal point. In fact, the discovery of the AHR greatly expanded the research efforts on dioxins and dioxin-related chemicals as well as the number of toxicologists and individuals in regulatory agencies that became involved in these efforts. New questions about the toxicity of these chemicals and possible relationships to the AHR needed to be addressed. Does the AHR mediate the toxicity of these chemicals? If so, is the induction of AHH or other drug metabolizing enzymes responsible for this toxicity? If not, then what other genes might be responsible? Do the different properties of the AHR among species have anything to do with the varying sensitivities of these species to toxic AHR ligands? Can the properties of the hAHR give us clues as to the relative sensitivity of people to these chemicals? How can our understanding of the properties of the AHR and the mechanisms by which it mediates toxicity and regulates intracellular processes be used to improve our risk assessment of these chemicals? Also, based on this knowledge, what might be the most sensitive toxic end points in people and thus the potentially most sensitive subpopulations?

1.6.2 Agonists Versus Antagonists—Or Somewhere In Between

In the early 1980s, Dr. Poland's laboratory reported that the toxicity of TCDD in various mouse strains segregated with the *Ah* locus and that the potency of various dioxins, halogenated biphenyls, and PAHs to bind to the AHR correlated with their ability to produce a variety of toxic end points and was structure dependent [123, 124]. In the subsequent 15 years, these data were extended by a number of laboratories. In particular, work by Dr. Steve Safe and his colleagues (first at the University of Guelph and later at Texas A&M University) increased the tools available by synthesizing a number of chlorinated dibenzo-*p*-dioxins, dibenzofurans, and biphenyls (PCBs) that were previously unavailable for research purposes. They and others demonstrated that, for the most part (except for those congeners more susceptible to metabolism), the SAR for AHR binding was closely related to the ability of the chemicals to produce a number of toxic end points in different laboratory animals [125–130]. These studies also extended our understanding of the structural requirements for ligand binding to the AHR (see further discussion of AHR ligands below).

The interest in SAR also led to the search for compounds that could bind to the AHR but act as AHR antagonists to block toxicity. In the 1980s, Dr. Safe and colleagues recognized that some of the chlorinated dibenzofurans and biphenyls had relatively high affinity for the AHR but were poor inducers of AHH activity, and furthermore could partially antagonize the response elicited by TCDD [131]. This finding led to the synthesis of additional compounds by the Safe lab that proved to be effective antagonists [132–134]. These compounds, as well as those later identified and/or synthesized by the labs of Drs. Safe, Gasiewicz, Perdew, and others [135–142], provided additional evidence for the role of the AHR in mediating toxicity, as well as valuable tools for exploring mechanisms by which the AHR acted to modulate cell functions. The potential of such compounds to block the toxicity caused by the inadvertent exposure of animals and humans to TCDD was also being explored. However, as research progressed on these compounds, it became clear that many factors determine their relative ability to block toxic effects mediated by the AHR. While most of the identified AHR antagonists appear to compete with TCDD for the same binding site, others may act as antagonists by several different mechanisms [143–148]. Furthermore, those that bind to the AHR may affect its activity in a very species-specific manner such that a compound may act more like an antagonist in one species but more like an agonist in another species [149, 150]. Importantly, the agonist versus antagonist activity of a particular compound may depend on relative affinity as well as intrinsic efficacy [151], coincident with an ability to produce a particular conformational change that may be dependent on the species-specific amino acid sequence within the ligand binding site [106, 150, 152]. Thus, while the development of antagonists provided valuable tools for research, their more practical applications to incidents of poisoning were hindered by the lack of understanding of the factors that regulated their ability to turn on or turn off AHR activity, the specificity of the chemicals for these actions, and the growing knowledge that the AHR may have some normal functions (see below).

1.6.3 The AHR Model as a Means to Approach Risk Assessment

It is important to note that evidence to date based on SAR and genetics supports the contention that the toxicity of TCDD and related chemicals is mediated through the AHR. Although over the years there have been some reports suggesting that certain toxic effects, particularly some immune system alterations, may be AHR independent [153–158], there is a general consensus that the AHR has an obligatory role in TCDD-induced toxicity, and this is an important factor in risk assessment. However, since TCDD is just one of a large number of structurally related contaminants that are widely distributed in the environment, generally in complex mixtures, the toxicity testing of all of them individually would be cost prohibitive. Recognition that many of them likely act through this same AHR-dependent mechanism led the international community in the late 1980s and early 1990s to devise an approach to risk assessment for these congeners based on a principle of "toxic equivalency factors" (TEFs) [159]. In 1990, Dr. Safe published a summary of the existing information on the relative potency of these chemicals and suggested TEF values for those chemicals that had dioxin-like activity [160]. Based on additional data over the years, these values have undergone several reevaluations [161]. Despite the general acceptance of the use of TEFs and validation of the approach by several labs [162–166], considerable controversy remains [167–172]. Today, it is widely recognized that the assigned TEF values are of necessity approximations based on incomplete data and may not accurately reflect additive or synergistic interactions or the actual metabolism of individual compounds within different tissues, considerations of different types of exposures (e.g., acute versus chronic), different toxic end points in different animal species, and the possible species-specific agonist versus antagonist activity of different ligands (as discussed above; see also Ref. 173). Although it is clear that knowledge gaps in the differences in biology among these chemicals need to be addressed to more accurately approach the risk assessment for xenobiotic AHR ligands, the concept that their toxicity is mediated by the AHR has nevertheless played, and will continue to play, a key role in driving the directions for this process.

1.6.4 More Evidence to Support the Model: Knockouts, Lots of Genes, and More Questions

As the evidence mounted that the toxicity of TCDD and related xenobiotics was mediated through the AHR and that this protein clearly functioned as a ligand-activated transcription factor, much research in the late 1980s became directed at determining the genes that were dysregulated by TCDD and associated with particular toxic responses. Early studies were focused on the "AH gene battery" that was initially identified and included those genes encoding CYP1A1, CYP1A2, NAD(P)H:quinone oxidoreductase, aldehyde dehydrogenase 3, UDP-glucuronosyltransferase 1A6, and glutathione S-transferase Ya [192]. Altered CYP expression may contribute to tissue damage and tumorigenesis, for example, through induced metabolism of some xenobiotics and DNA adduct formation, as well as oxidative DNA damage via the production of reactive oxygen species [3, 174–177]. The clearest association between TCDD-elicited induction of a *Cyp* gene and induced toxicity was made though the collaboration of the Andrew Smith lab at Leicester University, Dan Nebert (then at the University of Cincinnati), and Peter Sinclair at Vermont, when it was found that knocking out the expression of the *Cyp1a2* gene protects against uroporphyria and hepatic injury following treatment with TCDD [178]. In most cases, however, a clear association between induction of these enzymes and the species- and tissue-specific toxicity of TCDD has not been shown; indeed, it has been demonstrated, for example, that some tissues and species show a substantial induction of these without observable toxicity [179], and elimination of CYP1A1 expression does not ablate induced toxicity [180].

The 1990s brought new technology to the search for differentially expressed genes. Drs. Sutter and Greenlee, working at what was then called the Chemical Industry Institute of Toxicology (now the Hamner Institute for Health Sciences), first used differential hybridization to identify in human keratinocytes several genes regulated by the AHR that may play a role in TCDD-elicited alterations in inflammatory and differentiation processes [181]. Since that time the invention and use of such technologies as gene expression arrays has resulted in the identification of hundreds of TCDD-modified genes and signaling pathways in a variety of tissues [182–191]. Advances in cloning, sequencing, bioinformatics, biostatistics, transfection techniques, and chromatin analyses and the use of specific gene reporters have further allowed researchers from the labs of Dr. Puga at Cincinnati, Dr. Zacharewski at Michigan, and others to determine the presence of functional AHREs in regulatory regions with the objective to define which modified genes are actually directly responsive to the AHR and the interactions between the signaling pathways regulated by these genes [54, 56, 57, 61, 189, 192, 193].

Targeted gene disruption is an additional technology developed in the 1990s that has enabled another milestone in our understanding of the AHR's roles in both toxicity and normal physiology. In 1995, Frank Gonzalez and his group at NIH published the first report describing the successful creation of an *Ahr*-null allele (AHR-KO) mouse [194]. This was closely followed by publications from the Bradfield and Fujii-Kuriyama labs, also demonstrating the creation of KO mice based on targeting a different exon of the *Ahr* gene [195–197]. Studies using these mice verified that the presence of the AHR is necessary for susceptibility to a number of TCDD-induced effects, including lethality, hepatoxicity, immunotoxicity, teratogenesis, effects on developing prostate and uterus, and carcinogenicity [194, 196, 198–206]. These mice are also resistant to BP-induced carcinogenesis, UV radiation-induced inflammatory responses in the skin, and benzene-induced hematotoxicity [207–209].

These data coupled with our growing knowledge of the details of AHR structure and characteristics have given us a wealth of information. However, both despite and because of this knowledge, it has become clear that determining what specific molecular pathways designate a particular toxic outcome in response to TCDD is likely to be much more complex than what most of us working in this area might have believed. By the end of the 1990s, several features had become apparent. For example, although the presence of the AHR appears to be necessary for toxicity, it is not sufficient. Accumulating data show that the toxic and genomic responses to TCDD and related chemicals are clearly dose, cell, tissue, and species specific. These different molecular and biological responses may reflect differences in the developmental stage, state of cell cycling and differentiation, and the tissue environment surrounding the particular cell type responding to TCDD. In most cases, the embryo and fetus are much more susceptible to the toxicity of dioxin-like chemicals than adult animals (see Ref. 210 and Section 1.9.1). We also know that species and strain differences in responses likely reflect differences in AHR structure. As indicated above, certain mouse strains possess a mutation in the ligand binding domain that alters their affinity and thus sensitivity to certain dioxin-like xenobiotics [115]. Other data resulting from collaborations between Dr. Pohjanvirta's lab in Finland and that of Dr. Okey indicate striking differences in the transactivation domain of the AHR among species and strains that likely alter the AHR's ability to differentially interact with other transcription factors, coregulators, or corepressors to produce gene selectivity [120, 121, 211–213]. In addition, it is becoming more apparent that in most, if not all, cases, TCDD-induced toxicity results from the dysregulation of multiple genes and pathways rather than one specific gene. To date, only a few specific target genes and signaling pathways have been associated with particular end points of toxicity [214, 215]. As such, tissue-, species-, and developmental stage-specific differences in the regulation of one or several of these pathways are likely to make a huge difference in the ability of TCDD to elicit a biological response. Furthermore, it is becoming recognized that developmental, long-lasting, and even transgeneration effects of TCDD and related xenobiotics may be mediated by epigenetic mechanisms. As the 1990s progressed and ended, more and more researchers in this area have been focusing their attention on the mechanisms of very specific and sensitive *in vivo* responses to TCDD, recognizing that many *in vitro* systems may not accurately represent the *in vivo* situation and that defined mechanisms in one tissue may not be relevant in another tissue or even at a different stage of development. Many of the chapters in this book are a reflection of these efforts (see Fig. 1.2).

1.7 RECOGNIZING THE IMPORTANCE OF REGULATING AHR PRESENCE AND ACTIVITY

Since the discovery of the AHR, many publications have documented differences in its expression among a variety of tissues and cells and changes in its expression or activity in response to a number of agents and conditions (reviewed in Ref. 216). Although it was well recognized that these changes could potentially alter a tissue's toxic response to xenobiotic ligands, throughout most of the 1990s there was little understanding of the mechanisms that regulated AHR levels in tissues or its activity. In the past 15 years, several important discoveries have added much to our understanding of these mechanisms.

Several labs reported that, in many but not all cases, exposure to AHR agonists elicited a rapid depletion of AHR protein without an effect on AHR mRNA [217–220]. In 1999 and 2000, the labs of Drs. Pollenz, Whitelaw, and Ma

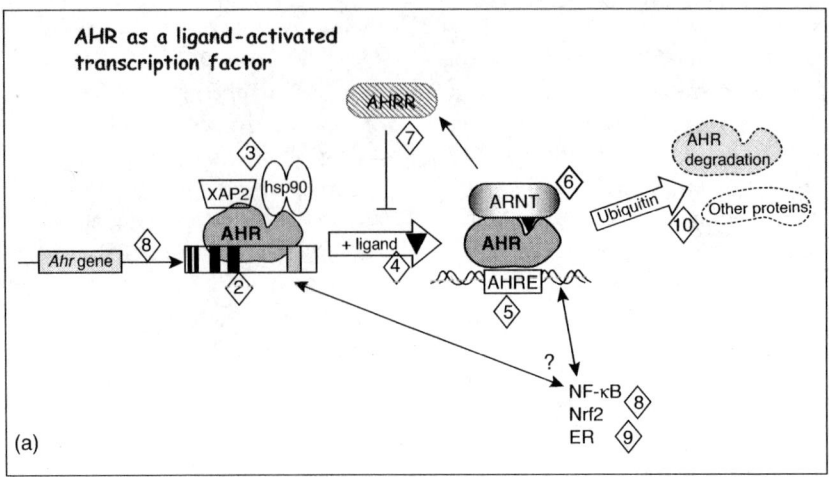
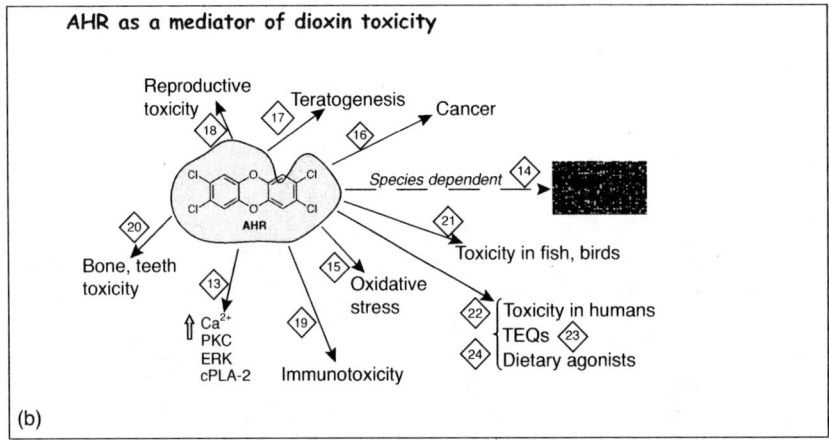
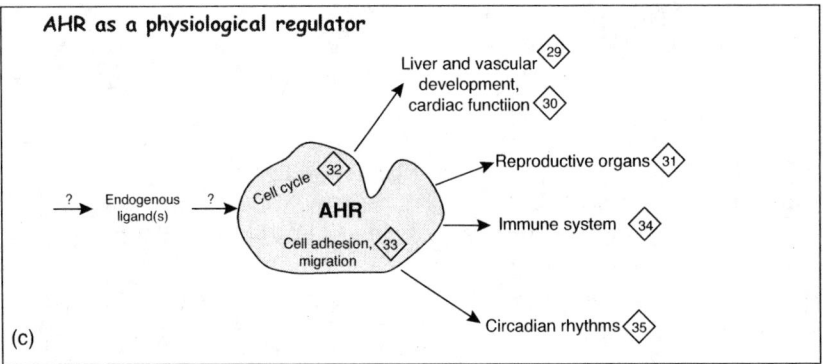

FIGURE 1.2 Overview of the multiple roles of AHR as covered in this book. Parts (a), (b), and (c) correspond to the sections of the book. Numbers in diamonds, for example, ⟨4⟩ indicate the chapter in which the indicated pathway, process, or feature is discussed.

independently demonstrated that this loss of AHR protein required nuclear uptake of the ligand-bound AHR, ubiquitination, and proteasome-mediated degradation [221–223]. Several subsequent studies have shown that this degradation depends on other cell-specific conditions and factors and may also occur independently of ligand [224–227]. Furthermore, most of these investigations were carried out using cultured cells, and it is not clear whether this induced degradation would occur in most tissues within an intact animal. Although, in theory, blocking receptor degradation may make a cell hyperresponsive to its own ligand [228], and indeed inhibiting proteasomal degradation has been shown to cause a "superinduction" of CYP1A1 by TCDD [229], it does not necessarily follow that all responses to any AHR ligand would be heightened. The importance of this pathway to control AHR protein levels and particular toxic or physiological responses to AHR ligands clearly needs to be examined in more detail.

In addition, the level of AHR expression has been found to change with cell density, state of cell differentiation, presence or absence of growth factors and hormones, and neoplastic transformation in the absence of known AHR ligands [216]. It seems likely that in some cases this may reflect differential regulation of the *Ahr* gene. The cloning of the *Ahr* gene from human, rat, and mouse in the early 1990s revealed the presence of many potential transcription factor binding sites [197, 230–232]. The presence of these sites could explain the observations made by many labs of tissue- and differentiation state-dependent expression, as well as the up- and downregulation of AHR expression by several factors such as IL-4, TGFβ, and pituitary hormones [216, 233–236]. In 2001, it was subsequently shown by Dr. Abel's group that TGFβ regulates *Ahr* expression through Smad binding sites within the 5′-flanking region [236]. However, for the most part, the functionality of other potential transcription factor binding sites in the *Ahr* gene remains largely unexplored. It would seem that this would be extremely fertile ground for further important investigations.

In 1999, a report from the Fujii-Kuriyama lab identified a novel protein that appeared to bind to the AHR and restrict AHR-dependent gene transcription [237]. Two years later, this group published details on the structure and expression of the gene that encoded for this bHLH–PAS protein, called the Ah receptor repressor (AHRR) [238]. The N-terminal region of the AHRR is similar to the AHR, but lacks the PAS-B domain that functions in ligand binding and hsp90 interaction. It is hypothesized that AHRR blocks AHR activity by dimerizing with ARNT and competing with the AHR–ARNT complexes for binding to AHRE sites. Notably, the *Ahrr* gene is directly upregulated by activated AHR [238], and the levels, as one might expect, are considerably lower in AHR-null allele mice [239]. However, the pattern of AHRR expression does not appear to always correlate directly with the ability to block the response to AHR ligands [239–241]. As such, it is possible that the AHRR may have additional functions, may repress AHR function by additional mechanisms, or may be very tissue specific and have some selectivity for particular AHR-responsive genes [242]. Recent reports suggest that AHRR may directly interact with ERα and suppress its transcriptional activity [243] and may act as a tumor suppressor for several types of human cancers [244]. Clearly, much more work is needed to understand the multiple functions of AHRR, especially since several studies suggest a linkage between some AHRR gene variants and reproductive disorders in humans (reviewed in Ref. 244).

Much evidence gathered over the past 30 years shows a clear tissue-specific developmental regulation of AHR expression [38, 245–250]. Our increased knowledge of gene regulatory processes also indicates that developmental gene expression is regulated, in part, by epigenetic mechanisms [251, 252]. A recent report from Dr. Klaassen's lab at the University of Kansas suggests that an enrichment of specific methylation patterns may regulate the ontogenic expression of AHR mRNA in mouse liver [253]. These data, along with the report that silencing of the *Ahr* gene may be associated with the progression of certain cancers [254–256], suggest that the epigenetic regulation of AHR expression is also an important area that needs to be explored.

These findings, and others summarized elsewhere [216], indicate that normal cellular processes regulate the level of AHR protein and its activity. It is safe to say that these are important for the organism's responses to a variety of environmental chemicals. Furthermore, it is increasingly apparent that endogenous AHR regulatory processes play important roles in cell growth, differentiation, and susceptibility to disease (see below). That this represents an area rich for further investigation is surely an understatement.

1.8 AHRE-DEPENDENT VERSUS AHRE-INDEPENDENT PATHWAYS

As the twenty-first century started, most evidence was consistent with the notion that the ligand-activated AHR modulates gene expression and intracellular signaling pathways through its ability to interact with ARNT and bind AHREs in responsive genes. However, it was (and still is) not unreasonable to hypothesize that a ligand-induced change in AHR conformation could also alter its ability to interact with other proteins to influence signaling pathways in a manner not dependent on the AHR interacting with DNA. Indeed, during the past 10–15 years, there have been several such reports suggesting AHRE-independent actions. These include interactions with c-Src kinase [257–260], NF-κB subunits [261–266], retinoblastoma protein [267–269], estrogen receptor [66, 270, 271], Nedd8 [272], Myb binding protein [273], Nrf2 [274, 275], p38 MAPK [276], β-catenin [277], E2F1 [278], and CD4K [279]. Furthermore, it is reasonable to speculate that the known interactions of the AHR with chaperone proteins in the AHR core complex [100] as well as basal transcription factors [280] could indirectly affect other signaling pathways. Since the middle 1980s, Fumio Matsumura at the University of California at Davis has proposed that nongenomic pathways, in particular those involved in inflammatory processes, may be mediated by the ligand-activated AHR [281]. This has been supported, in part, by a recent finding by the Perdew lab that the AHR can repress acute phase response gene expression without AHRE binding [282]. The finding published in 2007 that the AHR is a ligand-dependent E3 ubiquitin ligase [270] also has implications for affecting many signaling pathways.

Within the past 8 years, the Bradfield lab developed two mouse models that have mutations at the nuclear localization sequence or the DNA binding domain of the AHR, and the use of these suggests that many toxic responses to TCDD are absolutely dependent on the ability of the AHR to translocate to the nucleus and bind to AHREs [283, 284]. On the other hand, it is possible that some toxic end points not yet assessed in these models are mediated through mechanisms not dependent on AHRE binding. Furthermore, the AHR may affect a number of signaling pathways involved in normal physiological functions of the AHR, that is, not related to toxic end points, through non-AHRE binding mechanisms (see below). Relevant to this, comparison of the transcriptome in AHR-KO and wild-type mice implies that the AHR regulates distinct TCDD-dependent and TCDD-independent gene batteries [285], although it is not yet clear whether these differences are due to AHRE-dependent or AHRE-independent mechanisms. Certainly, future efforts should be directed at determining the role of non-AHRE-mediated mechanisms in AHR-dependent toxicity and in regulating normal cellular processes.

1.9 THE PHYSIOLOGICAL ROLE IS STILL ELUSIVE

Since discovery of the *Ah* locus and the AHR, researchers in the field have been struggling with questions surrounding the normal function of this protein. Is it just a xenobiotic sensor? What are its endogenous activators? Does it function to monitor changes in the tissue environment, including those brought about by xenobiotic exposures, and act to produce adaptive responses? Does the AHR have more fundamental roles in normal development and tissue differentiation?

As indicated above, much effort has been devoted to determining mechanisms of xenobiotic toxicity and risks to exposed human populations. However, there has always been an underlying interest in the role of the AHR in normal cellular processes and how this may lead us to a greater knowledge of events leading to human disease that may not necessarily be mediated by xenobiotic exposure. In fact, many of us have used TCDD as a "molecular probe" [27] to gather such information. As far back as the middle 1980s, Dan Nebert noted significant differences in the health, fertility, and life span of mouse strains differing at the *Ah* locus. For example, strains possessing a "high-affinity" receptor (for TCDD) have a longer life span than those possessing a "low-affinity" receptor [286, 287]. However, it has only been within the past 6–8 years that questions regarding the physiological function of the AHR have become the major focus of much of the research in the AHR field. This has been driven by the search for physiological AHR ligands and ligands that could have potential therapeutic uses and by the accumulating evidence of the importance of AHR in many cellular and tissue processes.

1.9.1 Are the Toxic Responses to Xenobiotic AHR Ligands Adaptive or a Reflection of Out-of-Control Physiological Responses That Normally Maintain Homeostasis?

It is clear from the above discussion (see Section 1.6), as well as the publication by Bock and Köhle [210], and discussion by a number of other investigators in this book and elsewhere that TCDD disrupts, directly or indirectly, a large number of tissue and cellular processes. These include, but certainly are not limited to, T-cell development, inflammatory responses [57, 288–296], cardiovascular development and function [190, 297–299], growth and regulation of reproductive tissues [300], pathways of cell differentiation [301–303], signals in cell cycle that regulate the balance between cell proliferation, death, senescence, and differentiation [268, 304–306], circadian rhythms [307–309], mesenchymal–epithelial interactions during development of tissues such as the palate [310] and prostate [311], cell trafficking and invasive tumor growth [312–315], neurological development and function [316–319], and bone development and maintenance [320–323]. Certainly, there is much overlap among several of these processes. However, in terms of the cell biology, some of the most consistent effects of TCDD-induced AHR activation are altered cell cycle, altered patterns of cellular proliferation and differentiation, and altered cell–cell communication. Based on this, it is reasonable to predict that the most sensitive tissues to TCDD would be those undergoing differentiation and proliferation, such as in the fetus or in the immune system even in adults. Furthermore, if the toxicity of xenobiotic AHR ligands is a reflection of dysregulated physiological functions, then the AHR is likely to have a normal role in those processes. Both of these generalizations appear to be appropriate. As indicated in Section 1.6, despite the challenging research path to decipher the exact molecular targets of TCDD–AHR and pathways leading to specific toxic phenotypes, a few have been identified, as recently reviewed [215]. Clearly, the dysregulated pathways are likely to be tissue and developmental stage specific and may involve multiple interacting pathways. Indeed, much of the available data on induced gene alterations indicates that this is the case [54, 56, 57, 189, 192, 212, 216, 324]. Many tools are now available to apply to this quest, so rapid expansion of our understanding of the molecular mechanisms of AHR-mediated toxicity is anticipated.

1.9.2 Step by Step, Chemical by Chemical, the Search for Endogenous AHR Ligands Continues

In 1984, a publication by Drs. Nebert, Eisen, and Hankinson asked the question of whether the AHR has "... specificity only for foreign chemicals" [325]. They concluded that the receptor may be "required for endogenous functions critical to life processes, as well as its function in the induction of drug metabolism." However, during the 1980s much of the effort was still directed to the toxicology associated with the AHR and the identification of xenobiotic AHR ligands (see Section 1.6.2). It really was not until the 1990s that many labs began an earnest search for other compounds, both dietary and physiological, that may serve as AHR agonists or antagonists. Many of those identified are dietary components as well as products and intermediates of normal metabolic pathways (reviewed in Refs 326 and 327). Of these, the flavonoids are the largest group of dietary ligands that have both agonist and antagonist activities [139, 326, 328]. However, the majority of these have AHR binding affinities that are orders of magnitude less than TCDD and potencies lower than might be expected for a physiological ligand.

About this same time, several indole-based compounds were identified that are the most promising of the putative physiological ligands. These include indigo and several photooxidation and/or metabolic products of tryptophan [327, 329–332]. The tryptophan derivatives are of particular interest. It was recognized as early as 1987 that several photooxidation products of tryptophan such as 6-formylindolo[3,2-*b*]carbazole (FICZ) bind to the AHR with very high affinity [333]. Subsequently, indolo[3,2-*b*] carbazole has been found in certain plants of the *Brassica* genus [334], and a putative secondary metabolite of tryptophan, 2-(1*H*-indole-3-carbonyl)-thiazole-4-carboxylic acid methyl ester, has been isolated from porcine lung [335]. Both have high affinity for the AHR [334, 336], although it is not clear whether the latter compound is actually produced in tissues or was a purification by-product. Several sulfoconjugates of FICZ have been found in human urine [337], strongly indicating the endogenous presence of tryptophan derivatives that could be AHR ligands. Although ultraviolet light and/or circadian rhythms may play a role in mediating AHR signaling by creating products of tryptophan, for better or worse, especially in the skin [338, 339], the Bradfield lab has demonstrated that both aspartate aminotransferase and D-amino acid oxidase can also generate AHR ligands from tryptophan [340–342]. Interestingly, indoleamine-2,3-dioxygenase (a major pathway of tryptophan metabolism notably in inflammation and in dendritic cells) is activated by TCDD, and its products, kynurenine and kynurenic acid, can activate AHR [343, 344]. Furthermore, AHR activation by kynurenine mediates FoxP3$^+$ Treg cell generation [344, 345].

Arachidonic acid derivatives are another class of endogenous substances that was discovered in the late 1990s to bind the AHR [346–349]. Tissue concentrations of these compounds can be as high as the μM range. Furthermore, the fact that many of these compounds are metabolized by cytochrome P450 isozymes regulated by the AHR, and are involved in inflammatory responses observed in TCDD-treated or AHR-KO mice, makes them attractive as *bona fide* endogenous AHR ligands [347, 350]. The ability of modified low-density lipoprotein to activate the AHR [351] could be mediated by products of arachidonic acid metabolism.

Although it is likely that one or more of these compounds, especially the tryptophan and arachidonic acid derivatives, will prove to be physiological ligands for the AHR, the definitive studies have yet to be performed. However, it is important to mention that the years of research on the AHR have provided convincing evidence of the existence, though not yet the identity, of endogenous ligands. For example, in the absence of known exogenous ligands, AHR-dependent regulation of the cell cycle [352, 353], activation of the AHR and induction of responsive genes under different conditions of cell culture [354–357], and cytoplasmic–nuclear shuttling of the AHR [358, 359] have been observed. There may be certain conditions during cell cycle, differentiation, or interactions with other cells in which endogenous ligands are produced. It is also possible that endogenous ligands may be generated by changes in the tissue environment that occur during tissue development, remodeling, repair, or during processes such as angiogenesis. This is particularly intriguing given that several members of the bHLH–PAS family are involved in sensing of changes in tissue environments [75]. However, although in most cases the transcriptional activity of the AHR is clearly ligand dependent, it is not yet known whether there are other pathways or intracellular signals, for example, posttranslational modifications such as phosphorylation, that may lead to AHR activation in the absence of ligand.

Given the ever-increasing knowledge of AHR physiology and ligands, there is also recent interest in whether any AHR ligands might have therapeutic uses or whether some therapeutic compounds already in use or under development may act, in part, by modulating AHR activity. In particular, several anti-inflammatory and antiallergic drugs have been found to have AHR agonist activity [360–363]. In support of the anti-inflammatory properties of some AHR ligands, the Perdew lab has shown that WAY-169916, a selective estrogen receptor modulator (SERM) with anti-inflammatory properties in models of rheumatoid arthritis, is also an AHR ligand. Furthermore, the suppression of acute phase response activity by this compound is dependent on AHR binding but not on ER- or AHRE-mediated gene expression [364]. The group went on to synthesize a derivative of WAY-169916 that lacked ER binding but retained high-affinity AHR binding as well as anti-inflammatory properties [365]. Very recently, groups at the Scripps Research Institute and Novartis Research Foundation, in collaboration with Gary Perdew and Mike Denison, found selective human AHR antagonists that were able to promote the expansion of human hematopoietic stem cells [366] and further suggested that this may facilitate the use of stem cell transplants. These observations are consistent with data published by the Gasiewicz and Kanno groups, indicating substantial effects of TCDD on hematopoietic stem cell characteristics and function [307, 367–370], as well as altered cycling of these cells from *Ahr*-null allele mice [368]. While together these data strongly suggest an important function of the AHR in the immune system, and particularly in hematopoietic stem cells, the exact signaling pathways involved have not yet been determined.

The interaction of some ligands with both the ER and AHR is of particular interest, especially since a number of studies, even from the 1970s, noted crosstalk between the AHR and ER signaling pathways. In 1978, a group led by Richard Kociba reported that the chronic dietary treatment of female rats with TCDD inhibited spontaneous mammary and uterine tumor formation [371]. Since that time, a number of investigators, mainly Steve Safe at Texas A&M, have observed crosstalk between these pathways at a number of different levels (reviewed in Ref. 372; see also Section 1.8). These studies eventually led Dr. Safe and his colleagues to develop several compounds that were selective aryl hydrocarbon receptor modulators (SARMs) inhibiting hormone-induced growth of ER-positive tumors, particularly breast cancer [373–376]. Recently, Richard Peterson, at the University of Wisconsin, showed that one of these compounds, 6-methyl-1,3,8-trichlorodibenzofuran, inhibited prostate tumor metastasis in a mouse model (TRAMP) that spontaneously develops prostate cancer [255]. In a variety of human breast cancer cell lines (ER-dependent and ER-independent) and in breast cancer stem-like cells, AHR activation inhibited invasive and metastatic activities [377, 378]. On the ER side, a recent publication from the McDonnell lab noted that 4-hydroxytamoxifen, an active metabolite of tamoxifen, directly binds to and modulates the transcriptional activity of the AHR [379]. Their studies further suggest that some of the effects originally attributed to the antiestrogenic action of tamoxifen may in fact be mediated through the AHR.

For purposes of defining AHR ligands that are toxic xenobiotics, endogenous mediators, or effective therapeutics, much information is likely to emerge from an analysis of the ligand binding domain. Although there is currently no information on three-dimensional structure for the AHR, in the past 10 years several groups have used mutational analysis and comparisons with structural data from homologous PAS-containing proteins to develop working models of the AHR LBD [380–382]. However, some of this modeling is based on the binding of xenobiotics such as TCDD, and it is not clear whether the parameters are also relevant for endogenous ligands. Furthermore, genes regulated by endogenous ligand-activated AHR may or may not overlap those regulated by xenobiotic ligand–AHR interactions. Indeed, some work suggests that the AHR regulates distinct TCDD-dependent and TCDD-independent gene batteries [285]. Recent work indicates that there may be some diversity in ligand selectivity designated by the human and mouse AHR [383, 384]. As such, modeling a rodent AHR may not accurately represent ligand activity toward the human AHR.

1.9.3 Knockout and Knock-In Show Glimpses of Physiological Functions

The development in the middle 1990s of several lines of mice lacking AHR [194, 196, 197] has proven to be another milestone in the quest to understand physiological functions of the AHR. Since that time, an extensive literature has developed in this area, complemented by the development of mice expressing a constitutively active [385, 386] or hypomorphic AHR [387], or with AHR that can be conditionally expressed [388]. Use of these models has suggested physiological roles of the AHR in angiogenesis and cardiovascular development and function [389–396], hematopoiesis and development and function of the immune system [57, 194, 263, 293, 294, 368, 370, 387, 397–402], melanogenesis [403], development of female reproductive tissues [404–409], mammary gland development [410], prostate development [411], maintenance of pregnancy [404], wound healing [412], tumorigenicity [413], and aging [414, 415]. However, use of *Ahr*-null mice, in which the AHR is lacking in all tissues, may lead to misinterpretations of primary versus secondary changes as a result of AHR loss. Using cell type-specific AHR deletion, for example, the Bradfield lab demonstrated that AHR within endothelial cells is necessary for normal development of liver vasculature, whereas AHR-mediated hepatotoxicity depends on expression of AHR in hepatocytes [388]. This group has likewise developed mice with analogous modifications of expression of the AHR chaperone protein, AIP [416, 417]. Cell type-specific effects of AIP deletion on vascular development and on TCDD-induced hepatotoxicity were consistent with the role of AIP in maintenance of AHR levels. Clearly, further study of the consequences of AHR loss in specific cell types will provide important new understanding of the physiology of the AHR.

1.10 SUMMARY AND FUTURE OF AHR RESEARCH

Less than 40 years ago, the AHR was a newly discovered binding protein associated with toxicity of a specific class of structurally related pollutants. Now it is additionally recognized as a ubiquitous transcription factor within a large family of related proteins that have important roles in development and cell signaling. Among the many significant milestones (Fig. 1.1) in the relatively short history of research focused on this protein, the following are surely highlights:

- discovery and initial characterization of the AHR protein (Fig. 1.2a);
- the understanding that the toxicity of a large class of xenobiotics is mediated by their binding to the AHR (Fig. 1.2b) provided an important tool for risk assessment of these chemicals;

- advances in biomedical research technology enabled elucidation of its activity as a ligand-activated transcription factor and revealed AHR to be a member of a large family of bHLH–PAS proteins that crosstalk with numerous other signaling pathways (Fig. 1.2a);
- further development of tools in molecular biology simplified the identification of AHR-mediated changes in gene expression and made possible the manipulation of the *Ahr* gene *in vitro* and *in vivo*.

From the extensive and detailed work reviewed throughout this book, it has became clear that not only is AHR a necessary mediator of essentially all toxic end points of TCDD, but it is also an important physiological signaling factor in many cells and tissues, notably during development (Fig. 1.2c).

Research interest in this protein and its functions will doubtless continue, and the spectrum of pathways that are discovered to have AHR involvement will doubtless further expand. Clearly the importance of AHR in toxicology will not wane, as identifying the key target genes and signaling pathways responsible for the diverse manifestations of toxicity of its xenobiotic ligands will be another major milestone in this history. However, possibly the most fruitful and intriguing directions for further study of AHR are expected to be in its basic biology. Determination of its endogenous ligand(s) and the regulators of its expression (temporally and spatially) will be important milestones not only for AHR research and toxicology, but also in our understanding of basic physiology and cell biology for which TCDD is an invaluable tool rather than just a toxic contaminant. Particular goals for further research include

- unraveling of the role(s) of AHR in normal development and how they may change during differentiation of various cell types;
- understanding in more detail the biology of AHR, including the critical target pathways impacted by its activation by toxic ligands, which may lead to development of therapeutic ligands or other means of perturbing pathways impacted by AHR-mediated toxicity or disease;
- investigating potential epigenetic mechanisms of AHR-mediated toxicity and endogenous signaling, especially in fetal development and differentiation;
- defining whether there are non-AHRE-dependent pathways of AHR response to either xenobiotic or endogenous signals;
- understanding which human diseases may have AHR-dependent etiology (besides or in addition to toxicity from xenobiotic ligands) or may be impacted by AHR-dependent pathways.

Perhaps in the not so distant future, another retrospective on AHR research will include milestones on the pathway toward such goals, as well as new insights from research directions that are currently unpredictable.

REFERENCES

1. Poland, A., Glover, E., and Kende, A. S. (1976). Stereospecific, high affinity binding of 2,3,7,8-tetrachlorodibenzo-*p*-dioxin by hepatic cytosol. *Journal of Biological Chemistry*, 251, 4936–4946.
2. Bradshaw, T. D. and Bell, D. R. (2009). Relevance of the aryl hydrocarbon receptor (AhR) for clinical toxicology. *Clinical Toxicology*, 47, 632–642.
3. Knerr, S. and Schrenk, D. (2006). Carcinogenicity of 2,3,7,8-tetrachlorodibenzo-*p*-dioxin in experimental models. *Molecular Nutrition & Food Research*, 50, 897–907.
4. Pohjanvirta, R. and Tuomisto, J. (1994). Short-term toxicity of 2,3,7,8-tetrachlorodibenzo-*p*-dioxin in laboratory animals: effects, mechanisms, and animal models. *Pharmacological Reviews*, 46, 483–549.
5. Schecter, A. and Gasiewicz, T. A. (2003). *Dioxins and Health*. Wiley, New York.
6. White, S. S. and Birnbaum, L. S. (2009). An overview of the effects of dioxins and dioxin-like compounds on vertebrates, as documented in human and ecological epidemiology. *Journal of Environmental Science and Health, Part C*, 27, 197–211.
7. Okey, A. B. (2007). An aryl hydrocarbon receptor odyssey to the shores of toxicology: the Deichmann Lecture, International Congress of Toxicology – XI. *Toxicological Sciences*, 98, 5–38.
8. Richardson, H. L., Stier, A. R., and Borsos-Natch-Nebel, E. (1952). Liver tumor inhibition and adrenal histologic responses in rats to which 3′-methyl-4-dimethyl-aminoazobenzene and 20-methylcholanthrene were simultaneously administered. *Cancer Research*, 12, 356–361.
9. Conney, A. H., Miller, E. C., and Miller, J. A. (1956). The metabolism of methylated amino-azo dyes. V. Evidence for induction of enzyme synthesis in the rat by 3-methylcholanthrene. *Cancer Research*, 16, 450–459.
10. Conney, A. H., Miller, E. C., and Miller, J. A. (1957). Substrate-induced synthesis and other properties of benzpyrene hydroxylase in rat liver. *Journal of Biological Chemistry*, 228, 753–766.
11. Conney, A. H., Gillette, J. R., Inscoe, J. K., Trams, E. R., and Posner, H. S. (1959). Induced synthesis of liver microsomal enzymes which metabolize foreign compounds. *Science 130*, 1478–1479.
12. Nebert, D. W. and Bausserman, L. L. (1970). Genetic differences in the extent of aryl hydrocarbon hydroxylase induction in mouse fetal cell cultures. *Journal of Biological Chemistry*, 245, 6373–6382.
13. Nebert, D. W. and Gelboin, H. V. (1969). The *in vivo* and *in vitro* induction of aryl hydrocarbon hydroxylase in mammalian cells of different species, tissues, strains, and developmental and hormonal status. *Archives of Biochemistry and Biophysics*, 134, 76–89.

14. Schmid, A., Elmer, I., and Tarnowski, G. S. (1969). Genetic determination of differential inflammatory reactivity and subcutaneous tumor susceptibility of AKR/J and C57BL/6J mice to 7,12-dimethylbenz[a]anthracene. *Cancer Research*, 29, 1585–1589.
15. Gielen, J. E., Goujon, F. M., and Nebert, D. W. (1972). Genetic regulation of aryl hydrocarbon hydroxylase induction. II. Simple Mendelian expression in mouse tissues in vivo. *Journal of Biological Chemistry*, 247, 1125–1137.
16. Nebert, D. W., Goujon, F. M., and Gielen, J. E. (1972). Aryl hydrocarbon hydroxylase induction by polycyclic hydrocarbons: simple autosomal dominant trait in the mouse. *Nature New Biology*, 236, 107–110.
17. Nebert, D. W., Negishi, M., Lang, M. A., Hjelmeland, L. M., and Eisen, H. J. (1982). The Ah locus, a multigene family necessary for survival in a chemically adverse environment: comparison with the immune system. *Advances in Genetics*, 21, 1–51.
18. Nebert, D. W. (1986). The 1986 Bernard B. Brodie award lecture. The genetic regulation of drug-metabolizing enzymes. *Drug Metabolism and Disposition*, 16, 1–7.
19. Poland, A. P., Smith, D., Metter, G., and Possick, P. (1971). A health survey of workers in a 2,4-D and 2,4,5-T plant with special attention to chloracne, porphyria cutanea tarda, and psychologic parameters. *Archives of Environmental Health*, 22, 316–327.
20. Kimmig, J. and Schulz, K. H. (1957). Occupational acne (so-called chloracne) due to chlorinated aromatic cyclic ethers. *Dermatologica*, 115, 540–546.
21. Schulz, K. H. (1957). Clinical and experimental studies on the etiology of chloracne. *Archives of Klinical and Experimental Dermatology*, 206, 589–596.
22. Poland, A. and Glover, E. (1973). 2,3,7,8-Tetrachlorodibenzo-p-dioxin: a potent inducer of δ-aminolevulinic acid synthetase. *Science*, 179, 476–477.
23. Poland, A. and Glover, E. (1973). Chlorinated dibenzo-p-dioxins: potent inducers of delta-aminolevulinic acid synthetase and aryl hydrocarbon hydroxylase. II. A study of the structure–activity relationship. *Molecular Pharmacology*, 9, 736–747.
24. Poland, A. and Glover, E. (1974). Comparison of 2,3,7,8-tetrachlorodibenzo-p-dioxin, a potent inducer of aryl hydrocarbon hydroxylase, with 3-methylcholanthrene. *Molecular Pharmacology*, 10, 349–359.
25. Poland, A., Glover, E., Robinson, J. R., and Nebert, D. W. (1974). Genetic expression of aryl hydrocarbon hydroxylase activity: induction of monooxygenase activities and cytochrome P1-450 formation by 2,3,7,8-tetrachlorodibenzo-p-dioxin in mice genetically "nonresponsive" to other aromatic hydrocarbons. *Journal of Biological Chemistry*, 249, 5599–5606.
26. Poland, A. and Glover, E. (1975). Genetic expression of aryl hydrocarbon hydroxylase by 2,3,7,8-tetrachlorodibenzo-p-dioxin: evidence for a receptor mutation in genetically non-responsive mice. *Molecular Pharmacology*, 11, 389–398.
27. Poland, A. and Kende, A. S. (1976). 2,3,7,8-Tetrachlorodibenzo-p-dioxin: environmental contaminant and molecular probe. *Federal Proceedings*, 35, 2404–2411.
28. Greenlee, W. F. and Poland, A. (1979). Nuclear uptake of 2,3,7,8-tetrachlorodibenzo-p-dioxin in C57BL/6J and DBA/2J mice. Role of the hepatic cytosol receptor protein. *Journal of Biological Chemistry*, 254, 9814–9821.
29. Okey, A. B., Bondy, G. P., Mason, M. E., Kahl, G. F., Eisen, H. J., Guenthner, T. M., and Nebert, D. W. (1979). Regulatory gene product of the Ah locus: characterization of the cytosolic inducer–receptor complex and evidence for its nuclear translocation. *Journal of Biological Chemistry*, 254, 11636–11648.
30. Perdew, G. H., Abbott, B., and Stanker, L. H. (1995). Production and characterization of monoclonal antibodies directed against the Ah receptor. *Hybridoma*, 14, 279–283.
31. Pollenz, R. S., Sattler, C. A., and Poland, A. (1994). The aryl hydrocarbon receptor and aryl hydrocarbon receptor nuclear translocator protein show distinct subcellular localizations in Hepa 1c1c7 cells by immunofluorescence microscopy. *Molecular Pharmacology*, 45, 428–438.
32. Carlstedt-Duke, J., Elstrom, G., Snochowski, M., Hogberg, B., and Gustafsson, J. A. (1978). Detection of the 2,3,7,8-tetrachlorodibenzo-p-dioxin (TCDD) receptor in rat liver by isoelectric focusing in polyacrylamide gels. *Toxicology Letters*, 2, 365–373.
33. Denison, M. S., Hamilton, J. W., and Wilkinson, C. F. (1985). Comparative studies of aryl hydrocarbon hyroxylase and the Ah receptor in non-mammalian species. *Comparative Biochemistry and Physiology, Part C*, 80, 319–324.
34. Denison, M. S., Vella, L. M., and Okey, A. B. (1986). Structure and function of the Ah receptor for 2,3,7,8-tetrachlorodibenzo-p-dioxin. Species differences in molecular properties of the receptors from mouse and rat hepatic cytosols. *Journal of Biological Chemistry*, 261, 3987–3995.
35. Denison, M. S. and Wilkinson, C. F. (1985). Identification of the Ah receptor in selected mammalian species and induction of aryl hydrocarbon hydroxylase. *European Journal of Biochemistry*, 147, 429–435.
36. Gasiewicz, T. A. and Bauman, P. A. (1987). Heterogeneity of the rat hepatic Ah receptor and evidence for transformation in vitro and in vivo. *Journal of Biological Chemistry*, 262, 2116–2120.
37. Gasiewicz, T. A. and Rucci, G. (1984). Cytosolic receptor for 2,3,7,8-tetrachlorodibenzo-p-dioxin. Evidence for a homologous nature among various mammalian species. *Molecular Pharmacology*, 26, 90–98.
38. Kahl, G. F., Freiderici, D. E., Bigelow, S. W., Okey, A. B., and Nebert, D. W. (1980). Ontogenic expression of regulatory and structural gene products associated with the Ah locus. Comparison of rat, mouse, rabbit and *Sigmoden hispedis*. *Developmental Pharmacology and Therapeutics*, 1, 137–162.
39. Manchester, D. K., Gordon, S. K., Golas, C. L., Roberts, E. A., and Okey, A. B. (1987). Ah receptor in human placenta: stabilization by molybdate and characterization of binding

of 2,3,7,8-tetrachlorodibenzo-*p*-dioxin, 3-methylcholanthrene, and benzo[*a*]pyrene. *Cancer Research*, 47, 4861–4868.

40. Okey, A. B., Bondy, G. P., Mason, M. E., Nebert, D. W., Forster-Gibson, C. J., Muncan, J., and Dufresne, M. J. (1980). Temperature-dependent cytosol-to-nucleus translocation of the Ah receptor for 2,3,7,8-tetrachlorodibenzo-*p*-dioxin in continuous cell culture lines. *Journal of Biological Chemistry*, 255, 11415–11422.

41. Okey, A. B., Vella, L. M., and Harper, P. A. (1989). Detection and characterization of a low affinity form of cytosolic Ah receptor in livers of mice nonresponsive to induction of cytochrome P1-450 by 3-methylcholanthrene. *Molecular Pharmacology*, 35, 823–830.

42. Roberts, E. A., Golas, C. L., and Okey, A. B. (1986). Ah receptor mediating induction of aryl hydrocarbon hydroxylase: detection in human lung by binding of 2,3,7,8-tetrachlorodibenzo-*p*-dioxin. *Cancer Research*, 46, 3739–3743.

43. Prokipcak, R. D. and Okey, A. B. (1988). Physicochemical characterization of the nuclear form of Ah receptor from mouse hepatoma cells exposed in culture to 2,3,7,8-tetrachlorodibenzo-*p*-dioxin. *Archives of Biochemistry and Biophysics*, 267, 811–828.

44. Hankinson, O. (1983). Dominant and recessive aryl hydrocarbon hydroxylase-deficient mutants of the mouse hepatoma line, Hepa-1, and assignment of the recessive mutants to three complementation groups. *Somatic and Cellular Genetics*, 9, 497–514.

45. Hankinson, O. (1979). Single-step selection of clones of a mouse hepatoma line deficient in aryl hydrocarbon hydroxylase. Proceedings of the National Academy of Sciences of the United States of America, 76, 373–376.

46. Legraverand, C., Hannah, R. R., Eisen, H. J., Owens, I. S., Nebert, D. W., and Hankinson, O. (1982). Regulatory gene product of the Ah locus. Characterization of receptor mutants among mouse hepatoma clones. *Journal of Biological Chemistry*, 257, 6402–6407.

47. Miller, A. G., Israel, D. I., and Whitlock, J. P., Jr. (1983). Biochemical and genetic analysis of variant mouse hepatoma cells defective in the induction of benzo(*a*)pyrene-metabolizing enzyme activity. *Journal of Biological Chemistry*, 258, 3523–3527.

48. Miller, A. G. and Whitlock, J. P., Jr. (1981). Novel variants in benzo(*a*)pyrene metabolism. Isolation by fluorescence-activated cell sorting. *Journal of Biological Chemistry*, 256, 2433–2437.

49. Hannah, R. R., Lund, J., Poellinger, L., Gillner, M., and Gustafsson, J. A. (1986). Characterization of the DNA-binding properties of the receptor for 2,3,7,8-tetrachlorodibenzo-*p*-dioxin. *European Journal of Biochemistry*, 156, 237–242.

50. Denison, M. S., Fisher, J. M., and Whitlock, J. P., Jr. (1988). The DNA recognition site for the dioxin–Ah receptor complex. Nucleotide sequence and functional analysis. *Journal of Biological Chemistry*, 263, 17221–17224.

51. Jones, P. B. C., Galeazzi, D. R., Fisher, J. M., and Whitlock, J. P., Jr. (1986). Control of cytochrome P_1-450 gene expression: analysis of a dioxin-responsive enhancer system. *Proceedings of the National Academy of Sciences of the United States of America*, 83, 2802–2806.

52. Whitlock, J. P., Jr. (1999). Induction of cytochrome P4501A1. *Annual Review of Pharmacology and Toxicology*, 39, 103–125.

53. Fujisawa-Sehara, A., Yamane, M., and Fujii-Kuriyama, Y. (1988). A DNA-binding factor specific for xenobiotic responsive elements of P-450c gene exists as a cryptic form in cytoplasm: its possible translocation to nucleus. *Proceedings of the National Academy of Sciences of the United States of America*, 85, 5859–5863.

54. Frericks, M., Burgoon, L. D., Zacharewski, T., and Esser, C. (2008). Promoter analysis of TCDD-inducible genes in a thymic epithelial cell line indicates the potential for cell-specific transcription factor cross-talk in the AhR response. *Toxicology and Applied Pharmacology*, 232, 268–279.

55. Gasiewicz, T. A., Henry, E. C., and Collins, L. L. (2008). Expression and activity of aryl hydrocarbon receptors in development and cancer. *Critical Reviews in Eukaryotic Gene Expression*, 18, 279–321.

56. Lee, K., Burgoon, L. D., Lamb, L., Dere, E., and Zacharewski, T. (2006). Identification and characterization of genes susceptible to transcriptional cross-talk between hypoxia and dioxin signaling cascades. *Chemical Research in Toxicology*, 19, 1284–1283.

57. Stevens, E. A., Mezrich, J. D., and Bradfield, C. A. (2009). The aryl hydrocarbon receptor: a perspective on potential roles in the immune system. *Immunology*, 127, 299–311.

58. Matikainen, T., Perez, G. I., Jurisicova, A., Pru, J. K., Schlezinger, J. J., Ryu, H.-Y., Laine, J., Sakai, T., Korsmeyer, S. J., Casper, R. F., Sherr, D. H., and Tilly, J. L. (2001). Aromatic hydrocarbon receptor-driven *Bax* gene expression is required for premature ovarian failure caused by biohazardous environmental chemicals. *Nature Genetics*, 28, 355–360.

59. Yao, E. F. and Denison, M. S. (1992). DNA sequence determinants for binding of transformed Ah receptor to a dioxin-responsive enhancer. *Biochemistry*, 31, 5060–5067.

60. Bank, P. A., Yao, E. F., Phelps, C. L., Harper, P. A., and Denison, M. S. (1992). Species-specific binding of transformed Ah receptor to a dioxin responsive transcriptional enhancer. *European Journal of Pharmacology*, 228, 85–94.

61. Sun, Y. V., Boverhof, D. R., Burgoon, L. D., Fielden, M. R., and Zacharewski, T. (2004). Comparative analysis of dioxin response elements in human, mouse and rat genomic sequences. *Nucleic Acids Research*, 32, 4512–4523.

62. Boutros, P. C., Moffat, I. D., Franc, M. A., Tijet, N., Tuomisto, J., Pohjanvirta, R., and Okey, A. B. (2004). Dioxin-responsive AHRE-II gene battery: identification by phylogenetic footprinting. *Biochemical and Biophysical Research Communications*, 321, 707–715.

63. Sogawa, K., Numayama-Tsuruta, K., Takahashi, T., Matsushita, N., Miura, C., Nikawa, J., Gotoh, O., Kikuchi, Y., and Fujii-Kuriyama, Y. (2004). A novel induction mechanism of the rat CYP1A2 gene mediated by Ah receptor–Arnt heterodimer. *Biochemical and Biophysical Research Communications*, 318, 746–755.

64. Gillesby, B., Santostefano, M., Porter, W., Wu, Z. F., Safe, S., and Zacharewski, T. (1997). Identification of a motif within the 5′-regulatory region on pS2 which is responsible for Ap1 binding and TCDD-mediated suppression. *Biochemistry*, 36, 6080–6089.

65. Krishnan, V., Porter, W., Santostefano, W., Wang, X., and Safe, S. (1995). Molecular mechanism of inhibition of estrogen-induced cathepsin D gene expression by 2,3,7,8-tetrachlorodibenzo-p-dioxin (TCDD) in MCF-7 cells. *Molecular and Cellular Biology*, 15, 6710–6719.

66. Ohtake, F., Takeyama, K., Matsumoto, T., Kitagawa, H., Yamamoto, Y., Nohara, K., Tohyama, C., Krust, A., Mimura, J., Chambon, P., Yanagisawa, J., Fujii-Kuriyama, Y., and Kato, S. (2003). Modulation of oestrogen receptor signaling by association with the activated dioxin receptor. *Nature*, 423, 545–550.

67. Porter, W., Wang, F., Duan, R., Qin, C., Castro-Rivera, E., Kim, K., and Safe, S. (2001). Transcriptional activation of heat shock protein 27 gene expression by 17β-estradiol and modulation by antiestrogens and aryl hydrocarbon receptor agonists. *Journal of Molecular Endocrinology*, 26, 31–42.

68. Jones, B. C. J., Galeazzi, D. R., Fisher, J. M., and Whitlock, J. P., Jr. (1985). Control of cytochrome P_1-450 gene expression by dioxin. *Science*, 227, 1499–1502.

69. Walsh, A. A., Tullis, K., Rice, R. H., and Denison, M. S. (1996). Identification of a novel *cis*-acting negative regulatory element affecting expression of the CYP1A1 gene in rat epidermal cells. *Journal of Biological Chemistry*, 271, 22746–22753.

70. Elferink, C. J., Gasiewicz, T. A., and Whitlock, J. P., Jr. (1990). Protein–DNA interactions at the dioxin-responsive enhancer. Evidence that the transformed Ah receptor is heteromeric. *Journal of Biological Chemistry*, 265, 20708–20712.

71. Gasiewicz, T. A., Elferink, C. J., and Henry, E. C. (1991). Characterization of multiple forms of the Ah receptor: recognition of a dioxin-responsive enhancer involves heteromer formation. *Biochemistry*, 30, 2909–2916.

72. Henry, E. C. and Gasiewicz, T. A. (1991). Inhibition and reconstitution of Ah receptor transformation *in vitro*: role and partial characterization of a cytosolic factor(s). *Archives of Biochemistry and Biophysics*, 288, 149–156.

73. Hoffman, E. C., Reyes, H., Chu, F. F., Sander, F., Conley, L. H., Brooks, B. A., and Hankinson, O. (1991). Identification of the Ah receptor nuclear translocator proteins (Arnt) as a component of the DNA binding form of the Ah receptor. *Science*, 252, 954–958.

74. Reyes, H., Reisz-Porszasz, S., and Hankinson, O. (1992). Identification of the Ah receptor nuclear translocator protein (Arnt) as a component of the DNA binding form of the Ah receptor. *Science*, 256, 1193–1195.

75. Gu, Y. Z., Hogenesch, J. B., and Bradfield, C. A. (2000). The PAS superfamily: sensors of environmental and developmental signals. *Annual Review of Pharmacology and Toxicology*, 40, 519–561.

76. Wang, G. L., Jiang, B. H., Rue, E. A., and Semenza, G. L. (1995). Hypoxia-inducible factor 1 is a basic-helix–loop–helix-PAS heterodimer regulated by cellular O_2 tension. *Proceedings of the National Academy of Sciences of the United States of America*, 92, 5510–5514.

77. Chan, W. K., Yao, G., Gu, Y. Z., and Bradfield, C. A. (1999). Cross-talk between the aryl hydrocarbon receptor and hypoxia inducible factor signaling pathways. Demonstration of competition and compensation. *Journal of Biological Chemistry*, 274, 12115–12123.

78. Schults, M. A., Timmermans, L., Godschalk, R. W., Theys, J., Wouters, B. G., van Schooten, F. J., and Chiu, R. K. (2010). Diminished carcinogen detoxification is a novel mechanism for hypoxia-inducible factor-1-mediated genetic instability. *Journal of Biological Chemistry*, 285, 14558–14564.

79. Siefert, A., Kaschinski, D. M., Tonack, S., Fisher, B., and Navarrente Santos, A. (2008). Significance of prolyl hydroxylase 2 in the interference of aryl hydrocarbon receptor and hypoxia-inducible factor-1 alpha signaling. *Chemical Research in Toxicology*, 21, 341–348.

80. Bradfield, C. A., Glover, E., and Poland, A. (1991). Purification and N-terminal amino acid sequence of the Ah receptor from the C57BL/6J mouse. *Molecular Pharmacology*, 39, 13–19.

81. Bradfield, C. A., Kende, A. S., and Poland, A. (1988). Kinetic and equilibrium studies of Ah receptor-ligand binding: use of [^{125}I]2-iodo-7,8-dibromodibenzo-p-dioxin. *Molecular Pharmacology*, 34, 229–237.

82. Perdew, G. H. and Poland, A. (1988). Purification of the Ah receptor from C57BL/6J mouse liver. *Journal of Biological Chemistry*, 263, 9848–9852.

83. Poland, A., Glover, E., and Bradfield, C. A. (1991). Characterization of polyclonal antibodies of the Ah receptor prepared by immunization with a synthetic peptide hapten. *Molecular Pharmacology*, 39, 20–26; Erratum, 39, 435.

84. Burbach, K. M., Poland, A., and Bradfield, C. A. (1992). Cloning of the Ah-receptor cDNA reveals a distinctive ligand-activated transcription factor. *Proceedings of the National Academy of Sciences of the United States of America*, 89, 8185–8189.

85. Ema, M., Sogawa, K., Watanabe, N., Chujoh, Y., Matsushita, N., Gotoh, O., Funae, Y., and Fujii-Kuriyama, Y. (1992). cDNA cloning and structure of mouse putative Ah receptor. *Biochemical and Biophysical Research Communications*, 184, 246–253.

86. Citri, Y., Colot, H. V., Jacquier, A. C., Yu, Q., Hall, J. C., Baltimore, D., and Rosbash, M. (1987). A family of unusually spliced biologically active transcripts encoded by a *Drosophila* clock gene. *Nature*, 326, 42–44.

87. Jackson, F. R., Bargiello, T. A., Yun, S. H., and Young, N. M. (1986). Product of per locus of *Drosophila* shares homology with proteoglycans. *Nature*, 320, 185–188.

88. Nambu, J. R., Lewis, J. O., Wharton, K. A., Jr. and Crews, S. T. (1991). The *Drosophila* single-minded gene encodes a helix–loop–helix protein that acts as a master regulator of CNS midline development. *Cell*, 67, 1157–1167.

89. Reddy, P., Jacquier, A. C., Abovich, N., Petersen, G., and Rosbash, M. (1986). The period clock locus of *D. melanogaster* codes for a proteoglycan. *Cell*, 46, 53–61.

90. Furness, S. G., Lees, M. J., and Whitelaw, M. L. (2007). The dioxin (aryl hydrocarbon) receptor as a model for adaptive responses of bHLH/PAS transcription factors. *FEBS Letters*, *581*, 3616–3625.

91. Antonsson, C., Whitelaw, M. L., McGuire, J., Gustafsson, J.-A., and Poellinger, L. (1995). Distinct roles of the molecular chaperone hsp90 in modulating dioxin receptor function via the basic helix–loop–helix and PAS domains. *Molecular and Cellular Biology*, *15*, 756–765.

92. Carver, L. A. and Bradfield, C. A. (1997). Ligand dependent interaction of the Ah receptor with a novel immunophilin homolog *in vivo*. *Journal of Biological Chemistry*, *272*, 11452–11456.

93. Carver, L. A., Jackiw, V., and Bradfield, C. A. (1994). The 90-kDa heat shock protein is essential for Ah receptor signaling in a yeast expression system. *Journal of Biological Chemistry*, *269*, 30109–30112.

94. Kazlauskas, A., Poellinger, L., and Pongratz, I. (1999). Evidence that the co-chaperone p23 regulates ligand responsiveness of the dioxin (aryl hydrocarbon) receptor. *Journal of Biological Chemistry*, *274*, 13519–13524.

95. LaPres, J. J., Glover, E., Dunham, E. E., Bunger, M. K., and Bradfield, C. A. (2000). ARA9 modifies agonist signaling through an increase in cytosolic aryl hydrocarbon receptor. *Journal of Biological Chemistry*, *275*, 6153–6159.

96. Ma, Q. and Whitlock, J. P., Jr. (1997). A novel cytoplasmic protein that interacts with the Ah receptor, contains tetratricopeptide repeat motifs, and augments the transcriptional response to 2,3,7,8-tetrachlorodibenzo-*p*-dioxin. *Journal of Biological Chemistry*, *272*, 8878–8884.

97. Meyer, B. K. and Perdew, G. H. (1999). Characterization of the AhR–hsp90–XAP2 core complex and the role of the immunophilin-related protein XAP2 in AhR stabilization. *Biochemistry*, *38*, 8907–8917.

98. Meyer, B. K., Pray-Grant, M., Vanden Heuvel, J. P., and Perdew, G. H. (1998). Hepatitis B virus X-associated protein 2 is a subunit of the unliganded aryl hydrocarbon receptor core complex and exhibits transcriptional enhancer activity. *Molecular and Cellular Biology*, *18*, 978–988.

99. Perdew, G. H. (1988). Association of the Ah receptor with 90-kDa heat shock protein. *Journal of Biological Chemistry*, *263*, 13802–13805.

100. Petrulis, J. R. and Perdew, G. H. (2002). The role of chaperone proteins in the aryl hydrocarbon receptor core complex. *Chemico-Biological Interactions*, *141*, 25–40.

101. Pongratz, I., Mason, G. G. F., and Poellinger, L. (1992). Dual roles of the 90 kDa heat shock protein hsp90 in modulating functional activities of the dioxin receptor. *Journal of Biological Chemistry*, *267*, 13728–13734.

102. Bacsi, S. G., Reisz-Porszasz, S., and Hankinson, O. (1995). Orientation of the heterodimeric aryl hydrocarbon (dioxin) receptor complex on its asymmetric DNA recognition sequence. *Molecular Pharmacology*, *47*, 432–438.

103. Dolwick, K. M., Swanson, H. I., and Bradfield, C. A. (1993). *In vitro* analysis of Ah receptor domains involved in ligand-activated DNA recognition. *Proceedings of the National Academy of Sciences of the United States of America*, *90*, 8566–8570.

104. Fukunaga, B. N., Probst, M. R., Reisz-Porszasz, S., and Hankinson, O. (1995). Identification of functional domains of the aryl hydrocarbon receptor. *Journal of Biological Chemistry*, *270*, 29270–29278.

105. Jain, S., Dolwick, K. M., Schmidt, J. V., and Bradfield, C. A. (1994). Potent transactivation domains of the Ah receptor and the Ah receptor nuclear translocator map to their carboxyl termini. *Journal of Biological Chemistry*, *269*, 31518–31524.

106. Pandini, A., Denison, M. S., Song, Y., Soshilov, A. A., and Bonati, L. (2007). Structural and functional characterization of the aryl hydrocarbon receptor ligand binding domain by homology modeling and mutational analysis. *Biochemistry*, *46*, 696–708.

107. Sogawa, K., Iwabuchi, K., Abe, H., and Fujii-Kuriyama, Y. (1995). Transcriptional activation domains of the Ah receptor and Ah receptor nuclear translocator. *Journal of Cancer Research and Clinical Oncology*, *121*, 612–620.

108. Swanson, H. I. and Yang, J. (1996). Mapping the protein/DNA contact sites of the Ah receptor and Ah receptor nuclear translocator. *Journal of Biological Chemistry*, *271*, 31657–31665.

109. Whitelaw, M. L., Gottlicher, M., Gustafsson, J. A., and Poellinger, L. (1993). Definition of a novel ligand binding domain of nuclear bHLH receptor: co-localization of ligand and hsp90 binding activities with the regulable inactivation domain of the dioxin receptor. *EMBO Journal*, *12*, 4169–4179.

110. Whitelaw, M. L., Gustafsson, J. A., and Poellinger, L. (1994). Identification of transactivation and repression functions of the dioxin receptor and its basic helix–loop–helix/PAS partner factor Arnt: inducible versus constitutive modes of regulation. *Molecular and Cellular Biology*, *14*, 8343–8355.

111. Poellinger, L., Lund, J., Gillner, M., Hansson, L. A., and Gustafsson, J.-A. (1983). Physicochemical characterization of specific and nonspecific polyaromatic hydrocarbon binders in rat and mouse liver cytosol. *Journal of Biological Chemistry*, *258*, 13535–13542.

112. Hahn, M. E., Poland, A., Glover, E., and Stegeman, J. J. (1994). Photoaffinity labeling of the Ah receptor: phylogenetic survey of diverse vertebrate and invertebrate species. *Archives of Biochemistry and Biophysics*, *310*, 218–228.

113. Poland, A. and Glover, E. (1990). Characterization and strain distribution pattern of the murine Ah receptor specified by the Ah^d and Ah^{b-3} alleles. *Molecular Pharmacology*, *38*, 306–312.

114. Poland, A. and Glover, E. (1987). Variation in the molecular mass of the Ah receptor among vertebrate species and strains of rats. *Biochemical and Biophysical Research Communications*, *146*, 1439–1449.

115. Ema, M., Ohe, N., Suzuki, M., Mimura, J., Sogawa, K., Ikawa, S., and Fujii-Kuriyama, Y. (1994). Dioxin binding activities of polymorphic forms of mouse and human arylhydrocarbon receptors. *Journal of Biological Chemistry*, *269*, 27337–27343.

116. Hahn, M. E. (1998). The aryl hydrocarbon receptor: a comparative perspective. *Comparative Biochemistry and Physiology, Part C*, *121*, 23–53.

117. Hahn, M. E. (2002). Aryl hydrocarbon receptors: diversity and evolution. *Chemico-Biological Interactions*, *141*, 131–160.

118. Hahn, M. E., Karchner, S. I., Evans, B. R., Franks, D. G., Merson, R. R., and Lapseritis, J. M. (2006). Unexpected diversity of aryl hydrocarbon receptors in non-mammalian vertebrates: insights from comparative genomics. *Journal of Experimental Zoology, Part A*, *305*, 693–706.

119. Korkalainen, M., Tuomisto, J., and Pohjanvirta, R. (2001). The AH receptor of the most dioxin-sensitive species, guinea pig, is highly homologous to the human AH receptor. *Biochemical and Biophysical Research Communications*, *285*, 1121–1129.

120. Korkalainen, M., Tuomisto, J., and Pohjanvirta, R. (2000). Restructured transactivation domain in hamster AH receptor. *Biochemical and Biophysical Research Communications*, *285*, 272–281.

121. Pohjanvirta, R., Wong, J. M., Li, W., Harper, P. A., Tuomisto, J., and Okey, A. B. (1998). Point mutation in intron sequence causes altered carboxyl-terminal structure in the aryl hydrocarbon receptor of the most 2,3,7,8-tetrachlorodibenzo-*p*-dioxin-resistant rat strain. *Molecular Pharmacology*, *54*, 86–93.

122. Dolwick, K. M., Schmidt, J. V., Carver, L. A., Swanson, H. I., and Bradfield, C. A. (1993). Cloning and expression of a human Ah receptor cDNA. *Molecular Pharmacology*, *44*, 911–917.

123. Poland, A. and Glover, E. (1980). 2,3,7,8-Tetrachlorodibenzo-*p*-dioxin:segregation of toxicity with the Ah locus. *Molecular Pharmacology*, *17*, 86–94.

124. Poland, A. and Knutson, J. C. (1982). 2,3,7,8-Tetrachlorodibenzo-*p*-dioxin and related halogenated aromatic hydrocarbons: examination of the mechanism of toxicity. *Annual Review of Pharmacology and Toxicology*, *22*, 517–554.

125. Bandiera, S., Sawyer, T., Romkes, M., Zmudzka, B., Safe, L., Mason, G. G., Keys, B., and Safe, S. (1984). Polychlorinated dibenzofurans (PCDFs): effects of structure on binding to the 2,3,7,8-TCDD cytosolic receptor protein, AHH induction and toxicity. *Toxicology 32*, 131–144.

126. Davis, D. and Safe, S. (1988). Immunosuppressive activities of polychlorinated dibenzofuran congeners: quantitative structure–activity relationships and interactive effects. *Toxicology and Applied Pharmacology*, *94*, 141–149.

127. Kerkvliet, N. I., Baecher-Steppan, L., Smith, B. B., Youngberg, J. A., Henderson, M. C., and Buhler, D. R. (1990). Role of the Ah locus in suppression of cytotoxic T lymphocyte activity by halogenated aromatic hydrocarbons (PCBs and TCDD): structure–activity relationships and effects in C57Bl/6 mice congenic at the Ah locus. *Fundamental and Applied Toxicology*, *14*, 532–541.

128. Mason, G. G., Farrell, K., Keys, B., Piskorska-Pliszczynska, J., Safe, L., and Safe, S. (1986). Polychlorinated dibenzo-*p*-dioxins: quantitative *in vitro* and *in vivo* structure–activity relationships. *Toxicology*, *41*, 21–31.

129. Romkes, M., Piskorska-Pliszczynska, J., Keys, B., Safe, S., and Fujita, T. (1987). Quantitative structure–activity relationships: analysis of interactions of 2,3,7,8-tetrachlorodibenzo-*p*-dioxin and 2-substituted analogues with rat, mouse, guinea pig, and hamster cytosolic receptor. *Cancer Research*, *47*, 5108–5111.

130. Safe, S., Bandiera, S., Sawyer, T., Zmudzka, B., Mason, G. G., Romkes, M., Denomme, M. A., Sparling, J., Okey, A. B., and Fujita, T. (1985). Effects of structure on binding to the 2,3,7,8-TCDD receptor protein and AHH induction: halogenated biphenyls. *Environmental Health Perspectives*, *61*, 21–33.

131. Keys, B., Piskorska-Pliszczynska, J., and Safe, S. (1986). Polychlorinated dibenzofurans as 2,3,7,8-TCDD antagonists: *in vitro* inhibition of monooxygenase enzyme induction. *Toxicology Letters*, *31*, 151–158.

132. Astroff, B. and Safe, S. (1989). 6-Substituted-1,3,8-trichlorodibenzofurans as 2,3,7,8-tetrachlorodibenzo-*p*-dioxin antagonists in the rat: structure activity relationships. *Toxicology 59*, 285–296.

133. Harris, M., Zacharewski, T., Astroff, B., and Safe, S. (1989). Partial antagonism of 2,3,7,8-tetrachlorodibenzo-*p*-dioxin-mediated induction of aryl hydrocarbon hydroxylase by 6-methyl-1,3,8-trichlorodibenzofuran: mechanistic studies. *Molecular Pharmacology*, *35*, 729–735.

134. Lu, Y.-F., Santostefano, M., Cunningham, B. D. M., Threadgill, M. D., and Safe, S. (1995). Identification of 3′-methoxy-4′-nitroflavone as a pure aryl hydrocarbon (Ah) receptor antagonist and evidence for more than one form of the nuclear Ah receptor in MCF-7 human breast cancer cells. *Archives of Biochemistry and Biophysics*, *316*, 470–477.

135. Casper, R. F., Quesne, M., Rogers, I. M., Shirota, T., Jolivet, A., Milgrom, E., and Savouret, J.-F. (1999). Resveratrol has antagonistic activity on the aryl hydrocarbon receptor: implications for prevention of dioxin toxicity. *Molecular Pharmacology*, *56*, 784–790.

136. Chen, G. and Bunce, N. J. (2003). Polybrominated diphenyl ethers as Ah receptor agonists and antagonists. *Toxicological Sciences*, *76*, 310–320.

137. Chen, I., Safe, S., and Bjeldanes, L. (1996). Indole-3-carbinol and diindolylmethane as aryl hydrocarbon (Ah) receptor agonists and antagonists in T47D human breast cancer cells. *Biochemical Pharmacology*, *51*, 1069–1076.

138. Fukuda, I., Kaneko, A., Nishiumi, S., Kawase, M., Nishikiori, R., Fujitake, N., and Ashida, H. (2009). Structure–activity relationships of anthraquinones on the suppression of DNA-binding activity of the aryl hydrocarbon receptor induced by 2,3,7,8-tetrachlorodibenzo-*p*-dioxin. *Journal of Bioscience and Bioengineering*, *107*, 296–300.

139. Gasiewicz, T. A., Kende, A. S., Rucci, G., Whitney, B., and Willey, J. (1996). Analysis of structural requirements for Ah receptor antagonist activity: ellipticines, flavones, and related compounds. *Biochemical Pharmacology*, *52*, 1787–1803.

140. Kim, S.-H., Henry, E. C., Kim, D.-K., Kim, Y.-H., Shin, K. J., Han, M. S., Lee, T. G., Kang, J.-K., Gasiewicz, T. A., Ryu, S. H., and Suh, P.-G. (2006). Novel compound 2-methyl-2*H*-pyrazole-3-carboxylic acid (2-methyl-4-*o*-tolylazo-phenyl)-

amide (CH-223191) prevents 2,3,7,8-TCDD-induced toxicity by antagonizing the aryl hydrocarbon receptor. *Molecular Pharmacology, 69,* 1871–1878.
141. Lu, Y.-F., Santostefano, M., Cunningham, B. D. M., Threadgill, M. D., and Safe, S. (1996). Substituted flavones as aryl hydrocarbon (Ah) receptor agonists and antagonists. *Biochemical Pharmacology, 51,* 1077–1087.
142. Puppala, D., Lee, H., Kim, K. B., and Swanson, H. I. (2008). Development of an aryl hydrocarbon receptor antagonist using the proteolysis-targeting chimeric molecules approach: a potential tool for chemoprevention. *Molecular Pharmacology, 73,* 1064–1071.
143. Gasiewicz, T. A. and Rucci, G. (1991). Alpha-naphthoflavone acts as an antagonist of 2,3,7,8-tetrachlorodibenzo-*p*-dioxin by forming an inactive complex with the Ah receptor. *Molecular Pharmacology, 40,* 607–612.
144. Henry, E. C., Kende, A. S., Rucci, G., Totleben, M. J., Willey, J., Dertinger, S. D., Pollenz, R. S., Jones, J. P., and Gasiewicz, T. A. (1999). Flavone antagonists bind competitively with 2,3,7,8-tetrachlorodibenzo-*p*-dioxin (TCDD) to the aryl hydrocarbon receptor but inhibit transformation. *Molecular Pharmacology, 55,* 716–725.
145. Nishiumi, S., Yoshida, K., and Ashida, H. (2007). Curcumin suppresses the transformation of an aryl hydrocarbon receptor through its phosphorylation. *Archives of Biochemistry and Biophysics, 466,* 267–273.
146. Palermo, C. M., Westlake, C. A., and Gasiewicz, T. A. (2005). Epigallocatechin gallate inhibits aryl hydrocarbon receptor gene transcription through an indirect mechanism involving binding to a 90 kDa heat shock protein. *Biochemistry 44,* 5041–5052.
147. Zhang, S., Qin, C., and Safe, S. (2003). Flavonoids as aryl hydrocarbon receptor agonists/antagonists: effects of structure and cell context. *Environmental Health Perspectives, 111,* 1877–1882.
148. Zhou, J.-G. and Gasiewicz, T. A. (2003). 3′-Methoxy-4′-nitroflavone, a reported aryl hydrocarbon receptor antagonist, enhances Cyp1a1 transcription by a dioxin responsive element-dependent mechanism. *Archives of Biochemistry and Biophysics, 416,* 68–80.
149. Aarts, J. M., Denison, M. S., Cox, M. A., Schalk, M. A., Garrison, P. M., Tullis, K., de Haan, L. H., and Brouwer, A. (1995). Species-specific antagonism of Ah receptor action by 2,2′,5,5′-tetrachloro- and 2,2′,3,3′,4,4′-hexachlorobiphenyl. *European Journal of Pharmacology, 293,* 463–474.
150. Henry, E. C. and Gasiewicz, T. A. (2008). Molecular determinants of species-specific agonist and antagonist activity of a substituted flavone towards the aryl hydrocarbon receptor. *Archives of Biochemistry and Biophysics, 472,* 77–88.
151. Hesterman, E. V., Stegeman, J. J., and Hahn, M. E. (2000). Relative contributions of affinity and intrinsic efficacy to aryl hydrocarbon receptor ligand potency. *Toxicology and Applied Pharmacology, 168,* 160–172.
152. Henry, E. C. and Gasiewicz, T. A. (2003). Agonist but not antagonist ligands induce conformational change in the mouse aryl hydrocarbon receptor as detected by partial proteolysis. *Molecular Pharmacology, 63,* 392–400.
153. Davis, D. and Safe, S. (1991). Halogenated aryl hydrocarbon-induced suppression of the *in vitro* plaque-forming cell response to sheep red blood cells is not dependent on the Ah receptor. *Immunopharmacology 21,* 183–190.
154. Kerkvliet, N. I., Steppan, L. B., Brauner, J. A., Deyo, J. A., Henderson, M. C., Tomar, R. S., and Buhler, D. R. (1990). Influence of the Ah locus on the humoral immunotoxicity of 2,3,7,8-tetrachlorodibenzo-*p*-dioxin (TCDD): evidence for Ah receptor-dependent and Ah receptor-independent mechanisms of immunosuppression. *Toxicology and Applied Pharmacology, 105,* 26–36.
155. Morris, D. L. and Holsapple, M. P. (1991). Effects of 2,3,7,8-tetrachlorodibenzo-*p*-dioxin (TCDD) on humoral immunity. 2. B cell activation. *Immunopharmacology 21,* 171–181.
156. Morris, D. L., Jordan, S. D., and Holsapple, M. P. (1991). Effects of 2,3,7,8-tetrachlorodibenzo-*p*-dioxin (TCDD) on humoral immunity. 1. Similarities to *Staphylococcus aureus* Cowan Strain I (SAC) in the *in vitro* T-dependent antibody response. *Immunopharmacology 21,* 159–169.
157. Morris, D. L., Snyder, N. K., Gokani, R. E., Blair, R. E., and Holsapple, M. P. (1992). Enhanced suppression of humoral immunity in DBA/2 mice following subchronic exposure to 2,3,7,8-tetrachlorodibenzo-*p*-dioxin (TCDD). *Toxicology and Applied Pharmacology, 112,* 128–132.
158. Tucker, A. N., Vore, S. J., and Luster, M. I. (1986). Suppression of B cell differentiation by 2,3,7,8-tetrachlorodibenzo-*p*-dioxin. *Molecular Pharmacology, 29,* 372–377.
159. NATO/CCMS (1988). Pilot study on international information exchange on dioxins and related compounds. International toxicity equivalent factors (I-TEF) method of risk assessment for complex mixtures of dioxins and related compounds. Report No. 176, pp. 1–26.
160. Safe, S. (1990). Polychlorinated biphenyls (PCBs), dibenzo-*p*-dioxins (PCDDs), dibenzofurans (PCDFs), and related compounds: environmental and mechanistic considerations which support the development of toxic equivalency factors (TEFs). *Critical Reviews in Toxicology, 21,* 51–88.
161. Van den Berg, M., Birnbaum, L. S., Denison, M. S., DeVito, M. J., Farland, W., Feeley, M., Fiedler, H., Hakansson, H., Hanberg, A., Haws, L., Rose, M., Safe, S., Schrenk, D., Tohyama, C., Tritscher, A., Tuomisto, J., Tysklind, M., Walker, N., and Peterson, R. E. (2006). The 2005 World Health Organization reevaluation of human and mammalian toxic equivalency factors for dioxins and dioxin-like compounds. *Toxicological Sciences, 93,* 223–241.
162. Birnbaum, L. S. and DeVito, M. J. (1995). Use of toxic equivalency factors for risk assessment for dioxins and related compounds. *Toxicology 105,* 391–401.
163. Gao, X., Son, D. S., Terranova, P. F., and Rozman, K. K. (1999). Toxic equivalency factors of polychlorinated dibenzo-*p*-dioxins in an ovulation model: validation of the toxic equivalency concept for one aspect of endocrine disruption. *Toxicology and Applied Pharmacology, 157,* 107–116.

164. Hamm, J. T., Chen, C. Y., and Birnbaum, L. S. (2003). A mixture of dioxins, furans, and non-*ortho* PCBs based on consensus toxic equivalency factors produces dioxin-like reproductive effects. *Toxicological Sciences, 74*, 182–191.

165. Hornung, M. W., Zabel, E. W., and Peterson, R. E. (1996). Additive interactions between pairs of polybrominated dibenzo-*p*-dioxin, dibenzofuran, and biphenyl congeners in a rainbow trout early life stage mortality bioassay. *Toxicology and Applied Pharmacology, 140*, 345–355.

166. Simanainen, U., Tuomisto, J. T., Tuomisto, J., and Viluksela, M. (2002). Structure–activity relationships and dose responses of polychlorinated dibenzo-*p*-dioxins for short-term effects in 2,3,7,8-tetrachlorodibenzo-*p*-dioxin-resistant and -sensitive rat strains. *Toxicology and Applied Pharmacology, 181*, 38–47.

167. Battershill, J. M. (1994). Review of the safety assessment of polychlorinated biphenyls (PCBs) with particular reference to reproductive toxicity. *Human and Experimental Toxicology, 13*, 581–597.

168. Finley, B. L., Conner, K. T., and Scott, P. K. (2003). The use of toxic equivalency factors in probabilistic risk assessment for dioxins, furans, and PCBs. *Journal of Toxicology and Environmental Health, Part A, 66*, 533–550.

169. Gray, M. N., Aylward, L. L., and Keenan, R. E. (2006). Relative cancer potencies of selected dioxin-like compounds on a body-burden basis: comparison to current toxic equivalency factors (TEFs). *Journal of Toxicology and Environmental Health, Part A, 69*, 907–917.

170. Pohjanvirta, R., Unkila, M., Linden, J., Tuomisto, J. T., and Tuomisto, J. (1995). Toxic equivalency factors do not predict the acute toxicities of dioxins in rats. *European Journal of Pharmacology, 293*, 341–353.

171. Starr, T. B., Greenlee, W. F., Neal, R. A., Poland, A., and Sutter, T. R. (1999). The trouble with TEFs. *Environmental Health Perspectives, 107*, A492–A493.

172. Van den Berg, M., Birnbaum, L. S., Bosveld, A. T. C., Brunstrom, B., Cook, P., Feeley, M., Giesy, J. P., Hanberg, A., Hasegawa, R., Kennedy, S. W., Kubiak, T., Larsen, J. C., van Leeuwen, F. X. R., Liem, A. K. D., Nolt, C., Peterson, R. E., Poellinger, L., Safe, S., Schrenk, D., Tillitt, D., Tysklind, M., Younes, M., Waern, F., and Zacharewski, T. (1998). Toxic equivalency factors (TEFs) for PCBs, PCDDs, PCDFs for humans and wildlife. *Environmental Health Perspectives, 106*, 775–792.

173. Howard, G. J., Schlizinger, J. J., Hahn, M. E., and Webster, T. F. (2010). Generalized concentration addition predicts joint effects of aryl hydrocarbon receptor agonists with partial agonists and competitive antagonists. *Environmental Health Perspectives, 118*, 666–672.

174. Goldstone, H. M. and Stegeman, J. J. (2006). Molecular mechanisms of 2,3,7,8-tetrachlorodibenzo-*p*-dioxin cardiovascular embryotoxicity. *Drug Metabolism Reviews, 38*, 261–289.

175. Nebert, D. W. and Dalton, T. P. (2006). The role of cytochrome P450 enzymes in endogenous signalling pathways and environmental carcinogenesis. *Nature Reviews Cancer, 6*, 947–960.

176. Nebert, D. W., Dalton, T. P., Okey, A. B., and Gonzalez, F. J. (2004). Role of aryl hydrocarbon receptor-mediated induction of the CYP1 enzymes in environmental toxicity and cancer. *Journal of Biological Chemistry, 279*, 23847–23850.

177. Reichard, J. F., Dalton, T. P., Shertzer, H. G., and Puga, A. (2006). Induction of oxidative stress responses by dioxin and other ligands of the aryl hydrocarbon receptor. *Dose Response, 1*, 306–331.

178. Smith, A. G., Clothier, B., Carthew, P., Childs, N. L., Sinclair, P. R., Nebert, D. W., and Dalton, T. P. (2001). Protection of the Cyp1a2(−/−) null mouse against uroporphyria and hepatic injury following exposure to 2,3,7,8-tetrachlorodibenzo-*p*-dioxin. *Toxicology and Applied Pharmacology, 173*, 89–98.

179. Gasiewicz, T. A., Rucci, G., Henry, E. C., and Baggs, R. B. (1986). Changes in hamster hepatic cytochrome P-450, ethoxycoumarin O-deethylase, and reduced NAD(P):menadione oxidoreductase following treatment with 2,3,7,8-tetrachlorodibenzo-*p*-dioxin. Partial dissociation of temporal and dose–response relationships from elicited toxicity. *Biochemical Pharmacology, 35*, 2737–2742.

180. Carney, S. A., Peterson, R. E., and Heideman, W. (2004). 2,3,7,8-Tetrachlorodibenzo-*p*-dioxin activation of the aryl hydrocarbon receptor/aryl hydrocarbon receptor nuclear translocator pathway causes developmental toxicity through a CYP1A1-independent mechanism in zebrafish. *Molecular Pharmacology, 66*, 512–521.

181. Sutter, T. R., Guzman, K., Dold, K. M., and Greenlee, W. F. (1991). Targets for dioxin: genes for plasminogen activator inhibtor-2 and interleukin-1 beta. *Science, 254*, 415–418.

182. Boverhof, D. R., Burgoon, L. D., Tashiro, C., Chittim, B., Harkema, J. R., Jump, D. B., and Zacharewski, T. R. (2005). Temporal and dose-dependent hepatic gene expression patterns in mice provide new insights into TCDD-mediated hepatotoxicity. *Toxicological Sciences, 85*, 1048–1063.

183. Boverhof, D. R., Burgoon, L. D., Tashiro, C., Sharratt, B., Chittim, B., Harkema, J. R., Mendrick, D. L., and Zacharewski, T. R. (2006). Comparative toxicogenomic analysis of the hepatotoxic effects of TCDD in Sprague Dawley rats and C57BL/6 mice. *Toxicological Sciences, 94*, 398–416.

184. Fletcher, N., Wahlstrom, D., Lundberg, R., Nilsson, C. B., Nilsson, K. C., Stockling, K., Hellmold, H., and Hakansson, H. (2005). 2,3,7,8-Tetrachlorodibenzo-*p*-dioxin (TCDD) alters the mRNA expression of critical genes associated with cholesterol metabolism, bile acid synthesis, and bile transport in rat liver: a microarray study. *Toxicology and Applied Pharmacology, 207*, 1–24.

185. Frueh, F. W., Hayashibara, K. C., Brown, P. O., and Whitlock, J. P., Jr. (2001). Use of cDNA microarrays to analyze dioxin-induced changes in human liver gene expression. *Toxicology Letters, 122*, 189–203.

186. Kim, S., Dere, E., Burgoon, L. D., Chang, C. C., and Zacharewski, T. R. (2009). Comparative analysis of AhR-mediated TCDD-elicited gene expression in human liver adult stem cells. *Toxicological Sciences, 112*, 229–244.

187. Laiosa, M. D., Mills, J. H., Lai, Z. W., Singh, K. P., Middleton, F. A., Gasiewicz, T. A., and Silverstone, A. E. (2010).

Identification of stage-specific gene modulation during early thymocyte development by whole-genome profiling analysis after aryl hydrocarbon receptor activation. *Molecular Pharmacology*, 77, 773–783.

188. Martinez, J. M., Afshari, C. A., Bushel, P. R., Masuda, A., Takahashi, T., and Walker, N. J. (2002). Differential toxicogenomic responses to 2,3,7,8-tetrachlorodibenzo-*p*-dioxin in malignant and nonmalignant human airway epithelial cells. *Toxicological Sciences*, 69, 409–423.

189. Puga, A., Maier, A., and Medvedovic, M. (2000). The transcriptional signature of dioxin in human hepatoma HepG2 cells. *Biochemical Pharmacology*, 60, 1129–1142.

190. Thackaberry, E. A., Nunez, B. A., Ivnitski-Steele, I. D., Friggins, M., and Walker, M. K. (2005). Effect of 2,3,7,8-tetrachlorodibenzo-*p*-dioxin on murine heart development: alteration in fetal and postnatal cardiac growth, and postnatal cardiac chronotropy. *Toxicological Sciences*, 88, 242–249.

191. Thomas, R. S., Rank, D. R., Penn, S. G., Zastrow, G. M., Hayes, K. R., Pande, K., Glover, E., Silander, T., Craven, M. W., Reddy, J. K., Jovanovich, S. B., and Bradfield, C. A. (2001). Identification of toxicologically predictive gene sets using cDNA microarrays. *Molecular Pharmacology*, 60, 1189–1194.

192. Johnson, C. D., Balagurunathan, Y., Tadesse, M. G., Falahatpisheh, M. H., Brun, M., Walker, M. K., Dougherty, E. R., and Ramos, K. S. (2004). Unraveling gene–gene interactions regulated by ligands of the aryl hydrocarbon receptor. *Environmental Health Perspectives*, 112, 403–412.

193. Sartor, M. A., Schnekenburger, M., Marlowe, J. L., Reichard, J. F., Wang, Y., Fan, Y., Ma, C., Karyala, S., Halbleib, D., Liu, X., Medvedovic, M., and Puga, A. (2009). Genomewide analysis of aryl hydrocarbon receptor binding targets reveals an extensive array of gene clusters that control morphogenetic and developmental programs. *Environmental Health Perspectives*, 117, 1139–1146.

194. Fernandez-Salguero, P. M., Pineau, T., Hilbert, D. M., McPhail, T., Lee, S. S. T., Kimura, S., Nebert, D. W., Rudikoff, S., Ward, J. M., and Gonzalez, F. J. (1995). Immune system impairment and hepatic fibrosis in mice lacking the dioxin-binding Ah receptor. *Science*, 268, 722–726.

195. Lahvis, G. P. and Bradfield, C. A. (1998). Ahr null alleles: distinctive or different? *Biochemical Pharmacology*, 56, 781–787.

196. Mimura, J., Yamashita, K., Nakamura, K., Morita, M., Takagi, T. N., Nakao, K., Ema, M., Sogawa, K., Yasuda, M., Katsuki, M., and Fujii-Kuriyama, Y. (1997). Loss of teratogenic response to 2,3,7,8-tetrachlorodibenzo-*p*-dioxin (TCDD) in mice lacking the Ah (dioxin) receptor. *Genes Cells*, 2, 645–654.

197. Schmidt, J. V., Huei-Ting, G., Reddy, J. K., Simon, M. C., and Bradfield, C. A. (1996). Characterization of a murine Ahr null allele: involvement of the Ah receptor in hepatic growth and development. *Proceedings of the National Academy of Sciences of the United States of America*, 93, 6731–6736.

198. Buchanan, D. L., Sato, T., Peterson, R. E., and Cooke, P. S. (2002). Antiestrogenic effects of 2,3,7,8-tetrachlorodibenzo-*p*-dioxin in mouse uterus: critical role of the aryl hydrocarbon receptor in stromal tissue. *Toxicological Sciences*, 57, 302–311.

199. Fernandez-Salguero, P. M., Hilbert, D. M., Rudikoff, S., Ward, J. M., and Gonzalez, F. J. (1996). Aryl-hydrocarbon receptor-deficient mice are resistant to 2,3,7,8-tetrachlorodibenzo-*p*-dioxin-induced toxicity. *Toxicology and Applied Pharmacology*, 140, 173–179.

200. Kerkvliet, N. I., Shepherd, D. M., and Baecher-Steppan, L. (2002). T lymphocytes are direct, aryl hydrocarbon receptor (AhR)-dependent targets of 2,3,7,8-tetrachlorodibenzo-*p*-dioxin (TCDD): AhR expression in both $CD4^+$ and $CD8^+$ T cells is necessary for full suppression of a cytotoxic T lymphocyte response by TCDD. *Toxicology and Applied Pharmacology*, 185, 146–152.

201. Ko, K., Moore, R. W., and Peterson, R. E. (2004). Aryl hydrocarbon receptors in urogenital sinus mesenchyme mediate the inhibition of prostatic epithelial bud formation by 2,3,7,8-tetrachlorodibenzo-*p*-dioxin. *Toxicology and Applied Pharmacology*, 96, 149–155.

202. Lin, T.-M., Ko, K., Moore, R. W., Buchanan, D. L., Cooke, P. S., and Peterson, R. E. (2001). Role of the aryl hydrocarbon receptor in the development of control and 2,3,7,8-tetrachlorodibenzo-*p*-dioxin-exposed male mice. *Journal of Toxicology and Environmental Health, Part A*, 64, 327–342.

203. Peters, J. M., Narotsky, M. G., Elizondo, G., Fernandez-Salguero, P. M., Gonzalez, F. J., and Abbott, B. (1999). Amelioration of TCDD-induced teratogenesis in aryl hydrocarbon receptor (AhR)-null mice. *Toxicological Sciences*, 47, 86–92.

204. Staples, J. E., Murante, F. G., Fiore, N. C., Gasiewicz, T. A., and Silverstone, A. E. (1998). Thymic alterations induced by 2,3,7,8-tetrachlorodibenzo-*p*-dioxin are strictly dependent on aryl hydrocarbon receptor activation in hemopoietic cells. *Journal of Immunology*, 160, 3844–3854; Erratum, 163, 1092.

205. Thurmond, T. S., Silverstone, A. E., Baggs, R. B., Quimby, F. W., Staples, J. E., and Gasiewicz, T. A. (1999). A chimeric aryl hydrocarbon receptor knockout mouse model indicates that aryl hydrocarbon receptor activation in hematopoietic cells contributes to the hepatic lesions induced by 2,3,7,8-tetrachlorodibenzo-*p*-dioxin. *Toxicology and Applied Pharmacology*, 158, 33–40.

206. Vorderstrasse, B. A., Steppan, L. B., Silverstone, A. E., and Kerkvliet, N. I. (2001). Aryl hydrocarbon receptor-deficient mice generate normal immune responses to model antigens and are resistant to TCDD-induced immune suppression. *Toxicology and Applied Pharmacology*, 171, 157–164.

207. Fritsche, E., Schafer, C., Calles, C., Bernsmann, T., Bernshausen, T., Wurm, M., Hubenthal, U., Cline, J. E., Hajimiranha, H., Schroeder, P., Klotz, L. O., Rannug, A., Furst, P., Hananberg, H., Abel, J., and Krutmann, J. (2007). Lightening up the UV response by identification of the aryl hydrocarbon receptor as a cytoplasmic target for ultraviolet B radiation. *Proceedings of the National Academy of Sciences of the United States of America*, 104, 8851–8856.

208. Shimizu, Y., Nakatsuru, Y., Ichinose, M., Takahashi, Y., Kume, H., Mimura, J., Fujii-Kuriyama, Y., and Ishikawa, T.

(2000). Benzo[a]pyrene carcinogenicity is lost in mice lacking the aryl hydrocarbon receptor. *Proceedings of the National Academy of Sciences of the United States of America*, 97, 779–782.

209. Yoon, B. I., Hirabayashi, Y., Kawasaki, Y., Kodama, Y., Kaneko, T., Kanno, J., Kim, D. Y., Fujii-Kuriyama, Y., and Inoue, T. (2002). Aryl hydrocarbon receptor mediates benzene-induced hematoxicity. *Toxicological Sciences*, 70, 150–156.

210. Bock, K. W. and Köhle, C. (2006). Ah receptor: dioxin-mediated toxic responses as hints to deregulated physiologic functions. *Biochemical Pharmacology*, 72, 393–404.

211. Boutros, P. C., Yao, C. Q., Watson, J. D., Wu, A. H., Moffat, I. D., Prokopec, S. D., Smith, A. B., Okey, A. B., and Pohjanvirta, R. (2011). Hepatic transcriptomic responses to TCDD in dioxin-sensitive and dioxin-resistant rats during the onset of toxicity. *Toxicology and Applied Pharmacology*, 251, 119–129.

212. Franc, M. A., Moffat, I. D., Boutros, P. C., Tuomisto, J. T., Tuomisto, J., Pohjanvirta, R., and Okey, A. B. (2008). Patterns of dioxin-altered mRNA expression in livers of dioxin-sensitive versus dioxin-resistant rats. *Archives of Toxicology*, 82, 809–830.

213. Moffat, I. D., Roblin, S., Harper, P. A., Okey, A. B., and Pohjanvirta, R. (2007). Aryl hydrocarbon receptor splice variants in the dioxin-resistant rat: tissue expression and transactivational activity. *Molecular Pharmacology*, 72, 956–966.

214. Xiong, K., Peterson, R. E., and Heideman, W. (2008). AHR activation by TCDD downregulates Sox9b expression producing jaw malformation in zebrafish embryos. *Molecular Pharmacology*, 74, 1544–1553.

215. Yoshioka, W., Peterson, R. E., and Tohyama, C. (2010). Molecular targets that link dioxin exposure to toxicity phenotypes. *Journal of Steroid Biochemistry and Molecular Biology*. doi: 10.1016/j.jsbmb.2010.12.005.

216. Harper, P. A., Riddick, D. S., and Okey, A. B. (2006). Regulating the regulator: factors that control levels and activity of the aryl hydrocarbon receptor. *Biochemical Pharmacology*, 72, 267–279.

217. Giannone, J. V., Li, W., Probst, M., and Okey, A. B. (1998). Prolonged depletion of AH receptor without alteration of receptor mRNA levels after treatment of cells in culture with 2,3,7,8-tetrachlorodibenzo-p-dioxin. *Biochemical Pharmacology*, 55, 489–497.

218. Huang, P., Rannug, A., Ahlbom, F., Hakansson, H., and Ceccatelli, S. (2000). Effect of 2,3,7,8-tetrachlorodibenzo-p-dioxin on the expression of cytochrome P4501A1, the aryl hydrocarbon receptor, and aryl hydrocarbon receptor nuclear translocator in rat brain and pituitary. *Toxicology and Applied Pharmacology*, 169, 159–167.

219. Pollenz, R. S. (1996). The aryl-hydrocarbon receptor, but not the aryl-hydrocarbon receptor nuclear translocator protein, is rapidly depleted in hepatic and nonhepatic culture cells exposed to 2,3,7,8-tetrachlorodibenzo-p-dioxin. *Molecular Pharmacology*, 49, 391–398.

220. Roman, B. L., Pollenz, R. S., and Peterson, R. E. (1998). Responsiveness of the adult male rat reproductive tract to 2,3,7,8-tetrachlorodibenzo-p-dioxin exposure: Ah receptor and ARNT expression, CYP1A1 induction, and Ah receptor down-regulation. *Toxicology and Applied Pharmacology*, 150, 228–239.

221. Davarios, N. A. and Pollenz, R. S. (1999). Aryl hydrocarbon receptor imported into nucleus following ligand binding is rapidly degraded via the cytoplasmic proteosome following nuclear export. *Journal of Biological Chemistry*, 274, 28708–28715.

222. Ma, Q. and Baldwin, K. T. (2000). 2,3,7,8-Tetrachlorodibenzo-p-dioxin-induced degradation of the aryl hydrocarbon receptor (AhR) by the ubiquitin-proteasome pathway. Role of the transcription activation and DNA binding of AhR. *Journal of Biological Chemistry*, 275, 8432–8438.

223. Roberts, B. J. and Whitelaw, M. L. (1999). Degradation of the basic helix–loop–helix/Per-ARNT-Sim homology domain dioxin receptor via the ubiquitin/proteasome pathway. *Journal of Biological Chemistry*, 274, 36351–36356.

224. Chen, S., Operana, T., Bonzo, J., Nguyen, N., and Tukey, R. H. (2005). ERK kinase inhibition stabilizes the aryl hydrocarbon receptor. Implications for transcriptional activation and protein degradation. *Journal of Biological Chemistry*, 280, 4350–4359.

225. Dunham, E. E., Stevens, E. A., Glover, E., and Bradfield, C. A. (2006). The aryl hydrocarbon receptor signaling pathway is modified through interactions with a Kelch protein. *Molecular Pharmacology*, 70, 8–15.

226. Pollenz, R. S. (2002). The mechanism of AH receptor protein down-regulation (degradation) and its impact on AH receptor-mediated gene regulation. *Chemico-Biological Interactions*, 141, 41–61.

227. Pollenz, R. S., Popat, J., and Dougherty, E. J. (2005). Role of the carboxy-terminal transactivation domain and active transcription in the ligand-induced and ligand-independent degradation of the mouse Ah[b-1] receptor. *Biochemical Pharmacology*, 70, 1623–1633.

228. Andersen, M. E. and Barton, H. A. (1999). Biological regulation of receptor-hormone complex concentrations in relation to dose–response assessments for endocrine-active compounds. *Toxicological Sciences*, 48, 38–50.

229. Ma, Q., Renzelli, A. J., Baldwin, K. T., and Antonini, J. M. (2000). Superinduction of CYP1A1 gene expression. Regulation of 2,3,7,8-tetrachlorodibenzo-p-dioxin-induced degradation of Ah receptor by cycloheximide. *Journal of Biological Chemistry*, 275, 12676–12683.

230. Eguchi, H., Hayashi, S., Watanabe, J., Gotoh, O., and Kawajiri, K. (1994). Molecular cloning of the human AH receptor gene promoter. *Biochemical and Biophysical Research Communications*, 203, 615–622.

231. Garrison, P. M. and Denison, M. S. (2000). Analysis of the murine AhR gene promoter. *Journal of Biochemistry and Molecular Toxicology*, 14, 1–10.

232. Mimura, J., Ema, M., Sogawa, K., Ikawa, S., and Fujii-Kuriyama, Y. (1994). A complete structure of the mouse Ah receptor gene. *Pharmacogenetics*, 4, 349–354.

233. Dohr, O., Sinning, R., Vogel, C. F. A., Munzel, P. A., and Abel, J. (1997). Effect of transforming growth factor-beta 1 on expression of aryl hydrocarbon receptor and genes of Ah gene battery: clues for independent down-regulation in A549 cells. *Molecular Pharmacology, 51*, 703–710.

234. Tanaka, G., Kanaji, S., Hirano, A., Arima, K., Shinagawa, A., Goda, C., Yasunaga, S., Ikizawa, K., Yanagihara, Y., Kubo, M., Fujii-Kuriyama, Y., Sugita, Y., Inokuchi, A., and Izuhara, K. (2005). Induction and activation of the aryl hydrocarbon receptor by IL-4 in B cells. *International Immunology 17*, 797–805.

235. Timsit, Y. E., Chia, F. S. C., Bhanthena, A., and Riddick, D. S. (2002). Aromatic hydrocarbon receptor expression and function in liver of hypophysectomized male rats. *Toxicology and Applied Pharmacology, 185*, 136–145.

236. Wolff, S., Harper, P. A., Wong, J. M. Y., Mostert, V., Wang, Y., and Abel, J. (2001). Cell-specific regulation of human aryl hydrocarbon receptor expression by transforming growth factor-β_1. *Molecular Pharmacology, 59*, 716–724.

237. Mimura, J., Ema, M., Sogawa, K., and Fujii-Kuriyama, Y. (1999). Identification of a novel mechanism of regulation of Ah (dioxin) receptor function. *Genes & Development, 13*, 20–25.

238. Baba, T., Mimura, J., Gradin, K., Kuroiwa, A., Watanabe, T., Matsuda, Y., Inazawa, J., Sogawa, K., and Fujii-Kuriyama, Y. (2001). Structure and expression of the Ah receptor repressor gene. *Journal of Biological Chemistry, 276*, 33101–33110.

239. Bernshausen, T., Jux, B., Esser, C., Abel, J., and Fritsche, E. (2006). Tissue distribution and function of the aryl hydrocarbon receptor repressor (AhRR) in C57BL/6 and aryl hydrocarbon receptor deficient mice. *Archives of Toxicology, 80*, 206–211.

240. Evans, B. R., Karchner, S. I., Allan, L. L., Pollenz, R. S., Tanguay, R. L., Jenny, M. J., Sherr, D. H., and Hahn, M. E. (2008). Repression of aryl hydrocarbon receptor (AHR) signaling by AHR repressor: role of DNA binding and competition for AHR nuclear translocator. *Molecular Pharmacology, 73*, 387–398.

241. Tsuchiya, Y., Nakajima, M., Itoh, S., Iwanari, M., and Yokoi, T. (2003). Expression of aryl hydrocarbon receptor repressor in normal human tissues and inducibility by polycyclic aromatic hydrocarbons in human tumor-derived cell lines. *Toxicological Sciences, 72*, 253–259.

242. Hahn, M. E., Allan, L. L., and Sherr, D. H. (2009). Regulation of constitutive and inducible AHR signaling: complex interactions involving the AHR repressor. *Biochemical Pharmacology, 77*, 485–497.

243. Kanno, Y., Takane, Y., Takizawa, Y., and Inouye, Y. (2008). Suppressive effect of aryl hydrocarbon receptor repressor on transcriptional activity of estrogen receptor alpha by protein–protein interaction in stably and transiently expressing cell lines. *Molecular and Cellular Endocrinology 291*, 87–94.

244. Zudaire, E., Cuesta, N., Murty, V., Woodson, K., Adams, L., Gonzalez, N., Martinez, A., Narayan, G., Kirsch, I., Franklin, W., Hirsch, F., Birrer, M., and Cuttitta, F. (2008). The aryl hydrocarbon receptor repressor is a putative tumor suppressor gene in multiple human cancers. *Journal of Clinical Investigation 118*, 640–650.

245. Abbott, B., Birnbaum, L. S., and Perdew, G. H. (1995). Developmental expression of two members of a new class of transcription factors. I. Expression of aryl hydrocarbon receptor in C57BL/6N mouse embryo. *Developmental Dynamics, 204*, 133–143.

246. Carlstedt-Duke, J., Elstrom, G., Hogberg, B., and Gustafsson, J. A. (1979). Ontogeny of the rat hepatic receptor for 2,3,7,8-tetrachlorodibenzo-*p*-dioxin and its endocrine dependence. *Cancer Research, 39*, 4653–4656.

247. Furness, S. G. B. and Whelan, F. (2009). The pleiotropy of dioxin toxicity: xenobiotic misappropriation of the aryl hydrocarbon receptor's alternative physiological roles. *Pharmacology & Therapeutics, 124*, 336–353.

248. Gasiewicz, T. A., Ness, W. C., and Rucci, G. (1984). Ontogeny of the cytosolic receptor for 2,3,7,8-tetrachlorodibenzo-*p*-dioxin in rat liver, lung, and thymus. *Biochemical and Biophysical Research Communications, 118*, 183–190.

249. Sommer, R. J., Sojka, K. M., Pollenz, R. S., Cooke, P. S., and Peterson, R. E. (1999). Ah receptor and ARNT protein and mRNA concentrations in rat prostate: effects of stage of development and 2,3,7,8-tetrachlorodibenzo-*p*-dioxin treatment. *Toxicology and Applied Pharmacology, 155*, 177–189.

250. Wu, Q., Ohsako, S., Baba, T., Miyamoto, K., and Tohyama, C. (2002). Effects of 2,3,7,8-tetrachlorodibenzo-*p*-dioxin (TCDD) on preimplantation embryos. *Toxicology, 174*, 119–129.

251. Jaenish, R. and Bird, A. (2003). Epigenetic regulation of gene expression: how the genome integrates intrinsic and environmental signals. *Nature Genetics, 33* (Suppl.) 245–254.

252. Kiefer, J. C. (2007). Epigenetics in development. *Developmental Dynamics, 236*, 1144–1156.

253. Cui, Y. J., Yeager, R. L., Zhong, X., and Klaassen, C. D. (2009). Ontogenic expression of hepatic Ahr mRNA is associated with histone H3K4 di-methylation during mouse liver development. *Toxicology Letters, 189*, 184–190.

254. Fan, Y., Bovin, G. P., Knudsen, E. S., Nebert, D. W., Xia, Y., and Puga, A. (2010). The aryl hydrocarbon receptor functions as a tumor suppressor of liver carcinogenesis. *Cancer Research, 70*, 212–220.

255. Fritz, W. A., Lin, T.-M., Cardiff, R. D., and Peterson, R. E. (2007). The aryl hydrocarbon receptor inhibits prostate carcinogenesis in TRAMP mice. *Carcinogenesis 28*, 497–505.

256. Mulero-Navarro, S., Carvajal-Gonzalez, J. M., Herranz, M., Ballestar, E., Fraga, M. F., Ropero, S., Esteller, M., and Fernandez-Salguero, P. M. (2006). The dioxin receptor is silenced by promoter hypermethylation in human acute lymphoblastic leukemia through inhibition of Sp1 binding. *Carcinogenesis, 27*, 1099–1104.

257. Blankenship, A. and Matsumura, F. (1997). 2,3,7,8-Tetrachlorodibenzo-*p*-dioxin-induced activation of a protein tyrosine kinase, pp60src, in murine hepatic cytosol using a cell-free system. *Molecular Pharmacology, 52*, 667–675.

258. Enan, E. and Matsumura, F. (1996). Identification of c-Src as the integral component of the cytosolic Ah receptor complex, transducing the signal of 2,3,7,8-tetrachlorodibenzo-p-dioxin (TCDD) through the protein phosphorylation pathway. *Biochemical Pharmacology*, 52, 1599–1612.

259. Köhle, C., Gschaidmeier, H., Lauth, D., Topell, S., Zitzer, H., and Bock, K. W. (1999). 2,3,7,8-Tetrachlorodibenzo-p-dioxin (TCDD)-mediated membrane translocation of c-Src protein kinase in liver WB-F344 cells. *Archives of Toxicology*, 73, 152–158.

260. Park, S., Dong, B., and Matsumura, F. (2007). Rapid activation of c-Src kinase by dioxin is mediated by the Cdc37–HSP90 complex as part of Ah receptor signaling in MCF10A cells. *Biochemistry 46*, 899–908.

261. Kim, D. W., Gazourian, L., Quadri, S. A., Romieu-Mourez, R., Sherr, D. H., and Sonenshein, G. E. (2000). The RelA NF-kappaB subunit and the aryl hydrocarbon receptor (AhR) cooperate to transactivate the c-myc promoter in mammary cells. *Oncogene 19*, 5498–5506.

262. Ruby, D. E., Leid, M., and Kerkvliet, N. I. (2002). 2,3,7,8-Tetrachlorodibenzo-p-dioxin suppresses tumor necrosis factor-alpha and anti-CD40-induced activation of NF-kappaB/Rel in dendritic cells: p50 homodimer activation is not affected. *Molecular Pharmacology*, 62, 722–728.

263. Thatcher, T. H., Maggirwar, S. B., Baglole, C. J., Lakatos, H. F., Gasiewicz, T. A., Phipps, R. P., and Sime, P. J. (2007). Aryl hydrocarbon receptor-deficient mice develop heightened inflammatory responses to cigarette smoke and endotoxin associated with rapid loss of the nuclear factor-κB component RelB. *American Journal of Physiology*, 170, 855–864.

264. Tian, Y., Ke, S., Denison, M. S., Rabson, A. B., and Gallo, M. A. (1999). Ah receptor and NF-κB interactions, a potential mechanism for dioxin toxicity. *Journal of Biological Chemistry*, 274, 510–515.

265. Vogel, C. F. A. and Matsumura, F. (2009). A new cross-talk between the aryl hydrocarbon receptor and RelB, a member of the NF-κB family. *Biochemical Pharmacology*, 77, 734–745.

266. Vogel, C. F. A., Sciullo, E., Li, W., Wong, P., Lazennec, G., and Matsumura, F. (2007). RelB, a new partner of aryl hydrocarbon receptor-mediated transcription. *Molecular Endocrinology*, 21, 2941–2955.

267. Elferink, C. J., Ge, N. L., and Levine, A. (2001). Maximal aryl hydrocarbon receptor activity depends on an interaction with the retinoblastoma protein. *Molecular Pharmacology*, 59, 664–673.

268. Ge, N. L. and Elferink, C. J. (1998). A direct interaction between the aryl hydrocarbon receptor and retinoblastoma protein. Linking dioxin signaling to the cell cycle. *Journal of Biological Chemistry*, 273, 22708–22713.

269. Puga, A., Barnes, S. J., Dalton, T. P., Chang, C., Knudsen, E. S., and Maier, M. A. (2000). Aromatic hydrocarbon receptor interaction with the retinoblastoma protein potentiates repression of E2F-dependent transcription and cell cycle arrest. *Journal of Biological Chemistry*, 275, 2943–2950.

270. Ohtake, F., Baba, A., Takada, I., Okada, M., Iwasaki, K., Miki, H., Takahashi, S., Kouzmenko, A., Nohara, K., Chiba, T., Fujii-Kuriyama, Y., and Kato, S. (2007). Dioxin receptor is a ligand-dependent E3 ubiquitin ligase. *Nature*, 446, 562–566.

271. Wormke, M., Stoner, M., Saville, B., Walker, K., Abdelrahim, M., Burghardt, R., and Safe, S. (2003). The aryl hydrocarbon receptor mediates degradation of estrogen receptor alpha through activation of proteasomes. *Molecular and Cellular Biology*, 23, 1843–1855.

272. Antenos, M., Casper, R. F., and Brown, T. J. (2002). Interaction with Nedd8, a ubiquitin-like protein, enhances the transcriptional activity of the aryl hydrocarbon receptor. *Journal of Biological Chemistry*, 277, 44028–44034.

273. Jones, L. C., Okino, S. T., Gonda, T. J., and Whitlock, J. P., Jr. (2002). Myb-binding protein 1a augments AhR-dependent gene expression. *Journal of Biological Chemistry*, 277, 22515–22519.

274. Ma, Q., Kinneer, K., Bi, Y., Chan, J. Y., and Kan, Y. W. (2004). Induction of murine NAD(P)H:quinone oxidoreductase by 2,3,7,8-tetrachlorodibenzo-p-dioxin requires the CNC (cap 'n' collar) basic leucine zipper transcription factor Nrf2 (nuclear factor erythroid 2-related factor 2): cross-interaction between AhR (aryl hydrocarbon receptor) and Nrf2 signal transduction. *Biochemical Journal*, 377, 205–213.

275. Miao, W., Hu, L., Scrivens, P. J., and Batist, G. (2005). Transcriptional regulation of NF-E2 p45-related factor (NRF2) expression by the aryl hydrocarbon receptor-xenobiotic response element signaling pathway: direct cross-talk between phase I and II drug-metabolizing enzymes. *Journal of Biological Chemistry*, 280, 20340–20348.

276. Weiss, C., Faust, D., Durk, H., Kolluri, S. K., Pelzer, A., Schneider, S., Dietrich, C., Oesch, F., and Gottlicher, M. (2005). TCDD induces c-jun expression via a novel Ah (dioxin) receptor-mediated p38-MAPK-dependent pathway. *Oncogene 24*, 4975–4983.

277. Kawajiri, K., Kobayashi, Y., Ohtake, F., Ikuta, T., Matsushima, Y., Mimura, Y., Pettersson, S., Pollenz, R. S., Sakaki, T., Hirokawa, T., Akiyama, T., Kurisumi, M., Poellinger, L., Kato, S., and Fujii-Kuriyama, Y. (2009). Aryl hydrocarbon receptor suppresses intestinal carcinogenesis in Apc$^{Min/+}$ mice with natural ligands. *Proceedings of the National Academy of Sciences of the United States of America*, 106, 13481–13486.

278. Marlowe, J. L., Fan, Y., Chang, X., Peng, L., Knudsen, E. S., Xia, Y., and Puga, A. (2008). The aryl hydrocarbon receptor binds to E2F1 and inhibits E2F1-induced apoptosis. *Molecular and Cellular Biology*, 19, 3263–3271.

279. Barhoover, M. S., Hall, J. M., Greenlee, W. F., and Thomas, R. S. (2010). Aryl hydrocarbon receptor regulates cell cycle progression in human breast cancer cells via a functional interaction with cyclin-dependent kinase 4. *Molecular Pharmacology*, 77, 195–201.

280. Hankinson, O. (2005). Role of coactivators in transcriptional activation by the aryl hydrocarbon receptor. *Archives of Biochemistry and Biophysics*, 433, 379–386.

281. Matsumura, F. (2009). The significance of the nongenomic pathway in mediating inflammatory signaling of the dioxin-

282. Patel, R. D., Murray, I. A., Flaveny, C. A., Kusnadi, A., and Perdew, G. H. (2009). Ah receptor represses acute phase response gene expression without binding to its cognate response element. *Laboratory Investigation, 89*, 695–707.

283. Bunger, M. K., Glover, E., Moran, S. M., Walisser, J. A., Lahvis, G. P., Hsu, E. L., and Bradfield, C. A. (2008). Abnormal liver development and resistance to 2,3,7,8-tetrachlorodibenzo-*p*-dioxin toxicity in mice carrying a mutation in the DNA binding domain of the aryl hydrocarbon receptor. *Toxicological Sciences, 106*, 83–92.

284. Bunger, M. K., Moran, S. M., Glover, E., Thomae, T. L., Lahvis, G. P., Lin, B. C., and Bradfield, C. A. (2003). Resistance to 2,3,7,8-tetrachlorodibenzo-*p*-dioxin toxicity and abnormal liver development in mice carrying a mutation in the nuclear localization sequence of the aryl hydrocarbon receptor. *Journal of Biological Chemistry, 278*, 17767–17774.

285. Tijet, N., Boutros, P. C., Moffat, I. D., Okey, A. B., Tuomisto, J., and Pohjanvirta (2006). Aryl hydrocarbon receptor regulates distinct dioxin-dependent and dioxin-independent gene batteries. *Molecular Pharmacology, 69*, 140–153.

286. Nebert, D. W. (1989). The Ah locus: genetic differences in toxicity, cancer, mutation and birth defects. *Critical Reviews in Toxicology, 20*, 153–174.

287. Nebert, D. W., Brown, D. D., Towne, D. W., and Eisen, H. J. (1984). Association of fertility, fitness, and longevity with the murine Ah locus among (C57BL/6N) (C3H/HeN) recombinant inbred mice. *Biology of Reproduction, 30*, 363–373.

288. Bankoti, J., Rase, B., Simones, T., and Shepherd, D. M. (2010). Functional and phenotypic effects of AhR activation in inflammatory dendritic cells. *Toxicology and Applied Pharmacology, 246*, 18–28.

289. Esser, C., Rannug, A., and Stockinger, B. (2009). The aryl hydrocarbon receptor in immunity. *Trends in Immunology 30*, 447–454.

290. Funatake, C. J., Ao, K., Suzuki, T., Yamamoto, M., Fujii-Kuriyama, Y., Kerkvliet, N. I., and Nohara, K. (2009). Expression of constitutively-active aryl hydrocarbon receptor in T-cells enhances the down-regulation of CD62L, but does not alter expression of CD25 or suppress the allogeneic CTL response. *Journal of Immunotoxicology 6*, 194–203.

291. Hauben, E., Gregory, S., Draghici, E., Migliavacca, B., Olivieri, S., Woisetschläger, M., and Roncarolo, M. G. (2008). Activation of the aryl hydrocarbon receptor promotes allograft-specific tolerance through direct and dendritic cell-mediated effects on regulatory T cells. *Blood, 112*, 1214–1222.

292. Head, J. L. and Lawrence, B. P. (2009). The aryl hydrocarbon receptor is a modulator of anti-viral immunity. *Biochemical Pharmacology, 77*, 642–653.

293. Kerkvliet, N. I. (2009). AHR-mediated immunomodulation: the role of altered gene transcription. *Biochemical Pharmacology, 77*, 746–760.

294. Kimura, A., Naka, T., Nohara, K., Fujii-Kuriyama, Y., and Kishimoto, T. (2008). Aryl hydrocarbon receptor regulates Stat1 activation and participates in the development of Th17 cells. Proceedings of the National Academy of Sciences of the United States of America, 105, 9721–9726.

295. Marshall, N. B., Vorachek, W. R., Steppan, L. B., Mourich, D. V., and Kerkvliet, N. I. (2008). Functional characterization and gene expression analysis of $CD4^+$ $CD25^+$ regulatory T cells generated in mice treated with 2,3,7,8-tetrachlorodibenzo-*p*-dioxin. *Journal of Immunology, 181*, 2382–2391.

296. Quintana, F. J., Basso, A. S., Iglesias, A. H., Korn, T., Farez, M. F., Bettelli, E., Caccamo, M., Oukka, M., and Weiner, H. L. (2008). Control of T_{reg} and T_H17 cell differentiation by the aryl hydrocarbon receptor. *Nature, 453*, 65–71.

297. Aragon, A. C., Kopf, P. G., Campen, M. J., Huwe, J. K., and Walker, M. K. (2008). In utero and lactational 2,3,7,8-tetrachlorodibenzo-*p*-dioxin exposure: effects on fetal and adult cardiac gene expression and adult cardiac and renal morphology. *Toxicological Sciences, 101*, 321–330.

298. Dalton, T. P., Kerzee, J. K., Wang, B., Miller, M., Dieter, M. Z., Lorenz, J. N., Shertzer, H. G., and Nebert, D. W. (2001). Dioxin exposure is an environmental risk factor for ischemic heart disease. *Cardiovascular Toxicology, 1*, 285–298.

299. Wang, Y., Fan, Y., and Puga, A. (2010). Dioxin exposure disrupts the differentiation of mouse embryonic stem cells into cardiomyocytes. *Toxicological Sciences, 115*, 225–237.

300. Hernandez-Ochoa, I., Karman, B. N., and Flaws, J. A. (2009). The role of the aryl hydrocarbon receptor in the female reproductive system. *Biochemical Pharmacology, 77*, 547–559.

301. Loertscher, J. A., Sattler, C. A., and Allen-Hoffman, B. L. (2001). 2,3,7,8-Tetrachlorodibenzo-*p*-dioxin alters the differentiation pattern of human keratinocytes in organotypic culture. *Toxicology and Applied Pharmacology, 175*, 121–129.

302. Panteleyev, A. A. and Bicker, D. R. (2006). Dioxin-induced chloracne: reconstructing the cellular and molecular mechanisms of a classic environmental disease. *Experimental Dermatology, 15*, 705–730.

303. Vorderstrasse, B. A., Fenton, S. E., Bohn, A. A., Cundiff, J. A., and Lawrence, B. P. (2004). A novel effect of dioxin: exposure during pregnancy severely impairs mammary gland differentiation. *Toxicological Sciences, 78*, 248–257.

304. Huang, G. and Elferink, C. J. (2005). Multiple mechanisms are involved in Ah receptor-mediated cell cycle arrest. *Molecular Pharmacology, 67*, 88–96.

305. Mitchell, K. A. and Elferink, C. J. (2009). Timing is everything: consequences of transient and sustained AhR activity. *Biochemical Pharmacology, 77*, 947–956.

306. Ray, S. S. and Swanson, H. I. (2009). Activation of the aryl hydrocarbon receptor by TCDD inhibits senescence: a tumor promoting event? *Biochemical Pharmacology, 77*, 681–688.

307. Garrett, R. W. and Gasiewicz, T. A. (2006). The aryl hydrocarbon receptor agonist 2,3,7,8-tetrachlorodibenzo-*p*-dioxin alters the circadian rhythms, quiescence, and expression of clock genes in murine hematopoietic stem and progenitor cells. *Molecular Pharmacology, 69*, 2076–2083.

308. Mukai, M., Lin, T.-M., Peterson, R. E., Cooke, P. S., and Tischkau, S. A. (2008). Behavioral rhythmicity of mice lacking AhR and attenuation of light-induced phase shift by

2,3,7,8-tetrachlorodibenzo-*p*-dioxin. *Journal of Biological Rhythms*, 23, 200–210.

309. Shimba, S. and Watabe, Y. (2009). Crosstalk between the AHR signaling pathway and circadian rhythm. *Biochemical Pharmacology*, 77, 560–565.

310. Abbott, B. D. and Birnbaum, L. S. (1991). TCDD exposure of human palatal shelves in organ culture alters the differentiation of medial epithelial cells. *Teratology*, 43, 119–132.

311. Vezina, C. M., Lin, T.-M., and Peterson, R. E. (2009). AHR signaling in prostate growth, morphogenesis, and disease. *Biochemical Pharmacology*, 77, 566–576.

312. Dietrich, C. and Kaina, B. (2010). The aryl hydrocarbon receptor (AhR) in the regulation of cell–cell contact and tumor growth. *Carcinogenesis*, 31, 1319–1328.

313. Fritz, W. A., Lin, T.-M., Safe, S., Moore, R. W., and Peterson, R. E. (2009). The selective aryl hydrocarbon receptor modulator 6-methyl-1,3,8,-trichlorodibenzofuran inhibits prostate tumor metastasis in TRAMP mice. *Biochemical Pharmacology*, 77, 1151–1160.

314. Kung, T., Murphy, K. A., and White, L. A. (2009). The aryl hydrocarbon receptor (AhR) pathway as a regulatory pathway for cell adhesion and matrix metabolism. *Biochemical Pharmacology*, 77, 536–546.

315. Villano, C. M., Murphy, K. A., Akintobi, A., and White, L. A. (2006). 2,3,7,8-Tetrachlorodibenzo-*p*-dioxin (TCDD) induces matrix metalloproteinase (MMP) expression and invasion in A2058 melanoma cells. *Toxicology and Applied Pharmacology*, 210, 212–224.

316. Birnbaum, L. S. (1995). Developmental effects of dioxins and related endocrine disrupting chemicals. *Toxicology Letters*, 82–83, 743–750.

317. Kakeyama, M. and Tohyama, C. (2003). Developmental neurotoxicity of dioxin and its related compounds. *Industrial Health*, 41, 215–230.

318. Nakajima, S., Saijo, Y., Kato, S., Sasaki, S., Uno, A., Kanagami, N., Hirakawa, H., Hori, T., Tobiishi, K., Todaka, T., Nakamura, Y., Yanagiya, S., Senogoku, Y., Iida, T., Sata, F., and Kishi, R. (2006). Effects of prenatal exposure to polychlorinated biphenyls and dioxins on mental and motor development in Japanese children at 6 months of age. *Environmental Health Perspectives*, 114, 147–152.

319. Williamson, M. A., Gasiewicz, T. A., and Opanashuk, L. A. (2005). Aryl hydrocarbon receptor expression and activity in cerebellar granule neuroblasts: implications for development and dioxin neurotoxicity. *Toxicological Sciences*, 83, 340–348.

320. Herlin, M., Kalantari, F., Stern, N., Sand, S., Larsson, S., Viluksela, M., Tuomisto, J. T., Tuomisto, J., Tuukkanen, J., Jamsa, T., Lind, P. M., and Hakansson, H. (2010). Quantitative characterization of changes in bone geometry, mineral density, and biomechanical properties in two rat strains with different Ah-receptor structures after long-term exposure to 2,3,7,8-tetrachlorodibenzo-*p*-dioxin. *Toxicology* 273, 1–11.

321. Hermsen, S. A., Larsson, S., Arima, A., Muneoka, A., Ihara, T., Sumida, H., Fukasato, T., Kubota, S., Yasuda, M., and Lind, P. M. (2008). In utero and lactational exposure to 2,3,7,8-tetrachlorodibenzo-*p*-dioxin (TCDD) affects bone tissue in rhesus monkeys. *Toxicology* 253, 147–152.

322. Korkalainen, M., Kallio, E., Olkku, A., Nelo, K., Ilvesaro, J., Tuukkanen, J., Mahonen, A., and Viluksela, M. (2009). Dioxins interfere with differentiation of osteoblasts and osteoclasts. *Bone*, 44, 1134–1142.

323. Wijheden, C., Brunnberg, S., Larsson, S., Lind, P. M., Andersson, G., and Hanberg, A. (2010). Transgenic mice with a constitutively active aryl hydrocarbon receptor display a gender-specific bone phenotype. *Toxicological Sciences*, 114, 48–58.

324. Haarmann-Stemmann, T., Bothe, H., and Abel, J. (2009). Growth factors, cytokines and their receptors as downstream targets of arylhydrocarbon receptor (AhR) signaling pathways. *Biochemical Pharmacology*, 77, 508–250.

325. Nebert, D. W., Eisen, H. J., and Hankinson, O. (1984). The Ah receptor: binding specific only for foreign chemicals? *Biochemical Pharmacology*, 33, 917–924.

326. Denison, M. S. and Nagy, S. R. (2003). Activation of the aryl hydrocarbon receptor by structurally diverse exogenous and endogenous chemicals. *Annual Review of Pharmacology and Toxicology*, 43, 309–334.

327. Nguyen, L. P. and Bradfield, C. A. (2008). The search for endogenous activators of the aryl hydrocarbon receptor. *Chemical Research in Toxicology*, 21, 102–116.

328. Ashida, H., Fukuda, I., Yamashita, T., and Kanazawa, K. (2000). Flavones and flavonols at dietary levels inhibit a transformation of aryl hydrocarbon receptor induced by dioxin. *FEBS Letters*, 476, 213–217.

329. Adachi, J., Mori, Y., Matsui, S., Takigami, H., Fujino, J., Kitagawa, H., Miller, C. A., Kato, T., Saeki, K., and Matsuda, T. (2001). Indirubin and indigo are potent aryl hydrocarbon receptor ligands present in human urine. *Journal of Biological Chemistry*, 276, 31475–31478.

330. Ciolino, H. P., Daschner, P. J., and Yeh, G. C. (1999). Dietary flavonols quercetin and kaempferol are ligands of the aryl hydrocarbon receptor that affect CYP1A1 transcription differentially. *Biochemical Journal*, 340, 715–722.

331. Heath-Pagliuso, S., Rogers, W. J., Tullis, K., Seidel, S. D., Cenijn, P. H., Brouwer, A., and Denison, M. S. (1998). Activation of the Ah receptor by tryptophan and tryptophan metabolites. *Biochemistry*, 37, 11508–11515.

332. Sinal, C. J. and Bend, J. R. (1997). Aryl hydrocarbon receptor-dependent induction of cyp1a1 by bilirubin in mouse hepatoma Hepa 1c1c7 cells. *Molecular Pharmacology*, 52, 590–599.

333. Rannug, A., Rannug, U., Rosenkranz, H. S., Winqvist, L., Westerholm, R., Agurell, E., and Grafstrom, A.-K. (1987). Certain photooxidized derivatives of tryptophan bind with very high affinity to the Ah receptor and are likely to be endogenous signal substances. *Journal of Biological Chemistry*, 262, 15422–15427.

334. Bjeldanes, L. F., Kim, J. Y., Grose, K. R., Bartholomew, J. C., and Bradfield, C. A. (1991). Aromatic hydrocarbon responsiveness-receptor agonists generated from indole-3-carbinol *in vitro* and *in vivo*: comparison with 2,3,7,8-tetrachlorodi-

benzo-*p*-dioxin. *Proceedings of the National Academy of Sciences of the United States of America, 88*, 9543–9547.

335. Song, J., Clagett-Dame, M., Peterson, R. E., Hahn, M. E., Westler, W. M., Sicinski, R. R., and DeLuca, H. F. (2002). A ligand for the aryl hydrocarbon receptor isolated from lung. *Proceedings of the National Academy of Sciences of the United States of America, 99*, 14694–14699.

336. Henry, E. C., Bemis, J. C., Henry, O., Kende, A. S., and Gasiewicz, T. A. (2006). A potential endogenous ligand for the aryl hydrocarbon receptor has potent agonist activity *in vitro* and *in vivo*. *Archives of Biochemistry and Biophysics, 450*, 67–77.

337. Wincent, E., Amini, N., Luecke, S., Glatt, H., Bergman, J., Crescenzi, C., Rannug, A., and Rannug, U. (2009). The suggested physiologic aryl hydrocarbon receptor activator and cytochrome P4501 substrate 6-formylindolo[3,2-*b*]carbazole is present in humans. *Journal of Biological Chemistry, 284*, 2690–2696.

338. Agostinis, P., Garmyn, M., and Van Laethem, A. (2007). The aryl hydrocarbon receptor: an illuminating effector of the UVB response. *Science STKE* pe49.

339. Mukai, M. and Tischkau, S. A. (2007). Effects of tryptophan photoproducts in the circadian timing system: searching for a physiological role for aryl hydrocarbon receptor. *Toxicological Sciences, 95*, 172–181.

340. Bittinger, M. A., Nguyen, L. P., and Bradfield, C. A. (2003). Aspartate aminotransferase generates proagonists of the aryl hydrocarbon receptor. *Molecular Pharmacology, 64*, 550–556.

341. Chowdhury, G., Dostalek, M., Hsu, E. L., Stec, D. F., Bradfield, C. A., and Guengerich, F. P. (2009). Structural identification of diindole agonists of the aryl hydrocarbon receptor derived from degradation of indole-3-pyruvic acid. *Chemical Research in Toxicology, 22*, 1905–1912.

342. Nguyen, L. P., Hsu, E. L., Chowdhury, G., Dostalek, M., Guengerich, F. P., and Bradfield, C. A. (2009). D-amino acid oxidase generates agonists of the aryl hydrocarbon receptor from D-tryptophan. *Chemical Research in Toxicology, 22*, 1897–1904.

343. DiNatale, B. C., Murray, I. A., Schroeder, J. C., Flaveny, C. A., Lahoti, T. S., Laurenzana, E. M., Omiecinski, C. J., and Perdew, G. H. (2010). Kynurenic acid is a potent endogenous aryl hydrocarbon receptor ligand that synergistically induces interleukin-6 in the presence of inflammatory signaling. *Toxicological Sciences, 115*, 89–97.

344. Mezrich, J. D., Fechner, J. H., Zhang, X., Johnson, B. P., Burlingham, W. J., and Bradfield, C. A. (2010). An interaction between kynurenine and the aryl hydrocarbon receptor can generate regulatory T cells. *Journal of Immunology, 185*, 3190–3198.

345. Nguyen, N. T., Kimura, A., Nakahama, T., Chinen, I., Masuda, K., Nohara, K., Fujii-Kuriyama, Y., and Kishimoto, T. (2010). Aryl hydrocarbon receptor negatively regulates dendritic cell immunogenicity via a kynurenine-dependent mechanism. *Proceedings of the National Academy of Sciences of the United States of America, 107*, 19961–19966.

346. Chiaro, C. R., Morales, J. L., Prabhu, K. S., and Perdew, G. H. (2008). Leukotriene A4 metabolites are endogenous ligands for the Ah receptor. *Biochemistry 47*, 8445–8455.

347. Chiaro, C. R., Patel, R. D., and Perdew, G. H. (2008). 12(*R*)-Hydroxy-5(*Z*),8(*Z*),10(*E*),14(*Z*)-eicosatetraenoic acid [12(*R*)-HETE], an arachadonic acid derivative, is an activator of the aryl hydrocarbon receptor. *Molecular Pharmacology, 74*, 1649–1656.

348. Schaldach, C. M., Riby, J., and Bjeldanes, L. F. (1999). Lipoxin A_4: a new class of ligand for the Ah receptor. *Biochemistry, 38*, 7594–7600.

349. Seidel, S. D., Winters, G. M., Rogers, W. J., Ziccardi, M. H., Li, V., Keser, B., and Denison, M. S. (2001). Activation of the Ah receptor signaling pathway by prostaglandins. *Journal of Biochemistry and Molecular Toxicology, 15*, 187–196.

350. Nebert, D. W. and Karp, C. L. (2008). Endogenous functions of the aryl hydrocarbon receptor (AHR): intersection of cytochrome P4501 (CYP1)-metabolized eicosanoids and AHR biology. *Journal of Biological Chemistry, 283*, 36061–36065.

351. McMillan, B. J. and Bradfield, C. A. (2007). The aryl hydrocarbon receptor is activated by modified low-density lipoprotein. *Proceedings of the National Academy of Sciences of the United States of America, 104*, 1412–1417.

352. Ma, Q. and Whitlock, J. P., Jr. (1996). The aromatic hydrocarbon receptor modulates the Hepa 1c1c7 cell cycle and differentiated state independently of dioxin. *Molecular and Cellular Biology, 16*, 2144–2150.

353. Weiss, C., Kolluri, S. K., Kiefer, F., and Gottlicher, M. (1996). Complementation of Ah receptor deficiency in hepatoma cells: negative feedback regulation and cell cycle control by the Ah receptor. *Experimental Cell Research, 226*, 154–163.

354. Chang, C.-Y. and Puga, A. (1998). Constitutive activation of the aromatic hydrocarbon receptor. *Molecular and Cellular Biology, 18*, 525–535.

355. Crawford, R. B., Holsapple, M. P., and Kaminski, N. E. (1997). Leukocyte activation induces aryl hydrocarbon receptor upregulation, DNA binding, and increased Cyp1a1 expression in the absence of exogenous ligand. *Molecular Pharmacology, 52*, 921–927.

356. Ikuta, T., Kobayashi, Y., and Kawajiri, K. (2004). Cell density regulates intracellular localization of aryl hydrocarbon receptor. *Journal of Biological Chemistry, 279*, 19209–19216.

357. Sadek, C. M. and Allen-Hoffmann, B. L. (1994). Suspension-mediated induction of Hepa 1c1c7 Cyp1a-1 expression is dependent on the Ah receptor signal transduction pathway. *Journal of Biological Chemistry, 269*, 31505–31509.

358. Santiago-Josefat, B., Pozo-Guisado, E., Mulero-Navarro, S., and Fernandez-Salguero, P. M. (2001). Proteasome inhibition induces nuclear translocation and transcriptional activation of the dioxin receptor in mouse embryo primary fibroblasts in the absence of xenobiotics. *Molecular and Cellular Biology, 21*, 1700–1709.

359. Singh, S., Nord, N., and Perdew, G. H. (1996). Characterization of the activated form of the aryl hydrocarbon receptor in

the nucleus of HeLa cells in the absence of endogenous ligand. *Archives of Biochemistry and Biophysics*, 329, 47–55.

360. Hu, W., Sorrentino, C., Denison, M. S., Kolaja, K., and Fielden, M. R. (2007). Induction of cyp1a1 is a nonspecific biomarker of aryl hydrocarbon receptor activation: results of a large scale screening of pharmaceuticals and toxicants *in vivo* and *in vitro*. *Molecular Pharmacology*, 71, 1475–1486.

361. Lawrence, B. P., Denison, M. S., Novak, H., Vorderstrasse, B. A., Harrer, N., Neruda, W., Reichel, C., and Woisetschläger, M. (2008). Activation of the aryl hydrocarbon receptor is essential for mediating the anti-inflammatory effects of a novel low-molecular-weight compound. *Blood*, 112, 1158–1165.

362. Morales, J. L., Krzeminski, J., Amin, S., and Perdew, G. H. (2008). Characterization of the antiallergic drugs 3-[2-(2-phenylethyl)benzoimidazole-4-yl]-3-hydroxypropanoic acid and ethyl 3-hydroxy-3-[2-(2-phenylethyl)benzoimidazol-4-yl]propanoate as full aryl hydrocarbon receptor agonists. *Chemical Research in Toxicology*, 21, 472–482.

363. Negishi, T., Kato, Y., Ooneda, O., Mimura, J., Takada, T., Mochizuki, H., Yamamoto, M., Fujii-Kuriyama, Y., and Furusako, S. (2005). Effects of aryl hydrocarbon receptor signaling on the modulation of Th1/Th2 balance. *Journal of Immunology*, 175, 7348–7356.

364. Murray, I. A., Morales, J. L., Flaveny, C. A., DiNatale, B. C., Chiaro, C., Gowdahalli, K., Amin, S., and Perdew, G. H. (2010). Evidence for ligand-mediated selective modulation of aryl hydrocarbon receptor activity. *Molecular Pharmacology*, 77, 247–254.

365. Murray, I. A., Krishnegowda, G., DiNatale, B. C., Flaveny, C. A., Chiaro, C., Lin, J.-M., Sharma, A. K., Amin, S., and Perdew, G. H. (2010). Development of a selective modulator of aryl hydrocarbon (Ah) receptor activity that exhibits anti-inflammatory properties. *Chemical Research in Toxicology*, 23, 955–966.

366. Boitano, A. E., Wang, J., Romeo, R., Bouchez, L. C., Parker, A. E., Sutton, S. E., Walker, J. R., Flaveny, C. A., Perdew, G. H., Denison, M. S., Schultz, P. G., and Cooke, M. P. (2010). Aryl hydrocarbon receptor antagonists promote the expansion of human hematopoietic stem cells. *Science*, 329, 1345–1348.

367. Sakai, R., Kajiume, T., Inoue, H., Kanno, R., Miyazaki, M., Ninomiya, Y., and Kanno, M. (2003). TCDD treatment eliminates the long-term reconstitution activity of hematopoietic stem cells. *Toxicological Sciences*, 72, 84–91.

368. Singh, K. P., Garrett, R. W., Casado, F. L., and Gasiewicz, T. A. (2011). Aryl hydrocarbon receptor-null allele mice have hematopoietic stem/progenitor cells with abnormal characteristics and functions. *Stem Cells and Development*, 20, 769–784.

369. Singh, K. P., Wyman, A., Casado, F. L., Garrett, R. W., and Gasiewicz, T. A. (2009). Treatment of mice with the Ah receptor agonist and human carcinogen dioxin results in altered numbers and function of hematopoietic stem cells. *Carcinogenesis* 30, 11–19.

370. Thurmond, T. S., Staples, J. E., Silverstone, A. E., and Gasiewicz, T. A. (2000). The aryl hydrocarbon receptor has a role in the *in vivo* maturation of murine bone marrow B lymphocytes and their response to 2,3,7,8-tetrachlorodibenzo-*p*-dioxin. *Toxicology and Applied Pharmacology*, 165, 227–236.

371. Kociba, R. J., Keyes, D. G., Beyer, J. E., Carreon, R. M., Wade, C. E., Dittenber, D. A., Kalnins, R. P., Frauson, L. E., Park, C. N., Barnard, S. D., Hummel, R. A., and Humiston, C. G. (1978). Results of a two-year chronic toxicity and oncogenicity study of 2,3,7,8-tetrachlorodibenzo-*p*-dioxin in rats. *Toxicology and Applied Pharmacology*, 46, 279–303.

372. Safe, S. and Wormke, M. (2003). Inhibitory aryl hydrocarbon receptor–estrogen receptor a cross-talk and mechanisms of action. *Chemical Research in Toxicology*, 16, 807–816.

373. McDougal, A., Wormke, M., Calvin, J., and Safe, S. (2001). Tamoxifen-induced antitumorigenic/antiestrogenic action synergized by a selective Ah receptor modulator. *Cancer Research*, 61, 3901–3907.

374. Safe, S. and McDougal, M. (2003). Mechanism of action and development of selective aryl hydrocarbon receptor modulators for treatment of hormone-dependent cancers. *International Journal of Oncology*, 20, 1123–1128.

375. Safe, S., Qin, C., and McDougal, A. (1999). Development of selective aryl hydrocarbon receptor modulators (SARMs) for treatment of breast cancer. *Expert Opinion on Drug Investigational Drugs*, 8, 1385–1396.

376. Zhang, S., Lei, P., Liu, X., Li, X., Walker, K., Kotha, L., Rowlands, C., and Safe, S. (2009). The aryl hydrocarbon receptor as a target for estrogen receptor-negative breast cancer chemotherapy. *Endocrine-Related Cancer*, 16, 835–844.

377. Hall, J. M., Barhoover, M. A., Kazmin, D., McDonnell, D. P., Greenlee, W. F., and Thomas, R. S. (2010). Activation of the aryl hydrocarbon receptor inhibits invasive and metastatic features of human breast cancer cells and promotes breast cancer cell differentiation. *Molecular Endocrinology*, 24, 359–369.

378. Prud'homme, G. J., Glinka, Y., Toulina, A., Ace, O., Venkateswaran, S., and Jothy, S. (2010). Breast cancer stem-like cells are inhibited by a non-toxic aryl hydrocarbon receptor agonist. *PLoS One*, 5, e13831.

379. DuSell, C. D., Nelson, E. R., Wittmann, B. M., Fretz, J. A., Kazmin, D., Thomas, R. S., Pike, J. W., and McDonnell, D. P. (2010). Regulation of aryl hydrocarbon receptor function by selective estrogen receptor modulators. *Molecular Endocrinology*, 24, 33–46.

380. Bisson, W. H., Koch, D. C., O'Donnell, E. F., Khalil, S. M., Kerkvliet, N. I., Tanguay, R. L., Abagyan, R., and Kolluri, S. K. (2009). Modeling of the aryl hydrocarbon receptor (AhR) ligand binding domain and its utility in virtual ligand screening to predict new AhR ligands. *Journal of Medicinal Chemistry*, 52, 5635–5641.

381. Pandini, A., Soshilov, A. A., Song, Y., Zhao, J., Bonati, L., and Denison, M. S. (2009). Detection of the TCDD binding-fingerprint within the Ah receptor ligand binding domain by structurally driven mutagenesis and functional analysis. *Biochemistry* 48, 5972–5983.

382. Procopio, M., Lahm, A., Tramontano, A., Bonati, L., and Pitea, D. (2002). A model for recognition of polychlorinated dibenzo-*p*-dioxins by the aryl hydrocarbon receptor. *European Journal of Biochemistry*, 269, 13–18.

383. Flaveny, C. A., Murray, I. A., Chiaro, C. R., and Perdew, G. H. (2009). Ligand selectivity and gene regulation by the human aryl hydrocarbon receptor in transgenic mice. *Molecular Pharmacology*, 75, 1412–1420.

384. Flaveny, C. A. and Perdew, G. H. (2009). Transgenic humanized AHR mouse reveals differences between human and mouse AHR ligand selectivity. *Molecular and Cellular Pharmacology*, 1, 119–123.

385. McGuire, J., Okamoto, K., Whitelaw, M. L., Tanaka, H., and Poellinger, L. (2001). Definition of a dioxin receptor mutant that is a constitutive activator of transcription: delineation of overlapping repression and ligand binding functions within the PAS domain. *Journal of Biological Chemistry*, 276, 41841–41849.

386. Tauchi, M., Hida, A., Negishi, T., Katsuoka, F., Noda, S., Mimura, J., Hosoya, T., Yanaka, A., Aburatani, H., Fujii-Kuriyama, Y., Motohashi, H., and Yamamoto, M. (2005). Constitutive expression of aryl hydrocarbon receptor in keratinocytes causes inflammatory skin lesions. *Molecular and Cellular Biology*, 25, 9360–9368.

387. Walisser, J. A., Bunger, M. K., Glover, E., and Bradfield, C. A. (2004). Gestational exposure of Ahr and Arnt hypomorphs to dioxin rescues vascular development. *Proceedings of the National Academy of Sciences of the United States of America*, 101, 16677–16682.

388. Walisser, J. A., Glover, E., Pande, K., Liss, A. L., and Bradfield, C. A. (2005). Aryl hydrocarbon receptor-dependent liver development and hepatotoxicity are mediated by different cell types. *Proceedings of the National Academy of Sciences of the United States of America*, 102, 17858–17863.

389. Harstad, E. B., Guite, C. A., Thomae, T. L., and Bradfield, C. A. (2006). Liver deformation in Ahr-null mice: evidence for aberrant hepatic perfusion in early development. *Molecular Pharmacology*, 69, 1534–1541.

390. Lahvis, G. P., Lindell, S. L., Thomas, R. S., McCuskey, R. S., Murphy, C., Glover, E., Bentz, M., Southard, J., and Bradfield, C. A. (2000). Portosystemic shunting and persistent fetal vascular structures in aryl hydrocarbon receptor-deficient mice. *Proceedings of the National Academy of Sciences of the United States of America*, 97, 10442–10447.

391. Lahvis, G. P., Pyzalski, R. W., Glover, E., Pitot, H. C., McElwee, H. K., and Bradfield, C. A. (2005). The aryl hydrocarbon receptor is required for developmental closure of the ductus venosus in the neonatal mouse. *Molecular Pharmacology*, 67, 714–720.

392. Lund, A. K., Goens, M. B., Kanagy, N. L., and Walker, M. K. (2003). Cardiac hypertrophy in aryl hydrocarbon receptor null mice is correlated with elevated angiotensin II, endothelin-1, and mean arterial blood pressure. *Toxicology and Applied Pharmacology*, 193, 177–187.

393. Roman, A. C., Carvajal-Gonzalez, J. M., Rico-Leo, E. M., and Fernandez-Salguero, P. M. (2009). Dioxin receptor deficiency impairs angiogenesis by a mechanism involving VEGF-A depletion in the endothelium and transforming growth factor-beta overexpression in the stroma. *Journal of Biological Chemistry*, 284, 25135–25148.

394. Thackaberry, E. A., Bedrick, E. J., Goens, M. B., Danielson, L., Lund, A. K., Gabaldon, D., Smith, S. M., and Walker, M. K. (2003). Insulin regulation in AhR-null mice: embryonic cardiac enlargement, neonatal macrosomia, and altered insulin regulation and response in pregnant and aging AhR-null females. *Toxicological Sciences*, 76, 407–417.

395. Vasquez, A., Atallah-Yunes, N., Smith, F. C., You, X., Chase, S. E., Silverstone, A. E., and Vikstrom, K. L. (2003). A role for the aryl hydrocarbon receptor in cardiac physiology and function as demonstrated by AhR knockout mice. *Cardiovascular Toxicology*, 3, 153–163.

396. Zhang, N., Agbor, L. N., Scott, J. A., Zalabowski, T., Elased, K. M., Trujillo, A., Duke, M. S., Wolf, V., Walsh, M. T., Born, J. L., Felton, L. A., Wang, J., Kanagy, N. L., and Walker, M. K. (2010). An activated renin-angiotensin system maintains normal blood pressure in aryl hydrocarbon receptor heterozygous mice but not in null mice. *Biochemical Pharmacology*, 80, 197–204.

397. Kimura, A., Naka, T., Nakamura, T., Chinen, I., Masuda, K., Nohara, K., Fujii-Kuriyama, Y., and Kishimoto, T. (2009). Aryl hydrocarbon receptor in combination with Stat1 regulates LPS-induced inflammatory responses. *Journal of Experimental Medicine*, 206, 2027–2035.

398. Lindsey, S. and Papoutsakis, E. T. (2011). The aryl hydrocarbon receptor (AHR) transcription factor regulates megakaryocytic polyploidization. *British Journal of Haematology*, 154, 469–484.

399. Rodriguez-Sosa, M., Elizondo, G., Lopez-Duran, R. M., Rivera, I., Gonzalez, F. J., and Vega, L. (2005). Over-production of IFN-γ and IL-12 in AhR-null mice. *FEBS Letters*, 579, 6403–6410.

400. Sekine, H., Mimura, J., Oshima, M., Okawa, H., Kanno, J., Igarashi, K., Gonzalez, F. J., Ikuta, T., Kawijiri, K., and Fujii-Kuriyama, Y. (2009). Hypersensitivity of AhR-deficient mice to LPS-induced septic shock. *Molecular and Cellular Biology*, 29, 6391–6400.

401. Shi, L. Z., Faith, N. G., Nakayama, Y., Suresh, M., Steinberg, H., and Czuprynski, C. J. (2007). The aryl hydrocarbon receptor is required for optimal resistance to *Listeria monocytogenes* infection in mice. *Journal of Immunology*, 179, 6952–6962.

402. Veldhoen, M., Hirata, K., Bauer, J., Dumoutier, L., Renauld, J. C., and Stockinger, B. (2008). The aryl hydrocarbon receptor links TH17-cell-mediated autoimmunity to environmental toxins. *Nature* 453, 106–109.

403. Jux, B., Kadow, S., Luecke, S., Rannug, A., Krutmann, J., and Esser, C. (2011). The aryl hydrocarbon receptor mediates UVB radiation-induced skin tanning. *Journal of Investigative Dermatology*, 131, 203–210.

404. Abbott, B. D., Schmid, J. E., Pitt, J. A., Buckalew, A. R., Wood, C. R., Held, G. A., and Diliberto, J. J. (1999). Adverse reproductive outcomes in the transgenic Ah receptor-deficient mouse. *Toxicology and Applied Pharmacology*, 155, 62–70.

405. Baba, T., Mimura, J., Nakamura, N., Harada, N., Yamamoto, M., Morohashi, K., and Fujii-Kuriyama, Y. (2005). Intrinsic function of the aryl hydrocarbon (dioxin) receptor as a key factor in female reproduction. *Molecular and Cellular Biology*, 25, 10040–10051.

406. Barnett, K. R., Tomic, D., Gupta, R. K., Babus, J. K., Roby, K. F., Terranova, P. F., and Flaws, J. A. (2007). The aryl hydrocarbon receptor is required for normal gonadotropin responsiveness in the mouse ovary. *Toxicology and Applied Pharmacology*, 223, 66–77.

407. Barnett, K. R., Tomic, D., Gupta, R. K., Miller, K. P., Meachum, S., Paulose, T., and Flaws, J. A. (2007). The aryl hydrocarbon receptor affects mouse ovarian follicle growth via mechanisms involving estradiol regulation and responsiveness. *Biology of Reproduction*, 76, 1062–1070.

408. Benedict, J. C., Lin, T.-M., Loeffler, I. K., Peterson, R. E., and Flaws, J. A. (2000). Physiological role of the aryl hydrocarbon receptor in mouse ovary development. *Toxicological Sciences*, 56, 382–388.

409. Robles, R., Morita, Y., Mann, K. K., Perez, G. I., Yang, S., Matikainen, T., Sherr, D. H., and Tilly, J. L. (2000). The aryl hydrocarbon receptor, a basic helix–loop–helix transcription factor of the PAS gene family, is required for normal ovarian germ cell dynamics in the mouse. *Endocrinology 141*, 450–453.

410. Hushka, L. J., Williams, J. S., and Greenlee, W. F. (1998). Characterization of 2,3,7,8-tetrachlorodibenzofuran-dependent suppression and AH receptor pathway gene expression in the developing mouse mammary gland. *Toxicology and Applied Pharmacology*, 152, 200–210.

411. Lin, T.-M., Ko, K., Moore, R. W., Simanainen, U., Oberley, T. D., and Peterson, R. E. (2002). Effects of aryl hydrocarbon receptor null mutation and *in utero* and lactational 2,3,7,8-tetrachlorodibenzo-*p*-dioxin exposure on prostate and seminal vesicle development in C57BL/6 mice. *Toxicological Sciences*, 68, 479–487.

412. Carvajal-Gonzalez, J. M., Roman, A. C., Cerezo-Guisado, M. I., Rico-Leo, E. M., Martin-Partido, G., and Fernandez-Salguero, P. M. (2009). Loss of dioxin-receptor expression accelerates wound healing *in vivo* by a mechanism involving TGFβ. *Journal of Cell Science*, 122, 1823–1833.

413. Mulero-Navarro, S., Pozo-Guisado, E., Perez-Mancera, P. A., Alvarez-Barrientos, A., Catalina-Fernandez, I., Hernandez-Nieto, E., Saenz-Santamaria, J., Martinez, N., Rojas, J. M., Sanchez-Garcia, I., and Fernandez-Salguero, P. M. (2005). Immortalized mouse mammary fibroblasts lacking dioxin receptor have impaired tumorigenicity in a subcutaneous mouse xenograft model. *Journal of Biological Chemistry*, 280, 28731–28741.

414. Fernandez-Salguero, P. M., Ward, J. M., Sundberg, J. P., and Gonzalez, F. J. (1997). Lesions of aryl-hydrocarbon receptor-deficient mice. *Veterinary Pathology*, 34, 605–614.

415. Hirabayashi, Y. and Inoue, T. (2009). Aryl hydrocarbon receptor biology and xenobiotic responses in hematopoietic progenitor cells. *Biochemical Pharmacology*, 77, 521–535.

416. Lin, B. C., Nguyen, N., Walisser, J. A., and Bradfield, C. A. (2008). A hypomorphic allele of aryl hydrocarbon receptor-associated protein-9 produces a phenocopy of the Ahr-null mouse. *Molecular Pharmacology*, 74, 1367–1371.

417. Nukaya, M., Lin, B. C., Glover, E., Moran, S. M., Kennedy, G. D., and Bradfield, C. A. (2010). The aryl hydrocarbon receptor-interacting protein (AIP) is required for dioxin-induced hepatotoxicity but not for the induction of the *Cyp1a1* and *Cyp1a2* genes. *Journal of Biological Chemistry*, 285, 35599–35605.

PART II

AHR AS A LIGAND-ACTIVATED TRANSCRIPTION FACTOR

2

OVERVIEW OF AHR FUNCTIONAL DOMAINS AND THE CLASSICAL AHR SIGNALING PATHWAY: INDUCTION OF DRUG METABOLIZING ENZYMES*

QIANG MA

2.1 INTRODUCTION

The aryl hydrocarbon receptor (AHR) is a ligand-activated receptor/transcription factor of the basic region helix–loop–helix–PER/ARNT/SIM homology (bHLH–PAS) family [1]. AHR was originally discovered in the studies of induction of cytochrome P4501A1 (CYP1A1). CYP1A1 catalyzes the oxidation reactions of polycyclic aromatic hydrocarbons (PAHs) in the biotransformation of the chemicals. Most notably, monooxygenation of carcinogenic PAHs, such as benzo[a]pyrene (B[a]p), leads to the formation of electrophilic ultimate carcinogens that covalently bind macromolecules to cause cancer and toxicity; induction of CYP1A1 through AHR is a necessary step in the metabolic activation of B[a]p [2, 3]. Later, AHR was found to regulate the induction of a number of drug metabolizing enzymes (DMEs) that, in addition to CYP1A1, include CYP1A2, CYP1B1, glutathione S-transferase (GST) A1, NAD(P)H: quinone oxidoreductase 1 (NQO1), UDP-glucuronosyltransferase (UGT) 1A, and aldehyde dehydrogenase (ALDH) 3A1 (Table 2.1). Induction of the enzymes is important in the metabolism, disposition, or bioactivation of many drugs and environmental chemicals as well as certain endogenous compounds. The importance of AHR in toxicology and environmental health was recognized as it was identified as the receptor to mediate the toxic effects of the ubiquitous environmental contaminant 2,3,7,8-tetrachlorodibenzo-p-dioxin (TCDD, dioxin) and related halogenated aromatic hydrocarbons (HAHs) [4–6]. In recent years, genetic models with altered AHR activities helped uncover multiple roles of AHR in development, physiology, disease, and evolution [7–14]. From the mechanistic point of view, the studies of induction of CYP1A1 and other DMEs have been the key to elucidating the identity and function of AHR in PAH carcinogenesis, dioxin toxicity, drug metabolism, and mammalian physiology. Such studies have also served as a model for analyzing how environmental chemicals impact on human health and, as such, have influenced many areas in biology and medicine including cancer research, toxicology, drug metabolism, pharmacology, and human risk assessment over several decades [1, 3].

2.2 INDUCTION OF CYP1A1 AND THE AHR BATTERY OF DMES

DMEs are a large, diverse group of promiscuous enzymes that are responsible for metabolizing xenobiotics and certain endobiotics in mammalian species [15]. Drug metabolism in general leads to the formation of metabolites with decreased activity and increased water solubility to facilitate detoxification and elimination of the chemicals from the body. Induction of DMEs is the principal mechanism by which the body regulates the metabolism and disposition of chemicals. Induction of DMEs is generally controlled by a specific, ligand-activated receptor/transcription factor known as the xenobiotic-activated receptor (XAR), such as AHR. In most cases, an XAR regulates the induction of a battery of DMEs to coordinate the metabolism of a chemical through multiple

---

*The findings and conclusions in this chapter are those of the authors and do not necessarily represent the views of the National Institute for Occupational Safety and Health.

The AH Receptor in Biology and Toxicology, First Edition. Edited by Raimo Pohjanvirta.
© 2012 John Wiley & Sons, Inc. Published 2012 by John Wiley & Sons, Inc.

TABLE 2.1 Major DMEs and Drug Transporters Inducible Through AHR

Gene Products	Reaction/Substrate	Response Element/Induction	References
CYP1A1	Oxidation/PAHs, etc.	DRE/highly inducible	17
CYP1A2	Oxidation/heterocyclic amines/amides, etc.	DRE/major human liver form of CYP1A	18, 19
CYP1B1	Oxidation/PAHs, etc.	DRE/extrahepatic induction	20, 21
CYP2S1	Oxidation/all-*trans* retinoic acid, carcinogens	DRE/extrahepatic induction	22
CYP2A5	Oxidation/nitrosamine, aflatoxin, and coumarin	DRE, ARE induction in liver, nasal mucosa	23, 24
GSTA1	GST conjugation/electrophilic chemical	DRE, ARE	25, 26
UGT1A1, 1A6, 1A9	Glucuronidation/bilirubin drugs	DRE, ARE	27–31
NQO1	Two-electron reduction/quinones	DRE, ARE	32
ALDH3A1	Dehydrogenation/retinaldehyde, aldehydes	DRE/highly inducible	33
MRP2, 3, 5, 6, 7	Transport/drugs and some endobiotics	DRE? ARE?	34

reactions by multiple DMEs [16]. Induction accelerates the metabolism and usually reduces the efficacy and toxicity of drugs and other chemicals, thereby maintaining the chemical homeostasis of cells in the presence of chemical challenge. Under certain circumstances, induction enhances the activity of a chemical, such as in the case of metabolic activation of B[*a*]p, or increases the propensity for undesirable chemical–chemical interaction if induction of a DME affects the disposition of other, coexposed chemicals.

The AHR battery of inducible DMEs consists of dozens of enzymes that catalyze a wide range of xenobiotic metabolism reactions. Induction of the enzymes requires AHR and most of the genes encoding the enzymes have been shown to contain one or several copies of a common DNA sequence called dioxin-responsive element (DRE) in the enhancer region to which AHR binds to increase the rate of transcription of the genes (Table 2.1) [17–34]. Among the DMEs, CYP1A1 is expressed in many tissues in animals and humans at a low level but is highly inducible (up to 10–100-fold). Although CYP1A1 is well known for its role in the metabolic activation of B[*a*]p and other carcinogenic PAHs to ultimate carcinogens, such as the *trans*-7,8-diol 9,10-epoxide of B[*a*]p, the enzyme induced in the small intestine and liver is critical for reducing the systemic exposure to B[*a*]p exposed through the digestive tract [35]. CYP1A2 is the major hepatic form of CYP1A enzymes and is inducible through AhR. The enzyme metabolizes a range of drugs, such as caffeine, melatonin, theophylline, and lidocaine, and is notable for its capacity to *N*-oxidize heterocyclic aromatic amines/amides to carcinogens. In addition, CYP1A2 binds TCDD in the liver and thereby modulates its kinetics when induced. CYP1B1 is highly inducible by AHR ligands and is involved in 7,12-dimethylbenz[*a*]anthracene-induced lymphomas and B[*a*]p-induced toxicity in the bone marrow. CYP2S1 catalyzes the oxidation of all-*trans* retinoic acid and certain carcinogens, whereas CYP2A5 oxygenates nitrosamines, aflatoxin, and coumarin. Both CYP2S1 and CYP2A5 are induced by TCDD and B[*a*]p. GSTA1, some UGTs (1A1, 1A6, and 1A9), and NQO1 are a group of DMEs that are induced by both AHR ligands and oxidant/electrophilic chemicals. ALDH3A1 oxidizes aldehydes, such as retinaldehyde, and is inducible via AHR. In recent studies, a group of drug transporters including the multidrug resistance-associated proteins (MRPs) 2, 3, 5, 6, and 7 were shown to be induced by AHR ligands to modulate the disposition and elimination of many exogenous and endogenous chemicals and metabolites in the liver and other tissues (Table 2.1).

2.3 DEFINING MAJOR COMPONENTS OF CYP1A1 INDUCTION

Early studies employed genetic, pharmacological, and molecular approaches to identify the major components necessary for the induction of CYP1A1 and establish the framework of the signal transduction pathway of induction.

The discovery of induction of CYP1A enzymes originates from the observation that PAHs induce their own metabolism. In the early 1950s, coadministration of 3′-methyl-4-dimethylaminoazobenzene (3′-Me-DAB), a potent hepatocarcinogen, with a low dose of 3-methylcholanthrene (3-MC), a carcinogenic PAH, was found to totally inhibit or delay the carcinogenic and hepatotoxic effects of 3′-Me-DAB [36]. Importantly, inhibition of aminoazo dye-induced liver cancer and cirrhosis by PAH correlated with induction of azo dye *N*-demethylase that catalyzes the hepatic *N*-demethylation and azo-link reduction of aminoazo dyes, metabolic pathways resulting in noncarcinogenic products [37]. These and subsequent studies demonstrated that PAH induces its own metabolism to protect animals from some chemical-induced cancer and toxicity and, thus, opened the gateway of research on induction of CYPs and other DMEs [38].

Characterization of induction of aryl hydrocarbon hydroxylase (AHH), which is measured as the enzymatic activity catalyzing B[*a*]p 3-hydroxylation, in inbred mice contributed significantly to the understanding of the mechanism of CYP1A1 induction. Some inbred mouse strains such as C57BL/6 (B6) are sensitive to induction by 3-MC, whereas other strains such as DBA/2 (D2) are resistant to the induction. Moreover, the sensitive phenotype segregates as a single autosomal dominant trait [39, 40]. The polymorphism of the genetic trait defines a genetic locus known as the aromatic hydrocarbon responsiveness or *Ah* locus. Ah^b represents the "Ah responsive" allele and Ah^d the "Ah nonresponsive" allele.

Among the inducers of CYP1A1, TCDD was found to be the most potent. TCDD and 3-MC produce parallel dose–response curves in AHH induction, but TCDD is 30,000 times more potent than 3-MC and induction by TCDD is persistent due to a long half-life of TCDD in the body [41]. Induction of AHH by TCDD is also observed in "nonresponsive" strains, albeit with a 10-fold higher ED_{50} compared to that in "responsive" strain (10 versus 1 nmol/kg body weight); induction in the heterozygous genotype (b/d) is intermediate between b/b and d/d homozygous [41]. Induction by TCDD and its congeners showed stereospecificity. Finally, radiolabeled TCDD binds to a cytoplasmic preparation of mouse liver in a saturable and reversible manner with a high affinity. These studies provided pharmacological evidence that the Ah locus product functions as an "induction receptor," hence the name "Ah receptor," which binds the inducers and mediates induction [41–43].

The mouse hepatoma cell line, Hepa1c1c7, is a powerful cell system for studying the mechanism of induction of CYP1A1 [43]. Induction in the cell is rapid and robust. Furthermore, variant hepatoma cells that are defective in the induction were obtained by using two independent approaches: (a) a single-step selection procedure, in which B[*a*]p induces CYP1A1 in wild-type, but not variant cells, leading to formation of toxic metabolites and cell death in wild-type cells and selective growth of variant cells; and (b) fluorescence-activated cell sorter (FACS), in which variant cells were selected because they exhibit lower or higher rates in the metabolism of B[*a*]p [44, 45]. Cell fusion analyses of the variant cells revealed that the variants fell into several complementation groups. The class I variant exhibits diminished binding to inducers and is defective in AHR, whereas the class II variant is defective in a protein termed Ah receptor nuclear translocator (ARNT) because the cells exhibit abnormal nuclear localization of ligand-activated AHR. The third class of variants lacks functional CYP1A1. The findings imply that multiple gene products contribute to the induction mechanism; AHR, ARNT, and functional CYP1A1 are major components required for AHH induction.

The identification of the transcription enhancer and promoter responsible for induction of CYP1A1 is a classical example of molecular approaches to defining the modular DNA response element required for inducible gene transcription [1]. Two assays were particularly useful in defining the ligand-dependent AHR–DNA interaction: the enhancer/promoter-driven reporter assay that allows the isolation of functional enhancer and promoter sequences of the gene, and the electrophoretic mobility shift assay that defines the consensus DNA sequences to which AHR and ARNT bind. Several copies of sequences in the enhancer region of CYP1A1 were identified as the DNA component specific for CYP1A1 induction and were designated as DRE [17]. The consensus DRE sequence has a core of 5′-TNGCGTG-3′ and the 5′-CGTG-3′ motif is required for inducible, AHR-dependent, and ARNT-dependent interactions at the enhancer. Protein–DNA interactions occur within the major groove of DNA and one AHR molecule binds to a DRE. The CYP1A1 promoter contains a TATA site and the promoter is silent in the absence of inducer [46]. These analyses revealed an enhancer-dependent gene transcription for CYP1A1 induction. These studies opened a new era of molecular studies of AHR for AHR–DNA interaction, chromatin modification, new target gene search, and, lately, the genomic response to TCDD and other AHR ligands [1].

2.4 MOLECULAR CHARACTERIZATION OF AHR AND ARNT

A photoaffinity ligand of AHR (2-azido-3-[^{125}I]iodo-7,8-dibromodibenzo-*p*-dioxin) was developed to covalently label AHR from the mouse liver and purify the protein to apparent homogeneity under denaturing conditions [47]. A partial, amino-terminal sequence was obtained from purified receptor. Cloning of the AHR cDNA (*b*-1 allele) using degenerate oligonucleotide probes derived from the peptide sequence revealed that the mouse AHR contains 805 amino acid residues with a structural organization that is representative of bHLH–PAS transcription factors (Fig. 2.1) [48]. The N-terminus contains a stretch of basic residues followed by a helix–loop–helix turn motif; the basic residues are critical for specific AHR–DRE interaction and the HLH motif mediates heterodimerization between AHR and ARNT. Following bHLH are two imperfect inverted repeats of 51-amino acid residues in length that were also found in PER—a circadian transcription factor, ARNT, and SIM—the *Drosophila* "single-minded" protein involved in neuronal development, hence the name "PAS". The PAS motif of AHR has multiple functions contributing to binding of AHR to Hsp90 and AIP—an immunophilin-like chaperone—in the cytoplasm [49], ligand binding, and AHR–ARNT interaction in the nucleus. The carboxyl half of AHR contains three separable transcription activation (TA) domains that are characteristic of acidic, serine/threonine-rich, and glutamine-rich TA motifs [50]. Immediately after the PAS-B region and before the TA domains, a short peptide of ∼81 residues was identified to have an inhibitory activity responsible for suppression of the TA activity of AHR in the absence of an agonist; the motif was named as the inhibitory domain (ID) [50].

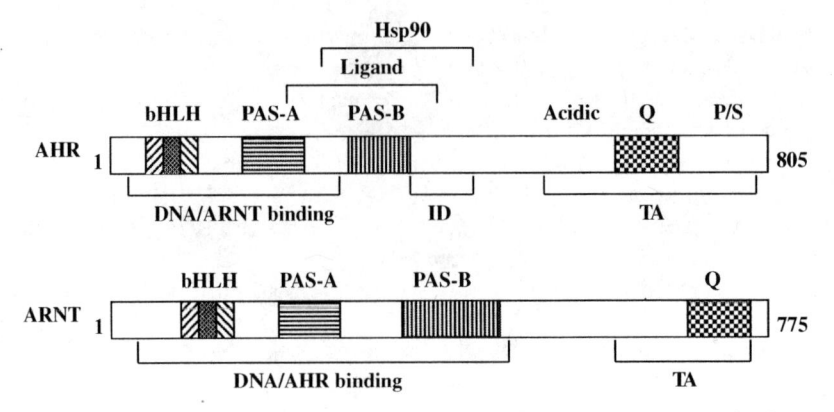

FIGURE 2.1 Domain structures of mouse AHR and ARNT. Numbers represent the first and last amino acid residues of mouse AHR (*b*-1 allele) and ARNT. bHLH, basic region helix–loop–helix; PAS, PER–ARNT–SIM homology; ID, inhibitory domain; Q, glutamine-rich; P/S, proline- and serine-rich.

ARNT cDNA was cloned by complementation of the class II variant cells in which liganded AHR fails to accumulate in the nucleus [51, 52]. ARNT is localized in the nucleus. Heterodimerization between AHR and ARNT in the nucleus is necessary, at least, for three major processes of induction: (a) nuclear enrichment of ligand-activated AHR; (b) binding of the AHR/ARNT dimer to DRE; and (c) activation of the transactivation activity of the AHR TA domains required for transcription of the endogenous *Cyp1a1* gene in intact cells. The structural organization of ARNT is similar to that of AHR (Fig. 2.1). The N-terminal region contains a bHLH domain and a nuclear localization signal sequence, followed by the PAS-A and PAS-B motifs. The C-terminal half contains a potent TA domain. In addition to AHR, ARNT dimerizes with the hypoxia-inducible factor 1α (HIF1α) to mediate the transcriptional response to low oxygen tension.

Heterodimerization of AHR and ARNT generates a functional complex with unique properties. The basic region of ARNT binds to the 5′-GTG-3′ half of the DRE, whereas the basic region of AHR binds the neighboring nucleotides. The HLH and PAS motifs of ARNT are involved in binding to AHR but not ligands. As noted above, the transactivation capacity of AHR is suppressed in the intact protein by its ID motif and is revealed *in vitro* when ID is deleted [50]. On the other hand, the transactivation activity of ARNT is constitutively expressed. However, reconstitution of induction of endogenous *Cyp1a1* in AHR- or ARNT-deficient variants cells revealed that the AHR TA domain but not the ARNT TA domain is required for induction [53]. Reconstitution is also achieved with a heterologous TA domain in place of AHR TA [54]. It is possible that the AHR/ARNT complex has versatile transactivation capacity *in vivo* to enable it to transcribe target genes in different regulatory contexts.

2.5 OTHER COMPONENTS OF INDUCTION

In addition to AHR, ARNT, and DRE, several protein and DNA factors were found to interact with the AHR pathway and modulate CYP1A1 induction at different steps.

The unliganded, cytoplasmic AHR exists in a complex that, in addition to AHR, contains two molecules of Hsp90 [55, 56] and a protein of ∼37 kDa known as the AHR-interacting protein (AIP) [49]; AIP was also named as ARA9 and Xap2 [57, 58]. Interaction between Hsp90 and AHR PAS-B motif may be necessary to create the ligand binding cavity. Reconstitution of AHR/ARNT/DRE-mediated gene induction in Hsp90-defective yeast strains showed that Hsp90 is required for induction by TCDD [55, 56]. The AIP protein is an immunophilin-like chaperone that contains three copies of tetratricopeptide repeats (TPR), a motif that mediates protein–protein interactions [49]. Overexpression of AIP increases the extent of CYP1A1 induction. Knockout of AIP in mice results in embryonic death, whereas reduced expression of the AIP protein is correlated with the appearance of vascular abnormality that is also observed in AHR-null mice, suggesting that AIP is essential for AHR signaling during development [59].

Unliganded AHR is a stable protein with a half-life ($t_{1/2}$) of ∼28 h, consistent with the observation that AHR-mediated induction of *Cyp1a1* mRNA does not require new protein synthesis. On the other hand, treatment with an agonist, such as TCDD or B[*a*]p, induces rapid turnover of the AHR protein to shorten its $t_{1/2}$ to 3 h [60]. Several lines of evidence implicate the ubiquitin 26S proteasome pathway in agonist-induced degradation of AHR: (a) the 26S proteasome inhibitors MG132 and lactacystin block TCDD-induced AHR degradation; (b) TCDD induces the polyubiquitination of AHR that is further enhanced by MG132; (c) degradation depends on the presence of E1, an enzyme required for the activation of ubiquitin for protein polyubiquitination; and (d) MG132 increases the amount of the AHR protein and AHR–DRE binding in the nucleus and augments CYP1A1 induction [60]. Thus, agonist-activated degradation of AHR likely occurs in the nucleus to downregulate nuclear AHR.

A prominent role of negative regulation of AHR in CYP1A1 induction is clearly demonstrated in the "superinduction" of CYP1A1, in which pre- or cotreatment of cells with a protein synthesis inhibitor, such as cycloheximide, in the presence of TCDD or other AHR agonists increases the induction to 5–10-fold higher than treatment with the agonist or cycloheximide alone, hence the term "superinduction" [61, 62]. Superinduction is mediated through AHR, occurs at the level of transcription, and is also observed with other AHR target genes [62, 63]. The findings implicate a negative regulatory mechanism in controlling the extent of CYP1A1 induction such that induction does not reach its maximal capacity but is adequate and flexible to the physiological need of the cell and can be terminated promptly, whereas a maximal induction may be detrimental to the cell.

Mechanistically, superinduction suggests that a cycloheximide-sensitive protein factor—either a short-lived or an inducible protein or both—is responsible for the suppression of induction. Two possible modes of action of the labile inhibitor have been raised: the protein interacts with a specific DNA sequence to inhibit CYP1A1 transcription; alternatively, the protein inhibits the AHR signal transduction to suppress induction. Although cis-acting inhibitory DNA sequences have been identified to inhibit constitutive and/or inducible Cyp1a1 transcription, evidence of binding of a labile protein to the DNA elements to inhibit induction is lacking. On the other hand, cycloheximide was found to totally block TCDD-induced ubiquitination and proteasomal degradation of AHR at concentrations at which it superinduces Cyp1a1. The findings suggest that a labile protein factor controls the agonist-induced AHR degradation and hence it was named AHR degradation promoting factor (ADPF) [62]; inhibition of the synthesis of the protein by cycloheximide removes such downregulation of AHR, resulting in the superinduction of the gene [64, 65]. The nature of the labile protein has not been elucidated.

The low level of constitutive mRNA expression of Cyp1a1 in most cells and tissues is in agreement with the observations that both the enhancer and the promoter of Cyp1a1 assume a nucleosomal configuration in which the DNA sequences associate with histones and other chromosomal proteins to form chromatin, which limits the access of transcription factors and DNA polymerase II to the promoter for transcription of the gene [46]. It is conceivable that heterodimerization of liganded AHR with ARNT in the nucleus enables the dimer to bind to DRE, during which considerable alteration of the chromatin structure at the enhancer occurs.

The binding triggers the recruitment of coactivators and chromatin-modifying complexes to open the chromatin at the promoter and recruit the general transcription machinery to the gene. Since the TA domain of AHR, but not that of ARNT, is required and sufficient for induction, it is postulated that the AHR TA is exposed in the AHR/ARNT/DRE complex and recruits the coactivators and chromatin-modifying complexes to the promoter to mediate induction [50, 53, 54].

2.6 THE CLASSICAL SIGNALING PATHWAY OF GENE INDUCTION BY AHR

The identification of AHR and ARNT as the principal protein components and DRE as the cis-acting DNA component for CYP1A1 induction, the characterization of the interactions between the DNA and protein components, and the cloning of other protein factors contributing to AHR signaling delineate the so-called "classical" signaling pathway of AHR, also known as the AHR/ARNT/DRE paradigm, for gene induction (Fig. 2.2). TCDD or other inducers bind to cytoplasmic AHR, which is in a complex with Hsp90 and AIP. Activation of the AHR by an agonist triggers a series of sequential signaling events that include dissociation of AHR from Hsp90 and AIP, translocation of liganded AHR into the nucleus, dimerization of AHR with ARNT, binding of the AHR/ARNT dimer to DREs, activation of the TA domain of AHR, and finally transcription of Cyp1a1. Transcription involves disruption of chromatin structures associated with the Cyp1a1 enhancer and promoter. The labile ADPF promotes the polyubiquitination of agonist-activated AHR in the nucleus, leading to rapid degradation of the AHR protein through the 26S proteasome pathway to negatively control and terminate the transcription by nuclear AHR.

2.7 EXPANDING ROLE OF THE AHR/ARNT/DRE PARADIGM IN AHR BIOLOGY

The AHR/ARNT/DRE paradigm of CYP1A1 induction described above has been useful for generating testable hypothesis for mechanistic studies on the induction of CYP1A1 and the AHR battery of DMEs. It has also been the key to understanding other biological responses to AHR ligands, such as TCDD toxicity, and the endogenous functions of AHR. In fact, many of these AHR functions appear to result from the modulation of the "classical" signaling pathway of AHR by other signaling molecules and pathways.

It is known that a subset of DMEs including GSTA1, some UGTs, and NQO1 are induced by both AHR agonists, such as dioxin and B[a]p, and oxidant/antioxidant/electrophilic chemicals, such as reactive oxygen species, transition

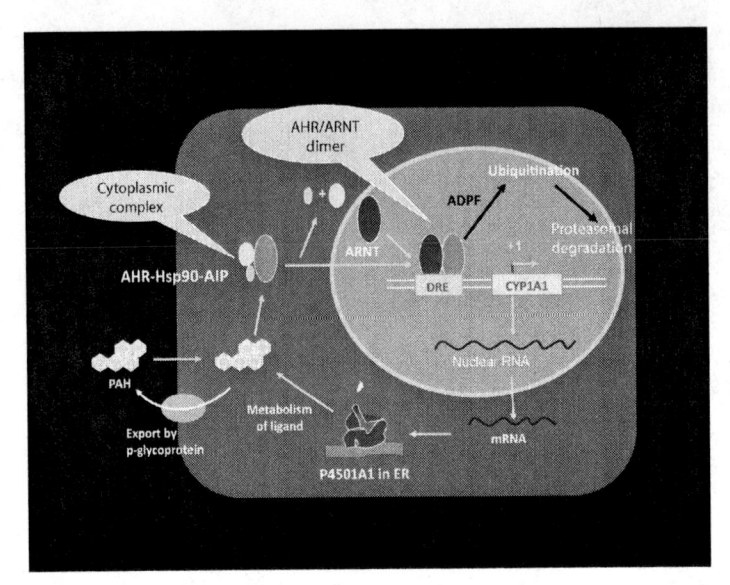

FIGURE 2.2 The "classical" pathway of AHR signaling for CYP1A1 induction. (See the color version of this figure in Color Plates section.)

metals, and natural and synthetic antioxidant agents (Table 2.1) [25–28, 30, 32, 66, 67]. In the former case, induction requires AHR, ARNT, and DRE, whereas Nrf2 and ARE dictate the basal expression and induction of the enzymes in the latter scenario. Interestingly, induction of the enzymes by AHR agonists also requires the presence of functional Nrf2, implicating a role of Nrf2 in AHR signal transduction for induction of the enzymes [68]. Recent studies revealed a growing list of genes inducible by both TCDD and antioxidants including drug transporters, such as MRPs [34]; whether AHR interacts with Nrf2 to govern the induction of these genes similarly to that of NQO1 is currently unclear. The mechanism by which Nrf2 affects the AHR pathway in the induction of DMEs and drug transporters remains to be elucidated. Conceptually, such interaction broadens the regulatory repertoire or forms a regulatory safety net in the metabolism of chemicals.

ARNT is recruited to dimerize with HIF1α in response to hypoxia [69]. The HIF1α/ARNT dimer binds to an enhancer containing the same core tetranucleotide motif 5′-CGTG-3′ that AHR/ARNT recognizes. The HIF1α/Arnt controls the induction of a number of glycolytic enzymes, erythropoietin, and genes important in vascular growth in response to hypoxia. For this reason, ARNT is also named as HIF1β. These findings raise the possibility that ARNT becomes a limiting factor for both AHR and HIF1α-mediated responses if both pathways are active in a cell. Indeed, it was shown that exposure to hypoxia reduces induction of AHR target genes by B[a]p in cultured cells; similar results were obtained with TCDD *in vivo* [70, 71]. ARNT was also found to serve as a coactivator of estrogen receptor beta (ERβ). Recruitment of ARNT to AHR by TCDD or to HIF1α by hypoxia reduces recruitment of ARNT to estradiol-regulated promoters [72].

The competition for ARNT between AHR and ERβ in the presence of TCDD may partially explain the antiestrogenic effect of TCDD.

Inflammation is a common component of disease and xenobiotic response. Inflammation may modulate AHR functions, such as DME induction, by way of modulating NF-κB function. Experimental results showed that AHR interacts with NF-κB in a ligand-dependent manner. AHR and NF-κB form a heterodimer to suppress gene transcription through the DRE or the NF-κB binding elements in reporter assays [73]. In mammary cells, AHR and the RelA NF-κB subunit cooperate to bind to NF-κB binding elements to induce c-myc expression [74]. These findings provide new opportunities to study the mechanism by which AHR mediates the immunotoxicity and tumor formation by TCDD.

Although AHR exhibits very high affinity and specificity toward some ligands such as TCDD, it also binds a wide range of structurally diverse chemicals with somewhat lower affinities [75], supporting the notion that AHR functions as a promiscuous xenobiotic-activated receptor. Omeprazole, bilirubin, and tryptophan metabolites, including indole-3-carbinol, indigo, and indirubin, are a few examples. Furthermore, some of the newly identified ligands of AHR, such as tryptophan photoproducts, lipoxin A4, and prostaglandin G2, are produced endogenously and, thus, likely serve as endogenous ligands of AHR for some of the physiological and developmental functions of AHR [75, 76].

Because AHR is a transcription factor, it is generally accepted that most, if not all, of the biological functions of AHR are mediated by modulating the expression of a range of AHR target genes in ligand-dependent manners similarly to the AHR/ARNT/DRE paradigm for CYP1A1 induction. However, the lack of knowledge on AHR target genes other

TABLE 2.2 Some AHR Target Genes Relevant to AHR Functions in Physiology and TCDD Toxicity

Gene Products	Potential Relevance to AHR Function	Effect of AHR on Expression	References
Bax	PAH-induced premature ovarian failure	DRE-dependent induction in oocytes	77
Cathepson D	Endocrine disruption by TCDD	DRE-dependent suppression of induction by estrogen in MCF-7 cells	78
AhRR	Negative regulation of AHR	DRE-dependent induction	79
c-myc	Tumorigenesis	DRE-dependent suppression of c-myc expression in mammary tumor cells	80
pS2	Endocrine disruption by TCDD	Suppression by TCDD	81
IgM μ gene	Inhibition of IgM production by TCDD	DRE-dependent suppression of IgMμ gene expression	82
IL-2	Thymus involution	DRE-dependent induction by TCDD in thymocytes	83
ILGFBP-1	Diabetic effect of TCDD	DRE-dependent induction by TCDD	84
$p27^{KIP1}$	TCDD-induced suppression of cell proliferation	Induction in developing thymus and hepatoma cells by TCDD	85
$p27^{CIP1}$	Inhibition of G(I) to S transition by TCDD	DRE-dependent suppression by TCDD	86
KLF2	Effect on early thymopoiesis	Premature induction of KLF2 in thymocytes (CD4, CD8, and CD3)	87
TiPARP	T-cell function	Induction by TCDD	88
MHCQ1b	Immunosuppression by TCDD	Suppression by TCDD	89
IL-6	Inhibition of dendritic cell-dependent T helper cell development	AHR-dependent suppression by antiallergic agent	90
IL-22	Enhancing T_H17 cell development and autoimmune pathology	Promoting IL-22 expression in T_H17 cell by AHR	14

than drug metabolizing enzymes has been a major challenge to this postulate for a long time. In recent years, many new target genes of AHR have been identified whose expressions clearly depend on the presence of functional AHR and ARNT, and in many cases, DREs or DRE-like DNA elements. Induction or suppression of the genes is often elicited by specific AHR ligands in specific tissues and cell types to account for many of the AHR functions, such as tissue-specific toxicities of TCDD. Table 2.2 provides a partial list of new AHR target genes; many of the genes are involved in critical physiological and pathological functions, including apoptosis, cell cycle control, oncogenesis, endocrine disruption, immune suppression, and anti-inflammatory response [14, 77–90]. As more of the AHR functions are being analyzed at the whole genome levels now and in the near future, it can be anticipated that this list of AHR target genes will expand rapidly.

2.8 CONCLUSIONS

In conclusion, numerous examples have been provided in which mechanisms involving new protein–protein and protein–DNA interactions, new AHR ligands that are either endogenous or exogenous, and new target genes with DRE or DRE-like elements interact with the "classical" AHR/ARNT/DRE pathway. It is foreseeable that these interactions not only provide mechanistic explanations to many of the species-, tissue-, cell type-, ligand-, and developmental stage-specific functions of AHR, but also reveal new directions for future studies of the biological and toxicological functions of AHR.

REFERENCES

1. Whitlock, J. P., Jr. (1999). Induction of cytochrome P4501A1. *Annual Reviews in Pharmacology and Toxicology*, 39, 103–125.
2. Conney, A. H. (1982). Induction of microsomal enzymes by foreign chemicals and carcinogenesis by polycyclic aromatic hydrocarbons: G. H. A. Clowes Memorial Lecture. *Cancer Research*, 42, 4875–4917.
3. Ma, Q. and Lu, A. Y. (2007). CYP1A induction and human risk assessment: an evolving tale of *in vitro* and *in vivo* studies. *Drug Metabolism and Disposition*, 35, 1009–1016.
4. Poland, A. and Knutson, J. C. (1982). 2,3,7,8-Tetrachlorodibenzo-*p*-dioxin and related halogenated aromatic hydrocarbons: examination of the mechanism of toxicity. *Annual Reviews in Pharmacology and Toxicology*, 22, 517–554.
5. Fernandez-Salguero, P. M., Hilbert, D. M., Rudikoff, S., Ward, J. M., and Gonzalez, F. J. (1996). Aryl-hydrocarbon receptor-deficient mice are resistant to 2,3,7,8-tetrachlorodibenzo-*p*-

dioxin-induced toxicity. *Toxicology and Applied Pharmacology*, 140, 173–179.

6. White, S. S. and Birnbaum, L. S. (2009). An overview of the effects of dioxins and dioxin-like compounds on vertebrates, as documented in human and ecological epidemiology. *Journal of Environmental Science and Health, Part C*, 27, 197–211.

7. Lahvis, G. P., Pyzalski, R. W., Glover, E., Pitot, H. C., McElwee, M. K., and Bradfield, C. A. (2005). The aryl hydrocarbon receptor is required for developmental closure of the ductus venosus in the neonatal mouse. *Molecular Pharmacology*, 67, 714–720.

8. Zaher, H., Fernandez-Salguero, P. M., Letterio, J., Sheikh, M. S., Fornace, A. J., Jr., Roberts, A. B., and Gonzalez, F. J. (1998). The involvement of aryl hydrocarbon receptor in the activation of transforming growth factor-beta and apoptosis. *Molecular Pharmacology*, 54, 313–321.

9. Bock, K. W. and Kohle, C. (2009). The mammalian aryl hydrocarbon (Ah) receptor: from mediator of dioxin toxicity toward physiological functions in skin and liver. *Biological Chemistry*, 390, 1225–1235.

10. Esser, C., Rannug, A., and Stockinger, B. (2009). The aryl hydrocarbon receptor in immunity. *Trends in Immunology*, 30, 447–454.

11. Singh, K. P., Casado, F. L., Opanashuk, L. A., and Gasiewicz, T. A. (2009). The aryl hydrocarbon receptor has a normal function in the regulation of hematopoietic and other stem/progenitor cell populations. *Biochemical Pharmacology*, 77, 577–587.

12. Ma, Q. and Whitlock, J. P., Jr. (1996). The aromatic hydrocarbon receptor modulates the Hepa 1c1c7 cell cycle and differentiated state independently of dioxin. *Molecular and Cellular Biology*, 16, 2144–2150.

13. Ma, Q. (2008). Ah receptor: xenobiotic response meets inflammation. *Blood*, 112, 928–929.

14. Veldhoen, M., Hirota, K., Westendorf, A. M., Buer, J., Dumoutier, L., Renauld, J. C., and Stockinger, B. (2008). The aryl hydrocarbon receptor links T_H17-cell-mediated autoimmunity to environmental toxins. *Nature*, 453, 106–109.

15. Ma, Q. and Lu, A. Y. H. (2008). The challenges of dealing with promiscuous drug-metabolizing enzymes, receptors and transporters. *Current Durg Metabolism*, 9, 374–383.

16. Ma, Q. (2008). Xenobiotic-activated receptors: from transcription to drug metabolism to disease. *Chemical Research in Toxicology*, 21, 1651–1671.

17. Denison, M. S., Fisher, J. M., and Whitlock, J. P., Jr. (1988). Inducible, receptor-dependent protein–DNA interactions at a dioxin-responsive transcriptional enhancer. *Proceedings of the National Academy of Sciences of the United States of America*, 85, 2528–2532.

18. Strom, D. K., Postlind, H., and Tukey, R. H. (1992). Characterization of the rabbit CYP1A1 and CYP1A2 genes: developmental and dioxin-inducible expression of rabbit liver P4501A1 and P4501A2. *Archives of Biochemistry and Biophysics*, 294, 707–716.

19. Nukaya, M., Moran, S., and Bradfield, C. A. (2009). The role of the dioxin-responsive element cluster between the Cyp1a1 and Cyp1a2 loci in aryl hydrocarbon receptor biology. *Proceedings of the National Academy of Sciences of the United States of America*, 106, 4923–4928.

20. Sutter, T. R., Tang, Y. M., Hayes, C. L., Wo, Y. Y., Jabs, E. W., Li, X., Yin, H., Cody, C. W., and Greenlee, W. F. (1994). Complete cDNA sequence of a human dioxin-inducible mRNA identifies a new gene subfamily of cytochrome P450 that maps to chromosome 2. *Journal of Biological Chemistry*, 269, 13092–13099.

21. Zhang, L., Savas, U., Alexander, D. L., and Jefcoate, C. R. (1998). Characterization of the mouse Cyp1B1 gene. Identification of an enhancer region that directs aryl hydrocarbon receptor-mediated constitutive and induced expression. *Journal of Biological Chemistry*, 273, 5174–5183.

22. Rivera, S. P., Wang, F., Saarikoski, S. T., Taylor, R. T., Chapman, B., Zhang, R., and Hankinson, O. (2007). A novel promoter element containing multiple overlapping xenobiotic and hypoxia response elements mediates induction of cytochrome P4502S1 by both dioxin and hypoxia. *Journal of Biological Chemistry*, 282, 10881–10893.

23. Arpiainen, S., Raffalli-Mathieu, F., Lang, M. A., Pelkonen, O., and Hakkola, J. (2005). Regulation of the Cyp2a5 gene involves an aryl hydrocarbon receptor-dependent pathway. *Molecular Pharmacology*, 67, 1325–1333.

24. Lamsa, V., Levonen, A. L., Leinoene, H., Yla-Herttuala, S., Yamamoto, M., and Hakkola, J. (2010). Cytochrome P450 2A5 constitutive expression and induction by heavy metals is dependent on redox-sensitive transcription factor Nrf2 in liver. *Chemical Research in Toxicology*, 23, 977–985.

25. Rushmore, T. H., King, R. G., Paulson, K. E., and Pickett, C. B. (1990). Regulation of glutathione S-transferase Ya subunit gene expression: identification of a unique xenobiotic-responsive element controlling inducible expression by planar aromatic compounds. *Proceedings of the National Academy of Sciences of the United States of America*, 87, 3826–3830.

26. Rushmore, T. H. and Pickett, C. B. (1990). Transcriptional regulation of the rat glutathione S-transferase Ya subunit gene. Characterization of a xenobiotic-responsive element controlling inducible expression by phenolic antioxidants. *Journal of Biological Chemistry*, 265, 14648–14653.

27. Yueh, M. F., Huang, Y. H., Hiller, A., Chen, S., Nguyen, N., and Tukey, R. H. (2003). Involvement of the xenobiotic response element (XRE) in Ah receptor-mediated induction of human UDP-glucuronosyltransferase 1A1. *Journal of Biological Chemistry*, 278, 15001–15006.

28. Yueh, M. F. and Tukey, R. H. (2007). Nrf2-Keap1 signaling pathway regulates human UGT1A1 expression *in vitro* and in transgenic UGT1 mice. *Journal of Biological Chemistry*, 282, 8749–8758.

29. Emi, Y., Ikushiro, S., and Iyanagi, T. (1996). Xenobiotic responsive element-mediated transcriptional activation in the UDP-glucuronosyltransferase family 1 gene complex. *Journal of Biological Chemistry*, 271, 3952–3958.

30. Munzel, P. A., Schmohl, S., Heel, H., Kalberer, K., Bock-Hennig, B. S., and Bock, K. W. (1999). Induction of human UDP glucuronosyltransferases (UGT1A6, UGT1A9, and UGT2B7) by *t*-butylhydroquinone and 2,3,7,8-tetrachlorodi-

benzo-*p*-dioxin in Caco-2 cells. *Drug Metabolism and Disposition*, 27, 569–573.

31. Munzel, P. A., Schmohl, S., Buckler, F., Jaehrling, J., Raschko, F. T., Kohle, C., and Bock, K. W. (2003). Contribution of the Ah receptor to the phenolic antioxidant-mediated expression of human and rat UDP-glucuronosyltransferase UGT1A6 in Caco-2 and rat hepatoma 5L cells. *Biochemical Pharmacology*, 66, 841–847.

32. Favreau, L. V. and Pickett, C. B. (1991). Transcriptional regulation of the rat NAD(P)H:quinone reductase gene. Identification of regulatory elements controlling basal level expression and inducible expression by planar aromatic compounds and phenolic antioxidants. *Journal of Biological Chemistry*, 266, 4556–4561.

33. Vasiliou, V., Reuter, S. F., Williams, S., Puga, A., and Nebert, D. W. (1999). Mouse cytosolic class 3 aldehyde dehydrogenase (Aldh3a1): gene structure and regulation of constitutive and dioxin-inducible expression. *Pharmacogenetics*, 9, 569–580.

34. Maher, J. M., Cheng, X., Slitt, A. L., Dieter, M. Z., and Klaassen, C. D. (2005). Induction of the multidrug resistance-associated protein family of transporters by chemical activators of receptor-mediated pathways in mouse liver. *Drug Metabolism and Disposition*, 33, 956–962.

35. Uno, S., Dalton, T. P., Dragin, N., Curran, C. P., Derkenne, S., Miller, M. L., Shertzer, H. G., Gonzalez, F. J., and Nebert, D. W. (2006). Oral benzo[*a*]pyrene in Cyp1 knockout mouse lines: CYP1A1 important in detoxication, *CYP1B1 metabolism required for immune damage independent of total-body burden and clearance rate. Molecular Pharmacology*, 69, 1103–1114.

36. Richardson, H. L., Stier, A. R., and Borsos-Nachtnebel, E. (1952). Liver tumor inhibition and adrenal histologic responses in rats to which 3′-methyl-4-dimethylaminoazobenzene and 20-methylcholanthrene were simultaneously administered. *Cancer Research*, 12, 356–361.

37. Conney, A. H., Miller, E. C., and Miller, J. A. (1956). The metabolism of methylated aminoazo dyes. V. Evidence for induction of enzyme synthesis in the rat by 3-methylcholanthrene. *Cancer Research*, 16, 450–459.

38. Conney, A. H. (2003). Induction of drug-metabolizing enzymes: a path to the discovery of multiple cytochromes P450. *Annual Review of Pharmacology and Toxicology*, 43, 1–30.

39. Gielen, J. E., Goujon, F. M., and Nebert, D. W. (1972). Genetic regulation of aryl hydrocarbon hydroxylase induction. II. Simple Mendelian expression in mouse tissues *in vivo*. *Journal of Biological Chemistry*, 247, 1125–1137.

40. Thomas, P. E., Kouri, R. E., and Hutton, J. J. (1972). The genetics of aryl hydrocarbon hydroxylase induction in mice: a single gene difference between C57BL-6J and DBA-2J. *Biochemical Genetics*, 6, 157–168.

41. Poland, A. and Glover, E. (1974). Comparison of 2,3,7,8-tetrachlorodibenzo-*p*-dioxin, a potent inducer of aryl hydrocarbon hydroxylase, with 3-methylcholanthrene. *Molecular Pharmacology*, 10, 349–359.

42. Poland, A. and Glover, E. (1975). Genetic expression of aryl hydrocarbon hydroxylase by 2,3,7,8-tetrachlorodibenzo-*p*-dioxin: evidence for a receptor mutation in genetically nonresponsive mice. *Molecular Pharmacology*, 11, 389–398.

43. Poland, A., Glover, E., and Kende, A. S. (1976). Stereospecific, high affinity binding of 2,3,7,8-tetrachlorodibenzo-*p*-dioxin by hepatic cytosol. Evidence that the binding species is receptor for induction of aryl hydrocarbon hydroxylase. *Journal of Biological Chemistry*, 251, 4936–4946.

44. Whitlock, J. P., Jr. (1990). Genetic and molecular aspects of 2,3,7,8-tetrachlorodibenzo-*p*-dioxin action. *Annual Review of Pharmacology and Toxicology*, 30, 251–277.

45. Hankinson, O. (1995). The aryl hydrocarbon receptor complex. *Annual Review of Pharmacology and Toxicology*, 35, 307–340.

46. Okino, S. T. and Whitlock, J. P., Jr. (1995). Dioxin induces localized, graded changes in chromatin structure: implications for Cyp1A1 gene transcription. *Molecular and Cellular Biology*, 15, 3714–3721.

47. Bradfield, C. A., Glover, E., and Poland, A. (1991). Purification and N-terminal amino acid sequence of the Ah receptor from the C57BL/6J mouse. *Molecular Pharmacology*, 39, 13–19.

48. Burbach, K. M., Poland, A., and Bradfield, C. A. (1992). Cloning of the Ah-receptor cDNA reveals a distinctive ligand-activated transcription factor. *Proceedings of the National Academy of Sciences of the United States of America*, 89, 8185–8189.

49. Ma, Q. and Whitlock, J. P., Jr. (1997). A novel cytoplasmic protein that interacts with the Ah receptor, contains tetratricopeptide repeat motifs, and augments the transcriptional response to 2,3,7,8-tetrachlorodibenzo-*p*-dioxin. *Journal of Biological Chemistry*, 272, 8878–8884.

50. Ma, Q., Dong, L., and Whitlock, J. P., Jr. (1995). Transcriptional activation by the mouse Ah receptor. Interplay between multiple stimulatory and inhibitory functions. *Journal of Biological Chemistry*, 270, 12697–12703.

51. Hoffman, E. C., Reyes, H., Chu, F. F., Sander, F., Conley, L. H., Brooks, B. A., and Hankinson, O. (1991). Cloning of a factor required for activity of the Ah (dioxin) receptor. *Science*, 252, 954–958.

52. Reyes, H., Reisz-Porszasz, S., and Hankinson, O. (1992). Identification of the Ah receptor nuclear translocator protein (Arnt) as a component of the DNA binding form of the Ah receptor. *Science*, 256, 1193–1195.

53. Ko, H. P., Okino, S. T., Ma, Q., and Whitlock, J. P., Jr. (1996). Dioxin-induced CYP1A1 transcription *in vivo*: the aromatic hydrocarbon receptor mediates transactivation, enhancer-promoter communication, and changes in chromatin structure. *Molecular and Cellular Biology*, 16, 430–436.

54. Ko, H. P., Okino, S. T., Ma, Q., and Whitlock, J. P., Jr. (1997). Transactivation domains facilitate promoter occupancy for the dioxin-inducible CYP1A1 gene *in vivo*. *Molecular and Cellular Biology*, 17, 3497–3507.

55. Carver, L. A., Jackiw, V., and Bradfield, C. A. (1994). The 90-kDa heat shock protein is essential for Ah receptor signaling in a yeast expression system. *Journal of Biological Chemistry*, 269, 30109–30112.

56. Whitelaw, M. L., McGuire, J., Picard, D., Gustafsson, J. A., and Poellinger, L. (1995). Heat shock protein hsp90 regulates

57. Carver, L. A. and Bradfield, C. A. (1997). Ligand-dependent interaction of the aryl hydrocarbon receptor with a novel immunophilin homolog *in vivo*. *Journal of Biological Chemistry*, 272, 11452–11456.
58. Meyer, B. K., Pray-Grant, M. G., Vanden Heuvel, J. P., and Perdew, G. H. (1998). Hepatitis B virus X-associated protein 2 is a subunit of the unliganded aryl hydrocarbon receptor core complex and exhibits transcriptional enhancer activity. *Molecular and Cellular Biology*, 18, 978–988.
59. Lin, B. C., Nguyen, L. P., Walisser, J. A., and Bradfield, C. A. (2008). A hypomorphic allele of aryl hydrocarbon receptor-associated protein-9 produces a phenocopy of the AHR-null mouse. *Molecular Pharmacology*, 74, 1367–1371.
60. Ma, Q. and Baldwin, K. T. (2000). 2,3,7,8-Tetrachlorodibenzo-*p*-dioxin-induced degradation of aryl hydrocarbon receptor (AhR) by the ubiquitin-proteasome pathway. Role of the transcription activation and DNA binding of AhR. *Journal of Biological Chemistry*, 275, 8432–8438.
61. Lusska, A., Wu, L., and Whitlock, J. P., Jr. (1992). Superinduction of CYP1A1 transcription by cycloheximide. Role of the DNA binding site for the liganded Ah receptor. *Journal of Biological Chemistry*, 267, 15146–15151.
62. Ma, Q., Renzelli, A. J., Baldwin, K. T., and Antonini, J. M. (2000). Superinduction of CYP1A1 gene expression. Regulation of 2,3,7,8-tetrachlorodibenzo-*p*-dioxin-induced degradation of Ah receptor by cycloheximide. *Journal of Biological Chemistry*, 275, 12676–12683.
63. Ma, Q. (2002). Induction and superinduction of 2,3,7,8-tetrachlorodibenzo-rho-dioxin-inducible poly(ADP-ribose) polymerase: role of the aryl hydrocarbon receptor/aryl hydrocarbon receptor nuclear translocator transcription activation domains and a labile transcription repressor. *Archives of Biochemistry and Biophysics*, 404, 309–316.
64. Ma, Q. and Baldwin, K. T. (2002). A cycloheximide-sensitive factor regulates TCDD-induced degradation of the aryl hydrocarbon receptor. *Chemosphere*, 46, 1491–1500.
65. Ma, Q. (2007). Aryl hydrocarbon receptor degradation-promoting factor (ADPF) and the control of the xenobiotic response. *Molecular Interventions*, 7, 133–137.
66. Talalay, P., Dinkova-Kostova, A. T., and Holtzclaw, W. D. (2003). Importance of phase 2 gene regulation in protection against electrophile and reactive oxygen toxicity and carcinogenesis. *Advances in Enzyme Regulation*, 43, 121–134.
67. Ma, Q. (2010). Transcriptional responses to oxidative stress: pathological and toxicological implications. *Pharmacological Therapy*, 125, 376–393.
68. Ma, Q., Kinneer, K., Bi, Y., Chan, J. Y., and Kan, Y. W. (2004). Induction of murine NAD(P)H:quinone oxidoreductase by 2,3,7,8-tetrachlorodibenzo-*p*-dioxin requires the CNC (cap 'n' collar) basic leucine zipper transcription factor Nrf2 (nuclear factor erythroid 2-related factor 2): cross-interaction between AhR (aryl hydrocarbon receptor) and Nrf2 signal transduction. *Biochemical Journal*, 377, 205–213.
69. Wang, G. L., Jiang, B. H., Rue, E. A., and Semenza, G. L. (1995). Hypoxia-inducible factor 1 is a basic-helix–loop–helix–PAS heterodimer regulated by cellular O_2 tension. *Proceedings of the National Academy of Sciences of the United States of America*, 92, 5510–5514.
70. Fleming, C. R., Billiard, S. M., and Di Giulio, R. T. (2009). Hypoxia inhibits induction of aryl hydrocarbon receptor activity in topminnow hepatocarcinoma cells in an ARNT-dependent manner. *Comparative Biochemistry and Physiology, Part C*, 150, 383–389.
71. Hofer, T., Pohjanvirta, R., Spielmann, P., Viluksela, M., Buchmann, D. P., Wenger, R. H., and Gassmann, M. (2004). Simultaneous exposure of rats to dioxin and carbon monoxide reduces the xenobiotic but not the hypoxic response. *Biological Chemistry*, 385, 291–294.
72. Ruegg, J., Swedenborg, E., Wahlstrom, D., Escande, A., Balaguer, P., Pettersson, K., and Pongratz, I. (2008). The transcription factor aryl hydrocarbon receptor nuclear translocator functions as an estrogen receptor beta-selective coactivator, and its recruitment to alternative pathways mediates antiestrogenic effects of dioxin. *Molecular Endocrinology*, 22, 304–316.
73. Ke, S., Rabson, A. B., Germino, J. F., Gallo, M. A, and Tian, Y. (2001). Mechanism of suppression of cytochrome P-450 1A1 expression by tumor necrosis factor-alpha and lipopolysaccharide. *Journal of Biological Chemistry*, 276, 39638–39644.
74. Kim, D. W., Gazourian, L., Quadri, S. A., Romieu-Mourez, R., Sherr, D. H., and Sonenshein, G. E. (2000). The RelA NF-kappaB subunit and the aryl hydrocarbon receptor (AhR) cooperate to transactivate the c-myc promoter in mammary cells. *Oncogene*, 19, 5498–5506.
75. Denison, M. S., and Nagy, S. R. (2003). Activation of the aryl hydrocarbon receptor by structurally diverse exogenous and endogenous chemicals. *Annual Review of Pharmacology and Toxicology*, 43, 309–334.
76. Nguyen, L. P. and Bradfield, C. A. (2008). The search for endogenous activators of the aryl hydrocarbon receptor. *Chemical Research in Toxicology*, 21, 102–116.
77. Matikainen, T., Perez, G. I., Jurisicova, A., Pru, J. K., Schlezinger, J. J., Ryu, H. Y., Laine, J., Sakai, T., Korsmeyer, S. J., Casper, R. F., Sherr, D. H., and Tilly, J. L. (2001). Aromatic hydrocarbon receptor-driven Bax gene expression is required for premature ovarian failure caused by biohazardous environmental chemicals. *Nature Genetics*, 28, 355–360.
78. Krishnan, V., Porter, W., Santostefano, M., Wang, X., and Safe, S. (1995). Molecular mechanism of inhibition of estrogen-induced cathepsin D gene expression by 2,3,7,8-tetrachlorodibenzo-*p*-dioxin (TCDD) in MCF-7 cells. *Molecular and Cellular Biology*, 15, 6710–6719.
79. Baba, T., Mimura, J., Gradin, K., Kuroiwa, A., Watanabe, T., Matsuda, Y., Inazawa, J., Sogawa, K., and Fujii-Kuriyama, Y. (2001). Structure and expression of the Ah receptor repressor gene. *Journal of Biological Chemistry*, 276, 33101–33110.
80. Yang, X., Liu, D., Murray, T. J., Mitchell, G. C., Hesterman, E. V., Karchner, S. I., Merson, R. R., Hahn, M. E., and Sherr, D. H. (2005). The aryl hydrocarbon receptor constitutively represses

c-myc transcription in human mammary tumor cells. *Oncogene*, 24, 7869–7881.
81. Gillesby, B. E., Stanostefano, M., Porter, W., Safe, S., Wu, Z. F., and Zacharewski, T. R. (1997). Identification of a motif within the 5′ regulatory region of pS2 which is responsible for AP-1 binding and TCDD-mediated suppression. *Biochemistry*, 36, 6080–6089.
82. Sulentic, C. E., Holsapple, M. P., and Kaminski, N. E. (2000). Putative link between transcriptional regulation of IgM expression by 2,3,7,8-tetrachlorodibenzo-p-dioxin and the aryl hydrocarbon receptor/dioxin-responsive enhancer signaling pathway. *Journal of Pharmacology and Experimental Therapeutics*, 295, 705–716.
83. Jeon, M. S. and Esser, C. (2000). The murine IL-2 promoter contains distal regulatory elements responsive to the Ah receptor, a member of the evolutionarily conserved bHLH–PAS transcription factor family. *Journal of Immunology*, 165, 6975–6983.
84. Marchand, A., Tomkiewicz, C., Marchandeau, J. P., Boitier, E., Barouki, R., and Garlatti, M. (2005). 2,3,7,8-Tetrachlorodibenzo-p-dioxin induces insulin-like growth factor binding protein-1 gene expression and counteracts the negative effect of insulin. *Molecular Pharmacology*, 67, 444–452.
85. Kolluri, S. K., Weiss, C., Koff, A., and Gottlicher, M. (1999). p27(Kip1) induction and inhibition of proliferation by the intracellular Ah receptor in developing thymus and hepatoma cells. *Genes & Development*, 13, 1742–1753.
86. Barnes-Ellerbe, S., Knudsen, and K. E., Puga, A. (2004). 2,3,7,8-Tetrachlorodibenzo-p-dioxin blocks androgen-dependent cell proliferation of LNCaP cells through modulation of pRB phosphorylation. *Molecular Pharmacology*, 66, 502–511.
87. McMillan, B. J., McMillan, S. N., Glover, E., and Bradfield, C. A. (2007). 2,3,7,8-Tetrachlorodibenzo-p-dioxin induces premature activation of the KLF2 regulon during thymocyte development. *Journal of Biological Chemistry*, 282, 12590–12597.
88. Ma, Q., Baldwin, K. T., Renzelli, A. J., McDaniel, A., and Dong, L. (2001). TCDD-inducible poly(ADP-ribose) polymerase: a novel response to 2,3,7,8-tetrachlorodibenzo-p-dioxin. *Biochemical and Biophysical Research Communications*, 289, 499–506.
89. Dong, L., Ma, Q., and Whitlock, J. P., Jr. (1997). Down-regulation of major histocompatibility complex Q1b gene expression by 2,3,7,8-tetrachlorodibenzo-p-dioxin. *Journal of Biological Chemistry*, 272, 29614–29619.
90. Lawrence, B. P., Denison, M. S., Novak, H., Vorderstrasse, B. A., Harrer, N., Neruda, W., Reichel, C., and Woisetschlager, M. (2008). Activation of the aryl hydrocarbon receptor is essential for mediating the anti-inflammatory effects of a novel low-molecular-weight compound. *Blood*, 112, 1158–1165.

3

ROLE OF CHAPERONE PROTEINS IN AHR FUNCTION

IAIN A. MURRAY AND GARY H. PERDEW

3.1 INTRODUCTION

The aryl hydrocarbon receptor (AHR) is a ligand-activated member of the basic helix–loop–helix/Per/Arnt/Sim (bHLH–PAS) family of transcription factors; members of this family are known to be regulated by external stimuli such as oxygen levels and light [1]. The AHR was first identified through the use of 2,3,7,8-[^{3}H]chlorodibenzo-p-dioxin binding to a specific protein in liver cytosol [2]. The AHR sedimented on sucrose density gradients as a 9S peak, which suggested that the AHR most likely existed in an oligomeric complex. This concept, combined with its inherent instability in cytosol, made purifying this receptor a challenge, which was eventually overcome through the use of a [^{125}I]-photoaffinity ligand and denaturing chromatographic techniques [3, 4]. This subsequently led to the cloning of the AHR and the production of antibodies against the AHR, important tools to facilitate detailed biochemical analysis of this receptor [5, 6].

3.2 COMPOSITION OF THE UNLIGANDED AHR CORE COMPLEX

The first step toward understanding the regulation of AHR function and protein composition of the 9S complex came in 1988 with evidence that the AHR exists bound to heat shock protein 90 (HSP90) [7, 8]. This discovery was facilitated by the production of HSP90 antibodies that can recognize HSP90 in large oligomeric complexes such as the AHR complex and similar to that observed previously with many steroid receptors (e.g., glucocorticoid receptor, estrogen receptor) [9]. Four years later, Perdew utilized a chemical cross-linking approach to define the molecular weight and number of proteins in the 9S AHR complex [10]. The receptor complex was found to be a tetrameric complex composed of the AHR and three proteins with relative molecular weights (M_r) of ~96, 88, and 46 kDa through examination of the M_r of partially cross-linked complexes after polyacrylamide gel electrophoresis. The identities of the 96 and 88 kDa proteins were later established to be hsp86 or hsp84, two isoforms of HSP90 in mice [11]. These data also demonstrate that HSP90 is directly bound to the AHR. These studies were all performed in cytosol isolated from Hepa 1 cells, a mouse-derived hepatoma cell line. It is quite possible that in other tissues or species the composition of the AHR core complex could vary and this possibility has not been adequately explored.

The specific domains involved in the interaction between the AHR and the HSP90 dimer have been mapped and this information offers insight into its functional importance. The AHR binds to the middle domain of HSP90 (amino acids 272–617), which also binds the estrogen receptor [12]. Although HSP90 binds to both bHLH and PAS B domains, this latter domain appears to contain the ligand binding domain [13–16]. The fact that two distinct AHR domains can stably bind to HSP90 may indicate that both HSP90 subunits come in contact with the AHR.

3.2.1 The Role of HSP90 in the AHR Complex

The molecular chaperone HSP90 is a critical protein in vertebrates as disruption of its expression is lethal. HSP90 has been examined at length as a key component of the folding machinery and as a stable chaperone utilized by steroid receptors [17]. The ability of HSP90 to stably interact with steroid receptors was established before the AHR was demonstrated to bind to HSP90. Studies with the estrogen,

---

The AH Receptor in Biology and Toxicology, First Edition. Edited by Raimo Pohjanvirta.
© 2012 John Wiley & Sons, Inc. Published 2012 by John Wiley & Sons, Inc.

glucocorticoid, and progesterone receptors revealed that these complexes were unstable in cytosolic preparations and required molybdate to stabilize and maintain ligand binding potential. Molybdate has been shown to stabilize a diverse group of client proteins bound to HSP90 [18]. This fact suggested that the effect of molybdate on steroid receptor complexes most likely involved binding to HSP90. It is believed that molybdate binds to the ATP binding site on HSP90 and induces a conformational change leading to a stabilized complex, although this concept has not been rigorously established [19]. The presence of HSP90 bound to the AHR in a 9S complex that in the presence of ligand treatment of cells leads to formation of a 4S complex in the nucleus appears to be a situation similar to that observed with the glucocorticoid receptor [20]. Indeed, prior to cloning of the AHR, it was believed that this receptor was most likely a member of the steroid receptor superfamily. Interestingly, the mouse AHR (Ah^b allele) complex does not require molybdate and is resistant to salt treatment, and thus the AHR/HSP90 complex may be the most stable HSP90 complex known. Several studies have demonstrated that HSP90 is required for expression of an AHR capable of binding ligand after initial synthesis [21–23]. However, once the complex is formed, there are considerable species differences in the stability of the HSP90/AHR complex, as well as the requirement of HSP90 in vitro to maintain a ligand binding conformation [24]. For example, in contrast to the mouse AHR, the specific binding of TCDD to the human AHR in cytosolic preparations required the presence of sodium molybdate [25, 26]. These results may suggest that the human AHR may differ in its behavior in vivo in terms of its ability to transform to its DNA binding form in the presence of various ligands.

The stability of the AHR in cells depends highly on the presence of HSP90 and indeed the AHR may not exist in the cell as a monomer. This is based on the fact that the AHR is highly sensitive to proteolytic turnover after disruption of HSP90 function established through the use of HSP90 inhibitors, such as geldanamycin [27, 28]. The AHR has been shown to have both nuclear localization and nuclear export sequences that are actively involved in mediating its cellular localization [29, 30]. In most cell types, the AHR appears to be predominantly in the cytoplasm. However, studies with nuclear export inhibitors suggest that, upon inhibition of nuclear export, the unliganded AHR largely accumulates in the nucleus within 1 h, thus demonstrating that the AHR undergoes dynamic nucleocytoplasmic shuttling [31–33]. The AHR appears to shuttle as a receptor/HSP90 complex as molybdate treatment of cells fails to block AHR nucleocyctoplasmic shuttling, but yet blocks AHR transformation to a DNA binding complex [34]. In addition, through the use of an AHR mutant constitutively present in the nucleus, the AHR was found to bind HSP90 [35]. The significance of ligand-independent nucleocytoplasmic shuttling of the AHR remains unclear; perhaps this process allows the AHR to more effectively respond to ligand exposure and also rapidly downregulates transcriptional activity in the absence of ligand.

Upon binding ligand, the AHR translocates from the cytoplasm to the nucleus (Fig. 3.1). This appears to be due to a conformational change in the AHR that enhances

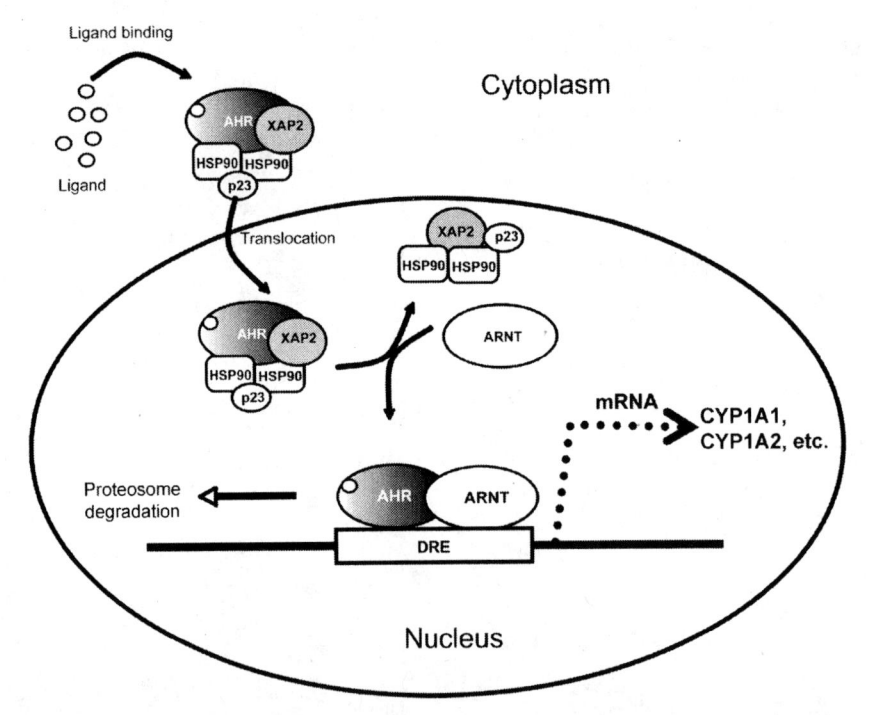

FIGURE 3.1 Model of ligand-mediated translocation of the AHR complex into the nucleus.

pendulin binding that facilitates active transport [32]. There is considerable evidence that the liganded AHR translocates into the nucleus bound to HSP90 [36, 37]. Once in the nucleus, ARNT most likely displaces HSP90, leading to the formation of the ARNT/AHR heterodimer [16, 38]. This process appears to be inefficient considering that only a small portion of the translocated receptor actually dimerizes with ARNT. The 6S nuclear form of the AHR has been shown to be a heterodimer bound to ARNT, which was confirmed through the use of cross-linking studies [10, 39]. The fate of HSP90 and other co-chaperone proteins displaced from the AHR in the nucleus is unknown. Another event that occurs to AHR/HSP90 complexes upon ligand binding is a rapid proteolytic turnover of the existing AHR pool [40–42]. Turnover of the AHR can occur both in the cytoplasm and in the nucleus, with the half-life of the AHR being considerably reduced in the nucleus [43, 44]. These results suggest a feedback loop that shortens the time frame of AHR transcriptional activity.

3.2.2 XAP2 a Co-Chaperone in the Unliganded AHR Complex

The current understanding and prevailing representation of the core cytoplasmic AHR complex consists of a minimal tetrameric organization of an AHR monomer associated with two molecules of the chaperone HSP90 and a single molecule of each of the co-chaperones XAP2 and p23 (Fig. 3.1). Here, we shall examine the evidence that has led to identification of XAP2 as a primary component of the minimal AHR complex.

Early studies investigating the composition of the unliganded cytoplasmic AHR complex revealed the presence of an unidentified subunit with an apparent molecular weight of 38 kDa that coprecipitated with the AHR from Hepa 1 cell extracts [11]. This previously uncharacterized protein appeared to be present in addition to the already characterized HSP90 subunits [7, 8, 45]. Subsequent work demonstrated that this unknown factor had the capacity to self-assemble into the 9S AHR/HSP90 complex without exogenous addition of TCDD in COS-1 cells transfected with mouse AHR, as determined by sucrose density fractionation studies [46]. The identity of this low molecular weight subunit was ultimately revealed, utilizing reverse genetics and yeast two-hybrid screening approaches by three research groups, to be a protein previously characterized as a repressor of the hepatitis B virus (HBV) encoded protein X (HBx). It was thus cloned and termed X-associated protein 2 (XAP2) [46], AHR-interacting protein (AIP) [47], or AHR-associated protein 9 (ARA9) [48], and less frequently the FK506 binding proteins FKBP16 or FKBP37. An examination of the amino acid sequence of XAP2 reveals a significant degree of homology to the immunosuppressive FKBP12, FKBPs51/52, and Cyp40, which represent members of the immunophilin family of co-chaperones (Fig. 3.2) [46–49]. Like the immunophilins, XAP2 has a discrete domain organization, minimally comprising an FK domain and a tetratricopeptide repeat (TPR) motif. The FK domain of XAP2 harbors a putative peptidylprolyl isomerase region that in other immunophilins exerts a catalytic function that assists in protein folding. However, an examination of these cells failed to identify any peptidylprolyl isomerase activity associated with XAP2 [48]. The carboxyl portion of XAP2 contains three TPR motifs that are highly conserved, with the most carboxyl TPR motif nearly 100% invariant with that found in FKBP52 [46]. As with the TPR domains found in other proteins, the XAP2 TPR regions, in combination with the extreme terminal sequences, adopt a α-helical conformation that generates a protein–protein interaction interface. It is this carboxyl-terminal TPR-containing domain that facilitates most, if not all, of XAP2 protein–protein interactions. The other two TPR domains of XAP2 are less homologous with those found in FKBP52, and it has been speculated that these are the specificity determinants that regulate binding of XAP2 and FKBP52 to their respective major client proteins. Interestingly, the ubiquitously expressed immunophilins were already characterized as components of *apo* 9S untransformed steroid nuclear hormone, for example, glucocorticoid (GR) and progesterone (PR) receptor complexes, suggesting that some common functionality may exist between XAP2 and that of immunophilins [50–52]. However, despite the marked degree of homology, XAP2 is not itself classified as an immunophilin because it lacks affinity for the immunosuppressive drugs FK506, cyclosporin A, or rapamycin [53].

The incorporation and self-assembly of XAP2 into a 9S complex with AHR and HSP90 raised the question as to the nature and organization of these factors in the complex. Previous studies utilizing a chemical cross-linking approach revealed that AHR directly binds to HSP90 [7, 8]. Although they were not fully characterized, additional cross-linking studies did reveal the presence of a protein species that in retrospect represented XAP2 in a complex with AHR [10]. However, it remained to be established whether XAP2 bound directly to AHR or indirectly through HSP90. *In vitro* translation studies performed in the absence of AHR expression indicate that XAP2 has little or no affinity for HSP90, which implied that the association of XAP2 into the complex is mediated not through HSP90 but through AHR [45, 46]. These data are in contrast to studies performed in cell culture, which demonstrate that XAP2 can associate with HSP90 [46, 47]. These opposing views have been reconciled through a hypothesis suggesting that in a cellular context formation of an XAP2:HSP90 complex is a tripartite, cooperative, and active process that requires the presence of AHR, such that XAP2 binding to AHR requires HSP90 and AHR stabilizes the binding of XAP2 to HSP90 that in turn enhances the limited stability of AHR [45]. Supporting evidence for this

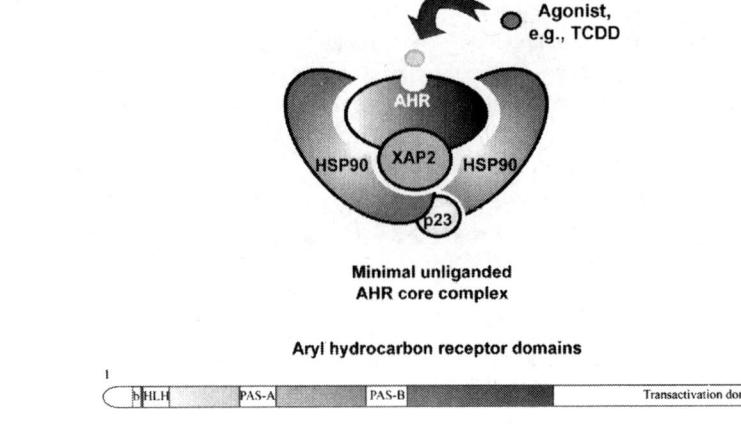

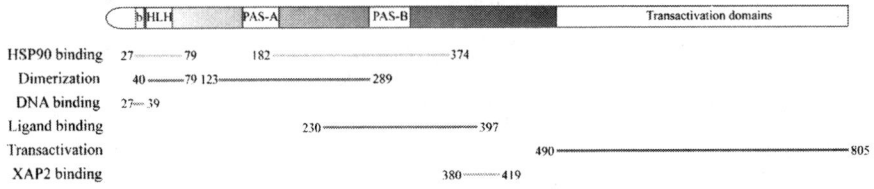

FIGURE 3.2 Composition, organization, and domains of the AHR involved in the assembly of the minimal core AHR complex. Experimental evidence indicates that in the ligand-free state the cytoplasmic AHR exists minimally as a core tetrameric complex consisting of an AHR monomer that physically interacts through separate binding domains with both the chaperone HSP90 (as a HSP90 dimer) and the co-chaperone XAP2. The minimal complex is completed by the incorporation of the co-chaperone p23, the binding of which is facilitated by HSP90. In addition to binding AHR, both XAP2 and HSP90 physically associate with each other while in the core complex. This tripartite cooperative interaction serves to stabilize the core complex. The various domains of the AHR that contribute to its major function are illustrated in the lower panel.

cooperative hypothesis is provided by data showing that the association of XAP2 with AHR is sensitive to the HSP90 disrupting agents, for example, geldanamycin and epigallocatechin gallate [45, 54]. It is likely that each factor in the complex induces a conformational change in the others that increases the net binding capacity of the complex as a whole. The nature of these structural changes is yet to be determined, primarily due to the fact that the AHR has so far proven to be refractory to crystallization.

The nature of the association between AHR/HSP90 and XAP2 has principally been deduced through mutational analysis of these factors. Truncation mutants of XAP2 have highlighted the critical importance of the carboxyl half of XAP2 for mediating binding to both AHR and HSP90, such that deletion of this region ablates binding of both AHR and HSP90 [53]. The TPR domains of numerous proteins have been demonstrated to facilitate protein–protein interactions, and thus the initial studies examining the incorporation of XAP2 into the core complex focused on the three TPR domains of XAP2. Mutation of the TPR repeats revealed that they are indeed critical for binding to both AHR and HSP90, with the most carboxy-terminal TPR being predominant. Perhaps not surprisingly, mutation of different amino acids within the TPR region has revealed unique sensitivities with regard to binding of AHR or HSP90, suggesting that while utilizing a common TPR region of XAP2, AHR and HSP90 appear to have subtly different binding interfaces. Furthermore, TPR mutations (e.g., K266A), which completely inhibit HSP90 binding to XAP2, diminish only XAP2 binding to AHR, which implies that the presence of HSP90 is not an absolute requirement for binding of XAP2 to AHR [45]. Supporting evidence is also provided by *in vitro* studies using cell extracts immunodepleted of HSP90, which similarly exhibit complex formation between AHR and XAP2 despite the absence of HSP90 [12]. Further subtle differences have also been identified within the extreme carboxyl-terminal end beyond the TPR motifs of XAP2, which lend credence to the notion that AHR and HSP90 bind with close proximity to XAP2. The last 32 amino acid residues of XAP2 have been demonstrated to be critical for the interaction of XAP2 with HSP90, since loss of HSP90 binding is observed upon deletion of these residues [12]. Conversely, removal or mutation of the last five residues has been shown through *in vitro* translation studies to facilitate both HSP90 and AHR binding to XAP2 [45].

As with XAP2 and HSP90, mutation and truncation studies proved instrumental in identifying which regions of AHR facilitate assembly of the minimal core complex. The association of AHR with HSP90 is mediated primarily through the amino-terminal half of AHR and is described elsewhere in this chapter. Removal of the first 130 amino acid residues, which encompass the DNA binding helix–loop–helix and ARNT dimerization domains, fails to disrupt XAP2 binding. However, a more expansive truncation spanning the DNA binding site, ARNT dimerization and part of the ligand binding domains up to residue 287, ablates the association with XAP2 [48]. Deletion of residues from the carboxyl-terminal transactivation domain up to amino acid 492 has little influence on XAP2 binding. However, increasing the length of the carboxyl-terminal deletion to 425 amino acid residues results in a complete loss of XAP2 binding [12, 48]. Thus, it has been deduced that the XAP2-interacting domain of AHR maps to the central portion of the receptor, spanning residues 380–419 [12]. The different binding sites for HSP90 and XAP2 perhaps indicate a lack of redundancy and distinct evolutionary adaption with regard to chaperone functions. The partial overlap between the regions that encompass ligand and XAP2 binding may contribute to the speculative increase in ligand sensitivity of AHR in the presence of XAP2. Additional studies have proposed that the interaction between AHR and XAP2 mediated by this central domain is primarily hydrophobic in nature by virtue of being insensitive to high salt concentrations and yet acutely sensitive to detergent, when examined in the context of coimmunoprecipitation [55].

3.2.3 Role of XAP2 in the AHR Core Complex

The identification and characterization of XAP2 as a primary component of the core unliganded cytoplasmic AHR complex comprising an AHR monomer/HSP90 dimer/p23 monomer [56] and at least one XAP2 molecule (Fig. 3.1) raises the following questions: What biological function does XAP2 serve in the complex? Does XAP2 impact AHR function in a regulated fashion or in a constitutive scaffolding role? Does XAP2 assist in ligand-mediated AHR activation or function to limit AHR transformation? What role does XAP2 play in the ligand-independent biological activities of AHR? Here, we shall attempt to address these questions and highlight the supporting evidence collected over the past 12 years since the initial identification of XAP2 as an AHR-interacting protein [46–48].

The first studies indicating a functional relevance of the association between XAP2 and AHR were derived from overexpression studies using the murine hepatoma clone Hepa 1c1c7 [57], which revealed that exogenous expression of XAP2 resulted in enhanced AHR transcriptional activity, as determined by a statistically significant threefold increase in *Cyp1a1* expression (the archetypal biomarker of ligand-mediated AHR activation) following exposure to TCDD [47]. This observation was confirmed in additional murine and human cell lines transiently transfected with AHR-responsive luciferase reporter constructs together with forced XAP2 expression [46]. Interpretation of such studies suggests that XAP2 has a positive influence on ligand-mediated AHR transcriptional activity. However, an XAP2 expression analysis of an array of cell lines and tissues from various species demonstrates that XAP2 is ubiquitously expressed, although at variable levels [46, 47]. Thus, these initial studies were performed in the context of AHR already in an endogenous core complex with XAP2. Interestingly, subsequent studies examining the effect of XAP2 on AHR transcriptional activity demonstrate that XAP2 can exert a repressive activity [33, 55, 58]. The role for XAP2 in modulating (positively or negatively) the ligand sensitivity and transcriptional responses of AHR is further supported by studies examining AHR and XAP2 homologs from less complex organisms, for example, *Caenorhabditis elegans* and *Drosophila melanogaster*. These AHR homologs demonstrate little or no ligand binding capacity, at least with typical AHR ligands, that is, polyaromatic and halogenated hydrocarbons such as 2,3,7,8-tetrachlorodibenzo-*p*-dioxin (TCDD), and have also been shown to lack XAP2 binding [45]. The inconsistency of the data regarding the effect of XAP2 on ligand-induced AHR activity perhaps underlies the different end points and experimental approaches employed, thus failing to observe the overall action of XAP2. The data may also be interpreted to suggest that two distinct populations of AHR complex exist, one population containing XAP2 and the other that does not, and that these two populations have separate biological activities. Supporting evidence for XAP2-independent AHR activity has been demonstrated. Cell culture-based studies in which XAP2 expression has been largely ablated or mutants of XAP2 deficient in AHR binding reveal that XAP2 is dispensable with regard to driving ligand-mediated AHR activation [59, 60]. In a cellular context, it is therefore possible that AHR can retain its activation potential in the absence of XAP2; however, due to the ubiquitous expression of XAP2, it seems unlikely that XAP2 will be limiting in a given tissue. Indeed, a transgenic mouse model was generated that overexpresses XAP2 in hepatocytes, and enhanced expression in this tissue was chosen due to relatively low levels of endogenous XAP2. Studies in this mouse revealed that overexpression of XAP2 had no effect on basal or ligand-induced AHR transcriptional activity [58]. As will be described later, XAP2 has the capacity to interact with multiple proteins, and some of which can influence AHR activity, such as phosphodiesterases PDE2A and PDE4A5. Therefore, it is conceivable that under different conditions these additional binding partners may contribute to the observed discrepancies. Notwithstanding, XAP2 clearly influences AHR-mediated transcriptional responses.

Early studies utilizing a yeast AHR signaling model revealed that in addition to increasing the ligand sensitivity of AHR to β-naphthoflavone, expression of XAP2 prompted a significant enhancement in the maximal transcriptional response [61]. Such data suggested that XAP2 has a stabilizing effect on the ligand-accessible form of AHR. Although this sensitization effect has been observed with other AHR ligands, a study on the effect of XAP2 in the context of different classes of ligands, for example, antagonists and novel selective AHR ligands, has yet to be undertaken. The increased ligand responsiveness in the presence of XAP2 has been attributed to enhanced receptor stability rather than a change in receptor conformation. This hypothesis was subsequently confirmed in mammalian systems, which demonstrated higher levels of cytoplasmic AHR in COS-1 cells transfected with XAP2 when compared to cells not transfected with XAP2 [12, 59, 62]. Furthermore, the increase in AHR protein elicited by XAP2 was demonstrated to be specific for XAP2, since immunophilins such as FKBP52 failed to mediate a positive influence on AHR levels [46]. Data exist that discredit the notion that XAP2 can increase ligand responsiveness through a conformational change [63]. AHR ligand responsiveness has been shown to exhibit a marked degree of species dependency, with mouse AHR being fivefold more sensitive than its human homolog to most but not all AHR agonists [25, 64]. However, chimeric mouse/human receptors fail to exchange ligand sensitivity, suggesting that ligand responsiveness is inherent to the receptor itself rather than a consequence of conformational changes facilitated by the XAP2 chaperone complex [63, 65]. Interestingly, the initial purification of AHR protein was hampered in part by its instability and acute sensitivity to heat stress. This sensitivity to temperature can be somewhat overcome by incorporation of XAP2 into the AHR core complex, supporting a role of XAP2 in stabilizing the AHR complex [62]. Further investigation has indicated that the stabilizing effect of XAP2, in combination with HSP90 and p23, may be due to XAP2 protecting AHR from proteolytic degradation through inhibition of AHR ubiquitination by E3 ligases, such as that observed with carboxyl terminus of Hsc70-interacting protein (CHIP), although the role of CHIP in AHR degradation has only been demonstrated in an *in vitro* context [66–68].

In addition to stabilizing the AHR, studies utilizing fluorescently tagged AHR identified XAP2 as a factor involved in the cytoplasmic retention of the AHR. Although previously considered primarily cytoplasmic in the unliganded state, the AHR has actually been shown to undergo dynamic nucleocytoplasmic shuttling in the absence of a ligand, resulting in a diffuse cytoplasmic and nuclear staining when visualized by immunofluorescence or the use of fluorescently tagged AHR constructs [33, 69]. Treatment with AHR ligands, for example, TCDD, prompts a marked and rapid redistribution of AHR into the nucleus, typically within 1 h, consistent with its role as a transcription factor. Intracellular movement of the AHR through constitutive nucleocytoplasmic shuttling or ligand-induced translocation is brought about through the action of a strong bipartite nuclear localization and export signals within the AHR [29, 70]. XAP2 overexpression studies in COS-1 cells have shown that XAP2 can markedly diminish the AHR translocation process, under both basal and ligand-stimulated conditions [33, 60, 65, 69]. These observations support the concept that XAP2 can alter AHR-mediated transcription. First, by limiting translocation, XAP2 can be classified as an inhibitor of AHR action. Conversely, the redistribution of unstimulated AHR from the nuclear compartment into the cytoplasm will result in elevated cytoplasmic levels of AHR, which is likely to increase the ligand sensitivity in any given cell. Despite these differences of interpretation, XAP2 clearly influences the localization of the AHR and a number of potential mechanisms have been proposed to explain this phenomenon. The immunophilins present in the glucocorticoid receptor have also been shown to influence receptor localization and transport through a mechanism that involves the peptidylprolyl isomerase domain of FKBP52, forming a transport complex with microtubules to facilitate ATP-dependent nuclear redistribution [71]. A complementary mechanism involving XAP2 and AHR has largely been discounted, due to the observation that chemical microtubule disruption fails to overtly influence ligand-induced AHR translocation [33]. It has also been speculated that binding of XAP2 to the AHR cytoplasmic complex may result in physical masking or a conformational change of the nuclear localization signal, thus preventing translocation. Such a mechanism seems unlikely, since an antibody directed against the nuclear localization signal is not blocked when AHR is bound to XAP2 [33]. Translocation of large molecular weight and multimeric complexes containing nuclear localization signals across the nuclear envelope is highly regulated, involving multiple adaptor proteins [72]. The AHR complex has been shown to interact with components of this transport machinery and suggest a mechanism by which XAP2 can influence AHR compartmentalization [33, 73]. Immunoprecipitation and GST pull-down studies using the mouse AHR demonstrate a physical interaction with importin β, a member of the karyopherin family of adaptor proteins. Importin α interacts with basic arginine-rich nuclear localization signals such as that carried by the AHR. Subsequent interaction of this complex with importin β and GDP-bound Ran facilitates transport across the nuclear envelope. Inclusion of XAP2 in the AHR complex appears to diminish the association of AHR, through its nuclear localization signal, with importin β and thus is likely to limit transport into the nucleus, which would explain the observed cytoplasmic retention of AHR [33]. The nature of the XAP2-mediated inhibition of importin β binding has yet to be elucidated, but, as mentioned earlier, it does not seem to involve blockade of the nuclear localization sequence.

In conclusion, XAP2 represents a somewhat enigmatic regulator of AHR function. First, XAP2 increases AHR protein stability, prompting an apparent enhancement of ligand sensitivity, which leads to increased AHR-mediated transcription of appropriate target genes following exposure to AHR agonists. However, incorporation of XAP2 into the

TABLE 3.1 XAP2-Interacting Proteins

Protein	Comments	References
Hepatitis B virus X protein	Interaction with XAP2 identified through yeast two-hybrid assay. XAP2 inhibits transactivation potential of HBx and may influence hepatitis B pathology	[76–78, 80, 81]
Heat shock protein 90	XAP2 undergoes cooperative binding to both AHR and HSP90 to form the mature and stable 9S cytoplasmic AHR core complex	[45–47]
Aryl hydrocarbon receptor	Association of XAP2 into the AHR complex with HSP90 and p23 stabilizes AHR and favors cytoplasmic retention of AHR to limit AHR-dependent transcription. Conversely, stabilization of AHR also appears to increase the ligand sensitivity of AHR, allowing greater transcription of AHR-dependent target genes	[12, 45–48, 53, 59]
Glucocorticoid receptor	Binding of XAP2 to GR requires HSP90 and serves to limit GR responsiveness to glucocorticoid ligands, possibly through stabilization and cytoplasmic retention	[84]
Peroxisome proliferator-activated receptor alpha	Similar to GR, the association of XAP2 with PPARα and HSP90 limits transactivation of PPARα target gene expression	[83]
Thyroid hormone receptor β1 (TRβ1)	XAP2 enhances the transcriptional activity of TRβ1 but not TRβ2 through an unidentified mechanism	[86, 87]
Phosphodiesterase 2A (PDE2A)	Interaction with XAP increases the phosphodiesterase activity of PDE2A and limits the cAMP-dependent translocation and activation of AHR	[97, 102]
Phosphodiesterase 4A5 (PDE4A5)	Association with XAP2 diminishes the phosphodiesterase activity of PDE4A5 and thus may elevate intracellular levels of cAMP	[98, 102]
Survivin (BRIC5)	Binding to XAP2 stabilizes survivin to elevate the cellular antiapoptotic threshold	[88, 102]
Epstein–Barr virus-encoded nuclear antigen-3	Association between EBNA-3 and XAP2 influences ligand-mediated activation of AHR	[82]

AHR complex results in a marked retention of AHR in the cytoplasmic compartment, probably through disruption of the association of AHR with the nuclear transport machinery. Such retention obviously sequesters AHR away from the site of its primary function as a transcription factor, that is, the nucleus. How these two opposing phenomena can be reconciled remains to be established, and it suggests that XAP2 regulation of AHR function is a highly ordered and complex process.

3.2.4 Insights from XAP2 Activity in Other Protein Complexes

Since the initial cloning of XAP2, its role in the cell has predominantly been studied and characterized in the context of its association with AHR; however, an increasing number of additional interactions have recently been demonstrated, which suggests that XAP2 has physiological roles independent of AHR. Evidence for such roles is provided by developmental expression studies, which illustrate that XAP2 expression precedes that of AHR and that XAP2 expression exhibits a more ubiquitous tissue expression profile than AHR [74]. Furthermore, ablation of XAP2 in homozygous *Xap2* null mice results in embryonic lethal phenotypes that are not matched by their *Ahr*-null counterparts [75]. These non-AHR XAP2 complexes and their biological implications are outlined in Table 3.1.

The discovery of XAP2 was the result of yeast two-hybrid assays performed to identify proteins that interact with the HBV X protein [76]. HBx is a promiscuous indirect transcriptional activator that appears to be required for viral replication and thus contributes to the pathology of HBV infection [77–80]. HBx is thought to promote cirrhosis of the liver and hepatocarcinoma in HBV-infected individuals through multiple mechanisms [81]. Despite the relatively low expression of XAP2 in hepatocytes, studies reveal that XAP2 can abolish HBx-mediated transactivation [76] and through this mechanism may provide some degree of protection against HBV replication and associated liver dysfunction. XAP2 is also reported to interact with another virally encoded protein, Epstein–Barr virus nuclear antigen-3 (EBNA-3) [82]. Although this association is reported to influence ligand-dependent activation of AHR, the effect of XAP2 on EBV-mediated lymphocyte pathology has not been established. XAP2 has been demonstrated to be a component of nuclear receptor complexes, for example, GR, peroxisome proliferator-activated receptor α (PPARα), and thyroid hormone receptors (TRβ). These interactions require

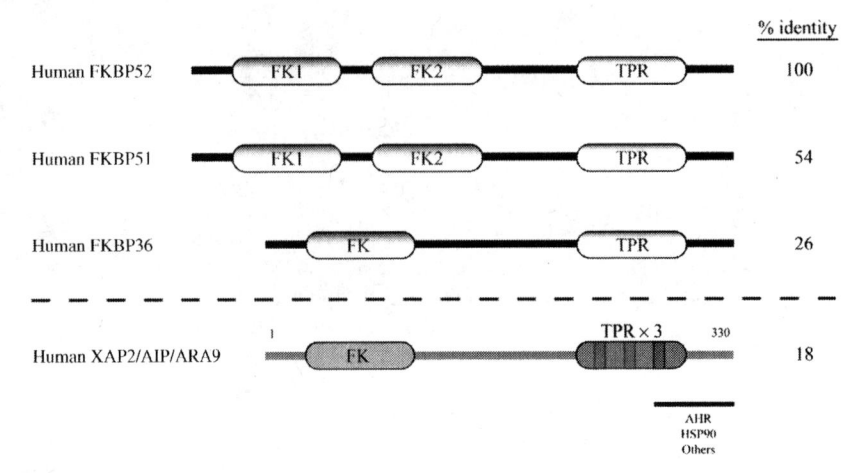

FIGURE 3.3 Homology between the co-chaperone XAP2 and the immunophilins. XAP2 shares a common domain structure with many of the FK506 binding immunophilin family of proteins consisting of an amino-terminal FK peptidylprolyl isomerase domain that does not exhibit FK506 binding or isomerase activity common to the immunophilins such as FKBP52. The carboxyl-terminal half of XAP2 encompasses $3 \times$ TPR protein–protein interaction domains, the most C-terminal of which facilitates its interaction with AHR and HSP90.

binding of XAP2 to HSP90 through the TPR domains and appear to limit responsiveness to ligand in the case of GR and PPAR [83, 84]. It is possible that in this context XAP2 may mimic the role of FKBP51, which is overexpressed in New World primates and in part contributes to their relative insensitivity to glucocorticoids [85]. Conversely, TRβ transcriptional activity is diminished in an isoform-specific fashion following siRNA-mediated knockdown of XAP2, with TRβ1 being influenced while TRβ2 activity is unaffected. This implies that XAP2 specifically stabilizes TRβ1, thus facilitating thyroid hormone-mediated transcription [86, 87].

In addition to influencing immediate ligand-induced transcriptional responses, evidence exists to suggest that XAP2 may also function as a determinate of cell fate through nontranscriptional modes of action. For example, XAP2 has been demonstrated to interact with survivin (*BRIC5*) through its TPR-containing C-terminal end [88, 89]. As a member of the inhibitor of the apoptosis family of proteins, survivin serves as a survival factor by inhibiting the proapoptotic activity of cleaved caspases 3 and 7. Interestingly, survivin is highly expressed in tumors and thus likely contributes to the malignant phenotype. The interaction between XAP2 and survivin appears to enhance the stability of survivin, rendering cells less sensitive to programmed cell death. Recently, XAP2 has been shown to associate *in vivo* with the tyrosine receptor kinase RET51 [89]. RET51 serves as the receptor for glial cell line-derived neurotrophic family of growth factors and has the capacity to transduce proapoptotic stimuli, resulting in cell death. The XAP2/RET51 association has been suggested to promote apoptosis by a mechanism involving the sequestration of XAP2 away from survivin, which then diminishes the stabilizing effect. Clearly, these interactions implicate XAP2 as factor that may impact tumor growth and progression and thus requires further investigation (Fig. 3.3). While no data are currently available, the action of XAP2 in determining apoptotic signaling may have particular relevance in the context of AHR. AHR is well characterized as a mediator of xenobiotic-induced carcinogenesis through the induction of *CYP1A1* and subsequent biotransformation of procarcinogens (e.g., benzo[*a*]pyrene) into genotoxic intermediates. Activation of AHR ultimately leads to dissociation of XAP2 from the AHR complex and the fate of liberated XAP2 remains to be established. It may be speculated that once XAP2 is liberated, it may associate with proteins such as RET51 or survivin and thus may contribute toward the tumorigenic properties of carcinogenic xenobiotics that bind to the AHR.

Recently, a genetic linkage has associated XAP2 with the occurrence of familial and sporadic isolated pituitary adenoma with XAP2 mutations (49 mutations characterized to date) found in 15–40% of cases [90–93]. In the majority of cases, the mutations resulted in truncated or nonsense expression of XAP2, with the loss of the TPR domain. Since the TPR domain is the primary site of protein–protein interaction for XAP2, it is likely that these functions of XAP2 are lost or diminished. It is unclear by which mechanism the loss of XAP2 function contributes to the specific development and progression of pituitary adenomas. *In vitro* studies examining the effect of pathogenic XAP2 mutations reveal no disruption in the context of RET51 binding. Similar studies examining the survivin protein have yet to be performed; therefore, the role of mutated XAP2 and its interaction with these cell fate factors with regard to pituitary pathogenesis remain to be established. Other potential mechanisms may involve the endocrine nature of the pituitary gland. As highlighted previously, XAP2 can specifically enhance TRβ1-mediated gene expression, with increased expression of hypothalamic TSH being the prime example. Also, the effect of XAP2 on diminished GR sensitivity may together promote an endocrine disrupting effect on the hypothalamic–pituitary–adrenal axis, thus providing a pro-tumorigenic hormonal environment. Clearly, XAP2 must be providing a tumor suppressor function, which is lost upon mutation, and thus additional studies are required to delineate the mechanisms by which wild-type XAP2 exerts its suppressor activity. Currently, no linkage analysis has been performed or detected with other nonpituitary tumors, but given the prevalence of hepatitis B-associated hepatic cancer and the interaction between HBx and XAP2, it is likely that this interaction has some impact on hepatic tumorigenesis. Unfortunately, XAP2 null animal models are not available to explore the role of XAP2 in carcinogenesis due to embryonic lethality [75]. However, the development of tissue-specific conditional knockout, mutant, or hypomorphic models may prove informative [94].

Another mechanism by which XAP2 may modulate cellular function is through its association with phosphodiesterases, enzymes that degrade the second messenger cyclic nucleotides cAMP and cGMP. Phosphodiesterases are ubiquitously expressed and there are numerous subtypes encoded by different genes with different modes of regulation and nucleotide specificity [95, 96]. Two-hybrid assays and subsequent GST pull-down studies have identified and validated two phosphodiesterases, namely, PDE2A and PDE4A5, that interact with XAP2 [97, 98]. Both PDE2A and PDE4A5 were shown to associate with XAP2 through the C-terminal TPR domain. Interestingly, despite sharing the same binding site on XAP2, the domains on the respective phosphodiesterase isotypes that facilitate the interaction are different, suggesting two independent modes of XAP2 binding [97, 98]. In the case of PDE4A5, the interaction is highly specific for XAP2 since there is no evidence for any association with the immunophilins, for example, FKBP52 [98]. Similarly, among the PDE4 isotypes, PDE4A5 appears to be the only one that complexes with XAP2. The functional significance of the XAP2–PDE4A5 interaction has yet to be elucidated in the context of AHR signaling, but formation of this complex has profound effects on PDE4A5 function. The phosphodiesterase catalytic activity of PDE4A5 is greatly diminished by 60% and its sensitivity to the phosphodiesterase inhibitor rolipram is enhanced when bound to XAP2 [98]. In addition, XAP2 attenuates the capacity of protein kinase A to phosphorylate PDE4A5. Data are limited regarding the physiological role of PDE4A5, but it has been implicated in sperm motility, apoptosis, inflammation, and adipocyte metabolism. Given the lack of information at this time regarding the role of PDE4A5, it is difficult to ascertain the precise physiological significance of its interaction with XAP2. Intriguingly, elevation of intracellular cAMP, as would occur through inhibition of PDE4A5, such as with rolipram or reduced phosphodiesterase activity, has been shown to attenuate inflammatory responses and is currently an active area of research. It could be speculated that some of the anti-inflammatory effects may arise from liberation of XAP2 from the AHR complex (which has also been shown to exhibit anti-inflammatory properties) following ligand activation. This free XAP2 might then subsequently bind to PDE4A5 and attenuate its catalytic activity, thus elevating cAMP. However, such a mechanism remains to be established. The second isotype of phosphodiesterase demonstrated to bind XAP2 is PDEA2. Similar to PDE4A5, PDEA2 binding is mediated through the XAP2 TPR domain; however, unlike PDE4A5, PDEA2 utilizes a different binding site that involves the GAF-B cyclic nucleotide binding site [97]. Interestingly, despite utilizing the GAF-B domain, binding of XAP2 does not attenuate the catalytic activity of PDEA2, which is in contrast to PDE4A5, suggesting that XAP2 interaction serves two distinct roles with regard to these phosphodiesterase isotypes. Unlike the PDE4A5 interaction, the binding of XAP2 to PDEA2 has been demonstrated to exert a functional effect on AHR signaling [97]. A number of groups have highlighted the phenomenon of ligand-independent AHR activation following exposure to cAMP or the adenylate cyclase activator forskolin, often with conflicting observations that are attributed to species, cell, or tissue differences [99–101]. Data indicate that cAMP or its analogs promote nuclear translocation of AHR and subsequent transcription of AHR target genes; however, forced expression of PDEA2 through transient transfection results in increased cytoplasmic retention of AHR under basal, TCDD-, and cAMP-induced states [97, 99, 102]. Under these conditions, the reduction in AHR activity is attributed to increased binding of XAP2 to PDEA2. Such data may indicate that PDEA2 forms a complex with AHR mediated through a mutual interaction with XAP2, an interaction that may be further stabilized through the involvement of HSP90. The presence of PDEA2 as a component of the AHR complex has been speculated to diminish the local cAMP concentration in the vicinity of AHR, thus decreasing its capacity to undergo nuclear translocation. Such a mechanism may provide scope for the subtle regulation of AHR activity or allow contextual activation or inhibition of AHR when multiple signaling stimuli are exerted (Fig. 3.4).

Thus, in conclusion, XAP2 appears to have evolved to be a versatile pleiotropic factor with the capacity to modulate diverse biological activities through numerous protein interactions. Evidence suggests that some of these interactions may influence the activity and biological functions of the most widely examined XAP2 binding partner the AHR. It is likely that further investigation will identify additional novel XAP2-interacting proteins and provide valuable information

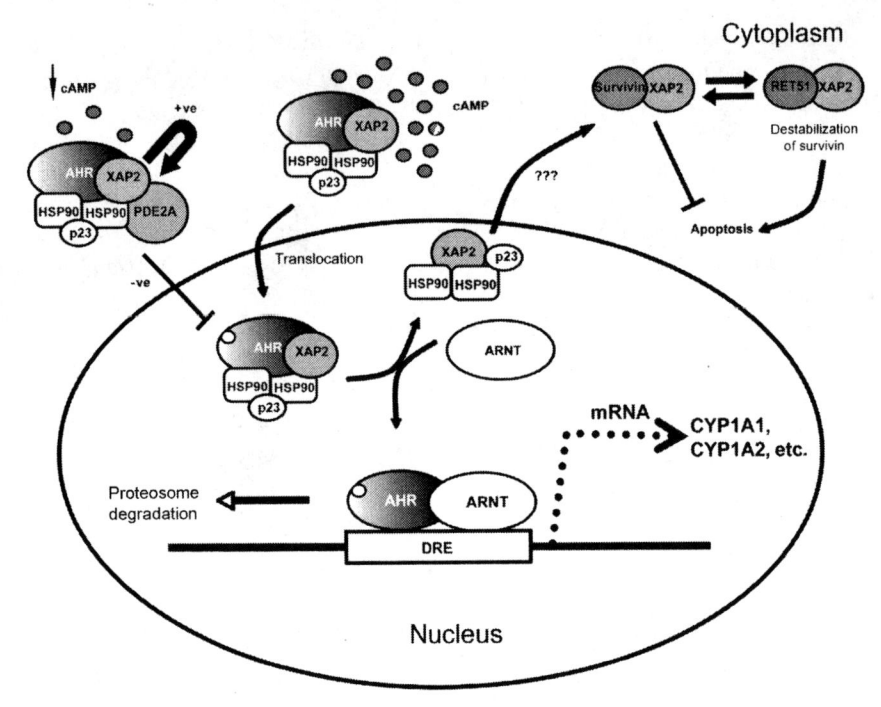

FIGURE 3.4 Potential role of XAP2 in influencing cAMP-dependent AHR activity and cellular apoptotic fate. XAP2 is reported to interact with specific isotypes of phosphodiesterase such as PDEA2. These enzymes catalyze the degradation of cyclic nucleotides, that is, cAMP and cGMP, limiting their signal transduction capacity. The association of PDEA2 into the AHR complex mediated by XAP2 thus may reduce the local concentration of cyclic nucleotides in the vicinity of AHR and diminish AHR translocation into the nucleus. The interaction of XAP2 with the antiapoptotic factor survivin has a stabilizing effect on this factor, thus raising the apoptotic threshold. Conversely, binding of XAP2 to RET51 sequesters XAP2 away from survivin facilitating its degradation and thus may influence cell fate in response to apoptotic stimuli.

regarding the potential crosstalk or integration of XAP2-dependent signaling pathways, thus enhancing the understanding of the diverse physiological roles of XAP2 and its binding partners.

3.3 PRESENCE OF OTHER PROTEINS IN THE AHR CORE COMPLEX

The HSP90 co-chaperone p23 has been shown to influence AHR activity [56]. p23 is a ubiquitous and highly conserved protein that has been found bound to a variety of HSP90 client protein complexes [103]. Studies *in vitro* with the glucocorticoid receptor suggest that p23 stabilizes the ATP-induced conformation of HSP90 and enhances its ability to bind ligand [104]. However, p23 does not appear to be required for the folding of a number of HSP90 client proteins, such as the androgen receptor [105]. Another possible function of p23 is the enhancement of ATP-dependent release of the client protein from HSP90 [106]. The ability of the AHR to bind ligand, dioxin-responsive elements and increase transcriptional activity have been found to be enhanced by p23 [56, 107, 108]. However, studies in p23 null mice revealed that the AHR in embryonic liver does not require p23 for efficient binding of ligand [109]. Also, an AHR mutant was generated that fails to bind XAP2, and this mutant maintains its ability to both bind ligand and mediate transcriptional activity [55]. *In vitro* studies examining the initial assembly of the AHR/HSP90 complex indicate that the AHR can transiently interact with p60 and Hip in a manner similar to that observed with other HSP90 chaperone proteins [110]. Clearly, additional studies are needed to firmly establish the role of HSP90 co-chaperones in AHR function.

3.4 CONCLUSIONS

The AHR is found in the cytoplasm of cells in a core complex composed of a dimer of HSP90 and XAP2. In addition, p23 appears to also play a role in modulating HSP90 and its function relative to AHR. This complex is in many ways similar to the progesterone and the glucocorticoid receptor, except instead of XAP2 in the receptor complex, these receptors contain FKBP51/52 or CYP40. The exact role of XAP2 in various protein complexes remains to be established, as well as its role in the AHR complex, which may be species specific. Also, whether cytoskeletal structures play a role in the transport of the AHR complex into the nucleus remains to be explored. Another aspect of AHR function that remains to be established is whether the AHR can be activated to a DNA binding species in the absence of ligand and the possible role chaperones would play in that process. A greater understanding of XAP2/HSP90 and their role in modulating AHR function should result in a better assessment of the biological function of this important receptor.

ACKNOWLEDGMENT

This work was supported by NIH grant ES04869.

REFERENCES

1. Gu, Y. Z., Hogenesch, J. B., and Bradfield, C. A. (2000). The PAS superfamily: sensors of environmental and developmental signals. *Annual Review of Pharmacology and Toxicology*, 40, 519–561.
2. Poland, A., Glover, E., and Kende, A. S. (1976). Stereospecific, high affinity binding of 2,3,7,8-tetrachlorodibenzo-*p*-dioxin by hepatic cytosol. Evidence that the binding species is receptor for induction of aryl hydrocarbon hydroxylase. *Journal of Biological Chemistry*, 251, 4936–4946.
3. Perdew, G. H. and Poland, A. (1988). Purification of the Ah receptor from C57BL/6J mouse liver. *Journal of biological chemistry*, 263, 9848–9852.
4. Bradfield, C. A., Glover, E., and Poland, A. (1991). Purification and N-terminal amino acid sequence of the Ah receptor from the C57BL/6J mouse. *Molecular Pharmacology*, 39, 13–19.
5. Poland, A., Glover, E., and Bradfield, C. A. (1991). Characterization of polyclonal antibodies to the Ah receptor prepared by immunization with a synthetic peptide hapten. *Molecular Pharmacology*, 39, 20–26.
6. Burbach, K. M., Poland, A., and Bradfield, C. A. (1992). Cloning of the Ah-receptor cDNA reveals a distinctive ligand-activated transcription factor. *Proceedings of the National Academy of Sciences of the United States of America*, 89, 8185–8189.
7. Perdew, G. H. (1988). Association of the Ah receptor with the 90-kDa heat shock protein. *Journal of Biological Chemistry*, 263, 13802–13805.
8. Denis, M., Cuthill, S., Wikstrom, A. C., Poellinger, L., and Gustafsson, J. A. (1988). Association of the dioxin receptor with the M_r 90,000 heat shock protein: a structural kinship with the glucocorticoid receptor. *Biochemical and Biophysical Research Communications*, 155, 801–807.
9. Perdew, G. H. and Whitelaw, M. L. (1991) Evidence that the 90-kDa heat shock protein (HSP90) exists in cytosol in heteromeric complexes containing HSP70 and three other proteins with M_r of 63,000, 56,000, and 50,000. *Journal of Biological Chemistry*, 266, 6708–6713.
10. Perdew, G. H. (1992). Chemical cross-linking of the cytosolic and nuclear forms of the Ah receptor in hepatoma cell line 1c1c7. *Biochemical and Biophysical Research Communications*, 182, 55–62.
11. Chen, H. S. and Perdew, G. H. (1994). Subunit composition of the heteromeric cytosolic aryl hydrocarbon receptor complex. *Journal of Biological Chemistry*, 269, 27554–27558.
12. Meyer, B. K. and Perdew, G. H. (1999). Characterization of the AhR-HSP90-XAP2 core complex and the role of the immunophilin-related protein XAP2 in AhR stabilization. *Biochemistry*, 38, 8907–8917.
13. Fukunaga, B. N., Probst, M. R., Reisz-Porszasz, S., and Hankinson, O. (1995). Identification of functional domains of the aryl hydrocarbon receptor. *Journal of Biological Chemistry*, 270, 29270–29278.
14. Perdew, G. H. and Bradfield, C. A. (1996). Mapping the 90 kDa heat shock protein binding region of the Ah receptor. *Biochemistry and Molecular Biology International*, 39, 589–593.
15. Antonsson, C., Whitelaw, M. L., McGuire, J., Gustafsson, J. A., and Poellinger, L. (1995). Distinct roles of the molecular chaperone HSP90 in modulating dioxin receptor function via the basic helix–loop–helix and PAS domains. *Molecular and Cellular Biology*, 15, 756–765.
16. Soshilov, A. and Denison, M. S. (2008). Role of the Per/Arnt/Sim domains in ligand-dependent transformation of the aryl hydrocarbon receptor. *Journal of Biological Chemistry*, 283, 32995–33005.
17. Pratt, W. B. and Toft, D. O. (1997). Steroid receptor interactions with heat shock protein and immunophilin chaperones. *Endocrine Reviews*, 18, 306–360.
18. Pratt, W. B. and Toft, D. O. (2003). Regulation of signaling protein function and trafficking by the HSP90/hsp70-based chaperone machinery. *Experimental Biology and Medicine (Maywood)*, 228, 111–133.
19. Csermely, P., Kajtar, J., Hollosi, M., Jalsovszky, G., Holly, S., Kahn, C. R., Gergely, P., Jr., Soti, C., Mihaly, K., and Somogyi, J. (1993). ATP induces a conformational change of the 90-kDa heat shock protein (HSP90). *Journal of Biological Chemistry*, 268, 1901–1907.
20. Cuthill, S., Poellinger, L., and Gustafsson, J. A. (1987). The receptor for 2,3,7,8-tetrachlorodibenzo-p-dioxin in the mouse hepatoma cell line Hepa 1c1c7. A comparison with the glucocorticoid receptor and the mouse and rat hepatic dioxin receptors. *Journal of Biological Chemistry*, 262, 3477–3481.
21. Carver, L. A., Jackiw, V., and Bradfield, C. A. (1994). The 90-kDa heat shock protein is essential for Ah receptor signaling in a yeast expression system. *Journal of Biological Chemistry*, 269, 30109–30112.
22. Pongratz, I., Mason, G. G., and Poellinger, L. (1992). Dual roles of the 90-kDa heat shock protein HSP90 in modulating functional activities of the dioxin receptor. Evidence that the dioxin receptor functionally belongs to a subclass of nuclear receptors which require HSP90 both for ligand binding activity and repression of intrinsic DNA binding activity. *Journal of Biological Chemistry* 267, 13728–13734.

23. Whitelaw, M. L., McGuire, J., Picard, D., Gustafsson, J. A., and Poellinger, L. (1995). Heat shock protein HSP90 regulates dioxin receptor function *in vivo*. *Proceedings of the National Academy of Sciences of the United States of America*, 92, 4437–4441.
24. Phelan, D. M., Brackney, W. R., and Denison, M. S. (1998). The Ah receptor can bind ligand in the absence of receptor-associated heat-shock protein 90. *Archives of Biochemistry and Biophysics*, 353, 47–54.
25. Manchester, D. K., Gordon, S. K., Golas, C. L., Roberts, E. A., and Okey, A. B. (1987). Ah receptor in human placenta: stabilization by molybdate and characterization of binding of 2,3,7,8-tetrachlorodibenzo-*p*-dioxin, 3-methylcholanthrene, and benzo(*a*)pyrene. *Cancer Research*, 47, 4861–4868.
26. Harper, P. A., Golas, C. L., and Okey, A. B. (1988). Characterization of the Ah receptor and aryl hydrocarbon hydroxylase induction by 2,3,7,8-tetrachlorodibenzo-p-dioxin and benz(a)anthracene in the human A431 squamous cell carcinoma line. *Cancer Research*, 48, 2388–2395.
27. Chen, H. S., Singh, S. S., and Perdew, G. H. (1997). The Ah receptor is a sensitive target of geldanamycin-induced protein turnover. *Archives of Biochemistry and Biophysics*, 348, 190–198.
28. Song, Z. and Pollenz, R. S. (2002). Ligand-dependent and independent modulation of aryl hydrocarbon receptor localization, degradation, and gene regulation. *Molecular Pharmacology*, 62, 806–816.
29. Ikuta, T., Eguchi, H., Tachibana, T., Yoneda, Y., and Kawajiri, K. (1998). Nuclear localization and export signals of the human aryl hydrocarbon receptor. *Journal of Biological Chemistry*, 273, 2895–2904.
30. Ikuta, T., Tachibana, T., Watanabe, J., Yoshida, M., Yoneda, Y., and Kawajiri, K. (2000). Nucleocytoplasmic shuttling of the aryl hydrocarbon receptor. *Journal of Biochemistry*, 127, 503–509.
31. Richter, C. A., Tillitt, D. E., and Hannink, M. (2001). Regulation of subcellular localization of the aryl hydrocarbon receptor (AhR). *Archives of Biochemistry and Biophysics*, 389, 207–217.
32. Kazlauskas, A., Sundstrom, S., Poellinger, L., and Pongratz, I. (2001). The HSP90 chaperone complex regulates intracellular localization of the dioxin receptor. *Molecular and Cellular Biology*, 21, 2594–2607.
33. Petrulis, J. R., Kusnadi, A., Ramadoss, P., Hollingshead, B., and Perdew, G. H. (2003). The HSP90 co-chaperone XAP2 alters importin beta recognition of the bipartite nuclear localization signal of the Ah receptor and represses transcriptional activity. *Journal of Biological Chemistry*, 278, 2677–2685.
34. Heid, S. E., Pollenz, R. S., and Swanson, H. I. (2000). Role of heat shock protein 90 dissociation in mediating agonist-induced activation of the aryl hydrocarbon receptor. *Molecular Pharmacology*, 57, 82–92.
35. Lees, M. J. and Whitelaw, M. L. (1999). Multiple roles of ligand in transforming the dioxin receptor to an active basic helix–loop–helix/PAS transcription factor complex with the nuclear protein Arnt. *Molecular and Cellular Biology*, 19, 5811–5822.
36. Perdew, G. H. (1991). Comparison of the nuclear and cytosolic forms of the Ah receptor from Hepa 1c1c7 cells: charge heterogeneity and ATP binding properties. *Archives of Biochemistry and Biophysics*, 291, 284–290.
37. Wilhelmsson, A., Cuthill, S., Denis, M., Wikstrom, A. C., Gustafsson, J. A., and Poellinger, L. (1990). The specific DNA binding activity of the dioxin receptor is modulated by the 90 kd heat shock protein. *EMBO journal*, 9, 69–76.
38. McGuire, J., Whitelaw, M. L., Pongratz, I., Gustafsson, J. A., and Poellinger, L. (1994). A cellular factor stimulates ligand-dependent release of HSP90 from the basic helix–loop–helix dioxin receptor. *Molecular and Cellular Biology*, 14, 2438–2446.
39. Probst, M. R., Reisz-Porszasz, S., Agbunag, R. V., Ong, M. S., and Hankinson, O. (1993). Role of the aryl hydrocarbon receptor nuclear translocator protein in aryl hydrocarbon (dioxin) receptor action. *Molecular Pharmacology*, 44, 511–518.
40. Pollenz, R. S. (2002). The mechanism of AH receptor protein down-regulation (degradation) and its impact on AH receptor-mediated gene regulation. *Chemico-Biological Interactions*, 141, 41–61.
41. Ma, Q. and Baldwin, K. T. (2000). 2,3,7,8-Tetrachlorodibenzo-*p*-dioxin-induced degradation of aryl hydrocarbon receptor (AhR) by the ubiquitin-proteasome pathway. Role of the transcription activation and DNA binding of AhR. *Journal of Biological Chemistry*, 275, 8432–8438.
42. Roberts, B. J. and Whitelaw, M. L. (1999). Degradation of the basic helix–loop–helix/Per–ARNT–Sim homology domain dioxin receptor via the ubiquitin/proteasome pathway. *Journal of Biological Chemistry*, 274, 36351–36356.
43. Davarinos, N. A. and Pollenz, R. S. (1999). Aryl hydrocarbon receptor imported into the nucleus following ligand binding is rapidly degraded via the cytosplasmic proteasome following nuclear export. *Journal of Biological Chemistry*, 274, 28708–28715.
44. Song, Z. and Pollenz, R. S. (2003). Functional analysis of murine aryl hydrocarbon (AH) receptors defective in nuclear import: impact on AH receptor degradation and gene regulation. *Molecular Pharmacology*, 63, 597–606.
45. Bell, D. R. and Poland, A. (2000). Binding of aryl hydrocarbon receptor (AhR) to AhR-interacting protein. The role of HSP90. *Journal of Biological Chemistry*, 275, 36407–36414.
46. Meyer, B. K., Pray-Grant, M. G., Vanden Heuvel, J. P., and Perdew, G. H. (1998). Hepatitis B virus X-associated protein 2 is a subunit of the unliganded aryl hydrocarbon receptor core complex and exhibits transcriptional enhancer activity. *Molecular and Cellular Biology*, 18, 978–988.
47. Ma, Q. and Whitlock, J. P., Jr. (1997). A novel cytoplasmic protein that interacts with the Ah receptor, contains tetratricopeptide repeat motifs, and augments the transcriptional response to 2,3,7,8-tetrachlorodibenzo-*p*-dioxin. *Journal of Biological Chemistry* 272, 8878–8884.

48. Carver, L. A., LaPres, J. J., Jain, S., Dunham, E. E., and Bradfield, C. A. (1998). Characterization of the Ah receptor-associated protein, ARA9. *Journal of Biological Chemistry*, 273, 33580–33587.

49. Schreiber, S. L. (1991). Chemistry and biology of the immunophilins and their immunosuppressive ligands. *Science*, 251, 283–287.

50. Tai, P. K. and Faber, L. E. (1985). Isolation of dissimilar components of the 8.5S nonactivated uterine progestin receptor. *Canadian Journal of Biochemistry and Cell Biology*, 63, 41–49.

51. Renoir, J. M., Radanyi, C., Faber, L. E., and Baulieu, E. E. (1990). The non-DNA-binding heterooligomeric form of mammalian steroid hormone receptors contains a HSP90-bound 59-kilodalton protein. *Journal of Biological Chemistry*, 265, 10740–10745.

52. Sanchez, E. R. (1990). Hsp56: a novel heat shock protein associated with untransformed steroid receptor complexes. *Journal of Biological Chemistry*, 265, 22067–22070.

53. Carver, L. A. and Bradfield, C. A. (1997). Ligand-dependent interaction of the aryl hydrocarbon receptor with a novel immunophilin homolog *in vivo*. *Journal of Biological Chemistry*, 272, 11452–11456.

54. Palermo, C. M., Westlake, C. A., and Gasiewicz, T. A. (2005). Epigallocatechin gallate inhibits aryl hydrocarbon receptor gene transcription through an indirect mechanism involving binding to a 90 kDa heat shock protein. *Biochemistry*, 44, 5041–5052.

55. Hollingshead, B. D., Petrulis, J. R., and Perdew, G. H. (2004). The aryl hydrocarbon (Ah) receptor transcriptional regulator hepatitis B virus X-associated protein 2 antagonizes p23 binding to Ah receptor-HSP90 complexes and is dispensable for receptor function. *Journal of Biological Chemistry*, 279, 45652–45661.

56. Kazlauskas, A., Poellinger, L., and Pongratz, I. (1999). Evidence that the co-chaperone p23 regulates ligand responsiveness of the dioxin (aryl hydrocarbon) receptor. *Journal of Biological Chemistry*, 274, 13519–13524.

57. Hankinson, O. (1979). Single-step selection of clones of a mouse hepatoma line deficient in aryl hydrocarbon hydroxylase. *Proceedings of the National Academy of Sciences of the United States of America*, 76, 373–376.

58. Hollingshead, B. D., Patel, R. D., and Perdew, G. H. (2006). Endogenous hepatic expression of the hepatitis B virus X-associated protein 2 is adequate for maximal association with aryl hydrocarbon receptor-90-kDa heat shock protein complexes. *Molecular Pharmacology*, 70, 2096–2107.

59. Meyer, B. K., Petrulis, J. R., and Perdew, G. H. (2000). Aryl hydrocarbon (Ah) receptor levels are selectively modulated by HSP90-associated immunophilin homolog XAP2. *Cell Stress & Chaperones*, 5, 243–254.

60. Pollenz, R. S. and Dougherty, E. J. (2005). Redefining the role of the endogenous XAP2 and C-terminal hsp70-interacting protein on the endogenous Ah receptors expressed in mouse and rat cell lines. *Journal of Biological Chemistry* 280, 33346–33356.

61. Miller, C. A., 3rd. (1997). Expression of the human aryl hydrocarbon receptor complex in yeast. Activation of transcription by indole compounds. *Journal of Biological Chemistry*, 272, 32824–32829.

62. LaPres, J. J., Glover, E., Dunham, E. E., Bunger, M. K., and Bradfield, C. A. (2000). ARA9 modifies agonist signaling through an increase in cytosolic aryl hydrocarbon receptor. *Journal of Biological Chemistry*, 275, 6153–6159.

63. Ramadoss, P., Petrulis, J. R., Hollingshead, B. D., Kusnadi, A., and Perdew, G. H. (2004). Divergent roles of hepatitis B virus X-associated protein 2 (XAP2) in human versus mouse Ah receptor complexes. *Biochemistry*, 43, 700–709.

64. Okey, A. B., Vella, L. M., and Harper, P. A. (1989). Detection and characterization of a low affinity form of cytosolic Ah receptor in livers of mice nonresponsive to induction of cytochrome P1-450 by 3-methylcholanthrene. *Molecular Pharmacology*, 35, 823–830.

65. Ramadoss, P. and Perdew, G. H. (2005). The transactivation domain of the Ah receptor is a key determinant of cellular localization and ligand-independent nucleocytoplasmic shuttling properties. *Biochemistry*, 44, 11148–11159.

66. Kazlauskas, A., Poellinger, L., and Pongratz, I. (2000). The immunophilin-like protein XAP2 regulates ubiquitination and subcellular localization of the dioxin receptor. *Journal of Biological Chemistry*, 275, 41317–41324.

67. Morales, J. L. and Perdew, G. H. (2007). Carboxyl terminus of hsc70-interacting protein (CHIP) can remodel mature aryl hydrocarbon receptor (AhR) complexes and mediate ubiquitination of both the AhR and the 90 kDa heat-shock protein (HSP90) *in vitro*. *Biochemistry*, 46, 610–621.

68. Lees, M. J., Peet, D. J., and Whitelaw, M. L. (2003). Defining the role for XAP2 in stabilization of the dioxin receptor. *Journal of Biological Chemistry*, 278, 35878–35888.

69. Petrulis, J. R. and Perdew, G. H. (2001). Monitoring nuclear import with GFP-variant fusion proteins in digitonin-permeabilized cells. *BioTechniques*, 31, 772–775.

70. Eguchi, H., Ikuta, T., Tachibana, T., Yoneda, Y., and Kawajiri, K. (1997). A nuclear localization signal of human aryl hydrocarbon receptor nuclear translocator/hypoxia-inducible factor 1β is a novel bipartite type recognized by the two components of nuclear pore-targeting complex. *Journal of Biological Chemistry*, 272, 17640–17647.

71. Galigniana, M. D., Radanyi, C., Renoir, J. M., Housley, P. R., and Pratt, W. B. (2001). Evidence that the peptidylprolyl isomerase domain of the HSP90-binding immunophilin FKBP52 is involved in both dynein interaction and glucocorticoid receptor movement to the nucleus. *Journal of Biological Chemistry*, 276, 14884–14889.

72. Pemberton, L. F. and Paschal, B. M. (2005). Mechanisms of receptor-mediated nuclear import and nuclear export. *Traffic (Copenhagen, Denmark)*, 6, 187–198.

73. Berg, P. and Pongratz, I. (2002). Two parallel pathways mediate cytoplasmic localization of the dioxin (aryl hydrocarbon) receptor. *Journal of Biological Chemistry*, 277, 32310–32319.

74. Jain, S., Maltepe, E., Lu, M. M., Simon, C., and Bradfield, C. A. (1998). Expression of ARNT, ARNT2, HIF1 alpha, HIF2

alpha and Ah receptor mRNAs in the developing mouse. *Mechanisms of Development*, 73, 117–123.

75. Lin, B. C., Sullivan, R., Lee, Y., Moran, S., Glover, E., and Bradfield, C. A. (2007). Deletion of the aryl hydrocarbon receptor-associated protein 9 leads to cardiac malformation and embryonic lethality. *Journal of Biological Chemistry*, 282, 35924–35932.

76. Kuzhandaivelu, N., Cong, Y. S., Inouye, C., Yang, W. M., and Seto, E. (1996). XAP2, a novel hepatitis B virus X-associated protein that inhibits X transactivation. *Nucleic Acids Research*, 24, 4741–4750.

77. Aufiero, B. and Schneider, R. J. (1990). The hepatitis B virus X-gene product trans-activates both RNA polymerase II and III promoters. *EMBO Journal*, 9, 497–504.

78. Lucito, R. and Schneider, R. J. (1992). Hepatitis B virus X protein activates transcription factor NF-κB without a requirement for protein kinase C. *Journal of Virology*, 66, 983–991.

79. Haviv, I., Vaizel, D., and Shaul, Y. (1995). The X protein of hepatitis B virus coactivates potent activation domains. *Molecular and Cellular Biology*, 15, 1079–1085.

80. Gearhart, T. L. and Bouchard, M. J. The hepatitis B virus X protein modulates hepatocyte proliferation pathways to stimulate viral replication. *Journal of Virology*, 84, 2675–2686.

81. Zhang, X., Zhang, H., and Ye, L. (2006). Effects of hepatitis B virus X protein on the development of liver cancer. *Journal of Laboratory And Clinical Medicine*, 147, 58–66.

82. Kashuba, E. V., Gradin, K., Isaguliants, M., Szekely, L., Poellinger, L., Klein, G., and Kazlauskas, A. (2006). Regulation of transactivation function of the aryl hydrocarbon receptor by the Epstein–Barr virus-encoded EBNA-3 protein. *Journal of Biological Chemistry*, 281, 1215–1223.

83. Sumanasekera, W. K., Tien, E. S., Turpey, R., Vanden Heuvel, J. P., and Perdew, G. H. (2003). Evidence that peroxisome proliferator-activated receptor alpha is complexed with the 90-kDa heat shock protein and the hepatitis virus B X-associated protein 2. *Journal of Biological Chemistry*, 278, 4467–4473.

84. Laenger, A., Lang-Rollin, I., Kozany, C., Zschocke, J., Zimmermann, N., Ruegg, J., Holsboer, F., Hausch, F., and Rein, T. (2009). XAP2 inhibits glucocorticoid receptor activity in mammalian cells. *FEBS Letters*, 583, 1493–1498.

85. Scammell, J. G., Denny, W. B., Valentine, D. L., and Smith, D. F. (2001). Overexpression of the FK506-binding immunophilin FKBP51 is the common cause of glucocorticoid resistance in three New World primates. *General and Comparative Endocrinology*, 124, 152–165.

86. Froidevaux, M. S., Berg, P., Seugnet, I., Decherf, S., Becker, N., Sachs, L. M., Bilesimo, P., Nygard, M., Pongratz, I., and Demeneix, B. A. (2006). The co-chaperone XAP2 is required for activation of hypothalamic thyrotropin-releasing hormone transcription *in vivo*. *EMBO Reports*, 7, 1035–1039.

87. Decherf, S., Hassani, Z., and Demeneix, B. A. (2008). *In vivo* siRNA delivery to the mouse hypothalamus shows a role of the co-chaperone XAP2 in regulating TRH transcription. *Methods in Molecular Biology*, 433, 355–366.

88. Kang, B. H. and Altieri, D. C. (2006). Regulation of survivin stability by the aryl hydrocarbon receptor-interacting protein. *Journal of Biological Chemistry*, 281, 24721–24727.

89. Vargiolu, M., Fusco, D., Kurelac, I., Dirnberger, D., Baumeister, R., Morra, I., Melcarne, A., Rimondini, R., Romeo, G., and Bonora, E. (2009). The tyrosine kinase receptor RET interacts *in vivo* with aryl hydrocarbon receptor-interacting protein to alter survivin availability. *Journal of Clinical Endocrinology and Metabolism*, 94, 2571–2578.

90. Ozfirat, Z. and Korbonits, M. AIP gene and familial isolated pituitary adenomas. *Molecular and Cellular Endocrinology*.

91. Cazabat, L., Libe, R., Perlemoine, K., Rene-Corail, F., Burnichon, N., Gimenez-Roqueplo, A. P., Dupasquier-Fediaevsky, L., Bertagna, X., Clauser, E., Chanson, P., Bertherat, J., and Raffin-Sanson, M. L. (2007). Germline inactivating mutations of the aryl hydrocarbon receptor-interacting protein gene in a large cohort of sporadic acromegaly: mutations are found in a subset of young patients with macroadenomas. *European Journal of Endocrinology/European Federation of Endocrine Societies*, 157, 1–8.

92. Igreja, S., Chahal, H. S., Akker, S. A., Gueorguiev, M., Popovic, V., Damjanovic, S., Burman, P., Wass, J. A., Quinton, R., Grossman, A. B., and Korbonits, M. (2009). Assessment of p27 (cyclin-dependent kinase inhibitor 1B) and aryl hydrocarbon receptor-interacting protein (AIP) genes in multiple endocrine neoplasia (MEN1) syndrome patients without any detectable MEN1 gene mutations. *Clinical Endocrinology*, 70, 259–264.

93. Leontiou, C. A., Gueorguiev, M., van der Spuy, J., Quinton, R., Lolli, F., Hassan, S., Chahal, H. S., Igreja, S. C., Jordan, S., Rowe, J., Stolbrink, M., Christian, H. C., Wray, J., Bishop-Bailey, D., Berney, D. M., Wass, J. A., Popovic, V., Ribeiro-Oliveira, A., Jr. Gadelha, M. R., Monson, J. P., Akker, S. A., Davis, J. R., Clayton, R. N., Yoshimoto, K., Iwata, T., Matsuno, A., Eguchi, K., Musat, M., Flanagan, D., Peters, G., Bolger, G. B., Chapple, J. P., Frohman, L. A., Grossman, A. B., and Korbonits, M. (2008). The role of the aryl hydrocarbon receptor-interacting protein gene in familial and sporadic pituitary adenomas. *Journal of Clinical Endocrinology and Metabolism*, 93, 2390–2401.

94. Lin, B. C., Nguyen, L. P., Walisser, J. A., and Bradfield, C. A. (2008). A hypomorphic allele of aryl hydrocarbon receptor-associated protein-9 produces a phenocopy of the AHR-null mouse. *Molecular Pharmacology*, 74, 1367–1371.

95. Beavo, J. A., Conti, M., and Heaslip, R. J. (1994). Multiple cyclic nucleotide phosphodiesterases. *Molecular Pharmacology*, 46, 399–405.

96. Conti, M. and Beavo, J. (2007). Biochemistry and physiology of cyclic nucleotide phosphodiesterases: essential components in cyclic nucleotide signaling. *Annual Review of Biochemistry*, 76, 481–511.

97. de Oliveira, S. K., Hoffmeister, M., Gambaryan, S., Muller-Esterl, W., Guimaraes, J. A., and Smolenski, A. P. (2007). Phosphodiesterase 2A forms a complex with the co-chaperone

XAP2 and regulates nuclear translocation of the aryl hydrocarbon receptor. *Journal of Biological Chemistry, 282,* 13656–13663.

98. Bolger, G. B., Peden, A. H., Steele, M. R., MacKenzie, C., McEwan, D. G., Wallace, D. A., Huston, E., Baillie, G. S., and Houslay, M. D. (2003). Attenuation of the activity of the cAMP-specific phosphodiesterase PDE4A5 by interaction with the immunophilin XAP2. *Journal of Biological Chemistry, 278,* 33351–33363.

99. Oesch-Bartlomowicz, B., Huelster, A., Wiss, O., Antoniou-Lipfert, P., Dietrich, C., Arand, M., Weiss, C., Bockamp, E., and Oesch, F. (2005). Aryl hydrocarbon receptor activation by cAMP vs. dioxin: divergent signaling pathways. *Proceedings of the National Academy of Sciences of the United States of America, 102,* 9218–9223.

100. Zheng, W., Brake, P. B., Bhattacharyya, K. K., Zhang, L., Zhao, D., and Jefcoate, C. R. (2003). Cell selective cAMP induction of rat CYP1B1 in adrenal and testis cells. Identification of a novel cAMP-responsive far upstream enhancer and a second Ah receptor-dependent mechanism. *Archives of Biochemistry and Biophysics, 416,* 53–67.

101. Zhang, Q. Y., He, W., Dunbar, D., and Kaminsky, L. (1997). Induction of CYP1A1 by β-naphthoflavone in IEC-18 rat intestinal epithelial cells and potentiation of induction by dibutyryl cAMP. *Biochemical and Biophysical Research Communications, 233,* 623–626.

102. de Oliveira, S. K. and Smolenski, A. (2009). Phosphodiesterases link the aryl hydrocarbon receptor complex to cyclic nucleotide signaling. *Biochemical Pharmacology, 77,* 723–733.

103. Felts, S. J. and Toft, D. O. (2003). p23, a simple protein with complex activities. *Cell Stress & Chaperones, 8,* 108–113.

104. Young, J. C., Moarefi, I., and Hartl, F. U. (2001). HSP90: a specialized but essential protein-folding tool. *Journal of Cell Biology, 154,* 267–273.

105. Freeman, B. C., Felts, S. J., Toft, D. O., and Yamamoto, K. R. (2000). The p23 molecular chaperones act at a late step in intracellular receptor action to differentially affect ligand efficacies. *Genes & Development, 14,* 422–434.

106. Young, J. C. and Hartl, F. U. (2000). Polypeptide release by HSP90 involves ATP hydrolysis and is enhanced by the co-chaperone p23. *EMBO Journal, 19,* 5930–5940.

107. Cox, M. B. and Miller, C. A., 3rd. (2002). The p23 co-chaperone facilitates dioxin receptor signaling in a yeast model system. *Toxicology Letters, 129,* 13–21.

108. Shetty, P. V., Bhagwat, B. Y., and Chan, W. K. (2003). P23 enhances the formation of the aryl hydrocarbon receptor–DNA complex. *Biochemical Pharmacology, 65,* 941–948.

109. Flaveny, C., Perdew, G. H., and Miller, C. A., 3rd. (2009). The aryl-hydrocarbon receptor does not require the p23 co-chaperone for ligand binding and target gene expression *in vivo*. *Toxicology Letters, 189,* 57–62.

110. Nair, S. C., Toran, E. J., Rimerman, R. A., Hjermstad, S., Smithgall, T. E., and Smith, D. F. (1996). A pathway of multi-chaperone interactions common to diverse regulatory proteins: estrogen receptor, Fes tyrosine kinase, heat shock transcription factor Hsf1, and the aryl hydrocarbon receptor. *Cell Stress & Chaperones, 1,* 237–250.

4

AHR LIGANDS: PROMISCUITY IN BINDING AND DIVERSITY IN RESPONSE

Danica DeGroot, Guochun He, Domenico Fraccalvieri, Laura Bonati, Allessandro Pandini, and Michael S. Denison

4.1 THE AH RECEPTOR

The Ah receptor (AHR) is a ligand-dependent basic helix–loop–helix–PER/ARNT/SIM (bHLH–PAS)-containing transcription factor that responds to exogenous and endogenous chemicals with the induction/repression of expression of a large battery of genes and production of a diverse spectrum of biological and toxic effects in a wide range of species and tissues [1–8]. More recently, the AHR has been shown to play a key regulatory role in a variety of endogenous developmental processes [7–9]. Induction of gene expression is one consistent AHR-dependent response and it was used as the model system to elucidate steps in the ligand-dependent AHR signaling pathway paradigm. Mechanistically, the inducing chemical ligand enters the cell and binds to the cytosolic AHR, which is present in a multiprotein complex containing two molecules of heat shock protein 90 (hsp90) [10], XAP2 [11], and co-chaperone p23 [12]. Following ligand binding, the AHR undergoes a conformational change, exposing a nuclear localization sequence (NLS) and the dimerization interface for the ARNT protein [13, 14]. The liganded AHR complex translocates into the nucleus [15, 16] and the AHR is released from its protein complex following its dimerization with ARNT [3, 14]. Formation of the liganded AHR:ARNT heterodimer converts the AHR complex into a high-affinity DNA binding form [3, 17] and binding of the complex to its specific DNA recognition site, the dioxin-responsive element (DRE), upstream of AHR-responsive genes leads to coactivator recruitment, chromatin rearrangement, increased promoter accessibility, and increased gene transcription [6, 17–19].

The identification and the characterization of AHR ligands have been an area of intense study for more than 30 years, following the discovery of the AHR in 1976 and demonstration of its ability to specifically bind 2,3,7,8-[^{3}H]tetrachlorodibenzo-p-dioxin (TCDD, dioxin) [20]. The best-characterized high-affinity ligands for the AHR include a variety of planar hydrophobic toxic halogenated aromatic hydrocarbons (HAHs), such as the polychlorinated dibenzo-p-dioxins, dibenzofurans, and biphenyls, and numerous polycyclic aromatic hydrocarbons (PAHs) and PAH-like chemicals, such as benzo(a)pyrene, 3-methylcholanthrene, and beta-naphthoflavone (BNF) [1, 2, 21]. However, a relatively large number of natural, endogenous and synthetic AHR agonists have also been identified in recent years whose structure and physicochemical characteristics are dramatically different from the prototypical HAH and PAH AHR ligands [21–23]. This suggests that the AHR has an extremely promiscuous ligand binding pocket. Interestingly, these studies have not only reported differences in the interaction of some ligands within the AHR ligand binding domain (LBD), but also observed ligand-specific differences in AHR-dependent gene expression. However, until very recently, the lack of understanding of the diversity in AHR ligand structure, the absence of a three-dimensional structure of the AHR LBD, and a lack of appreciation of ligand specificity in AHR response had limited our understanding of the mechanisms responsible for the spectrum of AHR-dependent responses. Accordingly, this chapter will describe recent developments and progress regarding the structural diversity of AHR ligands, the structure and function of the AHR LBD, and the diversity in response to different AHR

The AH Receptor in Biology and Toxicology, First Edition. Edited by Raimo Pohjanvirta.
© 2012 John Wiley & Sons, Inc. Published 2012 by John Wiley & Sons, Inc.

ligands. For a more in-depth description of AHR structure/function, AHR-dependent signal transduction, and biochemical and toxic effects of AHR ligands, the reader is referred to other chapters in this book and to many excellent published reviews [1–4, 8, 9].

4.2 AHR LIGANDS

The physicochemical characteristics necessary for binding of a chemical to the AHR have been examined by many laboratories for more than 30 years and have led to limited modeling of AHR ligand binding characteristics and to the identification of numerous AHR ligands. However, a greater diversity of novel AHR ligands and classes of ligands has been identified through the application of AHR-based bioassay approaches for chemical characterization and screening [24–26]. AHR ligands can be divided into two major categories: those that are planar and with characteristics similar to those of TCDD designated as "classical" AHR ligands, and those that have physicochemical/structural characteristics that dramatically diverge from that of TCDD and related chemicals designated as "nonclassical" AHR ligands. Given the extensive literature in this area, only representative examples of classical and nonclassical AHR ligands (agonists and antagonists) are presented in Fig. 4.1 and Table 4.1 in order to show the diversity in AHR ligands.

Classical AHR ligands represent the best-characterized high-affinity ligands for the AHR and they include the toxic HAHs, such as the polychlorinated and polybrominated dibenzo-p-dioxins, dibenzofurans, biphenyls and related compounds, and numerous PAHs and PAH-like chemicals [1, 2, 27]. These ligands are the ones responsible for producing the prototypical species- and tissue-specific AHR-dependent toxic responses and, with the exception of a few chemicals, members of this category are the highest affinity and most potent AHR ligands identified to date. While HAHs have a relatively high binding affinity for the AHR (in the pM to nM range), PAHs typically have a significantly lower affinity (in the high nM to μM range). Structure–activity relationship studies with a large number of classical AHR ligands suggest that the AHR binding pocket can accept planar ligands with maximal dimensions of $14 \text{ Å} \times 12 \text{ Å} \times 5 \text{ Å}$ and high-affinity binding appears to be critically dependent on key thermodynamic and electronic properties of the ligands [21, 27, 28–31]. While these modeling studies have some predictive applications for identification of new high-affinity AHR ligands, the constraints of these models still remain too restrictive, especially given current knowledge of AHR ligand diversity.

The greatest expansion in our knowledge about AHR ligands over the past 10 years has come from the identification and characterization of a relatively large number of natural, synthetic, and potentially endogenous AHR agonists and antagonists whose structure and physicochemical characteristics are dramatically different from those of the prototypical "classical" HAH and PAH ligands (reviewed in Refs 21–23) (Fig. 4.1 and Table 4.1). More recently, the identification of several endogenous lipids (HETE, PGG2, lipoxin A4, leukotriene metabolites, oxidized low-density lipoprotein [32–36]) and tryptophan metabolites

TABLE 4.1 Diversity of AHR Agonists and Antagonists

AHR Agonists (References)	AHR Antagonists (References)
Halogenated aromatic hydrocarbons	Halogenated aromatic hydrocarbons
Halogenated dibenzo-p-dioxins [2, 30]	MCDF [109, 149]
Halogenated dibenzofurans [2, 30]	Selected PCBs [83]
Halogenated biphenyls [2, 30]	Substituted flavones
Chlorinated naphthalenes [2, 30, 135]	Synthetic [89, 90, 143, 150, 151]
Polycyclic aromatic hydrocarbons	Natural [152–154]
PAHs [30, 135, 136]	Imidazoles and indoles
trans-Stilbenes [137]	SB203580 [155]
Hydroxylated benzo(a)pyrenes [81]	Rutaecarpine alkaloids [140]
Indoles	Diindolylmethane [156]
Indigo and indirubins [41, 138]	Ellipticine [140, 157]
Rutacarpine alkaloids [139]	Miscellaneous
Ellipticines [140]	SP600125 [158]
Various indoles [37–40, 141, 142]	ECGC and EGC [159]
Flavones/benzoflavones [30, 135, 140, 143]	CH223191 [90, 160]
Imidazoles and pyridines [42, 84, 85, 144]	7-Ketocholesterol [161]
Lipids and lipid metabolites [32–36]	Diflubenzuron [162]
Miscellaneous	Resveratrol [163]
Bilirubin [145]	SR1 [164]
2,3-Diaminotoluene [146]	
YH439 [44, 147]	
Carbaryl [111, 148]	

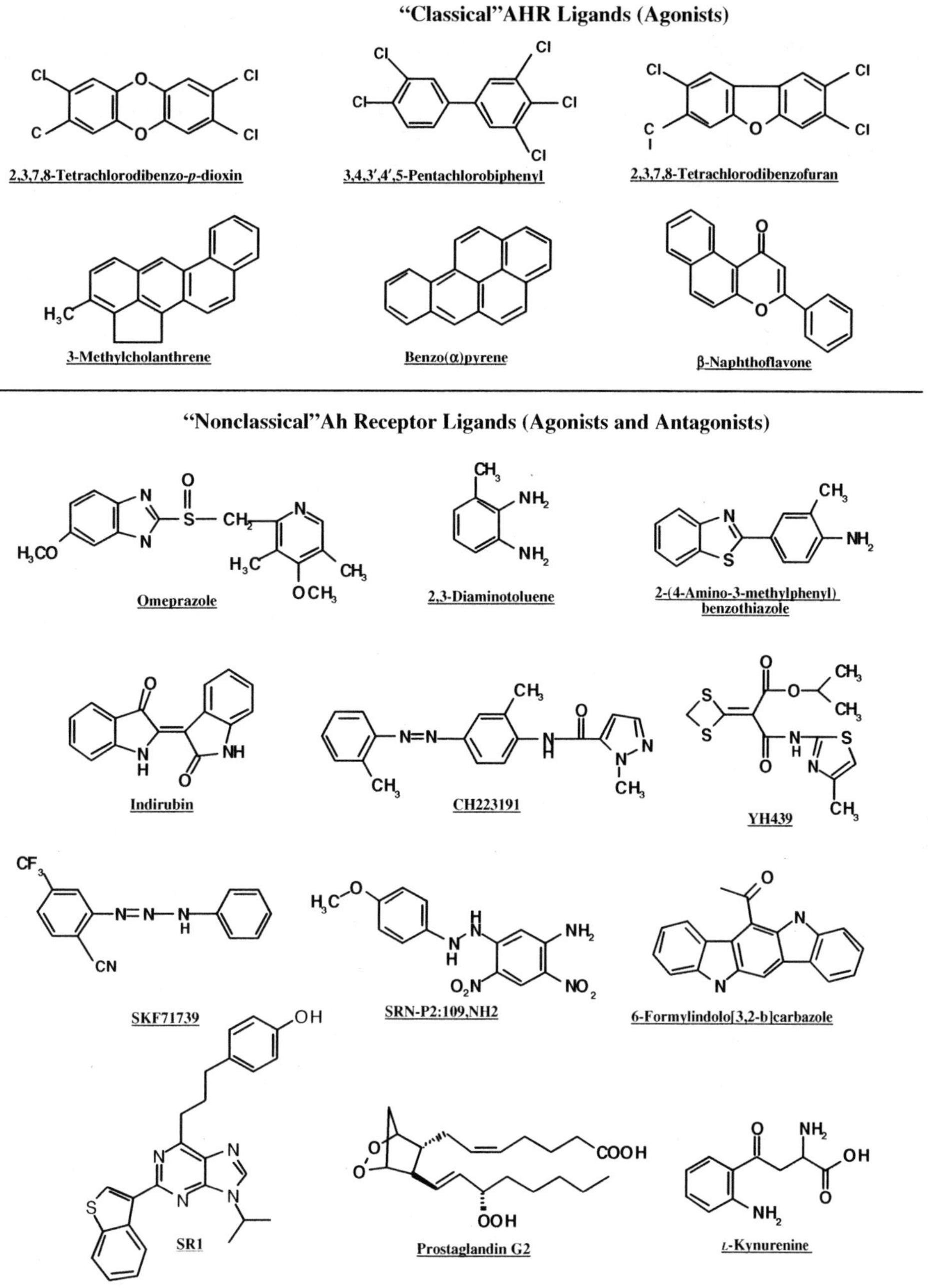

FIGURE 4.1 Representative structures of classical and nonclassical AHR ligands.

(kynurenines [37, 38]) as possible physiological ligands for the AHR has contributed to the known structural diversity of AHR ligands. Although the majority of these "nonclassical" AHR ligands/agonists are relatively low-affinity ligands and only moderately potent inducers of AHR-dependent gene expression (compared to TCDD), some potent agonists have been identified [39–41]. The identification of this striking structural diversity of these AHR ligands is important since it suggests that the AHR has an extremely promiscuous ligand binding site and demonstrates that the structural spectrum of synthetic and likely natural/endogenous AHR ligands is clearly much greater than originally thought. As described below, not only is there dramatic structural variation in the ligands that can bind to the AHR from a single species, but the ligand binding specificity of the AHR is also not identical between different species and this complicates studies and interpretation of the role of the AHR in chemical-specific effects within and between species. In addition, the identification of this promiscuity in AHR ligand structure supports a reevaluation of the currently accepted view of what makes a chemical an AHR ligand and what physicochemical characteristics are necessary for binding of a chemical within the

AHR LBD. Not surprisingly, previous quantitative structure–activity relationship (QSAR)-type studies attempting to include all known AHR ligands into a single binding model have been unsuccessful and suggest some diversity in ligand binding within the AHR LBD. Our recent collaborative studies with Mekenyan and coworkers [31] examining the relative binding activity/potency of more than 142 AHR agonists led to the development of a categorical common reactivity pattern (COREPA)-based structure–activity relationship model for predicting AHR ligands with different binding activities/potencies. These analyses suggested the existence of at least two distinct binding patterns for AHR agonists within the AHR LBD, one that was typified by that of TCDD and the other typified by coplanar polychlorinated biphenyls (PCBs), and that structurally diverse ligands exhibited a TCDD- or PCB-like binding pattern. These results and others described below are consistent with the idea that at least some AHR ligand promiscuity can be accounted for by differences in the binding of structurally diverse ligands within the AHR ligand binding pocket [42–44]. While the extreme structural diversity of AHR ligands is analogous to the ligand promiscuity reported for some members of the nuclear hormone receptors, such as the pregnane X receptor (PXR) [45, 46], a mechanistic understanding of AHR ligand diversity and the specific interactions of structurally diverse ligands with amino acids within the AHR LBD is lacking due to the absence of a 3D crystal/NMR structure of the AHR LBD.

4.3 AHR LIGAND BINDING DOMAIN

The AHR contains several defined protein domains that are responsible for each of its functional activities and each is affected by ligand binding. The N-terminal bHLH domain functions in DNA binding, hsp90 binding, and AHR:ARNT dimerization. The PAS domain in the central region of the AHR contains two structural repeats (PAS A and PAS B) that are involved in AHR/ARNT dimerization, and PAS B is also responsible for ligand and hsp90 binding [3–6]. Binding of hsp90 to the PAS B appears to direct the correct folding and maintenance of the high-affinity ligand binding conformation of AHR [47, 48] and ligand binding is proposed to produce a conformational alteration in AHR–hsp90 interactions in this region that unmasks both the N-terminal NLS [13] and the PAS A:ARNT dimerization interface [14]. XAP2 also interacts with the PAS B domain and contributes to the cytosolic localization of unliganded AHR [49, 50]. Inherent in the ligand-dependent AHR mechanism of action is a close association of the LBD with hsp90. While deletion of the LBD resulted in an hsp90-free, ligand-independent, and constitutively active AHR [14, 51], expression of an AHR fragment consisting of only the LBD (residues 230–421, 24% of the total murine AHR protein) produced an AHR–hsp90 complex that binds ligand (TCDD) with an affinity comparable to that of the wild-type AHR [52]. These results and the ability of several naturally occurring mutations within this region to alter AHR ligand binding affinity [53, 54] define the PAS B region as containing the AHR ligand binding pocket and responsible for the initial steps in ligand-dependent activation of the AHR. For a more detailed description of the AHR and ARNT domain structure, the reader is directed to several recent reviews ([3, 4, 55] and references therein) and other chapters in this book.

A molecular understanding of AHR ligand binding, ligand specificity, and the mechanism of ligand-dependent AHR activation requires detailed structural information about the AHR PAS B LBD. However, because of an inability to prepare sufficient amounts of purified AHR protein, no X-ray or NMR-determined structures of the liganded or unliganded AHR have been determined to date. However, once the first crystal structures of distant homologous proteins belonging to the PAS superfamily became available (the bacterial photoactive yellow protein (PYP) [56, 57], the PAS domain of the human potassium channel HERG [58], and the heme binding domain of the bacterial O_2 sensing FixL protein [59]), Bonati and coworkers [60] developed the first theoretical model for the LBD of the mouse AHR (mAHR) by applying homology modeling techniques. Despite the low level of sequence similarity, the structures of those three PAS domains showed highly conserved characteristics and FixL was found to be the best available template for homology modeling. The resulting model showed consistency with available experimental data, thus providing an initial framework to study the characteristics and functionality of the AHR LBD [27, 61]. The subsequent determination of the X-ray structure of the PAS domain of the fern photoreceptor Phy3 [62] improved the knowledge on this superfamily and highlighted the presence of different topologies of the ligand/cofactor binding cavity entrance in the known PAS structures. However, these first models of the mAHR LBD were not optimal due to the low degree of sequence similarity with the templates and the resulting uncertainty in the alignment of some regions. Although there was a high degree of confidence about the conserved fold characteristics of the initial modeled mAHR LBD, the resolution of the models was not sufficient to allow reliable detection and evaluation of more subtle details, such as loop lengths and secondary structure arrangements. Improvement in the model required new template structures of homologous proteins with significantly higher sequence identity to the AHR.

The availability of better template structures came from the impressive expansion of structural and functional information on PAS domains and a number of PAS structures derived by NMR spectroscopy or X-ray diffraction, including the N-terminal PAS domain of human PAS kinase (hPASK) [63], the C-terminal PAS B domain of human hypoxia-inducible factor 2α (HIF-2α) [64], the PAS B domain of the mouse nuclear coactivator NCoA1 [65], the N-terminal fragment of the *Drosophila* clock protein PERIOD (dPER) inclusive of two PAS repeats [66], and the PAS B domain of the human ARNT protein [67]. Accordingly, an improved homology model of

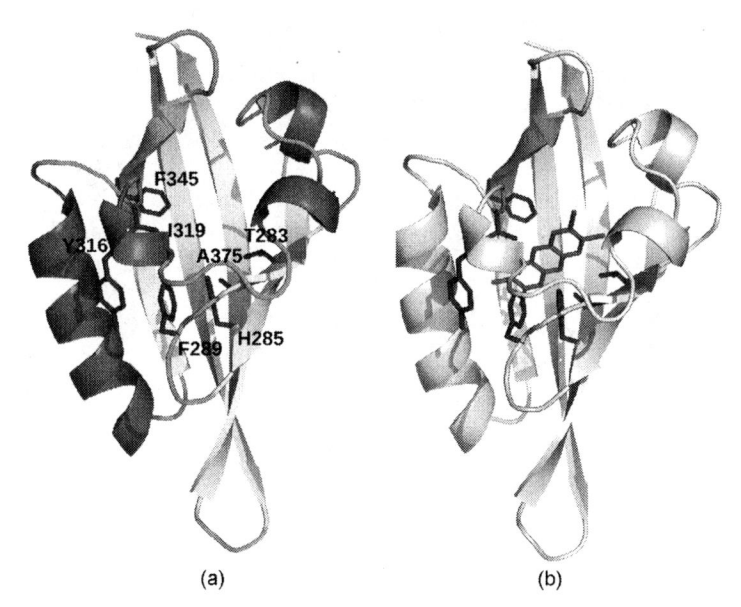

FIGURE 4.2 Homology model of the mouse AHR ligand binding domain and docking of TCDD in the binding cavity. (a) A cartoon representation of the modeled mAHR ligand binding domain with the "TCDD binding fingerprint" of conserved residues within the AHR ligand binding cavity necessary for optimal high-affinity TCDD binding [71]. (b) The lowest energy docked conformation of TCDD (blue sticks) within the binding cavity relative to the "binding fingerprint" residues. (See the color version of this figure in Color Plates section.)

the AHR LBD was built using the NMR-determined structures of the PAS B domains of HIF-2α [68] and ARNT [67] proteins as templates, given their higher degree of sequence identity and similarity with the AHR PAS B domain and their functional relationship to the AHR [69]. Site-directed mutagenesis, functional analysis, and ligand docking studies subsequently validated the improved AHR LBD model shown in Fig. 4.2 [70, 71]. More recently, other investigators have also reported development of comparable AHR LBD homology models, utilizing information derived from our LBD models, and have carried out comparable mutational analysis and ligand docking studies [43, 72–77].

Analysis of the mutagenesis results in the framework of the three-dimensional AHR LBD model identified several specific structural and chemical requirements for ligand binding. The structural analysis of the modeled domain indicated the presence of a buried cavity and the effects of point mutations in selected key residues on both TCDD binding and AHR functionality confirmed the role of the structural cavity as the site involved in ligand binding. While mutations of residues that point outside the domain do not affect AHR TCDD binding, mutations of internal residues reduce or eliminate TCDD binding and suggest that TCDD may bind close to Ala375 and His285, at the center of the cavity [70, 74]. These results provide an explanation for the lower ligand affinity of the AHR present in human and some mouse strains, since the presence of a valine residue in position 375 in the AHR in these species reduces the accessibility of the ligand within the LBD pocket. Further analysis of the cavity reveals additional insights into ligand binding. The surface of the cavity is characterized by a considerable aromatic character essential in stabilizing the TCDD binding, presumably by π–π interactions [71]. The cavity surface also contains a histidine and a threonine residue whose polarity seems to be necessary for specific binding to TCDD and a group of polar residues at both sides of the cavity surface that appears to contribute to TCDD stabilization probably through a network of weak interactions. The dramatic loss of AHR ligand binding with mutation of a residue in the central helical connector (I332) to the helix-breaking proline demonstrated the role of this structural element in maintaining the overall fold of the domain and the topology of the cavity. Finally, analysis of the LBD models of six mammalian AHRs with high affinity for TCDD revealed that the physicochemical characteristics of the modeled internal cavity spaces are well conserved in each. In addition, they allowed us to derive the complete list of residues needed for efficient TCDD binding and ligand-dependent transformation [71], inclusive of residues internal to all these cavities and conserved in all sequences. The role of each residue was validated by site-directed mutagenesis and AHR functional analysis [70, 71], allowing determination of the "TCDD binding fingerprint" (the group of seven residues in the AHR LBD whose characteristics are needed for optimal TCDD binding) (Fig. 4.2) [71]. The A375 and I319 residues exert a critical role in molecular recognition of TCDD due to their steric characteristics, F289, Y316, and F345 as a consequence of aromatic interactions, and T283 and H285 for electrostatic stabilization.

The residues and characteristics of the mouse AHR LBD important in ligand (TCDD) binding identified by mutagenesis and functional analysis were confirmed by our molecular docking simulations of TCDD within the AHR ligand binding pocket as shown in Fig. 4.2b [61]. More recently, utilization of aspects of our LBD modeling results for initial ligand docking analysis has been reported by several laboratories [72–77]. Overall, these analyses as well as future improvements in the AHR LBD model will certainly provide important insights into the mechanisms contributing to the observed structural diversity of AHR ligands and the mechanisms of ligand-dependent AHR signaling. While the current AHR LBD model has begun to increase our understanding of the molecular mechanisms of ligand binding, the structural diversity of AHR ligands presents some limitations with regard to detailed docking analysis. Initial examination of the improved AHR LBD model allowed determination of the internal cavity space of the ligand binding pocket by CASTp analysis [70]. While the AHR ligand binding pocket is sufficiently large to accommodate most ligands, several relatively large ligands that have been identified would not fit into the modeled cavity, implying that the AHR binding pocket must be larger than predicted by the current model. When we consider that the available PAS structures used for modeling do not contain bound hsp90, the proposed AHR LBD model would more appropriately describe the structural features of the ligand-bound hsp90-free activated form of AHR. Accordingly, we hypothesize that the buried cavity of the unliganded AHR LBD is actually made more accessible and significantly greater in size as a result of its association with hsp90, which binds to or within this region. Not only would this allow access and binding by ligands significantly larger than TCDD, but mechanistically one can also envision that ligand binding could disrupt AHR:hsp90 interactions at/within the ligand binding pocket, resulting in closure of the binding pocket around the ligand (as in our current LBD model), effectively trapping the ligand within the AHR LBD. This ligand binding scenario is consistent with current AHR ligand binding kinetics and provides a possible mechanism to explain the essentially "irreversible" binding nature of ligands (TCDD, BNF, and others) that has been reported by our laboratory and others [78–80].

4.3.1 Species-Specific Differences in Ligand Binding

While the current mouse AHR LBD model provides a framework for studies of ligand binding and activation of the AHR from other species, similar site-directed mutagenesis and functional analysis will be necessary to allow increased understanding of the mechanisms responsible for the observed species variation in ligand binding and ligand-dependent responsiveness. It is clear that some of this diversity can be attributed to differences in species-specific biochemical and physiological characteristics (particularly as they relate to differences in ligand pharmacokinetics, pharmacodynamics, and metabolism); however, although the binding specificity and rank order potency of HAH and PAH ligands are similar for AHRs among species and tissues, they are not identical and significant species differences do exist [81–83]. For example, although TCDD-inducible, AHR-dependent gene expression and TCDD binding of the mouse AHR were antagonized by several non-dioxin-like di-*ortho* PCBs, particularly 2,2′,5,5′-tetrachlorobiphenyl, it only partially antagonized activity of the rat AHR and had no effect on ligand binding or transcriptional activity of the human or guinea pig AHR [83]. Subsequent studies showed that although several benzimidazole drugs (omeprazole, thiabendazole, and lansoprazole) were unable to activate the AHR in mouse hepatoma (Hepa1c1c7) cells, they induce AHR-dependent gene expression in human hepatoma (HepG2) cells [84, 85]. Additional evidence for species-specific differences in ligand binding comes from studies using a series of PAHs [82] and single hydroxylated benzo[*a*]pyrene molecules [81], where significant differences in rank order potency of ligand binding were observed. Andersson and coworkers [86, 87] have demonstrated the ability of phenobarbital and gamma-aminobutyric acid (GABA) to bind to and activate the rainbow trout AHR, yet these chemicals are not known to bind to and/or activate the mammalian AHR. Additional studies observed that although mono-*ortho* PCBs were able to activate the human AHR, they were generally ineffective in activating rainbow trout and zebrafish AHRs [88].

While the majority of AHR LBD modeling has been done using the mouse AHR sequence, LBD models for rat, human, and zebrafish AHRs have also been reported [72, 74, 77] and comparison of the primary sequence of the AHR LBD from different species reveals some significant variation that could contribute to species differences in ligand response. Direct comparison of homology models of the LBD of AHRs from different species, such as that shown in Fig. 4.3, reveals amino acid differences that lie within the binding pocket and when combined with mutagenesis and functional analysis will help identify residues contributing to ligand differences among species. In fact, mutagenesis studies have already begun to identify several species-specific amino acids responsible for differences in ligand binding. 3′-Methoxy-4′-nitroflavone has been shown to be an antagonist/partial agonist of TCDD-induced AHR DRE binding and reporter gene induction in rat cells, yet it was a full agonist in guinea pig cells [89, 90], and this species variation appears to be due to a single amino acid difference in the AHR LBD of these two species (R355 in the mAHR and I360 in the guinea pig AHR) [91]. Similarly, Hahn and coworkers [92] demonstrated through site-directed mutagenesis studies that the 250-fold lower sensitivity of the tern AHR, compared to the chicken AHR, to TCDD-like chemicals was due to a two-amino acid difference in the tern LBD (Val325 and Ala381)

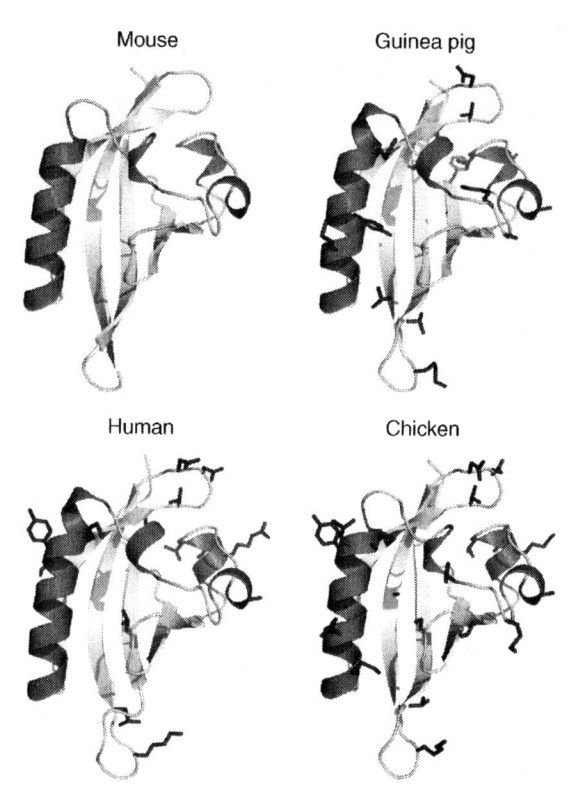

FIGURE 4.3 Cartoon representation of the modeled structures of the mouse, guinea pig, human, and chicken AHR ligand binding domains. Residues that are different from those in the mouse AHR ligand binding domain are indicated as blue sticks. (See the color version of this figure in Color Plates section.)

that was responsible for its reduced ligand binding affinity. Similarly, not only does the human AHR LBD exhibit sequence differences from the LBD of the high-affinity mouse AHR (specifically the modeled AHR^b allele), but it has a 10-fold lower affinity for TCDD and other ligands, and this lower affinity is due to the presence of a valine residue at position 381 (which is analogous to alanine at position 375 in the mouse AHR^b) [54, 93–95]. A valine at position 375 in the nonresponsive mouse AHR^d allele is responsible for its reduced ligand binding affinity compared to the mouse AHR^b allele [93] and mutation of alanine 375 to valine in the AHR^b allele significantly decreased its TCDD binding affinity [70, 93], demonstrating the critical nature of this amino acid. Interestingly, in a recent study using a transgenic mouse that expresses the human AHR protein in its liver, Perdew and coworkers [96] demonstrated that the human AHR exhibits a higher ligand binding affinity and responsiveness to two AHR ligands, indirubin and quercetin, than the endogenous mouse AHR^b allele, although the responsible amino acids remain to be determined. Not only do these studies further demonstrate that species differences in amino acids in the LBD contribute to ligand-specific differences in AHR responsiveness, but they also indicate that studies carried out in one species may not accurately predict the response in another. The availability of validated models of the AHR LBD from a variety of species will facilitate ongoing analyses of ligand-specific differences in AHR ligand binding and ligand responsiveness.

4.3.2 Ligand-Specific Differences in Ligand Binding

The AHR can bind and be activated or inhibited by a remarkably wide variety of structurally dissimilar compounds [21–23] and significant species-specific differences in ligand binding specificity and ligand-specific AHR functionality have been reported [21, 81–83]. Given the restrictions of the AHR LBD model, it remains an open question as to how the AHR LBD can accommodate such a diversity of ligand structures. The extreme structural diversity of AHR ligands is similar to the ligand promiscuity reported for some members of the nuclear hormone receptor superfamily, such as that of PXR [45, 46], and by analogy suggests that the promiscuity in AHR ligand structure also results from differential binding of ligands within the AHR ligand binding pocket. In fact, this hypothesis is supported by several recent studies. As described above, COREPA-based structure–activity relationship analysis of a large group of structurally diverse AHR agonists suggested the presence of at least two distinct binding patterns, one similar to TCDD and the other similar to coplanar PCBs [31]. More recently, we identified a novel ligand-selective AHR antagonist (CH223191) that preferentially inhibits the ability of TCDD and related HAHs to bind to the AHR, to stimulate AHR-dependent transformation, nuclear translocation and DNA binding and activation of AHR-dependent gene expression, but it has little inhibitory effect on the AHR agonist activity of BNF, PAHs, flavonoids, and indirubin [90]. While it is clear from previous [^{3}H]TCDD competitive binding studies that HAH and non-HAH AHR ligands can directly compete with each other for binding to the AHR ligand binding pocket and thus must have overlapping binding sites and common binding residues (reviewed in Ref. 25), CH223191 must interact with the AHR in such a way as to preferentially inhibit the binding of HAHs, but not non-HAH ligands. Similar to CH223191, the flavonoid antagonists 3′-methoxy-4′-nitroflavone and 6,2′,4,-trimethoxyflavone also exhibited ligand-specific antagonism, suggesting that all three compounds exert antagonism through a common mechanism. The preferential antagonism of HAHs by these three inhibitors is consistent with the hypothesis that there are significant differences in the binding of the HAHs and non-HAH AHR agonists within the ligand binding pocket and are supportive of at least two distinct classes of AHR agonists (based on their response to CH223191).

Differential binding of structurally diverse ligands within the AHR LBD are also indirectly supported by several site-directed mutagenesis/ligand binding analysis studies [42–44]. Mutation of a conserved tyrosine residue into phenylalanine (Y320F) in the human AHR LBD (comparable to residue 316 in the mouse AHR) reportedly resulted in the selective loss of binding and activation by several low-affinity non-HAH AHR ligands (i.e., 2-mercapto-5-methoxybenzimidazole, primaquine, and omeprazole), but not by the high-affinity ligand TCDD [42]. In contrast, insertion of a single mutation in a closely associated position into the mouse AHR LBD (F318L) reportedly produced a receptor that could be activated by the ligand 3-methylcholanthrene, but not by BNF or TCDD [43]. On the basis of our mouse AHR LBD homology model [70, 71], the mutations F318L and Y316F (Y320 in the human AHR) both reside in the same small alpha helix (Eα) present at the top of the ligand binding cavity. Interestingly, while the loss of the aromatic group in the side chain with the Y318L mutation eliminates binding by TCDD/BNF [43], the Y316F mutation, which retains an aromatic ring in this position, eliminates binding by low-affinity non-HAH ligands, but not TCDD [42]. These results, combined with our previous mutagenesis results [70, 71], indicate that the presence of an aromatic residue in these positions is critical for AHR ligand binding/functional activity and that changes in the specific residues within this helix differentially affect binding of HAH and non-HAH ligands. Another key residue, histidine 285, is contained within a central strand of the beta sheet (Aβ) of the LBD with its side chain pointing toward the center of the binding cavity, and previous studies [71] suggest that it plays a role in stabilizing TCDD binding through aromatic interactions. Mutation of histidine 285 to alanine in the LBD eliminated binding and activation of the AHR by TCDD, while mutation to phenylalanine reduced only the potency/affinity of TCDD, suggesting the importance of an aromatic group in this position as well [71]. Interestingly, the results of Whelan et al. [44] indicate that the presence of a tyrosine residue in this position allowed binding and activation of the AHR by the novel agonist YH439 but not TCDD, consistent with differences in the ability of these two agonists to bind within the ligand binding pocket and demonstrating the key nature of the aromatic side chain of this amino acid in binding specificity. Similarly, our site-directed mutagenesis and functional analysis reveals that a mAHR containing the double mutation F289L–R282A has a significantly enhanced transformation and DRE binding response to indirubin and indirubin-derivative ligands compared to wild-type mAHR (Fig. 4.4), indicating differential ligand binding to the wild-type and mutant AHRs. The effect of the specific mutations described above on the overall 3D structure of the AHR LBD and whether they result in an altered conformation of the LBD that contributes to or is responsible for these observed differential ligand-specific effects is not known. However, taken together, these studies support the hypothesis that the observed structural promiscuity of AHR ligands is derived, at least in part, from differential binding by ligands or classes of ligands within the AHR ligand binding pocket. Future mutagenesis, protein modeling, and docking analysis using the AHR LBD models and structurally diverse AHR ligands will provide further insights into the mechanisms responsible for AHR ligand promiscuity.

4.4 PROMISCUITY IN LIGAND-DEPENDENT AHR ACTIVATION

The structural diversity and the differential binding of AHR ligands also suggest the existence of selective modulators of the AHR similar to that reported for nuclear steroid hormone receptors. Previous studies have shown that the functional activity of nuclear hormone receptors can be altered in a ligand-selective manner and these functional changes appear to be directly related to ligand-specific changes in the overall structure of the receptor. This consequently impacts the specific proteins (i.e., coactivators) that interact with it and alters the overall biological response [97–100]. Similarly, given the ligand promiscuity of the AHR and differences in ligand binding, one can envision the existence of ligand-specific differences in AHR structure and function resulting from specific differences in the binding of ligands to the AHR (i.e., the existence of selective AHR modulators (SAHRMs)). In fact, several SAHRMs have already been identified based on their ability to selectively produce some AHR-dependent responses and not others (i.e., inhibition of inflammatory gene expression without induction of CYP1A1 [101, 102], ligand-selective inhibition of estrogen receptor action [103, 104], and differential gene expression). Although agonist-dependent changes in AHR protein structure have been observed [14], few studies have revealed any significant ligand-specific difference in AHR protein structure. Limited proteolysis studies have been used to demonstrate a difference in structure of the AHR when it is bound by an agonist or antagonist [105–107]. Interestingly, treatment of cells in culture with geldanamycin (an inhibitor that disrupts hsp90–AHR interactions) can produce an artificially transformed (i.e., activated) AHR that is structurally distinct from the liganded AHR [108]. This suggests that it is possible to produce distinct DNA binding forms of the AHR, although whether this can occur with different ligands remains to be demonstrated.

If the AHR can function similarly to the steroid hormone receptors and can assume ligand-specific conformations, then differential recruitment of coregulatory proteins and other members of the transcription complex would be predicted. Although the AHR antagonist (partial agonist) 3,3'-diindolylmethane (DIM) has been shown to be less efficient at recruiting RNA polymerase II to the cytochrome P4501A1 promoter in response to β-naphthoflavone treatment, no difference in recruitment was observed between DIM and TCDD. This may reflect a true ligand-specific difference between β-naphthoflavone and TCDD or may be the result of

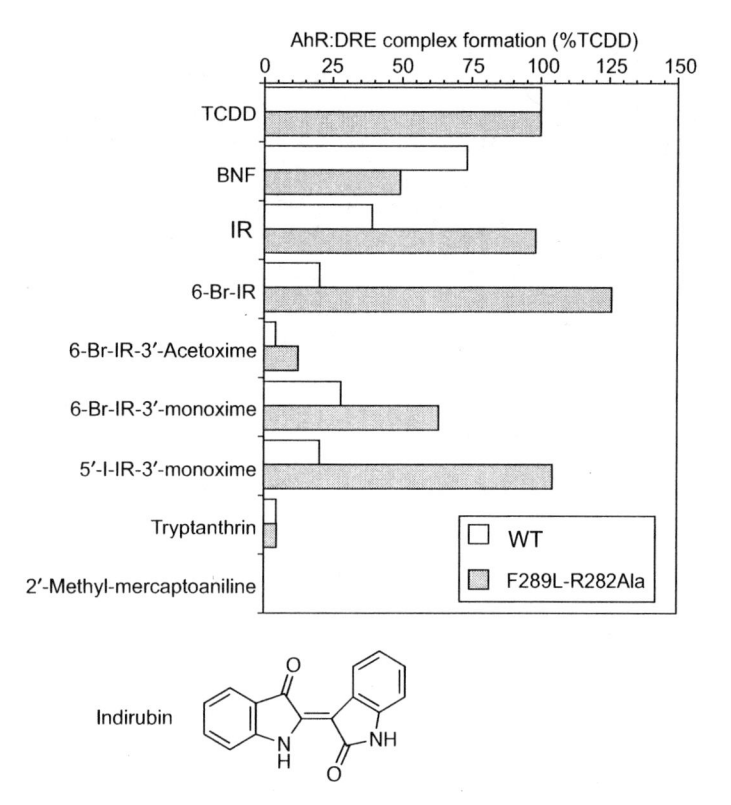

FIGURE 4.4 Mutation of phenylalanine 289 and arginine 282 within the LBD of the mAHR LBD differentially enhances the agonist activity of indirubin (structure shown) and indirubin derivatives to stimulate AHR transformation and DNA binding compared to TCDD. *In vitro* expressed wild-type AHR (white) or AHR containing the mutations R282A and F289L (black) and wt ARNT were incubated with DMSO (control), TCDD, indirubin (IR), indirubin derivative, tryptanthrin, or 2'-methyl-mercaptoaniline and ligand-dependent AHR:ARNT:DRE complex formation determined by gel retardation analysis [70, 71]. The amount of inducible AHR:ARNT:DRE complex formation was determined by phosphorimager analysis and the results expressed as the mean of at least triplicate analyses.

a cell line- or other tissue-specific difference [109, 110]. A decrease in the recruitment of the CREB binding protein (a histone acetyltransferase) has also been reported for AHRs bound by DIM or carbaryl as assessed by chromatin immunoprecipitation and yeast two-hybrid assays [110, 111]. More recently, ligand-specific differences in coactivator recruitment were observed for AHRs bound by different HAHs (i.e., polychlorinated dioxins, furans, or biphenyls) [103]. This result was surprising and raises the possibility that, even within a structurally related group of AHR ligands (agonists), minor physiochemical differences could produce a unique AHR conformation/functionality, recruit different coactivators, and result in differential gene transcription. Interestingly, the kinetics of AHR recruitment to DREs on its own target genes may also be ligand specific. BNF and DIM appear to cause oscillatory recruitment of the AHR to the CYP1A1 promoter and enhancer regions [110, 112], whereas TCDD does not [113, 114]. However, although oscillatory recruitment of the AHR to the CYP1A1 promoter in T47D cells can occur with the AHR ligands TCDD and 3-methylcholanthrene if the time course is extended to 4.5 h, these effects appear to be species- or cell line-specific as they were not observed in mouse hepatoma cells [115]. Taken together, although current evidence is limited, it does suggest the existence of some ligand-dependent differences in AHR structure/function that can contribute to the diversity in AHR response.

Assuming that different ligands can induce a unique AHR conformation, an additional possibility to consider is that ligand-dependent changes in AHR structure could lead to alterations in its nucleotide specificity of DNA binding. Classically, the ligand:AHR:ARNT complex has been shown to bind to the DRE, which contains the invariant sequence 5'-GCGTG-3'. This core sequence has been shown to be necessary, but not sufficient, for TCDD-induced AHR:ARNT complex binding and transcriptional activation and mutagenesis studies have identified that nucleotides adjacent to the core sequence are also required [116, 117]. Although our early targeted DNA mutagenesis failed to demonstrate any ligand-specific differences in AHR DNA binding [118], two studies have suggested that this can occur. Matikainen et al. [119] reported that two degenerate DREs present in the upstream regulatory region of the Bax gene were able to confer AHR-dependent responsiveness to a PAH metabolite (7,12-dimethylbenz(*a*)anthracene-3,4-dihydrodiol), but not to TCDD. However, a single nucleotide mutation that restored the consensus DRE sequence conferred TCDD responsiveness. More recently, Gouédard et al. [120] showed that a nonclassical DRE (4/5 consensus fit with the invariant DRE core sequence) present in the upstream region of the human paraoxonase 1 (Pon-1) gene was sufficient to allow the ligands 3-methylcholanthrene and quercetin, but not TCDD, to induce gene expression through an AHR-dependent mechanism. To our knowledge, these two reports stand alone in demonstrating ligand-specific differences in AHR DNA binding, and results from our laboratory have failed to demonstrate direct AHR binding to the Bax or Pon-1 "DRE-like" response element (DeGroot and Denison, unpublished observations). While it is well established that ligand-dependent transcription factors, such as the estrogen receptor, can bind to imperfect consensus sequences [121, 122], we are currently unaware of any ligand-activated transcription factor that binds to or stimulates gene expression from a completely unique consensus sequence based solely on the bound ligand. Thus, ligand-dependent variation in receptor response more likely resides in the proteins that can differentially interact with the AHR bound by distinct ligands rather than through an alteration of its DNA binding specificity.

Ligand-specific differences in AHR structure and function should also be predictive of differences in downstream biological or toxicological effects that cannot be explained by the inducing ligand's metabolic stability or affinity for the AHR. The SAHRMs are a group of AHR ligands with unique biological effects. However, based on its use in the literature, the exact definition of a SAHRM remains to be resolved. One definition, similar to that described for the selective estrogen receptor modulators (SERMs), stated that SAHRMs are compounds that produce species-, tissue-, and cell-specific agonist or antagonist activities [123, 124]. The compounds DIM and 6-methyl-1,3,8-trichlorodibenzofuran (MCDF) are classic SAHRMs having demonstrated AHR-dependent antiestrogenic activity *in vitro* and *in vivo* with minimal inducing effects on classical AHR-responsive genes, in contrast to TCDD and other full AHR agonists that induce AHR-dependent gene expression and are relatively potent antiestrogens [125, 126]. Ligand-specific differences between these compounds and TCDD have also been reported [127, 128]. Interestingly, Murray et al. [102, 103] presented evidence of novel SAHRMs, namely, WAY-169916 and SGA 360, that repress inflammatory gene expression in an AHR-dependent manner, yet produce no AHR:ARNT DNA binding or DRE-dependent transcriptional activation. Their results suggest that, while the AHR can be bound and activated by these compounds (with activation based on their ability to stimulate AHR nuclear translocation), the AHR must take on a unique protein conformation that not only prevents it from binding to ARNT, but also allows it to bind to other nuclear factors and repress expression of specific genes. The mechanism by which these novel AHR ligands (WAY-169916, SGA 360, DIM, and MCDF) can differentially affect the structure and function of the AHR is an exciting area of future research.

4.5 CONCLUSIONS

The AHR, unlike most ligand-dependent transcription factors, can be bound and activated by structurally diverse ligands and this promiscuity in ligand binding appears to be derived, at least in part, from differential ligand interactions with amino acids within the AHR ligand binding pocket. This diversity is clearly evidenced by comparison of the structure of these "nonclassical" ligands described in this chapter to that of TCDD and other "classical" AHR ligands. Not only is there a dramatic diversity in the structure of ligands for the AHR from a given species, but significant differences in ligand specificity also exist among species and these differences contribute to the diversity in response to AHR activators. The recently reported role of the AHR in a variety of endogenous physiological and developmental responses and the identification of the ability of numerous endogenous chemicals to bind to and activate the AHR strongly support the existence of multiple (structurally diverse) endogenous ligands. One can envision that cell- and tissue-specific endogenous ligands exist that can activate AHR-dependent gene expression to produce unique cell- or tissue-specific biological responses. The availability of validated theoretical structures of the AHR LBD from various species provides avenues in which to analyze species differences in ligand binding and the mechanisms of ligand-dependent activation of the AHR. More significantly is the rapidly expanding application of these LBD models for the analysis of AHR ligand binding by molecular docking approaches. These studies will greatly expand our understanding of the molecular mechanisms of AHR ligand binding and provide key insights into the unique structural diversity of AHR ligands among species. In addition, these models can be used to rapidly screen chemical libraries to identify novel AHR ligands and this will be an important area of future research. From a more practical perspective, the identification and the characterization of novel AHR agonists/antagonists become even more significant when one considers recent studies demonstrating involvement of the AHR in a variety of numerous clinically important human diseases [7, 8, 129–134]. Because of these associations and the diversity of AHR ligand structure and ligand-dependent responses, extensive screening efforts to identify and characterize novel AHR ligands as potential human therapeutic agents are already currently underway in numerous laboratories. The results of these studies are certain to yield new and interesting compounds, from both pharmacological and AHR research perspectives, as well as new insights into the endogenous role of this enigmatic receptor.

ACKNOWLEDGMENTS

We thank Grace Bedoian for her expert editorial assistance. Work in our laboratories on AHR ligands and the AHR ligand binding domain has been supported by the National Institutes of Environmental Health Sciences of the National Institutes of Health (ES012498, ES007685, Environmental Toxicology Training Grant ES07059 (to DD), and Superfund Research Grant (ES004699)), the California Agricultural Experiment Station, and the American taxpayers.

REFERENCES

1. Poland, A. and Knutson, J. C. (1982). 2,3,7,8-Tetrachlorodibenzo-*p*-dioxin and related halogenated aromatic hydrocarbons: examination of the mechanism of toxicity. *Annual Review of Pharmacology and Toxicology*, 22, 517–542.
2. Safe, S. (1990). Polychlorinated biphenyls (PCBs), dibenzo-*p*-dioxins (PCDDs), dibenzofurans (PCDFs), and related compounds: environmental and mechanistic considerations which support the development of toxic equivalency factors (TEFs). *Critical Review in Toxicology*, 21, 51–88.
3. Hankinson, O. (1995). The aryl hydrocarbon receptor complex. *Annual Review of Pharmacology and Toxicology*, 35, 307–340.
4. Ma, Q. (2001). Induction of CYP1A1. The AhR/DRE paradigm: transcription, receptor regulation, and expanding biological roles. *Current Drug Metabolism*, 2, 149–164.
5. Kewley, R. J., Whitelaw, M. L., and Chapman-Smith, A. (2004). The mammalian basic helix–loop–helix/PAS family of transcriptional regulators. *International Journal of Biochemistry and Cell Biology*, 36, 189–204.
6. Beischlag, T. V., Luis Morales, J., Hollingshead, B. D., and Perdew, G. H. (2008). The aryl hydrocarbon receptor complex and the control of gene expression. *Critical Review in Eurkaryotic Gene Expression*, 18, 207–250.
7. Bradshaw, T. D. and Bell, D. R. (2009). Relevance of the aryl hydrocarbon receptor (AhR) for clinical toxicology. *Clinical Toxicology*, 47, 632–642.
8. Furness, S. G. and Whelan, F. (2009). The pleiotropy of dioxin toxicity–xenobiotic misappropriation of the aryl hydrocarbon receptor's alternative physiological roles. *Pharmacological Therapeutics*, 124, 336–354.
9. McIntosh, B. E., Hogenesch, J. B., and Bradfield, C. A. (2010). Mammalian Per–Arnt–Sim proteins in environmental adaptation. *Annual Review of Physiology*, 72, 625–645.
10. Chen, H.-S. and Perdew, G. H. (1994). Subunit composition of the heteromeric cytosolic aryl hydrocarbon receptor complex. *Journal of Biological Chemistry*, 269, 27554–27558.
11. Meyer, B. K., Pray-Grant, M. G., Vanden Heuvel, J. P., and Perdew, G. H. (1998). Hepatitis B virus X-associated protein 2 is a subunit of the unliganded aryl hydrocarbon receptor core complex and exhibits transcriptional enhancer activity. *Molecular and Cellular Biology*, 18, 978–988.
12. Kazlauskas, A., Poellinger, L., and Pongratz, I. (1999). Evidence that the co-chaperone p23 regulates ligand responsiveness of the dioxin (aryl hydrocarbon) receptor. *Journal of Biological Chemistry*, 274, 13519–13524.
13. Ikuta, T., Eguchi, H., Tachibana, T., Yoneda, Y., and Kawajiri, K. (1998). Nuclear localization and export signals of the human aryl hydrocarbon receptor. *Journal of Biological Chemistry*, 273, 2895–2904.
14. Soshilov, A. A. and Denison, M. S. (2008). Role of the Per/Arnt/Sim domains in ligand-dependent transformation of the aryl hydrocarbon receptor. *Journal of Biological Chemistry*, 283, 32995–33005.
15. Hord, N. G. and Perdew, G. H. (1994). Physiochemical and immunochemical analysis of aryl hydrocarbon receptor nuclear translocator: characterization of two monoclonal antibodies to the aryl hydrocarbon receptor nuclear translocator. *Molecular Pharmacology*, 46, 618–624.
16. Pollenz, R. S., Sattler, C. A., and Poland, A. (1994). The aryl hydrocarbon receptor and aryl hydrocarbon receptor nuclear translocator protein show distinct subcellular localizations in Hepa 1c1c7 cells by immunofluorescence microscopy. *Molecular Pharmacology*, 45, 428–438.
17. Denison, M. S., Fisher, J. M., and Whitlock, J. P., Jr. (1988). The DNA recognition site for the dioxin–Ah receptor complex: nucleotide sequence and functional analysis. *Journal of Biological Chemistry*, 263, 17721–17724.
18. Carlson, D. B. and Perdew, G. H. (2002). A dynamic role for the Ah receptor in cell signaling? Insights from a diverse group of Ah receptor interacting proteins. *Journal of Biochemical and Molecular Toxicology*, 16, 317–325.
19. Hankinson, O. (2005). Role of coactivators in transcriptional activation by the aryl hydrocarbon receptor. *Archives of Biochemistry and Biophysics*, 433, 379–386.
20. Poland, A., Glover, E., and Kende, A. S. (1976). Stereospecific, high affinity binding of 2,3,7,8-tetrachlorodibenzo-*p*-dioxin by hepatic cytosol: evidence that the binding species is the receptor for induction of aryl hydrocarbon hydroxylase. *Journal of Biological Chemistry*, 251, 4936–4946.
21. Denison, M. S., Seidel, S. D., Rogers, W. J., Ziccardi, M., Winter, G. M., and Heath-Pagliuso, S. (1998). Natural and synthetic ligands for the Ah receptor. In *Molecular Biology Approaches to Toxicology* (Puga, A. and Wallace, K. B.,Eds). Taylor & Francis, Philadelphia, pp. 393–410.
22. Denison, M. S. and Nagy, S. R. (2003). Activation of the aryl hydrocarbon receptor by structurally diverse exogenous and endogenous chemicals. *Annual Review of Pharmacology and Toxicology*, 43, 309–334.

23. Nguyen, L. P. and Bradfield, C. A. (2008). The search for endogenous activators of the aryl hydrocarbon receptor. *Chemical Research in Toxicology, 21*, 102–106.
24. Seidel, S. D., Li, V., Winter, G. M., Rogers, W. J., Martinez, E. I., and Denison, M. S. (2000). Ah receptor-based chemical screening bioassays: application and limitations for the detection of Ah receptor agonists. *Toxicological Sciences, 55*, 107–115.
25. Denison, M. S., Zhao, B., Baston, D. S., Clark, G. C., Murata, H., and Han, D.-H. (2004). Recombinant cell bioassay systems for the detection and relative quantitation of halogenated dioxins and related chemicals. *Talanta, 63*, 1123–1133.
26. Nagy, S. R., Liu, G., Lam, K., and Denison, M. S. (2002). Identification of novel Ah receptor agonists using a high-throughput green fluorescent protein-based recombinant cell bioassay. *Biochemistry, 41*, 861–868.
27. Denison, M. S., Pandini, A., Nagy, S. R., Baldwin, E. P., and Bonati, L. (2002). Ligand binding and activation of the Ah receptor. *Chemico-Biological Interactions, 141*, 3–24.
28. Bonati, L., Fraschini, E., Lasagni, M., Modoni, E. P., and Pitea, D. (1995). A hypothesis on the mechanism of PCDD biological activity based on molecular electrostatic potential modeling. Part 2. *Journal of Molecular Structure (Theochem), 340*, 83–95.
29. Zhao, Y. Y., Tao, F. M., and Zeng, E. Y. (2008). Theoretical study of the quantitative structure–activity relationships for the toxicity of dibenzo-*p*-dioxins. *Chemosphere, 73*, 86–91.
30. Diao, J., Li, Y., Shi, S., Sun, Y., and Sun, Y. (2010). QSAR models for predicting toxicity of polychlorinated dibenzo-*p*-dioxins and dibenzofurans using quantum chemical descriptors. *Bulletin of Environmental Contamination and Toxicology, 85*, 109–115.
31. Petko, P. I., Rowlands, J. C., Budinsky, R., Zhao, B., Denison, M. S., and Mekenyan, O. (2010). Mechanism based common reactivity pattern (COREPA) modeling of AhR binding affinity. *SAR QSAR in Environmental Research, 21*, 187–214.
32. Schladach, C. N., Riby, J., and Bjeldanes, L. F. (1999). Lipoxin A4: a new class of ligand for the Ah receptor. *Biochemistry, 38*, 7594–7600.
33. Seidel, S. D., Winters, G. M., Rogers, W. J., Ziccardi, M. H., Li, V., Keser, B., and Denison, M. S. (2001). Activation of the Ah receptor signaling pathway by prostaglandins. *Journal of Biochemical and Molecular Toxicology, 15*, 187–196.
34. Chiaro, C. R., Patel, R. D., and Perdew, G. H, (2008). 12(*R*)-Hydroxy-5(*Z*),8(*Z*),10(*E*),14(*Z*)-eicosatetraenoic acid [12(*R*)-HETE], an arachidonic acid derivative, is an activator of the aryl hydrocarbon receptor. *Molecular Pharamacology, 74*, 1649–1656.
35. Chiaro, C. R., Morales, J. L., Prabhu, K. S., and Perdew, G. H. (2008). Leukotriene A4 metabolites are endogenous ligands for the Ah receptor. *Biochemistry, 47*, 8445–8455.
36. McMillan, B. J. and Bradfield, C. A. (2007). The aryl hydrocarbon receptor is activated by modified low-density lipoprotein. *Proceedings of the National Academy of Sciences of the United States of America, 104*, 1412–1417.
37. DiNatale, B. C., Murray, I. A., Schroeder, J. C., Flaveny, C. A., Lahoti, T. S., Laurenzana, E. M., Omiecinski, C. J., and Perdew, G. H. (2010). Kynurenic acid is a potent endogenous aryl hydrocarbon receptor ligand that synergistically induces interleukin-6 in the presence of inflammatory signaling. *Toxicological Sciences, 115*, 89–97.
38. Mezrich, J. D., Fechnew, J. H., Zhang, X., Johnson, B. P., Burlingham, W. J., and Bradfield, C. A. (2010). An interaction between kynurenine and the aryl hydrocarbon receptor can generate regulatory T cells. *Journal of Immunology, 185*, 3190–3198.
39. Bjeldanes, L. F., Kim, J.-L., Grose, K. R, Bartholomew, J. C., and Bradfield, C. A. (1991). Aromatic hydrocarbon responsiveness-receptor agonists generated from indole-3-carbinol *in vitro* and *in vivo*: comparisons with 2,3,7,8-tetrachlorodibenzo-*p*-dioxin. *Proceedings of the National Academy of Sciences of the United States of America, 88*, 9543–9547.
40. Rannug, A., Rannug, U., Rosenkranz, H. S., Winqvist, L., Westerholm, R., Agurell, E, and Grafstrom, A.-K. (1987). Certain photooxidized derivatives of tryptophan bind with very high affinity to the Ah receptor and are likely to be endogenous signal substances. *Journal of Biological Chemistry, 262*, 15422–15427.
41. Knockaert, M., Blondel, M., Bach, S., Leost, M., Elbi, C., Hager, G. L., Nagy, S. R., Han, D., Denison, M., French, M., Ryan, X. P., Magiatis, P., Polychronopoulos, P., Greengard, P., Skaltsounis, L., and Meijer, L. (2004). Independent actions on cyclin-dependent kinases and aryl hydrocarbon receptor mediate the antiproliferative effects of indirubins. *Oncogene, 23*, 4400–4412.
42. Backlund, M. and Ingelman-Sundberg, M. (2004). Different structural requirements of the ligand binding domain of the aryl hydrocarbon receptor for high- and low-affinity ligand binding and receptor activation. *Molecular Pharmacology, 65*, 416–425.
43. Goryo, K., Suzuki, A., Del Carpio, C. A., Siizaki, K., Kuriyama, E., Mikami, Y., Kinoshita, K., Yasumoto, K., Rannug, A., Miyamoto, A., Fujii-Kuriyama, Y., and Sogawa, K. (2007). Identification of amino aid residues in the Ah receptor involved in ligand binding. *Biochemical Biophysical Research Communications, 354*, 396–402.
44. Whelan, F., Hao, N., Furness, S. G., Whitelaw, M. L., and Chapman-Smith, A. (2010). Amino acid substitutions in the aryl hydrocarbon receptor (AhR) ligand binding domain reveal YH439 as a atypical AhR activator. *Molecular Pharmacology, 77*, 1037–1046.
45. Noy, N. (2007). Ligand specificity of nuclear hormone receptors: sifting through promiscuity. *Biochemistry, 46*, 13461–13467.
46. Ngan, C. H., Beglov, D., Rudnitskaya, A. N., Kozakov, D., Waxman, D. J., and Vajda, S. (2009). The structural basis of pregnane X receptor binding promiscuity. *Biochemistry, 48*, 11572–11581.
47. Carver, L. A., Jackiw, V., and Bradfield, C. A. (1994). The 90-kDa heat shock protein is essential for Ah receptor signaling in

a yeast expression system. *Journal of Biological Chemistry, 269,* 30109–30112.

48. Pongratz, I., Mason, G. G. F., and Poellinger, L. (1992). Dual roles of the 90-kDa heat shock protein hsp90 in modulating functional activities of the dioxin receptor. *Journal of Biological Chemistry, 267,* 13728–13734.

49. Petrulius, J. R., Hord, N. G., and Perdew, G. H. (2000). Subcellular location of the aryl hydrocarbon receptor is modulated by the immunophilin homolog hepatitis B virus X-associated protein 2. *Journal of Biological Chemistry, 275,* 37448–37453.

50. LaPres, J. J., Glover, E., Dunham, E. E., Bunger, M. K., and Bradfield, C. A. (2000). ARA9 modifies agonist signaling through an increase in cytosolic aryl hydrocarbon receptor. *Journal of Biological Chemistry, 275,* 6153–6159.

51. McGuire, J., Okamoto, K., Whitelaw, M. L., Tanaka, H., and Poellinger, L. (2001). Definition of a dioxin receptor mutant that is a constitutive activator of transcription: delineation of overlapping repression and ligand binding functions within the PAS domain. *Journal of Biological Chemistry, 276,* 41841–41849.

52. Coumailleau, P., Poellinger, L., Gustafsson, J.-A., and Whitelaw, M. L. (1995). Definition of a minimal domain of the dioxin receptor that is associated with hsp90 and maintains wild type ligand binding affinity and specificity. *Journal of Biological Chemistry, 270,* 25291–25300.

53. Ema, M., Ohe, N., Suzuki, M., Mimura, J., Sogawa, K., Ikawa, S., and Fujii-Kuriyama, Y. (1994). Dioxin binding activities of polymorphic forms of mouse and human aryl hydrocarbon receptors. *Journal of Biological Chemistry, 269,* 27337–27343.

54. Poland, A., Palen, D., and Glover, E. (1994). Analysis of the four alleles of the murine aryl hydrocarbon receptor. *Molecular Pharmacology, 46,* 915–921.

55. Rowlands, J. C. and Gustafsson, J.-A. (1997). Aryl hydrocarbon receptor-mediated signal transduction. *Critical Reviews in Toxicology, 27,* 109–134.

56. Borgstahl, G. E., Williams, D. R., and Getzoff, E. D. (1995). 1.4 Å structure of photoactive yellow protein, a cytosolic photoreceptor: unusual fold, active site, and chromophore. *Biochemistry, 34,* 6278–6287.

57. Dux, P., Rubinstenn, G., Vuister, G. W., Boelens, R., Mulder, F. A., Hard, K., Hoff, W. D., Kroon, A. R., Crielaard, W., Hellingwerf, K. J., and Kaptein, R. (1998). Solution structure and backbone dynamics of the photoactive yellow protein. *Biochemistry, 37,* 12689–12699.

58. Cabral, J. H. M., Lee, A., Cohen, S. L., Chait, B. T., Li, M., and Mackinnon, R. (1998). Crystal structure and functional analysis of the HERG potassium channel N terminus: a eukaryotic PAS domain. *Cell, 95,* 649–655.

59. Gong, W., Hao, B., Mansy, S. S., Gonzalez, G., Gilles-Gonzalez, M. A., and Chan, M. K. (1998). Structure of a biological oxygen sensor: a new mechanism for heme-driven signal transduction. *Proceedings of the National Academy of Sciences of the United States of America, 95,* 15177–15182.

60. Procopio, M., Lahm, A., Tramontano, A., Bonati, L., and Pitea, D. (2002). A model for recognition of polychlorinated dibenzo-*p*-dioxins by the aryl hydrocarbon receptor. *European Journal of Biochemistry, 269,* 13–18.

61. Murray, I. A., Reen, R. K., Leathery, N., Ramadoss, P., Bonati, L., Gonzalez, F. J., Peters, J. M., and Perdew, G. H. (2005). Evidence that ligand binding is a key determinant of Ah receptor-mediated transcriptional activity. *Archives of Biochemistry and Biophysics, 442,* 59–71.

62. Crosson, S. and Moffat, K. (2001). Structure of a flavin-binding plant photoreceptor domain: insights into light-mediated signal transduction. *Proceedings of the National Academy of Sciences of the United States of America, 98,* 2995–3000.

63. Amezcua, C., Harper, S., Rutter, J., and Gardner, K. (2002). Structure and interactions of PAS kinase N-terminal PAS domain: model for intramolecular kinase regulation. *Structure, 10,* 1349–1361.

64. Erbel, P. J., Card, P. B., Karakuzu, O., Bruick, R. K., and Gardner, K. H. (2003). Structural basis for PAS domain heterodimerization in the basic helix–loop–helix/PAS transcription factor hypoxia-inducible factor. *Proceedings of the National Academy of Sciences of the United States of America, 100,* 15504–15509.

65. Razeto, A., Ramakrishnan, V., Litterst, C. M., Giller, K., Griesinger, C., Carlomagno, T., Lakomek, N., Heimburg, T., Lodrini, M., Pfitzner, E., and Becker, S. (2004). Structure of the NCoA-1/Src-1 PAS-B domain bound to the LXXLL motif of the STAT6 transactivation domain. *Journal of Molecular Biology, 336,* 319–329.

66. Yildiz, O., Doi, M., Yujnovsky, I., Cardone, L., Berndt, A., Hennig, S., Schulze, S., Urbanke, C., Sassone-Corsi, P., and Wolf, E. (2005). Crystal structure and interactions of the PAS repeat region of the *Drosophila* clock protein PERIOD. *Molecular Cell, 17,* 69–82.

67. Card, P. B., Erbel, P. J., and Gardner, K. H. (2005). Structural basis of ARNT PAS-B dimerization: use of a common beta-sheet interface for hetero- and homodimerization. *Journal of Molecular Biology, 353,* 664–678.

68. Erbel, P. J., Card, P. B., Karakuzu, O., Bruick, R. K., and Gardner, K. H. (2003). Structural basis for PAS domain heterodimerization in the basic helix–loop–helix–PAS transcription factor hypoxia-inducible factor. *Proceedings of the National Academy of Sciences of the United States of America, 100,* 15504–15509.

69. Kewley, R. J., Whitelaw, M. L., and Chapman-Smith, A. (2004). The mammalian basic helix–loop–helix/PAS family of transcriptional regulators. *International Journal of Biochemistry and Cellular Biology, 36,* 189–204.

70. Pandini, A., Denison, M. S., Song, Y., Soshilov, A., and Bonati, L. (2007). Structural and functional characterization of the AhR ligand binding domain by homology modeling and mutational analysis. *Biochemistry, 23,* 696–708.

71. Pandini, A., Soshilov, A. A., Song, Y., Zhao, J., Bonati, L., and Denison, M. S. (2009). Detection of the TCDD binding fingerprint within the Ah receptor ligand binding domain by

structurally driven mutagenesis and functional analysis. *Biochemistry*, 48, 5972–5983.
72. Bisson, W. H., Koch, D. C., O'Donnell, E. F., Khalil, S. M., Kerkvliet, N. I., Tanguay, R. L., Abagyan, R., and Kolluri, S. K. (2009). Modeling of the aryl hydrocarbon receptor (AhR) ligand binding domain and its utility in virtual ligand screening to predict new AhR ligands. *Journal of Medicinal Chemistry*, 52, 5635–5641.
73. Murray, I. A., Flaveny, C. A., Chiaro, C. R., Sharma, A. K., Tanos, R. S., Schroeder, J. C., Amin, S. G., Bisson, W. H., Kolluri, S. K., and Perdew, G. H. (2010). Suppression of cytokine-mediated complement factor gene expression through selective activation of the Ah receptor with 3′,4′-dimethoxy-α-naphthoflavone. *Molecular Pharmacology*, 79, 508–519.
74. Wu, B., Zhang, Y., Kong, J., Zhang, X., and Cheng, S. (2009). In silico prediction of nuclear hormone receptors for organic pollutants by homology modeling and molecular docking. *Toxicology Letters*, 191, 69–73.
75. Olivero-Verbel, J., Cabaras-Montalvo, M., and Ortega-Uniga, C. (2010). Theoretical targets for TCDD: a bioinformatical approach. *Chemosphere*, 80, 1160–1166.
76. Jogalekar, A. S., Reiling, S., and Vaz, R. J. (2010). Identification of optimum computational protocols for modeling the aryl hydrocarbon receptor (AHR) and its interaction with ligands. *Bioorganic and Medicinal Chemistry Letters*, 20, 6616–6619.
77. Yoshikawa, E., Miyagi, S., Dedachi, K., Ishihara-Sugano, M., Itoh, S., and Kurita, N. (2010). Specific interactions between aryl hydrocarbon receptor and dioxin congeners: ab initio fragment molecular orbital calculations. *Journal of Molecular Graphics and Modelling*, 29, 197–205.
78. Bohonowych, J. E. and Denison, M. S. (2007). Persistent binding of ligands to the aryl hydrocarbon receptor. *Toxicological Sciences*, 98, 99–110.
79. Pertrulis, J. R. and Bunce, N. J. (2000). Competitive behavior in the interactive toxicology of halogenated aromatic compounds. *Journal of Biochemical Toxicology*, 14, 73–81.
80. Henry, E. C. and Gasiewicz, T. A. (1993). Transformation of the aryl hydrocarbon receptor to a DNA-binding form is accompanied by release of the 90 kDa heat-shock protein and increased affinity for 2,3,7,8-tetrachlorodibenzo-p-dioxin. *Biochemical Journal*, 294, 95–101.
81. Denison, M. S. and Wilkinson, C. F. (1985). Identification of the Ah receptor in selected mammalian species and its role in the induction of aryl hydrocarbon hydroxylase. *European Journal of Biochemistry*, 147, 429–435.
82. Denison, M. S., Vella, L. M., and Okey, A. B. (1986). Structure and function of the Ah receptor for 2,3,7,8-tetrachlorodibenzo-p-dioxin: species differences in molecular properties of the receptor from mouse and rat hepatic cytosol. *Journal of Biological Chemistry*, 261, 3987–3995.
83. Aarts, J. M. M. J. G., Denison, M. S., Cox, M. A., Schalk, M. A. C., Garrison, P. M., Tullis, K., de Haan, L. H. J., and Brouwer, A. (1996). Species-specific antagonism of Ah receptor action by 2,2′,5,5′-tetrachloro- and 2,2′,3,3′,4,4′-hexachlorobiphenyl. *European Journal of Pharamcology*, 293, 463–474.
84. Kikuchi, H., Kato, H., Mizuno, M., Hossain, A., Ikawa, S., Miyazaki, J., and Watanabe, M. (1996). Differences in inducibility of CYP1A1-mRNA by benzimidazole compounds between human and mouse cells: evidences of a human-specific signal transduction pathway for CYP1A1 induction. *Archives of Biochemistry and Biophysics*, 334, 235–240.
85. Shih, H., Pickwell, G. V., Guenette, D. K., Bilir, B., and Quattrochi, L. C. (1999). Species differences in hepatocyte induction of CYP1A1 and CYP1A2 by omeprazole. *Human and Experimental Toxicology*, 18, 95–105.
86. Sadar, M. D., Ash, R. Sundqvist, J., Olsson, P. E., and Andersson, T. B. (1996). Phenobarbital induction of CYP1A1 gene expression in a primary culture of rainbow trout hepatocytes. *Journal of Biological Chemistry*, 271, 17635–17643.
87. Sadar, M. D., Westlind, A., Blomstrand, F., and Andersson, T. B. (1996). Induction of CYP1A1 by GABA receptor ligands. *Biochemical and Biophysical Research Communications*, 229, 231–237.
88. Abnet, C. C., Tanguay, R. L., Heideman, W., and Peterson, R. E. (1999). Transactivation activity of human, zebrafish, and rainbow trout aryl hydrocarbon receptors expressed in COS-7 cells: greater insight into species differences in toxic potency of polychlorinated dibenzo-p-dioxin, dibenzofuran, and biphenyl congeners. *Toxicology and Applied Pharmacology*, 159, 41–51.
89. Zhou, J.-G., Henry, E. C., Palermo, C. M., Dertinger, S. D., and Gasiewicz, T. A. (2003). Species-specific transcriptional activity of synthetic flavonoids in guinea pig and mouse cells as a result of differential activation of the aryl hydrocarbon receptor to interact with dioxin-responsive elements. *Molecular Pharmacology*, 63, 915–924.
90. Zhao, B., Degroot, D., Hayashi, A., He, G., and Denison, M. S. (2010). CH223191 is a ligand-selective antagonist of the Ah (dioxin) receptor. *Toxicological Sciences*, 117, 393–340.
91. Henry, E. C. and Gasiewicz, T. A. (2008). Molecular determinants of species-specific agonist and antagonist activity of a substituted flavone towards the aryl hydrocarbon receptor. *Archives of Biochemistry and Biophysics*, 472, 77–88.
92. Karchner, S. I., Franks, D. G., Kennedy, S. W., and Hahn, M. E. (2006). The molecular basis for differential dioxin sensitivity in birds: role of the aryl hydrocarbon receptor. *Proceedings of the National Academy of Sciences of the United States of America*, 103, 6252–6257.
93. Okey, A. B., Boutros, P. C., Harper, and P. A. (2005). Polymorphisms of human nuclear receptors that control expression of drug-metabolizing enzymes. *Pharmacogenetics and Genomics*, 15, 371–379.
94. Ema, M., Ohe, N., Suzuki, M., Mimura, J., Sogawa, K., Ikawa, S., and Fujii-Kuriyama, Y. (1994). Dioxin binding activities of polymorphic forms of mouse and human arylhydrocarbon receptors. *Journal of Biological Chemistry*, 269, 27337–27343.
95. Ramadoss, P. and Perdew, G. H. (2004). Use of 2-azido-3-[^{125}I]iodo-7,8-dibromodibenzo-p-dioxin as a probe to determine the relative ligand affinity of human versus mouse aryl

hydrocarbon receptor in cultured cells. *Molecular Pharmcacology*, 66, 129–136.

96. Flaveny, C. A., Murray, I. A., Chiaro, C. R., and Perdew, G. H. (2009). Ligand selectivity and gene regulation by the human aryl hydrocarbon receptor in transgenic mice. *Molecular Pharmacology*, 75, 1412–1420.

97. Jeyakumar, M., Carlson, K.E., Gunther, J.R., Katzenellenbogen, J.A. (2011). Exploration of dimensions of estrogen potency: parsing ligand binding and coactivator binding affinities. *Journal of Biological Chemistry*, 286, 12971–12982.

98. Ellmann, S., Sticht, H., Thiel, F., Beckmann, M. W., Strick, R., and Strissel, P. L. (2009). Estrogen and progesterone receptors: from molecular structures to clinical targets. *Cell and Molecular Life Sciences*, 66, 2405–2426.

99. De Bosscher, K. (2010). Selective glucocorticoid receptor modulators. *Journal of Steroid Biochemistry and Molecular Biology*, 120, 96–104.

100. Norris, J. D., Joseph, J. D., Sherk, A. B., Juzumiene, D., Turnbull, P. S., Rafferty, S. W., Cui, H., Anderson, E., Fan, D., Dye, D. A., Deng, X., Kazmin, D., Chang, C. Y., Willson, T. M., and McDonnell, D. P. (2009). Differential presentation of protein interaction surfaces on the androgen receptor defines the pharmacological actions of bound ligands. *Chemical Biology*, 16, 452–460.

101. Murray, I. A., Krishnegowda, G., DiNatale, B. C., Flaveny, C., Chiaro, C., Lin, J. M., Sharma, A. K., Amin, S., and Perdew, G. H. (2010). Development of a selective modulator of aryl hydrocarbon (Ah) receptor activity that exhibits antiinflammatory properties. *Chemical Research in Toxicology*, 23, 955–966.

102. Murray, I. A., Morales, J. L., Flaveny, C. A., Dinatale, B. C., Chiaro, C., Gowdahalli, K., Amin, S., and Perdew, G. H. (2010). Evidence for ligand-mediated selective modulation of aryl hydrocarbon receptor activity. *Molecular Pharmacology*, 77, 247–254.

103. Zhang, S., Rowlands, C., and Safe, S. (2008). Ligand-dependent interactions of the Ah receptor with coactivators in a mammalian two-hybrid assay. *Toxicology and Applied Pharmacology*, 227, 196–206.

104. Zhang, S., Lei, P., Liu, X., Li, X., Walker, K., Kotha, L., Rowlands, C., and Safe, S. (2009). The aryl hydrocarbon receptor as a target for estrogen receptor-negative breast cancer chemotherapy. *Endocrine-Related Cancer*, 16, 835–844.

105. Henry, E. C. and Gasiewicz, T. A. (2003). Agonist but not antagonist ligands induce conformational change in the mouse aryl hydrocarbon receptor as detected by partial proteolysis. *Molecular Pharmacology*, 63, 392–400.

106. Kronenberg, S., Esser, C., and Carlberg, C. (2000). An aryl hydrocarbon receptor conformation acts as the functional core of nuclear dioxin signaling. *Nucleic Acids Research*, 28, 2286–2291.

107. Gerbal-Chaloin, S., Pichard-Garcia, L., Fabre, J. M., Sa-Cunha, A., Poellinger, L., Maurel, P., and Daujat-Chavanieu, M. (2006). Role of CYP3A4 in the regulation of the aryl hydrocarbon receptor by omeprazole sulphide. *Cell Signalling*, 18, 740–750.

108. Lees, M. J. and Whitelaw, M. L. (1999). Multiple roles of ligand in transforming the dioxin receptor to an active basic helix–loop–helix/PAS transcription factor complex with the nuclear protein Arnt. *Molecular and Cellular Biology*, 19, 5811–5822.

109. Fretland, A. J., Safe, S., and Hankinson, O. (2004). Lack of antagonism of 2,3,7,8-tetrachlorodibenzo-*p*-dioxin's (TCDDs) induction of cytochrome P4501A1 (CYP1A1) by the putative selective aryl hydrocarbon receptor modulator 6-alkyl-1,3,8-trichlorodibenzofuran (6-MCDF) in the mouse hepatoma cell line Hepa-1c1c7. *Chemico-Biological Interactions*, 150, 161–170.

110. Hestermann, E. V. and Brown, M. (2003). Agonist and chemopreventative ligands induce differential transcriptional cofactor recruitment by aryl hydrocarbon receptor. *Molecular and Cellular Biology*, 23, 7920–7925.

111. Boronat, S., Casado, S., Navas, J. M., and Pina, B. (2007). Modulation of aryl hydrocarbon receptor transactivation by carbaryl, a nonconventional ligand. *FEBS Journal*, 274, 3327–3339.

112. Wihlen, B., Ahmed, S., Inzunza, J., and Matthews, J. (2009). Estrogen receptor subtype- and promoter-specific modulation of aryl hydrocarbon receptor-dependent transcription. *Molecular Cancer Research*, 7, 977–986.

113. Matthews, J., Wihlen, B., Thomsen, J., and Gustafsson, J. A. (2005). Aryl hydrocarbon receptor-mediated transcription: ligand-dependent recruitment of estrogen receptor alpha to 2,3,7,8-tetrachlorodibenzo-*p*-dioxin-responsive promoters. *Molecular and Cellular Biology*, 25, 5317–5328.

114. Wang, S., Ge, K., Roeder, R. G., and Hankinson, O. (2004). Role of mediator in transcriptional activation by the aryl hydrocarbon receptor. *Journal of Biological Chemistry*, 279, 13593–13600.

115. Pansoy, A., Ahmed, S., Valen, E., Sandelin, A., and Matthews, J. (2010). 3-Methylcholanthrene induces differential recruitment of aryl hydrocarbon receptor to human promoters. *Toxicological Sciences*, 117, 90–100.

116. Yao, E. F. and Denison, M. S. (1992). DNA sequence determinants for binding of transformed Ah receptor to a dioxin-responsive enhancer. *Biochemistry*, 31, 5060–5067.

117. Lusska, A., Shen, E., and Whitlock, J. P., Jr. (1993). Protein–DNA interactions at a dioxin-responsive enhancer. Analysis of six bona fide DNA-binding sites for the liganded Ah receptor. *Journal of Biological Chemistry*, 268, 6575–6580.

118. Bank, P. A., Yao, E. F., Phelps, C. L., Harper, P. A., and Denison, M. S. (1992). Species-specific binding of transformed Ah receptor to a dioxin responsive transcriptional enhancer. *European Journal of Pharmacology*, 228, 85–94.

119. Matikainen, T., Perez, G. I., Jurisicova, A., Pru, J. K., Schlezinger, J. J., Ryu, H. Y., Laine, J., Sakai, T., Korsmeyer, S. J., Casper, R. F., Sherr, D. H., and Tilly, J. L. (2001). Aromatic hydrocarbon receptor-driven Bax gene expression is required for premature ovarian failure caused by biohazardous environmental chemicals. *Nature Genetics*, 28, 355–360.

120. Gouédard, C., Barouki, R., and Morel, Y. (2004). Dietary polyphenols increase paraoxonase 1 gene expression by an aryl hydrocarbon receptor-dependent mechanism. *Molecular and Cellular Biology*, 24, 5209–5222.

121. Gruber, C. J., Gruber, D M., Gruber, I. M., Wieser, F., and Huber, J. C. (2004). Anatomy of the estrogen response element. *Trends in Endocrinology and Metabolism*, 15, 73–78.
122. Mason, C. E., Shu, F. J., Wang, C., Session, R. M., Kallen, R. G., Sidell, N., Yu, T., Liu, M. H., Cheung, E., and Kallen, C. B. (2010). Location analysis for the estrogen receptor-alpha reveals binding to diverse ERE sequences and widespread binding within repetitive DNA elements. *Nucleic Acids Research*, 38, 2355–2368.
123. DuSell, C. D., Nelson, E. R., Wittmann, B. M., Fretz, J. A., Kazmin, D., Thomas, R. S., Pike, J. W., and McDonnell, D. P. (2010). Regulation of aryl hydrocarbon receptor function by selective estrogen receptor modulators. *Molecular Endocrinology*, 24, 33–46.
124. Safe, S. (2010). 3-Methylcholanthrene induces differential recruitment of aryl hydrocarbon receptor to human promoters. *Toxicological Sciences*, 117, 1–3.
125. Chen, I., McDougal, A., Wang, F., and Safe, S. (1998). Aryl hydrocarbon receptor-mediated antiestrogenic and antitumorigenic activity of diindolylmethane. *Carcinogenesis*, 19, 1631–1639.
126. McDougal, A., Wilson, C., and Safe, S. (1997). Inhibition of 7,12-dimethylbenz[a]anthracene-induced rat mammary tumor growth by aryl hydrocarbon receptor agonists. *Cancer Letters*, 120, 53–63.
127. Okino, S. T., Pookot, D., Basak, S., and Dahiya, R. (2009). Toxic and chemopreventive ligands preferentially activate distinct aryl hydrocarbon receptor pathways: implications for cancer prevention. *Cancer Prevention Research*, 2, 251–256.
128. Piskorska-Pliszczynska, J., Astroff, B., Zacharewski, T., Harris, M., Rosengren, R., Morrison, V., Safe, L., and Safe, S. (1991). Mechanism of action of 2,3,7,8-tetrachlorodibenzo-p-dioxin antagonists: characterization of 6-[^{125}I]methyl-8-iodo-1,3-dichlorodibenzofuran-Ah receptor complexes. *Archives of Biochemistry and Biophysics*, 284, 193–200.
129. Lawrence, B. P., Denison, M. S., Novak, H., Vorderstrasse, B. A., Harrer, N., Neruda, W., Reichel, C., and Woisetschläger, M. (2008). Activation of the aryl hydrocarbon receptor is essential for mediating the anti-inflammatory effects of a novel low-molecular-weight compound. *Blood*, 112, 1158–1165.
130. Peng, T. L., Chen, J., Mao, W., Liu. X., Tao, Y., Chen, L. Z., and Chen, M. H. (2009). Potential therapeutic significance of increased expression of aryl hydrocarbon receptor in human gastric cancer. *World Journal of Gastroenterology*, 15, 1719–1729.
131. Casado, F. L., Singh, K. P., and Gasiewicz, T. A. (2010). The aryl hydrocarbon receptor: regulation of hematopoiesis and involvement in the progression of blood diseases. *Blood Cells and Molecular Disease*, 44, 199–206.
132. Esser, C. and Krutmann, J. (2010). UV irradiation and skin pigmentation. Aryl hydrocarbon receptor—"new kid on the block". *Hautarzt*, 61, 561–566.
133. Hall, J., Barhoover, M. A., Kazmin, D., McDonnell, D. P., Greenlee, W. F., and Thomas, R. S. (2010). Activation of the aryl hydrocarbon receptor inhibits invasive and metastatic feature of human breast cancer cells and promotes breast cancer cell differentiation. *Molecular Endocrinology*, 24, 359–369.
134. Marshall, N. B. and Kerkvliet, N. I. (2010). Dioxin and immune regulation: emerging role of aryl hydrocarbon receptor in the generation of regulatory T cells. *Annals of the New York Academy of Sciences*, 1183, 25–37.
135. Behnisch, P. A., Hosoe, K., and Sakai, S.-I. (2003). Brominated dioxin-like compounds: *in vitro* assessment in comparison to classical dioxin-like compounds and other polyaromatic compounds. *Environment International*, 29, 861–877.
136. Gillner, M., Bergman, J., Cambillau, C., Alexandersson, M., Fernstrom, B., and Gustafsson, J.-A. (1993). Interactions of indolo[3,2-b]carbazoles and related polycyclic aromatic hydrocarbons with specific binding sites for 2,3,7,8-tetrachlorodibenzo-p-dioxin in rat liver. *Molecular Pharmacology*, 44, 336–345.
137. Bunce, N. J., Landers, J. P., Schneider, U. A., Safe, S. H., and Zacharewski, T. R. (1989). Chlorinated *trans* stilbenes: competitive binding to the Ah receptor, induction of cytochrome P-450 monooxygenase activity and partial 2,3,7,8-TCDD antagonism. *Toxicology and Environmental Chemistry*, 28, 217–229.
138. Adachi, J., Mori, Y., Matsui, S., Takigami, H., Fujino, J., Kitagawa, H., Miller, C. A., 3rd, Kato, T., Saeki, K., and Matsuda, T. (2001). Indirubin and indigo are potent aryl hydrocarbon receptor ligands present in human urine. *Journal of Biological Chemistry*, 276, 31475–31478.
139. Gillner, M., Bergman, J., Cambillau, C., and Gustafsson, J.-A. (1989). Interactions of rutaecarpine alkaloids with specific binding sites for 2,3,7,8-tetrachlorodibenzo-p-dioxin in rat liver. *Carcinogenesis*, 10, 651–654.
140. Gasiewicz, T. A., Kende, A. S., Rucci, G., Whitney, B., and Willey, J. J. (1996). Analysis of structural requirements for Ah receptor antagonist activity: ellipticines, flavones, and related compounds. *Biochemical Pharmacology*, 52, 1787–1803.
141. Heath-Pagliuso, S., Rogers, W. J., Tullis, K., Seidel, S. D., Cenijn, P. H., Brouwer, A., and Denison, M. S. (1998). Activation of the Ah receptor by tryptophan and tryptophan metabolites. *Biochemistry*, 37, 11508–11515.
142. Schroeder, J. C., Dinatale, B. C., Murray, I. A., Flaveny, C. A., Liu, Q., Laurenzana, E. M., Lin, J. M., Strom, S. C., Omiecinski, C. J., Amin, S., and Perdew, G. H. (2010). The uremic toxin 3-indoxyl sulfate is a potent endogenous agonist for the human aryl hydrocarbon receptor. *Biochemistry*, 49, 393–400.
143. Lu, Y.-F., Santostefano, M., Cunningham, B. D. M., Threadgill, M. D., and Safe, S. (1996). Substituted flavones as aryl hydrocarbon (Ah) receptor agonists and antagonists. *Biochemical Pharmacology*, 51, 1077–1087.
144. Kobayashi, Y., Matsuura, Y., Kotani, E., Fukuda, T., Aoyagi, T., Tobinaga, S., Yoshida, T., and Kuroiwa, Y. (1993). Structural requirements of the induction of hepatic microsomal cytochrome P450 by imidazole- and pyridine-containing compounds in rats. *Journal of Biochemistry*, 114, 697–701.
145. Phelan, D., Winter, G. M., Rogers, W. J., Lam, J. C., and Denison, M. S. (1998). Activation of the Ah receptor signal

transduction pathway by bilirubin and biliverdin. *Archives of Biochemistry and Biophysics*, 357, 155–163.
146. Cheung, Y.-L., Snelling, J., Mohammed, N. N. D., Gray, T. J. B., and Ioannides, C. (1996). Interaction with the aromatic hydrocarbon receptor, CYP1A induction, and mutagenicity of a series of diaminotoluenes: implications for their carcinogenicity. *Toxicology and Applied Pharmacology*, 139, 203–211.
147. Lee, I. J., Jeong, K. S., Roberts, B. J., Kallarakal, A. T., Fernandez-Salguero, P., Gonzalez, F. J., and Song, B. J. (1996). Transcriptional induction of the cytochrome P4501A1 gene by a thiazolium compound, YH439. *Molecular Pharmacology*, 49, 980–988.
148. Denison, M. S., Phelan, D., Winter, M. G., and Ziccardi, M. H. (1998). Carbaryl, a carbamate insecticide, is a ligand for the hepatic Ah (dioxin) receptor. *Toxicology and Applied Pharmacology*, 152, 406–414.
149. Merchant, M., Morrison, V., Santostefano, M., and Safe, S. (1992). Mechanism of action of aryl hydrocarbon receptor antagonists: inhibition of 2,3,7,8-tetrachlorodibenzo-*p*-dioxin induced CYP1A1 gene expression. *Archives of Biochemistry and Biophysics*, 298, 389–394.
150. Gasiewicz, T. A. and Rucci, G. (1991). Alpha-naphthoflavone acts as an antagonist of 2,3,7,8-tetrachlorodibenzo-*p*-dioxin by forming an inactive complex with the Ah receptor. *Molecular Pharmacology*, 40, 607–612.
151. Lee, J. E. and Safe, S. (2000). 3′,4′-Dimethoxyflavone as an aryl hydrocarbon receptor antagonist in human breast cancer cells. *Toxicological Sciences*, 58, 235–242.
152. Zhang, S., Qin, C., and Safe, S. H. (2003). Flavonoids as aryl hydrocarbon receptor agonists/antagonists: effects of structure and cell context. *Environmental Health Perspectives*, 111, 1877–1882.
153. Ashida, H. (2000). Suppressive effects of flavonoids on dioxin toxicity. *Biofactors*, 12, 201–206.
154. Hamada, M., Satsu, H., Natsume, Y., Nishiumi, S., Fukuda, I., Ashida, H., and Shimizu, M. (2006). TCDD-induced CYP1A1 expression, an index of dioxin toxicity, is suppressed by flavonoids permeating the human intestinal Caco-2 cell monolayers. *Journal of Agriculture and Food Chemistry*, 54, 8891–8898.
155. Shibazaki, M., Takeuchi, T., Ahmed, S., and Kikuchia, H. (2004). Blockage by SB203580 of Cyp1a1 induction by 2,3,7,8-tetrachlorodibenzo-*p*-dioxin, and the possible mechanism: possible involvement of the p38 mitogen-activated protein kinase pathway in shuttling of Ah receptor overexpressed in COS-7 cells. *Annals of the New York Academy of Sciences*, 1030, 275–281.
156. Chen, I., McDougal, A., Wang, F., and Safe, S. (1998). Aryl hydrocarbon receptor-mediated antiestrogenic and antitumorigenic activity of diindolylmethane. *Carcinogenesis*, 19, 1631–1639.
157. Fernandez, N., Roy, M., and Lesca, P. (1988). Binding characteristics of Ah receptors from rats and mice before and after separation from hepatic cytosols. 7-Hydroxyellipticine as a competitive antagonist of cytochrome P450 induction. *European Journal of Biochemistry*, 172, 585–589.
158. Joiakim, A., Mathieu, P. A., Palermo, C., Gasiewicz, T. A., and Reiners, J. J., Jr. (2003). The jun N-terminal kinase inhibitor SP600125 is a ligand and antagonist of the aryl hydrocarbon receptor. *Drug Metabolism and Disposition*, 31, 1279–1282.
159. Williams, S. N., Shih, S., Guenette, D. K., Brackney, W., Denison, M. S., Pickwell, G. V., and Quattrochi, L. C. (2000). Green tea flavonoids inhibit TCDD-induced CYP1A1 and CYP1A2 gene expression in human livers cells. *Chemico-Biological Interactions*, 28, 211–219.
160. Kim, S. H., Henry, E. C., Kim, D. K., Kim, Y. H., Shin, K. J., Han, M. S., Lee, T. G., Kang, J. K., Gasiewicz, T. A., Ryu, S. H., and Suh, P. G. (2006). Novel compound 2-methyl-2*H*-pyrazole-3-carboxylic acid (2-methyl-4-4-*o*-tolylazo-phenyl) amide (CH-223191) prevents 2,3,7,8-TCDD-induced toxicity by antagonizing the aryl hydrocarbon receptor. *Molecular Pharmacology*, 69, 1871–1878.
161. Savouret, J. F., Antenos, M., Quesne, M., Xu, J., Milgrom, E., and Casper, R. F. (2001). 7-Ketocholesterol is an endogenous modulator for the aryl hydrocarbon receptor. *Journal of Biological Chemistry*, 276, 3054–3059.
162. Ledirac, N., Delescluse, C., Lesca, P., Piechocki, M. P., Hines, R. N., de Sousa, G., Pralavorio, M., and Rahmani, R. (2000). Diflubenzuron, a benoyl-urea insecticide, is a potent inhibitor of TCDF-induced CYP1A1 expression in HepG2 cells. *Toxicology and Applied Pharmacology*, 164, 273–279.
163. Casper, R. F., Quesne, M., Rogers, I. M., Shirota, T., Jolivet, A., Milgrom, E., and Savouret, J. F. (1999). Resveratrol has antagonist activity on the aryl hydrocarbon receptor: implications for prevention of dioxin toxicity. *Molecular Pharmacology*, 56, 784–790.
164. Boitano, A. E., Wang, J., Romeo, R., Bouchez, L. C., Parker, A. E., Sutton, S. E., Walker, J. R., Flaveny, C. A., Perdew, G. H., Denison, M. S., Schultz, P. G., and Cooke, M. P. (2010). Aryl hydrocarbon receptor antagonists promote the expansion of human hematopoietic stem cells. *Science*, 329, 1345–1348.

5

DIOXIN RESPONSE ELEMENTS AND REGULATION OF GENE TRANSCRIPTION

HOLLIE SWANSON

5.1 THE INITIAL DISCOVERIES THAT SET THE STAGE: TCDD INDUCTION OF CYP1A1 REQUIRES SPECIFIC DNA SEQUENCES

The discovery that distinct DNA sequences, ultimately defined as sites recognized by the AHR/ARNT heterodimer, were required to mediate induction of TCDD (2,3,7,8-tetrachlorodibenzo-p-dioxin) of *Cyp1a1* (cytochrome p4501a1) transcription played a seminal role in establishing the aryl hydrocarbon receptor (AHR) as a unique DNA and ligand binding receptor. The pioneering work performed in the laboratory of Dr. James Whitlock [1] paralleled that of leaders of the rapidly maturing nuclear steroid receptor field as well as the biochemical/pharmacological [2, 3] and genetic [4] studies of the relatively nascent field of the AHR. Recognizing *Cyp1a1* as highly responsive to the effects of TCDD, Dr. Whitlock's group utilized promoter deletion analyses to identify a "TCDD-responsive" domain and subsequently an "AHR recognition site" that fulfilled the requirements of an enhancer element [1]. That is, it functioned in a position-independent manner, could "enhance" the activities of TATA-box promoters, and could act synergistically when present as multiple copies. An unexpected finding was that the consensus sequence, defined using DNA footprinting and gel shift analyses, was not a palindrome as was typical of the nuclear steroid receptors, but was asymmetric. Further analyses defined the core consensus as 5′-T/NGCGTG-3′, which has been described by various groups as AHRE (aryl hydrocarbon receptor enhancer) [5], DRE (dioxin response element) [1], or XRE (xenobiotic-responsive element) [6]. These initial discoveries established *Cyp1a1* as a prototypical TCDD/AHR target gene that has proven to be invaluable as a model to be used for understanding how the AHR can regulate genes [1, 7, 8]. The canonical, DRE-driven mechanisms as well as mechanisms that involve additional DNA sequences and interactions between AHR and various transcription factors will be the subject of discussion in this chapter.

5.2 THE AHR FORMS A DNA BINDING HETERODIMER WITH ARNT: TWO bHLH/PAS TRANSCRIPTION FACTORS

Prior to the initial discoveries that revealed the existence of the DREs, additional clues that aided our understanding of the DNA binding activities of the AHR involved biochemical observations. For example, it was found that administration of TCDD to cultured cells resulted in the nuclear uptake of a cytosolic receptor and its detergent-resistant binding to hepatic nuclei [2, 9]. Further, it was observed that the unliganded AHR existed in the cytosol as a "9S" form. The abbreviation "S" refers to Svedberg units, a measure of sedimentation rate of macromolecules obtained using ultracentrifugation techniques. The addition of TCDD transformed this 9S cytosolic form to a 6S nuclear form that was now capable of binding DNA. These biochemical properties proved to be similar to those of the nuclear hormone receptors and thus provoked the possibility that the AHR could be a family member of the steroids [10]. The 9S form of the AHR was found to contain the ubiquitous chaperone protein, HSP90, as well as additional chaperonins such as p23 and ARA9 (AHR associated protein 9) (also referred to as XAP2 (HBV X-associated protein 2) or AIP) [8]. UV cross-linking

---

The AH Receptor in Biology and Toxicology, First Edition. Edited by Raimo Pohjanvirta.
© 2012 John Wiley & Sons, Inc. Published 2012 by John Wiley & Sons, Inc.

revealed that the 6S form of the AHR interacted with the DRE as the ~100 kDa ligand binding AHR and a ~110 kDa protein that was ultimately identified as ARNT (aryl hydrocarbon receptor nuclear translocator) [11, 12]. Upon cloning, the AHR was identified as a member of the bHLH/PAS (basic helix–loop–helix–Per/Arnt/Sim domain) protein family and thus not a member of the steroid receptor superfamily [13]. The bHLH/PAS family now consists of 34 mammalian proteins that share a common 70-amino acid region [14]. At the functional level, the bHLH/PAS proteins can be classified into two classes [15, 16]. Within Class I are proteins such as the AHR that neither homodimerize nor heterodimerize with other Class I factors. Fulfillment of the gene transactivation properties of the Class I proteins (i.e., AHR) requires that they form a partnership with a Class II protein (i.e., ARNT). In contrast, Class II proteins can both homodimerize and heterodimerize, promiscuously bind multiple Class I proteins and thereby participate in multiple partnerships capable of regulating a wide variety of genes. Class I proteins are often induced by a signal spatiotemporally. Thus, temporal regulation of distinct subsets of genes can be achieved by the different combinations of the protein pairs formed and the availability (as determined by relative protein levels, posttranslational modifications, etc.) of each dimer partner. As many bHLH/PAS proteins, the presence of the AHR PAS domain restricts the number of possible partnerships that may be formed. Here, the AHR brings into play a property common to both bHLH/PAS and bHLH/LZ (basic helix–loop–helix leucine zipper) transcription families and uses its motif that lies immediately C-terminal to the HLH domain (i.e., its PAS domain) to dictate partner specificity. Unique capabilities of the PAS domain of the bHLH/PAS family members, however, are their environmental "sensing" activities and participation in general signal transduction activities. Most bHLH–PAS family members participate in gene regulation via two major routes [17]. Like the well-characterized canonical activities of the AHR and ARNT in their regulation of target genes such as *Cyp1a1*, they dimerize via their bHLH and PAS domains and bind their cognate DNA recognition sites following exposure to external stimuli. Alternatively, they indirectly participate in gene regulation by acting as transcription coactivators (or perhaps adaptor proteins) that do not bind DNA, but rather recruit and interact with general transcription factors including RNA polymerase II. The participation of the AHR in these latter activities is increasingly becoming more apparent.

The similarities and differences between the DNA binding properties of the AHR and the ARNT protein were revealed following deletion mapping and amino acid substitution analyses. The DNA binding activity of both the AHR and the ARNT is mediated primarily by their basic regions that lie within their N-termini [18–22]. They also participate in additional contacts with DNA that are contained within the PAS domain [23] and contribute toward DNA bending activities [24]. As with other bHLH proteins, DNA binding of the AHR/ARNT heterodimer appears to involve a change in the secondary structure of the basic region in which its conformation as a random coil transitions into an α helix as the AHR makes contact with the DNA [15]. ARNT makes direct contacts with the GCGTG site via amino acids H79, D83, R86, and R87 that lie within its basic region. With respect to the AHR, the primary nucleotide contacts are contained within amino acids 27–39 and in particular H38 and R39 [21, 22, 25]. While amino acids R14 and R15 of the AHR influence the ability of the AHR/ARNT heterodimer to interact with DNA, the role of these amino acids appears to be the maintenance of appropriate stability and perhaps tertiary structure of the DNA binding complex, rather than the participation in direct nucleotide contacts [26]. Thus, it is through these mechanisms that the AHR, stimulated by the presence of TCDD, enters the nucleus and forms a heterodimeric partnership with ARNT that allows each partner to participate in DNA binding by making direct contacts with specific nucleotides contained within the DRE.

5.3 THE AHR/ARNT HETERODIMER BINDS A CORE CONSENSUS SITE AND ADDITIONAL BINDING MOTIFS

Within the 5′-T/NGCGTG-3′ site, the AHR occupies the 5′-T/NGC half-site, whereas the ARNT occupies the 5′-GTG half-site [27, 28]. As a homodimer, ARNT preferentially binds the palindromic E-box CACGTG. Incorporation of these data into a classification scheme based on amino acid compositions of the basic regions coupled with DNA binding preferences placed ARNT as a member of binding group B and the AHR as a member of binding group C [29]. In addition to the core 5′-T/NGCGTG-3′ consensus site, sequences that lie immediately adjacent to the core have been consistently found to play a role in the DNA binding affinity/specificity of the AHR/ARNT heterodimer [27, 30, 31]. For example, extensive *in vitro* nucleotide substitution analyses performed by Yao and Denison demonstrated that these flanking nucleotides could significantly impact the DNA binding affinity of the AHR-containing DNA binding complex and thus extended the core to consist of the sequences **GCGTGNNA/TNNNC/G** [31]. Use of an oligonucleotide selection/amplification approach also identified preferences for flanking nucleotides and proposed that the optimal AHR/ARNT DNA recognition site was GGGNAT(C/T) GCGTGACNNCC [27]. Subsequent studies that utilized footprinting analyses of chimeric proteins indicate that these flanking nucleotides make direct contact with the bHLH regions of the AHR/ARNT heterodimer [24]. The PAS domain of the AHR also appears to be involved in contacts with the flanking nucleotides and in this manner contributes to the high-affinity DNA binding and stability of the

AHR/ARNT heterodimer by bending the DNA sequences and facilitating assembly of the enhanceosome complex.

It has been previously noted that the DRE sequence originally identified in the mouse *Cyp1a1* gene is most likely the response element required for the regulation of the majority of AHR target genes [10]. However, the mechanism by which the AHR regulates genes is becoming more complex. DNA sequences in addition to the DRE have also been found to be responsive to TCDD and require DNA binding of the AHR. Of these, the best characterized is that initially identified by Sogawa et al. [32] as an "XRE-II" site, which is located within the promoter of *Cyp1a2*. This site does not appear to involve the canonical AHR/ARNT/DNA binding interaction but is proposed to involve an as yet unidentified "adaptor" protein. Using phylogenetic footprinting and gene expression arrays, Boutros et al. [33] extended these observations by identifying 36 genes that harbored the XRE-II/AHRE-II site and were responsive to the presence of TCDD. In addition, a study that used a genome-wide search coupled with a southwestern chemistry-based enzyme-linked immunosorbent assay (to minimize nonspecific interactions) identified 77 AHR binding fragments that were confirmed to be TCDD responsive [34]. Unlike the canonical DRE, the vast majority of these sites were located distally from the promoter and were contained throughout the downstream sequences, including sites within the introns, exons, and coding sequences. Of these 77 fragments, 38 lacked the consensus GCGTG sequence. Taken together, these latter studies [33, 34] imply that DNA binding and gene regulation of the AHR involves at least two types of interactions. The first is the high-affinity DNA binding of the AHR/ARNT that regulates highly responsive AHR target genes. The second is the low-affinity DNA binding of the AHR/ARNT heterodimer (or the AHR complexed with other transcription factors as discussed below) that is represented by these degenerate sequences. Given the lower binding affinity involved in this latter mechanism, the genes regulated in this manner may be less responsive to the effects of the TCDD-activated AHR or may require additional input from an additional signaling pathway. AHR gene regulation involved in these two mechanisms will be discussed in more detail in the following sections.

5.4 REGULATION OF "CLASSIC" TCDD-INDUCIBLE GENES BY THE AHR

In the past decade, our understanding of how activation of the AHR by TCDD leads to altered gene expression has been greatly expanded by the use of numerous advanced technologies including promoter deletion, gene expression, and transcriptome analyses performed in a variety of cell types, tissues, experimental conditions, and species. Transcriptome analyses indicate that exposure to TCDD leads to the upregulation and downregulation of several hundred genes and that the regulation of many of these genes requires the presence of the AHR [35–37]. Within these studies, a few genes are consistently upregulated by TCDD, regardless of the species, tissues, cell types, or conditions examined (Table 5.1). The responsiveness of these genes to TCDD requires the presence of the AHR. The mechanisms by which the AHR is known to regulate these "classic" TCDD-inducible genes *Cyp1a1*, *Cyp1a2*, *Cyp1b1*, *Nrf2* (nuclear factor (erythroid-derived 2)-like 2) and *Tiparp* (TCDD-inducible poly(ADP-ribose) polymerase) are described in more detail below.

As the other CYP1 family members, CYP1A2 metabolizes many environmental procarcinogens and an as yet unidentified AHR ligand [5]. The TCDD-responsive element within the human *CYP1A2* was initially identified as a "cryptic" DRE that was bordered by nucleotides 2532 and 2423 [38] and within the rat as "AHRE II" that was located within nucleotides 2097 to 2073 [32]. Although these sites appeared to mediate the actions of the TCDD-induced AHR/ARNT complex, they were not well conserved among different species. More recent studies indicate that these so-called "cryptic" sites can play a role in the ability of TCDD to upregulate *Cyp1a2* transcription; however, this role is relatively minor to the primary site of induction that occurs within a "dioxin response element." The dioxin response element cluster lies proximal to the *Cyp1a1* promoter and is capable of regulating TCDD induction of both *Cyp1a1* and *Cyp1a2* and thereby acts as a bidirectional promoter [39, 40]. Thus, TCDD induction of *Cyp1a2* is dependent on the canonical AHR/ARNT/DRE activity that lies within the *Cyp1a1* promoter.

TABLE 5.1 AHR Target Genes Typically Upregulated by TCDD

Murine Gene	Number of Core DREs Present?	Number of AHRE-II Sites Present?	Regulated by DREs?	References
Cyp1a1	8	1	Yes, canonical	[30, 36]
Cyp1a2	1	1	Yes, distal	[32], [38], [39]
Cyp1b1	3	0	Yes, canonical	[42]
Nfe2l2 (*Nrf2*)	3	?	Yes, canonical	[46]
Tiparp	16	2	?	[36, 56]

CYP1B1 not only metabolizes environmental procarcinogens but also metabolizes endogenous estrogen [5]. While CYP1A1 and CYP1A2 exhibit low constitutive activity, that of CYP1B1 is considerably higher, albeit in a tissue-dependent manner. The mechanism by which TCDD induces *Cyp1b1* expression more closely resembles that of *Cyp1a1* in that it requires the canonical AHR/ARNT/DRE binding mechanism [41, 42]. A secondary mechanism by which TCDD can induce *Cyp1b1* expression involves the binding of an "anomalous complex" to a region that overlaps two canonical DREs and is composed of the AHR and as yet unidentified proteins [43]. Similar to that described for *Cyp1a1*, TCDD induces the binding of the AHR/ARNT heterodimer to multiple consensus DREs that lie within the *Cyp1b1* upstream promoter. In addition, maximal TCDD induction requires two SP1 sites contained within the basal promoter. However, despite the similarities in the DRE activities within the *Cyp1a1* and *Cyp1b1* promoters, there is evidence that the ability of TCDD to induce *Cyp1a1* and *Cyp1b1* can substantially vary in different cell lines [44]. The mechanism that underlies this latter phenomenon is not yet clear. Despite the presence of the AHR at the canonical DREs within both *Cyp1a1* and *Cyp1b1* promoters, their ability to respond to TCDD is dramatically different. Thus, it appears that mechanisms subsequent to the DNA binding of the AHR are at play.

NRF2, a member of the Cap-N-Collar transcription factor family, facilitates the antioxidant response via upregulation of its target genes [45]. TCDD induction of NRF2 expression involves three DREs sequences harbored within sites located both upstream of the promoter and within the first intron [46]. The presence of multiple copies of DREs within the promoters of the human, mouse, and rat *Nrf2* genes is well conserved. The observations that TCDD can induce the expression of both NRF2 and several NRF2 target genes (i.e., *Nqo1*, *Gsta1* (glutathione S-transferase A1), and *Ugt1* (UDP-glucuronosyltransferase 1)) [47] led to further investigations into the interplay that exists between AHR and NRF2 pathways [48, 49]. The current literature provides evidence for two possible mechanisms that underlie the crosstalk between AHR and NRF2. In the first mechanism, TCDD directly upregulates expression of NRF2, resulting in a subsequent upregulation in the expression of NRF2 target genes [50]. Hence, TCDD induction of NQO1 and other phase II metabolizing enzymes would be a secondary event. In the second mechanism, the TCDD-activated AHR/ARNT heterodimer and NRF2 engage in mutual binding to a composite DNA recognition site [48, 51, 52]. This mechanism is reminiscent of the interplay that exists between the AHR and other signaling pathways as will be discussed in Section 5.5. Since these mechanisms are not mutually exclusive, it is possible that they may both exert an effect in a given condition or cell type.

Poly(ADP-ribose) (PAR) polymerase enzymes catalyze poly(ADP-ribose) polymerization from donor NAD^+ molecules into target proteins. TCDD induction of *Tiparp* was discovered by Ma et al. [53], who employed differential display methodology to identify this novel protein that is expressed in a wide range of tissues. The specific function of TiPARP is not fully understood; however, it is possible that TCDD's induction of TiPARP expression may impact the apoptotic response. TCDD induction of *Tiparp* is consistently reported in a variety of cell types and conditions and ranges from 2- to 10-fold induction [36, 54, 55]. However, the mechanisms by which TCDD induces its expression have not been fully characterized. The limited evidence available thus far indicates that TCDD induction of *Tiparp* transcription involves both the AHR/ARNT heterodimer and an as yet unidentified transcription repressor [56].

5.5 GENE REGULATION BY THE AHR INVOLVES INTERACTIONS WITH NUMEROUS TRANSCRIPTION FACTORS

The AHR interacts with a number of transcription factors in addition to ARNT and in this mode can modulate a variety of biological responses. In these interactions, the AHR participates in crosstalk with other transcription factors. That is, the activity of the AHR correspondingly modulates the ability of other transcription factors to activate its target genes and vice versa. These activities of the AHR expand its realm of gene regulation beyond that of xenobiotic metabolism to include modulation of ER (estrogen receptor) signaling, cell cycle progression, and inflammatory response. This section describes only the best characterized crosstalk mechanisms involving the AHR. Additional interactions between the AHR and other transcription factors that may also represent crosstalk mechanisms of physiological importance do exist, but will not be discussed here. This includes interactions between AHR and HIF (hypoxia-inducible factor) [55, 57], BRCA1 (breast cancer type 1 susceptibility protein) [58], Epstein–Barr virus-encoded EBNA-3 protein [59], and BMAL1 (brain and muscle aryl hydrocarbon nuclear translocator-like protein) [60]. In their aggregate, the studies described below have cast light on the AHR as not only an activator of the transcription of genes such as the prototypical *Cyp1a1*, but also a repressor, coactivator, and perhaps corepressor.

5.5.1 The AHR and AHRR Interaction

Perhaps the most significant protein/protein interaction affecting the responsiveness of the AHR pathway is the interaction between the AHR and the AHRR (aryl hydrocarbon receptor repressor). The AHRR was first identified as an "AHR-like" protein that was initially hypothesized to compete with the AHR for binding to ARNT [61]. At the amino acid level, only the N-termini of the AHR and AHRR, consisting of the bHLH/PAS A, exhibit high sequence homology [62]. The motifs contained within the bHLH/PAS A region of the AHRR facilitate the heterodimerization between AHRR and ARNT and binding to the DREs; however, due to the lack of a discernible "AHR-like" PAS B domain, the AHRR is incapable of ligand binding. The AHRR is constitutively expressed in a number of cell types, albeit in an apparent tissue-specific manner. The ability of the AHRR to repress the AHR has been demonstrated in a variety of cultured cells [63]. Here, AHRR expression levels correlate with the TCDD responsiveness of at least *Cyp1a1* as a representative AHR target gene. With respect to the repressive activities of the AHRR, however, it is becoming more apparent that the mechanisms by which the AHRR represses the actions of the AHR are more complex than that originally proposed and may include at least two mechanisms [62]. In the first proposed mechanism, the AHRR heterodimerizes with ARNT but does not competitively displace AHR from the AHR/ARNT heterodimer. This AHRR/ARNT heterodimer can then bind DREs and recruit corepressor proteins. In the second proposed mechanism, the AHRR forms a DNA binding-independent and perhaps ARNT-independent complex that acts as a transrepressor. It should also be noted that the repressor activity of the AHRR is not limited to the actions of the AHR, as it can also repress the transcriptional activity of other transcription factors, such as ERα [64].

5.5.2 The AHR and ER Interaction

The well-noted "antiestrogenic" effects of TCDD and interplay between AHR and ER pathways arose from the early observations of Kociba et al. [65], who reported that administration of TCDD increased the incidences of liver tumors only in female rats. Subsequent studies performed in female rats demonstrated that administration of TCDD could counter the ability of 17β-estradiol to induce uterine wet weight, a well established *in vivo* marker of 17β-estradiol action [66]. Estrogen receptors α and β mediate the actions of 17β-estradiol in responsive tissues [67]. The role of ERα versus ERβ in 17β-estradiol gene regulation is dictated primarily by their relative tissue concentrations. For example, the high expression of ERα contributes to its predominate gene regulation in the uterus and liver. Similarly, high expression and predominant gene regulation by ERβ exist in the granulosa cells of the ovary. Our current literature indicates that the AHR/ER crosstalk occurs at multiple levels, some of which appear to occur in a cell-type and species-specific manner [68, 69]. The AHR directly interacts with not only ERα, but also ERβ [70] and ER-related proteins (COUP-TFI (chicken ovalbumin upstream promoter transcription factor 1) and ERRα1 (estrogen receptor related protein α1)) [71]. Furthermore, in its role as a DNA binding partner with ARNT, the AHR can interact with naturally occurring estrogen response elements (EREs) that are defined as "imperfect" as they fail to meet the requirements of consensus ERE DNA recognition sites [72]. As a consequence of its direct interaction with ERα, the AHR can significantly impact the ability of ERα to regulate genes and vice versa as defined by experiments performed in various cultured cells [8, 69]. For example, TCDD activation of the AHR leads to subsequent recruitment of ERα to AHR target genes [73]. Conversely, the AHR can be recruited to ERα target genes. It appears that the AHR and ERα engage in mutual coactivator activities that then alter the regulation of multiple target genes in multiple tissues. TCDD not only alters the genomic binding of ERα and thereby the regulation of a specific subset of ERα target genes but also increases the occupancy of a number of target gene promoters (some of which contain consensus DREs) by both AHR and ERα. Thus, the AHR modulates two subsets of E2-responsive genes, ERα target genes and AHR/ERα target genes. The extent to which the AHR modulates ERβ target genes is uncertain. The AHR can also engage in crosstalk relationships with other nuclear hormone receptors, including the androgen [74] and glucocorticoid receptors [75, 76]. However, it is not yet clear whether these crosstalk relationships are as extensive as that of the AHR and ERα and whether their crosstalk activities are mediated via similar underlying mechanisms.

5.5.3 The AHR and NF-κB Interaction

The NF-κB/Rel proteins (RelA, c-Rel, RelB, NF-κB1, and NF-κB2) are transcription factors that play important roles in immunity, cell survival/differentiation, and inflammatory response [77]. Specific triggers can activate distinct pathways that are defined as the classical and the alternative NF-κB pathway. The classical pathway is upregulated by cytokines, growth factors, and immune activation stimuli and involves nuclear translocation of RelA. The alternative pathway is activated by a small subset of TNF family members and by NF-κB2 and involves nuclear translocation of RelB. While RelA activation is often transient, RelB activation appears to be slower and more persistent. The AHR can engage in crosstalk with both classical and alternative NF-κB pathways and thereby modulate the proinflammatory response [78, 79]. Similar to that of the ER, the crosstalk between AHR and NF-κB is bidirectional with NF-κB activators (e.g., IL-1β (interleukin 1 beta), IL-6 (interleukin 6), TNFα (tumor necrosis factor alpha), and LPS (lipopolysaccharide)) suppressing the ability of the AHR to induce transcription of AHR target genes such as *Cyp1a1*. While the presence of AHR agonists impinges on the ability of NF-κB to exert its signal, these AHR-mediated actions appear to be agonist dependent. Both *in vitro* and *in vivo* studies indicate that sustained AHR activation that occurs with TCDD-like agonists *enhances* proinflammatory signaling [8, 78, 79]. However, transient AHR activation (such as that expected to arise from more rapidly metabolized and perhaps endogenous AHR agonists) *inhibits* acute NF-κB-mediated inflammatory responses [80].

At the mechanistic level, the AHR/NF-κB interplay reportedly involves a physical interaction with RelA (classical pathway) [81, 82], a physical interaction with RelB (alternative pathway) [79], an AHR-mediated inhibition of RelA nuclear translocation [82], and an AHR-mediated degradation of RelB [83]. It should also be noted that the AHR can also upregulate its own NF-κB-independent, proinflammatory response via the canonical AHR/ARNT/DRE mode of action [84]. Additional mechanisms by which the AHR can modulate the inflammatory response include its interaction with the STAT1 (signal transducer and activator of transcription family member 1) and STAT5 proteins [85] and participation in a complex that consists of both STAT1 and NF-κB1 (also termed p50) [86]. The possible cell-type, context-dependent nature of the interplay between AHR and NF-κB/STAT pathways is yet to be determined.

The interaction between AHR and RelB is a relatively recent discovery [79]. In this role, the AHR physically interacts with RelB and transcriptionally upregulates the expression of chemokines such as IL-8 via a recognition site termed the "RelB/AHR-responsive element." In addition to IL-8, genes that harbor "RelB/AHR response element" include B-cell activating factor of the tumor necrosis factor family (*BAFF*), B-lymphocyte chemoattractant (*BLC*), CC-chemokine ligand (*CCL1*), and transcription factor interferon γ response factor 3 (*IFR3*). In addition to binding the "RelB/AHR response element" within the IL-8 promoter, the RelB/AHR is capable of interacting with consensus NF-κB sites, albeit when the AHR is in an unliganded state. Further studies performed in disease relevant models that mirror appropriate *in vivo* inflammatory responses will be needed to fully appreciate the functional consequences of these AHR/NF-κB crosstalk mechanisms.

5.5.4 The AHR and RB/E2F1 Interaction

Attempts to understand the tumor promotion/cell proliferative effects of the TCDD-activated AHR pathway led to the discovery that the AHR interacts directly with cell cycle regulatory proteins and that these interactions are ligand dependent [87–90]. The initial discovery was that the AHR interacted directly with RB (retinoblastoma protein) [87, 88]. RB is a tumor suppressor protein that in conjunction with the E2F proteins regulates cell cycle progression [91, 92]. The E2F family is composed of two groups: transcriptional activators, such as E2F1, and transcriptional repressors, such as E2F4 and E2F5. When cell division is favorable, the active E2F1 complexes push the cell into S phase. However, during conditions of unfavorable cell division, the E2F1 proteins remain inactively bound to RB and the cells remain stalled in G1 phase. The integrity of the E2F1/RB complex depends on the phosphorylation levels of RB. In its hypophosphorylated state, the RB/E2F1 interaction is intact. However, hyperphosphorylation of RB disrupts the RB/E2F1 complex, releases E2F1, and allows E2F1 to upregulate its target genes and facilitate cell entry into S phase. The initial phosphorylation of RB is performed by the cell cycle kinase proteins, cyclin D1/cyclin-dependent kinase 4/cyclin-dependent kinase 6.

The current working model by which the TCDD-activated AHR may interfere with the RB/E2F1 cell cycle machinery has been proposed by Barhoover et al. [90]. In this model, first the AHR exists within the cyclin D1/cyclin-dependent 4 complex. Within this complex, the AHR acts as a scaffolding protein that recruits RB to the complex, thereby facilitating its phosphorylation by cyclin-dependent kinase 4. Second, the presence of TCDD removes the AHR from this scaffolding role and allows RB to remain in an unphosphorylated state and thus preserves the integrity of the RB/E2F1 complex. The ultimate impact of the TCDD-activated AHR is inhibition of E2F1 target gene and inhibition of the ability of the cells to enter S phase. However, an additional role of the AHR within this realm of the cell cycle has also been proposed [93]. Here, it has been found that not only does the AHR directly interact with RB and the cyclin D1/cyclin-dependent 4 complex, but it can also interact directly with E2F. Interestingly, the AHR/E2F1-containing complex binds the consensus E2F1 site that exists within the *Apaf1* promoter and exerts a proproliferative effect.

5.5.5 The AHR and SP1 Interaction

SP1 is a zinc finger binding protein that regulates thousands of genes, interacts with numerous transcription factors, and influences chromatin remodeling [94]. A role of SP1 in the transactivation activities of the AHR/ARNT heterodimer was initially noted in the early analyses of the *Cyp1a1* promoter [95, 96]. Here, it was observed that a GC box, a putative binding site of SP1, participated with the DREs to synergistically upregulate the *Cyp1a1* promoter. The synergistic relationship between DREs and GC box is mediated by a direct interaction between the PAS domain of the AHR and the zinc finger domain of SP1 [97]. The AHR/SP1 interaction may also come into play in the context of gene repression based on the observations that the liganded form of the AHR can interfere with the transactivation of the ERα/SP1 complex that regulates at least some ERα-responsive genes [98]. Furthermore, the SP1/AHR interplay has been shown to be involved in the regulation of a number of TCDD-inducible genes, including *Cyp1b1* [99], *Ahrr* [100], and *epiregulin* [101]. Given the extensive nature of gene regulation by SP1, it is highly likely that SP1 may participate in additional AHR-regulated genes.

5.6 DO ALL AHR ACTIVITIES REQUIRE DNA BINDING?

Given the considerable variety of mechanisms by which the AHR can modulate gene transcription, it is reasonable to question the extent to which DNA binding of the AHR mediates its activities. The first step in addressing this question has been made through the generation of a transgenic mouse that harbors an AHR protein that is incapable of binding DNA [102]. The results obtained using this mouse indicate that at least some of the developmental activities of the AHR (i.e., those pertaining to early liver development and vascular remodeling) as well as certain toxic end points of TCDD (i.e., hepatomegaly and thymic involution) require DNA binding of the AHR. Although additional studies using this type of approach are necessary, it appears that DNA binding of the AHR is required for some but not all of its gene regulatory actions. Similar to that previously proposed [103], the studies discussed in this chapter support the idea that the AHR can participate in at least five mechanisms that are capable of impacting gene transcription. In the first mechanism, the AHR engages in nonnuclear activities. This would include its activity as an E3 ubiquitin ligase [104]. In the second mechanism, the AHR, as a heterodimeric partner with ARNT, binds to a consensus DRE, thereby regulating genes such as *Cyp1a1* and *Cyp1b1* [1, 41, 42]. The remaining mechanisms involve interactions between the AHR and other transcription factors. In the third mechanism, the AHR, via its protein–protein interactions with another transcription factor, inhibits the ability of this transcription factor to regulate its target genes. This type of *sequestering* event is exemplified by the AHR/NF-κB (RelA) interaction [80–82]. In the fourth mechanism, the AHR as a monomer acts as a *coactivator* by interacting with another transcription factor that retains its ability to bind its cognate response element. This type of *coactivator* mechanism is illustrated in the activities of the AHR with ERα [73]. In the fifth mechanism, the AHR interacts with another transcription factor and this AHR-containing heterodimer then binds a unique recognition site. An example of this type of mechanism is provided by the AHR/RelB complex that interacts with the RelB/AHR response element [79].

5.7 SUMMARY

Our journey toward understanding how the AHR binds DNA to regulate its target genes was initiated in the study of *Cyp1a1* that is regulated by the consensus AHR/ARNT DNA recognition site, the DRE. This prototypical AHR target gene has proven to be an excellent paradigm used to study inducible regulation of drug metabolizing enzymes. As the AHR field has matured, this classic AHR/ARNT/DRE mechanism has been expanded to include AHR-mediated modulation of gene transcription that involves a variety of AHR or other transcription factor interactions and regulation of a wide spectrum of genes. It is possible that the AHR may regulate genes in a multitiered manner, with the classic AHR/ARNT/DRE mechanism representing the "first response"

system designed to eliminate the presence of external stimuli (i.e., xenobiotics and excessive concentrations of endogenous AHR agonists). The second tier of events, those mediated by the interactions between the AHR and other transcription factors, represents an "adaptive" response [14] that allows the biological system to cope with inappropriate agonist activation of the AHR. As we continue to make progress in this field, the physiological, toxicological, and pharmacological implications of these AHR-mediated events will be further clarified.

REFERENCES

1. Whitlock, J., Jr. (1999). Induction of cytochrome P4501A1. *Annual Review of Pharmacology and Toxicology, 39*, 103–125.
2. Okey, A., Bondy, G., Mason, M., Kahl, G., Eisen, H., Guenthner, T., and Nebert, D. (1979). Regulatory gene product of the Ah locus. Characterization of the cytosolic inducer–receptor complex and evidence for its nuclear translocation. *Journal of Biological Chemistry, 254*, 11636–11648.
3. Poland, A., Knutson, J., and Glover, E. (1985). Studies on the mechanism of action of halogenated aromatic hydrocarbons. *Clinical Physiology and Biochemistry, 3*, 147–154.
4. Bigelow, S. and Nebert, D. (1986) The murine aromatic hydrocarbon responsiveness locus: a comparison of receptor levels and several inducible enzyme activities among recombinant inbred lines. *Journal of Biochemical Toxicology, 1*, 1–14.
5. Nebert, D. and Dalton, T. (2006). The role of cytochrome P450 enzymes in endogenous signalling pathways and environmental carcinogenesis. *Nature Reviews Cancer, 6*, 947–960.
6. Fujii-Kuriyama, Y. and Mimura, J. (2005). Molecular mechanisms of AhR functions in the regulation of cytochrome P450 genes. *Biochemical and Biophysical Research Communications, 338*, 311–317.
7. Hankinson, O. (2005). Role of coactivators in transcriptional activation by the aryl hydrocarbon receptor. *Archives of Biochemistry and Biophysics, 433*, 379–386.
8. Beischlag, T., Luis Morales, J., Hollingshead, B., Perdew, G. (2008). The aryl hydrocarbon receptor complex and the control of gene expression. *Critital Reviews in Eukaryotic Gene Expression, 18*, 207–250.
9. Okey, A. Bondy, G., Mason, M., Nebert, D., Forster-Gibson, C., Muncan, J., and Dufresne, M. (1980). Temperature-dependent cytosol-to-nucleus translocation of the Ah receptor for 2,3,7,8-tetrachlorodibenzo-*p*-dioxin in continuous cell culture lines. *Journal of Biological Chemistry, 255*, 11415–11422.
10. Okey, A. (2007). An aryl hydrocarbon receptor odyssey to the shores of toxicology: the Deichmann Lecture, International Congress of Toxicology-XI. *Toxicological Sciences, 98*, 5–38.
11. Elferink, C., Gasiewicz, T., and Whitlock, J., Jr. (1990). Protein–DNA interactions at a dioxin-responsive enhancer. Evidence that the transformed Ah receptor is heteromeric. *Journal of Biological Chemistry, 265*, 20708–20712.

12. Swanson, H., Tullis, K., and Denison, M. (1993). Binding of transformed Ah receptor complex to a dioxin responsive transcriptional enhancer: evidence for two distinct heteromeric DNA-binding forms. *Biochemistry*, *32*, 12841–12849.
13. Burbach, K., Poland, A., and Bradfield, C. (1992). Cloning of the Ah-receptor cDNA reveals a distinctive ligand-activated transcription factor. *Proceedings of the National Academy of Sciences of the United States of America 89*, 8185–8189.
14. McIntosh, B., Hogenesch, J., and Bradfield, C. (2010). Mammalian Per–Arnt–Sim proteins in environmental adaptation. *Annual Review of Physiology*, *72*, 625–645.
15. Kewley, R., Whitelaw, M., and Chapman-Smith, A. (2004). The mammalian basic helix–loop–helix/PAS family of transcriptional regulators. *International Journal of Biochemistry and Cell Biology*, *36*, 189–204.
16. Amoutzias, G., Robertson, D., Van de Peer, Y., and Oliver, S. (2008). Choose your partners: dimerization in eukaryotic transcription factors. *Trends in Biochemical Sciences*, *33*, 220–229.
17. Partch, C. and Gardner, K. (2010) Coactivator recruitment: a new role for PAS domains in transcriptional regulation by the bHLH–PAS family. *Journal of Cellular Physiology*, *223*, 553–557.
18. Dolwick, K., Swanson, H., and Bradfield, C. (1993). *In vitro* analysis of Ah receptor domains involved in ligand-activated DNA recognition. *Proceedings of the National Academy of Sciences of the United States of America*, *90*, 8566–8570.
19. Reisz-Porszasz, S., Probst, M., Fukunaga, B., and Hankinson, O. (1994). Identification of functional domains of the aryl hydrocarbon receptor nuclear translocator protein (ARNT). *Molecular and Cellular Biology*, *14*, 6075–6086.
20. Bacsi, S. and Hankinson, O. (1996). Functional characterization of DNA-binding domains of the subunits of the heterodimeric aryl hydrocarbon receptor complex imputing novel and canonical basic helix–loop–helix protein–DNA interactions. *Journal of Biological Chemistry*, *271*, 8843–8850.
21. Dong, L., Ma, Q., and Whitlock, J., Jr. (1996). DNA binding by the heterodimeric Ah receptor. Relationship to dioxin-induced CYP1A1 transcription *in vivo*. *Journal of Biological Chemistry*, *271*, 7942–7948.
22. Swanson, H. and Yang, J. (1996). Mapping the protein/DNA contact sites of the Ah receptor and Ah receptor nuclear translocator. *Journal of Biological Chemistry 271*, 31657–31665.
23. Sun, W., Zhang, J., and Hankinson, O. (1997). A mutation in the aryl hydrocarbon receptor (AHR) in a cultured mammalian cell line identifies a novel region of AHR that affects DNA binding. *Journal of Biological Chemistry*, *272*, 31845–31854.
24. Chapman-Smith, A. and Whitelaw M. (2006). Novel DNA binding by a basic helix–loop–helix protein. The role of the dioxin receptor PAS domain. *Journal of Biological Chemistry*, *281*, 12535–12545.
25. Fukunaga, B. and Hankinson, O. (1996). Identification of a novel domain in the aryl hydrocarbon receptor required for DNA binding. *Journal of Biological Chemistry*, *271*, 3743–3749.
26. Wache, S., Hoagland, E., Zeigler, G., and Swanson, H. (2005). Role of arginine residues 14 and 15 in dictating DNA binding stability and transactivation of the aryl hydrocarbon receptor/aryl hydrocarbon receptor nuclear translocator heterodimer. *Gene Expression*, *12*, 231–243.
27. Swanson, H., Chan, W., and Bradfield, C. (1995). DNA binding specificities and pairing rules of the Ah receptor, ARNT, and SIM proteins. *Journal of Biological Chemistry*, *270*, 26292–26302.
28. Bacsi, S., Reisz-Porszasz, S., Hankinson, O. (1995). Orientation of the heterodimeric aryl hydrocarbon (dioxin) receptor complex on its asymmetric DNA recognition sequence. *Molecular Pharmacology*, *47*, 432–438.
29. Atchley, W. and Fitch, W. (1997). A natural classification of the basic helix–loop–helix class of transcription factors. *Proceedings of the National Academy of Sciences of the United States of America*, *94*, 5172–5176.
30. Denison, M., Fisher, J., and Whitlock, J., Jr. (1989). Protein–DNA interactions at recognition sites for the dioxin–Ah receptor complex. *Journal of Biological Chemistry*, *264*, 16478–16482.
31. Yao, E. and Denison, M. (1992). DNA sequence determinants for binding of transformed Ah receptor to a dioxin-responsive enhancer. *Biochemistry*, *31*, 5060–5067.
32. Sogawa, K., Numayama-Tsuruta, K., Takahashi, T., Matsushita, N., Miura, C., Nikawa, J., Gotoh, O., Kikuchi, Y., and Fujii-Kuriyama, Y. (2004). A novel induction mechanism of the rat CYP1A2 gene mediated by Ah receptor–Arnt heterodimer. *Biochemical and Biophysical Research Communications 318*, 746–755.
33. Boutros, P., Moffat, I., Franc, M., Tijet, N., Tuomisto, J., Pohjanvirta, R., and Okey, A. (2004). Dioxin-responsive AHRE-II gene battery: identification by phylogenetic footprinting. *Biochemical and Biophysical Research Communications*, *321*, 707–715.
34. Kinehara, M., Fukuda, I., Yoshida, K., and Ashida, H. (2008). High-throughput evaluation of aryl hydrocarbon receptor-binding sites selected via chromatin immunoprecipitation-based screening in Hepa-1c1c7 cells stimulated with 2,3,7,8-tetrachlorodibenzo-p-dioxin. *Genes & Genetic Systems*, *83*, 455–468.
35. Sun, Y., Boverhof, D., Burgoon, L., Fielden, M., and Zacharewski, T. (2004). Comparative analysis of dioxin response elements in human, mouse and rat genomic sequences. *Nucleic Acids Research*, *32*, 4512–4523.
36. Tijet, N., Boutros, P., Moffat, I., Okey, A., Tuomisto, J., and Pohjanvirta, R. (2006). Aryl hydrocarbon receptor regulates distinct dioxin-dependent and dioxin-independent gene batteries. *Molecular Pharmacology*, *69*, 140–153.
37. Boutros, P., Bielefeld, K., Pohjanvirta, R., and Harper, P. (2009). Dioxin-dependent and dioxin-independent gene batteries: comparison of liver and kidney in AHR-null mice. *Toxicological Sciences*, *112*, 245–256.
38. Quattrochi, L., Vu, T., and Tukey, R. (1994). The human CYP1A2 gene and induction by 3-methylcholanthrene. A region of DNA that supports AH-receptor binding and

promoter-specific induction. *Journal of Biological Chemistry, 269,* 6949–6954.

39. Nukaya, M., Moran, S., and Bradfield, C. (2009). The role of the dioxin-responsive element cluster between the Cyp1a1 and Cyp1a2 loci in aryl hydrocarbon receptor biology. *Proceedings of the National Academy of Sciences of the United States of America, 106,* 4923–4928.

40. Nukaya, M. and Bradfield, C., (2009). Conserved genomic structure of the Cyp1a1 and Cyp1a2 loci and their dioxin responsive elements cluster. *Biochemical Pharmacology, 77,* 654–659.

41. Eltom, S., Larsen, M., and Jefcoate, C. (1998). Expression of CYP1B1 but not CYP1A1 by primary cultured human mammary stromal fibroblasts constitutively and in response to dioxin exposure: role of the Ah receptor. *Carcinogenesis, 19,* 1437–1444.

42. Zhang, L., Savas, U., Alexander, D., and Jefcoate, C. (1998). Characterization of the mouse Cyp1B1 gene. Identification of an enhancer region that directs aryl hydrocarbon receptor-mediated constitutive and induced expression. *Journal of Biological Chemistry, 273,* 5174–5183.

43. Zhang, L., Zheng, W., and Jefcoate, C. (2003). Ah receptor regulation of mouse Cyp1B1 is additionally modulated by a second novel complex that forms at two AhR response elements. *Toxicology and Applied Pharmacology, 192,* 174–190.

44. Beedanagari, S., Taylor, R., and Hankinson, O. (2010). Differential regulation of the dioxin-induced Cyp1a1 and Cyp1b1 genes in mouse hepatoma and fibroblast cell lines. *Toxicology Letters, 194,* 26–33.

45. Osburn, W. and Kensler, T. (2008). Nrf2 signaling: an adaptive response pathway for protection against environmental toxic insults. *Mutation Research, 659,* 31–39.

46. Miao, W., Hu, L., Scrivens, P., and Batist, G. (2005). Transcriptional regulation of NF-E2 p45-related factor (NRF2) expression by the aryl hydrocarbon receptor-xenobiotic response element signaling pathway: direct cross-talk between phase I and II drug-metabolizing enzymes. *Journal of Biological Chemistry, 280,* 20340–20348.

47. Nebert, D., Roe, A., Dieter, M., Solis, W., Yang, Y., and Dalton, T. (2000). Role of the aromatic hydrocarbon receptor and [Ah] gene battery in the oxidative stress response, cell cycle control, and apoptosis. *Biochemical Pharmacology, 59,* 65–85.

48. Ma, Q., Kinneer, K., Bi, Y., Chan, J., and Kan, Y. (2004). Induction of murine NAD(P)H:quinone oxidoreductase by 2,3,7,8-tetrachlorodibenzo-p-dioxin requires the CNC (cap 'n' collar) basic leucine zipper transcription factor Nrf2 (nuclear factor erythroid 2-related factor 2): cross-interaction between AhR (aryl hydrocarbon receptor) and Nrf2 signal transduction. *Biochemical Journal, 377,* 205–213.

49. Kohle, C. and Bock, K. (2007). Coordinate regulation of Phase I and II xenobiotic metabolisms by the Ah receptor and Nrf2. *Biochemical Pharmacology, 73,* 1853–1862.

50. Yeager, R., Reisman, S., Aleksunes, L., and Klaassen, C. (2009). Introducing the "TCDD-inducible AhR-Nrf2 gene battery". *Toxicological Sciences, 111,* 238–246.

51. Munzel, P., Schmohl, S., Buckler, F., Jaehrling, J., Raschko, F., Kohle, C., and Bock, K. (2003). Contribution of the Ah receptor to the phenolic antioxidant-mediated expression of human and rat UDP-glucuronosyltransferase UGT1A6 in Caco-2 and rat hepatoma 5L cells. *Biochemical Pharmacology, 66,* 841–847.

52. Kalthoff, S., Ehmer, U., Freiberg, N., Manns, M., and Strassburg, C. (2010). Interaction between oxidative stress sensor Nrf2 and xenobiotic-activated aryl hydrocarbon receptor in the regulation of the human phase II detoxifying UDP-glucuronosyltransferase 1A10. *Journal of Biological Chemistry, 285,* 5993–6002.

53. Ma, Q., Baldwin, K., Renzelli, A., McDaniel, A., and Dong, L. (2001). TCDD-inducible poly(ADP-ribose) polymerase: a novel response to 2,3,7,8-tetrachlorodibenzo-p-dioxin. *Biochemical and Biophysical Research Communications, 289,* 499–506.

54. Boverhof, D., Burgoon, L., Tashiro, C., Chittim, B., Harkema, J., Jump, D., and Zacharewski, T. (2005). Temporal and dose-dependent hepatic gene expression patterns in mice provide new insights into TCDD-mediated hepatotoxicity. *Toxicological Sciences, 85,* 1048–1063.

55. Frericks, M., Burgoon, L., Zacharewski, T., and Esser, C. (2008). Promoter analysis of TCDD-inducible genes in a thymic epithelial cell line indicates the potential for cell-specific transcription factor crosstalk in the AhR response. *Toxicology and Applied Pharmacology, 232,* 268–279.

56. Ma, Q. (2002). Induction and superinduction of 2,3,7,8-tetrachlorodibenzo-rho-dioxin-inducible poly(ADP-ribose) polymerase: role of the aryl hydrocarbon receptor/aryl hydrocarbon receptor nuclear translocator transcription activation domains and a labile transcription repressor. *Archives of Biochemistry and Biophysics, 404,* 309–316.

57. Rivera, S., Wang, F., Saarikoski, S., Taylor, R., Chapman, B., Zhang, R., and Hankinson, O. (2007). A novel promoter element containing multiple overlapping xenobiotic and hypoxia response elements mediates induction of cytochrome P4502S1 by both dioxin and hypoxia. *Journal of Biological Chemistry, 282,* 10881–10893.

58. Kang, H., Kim, H., Cho, C., Hu, Y., Li, R., and Bae, I. (2008). BRCA1 transcriptional activity is enhanced by interactions between its AD1 domain and AhR. *Cancer Chemotherapy and Pharmacology, 62,* 965–975.

59. Kashuba, E., Gradin, K., Isaguliants, M., Szekely, L., Poellinger, L., Klein, G., and Kazlauskas, A. (2006). Regulation of transactivation function of the aryl hydrocarbon receptor by the Epstein–Barr virus-encoded EBNA-3 protein. *Journal of Biological Chemistry, 281,* 1215–1223.

60. Xu, C., Krager, S., Liao, D., and Tischkau, S. (2010). Disruption of CLOCK-BMAL1 transcriptional activity is responsible for aryl hydrocarbon receptor-mediated regulation of Period1 gene. *Toxicological Sciences, 115,* 98–108.

61. Mimura, J., Ema, M., Sogawa, K., and Fujii-Kuriyama, Y. (1999). Identification of a novel mechanism of regulation of Ah (dioxin) receptor function. *Genes & Development, 13,* 20–25.

62. Hahn, M., Allan, L., and Sherr, D. (2009). Regulation of constitutive and inducible AHR signaling: complex interactions involving the AHR repressor. *Biochemical Pharmacology*, 77, 485–497.
63. Haarmann-Stemmann, T., Bothe, H., Kohli, A., Sydlik, U., Abel, J., and Fritsche, E. (2007). Analysis of the transcriptional regulation and molecular function of the aryl hydrocarbon receptor repressor in human cell lines. *Drug Metabolism and Disposition*, 35, 2262–2269.
64. Kanno, Y., Takane, Y., Takizawa Y., and Inouye, Y. (2008). Suppressive effect of aryl hydrocarbon receptor repressor on transcriptional activity of estrogen receptor alpha by protein–protein interaction in stably and transiently expressing cell lines. *Molecular and Cellular Endocrinology*, 291, 87–94.
65. Kociba, R., Keyes, D., Beyer, J., Carreon, R., Wade, C., Dittenber, D., Kalnins, R., Frauson, L., Park, C., Barnard, S., Hummel, R., and Humiston, C. (1978). Results of a two-year chronic toxicity and oncogenicity study of 2,3,7,8-tetrachlorodibenzo-p-dioxin in rats. *Toxicology and Applied Pharmacology*, 46, 279–303.
66. Gallo, M., Hesse, E., Macdonald, G., and Umbreit, T. (1986). Interactive effects of estradiol and 2,3,7,8-tetrachlorodibenzo-p-dioxin on hepatic cytochrome P-450 and mouse uterus. *Toxicology Letters*, 32, 123–132.
67. Chang, E., Charn, T., Park, S., Helferich, W., Komm, B., Katzenellenbogen, J., and Katzenellenbogen, B. (2008). Estrogen receptors alpha and beta as determinants of gene expression: influence of ligand, dose, and chromatin binding. *Molecular Endocrinology*, 22, 1032–1043.
68. Safe, S. and Wormke, M. (2003). Inhibitory aryl hydrocarbon receptor-estrogen receptor alpha cross-talk and mechanisms of action. *Chemical Research in Toxicology*, 16, 807–816.
69. Matthews, J. and Gustafsson, J. (2006). Estrogen receptor and aryl hydrocarbon receptor signaling pathways. *Nuclear Receptor Signaling*, 4, e016.
70. Ohtake, F., Takeyama, K., Matsumoto, T., Kitagawa, H., Yamamoto, Y., Nohara, K., Tohyama, C., Krust, A., Mimura, J., Chambon, P., Yanagisawa, J., Fujii-Kuriyama, Y., and Kato, S. (2003). Modulation of oestrogen receptor signalling by association with the activated dioxin receptor. *Nature*, 423, 545–550.
71. Klinge, C., Kaur, K., and Swanson, H. (2000). The aryl hydrocarbon receptor interacts with estrogen receptor alpha and orphan receptors COUP-TFI and ERRα1. *Archives of Biochemistry and Biophysics* 373, 163–174.
72. Klinge, C., Bowers, J., Kulakosky, P., Kamboj, K., and Swanson, H. (1999). The aryl hydrocarbon receptor (AHR)/AHR nuclear translocator (ARNT) heterodimer interacts with naturally occurring estrogen response elements. *Molecular and Cellular Endocrinology*, 157, 105–119.
73. Ahmed, S., Valen, E., Sandelin, A., and Matthews, J. (2009). Dioxin increases the interaction between aryl hydrocarbon receptor and estrogen receptor alpha at human promoters. *Toxicological Sciences*, 111, 254–266.
74. Kollara, A. and Brown, T. (2010). Four and a half LIM domain 2 alters the impact of aryl hydrocarbon receptor on androgen receptor transcriptional activity. *Journal of Steroid Biochemistry and Molecular Biology*, 118, 51–58.
75. Wang, S., Liang, C., Liu, Y., Huang, M., Huang, S., Hong, W., and Su, J. (2009). Crosstalk between activated forms of the aryl hydrocarbon receptor and glucocorticoid receptor. *Toxicology*, 262, 87–97.
76. Bielefeld, K., Lee, C., and Riddick, D. (2008). Regulation of aryl hydrocarbon receptor expression and function by glucocorticoids in mouse hepatoma cells. *Drug Metabolism and Disposition*, 36, 543–551.
77. Vallabhapurapu, S. and Karin, M. (2009). Regulation and function of NF-κB transcription factors in the immune system. *Annual Reviews of Immunology*, 27, 693–733.
78. Tian, Y. (2009). Ah receptor and NF-κB interplay on the stage of epigenome. *Biochemical Pharmacology*, 77, 670–680.
79. Vogel, C. and Matsumura, F. (2009). A new cross-talk between the aryl hydrocarbon receptor and RelB, a member of the NF-κB family. *Biochemical Pharmacology*, 77, 734–745.
80. Patel, R., Murray, I., Flaveny, C., Kusnadi, A., and Perdew, G. (2009). Ah receptor represses acute-phase response gene expression without binding to its cognate response element. *Laboratory Investigations*, 89, 695–707.
81. Tian, Y., Ke, S., Denison, M., Rabson, A., and Gallo, M. (1999). Ah receptor and NF-κB interactions, a potential mechanism for dioxin toxicity. *Journal of Biological Chemistry*, 274, 510–515.
82. Ruby, C., Leid, M., and Kerkvliet, N. (2002). 2,3,7,8-Tetrachlorodibenzo-p-dioxin suppresses tumor necrosis factor-alpha and anti-CD40-induced activation of NF-κB/Rel in dendritic cells: p50 homodimer activation is not affected. *Molecular Pharmacology*, 62, 722–728.
83. Thatcher, T., Maggirwar, S., Baglole, C., Lakatos, H., Gasiewicz, T., Phipps, R., and Sime, P. (2007). Aryl hydrocarbon receptor-deficient mice develop heightened inflammatory responses to cigarette smoke and endotoxin associated with rapid loss of the nuclear factor-κB component RelB. *American Journal of Pathology*, 170, 855–864.
84. Hollingshead, B., Beischlag, T., Dinatale, B., Ramadoss, P., and Perdew, G. (2008). Inflammatory signaling and aryl hydrocarbon receptor mediate synergistic induction of interleukin 6 in MCF-7 cells. *Cancer Research*, 68, 3609–3617.
85. Kimura, A., Naka, T., Nohara, K., Fujii-Kuriyama, Y., and Kishimoto, T. (2008). Aryl hydrocarbon receptor regulates Stat1 activation and participates in the development of T_h17 cells. *Proceedings of the National Academy of Sciences of the United States of America*, 105, 9721–9726.
86. Kimura, A., Naka, T., Nakahama, T., Chinen, I., Masuda, K., Nohara, K., Fujii-Kuriyama, Y., and Kishimoto T. (2009). Aryl hydrocarbon receptor in combination with Stat1 regulates LPS-induced inflammatory responses. *Journal of Experimental Medicine*, 206, 2027–2035.
87. Ge, N. and Elferink, C. (1998). A direct interaction between the aryl hydrocarbon receptor and retinoblastoma protein. Linking dioxin signaling to the cell cycle. *Journal of Biological Chemistry*, 273, 22708–22713.

88. Puga, A., Barnes, S., Dalton, T., Chang, C., Knudsen, E., and Maier, M. (2000). Aromatic hydrocarbon receptor interaction with the retinoblastoma protein potentiates repression of E2F-dependent transcription and cell cycle arrest. *Journal of Biological Chemistry*, 275, 2943–2950.

89. Puga, A., Ma, C., and Marlowe, J. (2009). The aryl hydrocarbon receptor cross-talks with multiple signal transduction pathways. *Biochemical Pharmacology*, 77, 713–722.

90. Barhoover, M., Hall, J., Greenlee, W., and Thomas, R. (2010). Aryl hydrocarbon receptor regulates cell cycle progression in human breast cancer cells via a functional interaction with cyclin-dependent kinase 4. *Molecular Pharmacology*, 77, 195–201.

91. Chen, H., Tsai, S., and Leone, G. (2009). Emerging roles of E2Fs in cancer: an exit from cell cycle control. *Nature Reviews Cancer*, 9, 785–797.

92. Sun, A., Bagella, L., Tutton, S., Romano, G., and Giordano, A. (2007). From G0 to S phase: a view of the roles played by the retinoblastoma (Rb) family members in the Rb-E2F pathway. *Journal of Cellular Biochemistry*, 102, 1400–1404.

93. Watabe, Y., Nazuka, N., Tezuka, M., and Shimba, S. (2010). Aryl hydrocarbon receptor functions as a potent coactivator of E2F1-dependent transcription activity. *Biological & Pharmaceutical Bulletin*, 33, 389–397.

94. Tan, N. and Khachigian, L. (2009). Sp1 phosphorylation and its regulation of gene transcription. *Molecular and Cellular Biology*, 29, 2483–2488.

95. Fisher, J., Wu, L., Denison, M., and Whitlock, J., Jr. (1990). Organization and function of a dioxin-responsive enhancer. *Journal of Biological Chemistry*, 265, 9676–9681.

96. Fujii-Kuriyama, Y., Sogawa, K., Imataka, H., Yasumoto, K., Kikuchi, Y., and Fujisawa-Sehara, A. (1990). Transcriptional regulation of 3-methylcholanthrene-inducible P-450 gene responsible for metabolic activation of aromatic carcinogenes. *Princess Takamatsu Symposium*, 21, 165–175.

97. Kobayashi, A., Sogawa, K., and Fujii-Kuriyama, Y. (1996). Cooperative interaction between AhR/Arnt and Sp1 for the drug-inducible expression of CYP1A1 gene. *Journal of Biological Chemistry*, 271, 12310–12316.

98. Khan, S., Barhoumi, R., Burghardt, R., Liu, S., Kim, K., and Safe S. (2006). Molecular mechanism of inhibitory aryl hydrocarbon receptor-estrogen receptor/Sp1 cross talk in breast cancer cells. *Molecular Endocrinology*, 20, 2199–2214.

99. Tsuchiya, Y., Nakajima, M., and Yokoi, T. (2003). Critical enhancer region to which AhR/ARNT and Sp1 bind in the human CYP1B1 gene. *Journal of Biochemistry*, 133, 583–592.

100. Baba, T., Mimura, J., Gradin, K., Kuroiwa, A., Watanabe, T., Matsuda, Y., Inazawa, J., Sogawa, K., and Fujii-Kuriyama, Y. (2001). Structure and expression of the Ah receptor repressor gene. *Journal of Biological Chemistry*, 276, 33101–33110.

101. Patel, R., Kim, D., Peters, J., and Perdew, G. (2006). The aryl hydrocarbon receptor directly regulates expression of the potent mitogen epiregulin. *Toxicological Sciences*, 89, 75–82.

102. Bunger, M., Glover, E., Moran, S., Walisser, J., Lahvis, G., Hsu, E., and Bradfield, C. (2008). Abnormal liver development and resistance to 2,3,7,8-tetrachlorodibenzo-*p*-dioxin toxicity in mice carrying a mutation in the DNA-binding domain of the aryl hydrocarbon receptor. *Toxicological Sciences*, 106, 83–92.

103. Perdew, G. (2008). Ah receptor binding to its cognate response element is required for dioxin-mediated toxicity. *Toxicological Sciences*, 106, 301–303.

104. Ohtake, F., Fujii-Kuriyama, Y., and Kato, S. (2009). AhR acts as an E3 ubiquitin ligase to modulate steroid receptor functions. *Biochemical Pharmacology*, 77, 474–484.

6

THE AHR/ARNT DIMER AND TRANSCRIPTIONAL COACTIVATORS

OLIVER HANKINSON

The role of coactivators in AHR/ARNT-mediated transcriptional activation has been most studied for the *CYP1A1* gene. Agonist binding to the AHR leads to the association of the AHR/ARNT dimer (also called the aryl hydrocarbon receptor complex (AHRC)) with xenobiotic response elements (XREs) within the enhancer, which is located about 1 kb upstream of the transcriptional start site of the *CYP1A1* gene [1, 2]. In mouse hepatoma cells, the binding of the AHRC to the XREs of the *Cyp1a1* gene has been shown to stimulate local changes in chromatin structure at the enhancer and to induce chromatin modification and nucleosomal displacement at the proximal promoter region located just upstream of the transcriptional start site [3]. The displacement of the fixed nucleosome exposes a TATA sequence and allows promoter accessibility by the preinitiation complex and stabilization of RNA polymerase II (pol II) binding. The binding of activators and general transcription factors alone is insufficient to activate most genes [4]. Alterations in chromatin structure directed by the action of transcriptional coactivator proteins are also required. Transcriptional regulation by the AHRC is therefore dependent on the coordinated recruitment of coactivator proteins and general transcription factors.

I previously reviewed this area in 2005 [5]. This chapter will focus on developments that have occurred since that time. Prior to that date, most studies focused on the mouse *Cyp1a1* gene. More recently, the human *CYP1A1* gene and certain other AHRC-inducible genes have increasingly become the focus of study.

DNA in mammalian chromatin is highly compacted. The fundamental repeating unit of chromatin is the nucleosome, comprising 146 bp of DNA wound around a core of histone proteins, consisting of two molecules each of H2A, H2B, H3, and H4. Neighboring nucleosomes are associated via the H1 linker histone, which facilitates additional compaction. Further folding leads to the generation of the highly condensed chromosome. Coactivators function by remodeling, relocating, or dissociating nucleosomes, thereby relieving their repressive effect on transcription, by directly recruiting basal transcription factors and pol II to the promoter, and by activating RNA polymerase [6–8].

A large number of coactivator proteins have been identified [9, 10]. Certain coactivators covalently modify the protruding amino-terminal tails of the core histone proteins. Coactivators possessing histone acetyltransferase (HAT) activity transfer an acetyl group to lysine residues in histones. Histone deacetylases remove acetyl groups. Histone methyltransferases (HMT) add one to three methyl groups to individual lysine residues, or one or two methyl groups to particular arginine residues. Histone demethylases remove these groups. Histone tails can also be phosphorylated, ubiquitinated, sumoylated, and ADP ribosylated [11]. Acetylation and ubiquitination are generally associated with an enhancement of transcription, lysine methylation with either gene repression or gene activation, and arginine methylation with an active state of chromatin. However, the effects on gene expression can depend on which specific amino acids are modified. The covalent modifications appear to modulate gene transcription both by affecting the tightness with which DNA associates with the core histones and by providing docking sites for the recruitment of other proteins that subsequently regulate chromatin structure and activity. The "histone code" hypothesis proposes that the precise pattern of covalent histone modifications dictates downstream regulatory events [12–14].

The AH Receptor in Biology and Toxicology, First Edition. Edited by Raimo Pohjanvirta.
© 2012 John Wiley & Sons, Inc. Published 2012 by John Wiley & Sons, Inc.

Coactivators exist in multisubunit complexes that under appropriate conditions can be released intact from chromatin. Transcriptional activation appears to involve the ordered association and dissociation, and perhaps the cycling on and off, of such complexes.

Evidence for a coactivator role for a particular protein in AHRC-mediated induction of *CYP1A1* has been obtained via a number of different experimental approaches. A convincing identification of a coactivator for the AHRC requires several such approaches. The best evidence has been obtained using experimental approaches that address the endogenous *CYP1A1* gene in its normal chromosomal setting without the use of overproduced proteins. These lines of evidence include the following, each of which has limitations:

1. Knockdown of the protein with small inhibitory RNA (siRNAs) oligonucleotide or with short hairpin inhibiting RNAs (shRNAs), or elimination of the corresponding genes functionality by gene knockout technology, or diminution of the protein's activity via use of a dominant-negative construct reduces or eliminates induction of CYP1A1 mRNA by an AHR agonist [15]. Reservations to the interpretation of these types of experiments include the following: (a) induction of CYP1A1 may be unaffected by treatment with siRNAs or shRNAs to a *bona fide* coactivator if these reagents (as is usually the case) are not 100% efficient and residual levels of the putative coactivator support full induction of CYP1A1, (b) certain coactivators may function redundantly and loss of activity of all of them may be required to see an effect in CYP1A1 induction, (c) the cells may compensate physiologically for the loss of function of the coactivator (particularly relevant with gene knockouts and long-term cultures of cells containing shRNAs), or selection occurs for cells that have adapted to the loss of the coactivator (particularly relevant with shRNA-containing cells maintained in culture for a long time).
2. Chromatin immunoprecipitation (ChIP) analysis demonstrates that the protein is associated with the *CYP1A1* gene in its normal chromosomal setting *in vivo*. The evidence is particularly compelling when the degree of association of the protein is enhanced after AHR agonist treatment [15, 16]. It should be noted that since some coactivators may be buried within a complex and therefore completely or partially unavailable for antibody binding, the lack of a positive result in this assay even when using a ChIP verified antibody does not prelude the possibility that the protein is associated with the *CYP1A1* gene in chromatin.

Other experimental strategies, although being capable of providing support for the notion that a particular protein functions as a coactivator for CYP1A1 transcription, are less compelling than those described above, although these strategies can provide important mechanistic insights into the mode of action of the putative coactivator that are not readily obtainable by the above strategies. These experimental approaches include the following:

1. Demonstration that AHR agonist-dependent transcription of a transiently transfected reporter gene (e.g., luciferase) driven by the 5′ upstream region of the *CYP1A1* gene is increased by cotransfection of a cDNA for the putative coactivator. If stimulation of transcription occurs when synthetically synthesized concatemerized XREs are used instead of the complete *CYP1A1* upstream region to drive the reporter gene in the above assay, then this provides evidence that the stimulatory effect of the protein in question is mediated by the AHRC, as should be the case for a *bona fide* coactivator for this dimer [17].
2. Several experimental strategies have been used to provide evidence for the physical interaction of a putative coactivator with AHR and ARNT. These assays are particularly compelling if the interactions are shown to be stimulated by treatment with an AHR agonist. However, these assays involve the use of high concentrations of the relevant proteins *in vitro* or the analysis of transfected cDNAs (and thus overexpressed proteins) in cell culture. Since artificially high concentrations of the proteins could lead to interactions that do not occur in the cell, the use of such experiments to identify AHRC coactivators should be viewed with caution. Nevertheless, these types of experiments have been used to localize the probable interaction domains of AHR and ARNT for several previously identified coactivators. Examples of these strategies include the following:

   - Coimmunoprecipitation of the *in vitro* expressed putative coactivator with *in vitro* expressed AHR and ARNT. A related approach is to attach glutathione S-transferase (GST) to AHR or ARNT and to coprecipitate the putative coactivator and the GST fusion protein with glutathione Sepharose beads under nondenaturing conditions. The experiment can also be performed in the reciprocal fashion. These approaches are suited for the identification of the interaction sites on the coactivator and AHR or ARNT [17].
   - Analogous experiments can also be performed in cells cotransfected with cDNAs for AHR or ARNT and the putative coactivator. Usually both proteins are tagged with an epitope to facilitate immunoprecipitation and Western blot analyses. This approach has the advantage that the cells are present in a cellular setting [18].

- Another approach is to ascertain whether agonist treatment of cells leads to altered localization of the protein in the cell. This approach is most compelling if the protein is shown to relocalize to the same sites as AHR and ARNT [17].
- The yeast two-hybrid procedure has been used to identify the interaction domains of the proteins, and it has frequently been used to complement the results obtained with the *in vitro* coimmunoprecipitation procedure [17]. Although the above approaches cannot distinguish between a direct interaction or an indirect interaction of the two proteins, interaction of the putative coactivator with AHR or ARNT in the two-hybrid system is more indicative of a direct interaction, since yeast cells are less likely to harbor relevant intermediary proteins than mammalian cells.
- Fluorescence resonance energy transfer (FRET) can be used to detect the interaction of two proteins within cells. Since both proteins must be tagged with appropriate fluorophores, cDNAs for the relevant proteins must be cotransfected into cells (and the proteins are therefore likely to be overexpressed). A positive signal indicates that the two proteins are in very close contact and likely to be interacting directly. It appears that FRET has not been used for the identification or analysis of coactivators for the AHRC.

TABLE 6.1 Coactivators for the AHRC

Coactivators with HAT activity
Ncoa1 (nuclear receptor coactivator 1, also called SRC-1)
Ncoa2 (nuclear receptor coactivator 2, or SRC-2)
Ncoa3 (also called p/CIP or SRC-3)
p300
CBP (CREB binding protein)
ATPase-dependent histone modifier
BRG1 (Brahma-related gene 1)
Mediator subunit
Med1 (also called Med220 or mediator subunit of 220 kDa)
Other coactivators
RIP 140 (receptor-interacting protein 140)
Thyroid hormone receptor/retinoblastoma-interacting protein 230 (TRIP230)
Coiled-coil coactivator (CoCoA)
GRIP1-associated coactivator 63 (GAC63)
Estrogen receptor α (ERα)
Breast cancer susceptibility gene 1 (BRCA1)

6.1 NEWLY IDENTIFIED COACTIVATORS FOR THE AHRC

I will discuss those coactivators identified since my last review for which compelling evidence has been obtained (indicated in bold font in Table 6.1). This evidence consists, minimally, of a demonstration that (i) an AHR agonist stimulates the recruitment of the putative coactivator to the endogenous *CYP1A1* (or other AHRC-responsive) gene and that (ii) knockdown or knockout of the coactivator reduces induction of CYP1A1 by agonist. Thyroid hormone receptor/retinoblastoma-interacting protein 230 (TRIP230), coiled-coil coactivator (CoCoA), GRIP1-associated coactivator 63 (GAC63), estrogen receptor α (ERα), and breast cancer susceptibility gene 1 (BRCA1) fulfill these criteria.

TRIP230 was initially identified as coactivator of the thyroid hormone receptor (a member of the nuclear hormone receptor (NR) superfamily). It also binds the retinoblastoma protein (Rb). TRIP230's AHRC coactivator function appears to be mediated by its binding to ARNT. The interaction domain on ARNT was localized to its bHLH–PAS region [18] and the interaction domain on TRIP230 to a 200-amino acid segment toward the carboxy terminus distinct from its thyroid hormone receptor and Rb interaction domains. Later studies localized the ARNT interaction domain to the PASB region and the interaction domain on TRIP230 to an LXXLL as the nuclear receptor box within the above 200-amino acid region. Interestingly, TRIP230 was found to bind on the interface of the ARNT PASB domain located on the opposite side used to associate with the PASB domain on its hypoxia-inducible factor 2α (HIF-2α) dimerization partner [19]. Presumably, TRIP230 also binds to the equivalent interface of the PASB domain of AHR.

CoCoA, like TRIP230, is a coiled-coil protein. It was originally identified as a NR coactivator. It binds the bHLH–PAS domain of p160 coactivators (i.e., Ncoa1, Ncoa2, and Ncoa3) and acts as a secondary coactivator of NRs, interacting with them indirectly through the primary p160 coactivators. However, CoCoA binds directly to the bHLH–PAS domains of both AHR and ARNT and acts as a primary coactivator for the AHRC. CoCoA also interacts with the carboxy-terminal transactivation domain of AHR but not with that of ARNT [20]. CoCoA interacts with the same interface of the ARNT PASB domain as TRIP230 [19].

GAC63 is a coactivator for NRs. Like CoCoA it binds to the bHLH–PAS domain of p160 coactivators. Although GAC63 can interact with the ligand binding domain of some NRs directly, its coactivator function depends on the presence of a p160 coactivator, and thus like CoCoA it acts as a secondary coactivator for NR-mediated gene transcription. However, similar to CoCoA, GAC63 acts as a primary coactivator on AHRC-mediated transcription (i.e., its coactivator function is independent of the presence of p160 coactivators or other coactivators) and interacts with AHR through the latter's bHLH–PAS domain [21].

Several interesting matters emerge from the studies of TRIP230, CoCoA, and GAC63. (i) The transcription activation domains (TADs) of both AHR and ARNT were previously localized to the carboxy-terminal regions of both proteins in a conventional fashion by testing segments of the proteins for their ability to stimulate expression of a reporter gene in transient transfection assays. For example, in the experiments of Jain et al. [22], chimeras of a series of AHR and ARNT deletion mutants with the DNA binding region of the yeast Gal4 protein were cotransfected into cells along with a reporter gene under the control of an enhancer element recognized by Gal4, and only the carboxy-terminal regions of AHR and ARNT stimulated expression of the reporter gene. However, the coactivator functions of TRIP230, CoCoA, and GAC63 are all mediated through their interactions with the bHLH–PAS domains of ARNT and AHR (although CoCoA also binds the carboxy-terminal TAD of AHR). Furthermore, previous studies demonstrate that the interaction domain in ARNT for Ncoa1 is represented by helix 2 and is thus also located in the former's bHLH–PAS region [17]. Thus, although they are inactive in the conventional transcriptional activator domain assay, the bHLH–PAS domains of AHR and ARNT nevertheless appear to act as docking sites for coactivators and therefore could be considered as activation domains. In this regard, it is of interest that whereas the carboxy-terminal TAD of AHR is required for AHRC-dependent activation of CYP1A1 transcription, that of ARNT is largely dispensable [23]. (ii) TRIP230 and CoCoA both bind to the opposite side of the PASB domain required for interaction with HIF-2α (and presumably also AHR), thus allowing for simultaneous interaction of ARNT with one of these coactivators and a dimerization partner. Since TRIP230 and CoCoA bind to the same interface of ARNT's PASB domain, they probably mutually exclude each other's binding to ARNT. (iii) The AHRC and NRs apparently utilize these coactivators in a different fashion. Thus, CoCoA and GAC63 act as primary coactivators for the AHRC, but secondary coactivators for NRs.

The crosstalk between ERα and AHRC has been the subject of intense study. Most studies have focused on the effect of the AHRC on ERα and ERβ activity, motivated to a considerable degree by the extensive observations that TCDD and other ligands for the AHR negatively affect ER function. These studies will not be discussed here, but the focus will be on the reciprocal effect of ERα on AHRC function (and particularly on AHRC transcriptional activation). However, since this area has been reviewed in depth recently [24], and is also discussed in another chapter in this book, it is only briefly discussed here. A common observation is that simultaneous treatment with an AHR agonist and estradiol induces recruitment of ERα to the CYP1A1 gene (in Hepa-1 mouse cells and in the MCF-7 and T47D human breast cancer cell lines). In some studies, treatment with an AHR agonist on its own induced this recruitment. Contradictory results have been obtained concerning the effect of estradiol on the function of the AHRC. One group reported that estradiol did not affect the induction of CYP1A1 by an AHR agonist, while another group reported that estradiol inhibited it. Treatment of cells with an siRNA to ERα was reported not to affect induction of CYP1A1 MCF-7 cells, but to reduce it in HC11 mouse mammary cells. Furthermore, induction of Cyp1a1 by an AHR ligand was diminished in the liver of ERα knockout mice. Thus, ERα appeared to act as a repressor of AHRC function in some studies, to act as an enhancer of AHRC activity in others, and to behave neutrally in others [25–27]. These differential effects of ERα may depend on cell- and organism-specific differences. ERα interacts with both AHR and ARNT although in each case the interaction may be indirect [26, 28, 29]. Further studies are needed to define the role of ERα in AHRC-mediated transcription. Since ERα is expressed in only certain tissues, it could play a role in mediating gender- and tissue-specific effects on AHRC activity. BRCA1 is another protein having sex-associated activities that acts as a coactivator for the AHRC. BRAC1's effect is probably mediated by its interaction with ARNT [30].

It should be noted that the carboxy-terminal transactivation domains of the mouse and human AHR differ considerably in sequence and therefore could potentially recruit different coactivators [31]. However, such differences have not yet been observed.

6.2 KINETICS AND TOPOLOGY OF COACTIVATOR RECRUITMENTS TO THE CYP1A1 GENE

In some experimental systems, recruitments of AHR, pol II, and coactivator proteins at the enhancer of the CYP1A1 gene have been distinguished from recruitments at the promoter. The ability to demonstrate distinct recruitments to the enhancer and promoter probably depends on the extent of protein–protein cross-linking generated under the specific condition used in the ChIP assay. Since coactivators are probably present in multiprotein complexes that loop between the enhancer and the promoter, the localization of a coactivator to the enhancer rather than the promoter probably reflects its closer proximity to the enhancer in the multisubunit complex. In this regard, several studies have demonstrated that Ncoa1, Ncoa2, Ncoa3, p300, BRG1, Med1, TRIP230, and ERα are all recruited to the enhancer after dioxin treatment, while, as expected, the TATA binding protein (TBP) and pol II are recruited to the promoter [15–18, 32]. In some studies, the recruitment of coactivators is shown to occur in a defined sequence to the CYP1A1 gene. Thus, Matthews et al. [33] found that after dioxin treatment of MCF-7 cells, the recruitment of AHR and ARNT was followed by that of Ncoa2 and finally CBP. However, the AHR, Ncoa3, p300, and Med1 were recruited with indistinguishable kinetics to the Cyp1a1 enhancer in Hepa-1 cells [15]. Wang et al. [15] performed a ChIP reimmunoprecipitation experiment using sequential treatments with p300 and Med1 antibodies and concluded that these two coactivators can occupy the same enhancer element. However, Ohtake et al. [29] isolated AHR-associated proteins from human HeLa cells and showed that p300 and Med1 reside in different AHR-interacting multiprotein complexes. Thus, the observations of Wang et al. [15] may result from different multisubunit complexes residing on different XREs within a particular enhancer element at any given time. These observations point to a need for caution in interpreting the results of ChIP reimmunoprecipitation experiments. siRNA experiments indicate that the recruitments of p300 and BRG1 to the enhancer of the *CYP1A1* gene in MCF-7 cells occur independently of each other after TCDD treatment. Surprisingly, these experiments suggested that the HAT coactivator Ncoa2 and the ATPase-dependent histone modifying subunit Brahma/BRG1 are recruited together [16].

Several investigations have reported a cycling on and off of AHR and coactivators in the *CYP1A1* gene after AHR agonist treatment of cells, with a periodicity of 60–90 min [27, 34, 35], while other investigations have not confirmed these reports [15, 33]. An explanation for these differing observations has not yet been forthcoming, but they may be related to the differences in AHR agonist or cell type used in the studies.

Fluorescence recovery and photobleaching (FRAP) analysis has demonstrated that several (but not all) transcription factors in mammalian cells bind to their recognition sequences in chromatin only transiently (in some cases, in concert with associated coactivator proteins), with residence times varying from a few milliseconds to a few minutes [36]. The relationship between these transient associations and the cyclical binding observed by ChIP analysis is not clear. Such FRAP analysis has not been reported for the AHRC.

An important observation made with ChIP experiments is that ligand-dependent recruitment of AHR to the enhancer of the *CYP1A1* gene does not necessarily result in induction of CYP1A1 mRNA [37, 38]. This is consistent with developing data from ChIP-on-chip and ChIP-seq experiments, indicating that transcription factor binding to chromatin may not necessarily lead to activation of transcription of a neighboring gene. In one case, the mechanism responsible for the inability of AHR binding to elicit transcriptional activation has been determined. In this study, human HepG2 cells were found to be noninducible for CYP1B1 by the AHR agonist 2,3,7,8-tetrachlorodibenzo-*p*-dioxin (TCDD). Like the *CYP1A1* gene, the *CYP1B1* gene possesses an enhancer region containing several XREs approximately 1 kb upstream of its transcriptional start site. TCDD treatment induced recruitment of AHR to the enhancer but failed to induce recruitment of the TBP or RNA polymerase II to the promoter of *CYP1B1*. The lack of recruitment to the promoter was shown to be due to complete methylation of all the cytosine residues in the CpG dinucleotides residing in the "CpG island" located over the promoter. Presumably, this configuration of the promoter precludes recruitment of TBP and pol II. In contrast to the promoter, the enhancer of the *CYP1B1* gene in HepG2 cells is only partially methylated. In particular, cytosines in the CpG dinucleotides located in two XRE core sequences (5′-T/GNGCGTG-3′) within the enhancer are only about 30% methylated, thus permitting recruitment of AHR (whose binding is prevented by methylation of this cytosine) [39]. Most interestingly, in addition to AHR, TCDD treatment leads to recruitment of the HAT coactivators p300 and PCAF (p300/CBP-associated factor) to the *CYP1B1* enhancer in HepG2 cells. Since TBP and pol II (and presumably the general transcription factors) are not recruited to the *CYP1B1* promoter in these cells, recruitment of p300 and PCAF does not require their tethering to the promoter. Tethering at the enhancer is sufficient. Thus, although the multiunit coactivator complex containing p300 and PCAF is envisaged as bridging the gap between the enhancer and the promoter, and chromatin is envisaged as forming a loop between the enhancer and the promoter, stable association of coactivators with the enhancer can apparently occur in the absence of this looping process [40].

6.3 ROLES OF COACTIVATORS IN TRANSCRIPTIONAL ACTIVATION OF CYP1A1

AHR agonist treatment was found to lead to increases in acetylation of histone 3 at lysine 9 (H3K9), acetylations of H3K14, H4K5, and H4K16, trimethylation of histone 3 at lysine 4 (H3K4), phosphorylation of histone 3 at serine 10 (H3S10), and a decrease in dimethylation of histone 3 at lysine 4 (H3K4) at the *CYP1A1* gene [38, 40–42]. These changes are all associated with actively transcribed genes. Interestingly, Schnekenburger et al. [41] found that whereas the changes in acetylation at H3K9 and H4K16 and the changes in methylation at H3K4 occurred primarily at the promoter of the *Cyp1a1* gene in Hepa-1 cells, the changes in H4K16 acetylation and H3S10 phosphorylation occurred primarily at the enhancer. This suggests that these changes may affect different components of the transcriptional apparatus. p300 was shown to be required for TCDD induction of CYP1B1 in MCF-7 cells and to be required for the TCDD-induced acetylations of H3K9, H3K14, and trimethylation at H3K4 at the *CYP1B1* gene [40]. These acetylations are known to be catalyzed by p300 (as well as certain other HAT coactivators [11]) and therefore may represent direct targets of p300 in the *CYP1B1* gene. Besides these observations, little progress has been made in determining the roles of specific coactivators in covalent modifications of histones at AHR-responsive genes.

6.4 ROLE OF REPRESSORS

In the absence of an AHR agonist, histone deacetylase 1 (HDAC1) and DNA methyltransferase 1 (DNMT1) were found to be associated with the *Cyp1a1* promoter in Hepa-1 cells, and they appear to repress transcription of the gene. AHR agonist treatment leads to release of these proteins from the promoter. This suggests that a repressor complex containing these two proteins limits the transcription of the *Cyp1a1* gene under uninduced conditions [41].

6.5 FUTURE AREAS FOR RESEARCH

(i) Over 300 coactivators have been described for the 50 or so mammalian NRs [10]. Although many function only with certain members of the NR superfamily, it is likely that many more coactivators for the AHRC are waiting to be identified.

(ii) The expression of some coactivators can be modified by hormones, such as glucocorticoids [43]. However, a more prominent role in modifying coactivator activity is most likely played by posttranslational modifications of coactivator proteins, particularly phosphorylation [9, 10, 44], and these modifications can be triggered by exogenous agents and exhibit tissue specificity. Furthermore, many coactivators are overexpressed in particular types of cancer [45], and such overexpression could explain, at least in part, the elevated expression of CYP1A1 (and CYP1B1) that has been observed in a variety of cancer types [46].

(iii) Tissue-specific differences in expression or activity of coactivators may help explain the differential degree of induction of CYP1A1 (and other AHRC-responsive genes) observed in different organs and tissues and the different spectra of genes induced by the AHRC in different tissues [47–49].

(iv) ChIP-on-chip and microarray experiments have been interpreted as providing evidence that unliganded AHR can activate the transcription of certain genes, perhaps mediated via nucleocytoplasmic shuttling of the receptor [47, 49, 50]. It will be interesting to ascertain whether the same or different coactivators are used by the liganded and unliganded AHR. There is some evidence that AHR can modify gene expression in the absence of ARNT. Which coactivators are used in this case?

(v) There is some evidence that the AHRC can bind and activate transcription from the XRE-II sequence, which is different from the conventional XRE [51, 52]. Perhaps the AHRC utilizes different coactivators when driving transcription from the XRE-II rather than the XRE.

(vi) There is increasing evidence for the existence of selective aryl hydrocarbon receptor modulators (SAhRMs) that activate or repress only a portion of AHRC-responsive genes [53, 54]. The basis of the "abnormal" pattern of gene expression elicited by these compounds may reside in their differential ability to recruit coactivators. This is illustrated by the AHR ligand 3,3'-diindolylmethane that poorly induces CYP1A1 in MCF-7 cells and induces the recruitment of AHR and some coactivators but not others (e.g., CBP) to the *CYP1A1* gene [34]. This is a potentially fruitful area of research as SAhRMs may represent novel agents for the chemoprevention or even treatment of cancer.

ACKNOWLEDGMENTS

The relevant research in the author's laboratory is supported by NIH grant CA RO1CA28868. The author thanks Dr. Feng Wang for critically reading the manuscript and Mike Thompson for advice.

REFERENCES

1. Kress, S., Reichert, J., and Schwarz, M. (1998). Functional analysis of the human cytochrome P4501A1 (CYP1A1) gene enhancer. *European Journal of Biochemistry*, 258, 803–812.
2. Tsuchiya, Y., Nakajima, M., and Yokoi, T. (2003). Critical enhancer region to which AhR/ARNT and Sp1 bind in the human CYP1B1 gene. *Journal of Biochemistry*, 133, 583–592.
3. Okino, S. T. and Whitlock, J. P., Jr. (1995). Dioxin induces localized, graded changes in chromatin structure: implications for Cyp1A1 gene transcription. *Molecular and Cellular Biology*, 15, 3714–3721.
4. Kim, Y. J., Bjorklund, S., Li, Y., Sayre, M. H., and Kornberg, R. D. (1994). A multiprotein mediator of transcriptional activation and its interaction with the C-terminal repeat domain of RNA polymerase II. *Cell*, 77, 599–608.
5. Hankinson, O. (2005). Role of coactivators in transcriptional activation by the aryl hydrocarbon receptor. *Archives of Biochemistry and Biophysics*, 433, 379–386.
6. Goll, M. G. and Bestor, T. H. (2002). Histone modification and replacement in chromatin activation. *Genes & Development*, 16, 1739–1742.
7. Kraus, W. L. and Wong, J. (2002). Nuclear receptor-dependent transcription with chromatin. Is it all about enzymes? *European Journal of Biochemistry*, 269, 2275–2283.
8. Felsenfeld, G. and Groudine, M. (2003). Controlling the double helix. *Nature*, 421, 448–453.
9. Lonard, D. M. and O'Malley, B. W. (2005). Expanding functional diversity of the coactivators. *Trends in Biochemical Sciences*, 30, 126–132.
10. Han, S. J., Lonard, D. M., and O'Malley, B. W. (2009). Multi-modulation of nuclear receptor coactivators through posttranslational modifications. *Trends in Endocrinology and Metabolism*, 20, 8–15.

11. Kouzarides, T. (2007). Chromatin modifications and their function. *Cell, 128,* 693–705.
12. Fischle, W., Wang, Y., and Allis, C. D. (2003). Histone and chromatin cross-talk. *Current Opinion in Cell Biology, 15,* 172–183.
13. Hake, S. B., Xiao, A., and Allis, C. D. (2004). Linking the epigenetic 'language' of covalent histone modifications to cancer. *British Journal of Cancer, 90,* 761–769.
14. Turner, B. M. (2002). Cellular memory and the histone code. *Cell, 111,* 285–291.
15. Wang, S., Ge, K., Roeder, R. G., and Hankinson, O. (2004). Role of mediator in transcriptional activation by the aryl hydrocarbon receptor. *Journal of Biological Chemistry, 279,* 13593–13600.
16. Taylor, R. T., Wang, F., Hsu, E. L., and Hankinson, O. (2009). Roles of coactivator proteins in dioxin induction of CYP1A1 and CYP1B1 in human breast cancer cells. *Toxicological Sciences, 107,* 1–8.
17. Beischlag, T. V., Wang, S., Rose, D. W., Torchia, J., Reisz-Porszasz, S., Muhammad, K., Nelson, W. E., Probst, M. R., Rosenfeld, M. G., and Hankinson, O. (2002). Recruitment of the NCoA/SRC-1/p160 family of transcriptional coactivators by the aryl hydrocarbon receptor/aryl hydrocarbon receptor nuclear translocator complex. *Molecular and Cellular Biology, 22,* 4319–4333.
18. Beischlag, T. V., Taylor, R. T., Rose, D. W., Yoon, D., Chen, Y., Lee, W. H., Rosenfeld, M. G., and Hankinson, O. (2004). Recruitment of thyroid hormone receptor/retinoblastoma-interacting protein 230 by the aryl hydrocarbon receptor nuclear translocator is required for the transcriptional response to both dioxin and hypoxia. *Journal of Biological Chemistry, 279,* 54620–54628.
19. Partch, C. L., Card, P. B., Amezcua, C. A., and Gardner, K. H. (2009). Molecular basis of coiled coil coactivator recruitment by the aryl hydrocarbon receptor nuclear translocator (ARNT). *Journal of Biological Chemistry, 284,* 15184–15192.
20. Kim, J. H. and Stallcup, M. R. (2004). Role of the coiled-coil coactivator (CoCoA) in aryl hydrocarbon receptor-mediated transcription. *Journal of Biological Chemistry, 279,* 49842–49848.
21. Chen, Y. H., Beischlag, T. V., Kim, J. H., Perdew, G. H., and Stallcup, M. R. (2006). Role of GAC63 in transcriptional activation mediated by the aryl hydrocarbon receptor. *Journal of Biological Chemistry, 281,* 12242–12247.
22. Jain, S., Dolwick, K. M., Schmidt, J. V., and Bradfield, C. A. (1994). Potent transactivation domains of the Ah receptor and the Ah receptor nuclear translocator map to their carboxyl termini. *Journal of Biological Chemistry, 269,* 31518–31524.
23. Reisz-Porszasz, S., Probst, M. R., Fukunaga, B. N., and Hankinson, O. (1994). Identification of functional domains of the aryl hydrocarbon receptor nuclear translocator protein (ARNT). *Molecular and Cellular Biology, 14,* 6075–6086.
24. Beischlag, T. V., Luis Morales, J., Hollingshead, B. D., and Perdew, G. H. (2008). The aryl hydrocarbon receptor complex and the control of gene expression. *Critical Reviews in Eukaryotic Gene Expression, 18,* 207–250.
25. Matthews, J., Wihlen, B., Heldring, N., MacPherson, L., Helguero, L., Treuter, E., Haldosen, L. A., and Gustafsson, J. A. (2007). Co-planar 3,3′,4,4′,5-pentachlorinated biphenyl and non-co-planar 2,2′,4,6,6′-pentachlorinated biphenyl differentially induce recruitment of oestrogen receptor alpha to aryl hydrocarbon receptor target genes. *Biochemical Journal, 406,* 343–353.
26. Beischlag, T. V. and Perdew, G. H. (2005). ER alpha–AHR–ARNT protein–protein interactions mediate estradiol-dependent transrepression of dioxin-inducible gene transcription. *Journal of Biological Chemistry, 280,* 21607–21611.
27. Wihlen, B., Ahmed, S., Inzunza, J., and Matthews, J. (2009). Estrogen receptor subtype- and promoter-specific modulation of aryl hydrocarbon receptor-dependent transcription. *Molecular Cancer Research, 7,* 977–986.
28. Brunnberg, S., Pettersson, K., Rydin, E., Matthews, J., Hanberg, A., and Pongratz, I. (2003). The basic helix–loop–helix–PAS protein ARNT functions as a potent coactivator of estrogen receptor-dependent transcription. *Proceedings of the National Academy of Sciences of the United States of America, 100,* 6517–6522.
29. Ohtake, F., Baba, A., Takada, I., Okada, M., Iwasaki, K., Miki, H., Takahashi, S., Kouzmenko, A., Nohara, K., Chiba, T., Fujii-Kuriyama, Y., and Kato, S. (2007). Dioxin receptor is a ligand-dependent E3 ubiquitin ligase. *Nature, 446,* 562–566.
30. Kang, H. J., Kim, H. J., Kim, S. K., Barouki, R., Cho, C. H., Khanna, K. K., Rosen, E. M., and Bae, I. (2006). BRCA1 modulates xenobiotic stress-inducible gene expression by interacting with ARNT in human breast cancer cells. *Journal of Biological Chemistry, 281,* 14654–14662.
31. Flaveny, C., Reen, R. K., Kusnadi, A., and Perdew, G. H. (2008). The mouse and human Ah receptor differ in recognition of LXXLL motifs. *Archives of Biochemistry and Biophysics, 471,* 215–223.
32. Wang, S. and Hankinson, O. (2002). Functional involvement of the Brahma/SWI2-related gene 1 protein in cytochrome P4501A1 transcription mediated by the aryl hydrocarbon receptor complex. *Journal of Biological Chemistry, 277,* 11821–11827.
33. Matthews, J., Wihlen, B., Thomsen, J., and Gustafsson, J. A. (2005). Aryl hydrocarbon receptor-mediated transcription: ligand-dependent recruitment of estrogen receptor alpha to 2,3,7,8-tetrachlorodibenzo-p-dioxin-responsive promoters. *Molecular and Cellular Biology, 25,* 5317–5328.
34. Hestermann, E. V. and Brown, M. (2003). Agonist and chemopreventative ligands induce differential transcriptional cofactor recruitment by aryl hydrocarbon receptor. *Molecular and Cellular Biology, 23,* 7920–7925.
35. Pansoy, A., Ahmed, S., Valen, E., Sandelin, A., and Matthews, J. (2010). 3-methylcholanthrene induces differential recruitment of aryl hydrocarbon receptor to human promoters. *Toxicological Sciences, 117,* 90–100.
36. Hager, G. L., McNally, J. G., and Misteli, T. (2009). Transcription dynamics. *Molecular Cell, 35,* 741–753.
37. Yang, X., Solomon, S., Fraser, L. R., Trombino, A. F., Liu, D., Sonenshein, G. E., Hestermann, E. V., and Sherr, D. H. (2008).

Constitutive regulation of CYP1B1 by the aryl hydrocarbon receptor (AhR) in pre-malignant and malignant mammary tissue. *Journal of Cellular Biochemistry, 104*, 402–417.

38. Beedanagari, S. R., Taylor, R. T., and Hankinson, O. (2010). Differential regulation of the dioxin-induced Cyp1a1 and Cyp1b1 genes in mouse hepatoma and fibroblast cell lines. *Toxicology Letters, 194*, 26–33.

39. Shen, E. S. and Whitlock, J. P., Jr. (1989). The potential role of DNA methylation in the response to 2,3,7, 8-tetrachlorodibenzo-*p*-dioxin. *Journal of Biological Chemistry, 264*, 17754–17758.

40. Beedanagari, S. R., Taylor, R. T., Bui, P., Wang, F., Nickerson, -D. W., and Hankinson, O. (2010). Role of epigenetic mechanisms in differential regulation of the dioxin-inducible human *CYP1A1* and *CYP1B1* genes. *Molecular Pharmacology, 78*, 608–616.

41. Schnekenburger, M., Talaska, G., and Puga, A. (2007). Chromium cross-links histone deacetylase 1–DNA methyltransferase 1 complexes to chromatin, inhibiting histone-remodeling marks critical for transcriptional activation. *Molecular and Cellular Biology, 27*, 7089–7101.

42. Tian, Y. (2009). Ah receptor and NF-κB interplay on the stage of epigenome. *Biochemical Pharmacology, 77*, 670–680.

43. Dvorak, Z. and Pavek, P. (2010). Regulation of drug-metabolizing cytochrome P450 enzymes by glucocorticoids. *Drug Metabolism Reviews, 42*, 621–635.

44. Weigel, N. L. and Moore, N. L. (2007). Kinases and protein phosphorylation as regulators of steroid hormone action. *Nuclear Receptor Signaling, 5*, e005.

45. O'Malley, B. W. and Kumar, R. (2009). Nuclear receptor coregulators in cancer biology. *Cancer Research, 69*, 8217–8222.

46. Androutsopoulos, V. P., Tsatsakis, A. M., and Spandidos, D. A. (2009). Cytochrome P450 CYP1A1: wider roles in cancer progression and prevention. *BMC Cancer, 9*, 187.

47. Boutros, P. C., Bielefeld, K. A., Pohjanvirta, R., and Harper, P. A. (2009). Dioxin-dependent and dioxin-independent gene batteries: comparison of liver and kidney in AHR-null mice. *Toxicological Sciences, 112*, 245–256.

48. Kim, S., Dere, E., Burgoon, L. D., Chang, C. C., and Zacharewski, T. R. (2009). Comparative analysis of AhR-mediated TCDD-elicited gene expression in human liver adult stem cells. *Toxicological Sciences, 112*, 229–244.

49. Frericks, M., Meissner, M., and Esser, C. (2007). Microarray analysis of the AHR system: tissue-specific flexibility in signal and target genes. *Toxicology and Applied Pharmacology, 220*, 320–332.

50. Sartor, M. A., Schnekenburger, M., Marlowe, J. L., Reichard, J. F., Wang, Y., Fan, Y., Ma, C., Karyala, S., Halbleib, D., Liu, X., Medvedovic, M., and Puga, A. (2009). Genomewide analysis of aryl hydrocarbon receptor binding targets reveals an extensive array of gene clusters that control morphogenetic and developmental programs. *Environmental Health Perspectives, 117*, 1139–1146.

51. Sogawa, K., Numayama-Tsuruta, K., Takahashi, T., Matsushita, N., Miura, C., Nikawa, J., Gotoh, O., Kikuchi, Y., and Fujii-Kuriyama, Y. (2004). A novel induction mechanism of the rat CYP1A2 gene mediated by Ah receptor–Arnt heterodimer. *Biochemical and Biophysical Research Communications, 318*, 746–755.

52. Boutros, P. C., Moffat, I. D., Franc, M. A., Tijet, N., Tuomisto, J., Pohjanvirta, R., and Okey, A. B. (2004). Dioxin-responsive AHRE-II gene battery: identification by phylogenetic footprinting. *Biochemical and Biophysical Research Communications, 321*, 707–715.

53. Fritz, W. A., Lin, T. M., Safe, S., Moore, R. W., and Peterson, R. E. (2009). The selective aryl hydrocarbon receptor modulator 6-methyl-1,3,8-trichlorodibenzofuran inhibits prostate tumor metastasis in TRAMP mice. *Biochemical Pharmacology, 77*, 1151–1160.

54. Murray, I. A., Krishnegowda, G., DiNatale, B. C., Flaveny, C., Chiaro, C., Lin, J. M., Sharma, A. K., Amin, S., and Perdew, G. H. (2010). Development of a selective modulator of aryl hydrocarbon (Ah) receptor activity that exhibits anti-inflammatory properties. *Chemical Research in Toxicology, 23*, 955–966.

7

REGULATION OF AHR ACTIVITY BY THE AHR REPRESSOR (AHRR)

YOSHIAKI FUJII-KURIYAMA AND KANAME KAWAJIRI

7.1 INTRODUCTION

Upon activation by an AHR ligand or other cellular stimuli, AHR translocates from the cytoplasm into the nucleus where it heterodimerizes with ARNT and enhances the expression of many genes related to toxicology and physiology. AHR signaling can be downregulated by at least two different mechanisms [1, 2]. After entering the nucleus, activated AHR is degraded by the ubiquitin/proteasome pathway either in the nucleus or in the cytoplasm after being exported from the nucleus [3, 4]. Self-ubiquitination of AHR also reportedly occurs in the nucleus [5]. On the other hand, as other PAS signaling pathways, such as hypoxic signaling [6, 7] and diurnal rhythmicity [8], the AHR signaling pathway also mediates negative feedback regulation. The AHR repressor (AHRR), whose expression is upregulated by activated AHR, inhibits the transcriptional activity of AHR by competing with AHR to form a heterodimer with ARNT and by binding to the XRE sequence with the AHR/ARNT heterodimer [9]. AHRR is structurally highly similar to AHR in the N-terminal bHLH region, which allows AHRR to dimerize with ARNT and bind DNA [9], and has a potent transcription repression domain at its C-terminus [10] (Fig. 7.1).

The downregulation of AHR signal transduction by these two independent mechanisms indicates that it may be essential to prevent the overactivation of the transcriptional activity of AHR. Since the ubiquitination and subsequent degradation of AHR are discussed elsewhere in this book, we focus here on the role of AHRR in downregulating the activity of AHR.

7.2 STRUCTURE AND EVOLUTION OF AHRR

AHRR was first isolated as an AHR-related cDNA clone from a mouse intestine cDNA library [9]. Sequence analysis of the cDNA clone revealed that the encoded primary structure consists of 701 amino acids and that the N-terminal third of the protein (about 275 amino acids), which contains the bHLH domain and a portion of the PAS domain, shares high similarity with that of AHR; however, aside from this area of homology, these two proteins are significantly different [9] as shown in Fig. 7.1. Since the bHLH and PAS domains of AHR are known to mediate dimerization with ARNT and DNA recognition (basic region), it was reasonable to hypothesize that, similar to AHR, AHRR forms a heterodimer with ARNT that recognizes and binds the XRE sequence [9]. The absence of a PAS-B region, which has ligand binding activity, suggests that AHRR functions as a transcription factor independent of ligand binding. Although experimental data support this hypothesis, transient AHRR transfection and expression in cultured cells showed that AHRR had no transcriptional activity toward an XRE-driven reporter gene. However, when AHRR was cotransfected with an AHR expression vector, AHRR potently repressed the AHR-mediated enhanced expression of this reporter gene. Thus, the expressed protein was identified as the AHR repressor [9].

AHRR orthologs have been identified and characterized in several mammalian species, including mice [9], humans [11], rats [12], amphibians [13], and many types of bony fish, including tomcod [14], killifish [15], and zebrafish [16]. A phylogenetic tree comparing the AHRR and AHR from various animal species is shown in Fig. 7.2. Zebrafish have

The AH Receptor in Biology and Toxicology, First Edition. Edited by Raimo Pohjanvirta.
© 2012 John Wiley & Sons, Inc. Published 2012 by John Wiley & Sons, Inc.

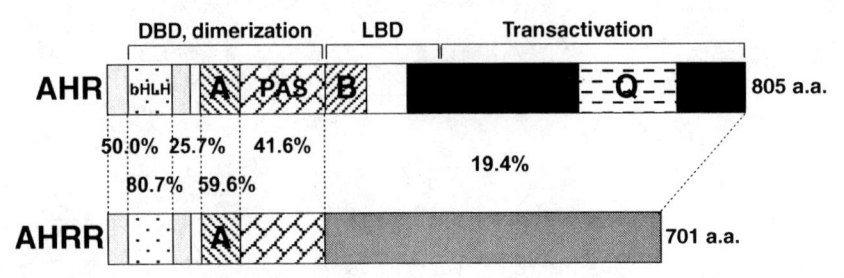

FIGURE 7.1 Domain structures of mouse AHR and AHRR. The percentages of identity between AHR and AHRR are indicated.

two AHRR paralogs that are considered to be co-orthologs of the mammalian AHRR based on phylogenetic analyses and conserved synteny [16]. Both AHRRs are induced similarly by TCDD, but experiments using antisense morpholino oligonucleotides (MOs) in zebrafish embryos and a zebrafish liver cell line (ZF-L) showed that these paralogs are functionally different [17]. Knockdown of AHRRb enhanced the TCDD-mediated induction of CYP1As, CYP1B1, and CYP1C1 and decreased the constitutive expression of Sox9b. Embryos that were microinjected with AHRRa MOs and treated with dimethyl sulfoxide (DMSO, a solvent) had developmental phenotypes that resembled those of TCDD-treated embryos, such as pericardial edema and lower jaw malformation, without affecting the constitutive expression of Sox9b. In contrast, these developmental phenotypes were not observed in DMSO-treated AHRRb morphants [17]. There have been no reports of AHRR in earlier diverging vertebrates or invertebrates. The genomes of earlier diverging animal species must be characterized before we can determine how the *AHRR* gene is distributed in the animal kingdom.

7.3 REGULATORY MECHANISMS OF AHRR EXPRESSION

In general, AHRR mRNA is constitutively expressed in various tissues in untreated animals, including mice [9, 18], rats [12], and humans [19], albeit at relatively low concentrations. However, upon AHR ligand stimulation, AHRR expression is markedly induced in a tissue-specific manner [9, 12, 18]. The isolated mouse *Ahrr* gene has no TATA or TATA-like sequence in the proximal promoter. However, as expected from the inducible expression of AHRR in response to an AHR ligand, the *Ahrr* gene has at least three copies of the XRE sequence in the promoter spanning 1.4 kb upstream of the transcription start site. All these XRE sequences are required for AHR-dependent transactivation, as assessed by a transient DNA transfection assay. Furthermore, these sites are conserved at the corresponding positions in the human *AHRR* gene located on chromosome 5 (5p15.3), a region syntenic to mouse chromosome 13 (13C2) [20] (Fig. 7.3). In the human gene, there are four XRE sequences in the first intron, and the XRE sequence at the +17,505 position was shown to function in the AHR-dependent signaling pathway [21]. The promoters of the human and mouse *AHRR* genes contain a region approximately −50 bp upstream of the transcription start site that contains three GC box sequences and one NF-κB site that partially overlap the XRE sequence (XRE1) at the −60 bp position. These GC box sequences are recognized and bound by Sp1 or Sp3 and are involved in constitutive expression of the *AHRR* gene [20]. Most importantly, these GC box sequences are required for both optimal ligand-induced

FIGURE 7.2 A phylogenetic tree of the AHR and AHRR proteins. Courtesy of Dr. O. Gotoh, Kyoto University.

```
     h CTCAGCTGCG CCCACCCCAC CTTGGGAGCT GTGTCTGGAA AAGAGACTTT TCCTGAGCCG
        **  *      *    *         *          *           *           *
-1400 m TCTAGGCATA GTTTACATTA AAAAAAAAAC AGATTAAAAT ACAATTAATC ATTTTTTTTA

     h ACAGCAGACA GCGAGTGCAA AGGCCCCAAG GCCCGCACGC CTTATTCAGA GAACAGCAGG
        *  *  *    *  *  *  ***   *         *  *****  * *    *    ****
-1340 m AAAAAACTGT GGTAGTGAGC CAGGTGGTGA CGGCACACGC CTTTAATCCC AACACTCAGG
                                                   XRE 3

     h AGGCGCGGGG CGGGTTGGGG GCGTCAGGGA GACGGGGTTT GACTGAGAAC CTGGGGCGC
        ** *      **         **  *       * **         *            *
-1280 m AGACAGGCAG GTGGATCTCT GAGTTCTAGG AAAGCCAGAG TTACACTGAG AAACTCTATC


     h AAT-ATTCTC TATAATTCGC CAA----AAA C------CAT GACT--CCTT TTACATTTTC
        *  ** **   *** ** **  **    ***  *       **    ***  **  *** **
-488  m ACTAATACTT TATGCTTTGC AAAGTATAAA CATACTGTAT CTTTAATCTT TTTTATTCTC

     h GAAACGACCT CATGGGGCAG GCGAGGGGGC ACACGCTCTC CTCACTTTAC CTGTTCGGGT
        ****  **   *   *  ***    *  ***  ***** **   ****    ****  *
-428  m CAAACAACTC TATAGCGCAA GAAAGGAGGA ACACGCTAAG TCTTGGTTAC CTATTCTGAT
                                         XRE 2

     h CTACACCGCC CGGCTCTAGG GCCGGCAGCT CAGCACAGCC GGCCGCGAGC GCTGGGGAGG
        ** **     *   *** *  **   **    **          **   ***   *    *
-368  m CT-CATTCTC TGGCCCCAGG G---GCCA-- -AGGGGCCCC AACCCACAGC TC-CGCAAAA


     h -CAGGGCAGC TGGGGCAGCC GGACCCGAGA CAGCCGGACC CGCGGGCGCC CTCTGCCGCC
         *        ****   **   **  **  *   *        *** ****   ******** *
-148  m TAAATATTGA AAGGGCGCGT GGCAGGCAGG GACAAGGATG TGCCGTCGCC CTCTGCCG-C

     h TCCCGGCCCT GGATGCTG-C GGGGCGGGGC GGGCGGCTCG CGTGCCGGGG TGGGGCAGTC
        *   ****   *  *** *  *  ****** *  ****** *  ***  **** *****    *
-89   m TT--TTCCCT TGCGTCTGCC GCTTCGGGCG GGGCGGTCG CGTGCTGGGG TGGGGCTTTC
                                        XRE 1                NF-κB

     h C--CTGGAG- CCGCCG-CGG GTTCCCGG
        *  **  **   *  *   *   **  **
-31   m CTTCTCTAGT CTGAGGCCAG GTCCCAGA
```

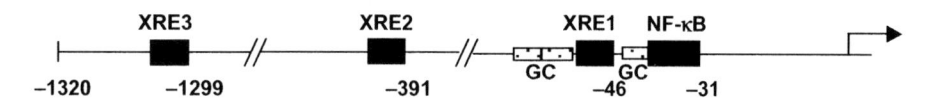

```
        XRE3              XRE2           XRE1    NF-κB
    ┣━━━■━━━━━━━//━━━━━━■━━━//━━━━━━[]■[]■━━━━━━▶
  -1320  -1299         -391           GC  GC
                                     -46  -31
```

FIGURE 7.3 Sequence comparison between the 5'-upstream regions of the human and mouse AHRR genes. The corresponding sequences in the human AHRR gene derived from the NCLB database are shown for comparison. Open boxes indicate the XRE and NF-κB binding sites and underlined sequences indicate the GC box sequences. The conserved nucleotides between human and mouse AHRR sequences are indicated by asterisks. h, human AHRR; m, mouse AHRR. A schematic representation of the promoter sequences of AHRR is also shown.

expression and basal expression of the mouse and human *AHRR* genes. *AHRR* gene expression is also activated by 12-*O*-tetradecanoylphorbol-13-acetate (TPA) through the NF-κB site in the promoter [20]. Interestingly, a kinetic induction of AHR and AHRR was observed in mouse bone marrow-derived macrophages after treating with Pam3CSK4 (Fig. 7.4), a TLR1/2 ligand that activates NF-κB (Mimura, J. and Fujii-Kuriyama, Y., unpublished observation), suggesting that the AHR/AHRR transcription regulation system of macrophages plays important roles in the protection against microbial infections. In this process, inducible AHRR expression depends on the presence of AHR without apparent exogenous AHR ligands. These observations suggest that highly promiscuous transcription factors regulate AHRR expression and that AHR overexpression is prevented by a feedback mechanism involving AHRR.

7.4 MOLECULAR MECHANISMS BY WHICH AHRR INHIBITS AHR ACTIVITY

It is important to elucidate the mechanisms by which AHRR inhibits the activity of AHR in order to fully understand AHR signaling in physiological processes such as cell proliferation [22] and estrous cycle [23]. As expected from the fact that AHRR lacks a PAS-B region, which functions as a ligand binding site for the activation and nuclear translocation of AHR, experiments using [^{3}H]TCDDD found no evidence that AHRR specifically binds to TCDD [15]. Upon synthesis, GFP-AHRR or native AHRR translocates into the nucleus and heterodimerizes with ARNT [9, 24]. AHRR contains both nuclear localization and nuclear export signals that mediate its nucleocytoplasmic shuttling, with equilibrium in favor of nuclear localization [24]. Although the precise

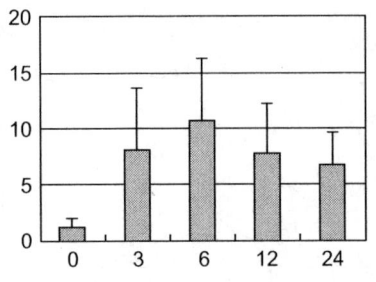

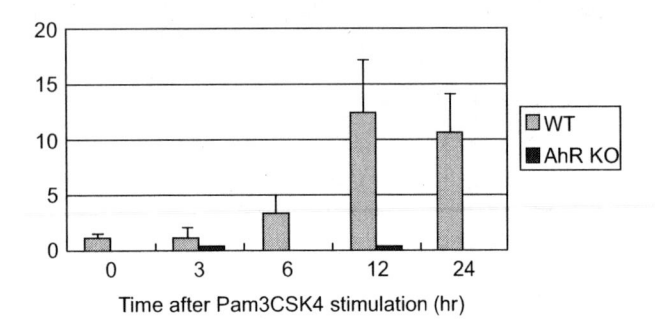

FIGURE 7.4 A kinetic analysis of AHR and AHRR induction in macrophages treated with Pam3CSK4. AHR and AHRR mRNA levels were measured by qRT-PCR in isolated macrophages at various time points after Pam3CSK4 treatment.

inhibitory mechanism of AHRR remains unknown, trichostatin A (TSA) significantly reverses the inhibitory activity of AHRR, suggesting that this process involves histone deacetylase (HDAC) activity. An analysis of the ability of AHRR to inhibit the expression of a reporter gene driven by a thymidine kinase promoter identified a core repressor domain at residues 555–701 whose inhibitory activity is also reversed by TSA treatment. On the other hand, the N-terminal sequence containing amino acids 1–342 shows little to no inhibitory activity in this assay [10]. In contrast, the N-terminal sequence was recently shown to inhibit the transactivation activity of the AHR/ARNT heterodimer [25]. It is not clear whether the inhibitory activity of the N-terminal sequence is TSA sensitive or is due to merely competitive binding of the N-terminal fragment (1–188 amino acids) of AHRR to ARNT with AHR.

To investigate how AHRR exhibits the TSA-sensitive inhibitory activity, a Cyto-Trap yeast two-hybrid screen was conducted with the C-terminal sequence of AHRR as the bait. This screen identified several binding proteins, including ankyrin repeat family A protein 2 (ANKRA2). This putative interaction is particularly interesting because ANKRA2 reportedly interacts with HDAC4 and HDAC5 [10]. Although a sequence comparison of the inhibitory domain of AHRR among various animal species did not reveal significant overall sequence similarity, three short segments that contain the consensus sequence for SUMOylation (small ubiquitin-like modifier) were distinctly conserved as shown in Fig. 7.5 [26]. Many transcription factors are SUMOylated, and in many cases SUMOylation enhances their repressive activity [27, 28]. An *in vivo* SUMOylation assay using COS-7 cells transfected with AHRR, SUMO-1, and Ubc9 expression plasmids showed that all three SUMOylation sites in AHRR are SUMOylated. Furthermore, a mutational analysis of the Lys to Arg residues in the SUMOylation sites revealed that all of the SUMOylations significantly contribute to the repressive activity of AHRR. Furthermore, SUMO modification of AHRR by Ubs9 and the SUMO substrate, SUMO-1, is markedly enhanced in the presence of the partner molecule, ARNT, and vice versa with the SUMOylation of ARNT. The PIAS family proteins, such as PIAS1, PIASxα, and PIASxβ, are known to function as SUMO E3 ligases, but are not required for AHRR or ARNT SUMOylation [26].

Because ARNT and AHRR interact with Ubc9 and the SUMOylation substrate, SUMO-1, it is reasonable to consider that AHRR and ARNT serve as the SUMO E3 ligase for each other. In this context, it is interesting to note that AHR serves as a ligand-dependent E3 ubiquitin ligase for proteasomal degradation of the estrogen receptor and androgen receptor [5]. Intriguingly, however, AHR appears to inhibit the SUMOylation of ARNT [26] because ARNT is not SUMOylated in the presence of AHR. These observations suggest that SUMOylation is intimately associated with the repressive activity of AHRR. We investigated how SUMOylation is associated with the repressive activity of AHRR and found that AHRR must be SUMOylated in order to interact with the corepressors ANKRA2, HDAC4, and HDAC5. Coimmunoprecipitation assays revealed that HDAC5 interacts with AHRR, directly or through ANKRA2, while HDAC4 binds AHRR only via ANKRA2. These interactions are markedly enhanced upon AHRR SUMOylation, while mutating the Lys to Arg residues in all three SUMOylation sites abolishes these interactions, indicating that AHRR/ARNT must be SUMOylated to form the repressor complex that includes ANKRA2, HDAC4, and HDAC5 [26]. Upon synthesis in the cytoplasm, AHRR is transported into the nucleus where it forms a heterodimer with ARNT, resulting in enhancement of mutual SUMOylation of AHRR and ARNT in the heterodimer complex. Thus, the SUMOylated AHRR–ARNT heterodimer becomes competent to recruit the corepressor components ANKRA2, HDAC4, and HDAC5 [26] (Fig. 7.6). However, it is still unknown whether this series of transcription repressor complexes forms at the XRE sequences in the target genes or in the nucleoplasm prior to DNA binding.

7.5 PHYSIOLOGICAL FUNCTIONS AND POLYMORPHISMS OF AHRR

The expression profiles of AHRR mRNA in various tissues and cells will likely provide useful information on the cell-

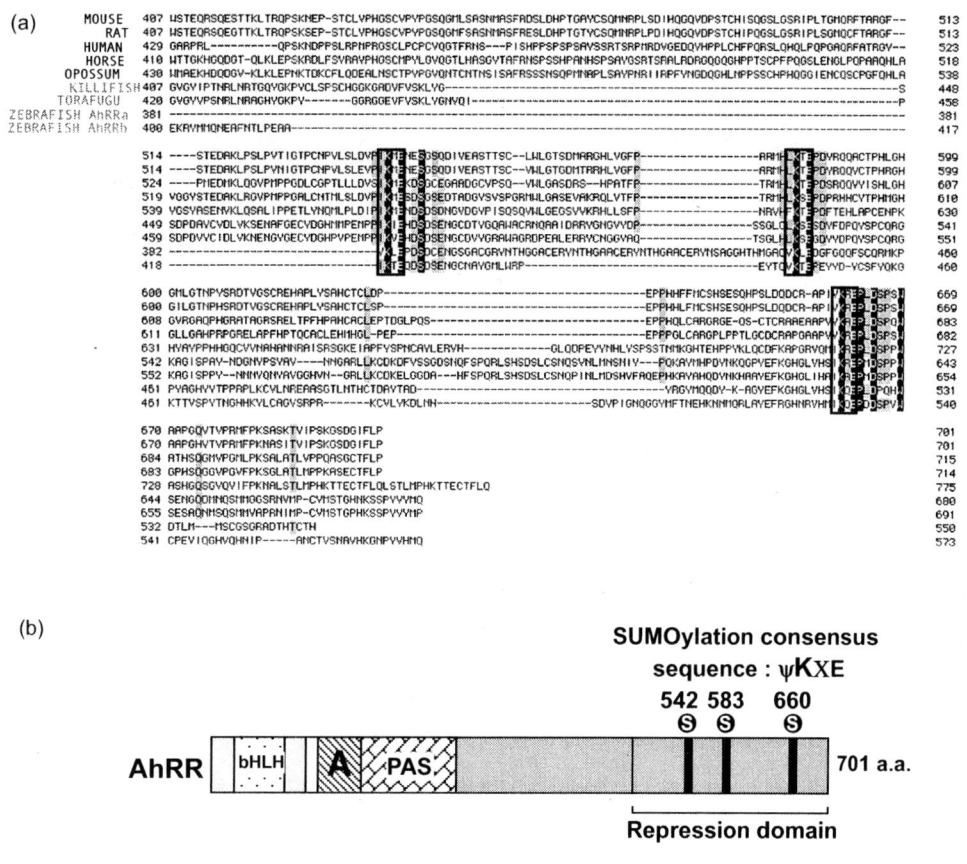

FIGURE 7.5 (a) Conserved SUMOylation sequences in AHRR from various animal species. (b) A schematic representation of mouse full-length AHRR. The characterized domains represented are the basic helix–loop–helix (bHLH), Per–ARNT–Sim (PAS), and repression domain. Three putative SUMOylation sites are located within the repression domain, and the target lysine residues are indicated.

and tissue-specific roles of the AHRR protein. RT-PCR analyses showed that HeLa cells (cervix uterus adenocarcinoma) have the highest AHRR mRNA levels among the seven examined cultured human cell lines and AHRR is relatively abundantly expressed in OMC-3 (ovarian carcinoma), NEC14 (testis embryonal carcinoma), HepG2 (hepatocellular carcinoma), LS-180 (colon carcinoma), HT-1197 (bladder carcinoma), ACHN (renal carcinoma), and A549 (lung carcinoma) cell lines in this order [19]. The AHRR mRNA expression levels may not precisely correlate with the protein levels and are not necessarily inversely related to the inducibility of AHR target genes, including *CYP1A1, 1A2, 1B1*, and *AHRR*, in response to TCDD and 3MC [2, 19]. There may be a translational control of AHRR mRNA in a cell- and tissue-specific manner [2]. HeLa cells are resistant to the induction of AHR target genes by TCDD treatment, but treating with AHRR siRNA or sodium butyrate, an HDAC inhibitor, restores their responsiveness to this induction [21]. In mouse embryonic fibroblasts (MEFs), AHRR and ANKRA2 siRNA treatment significantly enhances constitutive as well as inducible CYP1A1 expression in response to TCDD. These findings indicate that under normal conditions, a silent state of CYP1A1 gene expression is not only due to the absence of transcription activators, but also results from negative regulation by the AHRR/ARNT repressor complex containing ANKRA2, HDAC4, and HDAC5 [10].

In a primary cell culture of isolated human skin fibroblasts, CYP1A1 expression is nonresponsive to TCDD treatment and becomes responsive to dioxin when the cells are treated with TSA or AHRR siRNA [29, 30]. Taken together, these results indicate that in some cultured cells or tissues, AHRR functions as a repressor factor that silences AHR target genes. Consistent with these observations, CYP1A1 is more highly induced in the spleen, stomach, and skin of AHRR-null mice than wild-type mice in response to 3MC, while inducible CYP1A1 expression is not significantly different in other tissues, such as the liver, heart, and lung, between AHRR-null and wild-type mice [31]. The superinduction of CYP1A1 is not observed in all tissues in AHRR-null mice and does not appear to correlate with constitutive AHRR mRNA expression in the tissues of wild-type

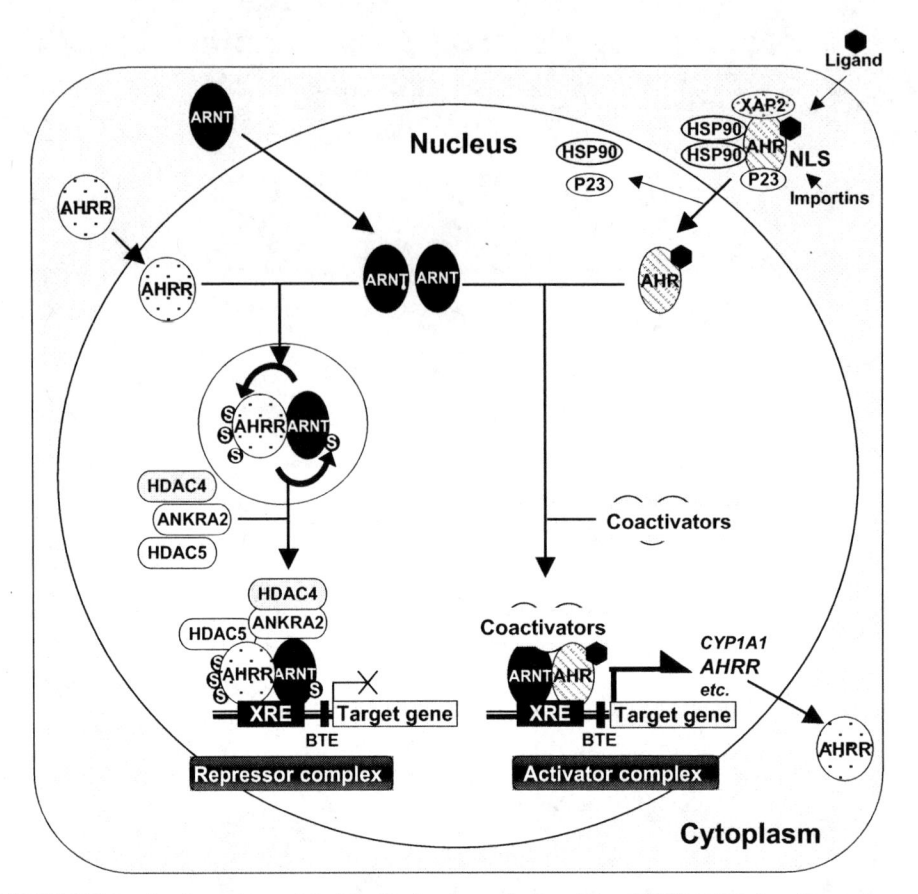

FIGURE 7.6 A schematic model of the feedback regulation of the AHR/AHRR signaling pathway. Activated AHR forms a heterodimer with ARNT and recruits coactivators, such as CBP/p300, to the promoter of target genes, including *AHRR*, to activate their expression. AHRR, in turn, forms a heterodimer with ARNT that leads to increased SUMOylation of both AHRR and ARNT. The SUMOylated heterodimer forms a repressor complex by recruiting the corepressors ANKRA2, HDAC4, and HDAC5.

mice [31]. As is often the case for mRNA expression, relative mRNA expression may not precisely correlate with the relative protein levels in various tissues.

In addition to negatively regulating the expression of drug metabolizing CYP1A1, AHRR is also thought to function as a tumor suppressor by regulating the expression of cell cycle-related genes such as *E2F*, *cyclin E1*, and *PCNA*. Variant cell lines stably expressing AHRR grow more slowly than the original MCF7 cells based on both cell number and cell proliferation, and the expression levels of E2F, cyclin E1 and PCNA are lower than those in the parental MCF7 cells [32]. Intriguingly, this inhibition of cell proliferation may be mediated through a direct interaction between AHRR and ERα in MCF7 cells. A ChIP analysis revealed that the AHRR/ERα complex is formed on the *cis*-element in the promoter of E2-responsive genes [33]. Consistent with the tumor suppressor function of AHRR, the human chromosomal location of AHRR (5p15.3) has been reported to be frequently deleted in a variety of tumors, such as cervical and testicular germ tumors, colorectal cancer, early-stage ovarian tumors, bladder cancer, esophageal cancer, and lung tumors [34]. Based on the frequent loss of heterozygosity of the *AHRR* gene locus, which is the chromosomal locus that contains the *AHRR* gene, this locus has been proposed to harbor at least one tumor suppressor gene. In human malignant tissues including colon, breast, lung, stomach, cervix, and ovary, AHRR mRNA expression is lost or markedly downregulated because of loss of the *AHRR* gene or DNA hypermethylation in its promoter region. siRNA-mediated knockdown of AHRR expression in a human lung cancer cell line enhances the tumorigenic properties of these cells, such as *in vitro* anchorage-dependent and -independent cell growth as well as *in vivo* cell growth in immunocompromised mice. On the other hand, ectopic expression of AHRR in tumor cells diminishes these tumorigenic properties and their angiogenic potential [34]. Increasing evidence indicates that AHR has a positive or negative effect on cell growth and proliferation in a cell-specific manner [35]. The roles of AHR in cell growth and proliferation should be elucidated considering the cell- and tissue-specific AHRR expression activated by AHR.

In association with high AHRR mRNA expression in hormone-dependent organs, such as the testis, ovary, and breast, it has been frequently reported that AHRR (Ala185-Pro) polymorphisms are linked to an increased risk of male and female reproductive abnormalities such as male infertility [36–38] and greater susceptibility to endometriosis [39–42]. However, these results are controversial and inconsistent in some cases. Since no published reports have evaluated the functional properties of the two variant AHRR proteins, it is difficult to understand the relationship between AHRR polymorphisms and reproductive abnormalities.

7.6 SUMMARY

Increasing evidence indicates that the AHR signaling pathway plays an important role in multiple physiological processes in addition to its role in pharmacology and toxicology, and these findings have furthered our understanding of the mechanisms that regulate AHR activity. We have summarized and discussed the structure and expression of AHRR and the molecular mechanisms by which it downregulates AHR activity, which constitutes a negative feedback loop in the AHR signaling pathway.

REFERENCES

1. Fujii-Kuriyama, Y. and Kawajiri, K. (2010). Molecular mechanisms of the physiological functions of the aryl hydrocarbon (dioxin) receptor, a multifunctional regulator that senses and responds to environmental stimuli. *Proceedings of the Japan Academy, Series B, 86*, 40–50.
2. Hahn, M. E., Lenka, L. A., and Serr, D. H. (2009). Regulation of constitutive and inducible AHR signaling: complex interactions involving the AHR repressor. *Biochemical Pharmacology, 77*, 485–497.
3. Davarinos, N. A. and Pollenz, R. S. (1999). Aryl hydrocarbon receptor imported into the nucleus following ligand binding is rapidly degraded via the cytoplasmic proteasome following nuclear export. *Journal of Biological Chemistry, 274*, 28704–28715.
4. Lee, M. J., Peet, D. J., and Whitelaw, M. (2003). Defining the role of XAP2 in stabilization of dioxin receptor. *Journal of Biological Chemistry, 278*, 35787–35888.
5. Ohtake, F., Baba, A., Takada, I., Okada, M., Iwasaki, K., Miki, H., Takahashi, S., Kouzmenko, A., Nohara, K., Chiba, T., Fujii-Kuriyama, Y., and Kato, S. (2007). Dioxin receptor is a ligand-dependent E3 ubiquitin ligase. *Nature, 446*, 562–566.
6. Makino, Y., Cao, R., Svensson, K., Bertilsson, G., Asman, M., Tanaka, H., Cao, Y., Berkenstam, A., and Poellinger, L. (2001). Inhibitory PAS domain is a negative regulator of hypoxia-inducible gene expression. *Nature, 414*, 550–554.
7. Yamashita, T., Ohneda, O., Nagano, M., Iemitsu, M., Makino, Y., Tanaka, H., Ohneda, K., Fujii-Kuriyama, Y., and Yamamoto, Y. (2008). Abnormal heart development and lung remodeling in mice lacking the hypoxia-inducible factor-related basic helix–loop–helix PAS protein NEPAS. *Molecular and Cellular Biology, 28*, 1285–1297.
8. Lee, C., Bae, K., and Edery, I. (1999). PER and TIM inhibit the DNA binding activity of a *Drosophila* CLOCK-CYC/dBMAL1 heterodimer without disrupting formation of the heterodimer: a basis circadian transcription. *Molecular and Cellular Biology, 19*, 5316–5325.
9. Mimura, J., Ema, M., Sogawa, K., and Fujii-Kuriyama, Y. (1999). Identification of a novel mechanism of regulation of Ah (dioxin) receptor function. *Genes & Development, 13*, 20–25.
10. Oshima, M., Mimura, J., Yamamoto, M., and Fujii-Kuriyama, Y. (2007). Molecular mechanism of transcriptional repression of AhR repressor involving ANKLA2, HDAC4 and HDAC5. *Biochemical and Biophysical Research Communications, 364*, 276–282.
11. Nagase, T., Kikuno, R., Nakayama, M., Hirosawa, M., and Ohara, O. (2000). Prediction of the coding sequences of unidentified human genes. XVIII. The complete sequences of 100 new cDNA clones from brain which code for large proteins in vitro. *DNA research, 7*, 273–281.
12. Korkalainen, M., Tuomisto, J., and Pohjanvirta, R. (2002). Primary structure and inducibility by 2,3,7, 8-tetrachlorodibenzo-*p*-dioxin (TCDD) of aryl hydrocarbon receptor repressor in a TCDD-sensitive and a TCDD-resistant rat strain. *Biochemical and Biophysical Research Communications, 315*, 123–131.
13. Zimmermann, A. L., King E. A., Dengler E., Scogin, S. R., and Powell, W. H. (2008). An aryl hydrocarbon receptor repressor from *Xenopus laevis*: function, expression, and role in dioxin responsiveness during frog development. *Toxicological Sciences, 104*, 124–134.
14. Roy, N. K., Courtenay, S. C., Chambers, R. C., and Wirgin, I. I. (2006). Characterization of the aryl hydrocarbon receptor repressor and comparison of its expression in Atlantic tomcod from resistant and sensitive populations. *Environmental Toxicological Chemistry, 25*, 560–571.
15. Karchner, S. I., Franks, D. G., Powell, W. H., and Hahn, M. E. (2002). Regulatory interactions among three members of the vertebrate aryl hydrocarbon receptor family: AHR repressor, AHR1 and AHR2. *Journal of Biological Chemistry, 277*, 6949–6959.
16. Evans, B. H., Karchner, S. I., Franks, D. G., and Hahn, M. E. (2005). Duplicate aryl hydrocarbon receptor repressor genes (*ahrr1* and *ahrr2*) in the zebrafish *Danio rerio*: structure, function, evolution, and AHR-dependent regulation *in vivo*. *Archives of Biochemistry and Biophysics, 441*, 151–167.
17. Jenny, M. J., Karchner, S. I., Franks, D. G., Woodin, B. R., Stegeman, J. J., and Hahn, M. E. (2008). Distinct roles of two zebrafish AHR repressors (AHRRa and AHRRb) in embryonic development and regulating the response to 2,3,7, 8-tetrachlorodibenzo-*p*-dioxin. *Toxicological Sciences, 110*, 426–441.
18. Bernshausen, T., Jux, B., Esser, C., Abel, J., and Frische E. (2006). Tissue distribution and function of the aryl hydrocarbon receptor repressor (AhRR) in C57BL/6 and aryl hydrocarbon receptor deficient mice. *Archives of Toxicology, 80*, 206–211.

19. Tsuchiya, Y., Nakajima, M., Itoh, S., Iwanari, M., and Yokoi, T. (2003). Expression of aryl hydrocarbon receptor repressor in normal human tissues and inducibility by polycyclic aromatic hydrocarbons in human tumor-derived cell lines. *Toxicological Sciences*, 72, 253–259.

20. Baba, T., Mimura, J., Gradin, K., Kuroiwa, A., Watanabe, T., Matsuda, Y., Inazawa, J., Sogawa, K., and Fujii-Kuriyama, Y. (2001). Structure and expression of the Ah receptor repressor gene. *Journal of Biological Chemistry*, 276, 33101–33110.

21. Haarmann-Stemmann, T., Bothe H., Kohli, A., Sydlik, U., Abel, J., and Fritsche E. (2007). Analysis of the transcriptional regulation and molecular function of the aryl hydrocarbon receptor repressor in human cell lines. *Drug Metabolism and Disposition*, 35, 2262–2269.

22. Puga, A., Ma, C., and Marlove, J. L. (2009). The aryl hydrocarbon receptor cross-talk with multiple signal transduction pathways. *Biochemical Pharmacology*, 77, 713–722.

23. Baba, T., Mimura, J., Nakamura, N., Harada, N., Yamamoto, M., Morohashi, K., and Fujii-Kuriyama, F. (2005). Intrinsic function of the aryl hydrocarbon (dioxin) receptor as a key factor in female reproduction. *Molecular and Cellular Biology*, 25, 10041–10051.

24. Kanno, Y., Miyama, Y., Takane, Y., Nakahama, T., and Inoue, Y. (2007). Identification of intracellular localization signals and of mechanisms underlining the nucleocytoplasmic shuttling of human aryl hydrocarbon receptor repressor. *Biochemical and Biophysical Research Communications*, 364, 1026–1031.

25. Evans, B. R., Karchner, S. I., Allan L. L., Pollenz, R. S., Tanguay, R. L., Jenny, M. J., Sherr, D. H., and Hahn, M. E. (2008). Repression of aryl hydrocarbon receptor (AHR) signaling by AHR repressor: role of DNA binding and competition for AHR nuclear translocator. *Molecular Pharmacology*, 73, 387–398.

26. Oshima, M., Mimura, J., Yamamoto, J., and Fujii-Kuriyama, Y. (2007). SUMO modification regulates the transcriptional repressor function of aryl hydrocarbon receptor repressor. *Journal of Biological Chemistry*, 284, 11017–11026.

27. Shallzi, A., Gaudillere, B., Slegmuller, J., Shirogane, T., Ge, Q., Yan, Y., Schulman, B., Harper, J. W., and Bonni, A. (2006). A calcium-regulated MEF2 sumoylation switch controls postsynaptic differentiation. *Science*, 311, 1012–1017.

28. Yang X. J. and Gregoire, S. (2006). A recurrent phospho-sumoylation switch in transcriptional repression and beyond. *Molecular Cell*, 23, 779–786.

29. Gradin, K., Toftgard, R., Poellinger L., and Berghard, A. (1999). Repression of dioxin signal transduction in fibroblast. *Journal of Biological Chemistry*, 274, 13511–13518.

30. Gradin, K., Mimura, J., Poellinger, L., and Fujii-Kuriyama, Y. (2002). Nonresponsiveness of normal human fibroblasts to dioxin is mediated by the aryl hydrocarbon receptor repressor. *The 14th international Symposium on Microsomes and Drug Oxidation Abstracts*, p. 170.

31. Hosoya T., Harada, N., Mimura, J., Motohashi, H., Takahashi, S., Nakajima, O., Morita, M., Kawauchi, S., Yamamoto, M., and Fujii-Kuriyama, Y. (2008). Inducibility of cytochrome P450 1A1 and chemical carcinogenesis by benzo[*a*]pyrene in AhR repressor-deficient mice. *Biochemical and Biophysical Research Communications*, 365, 562–567.

32. Kanno, Y., Takane, Y., Izawa, T., Nakahama, T., and Inoue, T. (2006). The inhibitory effect of aryl hydrocarbon receptor repressor (AhRR) on the growth of human breast cancer MCF-7 cells. *Biological and Pharmaceutical Bulletin*, 29, 1254–1257.

33. Kanno, Y., Takane, Y., Takizawa, Y., and Inoue, Y. (2008). Suppressive effect of aryl hydrocarbon receptor repressor on transcriptional activity of estrogen receptor alpha by protein–protein interaction in stably and transiently expressing cell lines. *Molecular and Cellular Endocrinology*, 291, 87–94.

34. Zudaire, E., Cuesta, N., Murty, V., Woodson, K., Adams, L., Gonzalez, N., Martinez, A., Narayan, G., Kirsch I., Franklin, W., Hirsch, F., Birrer, M., and Cuttitta, F. (2008). The aryl hydrocarbon receptor repressor is a putative tumor suppressor gene in multiple human cancers. *Journal of Clinical Investigation*, 118, 640–650.

35. Barouki, B., Coumoul, X., and Fernandez-Salguero A. M. (2007). The aryl hydrocarbon receptor, more than a xenobiotic-interacting protein. *FEBS Letters*, 581, 3608–3615.

36. Fujita, H., Kasai, R., Yoshihashi, H., Ogata, T., Tomita, M., Hasegawa, T., Takahashi, T., Matsuo, N., and Kosaki K. (2002). Characterization of the aryl hydrocarbon receptor repressor gene and association of its Pro185Ala polymorphism with micropenis. *Teratology*, 65, 10–18.

37. Watanabe, M., Sueoka, K., Sasagawa, I., Nakabayashi, A., Yoshimura, Y., and Ogata, T. (2004). Association of male infertility with Pro185Ala polymorphism in the aryl hydrocarbon receptor repressor gene: implication for the susceptibility to dioxin. *Fertility and Sterility*, 82, 1067–1071.

38. Merisalu, A., Punab, M., Altmae, S., Tiido, T., Peters, M., and Salmets, A. (2007). The contribution of genetic variations of aryl hydrocarbon receptor pathway genes to male fertility. *Fertility and Sterility*, 88, 854–859.

39. Watanabe, T., Imoto, I., Kosugi, Y., Mimura, J., Fujii, Y., Takayama, M., Sato, A., and Inazawa, J. (2001). *Journal of Human Genetics*, 46, 342–346.

40. Tsuchiya, M., Katoh, T., Motoyama, H., Sasaki, H., Tsugana, S., and Ikenoue, T. (2005). Analysis of the AhR, ARNT, and AhRR gene polymorphisms: genetic contribution to endometriosis susceptibility and severity. *Fertility and Sterility*, 88, 454–458.

41. Kim, S. H., Choi, Y. M., Lee, G. H., Hong, M. H., Lee, S. K., Lee, B. S., Kim, J. G., and Moon, S. Y. (2007). Association between susceptibility to advanced stage endometriosis and the genetic polymorphisms of aryl hydrocarbon receptor repressor and glutathione-S-transferase T1 genes. *Human Reproduction*, 22, 1866–1870.

42. Asada, H., Yagihashi, T., Furuya, M., Kosaki, K., Takahashi, T., and Yoshimura, Y. (2009). Association between patient age at the time of surgical treatment for endometriosis and aryl hydrocarbon receptor repressor polymorphism. *Fertility and Sterility*, 92, 1240–1242.

8

INFLUENCE OF HIF1α AND NRF2 SIGNALING ON AHR-MEDIATED GENE EXPRESSION, TOXICITY, AND BIOLOGICAL FUNCTIONS

THOMAS HAARMANN-STEMMANN AND JOSEF ABEL

8.1 INTRODUCTION

The aryl hydrocarbon receptor (AHR) is a key regulator of drug metabolism and the only known basic helix–loop–helix–Per/ARNT/Sim (bHLH/PAS) protein that is activated by low molecular weight compounds. The diverse spectrum of AHR-activating substances includes persistent organic pollutants (POP), especially dibenzo-*p*-dioxins and dibenzofurans, non-*ortho* polychlorinated biphenyls, and polycyclic aromatic hydrocarbons (PAH), as well as polyphenolic and alkaloid plant constituents, UVB-induced tryptophan photoproducts, and atmospheric ozone [1–3]. In the absence of a ligand, the AHR is located in the cytosol trapped in a multiprotein complex consisting of a heat shock protein 90 (HSP90) dimer, the hepatitis virus X-associated protein 2, the co-chaperone p23, and the tyrosine kinase c-src. Upon ligand binding, this multiprotein complex dissociates and the AHR translocates into the nucleus and dimerizes with the AHR nuclear translocator (ARNT). The resulting heterodimer then binds to xenobiotic-responsive elements (XRE) in gene promoters to stimulate their expression. The AHR gene battery encodes for several drug metabolizing enzymes as well as proteins involved in regulation of cell differentiation, proliferation, and apoptosis. Besides this direct transcriptional action, ligand binding of AHR leads to the release of c-src that subsequently phosphorylates the epidermal growth factor receptor (EGFR), a key event in the so-called nongenomic pathway of AHR signaling [4]. Downstream activation of mitogen-activated protein kinases (MAPK) results in gene induction of proinflammatory enzymes such as cyclooxygenase-2.

The majority of AHR/XRE-dependent drug metabolizing enzymes belong to phase I of xenobiotic metabolism, for example, cytochrome P450 (CYP) monooxygenases 1A1, 1A2, 1B1, and 2S1, paraoxonase 1, and aldehyde dehydrogenase 3. These enzymes introduce functional groups to lipophilic chemicals to enhance their water solubility. The resulting reactive phase I metabolites are then conjugated to hydrophilic substances, including glutathione, glucuronic acid, acetyl-CoA, and activated sulfate. Several phase II genes such as NAD(P)H:oxidoreductase (NQO1), several glutathione S-transferases (GST), and diverse UDP-glucuronosyltransferases (UGT) are also regulated in an AHR-dependent manner. Many of these phase II enzymes are transcriptionally coregulated by antioxidant-responsive elements (ARE) located adjacent to the XREs. These AREs are recognized by the transcription factor NF-E2 p45-regulated factor (Nrf2), the master regulator of the cellular antioxidant response [5]. Since exposure toward AHR binding substances such as dioxins, PAHs, flavonoids, and related polyphenols leads to the formation of reactive oxygen species (ROS) and electrophilic metabolites within the target cell or tissue, an interaction between AHR and Nrf2 signaling is most probably a crucial trigger for detoxification and cytoprotection.

TCDD and related POPs have been proven to cause distinctive skin disorders, hepatotoxicity, endocrine disruption, immune toxicity, and cancer, and a genetic knockout of

---

The AH Receptor in Biology and Toxicology, First Edition. Edited by Raimo Pohjanvirta.
© 2012 John Wiley & Sons, Inc. Published 2012 by John Wiley & Sons, Inc.

the AHR protects the animals against most of these toxic effects [6–8]. Besides this resistant phenotype toward dioxins and related chemicals, AHR-deficient mice also exhibit a markedly reduced liver weight, develop cardiac hypertrophy, have a reduced fertility and immunity, and display several alterations in the vascular system [9–12]. These phenotypic alterations provide evidence that the AHR is not only involved in coordination of drug metabolism but also has certain developmental functions. An important factor during embryonic development, especially for proper construction of the vascular system, is the hypoxia-inducible factor (HIF)-1α, a bHLH–PAS protein closely related to the AHR [13]. Hypoxia was shown to directly affect AHR-dependent gene regulation and thus may influence AHR function during both embryonic development and cellular defense against xenobiotics.

The broad spectrum of POP toxicity and the multiple abnormalities observed in AHR-deficient mice indicate that the AHR does not act in single player mode but instead closely crosstalks with several other cellular pathways including Nrf2 and HIF-1α signaling. The interactions of Nrf2 and HIF-1α with the AHR system and their relevance to AHR-dependent gene expression, toxicity, and developmental functions are discussed next.

8.2 THE AHR AND HIF-1α SIGNALING NETWORK

8.2.1 HIF-1α, ARNT, and Hypoxic Response

To maintain oxygen homeostasis, mammalian cells are equipped with cellular oxygen sensors that get activated in case of low oxygen partial pressure to induce an adaptive response within the cell and tissue. These oxygen sensors, termed hypoxia-inducible factors, are structurally related to the AHR and also belong to the bHLH–PAS protein family [13]. The most prominent one is HIF-1, a heterodimeric transcriptional activator consisting of two subunits, HIF-1α and ARNT (HIF-1β) [14]. While the ARNT protein is constitutively expressed and detectable in nearly every tissue [15], the HIF-1α subunit, although ubiquitously expressed on transcriptional level, is rapidly degraded by the 26S proteasome under normoxic conditions (Fig. 8.1). Precisely, in presence of oxygen the HIF-1α protein is modified by prolyl hydroxylation, which permits the binding of the von Hippel–Lindau tumor suppressor protein (pVHL), a recognition component of the E3 ubiquitin ligase complex [16]. In case of hypoxia, the protein level of the HIF-1α subunit increases rapidly via inhibition of ubiquitination and subsequent proteasomal degradation. Stabilized HIF-1α translocates into the nucleus and dimerizes with ARNT to bind to hypoxia-responsive elements (HRE) within the promoter of target genes to enforce their expression (Fig. 8.1). Among others, the hypoxia-sensitive gene battery includes erythro-

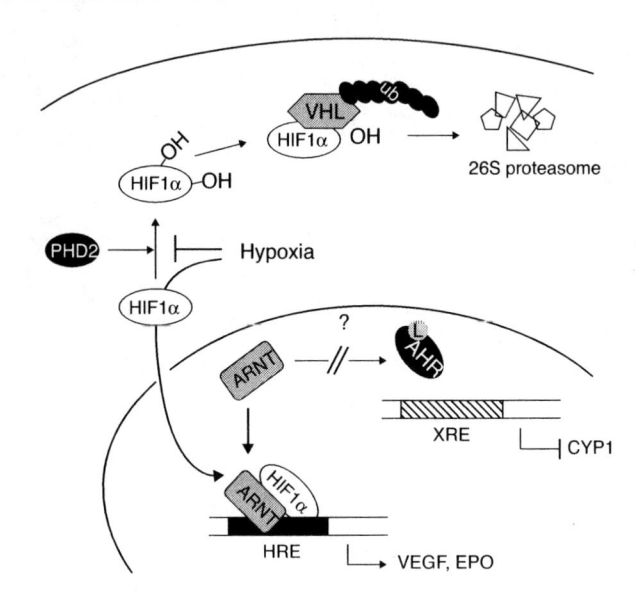

FIGURE 8.1 The HIF-1 signaling pathway. Under normoxic conditions, the HIF-1α protein is constantly hydroxylated by prolyl hydroxylases, a process that leads to the recognition and subsequent ubiquitination of HIF-1α by the VHL tumor suppressor protein, followed by proteasomal degradation. Hypoxic conditions lead to the stabilization of the HIF-1α protein, which rapidly enters the nucleus to dimerize with ARNT, bind to HRE in the promoters of target genes such as EPO or VEGF, and stimulate gene expression. For some cell types, it was postulated that hypoxia-activated HIF-1α sequesters the nuclear ARNT protein, which may probably result in a decreasing transcriptional activity of AHR.

poietin (EPO), vascular endothelial growth factor (VEGF), VEGF receptor Flt1 (VEGFR-1), heme oxygenase 1 (HO-1), nitric oxide synthase 2, and several enzymes involved in glucose transport and glycolysis [17, 18]. Besides HIF-1α, the closely related HIF-2α (endothelial PAS protein 1), HIF-3α, and inhibitory PAS protein, an alternative splice variant of HIF-3α, are also involved in regulation of the hypoxic response [19–21].

As a useful experimental tool, hypoxic conditions can be easily mimicked by using iron-chelating agents such as cobalt chloride ($CoCl_2$), nickel chloride, picolinic acid, or desferrioxamine (DFX) that inhibit the iron-dependent enzymatic activity of prolyl hydroxylases and thereby stabilize the HIF-1α protein. Besides its activation by hypoxia or iron chelators, the HIF-1 pathway can also be stimulated by growth factors and cytokines under normoxic conditions, for instance, by insulin, insulin-like growth factors (IGF), transforming growth factors (TGF), and interleukin (IL)-1β [22–25]. Whereas hypoxia is associated with a decreased proteasomal degradation of HIF-1α, these growth factors induce HIF-1α translation irrespective of hypoxia and thus overcome the oxygen sensor-mediated degradation under normoxia. In most cases, stimulation of receptor tyrosine kinases and subsequent activation of phosphatidylinositol 3-kinase (PI3K), MAPK, and downstream signal transduction are involved. A proper HIF-1 signaling is essential during embryonic development and for wound healing. In addition, HIF-1 signaling is a crucial trigger for cancer development and growth: Since hypoxic conditions often occur within fast growing tumors, HIF-1-induced neovascularization ensures proper blood supply and thus is indispensable for tumor survival [26].

8.2.2 Crosstalk of AHR and HIF-1α Signaling

The discovery that HIF-1α and AHR share ARNT as common dimerization partner raised the question whether a crosstalk between oxygen and dioxin signal transduction pathways exists. The probably most obvious connection of both pathways is that they share a common subset of target genes. Indeed, several reporter gene assays and gel mobility shift analyses brought a couple of genes to light that contain binding sequences for both AHR and HIF-1α within their promoter or enhancer region. For instance, transcription of EPO [27], breast cancer resistance protein [28, 29], IGF binding protein 1 [30, 31], or CYP2S1 [32] was shown to be under control of both signaling cascades. Global gene expression analyses in Hep3B cells revealed 767 and 430 genes that are modulated by $CoCl_2$ and TCDD, respectively [33]. Further analyses revealed an overlap of 33 genes, including *CYP1A1* and *carbonic anhydrase IX (CA9)*, that are probably sensitive to the activation of both signaling pathways. However, besides this indirect way of interference, the question arises whether AHR and HIF-1α crosstalk directly on signaling level. To answer this, several expression analyses and reporter gene studies using different human and rodent cell lines were performed. These investigations demonstrate the existence of such an interaction, at least on a certain level, since activation of one transcription factor was often found to be associated with a reduced activity of the other. For instance, an early study on human and murine hepatoma cells revealed that HIF-1α has a high affinity for ARNT, and consequently hypoxic activation of HIF-1α signaling inhibited both ligand-dependent DNA binding and transcriptional activity of the AHR [34]. In contrast, exposure to 2,3,7,8-tetrachlorodibenzofuran had no inhibiting effect on the expression of HIF-1α-dependent target genes such as EPO and aldolase A. The authors conclude that both factors compete for their common binding partner ARNT and that HIF-1α dominates over AHR signaling, as shown in Fig. 8.1. The hypothesis of competition was taken over by another study showing that a coexposure of Hepa 1 cells with TCDD and different concentrations of the iron chelators $CoCl_2$, DFX, and picolinic acid resulted in a dose-dependent inhibition of AHR/XRE-driven reporter gene activity [35]. As expected, a cotreatment with ferrous sulfate restored AHR-dependent luciferase expression, but, most interestingly, a coadministration of nitroarginine, an inhibitor of nitric oxide synthase, exhibited the same effect. This finding points to the idea that nitric oxide is an important mediator for the inhibition of AHR-dependent transcription, at least by iron chelators [36]. In contrast to these studies, Shearer and colleagues analyzed the ARNT protein level during hypoxia and revealed that the activated HIF-1α sequesters only 15% of the cellular ARNT pool. Moreover, gel mobility shift analyses and Western blots did not show any quantitative changes in TCDD-induced AHR/ARNT complex during hypoxia, a finding clearly being contrary to the competition hypothesis [37]. Another study has shown that the AHR-induced expression of aldehyde dehydrogenase 3 is downregulated during hypoxia even when ARNT is ectopically overexpressed, a finding that also counteracts the idea of ARNT as limiting factor [38]. In consistence with these results, a hypoxia-mediated downregulation was also observed for CYP2J2 and CYP3A4, two AHR-independent P450 enzymes [39, 40]. Finally, two studies were able to show a hypoxia-mediated downregulation of constitutive as well as AHR ligand-induced gene expression in HIF-1α-deficient cells and thus clearly refute the competition theory [41, 42]. However, cell-specific differences cannot be excluded. Interestingly, an analysis done in human endothelial cells revealed that hypoxia reduces constitutive CYP1A1 expression, which was probably due to a prior HIF-2α-mediated decrease in AHR expression [43]. Mechanistic investigations on a hypoxic rabbit model uncovered a possible involvement of the serum mediators IL-1β, IL-2, and interferon-γ (IFNγ), which are known to be secreted in response to hypoxia [44], in downregulation of CYP enzymes [45, 46]. These cytokines can induce NF-κB [47, 48], a transcription factor that was previously shown to repress AHR-dependent transcription in response to tumor necrosis factor-α and lipopolysaccharide exposure [49]. The activation of NF-κB by those inflammatory mediators blocked histone acetylation at the CYP1A1 TATA box, a process normally performed by CBP/p300 and related coactivators of the AHR/ARNT complex [50]. Accordingly, IL-1β and IFNγ were released upon both aseptic inflammation and viral infection to reduce CYP1A1 and CYP1A2 expression in rabbits [51]. Whether this mechanism is also responsible for the hypoxia-mediated inhibition of AHR-dependent gene expression is not clear. However, the early hypoxia-mediated repression was shown to be sensitive toward cycloheximide treatment, indicating that *de novo* protein synthesis of an inhibitory factor is required [52]. Longer lasting hypoxic conditions resulted in a reduced expression of AHR coactivator BRCA1, which may also account for a decrease in CYP1A1 transcription [52]. Other cofactors and coactivators such as HSP90 [53, 54], TRIP230 [55], SRC-1 [56, 57], CBP/p300 [56, 58], and BRG-1 [59, 60] are also commonly used by both AHR and HIF-1α and therefore represent possible limiting factors. Another regulatory player capable of suppressing CYP1 transcription is the AHR repressor (AHRR), a negative feedback inhibitor of AHR signaling [61, 62]. Even though no functional HRE motif has been identified so far, a putative NF-κB site was found in the promoter region of the human and mouse AHRR gene [63, 64]. Since both hypoxia and DFX are known to activate NF-κB [65], one could speculate that the AHRR mRNA increase upon hypoxic stimuli, as observed, for instance, in DFX-treated HepG2 cells (Fig. 8.2), is NF-κB dependent. However, in comparison to the CYP1A1 repression (paralleled by VEGF induction), the DFX-induced upregulation of AHRR transcription occurs timely delayed and thus indicates that both events occur independently (Fig. 8.2). In line with this observation, treatment of ZR75 breast cancer cells with $CoCl_2$ resulted in a repression of both TCDD-induced CYP1A1 and AHRR expression [52]. The authors conclude that AHRR is most likely not involved in the hypoxia-mediated inhibition of CYP1A1 expression. Noteworthy, a recent study identified a new splice variant of the human AHRR that lacks exon 8 and that was proven to efficiently repress AHR- as well as HIF-1α-driven reporter gene activity [66].

As mentioned above, hypoxia-mediated repression of AHR activity was also shown in rats and rabbits, pointing to the *in vivo* relevance of this crosstalk in rodents [46, 67]. In contrast, studies on human volunteers that were exposed to altitude-induced hypoxia corresponding to approximately 4500 m above sea level for a period of 2 or 7 days did not show any alterations in the metabolism of substrates for certain CYP enzymes, including the CYP1A2 substrates caffeine and theophylline [68, 69]. Since the CYP system is redundant regarding their substrate spectrum, these findings at least indicate that acute hypoxia has no clinically significant effect on overall CYP activity in humans.

8.2.3 AHR, HIF-1α, and ARNT in Mouse Development

The requirement of a functional HIF-1 signaling cascade for proper embryonic development was demonstrated by the generation of animals that lack HIF-1α or its binding partner ARNT. HIF-1α-deficient mice were not viable past gestation day 11 due to massive developmental defects [70]. The embryos had a markedly decreased VEGF expression and showed manifested defects in the neural tube and several cardiovascular malformations. In addition, the cephalic mesenchyme was shown to be affected by massive cell death. ARNT-deficient embryos died at embryonic day 10.5, showed reduced VEGF expression, and had basically the same phenotype as VEGF-deficient animals: a decreased vascularization of both yolk sac and solid tissues like the branchial arches [71, 72]. However, an investigation of a competing group did not observe the described abnormalities, but claimed a defective vascularization of the placenta to be responsible for midstage embryonic lethality [73]. It is discussed that this discrepancy in phenotype might be due to methodical differences in generation and breeding of the respective mouse strains. As expected, a conditional deletion

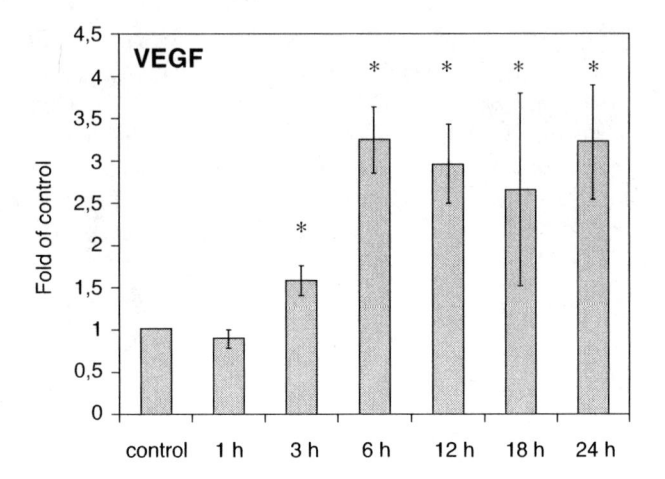

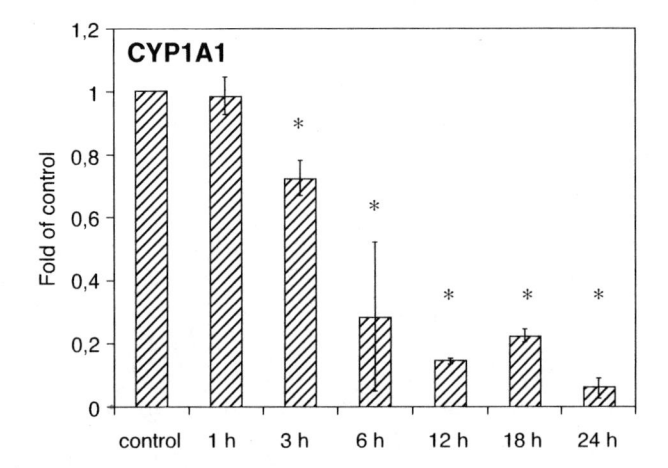

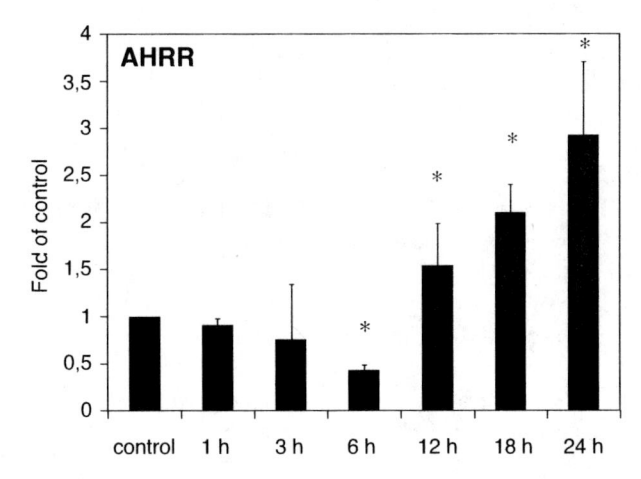

FIGURE 8.2 Chemical hypoxia-induced changes in gene expression in HepG2 cells. HepG2 cells were treated with 300 μM DFX for the indicated time points. Total RNA was isolated and cDNA was synthesized using M-MLV reverse transcriptase as described in Ref. 62. Subsequently, gene expression of VEGF, CYP1A1, and AHRR was measured by quantitative real-time PCR and normalized to β-actin. Gene expression is shown as fold of solvent (dH_2O) control; asterisk denotes significantly different from solvent control, $p \leq 0.05$.

of ARNT is accompanied by a loss of target gene induction by both AHR and HIF-1α [74]. Embryos that lack HIF-2α also developed severe vascular defects in yolk sac and did not survive past gestation day 12.5 [75]. The embryos had normal vasculogenesis but exhibited severe problems in remodeling the primary vascular network into a mature hierarchy pattern, for example, vessels failed to assemble into larger vessels during development or were fused inadequately.

In contrast to these lethal genetic knockouts, animals that lack a functional AHR gene are viable but show some developmental alterations that are probably related to disturbances in the maturing vascular system. Especially neonatal vascular structures that normally resolve in early development of wild-type mice are maintained in these mice. The best example is the reduced liver weight observed in AHR-deficient mice, which is due to an inappropriate closure of the *ductus venosus*, an essential shunt that bypasses the liver sinusoids during embryogenesis [76, 77]. Moreover, a persistent hyaloid artery, a fetal artery that is necessary for a proper blood supply of the developing eye, and alterations in vascular architecture of the kidneys were noticed [76]. These observations indicate that HIF-1α/ARNT induces the formation of shunts and vessels that are essential during certain development stages and that have to be resolved by AHR/ARNT when no longer needed. Furthermore, AHR-deficient mice exhibit cardiac enlargement due to an enhanced protein expression of HIF-1α and VEGF in the absence of myocardial hypoxia [12]. Recently, it was shown that AHR deficiency impairs angiogenesis by a complex mechanism that involves VEGF-A depletion in the endothelium paralleled by an overexpression of TGFβ in stromal myofibroblasts [78]. The relevance of these findings becomes evident by the observation that AHR-deficient mice lose their capacity to support the growth of melanoma tumors due to an inappropriate angiogenesis [79]. Accordingly, transgenic animals that express a constitutively active form of the AHR are cancer prone and develop tumors in different organs such as stomach and liver [80, 81]. Since mice that express a constitutively active AHR under control of the T-cell-specific CD2 promoter exhibited a reduced number of thymocytes and a suppressed expansion of immunization-induced T cells and B cells [82], a significant loss of immune function may contribute to cancer development in these animals.

8.2.4 TCDD Toxicity and Its Effects on Vasculogenesis and Angiogenesis

Numerous studies have shown that TCDD and related environmental pollutants can directly affect processes of neovascularization in certain organs and tissues [83]. The formation of new blood vessels may occur through either vasculogenesis, which means the *de novo* assembly of blood vessels from angioblast precursors, or angiogenesis, which stands for the formation of new capillary sprouts from preexisting vessels.

Studies on TCDD-exposed chicken embryos have shown that these animals have an abnormal cardiac vascularization that displays symptoms of cardiomyopathy and congestive heart failure. For instance, TCDD intoxication caused an increased heart weight, enlarged right and left ventricles, a thickened ventricular septum, and a thinner left ventricular wall [84, 85]. Moreover, dioxin decreases myocyte proliferation and increases cardiac apoptosis during coronary development [83]. The observed structural alterations correlate temporally and locally with the expression and activation status of AHR and ARNT, indicating that this heterodimer is responsible for these effects [84, 86]. Further experiments revealed a TCDD-induced reduction in cardiac hypoxia and a decreased cardiac expression of HIF-1α and VEGF [87]. This finding fits roughly with the phenotype of TCDD-treated mice and with the already mentioned observation in AHR-deficient mice, showing an enhanced cardiac expression of both genes [12, 88]. Early TCDD-mediated disturbances of endothelial tube formation and differentiation in chick were shown to be reversible by addition of exogenous recombinant VEGF or chemical mimicry of hypoxia [89]. Thus, TCDD disrupts proper heart development by both inhibiting the expression of VEGF and reducing the responsiveness of endothelial cells to this growth factor. Related antiangiogenic effects were observed in human umbilical vascular endothelial cells (HUVEC) upon exposure to different PAHs. Treatment of HUVECs with benzo(*a*)pyrene (BaP) in the presence of angiogenic factors reduces tube formation, cell migration, MAPK signaling, and integrin expression [90]. Accordingly, exposure of HUVECs toward 3-methylcholanthrene (3-MC) resulted in an enhanced induction of cell cycle arrest and decrease in cell permeability, adhesion, and tube formation [91]. Alterations in cell adhesion were due to a lower phosphorylation status of focal adhesion kinase, a key enzyme in integrin signaling. Most of these effects provoked by PAHs were reversible by addition of the AHR antagonist α-naphtoflavone (αNF), providing evidence for an involvement of AHR.

TCDD is also known to alter the vasculature of the placenta and to induce fetal death in several animal species [92]. Vascular remodeling in the rat placenta normally occurs during the late period of gestation, involves the VEGF/VEGFR as well as the angiopoietin (Ang)/Tie2 system, and is characterized by changes in cell shape and apoptotic removal of trophoblasts. The result is a volume increase in maternal and fetal blood in the placenta in order to ensure sufficient supply with oxygen and nutrients during late gestational stages. *In utero* TCDD exposure induces fetal death by suppressing the dilatation of sinusoids and fetal capillaries in the placenta through interference with the Ang/Tie2 and VEGF/VEGFR systems. Moreover, TCDD blocks trophoblast apoptosis and induces placental hypoxia as well as changes in glucose metabolism [93, 94]. Analyses done in a murine ischemia model exhibited enhanced VEGF expression and angiogenesis in AHR-deficient animals, which were probably due to an increased abundance and activity of the HIF-1α/ARNT heterodimer [95]. Oral administration of BaP to ischemic wild-type mice significantly inhibited ischemia-induced angiogenesis, an observation that was accompanied with decreasing levels of IL-6 and VEGF transcription [96]. As expected, the effects were markedly attenuated in ischemic AHR-deficient mice.

Taken together, TCDD can disturb angiogenesis and vasculogenesis during organ development, a finding implying that this compound, although classified by the International Agency for Research on Cancer (IARC) as a group I carcinogen [97], may probably exert inhibitory actions on tumor growth. Indeed, several studies noticed antitumorigenic effects of TCDD [98, 99]. TCDD was reported to repress the expression of hypoxia-induced G-protein-coupled receptor CXCR4, which is critically involved in the regulation of metastasis of many tumor types [100], as well as the hypoxia-driven expression of the Zn^{2+} binding metalloenzyme CA9, whose distribution in tumor tissues significantly correlates with a poor prognosis [101]. In addition, the antiangiogenic action of TCDD might be, at least to a certain extent, due to the inhibition of hypoxia-induced upregulation of VEGF. Therefore, the HIF-1α/VEGF axis moves into the focus of modern cancer therapy [102]. With respect to the search for and the development of new preventive compounds, natural substances, especially plant-derived polyphenols, represent an inexhaustible reservoir of putative anticancer drugs. Indeed, several flavonoids such as apigenin, chrysin, quercetin, or aminoflavone have been identified to disturb hypoxia signaling and subsequent VEGF expression in tumors and tumor-derived cell lines, respectively [103–106]. Interestingly, each of those flavonoids was previously reported to stimulate AHR signaling [106–108], pointing to the idea that these nontoxic substances act in a dioxin-like, antiangiogenic fashion. The respective flavonoids reduce the expression and stability of HIF-1α as well as VEGF expression. Especially the influence on HIF-1α protein stability is of interest, because a similar observation was made in A549 cells treated with the BaP metabolite BaP-3,6-dione [109]. Since HIF-1α stabilization is mediated via prolyl hydroxylation, the idea was that these compounds may lead to alterations in prolyl hydroxylase domain-containing protein 2 (PHD2) expression or activity. Both hypoxia and TCDD exposure were shown to enhance PHD2 expression, indicating that PHD2 expression is regulated in an AHR-dependent manner under normoxic conditions. However, TCDD treatment of hypoxic cells led to the destabilization of HIF-1α, which is accompanied by a reduction in PHD2 expression. Therefore, PHD2 seems to be not the mediator of TCDD-stimulated HIF-1α destabilization and obviously does not control the interference of AHR and HIF-1α signaling [110]. Whether TCDD and AHR-activating flavonoids modulate the expression or activity of other factors involved in HIF-1α stabilization or proteasomal degradation, for example, pVHL protein [16] or ARD protein acetyltransferase [111], remains to be elucidated. Regarding proteasomal degradation, it has to be mentioned that the AHR itself was identified as an integral part of a E3 ubiquitin ligase complex that targets β-catenin and several steroid hormone receptors (e.g., estrogen receptor-α) to the proteasome [112, 113]. Whether this process is involved in AHR-driven destabilization of HIF-1α has not been investigated so far. Noteworthy, another flavonoid and known AHR inhibitor [114], epigallocatechin-3-gallate (EGCG), represses HIF-1α and VEGF expression in human cervical carcinoma and hepatoma cells [115]. However, since EGCG is an HSP90 blocker [114], it probably inhibits AHR as well as HIF-1α signaling and thus may exert its antiangiogenic effects through a different mechanism.

In contrast to this inhibitory action of AHR on the HIF-1α/VEGF axis, several other studies exhibited an enhanced VEGF expression upon TCDD exposure, for instance, in U937 macrophages [116] or the cortical thymic epithelial cell line ET [117], pointing to an alternative, probably cell- or tissue-specific interference. In line with this findings, a siRNA-mediated knockdown of AHR was shown to be accompanied with a decrease in VEGF-A and VEGF-B mRNA levels [78], and an intraperitoneal injection of TCDD into C57BL/6 mice resulted in an abnormal vascularization in the eye, due to an enhanced AHR-dependent VEGF-A production [118]. An increased expression of VEGF and other angiogenic factors was also observed in a BaP feeding study in mice [119]. Since no functional XRE motifs were identified within the VEGF promoter so far, alternative signaling routes may connect AHR signaling and VEGF transcription and influence its direction. One possible link might be the EGFR. It is well described that EGFR (i) triggers VEGF-mediated angiogenesis in tumor cells [120] and (ii) is activated by the tyrosine kinase c-src upon ligand binding of the AHR [121, 122]. Hence, a modulation of AHR activity or expression may directly affect EGFR and downstream MAPK signaling and thus influence VEGF expression and subsequent vascularization. The discovery of an AHR-dependent regulation of early growth response factor 1 (EGR1), an immediate early gene, may serve as another explanation for AHR-mediated VEGF modulation [123]. The expression of EGR1 increases during hypoxia [124, 125] and EGR1 binds to the proximal region of the VEGF-A promoter and thus enforces HIF-1α-mediated VEGF-A expression [126]. In addition, it was shown that the PAH-mediated increase in VEGF expression in epidermal cells was mediated by the PI3K/Akt pathway and subsequent AP-1 activation, a HIF-1α-independent signaling route [127, 128]. From previous studies, it is already known that TCDD exposure increases both the expression of the AP-1 subunits c-fos and c-jun and the AP-1 transcriptional activity [129, 130].

In summary, TCDD intoxication may cause disorders and diseases in the vascular system, and thus the AHR is an important ruler of angiogenic factors and their regulators. Indeed, TCDD-exposed herbicide production workers were shown to have endothelial dysfunctions and impaired microvascular reactivities and do more often suffer from vascular diseases [131, 132]. Furthermore, a meta-analysis of 36 cohort studies on the mortality of dioxin-contaminated herbicide and chlorophenol production workers revealed an increase in the risk for circulatory diseases, especially ischemic heart disease [133]. Finally, a 20-year follow-up study on cancer incidence in the dioxin-contaminated Seveso population revealed an increased risk of lymphatic and hematopoietic tissue cancers in the most exposed zones (especially zone B) [134], again pointing to a putative interference with hypoxia-associated signaling.

8.3 THE AHR AND NRF2 SIGNALING NETWORK

8.3.1 Nrf2, Keap1, and Antioxidant Response

Oxidative stress is an imbalance between production and cellular detoxification of ROS that disturbs the redox state of the cell and can result in inflammation, cancer, cardiovascular diseases, kidney and liver diseases, and neurological disorders. The main source of ROS is the leakage of the mitochondrial respiratory chain as well as CYP-mediated oxidations. Oxidative stress leads to the activation of several transcription factors such as NF-κB, AP-1, β-catenin, peroxisome proliferator-activated receptor-γ, and Nrf2, the master regulator of the antioxidant response. Nrf2 belongs to the Cap-N-Collar family of basic region leucine zipper (CNC-bZIP) proteins and, as shown in Fig. 8.3, is bound to the kelch-like ECH-associated protein 1 (Keap1) in the cytoplasm under basal conditions [135, 136]. Keap1 is an adaptor protein that links Nrf2 to a cullin-3-dependent E3 ubiquitin ligase complex [137]. Thus, Keap1 targets Nrf2 to proteasomal degradation in the absence of oxidative stress. Upon occurrence of oxidative stressors such as ROS, reactive nitrogen species, heavy metals, lipid aldehydes, electrophilic xenobiotics, and their metabolites inside the cell, the Keap1/Nrf2 binding dissociates and Nrf2 translocates into the nucleus. This dissociation step is due to (i) the interaction of reactive cysteine residues in the Keap1 protein that form protein–protein cross-links in the presence of electrophils [138] and (ii) the activation of protein kinase C (PKC), PI3K, and MAPKs that phosphorylate and thereby stabilize the Nrf2 protein [139, 140]. The nuclear localized Nrf2 forms a heterodimer with a partner protein, usually small Maf

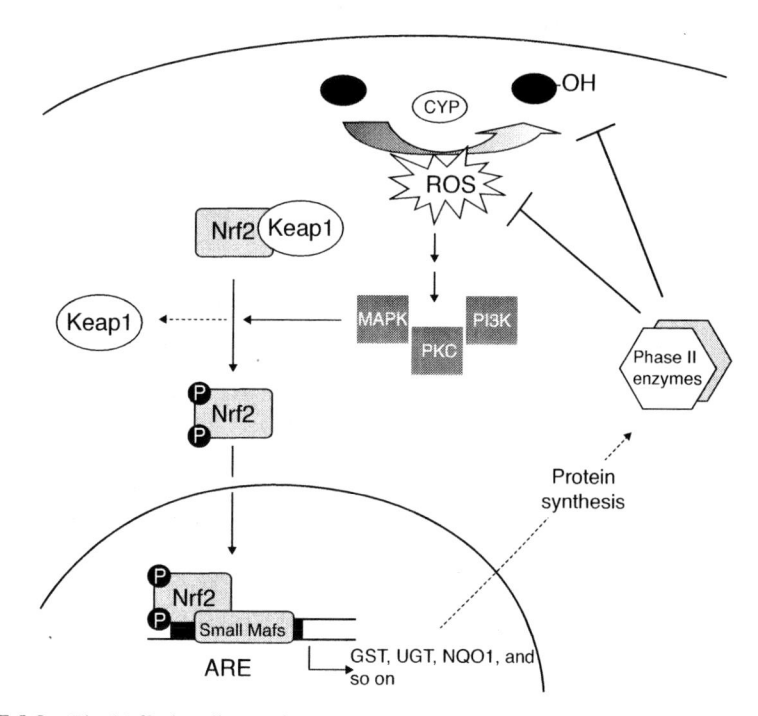

FIGURE 8.3 The Nrf2 signaling pathway. In unexposed cells, the Nrf2 protein is associated with Keap1 and thereby kept in the cytoplasm. Upon intracellular production of ROS or electrophilic metabolites, for instance, derived from CYP-mediated biotransformation processes, the binding of Keap1 to NRf2 is disturbed by electrophils and activated PI3K, MAPK, and PKC further stabilize the Nrf2 protein. Nrf2 enters the nucleus, binds to small Maf proteins, and recognizes AREs in the promoter sequences of target genes (e.g., *NQO1*) to stimulate their expression.

proteins, and binds to ARE in the promoter region of target genes to enhance their transcription (Fig. 8.3) [141]. Nrf2 directly induces the expression of cytoprotective genes, including NQO1, several UGT and GST isoforms, epoxide hydrolase, HO-1, multidrug resistance-associated proteins (MRPs), and others [5, 142]. In addition, Nrf2 acts in an autoregulatory manner and regulates its own expression through interaction with Nrf2 binding sites within its gene promoter [143].

Nrf2-deficient mice grow normally and are fertile [144], but are susceptible to oxidative stress and reactive electrophiles. Indeed, Nrf2-deficient animals are more prone to dextran sulfate sodium-induced colitis [145], cigarette smoke-induced emphysema [146], endotoxin-stimulated septic shock [147], butylated hydroxytoluene-provoked pulmonary diseases, [148] and BaP-induced carcinogenesis [149]. However, deficiency of both Nrf2 and Nrf1 results in early embryonic lethality due to oxidative stress, pointing to a functional redundancy [150]. The importance of a functional Nrf2 antioxidant response was proven by cullin-3 E3 ubiquitin ligase overexpression in breast cancer cells, which leads to Nrf2 depletion followed by cellular sensitization toward carcinogens, oxidative stress, and chemotherapeutics [151]. Newborn animals deficient in Keap1 display a normal phenotype but die within 3 weeks after birth most probably due to malnutrition. These pups develop hyperkeratotic lesions that constrict esophagus and forestomach and thus prevent sufficient milk uptake [152]. Interestingly, these severe effects were completely reversed in Keap1/Nrf2 double knockout mice, indicating that the constitutively active Nrf2 is responsible for the lethal phenotype of Keap1-null mice.

8.3.2 AHR and Nrf2: Evidences for Coupled Interactions

In the beginning of the 1990s, several investigations had shown that certain compounds, namely, TCDD, azo dyes, PAHs, and βNF are capable of inducing the expression and activity of both phase I carcinogen metabolizing enzymes and phase II detoxifying enzymes such as NQO1, UGT, and GST. In contrast to these bifunctional inducers, *tert*-butylhydroquinone, thiocarbamates, isothiocyanates, dithiolthiones, and diphenols were classified as monofunctional enzyme inducers, since they significantly enhance NQO1 or GST but not aryl hydroxylase (CYP1) activity [153, 154]. In an early study, Prochaska and Talalay used AHR-defective murine hepatoma cells to elucidate whether mono- and bifunctional inducers mechanistically differ from each other regarding NQO1 induction. Interestingly, the authors conclude that while monofunctional inducers stimulate NQO1 expression independently from AHR, bifunctional inducers require a competent AHR cascade as well as an additional unknown regulatory mechanism to enhance NQO1 function [153]. Meanwhile, it is well accepted that TCDD, PAHs, and other bifunctional inducers stimulate CYP1 enzyme activity that then leads to the formation of reactive metabolites and redox cycler and thus to oxidative stress (Fig. 8.4). This in turn activates the antioxidant signaling cascade with Nrf2 as master and commander. Besides this indirect way of crosstalk between AHR and Nrf2 signaling, the identification of responsive XREs and AREs in a couple of common phase II target genes, for example, NQO1 [155, 156], GST A2 [157, 158], and UGT1A1, 1A3, 1A4, 1A6, 1A7, 1A8, 1A9, and 1A10 [159–162], points to a putative protein–protein interaction between both transcription factors. Increasing evidence from several studies implies that AHR and Nrf2 have direct links (Fig. 8.4). Miao et al. demonstrated that Nrf2 gene expression is directly modulated by TCDD-mediated AHR activation [163]. Subsequent DNA sequence analyses of the mouse Nrf2 promoter revealed the existence of three TCDD-sensitive XRE-like motifs, whose functionality was confirmed by reporter gene studies, site-directed mutagenesis, and gel shift experiments. *Vice versa*, Kensler and colleagues revealed that constitutive AHR expression height depends on Nrf2 genotype [164]. Furthermore, pharmacological activation of Nrf2 by 1-[2-cyano-3,12-dioxooleana-1,9(11)-dien-28-oyl]imidazole resulted in an enhanced transcription of AHR, CYP1A1, and CYP1B1 in Nrf2-expressing but not in Nrf2-deficient mouse embryonic fibroblasts. By means of reporter gene analyses and chromatin immunoprecipitation assays, the authors showed that Nrf2 binds directly to one ARE located in the murine AHR gene promoter. The importance of this bidirectional transcriptional control in regulation of xenobiotic metabolizing enzymes has been demonstrated in several animal models. A recently published study aimed to investigate the impact of Nrf2 on constitutive expression of several phase I and phase II enzymes as well as on phase III drug transporters [165]. As expected, the expression and catalytic activity of NQO1 and GST A1 were significantly lower in the livers of Nrf2-deficient animals compared to their wild-type littermates. In addition, hepatic expression of the three CYP1 isoforms, CYP2B10, and several drug transporters including MRPs and solute carrier organic anion transporters was significantly decreased in Nrf2-null mice. Nrf2-deficient mice further exhibited reduced AHR mRNA and protein levels, pointing again to the importance of Nrf2 for AHR gene regulation. The expression of two other master regulators of drug metabolism, constitutive androstane receptor and pregnane X receptor, as well as of nuclear corepressor 1 and 2 was also significantly reduced in Nrf2-null mice, indicating that Nrf2 is a crucial regulator not only for AHR signaling but also for other drug metabolizing pathways [165]. The integrated function of AHR- and Nrf2-regulated enzymes in detoxification was investigated in AHR/Nrf2 compound null mutants exposed to the prototype AHR agonist 3-MC and the Nrf2 inducer butylated hydro-

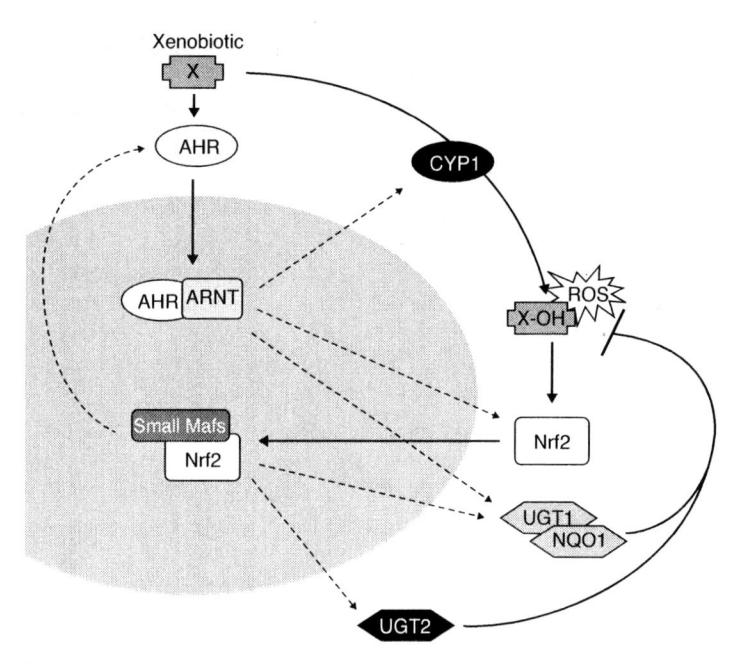

FIGURE 8.4 Crosstalk between AHR and Nrf2 signaling. Activation of the AHR pathway by a chemical ligand leads to the transcriptional induction of phase I enzymes (*CYP1*), phase II enzymes (*UGT1, NQO1*), *Nrf2*, and other target genes. The activating ligand gets oxidized by CYP1, a mechanism that can produce reactive metabolites and ROS, which in turn stabilize and activate Nrf2. Nrf2 further enhances transcription of several phase II enzymes, including AHR-independent proteins such as UGT2. The transcribed and translated phase II enzymes conjugate the electrophilic metabolites to hydrophilic substances and scavenge ROS. In addition, activation of Nrf2 is accompanied by a transcriptional increase in the AHR, probably to regenerate the cellular AHR pool for future xenobiotic threats. The dotted lines indicate processes of transcription and translation.

xyanisole (BHA) [166]. 3-MC-triggered gene induction of CYP1A1 and CYP1A2 as well as of NQO1 was completely abolished in the AHR/Nrf2 double mutants. After oral BHA administration, gene induction of GST Pi and NQO1 was significantly reduced in AHR/Nrf2 double mutant mice. The authors noticed a 45% and 67% decrease in NQO1 and GST Pi gene induction in AHR-deficient mice, respectively, pointing to the importance of a functional interaction of AHR and Nrf2 signaling pathways, especially in the context of metabolic biotransformation and toxicity of various drugs. An mRNA expression analysis of drug processing genes in wild-type and Nrf2-deficient mice treated with a high dose of TCDD revealed gene-specific effects. While the TCDD-triggered increase of CYP1A1 and UGT1A1 transcripts was similar in wild-type and Nrf2 knockout mice, the response of UGT1A5 and UGT1A9 toward dioxin was lost in Nrf2-null animals [167]. Exposure with TCDD resulted in gene induction of NQO1, UGT1A6, 2B34, 2B35, 2B36, and microsomal GST1 and GST A1, M1, M2, M3, M6, P2, and T2 in wild-type but not in Nrf2-deficient mice, demonstrating that Nrf2 is indispensable for TCDD-mediated induction of classical AHR battery genes in murine liver. Accordingly, TCDD-induced transcription was accompanied by an increased amount of nuclear Nrf2 [167].

8.3.3 Influence of AHR/Nrf2 Crosstalk on Metabolic Activation of PAHs

The extensive crosstalk of AHR and Nrf2 on transcriptional level implies a fundamental role of Nrf2 in AHR-triggered drug metabolism as well as metabolic activation of chemicals. It has been known for years that the AHR and downstream CYP1 monooxygenases play a crucial role in metabolic activation of procarcinogens relevant to human exposure. Accordingly, chemical inhibition of AHR signaling is thought to act protectively against DNA adduct formation of PAHs such as BaP or 7,12-dimethylbenzo(*a*)anthracene (DMBA). For instance, it was shown that treatment of human MCF-7 breast cancer cells with chrysoeriol and curcumin resulted in an inhibition of CYP1 activities and a reduced formation of BaP–DNA and DMBA–DNA adducts, respectively [168, 169]. Recent publications provide evidence that the Nrf2 system is of functional relevance for the chemopreventive properties of such plant polyphenols. Maru and coworkers pretreated mice orally with curcumin and subsequently exposed the mice to BaP [170]. In contrast to control mice, the BaP-stimulated induction of CYP1 expression and activity was markedly reduced in lungs and livers of curcumin-pretreated animals. In addition, curcumin altered BaP-induced phosphorylation, nuclear translocation, and DNA binding of AHR. This was accompanied by an increased protein expression and an enhanced nuclear localization of Nrf2, as well as by an increased Nrf2/ARE binding. Curcumin-mediated activation of Nrf2 was paralleled by increased GST and NQO1 gene expression and activity, resulting in an enhanced detoxification of BaP. The relevance of these findings regarding a potential protection against BaP-initiated carcinogenicity became evident by the observation that curcumin-exposed animals showed diminished DNA adduct formation, reduced oxidative damage, and less inflammatory markers [170]. Another study on female albino mice confirmed these findings. The animals were coexposed toward BaP and sulforaphane, a constituent of cruciferous vegetables [171]. Subsequent determination of AHR and CYP1 expression and activity suggests an antagonistic function of sulforaphane on the AHR system. Again, this was accompanied with an increased expression and activation of Nrf2 and downstream phase II enzymes, which probably contributes to the known anticarcinogenic action of sulforaphane [171, 172]. Mechanistic *in vitro* experiments performed in MCF-7 cells cotreated with DMBA and the anti-inflammatory agent 1-furan-2-yl-3-pyridin-2-yl-propenone (FPP-3) revealed that FPP-3-mediated reduction of DMBA genotoxicity is due to inhibition of nuclear AHR shuttling, CYP1 expression and activity, and a parallel stimulation of Nrf2 and its target enzymes NQO1 and GST [173]. Inhibitor studies indicate that the stimulating effect of FPP-3 on NQO1 and GST enzymes is mediated by PKC and p38 MAPK, pointing to the involvement of the classical Nrf2 cascade (Fig. 8.3).

Although chemopreventive agents may act by inhibiting AHR-triggered CYP1 metabolizing enzymes or by inducing Nrf2-dependent phase II detoxifying enzymes, these recently published studies imply that the most promising candidates are probably those that do both simultaneously.

8.4 CONCLUSION

As outlined in this chapter, multiple hierarchical levels and numerous signaling routes exist by which the AHR communicates and interferes with HIF-1α and Nrf2. Recently published studies also provide evidence for an active crosstalk between Nrf2 and HIF-1α during ethanol-induced oxidative stress [174], hypoxia [175, 176], and tumor angiogenesis [177], a fact that further complicates the unravelling of the AHR/Nrf2 and AHR/HIF-1α networks. However, both the AHR/HIF-1α and the AHR/Nrf2 axis represent suitable targets for the development of new chemotherapeutic and chemopreventive strategies. A combined chemical modulation of AHR and Nrf2 signaling is probably a good tool to protect tissues, in particular those frequently exposed to environmentally, dietary-, or lifestyle-derived AHR ligands such as liver, lung, skin, or GI tract, against the metabolic activation and thus tumor-initiating potency of procarcinogenic chemicals. Although cellular coexposure scenarios clearly point to the idea that the HIF-1α-mediated hypoxic response dominates over AHR signaling and thus may affect biotransformation of drugs and chemicals, the overall relevance of this "master–slave" relationship seems to be negligible for humans. Instead, AHR agonistic plant polyphenols seem to exert antiangiogenic effects via suppression of HIF-1α-dependent gene expression (especially VEGF), which might be a useful therapeutic approach to block oxygen supply and thereby growth and progression of tumors.

Future mechanistic investigations will shed light on detailed molecular facets of these crosstalks and moreover elucidate the underlying mechanisms of often observed cell and tissue specificity. Other highly involved factors and pathways have to be discovered to completely understand the fascinating biology of AHR and its crucial role during development, drug metabolism, and POP toxicity.

REFERENCES

1. Abel, J. and Haarmann-Stemmann, T. (2010). An introduction to the molecular basics of aryl hydrocarbon receptor biology. *Biological Chemistry*, 391, 1235–1248.
2. Afaq, F., Zaid, M. A., Pelle, E., Khan, N., Syed, D. N., Matsui, M. S., Maes, D., and Mukhtar, H. (2009). Aryl hydrocarbon receptor is an ozone sensor in human skin. *Journal of Investigative Dermatology*, 129, 2396–2403.
3. Denison, M. S. and Nagy, S. R. (2003). Activation of the aryl hydrocarbon receptor by structurally diverse exogenous and endogenous chemicals. *Annual Review of Pharmacology and Toxicology*, 43, 309–334.
4. Haarmann-Stemmann, T., Bothe, H., and Abel, J. (2009). Growth factors, cytokines and their receptors as downstream targets of arylhydrocarbon receptor (AhR) signaling pathways. *Biochemical Pharmacology*, 77, 508–520.
5. Osburn, W. O. and Kensler, T. W. (2008). Nrf2 signaling: an adaptive response pathway for protection against environmental toxic insults. *Mutation Research*, 659, 31–39.
6. Fernandez-Salguero, P. M., Hilbert, D. M., Rudikoff, S., Ward, J. M., and Gonzalez, F. J. (1996). Aryl-hydrocarbon receptor-deficient mice are resistant to 2,3,7, 8-tetrachlorodibenzo-p-dioxin-induced toxicity. *Toxicology and Applied Pharmacology*, 140, 173–179.
7. Mimura, J., Yamashita, K., Nakamura, K., Morita, M., Takagi, T. N., Nakao, K., Ema, M., Sogawa, K., Yasuda, M., Katsuki, M., and Fujii-Kuriyama, Y. (1997). Loss of teratogenic response to 2,3,7, 8-tetrachlorodibenzo-p-dioxin (TCDD) in mice lacking the Ah (dioxin) receptor. *Genes to Cells*, 2, 645–654.
8. Shimizu, Y., Nakatsuru, Y., Ichinose, M., Takahashi, Y., Kume, H., Mimura, J., Fujii-Kuriyama, Y., and Ishikawa, T. (2000). Benzo[*a*]pyrene carcinogenicity is lost in mice lacking the aryl hydrocarbon receptor. *Proceedings of the National Academy of Sciences of the United States of America*, 97, 779–782.

9. Fernandez-Salguero, P. M., Ward, J. M., Sundberg, J. P., and Gonzalez, F. J. (1997). Lesions of aryl-hydrocarbon receptor-deficient mice. *Veterinary Pathology*, 34, 605–614.
10. Schmidt, J. V., Su, G. H., Reddy, J. K., Simon, M. C., and Bradfield, C. A. (1996). Characterization of a murine Ahr null allele: involvement of the Ah receptor in hepatic growth and development. *Proceedings of the National Academy of Sciences of the United States of America*, 93, 6731–6736.
11. Sekine, H., Mimura, J., Oshima, M., Okawa, H., Kanno, J., Igarashi, K., Gonzalez, F. J., Ikuta, T., Kawajiri, K., and Fujii-Kuriyama, Y. (2009). Hypersensitivity of aryl hydrocarbon receptor-deficient mice to lipopolysaccharide-induced septic shock. *Molecular and Cellular Biology*, 29, 6391–6400.
12. Thackaberry, E. A., Gabaldon, D. M., Walker, M. K., and Smith, S. M. (2002). Aryl hydrocarbon receptor null mice develop cardiac hypertrophy and increased hypoxia-inducible factor-1α in the absence of cardiac hypoxia. *Cardiovascular Toxicology*, 2, 263–274.
13. Gu, Y. Z., Hogenesch, J. B., and Bradfield, C. A. (2000). The PAS superfamily: sensors of environmental and developmental signals. *Annual Review of Pharmacology and Toxicology*, 40, 519–561.
14. Wang, G. L., Jiang, B. H., Rue, E. A., and Semenza, G. L. (1995). Hypoxia-inducible factor 1 is a basic-helix–loop–helix–PAS heterodimer regulated by cellular O_2 tension. *Proceedings of the National Academy of Sciences of the United States of America*, 92, 5510–5514.
15. Abbott, B. D. and Probst, M. R. (1995). Developmental expression of two members of a new class of transcription factors: II. Expression of aryl hydrocarbon receptor nuclear translocator in the C57BL/6N mouse embryo. *Developmental Dynamics*, 204, 144–155.
16. Maxwell, P. H., Wiesener, M. S., Chang, G. W., Clifford, S. C., Vaux, E. C., Cockman, M. E., Wykoff, C. C., Pugh, C. W., Maher, E. R., and Ratcliffe, P. J. (1999). The tumour suppressor protein VHL targets hypoxia-inducible factors for oxygen-dependent proteolysis. *Nature*, 399, 271–275.
17. Lee, J. W., Bae, S. H., Jeong, J. W., Kim, S. H., and Kim, K. W. (2004). Hypoxia-inducible factor (HIF-1)α: its protein stability and biological functions. *Experimental Molecular Medicine*, 36, 1–12.
18. Semenza, G. L. (2000). HIF-1 and human disease: one highly involved factor. *Genes & Development*, 14, 1983–1991.
19. Ema, M., Taya, S., Yokotani, N., Sogawa, K., Matsuda, Y., and Fujii-Kuriyama, Y. (1997). A novel bHLH–PAS factor with close sequence similarity to hypoxia-inducible factor 1α regulates the VEGF expression and is potentially involved in lung and vascular development. *Proceedings of the National Academy of Sciences of the United States of America*, 94, 4273–4278.
20. Gu, Y. Z., Moran, S. M., Hogenesch, J. B., Wartman, L., and Bradfield, C. A. (1998). Molecular characterization and chromosomal localization of a third alpha-class hypoxia inducible factor subunit, HIF3α. *Gene Expression*, 7, 205–213.
21. Makino, Y., Kanopka, A., Wilson, W. J., Tanaka, H., and Poellinger, L. (2002). Inhibitory PAS domain protein (IPAS) is a hypoxia-inducible splicing variant of the hypoxia-inducible factor-3α locus. *Journal of Biological Chemistry*, 277, 32405–32408.
22. Fukuda, R., Hirota, K., Fan, F., Jung, Y. D., Ellis, L. M., and Semenza, G. L. (2002). Insulin-like growth factor 1 induces hypoxia-inducible factor 1-mediated vascular endothelial growth factor expression, which is dependent on MAP kinase and phosphatidylinositol 3-kinase signaling in colon cancer cells. *Journal of Biological Chemistry*, 277, 38205–38211.
23. Gorlach, A., Diebold, I., Schini-Kerth, V. B., Berchner-Pfannschmidt, U., Roth, U., Brandes, R. P., Kietzmann, T., and Busse, R. (2001). Thrombin activates the hypoxia-inducible factor-1 signaling pathway in vascular smooth muscle cells: role of the p22(phox)-containing NADPH oxidase. *Circulation Research*, 89, 47–54.
24. Stiehl, D. P., Jelkmann, W., Wenger, R. H., and Hellwig-Burgel, T. (2002). Normoxic induction of the hypoxia-inducible factor 1α by insulin and interleukin-1β involves the phosphatidylinositol 3-kinase pathway. *FEBS Letters*, 512, 157–162.
25. Treins, C., Giorgetti-Peraldi, S., Murdaca, J., Semenza, G. L., and Van Obberghen, E. (2002). Insulin stimulates hypoxia-inducible factor 1 through a phosphatidylinositol 3-kinase/target of rapamycin-dependent signaling pathway. *Journal of Biological Chemistry*, 277, 27975–27981.
26. Ryan, H. E., Lo, J., and Johnson, R. S. (1998). HIF-1α is required for solid tumor formation and embryonic vascularization. *EMBO Journal*, 17, 3005–3015.
27. Chan, W. K., Yao, G., Gu, Y. Z., and Bradfield, C. A. (1999). Cross-talk between the aryl hydrocarbon receptor and hypoxia inducible factor signaling pathways. Demonstration of competition and compensation. *Journal of Biological Chemistry*, 274, 12115–12123.
28. Krishnamurthy, P., Ross, D. D., Nakanishi, T., Bailey-Dell, K., Zhou, S., Mercer, K. E., Sarkadi, B., Sorrentino, B. P., and Schuetz, J. D. (2004). The stem cell marker Bcrp/ABCG2 enhances hypoxic cell survival through interactions with heme. *Journal of Biological Chemistry*, 279, 24218–24225.
29. Tan, K. P., Wang, B., Yang, M., Boutros, P. C., Macaulay, J., Xu, H., Chuang, A. I., Kosuge, K., Yamamoto, M., Takahashi, S., Wu, A. M., Ross, D. D., Harper, P. A., and Ito, S. (2010). Aryl hydrocarbon receptor is a transcriptional activator of the human breast cancer resistance protein (BCRP/ABCG2). *Molecular Pharmacology*, 78, 175–185.
30. Marchand, A., Tomkiewicz, C., Marchandeau, J. P., Boitier, E., Barouki, R., and Garlatti, M. (2005). 2,3,7,8-Tetrachlorodibenzo-*p*-dioxin induces insulin-like growth factor binding protein-1 gene expression and counteracts the negative effect of insulin. *Molecular Pharmacology*, 67, 444–452.
31. Tazuke, S. I., Mazure, N. M., Sugawara, J., Carland, G., Faessen, G. H., Suen, L. F., Irwin, J. C., Powell, D. R., Giaccia, A. J., and Giudice, L. C. (1998). Hypoxia stimulates insulin-like growth factor binding protein 1 (IGFBP-1) gene expression in HepG2 cells: a possible model for IGFBP-1 expression in fetal hypoxia. *Proceedings of the National Academy of Sciences of the United States of America*, 95, 10188–10193.

32. Rivera, S. P., Wang, F., Saarikoski, S. T., Taylor, R. T., Chapman, B., Zhang, R., and Hankinson, O. (2007). A novel promoter element containing multiple overlapping xenobiotic and hypoxia response elements mediates induction of cytochrome P4502S1 by both dioxin and hypoxia. *Journal of Biological Chemistry*, 282, 10881–10893.

33. Lee, K. A., Burgoon, L. D., Lamb, L., Dere, E., Zacharewski, T. R., Hogenesch, J. B., and LaPres, J. J. (2006). Identification and characterization of genes susceptible to transcriptional cross-talk between the hypoxia and dioxin signaling cascades. *Chemical Research in Toxicology*, 19, 1284–1293.

34. Gradin, K., Takasaki, C., Fujii-Kuriyama, Y., and Sogawa, K. (2002). The transcriptional activation function of the HIF-like factor requires phosphorylation at a conserved threonine. *Journal of Biological Chemistry*, 277, 23508–23514.

35. Kim, J. E. and Sheen, Y. Y. (2000). Inhibition of 2,3,7, 8-tetrachlorodibenzo-*p*-dioxin (TCDD)-stimulated Cyp1a1 promoter activity by hypoxic agents. *Biochemical Pharmacology*, 59, 1549–1556.

36. Kim, J. E. and Sheen, Y. Y. (2000). Nitric oxide inhibits dioxin action for the stimulation of Cyp1a1 promoter activity. *Biological & Pharmaceutical Bulletin*, 23, 575–580.

37. Pollenz, R. S., Davarinos, N. A., and Shearer, T. P. (1999). Analysis of aryl hydrocarbon receptor-mediated signaling during physiological hypoxia reveals lack of competition for the aryl hydrocarbon nuclear translocator transcription factor. *Molecular Pharmacology*, 56, 1127–1137.

38. Reisdorph, R. and Lindahl, R. (2001). Aldehyde dehydrogenase 3 gene regulation: studies on constitutive and hypoxia-modulated expression. *Chemico-Biological Interactions*, 130–132, 227–233.

39. Legendre, C., Hori, T., Loyer, P., Aninat, C., Ishida, S., Glaise, D., Lucas-Clerc, C., Boudjema, K., Guguen-Guillouzo, C., Corlu, A., and Morel, F. (2009). Drug-metabolising enzymes are down-regulated by hypoxia in differentiated human hepatoma HepaRG cells: HIF-1α involvement in CYP3A4 repression. *European Journal of Cancer*, 45, 2882–2892.

40. Marden, N. Y., Fiala-Beer, E., Xiang, S. H., and Murray, M. (2003). Role of activator protein-1 in the down-regulation of the human CYP2J2 gene in hypoxia. *Biochemical Journal*, 373, 669–680.

41. Allen, J. W., Johnson, R. S., and Bhatia, S. N. (2005). Hypoxic inhibition of 3-methylcholanthrene-induced CYP1A1 expression is independent of HIF-1α. *Toxicology Letters*, 155, 151–159.

42. Davidson, T., Salnikow, K., and Costa, M. (2003). Hypoxia inducible factor-1α-independent suppression of aryl hydrocarbon receptor-regulated genes by nickel. *Molecular Pharmacology*, 64, 1485–1493.

43. Zhang, N. and Walker, M. K. (2007). Crosstalk between the aryl hydrocarbon receptor and hypoxia on the constitutive expression of cytochrome P4501A1 mRNA. *Cardiovascular Toxicology*, 7, 282–290.

44. Naldini, A., Carraro, F., Silvestri, S., and Bocci, V. (1997). Hypoxia affects cytokine production and proliferative responses by human peripheral mononuclear cells. *Journal of Cellular Physiology*, 173, 335–342.

45. Fradette, C., Bleau, A. M., Pichette, V., Chauret, N., and du Souich, P. (2002). Hypoxia-induced down-regulation of CYP1A1/1A2 and up-regulation of CYP3A6 involves serum mediators. *British Journal of Pharmacology*, 137, 881–891.

46. Fradette, C., Batonga, J., Teng, S., Piquette-Miller, M., and du Souich, P. (2007). Animal models of acute moderate hypoxia are associated with a down-regulation of CYP1A1, 1A2, 2B4, 2C5, and 2C16 and up-regulation of CYP3A6 and P-glycoprotein in liver. *Drug Metabolism and Disposition*, 35, 765–771.

47. Pahl, H. L. (1999). Activators and target genes of Rel/NF-κB transcription factors. *Oncogene*, 18, 6853–6866.

48. Hiscott, J., Kwon, H., and Genin, P. (2001). Hostile takeovers: viral appropriation of the NF-κB pathway. *Journal of Clinical Investigation*, 107, 143–151.

49. Tian, Y., Ke, S., Denison, M. S., Rabson, A. B., and Gallo, M. A. (1999). Ah receptor and NF-κB interactions, a potential mechanism for dioxin toxicity. *Journal of Biological Chemistry*, 274, 510–515.

50. Ke, S., Rabson, A. B., Germino, J. F., Gallo, M. A., and Tian, Y. (2001). Mechanism of suppression of cytochrome P-450 1A1 expression by tumor necrosis factor-alpha and lipopolysaccharide. *Journal of Biological Chemistry*, 276, 39638–39644.

51. Bleau, A. M., Maurel, P., Pichette, V., Leblond, F., and du Souich, P. (2003). Interleukin-1β, interleukin-6, tumour necrosis factor-alpha and interferon-gamma released by a viral infection and an aseptic inflammation reduce CYP1A1, 1A2 and 3A6 expression in rabbit hepatocytes. *European Journal of Pharmacology*, 473, 197–206.

52. Khan, S., Liu, S., Stoner, M., and Safe, S. (2007). Cobaltous chloride and hypoxia inhibit aryl hydrocarbon receptor-mediated responses in breast cancer cells. *Toxicology and Applied Pharmacology*, 223, 28–38.

53. Denis, M., Cuthill, S., Wikstrom, A. C., Poellinger, L., and Gustafsson, J. A. (1988). Association of the dioxin receptor with the M_r 90, 000 heat shock protein: a structural kinship with the glucocorticoid receptor. *Biochemical and Biophysical Research Communications*, 155, 801–807.

54. Minet, E., Mottet, D., Michel, G., Roland, I., Raes, M., Remacle, J., and Michiels, C. (1999). Hypoxia-induced activation of HIF-1: role of HIF-1α–Hsp90 interaction. *FEBS Letters*, 460, 251–256.

55. Beischlag, T. V., Taylor, R. T., Rose, D. W., Yoon, D., Chen, Y., Lee, W. H., Rosenfeld, M. G., and Hankinson, O. (2004). Recruitment of thyroid hormone receptor/retinoblastoma-interacting protein 230 by the aryl hydrocarbon receptor nuclear translocator is required for the transcriptional response to both dioxin and hypoxia. *Journal of Biological Chemistry*, 279, 54620–54628.

56. Carrero, P., Okamoto, K., Coumailleau, P., O'Brien, S., Tanaka, H., and Poellinger, L. (2000). Redox-regulated recruitment of the transcriptional coactivators CREB-binding protein and SRC-1 to hypoxia-inducible factor 1α. *Molecular and Cellular Biology*, 20, 402–415.

57. Kumar, M. B. and Perdew, G. H. (1999). Nuclear receptor coactivator SRC-1 interacts with the Q-rich subdomain of the AhR and modulates its transactivation potential. *Gene Expression*, *8*, 273–286.
58. Kobayashi, A., Numayama-Tsuruta, K., Sogawa, K., and Fujii-Kuriyama, Y. (1997). CBP/p300 functions as a possible transcriptional coactivator of Ah receptor nuclear translocator (Arnt). *Journal of Biochemistry*, *122*, 703–710.
59. Wang, S. and Hankinson, O. (2002). Functional involvement of the Brahma/SWI2-related gene 1 protein in cytochrome P4501A1 transcription mediated by the aryl hydrocarbon receptor complex. *Journal of Biological Chemistry*, *277*, 11821–11827.
60. Wang, F., Zhang, R., Beischlag, T. V., Muchardt, C., Yaniv, M., and Hankinson, O. (2004). Roles of Brahma and Brahma/SWI2-related gene 1 in hypoxic induction of the erythropoietin gene. *Journal of Biological Chemistry*, *279*, 46733–46741.
61. Haarmann-Stemmann, T. and Abel, J. (2006). The arylhydrocarbon receptor repressor (AhRR): structure, expression, and function. *Biological Chemistry*, *387*, 1195–1199.
62. Haarmann-Stemmann, T., Bothe, H., Kohli, A., Sydlik, U., Abel, J., and Fritsche, E. (2007). Analysis of the transcriptional regulation and molecular function of the aryl hydrocarbon receptor repressor in human cell lines. *Drug Metabolism and Disposition*, *35*, 2262–2269.
63. Baba, T., Mimura, J., Gradin, K., Kuroiwa, A., Watanabe, T., Matsuda, Y., Inazawa, J., Sogawa, K., and Fujii-Kuriyama, Y. (2001). Structure and expression of the Ah receptor repressor gene. *Journal of Biological Chemistry*, *276*, 33101–33110.
64. Cauchi, S., Stucker, I., Cenee, S., Kremers, P., Beaune, P., and Massaad-Massade, L. (2003). Structure and polymorphisms of human aryl hydrocarbon receptor repressor (AhRR) gene in a French population: relationship with CYP1A1 inducibility and lung cancer. *Pharmacogenetics*, *13*, 339–347.
65. Jeong, H. J., Chung, H. S., Lee, B. R., Kim, S. J., Yoo, S. J., Hong, S. H., and Kim, H. M. (2003). Expression of proinflammatory cytokines via HIF-1α and NF-κB activation on desferrioxamine-stimulated HMC-1 cells. *Biochemical and Biophysical Research Communications*, *306*, 805–811.
66. Karchner, S. I., Jenny, M. J., Tarrant, A. M., Evans, B. R., Kang, H. J., Bae, I., Sherr, D. H., and Hahn, M. E. (2009). The active form of human aryl hydrocarbon receptor (AHR) repressor lacks exon 8, and its Pro 185 and Ala 185 variants repress both AHR and hypoxia-inducible factor. *Molecular and Cellular Biology*, *29*, 3465–3477.
67. Hofer, T., Pohjanvirta, R., Spielmann, P., Viluksela, M., Buchmann, D. P., Wenger, R. H., and Gassmann, M. (2004). Simultaneous exposure of rats to dioxin and carbon monoxide reduces the xenobiotic but not the hypoxic response. *Biological Chemistry*, *385*, 291–294.
68. Jurgens, G., Christensen, H. R., Brosen, K., Sonne, J., Loft, S., and Olsen, N. V. (2002). Acute hypoxia and cytochrome P450-mediated hepatic drug metabolism in humans. *Clinical Pharmacology and Therapeutics*, *71*, 214–220.
69. Streit, M., Goggelmann, C., Dehnert, C., Burhenne, J., Riedel, K. D., Menold, E., Mikus, G., Bartsch, P., and Haefeli, W. E. (2005). Cytochrome P450 enzyme-mediated drug metabolism at exposure to acute hypoxia (corresponding to an altitude of 4, 500 m). *European Journal of Clinical Pharmacology*, *61*, 39–46.
70. Iyer, N. V., Kotch, L. E., Agani, F., Leung, S. W., Laughner, E., Wenger, R. H., Gassmann, M., Gearhart, J. D., Lawler, A. M., Yu, A. Y., and Semenza, G. L. (1998). Cellular and developmental control of O_2 homeostasis by hypoxia-inducible factor 1α. *Genes & Development*, *12*, 149–162.
71. Carmeliet, P., Ferreira, V., Breier, G., Pollefeyt, S., Kieckens, L., Gertsenstein, M., Fahrig, M., Vandenhoeck, A., Harpal, K., Eberhardt, C., Declercq, C., Pawling, J., Moons, L., Collen, D., Risau, W., and Nagy, A. (1996). Abnormal blood vessel development and lethality in embryos lacking a single VEGF allele. *Nature*, *380*, 435–439.
72. Maltepe, E., Schmidt, J. V., Baunoch, D., Bradfield, C. A., and Simon, M. C. (1997). Abnormal angiogenesis and responses to glucose and oxygen deprivation in mice lacking the protein ARNT. *Nature*, *386*, 403–407.
73. Kozak, K. R., Abbott, B., and Hankinson, O. (1997). ARNT-deficient mice and placental differentiation. *Developmental Biology*, *191*, 297–305.
74. Tomita, S., Sinal, C. J., Yim, S. H., and Gonzalez, F. J. (2000). Conditional disruption of the aryl hydrocarbon receptor nuclear translocator (Arnt) gene leads to loss of target gene induction by the aryl hydrocarbon receptor and hypoxia-inducible factor 1α. *Molecular Endocrinology*, *14*, 1674–1681.
75. Peng, J., Zhang, L., Drysdale, L., and Fong, G. H. (2000). The transcription factor EPAS-1/hypoxia-inducible factor 2α plays an important role in vascular remodeling. *Proceedings of the National Academy of Sciences of the United States of America*, *97*, 8386–8391.
76. Lahvis, G. P., Lindell, S. L., Thomas, R. S., McCuskey, R. S., Murphy, C., Glover, E., Bentz, M., Southard, J., and Bradfield, C. A. (2000). Portosystemic shunting and persistent fetal vascular structures in aryl hydrocarbon receptor-deficient mice. *Proceedings of the National Academy of Sciences of the United States of America*, *97*, 10442–10447.
77. Lahvis, G. P., Pyzalski, R. W., Glover, E., Pitot, H. C., McElwee, M. K., and Bradfield, C. A. (2005). The aryl hydrocarbon receptor is required for developmental closure of the ductus venosus in the neonatal mouse. *Molecular Pharmacology*, *67*, 714–720.
78. Roman, A. C., Carvajal-Gonzalez, J. M., Rico-Leo, E. M., and Fernandez-Salguero, P. M. (2009). Dioxin receptor deficiency impairs angiogenesis by a mechanism involving VEGF-A depletion in the endothelium and transforming growth factor-beta overexpression in the stroma. *Journal of Biological Chemistry*, *284*, 25135–25148.
79. Mulero-Navarro, S., Pozo-Guisado, E., Perez-Mancera, P. A., varez-Barrientos, A., Catalina-Fernandez, I., Hernandez-Nieto, E., Saenz-Santamaria, J., Martinez, N., Rojas, J. M., Sanchez-Garcia, I., and Fernandez-Salguero, P. M. (2005).

Immortalized mouse mammary fibroblasts lacking dioxin receptor have impaired tumorigenicity in a subcutaneous mouse xenograft model. *Journal of Biological Chemistry, 280,* 28731–28741.

80. Andersson, P., McGuire, J., Rubio, C., Gradin, K., Whitelaw, M. L., Pettersson, S., Hanberg, A., and Poellinger, L. (2002). A constitutively active dioxin/aryl hydrocarbon receptor induces stomach tumors. *Proceedings of the National Academy of Sciences of the United States of America, 99,* 9990–9995.

81. Moennikes, O., Loeppen, S., Buchmann, A., Andersson, P., Ittrich, C., Poellinger, L., and Schwarz, M. (2004). A constitutively active dioxin/aryl hydrocarbon receptor promotes hepatocarcinogenesis in mice. *Cancer Research, 64,* 4707–4710.

82. Nohara, K., Pan, X., Tsukumo, S., Hida, A., Ito, T., Nagai, H., Inouye, K., Motohashi, H., Yamamoto, M., Fujii-Kuriyama, Y., and Tohyama, C. (2005). Constitutively active aryl hydrocarbon receptor expressed specifically in T-lineage cells causes thymus involution and suppresses the immunization-induced increase in splenocytes. *Journal of Immunology, 174,* 2770–2777.

83. Ivnitski-Steele, I. and Walker, M. K. (2005). Inhibition of neovascularization by environmental agents. *Cardiovascular Toxicology, 5,* 215–226.

84. Walker, M. K., Pollenz, R. S., and Smith, S. M. (1997). Expression of the aryl hydrocarbon receptor (AhR) and AhR nuclear translocator during chick cardiogenesis is consistent with 2,3,7,8-tetrachlorodibenzo-p-dioxin-induced heart defects. *Toxicology and Applied Pharmacology, 143,* 407–419.

85. Walker, M. K. and Catron, T. F. (2000). Characterization of cardiotoxicity induced by 2,3,7,8-tetrachlorodibenzo-p-dioxin and related chemicals during early chick embryo development. *Toxicology and Applied Pharmacology, 167,* 210–221.

86. Heid, S. E., Walker, M. K., and Swanson, H. I. (2001). Correlation of cardiotoxicity mediated by halogenated aromatic hydrocarbons to aryl hydrocarbon receptor activation. *Toxicological Sciences, 61,* 187–196.

87. Ivnitski-Steele, I. D., Sanchez, A., and Walker, M. K. (2004). 2,3,7,8-Tetrachlorodibenzo-p-dioxin reduces myocardial hypoxia and vascular endothelial growth factor expression during chick embryo development. *Birth Defects Research, Part A: Clinical and Molecular Teratology, 70,* 51–58.

88. Thackaberry, E. A., Nunez, B. A., Ivnitski-Steele, I. D., Friggins, M., and Walker, M. K. (2005). Effect of 2,3,7,8-tetrachlorodibenzo-p-dioxin on murine heart development: alteration in fetal and postnatal cardiac growth, and postnatal cardiac chronotropy. *Toxicological Sciences, 88,* 242–249.

89. Ivnitski-Steele, I. D. and Walker, M. K. (2003). Vascular endothelial growth factor rescues 2,3,7,8-tetrachlorodibenzo-p-dioxin inhibition of coronary vasculogenesis. *Birth Defects Research, Part A: Clinical and Molecular Teratology, 67,* 496–503.

90. Li, C. H., Cheng, Y. W., Hsu, Y. T., Hsu, Y. J., Liao, P. L., and Kang, J. J. (2010). Benzo[a]pyrene inhibits angiogenic factors avb3 integrin expression, neovasculogenesis, and angiogenesis in human umbilical vein endothelial cells. *Toxicological Sciences, 118,* 544–553.

91. Juan, S. H., Lee, J. L., Ho, P. Y., Lee, Y. H., and Lee, W. S. (2006). Antiproliferative and antiangiogenic effects of 3-methylcholanthrene, an aryl-hydrocarbon receptor agonist, in human umbilical vascular endothelial cells. *European Journal of Pharmacology, 530,* 1–8.

92. Ishimura, R., Kawakami, T., Ohsako, S., and Tohyama, C. (2009). Dioxin-induced toxicity on vascular remodeling of the placenta. *Biochemical Pharmacology, 77,* 660–669.

93. Ishimura, R., Ohsako, S., Kawakami, T., Sakaue, M., Aoki, Y., and Tohyama, C. (2002). Altered protein profile and possible hypoxia in the placenta of 2,3,7,8-tetrachlorodibenzo-p-dioxin-exposed rats. *Toxicology and Applied Pharmacology, 185,* 197–206.

94. Ishimura, R., Kawakami, T., Ohsako, S., Nohara, K., and Tohyama, C. (2006). Suppressive effect of 2,3,7,8-tetrachlorodibenzo-p-dioxin on vascular remodeling that takes place in the normal labyrinth zone of rat placenta during late gestation. *Toxicological Sciences, 91,* 265–274.

95. Ichihara, S., Yamada, Y., Ichihara, G., Nakajima, T., Li, P., Kondo, T., Gonzalez, F. J., and Murohara, T. (2007). A role for the aryl hydrocarbon receptor in regulation of ischemia-induced angiogenesis. *Arteriosclerosis, Thrombosis and Vascular Biology, 27,* 1297–1304.

96. Ichihara, S., Yamada, Y., Gonzalez, F. J., Nakajima, T., Murohara, T., and Ichihara, G. (2009). Inhibition of ischemia-induced angiogenesis by benzo[a]pyrene in a manner dependent on the aryl hydrocarbon receptor. *Biochemical and Biophysical Research Communications, 381,* 44–49.

97. IARC (1997). Polychlorinated dibenzo-*para*-dioxins and polychlorinated dibenzofurans. *IARC Monographs on the Evaluation of Carcinogenic Risks to Humans, 69,* 33–334.

98. Holcomb, M. and Safe, S. (1994). Inhibition of 7,12-dimethylbenzanthracene-induced rat mammary tumor growth by 2,3,7,8-tetrachlorodibenzo-p-dioxin. *Cancer Letters, 82,* 43–47.

99. Kociba, R. J., Keyes, D. G., Beyer, J. E., Carreon, R. M., Wade, C. E., Dittenber, D. A., Kalnins, R. P., Frauson, L. E., Park, C. N., Barnard, S. D., Hummel, R. A., and Humiston, C. G. (1978). Results of a two-year chronic toxicity and oncogenicity study of 2,3,7,8-tetrachlorodibenzo-p-dioxin in rats. *Toxicology and Applied Pharmacology, 46,* 279–303.

100. Hsu, E. L., Yoon, D., Choi, H. H., Wang, F., Taylor, R. T., Chen, N., Zhang, R., Hankinson, O. (2007). A proposed mechanism for the protective effect of dioxin against breast cancer. *Toxicological Sciences, 98,* 436–444.

101. Takacova, M., Holotnakova, T., Vondracek, J., Machala, M., Pencikova, K., Gradin, K., Poellinger, L., Pastorek, J., Pastorekova, S., and Kopacek, J. (2009). Role of aryl hydrocarbon receptor in modulation of the expression of the hypoxia marker carbonic anhydrase IX. *Biochemical Journal, 419,* 419–425.

102. Semenza, G. L. (2003). Targeting HIF-1 for cancer therapy. *Nature Reviews Cancer, 3,* 721–732.

103. Fang, J., Zhou, Q., Liu, L. Z., Xia, C., Hu, X., Shi, X., and Jiang, B. H. (2007). Apigenin inhibits tumor angiogenesis through decreasing HIF-1α and VEGF expression. *Carcinogenesis*, 28, 858–864.

104. Fu, B., Xue, J., Li, Z., Shi, X., Jiang, B. H., and Fang, J. (2007). Chrysin inhibits expression of hypoxia-inducible factor-1alpha through reducing hypoxia-inducible factor-1α stability and inhibiting its protein synthesis. *Molecular Cancer Therapy*, 6, 220–226.

105. Oh, S. J., Kim, O., Lee, J. S., Kim, J. A., Kim, M. R., Choi, H. S., Shim, J. H., Kang, K. W., and Kim, Y. C. (2010). Inhibition of angiogenesis by quercetin in tamoxifen-resistant breast cancer cells. *Food Chemistry and Toxicology*, 48, 3227–3234.

106. Terzuoli, E., Puppo, M., Rapisarda, A., Uranchimeg, B., Cao, L., Burger, A. M., Ziche, M., and Melillo, G. (2010). Aminoflavone, a ligand of the aryl hydrocarbon receptor, inhibits HIF-1α expression in an AhR-independent fashion. *Cancer Research*, 70, 6837–6848.

107. Ciolino, H. P., Daschner, P. J., and Yeh, G. C. (1999). Dietary flavonols quercetin and kaempferol are ligands of the aryl hydrocarbon receptor that affect CYP1A1 transcription differentially. *Biochemical Journal*, 340, 715–722.

108. Zhang, S., Qin, C., and Safe, S. H. (2003). Flavonoids as aryl hydrocarbon receptor agonists/antagonists: effects of structure and cell context. *Environmental Health Perspectives*, 111, 1877–1882.

109. Li, Z. D., Liu, L. Z., Shi, X., Fang, J., Jiang, B. H. (2007). Benzo[a]pyrene-3,6-dione inhibited VEGF expression through inducing HIF-1α degradation. *Biochemical and Biophysical Research Communications*, 357, 517–523.

110. Seifert, A., Katschinski, D. M., Tonack, S., Fischer, B., and Navarrete, S. A. (2008). Significance of prolyl hydroxylase 2 in the interference of aryl hydrocarbon receptor and hypoxia-inducible factor-1 alpha signaling. *Chemical Research in Toxicology*, 21, 341–348.

111. Jeong, J. W., Bae, M. K., Ahn, M. Y., Kim, S. H., Sohn, T. K., Bae, M. H., Yoo, M. A., Song, E. J., Lee, K. J., and Kim, K. W. (2002). Regulation and destabilization of HIF-1α by ARD1-mediated acetylation. *Cell*, 111, 709–720.

112. Kawajiri, K., Kobayashi, Y., Ohtake, F., Ikuta, T., Matsushima, Y., Mimura, J., Pettersson, S., Pollenz, R. S., Sakaki, T., Hirokawa, T., Akiyama, T., Kurosumi, M., Poellinger, L., Kato, S., and Fujii-Kuriyama, Y. (2009). Aryl hydrocarbon receptor suppresses intestinal carcinogenesis in ApcMin/+ mice with natural ligands. *Proceedings of the National Academy of Sciences of the United States of America*, 106, 13481–13486.

113. Ohtake, F., Baba, A., Takada, I., Okada, M., Iwasaki, K., Miki, H., Takahashi, S., Kouzmenko, A., Nohara, K., Chiba, T., Fujii-Kuriyama, and Y., Kato, S. (2007). Dioxin receptor is a ligand-dependent E3 ubiquitin ligase. *Nature*, 446, 562–566.

114. Palermo, C. M., Westlake, C. A., and Gasiewicz, T. A. (2005). Epigallocatechin gallate inhibits aryl hydrocarbon receptor gene transcription through an indirect mechanism involving binding to a 90 kDa heat shock protein. *Biochemistry*, 44, 5041–5052.

115. Zhang, Q., Tang, X., Lu, Q., Zhang, Z., Rao, J., and Le, A. D. (2006). Green tea extract and (−)-epigallocatechin-3-gallate inhibit hypoxia- and serum-induced HIF-1α protein accumulation and VEGF expression in human cervical carcinoma and hepatoma cells. *Molecular Cancer Therapeutics*, 5, 1227–1238.

116. Sciullo, E. M., Vogel, C. F., Wu, D., Murakami, A., Ohigashi, H., and Matsumura, F. (2010). Effects of selected food phytochemicals in reducing the toxic actions of TCDD and p,p′-DDT in U937 macrophages. *Archives of Toxicology*, 84, 957–966.

117. Frericks, M., Burgoon, L. D., Zacharewski, T. R., and Esser, C. (2008). Promoter analysis of TCDD-inducible genes in a thymic epithelial cell line indicates the potential for cell-specific transcription factor crosstalk in the AhR response. *Toxicology and Applied Pharmacology*, 232, 268–279.

118. Takeuchi, A., Takeuchi, M., Oikawa, K., Sonoda, K. H., Usui, Y., Okunuki, Y., Takeda, A., Oshima, Y., Yoshida, K., Usui, M., Goto, H., and Kuroda, M. (2009). Effects of dioxin on vascular endothelial growth factor (VEGF) production in the retina associated with choroidal neovascularization. *Investigative Ophthalmology & Visual Science*, 50, 3410–3416.

119. Bandi, N., Ayalasomayajula, S. P., Dhanda, D. S., Iwakawa, J., Cheng, P. W., and Kompella, U. B. (2005). Intratracheal budesonide-poly(lactide-co-glycolide) microparticles reduce oxidative stress, VEGF expression, and vascular leakage in a benzo(a)pyrene-fed mouse model. *Journal of Pharmacy and Pharmacology*, 57, 851–860.

120. Petit, A. M., Rak, J., Hung, M. C., Rockwell, P., Goldstein, N., Fendly, B., and Kerbel, R. S. (1997). Neutralizing antibodies against epidermal growth factor and ErbB-2/neu receptor tyrosine kinases down-regulate vascular endothelial growth factor production by tumor cells *in vitro* and *in vivo*: angiogenic implications for signal transduction therapy of solid tumors. *American Journal of Pathology*, 151, 1523–1530.

121. Fritsche, E., Schafer, C., Calles, C., Bernsmann, T., Bernshausen, T., Wurm, M., Hubenthal, U., Cline, J. E., Hajimiragha, H., Schroeder, P., Klotz, L. O., Rannug, A., Furst, P., Hanenberg, H., Abel, J., and Krutmann, J. (2007). Lightening up the UV response by identification of the arylhydrocarbon receptor as a cytoplasmatic target for ultraviolet B radiation. *Proceedings of the National Academy of Sciences of the United States of America*, 104, 8851–8856.

122. Köhle, C., Gschaidmeier, H., Lauth, D., Topell, S., Zitzer, H., and Bock, K. W. (1999). 2,3,7,8-Tetrachlorodibenzo-p-dioxin (TCDD)-mediated membrane translocation of c-Src protein kinase in liver WB-F344 cells. *Archives of Toxicology*, 73, 152–158.

123. Martinez, J. M., Baek, S. J., Mays, D. M., Tithof, P. K., Eling, T. E., and Walker, N. J. (2004). EGR1 is a novel target for AhR agonists in human lung epithelial cells. *Toxicological Sciences*, 82, 429–435.

124. Lo, L. W., Cheng, J. J., Chiu, J. J., Wung, B. S., Liu, Y. C., Wang, D. L. (2001). Endothelial exposure to hypoxia induces Egr-1 expression involving PKCα-mediated Ras/Raf-1/ERK1/2 pathway. *Journal of Cellular Physiology*, 188, 304–312.

125. Yan, S. F., Lu, J., Zou, Y. S., Soh-Won, J., Cohen, D. M., Buttrick, P. M., Cooper, D. R., Steinberg, S. F., Mackman, N., Pinsky, D. J., and Stern, D. M. (1999). Hypoxia-associated induction of early growth response-1 gene expression. *Journal of Biological Chemistry*, 274, 15030–15040.

126. Shimoyamada, H., Yazawa, T., Sato, H., Okudela, K., Ishii, J., Sakaeda, M., Kashiwagi, K., Suzuki, T., Mitsui, H., Woo, T., Tajiri, M., Ohmori, T., Ogura, T., Masuda, M., Oshiro, H., and Kitamura, H. (2010). Early growth response-1 induces and enhances vascular endothelial growth factor-A expression in lung cancer cells. *American Journal of Pathology*, 177, 70–83.

127. Ding, J., Li, J., Chen, J., Chen, H., Ouyang, W., Zhang, R., Xue, C., Zhang, D., Amin, S., Desai, D., and Huang, C. (2006). Effects of polycyclic aromatic hydrocarbons (PAHs) on vascular endothelial growth factor induction through phosphatidylinositol 3-kinase/AP-1-dependent, HIF-1α-independent pathway. *Journal of Biological Chemistry*, 281, 9093–9100.

128. Huang, C., Li, J., Song, L., Zhang, D., Tong, Q., Ding, M., Bowman, L., Aziz, R., and Stoner, G. D. (2006). Black raspberry extracts inhibit benzo(*a*)pyrene diol-epoxide-induced activator protein 1 activation and VEGF transcription by targeting the phosphotidylinositol 3-kinase/Akt pathway. *Cancer Research*, 66, 581–587.

129. Puga, A., Nebert, D. W., and Carrier, F. (1992). Dioxin induces expression of c-fos and c-jun proto-oncogenes and a large increase in transcription factor AP-1. *DNA and Cell Biology*, 11, 269–281.

130. Puga, A., Barnes, S. J., Chang, C., Zhu, H., Nephew, K. P., Khan, S. A., and Shertzer, H. G. (2000). Activation of transcription factors activator protein-1 and nuclear factor-κB by 2,3,7,8-tetrachlorodibenzo-*p*-dioxin. *Biochemical Pharmacology*, 59, 997–1005.

131. Pelclova, D., Prazny, M., Skrha, J., Fenclova, Z., Kalousova, M., Urban, P., Navratil, T., Senholdova, Z., and Smerhovsky, Z. (2007). 2,3,7,8-TCDD exposure, endothelial dysfunction and impaired microvascular reactivity. *Human and Experimental Toxicology*, 26, 705–713.

132. Pelclova, D., Fenclova, Z., Urban, P., Ridzon, P., Preiss, J., Kupka, K., Malik, J., Dubska, Z., and Navratil, T. (2009). Chronic health impairment due to 2,3,7,8-tetrachloro-dibenzo-*p*-dioxin exposure. *Neuroendocrinology Letters*, 30 (Suppl. 1), 219–224.

133. Vena, J., Boffetta, P., Becher, H., Benn, T., Bueno-de-Mesquita, H. B., Coggon, D., Colin, D., Flesch-Janys, D., Green, L., Kauppinen, T., Littorin, M., Lynge, E., Mathews, J. D., Neuberger, M., Pearce, N., Pesatori, A. C., Saracci, R., Steenland, K., and Kogevinas, M. (1998). Exposure to dioxin and nonneoplastic mortality in the expanded IARC international cohort study of phenoxy herbicide and chlorophenol production workers and sprayers. *Environmental Health Perspectives*, 106 (Suppl. 2), 645–653.

134. Pesatori, A. C., Consonni, D., Rubagotti, M., Grillo, P., and Bertazzi, P. A. (2009). Cancer incidence in the population exposed to dioxin after the "Seveso accident": twenty years of follow-up. *Environmental Health*, 8, 39.

135. Moi, P., Chan, K., Asunis, I., Cao, A., and Kan, Y. W. (1994). Isolation of NF-E2-related factor 2 (Nrf2), a NF-E2-like basic leucine zipper transcriptional activator that binds to the tandem NF-E2/AP1 repeat of the beta-globin locus control region. *Proceedings of the National Academy of Sciences of the United States of America*, 91, 9926–9930.

136. Itoh, K., Wakabayashi, N., Katoh, Y., Ishii, T., Igarashi, K., Engel, J. D., and Yamamoto, M. (1999). Keap1 represses nuclear activation of antioxidant responsive elements by Nrf2 through binding to the amino-terminal Neh2 domain. *Genes & Development*, 13, 76–86.

137. Kobayashi, A., Kang, M. I., Okawa, H., Ohtsuji, M., Zenke, Y., Chiba, T., Igarashi, K., and Yamamoto, M. (2004). Oxidative stress sensor Keap1 functions as an adaptor for Cul3-based E3 ligase to regulate proteasomal degradation of Nrf2. *Molecular and Cellular Biology*, 24, 7130–7139.

138. Yamamoto, T., Suzuki, T., Kobayashi, A., Wakabayashi, J., Maher, J., Motohashi, H., and Yamamoto, M. (2008). Physiological significance of reactive cysteine residues of Keap1 in determining Nrf2 activity. *Molecular and Cellular Biology*, 28, 2758–2770.

139. Nakaso, K., Yano, H., Fukuhara, Y., Takeshima, T., Wada-Isoe, K., and Nakashima, K. (2003). PI3K is a key molecule in the Nrf2-mediated regulation of antioxidative proteins by hemin in human neuroblastoma cells. *FEBS Letters*, 546, 181–184.

140. Huang, H. C., Nguyen, T., and Pickett, C. B. (2002). Phosphorylation of Nrf2 at Ser-40 by protein kinase C regulates antioxidant response element-mediated transcription. *Journal of Biological Chemistry*, 277, 42769–42774.

141. Itoh, K., Chiba, T., Takahashi, S., Ishii, T., Igarashi, K., Katoh, Y., Oyake, T., Hayashi, N., Satoh, K., Hatayama, I., Yamamoto, M., and Nabeshima, Y. (1997). An Nrf2/small Maf heterodimer mediates the induction of phase II detoxifying enzyme genes through antioxidant response elements. *Biochemical and Biophysical Research Communications*, 236, 313–322.

142. Lee, J. S. and Surh, Y. J. (2005). Nrf2 as a novel molecular target for chemoprevention. *Cancer Letters*, 224, 171–184.

143. Kwak, M. K., Itoh, K., Yamamoto, M., and Kensler, T. W. (2002). Enhanced expression of the transcription factor Nrf2 by cancer chemopreventive agents: role of antioxidant response element-like sequences in the nrf2 promoter. *Molecular and Cellular Biology*, 22, 2883–2892.

144. Chan, K., Lu, R., Chang, J., and Kan, Y. (1996). NRF2, a member of the NFE2 family of transcription factors, is not essential for murine erythropoiesis, growth, and development. *Proceedings of the National Academy of Sciences of the United States of America*, 93, 13943–13948.

145. Khor, T. O., Huang, M. T., Kwon, K. H., Chan, J. Y., Reddy, B. S., and Kong, A. N. (2006). Nrf2-deficient mice have an increased susceptibility to dextran sulfate sodium-induced colitis. *Cancer Research*, 66, 11580–11584.

146. Rangasamy, T., Cho, C. Y., Thimmulappa, R. K., Zhen, L., Srisuma, S. S., Kensler, T. W., Yamamoto, M., Petrache, I., Tuder, R. M., and Biswal, S. (2004). Genetic ablation of Nrf2 enhances susceptibility to cigarette smoke-induced emphysema in mice. *Journal of Clinical Investigation*, 114, 1248–1259.

147. Thimmulappa, R. K., Lee, H., Rangasamy, T., Reddy, S. P., Yamamoto, M., Kensler, T. W., and Biswal, S. (2006). Nrf2 is a critical regulator of the innate immune response and survival during experimental sepsis. *Journal of Clinical Investigation, 116,* 984–995.

148. Chan, K. and Kan, Y. W. (1999). Nrf2 is essential for protection against acute pulmonary injury in mice. *Proceedings of the National Academy of Sciences of the United States of America, 96,* 12731–12736.

149. Ramos-Gomez, M., Kwak, M. K., Dolan, P. M., Itoh, K., Yamamoto, M., Talalay, P., and Kensler, T. W. (2001). Sensitivity to carcinogenesis is increased and chemoprotective efficacy of enzyme inducers is lost in nrf2 transcription factor-deficient mice. *Proceedings of the National Academy of Sciences of the United States of America, 98,* 3410–3415.

150. Leung, L., Kwong, M., Hou, S., Lee, C., and Chan, J. Y. (2003). Deficiency of the Nrf1 and Nrf2 transcription factors results in early embryonic lethality and severe oxidative stress. *Journal of Biological Chemistry, 278,* 48021–48029.

151. Loignon, M., Miao, W., Hu, L., Bier, A., Bismar, T. A., Scrivens, P. J., Mann, K., Basik, M., Bouchard, A., Fiset, P. O., Batist, Z., and Batist, G. (2009). Cul3 overexpression depletes Nrf2 in breast cancer and is associated with sensitivity to carcinogens, to oxidative stress, and to chemotherapy. *Molecular Cancer Therapy, 8,* 2432–2440.

152. Wakabayashi, N., Itoh, K., Wakabayashi, J., Motohashi, H., Noda, S., Takahashi, S., Imakado, S., Kotsuji, T., Otsuka, F., Roop, D. R., Harada, T., Engel, J. D., and Yamamoto, M. (2003). Keap1-null mutation leads to postnatal lethality due to constitutive Nrf2 activation. *Nature Genetics, 35,* 238–245.

153. Prochaska, H. J. and Talalay, P. (1988). Regulatory mechanisms of monofunctional and bifunctional anticarcinogenic enzyme inducers in murine liver. *Cancer Research, 48,* 4776–4782.

154. Holtzclaw, W. D. Dinkova-Kostova, A. T., and Talalay, P. (2004). Protection against electrophile and oxidative stress by induction of phase 2 genes: the quest for the elusive sensor that responds to inducers. *Advances in Enzyme Regulation, 44,* 335–367.

155. Nioi, P., McMahon, M., Itoh, K., Yamamoto, M., and Hayes, J. D. (2003). Identification of a novel Nrf2-regulated antioxidant response element (ARE) in the mouse NAD(P)H:quinone oxidoreductase 1 gene: reassessment of the ARE consensus sequence. *Biochemical Journal, 374,* 337–348.

156. Jaiswal, A. K. (1991). Human NAD(P)H:quinone oxidoreductase (NQO1) gene structure and induction by dioxin. *Biochemistry, 30,* 10647–10653.

157. Hayes, J. D., Flanagan, J. U., and Jowsey, I. R. (2005). Glutathione transferases. *Annual Review of Pharmacology and Toxicology, 45,* 51–88.

158. Nguyen, T., Sherratt, P. J., and Pickett, C. B. (2003). Regulatory mechanisms controlling gene expression mediated by the antioxidant response element. *Annual Review of Pharmacology and Toxicology, 43,* 233–260.

159. Kalthoff, S., Ehmer, U., Freiberg, N., Manns, M. P., and Strassburg, C. P. (2010). Interaction between oxidative stress sensor Nrf2 and xenobiotic-activated aryl hydrocarbon receptor in the regulation of the human phase II detoxifying UDP-glucuronosyltransferase 1A10. *Journal of Biological Chemistry, 285,* 5993–6002.

160. Kalthoff, S., Ehmer, U., Freiberg, N., Manns, M. P., and Strassburg, C. P. (2010). Coffee induces expression of glucuronosyltransferases by the aryl hydrocarbon receptor and Nrf2 in liver and stomach. *Gastroenterology, 139,* 1699–1710.

161. Munzel, P. A., Schmohl, S., Buckler, F., Jaehrling, J., Raschko, F. T., Kohle, C., and Bock, K. W. (2003). Contribution of the Ah receptor to the phenolic antioxidant-mediated expression of human and rat UDP-glucuronosyltransferase UGT1A6 in Caco-2 and rat hepatoma 5L cells. *Biochemical Pharmacology, 66,* 841–847.

162. Yueh, M. F., Huang, Y. H., Hiller, A., Chen, S., Nguyen, N., and Tukey, R. H. (2003). Involvement of the xenobiotic response element (XRE) in Ah receptor-mediated induction of human UDP-glucuronosyltransferase 1A1. *Journal of Biological Chemistry, 278,* 15001–15006.

163. Miao, W., Hu, L., Scrivens, P. J., and Batist, G. (2005). Transcriptional regulation of NF-E2 p45-related factor (NRF2) expression by the aryl hydrocarbon receptor-xenobiotic response element signaling pathway: direct cross-talk between phase I and II drug-metabolizing enzymes. *Journal of Biological Chemistry, 280,* 20340–20348.

164. Shin, S., Wakabayashi, N., Misra, V., Biswal, S., Lee, G. H., Agoston, E. S., Yamamoto, M., and Kensler, T. W. (2007). NRF2 modulates aryl hydrocarbon receptor signaling: influence on adipogenesis. *Molecular and Cellular Biology, 27,* 7188–7197.

165. Anwar-Mohamed, A., Degenhardt, O. S., Gendy, M. A., Seubert, J. M., Kleeberger, S. R., and El-Kadi, A. O. (2011) The effect of Nrf2 knockout on the constitutive expression of drug metabolizing enzymes and transporters in C57Bl/6 mice livers. *Toxicology In Vitro, 25,* 785–795.

166. Noda, S., Harada, N., Hida, A., Fujii-Kuriyama, Y., Motohashi, H., and Yamamoto, M. (2003). Gene expression of detoxifying enzymes in AhR and Nrf2 compound null mutant mouse. *Biochemical and Biophysical Research Communications, 303,* 105–111.

167. Yeager, R. L., Reisman, S. A., Aleksunes, L. M., and Klaassen, C. D. (2009). Introducing the "TCDD-inducible AhR-Nrf2 gene battery". *Toxicological Sciences, 111,* 238–246.

168. Ciolino, H. P., Daschner, P. J., Wang, T. T., and Yeh, G. C. (1998). Effect of curcumin on the aryl hydrocarbon receptor and cytochrome P450 1A1 in MCF-7 human breast carcinoma cells. *Biochemical Pharmacology, 56,* 197–206.

169. Takemura, H., Nagayoshi, H., Matsuda, T., Sakakibara, H., Morita, M., Matsui, A., Ohura, T., and Shimoi, K. (2010). Inhibitory effects of chrysoeriol on DNA adduct formation with benzo[a]pyrene in MCF-7 breast cancer cells. *Toxicology, 274,* 42–48.

170. Garg, R., Gupta, S., and Maru, G. B. (2008). Dietary curcumin modulates transcriptional regulators of phase I and phase II enzymes in benzo[a]pyrene-treated mice: mechanism of its anti-initiating action. *Carcinogenesis, 29,* 1022–1032.

171. Kalpana Deepa, P. D., Gayathri, R., and Sakthisekaran, D. (2010). Role of sulforaphane in the anti-initiating mechanism of lung carcinogenesis *in vivo* by modulating the metabolic activation and detoxification of benzo(*a*)pyrene. *Biomedicine and Pharmacotherapy*, 65, 9–16.

172. Zhang, Y., Kensler, T. W., Cho, C. G., Posner, G. H., and Talalay, P. (1994). Anticarcinogenic activities of sulforaphane and structurally related synthetic norbornyl isothiocyanates. *Proceedings of the National Academy of Sciences of the United States of America*, 91, 3147–3150.

173. Hwang, Y. P., Han, E. H., Choi, J. H., Kim, H. G., Lee, K. J., Jeong, T. C., Lee, E. S., and Jeong, H. G. (2008). Chemopreventive effects of furan-2-yl-3-pyridin-2-yl-propenone against 7,12-dimethylbenz[a]anthracene-inducible genotoxicity. *Toxicology and Applied Pharmacology*, 228, 343–350.

174. Yeligar, S. M., Machida, K., and Kalra, V. K. (2010). Ethanol-induced HO-1 and NQO1 are differentially regulated by HIF-1α and Nrf2 to attenuate inflammatory cytokine expression. *Journal of Biological Chemistry*, 285, 35359–35373.

175. Malec, V., Gottschald, O. R., Li, S., Rose, F., Seeger, W., and Hanze, J. (2010). HIF-1α signaling is augmented during intermittent hypoxia by induction of the Nrf2 pathway in NOX1-expressing adenocarcinoma A549 cells. *Free Radical Biology and Medicine*, 48, 1626–1635.

176. Loboda, A., Stachurska, A., Florczyk, U., Rudnicka, D., Jazwa, A., Wegrzyn, J., Kozakowska, M., Stalinska, K., Poellinger, L., Levonen, A. L., Yla-Herttuala, S., Jozkowicz, A., and Dulak, J. (2009). HIF-1 induction attenuates Nrf2-dependent IL-8 expression in human endothelial cells. *Antioxidants & Redox Signaling*, 11, 1501–1517.

177. Kim, T. H., Hur, E. G., Kang, S. J., Kim, J. A., Thapa, D., Lee, Y. M., Ku, S. K., Jung, Y., and Kwak, M. K. (2011). NRF2 blockade suppresses colon tumor angiogenesis by inhibiting hypoxia-induced activation of HIF-1α. *Cancer Research*, 71, 2260–2275.

9

FUNCTIONAL INTERACTIONS OF AHR WITH OTHER RECEPTORS

Sara Brunnberg, Elin Swedenborg, and Jan-Åke Gustafsson

9.1 INTRODUCTION

Several lines of evidence suggest that the AHR is a multifunctional protein and plays a fundamental role in normal physiology, beyond the response to xenobiotics. Also, the AHR signaling pathway converges with several other important signal transduction pathways, thus modulating the cellular responses. For instance, AHR has been shown to play an important role in the control of autoimmune disease such as multiple sclerosis (MS) [1, 2]. In addition, AHR is involved in cell cycle regulation and in crosstalk with MAP kinases (described elsewhere in this book and also reviewed in Ref. 3). In this chapter, we will focus on the convergence of the AHR and the nuclear receptor (NR)-mediated signaling pathways and the implications of this crosstalk. First, we will attempt to summarize the interactions between estrogen receptors (ERs) and AHR. Second, an overview of other nuclear receptors' crosstalk with AHR, or indications thereof, will be provided.

9.2 NUCLEAR RECEPTORS

Nuclear receptors and basic helix–loop–helix–PER/ARNT/SIM (bHLH–PAS) proteins represent two important superfamilies of transcription factors. NRs, such as estrogen, glucocorticoid (GC), progesterone, and androgen receptors, control vital processes such as reproduction, metabolism, and development (reviewed in Refs 4 and 5).

The nuclear receptors are constructed as modular proteins where each domain can function separately from the others (Fig. 9.1). The major domains are as follows:

- The transactivation domain in the amino terminus, which can be of variable length and is involved in interactions with coactivators and corepressors
- The DNA binding domain, containing the two zinc finger motifs characteristic of the NR superfamily
- The carboxy-terminal domain, which has a ligand-regulated activation function (AF-2) and is central for transcriptional activity

9.3 THE ESTROGEN RECEPTORS

The physiological actions of estrogenic hormones are mediated by the two estrogen receptor isoforms ERα and ERβ that belong to the NR superfamily. Although these receptors are predominantly activated by endogenous hormones such as 17β-estradiol (E2), they may also bind to and become activated by a wide range of natural and synthetic compounds, such as dietary substances, pharmaceuticals, and certain environmental pollutants.

The ligand binding domains of the ER isoforms, which become activated by ligand, are only 56% identical at the amino acid level [6]. This difference provides a basis for differential responses to certain ligands, that is, isoform-selective ligands [7]. Both ERs display two transactivation domains, AF-1 and AF-2, and synergy between these two functions leads to full transcriptional activity. However, whereas the ligand-dependent AF-2 of the ER subtypes are equally potent, the ERβ AF-1 appears weaker, and thus ERβ activity relies more on the AF-2 function [8].

Much like AHR, ERs exist as multiprotein complexes together with various coregulator proteins. The coregulators

---

The AH Receptor in Biology and Toxicology, First Edition. Edited by Raimo Pohjanvirta.
© 2012 John Wiley & Sons, Inc. Published 2012 by John Wiley & Sons, Inc.

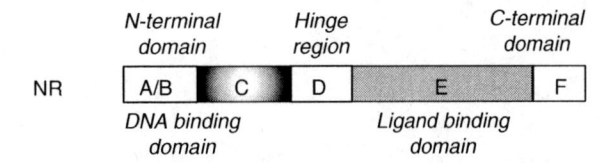

FIGURE 9.1 NRs have a modular structure consisting of six functional domains labeled A–F. The main domains and their functions are A/B, the transactivation domain in the amino terminus that can be of variable length and is involved in interactions with coactivators and corepressors; C, the central DNA binding domain that contains the two zinc finger motifs characteristic of the NR superfamily; and E, the carboxy-terminal domain that has a ligand-regulated activation function (AF-2) and is central to transcriptional activity.

can stimulate or inhibit transcription; hence, they are referred to as *coactivators* (e.g., steroid receptor coactivator 1 (SRC1) and transcriptional intermediary factor 2 (TIF2)) or *corepressors* (e.g., nuclear receptor corepressor 1 (NCoR1) and silencing mediator of retinoid and thyroid hormone receptors (SMRT)). The specific set of coregulators in the cell or tissue in the presence of a certain ligand determines the final outcome of the ERs' transcriptional response.

The transcriptional activity of the ERs is generally classified into two separate modes of action, referred to as genomic and nongenomic signaling (Fig. 9.2). These modes are characterized by:

1. *Genomic*: Classic ER signaling, where the ER dimer, upon ligand activation, *directly* binds to its recognition sequence in the promoter of target genes. This signaling may also take place *indirectly*, through association with other transcription factors such as AP-1 and Sp1. This occurs at AP-1 and Sp1 binding sites in the regulatory regions of target genes.

2. *Nongenomic*: This signaling is characterized by rapid effects that cannot be explained by an increase in transcription and subsequent protein translation. It has been reported to involve membrane localized estrogen receptors as well as cytoplasmic signal transduction molecules such as tyrosine kinases and STATs [9, 10].

Besides the intracellular receptors, there are numerous reports of membrane proteins mediating effects of steroid hormones. For instance, both progesterone and testosterone appear to have such receptors [11–13]. In addition, in the pancreas, a nonclassical membrane estrogen receptor has been described as a mediator of rapid nongenomic effects of estrogens and xenoestrogens [14]. The receptor appears unrelated to the ERs. Instead, it binds E2 and xenoestrogens, such as diethylstilbestrol and bisphenol A, as well as neurotransmitters, such as dopamine and epinephrine. Moreover, it appears insensitive to classical ER antagonists such as ICI182,780 and there is evidence indicating that it might be a G-protein-coupled type of receptor [15].

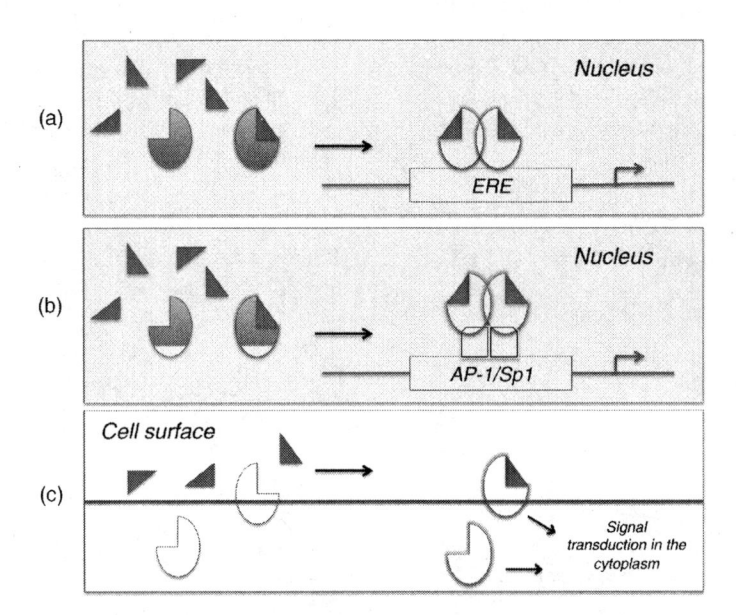

FIGURE 9.2 ER signaling: different modes of action. The classical, *genomic* ER signaling occurs through direct binding of ligand-bound ER dimers to ERE sequences in the regulatory regions of estrogen-responsive genes (a). Another type of *genomic* activity occurs through indirect association with promoters by protein–protein interactions with other transcription factors, such as AP-1 or Sp1 (b). The second mode of action, referred to as *nongenomic* activity, has been shown to involve interactions with cytoplasmic signal transduction proteins, such as kinases (c). (See the color version of this figure in Color Plates section.)

In addition to the above-mentioned mechanisms, rapid effects of 17β-estradiol have been studied in isolated pancreatic β cells by using patch clamp technique [16]. Findings from this study suggest that ERβ can potentiate the effects elicited by E2 in β cells by activating the membrane guanylate cyclase A receptor (GC-A), which in turn leads to glucose-stimulated insulin secretion.

Another G-protein-coupled estrogen receptor is GPR30 (also known as GPER; reviewed in Ref. 17). This transmembrane receptor appears to have both distinct and redundant functions to those of the nuclear ERs in a wide range of tissues and cell types. Still, much remains to be investigated about GPR30 and its physiological role in estrogen biology.

9.4 INTERACTIONS BETWEEN AHR AND ER

9.4.1 *In Vivo* Findings

Several studies have suggested an intricate physiological relationship between AHR and estrogen receptor systems. For instance, the antiestrogenic influence of TCDD on ER signaling is well documented, such as suppression of estrogenic responses in rodent uterus and mammary glands as well as in human breast cancer cells. Inhibitory AHR–ER crosstalk has been shown both *in vitro* and *in vivo* (reviewed in Refs 18 and 19).

Aiming to elucidate the physiological role of AHR, besides the xenobiotic response, AHR knockout mice have been produced by three independent research groups [20–22]. Analysis of the phenotypes of these mice has provided significant evidence for a role of AHR in the female reproductive tract. For instance, it was demonstrated that female AHR-null mice have abnormal phenotypes in the reproductive tissues and display reduced reproductive success such as small litter sizes and subfertility [23].

In a long-term (2 years) chronic exposure study of TCDD in rats, it was demonstrated that the incidence of both mammary and uterine tumors, two types of estrogen-dependent tumors, was significantly decreased [24]. These indications of tissue-specific AHR–ER crosstalk led to several studies in rodent models, investigating the possible effects of TCDD on the reproductive systems. In addition, it has been reported that women accidentally exposed to TCDD following the plant explosion in Italy (Seveso, 1976) were less prone to develop breast cancer than nonexposed control groups [25].

Many adverse or inhibitory effects of TCDD on estrogen-regulated processes in the reproductive tract of mice and rats have been observed. For instance, inhibition of uterine wet weight increase, impaired progesterone receptor (PR) binding, and altered c-fos mRNA levels have been reported in dioxin-treated rodents [26–28]. The actual involvement of the AHR in these effects is implicated by the fact that the antiestrogenic potencies of tested aromatic hydrocarbon compounds coincide with their binding affinities to AHR [29]. In addition, using AHRKO mice, Buchanan et al. have demonstrated that TCDD effects on uterine epithelial mitogenic activity and secretory function are clearly dependent on stromal AHR [30].

9.5 AHR–ER CROSSTALK: PROPOSED MOLECULAR MECHANISMS

Several molecular mechanisms of crosstalk between AHR and ER pathways have been suggested (see Fig. 9.3 for a schematic overview). It is still not clear which of these mechanisms are predominant, or if there exist, for instance, tissue- or cell-specific differences.

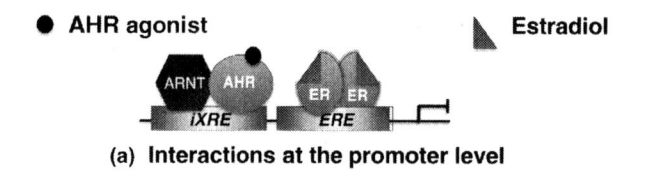

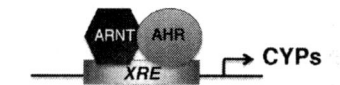

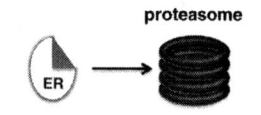

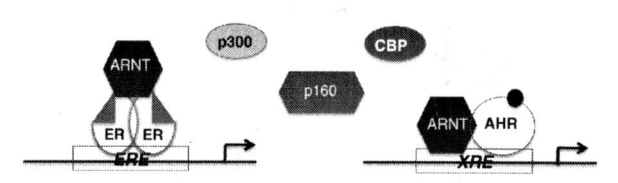

FIGURE 9.3 ER–AHR crosstalk mechanisms. (a) ER transactivation is repressed by the binding of liganded AHR to XREs, which blocks the ER binding or formation of the ERα/Sp1 or ERα/AP-1 complex. (b) AHR ligands can activate the CYP1A1 and 1A2 enzymatic activities. Given that circulating estrogens are metabolized in the liver, predominantly by 1A1 and 1A2, exposure to AHR agonists could thus lead to altered hormonal balance. (c) TCDD has been shown to enhance degradation of ERα by targeting it to the proteasome, thus lowering the cellular levels of ER. (d) Simultaneous induction of ER and AHR leads to competition for common regulatory factors or coactivators, thus inhibiting full transcriptional activity of the receptors. (See the color version of this figure in Color Plates section.)

9.5.1 Interactions at the Promoter Level

Xenobiotic response elements (XREs) have been identified in the regulatory regions of classical estrogen target genes such as *pS2, cathepsin D*, and *c-fos* [31–33]. Here, formation of ERα/Sp1 or ERα/AP-1 complexes is inhibited, and ER has decreased access to its binding sites. Thus, ER transcription is suppressed. Competition for binding sites, that is, steric hindrance caused by AHR complexes associated with XREs located adjacent to or overlapping with EREs has also been demonstrated [34] (see Fig. 9.3a).

9.5.2 Metabolism of Circulating Hormones

AHR agonists such as dioxin activate the enzymatic activities of CYP1A1 and 1A2, which are typical AHR target genes. Once activated, these enzymes are responsible for breaking down circulating estrogens. AHR also regulates expression of the ovarian P450 aromatase, CYP19, a key enzyme in estrogen synthesis [23]. Thus, activation of the AHR pathway may perturb estrogen levels [35] (see Fig. 9.3b).

9.5.3 Accelerated Turnover of the ER(s)

It has been shown that the AHR, once activated by ligand, can increase the degradation of ERα through activation of the proteasome [36]. This mechanism seems ligand dependent as TCDD exerts a stronger effect than other weaker ligands such as benzo[*a*]pyrene. Moreover, AHR has been shown to be part of a cullin 4B ubiquitin ligase complex, which is ligand dependent [37] (see Fig. 9.3c).

9.5.4 Competition for Common Coregulators

Several studies have reported that AHR and ER interact with common nuclear cofactors, both coactivators and corepressors. For instance, SMRT, SRC1, and RIP140 are important ER cofactors that have been shown to be recruited also by the AHR [38–42]. Thus, this may constitute yet another mechanism by which crosstalk may occur between AHR and ER pathways, when they are simultaneously activated (see Fig. 9.3d).

Our group has investigated the role of the AHR partner protein ARNT as a modulator of the estrogen receptors. We have demonstrated that, in the presence of estradiol, ARNT is recruited to estrogen-responsive promoters. This function is independent of AHR and leads to increased transcriptional activity of both ERs. Both the ER ligand binding domain and the C-terminal part of ARNT are required for the interaction, and one theory is that ARNT mediates or stabilizes the interaction between ER and cofactor p300 [42, 43]. P300 has a strong histone methyltransferase activity and has been reported to mediate synergy between the two ER transactivation domains, AF-1 and AF-2, thus enhancing ER transcriptional capacity [44, 45].

9.6 AHR AGONISTS MODULATE ER SIGNALING DIFFERENTLY

The complex interplay between AHR and ER pathways is apparent at different levels. Some data suggest a role for ERα in AHR gene regulation. For instance, AHR has been shown to recruit ERα to AHR target genes such as *CYP1A1*, and the association was further increased by estradiol addition [46, 47]. The mechanism by which ERα modulates AHR-regulated transcription is unclear and could involve, for instance, protein–protein interactions between ER and AHR and influence of additional coregulators.

The antiestrogenic effects of TCDD on ER signaling have been shown both *in vitro* and *in vivo* (reviewed in Refs 18 and 19). In contrast to the TCDD effects, some results show that another prototypical AHR agonist, 3-methylcholanthrene (3-MC), may actually exert estrogenic activities in certain cell types. It has been demonstrated that this phenomenon is cell type specific and results from biotransformation of the parent compound [48]. Also, reports have shown that 3-MC can activate ERα in MCF-7 cells and that estrogenic responses can be detected in both AHR$^{+/+}$ and AHR$^{-/-}$ mice, demonstrating AHR independence [49, 50]. As an alternative model, it has been proposed that estrogenic effects of 3-MC could be due to recruitment of liganded AHR to the ERα, AHR thus acting as a coactivator of ER. [51]. It is possible that the discrepancies between these experimental results could be explained by different 3-MC concentrations used in the various studies.

ER subtype- and cell-specific differences in the ER–AHR crosstalk have also been reported. For instance, knockdown of ERβ in mouse mammary cells (HC11) potentiated the AHR-regulated Cyp1a1 expression, whereas ERα knockdown had an inhibitory effect. In contrast, no effect of ERα downregulation on CYP1A1 expression could be measured in human breast cancer MCF-7 cells [52].

In addition, data show that the antiestrogenic effects of TCDD are much stronger on ERβ than on ERα-mediated activities, indicating subtype-selective mechanisms [43]. Moreover, TCDD inhibits ERα-mediated upregulation of ERβ mRNA in breast cancer cells, suggesting that TCDD affects ERβ activity more severely than that of ERα [53].

9.7 OTHER NUCLEAR RECEPTORS: INTERACTIONS WITH AHR

9.7.1 Androgen Receptor

Androgens are important for the development and maintenance of the male reproductive organs, including prostate, seminal vesicles, epididymis, and testis. The main endogenous androgenic hormone is testosterone and it is produced primarily in the testis and, to some extent, in the adrenals.

Testosterone can be further metabolized to dihydrotestosterone (DHT), a ligand showing higher affinity to AR than testosterone.

Androgens function through the androgen receptor (AR), which is a member of the steroid hormone receptor group within the nuclear receptor family. To regulate gene expression, AR functions by a mechanism similar to that previously described for ER: androgens activate the AR, which is then translocated to the cell nucleus. Here, ligand-bound AR homodimerizes and binds to chromatin at specific androgen response elements (AREs) in the regulatory regions of target genes.

Male reproductive organs are considered to be the most susceptible target tissues for *in utero* and lactational exposure to TCDD in laboratory animals. In cell cultures, AHR, activated by ligand or as a constitutively active mutant, represses AR function, as demonstrated by decreased androgen-induced prostate-specific antigen (PSA) mRNA and protein levels following exposure to TCDD. AR mRNA levels, however, remain stable [54, 55]. Overexpression of AHR/ARNT, in the absence of TCDD, also downregulated the transcriptional activity of ligand-bound AR [56]. Therefore, some of the effects observed in male laboratory animals may come as a consequence of a potential interference between AHR and AR pathways (Fig. 9.4). *In vitro* studies further support this notion, and this crosstalk appears to be cell or context specific [57].

Since AR and ER show functional similarities, it is not surprising that AHR has been shown to interact with AR in similar ways as previously described for ER. Suggested AHR–AR crosstalk mechanisms are exemplified next.

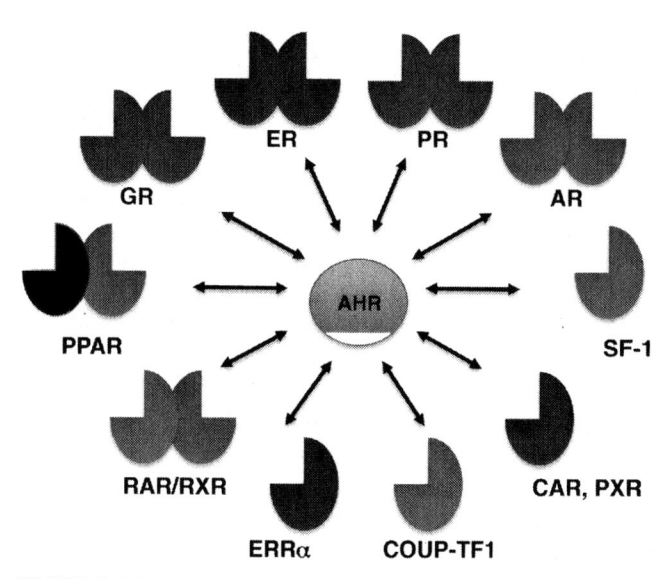

FIGURE 9.4 Overview of AHR–NR crosstalk. There are many indications of the AHR pathway converging with various NR signaling pathways. Some are described in the chapter. (See the color version of this figure in Color Plates section.)

9.7.1.1 Proteosomal Degradation 3-MC-activated AHR has been shown to decrease AR protein levels by promoting ubiquitination and proteosomal degradation by assembling a ubiquitin ligase complex called CUL4B [37]. Thus, AHR agonists such as TCDD may modulate the androgen response by controlling AR protein levels.

9.7.1.2 Disruption of Hormone Levels There exist conflicting data whether TCDD exposure decreases circulating or cellular androgen levels [58–62]. Ligand-activated AHR has been shown, in some studies, to suppress the levels of testicular cytochrome P450 side chain cleavage (P450scc). This enzyme converts cholesterol into pregnenolone and its catalytic activity is also decreased, which may result in decreased testosterone levels [63, 64]. The catalytic activity of P450scc may be repressed by a decrease in the mobilization of the substrate cholesterol to P450scc [65]. Steroidogenic acute regulatory protein (StAR) is responsible for the intramitochondrial movement of cholesterol. The expression of this enzyme is unaffected or downregulated by TCDD exposure [66, 67].

9.7.1.3 Competition Between Common Coactivators Four and a half LIM domain 2 (FHL2) is a multifunctional protein that can act as a nuclear receptor coregulator. FLH2 interacts with and coactivates the transcriptional activity of AR, similarly to the well-characterized NR coactivator SRC1. In addition, FLH2 acts as a coregulator of AHR independent of ligand in a cell-specific manner by increasing its activity in MCF-7 and PC-3 cells or decreasing its activity in T47D and LNCaP cells [68, 69]. Also, another coactivator, nuclear receptor coactivator 4 (NCOA4), had similar effects [70]. The increased activity of AR was repressed in the presence of TCDD in PC-3 cells. Thus, the authors propose a model for AR–AHR crosstalk involving sequestration of a common coactivator, explaining the antiandrogenic properties of TCDD. AHR significantly repressed the ligand-activated transcriptional activity of AR independent of the presence of an AHRE within the regulatory region of the AR target gene PSA. This indicates that AHR may inhibit AR activity through squelching mechanisms [56]. TCDD induces expression of *c-fos* and *c-jun* genes, and, consequently, the transcription factor AP-1 [71]. This could lead to an increased interaction of AR and AP-1, which may result in an inhibition of the binding of AR to ARE in the transcription regulatory region of target genes such as *PSA* [72].

In line with ER data, studies show that AHR may increase the transcriptional activity of AR. AHR, independent of ligand, increased the transcriptional activity of DHT-bound AR in a PSA promotor–luciferase reporter [69]. AR appears to interact with both AHR and ARNT at the *PSA* promoter in LNCaP cells. In contrast, however, to ER, ARNT did not coactivate the transcriptional activity of AR in this system.

In addition, the signaling pathways of AR and AHR appear to interact reciprocally to modify each other's transcriptional activity. Ligand-activated AR inhibits the transcriptional activity of liganded AHR in human prostate cancer LNCaP cells [54, 73]. These effects were reversed by AR knockdown by siRNA [73].

9.7.2 Glucocorticoid Receptor

Toxicity studies of TCDD exposure also indicate interactions of AHR with the glucocorticoid receptor (GR) (Fig. 9.4). The adrenal cortex produces glucocorticoids, which are important endocrine regulators of a wide range of physiological systems ranging from respiratory development and immune function to responses to stress. GC binds to and activates the GR that signals in a similar fashion as the other NRs. However, GR also shares certain features common to the AHR. In fact, GR was previously used as an analog for the studies of AHR. Both receptors reside in the cytoplasm in a heteromeric complex of several proteins including HSP90 and p23. Upon ligand binding, GR homodimerizes and binds to specific glucocorticoid response elements (GRE) within promoters of target genes. Furthermore, GR action also relies on its direct protein–protein interaction capabilities with other transcriptional regulators bound to DNA at their own high-affinity binding sites. Such interactions give rise to the subsequent control of distinct subsets of target genes [74].

The interactions of GR and AHR are complex and *in vivo* studies show that GC treatment could attenuate acute toxicity of TCDD [75] or aggravate its toxic effects such as cleft palate [76, 77]. Cotreatment of TCDD and GC synergistically increased CYP1A1 activity in the rat liver [78]. Moreover, gestational GC treatment increased mRNA and protein levels of AHR in craniofacial tissue of mice, in line with aggravated cleft palate [79]. On the other hand, GC withdrawal by adrenalectomy depleted the AHR protein levels in rat liver by 50–60%. This reduction was not rescued by acute treatment of dexamethasone (DEX), a synthetic GR agonist [80].

Numerous studies conducted *in vitro* have given rather conflicting results probably caused by species differences. GR activated by DEX increased mRNA and protein levels of AHR in rodent-derived cell cultures [81]. In line with this, DEX potentiated AHR-mediated increase of mRNA and protein levels as well as activity of CYP1A1 in rat-derived cells in a GR-dependent manner [82–86]. In fact, DEX alone also increased CYP1A1 expression and activity [87, 88]. This effect is most likely due to functional GRE(s) within the first intron of rat CYP1A1 gene overlapping with AHRE. This overlap might cause potentiation or repression of gene expression, depending on cellular context [83]. In contrast, in human-derived cells, DEX inhibits AHR transactivation and GR antagonist reverses the effects [82, 87, 89, 90]. This species difference could be explained by the presence of two putative GREs in the rat AHR promoter, whereas there are no identified functional GREs in the human AHR promoter [82]. This is further supported by the observations that the protein synthesis inhibitor cycloheximide could block the DEX-induced stimulation of CYP1A1 expression, suggesting that the enhancing effect of DEX via GR on TCDD target genes requires the synthesis of an additional protein, such as AHR [82]. However, it has also been reported that DEX inhibits induction of human CYP1A1 in a GR-independent manner [91]. In conclusion, crosstalk between AHR and GR has been well documented but the mechanisms involved still remain unclear.

9.7.3 Retinoic Acid Receptors

Retinoids regulate various important biological functions from embryogenesis, growth, and reproduction to vision and maintenance of numerous epithelial tissues [92]. The two most active forms of retinoids are retinol (vitamin A) and retinoic acid, the latter existing as three different types, all-*trans* retinoic acid (atRA), 9-*cis*-retinoic acid, and 13-*cis*-retinoic acid [93]. The effects of retinoids are mediated by the retinoic acid receptors (RARs). These receptors are members of the subfamily I of the nuclear receptor family. RAR heterodimerizes with the promiscuous retinoid X receptor (RXR), which belongs to another group of the nuclear receptor family. Both RAR and RXR exist in three different subtypes, and with numerous isoforms (extensively reviewed in Ref. [94]). AtRA binds selectively to RAR, while 9-*cis*-RA binds to both RARs and RXRs [93].

In the absence of an RAR agonist, the RAR/RXR heterodimers are bound to the retinoic acid response element (RARE) and recruit to this complex SMRT or other nuclear receptor corepressors. These corepressors function by recruiting other complexes containing histone deacetylases to the promoter. Upon ligand binding, the RAR/RXR heterodimers are released from this complex and interact with coactivator complexes (SRC1/TIF2/RAC3 and CBP/p300) that stimulate transactivation [92].

There exists a vast literature on the impact of TCDD on vitamin A homeostasis (reviewed in Refs 95 and 92) (Fig. 9.4). Several of the effects of TCDD exposure observed in laboratory animals resemble those of vitamin A deficiency, including impaired growth, impaired immune function, lesions of epithelial linings, and an increase in the levels of kidney vitamin A. Some of the adverse effects of TCDD exposure may be partly reversed by coadministration of vitamin A. On the other hand, excess of vitamin A during gestation causes severe developmental defects in mice, for example, cleft palate, in a similar manner as exposure to TCDD. Furthermore, TCDD and atRA exposure *in utero* synergistically provokes cleft palate in the offspring [95].

Reported *in vitro* data diverge although it is clear that the two receptor families reciprocally interfere with each other's signaling pathways. In a human keratinocyte cell line, TCDD repressed the atRA-induced release of TGFβ and decreased the binding of atRA to RARs. However, the mRNA levels of RARα and RARγ were unchanged [96]. TCDD also inhibited the expression of other atRA target genes, such as *RDH9* and *CRABPII* [97, 98]. Furthermore, in SCC-4 cells, TCDD suppressed RA induction of both mRNA level and enzyme activity of the RA target gene *transglutaminase* [99, 100]. The suppressive effect of TCDD on transglutaminase mRNA expression was abolished when the cells were cotreated with a histone deacetylase inhibitor, indicating the presence of a corepressor [100]. The corepressor SMRT inhibited TCDD increased activity of a CYP1A1-driven promoter construct in MCF-7 cells [38, 101]. However, SMRT enhanced the TCDD–AHR activity on a dioxin-responsive fragment from the mouse *Cyp1a1* regulatory region [101].

TCDD has also been reported to increase the transactivation of a transiently expressed RARE reporter construct in MCF-7 cells [102]. Coexposure of atRA and TCDD further increased the activation of the reporter construct [102]. In the presence of an AHR or RARα antagonist, the effects were suppressed, implying that both receptors are involved in this additive transactivation. The authors propose that the underlying mechanism could be that AHR acts through sequestering SMRT. In keratinocytes, the mRNA expression of matrix metalloproteinase 1 (MMP-1) and PAI-2 was induced by TCDD and coexposure of atRA further increased this response [103]. The presence of the AHR antagonist α-naphthoflavone diminished this effect, indicating that the AHR pathway is involved. The mechanism behind this was shown to involve an increase in MMP-1 promoter activity mainly dependent on two proximal AP-1 sites present in the promoter, as well as posttranscriptional effects such as augmented mRNA stability [103].

In addition, retinoids have been shown to inhibit AHR-mediated transcription. Several mechanisms may be involved in CYP1A1 expression, which is repressed in the presence of atRA [104–107]. AtRA seems to abolish TCDD-induced binding of AHR to the CYP1A1 XRE sequence in keratinocytes [103]. Moreover, in the presence of atRA, TCDD-activated AHR failed to transactivate an XRE-driven luciferase construct [102]. In colorectal adenocarcinoma cells, RA alone or in combination with 3-MC suppressed both the mRNA and protein levels of AHR and the CYP1A1 enzymatic activity [107]. However, in the presence of trichostatin A, a histone deacetylase inhibitor, RA instead increased the expression of CYP1A1 mRNA levels, indicating involvement of a corepressor in the repressive effect of RA on CYP1A1 induction by 3-MC. Cotransfection of the corepressor SMRT inhibited the XRE-driven reporter and SMRT was also coimmunoprecipitated with AHR in the presence of 3-MC and atRA [107]. Thus, although it appears that there is a physical and functional interaction between AHR and SMRT, it is still unclear whether this leads to inhibition or activation of AHR [38, 101].

The synthetic retinoid AGN 193109, an RAR pan-antagonist, induces the levels of CYP1A1 mRNA, protein, and enzymatic activity in Hepa 1c1c7 cells as well as the Cyp1a1 mRNA levels in mouse embryos. This effect seems to be mediated by an AHR/ARNT-dependent pathway and not by the RAR/RXR and was not observed in the presence of atRA [108].

In conclusion, it seems as if TCDD may give rise to both effects resembling those of vitamin A deficiency and, conversely, symptoms similar to those characteristic of vitamin A excess. The mechanisms behind the effects of TCDD on the vitamin A system need further exploration.

9.7.4 Peroxisome Proliferator-Activated Receptors

One of the hallmarks of acute exposure to TCDD in many species is a condition known as wasting syndrome. Animals exposed to TCDD exhibit prolonged weight loss due to depletion of both lean and adipose tissue mass. Biochemical changes such as inhibition of glucose transport, lipoprotein lipase activity, and fatty acid synthesis are observed in adipose tissue after TCDD exposure and are associated with serum hyperlipidemia [109–112]. Intriguingly, in the liver of AHR-null mice, droplets of fat are microscopically visible in the cells over the first 2 weeks of life. This condition is known as fatty metamorphosis and suggests perturbation of hepatic lipid metabolism [21]. In addition, TCDD exposure suppresses the conversion of the preadipocyte cell line, 3T3-L1, the multipotential fibroblast C3H10T1/2 (10T1/2) cell line, and primary mouse embryo fibroblasts (MEFs) into mature adipocytes [113–116]. Taken together, these results suggest extensive crosstalk between the AHR and the pathways involved in metabolism (Fig. 9.4).

Peroxisome proliferator-activated receptors (PPARs) are a subgroup of the nuclear receptor superfamily and key regulators of lipid metabolism. Several observations suggest a possible intersection of AHR and PPARs. Three PPAR isotypes, α, γ, and δ, have been identified in mammalian species. PPARα is an important regulator of energy homeostasis. PPARα activates fatty acid catabolism and stimulates gluconeogenesis and cholesterol catabolism in the liver. The function of PPARδ remains elusive, but it appears to be a potent inhibitor of ligand-induced transcriptional activity of PPARα and PPARγ. PPARγ exists in several isoforms, three in humans and two in mouse. PPARγ is important for adipose tissue differentiation, is required for the survival of differentiated adipocytes, and is also involved in glucose metabolism [117, 118]. PPARs function as heterodimers with the RXR, another member of the nuclear receptor superfamily described above. The heterodimeric complexes signal through binding to a specific peroxisome proliferator response element in target genes. Upon activation by their respective ligands, PPARs regulate genes involved in metabolism of lipids or carbohydrates, cell proliferation, cell differentiation, and inflammation. The endogenous ligands identified for the PPARs include long-chain fatty acids, polyunsaturated fatty acids such as linoleic and arachidonic acid, saturated fatty acids, and eicosanoids [118, 119]. There exist several synthetic ligands for the PPARs used as pharmaceuticals, for instance, thiazolidinediones, which activate PPARγ and are antidiabetic [120].

9.7.5 AHR and PPARα

A synthetic PPARα agonist, WY, increased the mRNA and protein expression of AHR in human colorectal adenocarcinoma (Caco-2) cells in a PPARα-dependent manner [121]. A putative PPRE site is located within the promoter of the *AHR* gene and mutation of this sequence repressed the *AHR* promoter activation by WY in transient transfections [122]. WY was also shown to potentiate 3-MC-induced CYP1A1 expression, perhaps due to increased AHR protein levels. On the other hand, WY also induced CYP1A1 expression and activity in an AHR-independent manner, probably by binding of PPARα to PPRE sites located within the promoter of *CYP1A1*. 3-MC failed to modify PPARα expression in this study [121].

9.7.6 AHR and PPARγ

AHR and PPARγ appear to interact during adipocyte differentiation and adipogenesis. For instance, it has been shown that TCDD exposure downregulates PPARγ2 mRNA during differentiation of isolated preadipocytes and SCD-1 and C/EBPα mRNA in MEFs [115, 123]. However, TCDD did not affect the mRNA levels of RXRα and RARα [124, 125]. The inhibitory effect of TCDD on adipocyte differentiation was AHR dependent, as judged by the effects of the AHR antagonist α-naphthoflavone in preadipocytes from AHR null animals as well as from transient transfection of dominant negative mutant of AHR or AHR antisense [126, 127].

Addition of TCDD before or concurrent with hormonal induction suppressed PPARγ mRNA and adipogenesis [116]. This effect of TCDD treatment was absent in AHR$^{-/-}$ MEFs, thus establishing the role of AHR in hormone-induced adipogenesis. Furthermore, overexpression of AHR inhibits adipose differentiation in 3T3-L1 cells independent of dioxin [127]. This inhibition was restored by exposing the 3T3-L1 cells to PPARγ ligands.

The suppression of PPARγ1 and subsequent adipogenesis by AHR activation involves the MEK/ERK pathway and appears to depend on cooperation with EGF activation (of the MEK/ERK pathway) [125, 127–129]. When 3T3-L1 adipocytes were exposed to TCDD or modified with vectors that lead to increased expression of the AHR, phosphorylation and activation of ERK1/2 and JNK was detected [125, 127, 130]. MEK and JNK inhibitors reversed the suppression of PPARγ and adipogenesis [125, 127]. C-Src is a tyrosine kinase associated with the AHR that triggers several phosphorylation cascades upon activation with AHR agonists [123, 131]. Interestingly, low levels of a selective Src kinase inhibitor completely prevented the inhibitory effect of the TCDD/EGF combination, suggesting the involvement of c-src [132]. Also, adipocyte cell lines exposed to TCDD showed increased expression and secretion of TNFα, which subsequently could lead to repression of PPARγ activity [120, 130, 133].

9.7.7 Progesterone Receptor

In addition to other steroid receptors, PR is also involved in crosstalk with AHR (Fig. 9.4). The human PR exists in two isoforms, PR-A and PR-B, that can mediate progestin-activated gene transcription at specific progestin-responsive elements (PREs). The human PR-B contains an additional 164 amino acids at its N-terminus that are absent in the hPR-A. Both isoforms can act as potent repressors of AHR transcriptional activity. hPR-B represses transcription of the AHR–ARNT complex in its agonist-bound and transcriptionally active conformation. In contrast, hPR-A-mediated repression requires neither a transcriptionally active conformation nor interaction with the cognate DNA element [134]. Exposure to TCDD has also been shown to decrease progesterone production in luteinized granulosa cells, indicating yet another PR–AHR crosstalk mechanism [135].

9.8 BRIEF OVERVIEW OF OTHER NUCLEAR RECEPTORS

9.8.1 Thyroid Hormone Receptor

TCDD is known to disrupt the thyroid hormone system primarily by changing thyroid hormone turnover. This is, however, independent of an interaction with thyroid hormone receptors (TRs). Instead, ligand-activated AHR transcriptionally regulates UDP-glucuronosyltransferase (UDPGT) that glucuronidates thyroxine (T4) in rodent liver. The increase in enzyme levels results in reduced circulating T4 levels by accelerated biliary excretion of T4 glucuronides [136].

9.8.2 Constitutive Androstane Receptor and Pregnane X Receptor

Constitutive androstane receptor (CAR) and pregnane X receptor (PXR) are xenobiotic sensors involved in the regulation of phase I and phase II enzymes, such as CYP2B and CYP3A, and glucuronosyltransferases [137]. PXR and CAR bind not only multiple ligands including drugs such as phenobarbital (PB) and rifampicin but also endogenous steroid hormones [137]. Recently, AHR activation by β-naphthoflavone was shown to increase CAR mRNA in mouse liver, correlating with an increase in transcriptional activity of CAR [138]. Rifampicin-activated PXR increased the expression and activity of CYP3A, which was followed by an increased metabolism of omeprazole sulfide, an antagonist to the AHR agonist, omeprazole [139]. These studies address potential crosstalk of AHR with xenobiotic binding nuclear receptors (Fig. 9.4).

9.8.3 Adrenal 4 Binding Protein/Steroidogenic Factor 1

The adrenal 4 binding protein/steroidogenic factor 1 (A4BP/SF-1 or SF-1) is an orphan receptor that controls sexual development in the embryo and at puberty. This transcription factor regulates the transcription of key genes involved in sexual development and reproduction. Mutations in the *SF-1* (*NR5A1*) gene are associated with ovarian insufficiency and can lead to intersex genitals, absence of puberty, and infertility.

Ligand-activated AHR has been demonstrated to physically interact with SF-1 [23]. Formation of this complex, AHR/ARNT and SF-1, synergistically activates *Cyp19* gene transcription in mouse ovarian granulosa cells by an interaction between XRE and Ad4-containing sequences within the *Cyp19* promoter region [23]. Thus, AHR may be a ligand-dependent regulator of SF-1 activity (Fig. 9.4).

9.8.4 The Estrogen-Related Receptor α and COUP-TFI

The estrogen-related receptor α (ERRα), an orphan receptor also known as NR3B1, is closely related to ERα but does not bind estradiol or to the same DNA elements as ERα. ERRα has been demonstrated to control ERα activities in certain tissues, and these two proteins also regulate many of the same genes, for example, lactoferrin [140]. ERRα knockout mice display impaired fat metabolism and absorption.

TCDD-activated AHR has been reported to physically interact with both the ERRα and the orphan receptor chicken ovalbumin upstream promoter transcription factor 1 (COUP-TF1) *in vitro* [141] (Fig. 9.4). COUP-TF1 and 2 are well-characterized orphan receptors involved in the regulation of organogenesis, neurogenesis, and cellular differentiation during embryonic development. There are also indications for a role in retinoid signaling [142]. In the context of AHR interactions, COUP-TF1 binds to consensus XREs, suggesting a putative role as competitor of AHR DNA binding activity [141]. However, to date the physiological importance of these findings remains elusive.

9.9 SUMMARY

Recent advances in the field of AHR research indicate that the AHR is a multifunctional transcription factor involved in many more physiological processes than first believed. Its molecular interactions with other vital transcription factors, such as the nuclear receptors, are yet to be fully elucidated. However, it is clear that this type of crosstalk is extensive and has a strong biological significance.

Common for the studies reviewed here is that there seem to be significant species- and cell context-specific differences regarding the mechanisms and the impact of the interactions. This may be critical when translating the results to other experimental systems.

The ER–AHR inhibitory crosstalk is complex and has been the subject of many studies. One outcome from this research was the development of selective aryl hydrocarbon modulators (SAHRMs), which were suggested to be used in, for example, the treatment of hormonal cancers, such as breast and prostate cancer. The molecular basis for these pharmaceutical candidates is the negative crosstalk between AHR and ER and AHR and AR [143]. Moreover, since the AHR seems to influence the PPARγ, which is a key player in adipogenesis, one might speculate about the possibility of also targeting this pathway by developing AHR modulators. These examples underline the importance of thorough molecular analysis of the mechanisms involved in AHR signaling and its convergence with other receptor-mediated signaling pathways.

ACKNOWLEDGMENT

This work was supported by a grant from the Swedish Cancer Society.

REFERENCES

1. Zhang, L., Ma, J., Takeuchi, M., Usui, Y., Hattori, T., Okunuki, Y, Yamakawa, N., Kezuka, T., Kuroda, M., and Goto, H. (2009). Activation of aryl hydrocarbon receptor suppresses experimental autoimmune uveoretinitis by inducing differentiation of regulatory T cells. *Investigative Ophthalmology & Visual Science*, *51*, 2109–2117.

2. Quintana, F. J., Basso, A. S., Iglesias, A. H., Korn, T., Farez, M. F., Bettelli, E., Caccamo, M., Oukka, M., and Weiner, H. L. (2008). Control of T_{reg} and T_H17 cell differentiation by the aryl hydrocarbon receptor. *Nature*, *453*, 65–71.

3. Puga, A., Ma, C., and Marlowe, J. L. (2009). The aryl hydrocarbon receptor cross-talks with multiple signal transduction pathways. *Biochemical Pharmacology*, *77*, 713–722.

4. Heldring, N., Pike, A., Andersson, S., Matthews, J., Cheng, G., Hartman, J., Tujague, M., Strom, A., Treuter, E., Warner, M.,

and Gustafsson, J.-A. (2007). Estrogen receptors: how do they signal and what are their targets. *Physiological Reviews*, 87, 905–931.

5. Gronemeyer, H., Gustafsson, J.-A., and Laudet, V. (2004). Principles for modulation of the nuclear receptor superfamily. *Nature Reviews Drug Discovery*, 3, 950–964.

6. Dahlman-Wright, K., Cavailles, V., Fuqua, S. A., Jordan, V. C., Katzenellenbogen, J.-A., Korach, K.-S., Maggi, A., Muramatsu, M., Parker, M.-G., and Gustafsson, J.-A. (2006). International Union of Pharmacology. LXIV. Estrogen receptors. *Pharmacological Reviews*, 58, 773–781.

7. Barkhem, T., Carlsson, B., Nilsson, Y., Enmark, E., Gustafsson, J., and Nilsson, S. (1998). Differential response of estrogen receptor alpha and estrogen receptor beta to partial estrogen agonists/antagonists. *Molecular Pharmacology*, 54, 105–112.

8. Delaunay, F., Pettersson, K., Tujague, M., and Gustafsson, J. A. (2000). Functional differences between the amino-terminal domains of estrogen receptors alpha and beta. *Molecular Pharmacology*, 58, 584–590.

9. Razandi, M., Pedram, A., Merchenthaler, I., Greene, G. L., and Levin, E. R. (2004). Plasma membrane estrogen receptors exist and functions as dimers. *Molecular Endocrinology*, 18, 2854–2865.

10. Wong, C. W., McNally, C., Nickbarg, E., Komm, B. S., and Cheskis, B. J. (2002). Estrogen receptor-interacting protein that modulates its nongenomic activity-crosstalk with Src/Erk phosphorylation cascade. *Proceedings of the National Academy of Sciences of the United States of America*, 99, 14783–14788.

11. Falkenstein, E., Heck, M., Gerdes, D., Grube, D., Christ, M., Weigel, M., Buddhikot, M., Meizel, S., and Wehling, M. (1999). Specific progesterone binding to a membrane protein and related nongenomic effects on Ca^{2+}-fluxes in sperm. *Endocrinology*, 140, 5999–6002.

12. Gerdes, D., Wehling, M., Leube, B., and Falkenstein, E. (1998). Cloning and tissue expression of two putative steroid membrane receptors. *Biological Chemistry*, 379, 907–911.

13. Benten, W. P., Lieberherr, M., Stamm, O., Wrehlke, C., Guo, Z., and Wunderlich, F. (1999). Testosterone signaling through internalizable surface receptors in androgen receptor-free macrophages. *Molecular Biology of the Cell*, 10, 3113–3123.

14. Nadal, A., Ropero, A. B., Laribi, O., Maillet, M., Fuentes, E., and Soria B. (2000). Nongenomic actions of estrogens and xenoestrogens by binding at a plasma membrane receptor unrelated to estrogen receptor alpha and estrogen receptor beta. *Proceedings of the National Academy of Sciences of the United States of America*, 97, 11603–11608.

15. Ropero, A. B., Soria, B., and Nadal A. (2002). A nonclassical estrogen membrane receptor triggers rapid differential actions in the endocrine pancreas. *Molecular Endocrinology*, 16, 497–505.

16. Soriano, S., Ropero, A. B., Alonso-Magdalena, P., Ripoll, C., Quesada, I., Gassner, B., Kuhn, M., Gustafsson, J. A., and Nadal A. (2009). Rapid regulation of K(ATP) channel activity by 17β-estradiol in pancreatic β-cells involves the estrogen receptor β and the atrial natriuretic peptide receptor. *Molecular Endocrinology*, 23, 1973–1982.

17. Maggiolini, M. and Picard D. (2010). The unfolding stories of GPR30, a new membrane-bound estrogen receptor. *Journal of Endocrinology*, 204, 105–114.

18. Safe, S., Wormke, M., and Samudio, I. (2000). Mechanisms of inhibitory aryl hydrocarbon receptor-estrogen receptor cross-talk in human breast cancer cells. *Journal of Mammary Gland Biology and Neoplasia*, 5, 295–306.

19. Safe, S. and Wormke, M. (2003). Inhibitory aryl hydrocarbon receptor-estrogen receptor alpha cross-talk and mechanisms of action. *Chemical Research in Toxicology*, 16, 807–816.

20. Fernandez-Salguero, P., Pineau, T., Hilbert, D. M., McPhail, T., Lee, S. S., Kimura, S., Nebert, D. W., Rudikoff, S., Ward, J. M., and Gonzalez F. J. (1995). Immune system impairment and hepatic fibrosis in mice lacking the dioxin-binding Ah receptor. *Science*, 268, 722–726.

21. Schmidt, J. V., Su, G. H., Reddy, J. K., Simon, M. C., and Bradfield C. A. (1996). Characterization of a murine Ahr null allele: involvement of the Ah receptor in hepatic growth and development. *Proceedings of the National Academy of Sciences of the United States of America*, 93, 6731–6736.

22. Mimura, J., Yamashita, K., Nakamura, K., Morita, M., Takagi, T. N., Nakao, K., Ema, M., Sogawa, K., Yasuda, M., Katsuki, M., and Fujii-Kuriyama, Y. (1997). Loss of teratogenic response to 2,3,7,8-tetrachlorodibenzo-p-dioxin (TCDD) in mice lacking the Ah (dioxin) receptor. *Genes to Cells: Devoted to Molecular & Cellular Mechanisms*, 2, 645–654.

23. Baba, T., Mimura, J., Nakamura, N., Harada, N., Yamamoto, M., Morohashi, K., and Fujii-Kuriyama Y. (2005). Intrinsic function of the aryl hydrocarbon (dioxin) receptor as a key factor in female reproduction. *Molecular and Cellular Biology*, 25, 10040–10051.

24. Kociba, R. J., Keyes, D. G., Beyer, J. E, Carreon, R. M., Wade, C. E., Dittenber, D. A., Kalnins, R. P., Frauson, L. E., Park, C. N., Barnard, S. D., Hummel, R. A., and Humiston, C. G. (1978). Results of a two-year chronic toxicity and oncogenicity study of 2,3,7,8-tetrachlorodibenzo-p-dioxin in rats. *Toxicology and Applied Pharmacology*, 46, 279–303.

25. Bertazzi, P. A., Consonni, D., Bachetti, D., Rubagotti, M., Baccarelli, A., Zocchetti, C., and Pesatori A. C. (2001). Health effects of dioxin exposure: a 20-year mortality study. *American Journal of Epidemiology*, 153, 1031–1044.

26. Romkes, M. and Safe, S. (1988). Comparative activities of 2,3,7,8-tetrachlorodibenzo-p-dioxin and progesterone as antiestrogens in the female rat uterus. *Toxicology and Applied Pharmacology*, 92, 368–380.

27. Astroff, B. and Safe, S. (1990). 2,3,7,8-Tetrachlorodibenzo-p-dioxin as an antiestrogen: effect on rat uterine peroxidase activity. *Biochemical Pharmacology*, 39, 485–488.

28. Astroff, B. and Safe, S. (1991). 6-Alkyl-1,3,8-trichlorodibenzofurans as antiestrogens in female Sprague-Dawley rats. *Toxicology*, 69, 187–197.

29. Narasimhan, T. R., Craig, A., Arellano, L., Harper, N., Howie, L., Menache, M., Birnbaum, L., and Safe, S.

(1994). Relative sensitivities of 2,3,7,8-tetrachlorodibenzo-*p*-dioxin-induced Cyp1a-1 and Cyp1a-2 gene expression and immunotoxicity in female B6C3F1 mice. *Fundamental and Applied Toxicology*, 23, 598–607.

30. Buchanan, D. L., Sato, T., Peterson, R. E., and Cooke P. S. (2000). Antiestrogenic effects of 2,3,7,8-tetrachlorodibenzo-*p*-dioxin in mouse uterus: critical role of the aryl hydrocarbon receptor in stromal tissue. *Toxicological Sciences*, 57, 302–311.

31. Duan, R., Porter, W., Samudio, I., Vyhlidal, C., Kladde, M., and Safe, S. (1999). Transcriptional activation of c-fos protooncogene by 17β-estradiol: mechanism of aryl hydrocarbon receptor-mediated inhibition. *Molecular Endocrinology*, 13, 1511–1521.

32. Astroff, B., Eldridge, B., and Safe, S. (1991). Inhibition of the 17β-estradiol-induced and constitutive expression of the cellular protooncogene c-fos by 2,3,7,8-tetrachlorodibenzo-*p*-dioxin (TCDD) in the female rat uterus. *Toxicology Letters*, 56, 305–315.

33. Krishnan, V., Porter, W., Santostefano, M., Wang, X., and Safe, S. (1995). Molecular mechanism of inhibition of estrogen-induced cathepsin D gene expression by 2,3,7,8-tetrachlorodibenzo-*p*-dioxin (TCDD) in MCF-7 cells. *Molecular and Cellular Biology*, 15, 6710–6719.

34. Klinge, C. M., Bowers, J. L., Kulakosky, P. C., Kamboj, K. K., and Swanson, H. I. (1999). The aryl hydrocarbon receptor (AHR)/AHR nuclear translocator (ARNT) heterodimer interacts with naturally occurring estrogen response elements. *Molecular and Cellular Endocrinology*, 157, 105–119.

35. Takemoto, K., Nakajima, M., Fujiki, Y., Katoh, M., Gonzalez, F. J., and Yokoi, T. (2004). Role of the aryl hydrocarbon receptor and Cyp1b1 in the antiestrogenic activity of 2,3,7,8-tetrachlorodibenzo-*p*-dioxin. *Archives of Toxicology*, 78, 309–315.

36. Wormke, M., Stoner, M., Saville, B., Walker, K., Abdelrahim, M., Burghardt, R., and Safe, S. (2003). The aryl hydrocarbon receptor mediates degradation of estrogen receptor α through activation of proteasomes. *Molecular and Cellular Biology*, 23, 1843–1855.

37. Ohtake, F., Baba, A., Takada, I., Okada, M., Iwasaki, K., Miki, H., Takahashi, S., Kouzmenko, A., Nohara, K., Chiba, T., Fujii-Kuriyama, Y., and Kato, S. (2007). Dioxin receptor is a ligand-dependent E3 ubiquitin ligase. *Nature*, 446, 562–566.

38. Nguyen, T. A., Hoivik, D., Lee, J. E., and Safe, S. (1999). Interactions of nuclear receptor coactivator/corepressor proteins with the aryl hydrocarbon receptor complex. *Archives of Biochemistry and Biophysics*, 367, 250–257.

39. Kumar, M. B., Tarpey, R. W., and Perdew, G. H. (1999). Differential recruitment of coactivator RIP140 by Ah and estrogen receptors. Absence of a role for LXXLL motifs. *Journal of Biological Chemistry*, 274, 22155–22164.

40. Kumar, M. B. and Perdew, G. H. (1999). Nuclear receptor coactivator SRC-1 interacts with the Q-rich subdomain of the AhR and modulates its transactivation potential. *Gene Expression*, 8, 273–286.

41. Beischlag, T. V., Wang, S., Rose, D. W., Torchia, J., Reisz-Porszasz, S., Muhammad, K., Nelson, W. E., Probst, M. R., Rosenfeld, M. G., and Hankinson, O. (2002). Recruitment of the NCoA/SRC-1/p160 family of transcriptional coactivators by the aryl hydrocarbon receptor/aryl hydrocarbon receptor nuclear translocator complex. *Molecular and Cellular Biology*, 22, 4319–4133.

42. Brunnberg, S., Pettersson, K., Rydin, E., Matthews, J., Hanberg, A., and Pongratz, I. (2003). The basic helix–loop–helix–PAS protein ARNT functions as a potent coactivator of estrogen receptor-dependent transcription. *Proceedings of the National Academy of Sciences of the United States of America*, 100, 6517–6522.

43. Ruegg, J., Swedenborg, E., Wahlstrom, D., Escande, A., Balaguer, P., Pettersson, K., and Pongratz, I. (2008). The transcription factor aryl hydrocarbon receptor nuclear translocator functions as an estrogen receptor beta-selective coactivator, and its recruitment to alternative pathways mediates antiestrogenic effects of dioxin. *Molecular Endocrinology*, 22, 304–316.

44. Kobayashi, Y., Kitamoto, T., Masuhiro, Y., Watanabe, M., Kase, T., Metzger, D., Yanagisawa, J., and Kato, S. (2000). p300 mediates functional synergism between AF-1 and AF-2 of estrogen receptor alpha and beta by interacting directly with the N-terminal A/B domains. *Journal of Biological Chemistry*, 275, 15645–15651.

45. Kobayashi, A., Numayama-Tsuruta, K., Sogawa, K., and Fujii-Kuriyama, Y. (1997). CBP/p300 functions as a possible transcriptional coactivator of Ah receptor nuclear translocator (Arnt). *Journal of Biochemistry*, 122, 703–710.

46. Matthews, J., Wihlen, B., Thomsen, J., and Gustafsson, J. A. (2005). Aryl hydrocarbon receptor-mediated transcription: ligand-dependent recruitment of estrogen receptor alpha to 2,3,7,8-tetrachlorodibenzo-*p*-dioxin-responsive promoters. *Molecular and Cellular Biology*, 25, 5317–5328.

47. Matthews, J., Wihlen, B., Heldring, N., MacPherson, L., Helguero, L., Treuter, E., Haldosen, L. A., and Gustafsson, J. A. (2007). Co-planar 3,3′,4,4′,5-pentachlorinated biphenyl and non-co-planar 2,2′,4,6,6′-pentachlorinated biphenyl differentially induce recruitment of oestrogen receptor alpha to aryl hydrocarbon receptor target genes. *Biochemical Journal*, 406, 343–353.

48. Swedenborg, E., Ruegg, J., Hillenweck, A., Rehnmark, S., Faulds, M. H., Zalko, D., Pongratz, I., and Pettersson, K. (2008). 3-Methylcholanthrene displays dual effects on estrogen receptor (ER) alpha and ER beta signaling in a cell-type specific fashion. *Molecular Pharmacology*, 73, 575–586.

49. Abdelrahim, M., Ariazi, E., Kim, K., Khan, S., Barhoumi, R., Burghardt, R., Liu, S., Hill, D., Finnell, R., Wlodarczyk, B., Jordan, V. C., and Safe, S. (2006). 3-Methylcholanthrene and other aryl hydrocarbon receptor agonists directly activate estrogen receptor alpha. *Cancer Research*, 66, 2459–2467.

50. Shipley, J. M. and Waxman, D. J. (2006). Aryl hydrocarbon receptor-independent activation of estrogen receptor-dependent transcription by 3-methylcholanthrene. *Toxicology and Applied Pharmacology*, 213, 87–97.

51. Ohtake, F., Takeyama, K., Matsumoto, T., Kitagawa, H., Yamamoto, Y., Nohara, K., Tohyama, C., Krust, A., Mimura, J., Chambon, P., Yanagisawa, J., Fujii-Kuriyama, Y., and Kato, S. (2003). Modulation of oestrogen receptor signalling by association with the activated dioxin receptor. *Nature*, 423, 545–550.

52. Wihlen, B., Ahmed, S., Inzunza, J., and Matthews, J. (2009). Estrogen receptor subtype- and promoter-specific modulation of aryl hydrocarbon receptor-dependent transcription. *Molecular Cancer Research*, 7, 977–986.

53. Kietz, S., Thomsen, J. S., Matthews, J., Pettersson, K., Strom, A., and Gustafsson, J. A. (2004). The Ah receptor inhibits estrogen-induced estrogen receptor beta in breast cancer cells. *Biochemical and Biophysical Research Communications*, 320, 76–82.

54. Jana, N. R., Sarkar, S., Ishizuka, M., Yonemoto, J., Tohyama, C., and Sone, H. (1999). Cross-talk between 2,3,7,8-tetrachlorodibenzo-*p*-dioxin and testosterone signal transduction pathways in LNCaP prostate cancer cells. *Biochemical and Biophysical Research Communications*, 256, 462–468.

55. Ohtake, F., Baba, A., Fujii-Kuriyama, Y., and Kato, S. (2008). Intrinsic AhR function underlies cross-talk of dioxins with sex hormone signalings. *Biochemical and Biophysical Research Communications*, 370, 541–546.

56. Barnes-Ellerbe, S., Knudsen, K. E., and Puga, A. (2004). 2,3,7,8-Tetrachlorodibenzo-*p*-dioxin blocks androgen-dependent cell proliferation of LNCaP cells through modulation of pRB phosphorylation. *Molecular Pharmacology*, 66, 502–511.

57. Vezina, C. M., Lin, T. M., and Peterson, R. E. (2009). AHR signaling in prostate growth, morphogenesis, and disease. *Biochemical Pharmacology*, 77, 566–576.

58. Johnson, L., Dickerson, R., Safe, S. H., Nyberg, C. L., Lewis, R. P., and Welsh, Jr., T. H. (1992). Reduced Leydig cell volume and function in adult rats exposed to 2,3,7,8-tetrachlorodibenzo-*p*-dioxin without a significant effect on spermatogenesis. *Toxicology*, 76, 103–118.

59. Adamsson, A., Simanainen, U., Viluksela, M., Paranko, J., and Toppari, J. (2009). The effects of 2,3,7,8-tetrachlorodibenzo-*p*-dioxin on fetal male rat steroidogenesis. *International Journal of Andrology*, 32, 575–585.

60. Ohsako, S., Miyabara, Y., Nishimura, N., Kurosawa, S., Sakaue, M., Ishimura, R., Sato, M., Takeda, K., Aoki, Y., Sone, H., Tohyama, C., and Yonemoto Y. (2001). Maternal exposure to a low dose of 2,3,7,8-tetrachlorodibenzo-*p*-dioxin (TCDD) suppressed the development of reproductive organs of male rats: dose-dependent increase of mRNA levels of 5α-reductase type 2 in contrast to decrease of androgen receptor in the pubertal ventral prostate. *Toxicological Sciences*, 60, 132–143.

61. Roman, B. L., Pollenz, R. S., and Peterson, R. E. (1998). Responsiveness of the adult male rat reproductive tract to 2,3,7,8-tetrachlorodibenzo-*p*-dioxin exposure: Ah receptor and ARNT expression, CYP1A1 induction, and Ah receptor down-regulation. *Toxicology and Applied Pharmacology*, 150, 228–239.

62. Roman, B. L., Sommer, R. J., Shinomiya, K., and Peterson, R. E. (1995). In utero and lactational exposure of the male rat to 2,3,7,8-tetrachlorodibenzo-*p*-dioxin: impaired prostate growth and development without inhibited androgen production. *Toxicology and Applied Pharmacology*, 134, 241–250.

63. Fukuzawa, N. H., Ohsako, S., Wu, Q., Sakaue, M., Fujii-Kuriyama, Y., Baba, T., and Tohyama, C. (2004). Testicular cytochrome P450scc and LHR as possible targets of 2,3,7,8-tetrachlorodibenzo-*p*-dioxin (TCDD) in the mouse. *Molecular and Cellular Endocrinology*, 221, 87–96.

64. Ruangwises, S., Bestervelt, L. L., Piper, D. W., Nolan, C. J., and Piper, W. N. (1991). Human chorionic gonadotropin treatment prevents depressed 17α-hydroxylase/C17-20 lyase activities and serum testosterone concentrations in 2,3,7,8-tetrachlorodibenzo-*p*-dioxin-treated rats. *Biology of Reproduction*, 45, 143–150.

65. DiBartolomeis, M. J., Moore, R. W., Peterson, R. E., Christian, R. E., and Jefcoate, C. R. (1987). Altered regulation of adrenal steroidogenesis in 2,3,7,8-tetrachlorodibenzo-*p*-dioxin-treated rats. *Biochemical Pharmacology*, 36, 59–67.

66. Pesonen, S. A., Haavisto, T. E., Viluksela, M., Toppari, J., and Paranko, J. (2006). Effects of in utero and lactational exposure to 2,3,7,8-tetrachlorodibenzo-*p*-dioxin (TCDD) on rat follicular steroidogenesis. *Reproductive Toxicology*, 22, 521–528.

67. Mutoh, J., Taketoh, J., Okamura, K., Kagawa, T., Ishida, T., Ishii, Y., and Yamada, H. (2006). Fetal pituitary gonadotropin as an initial target of dioxin in its impairment of cholesterol transportation and steroidogenesis in rats. *Endocrinology*, 147, 927–936.

68. Kollara, A. and Brown, T. J. (2009). Modulation of aryl hydrocarbon receptor activity by four and a half LIM domain 2. *International Journal of Biochemistry and Cell Biology*, 41, 1182–1188.

69. Kollara, A. and Brown, T. J. (2010). Four and a half LIM domain 2 alters the impact of aryl hydrocarbon receptor on androgen receptor transcriptional activity. *Journal of Steroid Biochemistry and Molecular Biology*, 118, 51–58.

70. Kollara, A. and Brown, T. J. (2006). Functional interaction of nuclear receptor coactivator 4 with aryl hydrocarbon receptor. *Biochemical and Biophysical Research Communications*, 346, 526–534.

71. Puga, A., Nebert, D. W., and Carrier, F. (1992). Dioxin induces expression of c-fos and c-jun proto-oncogenes and a large increase in transcription factor AP-1. *DNA and Cell Biology*, 11, 269–281.

72. Kizu, R., Okamura, K., Toriba, A., Kakishima, H., Mizokami, A., Burnstein, K. L., and Hayakawa, K. (2003). A role of aryl hydrocarbon receptor in the antiandrogenic effects of polycyclic aromatic hydrocarbons in LNCaP human prostate carcinoma cells. *Archives of Toxicology*, 77, 335–343.

73. Sanada, N., Gotoh, Y., Shimazawa, R., Klinge, C. M., and Kizu, R. (2009). Repression of activated aryl hydrocarbon receptor-induced transcriptional activation by 5α-dihydrotestosterone in human prostate cancer LNCaP and human breast

cancer T47D cells. *Journal of Pharmacological Sciences, 109*, 380–387.

74. Vegiopoulos, A. and Herzig, S. (2007). Glucocorticoids, metabolism and metabolic diseases. *Molecular and Cellular Endocrinology, 275*, 43–61.

75. Gorski, J. R., Lebofsky, M., and Rozman, K. (1988). Corticosterone decreases toxicity of 2,3,7,8-tetrachlorodibenzo-*p*-dioxin (TCDD) in hypophysectomized rats. *Journal of Toxicology and Environmental Health, 25*, 349–360.

76. Abbott, B. D. (1995). Review of the interaction between TCDD and glucocorticoids in embryonic palate. *Toxicology, 105*, 365–373.

77. Birnbaum, L. S., Harris, M. W., Miller, C. P., Pratt, R. M., and Lamb, J. C. (1986). Synergistic interaction of 2,3,7,8,-tetrachlorodibenzo-*p*-dioxin and hydrocortisone in the induction of cleft palate in mice. *Teratology, 33*, 29–35.

78. Taylor, M. J., Lucier, G. W., Mahler, J. F., Thompson, M., Lockhart, A. C., and Clark, G. C. (1992). Inhibition of acute TCDD toxicity by treatment with anti-tumor necrosis factor antibody or dexamethasone. *Toxicology and Applied Pharmacology, 117*, 126–132.

79. Abbott, B. D., Perdew, G. H., Buckalew, A. R., and Birnbaum, L. S. (1994). Interactive regulation of Ah and glucocorticoid receptors in the synergistic induction of cleft palate by 2,3,7,8-tetrachlorodibenzo-*p*-dioxin and hydrocortisone. *Toxicology and Applied Pharmacology, 128*, 138–150.

80. Mullen Grey, A. K. and Riddick, D. S. (2009). Glucocorticoid and adrenalectomy effects on the rat aryl hydrocarbon receptor pathway depend on the dosing regimen and post-surgical time. *Chemico-Biological Interactions, 182*, 148–158.

81. Bielefeld, K. A., Lee, C., and Riddick, D. S. (2008). Regulation of aryl hydrocarbon receptor expression and function by glucocorticoids in mouse hepatoma cells. *Drug Metabolism and Disposition, 36*, 543–551.

82. Sonneveld, E., Jonas, A., Meijer, O. C., Brouwer, A., and van der Burg, B. (2007). Glucocorticoid-enhanced expression of dioxin target genes through regulation of the rat aryl hydrocarbon receptor. *Toxicological Sciences, 99*, 455–469.

83. Mathis, J. M., Houser, W. H., Bresnick, E., Cidlowski, J. A., Hines, R. N., Prough, R. A., and Simpson, E. R. (1989). Glucocorticoid regulation of the rat cytochrome P450c (P450IA1) gene: receptor binding within intron I. *Archives of Biochemistry and Biophysics, 269*, 93–105.

84. Sherratt, A. J., Banet, D. E., and Prough, R. A. (1990). Glucocorticoid regulation of polycyclic aromatic hydrocarbon induction of cytochrome P450IA1, glutathione S-transferases, and NAD(P)H:quinone oxidoreductase in cultured fetal rat hepatocytes. *Molecular Pharmacology, 37*, 198–205.

85. Pinaire, J. A., Xiao, G. H., Falkner, K. C., and Prough, R. A. (2004). Regulation of NAD(P)H:quininone oxidoreductase by glucocorticoids. *Toxicology and Applied Pharmacology, 199*, 344–353.

86. Xiao, G. H., Pinaire, J. A., Rodrigues, A. D., and Prough, R. A. (1995). Regulation of the Ah gene battery via Ah receptor-dependent and independent processes in cultured adult rat hepatocytes. *Drug Metabolism and Disposition, 23*, 642–650.

87. Dvorak, Z., Vrzal, R., Pavek, P., and Ulrichova, J. (2008). An evidence for regulatory cross-talk between aryl hydrocarbon receptor and glucocorticoid receptor in HepG2 cells. *Physiological Research, 57*, 427–435.

88. Lai, K. P., Wong, M. H., and Wong, C. K. (2004). Modulation of AhR-mediated CYP1A1 mRNA and EROD activities by 17β-estradiol and dexamethasone in TCDD-induced H411E cells. *Toxicological Sciences, 78*, 41–49.

89. Wang, S. H., Liang, C. T., Liu, Y. W., Huang, M. C., Huang, S. C., Hong, W. F., and Su, J. G. (2009). Crosstalk between activated forms of the aryl hydrocarbon receptor and glucocorticoid receptor. *Toxicology, 262*, 87–97.

90. Vrzal, R., Stejskalova, L., Monostory, K., Maurel, P., Bachleda, P., Pavek, P., and Dvorak, Z. (2009). Dexamethasone controls aryl hydrocarbon receptor (AhR)-mediated CYP1A1 and CYP1A2 expression and activity in primary cultures of human hepatocytes. *Chemico-Biological Interactions, 179*, 288–296.

91. Monostory, K., Kohalmy, K., Prough, R. A., Kobori, L., and Vereczkey, L. (2005). The effect of synthetic glucocorticoid, dexamethasone on CYP1A1 inducibility in adult rat and human hepatocytes. *FEBS Letters, 579*, 229–235.

92. Murphy, K. A., Quadro, L., and White, L. A. (2007). The intersection between the aryl hydrocarbon receptor (AhR)- and retinoic acid-signaling pathways. *Vitamins & Hormones, 75*, 33–67.

93. Janosek, J., Hilscherova, K., Blaha, L., and Holoubek, I. (2006). Environmental xenobiotics and nuclear receptors—interactions, effects and *in vitro* assessment. *Toxicology In Vitro, 20*, 18–37.

94. Germain, P., Chambon, P., Eichele, G., Evans, R. M., Lazar, M. A., Leid, M., De Lera, A. R., Lotan, R., Mangelsdorf, D. J., and Gronemeyer, H. (2006). International Union of Pharmacology. LX. Retinoic acid receptors. *Pharmacological Reviews, 58*, 712–725.

95. Nilsson, C. B. and Håkansson, H. (2002). The retinoid signaling system—a target in dioxin toxicity. *Critical Reviews in Toxicology, 32*, 211–232.

96. Lorick, K. L., Toscano, D. L., and Toscano, Jr. W. A. (1998). 2,3,7,8-Tetrachlorodibenzo-*p*-dioxin alters retinoic acid receptor function in human keratinocytes. *Biochemical and Biophysical Research Communications, 243*, 749–752.

97. Tijet, N., Boutros, P. C., Moffat, I. D., Okey, A. B., Tuomisto, J., and Pohjanvirta, R. (2006). Aryl hydrocarbon receptor regulates distinct dioxin-dependent and dioxin-independent gene batteries. *Molecular Pharmacology, 69*, 140–153.

98. Weston, W. M., Nugent, P., and Greene, R. M. (1995). Inhibition of retinoic-acid-induced gene expression by 2,3,7,8-tetrachlorodibenzo-*p*-dioxin. *Biochemical and Biophysical Research Communications, 207*, 690–694.

99. Rubin, A. L. and Rice, R. H. (1988). 2,3,7,8-Tetrachlorodibenzo-*p*-dioxin and polycyclic aromatic hydrocarbons suppress retinoid-induced tissue transglutaminase in SCC-4 cultured human squamous carcinoma cells. *Carcinogenesis, 9*, 1067–1070.

100. Krig, S. R. and Rice, R. H. (2000). TCDD suppression of tissue transglutaminase stimulation by retinoids in malignant human keratinocytes. *Toxicological Sciences*, 56, 357–364.

101. Rushing, S. R. and Denison, M. S. (2002). The silencing mediator of retinoic acid and thyroid hormone receptors can interact with the aryl hydrocarbon (Ah) receptor but fails to repress Ah receptor-dependent gene expression. *Archives of Biochemistry and Biophysics*, 403, 189–201.

102. Widerak, M., Ghoneim, C., Dumontier, M. F., Quesne, M., Corvol M. T., and Savouret, J. F. (2006). The aryl hydrocarbon receptor activates the retinoic acid receptor alpha through SMRT antagonism. *Biochimie*, 88, 387–397.

103. Murphy, K. A., Villano, C. M., Dorn, R., and White, L. A. (2004). Interaction between the aryl hydrocarbon receptor and retinoic acid pathways increases matrix metalloproteinase-1 expression in keratinocytes. *Journal of Biological Chemistry*, 279, 25284–25293.

104. Hoegberg, P., Schmidt, C. K., Fletcher, N., Nilsson, C. B., Trossvik, C., Gerlienke Schuur, A., Brouwer, A., Nau, H., Ghyselinck, N. B., Chambon, P., and Håkansson, H. (2005). Retinoid status and responsiveness to 2,3,7,8-tetrachlorodibenzo-*p*-dioxin (TCDD) in mice lacking retinoid binding protein or retinoid receptor forms. *Chemico-Biological Interactions*, 156, 25–39.

105. Yamazaki, H. and Shimada, T. (1999). Effects of arachidonic acid, prostaglandins, retinol, retinoic acid and cholecalciferol on xenobiotic oxidations catalysed by human cytochrome P450 enzymes. *Xenobiotica*, 29, 231–241.

106. Wanner, R., Brommer, S., Czarnetzki, B. M., and Rosenbach T. (1995). The differentiation-related upregulation of aryl hydrocarbon receptor transcript levels is suppressed by retinoic acid. *Biochemical and Biophysical Research Communications*, 209, 706–711.

107. Fallone, F., Villard, P. H., Seree, E., Rimet, O., Nguyen, Q. B., Bourgarel-Rey, V., Fouchier, F., Barra, Y., Durand, A., and Lacarelle, B. (2004). Retinoids repress Ah receptor CYP1A1 induction pathway through the SMRT corepressor. *Biochemical and Biophysical Research Communications*, 322, 551–556.

108. Soprano, D. R., Gambone, C. J., Sheikh, S. N., Gabriel, J. L., Chandraratna, R. A., Soprano, K. J., and Kochhar, D. M. (2001). The synthetic retinoid AGN 193109 but not retinoic acid elevates CYP1A1 levels in mouse embryos and Hepa-1c1c7 cells. *Toxicology and Applied Pharmacology*, 174, 153–159.

109. Lakshman, M. R., Chirtel, S. J., Chambers, L. L., and Coutlakis, P. J. (1989). Effects of 2,3,7,8-tetrachlorodibenzo-*p*-dioxin on lipid synthesis and lipogenic enzymes in the rat. *Journal of Pharmacology and Experimental Therapeutics*, 248, 62–66.

110. Liu, P. C. and Matsumura, F. (1995). Differential effects of 2,3,7,8-tetrachlorodibenzo-*p*-dioxin on the "adipose- type" and "brain-type" glucose transporters in mice. *Molecular Pharmacology*, 47, 65–73.

111. Enan, E., Liu, P. C., and Matsumura, F. (1992). TCDD (2,3,7,8-tetrachlorodibenzo-*p*-dioxin) causes reduction in glucose uptake through glucose transporters on the plasma membrane of the guinea pig adipocyte. *Journal of Environmental Science and Health, Part B*, 27, 495–510.

112. Brewster, D. W. and Matsumura, F. (1984). TCDD (2,3,7,8-tetrachlorodibenzo-*p*-dioxin) reduces lipoprotein lipase activity in the adipose tissue of the guinea pig. *Biochemical and Biophysical Research Communications*, 122, 810–817.

113. Shimba, S., Todoroki, K., Aoyagi, T., and Tezuka, M. (1998). Depletion of arylhydrocarbon receptor during adipose differentiation in 3T3-L1 cells. *Biochemical and Biophysical Research Communications*, 249, 131–137.

114. Phillips, M., Enan, E., Liu, P. C., and Matsumura, F. (1995). Inhibition of 3T3-L1 adipose differentiation by 2,3,7,8-tetrachlorodibenzo-*p*-dioxin. *Journal of Cell Science*, 108(Pt 1), 395–402.

115. Brodie, A. E., Manning, V. A., and Hu, C. Y. (1996). Inhibitors of preadipocyte differentiation induce COUP-TF binding to a PPAR/RXR binding sequence. *Biochemical and Biophysical Research Communications*, 228, 655–661.

116. Alexander, D. L., Ganem, L. G., Fernandez-Salguero, P., Gonzalez, F., and Jefcoate, C. R. (1998). Aryl-hydrocarbon receptor is an inhibitory regulator of lipid synthesis and of commitment to adipogenesis. *Journal of Cell Science*, 111 (Pt 22), 3311–3322.

117. Michalik, L., Auwerx, J., Berger, J. P., Chatterjee, V. K., Glass, C. K., Gonzalez, F. J., Grimaldi, P. A., Kadowaki, T., Lazar, M. A., O'Rahilly, S., Palmer, C. N., Plutzky, J., Reddy, J. K., Spiegelman, B. M., Staels, B., and Wahli, W. (2006). International Union of Pharmacology. LXI. Peroxisome proliferator-activated receptors. *Pharmacological Reviews*, 58, 726–741.

118. Abbott, B. D. (2009). Review of the expression of peroxisome proliferator-activated receptors alpha (PPAR alpha), beta (PPAR beta), and gamma (PPAR gamma) in rodent and human development. *Reproductive Toxicology*, 27, 246–257.

119. Gottlicher, M., Widmark, E., Li, Q., and Gustafsson, J. Å. (1992). Fatty acids activate a chimera of the clofibric acid-activated receptor and the glucocorticoid receptor. *Proceedings of the National Academy of Sciences of the United States of America*, 89, 4653–4657.

120. Remillard, R. B. and Bunce, N. J. (2002). Linking dioxins to diabetes: epidemiology and biologic plausibility. *Environmental Health Perspectives*, 110, 853–858.

121. Fallone, F., Villard, P. H., Decome, L., Seree, E., Meo, M., Chacon, C., Durand, A., Barra, Y., and Lacarelle, B. (2005). PPARα activation potentiates AhR-induced CYP1A1 expression. *Toxicology*, 216, 122–128.

122. Villard, P. H., Caverni, S., Baanannou, A., Khalil, A., Martin, P. G., Penel, C., Pineau, T., Seree, E., and Barra, Y. (2007). PPARα transcriptionally induces AhR expression in Caco-2, but represses AhR pro-inflammatory effects. *Biochemical and Biophysical Research Communications*, 364, 896–901.

123. Vogel, C. F. and Matsumura, F. (2003). Interaction of 2,3,7,8-tetrachlorodibenzo-*p*-dioxin (TCDD) with induced adipocyte differentiation in mouse embryonic fibroblasts (MEFs)

involves tyrosine kinase c-Src. *Biochemical Pharmacology*, 66, 1231–1244.
124. Chen, C. L., Brodie, A. E., and Hu, C. Y. (1997). CCAAT/enhancer-binding protein beta is not affected by tetrachlorodibenzo-*p*-dioxin (TCDD) inhibition of 3T3-L1 preadipocyte differentiation. *Obesity Research*, 5, 146–152.
125. Hanlon, P. R., Ganem, L. G., Cho, Y. C., Yamamoto, M., and Jefcoate, C. R. (2003). AhR- and ERK-dependent pathways function synergistically to mediate 2,3,7,8-tetrachlorodibenzo-*p*-dioxin suppression of peroxisome proliferator-activated receptor-γ1 expression and subsequent adipocyte differentiation. *Toxicology and Applied Pharmacology*, 189, 11–27.
126. Shimada, T., Hiramatsu, N., Hayakawa, K., Takahashi, S., Kasai, A., Tagawa, Y., Mukai, M., Yao, J., Fujii-Kuriyama, Y., and Kitamura, M. (2009). Dual suppression of adipogenesis by cigarette smoke through activation of the aryl hydrocarbon receptor and induction of endoplasmic reticulum stress. *American Journal of Physiology: Endocrinology and Metabolism*, 296, E721–E730.
127. Shimba, S., Wada, T., and Tezuka, M. (2001). Arylhydrocarbon receptor (AhR) is involved in negative regulation of adipose differentiation in 3T3-L1 cells: AhR inhibits adipose differentiation independently of dioxin. *Journal of Cell Science*, 114(Pt 15), 2809–2817.
128. Hanlon, P. R., Cimafranca, M. A., Liu, X., Cho, Y. C., and Jefcoate, C. R. (2005). Microarray analysis of early adipogenesis in C3H10T1/2 cells: cooperative inhibitory effects of growth factors and 2,3,7,8-tetrachlorodibenzo-*p*-dioxin. *Toxicology and Applied Pharmacology*, 207, 39–58.
129. Liu, P. C. and Matsumura, F. (2006). TCDD suppresses insulin-responsive glucose transporter (GLUT-4) gene expression through C/EBP nuclear transcription factors in 3T3-L1 adipocytes. *Journal of Biochemical and Molecular Toxicology*, 20, 79–87.
130. Nishiumi, S., Yoshida, M., Azuma, T., Yoshida, K., and Ashida, H. (2010). 2,3,7,8-tetrachlorodibenzo-*p*-dioxin impairs an insulin signaling pathway through the induction of tumor necrosis factor-alpha in adipocytes. *Toxicological Sciences* 115, 482–491.
131. Hoelper, P., Faust, D., Oesch, F., and Dietrich, C. (2005). Evaluation of the role of c-Src and ERK in TCDD-dependent release from contact-inhibition in WB-F344 cells. *Archives of Toxicology*, 79, 201–207.
132. Hankinson, O. (2009). Repression of aryl hydrocarbon receptor transcriptional activity by epidermal growth factor. *Molecular Interventions*, 9, 116–118.
133. Kern, P. A., Fishman, R. B., Song, W., Brown, A. D., and Fonseca, V. (2002). The effect of 2,3,7,8-tetrachlorodibenzo-*p*-dioxin (TCDD) on oxidative enzymes in adipocytes and liver. *Toxicology*, 171, 117–125.
134. Kuil, C. W., Brouwer, A., van der Saag, P. T., and van der Burg, B. (1998). Interference between progesterone and dioxin signal transduction pathways. Different mechanisms are involved in repression by the progesterone receptor A and B isoforms. *Journal of Biological Chemistry*, 273, 8829–8834.
135. Enan, E., Lasley, B., Stewart, D., Overstreet, J., and Vandevoort, C. A. (1996). 2,3,7,8-Tetrachlorodibenzo-*p*-dioxin (TCDD) modulates function of human luteinizing granulosa cells via cAMP signaling and early reduction of glucose transporting activity. *Reproductive Toxicology*, 10, 191–198.
136. Okino, S. T. and Whitlock, Jr. J. P. (2000). The aromatic hydrocarbon receptor, transcription, and endocrine aspects of dioxin action. *Vitamins & Hormones*, 59, 241–264.
137. Timsit, Y. E. and Negishi, M. (2007). CAR and PXR: the xenobiotic-sensing receptors. *Steroids*, 72, 231–246.
138. Patel, R. D., Hollingshead, B. D., Omiecinski, C. J., and Perdew, G. H. (2007). Aryl-hydrocarbon receptor activation regulates constitutive androstane receptor levels in murine and human liver. *Hepatology*, 46, 209–218.
139. Gerbal-Chaloin, S., Pichard-Garcia, L., Fabre, A. Sa-Cunha, J. M., Poellinger, L., Maurel, P., and Daujat-Chavanieu, M. (2006). Role of CYP3A4 in the regulation of the aryl hydrocarbon receptor by omeprazole sulphide. *Cell Signal*, 18, 740–750.
140. Vanacker, J. M., Bonnelye, E., Chopin-Delannoy, S., Delmarre, C., Cavailles, V., and Laudet, V. (1999). Transcriptional activities of the orphan nuclear receptor ERR alpha (estrogen receptor-related receptor-alpha). *Molecular Endocrinology*, 13, 764–773.
141. Klinge, C. M., Kaur, K., and Swanson, H. I. (2000). The aryl hydrocarbon receptor interacts with estrogen receptor alpha and orphan receptors COUP-TFI and ERRα1. *Archives of Biochemistry and Biophysics*, 373, 163–174.
142. Brubaker, K., McMillan, M., Neuman, T., and Nornes, H. O. (1996). All-*trans* retinoic acid affects the expression of orphan receptors COUP-TF I and COUP-TF II in the developing neural tube. *Brain Research. Developmental Brain Research*, 93, 198–202.
143. Safe, S. and McDougal, A. (2002). Mechanism of action and development of selective aryl hydrocarbon receptor modulators for treatment of hormone-dependent cancers (Review). *International Journal of Oncology*, 20, 1123–1128.

10

THE E3 UBIQUITIN LIGASE ACTIVITY OF TRANSCRIPTION FACTOR AHR PERMITS NONGENOMIC REGULATION OF BIOLOGICAL PATHWAYS

Fumiaki Ohtake and Shigeaki Kato

The aryl hydrocarbon receptor (AHR) is a ligand-dependent transcription factor belonging to the basic helix–loop–helix/Per–Arnt–Sim (bHLH/PAS) family of proteins. Accumulating evidence suggests that direct regulation of target genes might not fully explain the wide variety of actions of AHR ligands. Recently, it was shown that AHR acts as a ligand-dependent ubiquitin ligase, regulating protein degradation through the ubiquitin–proteasome system. AHR assembles a CUL4B-based ubiquitin ligase, $CUL4B^{AHR}$, promoting degradation of estrogen (ER) and androgen (AR) receptors and β-catenin. Turnover of AHR itself is also tightly regulated both in the nucleus and in the cytosol. In this chapter, we will review the molecular mechanisms and biological significance of a nongenomic, ubiquitin-related function of AHR as well as regulation of AHR protein turnover. The ligand-dependent properties of AHR ubiquitin ligase are also compared to other signal-dependent ubiquitin ligases.

10.1 INTRODUCTION: MODE OF AHR ACTION BEYOND THAT OF A TRANSCRIPTION FACTOR

Environmental stresses, including UV, oxidative stress, infection, and toxic chemicals, are threats to organisms. Therefore, organisms have evolved mechanisms for sensing and protecting themselves against such stresses. To achieve stress-dependent control of the expression of specific proteins, organisms generally utilize two systems: transcriptional/translational regulation as a genomic pathway and proteolysis regulation as a nongenomic pathway.

Among environmental stresses, dioxin-type environmental contaminants, such as 2,3,7,8-tetrachlorodibenzo-p-dioxin (TCDD), exert a variety of toxic effects [1]. The AHR is a receptor for dioxins in vertebrates. Extensive studies including those using AHR-deficient mice have shown that most of the toxic effects of dioxins are mediated through AHR.

AHR is a ligand-dependent transcription factor belonging to the basic bHLH/PAS family [2–5]. The transcriptional activity of AHR is regulated by direct binding of its ligands [6, 7]. Although the nuclear receptor superfamily of proteins is regulated by its cognate ligands in a similar manner, ligand-dependent regulation of other proteins in the bHLH/PAS family has not yet been reported. Thus, it is an intriguing question whether AHR has endogenous ligand(s) for its biological functions. Ligand binding to the PAS-B region of AHR presumably induces conformational changes and subsequent translocation of the AHR complex to the nucleus. AHR then dimerizes with the AHR nuclear translocator (Arnt) in the nucleus after dissociating from the chaperone complex, recognizes the xenobiotic-responsive element (XRE), and recruits coactivators such as the histone acetyltransferases p300/CBP and SRC-1, chromatin remodeling factor Brg1, and the mediator (DRIP/TRAP) complex on the responsive promoters [2–4]. The AHR/Arnt heterodimer induces the expression of target genes such as CYP1A1 and CYP1A2 [1].

The AH Receptor in Biology and Toxicology, First Edition. Edited by Raimo Pohjanvirta.
© 2012 John Wiley & Sons, Inc. Published 2012 by John Wiley & Sons, Inc.

The responses of AHR's direct target genes do not fully explain AHR's toxicological and physiological effects [8, 9]. Accumulating evidence suggests that the AHR achieves its regulatory functions by modulating the functions of its interacting proteins [5, 10], including estrogen (ERα and ERβ) [11–16] and androgen [15, 16] receptors. The liganded AHR has been shown to promote the ubiquitination and proteasomal degradation of ERs and AR by assembling a ubiquitin ligase complex CUL4B^{AHR} [15, 16]. Thus, some of the actions of dioxins appear to be mediated by a nongenomic, proteolytic pathway through the AHR. Here, we summarize a novel role for the AHR as a component of an E3 ubiquitin ligase complex that mediates accelerated degradation of substrate proteins through the proteasome.

10.2 THE UBIQUITIN–PROTEASOME SYSTEM IN THE REGULATION OF AHR TURNOVER

10.2.1 An Overview of the Ubiquitin–Proteasome System in the Context of Dioxin Stressors

In metazoans, the transcriptional regulatory system and the ubiquitin–proteasome system are the two major target-selective systems that control intracellular protein levels in response to physiological demands (Fig. 10.1a). Although the transcriptional regulatory system is targeted by environmental fat-soluble ligands, the involvement of the ubiquitin–proteasome system in the presence of environmental toxins remains largely unknown. In these systems, target selectivity depends on recognition of specific DNA elements by sequence-specific transcription factors [17–19] and recognition of degradation substrates by E3 ubiquitin ligases [20–23]. The transcription factors and E3 ubiquitin ligases primarily serve as specific adapters to recruit enzymes such as transcriptional coregulators and ubiquitin-conjugating enzymes (E2), respectively, to appropriate targets. The transcriptional coregulator(s) and E2 regulate site-specific posttranslational modifications to promote transcription and protein degradation, respectively (Fig. 10.1b). Considering the functional analogy of E3 ubiquitin ligase and transcription factors, it is possible that E3 ubiquitin ligase also serves as a target of environmental toxins.

The ubiquitin–proteasome system plays a pivotal role in cellular homeostasis [20–23]. Ubiquitin is a 76-amino acid polypeptide that is highly conserved among eukaryotes. Ubiquitin is covalently attached to lysine (Lys) residues of substrate proteins. Ubiquitination of proteins is catalyzed by sequential reactions involving ubiquitin-activating enzyme (E1), ubiquitin-conjugating enzyme (E2), and ubiquitin

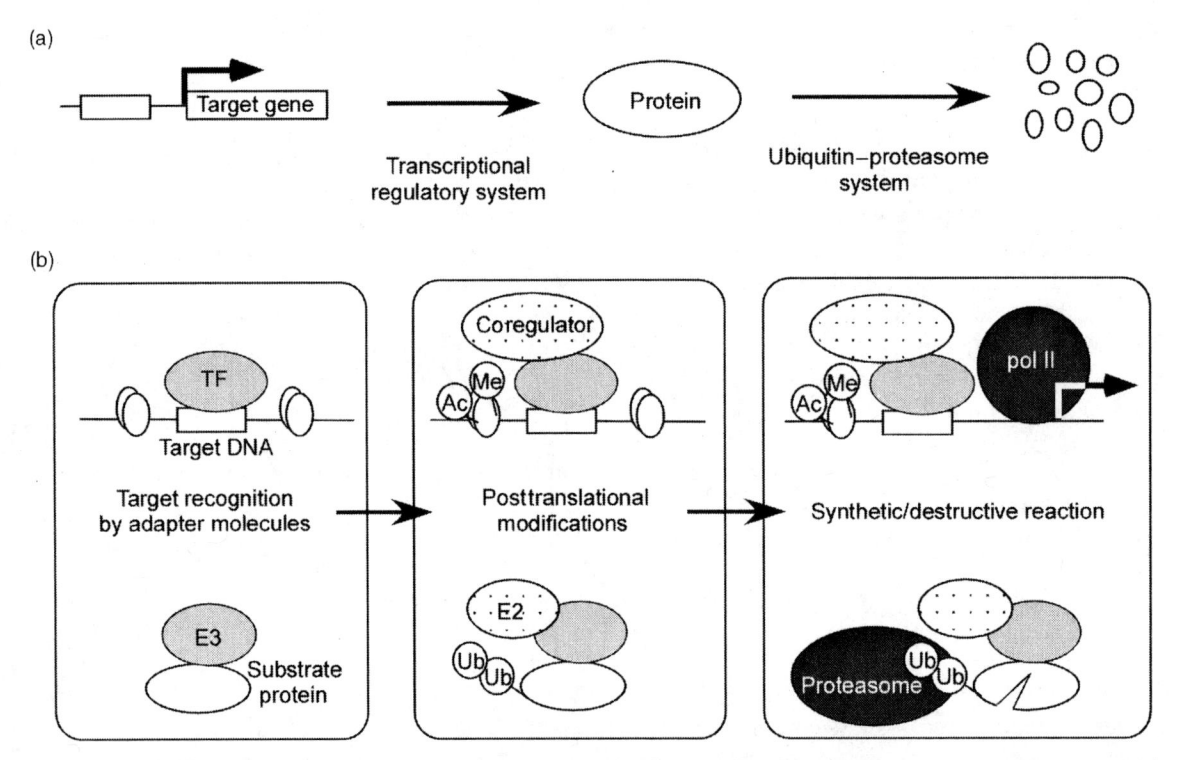

FIGURE 10.1 The ubiquitin–proteasome system. (a) The transcriptional regulatory system and the ubiquitin–proteasome system are two major target-selective systems that control intracellular protein levels. (b) The transcription factors and E3 ubiquitin ligases primarily serve as target-specifying adapters in these systems. Ub, ubiquitin; P, phosphorylated serine/threonine; Ac, acetylated lysine; Me, methylated lysine; pol II, RNA polymerase II.

protein ligase (E3). Ubiquitin is conjugated as one molecule (monoubiquitination) or as a tandem polymer (polyubiquitination). Polyubiquitination can occur at any of the seven lysine residues in the ubiquitin molecule, but Lys48, Lys63, and Lys11 are considered to be the major polyubiquitination sites [24]. The Lys48-linked polyubiquitin chain is then recognized by the 26S proteasome for subsequent proteolysis. On the other hand, the Lys63-linked polyubiquitin chain regulates proteasome-independent pathways such as signal transduction and DNA repair, whereas the Lys11-linked chain reportedly functions as a proteasomal signal [24].

10.2.2 The Diversity of E3 Ubiquitin Ligase

Among E1, E2, and E3 enzymes, the E3 ubiquitin ligases are the most diverse and are critical for the determination of substrate specificity. E3 acts as a bridge between E2 and the substrate, maintaining the appropriate distance. E2 then conjugates ubiquitin to the substrate [20–23]. Of the RING-type E3s, the largest class consists of the cullin–RING ubiquitin ligases (CRLs) [22, 23, 25–27]. CRLs are multi-subunit complexes that typically include a cullin (CUL1, 2, 3, 4A, 4B, or 5) subunit, a RING finger protein Rbx1 or Rbx2, and a substrate recognition subunit (Fig. 10.2a). Cullin serves as a scaffold protein, binding to the adapter protein at its N-terminus while binding to Rbx1 at its C-terminus [23]. Rbx1 binds to E2 enzymes through the RING finger to support efficient conjugation of ubiquitin to the substrates. Neddylation of cullins is critical for their activity. Their diverse substrate recognition subunits enable CRLs to target numerous substrates. The best-characterized CRLs are the Skp1–CUL1–F-box (SCF) complexes [28]. In a SCF complex, the F-box protein functions as a substrate recognition subunit by binding to the bridging factor, Skp1, which is bound to the N-terminal region of CUL1. F-box proteins and other types of substrate recognition subunits serve as specific adapters for target substrates (Fig. 10.2a).

Both the ubiquitination and the proteasomal degradation of proteins are primarily regulated through targeted posttranslational modifications of their substrates (Fig. 10.2b). Many cell cycle regulators are phosphorylated at a specific time, which induces association with their cognate ubiquitin

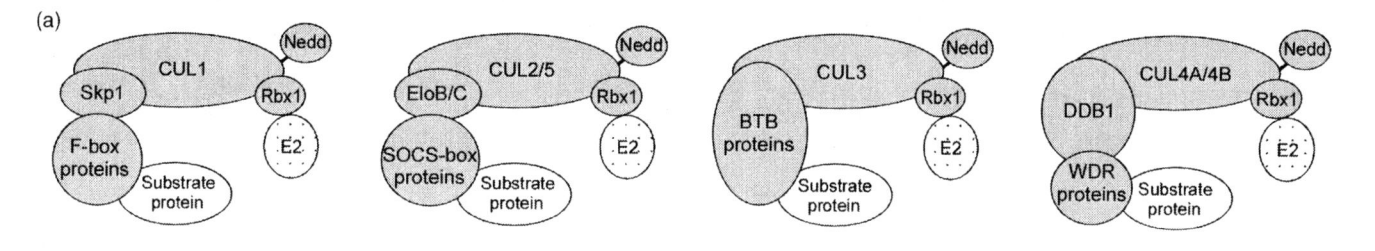

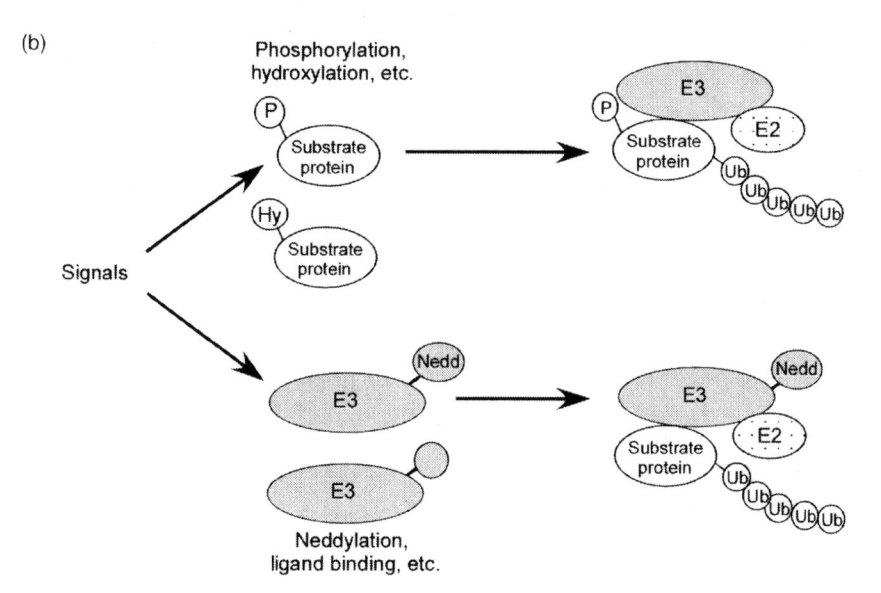

FIGURE 10.2 The cullin–RING ubiquitin ligases. (a) Complex assembly of CRLs through cullins as scaffold proteins. Cullins associate with different adapter components that specifically recognize various substrates. (b) The signal-sensing mechanism in the ubiquitin–proteasome system. In general, it is believed that posttranslational modifications for substrates are trigger of ubiquitination. Recent evidence suggests that ubiquitin ligase can also sense extracellular signals to regulate ubiquitination.

ligases for their proper destruction [21]. In this view, activity of ubiquitin ligase is considered to be constitutive. In the past decade, however, many examples were reported in which ubiquitin ligases directly sensed extracellular signals and thereby regulated protein turnover (Fig. 10.2b). Damaged DNA binding protein 2 (DDB2) and Cockayne syndrome protein A (CSA) both assemble CUL4-based ubiquitin ligase [29]. The cellular location and activities of $CUL4^{DDB2}$ and $CUL4^{CSA}$ are regulated by UV radiation through neddylation and deneddylation of CUL4. This discovery suggests that ubiquitin ligase itself can sense extracellular signaling. Several types of signal-dependent ubiquitin ligases have been identified, such as Keap1 [30] and CRBN [31], as described later.

Note that given its molecular mechanism, any protein binding to the core components of the CRL E3 complex could potentially act in a manner similar to substrate recognition subunits. More interestingly, F-box proteins and other types of substrate recognition subunits are rapidly degraded through an autocatalytic mechanism once they are integrated into the CRL core complexes [25]. In this way, CRLs can efficiently ubiquitinate different substrates by associating with different substrate recognition subunits. This raises the possibility that F-box (and similar) proteins act as substrates or as adapter components, as in the case of DDB2 in the CUL4-based CRL complex [32–37]. As described later, AHR also undergoes ligand-dependent ubiquitination and proteasomal degradation. This property of AHR is very similar to autodegradation of F-box proteins.

10.2.3 Regulation of Proteasomal Turnover of AHR

The level of AHR protein is tightly regulated by the ubiquitin–proteasome system in response to local processes, such as ligand binding, phosphorylation, and localization [38]. Unliganded AHR associates with p23/XAP2/Hsp90 and is thereby protected from ubiquitination and degradation (Fig. 10.3). The C-terminus of Hsp70-interacting protein (CHIP), a quality-controlling ubiquitin ligase that associates with the chaperone complex, reportedly promotes degradation of AHR [39]. Since both CHIP and unliganded AHR are mainly located in the cytosol, the degradation of AHR through CHIP is likely to occur in the cytosol [39] (Fig. 10.3).

Upon ligand binding, AHR dissociates from the p23/XAP2/Hsp90 complex and translocates into the nucleus [8]. In the nucleus, AHR interacts with CUL4B and the other components of the CUL4B ubiquitin ligase, including DDB1, TBL3, and Rbx1 [15]. This CUL4B complex promotes degradation of AHR. As discussed later, the CUL4B-mediated degradation of AHR involves an autocatalytic mechanism and the assembly of the $CUL4B^{AHR}$ complex. Immunostaining revealed that CUL4B is primarily located in

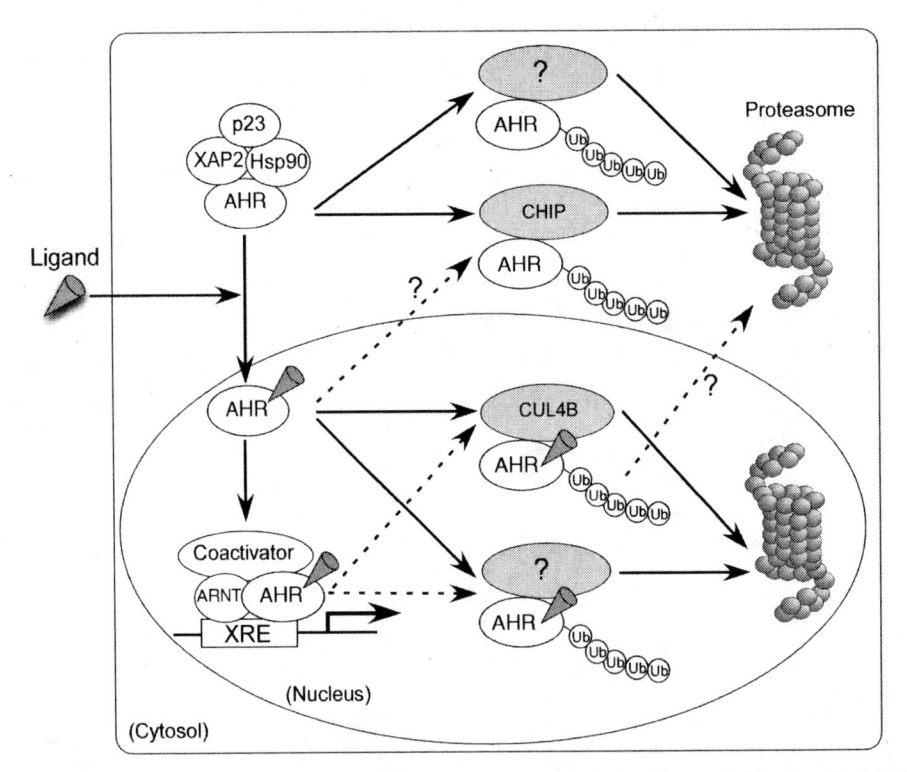

FIGURE 10.3 Proteasomal degradation of AHR in the nucleus and cytosol. The level of AHR protein is tightly regulated by the ubiquitin–proteasome system in response to local processes, such as ligand binding, phosphorylation, and localization. AHR ubiquitination occurs both in the nucleus and in the cytosol.

the nucleus [15]. Experiments using siRNAs revealed that a portion of the ligand-dependent ubiquitination and degradation of AHR in the nucleus is mediated through CUL4B. However, since degradation of AHR was not fully canceled by siCUL4B [15], there appear to be other (unknown) ubiquitin ligases for nuclear AHR degradation (Fig. 10.3). One possibility is that CHIP is also involved in nuclear AHR degradation. The location of AHR degradation is also unclear. Given that proteasome is located both in the nucleus and in the cytosol, it is possible that liganded AHR is degraded by the proteasome in the nucleus. Supporting this idea, we have identified proteasome as a nuclear AHR-associated complex [15]. On the other hand, it was reported that leptomycin B (LMB), an inhibitor of nuclear export, canceled ligand-dependent degradation of AHR in HepG2 cells [40].

Ubiquitination and degradation of AHR is in general accelerated by agonists, but inhibited by antagonists [41]. AHR agonists, such as TCDD and 3-methylcholanthrene (3MC), induce prompt degradation of AHR, which is inhibited by proteasomal inhibitors. Reciprocally, AHR antagonists such as α-naphthoflavone (αNF) repress the degradation. Thus, it appears that degradation of AHR is coupled with transcriptional activity. Studies using AHR derivatives expressed in cell lines reveal that the C-terminal transactivation domain and N-terminal DNA binding domain were required for efficient turnover of AHR [41].

Such transcription-coupled ubiquitination and degradation are commonly observed with several classes of transcription factors, such as ER, RAR, and PPAR. Ubiquitination and degradation of ERα are also enhanced by agonists but repressed by antagonists [42]. The biological significance of activity-coupled degradation is not fully understood, although several hypotheses have been proposed. One is that receptor degradation shortens its prolonged activity. Prompt degradation of ERβ upon ligand binding might be required for transcriptional termination after ligand depletion [43]. Another view is that cyclic recruitment and clearance of receptor by degradation are important for proper transcriptional regulation. Studies using chromatin immunoprecipitation (ChIP) analyses revealed cyclical promoter recruitment and dissociation of ERα and proteasomal subunits after ligand stimulation [44, 45].

10.3 AHR IS A LIGAND-DEPENDENT E3 UBIQUITIN LIGASE

10.3.1 Molecular Mechanisms of Crosstalk Between AHR and Estrogen or Androgen Receptor

The ubiquitin ligase activity of AHR was found in the course of studying signal crosstalk between AHR and sex hormone receptors. Dioxins exert both estrogen- and androgen-related effects [1, 10, 46–53] (Fig. 10.4). Dioxins have well-described antiestrogenic effects, such as the inhibition of estrogen-induced uterine enlargement, MCF-7 cell growth, and target gene induction [47, 49]. However, there is also evidence to the contrary. Dioxins have estrogenic effects, including the stimulation of uterine enlargement [50], induction of estrogen-responsive genes such as *VEGF*, *c-fos*, and *TERT*, and a pattern similar to estrogen in transcriptional regulation [46]. In addition, AHR-deficient mice exhibit impaired ovarian follicle maturation [54]. Using AHR-deficient cells, the importance of AHR in the proliferation of mammary cells has been confirmed [55]. These findings suggest that AHR, activated by its ligand, might modulate the estrogen signaling pathway. Similarly, dioxins exert both androgenic and antiandrogenic effects on prostate development in an age-specific manner [48]. As is true for other crosstalk pathways [19], the AHR appears to modulate estrogen/androgen signaling both positively and negatively depending on the cellular context.

The molecular mechanisms by which AHR modulates ERα have been extensively studied, and both direct and indirect regulatory mechanisms have been proposed. First, TCDD/AHR increases or decreases estrogen levels through an indirect mechanism [10, 56]. TCDD promotes the clearance of estrogen, thereby repressing ER transcriptional activity [56]. AHR-deficient mice have decreased estrogen production due to impaired induction of aromatase (CYP19) gene expression [54]. Another indirect mechanism involves AHR's and ER's competitive DNA binding on responsive promoters [10]. AHR and ER, each bound to its own target promoter, recruit transcriptional coregulators in a competitive manner. This mechanism might be limited to specific genes and conditions since not all the estrogen-responsive promoters contain XRE.

10.3.2 Direct Association of AHR with the Estrogen or the Androgen Receptor

In addition to the indirect mechanisms described above, direct association of AHR with ERs has been independently reported. Ligand-activated AHR/Arnt associates with ERα and ERβ through the N-terminal A/B region within ERs [11–15] (Fig. 10.4). Through this association, liganded AHR potentiates the transactivation function of 17β-estradiol (E_2)-unbound ERα, while it represses E_2-bound ERα-mediated transcription upon the estrogen-responsive element (ERE) [11]. The interaction of AHR/ER is induced by different AHR ligands, such as TCDD, 3MC, and β-naphthoflavone ((βNF). The activation of AHR is thought to be sufficient for interaction with ERα, as a constitutively active form of AHR [6] modulates ERα function in the absence of AHR ligand [16]. These results suggest that the crosstalk of AHR with ER is initiated primarily through stimulation of AHR. Supporting this, ERα is predominantly

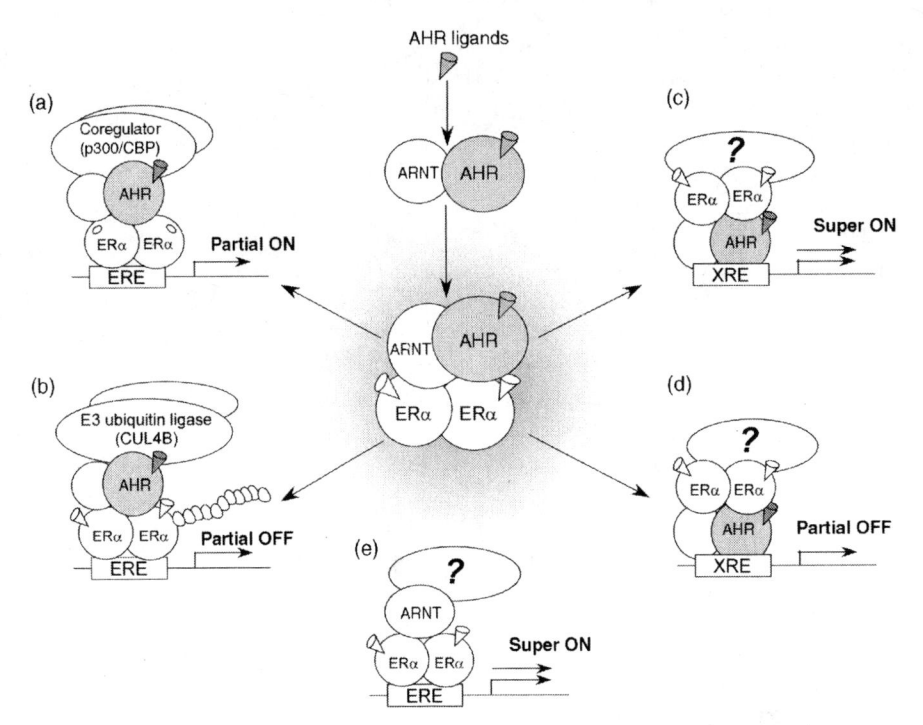

FIGURE 10.4 Crosstalk of AHR with ERα through direct association. Ligand-bound AHR directly associates with the estrogen (ERα, ERβ)) or the androgen receptor in the nucleus. This association leads to different types of crosstalks between AHR and ERs/AR, such as recruitment of transcriptional coactivators and formation of an E3 ubiquitin ligase complex, as illustrated. Crosstalks on both AHR and ER target genes have been reported.

located in the nucleus, whereas AHR translocates to the nucleus upon ligand stimulation. The association of AHR/ERα has been shown by *in vitro* [57] and *in vivo* techniques and by biochemical methods [15]. Moreover, AHR/ERα crosstalk in the transcriptional regulation of ERα-responsive genes is abolished in AHR-deficient mice [4, 54], confirming the specificity of the molecular pathway *in vivo* [11]. Reciprocally, E_2-bound ERα associates with XRE-bound AHR to potentiate [12] or repress [13] AHR-mediated transcription. Considered together, the AHR/ERα complex might bind to XRE or ERE through the attachment functions of AHR or ERα, respectively. Alternatively, different complex subtypes that contain AHR/ERα might control promoter selectivity (Fig. 10.4). Reflecting this functional crosstalk, Arnt also acts as a coregulator for both ERα and ERβ [58]. A genome-wide microarray analysis revealed that AHR and ER signals converge in target gene expression [46]. In addition, recent unbiased proteomic analysis also identified functional interaction between AHR and ER [59].

The proposed mechanism of AHR/ER association appears to be a reasonable explanation for dioxin/estrogen crosstalk. First, this mechanism explains the functional AHR/ER crosstalk irrespective of differences in target gene promoters. Second, ligand-dependent AHR/ER association could result in a rapid cellular response to dioxins in terms of ER activity. The responses of ER transcriptional activity to AHR ligands are observed within a few hours in cultured cells as well as in mice, which supports the existence of direct crosstalk mechanisms. Third, variations in AHR/ER-containing coregulator complexes might result in biphasic responses of AHR/ER crosstalk. Complexes containing different classes of transcription factors can recruit coregulator complexes distinct from their cognate associating complexes [19]. Consequently, it is possible that the AHR/ER complex, acting as a functional unit, might recruit different types of complexes depending on the cellular context. A current area of interest is the identification of the molecular determinants by which the activity of the AHR/ER complex is controlled. Recently, it was shown that constitutively active AHR also regulates the transcriptional activity of ER and AR, even in the absence of any exogenous ligand [16, 60]. This suggests that crosstalk with ER/AR is an intrinsic function of AHR and is not attributable to ligand-specific properties. Note that a naturally occurring AHR ligand, indole-3-carbinol, also induces AHR–ER crosstalk. Taken together, molecular dissection of signal crosstalk between AHR and other classes of transcription factors appears to be the key to identifying novel modes of AHR function beyond direct induction of target genes.

10.3.3 Ubiquitin Ligase Activity of AHR

As described previously, AHR regulates functions of ER/AR through direct association. However, the molecular mechanism for AHR-mediated repression of ER/AR remains to be addressed. In the crosstalk pathway, AHR has well-described effects on the transcriptional regulatory system. TCDD also decreases the uterine ERα protein level in the rat [61], suggesting that AHR might also be involved in the control of protein stability. Somewhat unexpectedly, our own ChIP analysis showed that ligand-bound AHR does not block coactivator recruitment of liganded ERα. In addition, repression of ERα transcriptional activity by AHR is not observed when ERα is overexpressed in transient reporter assays. These observations imply that ligand-activated AHR has an additional molecular role beyond transcriptional regulation, at least in the modulation of sex hormone signaling.

Exploring the functions of AHR in sex hormone signaling, we found that upon activation of AHR (by binding of AHR ligands such as 3MC and βNF), as well as by expression of constitutively active AHR, endogenous protein levels of ERα, ERβ, and AR were drastically decreased without alterations in mRNA levels [16]. Since AHR and ERα proteins that are bound to their cognate ligands are ubiquitinated for proteasome-mediated degradation [41, 62–66], we tested whether the functional modulation of ERs and AR by activated AHR was related to this degradation system. 3MC-enhanced degradation of sex steroid receptors is attenuated in the presence of the proteasome inhibitor MG132, and 3MC-enhanced polyubiquitination of ERα is consistently observed irrespective of E2 binding. MG132 treatment abrogates the transcriptional modulation of liganded sex steroid receptor function by activated AHR. This indicates that the ubiquitin–proteasome system mediates the repressive AHR–ER crosstalk pathway (Fig. 10.5).

The following experiments provided evidence that AHR acts as an E3 ubiquitin ligase component. First, FLAG–AHR immunoprecipitated complexes exert a self-ubiquitination activity in an E1/E2 enzyme-dependent manner *in vitro* [16]. Second, 3MC-dependent recognition of ER and AR by AHR [11] appears to induce ubiquitination of ER/AR. Third, degradation of AHR itself is accelerated upon activation of degradation of sex steroid receptors, which is a typical sign of self-ubiquitination of the E3 component [25]. Taken together, these properties of AHR resemble those of classical adapter components of the E3 ubiquitin ligase complex, such as F-box proteins in the SCF complex [21, 25], DDB2/CSA in the CUL4A complex [32–36], and VHL in the CUL2

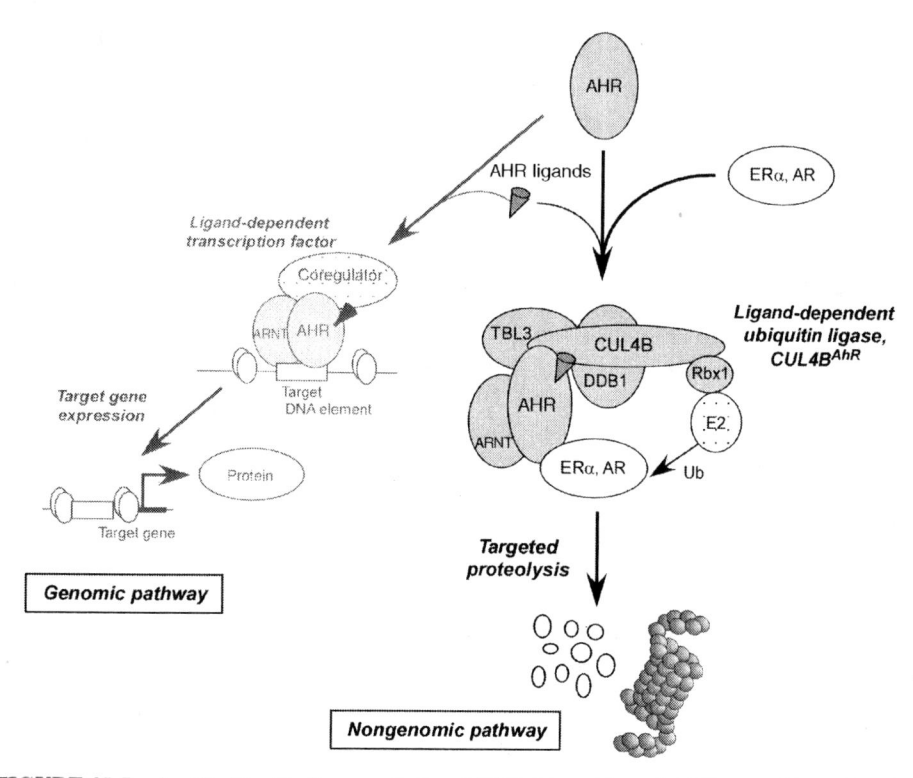

FIGURE 10.5 An E3 ubiquitin ligase activity of AHR. Ligand-bound AHR assembles a CUL4B-based atypical E3 ubiquitin ligase complex, CUL4B^{AHR}, to mediate a nongenomic signaling pathway of fat-soluble ligands. AHR serves as a ligand-dependent ubiquitin ligase as well as a transcription factor.

complex [67]. Actually, AHR possesses biochemical properties similar to those of Skp2, the first identified F-box protein [28]. Therefore, we reasoned that activated AHR might serve as an E3 ubiquitin ligase component.

10.3.4 AHR Assembles a CUL4B-Based Ubiquitin Ligase

To prove this hypothesis, an AHR-associating ubiquitin ligase complex was biochemically purified [68] from HeLa cells. This complex includes cullin 4B (CUL4B) [21, 69], damaged DNA binding protein 1 (DDB1) [29, 70], and Rbx1 [21], together with subunits of the 19S regulatory particle (19S RP) of the 26S proteasome as well as Arnt and transducin beta-like 3 (TBL3) (Fig. 10.5). The core complex appears to constitute a CRL-type E3 ligase and therefore is referred to as CUL4B^{AHR}. Although the typical CUL4B-type CRL complex contains substrate recognition components having a WDXR/DWD motif [32–36], no such component has been identified in this complex. Recent reports also show that other CUL4-based components, such as cereblon (CRBN), do not have the WDXR/DWD motif [31]. AHR directly interacts with the N-terminal region of CL4B in GST pull-down assays. Together with the direct interaction of AHR with ER, it appears that AHR might act as a substrate recognition component in the CUL4B^{AHR} complex. Using an *in vitro* reconstituted ubiquitination assay, the E3 ubiquitin ligase activity of CUL4B^{AHR} for ERα is dependent only on 3MC and not on E$_2$. This suggests that CUL4B^{AHR} has the unique property of being able to respond to ligand signals by complex assembly and ubiquitin ligase activity. The importance of the CUL4B^{AHR} components for the promotion of ERα ubiquitination and degradation has been demonstrated in knockdown experiments. Degradation of ERα or AR in the uterus and prostate was induced by treatment with AHR ligands. Such degradation of ERα or AR is not seen in AHR-deficient mice [4, 54]. These data confirm that the AHR has E3 ubiquitin ligase activity *in vivo*. The antiestrogenic effects of AHR ligands on estrogen-dependent uterine cell proliferation [11] appear to be mediated by the E3 ubiquitin ligase activity of AHR, since knockdown of CUL4B abolishes the antiproliferative effect of AHR ligand in primary uterine cells (Fig. 10.5).

Taken together, these results indicate that liganded AHR assembles a CUL4B-based ubiquitin ligase complex to target ER and AR for proteasomal degradation and repress their activity. Thus, it appears that AHR acts both as a transcription factor and as a ubiquitin ligase to mediate different signaling pathways. In addition to typical AHR ligands (TCDD, βNF, and 3MC), constitutively active mutations of AHR also induce ubiquitin ligase activity of CUL4B^{AHR} [16]. A naturally occurring AHR ligand, indole-3-carbinol, was recently shown to activate AHR-mediated ubiquitination and degradation of ER [71]. Thus, the ubiquitin ligase activity of AHR might mediate a part of the nongenomic function of AHR.

10.3.5 β-Catenin as a Substrate for E3 Ubiquitin Ligase Activity of AHR

The finding that AHR assembles a ubiquitin ligase complex suggests that the E3 ubiquitin ligase activity of AHR and the transcriptional activity of AHR are responsible for a distinct set of biological events induced by AHR ligands. As substrate-specific adapters of ubiquitin ligase complexes are capable of recognizing a number of proteins, identification of other CUL4B^{AHR} substrate proteins might reveal new molecular links between AHR-mediated signaling and other signaling pathways and cellular events.

Given that AHR-interacting proteins might also serve as degradation substrates for AHR, it is of interest that AHR interacts with various transcription factors [5], such as Rb/E2F1 [72], SF1/Ad4BP [54], and NF-κB [73], to modulate their functions. AHR has recently been shown to regulate the differentiation of T$_h$17 and T$_{reg}$ cells [74–76]. This might be mediated by functional interaction with STAT1 [76]. AHR also modulates the function of transcription factors [77] such as GR and RAR [78, 79]. Considering the evolutionary conservation of AHR, it is likely that the intrinsic function of AHR is to mediate the signal transduction of endogenous ligands in crosstalk pathways.

As one such substrate for AHR E3 ligase, β-catenin was recently found to be ubiquitinated by activated AHR [80] (Fig. 10.5). β-catenin is a transcription factor downstream from the Wnt signaling pathway. β-catenin has many profound functions in various tissues, including proliferation of colon/intestine cells. β-catenin is generally degraded through the APC/axin system in the absence of Wnt signaling. Thus, mutations in the *APC* gene cause colon tumors in humans and mice. It was found that activated AHR causes proteasomal degradation of β-catenin in colon tumor cell lines. Ligand-activated AHR associates with β-catenin and promotes its ubiquitination in a CUL4B-dependent manner. AHR represses the transcriptional activity of β-catenin in a CUL4B-dependent, but APC-independent, manner [80]. Reflecting these observations, AHR-deficient mice frequently develop colon tumors with abnormal accumulation of β-catenin protein. Administration of AHR ligands efficiently suppressed colon cancer in *APC*$^{Min/+}$ mice, an established mouse model of familial adenomatous polyposis. These findings suggest that β-catenin is another degradation substrate for CUL4B^{AHR} ubiquitin ligase (Fig. 10.6). Interestingly, ER, AR, and β-catenin promote cellular proliferation in their target tissues. Therefore, one biological role of the ubiquitin ligase function of AHR appears to be antiproliferative regulation through degradation of those transcription factors. This raises the possibility that ubiquitin ligase function-selective AHR ligands might be useful in cancer therapy.

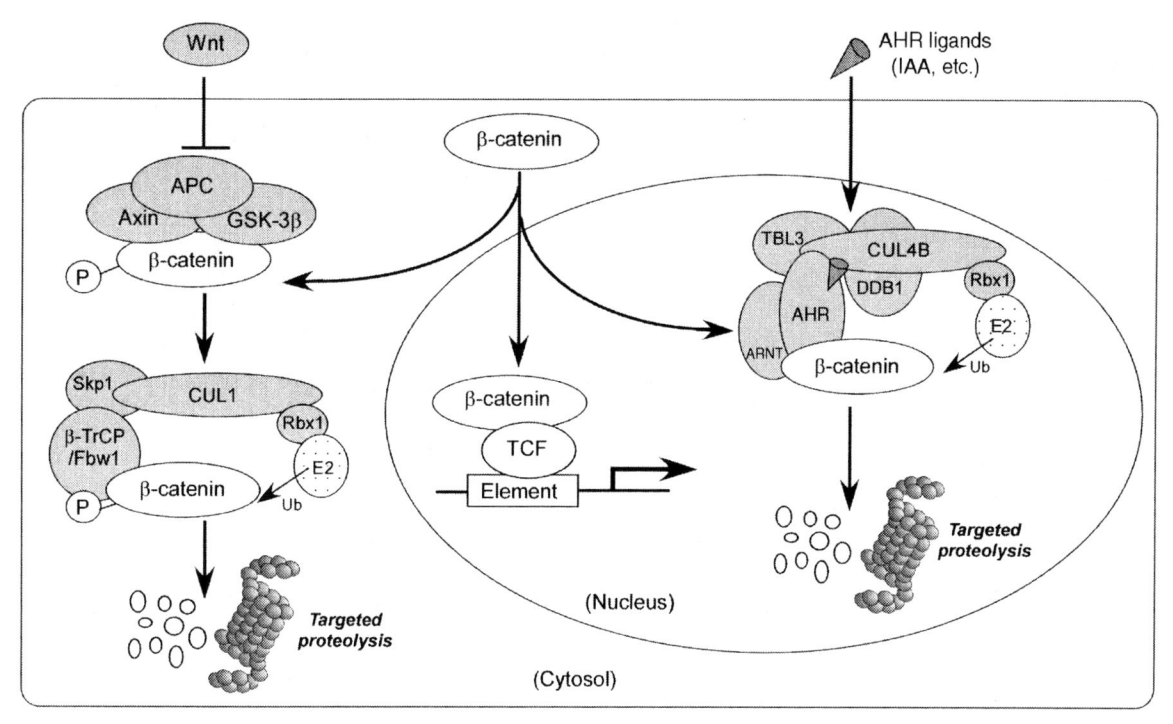

FIGURE 10.6 β-Catenin is a degradation substrate for AHR. Activated AHR causes proteasomal degradation of β-catenin in colon tumor cell lines. Ligand-activated AHR associates with β-catenin and promotes its ubiquitination in a CUL4B-dependent manner.

10.4 AHR AS A SIGNAL-SENSING UBIQUITIN LIGASE COMPLEX

10.4.1 A Nongenomic, Proteolytic Pathway for AHR Signaling

Although it is well established that AHR is a key factor in mediating the adverse effects of dioxin-type compounds [2–4], the underlying mechanisms for this remain elusive. The putative functions of the previously identified target genes for AHR appear unlikely to fully explain the diverse range of biological actions of AHR ligands [5]. The discovery of CUL4B^{AHR} suggests that the adverse effects of AHR ligands on sex hormone signaling are, at least in part, attributable to the enhanced degradation of sex steroid receptors through the E3 ubiquitin ligase activity of AHR [15, 16] (Figs 10.5 and 10.7). Target selectivity of the transcriptional regulatory system and the ubiquitin–proteasome system depends on specificity conferred by sequence-specific transcription factors and E3 ubiquitin ligases (Fig. 10.1b). To date, however, no single factor has been shown to function as a specificity factor in both the target selection systems. Therefore, AHR is the first sequence-specific transcription factor identified that acts as an E3 ubiquitin ligase that also targets substrates for accelerated protein degradation. It is possible that other transcription factors, such as nuclear receptors, also function as E3 ubiquitin ligase components in some cellular contexts. Fat-soluble ligands for nuclear receptors are reported to have "nongenomic" actions independent of transcriptional regulation. Considered together, ubiquitin ligase-based signaling mechanisms might be involved in nongenomic actions of various fat-soluble ligands.

10.4.2 Ubiquitin Ligase-Based Sensing of Environmental Stresses

What is the biological significance of the ubiquitin ligase activity of AHR? It is plausible that activation of atypical E3 complexes might provide a mechanism by which physiological sensors of environmental stresses could respond to those stresses (Fig. 10.7). Supporting this, Hsp70 acts as an atypical substrate-specific adapter within the CHIP E3 complex in response to heat shock stress [81]. Hsp70 interacts with misfolded proteins and promotes their degradation. It later undergoes autocatalytic degradation through CHIP [81]. In response to DNA damage, an atypical E3 complex alters the stability of TIP60, which in turn regulates ataxia telangiectasia mutated (ATM) activation in DNA repair [82]. Activating transcription factor-2 (ATF2) promotes the degradation of TIP60 by assembling a CUL3-based complex under nonstressed conditions. ATF2 dissociates from TIP60 in response to ionizing radiation (IR), resulting in enhanced TIP60 stability and activity [82]. Functional regulation of E3 components is also seen with the CUL3-based component

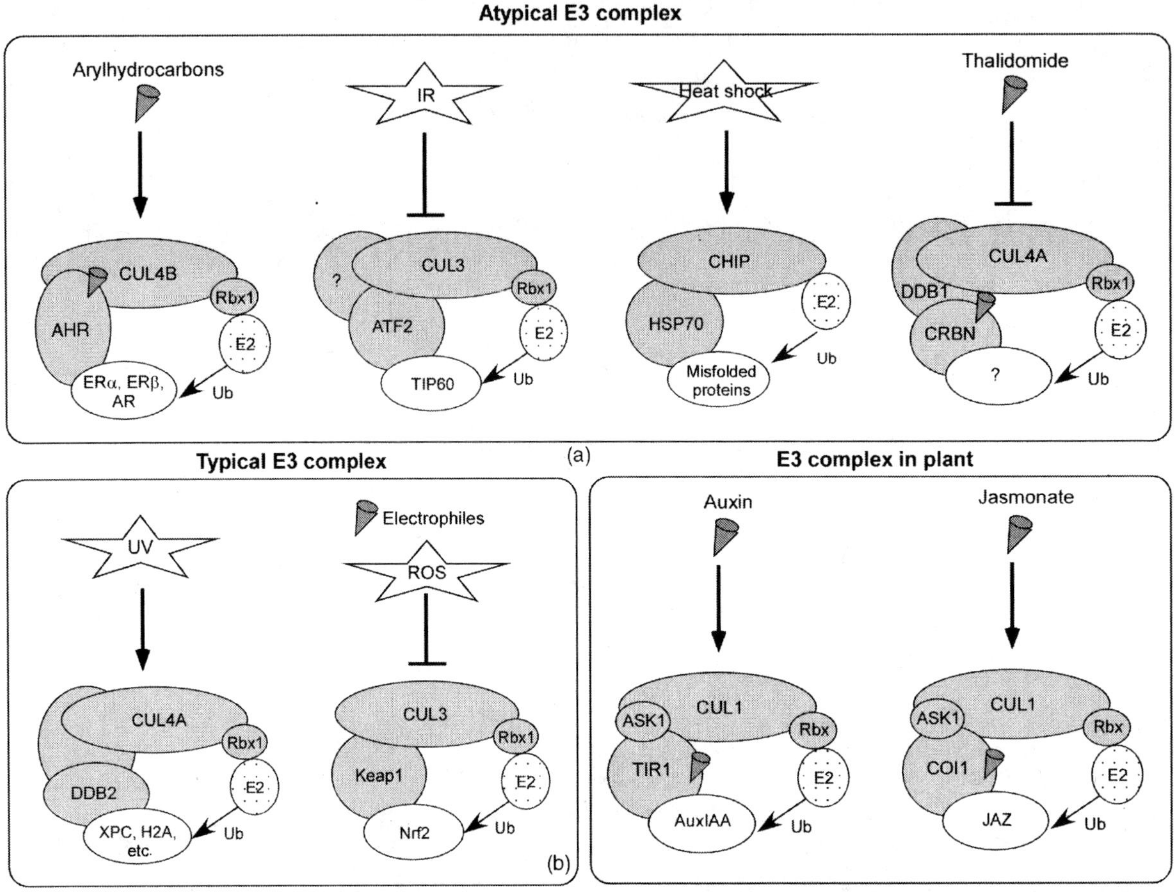

FIGURE 10.7 Atypical E3 complexes as sensors for environmental stresses. Several examples of E3 ubiquitin ligase-based sensing of environmental stresses are illustrated. (a) Signal-responsive factors serve as atypical components of E3 complexes. (b) Canonical E3 components with conserved signature motifs act as signal-responsive factors. E3s composed of plant hormone receptors are also shown.

Keap1 in the oxidative stress response [83] and CUL4A-based components DDB2 and CSA in the DNA damage response [29]. Interestingly, several viruses utilize such CRL-type ubiquitin ligases for infection. Human immunodeficiency virus-1 (HIV-1)-derived Vif protein acts as a substrate-specific component of CUL5 ligase to promote degradation of APOBEC3G protein [84]. Considered together, E3 components that respond to environmental stress might be more diverse than initially believed. It is possible that CUL4B^{AHR} might crosstalk with these stress-responsive E3 ligases to modulate their functions. As WDXR/DWD motif-containing components, including DDB2 and CSA, also bind to CUL4B [29], it is possible that AHR might associate or interfere with these CRL subunits.

More recently, it was reported that thalidomide, a teratogenic chemical, acts by binding to and inhibiting CRBN, a substrate-specific component of CUL4A-based ubiquitin ligase [31] (Fig. 10.7). By purification of thalidomide-associated proteins, CRBN was identified together with DDB1, a component of the CUL4A/CUL4B complex. Thalidomide was shown to inhibit E3 activity of the CUL4A^{CRBN} complex *in vitro* and *in vivo*. Moreover, the teratogenic activity of thalidomide in zebrafish was abolished by overexpression of a CRBN mutant that did not bind thalidomide. Thus, it is noteworthy that both dioxins and thalidomide, two prototypical toxins that originate from human industrial activity, exert their effects in part through CUL4-based ubiquitin ligase.

Finally, the species differences in the ligand-dependent ubiquitin system should be described. Whether endogenous small molecules directly regulate the ubiquitin ligase in animals has not been addressed, since endogenous ligands for AHR, Keap1, and CRBN have not been established. In plants, on the other hand, several hormones such as auxin and jasmonate are ligands for ubiquitin ligase. Auxin or indole-3-acetic acid (IAA), a growth hormone for plants, binds to TIR1, a substrate specificity component of CUL1 ubiquitin ligase [85, 86] (Fig. 10.7). Structural studies show that auxin binding promotes association of TIR1 with ubiquitinated substrates [87]. It is interesting that IAA is also produced in the intestines of mammals in the presence of microbes and can act as an agonist for AHR to induce ubiquitin ligase activity [80] (Fig. 10.6). Thus, IAA appears to activate different ubiquitin ligases in animals and plants. Another endogenous hormone for ubiquitin ligase in plants is jasmonate, a kind of immune system hormone produced upon injury. Jasmonate binds to and activates ubiquitin ligase COI1, which is also a substrate-specific component of CUL1 ubiquitin ligase [88, 89] (Fig. 10.7). Thus, ubiquitin ligase-based sensing of environmental signals utilizes diverse mechanisms among different species.

10.5 CONCLUSION

Cellular responses to dioxins cannot be fully explained by the transcriptional activity of the AHR on direct target genes. Here, we describe recent findings that AHR acts as a novel ligand-dependent E3 ubiquitin ligase that regulates target-specific protein destruction. This activity constitutes a nongenomic pathway for AHR ligands. The ubiquitin ligase activity of AHR and the transcriptional activity of AHR appear to be responsible for a distinct set of biological events induced by AHR ligands. In this regard, whether yet to be identified endogenous AHR ligands activate the ubiquitin ligase activity of AHR is an interesting area to be explored. Characterization of this new molecular aspect of AHR function, especially identification of the degradation substrates for AHR, might lead to a greater understanding of the diverse biological actions induced by endogenous and exogenous AHR ligands. Dissecting the molecular mechanism of ubiquitin ligase activity of AHR in comparison to other CRLs might shed light on the ubiquitin ligase-based stress-sensing mechanism.

ACKNOWLEDGMENTS

This work was supported in part by priority areas from the Ministry of Education, Culture, Sports, and Science and Technology (to F.O. and S.K.).

REFERENCES

1. Bock, K. W. (1994). Aryl hydrocarbon or dioxin receptor: biologic and toxic responses. *Reviews of Physiology, Biochemistry and Pharmacology, 125*, 1–42.
2. Poellinger, L. (2000). Mechanistic aspects—the dioxin (aryl hydrocarbon) receptor. *Food Additives and Contaminants, 17*, 261–266.
3. Hankinson, O. (1995). The aryl hydrocarbon receptor complex. *Annual Reviews of Pharmacology and Toxicology, 35*, 307–340.
4. Mimura, J. and Fujii-Kuriyama, Y. (2003). Functional role of AhR in the expression of toxic effects by TCDD. *Biochimical Biophysical Acta, 1619*, 263–268.
5. Matsumura, F. and Vogel, C. F. (2006). Evidence supporting the hypothesis that one of the main functions of the aryl hydrocarbon receptor is mediation of cell stress responses. *Biological Chemistry, 387*, 1189–1194.
6. Andersson, P., McGuire, J., Rubio, C., Gradin, K., et al. (2002). A constitutively active dioxin/aryl hydrocarbon receptor induces stomach tumors. *Proceedings of the National Academy of Sciences of the United States, of America. 99*, 9990–9995.
7. Gu, Y. Z., Hogenesch, J. B., and Bradfield, C. A. (2000). The PAS superfamily: sensors of environmental and developmental signals. *Annual Reviews of Pharmacology and Toxicology, 40*, 519–561.
8. Beischlag, T. V., Luis Morales, J., Hollingshead, B. D., and Perdew, G. H. (2008). The aryl hydrocarbon receptor complex and the control of gene expression. *Critical Reviews in Eukaryotic Gene Expression, 18*, 207–250.
9. Puga, A., Ma, C., and Marlowe, J. L. (2009). The aryl hydrocarbon receptor cross-talks with multiple signal transduction pathways. *Biochemical Pharmacology, 77*, 713–722.
10. Carlson, D. B. and Perdew, G. H. (2002). A dynamic role for the Ah receptor in cell signaling? Insights from a diverse group of Ah receptor interacting proteins. *Journal of Biochemical and Molecular Toxicology, 16*, 317–325.
11. Ohtake, F., Takeyama, K., Matsumoto, T., Kitagawa, H., et al. (2003). Modulation of oestrogen receptor signalling by association with the activated dioxin receptor. *Nature, 423*, 545–550.
12. Matthews, J., Wihlen, B., Thomsen, J., and Gustafsson, J. A. (2005). Aryl hydrocarbon receptor-mediated transcription: ligand-dependent recruitment of estrogen receptor alpha to 2,3,7 8-tetrachlorodibenzo-p-dioxin-responsive promoters. *Molecular and Cellular Biology, 25*, 5317–5328.
13. Beischlag, T. V. and Perdew, G. H. (2005). ERα–AHR–ARNT protein–protein interactions mediate estradiol-dependent transrepression of dioxin-inducible gene transcription. *Journal of Biological Chemistry, 280*, 21607–21611.
14. Wormke, M., Stoner, M., Saville, B., Walker, K., et al., The aryl hydrocarbon receptor mediates degradation of estrogen receptor alpha through activation of proteasomes. *Molecular and Cellular Biology, 23*, 1843–1855.
15. Ohtake, F., Baba, A., Takada, I., Okada, M., et al. (2007). Dioxin receptor is a ligand-dependent E3 ubiquitin ligase. *Nature, 446*, 562–566.
16. Ohtake, F., Baba, A., Fujii-Kuriyama, Y., and Kato, S. (2008). Intrinsic AhR function underlies cross-talk of dioxins with sex hormone signalings. *Biochemical and Biophysical Research Communications, 370*, 541–546.
17. McKenna, N. J. and O'Malley, B. W. (2002). Combinatorial control of gene expression by nuclear receptors and coregulators. *Cell, 108*, 465–474.

18. Mangelsdorf, D. J., Thummel, C., Beato, M., Herrlich, P., et al. (1995). The nuclear receptor superfamily: the second decade. *Cell, 83*, 835–839.
19. Rosenfeld, M. G., Lunyak, V. V., and Glass, C. K. (2006). Sensors and signals: a coactivator/corepressor/epigenetic code for integrating signal-dependent programs of transcriptional response. *Genes & Development, 20*, 1405–1428.
20. Hershko, A. and Ciechanover, A. (1998). The ubiquitin system. *Annual Review of Biochemistry, 67*, 425–479.
21. Deshaies, R. J. (1999). SCF and Cullin/Ring H2-based ubiquitin ligases. *Annual Review of Cell and Developmental Biology, 15*, 435–467.
22. Weissman, A. M. (2001). Themes and variations on ubiquitylation. *Nature Reviews Molecular Cell Biology, 2*, 169–178.
23. Zheng, N., Schulman, B. A., Song, L., Miller, J. J., et al. (2002). Structure of the Cul1–Rbx1–Skp1–F boxSkp2 SCF ubiquitin ligase complex. *Nature, 416*, 703–709.
24. Ikeda, F. and Dikic, I. (2008). Atypical ubiquitin chains: new molecular signals 'Protein Modifications: Beyond the Usual Suspects' review series. *EMBO Reports, 9*, 536–542.
25. Galan, J. M. and Peter, M. (1999). Ubiquitin-dependent degradation of multiple F-box proteins by an autocatalytic mechanism. *Proceedings of the National Academy of Sciences of the United States, of America, 96*, 9124–9129.
26. Ivan, M. and Kaelin, W. G., Jr. (2001). The von Hippel–Lindau tumor suppressor protein. *Current Opinion in Genetics & Development, 11*, 27–34.
27. Jaakkola, P., Mole, D. R., Tian, Y. M., Wilson, M. I., et al. (2001). Targeting of HIFα to the von Hippel–Lindau ubiquitylation complex by O_2-regulated prolyl hydroxylation. *Science, 292*, 468–472.
28. Bai, C., Sen, P., Hofmann, K., Ma, L., et al. (1996). SKP1 connects cell cycle regulators to the ubiquitin proteolysis machinery through a novel motif, the F-box. *Cell, 86*, 263–274.
29. Groisman, R., Polanowska, J., Kuraoka, I., Sawada, J., et al. (2003). The ubiquitin ligase activity in the DDB2 and CSA complexes is differentially regulated by the COP9 signalosome in response to DNA damage. *Cell, 113*, 357–367.
30. Kobayashi, A., Kang, M. I., Okawa, H., Ohtsuji, M., et al. (2004). Oxidative stress sensor Keap1 functions as an adaptor for Cul3-based E3 ligase to regulate proteasomal degradation of Nrf2. *Molecular and Cellular Biology, 24*, 7130–7139.
31. Ito, T., Ando, H., Suzuki, T., Ogura, T., et al. (2010). Identification of a primary target of thalidomide teratogenicity. *Science, 327*, 1345–1350.
32. Angers, S., Li, T., Yi, X., MacCoss, M. J., et al. (2006). Molecular architecture and assembly of the DDB1–CUL4A ubiquitin ligase machinery. *Nature, 443*, 590–593.
33. Jin, J., Arias, E. E., Chen, J., Harper, J. W., and Walter, J. C. (2006). A family of diverse Cul4-Ddb1-interacting proteins includes Cdt2, which is required for S phase destruction of the replication factor Cdt1. *Molecular Cell, 23*, 709–721.
34. Higa, L. A., Wu, M., Ye, T., Kobayashi, R., et al. (2006). CUL4-DDB1 ubiquitin ligase interacts with multiple WD40-repeat proteins and regulates histone methylation. *Nature Cell Biology, 8*, 1277–1283.
35. He, Y. J., McCall, C. M., Hu, J., Zeng, Y., and Xiong, Y. (2006). DDB1 functions as a linker to recruit receptor WD40 proteins to CUL4-ROC1 ubiquitin ligases. *Genes & Development, 20*, 2949–2954.
36. Wang, H., Zhai, L., Xu, J., Joo, H. Y., et al. (2006). Histone H3 and H4 ubiquitylation by the CUL4-DDB-ROC1 ubiquitin ligase facilitates cellular response to DNA damage. *Molecular Cell, 22*, 383–394.
37. Matsuda, N., Azuma, K., Saijo, M., Iemura, S., et al. (2005). DDB2, the xeroderma pigmentosum group E gene product, is directly ubiquitylated by Cullin 4A-based ubiquitin ligase complex. *DNA Repair (Amst), 4*, 537–545.
38. Ma, Q. (2007). Aryl hydrocarbon receptor degradation-promoting factor (ADPF) and the control of the xenobiotic response. *Molecular Interventions, 7*, 133–137.
39. Morales, J. L., and Perdew, G. H. (2007). Carboxyl terminus of hsc70-interacting protein (CHIP) can remodel mature aryl hydrocarbon receptor (AhR) complexes and mediate ubiquitination of both the AhR and the 90 kDa heat-shock protein (hsp90) in vitro. *Biochemistry, 46*, 610–621.
40. Pollenz, R. S. (2002). The mechanism of AH receptor protein down-regulation (degradation) and its impact on AH receptor-mediated gene regulation. *Chemico-Biological Interactions, 141*, 41–61.
41. Ma, Q., and Baldwin, K. T. (2000). 2,3,7,8-Tetrachlorodibenzo-*p*-dioxin-induced degradation of aryl hydrocarbon receptor (AhR) by the ubiquitin–proteasome pathway. Role of the transcription activation and DNA binding of AhR. *Journal of Biological Chemistry, 275*, 8432–8438.
42. Nawaz, Z., Lonard, D. M., Dennis, A. P., Smith, C. L., and O'Malley, B. W. (1999). Proteasome-dependent degradation of the human estrogen receptor. *Proceedings of the National Academy of Sciences of the United States, of America 96*, 1858–1862.
43. Tateishi, Y., Sonoo, R., Sekiya, Y., Sunahara, N., et al. (2006). Turning off estrogen receptor beta-mediated transcription requires estrogen-dependent receptor proteolysis. *Molecular and Cellular Biology, 26*, 7966–7976.
44. Shang, Y., Hu, X., DiRenzo, J., Lazar, M. A., and Brown, M. (2000). Cofactor dynamics and sufficiency in estrogen receptor-regulated transcription. *Cell, 103*, 843–852.
45. Metivier, R., Penot, G., Hubner, M. R., Reid, G., et al. (2003). Estrogen receptor-alpha directs ordered, cyclical, and combinatorial recruitment of cofactors on a natural target promoter. *Cell, 115*, 751–763.
46. Boverhof, D. R., Kwekel, J. C., Humes, D. G., Burgoon, L. D., and Zacharewski, T. R. (2006). Dioxin induces an estrogen-like, estrogen receptor-dependent gene expression response in the murine uterus. *Molecular Pharmacology, 69*, 1599–1606.
47. Boverhof, D. R., Burgoon, L. D., Williams, K. J., and Zacharewski, T. R. (2008). Inhibition of estrogen-mediated uterine gene expression responses by dioxin. *Molecular Pharmacology, 73*, 82–93.

48. Lin, T. M., Ko, K., Moore, R. W., Simanainen, U., et al. (2002). Effects of aryl hydrocarbon receptor null mutation and in utero and lactational 2,3,7 8-tetrachlorodibenzo-*p*-dioxin exposure on prostate and seminal vesicle development in C57BL/6 mice. *Toxicological Sciences*, 68, 479–487.

49. Astroff, B., Eldridge, B., and Safe, S. (1991). Inhibition of the 17β-estradiol-induced and constitutive expression of the cellular protooncogene c-fos by 2,3,7,8-tetrachlorodibenzo-*p*-dioxin (TCDD) in the female rat uterus. *Toxicological Letters*, 56, 305–315.

50. Brauze, D., Crow, J. S., and Malejka-Giganti, D. (1997). Modulation by beta-naphthoflavone of ovarian hormone dependent responses in rat uterus and liver *in vivo*. *Canadian Journal of Physiology and Pharmacology*, 75, 1022–1029.

51. Brown, N. M., Manzolillo, P. A., Zhang, J. X., Wang, J., and Lamartiniere, C. A. (1998). Prenatal TCDD and predisposition to mammary cancer in the rat. *Carcinogenesis*, 19, 1623–1629.

52. Cummings, A. M., Metcalf, J. L., and Birnbaum, L. (1996). Promotion of endometriosis by 2,3,7,8-tetrachlorodibenzo-*p*-dioxin in rats and mice: time–dose dependence and species comparison. *Toxicology and Applied Pharmacology*, 138, 131–139.

53. Cummings, A. M., Hedge, J. M., and Birnbaum, L. S. (1999). Effect of prenatal exposure to TCDD on the promotion of endometriotic lesion growth by TCDD in adult female rats and mice. *Toxicological Sciences*, 52, 45–49.

54. Baba, T., Mimura, J., Nakamura, N., Harada, N., et al. (2005). Intrinsic function of the aryl hydrocarbon (dioxin) receptor as a key factor in female reproduction. *Molecular and Cellular Biology*, 25, 10040–10051.

55. Mulero-Navarro, S., Pozo-Guisado, E., Perez-Mancera, P. A., Alvarez-Barrientos, A., et al. (2005). Immortalized mouse mammary fibroblasts lacking dioxin receptor have impaired tumorigenicity in a subcutaneous mouse xenograft model. *Journal of Biological Chemistry*, 280, 28731–28741.

56. Spink, D. C., and Lincoln, D. W.2nd, Dickerman, H. W., and Gierthy, J. F. (1990). 7,8-Tetrachlorodibenzo-*p*-dioxin causes an extensive alteration of 17β-estradiol metabolism in MCF-7 breast tumor cells. *Proceedings of the National Academy of Sciences of the United States, of America*, 87, 6917–6921.

57. Klinge, C. M., Kaur, K., and Swanson, H. I. (2000). The aryl hydrocarbon receptor interacts with estrogen receptor alpha and orphan receptors COUP-TFI and ERRα1. *Archives of Biochemistry and Biophysics*, 373, 163–174.

58. Brunnberg, S., Pettersson, K., Rydin, E., Matthews, J., et al. (2003). The basic helix–loop–helix–PAS protein ARNT functions as a potent coactivator of estrogen receptor-dependent transcription. *Proceedings of the National Academy of Sciences of the United States, of America*, 100, 6517–6522.

59. Ravasi, T., Suzuki, H., Cannistraci, C. V., Katayama, S., et al. (2008). An atlas of combinatorial transcriptional regulation in mouse and man. *Cell*, 140, 744–752.

60. Ohtake, F., Fujii-Kuriyama, Y., and Kato, S. (2009). AhR acts as an E3 ubiquitin ligase to modulate steroid receptor functions. *Biochemical Pharmacology*, 77, 474–484.

61. Medlock, K. L., Lyttle, C. R., Kelepouris, N., Newman, E. D., and Sheehan, D. M. (1991). Estradiol down-regulation of the rat uterine estrogen receptor. *Proceedings of the Society for Experimental Biology and Medicine*, 196, 293–300.

62. Lonard, D. M., Nawaz, Z., Smith, C. L., and O'Malley, B. W. (2000). The 26S proteasome is required for estrogen receptor-alpha and coactivator turnover and for efficient estrogen receptor-alpha transactivation. *Molecular Cell*, 5, 939–948.

63. Roberts, B. J., and Whitelaw, M. L. (1999). Degradation of the basic helix–loop–helix/Per–ARNT–Sim homology domain dioxin receptor via the ubiquitin/proteasome pathway. *Journal of Biological Chemistry*, 274, 36351–36356.

64. LaPres, J. J., Glover, E., Dunham, E. E., Bunger, M. K., and Bradfield, C. A. (2000). ARA9 modifies agonist signaling through an increase in cytosolic aryl hydrocarbon receptor. *Journal of Biological Chemistry*, 275, 6153–6159.

65. Petrulis, J. R., Hord, N. G., and Perdew, G. H. (2000). Subcellular localization of the aryl hydrocarbon receptor is modulated by the immunophilin homolog hepatitis B virus X-associated protein 2. *Journal of Biological Chemistry*, 275, 37448–37453.

66. Perissi, V., Aggarwal, A., Glass, C. K., Rose, D. W., and Rosenfeld, M. G. (2004). A corepressor/coactivator exchange complex required for transcriptional activation by nuclear receptors and other regulated transcription factors. *Cell*, 116, 511–526.

67. Maxwell, P. H., Wiesener, M. S., Chang, G. W., Clifford, S. C., et al. (1999). The tumour suppressor protein VHL targets hypoxia-inducible factors for oxygen-dependent proteolysis. *Nature*, 399, 271–275.

68. Yanagisawa, J., Kitagawa, H., Yanagida, M., Wada, O., et al. (2002). Nuclear receptor function requires a TFTC-type histone acetyl transferase complex. *Molecular Cell*, 9, 553–562.

69. Zhong, W., Feng, H., Santiago, F. E., and Kipreos, E. T. (2003). CUL-4 ubiquitin ligase maintains genome stability by restraining DNA-replication licensing. *Nature*, 423, 885–889.

70. Wertz, I. E., O'Rourke, K. M., Zhang, Z., Dornan, D., et al. (2004). Human de-etiolated-1 regulates c-Jun by assembling a CUL4A ubiquitin ligase. *Science*, 303, 1371–1374.

71. Marconett, C. N., Sundar, S. N., Poindexter, K. M., Stueve, T. R., et al. (2010). Indole-3-carbinol triggers aryl hydrocarbon receptor-dependent estrogen receptor (ER)α protein degradation in breast cancer cells disrupting an ERα-GATA3 transcriptional cross-regulatory loop. *Molecular Biology of the Cell*, 21, 1166–1177.

72. Puga, A., Barnes, S. J., Dalton, T. P., Chang, C., et al. (2000). Aromatic hydrocarbon receptor interaction with the retinoblastoma protein potentiates repression of E2F-dependent transcription and cell cycle arrest. *Journal of Biological Chemistry*, 275, 2943–2950.

73. Vogel, C. F., Sciullo, E., Li, W., Wong, P., et al. (2007). RelB, a New Partner of Aryl Hydrocarbon Receptor-Mediated Transcription. *Molecular Endocrinology*, 21, 2941–2955.

74. Quintana, F. J., Basso, A. S., Iglesias, A. H., Korn, T., et al. (2008). Control of T(reg) and T(H)17 cell differentiation by the aryl hydrocarbon receptor. *Nature*, 453, 65–71.

75. Veldhoen, M., Hirota, K., Westendorf, A. M., Buer, J., et al. (2008). The aryl hydrocarbon receptor links TH17-cell-mediated autoimmunity to environmental toxins. *Nature, 453*, 106–109.
76. Kimura, A., Naka, T., Nohara, K., Fujii-Kuriyama, Y., and Kishimoto, T. (2008). Aryl hydrocarbon receptor regulates Stat1 activation and participates in the development of Th17 cells. *Proceedings of the National Academy of Sciences of the United States, of America, 105*, 9721–9726.
77. Liu, P. C., Dunlap, D. Y., and Matsumura, F. (1998). Suppression of C/EBPα and induction of C/EBPβ by 2,3,7,8-tetrachlorodibenzo-*p*-dioxin in mouse adipose tissue and liver. *Biochemical Pharmacology, 55*, 1647–1655.
78. Celander, M., Weisbrod, R., and Stegeman, J. J. (1997). Glucocorticoid potentiation of cytochrome P4501A1 induction by 2,3,7, 8-tetrachlorodibenzo-*p*-dioxin in porcine and human endothelial cells in culture. *Biochemical and Biophysical Research Communications, 232*, 749–753.
79. Lorick, K. L., Toscano, D. L., and Toscano, W. A., Jr. (1998). 2,3,7,8-Tetrachlorodibenzo-*p*-dioxin alters retinoic acid receptor function in human keratinocytes. *Biochemical and Biophysical Research Communications, 243*, 749–752.
80. Kawajiri, K., Kobayashi, Y., Ohtake, F., Ikuta, T., et al. (2009). Aryl hydrocarbon receptor suppresses intestinal carcinogenesis in Apc$^{Min/+}$ mice with natural ligands. *Proceedings of the National Academy of Sciences of the United States, of America, 106*, 13481–13486.
81. Qian, S. B., McDonough, H., Boellmann, F., Cyr, D. M., and Patterson, C. (2006). CHIP-mediated stress recovery by sequential ubiquitination of substrates and Hsp70. *Nature, 440*, 551–555.
82. Bhoumik, A., Singha, N., O'Connell, M. J., and Ronai, Z. A. (2008). Regulation of TIP60 by ATF2 modulates ATM activation. *Journal of Biological Chemistry, 283*, 17605–17614.
83. Kobayashi, A., Kang, M. I., Watai, Y., Tong, K. I., et al. (2006). Oxidative and electrophilic stresses activate Nrf2 through inhibition of ubiquitination activity of Keap1. *Molecular and Cellular Biology, 26*, 221–229.
84. Yu, X., Yu, Y., Liu, B., Luo, K., et al. (2003). Induction of APOBEC3G ubiquitination and degradation by an HIV-1 Vif-Cul5-SCF complex. *Science, 302*, 1056–1060.
85. Kepinski, S., and Leyser, O. (2005). The Arabidopsis F-box protein TIR1 is an auxin receptor. *Nature, 435*, 446–451.
86. Dharmasiri, N., Dharmasiri, S., and Estelle, M. (2005). The F-box protein TIR1 is an auxin receptor. *Nature, 435*, 441–445.
87. Tan, X., Calderon-Villalobos, L. I., Sharon, M., Zheng, C., et al. (2007). Mechanism of auxin perception by the TIR1 ubiquitin ligase. *Nature, 446*, 640–645.
88. Chini, A., Fonseca, S., Fernandez, G., Adie, B., et al. (2007). The JAZ family of repressors is the missing link in jasmonate signalling. *Nature, 448*, 666–671.
89. Thines, B., Katsir, L., Melotto, M., Niu, Y., et al. (2007). JAZ repressor proteins are targets of the SCF(COI1) complex during jasmonate signalling. *Nature, 448*, 661–665.

11

EPIGENETIC MECHANISMS IN AHR FUNCTION

CHIA-I KO AND ALVARO PUGA

In the first part of this chapter, we will briefly discuss mechanisms of epigenetic modification, emphasizing those epigenetic changes that take place during development and their reversibility. We will focus on aspects of zygotic genome activation, *in vitro* ES cell differentiation, and environmental effects on the epigenome that we believe are more closely related to the known functions of the aryl hydrocarbon receptor and to its dual role during development and during adaptive and toxic responses to environmental injury. In the second part of this chapter, we will discuss the specifics of epigenetic mechanisms associated with the Ah receptor in the context of its role in regulation of gene expression.

11.1 EPIGENETICS, CELLULAR MEMORY, AND DEVELOPMENT

11.1.1 Epigenetic Inheritance

To ensure the ontogenic divergence of the different cell types of an organism, the differentiation of a multicellular eukaryote requires a mechanism that would control differential gene expression in the different cellular lineages. Even though all cells of an organism have identical genomic sequences, different cell types and tissues have their own specific patterns of gene expression, such that genes become activated or repressed according to their cellular functions and the physiological state of their corresponding tissues. Aside from physical changes in DNA (i.e., mutations and genetic recombination), accessibility to the genetic material is key to regulate transcription initiation and gene expression, which, at a molecular level, require binding of the RNA polymerase II complex to the promoter region of any genes that are to become transcriptionally active. Unlike the naked DNA chromosomes of prokaryotes, eukaryotic DNA is complexed with histones and organized into a structure termed chromatin, of which the basic unit, the nucleosome, consists of 146 base pairs of DNA wrapped around an octamer of histone proteins, including a tetramer of histones H3 and H4 and two dimers of histones H2A and H2B. With the addition of the linker histone H1, chromatin undergoes compaction and forms the chromosome. *In vivo*, transcriptionally active chromatin resolves into a 30 nm helical fiber, the so-called euchromatin, which is transcriptionally open. Further condensation into 60–130 nm fibers, or higher complexity structures, forms the transcriptionally inactive heterochromatin.

Animal development and cellular differentiation are characterized by the gradual loss of the cellular pluripotency of the embryonic cell and the gain of tissue- or cell-type specificity. *In vitro* embryonic stem (ES) cells are pluripotent cells derived from the inner cell mass of the early blastocyst. Differentiation-specific genes are transcriptionally repressed in ES cells, partly as a result of changes in the methylation status of their DNA and partly due to histone modifications. Upon fertilization, the bulk of the mammalian genome, excluding imprinted genes and the inactivated X chromosome, goes through a brief period of DNA demethylation. At implantation, *de novo* methylation starts from the cells in the inner cell mass and propagates quickly to the whole embryo, affecting the totality of the DNA with the exception of the CpG islands, which are not methylated at this point. At this time, genes with CpG-rich promoters are marked for a bivalent state of expression or repression by trimethylation of lysine 4 and lysine 27 residues in histone H3. This pluripotent state keeps these genes poised to be expressed or repressed upon cell fate commitment [1]. Trimethylation at lysine 4 (H3K4me3) is catalyzed by Trithorax (Trx) and

The AH Receptor in Biology and Toxicology, First Edition. Edited by Raimo Pohjanvirta.
© 2012 John Wiley & Sons, Inc. Published 2012 by John Wiley & Sons, Inc.

trimethylation at lysine 27 (H3K27me3) by Polycomb group (PcG) histone methyltransferase proteins, which mediate mitotic inheritance of lineage-specific gene expression programs and have key developmental functions [2]. PcG proteins target not only differentiation-specific genes in ES cells, but also lineage-committed genes in progenitor cells [3]. In genes that are stably repressed, trimethylation of H3K27 is quickly followed by DNA methylation, supporting the conclusion that H3K27me3 is a transient signal for gene repression [4]. In addition to methylation, histone tail amino acids can also be marked by other posttranslational modifications, including acetylation, phosphorylation, ADP ribosylation, sumoylation, and ubiquitination, providing a "histone code" for each gene that, together with DNA methylation and chromatin rearrangements, constitutes a cell- and tissue-type-specific repertoire of instructions to direct and maintain gene expression through the generations. This conservation and inheritance of information without change of gene sequence or structure is referred to as "cellular memory" or "epigenetic inheritance." Although the concept of epigenetic regulation of gene expression is widely accepted, there is much yet to be learned about the specific mechanisms that determine how genes and chromatin domains reproduce these patterns of instructions through DNA replication and how the instructions are transmitted to the next generation. The reader is directed to an excellent review in the recent literature that discusses this subject [5].

11.1.1.1 Epigenetic Reprogramming The successful correction of a genetic defect by transfer of nuclei derived from ES cells has made it possible to entertain the prospects of patient-specific nuclear transfer for cellular therapy in human disease [6]. Numerous nuclear transfer attempts in animal cloning have shown the reversibility of epigenetic programming; that is, the fact that the epigenetic instructions acquired during development of terminally differentiated cells can be erased. This concept has led to the hypothesis that nuclear transfer from somatic cells to oocytes should be able to direct normal embryonic development, since all cell types carry the same genetic information. This hypothesis, however, was incorrect, and most clones generated in this fashion could not complete embryonic development, and the few embryos that developed to term died shortly after birth or developed serious developmental defects, mainly respiratory distress and circulatory problems, leading to death, and oversized fetuses, likely associated with a dysfunctional placenta [7]. These problems depended on the donor, not the recipient cells [8, 9], indicating that the cellular memory of the donor nuclei bear a set of gene expression instructions that cause inadequate reprogramming of the recipient cells.

Similar experiments conducted with ES cell nuclei met with more success, although the transfer rate was low compared to the rate with somatic cells. The reason for this low rate was that the ES cell nuclei are in the S phase of the cell cycle and are not synchronized with the enucleated oocytes. However, embryos generated with transfer of ES cell G1 nuclei developed to term with a much higher efficiency than those from somatic donor nuclei [7]. These observations suggest that the ES epigenome is more plastic than the epigenome of differentiated cells and can reprogram a cell into a state closer to that of a normal embryo at an early stage.

Stable transfection of the genes coding for the pluripotent transcription factors OCT3/4, SOX2, c-MYC, and KLF4 resulted in the successful reprogramming of somatic cells of many types. These cells are the so-called iPS, or induced pluripotent stem cells, first reported by Takahashi and Yamanaka [10]. It seems that in these cells, reprogramming occurs by a stepwise reversal process, in which histone modifications take place first, followed by DNA demethylation [11].

11.1.2 Mechanisms of Epigenetic Inheritance

11.1.2.1 High-Order Chromatin Structure Arrangement: Heterochromatin Epigenetic modifications comprise all the changes that can alter chromatin structure and gene expression without changing the DNA sequence. These changes include DNA methylation, histone posttranslational modifications, and small interfering RNAs. DNA methylation is often a form of long-term silencing associated with limited, if any, reversibility. On the other hand, histone posttranslational modifications provide the chromatin with a spatiotemporal signature that can be more readily reversed according to cellular status. Both DNA methylation and histone modifications cause rearrangements of chromatin structure that, when gene repression is required, reorganizes itself into a heterochromatic structure. Conversely, a heterochromatic environment promotes an inhibitory effect on gene expression.

"Constitutive" heterochromatin is a tightly packed, highly condensed, and deeply staining portion of the genomic DNA, commonly localized to the periphery of the nucleus. Compared to euchromatin, which contains a high density of expressed genes and is dispersed during interphase, heterochromatin consists mainly of repetitive sequences that form centromeres and telomeres; any genes present in these sequences are very poorly expressed, if at all [12]. Three epigenetic features are characteristic of heterochromatin: histone hypoacetylation, histone H3K9 methylation, and DNA hypermethylation; these three are also chromatin modifications associated with silencing of euchromatin genes. It has been proposed that once a genomic region has been targeted for silencing by acquisition of one or more epigenetic marks, the silent state could be propagated to other chromatin regions through (1) positive signaling between epigenetic marks; (2) positive feedback of epigenetic generation of marks; and (3) recruitment of other chromatin modification enzymes, such as histone methyltransferases [13].

The term "facultative heterochromatin" or "heterochromatin-like" has appeared more recently in the literature. Unlike constitutive heterochromatin, facultative heterochromatin is transcriptionally silent, but can be converted to euchromatin depending on the differentiated state of the cell [14]. The conversion between euchromatin and facultative heterochromatin may depend on the incorporation of certain histone H1 isotypes, as is the case during silencing of the regulatory gene *MYOD* during muscle cell differentiation, which requires recruitment of histone H1b by the homeotic protein Msx1 [15].

The role of the heterochromatin environment on gene expression was first observed while studying eye color changes in X-ray-irradiated *Drosophila* flies during development and was termed "position effect variegation (PEV)" [16]. Irradiation caused the translocation of a euchromatic gene from its normal chromosome position to an adjacent heterochromatic location and resulted in a mutant phenotype. Extensive studies in *Drosophila* led to the identification of the gene interactions involved in PEV. Critical for gene silencing is the interaction between two proteins, one a chromodomain-bearing heterochromatin protein 1 (HP1), product of the *Drosophila Su(var)2-5* gene, and the other a SET domain (suppressor of variegation, enhancer of zeste, Trithorax)-containing histone methyltransferase product of the *Drosophila Su(var)3-9* gene. Chromodomain motifs are also found in members of the Polycomb group gene family and HP1 proteins bind specifically to methylated chromatin on H3K9 [17]. In turn, the SET domain-bearing histone methyltransferases are conserved through evolution and have H3K9-specific catalytic activity [18, 19].

11.1.2.2 DNA Methylation: Long-Term Silencing
DNA methylation is not often a primary event in gene silencing, but once it has taken place, it has a long-term effect, keeping genes stably repressed. The enzyme DNA methyltransferase (DNMT) is responsible for the methylation of cytosine residues in CpG dimers. Two types of DNA methyltransferases are found: DNMT1 is a maintenance methyltransferase, with preference for complementary pairs of CpG dimers in which one of the two cytosines is already methylated; DNMT3a and 3b are *de novo* methyltransferases. In mammals, "CpG islands" refer to genomic regions rich in CpG clusters, which often form part of promoter regions of genes closely associated with differentiation and development. CpG islands are *de novo* methylated at the moment of implantation during early embryonic development. Methylated CpG residues repress gene expression by preventing the binding of transcriptional activators or by recruiting the methylated CpG binding proteins MeCP1 and MeCP2 that promote repression by as yet unknown mechanisms [20]. DNA methylation and histone modifications could be coordinated through protein interactions, since both methylated DNA and MeCP2 are able to recruit histone deacetylases [21]; conversely, histone-modifying enzymes can recruit DNMT3a and DNMT3b.

The first step in the stable repression of a gene is to switch off its expression. To do this, transcription is first turned off by the binding of repressors to the promoter region, then histone modifications take place that induce further chromatin rearrangements, and finally *de novo* DNMT3s are recruited to the gene promoter, methylate the CpGs, and cause the stable repression of the gene. A classical example of this sequence of events is the silencing of the pluripotent genes *Oct3/4* and *Nanog* during differentiation, which are transcriptionally switched off and methylated by recruitment of DNMT3a and DNMT3b by the G9a histone methyltransferase [22]. A similar series of events was found for the silencing of repetitive sequences, which are methylated by DNMT3a and DNMT3b after being recruited by the SET-containing histone methyltransferase SUV39H, the human homolog of the *Drosophila* SU(VAR)3-9 protein [23].

Random inactivation of a single X chromosome in female somatic cells ensures that XX and XY individuals have equal expression dosage of genes located in the X chromosome. Coating of inactivated X (Xi) by noncoding RNA *Xist* (X-inactive-specific transcript) in *cis* triggers a number of chromosomal modifications, including histone hypoacetylation and lysine methylation on histones H3 and H4 [24]. Incorporation of the macroH2A1 histone subtype and DNA methylation further convert the Xi into a heterochromatic state [25]. Xi heterochromatin forms a particular structure, named the Barr body, that often localizes adjacent to the nuclear envelope and remains visible through interphase.

11.1.2.3 Polycomb Group-Mediated Gene Repression
PcG proteins were initially discovered as chromatin binding factors that ensured the appropriate expression of *Drosophila hox* genes. These proteins are critical for maintaining ES cell pluripotency because they repress the expression of transcription factors essential for differentiation [26, 27]. Studies both in mammals and in *Drosophila* show that most PcG target genes encode transcription factors, morphogens, and growth factor receptors [26–28]. Detailed mechanisms of this repression system are still unknown, although genetic and genome-wide studies in *Drosophila* and mammals provide clues as to their biological function.

Three Polycomb repressive complexes (PRC), PRC1, PRC2, and PhoRC, have been identified. In *Drosophila*, the core of the PRC1 complex consists of four PcG proteins, namely, Polycomb (Pc), Posterior sex combs (Psc), Polyhomeotic (Ph), and Ring (also known as sex comb extra). The chromodomain of the Pc protein binds to DNA sequences *via* its affinity for H3K27 trimethylation (H3K27me3) [29]. The mammalian homologs of these four proteins also form functional complexes [30]. Heterodimers of RING1B and BMI1 (B lymphoma Mo-MLV insertion region 1, the mammalian homolog of *Drosophila* Psc) have been shown to directly catalyze histone H2A ubiquitination, which may be an important step in PcG-mediated repression [30]. The PRC2 complex of *Drosophila* also contains four core proteins, namely, enhancer of zeste (E(z)), extra sex combs (Esc), suppressor of zeste 12 (Su(z)12), and Nurf55. The SET domain of the E(z) protein has histone H3K27 methyltransferase activity, which is only catalytic when assembled within the complex [31]; EZH2 is the homolog protein for this catalytic function in mammals. The third complex in *Drosophila* is PhoRC, which contains pleiohomeotic (Pho) and *Scm*-related gene containing four MBT domains (Sfmbt). The function of this complex and the existence of mammalian homologs are still uncertain. The mammalian PcG may be much more complex than its *Drosophila* counterpart, since there are a large number of isoforms of each of the mammalian proteins. For example, chromobox binding homologs (CBX) 2, 4, 6, 7, and 8 are human homologs of the *Drosophila* Pc protein; EZH2 and EZH1 are homologs of E(z). Multiple versions of PRC1 and PRC2 may confer different functions and may regulate distinct subsets of genes by different mechanisms.

A Polycomb response element (PRE) in DNA has only been identified in *Drosophila* [32] and has not been defined as a conserved sequence motif, but as many enhancer domains includes several conserved short motifs. ChIP-on-chip experiments in *Drosophila* reveal the colocalization of PRC1 and PRC2 to a well-defined region (known or presumptive PREs) unlike the much broader region covered by H3K27me3 on silenced genes, often involving many kilobases [33]. The mammalian PRE has not yet been defined, but unlike in flies, binding and distribution of PRC1 and PRC2 proteins are coextensive with the H3K27me3 binding domains [27, 34], with 90% of the binding located in the proximal promoter region of the controlled genes.

Binding of PRC to DNA may be the result of recruitment by transcription factors. In mouse myoblasts, knockdown of the YY1 protein (Yin Yang 1, a homolog of the *Drosophila* Pho protein) disassembles EZH2 and blocks H3K27 trimethylation [35]. Similarly, OCT4 knockdown in ES cells blocks PRC2 DNA binding [36]. Recent ChIP-seq data show that 88% of PRC2 binding DNA sequences in mouse ES cells are annotated CpG islands [37], and thus CpG binding proteins may also be good candidates for PcG proteins recruitment. In *Drosophila*, binding of PRC1 on H3K27me3 might facilitate the contact of PcG complexes to target gene sequences by loop formation [33]. PcG silencing does not restrict the access of transcription activators or RNA polymerase II complexes to DNA, but directly interferes with the activity of the transcriptional machinery [38, 39]. A similar silencing phenomenon is also observed in mouse ES cells in which histone H2A ubiquitination on lysine 119 (H2AK119ub1) by Ring1 impedes RNA polymerase II elongation but not its entry onto the promoter [40].

All PcG target genes are also subject to regulation by Trithorax group proteins, which provide a regulatory system antagonistic to PcG-mediated gene repression. Trx proteins are also chromatin binding factors that ensure appropriate expression of *hox* genes in *Drosophila* by activating their transcription. The mammalian Trx homologs have a SET domain with H3K4 methyltransferase activity [41, 42]. H3K4 trimethylation (H3K4me3) was long considered to be a histone mark for transcription activation, involving transcriptional maintenance rather than initiation, since the trimethylation of K4 first appears after the assembly of the preinitiation complex [39]. Genetic studies in *Drosophila* suggest that the biological function of Trx proteins is to prevent PcG-mediated silencing rather than simple gene activation [43, 44]. Alternatively, transcription factors could recognize H3K4me3 and stimulate gene activation-associated chromatin remodeling, as is the case of the nucleosome remodeling factor (NURF) complex in plants, a homeodomain (PHD) finger protein that couples NURF with H3K4me3 and recruits ATP-dependent chromatin remodeling complexes to maintain expression of *hox* genes [45].

Repression by a PcG-mediated mechanism instead of DNA methylation may be a more efficient and plastic way to keep genes poised for reactivation during development. Indeed, it is not very feasible to remove methyl groups from cytosine residues, since they are chemically very stable. No DNA demethylases have been reported, although recently two proteins, JMJD3 and UTX, have been identified as H3K27-specific demethylases [46–49]. Evidence that ES cells can be generated from $Ezh2^{-/-}$ embryo suggests that PcG proteins are not required for ES self-renewal, but rather to maintain ES identity by keeping differentiation genes in a state of developmental standby [50]. PcG-mediated repression may be a reversible system set to respond to developmental stimuli, with far more dynamic flexibility than DNA methylation.

11.1.2.4 Histone Code Amino acids on the amino-terminal end of histones are subject to variable posttranslational modifications (PTMs), including acetylation, phosphorylation, methylation, ubiquitination, and sumoylation. Acetylation of lysine 14 or lysine 9 in histone H3 (H3K14ac, H3K9ac), methylation of arginine 17 on histone H3 (H3R17me), and phosphoacetylation on histone H3 at serine 10/lysine 14 (H3S10p–K14ac) are PTMs generally associated with gene activation [51–53], whereas methylation on H3K9 and H3K27 are associated with HP1 binding, heterochromatin formation, and PcG-mediated gene repression. The protein module that interacts with acetylated lysine on these histone tails [54, 55] is the bromodomain, which is often present in transcriptional activators containing histone acetyltransferase (HAT) activity, such as p300, GCN5, TAFII250, and SWI/SNF chromatin remodeling complexes [56–58]. Methylated lysines could be recognized by chromodomain proteins, such as HP1 [17], chromodomain-containing histone methyltransferases, such as SU(VAR)3-9 and CHD [18, 59], and SET domain proteins with histone methyltransferase activity, such as G9a [22]. Identification of novel histone demethylases and analyses of their biological functions are areas of active research, since erasing and establishing acetyl and methyl marks are important for development, cellular differentiation, and disease [60]. Moreover, combinatorial interactions of different modifications can bring forth synergistic or antagonistic effects by recruiting alternative chromatin-associated proteins. Assembly of histone marks constitutes the critical epigenetic information specific for the differentiation and the developmental state of cells.

11.1.3 Developmental Epigenomics

Many recent findings underscore the significance of epigenetic modifications during development; the following sections briefly describe two examples of such findings in mammals: (1) epigenomic reprogramming before embryonic development and (2) epigenetic signature switching in ES cells during *in vitro* differentiation.

11.1.3.1 The Onset of Gene Expression in the Zygote

The mammalian genome undergoes two steps of epigenomic reprogramming mainly through changes of methylation patterns. The first step takes place during gametogenesis and the second at the time of embryo implantation. Reprogramming during gametogenesis results in epigenetic asymmetry between the parental genomes, a phenomenon crucial for the establishment of imprinting patterns [61]. Reprogramming during implantation is essential to initiate early embryo gene expression and lineage development and puts an end to embryo totipotency [62].

The difference between maternal and paternal genomes is manifested by the differential expression of imprinted genes during development. Epigenetic reprogramming during gametogenesis is first characterized by the erasure of methylation marks in the DNA of imprinted genes in primordial germ cells [63]. De novo methylation and establishment of maternal-specific imprints are important for oocyte maturation [64]. Besides *de novo* methylation, histone deacetylation and displacement of histones by protamines are important epigenomic reprogramming events during spermatogenesis [65]. Upon fertilization, the paternal genome is quickly demethylated, exchanges protamines for histones, and goes through extensive histone modifications. In contrast, the maternal genome remains relatively static at that time [62]. Differences in histone H3 marks between parental pronuclei may explain paternal-specific active demethylation and maternal-specific protection [66]. Embryonic *de novo* methylation begins in the inner cell mass of blastocysts and propagates to the entire embryo body, whereas a low level of methylation is maintained in the lineages giving rise to the extraembryonic structures. Recent studies also show an asymmetric H3K4/H3K27 methylation pattern on the promoters of developmental regulators between embryonic and extraembryonic lineages, which may be indicative of a distinctive cell fate commitment [67].

11.1.3.2 In Vitro **Differentiation from Embryonic Stem Cells**

Pluripotent ES cells constitute early progenitor cell populations difficult to acquire *in vivo*; directed *in vitro* differentiation of these cells may generate enough differentiated cells invaluable for transplantation and regeneration purposes in many diseases, although obtaining large amounts of a pure population of cells of a specific lineage is not often feasible. Different epigenetic modifications mark ES and differentiated cells, constituting a critical element in their characterization.

Genetic studies, especially in *Drosophila*, led to an understanding of the significance of PcG/Trx proteins in vertebrate development. To a great extent, the correct expression pattern of PcG/Trx proteins ensures proper cell fate commitment. Hence, the chromatin landmarks of the PcG/Trx proteins would have specific patterns depending on the differentiation state of the cell. In fact, genome-wide mapping of H3K4me3 and H3K27me3 revealed a profile in ES cells that may be a key biological gene expression characteristic [1]. Histone modification patterns of high-CpG promoter regions appear to be closely related to the nature of the corresponding genes. For the most part, the monovalent H3K4me3 modification marked promoters of housekeeping genes, whereas H3K4me3/H3K27me3 bivalent modifications marked promoters of genes associated with developmental factors, morphogens, and cell surface receptors. Consistent with the function of the corresponding genes, most H3K4me3-marked promoters retain that chromatin signature in the committed cells, while the majority of promoters with bivalent marks resolve into monovalent states, as H3K4me3 or H3K27me3, at which point the expression of genes with H3K4me3 monovalent marks is induced, whereas the expression of genes with H3K27me3 monovalent and H3K4me3/H3K27me3 bivalent marks is repressed or very low. These observations indicate that the plasticity of ES cells is built upon specific histone modification patterns that temporally repress expression of key developmental factors. In contrast to the silencing of imprinted genes by DNA methylation, PcG/Trx-mediated repression and activation represent a rapid reversible mechanism to keep functional genes poised for regulation.

11.1.4 Environmental Impact on Epigenetic Inheritance

Increasing experimental evidence indicates that in addition to genetic mechanisms, environmental factors may promote disease incidence in humans by interfering with the epigenetic mechanisms of gene regulation. The epigenome can be modulated by environmental agents, nutritional states, and behavior at the same time that it may integrate these environmental signals into chromatin remodeling processes that derail the normal function of the genome. Complex human diseases, such as cancer, heart disease, and type 2 diabetes, are suspected of having a strong environmental component superimposed onto genetic causes [68]. To complete appropriate cellular development and differentiation, it appears that the epigenome has to be fully erased and reestablished between generations. However, there have been numerous reports of incomplete erasure of parental marks at specific genes, which might be associated with the epigenetic inheritance of disease. Although the existence of such a phenomenon is still controversial, there is increasing evidence that it does occur in rodents and humans [69]. If so, environmental factors would affect not only those individuals directly exposed to the specific agent but also their offspring, providing a potential mechanism for the transgenerational transmission of the effects of environmental factors. These "transgenerational effects" cause offspring, with or without exposure to the same environmental factor, to have the same pathological phenotype as their parents or grandparents; in a most Lamarckian sense, the offspring inherit the traits that their ancestors acquired [70–72].

The first observation of an environmental effect *via* epigenetic changes was the finding of an elevated risk of breast and ovarian cancer in girls of women treated with diethylstilbestrol [73, 74]. These abnormalities were probably due to aberrant DNA methylation of genes that control uterine development [75]. Aberrant DNA methylation patterns also appear in offspring of rats exposed to the endocrine disruptor chemicals vinclozolin and methoxychlor [76]. Exposure to sexual hormone mimetics during the period of sex determination may cause infertility and increased risk of cancer associated with reproductive tract and organs [72]. Exposure during crucial developmental windows can interfere with the differentiation program, as is the case of reduced heart-to-body weight observed in fetal mice exposed *in utero* to TCDD during cardiovascular development [77, 78].

The nutritional state is also believed to cause pathological phenotypes *via* transgenerational epigenetic changes. Babies from women subjected to severe food restriction during the last trimester of pregnancy have lower birth weight, a trait that is also observed in the subsequent generation, despite normal diet conditions [79]. Insufficient food supply during the middle childhood of paternal grandparents was also linked to the mortality of the grandchildren [80], while a high-fat diet increased body length and reduced insulin sensitivity for at least two generations subsequent to the one initially exposed [81].

11.2 THE AHR AND ITS ROLE IN DEVELOPMENT

In the second part of this chapter, we will describe the involvement of the Ah receptor in development, the increasing body of data addressing the epigenetic mechanisms operative in AHR functions, and how these functions may be subverted by exposure to AHR agonists. We will address recent work dealing with the genome-wide identification of AHR target genes and the crosstalk of AHR signaling with other developmental pathways. To close, we will discuss AHR expression and possible functions during vertebrate development.

We will use throughout the abbreviation "AHR" in accordance with the rules of genetic nomenclature as the acronym for the aromatic hydrocarbon receptor protein. The AHR mediates the teratogenic and carcinogenic effects of environmental planar aromatic hydrocarbons, among which the most potent xenobiotic agent, 2,3,7,8-tetrachlorodibenzo-*p*-dioxin (TCDD), is the causative agent of chloracne, a long-lasting hyperkeratotic skin disease in humans [82]. The vertebrate AHR is a ligand-activated transcription factor that belongs to a protein family characterized by the presence of a basic helix–loop–helix/Per–ARNT–Sim (bHLH/PAS) domain [83, 84]. The unliganded AHR resides in the cytosol, where it forms a protein complex with two molecules of the 90 kDa heat shock protein hsp90, a p23 protein and the immunophilin homolog XAP2 [85–87]. Upon ligand activation, the AHR complex translocates into the nucleus, where the AHR dissociates from its cytosolic partner proteins and heterodimerizes with the Ah receptor nuclear translocator (ARNT), which is also a member of the PAS protein family [88]. The AHR–ARNT heterodimer binds to the AHR-, dioxin- or xenobiotic-responsive element (AhRE, DRE, or XRE) core consensus 5′-TNGCGTG-3′ located on the promoter region of AHR target genes [89] and recruits transcriptional coactivators and chromatin remodeling proteins that cooperate in the induction of AHR target gene expression [90]. Well-known AHR target genes are drug metabolizing enzymes belonging to the cytochrome P450 CYP1 family [91].

AHR-deficient mice are resistant to dioxin toxicity [92, 93] and show multiple organ defects, including reduced liver size, cardiovascular defects, diminished reproductive capabilities, immunosuppression, and epidermal hyperplasia [94–96], suggesting that the AHR has physiological functions other than detoxification and, albeit indirectly, arguing in favor of the existence of AHR endogenous ligands. Activation of detoxification pathways may be an evolutionary adaptation of the vertebrate AHR because invertebrate AHR homologs do not bind exogenous ligands and their functions are more closely related to embryonic development and differentiation. Spineless (ss), the *Drosophila* AHR homolog, is involved in nerve cell differentiation and development of distal segments of leg and antenna [97]. AHR-1, the AHR homolog in the nematode *Caenorhabditis elegans*, regulates the cell fate specification of GABAergic motor neurons [98]. Interestingly, differentiation of GABAergic neurons in the ventral telencephalon also appears to be a function of the murine AHR [99].

In support of a role for AHR in normal physiology, $Ahr^{-/-}$ mice and their wild-type counterparts show widely divergent hepatic gene expression profiles [100]. Sustained AHR activation in mice can also be harmful. Female mice with constitutively active AHR show a ductile bone phenotype, increased weight of multiple organs, atopic dermatitis, immune involution, and digestive tract cancer [101–106], suggesting that an adequate level of AHR activity may be essential for normal development, but that too much may be deleterious. Indeed, studies in cultured cell lines have shown that the AHR has an endogenous role in maintenance of a balance between proliferation and apoptosis, with cell phenotype being a critical determinant of this AHR function [107].

11.2.1 Ancestral AHR Functions

Emerging evidence implicates the AHR not only in response to xenobiotic intoxication and detoxification, but also in developmental and cell proliferation processes. Beyond its role as a mediator of xenobiotic toxicity, the normal role of the AHR in a number of biological processes is just beginning to be recognized and implicates its target genes and signaling pathways in fundamental cell regulatory pathways and developmental processes. The AHR is dispensable but has an important function in controlling the balance among processes involved in cell proliferation, death, and differentiation that, if deregulated, contribute to tumor initiation, promotion, and progression that ultimately lead to malignant tumor formation. In addition, the AHR is involved in regulating many signaling pathways, leading to regulatory alteration of gene expression in a multiplicity of biological processes [108].

11.2.1.1 Endogenous Functions of the AHR Studies with $Ahr^{-/-}$ mice and $Ahr^{-/-}$ cells derived from these mice suggest that the mammalian AHR is involved in a large variety of physiological processes. AHR knockout mice have a reduced liver size and mild portal fibrosis [94, 95] and show *patent ductus venosus*, a rare birth disorder known to happen in humans characterized by the failure to close the fetal hepatic vascular shunt after birth [109]. Failure to close this shunt allows a significant level of blood flow to bypass the portal circulation, causing toxicity and necrosis at the liver periphery. Gestational exposure of AHR hypomorphic mice to dioxin can rescue this vascular defect, suggesting a direct involvement of the AHR in normal vasculogenesis, including the proper processing of the mechanisms responsible for embryonic *ductus venosus* closure [110]. Vascular abnormalities are not limited to the liver, but have also been observed in other organs, particularly eyes and kidneys, of newborn $Ahr^{-/-}$ mice [96]. Consistent with the AHR having also a role in protecting the vascular endothelium against fluid shear stress, AHR-null mice show increased serum levels of oxidized low-density lipoprotein (oxLDL) [111] and have increased levels of liver retinoids, possibly as a result of loss of Cyp2C39 activity, with subsequent decreased metabolism of retinoid acid into 4-hydroxyretinoid acid [112].

Work from our own and from other laboratories has led to the concept that the AHR may function as a tumor suppressor gene that becomes silenced during the process of tumor formation. Epigenetic *AHR* silencing by promoter hypermethylation has recently been reported in a significant number of human acute lymphoblastic leukemia cases [113], whereas *Ahr* repression was also substantial in DEN-induced liver tumors of mice, albeit in the absence of any significant promoter hypermethylation [114]. In addition, AHR-null TRAMP (transgenic adenocarcinoma of the mouse prostate) mice develop prostate tumors with greater frequency than $Ahr^{+/+}$ TRAMP mice, suggesting that the AHR possesses tumor suppressor properties [115]. To determine whether the mouse *Ahr* gene was a tumor suppressor gene *in vivo*, we examined its role on liver tumorigenesis induced by diethylnitrosamine, a hepatic carcinogen that is not an AHR ligand. In mice administered a single i.p. injection of diethylnitrosamine, AHR antagonized liver tumor formation and growth by regulating cell proliferation, inflammatory cytokine expression, and DNA damage, which were significantly elevated in the livers of control, and more so, of diethylnitrosamine-exposed $Ahr^{-/-}$ mice. $Ahr^{-/-}$ hepatocytes also showed significantly higher numbers of 4N cells, increased expression of proliferative markers, and repression of tumor suppressor genes, supporting the conclusion that in its basal state in the absence of a xenobiotic ligand, the *Ahr* gene functions as a tumor suppressor gene and that its silencing may be associated with cancer progression [116]. Consistent with this role as a tumor suppressor, approximately one-third of human acute lymphoblastic leukemia patients and several leukemia-derived cell lines showed hypermethylation of *AHR* promoter CpG islands, which impaired binding of the transcription factor Sp1 and silenced gene expression [113].

Cardiac hypertrophy is an early phenotype of AHR knockout mice [117]. Studies from the Walker lab found that AHR knockout mice develop elevated arterial pressure and showed that these mice have increased levels of angiotensin II, progressing toward overt cardiac hypertrophy [118]. Suggestive of an epigenetic pathogenic mechanism, the cardiac enlargement was dependent on the maternal phenotype, since neonates born from AHR-null female had increased heart weight regardless of their genotype [119]. Possibly as a result of dysregulated estradiol synthesis within follicles and lack of normal levels of aromatase [120, 121], $Ahr^{-/-}$ female mice have reduced fertility and difficulty maintaining conceptuses through successive pregnancies, whereas their $Ahr^{-/-}$ pups show poor survival during lactation and after weaning [122]. Interestingly, *ahr-1* mutant *C. elegans* nematodes display deficits in growth and development including a reduced number of eggs laid and a higher proportion of dead embryos, suggesting that developmental aspects of AHR functions may be evolutionarily conserved [123].

In the adult immune system, the AHR is responsible for maintaining the balance between regulatory T_{reg} and helper T_h17 cells, which is a crucial determinant of an effective immune response and self-tolerance. $Ahr^{-/-}$ mice show numerous hematopoietic defects and an enlarged spleen [94, 124] and their naive T cells lose the ability to differentiate into T_h17 cells when induced by TGFβ and IL6 [125]. These mice also have a reduced number of T_h17 cells upon induction of experimental allergic encephalomyelitis. The dorsal skin of $Ahr^{-/-}$ mice shows severe epidermal hyperplasia accompanied by hyperkeratosis, acanthosis, and dermal fibrosis [95], suggesting that AHR reduces the inflammatory responses in the skin. Indeed, recent work has shown that in the early phase of wound healing the wound area of $Ahr^{-/-}$ mice decreases faster than in their wild-type counterparts, consistent with the finding that $Ahr^{-/-}$ dermal fibroblasts secrete higher levels of active TGFβ that increases keratinocyte migration for faster wound healing [126].

Mice that express a transgenic constitutively active AHR protein were first made by insertion of a truncated AHR cDNA in which the ligand binding domain had been removed. Surprisingly, these mice developed spontaneous stomach tumors [104] and were more likely to develop diethylnitrosamine-induced liver tumors than the wild-type mice [105]. Recent work in these mice has found a gender-specific bone phenotype, with transgenic females, but not males, showing increased osteoclast activity with subsequent elevation of the resorption index [106]. These mice also show atopic dermatitis accompanied by inflammation and immunological imbalance [103], probably because sustained AHR expression also causes thymus involution and suppresses immunization-induced splenocyte proliferation [127] but not Th2 cytokine production [102]. Constitutively active AHR overexpression in MCF-7 human breast cancer cells showed remodeled cytoskeleton, increased interaction with extracellular matrix, and loosening of cell–cell contacts and were correlated with activation of Jun N-terminal kinase [128].

It would seem that a precise homeostatic dose of AHR expression and activity is critical for the maintenance of normal physiological processes, functioning in cell cycle regulation as a tumor suppressor. In general terms, lack of AHR disturbs normal development and results in anomalies in hepatic, cardiovascular, reproductive, and immune systems; on the other hand, excessive AHR expression is carcinogenic.

11.2.1.2 Developmental Consequences of TCDD Exposure
TCDD is the most potent xenobiotic AHR ligand and can induce persistent receptor activation. Its high binding affinity in the midpicomolar range and its metabolic stability allow it to bioaccumulate in exposed organisms. Epidemiological studies in accidentally exposed populations have established a link between exposure to high doses of TCDD and certain types of cancer [129] and cardiovascular disease [130]. In humans, clinical manifestations after dioxin exposure include cancer, developmental abnormalities, reproductive and endocrine pathologies, liver damage, chloracne, and others [131]. Perhaps the best-studied consequence of exposure, chloracne appears to result from AHR-mediated epithelial differentiation from skin stem cells [132] and accelerated keratinocyte terminal differentiation [133]. Lipid accumulation, parenchymal cell necrosis, and infiltration of inflammatory cells were observed in livers of rats exposed to TCDD [134]. In mice, TCDD exposure during embryogenesis causes developmental abnormalities including hydronephrosis and cleft palate [135].

Several studies in mice also demonstrated effects of TCDD on the immune system, including increased inflammatory responses [136], suppression of antigen-specific antibody production [137], and arrest of thymocyte proliferation [138]. Consistent with the role of AHR in the establishment of a balance between T_{reg} and T_h17 cells, TCDD was shown to provide an early signal to activate T-cell proliferation and at the same time block T-cell survival [139], to induce T_{reg} proliferation accompanied by upregulation of TGFβ3, Blimp1, and granzyme B expression [140], and to increase the number of T_h17 cells and their production of cytokines [141].

Early work from our lab in $Apoe^{-/-}$ hyperlipidemic mice showed that TCDD exposure increased the blood levels of triglycerides and urinary excretion of vasoactive eicosanoids, leading to an increase in low-density lipoprotein, earlier onset and greater severity of atherosclerotic lesions, and dysregulation of blood pressure [142]. Extensive work from the Walker lab has uncovered the major impact that TCDD exposure has on ischemic heart disease and cardiac dysfunction. *In utero* exposure to dioxin was found to cause a decrease in fetal heart-to-body ratio and reduced cardiomyocyte proliferation, which translated into postnatal cardiac hypertrophy, showing that the fetal heart is a sensitive target of dioxin [77]. Subsequent studies showed that TCDD exposure during fetal development causes changes in expression of genes involved in cardiac extracellular matrix remodeling, such as matrix metalloproteinases 9 and 13 and preproendothelin-1, cardiac hypertrophy, such as atrial natriuretic factor, β-myosin heavy chain, and osteopontin, and AHR activation, such as *Cyp1a1* and *Ahrr*. All these changes, except for the upregulation of preproET-1, remained induced in the heart of adult offspring [78].

11.2.2 AHR Target Genes and Signaling Pathways

Two broad groups of AHR target genes are generally recognized based on their function [143, 144]. On the one hand, AHR targets a number of genes coding for enzymes involved in ligand metabolism, including the cytochrome P450s Cyp1A1, Cyp1A2, Cyp1B1, Cyp2A5, Cyp2S1, and the aldehyde dehydrogenase 3A1 (ALDH3A1), glutathione S-transferase Ya (GSTA), NAD(P)H–quinone oxidoreductase 1 (NQO1), UDP-glucuronosyltransferase 1A1 (UGT1A1), and UGT1A6 [108]. On the other hand, AHR regulates the expression of genes coding for proteins controlling cell proliferation, apoptosis, differentiation, and inflammation. Among these targets are included the genes coding for many factors involved in cell cycle progression, differentiation, and inflammation [108, 143], which in turn connect the AHR with other signaling pathways, such as those directed by NF-κB, TGFβ, epithelial–mesenchymal transition, and nuclear receptor signaling.

AHR signaling alters crucial aspects of matrix metabolism by regulating the expression of many genes coding for matrix metalloproteinases and other aspects of extracellular matrix (ECM) function, critical for physiological processes that require tissue remodeling and important mediators of cell signaling at the cell membrane [145]. Among the ECM genes that AHR regulates, those coding for the three TGFβ proteins are possibly the best understood. Binding of TGFβ to type II TGFβ receptors with serine–threonine kinase activity triggers TGFβ signaling by phosphorylating type I receptors and inducing downstream intracellular signaling Smad proteins [146]. TGFβ plays a major role in cellular homeostasis, and perturbation of TGFβ signaling is responsible for several diseases in humans [147–149]. TGFβ is secreted as a latent form from the producing cells into the ECM, where it binds to the propeptide LAP and to the latent TGFβ binding protein (LTBP) [150, 151]. $Ltbp1^{-/-}$ mice show altered septation of cardiac outflow tract [152], whereas $Ltbp3^{-/-}$ mice show bone dysfunction, including osteosclerosis and osteoarthritis [153], and $Ltbp4^{-/-}$ was associated with cardiomyopathy, defective lung development, and colorectal cancer [154].

Studies with AHR-deficient mice have shown a role for the AHR in repression of TGFβ expression both in vivo and in vitro. Livers from $Ahr^{-/-}$ mice and primary embryo fibroblasts (MEF) from the same mice showed increased secretion and expression of latent and active TGFβ and reduced cell proliferation levels [155, 156], possibly as a result of their LTBP-1 overexpression [157]. Wild-type MEF express low constitutive levels of LTBP-1 through the AHR/ARNT-dependent cross-linking of histone deacetylase 2 (HDAC2) to the promoter chromatin of the $Ltbp1$ gene, thus preventing the recruitment of the histone acetyltransferase coactivator CBP/p300 [158]. In $Ahr^{-/-}$ MEF, decreased recruitment of HDAC2 and increased binding of pCREB to the $Ltbp1$ promoter result in its overexpression. Consistent with AHR controlling constitutive LTBP-1 expression, TGFβ3 and various proteins involved in TGFβ signaling have been shown to be overexpressed in smooth muscle cells from aorta of $Ahr^{-/-}$ mice [159].

The estrogen receptors are likely partners that interact with the AHR in ECM functions such as mammary gland, reproductive tract, and vasculature remodeling [160–162]. Several estradiol-activated genes, including cytokeratin 18, $Tgf\alpha$, $Hsp27$, and $Cfos$, could also be repressed by TCDD treatment [163]. ERα–AHR interactions were recently demonstrated in MCF-7 human breast cancer cells with or without ligand activation [164–166]. ER-mediated mammary gland development is AHR signaling dependent since high levels of AHR protein were observed in mammary glands during development and $Ahr^{-/-}$ mice mammary glands had fewer terminal end buds and altered shapes [167, 168].

Unrelated to its interactions with steroid hormone receptors in other tissues, the AHR has a central signaling role in prostate growth, morphogenesis, and disease at various stages of development and during normal and aberrant prostate growth, regulating timing and patterning of prostate development. Inappropriate activation of AHR signaling by TCDD exposure during development disrupts the balance of these signals, impairs prostate morphogenesis, and has an imprinting effect on the developing prostate that predisposes to prostate disease and sets prostate aging in adulthood. Interaction with several signaling pathways, of which WNT, FGF10, and VEGF are key, seems to be responsible for the effect of the AHR–TCDD axis on differentiation, proliferation, and angiogenesis in the prostate [169]. Interestingly, AHR–WNT crosstalk has been shown to be critical in zebrafish tissue regeneration [170] and in neuronal development in the Baltic salmon [171].

During the past 10 years, evidence indicative of a close interaction between AHR and NF-κB pathways suggests that crosstalk between these two transcription factors may regulate physiological and pathological processes, including immune and inflammatory responses, carcinogenesis, and alterations of xenobiotic metabolism and disposition [172]. Studies in mice showed that nonspecific stimulation of the immune system of the dam protected the embryos against TCDD-induced cleft palate [173], whereas synergistic induction of cleft palate was observed in mice that received simultaneous NF-κB suppression and TCDD-dependent AHR activation [174]. Direct interactions between AHR and the RelA subunit of NF-κB have been found to be mutually repressive [175] as well as to cooperate in the transactivation of the c-Myc promoter [176]. There has been no satisfactory resolution to this discrepancy, but NF-κB and AHR share similar coregulator pool [158, 177] and hence may compete for coactivators. In addition, NF-κB activation was found to block AHR-dependent gene transactivation by inhibiting histone acetylation without interfering with AHR binding to the DNA [178]. Interactions between NF-κB and AHR may also happen at the transcriptional elongation level; TNFα-activated NF-κB was shown to inhibit TCDD-induced phosphorylation of serine 2 in the C-terminal domain of RNA polymerase II in mouse liver cells [179].

11.2.3 Regulation of AHR Expression

The AHR has been identified in all vertebrate species examined so far; their interspecies sequence similarity strongly indicates that the receptor structure has been well conserved during evolution [180]. The AHR protein is composed of two PAS (Per–ARNT–Sim) domains (PAS-A and PAS-B), a bHLH (basic region–helix–loop–helix) domain at its N-terminal end and a transactivation domain with a glutamine-rich subdomain located at the C-terminal end [181] responsible for protein–protein interactions with components of the basal transcription machinery and specific coregulators [182, 183]. DNA binding takes place within the bHLH sequence and heterodimerization with ARNT via the PAS-A and bHLH domains further strengthens the DNA binding affinity of the complex [184]. Amino acids 230–421, comprising the PAS-B domain, constitute the minimal ligand binding region [185, 186]. Different mouse strains show great differences in ligand binding affinity, resulting from amino acid differences in the PAS-B domain. Four alleles have been identified in mouse, of which the Ahr^d allele, which encodes an AHR protein with low ligand affinity, differs from the Ahr^b allele, which encodes an AHR protein with high ligand affinity by a valine to alanine substitution at residue 375 [187, 188].

The mammalian AHR gene promoter lacks TATA and CCAAT boxes, but it has multiple GC-rich sites, associated with transcription factor Sp1 binding [189–191], a characteristic often found in housekeeping genes. Many other transcription factor binding sites have been identified in the 5′-flanking region [191, 192].

AHR expression can be detected in mouse embryos as early as the preimplantation stage [193]. RT-PCR analysis revealed a high level of AHR mRNA at the one-cell stage, a decrease at the two- to eight-cell stages, followed by an increase at the blastocyst stage [194]. This expression pattern parallels the epigenetic reprogramming of gene expression during early mammalian development, suggesting that the AHR may be implicated in early cell differentiation. Such a role for the AHR is supported by the fact that embryos cultured in the presence of antisense AHR oligodeoxynucleotides have fewer cell numbers and develop damaged blastocysts [193]. Other studies investigating AHR expression during embryo gestation revealed the tissue-specific expression of AHR as early as embryonic day 9.5 (E9.5). By E10.5, AHR was detected in neuroepithelium, branchial arches, heart, somites, and liver [195, 196]. These studies also showed an age-specific AHR expression pattern with expression being maximal at E10.5–11.5 in the heart, decreasing with age thereafter and again suggesting an AHR involvement in early development. Interestingly, mice embryos become responsive to TCDD treatment only at the blastocyst stage, as shown by the induction of CYP1A1 mRNA [194]. Furthermore, exposure of preimplantation embryos to TCDD tended to decrease the expression levels of the imprinted genes H19 and Igf2 (insulin-like growth factor 2 gene) [197]. Studies at later embryonic times showed constitutive and TCDD-inducible tissue-specific expression of CYP1A1 during development, including heart and hindbrain (E8.5–14.5), kidney (E9.5–14.5), and muscle (E13.5–14.5) [198]. These data suggest the presence of endogenous ligand(s) at specific embryo stages, with embryos sensitive to xenobiotic impact during specific developmental windows.

11.2.4 Epigenetic Regulation of AHR Expression and Function

The mechanisms that regulate AHR expression at the transcriptional level are still largely unknown, although expression data during early embryonic and postnatal development have revealed various elements of epigenetic regulation. It was first observed that constitutive AHR expression in culture was suppressed by the histone deacetylase 1 (HDAC1) inhibitors n-butyrate and trichostatin A (TSA), indicating that histone acetylation was critical for the transcriptional activation of the Ahr promoter [199]. More recent developmental studies in mouse liver showed a doubling in Ahr mRNA expression between gestational day 19 and postnatal day 45 [200]. This modest hepatic upregulation did not change significantly the extent of Ahr promoter methylation or associated trimethylation of H3K27, but correlated with an increase of H3K4me2, although the meaning of this increase could not be assessed properly since neither the monomethylated nor the trimethylated forms were measured [200]. Consistent with the possible function in early development and differentiation discussed earlier, the Ahr promoter in pluripotent mouse stem cells is marked by H3K4me3/H3K27me3 bivalent domains [1].

Early in vitro work with AHR binding sequence motifs methylated at the cytosine residues showed that enhancer function, as demonstrated by AHR binding in electrophoretic mobility assays, was inhibited by methylation, suggesting that the in vivo response to AHR activation by ligand could be diminished if the cognate binding site was methylated [201]. This was in fact the case, and cell-type-specific transcriptional induction of the rabbit Cyp1a1 gene was found to be dependent on DNA methylation, with silent lung cells showing extensive hypermethylation, absent in kidney cells [202]. In contrast, hypomethylation of the promoter/enhancer region of the human CYP1B1 gene was found in human prostate cancer tissue but not in benign prostatic hyperplasia and was associated with CYP1B1 overexpression and risk of prostate cancer [203]. Recently, the methylation status of CpG islands in the 5′-flanking region of the CYP1A1 and CYP1B1 genes in 7 colorectal cancer cell lines and 40 primary colorectal cancers was probed to determine whether epigenetic mechanisms were responsible for the expression differences of these two AHR-regulated genes. In contrast to the hypomethylation observed in prostate cancer, a significant level of hypermethylation was observed in the *CYP1B1* but not in the *CYP1A1* promoter, with no change in AHR or ARNT expression [204].

Epigenetic mechanisms operative at the promoter of AHR target genes play a fundamental role in the control of their expression by the AHR, a property that has fundamental consequences on the balance between CYP1A1 and CYP1B1 expression, which is maintained by a diversity of epigenetic mechanisms that do not seem to have a common denominator for either gene. Thus, in the human breast cancer cell line MCF-7, *CYP1A1* is inducible to a greater degree than *CYP1B1* by dioxin activation of the AHR, concomitantly with its recruitment to the enhancer but not to the proximal promoter region of these genes; the converse is true for RNA polymerase II. Several coactivators are recruited with equal efficiency to the 5′-flanking regions of both genes, except for the BRM/BRG-1 subunit of the nucleosome remodeling factor, which is consistently recruited to *CYP1A1* more efficiently than to *CYP1B1*, suggesting that nucleosome remodeling is a significant determinant of the AHR-dependent regulation of some of its target genes, but not others [205]. In contrast to MCF-7 cells, only CYP1A1 is inducible in human hepatoma HepG2 cells because partial methylation of the *CYP1B1* enhancer and full methylation of its proximal promoter block RNA polymerase II and TATA binding protein recruitment, the establishment of acetylation marks in histones H3 and H4 and methylation marks in H4 [206].

Unlike MCF-7 cells, the mouse hepatoma cell line Hepa 1c1c7 expresses high dioxin-inducible levels of CYP1A1 mRNA but low or little CYP1B1, whereas the opposite is the case in C3H10T1/2 mouse embryonic fibroblast cell line. In this case, it appears that the *Cyp1a1* repression in fibroblasts is due to DNA hypermethylation and histone modifications, since it can be reversed by DNMT1 inhibitor 5-aza-2′-deoxycytidine and by treatment with the HDAC1 inhibitor TSA. In contrast, *Cyp1b1* repression could not be overcome by either of these treatments, suggesting that epigenetics plays a minor role in the silencing of *Cyp1b1* in Hepa 1c1c7 cells [207]. Differences in HDAC levels associated with the promoters of the human and the mouse *CYP1A1* gene have also been proposed to account for the difference in expression levels of these two genes in human HepG2 relative to mouse Hepa 1c1c7 hepatoma cells [208].

Diverse epigenetic mechanisms regulate a concerted series of chromatin remodeling events responsible for the initiation, completion, and sustained continuation of the transcriptional regulatory processes responsible for AHR-dependent gene transactivation. The objective of remodeling chromatin is to make it more accessible to transcription factors, coregulators, and the transcription machinery itself by relaxing the chromatin, repositioning the nucleosomes, and facilitating the recruitment of the RNA polymerase II transcription initiation complex. The 5′-flanking region of the mouse *Cyp1a1* gene was used as the target to study chromatin recruitment of the AHR and of the multiple factors and modified histones associated with gene transcription. Both coactivators and corepressors of transcription were shown in early work to bind to the AHR–ARNT complex. The estrogen receptor-associated protein 140 (ERAP 140) coactivator and SMRT, the silencing mediator for retinoic acid and thyroid hormone receptor, were both found to coimmunoprecipitate with AHR–ARNT and to enhance or inhibit, respectively, AHR-mediated gene expression [209]. Early work from the Hankinson laboratory showed that a number of members of the NCoA/SRC-1 family of transcriptional coactivators were recruited to the AHR–ARNT complex and were capable of independently enhancing TCDD-dependent gene induction. These coactivators, which have intrinsic histone acetyltransferase activity, could also recruit p300/CEBP, also with HAT activity, to the AHR transcriptional complex [210].

Other epigenetic mechanisms that do not involve histone acetylation are performed by the mammalian SWI/SNF ATP-dependent chromatin remodeling complexes, which contain a number of different protein subunits, of which the Brahma-related gene 1 (BRG-1) or Brahma itself perform the essential ATPase function. Chromatin remodeling by BRG-1 was found to potentiate TCDD-dependent AHR–ARNT-mediated transactivation by associating with the enhancer region of the 5′-flanking domain of AHR target genes and with the Q-rich domain of the AHR [211]. The TRAP/DRIP/ARC/mediator complex, known to provide cofactors essential for *in vitro* transcription, was also found to be able to associate with the AHR and with p300 at the enhancer of the *Cyp1a1* gene and to increase AHR-dependent transcription [212].

The successful conclusion of the initial steps to activate transcription is the recruitment of RNA polymerase II and all its associated cofactors. Treatment of MCF-7 cells with β-naphthoflavone, a common AHR agonist, led to rapid promoter association of the AHR, p160 coactivators, and p300 HAT, followed immediately by RNA polymerase II recruitment, whereas 3,3′-diindolylmethane, a selective AHR modulator ligand, failed to recruit RNA polymerase II or induce histone acetylation. Interestingly, promoter association of all these components was maximal at 30 min after treatment and cycled on and off the promoter with a 60 min periodicity [213]. In contrast, treatment of mouse hepatoma Hepa 1c1c7 cells with benzo[*a*]pyrene led to maximal AHR and RNA polymerase II promoter recruitment after 90 min and did not cycle [214]. Possibly, ligand, species, or tissue of origin differences account for the different kinetics of activation. Early work from our laboratory showed that ligand-mediated recruitment of AHR/ARNT complexes to the *Cyp1a1* enhancer domain and concomitant recruitment of p300 and gene induction required the removal of constitutively bound HDAC1 from the proximal promoter domain [215]. In a follow-up to that work, we observed that the bound HDAC1 was complexed to DNMT1 in the *Cyp1a1* promoter of uninduced cells and that the complex prevented the entry of RNA polymerase II into the proximal promoter and the establishment of histone acetylation marks specific to the induced state, without interfering with the kinetics of AHR–ARNT DNA binding at the enhancer domain. HDAC1 and DNMT1 inhibitors or depletion of HDAC1 or DNMT1 with siRNAs decreased the interaction of these proteins with the *Cyp1a1* promoter but were not sufficient to induce gene expression, although they allowed hyperacetylation marks at H3K14 and H4K16 to be established at levels similar to those found in B[*a*]P-induced cells, suggesting that the *Cyp1a1* gene is under active repression before its ligand-dependent induction. In addition to HDAC1–DNMT1 removal, *Cyp1a1* induction is associated with modification of specific chromatin marks, including hyperacetylation of histone H3K14 and H4K16, trimethylation of histone H3K4, and phosphorylation of H3S10. These results show that by blocking modification of histone marks, HDAC1 plays a central role in *Cyp1a1* expression and that its removal is a necessary but not sufficient condition for *Cyp1a1* induction [214, 216].

Not only is chromosome remodeling a determinant of AHR function, but there is growing evidence that the converse is also true and that AHR activity can reshape chromatin structure, including chromosome positioning. Thus, in primary human keratinocytes, TCDD-mediated AHR activation attenuates senescence and represses the transcription of the tumor suppressors p[16INK4a] and p53 by promoter methylation. As a consequence, TCDD treatment alone is sufficient to immortalize normal human keratinocytes, suggesting possible mechanisms by which TCDD may contribute to human malignancies [217]. Although no direct evidence of DNA methylation changes was collected in this work, a mixture of three non-*ortho* PCBs, six PCDDs, and seven PCDFs caused a significant decrease in *Dnmt1* expression in rat liver and hypothalamus and a significant elevation in brain, supporting the possibility that exposure to these environmental mixture of AHR agonists could lead to effects resulting from changes in DNA methylation [218]. Interestingly, data from our laboratory using several coplanar and noncoplanar PCBs with different toxic equivalence factors (TEF) show significant differences in histone mark establishment and kinetics of AHR recruitment that correlate with the TEF of each compound (Ovesen, Schnekenburger, and Puga, manuscript submitted for publication). Another PCB, the 3,3′,4,4′-pentachlorobiphenyl, was responsible for differential gene expression patterns between rat glioma and hepatoma cells, associated with tissue-selective histone deacetylase inhibition in glioma, but not in hepatoma cells [219].

In the interphase nucleus, each chromosome occupies its own position, known as its "territory," which is tissue specific and correlated with differentiation state. Activation of the AHR by dioxin was also shown to affect chromosome territory distribution. In human preadipocyte cells, treatment with dioxin was shown to enlarge the minimum distance between chromosome 12 and chromosome 16, which might have important consequences for the mechanisms of epigenetic control of interchromosomal interactions [220]. Epigenetic mechanisms have also been proposed to explain the apparent crosstalk of AHR and NF-κB signaling pathways. Studies on modifications in histone H4 have revealed an interplay between arginine acetylation and methylation marks at H4R3 and shown that TCDD treatment causes an increase in H4K5 acetylation and a reduction in H4R3 methylation marks, while TNFα treatment has the opposite effect, suggesting that protein arginine methyltransferase 1 (PRMT1), responsible for the methylation mark, is a coregulator of AHR-dependent gene activation, potentially explaining the seemingly diametrically opposed dynamic interplay between AHR and NF-κB interactions [172].

11.3 CONCLUSION

Traditionally, the biological consequences of Ah receptor activation have been ascribed to the toxicity of its ligands; however, it appears that the ancestral function of this receptor is not involved in the response to toxic agents, but is the regulation of specific aspects of embryonic development, having acquired the ability to bind xenobiotic compounds only during vertebrate evolution [180]. Hence, it is not unexpected to find that the AHR has diverse developmental functions. In this regard, the histone marks elicited by the binding of AHR to chromatin are the same modifications associated with the bivalent developmental marks in differentiating ES cells (see above), which cluster within regions enriched for genes encoding developmentally important homeobox transcription factors. These chromatin modifications induced by AHR may play an important role in the activation (or silencing) of developmentally regulated genes, critical for the concerted progress of differentiation programs, opening up the interesting possibility that exposure to AHR ligands during development may derail developmental patterns.

We expect (and hope) that knowledge of the range of genes directly regulated by the AHR would facilitate integration at a systems biology level of the complex physiological mechanisms that this receptor regulates both endogenously and in response to toxic ligands. Mapping the network of AHR binding targets in the mouse genome through a multipronged approach involving ChIP/chip and global gene expression signatures led to the finding that the naïve receptor bound in unstimulated cells to an extensive array of gene clusters with functions in regulation of gene expression, differentiation, and pattern specification, connecting multiple morphogenetic and developmental programs. Activation by either of two ligands, TCDD or B[a]P, displaced the receptor from some of these targets toward sites in the promoters of xenobiotic metabolism genes. Three unexpected findings stood out from the analysis of these results. First was the large number of gene promoter regions that showed significant AHR binding in naïve, unstimulated cells, suggestive of a physiological role for the receptor in the absence of an exogenous ligand; second was the large proportion of top-ranked genes in naïve cells that have transcriptional regulatory functions, including homeodomain genes with a morphogenetic role in development; third was the large fraction of top-ranked AHR binding genes in ligand-treated cells that responded to one ligand but not the other and *vice versa* [221]. Diversity of ligand responses and possibly of species-specific responses was made more evident in a similar recent genome-wide analysis of T47D human breast cancer cells treated with 3-methylcholanthrene [222], which shared very few common genes with our analyses of mouse hepatoma cells.

It would be reasonable to conclude that the AHR possesses a set of endogenous functions that are independent of its activation as a transcription factor by xenobiotic ligands and that these functions may perform a regulatory role during embryonic development, suggesting that gene–environment interactions during fetal life may be potential triggers of developmental abnormalities.

One of the most critical issues of environmental health research today is whether environmental exposures during development are critical for environmental disease susceptibility in the adult, and if so, what are the causative mechanistic connections between environmental exposure and disease. The findings discussed in this chapter may have serious implications on the etiology of diseases caused by AHR ligands, since any observed changes in gene regulation resulting from exposure might be not due to new regulatory units being recognized by the activated Ah receptor, but due to loss of regulation of other units preexisting in the unexposed organism. Further studies in this area will provide a unique resource to advance our ability to establish needed gene–environment connections between environmental disease and disturbances of developmental gene regulation patterns at the level of the whole genome. The molecular basis of adult environmental disease might be rooted in exposures occurring during fetal life; if this were the case, understanding adult environmental diseases will require the synergistic interaction between toxicology and developmental biology.

ACKNOWLEDGMENTS

The research in the authors' laboratory is supported by National Institutes of Environmental Health Sciences grants R01 ES06273, R01 ES10807, and P30 ES06096. C. K. was supported in part by Recherches Scientifiques Luxembourg asbl.

REFERENCES

1. Mikkelsen, T. S., Ku, M., Jaffe, D. B., Issac, B., Lieberman, E., Giannoukos, G., Alvarez, P., Brockman, W., Kim, T. K., Koche, R. P., Lee, W., Mendenhall, E., O'Donovan, A., Presser, A., Russ, C., Xie, X., Meissner, A., Wernig, M., Jaenisch, R., Nusbaum, C., Lander, E. S., and Bernstein, B. E. (2007). Genome-wide maps of chromatin state in pluripotent and lineage-committed cells. *Nature*, 448, 553–560.
2. Ringrose, L. and Paro, R. (2004). Epigenetic regulation of cellular memory by the Polycomb and Trithorax group proteins. *Annual Review of Genetics*, 38, 413–443.
3. Mohn, F., Weber, M., Rebhan, M., Roloff, T. C., Richter, J., Stadler, M. B., Bibel, M., and Schubeler, D. (2008). Lineage-specific polycomb targets and *de novo* DNA methylation define restriction and potential of neuronal progenitors. *Molecular Cell*, 30, 755–766.
4. Gal-Yam, E. N., Egger, G., Iniguez, L., Holster, H., Einarsson, S., Zhang, X., Lin, J. C., Liang, G., Jones, P. A., and Tanay, A. (2008). Frequent switching of Polycomb repressive marks and DNA hypermethylation in the PC3 prostate cancer cell line. *Proceedings of the National Academy of Sciences of the United States of America*, 105, 12979–12984.
5. Margueron, R. and Reinberg, D. (2010). Chromatin structure and the inheritance of epigenetic information. *Nature Reviews in Genetics*, 11, 285–296.
6. Rideout, W. M., III, Hochedlinger, K., Kyba, M., Daley, G. Q., and Jaenisch, R. (2002). Correction of a genetic defect by nuclear transplantation and combined cell and gene therapy. *Cell*, 109, 17–27.
7. Eggan, K., Akutsu, H., Loring, J., Jackson-Grusby, L., Klemm, M., Rideout, W. M., III, Yanagimachi, R., and Jaenisch, R. (2001). Hybrid vigor, fetal overgrowth, and viability of mice derived by nuclear cloning and tetraploid embryo complementation. *Proceedings of the National Academy of Sciences of the United States of America*, 98, 6209–6214.
8. Ng, R. K. and Gurdon, J. B. (2005). Epigenetic memory of active gene transcription is inherited through somatic cell nuclear transfer. *Proceedings of the National Academy of Sciences of the United States of America*, 102, 1957–1962.
9. Kohda, T., Inoue, K., Ogonuki, N., Miki, H., Naruse, M., Kaneko-Ishino, T., Ogura, A., and Ishino, F. (2005). Variation in gene expression and aberrantly regulated chromosome regions in cloned mice. *Biology of Reproduction*, 73, 1302–1311.
10. Takahashi, K. and Yamanaka, S. (2006). Induction of pluripotent stem cells from mouse embryonic and adult fibroblast cultures by defined factors. *Cell*, 126, 663–676.
11. Mikkelsen, T. S., Hanna, J., Zhang, X., Ku, M., Wernig, M., Schorderet, P., Bernstein, B. E., Jaenisch, R., Lander, E. S.,

Meissner, A. (2008). Dissecting direct reprogramming through integrative genomic analysis. *Nature, 454*, 49–55.

12. Henikoff, S. (2000). Heterochromatin function in complex genomes. *Biochimica et Biophysica Acta, 1470*, O1–O8.
13. Richards, E. J. and Elgin, S. C. (2002). Epigenetic codes for heterochromatin formation and silencing: rounding up the usual suspects. *Cell, 108*, 489–500.
14. Trojer, P. and Reinberg, D. (2007). Facultative heterochromatin: is there a distinctive molecular signature? *Molecular Cell, 28*, 1–13.
15. Lee, H., Habas, R., and Abate-Shen, C. (2004). MSX1 cooperates with histone H1b for inhibition of transcription and myogenesis. *Science, 304*, 1675–1678.
16. Muller, H. J. and Altenburg, E. (1930). The frequency of translocations produced by X-rays in *Drosophila. Genetics, 15*, 283–311.
17. Bannister, A. J., Zegerman, P., Partridge, J. F., Miska, E. A., Thomas, J. O., Allshire, R. C., and Kouzarides, T. (2001). Selective recognition of methylated lysine 9 on histone H3 by the HP1 chromo domain. *Nature, 410*, 120–124.
18. Kouzarides, T. (2007). Chromatin modifications and their function. *Cell, 128*, 693–705.
19. Schotta, G., Ebert, A., and Reuter, G. (2003). SU(VAR)3-9 is a conserved key function in heterochromatic gene silencing. *Genetica, 117*, 149–158.
20. Ferguson-Smith, A. C. and Surani, M. A. (2001). Imprinting and the epigenetic asymmetry between parental genomes. *Science, 293*, 1086–1089.
21. Jones, P. L., Veenstra, G. J., Wade, P. A., Vermaak, D., Kass, S. U., Landsberger, N., Strouboulis, J., and Wolffe, A. P. (1998). Methylated DNA and MeCP2 recruit histone deacetylase to repress transcription. *Nature Genetics, 19*, 187–191.
22. Epsztejn-Litman, S., Feldman, N., Abu-Remaileh, M., Shufaro, Y., Gerson, A., Ueda, J., Deplus, R., Fuks, F., Shinkai, Y., Cedar, H., and Bergman, Y. (2008). De novo DNA methylation promoted by G9a prevents reprogramming of embryonically silenced genes. *Nature Structural & Molecular Biology, 15*, 1176–1183.
23. Lehnertz, B., Ueda, Y., Derijck, A. A., Braunschweig, U., Perez-Burgos, L., Kubicek, S., Chen, T., Li, E., Jenuwein, T., and Peters, A. H. (2003). Suv39h-mediated histone H3 lysine 9 methylation directs DNA methylation to major satellite repeats at pericentric heterochromatin. *Current Biology, 13*, 1192–1200.
24. Heard, E. (2005). Delving into the diversity of facultative heterochromatin: the epigenetics of the inactive X chromosome. *Current Opinions in Genetics and Development, 15*, 482–489.
25. Costanzi, C. and Pehrson, J. R. (1998). Histone macroH2A1 is concentrated in the inactive X chromosome of female mammals. *Nature, 393*, 599–601.
26. Bernstein, B. E., Mikkelsen, T. S., Xie, X., Kamal, M., Huebert, D. J., Cuff, J., Fry, B., Meissner, A., Wernig, M., Plath, K., Jaenisch, R., Wagschal, A., Feil, R., Schreiber, S. L., and Lander, E. S. (2006). A bivalent chromatin structure marks key developmental genes in embryonic stem cells. *Cell, 125*, 315–326.
27. Boyer, L. A., Plath, K., Zeitlinger, J., Brambrink, T., Medeiros, L. A., Lee, T. I., Levine, S. S., Wernig, M., Tajonar, A., Ray, M. K., Bell, G. W., Otte, A. P., Vidal, M., Gifford, D. K., Young, R. A., and Jaenisch, R. (2006). Polycomb complexes repress developmental regulators in murine embryonic stem cells. *Nature, 441*, 349–353.
28. Schwartz, Y. B., Kahn, T. G., Nix, D. A., Li, X. Y., Bourgon, R., Biggin, M., and Pirrotta, V. (2006). Genome-wide analysis of Polycomb targets in *Drosophila melanogaster. Nature Genetics, 38*, 700–705.
29. Min, J., Zhang, Y., and Xu, R. M. (2003). Structural basis for specific binding of Polycomb chromodomain to histone H3 methylated at Lys 27. *Genes & Development, 17*, 1823–1828.
30. Levine, S. S., Weiss, A., Erdjument-Bromage, H., Shao, Z., Tempst, P., and Kingston, R. E. (2002). The core of the polycomb repressive complex is compositionally and functionally conserved in flies and humans. *Molecular and Cellular Biology, 22*, 6070–6078.
31. Muller, J., Hart, C. M., Francis, N. J., Vargas, M. L., Sengupta, A., Wild, B., Miller, E. L., O'Connor, M. B., Kingston, R. E., and Simon, J. A. (2002). Histone methyltransferase activity of a *Drosophila* Polycomb group repressor complex. *Cell, 111*, 197–208.
32. Muller, J. and Kassis, J. A. (2006). Polycomb response elements and targeting of Polycomb group proteins in *Drosophila. Current Opinions in Genetics and Development, 16*, 476–484.
33. Kahn, T. G., Schwartz, Y. B., Dellino, G. I., and Pirrotta, V. (2006). Polycomb complexes and the propagation of the methylation mark at the *Drosophila* ubx gene. *Journal of Biological Chemistry, 281*, 29064–29075.
34. Bracken, A. P., Dietrich, N., Pasini, D., Hansen, K. H., and Helin, K. (2006). Genome-wide mapping of Polycomb target genes unravels their roles in cell fate transitions. *Genes & Development, 20*, 1123–1136.
35. Caretti, G., Di, P. M., Micales, B., Lyons, G. E., and Sartorelli, V. (2004). The Polycomb Ezh2 methyltransferase regulates muscle gene expression and skeletal muscle differentiation. *Genes & Development, 18*, 2627–2638.
36. Endoh, M., Endo, T. A., Endoh, T., Fujimura, Y., Ohara, O., Toyoda, T., Otte, A. P., Okano, M., Brockdorff, N., Vidal, M., and Koseki, H. (2008). Polycomb group proteins Ring1A/B are functionally linked to the core transcriptional regulatory circuitry to maintain ES cell identity. *Development, 135*, 1513–1524.
37. Ku, M., Koche, R. P., Rheinbay, E., Mendenhall, E. M., Endoh, M., Mikkelsen, T. S., Presser, A., Nusbaum, C., Xie, X., Chi, A. S., Adli, M., Kasif, S., Ptaszek, L. M., Cowan, C. A., Lander, E. S., Koseki, H., and Bernstein, B. E. (2008). Genomewide analysis of PRC1 and PRC2 occupancy identifies two classes of bivalent domains. *PLoS Genetics, 4*, e1000242.
38. Dellino, G. I., Schwartz, Y. B., Farkas, G., McCabe, D., Elgin, S. C., and Pirrotta, V. (2004). Polycomb silencing blocks transcription initiation. *Molecular Cell, 13*, 887–893.
39. Pavri, R., Zhu, B., Li, G., Trojer, P., Mandal, S., Shilatifard, A., and Reinberg, D. (2006). Histone H2B monoubiquitination

functions cooperatively with FACT to regulate elongation by RNA polymerase II. *Cell*, *125*, 703–717.

40. Stock, J. K., Giadrossi, S., Casanova, M., Brookes, E., Vidal, M., Koseki, H., Brockdorff, N., Fisher, A. G., and Pombo, A. (2007). Ring1-mediated ubiquitination of H2A restrains poised RNA polymerase II at bivalent genes in mouse ES cells. *Nature Cell Biology*, *9*, 1428–1435.

41. Steward, M. M., Lee, J. S., O'Donovan, A., Wyatt, M., Bernstein, B. E., and Shilatifard, A. (2006). Molecular regulation of H3K4 trimethylation by ASH2L, a shared subunit of MLL complexes. *Nature Structural & Molecular Biology*, *13*, 852–854.

42. Guenther, M. G., Jenner, R. G., Chevalier, B., Nakamura, T., Croce, C. M., Canaani, E., and Young, R. A. (2005). Global and Hox-specific roles for the MLL1 methyltransferase. *Proceedings of the National Academy of Sciences of the United States of America*, *102*, 8603–8608.

43. Poux, S., Horard, B., Sigrist, C. J., and Pirrotta, V. (2002). The *Drosophila* trithorax protein is a coactivator required to prevent re-establishment of polycomb silencing. *Development*, *129*, 2483–2493.

44. Klymenko, T. and Muller, J. (2004). The histone methyltransferases Trithorax and Ash1 prevent transcriptional silencing by Polycomb group proteins. *EMBO Reproduction*, *5*, 373–377.

45. Wysocka, J., Swigut, T., Xiao, H., Milne, T. A., Kwon, S. Y., Landry, J., Kauer, M., Tackett, A. J., Chait, B. T., Badenhorst, P., Wu, C., and Allis, C. D. (2006). A PHD finger of NURF couples histone H3 lysine 4 trimethylation with chromatin remodelling. *Nature*, *442*, 86–90.

46. De, S. F., Totaro, M. G., Prosperini, E., Notarbartolo, S., Testa, G., and Natoli, G. (2007). The histone H3 lysine-27 demethylase Jmjd3 links inflammation to inhibition of polycomb-mediated gene silencing. *Cell*, *130*, 1083–1094.

47. Lan, F., Bayliss, P. E., Rinn, J. L., Whetstine, J. R., Wang, J. K., Chen, S., Iwase, S., Alpatov, R., Issaeva, I., Canaani, E., Roberts, T. M., Chang, H. Y., and Shi, Y. (2007). A histone H3 lysine 27 demethylase regulates animal posterior development. *Nature*, *449*, 689–694.

48. Lee, M. G., Villa, R., Trojer, P., Norman, J., Yan, K. P., Reinberg, D., Di, C. L., and Shiekhattar, R. (2007). Demethylation of H3K27 regulates polycomb recruitment and H2A ubiquitination. *Science*, *318*, 447–450.

49. Agger, K., Cloos, P. A., Christensen, J., Pasini, D., Rose, S., Rappsilber, J., Issaeva, I., Canaani, E., Salcini, A. E., and Helin, K. (2007). UTX and JMJD3 are histone H3K27 demethylases involved in HOX gene regulation and development. *Nature*, *449*, 731–734.

50. Shen, X., Liu, Y., Hsu, Y. J., Fujiwara, Y., Kim, J., Mao, X., Yuan, G. C., and Orkin, S. H. (2008). EZH1 mediates methylation on histone H3 lysine 27 and complements EZH2 in maintaining stem cell identity and executing pluripotency. *Molecular Cell*, *32*, 491–502.

51. Nicholson, J. M., Wood, C. M., Reynolds, C. D., Brown, A., Lambert, S. J., Chantalat, L., and Baldwin, J. P. (2004). Histone structures: targets for modifications by molecular assemblies. *Annals of the New York Academy of Science*, *1030*, 644–655.

52. Valls, E., Sanchez-Molina, S., and Martinez-Balbas, M. A. (2005). Role of histone modifications in marking and activating genes through mitosis. *Journal of Biological Chemistry*, *280*, 42592–42600.

53. Clayton, A. L. and Mahadevan, L. C. (2003). MAP kinase-mediated phosphoacetylation of histone H3 and inducible gene regulation. *FEBS Letters*, *546*, 51–58.

54. Mujtaba, S., Zeng, L., and Zhou, M. M. (2007). Structure and acetyl-lysine recognition of the bromodomain. *Oncogene*, *26*, 5521–5527.

55. Sanchez, R. and Zhou, M. M. (2009). The role of human bromodomains in chromatin biology and gene transcription. *Current Opinions in Drug Discovery & Developments*, *12*, 659–665.

56. Hassan, A. H., Neely, K. E., and Workman, J. L. (2001). Histone acetyltransferase complexes stabilize SWI/SNF binding to promoter nucleosomes. *Cell*, *104*, 817–827.

57. Li, S. and Shogren-Knaak, M. A. (2009). The Gcn5 bromodomain of the SAGA complex facilitates cooperative and cross-tail acetylation of nucleosomes. *Journal of Biological Chemistry*, *284*, 9411–9417.

58. Chen, J., and Ghazawi, F. M., Li, Q. (2010). Interplay of bromodomain and histone acetylation in the regulation of p300-dependent genes. *Epigenetics*, *5*, 509–515.

59. Pampal, A. (2010). CHARGE: an association or a syndrome? *International Journal of Pediatric Otorhinolaryngology*, *74*, 719–722.

60. Cloos, P. A., Christensen, J., Agger, K., and Helin, K. (2008). Erasing the methyl mark: histone demethylases at the center of cellular differentiation and disease. *Genes & Development*, *22*, 1115–1140.

61. Reik, W., Santos, F., Mitsuya, K., Morgan, H., and Dean, W. (2003). Epigenetic asymmetry in the mammalian zygote and early embryo: relationship to lineage commitment? *Philosophical Transactions of the Royal Society of London B, Biological Sciences*, *358*, 1403–1409.

62. Morgan, H. D., Santos, F., Green, K., Dean, W., and Reik, W. (2005). Epigenetic reprogramming in mammals. *Human Molecular Genetics*, *14*(Spec. No. 1) R47–R58.

63. Lee, J., Inoue, K., Ono, R., Ogonuki, N., Kohda, T., Kaneko-Ishino, T., Ogura, A., and Ishino, F. (2002). Erasing genomic imprinting memory in mouse clone embryos produced from day 11.5 primordial germ cells. *Development*, *129*, 1807–1817.

64. Hata, K., Okano, M., Lei, H., and Li, E. (2002). Dnmt3L cooperates with the Dnmt3 family of de novo DNA methyltransferases to establish maternal imprints in mice. *Development*, *129*, 1983–1993.

65. Hazzouri, M., Pivot-Pajot, C., Faure, A. K., Usson, Y., Pelletier, R., Sele, B., Khochbin, S., and Rousseaux, S. (2000). Regulated hyperacetylation of core histones during mouse spermatogenesis: involvement of histone deacetylases. *European Journal of Cell Biology*, *79*, 950–960.

66. Santos, F., Peters, A. H., Otte, A. P., Reik, W., and Dean, W. (2005). Dynamic chromatin modifications characterise the

first cell cycle in mouse embryos. *Developmental Biology*, 280, 225–236.

67. Dahl, J. A., Reiner, A. H., Klungland, A., Wakayama, T., and Collas, P. (2010). Histone H3 lysine 27 methylation asymmetry on developmentally-regulated promoters distinguish the first two lineages in mouse preimplantation embryos. *PLoS One*, 5, e9150.

68. Liu, L., Li, Y., and Tollefsbol, T. O. (2008). Gene–environment interactions and epigenetic basis of human diseases. *Current Issues in Molecular Biology*, 10, 25–36.

69. Franklin, T. B. and Mansuy, I. M. (2010). Epigenetic inheritance in mammals: evidence for the impact of adverse environmental effects. *Neurobiology of Disease*, 39, 61–65.

70. Nadeau, J. H. (2009). Transgenerational genetic effects on phenotypic variation and disease risk. *Human Molecular Genetics*, 18, R202–R210.

71. Skinner, M. K., Manikkam, M., and Guerrero-Bosagna, C. (2010). Epigenetic transgenerational actions of environmental factors in disease etiology. *Trends in Endocrinology and Metabolism*, 21, 214–222.

72. Guerrero-Bosagna, C. M. and Skinner, M. K. (2009). Epigenetic transgenerational effects of endocrine disruptors on male reproduction. *Seminars in Reproductive Medicine*, 27, 403–408.

73. Palmer, J. R., Wise, L. A., Hatch, E. E., Troisi, R., Titus-Ernstoff, L., Strohsnitter, W., Kaufman, R., Herbst, A. L., Noller, K. L., Hyer, M., and Hoover, R. N. (2006). Prenatal diethylstilbestrol exposure and risk of breast cancer. *Cancer and Epidemiology Biomarkers and Prevention*, 15, 1509–1514.

74. Titus-Ernstoff, L., Troisi, R., Hatch, E. E., Hyer, M., Wise, L. A., Palmer, J. R., Kaufman, R., Adam, E., Noller, K., Herbst, A. L., Strohsnitter, W., Cole, B. F., Hartge, P., and Hoover, R. N. (2008). Offspring of women exposed *in utero* to diethylstilbestrol (DES): a preliminary report of benign and malignant pathology in the third generation. *Epidemiology*, 19, 251–257.

75. Bromer, J. G., Wu, J., Zhou, Y., and Taylor, H. S. (2009). Hypermethylation of homeobox A10 by *in utero* diethylstilbestrol exposure: an epigenetic mechanism for altered developmental programming. *Endocrinology*, 150, 3376–3382.

76. Anway, M. D., Cupp, A. S., Uzumcu, M., and Skinner, M. K. (2005). Epigenetic transgenerational actions of endocrine disruptors and male fertility. *Science*, 308, 1466–1469.

77. Thackaberry, E. A., Nunez, B. A., Ivnitski-Steele, I. D., Friggins, M., and Walker, M. K. (2005). Effect of 2,3,7,8-tetrachlorodibenzo-*p*-dioxin on murine heart development: alteration in fetal and postnatal cardiac growth, and postnatal cardiac chronotropy. *Toxicological Sciences*, 88, 242–249.

78. Aragon, A. C., Kopf, P. G., Campen, M. J., Huwe, J. K., and Walker, M. K. (2008). *In utero* and lactational 2,3,7,8-tetrachlorodibenzo-*p*-dioxin exposure: effects on fetal and adult cardiac gene expression and adult cardiac and renal morphology. *Toxicological Sciences*, 101, 321–330.

79. Susser, M. and Stein, Z. (1994). Timing in prenatal nutrition: a reprise of the Dutch Famine Study. *Nutrition Reviews*, 52, 84–94.

80. Pembrey, M. E., Bygren, L. O., Kaati, G., Edvinsson, S., Northstone, K., Sjostrom, M., and Golding, J. (2006). Sex-specific, male-line transgenerational responses in humans. *European Journal of Human Genetics*, 14, 159–166.

81. Dunn, G. A. and Bale, T. L. (2009). Maternal high-fat diet promotes body length increases and insulin insensitivity in second-generation mice. *Endocrinology*, 150, 4999–5009.

82. Zugerman, C. (1990). Chloracne. Clinical manifestations and etiology. *Dermatological Clinic*, 8, 209–213.

83. Burbach, K. M., Poland, A., and Bradfield, C. A. (1992). Cloning of the Ah-receptor cDNA reveals a distinctive ligand-activated transcription factor. *Proceedings of the National Academy of Sciences of the United States of America*, 89, 8185–8189.

84. Ema, M., Sogawa, K., Watanabe, N., Chujoh, Y., Matsushita, N., Gotoh, O., Funae, Y., and Fujii-Kuriyama, Y. (1992). cDNA cloning and structure of mouse putative Ah receptor. *Biochemical and Biophysical Research Communications*, 184, 246–253.

85. Perdew, G. H. (1988). Association of the Ah receptor with the 90-kDa heat shock protein. *Journal of Biological Chemistry*, 263, 13802–13805.

86. Carver, L. A. and Bradfield, C. A. (1997). Ligand-dependent interaction of the aryl hydrocarbon receptor with a novel immunophilin homolog *in vivo*. *Journal of Biological Chemistry*, 272, 11452–11456.

87. Kazlauskas, A., Poellinger, L., and Pongratz, I. (1999). Evidence that the co-chaperone p23 regulates ligand responsiveness of the dioxin (aryl hydrocarbon) receptor. *Journal of Biological Chemistry*, 274, 13519–13524.

88. Lees, M. J. and Whitelaw, M. L. (1999). Multiple roles of ligand in transforming the dioxin receptor to an active basic helix–loop–helix/PAS transcription factor complex with the nuclear protein Arnt. *Molecular and Cellular Biology*, 19, 5811–5822.

89. Fujisawa-Sehara, A., Sogawa, K., Yamane, M., and Fujii-Kuriyama, Y. (1987). Characterization of xenobiotic responsive elements upstream from the drug-metabolizing cytochrome P-450c gene: a similarity to glucocorticoid regulatory elements. *Nucleic Acids Research*, 15, 4179–4191.

90. Hankinson, O. (2005). Role of coactivators in transcriptional activation by the aryl hydrocarbon receptor. *Archives in Biochemistry and Biophysics*, 433, 379–386.

91. Nebert, D. W., Dalton, T. P., Okey, A. B., and Gonzalez, F. J. (2004). Role of aryl hydrocarbon receptor-mediated induction of the CYP1 enzymes in environmental toxicity and cancer. *Journal of Biological Chemistry*, 279, 23847–23850.

92. Fernandez-Salguero, P. M., Hilbert, D. M., Rudikoff, S., Ward, J. M., and Gonzalez, F. J. (1996). Aryl-hydrocarbon receptor-deficient mice are resistant to 2,3,7,8-tetrachlorodibenzo-*p*-dioxin-induced toxicity. *Toxicology and Applied Pharmacology*, 140, 173–179.

93. Shimizu, Y., Nakatsuru, Y., Ichinose, M., Takahashi, Y., Kume, H., Mimura, J., Fujii-Kuriyama, Y., and Ishikawa, T. (2000). Benzo[*a*]pyrene carcinogenicity is lost in mice lacking the aryl hydrocarbon receptor. *Proceedings of the National*

Academy of Sciences of the United States of America, 97, 779–782.

94. Fernandez-Salguero, P., Pineau, T., Hilbert, D. M., McPhail, T., Lee, S. S., Kimura, S., Nebert, D. W., Rudikoff, S., Ward, J. M., and Gonzalez, F. J. (1995). Immune system impairment and hepatic fibrosis in mice lacking the dioxin-binding Ah receptor. *Science*, 268, 722–726.

95. Fernandez-Salguero, P. M., Ward, J. M., Sundberg, J. P., and Gonzalez, F. J. (1997). Lesions of aryl-hydrocarbon receptor-deficient mice. *Veterinary Pathology*, 34, 605–614.

96. Lahvis, G. P., Lindell, S. L., Thomas, R. S., McCuskey, R. S., Murphy, C., Glover, E., Bentz, M., Southard, J., and Bradfield, C. A. (2000). Portosystemic shunting and persistent fetal vascular structures in aryl hydrocarbon receptor-deficient mice. *Proceedings of the National Academy of Sciences of the United States of America*, 97, 10442–10447.

97. Emmons, R. B., Duncan, D., Estes, P. A., Kiefel, P., Mosher, J. T., Sonnenfeld, M., Ward, M. P., Duncan, I., and Crews, S. T. (1999). The spineless-aristapedia and tango bHLH–PAS proteins interact to control antennal and tarsal development in *Drosophila*. *Development*, 126, 3937–3945.

98. Huang, X., Powell-Coffman, J. A., and Jin, Y. (2004). The AHR-1 aryl hydrocarbon receptor and its co-factor the AHA-1 aryl hydrocarbon receptor nuclear translocator specify GABAergic neuron cell fate in *C. elegans*. *Development*, 131, 819–828.

99. Gohlke, J. M., Stockton, P. S., Sieber, S., Foley, J., and Portier, C. J. (2009). AhR-mediated gene expression in the developing mouse telencephalon. *Reproductive Toxicology*, 28, 321–328.

100. Tijet, N., Boutros, P. C., Moffat, I. D., Okey, A. B., Tuomisto, J., and Pohjanvirta, R. (2006). Aryl hydrocarbon receptor regulates distinct dioxin-dependent and dioxin-independent gene batteries. *Molecular Pharmacology*, 69, 140–153.

101. Brunnberg, S., Andersson, P., Lindstam, M., Paulson, I., Poellinger, L., and Hanberg, A. (2006). The constitutively active Ah receptor (CA-Ahr) mouse as a potential model for dioxin exposure—effects in vital organs. *Toxicology*, 224, 191–201.

102. Nohara, K., Suzuki, T., Ao, K., Murai, H., Miyamoto, Y., Inouye, K., Pan, X., Motohashi, H., Fujii-Kuriyama, Y., Yamamoto, M., and Tohyama, C. (2009). Constitutively active aryl hydrocarbon receptor expressed in T cells increases immunization-induced IFN-γ production in mice but does not suppress T_h2-cytokine production or antibody production. *International Immunology*, 21, 769–777.

103. Tauchi, M., Hida, A., Negishi, T., Katsuoka, F., Noda, S., Mimura, J., Hosoya, T., Yanaka, A., Aburatani, H., Fujii-Kuriyama, Y., Motohashi, H., and Yamamoto, M. (2005). Constitutive expression of aryl hydrocarbon receptor in keratinocytes causes inflammatory skin lesions. *Molecular and Cellular Biology*, 25, 9360–9368.

104. Andersson, P., McGuire, J., Rubio, C., Gradin, K., Whitelaw, M. L., Pettersson, S., Hanberg, A., and Poellinger, L. (2002). A constitutively active dioxin/aryl hydrocarbon receptor induces stomach tumors. *Proceedings of the National Academy of Sciences of the United States of America*, 99, 9990–9995.

105. Moennikes, O., Loeppen, S., Buchmann, A., Andersson, P., Ittrich, C., Poellinger, L., and Schwarz, M. (2004). A constitutively active dioxin/aryl hydrocarbon receptor promotes hepatocarcinogenesis in mice. *Cancer Research*, 64, 4707–4710.

106. Wejheden, C., Brunnberg, S., Larsson, S., Lind, P. M., Andersson, G., and Hanberg, A. (2010). Transgenic mice with a constitutively active aryl hydrocarbon receptor display a gender-specific bone phenotype. *Toxicological Sciences*, 114, 48–58.

107. Barouki, R., Coumoul, X., and Fernandez-Salguero, P. M. (2007). The aryl hydrocarbon receptor, more than a xenobiotic-interacting protein. *FEBS Letters*, 581, 3608–3615.

108. Gasiewicz, T. A., Henry, E. C., and Collins, L. L. (2008). Expression and activity of aryl hydrocarbon receptors in development and cancer. *Critical Reviews in Eukaryotic Gene Expression*, 18, 279–321.

109. Harstad, E. B., Guite, C. A., Thomae, T. L., and Bradfield, C. A. (2006). Liver deformation in Ahr-null mice: evidence for aberrant hepatic perfusion in early development. *Molecular Pharmacology*, 69, 1534–1541.

110. Walisser, J. A., Bunger, M. K., Glover, E., and Bradfield, C. A. (2004). Gestational exposure of Ahr and Arnt hypomorphs to dioxin rescues vascular development. *Proceedings of the National Academy of Sciences of the United States of America*, 101, 16677–16682.

111. McMillan, B. J. and Bradfield, C. A. (2007). The aryl hydrocarbon receptor is activated by modified low-density lipoprotein. *Proceedings of the National Academy of Sciences of the United States of America*, 104, 1412–1417.

112. Andreola, F., Hayhurst, G. P., Luo, G., Ferguson, S. S., Gonzalez, F. J., Goldstein, J. A., and De Luca, L. M. (2004). Mouse liver CYP2C39 is a novel retinoic acid 4-hydroxylase. Its downregulation offers a molecular basis for liver retinoid accumulation and fibrosis in aryl hydrocarbon receptor-null mice. *Journal of Biological Chemistry*, 279, 3434–3438.

113. Mulero-Navarro, S., Carvajal-Gonzalez, J. M., Herranz, M., Ballestar, E., Fraga, M. F., Ropero, S., Esteller, M., and Fernandez-Salguero, P. M. (2006). The dioxin receptor is silenced by promoter hypermethylation in human acute lymphoblastic leukemia through inhibition of Sp1 binding. *Carcinogenesis*, 27, 1099–1104.

114. Peng, L., Mayhew, C. N., Schnekenburger, M., Knudsen, E. S., and Puga, A. (2008). Repression of Ah receptor and induction of transforming growth factor-beta genes in DEN-induced mouse liver tumors. *Toxicology*, 246, 242–247.

115. Fritz, W. A., Lin, T. M., Safe, S., Moore, R. W., and Peterson, R. E. (2009). The selective aryl hydrocarbon receptor modulator 6-methyl-1,3,8-trichlorodibenzofuran inhibits prostate tumor metastasis in TRAMP mice. *Biochemical Pharmacology*, 77, 1151–1160.

116. Fan, Y., Boivin, G. P., Knudsen, E. S., Nebert, D. W., Xia, Y., and Puga, A. (2010). The aryl hydrocarbon receptor functions as a tumor suppressor of liver carcinogenesis. *Cancer Research*, 70, 212–220.

117. Vasquez, A., Atallah-Yunes, N., Smith, F. C., You, X., Chase, S. E., Silverstone, A. E., and Vikstrom, K. L. (2003). A role for the aryl hydrocarbon receptor in cardiac physiology and function as demonstrated by AhR knockout mice. *Cardiovascular Toxicology*, 3, 153–163.

118. Lund, A. K., Goens, M. B., Kanagy, N. L., and Walker, M. K. (2003). Cardiac hypertrophy in aryl hydrocarbon receptor null mice is correlated with elevated angiotensin II, endothelin-1, and mean arterial blood pressure. *Toxicology and Applied Pharmacology*, 193, 177–187.

119. Thackaberry, E. A., Bedrick, E. J., Goens, M. B., Danielson, L., Lund, A. K., Gabaldon, D., Smith, S. M., and Walker, M. K. (2003). Insulin regulation in AhR-null mice: embryonic cardiac enlargement, neonatal macrosomia, and altered insulin regulation and response in pregnant and aging AhR-null females. *Toxicological Sciences*, 76, 407–417.

120. Baba, T., Mimura, J., Nakamura, N., Harada, N., Yamamoto, M., Morohashi, K., and Fujii-Kuriyama, Y. (2005). Intrinsic function of the aryl hydrocarbon (dioxin) receptor as a key factor in female reproduction. *Molecular and Cellular Biology*, 25, 10040–10051.

121. Barnett, K. R., Tomic, D., Gupta, R. K., Miller, K. P., Meachum, S., Paulose, T., and Flaws, J. A. (2007). The aryl hydrocarbon receptor affects mouse ovarian follicle growth via mechanisms involving estradiol regulation and responsiveness. *Biology of Reproduction*, 76, 1062–1070.

122. Abbott, B. D., Schmid, J. E., Pitt, J. A., Buckalew, A. R., Wood, C. R., Held, G. A., and Diliberto, J. J. (1999). Adverse reproductive outcomes in the transgenic Ah receptor-deficient mouse. *Toxicology and Applied Pharmacology*, 155, 62–70.

123. Aarnio, V., Storvik, M., Lehtonen, M., Asikainen, S., Reisner, K., Callaway, J., Rudgalvyte, M., Lakso, M., and Wong, G. (2010). Fatty acid composition and gene expression profiles are altered in aryl hydrocarbon receptor-1 mutant *Caenorhabditis elegans*. *Comparative Biochemistry and Physiology, Part C*, 151, 318–324.

124. Schmidt, J. V., Su, G. H., Reddy, J. K., Simon, M. C., and Bradfield, C. A. (1996). Characterization of a murine Ahr null allele: involvement of the Ah receptor in hepatic growth and development. *Proceedings of the National Academy of Sciences of the United States of America*, 93, 6731–6736.

125. Kimura, A., Naka, T., Nohara, K., Fujii-Kuriyama, Y., and Kishimoto, T. (2008). Aryl hydrocarbon receptor regulates Stat1 activation and participates in the development of T_h17 cells. *Proceedings of the National Academy of Sciences of the United States of America*, 105, 9721–9726.

126. Carvajal-Gonzalez, J. M., Roman, A. C., Cerezo-Guisado, M. I., Rico-Leo, E. M., Martin-Partido, G., and Fernandez-Salguero, P. M. (2009). Loss of dioxin-receptor expression accelerates wound healing *in vivo* by a mechanism involving TGFβ. *Journal of Cell Science*, 122, 1823–1833.

127. Nohara, K., Pan, X., Tsukumo, S., Hida, A., Ito, T., Nagai, H., Inouye, K., Motohashi, H., Yamamoto, M., Fujii-Kuriyama, Y., and Tohyama, C. (2005). Constitutively active aryl hydrocarbon receptor expressed specifically in T-lineage cells causes thymus involution and suppresses the immunization-induced increase in splenocytes. *Journal of Immunology*, 174, 2770–2777.

128. Diry, M., Tomkiewicz, C., Koehle, C., Coumoul, X., Bock, K. W., Barouki, R., and Transy, C. (2006). Activation of the dioxin/aryl hydrocarbon receptor (AhR) modulates cell plasticity through a JNK-dependent mechanism. *Oncogene*, 25, 5570–5574.

129. Bertazzi, A., Pesatori, A. C., Consonni, D., Tironi, A., Landi, M. T., and Zocchetti, C. (1993). Cancer incidence in a population accidentally exposed to 2,3,7,8-tetrachlorodibenzo-para-dioxin. *Epidemiology*, 4, 398–406.

130. Steenland, K., Piacitelli, L., Deddens, J., Fingerhut, M., and Chang, L. I. (1999). Cancer, heart disease, and diabetes in workers exposed to 2,3,7,8-tetrachlorodibenzo-*p*-dioxin. *Journal of the National Cancer Institute*, 91, 779–786.

131. Schecter, A., Birnbaum, L., Ryan, J. J., and Constable, J. D. (2006). Dioxins: an overview. *Environmental Research*, 101, 419–428.

132. Panteleyev, A. A. and Bickers, D. R. (2006). Dioxin-induced chloracne—reconstructing the cellular and molecular mechanisms of a classic environmental disease. *Experimental Dermatology*, 15, 705–730.

133. Loertscher, J. A., Lin, T. M., Peterson, R. E., and Allen-Hoffmann, B. L. (2002). In utero exposure to 2,3,7,8-tetrachlorodibenzo-*p*-dioxin causes accelerated terminal differentiation in fetal mouse skin. *Toxicological Sciences*, 68, 465–472.

134. Poland, A. and Knutson, J. C. (1982). 2,3,7,8-Tetrachlorodibenzo-*p*-dioxin and related halogenated aromatic hydrocarbons: examination of the mechanism of toxicity. *Annual Review of Pharmacology and Toxicology*, 22, 517–554.

135. Birnbaum, L. S. (1995). Developmental effects of dioxins. *Environmental Health Perspectives*, 103(Suppl. 7), 89–94.

136. Kerkvliet, N. I. (1995). Immunological effects of chlorinated dibenzo-*p*-dioxins. *Environmental Health Perspectives*, 103 (Suppl. 9), 47–53.

137. Ito, T., Inouye, K., Fujimaki, H., Tohyama, C., and Nohara, K. (2002). Mechanism of TCDD-induced suppression of antibody production: effect on T cell-derived cytokine production in the primary immune reaction of mice. *Toxicological Sciences*, 70, 46–54.

138. Laiosa, M. D., Wyman, A., Murante, F. G., Fiore, N. C., Staples, J. E., Gasiewicz, T. A., and Silverstone, A. E. (2003). Cell proliferation arrest within intrathymic lymphocyte progenitor cells causes thymic atrophy mediated by the aryl hydrocarbon receptor. *Journal of Immunology*, 171, 4582–4591.

139. Funatake, C. J., Dearstyne, E. A., Steppan, L. B., Shepherd, D. M., Spanjaard, E. S., Marshak-Rothstein, A., and Kerkvliet, N. I. (2004). Early consequences of 2,3,7,8-tetrachlorodibenzo-*p*-dioxin exposure on the activation and survival of antigen-specific T cells. *Toxicological Sciences*, 82, 129–142.

140. Marshall, N. B., Vorachek, W. R., Steppan, L. B., Mourich, D. V., and Kerkvliet, N. I. (2008). Functional characterization and gene expression analysis of CD4+ CD25+ regulatory T cells generated in mice treated with 2,3,7,8-tetrachlorodibenzo-*p*-dioxin. *Journal of Immunology*, 181, 2382–2391.

141. Veldhoen, M., Hirota, K., Westendorf, A. M., Buer, J., Dumoutier, L., Renauld, J. C., and Stockinger, B. (2008). The aryl hydrocarbon receptor links T_H17-cell-mediated autoimmunity to environmental toxins. *Nature, 453*, 106–109.

142. Dalton, T. P., Kerzee, J. K., Wang, B., Miller, M., Dieter, M. Z., Lorenz, J. N., Shertzer, H. G., Nerbert, D. W., and Puga, A. (2001). Dioxin exposure is an environmental risk factor for ischemic heart disease. *Cardiovascular Toxicology, 1*, 285–298.

143. Bock, K. W. and Kohle, C. (2009). The mammalian aryl hydrocarbon (Ah) receptor: from mediator of dioxin toxicity toward physiological functions in skin and liver. *Biological Chemistry, 390*, 1225–1235.

144. Kawajiri, K. and Fujii-Kuriyama, Y. (2007). Cytochrome P450 gene regulation and physiological functions mediated by the aryl hydrocarbon receptor. *Archives of Biochemistry and Biophysics, 464*, 207–212.

145. Hillegass, J. M., Murphy, K. A., Villano, C. M., and White, L. A. (2006). The impact of aryl hydrocarbon receptor signaling on matrix metabolism: implications for development and disease. *Biological Chemistry, 387*, 1159–1173.

146. Massague, J. and Gomis, R. R. (2006). The logic of TGFβ signaling. *FEBS Letters, 580*, 2811–2820.

147. Neptune, E. R., Frischmeyer, P. A., Arking, D. E., Myers, L., Bunton, T. E., Gayraud, B., Ramirez, F., Sakai, L. Y., and Dietz, H. C. (2003). Dysregulation of TGF-beta activation contributes to pathogenesis in Marfan syndrome. *Nature Genetics, 33*, 407–411.

148. ten Dijke, P. and Hill, C. S. (2004). New insights into TGF-beta-Smad signalling. *Trends in Biochemical Sciences, 29*, 265–273.

149. Siegel, P. M. and Massague, J. (2003). Cytostatic and apoptotic actions of TGF-beta in homeostasis and cancer. *Nature Reviews in Cancer, 3*, 807–821.

150. Annes, J. P., Munger, J. S., and Rifkin, D. B. (2003). Making sense of latent TGFβ activation. *Journal of Cell Science, 116*, 217–224.

151. Saharinen, J., Hyytiainen, M., Taipale, J., and Keski-Oja, J. (1999). Latent transforming growth factor-beta binding proteins (LTBPs)—structural extracellular matrix proteins for targeting TGF-beta action. *Cytokine and Growth Factor Reviews, 10*, 99–117.

152. Todorovic, V., Frendewey, D., Gutstein, D. E., Chen, Y., Freyer, L., Finnegan, E., Liu, F., Murphy, A., Valenzuela, D., Yancopoulos, G., and Rifkin, D. B. (2007). Long form of latent TGF-beta binding protein 1 (Ltbp1L) is essential for cardiac outflow tract septation and remodeling. *Development, 134*, 3723–3732.

153. Dabovic, B., Chen, Y., Colarossi, C., Zambuto, L., Obata, H., and Rifkin, D. B. (2002). Bone defects in latent TGF-beta binding protein (Ltbp)-3 null mice: a role for Ltbp in TGF-beta presentation. *Journal of Endocrinology, 175*, 129–141.

154. Sterner-Kock, A., Thorey, I. S., Koli, K., Wempe, F., Otte, J., Bangsow, T., Kuhlmeier, K., Kirchner, T., Jin, S., Keski-Oja, J., and von Melchner, H. (2002). Disruption of the gene encoding the latent transforming growth factor-beta binding protein 4 (LTBP-4) causes abnormal lung development, cardiomyopathy, and colorectal cancer. *Genes & Development, 16*, 2264–2273.

155. Elizondo, G., Fernandez-Salguero, P., Sheikh, M. S., Kim, G. Y., Fornace, A. J., Lee, K. S., and Gonzalez, F. J. (2000). Altered cell cycle control at the G(2)/M phases in aryl hydrocarbon receptor-null embryo fibroblast. *Molecular Pharmacology, 57*, 1056–1063.

156. Chang, X., Fan, Y., Karyala, S., Schwemberger, S., Tomlinson, C. R., Sartor, M. A., and Puga, A. (2007). Ligand-independent regulation of transforming growth factor β1 expression and cell cycle progression by the aryl hydrocarbon receptor. *Molecular and Cellular Biology, 27*, 6127–6139.

157. Santiago-Josefat, B., Mulero-Navarro, S., Dallas, S. L., and Fernandez-Salguero, P. M. (2004). Overexpression of latent transforming growth factor-beta binding protein 1 (LTBP-1) in dioxin receptor-null mouse embryo fibroblasts. *Journal of Cell Science, 117*, 849–859.

158. Gomez-Duran, A., Ballestar, E., Carvajal-Gonzalez, J. M., Marlowe, J. L., Puga, A., Esteller, M., and Fernandez-Salguero, P. M. (2008). Recruitment of CREB1 and histone deacetylase 2 (HDAC2) to the mouse Ltbp-1 promoter regulates its constitutive expression in a dioxin receptor-dependent manner. *Journal of Molecular Biology, 380*, 1–16.

159. Guo, J., Sartor, M., Karyala, S., Medvedovic, M., Kann, S., Puga, A., Ryan, P., and Tomlinson, C. R. (2004). Expression of genes in the TGF-beta signaling pathway is significantly deregulated in smooth muscle cells from aorta of aryl hydrocarbon receptor knockout mice. *Toxicology and Applied Pharmacology, 194*, 79–89.

160. Brouchet, L., Krust, A., Dupont, S., Chambon, P., Bayard, F., and Arnal, J. F. (2001). Estradiol accelerates reendothelialization in mouse carotid artery through estrogen receptor-alpha but not estrogen receptor-beta. *Circulation, 103*, 423–428.

161. Haslam, S. Z. and Woodward, T. L. (2001). Reciprocal regulation of extracellular matrix proteins and ovarian steroid activity in the mammary gland. *Breast Cancer Research, 3*, 365–372.

162. Cox, D. A. and Helvering, L. M. (2006). Extracellular matrix integrity: a possible mechanism for differential clinical effects among selective estrogen receptor modulators and estrogens? *Molecular and Cellular Endocrinology, 247*, 53–59.

163. Safe, S., Wormke, M., and Samudio, I. (2000). Mechanisms of inhibitory aryl hydrocarbon receptor-estrogen receptor crosstalk in human breast cancer cells. *Journal of Mammary Gland Biology and Neoplasia, 5*, 295–306.

164. Beischlag, T. V. and Perdew, G. H. (2005). ERα–AHR–ARNT protein–protein interactions mediate estradiol-dependent transrepression of dioxin-inducible gene transcription. *Journal of Biological Chemistry, 280*, 21607–21611.

165. Hockings, J. K., Thorne, P. A., Kemp, M. Q., Morgan, S. S., Selmin, O., and Romagnolo, D. F. (2006). The ligand status of the aromatic hydrocarbon receptor modulates transcriptional activation of BRCA-1 promoter by estrogen. *Cancer Research, 66*, 2224–2232.

166. Ohtake, F., Takeyama, K., Matsumoto, T., Kitagawa, H., Yamamoto, Y., Nohara, K., Tohyama, C., Krust, A.,

Mimura, J., Chambon, P., Yanagisawa, J., Fujii-Kuriyama, Y., and Kato, S. (2003). Modulation of oestrogen receptor signalling by association with the activated dioxin receptor. *Nature, 423*, 545–550.

167. Bocchinfuso, W. P. and Korach, K. S. (1997). Mammary gland development and tumorigenesis in estrogen receptor knockout mice. *Journal of Mammary Gland Biology and Neoplasia, 2*, 323–334.

168. Hushka, L. J., Williams, J. S., and Greenlee, W. F. (1998). Characterization of 2,3,7,8-tetrachlorodibenzofuran-dependent suppression and AH receptor pathway gene expression in the developing mouse mammary gland. *Toxicology and Applied Pharmacology, 152*, 200–210.

169. Vezina, C. M., Lin, T. M., and Peterson, R. E. (2009). AHR signaling in prostate growth, morphogenesis, and disease. *Biochemical Pharmacology, 77*, 566–576.

170. Mathew, L. K., Sengupta, S. S., Ladu, J., Andreasen, E. A., and Tanguay, R. L. (2008). Crosstalk between AHR and Wnt signaling through R-Spondin1 impairs tissue regeneration in zebrafish. *FASEB Journal, 22*, 3087–3096.

171. Vuori, K. A., Nordlund, E., Kallio, J., Salakoski, T., and Nikinmaa, M. (2008). Tissue-specific expression of aryl hydrocarbon receptor and putative developmental regulatory modules in Baltic salmon yolk-sac fry. *Aquatic Toxicology, 87*, 19–27.

172. Tian, Y. (2009). Ah receptor and NF-κB interplay on the stage of epigenome. *Biochemical Pharmacology, 77*, 670–680.

173. Holladay, S. D., Sharova, L. V., Punareewattana, K., Hrubec, T. C., Gogal, R. M., Jr., Prater, M. R., and Sharov, A. A. (2002). Maternal immune stimulation in mice decreases fetal malformations caused by teratogens. *International Immunopharmacology, 2*, 325–332.

174. Abbott, B. D. (1995). Review of the interaction between TCDD and glucocorticoids in embryonic palate. *Toxicology, 105*, 365–373.

175. Tian, Y., Ke, S., Denison, M. S., Rabson, A. B., and Gallo, M. A. (1999). Ah receptor and NF-κB interactions, a potential mechanism for dioxin toxicity. *Journal of Biological Chemistry, 274*, 510–515.

176. Kim, D. W., Gazourian, L., Quadri, S. A., Romieu-Mourez, R., Sherr, D. H., and Sonenshein, G. E. (2000). The RelA NF-κB subunit and the aryl hydrocarbon receptor (AhR) cooperate to transactivate the c-myc promoter in mammary cells. *Oncogene, 19*, 5498–5506.

177. Sheppard, K. A., Rose, D. W., Haque, Z. K., Kurokawa, R., McInerney, E., Westin, S., Thanos, D., Rosenfeld, M. G., Glass, C. K., and Collins, T. (1999). Transcriptional activation by NF-κB requires multiple coactivators. *Molecular and Cellular Biology, 19*, 6367–6378.

178. Ke, S., Rabson, A. B., Germino, J. F., Gallo, M. A., and Tian, Y. (2001). Mechanism of suppression of cytochrome P-450 1A1 expression by tumor necrosis factor-alpha and lipopolysaccharide. *Journal of Biological Chemistry, 276*, 39638–39644.

179. Tian, Y., Ke, S., Chen, M., and Sheng, T. (2003). Interactions between the aryl hydrocarbon receptor and P-TEFb. Sequential recruitment of transcription factors and differential phosphorylation of C-terminal domain of RNA polymerase II at cyp1a1 promoter. *Journal of Biological Chemistry, 278*, 44041–44048.

180. Hahn, M. E. (2002). Aryl hydrocarbon receptors: diversity and evolution. *Chemico-Biological Interactions, 141*, 131–160.

181. Fukunaga, B. N., Probst, M. R., Reisz-Porszasz, S., and Hankinson, O. (1995). Identification of functional domains of the aryl hydrocarbon receptor. *Journal of Biological Chemistry, 270*, 29270–29278.

182. Kumar, M. B., Ramadoss, P., Reen, R. K., Vanden Heuvel, J. P., and Perdew, G. H. (2001). The Q-rich subdomain of the human Ah receptor transactivation domain is required for dioxin-mediated transcriptional activity. *Journal of Biological Chemistry, 276*, 42302–42310.

183. Watt, K., Jess, T. J., Kelly, S. M., Price, N. C., and McEwan, I. J. (2005). Induced alpha-helix structure in the aryl hydrocarbon receptor transactivation domain modulates protein–protein interactions. *Biochemistry, 44*, 734–743.

184. Chapman-Smith, A. and Whitelaw, M. L. (2006). Novel DNA binding by a basic helix–loop–helix protein. The role of the dioxin receptor PAS domain. *Journal of Biological Chemistry, 281*, 12535–12545.

185. Coumailleau, P., Poellinger, L., Gustafsson, J. A., and Whitelaw, M. L. (1995). Definition of a minimal domain of the dioxin receptor that is associated with Hsp90 and maintains wild type ligand binding affinity and specificity. *Journal of Biological Chemistry, 270*, 25291–25300.

186. Kudo, K., Takeuchi, T., Murakami, Y., Ebina, M., and Kikuchi, H. (2009). Characterization of the region of the aryl hydrocarbon receptor required for ligand dependency of transactivation using chimeric receptor between *Drosophila* and *Mus musculus*. *Biochimica et Biophysica Acta, 1789*, 477–486.

187. Poland, A., Palen, D., and Glover, E. (1994). Analysis of the four alleles of the murine aryl hydrocarbon receptor. *Molecular Pharmacology, 46*, 915–921.

188. Chang, C., Smith, D. R., Prasad, V. S., Sidman, C. L., Nebert, D. W., and Puga, A. (1993). Ten nucleotide differences, five of which cause amino acid changes, are associated with the Ah receptor locus polymorphism of C57BL/6 and DBA/2 mice. *Pharmacogenetics, 3*, 312–321.

189. Fitzgerald, C. T., Nebert, D. W., and Puga, A. (1998). Regulation of mouse Ah receptor (Ahr) gene basal expression by members of the Sp family of transcription factors. *DNA and Cellular Biology, 17*, 811–822.

190. Eguchi, H., Hayashi, S., Watanabe, J., Gotoh, O., and Kawajiri, K. (1994). Molecular cloning of the human AH receptor gene promoter. *Biochemical and Biophysical Research Communications, 203*, 615–622.

191. Schmidt, J. V., Carver, L. A., and Bradfield, C. A. (1993). Molecular characterization of the murine Ahr gene. Organization, promoter analysis, and chromosomal assignment. *Journal of Biological Chemistry, 268*, 22203–22209.

192. Harper, P. A., Riddick, D. S., and Okey, A. B. (2006). Regulating the regulator: factors that control levels and activity of the aryl hydrocarbon receptor. *Biochemical Pharmacology, 72*, 267–279.

193. Peters, J. M. and Wiley, L. M. (1995). Evidence that murine preimplantation embryos express aryl hydrocarbon receptor. *Toxicology and Applied Pharmacology*, 134, 214–221.

194. Wu, Q., Ohsako, S., Baba, T., Miyamoto, K., and Tohyama, C. (2002). Effects of 2,3,7,8-tetrachlorodibenzo-p-dioxin (TCDD) on preimplantation mouse embryos. *Toxicology*, 174, 119–129.

195. Jain, S., Maltepe, E., Lu, M. M., Simon, C., and Bradfield, C. A. (1998). Expression of ARNT, ARNT2, HIF1α, HIF2α and Ah receptor mRNAs in the developing mouse. *Mechanisms of Development*, 73, 117–123.

196. Abbott, B. D., Birnbaum, L. S., and Perdew, G. H. (1995). Developmental expression of two members of a new class of transcription factors: I. Expression of aryl hydrocarbon receptor in the C57BL/6N mouse embryo. *Developmental Dynamics*, 204, 133–143.

197. Wu, Q., Ohsako, S., Ishimura, R., Suzuki, J. S., and Tohyama, C. (2004). Exposure of mouse preimplantation embryos to 2,3,7,8-tetrachlorodibenzo-p-dioxin (TCDD) alters the methylation status of imprinted genes H19 and Igf2. *Biology of Reproduction*, 70, 1790–1797.

198. Campbell, S. J., Henderson, C. J., Anthony, D. C., Davidson, D., Clark, A. J., and Wolf, C. R. (2005). The murine Cyp1a1 gene is expressed in a restricted spatial and temporal pattern during embryonic development. *Journal of Biological Chemistry*, 280, 5828–5835.

199. Garrison, P. M. and Denison, M. S. (2000). Analysis of the murine AhR gene promoter. *Journal of Biochemical and Molecular Toxicology*, 14, 1–10.

200. Cui, Y. J., Yeager, R. L., Zhong, X. B., and Klaassen, C. D. (2009). Ontogenic expression of hepatic Ahr mRNA is associated with histone H3K4 di-methylation during mouse liver development. *Toxicological Letters*, 189, 184–190.

201. Shen, E. S. and Whitlock, J. P., Jr., (1989). The potential role of DNA methylation in the response to 2,3,7,8-tetrachlorodibenzo-p-dioxin. *Journal of Biological Chemistry*, 264, 17754–17758.

202. Takahashi, Y., Suzuki, C., and Kamataki, T. (1998). Silencing of CYP1A1 expression in rabbits by DNA methylation. *Biochemical and Biophysical Research Communications*, 247, 383–386.

203. Tokizane, T., Shiina, H., Igawa, M., Enokida, H., Urakami, S., Kawakami, T., Ogishima, T., Okino, S. T., Li, L. C., Tanaka, Y., Nonomura, N., Okuyama, A., and Dahiya, R. (2005). Cytochrome P450 1B1 is overexpressed and regulated by hypomethylation in prostate cancer. *Clinics in Cancer Research*, 11, 5793–5801.

204. Habano, W., Gamo, T., Sugai, T., Otsuka, K., Wakabayashi, G., and Ozawa, S. (2009). CYP1B1, but not CYP1A1, is downregulated by promoter methylation in colorectal cancers. *International Journal of Oncology*, 34, 1085–1091.

205. Taylor, R. T., Wang, F., Hsu, E. L., and Hankinson, O. (2009). Roles of coactivator proteins in dioxin induction of CYP1A1 and CYP1B1 in human breast cancer cells. *Toxicological Sciences*, 107, 1–8.

206. Beedanagari, S. R., Taylor, R. T., Bui, P., Wang, F., Nickerson, D. W., and Hankinson, O. (2010). Role of epigenetic mechanisms in differential regulation of the dioxin-inducible human Cyp1a1 and Cyp1b1 genes. *Molecular Pharmacology*, 78, 608–616.

207. Beedanagari, S. R., Taylor, R. T., and Hankinson, O. (2010). Differential regulation of the dioxin-induced Cyp1a1 and Cyp1b1 genes in mouse hepatoma and fibroblast cell lines. *Toxicological Letters*, 194, 26–33.

208. Suzuki, T. and Nohara, K. (2007). Regulatory factors involved in species-specific modulation of arylhydrocarbon receptor (AhR)-dependent gene expression in humans and mice. *Journal of Biochemistry*, 142, 443–452.

209. Nguyen, T. A., Hoivik, D., Lee, J. E., and Safe, S. (1999). Interactions of nuclear receptor coactivator/corepressor proteins with the aryl hydrocarbon receptor complex. *Archives of Biochemistry and Biophysics*, 367, 250–257.

210. Beischlag, T. V., Wang, S., Rose, D. W., Torchia, J., Reisz-Porszasz, S., Muhammad, K., Nelson, W. E., Probst, M. R., Rosenfeld, M. G., and Hankinson, O. (2002). Recruitment of the NCoA/SRC-1/p160 family of transcriptional coactivators by the aryl hydrocarbon receptor/aryl hydrocarbon receptor nuclear translocator complex. *Molecular and Cellular Biology*, 22, 4319–4333.

211. Wang, S. and Hankinson, O. (2002). Functional involvement of the Brahma/SWI2-related gene 1 protein in cytochrome P4501A1 transcription mediated by the aryl hydrocarbon receptor complex. *Journal of Biological Chemistry*, 277, 11821–11827.

212. Wang, S., Ge, K., Roeder, R. G., and Hankinson, O. (2004). Role of mediator in transcriptional activation by the aryl hydrocarbon receptor. *Journal of Biological Chemistry*, 279, 13593–13600.

213. Hestermann, E. V. and Brown, M. (2003). Agonist and chemopreventative ligands induce differential transcriptional cofactor recruitment by aryl hydrocarbon receptor. *Molecular and Cellular Biology*, 23, 7920–7925.

214. Schnekenburger, M., Peng, L., and Puga, A. (2007). HDAC1 bound to the Cyp1a1 promoter blocks histone acetylation associated with Ah receptor-mediated transactivation. *Biochimica et Biophysica Acta*, 1769, 569–578.

215. Wei, Y. D., Tepperman, K., Huang, M. Y., Sartor, M. A., and Puga, A. (2004). Chromium inhibits transcription from polycyclic aromatic hydrocarbon-inducible promoters by blocking the release of histone deacetylase and preventing the binding of p300 to chromatin. *Journal of Biological Chemistry*, 279, 4110–4119.

216. Schnekenburger, M., Talaska, G., and Puga, A. (2007). Chromium cross-links histone deacetylase 1-DNA methyltransferase 1 complexes to chromatin, inhibiting histone-remodeling marks critical for transcriptional activation. *Molecular and Cellular Biology*, 27, 7089–7101.

217. Ray, S. S. and Swanson, H. I. (2004). Dioxin-induced immortalization of normal human keratinocytes and silencing of p53 and p16INK4a. *Journal of Biological Chemistry*, 279, 27187–27193.

218. Desaulniers, D., Xiao, G. H., Leingartner, K., Chu, I., Musicki, B., and Tsang, B. K. (2005). Comparisons of brain, uterus, and

liver mRNA expression for cytochrome p450s, DNA methyltransferase-1, and catechol-*o*-methyltransferase in prepubertal female Sprague-Dawley rats exposed to a mixture of aryl hydrocarbon receptor agonists. *Toxicological Sciences, 86*, 175–184.

219. Maier, M. S., Legare, M. E., and Hanneman, W. H. (2007). The aryl hydrocarbon receptor agonist 3,3′,4,4′,5-pentachlorobiphenyl induces distinct patterns of gene expression between hepatoma and glioma cells: chromatin remodeling as a mechanism for selective effects. *Neurotoxicology, 28*, 594–612.

220. Oikawa, K., Yoshida, K., Takanashi, M., Tanabe, H., Kiyuna, T., Ogura, M., Saito, A., Umezawa, A., and Kuroda, M. (2008). Dioxin interferes in chromosomal positioning through the aryl hydrocarbon receptor. *Biochemical and Biophysical Research Communications, 374*, 361–364.

221. Sartor, M. A., Schnekenburger, M., Marlowe, J. L., Reichard, J. F., Wang, Y., Fan, Y., Ma, C., Karyala, S., Halbleib, D., Liu, X., Medvedovic, M., and Puga, A. (2009). Genomewide analysis of aryl hydrocarbon receptor binding targets reveals an extensive array of gene clusters that control morphogenetic and developmental programs. *Environmental Health Perspectives, 117*, 1139–1146.

222. Pansoy, A., Ahmed, S., Valen, E., Sandelin, A., and Matthews, J. (2010). 3-Methylcholanthrene induces differential recruitment of aryl hydrocarbon receptor to human promoters. *Toxicological Sciences, 117*, 90–100.

PART III

AHR AS A MEDIATOR OF XENOBIOTIC TOXICITIES: DIOXINS AS A KEY EXAMPLE

12

ROLE OF THE AHR AND ITS STRUCTURE IN TCDD TOXICITY

Raimo Pohjanvirta, Merja Korkalainen, Ivy D. Moffat, Paul C. Boutros, and Allan B. Okey

12.1 INTRODUCTION

The aryl hydrocarbon receptor (AHR) was initially found to be intimately involved in the induction of CYP1A1 (at that time called cytochrome P450c)-associated monooxygenase activities, particularly aryl hydrocarbon hydroxylase (AHH). Early studies in the 1960s had established that the polycyclic aromatic hydrocarbon compounds 3-methylcholanthrene (3-MC) and benzo[a]pyrene were able to cause a marked augmentation of AHH activity in C57BL/6 mice but not in DBA/2 mice. In the 1970s, Alan Poland and coworkers first showed that 2,3,7,8-tetrachlorodibenzo-p-dioxin (TCDD) is about 30,000 times more potent than 3-MC as an inducer of AHH and that TCDD could induce AHH even in so-called "nonresponsive" DBA/2 mice. Later Poland and coworkers, using ^{3}H-labeled TCDD, discovered that there is a stereospecific, saturable, high-affinity binding molecule in liver cytosol for this compound, thus proving the existence of a specific receptor [1, 2] (see also Chapter 1 of this book). Subsequent to this discovery, CYP1A1 induction was utilized in numerous laboratories as the model response in studies to determine the molecular mechanisms of AHR signaling. Induction of cytochrome P450-associated drug metabolizing enzyme activities is basically a beneficial adaptive response to xenobiotic exposure rather than a harmful response [3, 4]. Although first hints of the toxicity of TCDD were detected as early as 1957 [5], it was still unclear to what extent these involved the AHR and whether the findings made on the CYP1A1 induction mechanism would be directly relevant to understanding the mechanism of TCDD's toxic effects as well.

12.2 TOXICITY OF TCDD

Since there are several comprehensive reviews on the topic available in the literature [6–8], the toxic effects of TCDD will only be generally outlined here. On the basis of its acute lethality to guinea pigs (LD_{50} ~1 μg/kg), TCDD is the most toxic synthetic compound known [9]. However, there are exceptionally wide differences in sensitivity among laboratory animals; the most resistant species, the hamster, can tolerate over 1000-fold higher doses than guinea pig [10]. Variation in sensitivity of 1000-fold or greater magnitude also exists within species: as a striking example, the LD_{50} values for two strains of rat, Long–Evans (*Turku/AB*; L-E) and Han/Wistar (*Kuopio*; H/W), are about 10–20 and >9600 μg/kg, respectively [11, 12]. In addition to genetic background, other factors such as gender and age influence sensitivity to TCDD toxicity. For example, in TCDD-sensitive mouse strains, the LD_{50} value for male mice can be over 10-fold lower than that for females, whereas in TCDD-sensitive rat strains, male rats have about twofold higher LD_{50} values than females [11, 13] (and unpublished data). Newborn rats seem to be more susceptible to TCDD toxicity than adults for variable periods (1–3 weeks) after birth [14]. A highly characteristic feature of TCDD is the delayed emergence of its acute toxicity. Even at exceptionally high doses, exposed animals do not succumb until 1–8 weeks after dosing. Death is usually, particularly in rats and guinea pigs, preceded by a dramatic (up to >50%) body weight loss, the wasting syndrome. It appears that TCDD is capable of adjusting the putative central body weight set point to a lower level; thus, sublethal doses can persistently or even permanently reduce the size of the animals (for a recent review, see Ref. 15).

The AH Receptor in Biology and Toxicology, First Edition. Edited by Raimo Pohjanvirta.
© 2012 John Wiley & Sons, Inc. Published 2012 by John Wiley & Sons, Inc.

The hypophagia-based body weight loss is a principal reason for death of TCDD-exposed animals, as testified by a similar life span in control rats pair-fed to their lethally TCDD-treated counterparts [16, 17]. Nevertheless, additional factors seem to be involved because prevention of wasting by intravenous nutrition does not protect from TCDD lethality [18].

TCDD is immunotoxic (see Chapter 19) and a potent endocrine disruptor (Table 12.1). In rats, it decreases circulating levels of testosterone, thyroxine, melatonin, corticosterone (peak levels late in the light phase), β-endorphin, and insulin, while increasing levels of corticosterone (early in the light phase), adrenocorticotropin (ACTH), and thyrotropin (TSH). Some of these impacts vary among species; they may also not all be direct responses. The underlying biochemical mechanisms include accelerated elimination due to enzyme induction (thyroxine, melatonin), impaired biosynthesis (testosterone, corticosterone), and compensatory reactions (TSH) [19–23]. In addition to altering plasma levels of hormones, TCDD may alter the concentration or function of their receptors (Chapter 9). Likewise, TCDD is known to

TABLE 12.1 Major Toxic Effects of TCDD

Impact	Target Cell, Tissue, or Mechanism	Examples of Species Affected[a]	References
Acute lethality	Pathogenesis poorly understood	Guinea pig, rat, mouse	[9, 114]
Wasting syndrome	Central regulation of body weight?	Rat, guinea pig, mouse	[115, 116]
Endocrine disturbances			
Testosterone ↓	Impeded testicular biosynthesis	Rat	[117, 118]
Thyroxine ↓	Accelerated hepatic metabolism	Rat, mouse	[119–121]
Insulin ↓	Impaired pancreatic secretion	Rat	[122, 123]
Corticosterone ↓↑	Diurnal phase-dependent modulation	Rat	[19]
ACTH ↑		Rat	[124]
TSH ↑		Rat	[125]
β-Endorphin ↓	Altered pituitary conversion of POMC?	Rat	[126]
Melatonin ↓	Accelerated extrahepatic metabolism	Rat	[20]
Receptor modulation	Variable interactions	Rat, mouse	Chapter 9
Hypoglycemia	Inhibited gluconeogenesis	Rat, mouse	[17, 121, 127]
Hypercholesterolemia	Reduced hepatic metabolism and elimination	Rat	[128, 129]
Elevation of plasma FFA	Mobilization of peripheral fat	Rat	[128, 130]
Porphyria	Inhibition of uroporphyrinogen decarboxylase	Mouse, rat, human	[131–133]
Jaundice	Impaired biliary transport of bilirubin conjugates?	Rat	[12, 134, 135]
Accumulation of biliverdin	Enhanced hepatic biliverdin production?	Rat	[24]
Altered vitamin A homeostasis	Reduced hepatic esterification of retinol	Guinea pig, rat, mouse	[136, 137]
Enhanced oxidative stress	Mitochondria, microsomes, macrophages	Rat, mouse	Chapter 15
Skin lesions	Keratinocytes and sebaceous glands	Human, monkey, rabbit, mouse	[138–140]
Liver lesions	Hepatocytes and bile ducts; other liver cells?	Mouse, rat, rabbit	[141–143]
Testis lesions	Spermatozoa, Leydig cells, Sertoli cells	Rat, mouse, guinea pig, monkey	[114, 144, 145]
Epithelial and mucosal lesions	Urinary tract, stomach, intestine	Monkey, guinea pig, hamster	[7]
Cardiovascular lesions	Cardiomyocytes; endothelial cells?	Rat	[146]
Tooth lesions	Amelo- and odontoblasts	Rat	Chapter 20
Bone lesions	Osteoblasts; osteoclasts?	Rat	Chapter 20
Myelotoxicity	Hemopoietic progenitor cells	Mouse	[147]
Ascites and general edema	Hypoproteinemia?	Mouse	[114]
Altered levels or function of			
Cytokines and chemokines	For example, TNFα, IL-1β, IL-6, IL-17	Mouse, rat	[148–151]
Growth factors or receptors	EGFR, TGFα, TGFβ	Rat, mouse	[152, 153]
Immunotoxicity	B and T cells, NK cells, dendritic cells	Mouse, rat	Chapter 19
	Thymus atrophy	Mouse, rat, guinea pig	[114, 128]
Developmental toxicity	Hard palate and kidney	Mouse	Chapter 17
	Reproductive tissues and sexual behavior	Rat	Chapters 18 and 31
	Glutamatergic neurons and learning	Rat, monkey	[154–156]
	Molar teeth	Rat	[27]
	Heart	Mouse	[157]
Carcinogenicity	Liver	Rat, mouse	Chapter 16
	Facial skin	Hamster	[158]

[a] Strain differences occur in rats and mice.

interfere with the effects of various cytokines, chemokines, and growth factors or their receptors. Morphologically, some of the major targets for the acute toxicity of TCDD in adult animals include liver, thymus, testes, intestine, teeth, bone (Chapter 20), and urinary tract. However, most of these are highly dependent on which species is studied; only thymic atrophy is an almost invariable finding in mammals (see Ref. 7). TCDD toxicity affects mainly epithelial cells. In clinical chemistry values, typical changes at acutely toxic doses in rats include elevations in plasma cholesterol and free fatty acids (FFA) and a decrease in plasma glucose. Furthermore, TCDD derails heme metabolism leading to porphyria, jaundice, and accumulation of biliverdin and its metabolites in liver [6, 24]. This exceptional diversity of toxic impacts is probably the main reason for the surprising fact that after over 30 years of extensive research, the critical target tissue for the acute lethal toxicity of TCDD is still to be established.

TCDD is carcinogenic in all species tested; the targets again depend on species (Chapter 16). It is also teratogenic. In mice, two teratogenic outcomes are exquisitely typical of TCDD: hydronephrosis and cleft palate (Chapter 17). Rats are less prone to exhibit these effects. In rats, alterations in reproductive organs and in male sexual behavior as well as missing molar teeth are the most sensitive developmental disturbances [25–27] (see also Chapter 18). It is noteworthy that at the embryonal/fetal stage, the differences among species and strains in sensitivity to TCDD lethality are much smaller than in adulthood [28]. The major toxicities of TCDD are summed up in Table 12.1.

12.2.1 Indispensability of the AHR for TCDD Toxicity

Initial evidence for a crucial role of the AHR in TCDD toxicity emerged from two sources. First, it was found that the toxicity of dioxin congeners correlates with their binding affinity to the AHR [29, 30]. Second, sucrose density gradient assays revealed about a 10-fold difference in the binding affinity of the AHR to TCDD between C57BL/6J and DBA/2J mouse strains, concurrent with a sensitivity difference of a similar magnitude between them (as well as in C57BL/6J mice congenic at the *Ah* locus) for most of the toxic and biochemical impacts caused by TCDD (see below). The most definitive evidence, however, became available only with the advent of AHR-deficient mice. These were independently generated by three laboratories, and the three models consistently showed that the AHR is required for all major toxicities analyzed so far, including acute lethality (follow-up period 28 days), thymic atrophy, main features of the liver lesion (although scattered single-cell necrosis were recorded in *Ahr*-null mice), teratogenicity (cleft palate and hydronephrosis), developmental toxicity to male reproductive organs, reduced plasma thyroxine levels, disrupted vitamin A homeostasis, porphyria, and immune toxicity (humoral and cell-mediated immune reactions) [31–36].

Interestingly, heterozygous (+/-) mice did not deviate from wild-type mice in their sensitivity to TCDD-induced fetal hydronephrosis (which emerges at lower doses than cleft palate) but were less responsive than their wild-type counterparts to TCDD-induced cleft palate and immune toxicity [32, 33]. In another study, utilizing chimeras created by hematopoietic reconstitution of irradiated mice with cells from AHR wild-type or AHR-deficient mice, the presence of AHR in hepatic parenchyma alone was found to be sufficient for TCDD induction of hepatic necrosis, whereas its presence in hematopoietic cells was necessary for the inflammatory response (predominantly mononuclear) to hepatic lesions elicited by 30 μg/kg TCDD at 10 days [37]. In at least partial support of these findings, mice made AHR deficient in hepatocytes alone by the *Cre-lox* technology were largely protected against TCDD-provoked (100 μg/kg once a week for 4 weeks) hydropic degeneration and pyogranuloma formation in the liver at 7 days after the last treatment; their microsomal enzyme induction response was also attenuated (CYP1A1 and CYP1B1) or abolished (CYP1A2) [38]. Strictly speaking, all these data obtained with AHR knockouts apply to mouse and are not necessarily representative of other species because the impacts of TCDD vary considerably among species; for example, there is only a subtle overlap between mice and rats in hepatic gene expression profiles after TCDD treatment (Chapter 14). As soon as targeted gene disruption becomes a widely accessible option in rats [39], it will be important to verify that the AHR is essential to TCDD toxicity not only in mice but also in rats.

While the studies cited above confirm the essentiality of the AHR in the major toxicities of TCDD in mice, they do not prove that the molecular mechanism by which the AHR exerts its deleterious effects is equivalent to that for CYP1A1 induction. This question was elegantly addressed in Chris Bradfield's laboratory by generating transgenic mice unable to pass through a series of critical steps along the AHR signaling pathway (Fig. 12.1). One of their models harbored a mutated nuclear localization signal in the AHR protein that therefore could not translocate into the nucleus upon activation by ligand binding (the mutation also abolished the DRE binding ability). These mice were refractory to TCDD-induced hepatomegaly and thymic atrophy. Moreover, cultured palates from the transgenic mice fused normally when exposed to TCDD, whereas those from wild-type and heterozygous mice failed to do so [40]. In another model, the AHR was modified such that ligand binding, translocation, and heterodimerization with ARNT occurred normally, but DRE binding was prevented by addition of a single amino acid into the DNA binding domain. These mice were again nonresponsive to all impacts of TCDD examined: hepatomegaly and liver steatosis, teratogenicity (cleft palate and hydronephrosis), and thymic atrophy [41]. Their third model exhibited very low level of ARNT protein expression (~10% of wild-type mice). This manipulation resulted in

```
Cytoplasm
        TCDD ─┐ ┌─ AHR
              └─┘
           Ligand binding
                │
           Transformation
                │
              Trans-
         ─────────────────
              location
                   ┌─ ARNT
           Heterodimerization
Nucleus         │
           DNA binding
                │
           Transactivation
```

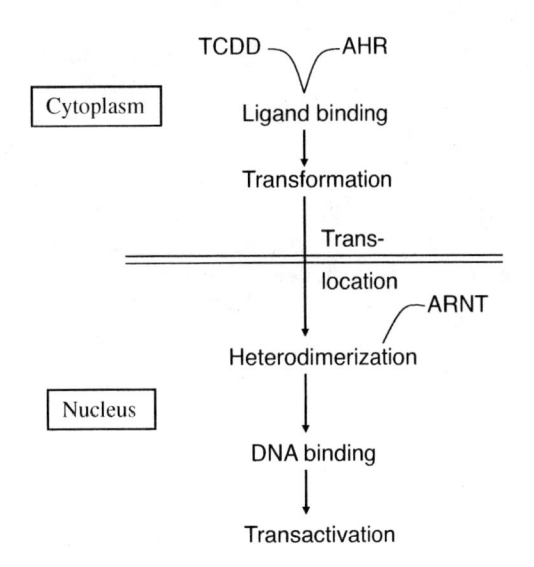

FIGURE 12.1 Key steps in the canonical signaling pathway of the AHR in cells.

a substantial attenuation of thymic atrophy and hepatotoxicity after TCDD treatment. Concomitantly, CYP1A1 induction remained unaffected [42], suggesting that the reduced AHR signaling capacity sufficed for this response but not for liver or thymic toxicity. Most recently, mice rendered ARNT deficient in hepatocytes alone by the *Cre*-lox methodology were found to be nonresponsive to TCDD-provoked liver toxicity (hepatomegaly, steatosis, hydropic degeneration, inflammation) and to hepatic induction of drug metabolizing enzymes by TCDD [43].

As to animals not exposed to TCDD, it is important to note that the mice defective in nuclear localization or DRE binding of the AHR, as well as the globally ARNT hypomorphic mice, phenotypically resembled AHR-deficient mice and exhibited one of the most prominent features of AHR knockouts, that is, patent *ductus venosus* in liver. In subsequent studies, activation of the canonical AHR signaling pathway by TCDD administration was actually shown to rescue the ARNT hypomorphic animals from patent *ductus venosus* [44]. Thus, both the major toxicities accruing from a persistent activation of the AHR and the physiological actions of the AHR in ontogeny appear to mainly utilize one and the same chief molecular mechanism. However, this generalization should be taken with caution at the moment because only a handful of toxic end points have so far been assessed with the mouse models. In a given tissue, the AHR signaling pathway may also be at work in different cells during development and under TCDD intoxication: studies with conditional AHR knockout mice (produced by the *Cre*-lox technology) have indicated that AHR signaling in endothelial/hematopoietic cells is necessary for developmental closure of the *ductus venosus*, whereas AHR signaling in hepatocytes is necessary to generate adaptive and toxic responses of the liver in response to dioxin exposure [38]. In any case, it is conceptually important to recognize that the major deleterious impacts of the exquisitely toxic environmental compound, TCDD, ultimately represent responses elicited by an endogenous protein that has only been rendered overactive by this xenobiotic.

One of the effects of TCDD, suggested by *in vitro* studies on cell lines to be mediated by a noncanonical (nongenomic) AHR pathway, is the induction of cyclooxygenase 2 (COX2) [45–50]. It was recently reported that hydronephrosis, caused in mouse pups by lactational exposure to TCDD, is critically dependent on COX2 [51]. However, mice carrying an AHR with abolished DRE binding ability did not display any cases of hydronephrosis after *in utero* exposure to TCDD in the face of 100% prevalence of the malformation in wild-type animals [41]. Hence, either this lesion evolves by different mechanisms, depending on developmental stage, or, alternatively, the induction of COX2 by TCDD occurs via the classical pathway in developing mice.

12.2.2 AHR Primary Structure as a Determinant of Sensitivity to TCDD Toxicity

Similar to the formal family of nuclear receptors, the AHR is composed of distinct functional domains. Contrary to the nuclear receptors, however, almost all the domains of functional significance are clustered within the amino-terminal half of the AHR (see Fig. 3.3). These include the ligand binding domain (LBD). Minimally, it harbors the PAS B motif along with flanking amino acid residues on both sides [52]. The LBD partially overlaps one of the two sites where HSP90 makes contact with the AHR. By this means, the HSP90 chaperone maintains the AHR in a configuration that is favorable for ligand binding [53]. This site is also essential for retention of the ligand-free receptor in the cytosol [54]. The molecules accepted by the LBD to bind are usually planar and hydrophobic although chemicals with a wide variety of structures can act as AHR ligands [55] (see Chapter 4). At the C-terminal end of the AHR, there is only a large transactivation domain (TAD) that harbors smaller, synergistically interacting subunits [56, 57]. Biochemically, these comprise acidic, glutamine-rich (Q-rich), and proline/serine/threonine-rich (P-S-T-rich) subregions [56, 57]. Of these subdomains, the Q-rich subregion appears to possess the greatest transactivation potency, although there is variation among AHRs of various species and among data from different experimental settings [57–60]. Likewise, the function of the acidic subdomain may vary across species. In the case of the mouse AHR, the acidic subdomain was able to transactivate the *Cyp1a1* gene, whereas in human AHR it did not enhance CYP1A1 activity when tested in BP8 rat hepatoma cells devoid of native AHR [58, 60]. The P-S-T-rich subdomain may have a repressive function on other subdomains, at least in humans [60].

12.3 LIGAND BINDING DOMAIN

That variations in LBD structure can influence dioxin sensitivity was initially demonstrated with inbred C57BL/6 and DBA/2 mouse strains, which constitute the oldest animal model for genetic studies of TCDD toxicity. More recently, these two strains were exploited to develop congenic C57BL/6J mouse strains, differing only at the *Ahr* locus, for this purpose. In these models, both sensitive and resistant strains exhibit largely an identical spectrum of toxicities, but the resistant mice (DBA/2 or C57BL/6J$^{d/d}$) require a 10–20-fold higher dose of TCDD. This pattern applies to impacts of TCDD such as acute lethality, hepatomegaly, and histopathological changes in liver, thymic atrophy, elevation of plasma corticosterone, reductions in plasma glucose, cholesterol, and triglycerides, suppression of the cytotoxic T-cell and the splenic plaque-forming cell responses, porphyria, teratogenicity (cleft palate and hydronephrosis), myelotoxicity, endotoxin hypersensitivity, aggravated oxidative stress, and decreased binding capacities of hepatic receptors for epidermal growth factor and estrogen [61–71]. Furthermore, a single-dose study revealed that hepatic tumor promotion by TCDD was also dependent on the AHR in this animal model [72]. By a modified sucrose density gradient analysis, Okey et al. discovered in the late 1980s that the binding affinity of the DBA/2 mouse hepatic AHR is about 10-fold lower than that of the C57BL/6 mouse and that receptor density is twice as high in C57BL/6 as in the DBA/2 strain [73]. Subsequent molecular biological studies disclosed several nonsynonymous single-nucleotide polymorphisms in the b-1 (C57BL/6) and d (DBA/2) alleles of the AHR [74]. Finally, Poland et al. [75] tracked down the critical polymorphism for AHR function to the codon encoding amino acid 375. In the b-1 allele, this is alanine but in the d allele it is converted to valine, accounting for the difference in the binding affinities of the two receptor forms.

Variation in the primary structure of the LBD of the AHR seems to be quite common in nature and underlies some of the between-species and within-species differences reported in TCDD sensitivity. Amphibians appear to be resilient to TCDD lethality [76, 77]. The model amphibian species, *Xenopus laevis*, possesses two AHRs encoded by distinct genes; both of these receptor isoforms bind TCDD with at least 20-fold lower affinity than the mouse AHR^{b-1} protein [78]. Birds are also endowed with two AHRs: AHR1 and AHR2. AHR1 is probably the dominant subtype [79] (see also Chapter 21). Among bird species, AHR binding affinity correlates with TCDD susceptibility, and two amino acids corresponding to Ile324 and Ser380 in the AHR1 of the highly TCDD-sensitive chicken (LD$_{50}$ 0.18 µg/kg by egg injection) play leading roles [80]. Interestingly, the latter position (380) is equivalent to the critical amino acid 375 in mouse AHR. At a position corresponding to 324 in chicken, the mouse has also isoleucine, and the mouse AHR^{b-1} combination, Ile–Ala, is present in birds lagging behind the Ile–Ser genotype but outperforming the third common avian combination, Val–Ala, in terms of TCDD binding avidity [80].

It is important to note that, based on the data described above from studies with inbred mice, a low binding affinity of TCDD, due to unfavorable structure of the AHR LBD, confers indiscriminate protection against all major toxicities triggered by this compound. Pertinent to risk assessment of dioxins, the equivalent amino acid in the human AHR (AA 381) is the same as in the insensitive mouse AHR^d isoform (valine). As might be expected, the human AHR displays lower binding affinity to TCDD compared to the product of the mouse b-1 allele [81, 82]. However, it should be kept in mind that for some AHR agonists such as indirubin the human receptor may show superior binding ability compared to its murine b-1 counterpart [83].

12.4 TRANSACTIVATION DOMAIN

In two animal models of exceptional resistance to the toxicity of TCDD-activated AHR, that is, hamsters and H/W rats, the LBD of the AHR is structurally unaltered and functionally intact [84–87]. However, they both have AHRs in which the C-terminal TAD of the AHR deviates from that occurring in TCDD-sensitive animals (Fig. 12.2). In hamster AHR, the glutamine-rich subregion is greatly expanded due to insertion of numerous microsatellite-like stretches into the coding region of the DNA. These encode for Glu, thereby doubling the number of Glu residues in this subdomain compared to the AHRs in other mammals (49, 28, 27, and 23 in hamster, mouse, rat, and guinea pig AHR, respectively) [85, 88]. In the H/W rat *AHR* gene, a G → A point mutation in the first nucleotide of the last intron disrupts the exon/intron boundary, resulting in altered splicing. As a result, three mRNAs are produced (two insertions and one deletion). Because both insertions will give rise to an identical protein, H/W rats express two AHR protein isoforms (Fig. 12.2): one (arising from the deletion) that lacks 43 AAs around the middle point of the P-S-T-rich subdomain of the TAD and the other (arising from the insertions) that lacks the last 45 AAs but has integration of 7 novel amino acids at its C-terminal end [86]. Intriguingly, the phenotypes of these two animal models, hamsters and H/W rats, have some common features. While both are extremely resistant to the acute lethality of TCDD (LD$_{50}$ values about 5000 for hamsters and >9600 µg/kg for H/W rats), they still avidly display the hallmark adaptive response, CYP1A1 induction [89–91]. In both cases, the resistance is also strictly age dependent, with the difference from sensitive species/strains shrinking down to some 10-fold for prenatal developmental stages [28, 92]. The response-selective sensitivity of these models thus distinguishes them from the models based on differential ligand

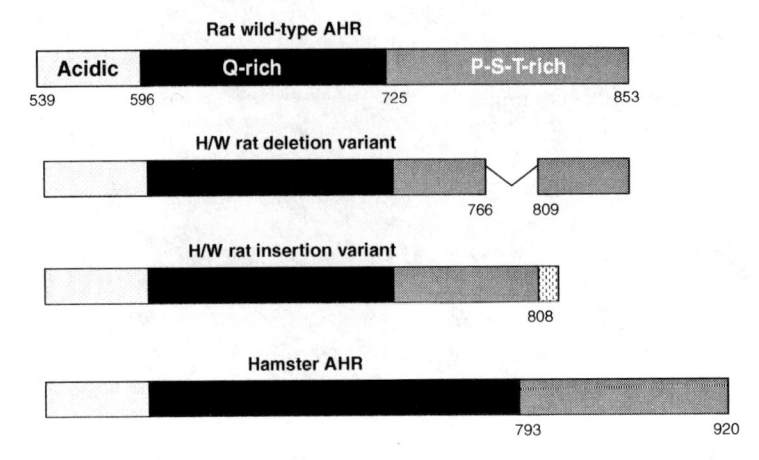

FIGURE 12.2 TAD structure in the AHRs of H/W rats and hamsters. The three major subdomains are shown. For comparison, the rat wild-type AHR is shown on top. The numbers refer to the approximate locations (amino acids) in the full-length AHR.

binding properties of the AHR. H/W rats have been studied in this context in more detail than hamsters and the data gathered from them on the involvement of the AHR in this phenomenon will be briefly outlined next.

Early studies in H/W and a highly TCDD-sensitive rat strain, L-E, uncovered that the difference in their vulnerability to the acute lethality of TCDD (1) did not appear to stem from toxicokinetic factors [93]; (2) was specific to dioxin-like compounds, since it vanished when the strains were tested with perfluorodecanoic acid, which also brings about a wasting syndrome but does not act through the AHR [94]; and (3) correlated with AHR binding affinity among dioxins, being ~1000 for TCDD (relative AHR binding affinity 1), 280 for 1,2,3,7,8-pentachlorodibenzo-p-dioxin (0.5–0.8), 20 for 1,2,3,4,7,8-hexachlorodibenzo-p-dioxin (0.04–0.3), and 7 for 1,2,3,4,6,7,8-heptachlorodibenzo-p-dioxin (predicted ~0.003) [11,12, 95–98]. In genetic crosses between the two strains, TCDD resistance was inherited as an autosomal dominant trait with 1–2 genes participating in its mediation [99]. Collectively, these findings pointed to involvement of the AHR in the foundations of the strain difference. The advent of AHR antibody later reinforced this notion because the apparent molecular mass of the AHR, as assessed by Western blot assays, proved to be smaller (~98 kDa) in H/W rats than in TCDD-sensitive rat strains (106 kDa); the small receptor protein also exhibited peculiar vulnerability to metal oxyanions and to buffers with high ionic strength, suggesting modified association with HSP90 [87]. The AHR-specific antibody helped verify the earlier genetic findings and was a key tool in establishing that the principal reason for H/W resistance lies in their *AHR* gene [100]. Molecular cloning of the AHRs from H/W and L-E rats then revealed the existence in H/W rats of the splice variants described above. (Both protein variants also contain a conservative Val → Ala substitution at position 497. This occurs in a hypervariable region and does not affect the predicted secondary structure of the proteins [101].) Based on mRNA analyses, the predominant AHR isoform in H/W rats turned out to be the insertion-derived receptor protein accounting for some 85% of total AHR mRNA expression in all tissues examined, independent of TCDD treatment status [101]. Two other major players exist in the AHR signaling pathway: ARNT (and ARNT2) and AHRR. No divergences that could readily explain the TCDD susceptibility difference between H/W and L-E rats were found in the primary structures or tissue expression levels of ARNT, ARNT2, or AHRR [102, 103].

As already mentioned, H/W rats are not refractory to all effects of TCDD. In addition to induction of drug metabolizing enzymes—including CYP1A1, CYP1A2, CYP1B1, ALDH3A1, NQO1, and UGT1A1—they respond nearly as effectively as L-E rats in terms of sensitivity and magnitude of response to such biochemical and toxic impacts of TCDD as thymic atrophy, fetolethality, disrupted vitamin A status, some endocrine disturbances, and changes in certain clinical chemistry values (Table 12.2). These have been dubbed "Type I" effects and they mainly encompass low-dose impacts (but are not confined to them: for example, incisor tooth defect is a typical high-dose effect). On the other hand, some other effects require considerably larger doses in H/W than L-E rats to be elicited or (the usual case) are drastically reduced in the magnitude of response. These "Type II" effects typically emerge at high doses of TCDD and include acute lethality, wasting syndrome, liver toxicity, changes in plasma biochemistry, and tumor promotion (Fig. 12.3 and Table 12.2). Teratogenicity constitutes a special case as it is triggered by similar doses in both strains, but the manifestations are disparate: although cleft palate is the predominant malformation in L-E rats, it has not been recorded in H/W rats. Instead, they exhibit hydronephrosis and intestinal hemorrhages [92].

TABLE 12.2 Sensitivity of H/W Rats in Comparison with L-E Rats to Various Impacts of TCDD

Type I Effects (Similar Responsiveness)	Type II Effects (Dissimilar Responsiveness)	References
Induction of Phase I and II enzymes	Acute lethality	[11, 90, 91, 159]
Thymus atrophy	Wasting syndrome	[12, 128]
Fetolethality	Liver toxicity	[92, 128]
Diminished brain inositol levels	Enhanced brain serotonin turnover	[160, 161]
Derailed vitamin A status	Lipid peroxidation in liver	[159, 162]
Hypercholesterolemia	Hyperbilirubinemia	[12, 98, 128]
Reduced plasma thyroxine levels	Elevated plasma FFA levels	[98, 128, 130]
Reduced plasma melatonin levels	Reduced plasma β-endorphin levels	[126, 163, 164]
Incisor tooth lesion	Changes in plasma amino acids	[98, 165]
Atrophy of the pituitary gland	Decreased liver glycogen concentration	[98, 126, 165]
Augmented neophobia	Accumulation of biliverdin in liver	[166, 167]
Diminished placental weight	Tumor promotion	[92, 113]
(Teratogenicity)	(Teratogenicity)	[92]

As assessed by a reporter gene assay employing Gal4–AHR TAD chimeras *in vitro*, the intrinsic transactivation activity of the H/W rat deletion (DEL) variant proved to be three times as high as that for the rat wild-type (rWT) AHR TAD. In contrast, the activity of the insertion (INS) variant was some 20% lower compared to that of the rWT TAD [101]. Secondary structure modeling revealed that in the INS variant, the altered amino acid sequence increases both α-helical content and hydrophobicity of the TAD terminus compared to rWT AHR. Overall, the modeling suggested that the TAD terminus of the insertion variant protein adopts a conformation that is less accessible to interactions with other proteins [101]. Interestingly, in the INS variant TAD, there is loss of a predicted Src kinase substrate motif and a second such motif adjacent to an inserted α-helix may be obstructed. Mice lacking c-Src kinase activity are resistant to lethality of TCDD but are fully responsive to CYP1A1 induction as are H/W rats [104–106].

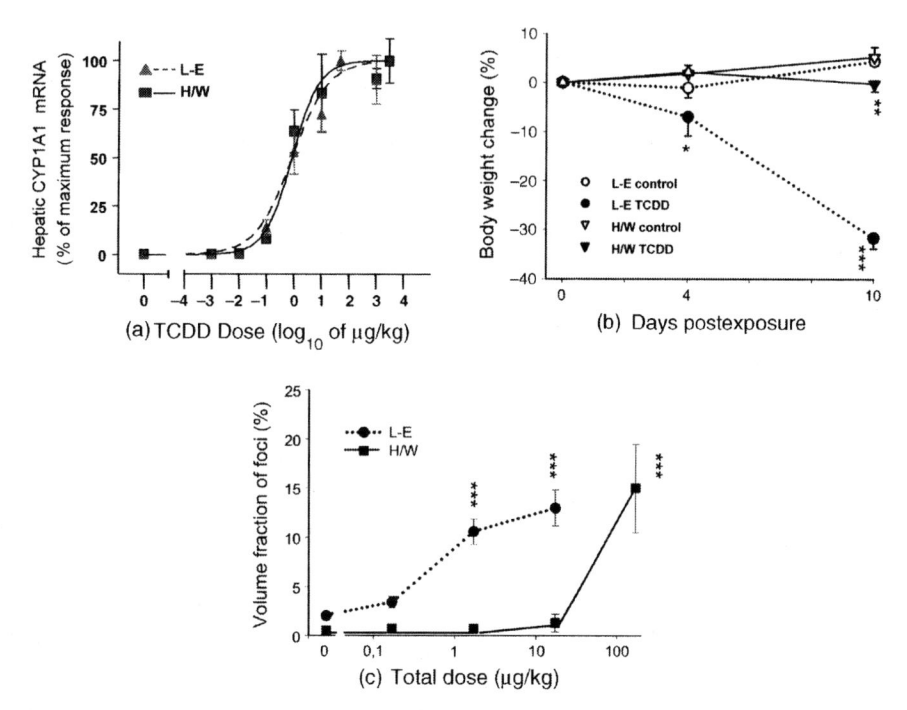

FIGURE 12.3 Examples of Type I and Type II effects of TCDD in L-E and H/W rats. (a) Dose–response for CYP1A1 induction 19 h after TCDD exposure (real-time RT-qPCR assay on mRNA levels; slightly modified from Ref. 110). (b) Body weight change (as % of initial weight) 4 and 10 days after exposure to 100 µg/kg TCDD by gavage. (c) Volume fraction of hepatic preneoplastic foci in a tumor promotion study (modified from Ref. 113).

To gain further insight into the actions of the rat AHR isoforms *in vivo*, they were transferred by nuclear microinjection into C57BL/6 mice, and the construct-positive individuals were then mated with AHR-deficient mice to eliminate the mouse innate AHR. By this means, a single line was obtained for each of the three rat AHR isoforms. All three lines expressed the construct globally, with the expression levels being comparable to, or exceeding, those of mouse AHR in C57BL/6 mice. When challenged with low doses of TCDD, all three lines responded by induction of the characteristic drug metabolizing enzymes CYP1A1, CYP1A2, and CYP1B1. The induction was fairly similar in magnitude across the lines and reflected the expression levels of the AHR transgene constructs. By contrast, responsiveness to the acute lethality of TCDD differed starkly among the lines with LD_{50} values being between 25 and 125 µg/kg for mice expressing the rWT variant, ~250 µg/kg for mice expressing the DEL variant, and >500 µg/kg for mice expressing the INS variant AHR. The highest dose tested was 500 µg/kg; no deaths were recorded at this dose level in either gender of INS mice [107] (unpublished data). A conspicuous facet distinguishing the TCDD-sensitive transgenic mice from all other laboratory animals studied previously was the exceptionally early emergence of mortality: in rWT mice, the first deaths took place as early as 4 days after TCDD exposure and appeared to be due to massive necrosis of the liver. In a time course experiment, necrotic changes in liver were detectable already on the first day postexposure. In mice that survived this early phase of toxicity, the necrotic foci substantially diminished or entirely disappeared by day 6, being replaced by degenerative alterations in hepatocytes and infiltration of inflammatory cells (unpublished data). Overall, transgenic mice expressing the rWT receptor were more susceptible to acute toxicity of TCDD than were their native C57BL/6 controls. This was also reflected in a swifter body weight loss: by day 4 (males) or 6 (females), the rWT had lost ~15% of their initial body weight, while in C57BL/6 mice, even at supralethal doses, the decrease was only a few percent at these time points. An intriguing finding was that in INS mice, TCDD treatment actually tended to augment body weight gain [107] (unpublished data).

As shown in Table 12.2, thymus atrophy belongs to Type I effects in the L-E and H/W rat model. Therefore, it was of interest that despite the fact that INS mice express more AHR mRNA in thymus than rWT mice, thymus atrophied sooner and to a larger degree in rWT than in INS mice. It is conceivable that the meager expression (~15% of total AHR mRNA) of the DEL variant in H/W rats with its high intrinsic transactivation activity is able to modulate their responses to TCDD. By and large, notwithstanding the induction of drug metabolizing enzymes, the INS mice have failed to markedly exhibit any toxic response assessed so far (e.g., endocrine effects have not yet been looked at). Another difference from the parent rat model was found when the transgenic mice were crossed with another (rWT × INS) or with C57BL/6 mice (rWT × C57BL/6; INS × C57BL/6). Whereas resistance to the acute lethality of TCDD is inherited as a dominant trait in rats, mice that carried the rWT cDNA construct in their genome proved to be highly sensitive to TCDD toxicity, irrespective of the type of the other AHR isoform they were carrying (whether it was the INS cDNA construct or the native mouse AHR gene). In other words, the presence of one rat INS allele in mice in which the other allele encodes wild-type rat AHR did not confer a "dominant-negative" protection from TCDD toxicity in the transgenic mouse model. This is in contrast to the response in rats where one copy of an H/W-derived AHR allele is sufficient to protect from toxicity when the other allele is rat wild type. One possible explanation for this discrepancy might be excessively lavish expression of the rWT construct in a critical target tissue. That AHR expression levels can indeed potentially be a crucial limiting factor for TCDD responsiveness was attested to by the degree of testicular induction of CYP1A1 in these transgenic mouse lines. In all three lines, testicular AHR expression exceeded that of normal wild-type C57BL/6 mice by some 30-fold, and this was accompanied by a similarly enhanced CYP1A1 induction [107]. In mouse testis, clearly there does not appear to be any AHR reserve ("spare receptors") in terms of this adaptive response. However, in all other tissues and organs analyzed hitherto, the departures from the native levels have been much smaller, and this also applies to differences among the transgenic lines themselves.

Similar to the INS variant AHR, α-helical content is predicted to be increased in the TAD of hamster AHR, albeit in a different subregion [101]. Unfortunately, the hamster AHR has not yet been explored any further. It is also noteworthy that the AHR encoded by the mouse d allele is 43 AAs longer at the C-terminus than that generated by the TCDD-responsive b-1 allele [74], but the functional significance of this difference is unknown at present.

The molecular details of how polymorphisms in AHR TAD structure result in such a selective interference on TCDD responses are still obscure. Folding of the AHR protein is likely to play an important role. Remodeled folding could lead to accelerated degradation of the AHR. However, studies in H/W and L-E rats with high single doses of TCDD or low repeated dosing regimens do not lend support to this possibility [108, 109]. In a time course study in the transgenic mouse lines expressing rat AHR variants, TCDD exposure rapidly and persistently (days 1–6) reduced hepatic AHR protein concentrations. Although the decrease was most prominent in rWT mice, it also occurred in DEL and INS mice; thus, it is hard to attribute differences in toxic response to differences among the AHR genotypes by alterations in AHR levels caused by TCDD treatment (unpublished data). A tenable hypothesis is that the structural alterations affect the interactions of the AHR with transcriptional coregulators that make contact with the AHR mainly at the TAD (see Chapter 5). Structural modifications in this region might render the AHR less suited to interplay with the coregulators (or even general transcription factors) overall. Recent microarray analyses provide some support to this hypothesis by revealing that fewer hepatic genes alter their expression after TCDD treatment in H/W than in L-E rats [110]. On the other hand, *in vitro* studies have implied that for CYP1A1 induction, the TAD of the AHR is crucial while that of ARNT is dispensable [111]; yet, this is typically a Type I response. Since the dissimilar responses in most cases occur only at high doses of TCDD, they might need the recruitment of a different set of coregulators (possibly having a relatively low affinity to the AHR and thus being more finicky with regard to its conformation) than the low-dose responses including CYP1A1 induction. Mediation of TCDD toxicity may also involve crosstalk with other signaling pathways that may be affected by TAD structure. The loss of a predicted Src kinase substrate motif in INS TAD is already mentioned above. Another candidate to consider is estrogen receptor. For example, in C57BL/6 mice, female animals can be over 10-fold less sensitive to the acute lethality of TCDD compared to males, but coexpression of rWT construct with their own murine AHR abolishes female resistance [107] (unpublished data). The AHR and the estrogen receptor have been shown to interact with one another in a number of ways (Chapters 9 and 10). Even isolated subdomains of human AHR TAD were found to interfere with estrogen receptor action, with the P-S-T-rich subdomain being the most potent in this regard [112]; this suggests potential diversity in outcomes depending on TAD structure.

Could these findings be directly utilized in risk assessment of dioxins and other AHR ligands—that is, have we already reached a position where we can predict the sensitivity of a given species (or a subpopulation within a species) to AHR-mediated toxicities based on the structure of its AHR alone? Unfortunately, this is not yet the case. Although certain structural features in the LBD are irrefutably associated with dioxin resistance (foremost the Ala versus Val variation), this information seems to be of limited value in the prediction of the AHR activating potential of other AHR ligands (e.g., indirubin). The functional consequences of alterations in the TAD are even less predictable, and much more work needs to be done before the roles of TAD subdomains (and key amino acids therein) in AHR responsiveness will be understood. It should also be kept in mind that all the animal models dealt with in this chapter ultimately represent models of dioxin *resistance*. For example, no such structural modifications of the AHR have so far been discovered in rats or mice that would *enhance* TCDD susceptibility over that associated with their wild-type receptor forms. In the species model of guinea pig versus hamster, the receptor that structurally deviates from others is again that of the resistant species, the hamster. The AHR of the most TCDD-susceptible mammal, guinea pig, does not contain any striking characteristics that would distinguish it from the AHR of most other species [88]. In fact, its closest homolog appears to be the human AHR, yet all the available evidence implies that humans belong to fairly TCDD-insensitive species.

12.5 CONCLUSIONS

The canonical AHR signaling pathway was originally uncovered in the context of CYP1A1 induction by TCDD. However, the evidence gathered subsequently, in particular from studies in AHR-deficient mice as well as from transgenic mice in which individual key steps along this pathway have been rendered nonfunctional, coherently indicates that all major forms of TCDD toxicity also employ the same fundamental AHR mechanism as CYP1A1 regulation. In agreement with this view, alterations of single critical amino acids in the LBD of the AHR tend to affect indiscriminately the toxic impacts of TCDD. In contrast, polymorphisms in the TAD of the AHR that exist in animal models of TCDD resistance seem to result in modifications in sensitivity that are highly dependent on which response or end point is being measured. However, this phenomenon has so far been examined in fair detail only in the H/W rat model. At present, our understanding of how AHR structure and AHR polymorphisms affect toxic responses is not sufficient to employ receptor structure per se as a major predictor of susceptibility in assessment of risk of dioxin-like chemicals to humans or other vertebrate species.

ACKNOWLEDGMENTS

This work was financially supported by grants from the Academy of Finland (grant number 123345; RP) and from the Canadian Institutes of Health Research (grant number MOP-57903; ABO and PC).

REFERENCES

1. Poland, A. and Glover, E. (1974). Comparison of 2,3,7,8-tetrachlorodibenzo-*p*-dioxin, a potent inducer of aryl hydrocarbon hydroxylase, with 3-methylcholanthrene. *Molecular Pharmacology*, 10, 349–359.
2. Poland, A., Glover, E., and Kende, A. S. (1976). Stereospecific, high affinity binding of 2,3,7,8-tetrachlorodibenzo-*p*-dioxin by hepatic cytosol. Evidence that the binding species is receptor for induction of aryl hydrocarbon hydroxylase. *Journal of Biological Chemistry*, 251, 4936–4946.
3. McIntosh, B. E., Hogenesch, J. B., and Bradfield, C. A. (2010). Mammalian Per–Arnt–Sim proteins in environmental adaptation. *Annual Review of Physiology*, 72, 625–645.

4. Nebert, D. W., Dalton, T. P., Okey, A. B., and Gonzalez, F. J. (2004). Role of aryl hydrocarbon receptor-mediated induction of the CYP1 enzymes in environmental toxicity and cancer. *Journal of Biological Chemistry*, 279, 23847–23850.

5. Kimmig, J. and Schulz, K. H. (1957). Chlorierte aromatische zyklische Äther als Ursache der sogenannten Chlorakne. *Natürwissenschaften*, 44, 337–338.

6. Pohjanvirta, R. and Tuomisto, J. (1994). Short-term toxicity of 2,3,7,8-tetrachlorodibenzo-p-dioxin in laboratory animals: effects, mechanisms, and animal models. *Pharmacological Reviews*, 46, 483–549.

7. Poland, A. and Knutson, J. C. (1982). 2,3,7,8-Tetrachlorodibenzo-p-dioxin and related halogenated aromatic hydrocarbons: examination of the mechanism of toxicity. *Annual Review of Pharmacology and Toxicology*, 22, 517–554.

8. Birnbaum, L. S. and Tuomisto, J. (2000). Non-carcinogenic effects of TCDD in animals. *Food Additives and Contaminants*, 17, 275–288.

9. Schwetz, B. A., Norris, J. M., Sparschu, G. L., Rowe, U. K., Gehring, P. J., Emerson, J. L., and Gerbig, C. G. (1973). Toxicology of chlorinated dibenzo-p-dioxins. *Environmental Health Perspectives*, 5, 87–99.

10. Henck, J. M., New, M. A., Kociba, R. J., and Rao, K. S. (1981). 2,3,7,8-Tetrachlorodibenzo-p-dioxin: acute oral toxicity in hamsters. *Toxicology and Applied Pharmacology*, 59, 405–407.

11. Pohjanvirta, R., Unkila, M., and Tuomisto, J. (1993). Comparative acute lethality of 2,3,7,8-tetrachlorodibenzo-p-dioxin (TCDD), 1,2,3,7,8-pentachlorodibenzo-p-dioxin and 1,2,3,4,7,8-hexachlorodibenzo-p-dioxin in the most TCDD-susceptible and the most TCDD-resistant rat strain. *Pharmacology & Toxicology*, 73, 52–56.

12. Unkila, M., Pohjanvirta, R., MacDonald, E., Tuomisto, J. T., and Tuomisto, J. (1994). Dose response and time course of alterations in tryptophan metabolism by 2,3,7,8-tetrachlorodibenzo-p-dioxin (TCDD) in the most TCDD-susceptible and the most TCDD-resistant rat strain: relationship with TCDD lethality. *Toxicology and Applied Pharmacology*, 128, 280–292.

13. Pohjanvirta, R., Boutros, P. C., Moffat, I. D., Linden, J., Wendelin, D., and Okey, A. B. (2008). Genome-wide effects of acute progressive feed restriction in liver and white adipose tissue. *Toxicology and Applied Pharmacology*, 230, 41–56.

14. Simanainen, U., Tuomisto, J. T., Pohjanvirta, R., Syrjala, P., Tuomisto, J., and Viluksela, M. (2004). Postnatal development of resistance to short-term high-dose toxic effects of 2,3,7,8-tetrachlorodibenzo-p-dioxin in TCDD-resistant and -semiresistant rats. *Toxicology and Applied Pharmacology*, 196, 11–19.

15. Linden, J., Lensu, S., Tuomisto, J., and Pohjanvirta, R. (2010). Dioxins, the aryl hydrocarbon receptor and the central regulation of energy balance. *Frontiers in Neuroendocrinology*, 31, 452–478.

16. Christian, B. J., Inhorn, S. L., and Peterson, R. E. (1986). Relationship of the wasting syndrome to lethality in rats treated with 2,3,7,8-tetrachlorodibenzo-p-dioxin. *Toxicology and Applied Pharmacology*, 82, 239–255.

17. Weber, L. W., Lebofsky, M., Stahl, B. U., Gorski, J. R., Muzi, G., and Rozman, K. (1991). Reduced activities of key enzymes of gluconeogenesis as possible cause of acute toxicity of 2,3,7,8-tetrachlorodibenzo-p-dioxin (TCDD) in rats. *Toxicology*, 66, 133–144.

18. Gasiewicz, T. A., Holscher, M. A., and Neal, R. A. (1980). The effect of total parenteral nutrition on the toxicity of 2,3,7,8-tetrachlorodibenzo-p-dioxin in the rat. *Toxicology and Applied Pharmacology*, 54, 469–488.

19. DiBartolomeis, M. J., Moore, R. W., Peterson, R. E., Christian, B. J., and Jefcoate, C. R. (1987). Altered regulation of adrenal steroidogenesis in 2,3,7,8-tetrachlorodibenzo-p-dioxin-treated rats. *Biochemical Pharmacology*, 36, 59–67.

20. Pohjanvirta, R., Laitinen, J., Vakkuri, O., Linden, J., Kokkola, T., Unkila, M., and Tuomisto, J. (1996). Mechanism by which 2,3,7,8-tetrachlorodibenzo-p-dioxin (TCDD) reduces circulating melatonin levels in the rat. *Toxicology*, 107, 85–97.

21. Moore, R. W., Jefcoate, C. R., and Peterson, R. E. (1991). 2,3,7,8-Tetrachlorodibenzo-p-dioxin inhibits steroidogenesis in the rat testis by inhibiting the mobilization of cholesterol to cytochrome P450scc. *Toxicology and Applied Pharmacology*, 109, 85–97.

22. Henry, E. C. and Gasiewicz, T. A. (1987). Changes in thyroid hormones and thyroxine glucuronidation in hamsters compared with rats following treatment with 2,3,7,8-tetrachlorodibenzo-p-dioxin. *Toxicology and Applied Pharmacology*, 89, 165–174.

23. Kohn, M. C. (2000). Effects of TCDD on thyroid hormone homeostasis in the rat. *Drug and Chemical Toxicology*, 23, 259–277.

24. Niittynen, M., Tuomisto, J. T., Auriola, S., Pohjanvirta, R., Syrjala, P., Simanainen, U., Viluksela, M., and Tuomisto, J. (2003). 2,3,7,8-Tetrachlorodibenzo-p-dioxin (TCDD)-induced accumulation of biliverdin and hepatic peliosis in rats. *Toxicological Sciences*, 71, 112–123.

25. Mably, T. A., Moore, R. W., and Peterson, R. E. (1992). In utero and lactational exposure of male rats to 2,3,7,8-tetrachlorodibenzo-p-dioxin. 1. Effects on androgenic status. *Toxicology and Applied Pharmacology*, 114, 97–107.

26. Mably, T. A., Moore, R. W., Goy, R. W., and Peterson, R. E. (1992). In utero and lactational exposure of male rats to 2,3,7,8-tetrachlorodibenzo-p-dioxin. 2. Effects on sexual behavior and the regulation of luteinizing hormone secretion in adulthood. *Toxicology and Applied Pharmacology*, 114, 108–117.

27. Kattainen, H., Tuukkanen, J., Simanainen, U., Tuomisto, J. T., Kovero, O., Lukinmaa, P. L., Alaluusua, S., Tuomisto, J., and Viluksela, M. (2001). In utero/lactational 2,3,7,8-tetrachlorodibenzo-p-dioxin exposure impairs molar tooth development in rats. *Toxicology and Applied Pharmacology*, 174, 216–224.

28. Kransler, K. M., McGarrigle, B. P., and Olson, J. R. (2007). Comparative developmental toxicity of 2,3,7,8-tetrachlorodibenzo-p-dioxin in the hamster, rat and guinea pig. *Toxicology*, 229, 214–225.

29. Poland, A. and Knutson, J. C. (1982). 2,3,7,8-Tetrachlorodibenzo-*p*-dioxin and related halogenated aromatic hydrocarbons: examination of the mechanism of toxicity. *Annual Review of Pharmacology and Toxicology*, 22, 517–554.

30. Safe, S. H. (1986). Comparative toxicology and mechanism of action of polychlorinated dibenzo-*p*-dioxins and dibenzofurans. *Annual Review of Pharmacology and Toxicology*, 26, 371–399.

31. Fernandez-Salguero, P. M., Hilbert, D. M., Rudikoff, S., Ward, J. M., and Gonzalez, F. J. (1996). Aryl-hydrocarbon receptor-deficient mice are resistant to 2,3,7,8-tetrachlorodibenzo-*p*-dioxin-induced toxicity. *Toxicology and Applied Pharmacology*, 140, 173–179.

32. Mimura, J., Yamashita, K., Nakamura, K., Morita, M., Takagi, T. N., Nakao, K., Ema, M., Sogawa, K., Yasuda, M., Katsuki, M., and Fujii-Kuriyama, Y. (1997). Loss of teratogenic response to 2,3,7,8-tetrachlorodibenzo-*p*-dioxin (TCDD) in mice lacking the Ah (dioxin) receptor. *Genes to Cells*, 2, 645–654.

33. Vorderstrasse, B. A., Steppan, L. B., Silverstone, A. E., and Kerkvliet, N. I. (2001). Aryl hydrocarbon receptor-deficient mice generate normal immune responses to model antigens and are resistant to TCDD-induced immune suppression. *Toxicology and Applied Pharmacology*, 171, 157–164.

34. Lin, T. M., Ko, K., Moore, R. W., Simanainen, U., Oberley, T. D., and Peterson, R. E. (2002). Effects of aryl hydrocarbon receptor null mutation and *in utero* and lactational 2,3,7,8-tetrachlorodibenzo-*p*-dioxin exposure on prostate and seminal vesicle development in C57BL/6 mice. *Toxicological Sciences*, 68, 479–487.

35. Nishimura, N., Yonemoto, J., Miyabara, Y., Fujii-Kuriyama, Y., and Tohyama, C. (2005). Altered thyroxin and retinoid metabolic response to 2,3,7,8-tetrachlorodibenzo-*p*-dioxin in aryl hydrocarbon receptor-null mice. *Archives of Toxicology*, 79, 260–267.

36. Davies, R., Clothier, B., Robinson, S. W., Edwards, R. E., Greaves, P., Luo, J., Gant, T. W., Chernova, T., and Smith, A. G. (2008). Essential role of the AH receptor in the dysfunction of heme metabolism induced by 2,3,7,8-tetrachlorodibenzo-*p*-dioxin. *Chemical Research in Toxicology*, 21, 330–340.

37. Thurmond, T. S., Silverstone, A. E., Baggs, R. B., Quimby, F. W., Staples, J. E., and Gasiewicz, T. A. (1999). A chimeric aryl hydrocarbon receptor knockout mouse model indicates that aryl hydrocarbon receptor activation in hematopoietic cells contributes to the hepatic lesions induced by 2,3,7,8-tetrachlorodibenzo-*p*-dioxin. *Toxicology and Applied Pharmacology*, 158, 33–40.

38. Walisser, J. A., Glover, E., Pande, K., Liss, A. L., and Bradfield, C. A. (2005). Aryl hydrocarbon receptor-dependent liver development and hepatotoxicity are mediated by different cell types. *Proceedings of the National Academy of Sciences of the United States of America*, 102, 17858–17863.

39. Kawamata, M. and Ochiya, T. (2010). Generation of genetically modified rats from embryonic stem cells. *Proceedings of the National Academy of Sciences of the United States of America*, 107, 14223–14228.

40. Bunger, M. K., Moran, S. M., Glover, E., Thomae, T. L., Lahvis, G. P., Lin, B. C., and Bradfield, C. A. (2003). Resistance to 2,3,7,8-tetrachlorodibenzo-*p*-dioxin toxicity and abnormal liver development in mice carrying a mutation in the nuclear localization sequence of the aryl hydrocarbon receptor. *Journal of Biological Chemistry*, 278, 17767–17774.

41. Bunger, M. K., Glover, E., Moran, S. M., Walisser, J. A., Lahvis, G. P., Hsu, E. L., and Bradfield, C. A. (2008). Abnormal liver development and resistance to 2,3,7,8-tetrachlorodibenzo-*p*-dioxin toxicity in mice carrying a mutation in the DNA-binding domain of the aryl hydrocarbon receptor. *Toxicological Sciences*, 106, 83–92.

42. Walisser, J. A., Bunger, M. K., Glover, E., Harstad, E. B., and Bradfield, C. A. (2004). Patent ductus venosus and dioxin resistance in mice harboring a hypomorphic Arnt allele. *Journal of Biological Chemistry*, 279, 16326–16331.

43. Nukaya, M., Walisser, J. A., Moran, S. M., Kennedy, G. D., and Bradfield, C. A. (2010). Aryl hydrocarbon receptor nuclear translocator in hepatocytes is required for aryl hydrocarbon receptor-mediated adaptive and toxic responses in liver. *Toxicological Sciences*, 118, 554–563.

44. Walisser, J. A., Bunger, M. K., Glover, E., and Bradfield, C. A. (2004). Gestational exposure of Ahr and Arnt hypomorphs to dioxin rescues vascular development. *Proceedings of the National Academy of Sciences of the United States of America*, 101, 16677–16682.

45. Sciullo, E. M., Dong, B., Vogel, C. F., and Matsumura, F. (2009). Characterization of the pattern of the nongenomic signaling pathway through which TCDD-induces early inflammatory responses in U937 human macrophages. *Chemosphere*, 74, 1531–1537.

46. Sciullo, E. M., Vogel, C. F., Li, W., and Matsumura, F. (2008). Initial and extended inflammatory messages of the nongenomic signaling pathway of the TCDD-activated Ah receptor in U937 macrophages. *Archives of Biochemistry and Biophysics*, 480, 143–155.

47. Dong, B. and Matsumura, F. (2009). The conversion of rapid TCCD nongenomic signals to persistent inflammatory effects via select protein kinases in MCF10A cells. *Molecular Endocrinology (Baltimore, Md.)* 23, 549–558.

48. Dong, B. and Matsumura, F. (2008). Roles of cytosolic phospholipase A2 and Src kinase in the early action of 2,3,7,8-tetrachlorodibenzo-*p*-dioxin through a nongenomic pathway in MCF10A cells. *Molecular Pharmacology*, 74, 255–263.

49. Dong, B., Nishimura, N., Vogel, C. F., Tohyama, C., and Matsumura, F. (2010). TCDD-induced cyclooxygenase-2 expression is mediated by the nongenomic pathway in mouse MMDD1 macula densa cells and kidneys. *Biochemical Pharmacology*, 79, 487–497.

50. Matsumura, F. (2009). The significance of the nongenomic pathway in mediating inflammatory signaling of the dioxin-activated Ah receptor to cause toxic effects. *Biochemical Pharmacology*, 77, 608–626.

51. Nishimura, N., Matsumura, F., Vogel, C. F., Nishimura, H., Yonemoto, J., Yoshioka, W., and Tohyama, C. (2008). Critical role of cyclooxygenase-2 activation in pathogenesis of hydronephrosis caused by lactational exposure of mice to dioxin. *Toxicology and Applied Pharmacology*, 231, 374–383.

52. Whitelaw, M. L., Gottlicher, M., Gustafsson, J. A., and Poellinger, L. (1993). Definition of a novel ligand binding domain of a nuclear bHLH receptor: co-localization of ligand and hsp90 binding activities within the regulable inactivation domain of the dioxin receptor. *EMBO Journal*, 12, 4169–4179.

53. Pongratz, I., Mason, G. G., and Poellinger, L. (1992). Dual roles of the 90-kDa heat shock protein hsp90 in modulating functional activities of the dioxin receptor. Evidence that the dioxin receptor functionally belongs to a subclass of nuclear receptors which require hsp90 both for ligand binding activity and repression of intrinsic DNA binding activity. *Journal of Biological Chemistry*, 267, 13728–13734.

54. Kazlauskas, A., Sundstrom, S., Poellinger, L., and Pongratz, I. (2001). The hsp90 chaperone complex regulates intracellular localization of the dioxin receptor. *Molecular and Cellular Biology*, 21, 2594–2607.

55. Denison, M. S. and Nagy, S. R. (2003). Activation of the aryl hydrocarbon receptor by structurally diverse exogenous and endogenous chemicals. *Annual Review of Pharmacology and Toxicology*, 43, 309–334.

56. Rowlands, J. C., McEwan, I. J., and Gustafsson, J. A. (1996). Trans-activation by the human aryl hydrocarbon receptor and aryl hydrocarbon receptor nuclear translocator proteins: direct interactions with basal transcription factors. *Molecular Pharmacology*, 50, 538–548.

57. Ma, Q., Dong, L., and Whitlock, J. P., Jr. (1995). Transcriptional activation by the mouse Ah receptor. Interplay between multiple stimulatory and inhibitory functions. *Journal of Biological Chemistry*, 270, 12697–12703.

58. Ko, H. P., Okino, S. T., Ma, Q., and Whitlock, J. P., Jr. (1997). Transactivation domains facilitate promoter occupancy for the dioxin-inducible CYP1A1 gene *in vivo*. *Molecular and Cellular Biology*, 17, 3497–3507.

59. Jain, S., Dolwick, K. M., Schmidt, J. V., and Bradfield, C. A. (1994). Potent transactivation domains of the Ah receptor and the Ah receptor nuclear translocator map to their carboxyl termini. *Journal of Biological Chemistry*, 269, 31518–31524.

60. Kumar, M. B., Ramadoss, P., Reen, R. K., Vanden Heuvel, J. P., and Perdew, G. H. (2001). The Q-rich subdomain of the human Ah receptor transactivation domain is required for dioxin-mediated transcriptional activity. *Journal of Biological Chemistry*, 276, 42302–42310.

61. Chapman, D. E. and Schiller, C. M. (1985). Dose-related effects of 2,3,7,8-tetrachlorodibenzo-p-dioxin (TCDD) in C57BL/6J and DBA/2J mice. *Toxicology and Applied Pharmacology*, 78, 147–157.

62. Birnbaum, L. S., McDonald, M. M., Blair, P. C., Clark, A. M., and Harris, M. W. (1990). Differential toxicity of 2,3,7,8-tetrachlorodibenzo-p-dioxin (TCDD) in C57BL/6J mice congenic at the Ah Locus. *Fundamental and Applied Toxicology*, 15, 186–200.

63. Kerkvliet, N. I., Baecher-Steppan, L., Smith, B. B., Youngberg, J. A., Henderson, M. C., and Buhler, D. R. (1990). Role of the Ah locus in suppression of cytotoxic T lymphocyte activity by halogenated aromatic hydrocarbons (PCBs and TCDD): structure–activity relationships and effects in C57Bl/6 mice congenic at the Ah locus. *Fundamental and Applied Toxicology*, 14, 532–541.

64. Davies, R., Clothier, B., Robinson, S. W., Edwards, R. E., Greaves, P., Luo, J., Gant, T. W., Chernova, T., and Smith, A. G. (2008). Essential role of the AH receptor in the dysfunction of heme metabolism induced by 2,3,7,8-tetrachlorodibenzo-p-dioxin. *Chemical Research in Toxicology*, 21, 330–340.

65. Lin, F. H., Clark, G., Birnbaum, L. S., Lucier, G. W., and Goldstein, J. A. (1991). Influence of the Ah locus on the effects of 2,3,7,8-tetrachlorodibenzo-p-dioxin on the hepatic epidermal growth factor receptor. *Molecular Pharmacology*, 39, 307–313.

66. Lin, F. H., Stohs, S. J., Birnbaum, L. S., Clark, G., Lucier, G. W., and Goldstein, J. A. (1991). The effects of 2,3,7,8-tetrachlorodibenzo-p-dioxin (TCDD) on the hepatic estrogen and glucocorticoid receptors in congenic strains of Ah responsive and Ah nonresponsive C57BL/6J mice. *Toxicology and Applied Pharmacology*, 108, 129–139.

67. Poland, A. and Glover, E. (1980). 2,3,7,8-Tetrachlorodibenzo-p-dioxin: segregation of toxicity with the Ah locus. *Molecular Pharmacology*, 17, 86–94.

68. Hong, L. H., McKinney, J. D., and Luster, M. I. (1987). Modulation of 2,3,7,8-tetrachlorodibenzo-p-dioxin (TCDD)-mediated myelotoxicity by thyroid hormones. *Biochemical Pharmacology*, 36, 1361–1365.

69. Mohammadpour, H., Murray, W. J., and Stohs, S. J. (1988). 2,3,7,8-Tetrachlorodibenzo-p-dioxin (TCDD)-induced lipid peroxidation in genetically responsive and non-responsive mice. *Archives of Environmental Contamination and Toxicology*, 17, 645–650.

70. Clark, G. C., Taylor, M. J., Tritscher, A. M., and Lucier, G. W. (1991). Tumor necrosis factor involvement in 2,3,7,8-tetrachlorodibenzo-p-dioxin-mediated endotoxin hypersensitivity in C57BL/6J mice congenic at the Ah locus. *Toxicology and Applied Pharmacology*, 111, 422–431.

71. Harper, N., Connor, K., and Safe, S. (1993). Immunotoxic potencies of polychlorinated biphenyl (PCB), dibenzofuran (PCDF) and dibenzo-p-dioxin (PCDD) congeners in C57BL/6 and DBA/2 mice. *Toxicology*, 80, 217–227.

72. Beebe, L. E., Fornwald, L. W., Diwan, B. A., Anver, M. R., and Anderson, L. M. (1995). Promotion of N-nitrosodiethylamine-initiated hepatocellular tumors and hepatoblastomas by 2,3,7,8-tetrachlorodibenzo-p-dioxin or Aroclor 1254 in C57BL/6, DBA/2, and B6D2F1 mice. *Cancer Research*, 55, 4875–4880.

73. Okey, A. B., Vella, L. M., and Harper, P. A. (1989). Detection and characterization of a low affinity form of cytosolic Ah receptor in livers of mice nonresponsive to induction of

cytochrome P1-450 by 3-methylcholanthrene. *Molecular Pharmacology*, 35, 823–830.

74. Chang, C., Smith, D. R., Prasad, V. S., Sidman, C. L., Nebert, D. W., and Puga, A. (1993). Ten nucleotide differences, five of which cause amino acid changes, are associated with the Ah receptor locus polymorphism of C57BL/6 and DBA/2 mice. *Pharmacogenetics*, 3, 312–321.

75. Poland, A., Palen, D., and Glover, E. (1994). Analysis of the four alleles of the murine aryl hydrocarbon receptor. *Molecular Pharmacology*, 46, 915–921.

76. Beatty, P. W., Holscher, M. A., and Neal, R. A. (1976). Toxicity of 2,3,7,8-tetrachlorodibenzo-*p*-dioxin in larval and adult forms of *Rana catesbeiana*. *Bulletin of Environmental Contamination and Toxicology*, 16, 578–581.

77. Jung, R. E. and Walker, M. K. (1997). Effects of 2,3,7,8-tetrachlorodibenzo-*p*-dioxin (TCDD) on development of anuran amphibians. *Environmental Toxicology and Chemistry*, 16, 230–240.

78. Lavine, J. A., Rowatt, A. J., Klimova, T., Whitington, A. J., Dengler, E., Beck, C., and Powell, W. H. (2005). Aryl hydrocarbon receptors in the frog *Xenopus laevis*: two AhR1 paralogs exhibit low affinity for 2,3,7,8-tetrachlorodibenzo-*p*-dioxin (TCDD). *Toxicological Sciences*, 88, 60–72.

79. Yasui, T., Kim, E. Y., Iwata, H., Franks, D. G., Karchner, S. I., Hahn, M. E., and Tanabe, S. (2007). Functional characterization and evolutionary history of two aryl hydrocarbon receptor isoforms (AhR1 and AhR2) from avian species. *Toxicological Sciences*, 99, 101–117.

80. Head, J. A., Hahn, M. E., and Kennedy, S. W. (2008). Key amino acids in the aryl hydrocarbon receptor predict dioxin sensitivity in avian species. *Environmental Science & Technology*, 42, 7535–7541.

81. Manchester, D. K., Gordon, S. K., Golas, C. L., Roberts, E. A., and Okey, A. B. (1987). Ah receptor in human placenta: stabilization by molybdate and characterization of binding of 2,3,7,8-tetrachlorodibenzo-*p*-dioxin, 3-methylcholanthrene, and benzo(*a*)pyrene. *Cancer Research*, 47, 4861–4868.

82. Ramadoss, P. and Perdew, G. H. (2004). Use of 2-azido-3-[^{125}I]iodo-7,8-dibromodibenzo-*p*-dioxin as a probe to determine the relative ligand affinity of human versus mouse aryl hydrocarbon receptor in cultured cells. *Molecular Pharmacology*, 66, 129–136.

83. Flaveny, C. A., Murray, I. A., Chiaro, C. R., and Perdew, G. H. (2009). Ligand selectivity and gene regulation by the human aryl hydrocarbon receptor in transgenic mice. *Molecular Pharmacology*, 75, 1412–1420.

84. Denison, M. S. and Wilkinson, C. F. (1985). Identification of the Ah receptor in selected mammalian species and induction of aryl hydrocarbon hydroxylase. *European Journal of Biochemistry*, 147, 429–435.

85. Korkalainen, M., Tuomisto, J., and Pohjanvirta, R. (2000). Restructured transactivation domain in hamster AH receptor. *Biochemical and Biophysical Research Communications*, 273, 272–281.

86. Pohjanvirta, R., Wong, J. M., Li, W., Harper, P. A., Tuomisto, J., and Okey, A. B. (1998). Point mutation in intron sequence causes altered carboxyl-terminal structure in the aryl hydrocarbon receptor of the most 2,3,7,8-tetrachlorodibenzo-*p*-dioxin-resistant rat strain. *Molecular Pharmacology*, 54, 86–93.

87. Pohjanvirta, R., Viluksela, M., Tuomisto, J. T., Unkila, M., Karasinska, J., Franc, M. A., Holowenko, M., Giannone, J. V., Harper, P. A., Tuomisto, J., and Okey, A. B. (1999). Physicochemical differences in the AH receptors of the most TCDD-susceptible and the most TCDD-resistant rat strains. *Toxicology and Applied Pharmacology*, 155, 82–95.

88. Korkalainen, M., Tuomisto, J., and Pohjanvirta, R. (2001). The AH receptor of the most dioxin-sensitive species, guinea pig, is highly homologous to the human AH receptor. *Biochemical and Biophysical Research Communications*, 285, 1121–1129.

89. Gasiewicz, T. A., Rucci, G., Henry, E. C., and Baggs, R. B. (1986). Changes in hamster hepatic cytochrome P-450, ethoxycoumarin *O*-deethylase, and reduced NAD(P): menadione oxidoreductase following treatment with 2,3,7,8-tetrachlorodibenzo-*p*-dioxin. Partial dissociation of temporal and dose–response relationships from elicited toxicity. *Biochemical Pharmacology*, 35, 2737–2742.

90. Pohjanvirta, R., Juvonen, R., Karenlampi, S., Raunio, H., and Tuomisto, J. (1988). Hepatic Ah-receptor levels and the effect of 2,3,7,8-tetrachlorodibenzo-*p*-dioxin (TCDD) on hepatic microsomal monooxygenase activities in a TCDD-susceptible and -resistant rat strain. *Toxicology and Applied Pharmacology*, 92, 131–140.

91. Unkila, M., Pohjanvirta, R., Honkakoski, P., Torronen, R., and Tuomisto, J. (1993). 2,3,7,8-Tetrachlorodibenzo-*p*-dioxin (TCDD) induced ethoxyresorufin-O-deethylase (EROD) and aldehyde dehydrogenase (ALDH3) activities in the brain and liver. A comparison between the most TCDD-susceptible and the most TCDD-resistant rat strain. *Biochemical Pharmacology*, 46, 651–659.

92. Huuskonen, H., Unkila, M., Pohjanvirta, R., and Tuomisto, J. (1994). Developmental toxicity of 2,3,7,8-tetrachlorodibenzo-*p*-dioxin (TCDD) in the most TCDD-resistant and -susceptible rat strains. *Toxicology and Applied Pharmacology*, 124, 174–180.

93. Pohjanvirta, R., Vartiainen, T., Uusi-Rauva, A., Monkkonen, J., and Tuomisto, J. (1990). Tissue distribution, metabolism, and excretion of ^{14}C-TCDD in a TCDD-susceptible and a TCDD-resistant rat strain. *Pharmacology & Toxicology*, 66, 93–100.

94. Pohjanvirta, R. (1991). Studies on the mechanism of acute toxicity of 2,3,7,8-tetrachlorodibenzo-*p*-dioxin (TCDD) in rats. Thesis, Publications of the National Public Health Institute (NPHI) A1.

95. Safe, S. (1990). Polychlorinated biphenyls (PCBs), dibenzo-*p*-dioxins (PCDDs), dibenzofurans (PCDFs), and related compounds: environmental and mechanistic considerations which support the development of toxic equivalency factors (TEFs). *Critical Reviews in Toxicology*, 21, 51–88.

96. Mekenyan, O. G., Veith, G. D., Call, D. J., and Ankley, G. T. (1996). A QSAR evaluation of Ah receptor binding of

halogenated aromatic xenobiotics. *Environmental Health Perspectives*, 104, 1302–1310.

97. Tuppurainen, K. and Ruuskanen, J. (2000). Electronic eigenvalue (EEVA): a new QSAR/QSPR descriptor for electronic substituent effects based on molecular orbital energies. A QSAR approach to the Ah receptor binding affinity of polychlorinated biphenyls (PCBs), dibenzo-*p*-dioxins (PCDDs) and dibenzofurans (PCDFs). *Chemosphere*, 41, 843–848.

98. Pohjanvirta, R., Unkila, M., Linden, J., Tuomisto, J. T., and Tuomisto, J. (1995). Toxic equivalency factors do not predict the acute toxicities of dioxins in rats. *European Journal of Pharmacology*, 293, 341–353.

99. Pohjanvirta, R. (1990). TCDD resistance is inherited as an autosomal dominant trait in the rat. *Toxicology Letters*, 50, 49–56.

100. Tuomisto, J. T., Viluksela, M., Pohjanvirta, R., and Tuomisto, J. (1999). The AH receptor and a novel gene determine acute toxic responses to TCDD: segregation of the resistant alleles to different rat lines. *Toxicology and Applied Pharmacology*, 155, 71–81.

101. Moffat, I. D., Roblin, S., Harper, P. A., Okey, A. B., and Pohjanvirta, R. (2007). Aryl hydrocarbon receptor splice variants in the dioxin-resistant rat: tissue expression and transactivational activity. *Molecular Pharmacology*, 72, 956–966.

102. Korkalainen, M., Tuomisto, J., and Pohjanvirta, R. (2003). Identification of novel splice variants of ARNT and ARNT2 in the rat. *Biochemical and Biophysical Research Communications*, 303, 1095–1100.

103. Korkalainen, M., Tuomisto, J., and Pohjanvirta, R. (2004). Primary structure and inducibility by 2,3,7,8-tetrachlorodibenzo-*p*-dioxin (TCDD) of aryl hydrocarbon receptor repressor in a TCDD-sensitive and a TCDD-resistant rat strain. *Biochemical and Biophysical Research Communications*, 315, 123–131.

104. Matsumura, F., Enan, E., Dunlap, D. Y., Pinkerton, K. E., and Peake, J. (1997). Altered *in vivo* toxicity of 2,3,7,8-tetrachlorodibenzo-*p*-dioxin (TCDD) in C-SRC deficient mice. *Biochemical Pharmacology*, 53, 1397–1404.

105. Enan, E., Dunlap, D. Y., and Matsumura, F. (1998). Use of c-Src and c-Fos knockout mice for the studies on the role of c-Src kinase signaling in the expression of toxicity of TCDD. *Journal of Biochemical and Molecular Toxicology*, 12, 263–274.

106. Dunlap, D. Y., Ikeda, I., Nagashima, H., Vogel, C. F., and Matsumura, F. (2002). Effects of src-deficiency on the expression of *in vivo* toxicity of TCDD in a strain of c-src knockout mice procured through six generations of backcrossings to C57BL/6 mice. *Toxicology*, 172, 125–141.

107. Pohjanvirta, R. (2009). Transgenic mouse lines expressing rat AH receptor variants—a new animal model for research on AH receptor function and dioxin toxicity mechanisms. *Toxicology and Applied Pharmacology*, 236, 166–182.

108. Franc, M. A., Pohjanvirta, R., Tuomisto, J., and Okey, A. B. (2001). *In vivo* up-regulation of aryl hydrocarbon receptor expression by 2,3,7,8-tetrachlorodibenzo-*p*-dioxin (TCDD) in a dioxin-resistant rat model. *Biochemical Pharmacology*, 62, 1565–1578.

109. Franc, M. A., Pohjanvirta, R., Tuomisto, J., and Okey, A. B. (2001). Persistent, low-dose 2,3,7,8-tetrachlorodibenzo-*p*-dioxin exposure: effect on aryl hydrocarbon receptor expression in a dioxin-resistance model. *Toxicology and Applied Pharmacology*, 175, 43–53.

110. Franc, M. A., Moffat, I. D., Boutros, P. C., Tuomisto, J. T., Tuomisto, J., Pohjanvirta, R., and Okey, A. B. (2008). Patterns of dioxin-altered mRNA expression in livers of dioxin-sensitive versus dioxin-resistant rats. *Archives of Toxicology*, 82, 809–830.

111. Ko, H. P., Okino, S. T., Ma, Q., and Whitlock, J. P., Jr. (1996). Dioxin-induced CYP1A1 transcription *in vivo*: the aromatic hydrocarbon receptor mediates transactivation, enhancer-promoter communication, and changes in chromatin structure. *Molecular and Cellular Biology*, 16, 430–436.

112. Reen, R. K., Cadwallader, A., and Perdew, G. H. (2002). The subdomains of the transactivation domain of the aryl hydrocarbon receptor (AhR) inhibit AhR and estrogen receptor transcriptional activity. *Archives of Biochemistry and Biophysics*, 408, 93–102.

113. Viluksela, M., Bager, Y., Tuomisto, J. T., Scheu, G., Unkila, M., Pohjanvirta, R., Flodstrom, S., Kosma, V. M., Maki-Paakkanen, J., Vartiainen, T., Klimm, C., Schramm, K. W., Warngard, L., and Tuomisto, J. (2000). Liver tumor-promoting activity of 2,3,7,8-tetrachlorodibenzo-*p*-dioxin (TCDD) in TCDD-sensitive and TCDD-resistant rat strains. *Cancer Research*, 60, 6911–6920.

114. McConnell, E. E., Moore, J. A., Haseman, J. K., and Harris, M. W. (1978). The comparative toxicity of chlorinated dibenzo-*p*-dioxins in mice and guinea pigs. *Toxicology and Applied Pharmacology*, 44, 335–356.

115. Seefeld, M. D., Corbett, S. W., Keesey, R. E., and Peterson, R. E. (1984). Characterization of the wasting syndrome in rats treated with 2,3,7,8-tetrachlorodibenzo-*p*-dioxin. *Toxicology and Applied Pharmacology*, 73, 311–322.

116. Kelling, C. K., Christian, B. J., Inhorn, S. L., and Peterson, R. E. (1985). Hypophagia-induced weight loss in mice, rats, and guinea pigs treated with 2,3,7,8-tetrachlorodibenzo-*p*-dioxin. *Fundamental and Applied Toxicology*, 5, 700–712.

117. Moore, R. W., Potter, C. L., Theobald, H. M., Robinson, J. A., and Peterson, R. E. (1985). Androgenic deficiency in male rats treated with 2,3,7,8-tetrachlorodibenzo-*p*-dioxin. *Toxicology and Applied Pharmacology*, 79, 99–111.

118. Kleeman, J. M., Moore, R. W., and Peterson, R. E. (1990). Inhibition of testicular steroidogenesis in 2,3,7,8-tetrachlorodibenzo-*p*-dioxin-treated rats: evidence that the key lesion occurs prior to or during pregnenolone formation. *Toxicology and Applied Pharmacology*, 106, 112–125.

119. Gorski, J. R. and Rozman, K. (1987). Dose–response and time course of hypothyroxinemia and hypoinsulinemia and characterization of insulin hypersensitivity in 2,3,7,8-tetrachlorodibenzo-*p*-dioxin (TCDD)-treated rats. *Toxicology*, 44, 297–307.

120. Bastomsky, C. H. (1977). Enhanced thyroxine metabolism and high uptake goiters in rats after a single dose of 2,3,7,8-tetrachlorodibenzo-*p*-dioxin. *Endocrinology*, 101, 292–296.

121. Weber, L. W., Lebofsky, M., Stahl, B. U., Smith, S., and Rozman, K. K. (1995). Correlation between toxicity and effects on intermediary metabolism in 2,3,7,8-tetrachlorodibenzo-*p*-dioxin-treated male C57BL/6J and DBA/2J mice. *Toxicology and Applied Pharmacology, 131*, 155–162.

122. Gorski, J. R., Muzi, G., Weber, L. W., Pereira, D. W., Arceo, R. J., Iatropoulos, M. J., and Rozman, K. (1988). Some endocrine and morphological aspects of the acute toxicity of 2,3,7,8-tetrachlorodibenzo-*p*-dioxin (TCDD). *Toxicologic Pathology, 16*, 313–320.

123. Novelli, M., Piaggi, S., and De Tata, V. (2005). 2,3,7,8-Tetrachlorodibenzo-*p*-dioxin-induced impairment of glucose-stimulated insulin secretion in isolated rat pancreatic islets. *Toxicology Letters, 156*, 307–314.

124. Bestervelt, L. L., Cai, Y., Piper, D. W., Nolan, C. J., Pitt, J. A., and Piper, W. N. (1993). TCDD alters pituitary–adrenal function. I: Adrenal responsiveness to exogenous ACTH. *Neurotoxicology and Teratology, 15*, 365–367.

125. Potter, C. L., Moore, R. W., Inhorn, S. L., Hagen, T. C., and Peterson, R. E. (1986). Thyroid status and thermogenesis in rats treated with 2,3,7,8-tetrachlorodibenzo-*p*-dioxin. *Toxicology and Applied Pharmacology, 84*, 45–55.

126. Pohjanvirta, R., Unkila, M., Tuomisto, J. T., Vuolteenaho, O., Leppaluoto, J., and Tuomisto, J. (1993). Effect of 2,3,7,8-tetrachlorodibenzo-*p*-dioxin (TCDD) on plasma and tissue beta-endorphin-like immunoreactivity in the most TCDD-susceptible and the most TCDD-resistant rat strain. *Life Sciences, 53*, 1479–1487.

127. Potter, C. L., Sipes, I. G., and Russell, D. H. (1983). Hypothyroxinemia and hypothermia in rats in response to 2,3,7,8-tetrachlorodibenzo-*p*-dioxin administration. *Toxicology and Applied Pharmacology, 69*, 89–95.

128. Pohjanvirta, R., Kulju, T., Morselt, A. F., Tuominen, R., Juvonen, R., Rozman, K., Mannisto, P., Collan, Y., Sainio, E. L., and Tuomisto, J. (1989). Target tissue morphology and serum biochemistry following 2,3,7,8-tetrachlorodibenzo-*p*-dioxin (TCDD) exposure in a TCDD-susceptible and a TCDD-resistant rat strain. *Fundamental and Applied Toxicology, 12*, 698–712.

129. Fletcher, N., Wahlstrom, D., Lundberg, R., Nilsson, C. B., Nilsson, K. C., Stockling, K., Hellmold, H., and Hakansson, H. (2005). 2,3,7,8-Tetrachlorodibenzo-*p*-dioxin (TCDD) alters the mRNA expression of critical genes associated with cholesterol metabolism, bile acid biosynthesis, and bile transport in rat liver: a microarray study. *Toxicology and Applied Pharmacology, 207*, 1–24.

130. Simanainen, U., Tuomisto, J. T., Tuomisto, J., and Viluksela, M. (2003). Dose–response analysis of short-term effects of 2,3,7, 8-tetrachlorodibenzo-*p*-dioxin in three differentially susceptible rat lines. *Toxicology and Applied Pharmacology, 187*, 128–136.

131. Cantoni, L., Rizzardini, M., Graziani, A., Carugo, C., and Garattini, S. (1987). Effects of chlorinated organics on intermediates in the heme pathway and on uroporphyrinogen decarboxylase. *Annals of the New York Academy of Sciences, 514*, 128–140.

132. Cantoni, L., Salmona, M., and Rizzardini, M. (1981). Porphyrogenic effect of chronic treatment with 2,3,7,8-tetrachlorodibenzo-*p*-dioxin in female rats. Dose–effect relationship following urinary excretion of porphyrins. *Toxicology and Applied Pharmacology, 57*, 156–163.

133. Pelclova, D., Urban, P., Preiss, J., Lukas, E., Fenclova, Z., Navratil, T., Dubska, Z., and Senholdova, Z. (2006). Adverse health effects in humans exposed to 2,3,7,8-tetrachlorodibenzo-*p*-dioxin (TCDD). *Reviews on Environmental Health, 21*, 119–138.

134. Zinkl, J. G., Vos, J. G., Moore, J. A., and Gupta, B. N. (1973). Hematologic and clinical chemistry effects of 2,3,7,8-tetrachlorodibenzo-*p*-dioxin in laboratory animals. *Environmental Health Perspectives, 5*, 111–118.

135. Yang, K. H., Yoo, B. S., and Choe, S. Y. (1983). Effects of halogenated dibenzo-*p*-dioxins on plasma disappearance and biliary excretion of ouabain in rats. *Toxicology Letters, 15*, 259–264.

136. Fletcher, N., Hanberg, A., and Hakansson, H. (2001). Hepatic vitamin A depletion is a sensitive marker of 2,3,7,8-tetrachlorodibenzo-*p*-dioxin (TCDD) exposure in four rodent species. *Toxicological Sciences, 62*, 166–175.

137. Yang, Y. M., Huang, D. Y., Liu, G. F., Zhong, J. C., Du, K., Li, Y. F., and Song, X. H. (2005). Effects of 2,3,7,8-tetrachlorodibenzo-*p*-dioxin on vitamin A metabolism in mice. *Journal of Biochemical and Molecular Toxicology, 19*, 327–335.

138. Passarini, B., Infusino, S. D., and Kasapi, E. (2010). Chloracne: still cause for concern. *Dermatology (Basel, Switzerland), 221*, 63–70.

139. McNulty, W. P., Pomerantz, I., and Farrell, T. (1981). Chronic toxicity of 2,3,7,8-tetrachlorodibenzofuran for rhesus macaques. *Food and Cosmetics Toxicology, 19*, 57–65.

140. Vos, J. G., Van Leeuwen, F.X., and de Jong, P. (1982). Acnegenic activity of 3-methylcholanthrene and benzo[*a*]pyrene, and a comparative study with 2,3,7,8-tetrachlorodibenzo-*p*-dioxin in the rabbit and hairless mouse. *Toxicology, 23*, 187–196.

141. Jones, G. and Greig, J. B. (1975). Pathological changes in the liver of mice given 2,3,7,8-tetrachlorodibenzo-*p*-dioxin. *Experientia, 31*, 1315–1317.

142. Jones, G. and Butler, W. H. (1974). A morphological study of the liver lesion induced by 2,3,7,8-tetrachlorodibenzo-*p*-dioxin in rats. *Journal of Pathology, 112*, 93–97.

143. Vos, J. G. and Beems, R. B. (1971). Dermal toxicity studies of technical polychlorinated biphenyls and fractions thereof in rabbits. *Toxicology and Applied Pharmacology, 19*, 617–633.

144. Chahoud, I., Hartmann, J., Rune, G. M., and Neubert, D. (1992). Reproductive toxicity and toxicokinetics of 2,3,7,8-tetrachlorodibenzo-*p*-dioxin. 3. Effects of single doses on the testis of male rats. *Archives of Toxicology, 66*, 567–572.

145. Johnson, L., Dickerson, R., Safe, S. H., Nyberg, C. L., Lewis, R. P., and Welsh, T. H., Jr. (1992). Reduced Leydig cell volume and function in adult rats exposed to 2,3,7,8-tetrachlorodibenzo-*p*-dioxin without a significant effect on spermatogenesis. *Toxicology, 76*, 103–118.

146. Jokinen, M. P., Walker, N. J., Brix, A. E., Sells, D. M., Haseman, J. K., and Nyska, A. (2003). Increase in cardiovascular pathology in female Sprague-Dawley rats following chronic treatment with 2,3,7,8-tetrachlorodibenzo-p-dioxin and 3,3′,4,4′,5-pentachlorobiphenyl. *Cardiovascular Toxicology*, 3, 299–310.

147. Murante, F. G. and Gasiewicz, T. A. (2000). Hemopoietic progenitor cells are sensitive targets of 2,3,7,8-tetrachlorodibenzo-p-dioxin in C57BL/6J mice. *Toxicological Sciences*, 54, 374–383.

148. Moos, A. B., Baecher-Steppan, L., and Kerkvliet, N. I. (1994). Acute inflammatory response to sheep red blood cells in mice treated with 2,3,7,8-tetrachlorodibenzo-p-dioxin: the role of proinflammatory cytokines, IL-1 and TNF. *Toxicology and Applied Pharmacology*, 127, 331–335.

149. Fan, F., Yan, B., Wood, G., Viluksela, M., and Rozman, K. K. (1997). Cytokines (IL-1β and TNFα) in relation to biochemical and immunological effects of 2,3,7,8-tetrachlorodibenzo-p-dioxin (TCDD) in rats. *Toxicology*, 116, 9–16.

150. Kim, H. J., Jeong, K. S., Park, S. J., Cho, S. W., Son, H. Y., Kim, S. R., Kim, S. H., An, M. Y., and Ryu, S. Y. (2003). Effects of benzo[alpha]pyrene, 2-bromopropane, phenol and 2,3,7,8-tetrachlorodibenzo-p-dioxin on IL-6 production in mice after single or repeated exposure. *In Vivo* (Athens, Greece), 17, 269–275.

151. Quintana, F. J., Basso, A. S., Iglesias, A. H., Korn, T., Farez, M. F., Bettelli, E., Caccamo, M., Oukka, M., and Weiner, H. L. (2008). Control of T_{reg} and T_H17 cell differentiation by the aryl hydrocarbon receptor. *Nature*, 453, 65–71.

152. Abbott, B. D. and Birnbaum, L. S. (1990). TCDD-induced altered expression of growth factors may have a role in producing cleft palate and enhancing the incidence of clefts after coadministration of retinoic acid and TCDD. *Toxicology and Applied Pharmacology*, 106, 418–432.

153. Dohr, O., Vogel, C., and Abel, J. (1994). Modulation of growth factor expression by 2,3,7,8-tetrachlorodibenzo-p-dioxin. *Experimental and Clinical Immunogenetics*, 11, 142–148.

154. Kakeyama, M., Sone, H., and Tohyama, C. (2001). Changes in expression of NMDA receptor subunit mRNA by perinatal exposure to dioxin. *Neuroreport*, 12, 4009–4012.

155. Widholm, J. J., Seo, B. W., Strupp, B. J., Seegal, R. F., and Schantz, S. L. (2003). Effects of perinatal exposure to 2,3,7,8-tetrachlorodibenzo-p-dioxin on spatial and visual reversal learning in rats. *Neurotoxicology and Teratology*, 25, 459–471.

156. Schantz, S. L. and Bowman, R. E. (1989). Learning in monkeys exposed perinatally to 2,3,7,8-tetrachlorodibenzo-p-dioxin (TCDD). *Neurotoxicology and Teratology*, 11, 13–19.

157. Aragon, A. C., Kopf, P. G., Campen, M. J., Huwe, J. K., and Walker, M. K. (2008). *In utero* and lactational 2,3,7,8-tetrachlorodibenzo-p-dioxin exposure: effects on fetal and adult cardiac gene expression and adult cardiac and renal morphology. *Toxicological Sciences*, 101, 321–330.

158. Rao, M. S., Subbarao, V., Prasad, J. D., and Scarpelli, D. G. (1988). Carcinogenicity of 2,3,7,8-tetrachlorodibenzo-p-dioxin in the Syrian golden hamster. *Carcinogenesis*, 9, 1677–1679.

159. Pohjanvirta, R., Hakansson, H., Juvonen, R., and Tuomisto, J. (1990). Effects of TCDD on vitamin A status and liver microsomal enzyme activities in a TCDD-susceptible and a TCDD-resistant rat strain. *Food and Chemical Toxicology*, 28, 197–203.

160. Pohjanvirta, R., Hirvonen, M. R., Unkila, M., Savolainen, K., and Tuomisto, J. (1994). TCDD decreases brain inositol concentrations in the rat. *Toxicology Letters*, 70, 363–372.

161. Unkila, M., Pohjanvirta, R., MacDonald, E., and Tuomisto, J. (1993). Differential effect of TCDD on brain serotonin metabolism in a TCDD-susceptible and a TCDD-resistant rat strain. *Chemosphere*, 27, 401–406.

162. Pohjanvirta, R., Sankari, S., Kulju, T., Naukkarinen, A., Ylinen, M., and Tuomisto, J. (1990). Studies on the role of lipid peroxidation in the acute toxicity of TCDD in rats. *Pharmacology & Toxicology*, 66, 399–408.

163. Pohjanvirta, R., Tuomisto, J., Linden, J., and Laitinen, J. (1989). TCDD reduces serum melatonin levels in Long–Evans rats. *Pharmacology & Toxicology*, 65, 239–240.

164. Linden, J., Pohjanvirta, R., Rahko, T., and Tuomisto, J. (1991). TCDD decreases rapidly and persistently serum melatonin concentration without morphologically affecting the pineal gland in TCDD-resistant Han/Wistar rats. *Pharmacology & Toxicology*, 69, 427–432.

165. Viluksela, M., Unkila, M., Pohjanvirta, R., Tuomisto, J. T., Stahl, B. U., Rozman, K. K., and Tuomisto, J. (1999). Effects of 2,3,7,8-tetrachlorodibenzo-p-dioxin (TCDD) on liver phosphoenolpyruvate carboxykinase (PEPCK) activity, glucose homeostasis and plasma amino acid concentrations in the most TCDD-susceptible and the most TCDD-resistant rat strains. *Archives of Toxicology*, 73, 323–336.

166. Tuomisto, J. T., Viluksela, M., Pohjanvirta, R., and Tuomisto, J. (2000). Changes in food intake and food selection in rats after 2,3,7,8-tetrachlorodibenzo-p-dioxin (TCDD) exposure. *Pharmacology, Biochemistry, and Behavior*, 65, 381–387.

167. Niittynen, M., Tuomisto, J. T., Auriola, S., Pohjanvirta, R., Syrjala, P., Simanainen, U., Viluksela, M., and Tuomisto, J. (2003). 2,3,7,8-Tetrachlorodibenzo-p-dioxin (TCDD)-induced accumulation of biliverdin and hepatic peliosis in rats. *Toxicological Sciences*, 71, 112–123.

13

NONGENOMIC ROUTE OF ACTION OF TCDD: IDENTITY, CHARACTERISTICS, AND TOXICOLOGICAL SIGNIFICANCE

FUMIO MATSUMURA

13.1 INTRODUCTION

The history of the studies on toxic actions of dioxin has been dominated by the discovery of its high potency to activate the AH receptor (AHR) to induce many drug metabolizing enzymes, which has helped to construct the initial theory of its toxic action, based on this remarkable phenomenon of the dioxin-activated AHR specifically inducing its target gene expressions such as cytochrome P450s. Among them particularly known is the gene encoding CYP1A1 protein. This theory, hereafter referred to as "the classical pathway," has contributed greatly to this field of science by providing a clear rationale and a logical approach to explore the molecular basis of the action of the family of dioxin-type chemicals, particularly that of TCDD, its most potent member of this group of environmental pollutants; so much so that now most of people, including many toxicologists, consider that induction of those drug metabolizing enzymes itself represents the basic mechanism of toxic actions of dioxins. So the question is whether this model provides all the answers needed to know about toxic actions of dioxins or not?

Well, the short answer is no, as there are several major problems in explaining many toxic actions of TCDD (see also Refs 1 and 2). First, let us consider this question from the specific viewpoint of evolutionary perspectives of the AHR. A simplest question one could ask on this role of AHR is "if the sole purpose of having this receptor is to mediate the toxicity of dioxins and how could such a receptor being kept by most of vertebrate species during the process of evolution, since no animal would have kept this receptor just to poison themselves?" One could argue that its power to activate those drug metabolizing enzymes must have helped them to survive. It may be so in the case of the animals encountering some degradable toxicants that activate AHR. However, TCDD itself is not readily degraded by those enzymes, particularly in humans. Thus, as far as TCDD and other very stable dioxin-type chemicals are concerned, induction of those "detoxification enzymes" does not help those animals to gain any survival advantages over those who do not have this receptor. The above question naturally leads to the next question: "What is then the major biological advantage of having AHR?" The logical answer, expressed in a most simplified manner, is that AHR must help those animals, individual organs, and cells to survive when they are exposed to these types of toxic chemicals that activate this receptor. Certainly the above answer may appear to be illogical: that is, how could the major survival advantage offered by TCDD-activated AHR result in toxic effects to the affected animals at the same time? This is an extremely important point, and it must be clearly explained here before proceeding to full descriptions on the nongenomic signaling pathway of TCDD.

In this chapter, one of the major roles of the ligand-activated AHR is being envisioned to assist animal survival by its ability to rapidly activate "cell stress response reactions." These reactions include many types of defensive, cell protecting measures such as activation of inflammation, resistance to apoptosis, and immune counteractive measures (Fig. 13.1). For the sake of clarity, let us cite here one simplest example of inflammation as a beneficial stress response. Acute inflammation has been known for some time to provide the benefits to the affected animals such as those damaged by

---

The AH Receptor in Biology and Toxicology, First Edition. Edited by Raimo Pohjanvirta.
© 2012 John Wiley & Sons, Inc. Published 2012 by John Wiley & Sons, Inc.

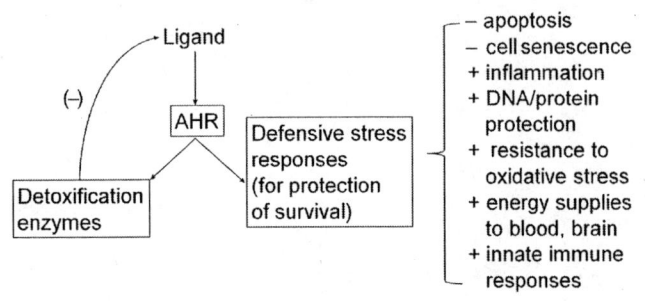

FIGURE 13.1 Dual missions of ligand-activated AHR. In this hypothesis-generated scheme, induction of detoxification enzymes by the ligand-activated AHR through the classical action pathway is envisioned mostly as the response of its target cells to metabolically degrade toxic ligands. In contrast, elicitation of defensive stress responses induced by the activated AHR at the same time is viewed as the prosurvival reactions that are mostly carried out through the nongenomic pathway. For the sake of clarity, the scheme shown here is simplified. In reality, some of the end results of activation of the classical pathway could also assist in defensive stress responses (e.g., induction of IGFBP and perhaps that of PAI-2 assisting expression of apoptosis resistance), and those of activation of the nongenomic could contribute to detoxification of toxic ligands. Note that not all of the "defensive stress responses" listed on the right have been firmly established in the case of TCDD-activated AHR signaling.

physical forces such as wounds by helping them to heal quicker than those cells not showing inflammation or providing a hostile environment to invading microorganisms and parasites. The source of confusion may be that currently most people do not view any type of inflammation, acute or chronic, to be beneficial at all. In contrast, most of scientists now accept the view that controlled short-term inflammatory responses are indeed beneficial in defending those animals from invading microorganisms [3]. Actually it is chronic inflammation that is becoming a great concern for health scientists because of its frequent association with a number of human diseases, such as rheumatoid arthritis, inflammatory bowel symptoms, asthma and other major lung diseases (such as chronic obstructive pulmonary disease), several types of cancers, diabetes, and brain diseases that include Alzheimer's disease.

The above brief introduction to this chapter has been intentionally made oversimplified so as to help initial understanding of the basic concept on stress response mechanisms, which provides the major rationale for this chapter. Certainly there are more complex cases where the risks and benefits of inflammatory responses including some undesirable or even life threatening effects of acute inflammation do exists. Those readers wishing to know more detailed information are advised to read more about the evolutionally meaning of inflammation and other cellular cell stress response mechanisms among animals [4–6].

13.2 WHAT IS "NONGENOMIC" SIGNALING?

The term, "nongenomic pathway" has been coined originally to explain the multiple actions of steroid hormones through their receptors (e.g., Ref. 7 for vitamin D3). A quick search in the PubMed using the keyword combination of "nongenomic" and "receptor" turns up 743 hits, whereas the use of "nongenomic" and "steroid receptor" gives 552 hits, indicating that the events covered by the term "nongenomic" are mostly those dealing with the steroid receptors. It is important to point out that, in the case of steroid receptor signaling, nongenomic actions of a given ligand are initiated by its binding to the portion of the relevant receptor located on the plasma membrane, not its usual locations such as cytosol or nucleus. According to Vasudevan and Pfaff [8], "Hormonal ligands for the nuclear receptor superfamily have at least two interacting mechanisms of action: (1) classical transcriptional regulation of target genes (genomic mechanisms) and (2) nongenomic actions that are initiated at the cell membrane, which activate protein kinases, one of which could activate the steroid receptor and thereby impact transcriptional processes as well." Although transcriptional mechanisms are increasingly well understood, membrane-initiated actions of these ligands are still incompletely understood. Historically, this has led to a considerable divergence of thoughts in the molecular endocrine field. There is evidence, however, that the membrane-limited actions of hormones, particularly estrogens, involve the rapid activation of kinases and the release of calcium. Membrane actions of estrogens, which activate these rapid signaling cascades, can also potentiate nuclear transcription. These signaling cascades may occur in parallel or in series along with the events taking place through the classical route of action of estrogens, but subsequently the signaling activities through these two pathways eventually converge at the level of modification of transcriptionally relevant molecules such as nuclear receptors and coactivators. In addition, other hormones or neurotransmitters may also activate cascades of crosstalks with other steroid receptor-mediated transcriptional activities. "The idea of synergistic coupling between membrane-initiated and genomic actions of hormones fundamentally revises the paradigms of cell signaling in neuroendocrinology" (a direct quote from Ref. 8). Another key statement in this article is: "the rapid activation of kinases and the release of Ca^{2+}, the former leading to the activation of the rapid signaling cascade, which results in stimulation of nuclear transcription." There is no question about the importance of the contribution made by many endocrinologists to clearly delineate the nongenomic signaling pathway apart from that of the classical signaling, which has greatly helped in the understanding of the divergent signaling mechanisms of steroids and provided the clear explanation on the necessity of having two separate pathways in achieving the mission of those steroid hormones (see Refs 9 and 10).

Having explained the origin of the term "nongenomic pathway" (abbreviated as NGP hereafter), on the other hand, it is also important to guard against the expectation of a perfect parallelism between the process of nongenomic signaling of the steroid receptor and that of the AHR, which should become apparent in the subsequent descriptions in this chapter.

13.3 THE VIEW ON THE ROLE OF THE NONGENOMIC PATHWAY WITHIN THE PARADIGM OF MULTIPLE TOXIC SIGNALING OF TCDD

The important premise in identifying any novel pathway is the clear understanding of its defining features that makes it truly distinct from others and clarifying the limits of the scope of signaling cascade of this newly found pathway that constitutes the downstream boundary beyond which its signaling starts becoming interacting with those generated from other pathways. For the sake of simplifying the picture, only three major pathways of the toxic actions of TCDD are envisioned in this chapter: (1) the classical pathway, (2) the nongenomic pathway, and (3) the nuclear crosstalk pathways, involving AHR [11]. Note that unlike the case of steroid receptors, the term "genomic" pathway, as opposed to the nongenomic one, is not used here, since once the nongenomic signaling reaches the nucleus and starts influencing gene expressions, this pathway is no longer considered purely "nongenomic." In a similar line of logic, signaling initially triggered by the classical route of action of TCDD further induces nuclear crosstalk, in addition to the direct activation of the target genes such as CYP1A1. Thus, interactions among the signaling of these three major pathways, and the eventual integration of their signaling, make it imperative to start with the simplified models.

13.4 AN EXAMPLE OF "NONGENOMIC SIGNALING" TRIGGERED BY A CANCER PROMOTER THAT IS SIMILAR TO THAT OF TCDD

The discovery of cancer promoting actions of phorbol diesters, particularly that of TPA (12-*O*-tetradecanoylphorbol-13-acetate) or PMA, has drastically altered the thinking of the entire field of carcinogenic process [12]. Namely, in addition to already known chemical carcinogens that cause genomic changes such as direct DNA damage and formation of DNA adducts, TPA and other phorbol diesters were found to act as cancer promoters that could act through the nongenomic route of action. In brief, TPA has been found to bind to its putative receptor, which was later identified to be protein kinase C (PKC), and to rapidly trigger a nongenomic signaling, which is mediated by cascades of signal transduction that are sequentially transmitted through protein kinases and phosphatases [13]. This nongenomic signaling of TPA, initially occurring on the plasma membrane, is transmitted through the cytosol to the final messengers such as jun and fos protein through their phosphorylation by their specific kinases (i.e., ERK–Elk and JNK, respectively), which promote their translocation into the nucleus. Once they are in the nucleus, these phosphorylated proteins (now called nuclear transcription factors) activate their target genes, of which promoter regions contain the specific "response element" called TRE (the TPA response element or now designated as the activator protein 1 (AP-1) response element) and thereby trigger cell mitotic responses. This TPA-induced signal transduction pathway has been well researched and is now well established [14]. Although the realization that the above process of nongenomic signaling of TPA is closely accompanied with elicitation of cellular stress responses was made rather early [15], the importance of such parallel actions of TPA in causing inflammation and oxidative stress [16] in carcinogenesis was largely ignored until recently when a host of publications on the favorable effects of anti-inflammatory or antioxidative agents against TPA-induced cancer promotion started to appear [17, 18]. As for the reason why TPA activates PKC, it has been identified to be due to its structural similarities to 1,2-diacylglycerol (DAG), which serves as the natural endogenous ligand for TPA receptor (i.e., PKC). According to Bell et al. [14], their activation model suggests that PKC binds to Ca^{2+} and four phosphatidylserine (PS) carboxyl groups to form a surface-bound, "primed" but inactive complex. DAG, which is produced by activation of phospholipase C, binds to the complex of the four PS carboxyl groups, the Ca^{2+}, and the PKC through three bonds, two to ester carbonyls and one to the 3-hydroxyl moiety, indicating that DAG is the initial trigger of the PKC cascade.

The most important lessons learned from this paradigm of nongenomic signaling of TPA are as follows: (a) both Ca^{2+} and lipids produced by phospholipases could act as the second messenger of this type of nongenomic signaling, (b) several protein kinases and phosphatases play the role of major transducers of such signaling through their ability to phosphorylate their specific downstream substrate protein and thereby activate to relay the message to the next carrier, and (c) the final act of the phosphorylated "anchor carrier" is to translocate into the nucleus, where it directly binds to its specific response element or forms protein complexes with other active nuclear transcription factors that bind to different response elements and thereby indirectly activate their target gene expressions. These final acts of phosphorylating (and therefore functionally activating) the target nuclear proteins are naturally viewed as "genomic" events inasmuch as they directly or indirectly influence the expression of their target genes.

13.5 SPECIFIC FEATURES OF A TYPICAL "NONGENOMIC" SIGNALING GENERATED BY TCDD THROUGH ACTIVATION OF AHR IN ONE TYPE OF CELL

Before getting into the details of nongenomic pathways mediating toxic actions of TCDD, it is necessary to caution the reader not to expect a consistent pattern of nongenomic signaling that could be uniformly applied to many different types of cells or tissues, as in the case of signaling of TCDD to induce CYP1A1 induction through the classical action pathway. As will be shown later, the nature of protein kinase-mediated nongenomic signaling varies greatly among different types of cells and animal species, depending on the roles played by each type of cells or tissues in defending themselves from the conceived attacks from the invading pathogens or toxic chemicals. Thus, one must be careful not to automatically assume the wide applicability of the nongenomic signaling pathway found only in one type of cells, even within the same species. Having warned of this inherent complexity of the nongenomic signaling, on the other hand, it must be also mentioned that there are ways to help understand the general pattern of cell–tissue responses to stress through this route of signaling, once one grasps the concept of tissue-specific roles of stress responses. That is, there are some stereotypic stress response patterns exhibited by several major tissues that can be recognized [19–21]. Nevertheless, it appears to be wise to first start with one concrete example of the NGP pathway in order to introduce the basic concept of the nongenomic signaling of TCDD. For this reason, the pertinent information on the NGP signaling gathered from studies on MCF10A (an immortalized but otherwise a normal mammary epithelial cell line) will be given below as a starter.

The most noticeable early event, occurring as a result of MCF10A cells exposure to TCDD, is rapid activation of Src tyrosine kinase that takes place within 5–15 min [22], which is accompanied with equally rapid activation of ERK1 and ERK2 (assessed by double phosphorylation on these two proteins) that takes place within 10–15 min. Such an action of TCDD on ERK was totally inhibited by PP-2, a specific inhibitor of Src tyrosine kinase (Fig. 13.2), indicating that Src-mediated activation of AP-1-dependent mitogenic signaling has already taken place by this action of TCDD, just as in the case of the action of TPA to induce rapid responses from the affected cells as described above.

Knowing that 15 min is too short for TCDD-activated AHR to accomplish induction of its DRE-based target genes (such as CYP1A1), and the importance of Src kinase in the process of carcinogenic transformation of mammary epithelial cells, we decided to investigate this phenomenon in more detail. It was found that this action of TCDD is accompanied with immediate suppression of signaling of insulin, which was confirmed in our subsequent work [23], resulting in inhibition of cell proliferation that was induced by exogenous insulin in MCF10A cells. Furthermore, in addition to insulin signaling, this action of TCDD to activate Src kinase is intimately coupled to activation of EGF receptor signaling, since preincubation of MCF10A cells with an antibody against TGFα clearly suppressed this action of TCDD, basing on the knowledge that signaling of EGF receptor is blocked by this antibody. As for the significance of Src kinase activation in terms of phenotypic changes in MCF10A cells, it was determined that one of the major consequences of this action of TCDD is to provide an effective protection of those cells from stress-induced apoptosis [24]. Judging by the effectiveness of PP-2 (a specific inhibitor of Src kinase) and other means to block the action of Src kinase in antagonizing the antiapoptotic action of TCDD, the essential role of Src kinase in conferring this cell survival ability has been confirmed. At the same time, it was noted that such an action of TCDD closely resembles a similar antiapoptotic action of EGF or TGFα on this cell line.

Although the above series of studies has clearly indicated the essential role of Src kinase in mediating the early action of TCDD in this cell line, no clue has emerged on how Src kinase is activated, or what the precise role of Src kinase in the nongenomic signaling of TCDD-activated AHR is at that time. An important clue, which has led us to the existence of a nongenomic pathway, was the discovery of rapid action of TCDD to cause elevation of the enzymatic activity of cytosolic phospholipase A2 (cPLA2) [25]. This effect of TCDD was discovered by the initial observation that treatment of

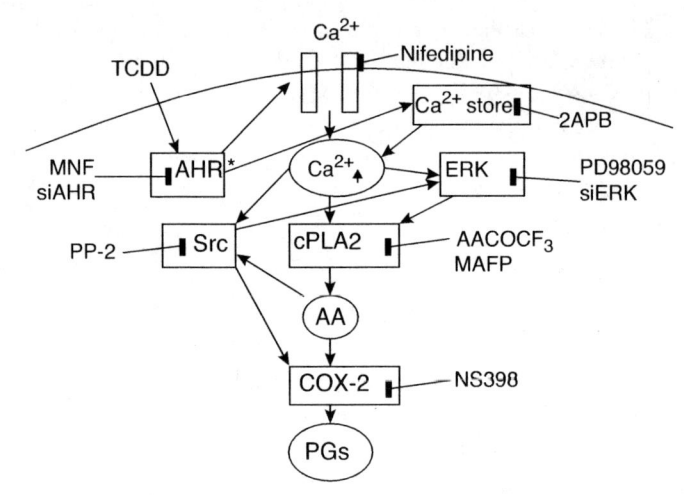

FIGURE 13.2 Simplified diagram representing the rapid action of TCDD in MCF10A cells, occurring within 1 h, which is mediated through the nongenomic pathway [28]. In this scheme, the initial action of the TCDD-activated AHR is to increase the intracellular concentration of Ca^{2+} from two different sources, which activates Ca^{2+}-dependent cPLA2, leading to increased production of AA and further activation of Src and ERK kinases. Elucidation of this rapidly reacting nongenomic signaling cascade has been greatly helped by the use of specific inhibitors, each affecting one key component of the signal transduction pathway.

MCF10A cells with TCDD caused rapid release of arachidonic acid (AA) within 15 min. Interestingly, this process was not inhibited by PP-2, a specific inhibitor of Src tyrosine kinase, but suppressed by MAFP, a specific inhibitor of cPLA2, or by MNF, an antagonist of AHR. This finding indicates the possibility that activation of cPLA2 acts as the upstream mediator of Src activation in the scheme of nongenomic signaling of TCDD in this cell line. This possibility was checked by studying the direct effect of exogenous AA on Src. The result of this line of approach unambiguously showed that AA does indeed activate Src function as attested to by the action of exogenously introduced AA to cause translocation of Src protein from cytosol to the plasma membrane within 10 min. In a similar manner, we could also demonstrate that both Src and AA could directly activate Cox-2 expression. cPLA2 is a cytosolic phospholipase known to be activated (with an exception of group IVD isoform) by the increase in intracellular Ca^{2+} (expressed as $[Ca^{2+}]_i$) that becomes accessible to cytosolic proteins. Knowing that its enzymatic activity is stimulated by A23187 (a calcium ionophore) and is inhibited by EGTA-AM (a Ca^{2+} chelator that penetrates into cells), it was hypothesized that the upstream event of cPLA2 activation is likely due to the increase in $[Ca^{2+}]_i$. To identify the source of this increase in $[Ca^{2+}]_i$, we tested many specific inhibitors of Ca^{2+} transport and found that nifedipine and 2APB are two of the most effective inhibitors in terms of suppressing the action of TCDD to activate cPLA2. The former is known to block the entry of extracellular Ca^{2+} into cells through the Ca^{2+} channel on the plasma membrane and the latter is known for its specific action to block Ca^{2+} release from its intracellular storage site (i.e., Ca^{2+} store) in endoplasmic reticulum. Thus, surprisingly TCDD appears to mobilize Ca^{2+} from two distinctly different sources of Ca^{2+} that are available in MCF10A cells. The overall scheme of the nongenomic pathway found in this cell line has been summarized and shown in Fig. 13.2. It must be added that our overall goal to recognize the nongenomic pathway, apart from the classical pathway of action of TCDD, was helped in the above studies on very early nongenomic actions of TCDD (i.e., taking place within the initial 30 min of action) that took advantage of the slower time course of the classical pathway. To relate the above findings to the eventual toxic actions of TCDD, however, it was necessary to conduct further in-depth studies on the downstream signal transduction processes occurring following the above initial triggering events.

In the subsequent study on the same cell line [26], we have addressed this question on the process of propagation of the initial nongenomic responses to TCDD exposure, leading to the longer term inflammatory status of those affected cells. This is not a trivial subject because of the transient nature of the initial Ca^{2+} signaling as originally shown by Puga et al. [27] in Hepa 1 cells exposed to TCDD. As in the case of many cell stress responses mediated by Ca^{2+} signaling, the initial rapid rise in $[Ca^{2+}]_i$ is inevitably compensated by negative feedback, once the initial signaling is processed, so as to help those cells to return to their normal resting state. Another good example is the case of TCDD-induced rapid activation of ERK1 and ERK2 [22] that reached its peak in 15 min, but by 60 min this effect of TCDD totally disappeared. From the viewpoint of toxicology, the transient nature of such initial signaling creates a big challenge, since for toxicologists the key consideration must be that such an early nongenomic response of cells would not mean much, if one cannot logically link it to any of the long-term toxic effects of TCDD *in vivo*. For this reason, we first studied a longer time course (0–72 h) of TCDD-induced changes in expression of select markers that were already identified in our previous study (note that in both cases TCDD was given as a single treatment at the beginning) in MCF10A cells [26]. We noted that the time-dependent pattern of CYP1A1 mRNA induction showed an initial rise, starting around 0.5–1 h, peaking by 2 h, and thereafter showing gradual decline during the next 70 h. In contrast, that of Cox-2 showed a rapid initial rise occurring around 20–30 min, reaching the plateau, and being continuously maintained at that level till the end of the experiment, 72 h. Some notable markers such as metalloproteinase-2 (MMP-2), colony-stimulating factor-1 (CSF-1), and aromatase (CYP19) showed a totally different pattern in that their expression kept increasing so that at the end of this experiment (i.e., 72 h), the level of their mRNA expression reached the highest point. Clearly, under this test condition, the signaling of the TCDD-activated AHR is continuously processed, propagated, and, in some cases, even enhanced (e.g., MMP-2, CSF-1, and aromatase in the case of TCDD-affected MCF10A cells).

To determine the basic cause for such long-lasting effects of TCDD, we hypothesized that some powerful protein kinases are involved somewhere in the process of propagating the initial nongenomic message of the TCDD-activated AHR, based on our previous observation made on the persistent activation of protein kinases in the liver of TCDD-exposed rats [28]. Three classes of protein kinases found to be elevated by the action of TCDD throughout this 72 h test period were tyrosine kinases (particularly Src kinase), cAMP-dependent protein kinases (PKAs), and protein kinase C. Of these, the influence of the first two groups of kinases appeared to be most persistent in MCF10A cells, and therefore we decided to study their effects on the expression of select markers by using class-specific inhibitors of those protein kinases (i.e., PP-2 for Src kinase and H89 for PKAs) [26]. Briefly, H89 was found to be very effective in antagonizing the 72 h effects of TCDD on Cox-2 and aromatase mRNA expression, whereas PP-2 was more effective than H89 in suppressing the effect of TCDD to upregulate MMP-2 and CSF-1 (Fig. 13.3). None of these protein kinase inhibitors showed any effects on CYP1A1 expression. Just to ascertain that these four marker expressions are clearly

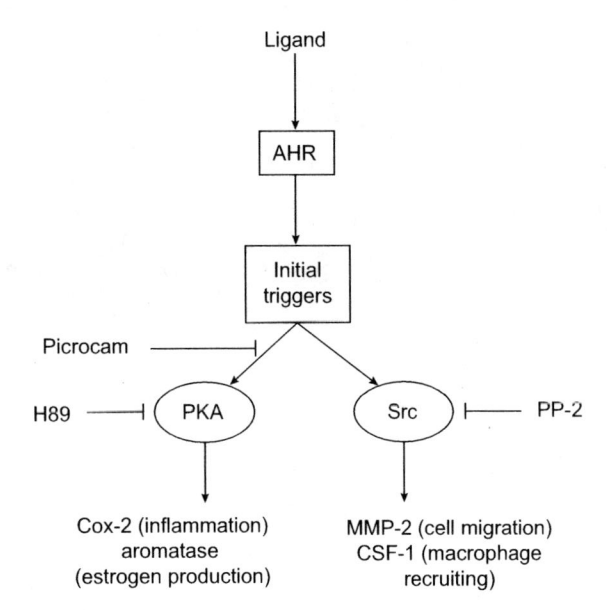

FIGURE 13.3 Propagation of the initial nongenomic signaling through activation of protein kinases in MCF10A cells. In this cell line, TCDD-induced transient nongenomic signaling occurring during the initial 30 min is propagated by sustained activation of two major protein kinases, PKA and Src, during the test period of 72 h [26]. The PKA-mediated signal propagation supports the mRNA expression of Cox-2 and aromatase (CYP19), and that supported by Src kinase, on the other hand, upregulates MMP-2 and CSF-1.

representing the long-term nongenomic signaling of TCDD, rather than that is mediated through the classical action of TCDD, we tested the effectiveness of two siRNA preparations, one against AHR and the other against ARNT, on their expressions. The results clearly showed that both siRNA preparations were quite effective in suppressing the action of TCDD to induce CYP1A1, but only that against AHR was effective in suppressing the expression of those nongenomic marker expressions, whereas that against ARNT showed no effect at all on their expressions. These results start revealing the divergence of protein kinase-mediated signaling at later stage of action of TCDD in this cell line. That is, long-term upregulation of one group of markers, represented by Cox-2 and aromatase in this study, appears to be mediated mainly by PKA, and another group (e.g., MMP-2 and CSF-1) is supported mainly by Src tyrosine kinase, probably through its activation of the AP-1-mediated nongenomic signaling. Interestingly, such an effect of TCDD to cause sustained upregulation of PKA enzymatic activities could be inhibited by piroxicam, a persistent Cox-2 inhibitor, which also suppressed 24 h action of TCDD to cause upregulation of Cox-2 and aromatase, but not that of MMP-2 and CSF-1, supporting our notion on the overall pattern of protein kinase-based propagation of the initial nongenomic signaling. Another piece of information important in the understanding of the nongenomic origin of this kinase-based propagation of the initial message is the effectiveness of pretreatments of MCF10A cells with MAFP (an inhibitor of cPLA2) or EGTA-AM (an intracellular chelator of Ca^{2+}) in terms of preventing the long-term action of TCDD (in this case 24 h, since the effect of EGTA-AM cannot be sustained too long) from inducing the expression of all of those markers except that for CYP1A1, indicating unambiguously that those protein kinase-mediated, long-term effects of TCDD on those nongenomic marker expressions are the actual extension of the original nongenomic signaling that is triggered by $[Ca^{2+}]_i$ and cPLA2 in this cell line. The apparent existence of at least two major branches of the nongenomic pathway in this cell line does indeed bring up an interesting question: Are there two independent mechanisms of turning on each branch, or do those two groups of genes just happen to have highly active response elements on their promoters that are being bound by PKA-activated or Src-activated nuclear transcription factors? More research efforts are needed to answer this question.

13.5.1 Variations Observed in Terms of Nongenomic Responses to TCDD Among Different Types of Isolated Cells in Culture

As mentioned above, one of the prerequisites in grasping the concept of inflammatory responses of cells to environmental stressors is to first accept the reality that their patterns differ greatly among different types of cells. This is particularly apparent in cases of systemic inflammation, such as the ones induced by dioxin-type pollutants, which involves coordination of several different types of cells distributed among various organs and tissues. As indicated already, basically most cases of acute inflammation are considered to be aimed at protecting those cells that are injured or facing the invasion of pathogenic microorganisms and therefore are considered to be basically beneficial to the organism affected [29]. However, it is chronic inflammation that is increasingly becoming major health threats in recent years [3]. Knowing the chemical stability and the persistence of TCDD in animal bodies, it is not surprising that most of its inflammatory effects would become chronic, particularly in those tissues known to accumulate TCDD such as adipose tissues and liver. Thus, in order to provide full understanding on this aspect of toxic action of TCDD, much more extensive data must become available than what we have at hand currently. The following are, therefore, only the initial glimpse on this subject based on the small number of existing publications.

A simplified picture of the major differences in terms of TCDD-induced upregulation of major nongenomic markers among five different types of cells is shown in Table 13.1 (note that only the markers that are already upregulated by 3 h action of TCDD are shown here). For instance, the conspicuous absence of changes in Src expression in U937 macrophages is immediately noticeable. In addition, the lack of

TABLE 13.1 Cellular Specificity of Rapid Inflammatory Responses to TCDD Exposure (>3 h)

	Macrophages (U937) [30]	Mammary Epithelial (MCF10A) [25, 26]	Adipocytes (3T3-L1 [31] and Human Stem Cell Derived [32])	Kidney Tubular Epithelial (MMDD1) [33]	Hepatoma c cells (HepG2)[a]
$[Ca^{2+}]_i$	+	+	+	+	+
cPLA2	+	+	+	+	+
Src	−	+	+	+	+
Cox-2	+	+	+	+	+
TNFα	+	−	+	−	−
IL-1β	+	+	−	−	−

[a] Li, W. and Matsumura, F., unpublished data.

effect of TCDD on mRNA expression of TNFα and IL-1β in HepG2 human hepatoma cells or mouse macula densa kidney tubular epithelial cells also helps to illustrate the difference between these cells and other types of cells.

It is known that inflammatory responses of the target tissues, in many cases, require the participation of other types of cells such as macrophages, neutrophils, and other hematopoietic cells to fully express their potentials to express inflammatory responses. Therefore, those cells studied *in vitro* in isolation from any other inflammation-mediating cells may not fully respond to TCDD. Yet, it is also important to learn about the direct effects of TCDD on its target cells in isolation from other inflammation assisting cells. Regarding the basic characteristics of TCDD-induced NGP, one could at least try to deduce the minimum common denominators among all types of cells such as $[Ca^{2+}]_i$, cPLA2, and Cox-2 from this set of data, to recognize the general pattern of the NGP activated by TCDD among different types of cells, at this early stage of action of TCDD.

The scope of variations among different types of cells starts expanding tremendously once we consider the effects of TCDD beyond this 3 h limit. Certainly, part of such differences is due to the specificity of cellular functions. For instance, effects of TCDD to downregulate leptin and adiponectin can only be seen in adipocytes, and similarly its effect on ion exchanging systems such as NaKCC2 and ROMK can only be found in kidney tubular epithelial cells, such as MMDD1 cells. So variations are the inherent nature of these types of cellular responses that result from various environmental stressors affecting many different types of cells. A more important question should be whether one can find common mechanistic principles that govern all these inflammatory/nongenomic signaling activities among different cells despite tremendous variations in their responses.

Although, at this stage of early development of this field of science, not enough data have been obtained through systematic study approaches, there are several pertinent observations that could be utilized to make attempts to construct a hypothetical framework for the future studies on this subject, which will be described next.

13.5.2 Propagation of Initial Nongenomic Signaling Triggered by the Ligand-Activated AHR

It was initially reported by Bombick et al. [34] that a single treatment of male rats and guinea pigs with low doses of TCDD *in vivo* causes sustained elevation of protein kinases in hepatic plasma membrane preparations lasting 20–40 days of the experimental period. The types of kinases in that study included those specifically stimulated by exogenous cAMP and PKAs and by Ca^{2+} and diacylglycerol, PKCs. Subsequently, Bombick and Matsumura [35] reported that TCDD also causes long-lasting activation of Src tyrosine kinase in hepatic plasma membrane preparations from liver from treated guinea pigs, rats, and mice, which could be detected after 25 days from the time of the single injection of TCDD. It must be added that the reason why the plasma membrane was chosen at that time as the material for the above studies was that Src-type protein kinases, which normally reside in cytosol as an inactivated state, translocate to the plasma membrane upon their activation. Madhukar et al. [36] further reported that TCDD induces long-lasting activation of the EGF receptor on the guinea pig liver plasma membrane. The EGF receptor (EGFR), which is already associated with its own protein tyrosine kinase, recruits Src kinase upon its activation in the plasma membrane. The membrane preparations obtained from TCDD-treated (again single dosing at the beginning) male rats clearly exhibit the activated state of EGFR during the entire test period of 20 days.

These earlier observations have shown that at least some types of protein kinases are chronically elevated as the result of single administration of TCDD exposure *in vivo* for extraordinarily long periods in some tissues. Since then a number of papers have been published indicating that indeed elevated protein kinase activities can be observed in animals treated with TCDD *in vivo* or in cultured cells exposed to TCDD. The first group of protein kinases that have attracted the attention of scientists other than our own laboratory was protein tyrosine kinases, particularly those associated with Src kinase and EGFR-associated tyrosine kinases. For instance, the group of scientists at EPA and NIEHS at Research Triangle Park have assessed the significance of those tyrosine kinases in carcinogenic action of TCDD in rat liver [37, 38]. Their main conclusion was that the effect of TCDD on the EGFR, which is accompanied by its phosphorylation, is a sensitive and reliable marker for liver carcinogenesis. Abbott and Birnbaum [39] exposed mouse embryos to TCDD at gestation day 10 or 12 and assessed EGFR status in medial epithelial cells of palatal shelves in relation to cleft palate formation and found that the expression of EGFR continues till PD 16, a critical timing in pathogenesis. Later Abbott et al. [40] have established unequivocally that the EGFR plays an indispensable role in TCDD-induced cleft palate formation in mice. The information related to our own work addressing the significance of Src kinase in mediating the pathogenesis of the "wasting syndrome" in vivo has already been reviewed [41]. Briefly, the key finding was the significant reduction in the manifestation of "wasting syndrome" induced by TCDD in the src$^{-/-}$ mice, compared to its genetically matched src$^{+/+}$ C57 BL/6 mice, judging by several well-accepted hallmark parameters such as accumulation of triglycerides and reduction in glycogen storage in liver, reduction in the weight gain, loss of fat mass, and serum hyperlipidemia. An additional observation that a similar pattern of reduction of those wasting effects of TCDD in wa1/wa1 mice, a strain with a mutant and therefore a defective TGFα gene, that confers impaired EGFR responses compared to the matched wild-type mice [42] has provided a logical link between the activation of Src kinase and the elevation of the EGFR signaling.

As for in vitro studies on this action of TCDD, Köhle et al. [43] have treated confluent WBF344 hepatocytes with TCDD (1 nM) and found that it indeed promotes translocation of the cytosolic form of Src protein into the plasma membrane within 20–60 min, a phenomenon well known to represent functional activation of this tyrosine kinase. At the same time, this treatment increased tyrosine phosphorylation on the EGF receptor under the same test condition. Since then there have been a number of publications reporting observations on the appearance of Src kinase in TCDD-exposed animals or in isolated cells. A quick search on PubMed shows 45 hits (using keywords of TCDD Src), indicating at least this subject has been attracting the attention of a number of research groups. A more recent review on this subject centered on growth factors, cytokines, and their receptors as downstream targets of AHR has been published by Haarmann-Stemmann et al. [44].

As for the effect of TCDD on PKC in vivo, it was initially reported by Bombick et al. [34] that a single treatment of male rats (at 25 μg/kg, i.p.) and guinea pigs (1 μg/kg i.p.) induced increased enzymatic activities of PKC in the hepatic plasma membrane preparation to the levels of 2.4- and 1.9-fold of control value, respectively, on posttreatment day 10, indicating that TCDD also causes long-lasting upregulation of PKC in these animal species under their test condition. In a subsequent study, the same group [45] showed that a single i.p. injection of TCDD (30 μg/kg) to strains of mice caused a time-dependent rise in enzymatic activities (as assessed by phosphorylation on histone) of both tyrosine kinases and PKC in mouse thymus, but not PKA. The time sequence of their activation showed that tyrosine kinases were activated immediately after administration of TCDD, whereas that of PKC showed an initial decrease that was eventually followed by a steep upregulation starting 12 h after the initial TCDD treatment in this test system. Bagchi et al. [46] reported that female Sprague Dawley rats treated with 50 μg/kg of TCDD show elevated activity of PKC both in liver (2.2-fold) and in brain (2.6-fold, both compared to control) after 24 h. Weber et al. [47] treated female rats with 10 μg/kg of TCDD in vivo and found increased activity of PKC in randomly cycling smooth vascular muscle cells that were isolated and cultured. All these in vivo studies showed that elevated PKC activities, all assessed as the sum of all PKC kinase activities that were stimulated by DAG and Ca^{2+}, were sustained at least in some of the TCDD target tissues for relatively long periods. In more recent years, studies on PKC have advanced enough to address the effects of TCDD on specific isoforms of PKC [48, 49], revealing intricate tissue-specific PKC response systems to the action of TCDD that are specifically mediating each type of cell responses, depending on the individual missions of those target cells.

Additional information supporting the indispensable role of PKC in aiding propagation of NGP signaling still mainly comes from studies on the effectiveness of class-specific PKC inhibitors in blocking the action of TCDD through the NGP. Wölfle et al. [50] have shown that the effect of TCDD, unlike the rapid and transient action of TPA on PKC, is characterized by the steady rise in the level of PKC activity in isolated rat hepatocytes, reaching an initial plateau by 3 h and continuing the process of steadily increasing till 48 h (i.e., the end of experiment). In this study, H7, a class-specific PKC inhibitor, was used to block the action of TPA-induced PKC. Li and Matsumura [31] have shown that a short-term (1 h) action of TCDD to induce mRNA expressions of interleukin-8 (IL-8) and monocyte chemoattractant protein-1 (MCP-1) in 3T3-L1 adipocytes through the NGP is completely abolished by the action of H7 or by the pretreatment of adipocytes with TPA (for 1 h), although neither treatment affected the action of TCDD to upregulate Cox-2 expression. It must be added that the above information on PKC is meant to come up with the circumstantial evidence, supporting the hypothesis that long-term activation of PKC could play significant roles as one of the mediators of propagation of the NGP signaling. To firmly establish such roles of PKC, much more concrete experimental data directly proving this hypothesis are urgently needed. In that case, the key experiment needed is to elucidate the mechanism through which the initial transient signaling of the NGP such as the increase in $[Ca^{2+}]_i$ (such as rapid activation of PKCα in most types

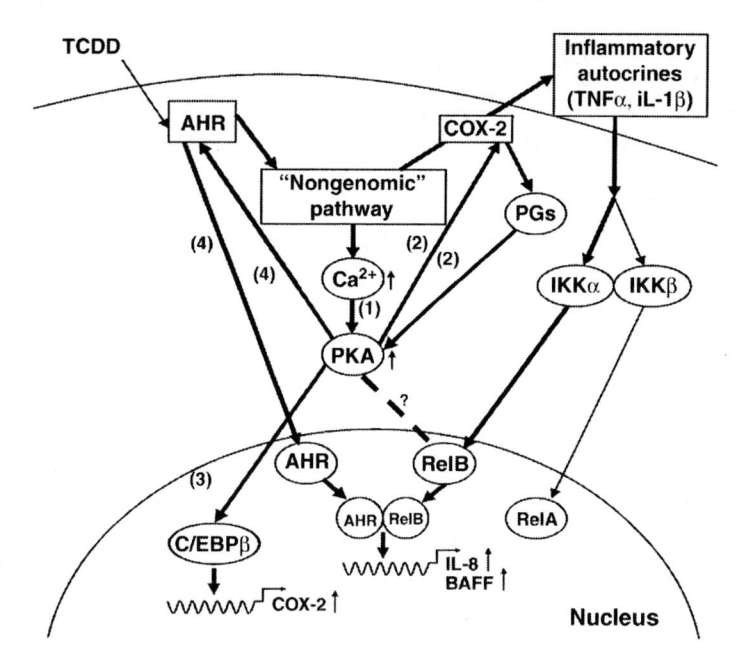

FIGURE 13.4 A hypothetical scheme through which PKA participates in propagation of nongenomic signaling of the ligand-activated AHR through multiple steps. Under this scheme, the transient initial Ca^{2+} signaling induced by TCDD is converted into more stable and long-lasting inflammatory effects by the signal amplifying functions of PKA: (1) PKA is initially activated by Ca^{2+} that leads to (2) activation of Cox-2 enzymes and subsequent increase in production of prostaglandins (PGs), (3) causing eventual autoamplification of PKA activity and thus amplifying PKA increases the C/EBPβ gene expression, leading to sustained increase in the Cox-2 gene and protein expression to further maintain the elevated level of Cox-2 activation. Furthermore, activated PKA also participates in increasing (4) translocation of the AHR monomer into the nucleus.

of cells) is converted into long-lasting activation of PKC in a given type of cells that are relevant to toxic effects of TCDD in that species.

In contrast to the role of PKC, the contribution of PKA in mediating propagation of the NGP signaling is more compelling. It appears that upon the increase in $[Ca^{2+}]_i$, PKA is rapidly activated in certain cells [51]. Therefore, there is the possibility that the initial rise in PKA in TCDD-affected cells is mediated through this route. However, knowing that the initial signaling of Ca^{2+} does not last too long, the question one must address is how this type of transient signaling is converted into sustained activation of PKA in TCDD-treated cells. Earlier in this chapter, the mechanism of $[Ca^{2+}]_i$-induced initial rise in cPLA2 activity causes the rise in the level of cellular AA, leading to activation of Cox-2 enzymatic activities. Given that Cox-2 is one of the most important early transducers of the NGP signaling of TCDD that is consistently found in several different types of cells as shown above as well as in a recent review from this laboratory [52], the critical role of this enzyme in regulating PKA activity must be considered next. It is well known that prostaglandins produced by Cox-2 are capable of stimulating PKA. Therefore, this action of Cox-2 may be regarded as the second autoamplification process in TCDD-affected cells to prolong the state of activation of PKA (note that the initial PKA activation by Ca^{2+} is considered as the first amplification process) (Fig. 13.4). Probably the most concrete mechanistic evidence supporting the important role of PKA on the action of TCDD to cause gene activation mechanisms to finally stabilize this amplification of the inflammatory signaling has been the work of Vogel et al. [53], who could show that TCDD-induced upregulation of C/EBPβ in 10T1/2 and Hepa 1c1c7 cells solely depends on its action to induce PKA activity. Further analysis using mutated constructs of the C/EBPβ gene promoter demonstrated that activation of the C/EBPβ gene is mediated through the cAMP response element binding protein (CREB) sites located close to the TATA box of the C/EBPβ gene. The protein kinase A (PKA) inhibitor H89 completely blocks this TCDD-dependent effect on C/EBPβ promoter activity, indicating that TCDD activates CREB binding via a cAMP/PKA pathway, which is supported by the increased cAMP level and PKA activity observed after TCDD treatment. C/EBPβ is one of the key nuclear transcription factors that are capable of activating Cox-2 gene expression as well as many other inflammation-mediating genes as shown by many scientists already. Thus, the information that PKA plays a pivotal role in TCDD-induced upregulation of C/EBPβ has provided the necessary evidence how this type of gene activation steps acts as the third process of the autoamplification of the initial transient signaling of the NGP and thereby acts as the important mechanism of stabilization of the initial NGP signaling in TCDD-affected cells now that the initial message has been finally transduced into stable expressions of many genes that are functionally activated by C/EBPβ protein in the nucleus. One example of the transduction process of the TCDD-induced NGP signaling through sustained activation of protein kinases is discussed next (see Fig. 13.4).

Having shown that PKA activation is one of the key amplifiers of the nongenomic signaling of TCDD at least in several types of cells, the next question is how PKA is actually being activated in one type of cell. It was shown in our study on the effect of TCDD on MCF10A cells [26] that TCDD causes continuous activation of the enzymatic activity of PKA during the test period of 24 h and that this effect of TCDD could be significantly inhibited by picrocam (approximately 82% inhibition, assessed after 6 h of action of TCDD), a persistent Cox-2 inhibitor, indicating that the majority of TCDD-induced PKA activity is likely linked to activation of Cox-2. On the other hand, in the presence of picrocam, TCDD caused a modest upregulation of PKA (approximately 18% at this time point, or as much as 50% of the total after 2 h), indicating that there is a portion of PKA that is not dependent on Cox-2 activation. Judging by the rapid action of TCDD to activate this portion of PKA within 30 min, and because by such an action TCDD is completely inhibited by H89 (a specific inhibitor of PKA), this initial action could be viewed as one of the triggering events to activate this inflammatory signaling. The important question being addressed here is whether PKA plays any role in initiating the triggering event itself. In the above work [26], it was revealed, by using the electrophoretic mobility shift assay (EMSA), that activation of nuclear protein binding to the labeled C/EBP response element oligonucleotide clearly increased within the initial 30 min of the action of TCDD, meaning that PKA appears to serve as one of the initiating factors as well. Nevertheless, the observation that this Cox-2 inhibitor showed a profound suppressive effect on PKA activity indicates that the later activated portion of PKA activity is likely a downstream event of initial Cox-2 activation, which takes place within 2 h in MCF10A cells. Accordingly, the most likely mediator of activation of PKA are prostaglandins produced by Cox-2 as shown by a number of scientists, including a report provided by McKenzie et al. [54], who showed such an action of PGE2 on PKA in normal mammary epithelial cells. Since TCDD-induced upregulation of both Cox-2 and aromatase (CYP19) mRNA expression after 72 h is significantly antagonized by both picrocam and H89 [26], it is safe to conclude that this route of action is providing the long-term, sustained action of TCDD on the expression of those inflammation markers in this cell line. Since upregulation of aromatase expression starts rather late (not observed after 6 h, clearly recognized after 24 h, and kept increasing till 72 h, which was the end of that experiment), this observation also supports the above conclusion. Also, it is apparent that the role of PKA in this case is to serve as the third autoamplification mediator, since it is well known that PKA itself is also known as one of the major activators of Cox-2 gene expression [53].

In the work of Li and Matsumura [31] on early action of TCDD (for 1 h) on 3T3-L1 adipocytes, it was reported that H89 preincubation clearly suppresses the action of TCDD to induce mRNA expression of both IL-8 (KC in mice) and VEGF. Since exogenously added forskolin, which is known to activate PKA, clearly elevated these two markers as well, in a parallel series of tests on 3T3-L1 adipocytes, this observation also provides support for the role of PKA in facilitating the early actions of TCDD in adipocytes, although much more work would be needed to firmly establish this role of PKA. Certainly, the molecular mechanism of rapid activation of PKA upon TCDD binding to AHR in cytosol is of great interest to several groups of scientists who are really specialized in this subject. There are two very insightful review papers on this subject, one of which has been published in a 2009 special issue of *Biochemical Pharmacology*: for example, by Barbara Oesch-Bartlomowicz and Franz Oesch [55] and by Simone Kobe de Oliveira and Albert Smolenski [56].

Another type of protein kinase potentially contributing nongenomic action of TCDD is CK2 (formerly casein kinase II), which has not attracted too much attention so far, but could play an important role in TCDD-induced cellular inflammation. CK2 is an ubiquitous serine/threonine kinase that is known to play important roles in mediating signaling of cellular stress responses, such as activation of NF-κB signaling in cytosol and nucleus, modulating inflammatory responses requiring nuclear coordination, and influencing the structure and function of chromatin and nuclear matrix through posttranslational modifications of various proteins in these dynamic structures [57]. It was initially reported by Enan and Matsumura [58], who studied TCDD-induced changes in protein phosphorylation activities in the nucleus of adipose tissues of guinea pigs, that among Mn^{2+}-stimulated protein kinases, CK2 was found to be the most predominant type of nuclear protein phosphorylating activity that was affected by TCDD: for example, approximately 60% of the total nuclear protein kinase activity was due to the heparin-sensitive casein kinase II (CK2). The level of CK2 activity in the nuclear protein preparation from adipose tissue of TCDD-treated guinea pigs (1 μg/kg) was only 35% of the control value after 24 h, indicating its significant decrease caused by this treatment, although in this particular study no efforts were made to assess the time course of action of TCDD to cause suppression of CK2 in the nucleus. This phenomenon was subsequently studied in more detail both *in vivo* and *in vitro* by Ashida et al. [59]. First, they found that TCDD (1 μg/kg, i.p. single dose), given to guinea pigs, causes suppression of CK2 in the nucleus obtained from liver. To study the biochemical events taking place at the early stage of the action of TCDD, a short-term *in vitro* model system was established using explants of guinea pig liver tissue maintained in tissue culture medium. It was found that TCDD causes a rapid reduction in the activity of nuclear CK2 with an accompanying increase in the cytosol, indicating that such an action of TCDD might be caused by rapid translocation of CK2 protein from the nucleus to cytosol, since this protein is known to be rapidly shuttled between nucleus and cytosol. Such changes in protein phosphorylation activities were also accompanied by an increase in the DNA binding activity of AP-1. As for the meaning of CK2 activation, most reviews support the notion that CK2 is essentially a proinflammatory factor by virtue of its ability to stimulate rapid turnover of IKBα and phosphorylate p65 (RelA) protein in cytosol and thereby activate this important proinflammation mediator, which in turn induces rapid accumulation of RelA/NF-κB in the nucleus [60]. Once in the nucleus, CK2 is also involved in coordinating cellular stress responses: for example, controlling the activities of histone deacetylase 2 (HDAC2), which mediates the repression of proinflammatory genes by deacetylating core histones supporting RelA/p65. This histone deacetylating enzyme is known to be phosphorylated by CK2, upon cellular stress induced by cigarette smoking and oxidative stress [61], which promotes degradation of HDAC2 and thereby accelerates cell inflammatory responses. Thus, the initial increase in CK2 in cytosol [59] and its gradual decrease in the nucleus by the action of TCDD could represent its initial proinflammatory action and the late negative feedback to suppress inflammation through the nuclear interactions, respectively, although much more work would be needed to prove or disprove this possibility.

It must be added, on the other hand, that other than those recent works in which the propagation of the NGP signaling has been carefully separated from that of the classical action mechanism [52], previous observations on TCDD-induced changes in protein kinases (published before 2000), particularly those conducted *in vivo*, were not specifically conducted for the purpose of establishing the NGP per se. Thus, in those cases one cannot totally rule out the possibility of the involvement of the classical route of signaling in the process of activating and maintaining those kinases. Nevertheless, it is clear that those processes, being involved in eliciting typical proinflammatory signaling through protein kinases that are capable of directly activating key nuclear transcription factors such as AP-1, CREB, C/EBP, and NF-κB proteins, are not directly activated by the classical action route of TCDD, since the routes of protein phosphorylation-induced activation of these nuclear transcription factors are carried out through the well-established nongenomic signal transduction pathways that are mediated by a sequential activation of a series of protein kinases and phosphatases, not through any direct gene activation mechanisms. Thus, the possibility of any significant contributions of the classical route of action of AHR signaling on those genes activated by any of the above response elements is very low.

13.6 EXAMPLES OF THE NUCLEAR EVENT THAT IS DIRECTLY RELATED TO THE INITIAL NGP SIGNALING

Having described the possibility of significant roles of selected group of protein kinases in propagating and stabilizing the initial NGP signaling mostly *in vitro*, it would be helpful to present at this stage some concrete examples of the active role of a specific protein kinase in propagating the NGP signaling particularly into the nucleus in terms of activating a specific gene or genes that are directly related to inflammation. TCDD-induced activation of the IL-8 gene expression is one such example, and therefore it will be described here in detail.

Vogel et al. [62] studied the expression of two inflammation assisting chemokines, KC (mouse keratinocyte chemoattractant, a gene homologous with human IL-8) and MCP-1, in different organs of mice after exposure to TCDD and found that TCDD exposure led to an early and clear-cut induction of KC in liver and spleen on day 1, which was sustained over the period of 10 days. The level of MCP-1 mRNA was also induced by TCDD on day 1 in spleen, lung, kidney, and liver, which was further increased at day 7. The increase of KC and MCP-1 expressions at day 7 in liver, thymus, kidney, adipose, and heart was associated with elevated levels of expression of the macrophage marker F4/80, indicating the infiltration of macrophages in these organs. Induction of KC requires a functional AHR, since mice with a mutation in the AHR nuclear localization domain (AHR^{nls}) were found to be resistant to TCDD-induced expression of KC. A subsequent study by Li and Matsumura [31] showed that when 3T3-L1 adipocytes were treated with TCDD, its early effect on adipocytes was found to be predominantly on inflammation, judging by significant induction of COX-2 and KC (IL-8), which was accompanied by upregulation of C/EBPβ. Previously, TCDD-induced activation of C/EBPβ in 3T3-L1 adipocytes was already documented by Liu and Matsumura [63], who concluded that this event is most likely the cause for its action of TCDD to suppress the expression of insulin-sensitive glucose transporter 4 (Glut 4) expression, eventually leading to prevention of hormone-induced differentiation of 3T3-L1 cells to adipocytes. With such fundamental information on the importance of PKA in mediating the action of TCDD, Vogel et al. [64] studied the details of the mechanism of action of TCDD to rapidly induce IL-8 expression in U937 human macrophages. In this cell line, TCDD-induced activation of IL-8 mRNA expression starts as early as 1 h and keeps increasing till 52 h, the end of experiment. Not surprisingly, this action was very similar to the pattern induced by forskolin, a phosphodiesterase inhibitor known to activate PKA. Furthermore, this action of TCDD was abolished when cells were pretreated with H89, a specific inhibitor of PKA, again confirming the importance of PKA in this action of TCDD. The result of the IL-8 gene promoter deletion analysis indicated that the most important region needed for full activation of this gene resides in the short fragment situated within the −1 to −50 bp immediately upstream of the starting site. Using mutation analyses on its core response element sequence 5′-AGATGAGGGTGCATAG-3′, they showed that any modification to simulate the DRE sequence (e.g., changing its central sequence, GGGTG to GCGTG) totally abolished its IL-8 activating ability. Eventually it was shown that the major proteins that bind to this site were a heterodimer consisting of RelB and AHR proteins. This site was named as RelBAHRE, which was found in the promoting regions of several chemokines known to mediate inflammatory response of cells, particularly immune-related ones [64]. The existence of this heterodimer in the nuclear fraction from U937 macrophages has been established through rigorous confirmatory tests that included electrophoretic gel mobility shift assay (EMSA), supershift assay, immunocoprecipitation, siRNA effect assays, and plasmid-based AHR and RelB overexpression assays. It is noteworthy that there was no sign of the involvement of ARNT protein in this process of forming the RelB:AHR complex, nor was its binding activity to RelBAHRE, indicating a clear separation of this event from the known classical action pathway of TCDD. Although this particular study did not directly address the role of PKA in promoting IL-8 induction, judging by the significant increase in the total quantity of the RelB:AHR protein complex induced by forskolin (a stimulator of PKA) and by the antagonistic action of H89 to cause its decrease, the most likely possibility is that PKA has induced the nuclear transport of the monomer of the AHR protein as pointed out by Oesch-Bartlomowicz and Oesch [55].

RelB (also called p67) is an important member of the family of NF-κB proteins, which are one of the key groups of inflammation-mediating nuclear transcription factors. Perhaps the most important contributions of the above finding from the viewpoint of relating to the central theme of this chapter are as follows: (a) the entire process of this PKA triggered activation of IL-8 gene is processed through the NGP, (b) it represents one concrete example of AHR directly affecting at least one route of the NF-κB-mediated inflammatory cascade, and (c) it offers a new paradigm of AHR crosstalk in the nucleus, meaning that this work establishes a role of the AHR monomer in nuclear crosstalks, in addition to the already established cases of nuclear crosstalk activities involving the AHR:ARNT dimer with other types of nuclear transcription factors [65, 66]. The final question of this section is on the possible functional significance of such a heterodimer formation. The interested readers are recommended to read the review paper solely focused on this specific subject [67], although a brief description on one of the concrete examples of its possible physiological roles may help the reader. In this case, the RelB:AHR dimer has been found to play a critical role in the process of maturation of antigen-presenting dendritic cells (DC). Here, it was reported by Vogel et al. [68] that activation of AHR by TCDD, through its complex formation with RelB, induces indoleamine 2,3-dioxygenase 1 (IDO1) and indoleamine 2,3-dioxygenase-like protein (IDO2) in a model dendritic cells derived from U937 human macrophages. Induction of IDO1 and IDO2 was also found in lung and spleen associated with an increase in the regulatory T cells (T_{reg}) as judged by the increased expression of marker Foxp3 in spleen of TCDD-treated C57BL/6 mice. T_{reg} cells are key players in regulating the quality of T cells to prevent the process of T-cell differentiation from becoming out of control (such as not allowing the development of aberrant T cells that cause autoimmune diseases and the failure of recognizing animals own embryos). The above finding confirms the report by Quintana et al. [69] that AHR activated by its ligand TCDD induces functional T_{reg} cells that suppressed experimental autoimmune encephalomyelitis, by showing that the activated AHR indeed serves as a regulator of T_{reg} and $T_H 17$ cell differentiation in mice with an additional information that the role of the antigen-presenting dendritic cells in promoting T-cell differentiation is mediated by the AHR:RelB heterodimer.

13.6.1 An Example of the Significant Contribution of the Nongenomic Action Route of TCDD to the Process of the Development of a Well-Defined Toxic Syndrome *In Vivo*

An appropriate question one could ask at this stage of the development of this field of science on the NGP is whether there are concrete examples of the NGP significantly contributing to the pathogenesis of any of the hallmark toxicities of dioxin-like chemicals. Previously, two such best examples on the importance of the NGP have been the case of src knockout mice not showing full effects of TCDD to induce the "wasting syndrome," compared to the matched wild-type mice [70], and a similar but less profound case of the reduction in this action of TCDD in a TGFα-mutant, wa-1/wa-1 mice [42]. In addition, in the case of the effect of src deficiency causing attenuation of the effect of TCDD on reduction in adipose tissues, Vogel and Matsumura [71] have shown that mouse embryonic fibroblasts (MEF) isolated from src knockout mice and cultured respond to the action of TCDD *in vitro* in a manner fundamentally different from the equivalent MEF isolated from the genetically matched wild-type ($src^{+/+}$) mice. To be more specific, when those cells were artificially induced to differentiate into adipocytes, both MEF lines responded in similar manners by showing lipid accumulation. However, TCDD clearly suppressed differentiation of src$^{+/+}$ MEF, but not src$^{-/-}$ MEF, as measured by TCDD-induced changes in the level of accumulation of triglycerides that are associated with increased expression of adipocyte differentiation-specific genes. Further studies revealed that TCDD indeed induced C/EBPβ and C/EBPδ mRNA expression and DNA binding activity in a time- and dose-dependent manner in MEF$^{+/+}$ to inhibit their adipogenesis, but not in MEF$^{-/-}$. In contrast, the actions of TCDD to induce cytochrome P450s, CYP1A1 and CYP1B1, were comparable in both MEF$^{+/+}$ and MEF$^{-/-}$ strains of mice, indicating that the above interstrain difference is not due to any abnormality of the classical action pathway in MEF from the src knockout mice. These data indicated that suppression of differentiation by TCDD in MEF requires a functional Src kinase activity and induced levels of C/EBPβ and C/EBPδ play a pivotal role in mediating this action of TCDD; that is, this specific effect of TCDD is clearly mediated through the NGP.

While there is no question about the importance of both Src kinase and EGF receptor in the nongenomic signaling of TCDD *in vivo*, the relationship between these two protein kinase-related factors to the inflammatory response of the TCDD-affected cells has not been totally clarified yet. Thus, it appears to be prudent to cite at least one clear-cut case of the inflammatory action of TCDD mainly through the NGP causing the pathogenesis of a definite toxic syndrome that would be accepted by many toxicologists.

Hydronephrosis is one such syndrome observed to occur among TCDD-exposed developing mouse neonates that has been well documented [72]. This syndrome is characterized by a marked dilatation of the pelvis and calyces and thinning of the renal parenchyma and is considered to be caused mainly by anatomical obstruction, most commonly found at the ureteropelvic junction of the ureter, but in some cases even by the abnormal peristalsis of the ureter without any physical blocking. In humans, furthermore, the urine volume that overwhelms the capacity of transfer of urine from the kidney to the bladder is known to contribute to hydronephrosis induction in the absence of anatomical obstruction. As for the action of TCDD on mouse neonates, it was originally reported by Moore et al. [73] that mouse embryos exposed to TCDD through treatment on their mothers (single treatment at 3 μg/kg) at the critical window of gestation days 10 through 13 develop into neonates that show very high incidence (95%) of hydronephrosis. Later it was shown, however, that the lactational exposure of newborn mice to TCDD was also reported to induce hydronephrosis [74], indicating that this action of TCDD does not totally depend on the embryonic developmental stages to manifest.

A collaborative study evolved between Dr Chiharu Tohyama's research team (at that time at the National Institute of Environmental Studies in Tsukuba, Japan) and our laboratory on the mechanism of TCDD-induced pathogenesis of hydronephrosis has been focused on the inflammatory effect of TCDD on the kidney of those neonates postnatally exposed to TCDD through their mothers' milk [75]. It was found that in the kidneys of these pups, the expressions of a battery of inflammatory cytokines including MCP-1, tumor necrosis factor α (TNFα), and IL-1β were upregulated as early as postnatal day (PND) 7. The amounts of Cox-2 mRNA and protein expressions, as well as that of prostaglandin E_2 (PGE_2), were conspicuously upregulated in an AHR-dependent manner in the TCDD-induced hydronephrotic kidney, with a subsequent downregulation of the gene expressions of Na^+ and K^+ transporters, NKCC2 and ROMK. These findings helped us to concentrate on the role of Cox-2 in mediating inflammation of kidney leading to hydronephrotic conditions, which has already been suggested by Seibert et al. [76]. *In situ* hybridization of kidney slices with a specific antibody against Cox-2 protein indicated that TCDD-induced increase in the protein expression of this enzyme is confined to the epithelial layers surrounding the inner walls of kidney tubules, particularly in distal tubules and macula densa cells of the kidney that are active in exchanging ions such as Na^+, K^+, and Cl^- through NKCC2 and ROMK as well as other mechanisms [75]. Furthermore, TCDD-exposed mouse pups showed an increased production of PGE_2 and a decrease in the levels of NKCC2 and ROMK mRNAs on PND 7. This observation seems to be analogous to the signs of hydronephrosis expressed in Bartter's syndrome, known to have a mutation in one of at least four genes encoding membrane proteins of the nephron segment, including NKCC2 and ROMK. The final proof that Cox-2 plays the pivotal role in mediating this action of TCDD comes from the effectiveness of Cox-2 inhibition in eliminating this syndrome; that is, daily administration of a COX-2-selective inhibitor, indomethacin *N*-octylamide, to newborns until PND 7 completely abrogated the TCDD-induced PGE_2 synthesis and gene expressions of inflammatory cytokines and electrolyte transporters and eventually prevented the onset of hydronephrosis. A bonus finding from this study was that this way of inducing hydronephrosis through the action of TCDD to activate Cox-2 in mouse neonates is not accompanied with overt physical blocking of the ureter, indicating that this increased Cox-2-mediated ion exchanging activity occurring in the tubular epithelial cells alone could account for the pathogenesis of hydronephrosis in this mouse model.

The involvement of the NGP in the above action of TCDD to activate Cox-2 in the kidney tubular epithelial cells was further studied in MMDD1, murine macula densa cells in culture [33]. Briefly, we found in this cell line that induction of Cox-2 by TCDD is accompanied with a rapid increase in the enzymatic activity of cPLA2 as well as activation of protein kinases. Calcium serves as a trigger for such an action of TCDD in this cell line. These observations indicate that the basic mode of action of TCDD to induce the rapid inflammatory response in MMDD1 is remarkably similar to those mediated by the nongenomic pathway of the AHR found in other types of cells (Table 13.1). Such an action of TCDD to induce Cox-2 in MMDD1 was not affected by "DRE decoy" oligonucleotide treatment or by introduction of a mutation on the DRE site of Cox-2 promoter in a *Cox-2* gene expression as assessed by a reporter assay based on luciferase activity, suggesting that this route of action of TCDD clearly depends on the NGP.

The significance of the above discovery is that such a definable and well-recognized inflammation inducing enzyme as Cox-2 serves as the direct target of toxic action of this chemical through the NGP. Knowing the importance of Cox-2 in mediating inflammatory responses in many different tissues, it is possible that there could be additional toxic end points resulting from such an action of TCDD on this enzyme, depending on the critical roles of this enzyme in the vital functions in those organs.

13.6.2 Critical Examination of the Roles of NGP in Comparison to Other Pathways of Toxic Signaling of TCDD

Although in this chapter many examples have been cited to support the existence and the importance of the NGP, it must be cautioned that one should not overinterpret those examples to start assuming that the NGP is a pathway that stands alone without the participation of any other pathways in helping those affected cells to manifest many of toxic effects of TCDD. In other words, in many cases one should not overlook the possibility of participation of other types of action pathways, in addition to that of the NGP, in producing the final toxic outcomes, depending on each species studied. For instance, in the above case of manifestation of hydronephrosis, we are aware of a subtle but definite effect of TCDD on the epithelial cells of ureter themselves, which, in severe cases, could also lead to the eventual blockage of urinary clearance through the ureter itself and thereby exacerbates the symptom of hydronephrosis. Whether this action of TCDD on the ureter, apart from the ion exchanging activities in the tubular epithelial cells, is mediated by the NGP, the classical pathway, or any other routes of action has not been totally elucidated yet. Another good example illustrating the above principle of possible interactions and integration among different pathways is the case of species variations observed among the various target genes that have been affected by TCDD. Even in the case of Cox-2, there are some possibilities that this gene could be also activated by TCDD through the classical pathway depending on the tissues and the species studied. For instance, it was shown by Yang and Bleich [77] that, in rat pancreatic β cells, TCDD causes upregulation of Cox-2 through the increase in nuclear protein binding not only to CRE and C/EBP, as shown above in kidney epithelial cells, but also to DRE (i.e., through the classical route of action), and each response site contributing significantly to the final outcome of upregulation of rat *Cox-2* gene in this type of cell. Even in the case of cPLA2 that has been shown to be a key mediator of the NGP signaling in our studies, it was reported recently by Kinehara et al. [78] that the TCDD-activated AHR binds to the second intron of the Pla2g4a gene, which encodes cytosolic phospholipase A(2)α (i.e., cPLA2α), in mouse hepatoma Hepa 1c1c7 cells, suggesting that Pla2g4a appears to be one of the target genes of the AHR through its classical action pathway. cPLA2α is one of the six major subtypes of cPLA2 that is expressed in both human and mouse. This cytosolic enzyme definitely requires Ca^{2+} to be activated, and, therefore, the above finding suggests that in Hepa 1c1c cells, TCDD could utilize both these pathways (i.e., NGP and the classical action pathway) to activate this enzyme. This type of redundancy in terms of having more than one way to elicit cell stress responses is known in the fields of stress responses and signaling studies on hormone receptors [79, 80].

While the mention of these multiple routes of actions of TCDD might be taken as the negative evidence against the validity of the NGP as one of the major pathways of toxic action of TCDD, this way of viewing each of the major pathways in relation to other actions of TCDD actually helps to illustrate at least one important principle that nongenomic signaling, no matter how it started independently of any other pathways, must be eventually integrated with other forces, including those generated through the classical action route, to produce most of those major toxic syndromes. In other words, the significance of describing the nongenomic pathway in isolation from any other routes as shown above actually serves several purposes: (a) to correct the past assumption that the classical model is sufficient in explaining most of the toxic actions of TCDD, (b) to offer the means to grasp the concept of the significant roles of cell stress response, particularly inflammatory reactions, in mediating the toxicity of TCDD mainly through this route of action, (c) to illustrate the important roles of protein kinases and phosphatases in toxic signaling of TCDD, and (d) to integrate many of valuable contributions made by many toxicologists who have reported actions of TCDD on signal transduction pathways such as those initiated by growth factors [81] and hormone receptors [82] as well as observations on the action of TCDD on various protein kinases and phosphatases [11], calcium signaling, and cell stress responses such as inflammation [27, 83], apoptosis resistance, cell senescence [84], and so on from the specific viewpoint of the nongenomic route of cell regulation in the action of the ligand-activated AHR to add a novel perspective that is different from the existing paradigms. An added bonus could be the new prospect of ameliorating some of toxic symptoms through the use of anti-inflammatory agents and protein kinase inhibitors and help to eventually construct a new strategy of therapeutic approaches based on the nongenomic routes of toxic signaling of this class of enigmatic environmental pollutants. The rationale of this optimistic prediction is that, as in the case of most cellular stress responses mediated through signal transduction pathways, it is possible, at least in theory, to block the whole cascade of events by preventing the initial triggering step from taking place. This might be the reason for the use of src knockout mice in assessing the effect of TCDD to cause the wasting syndrome [70] or the use of dexamethasone in reducing the incidence of TCDD-induced wasting and the subsequent mortality in mice [85].

13.7 CONCLUSION

Survival under stressful environments is the most fundamental requirement for the evolutionary success of any given species. Thus, during the course of evolution, many modes of stress response mechanisms have been developed. Good examples are those aimed against invading pathogenic organisms through the development of innate immune responses and the Toll-like receptors [86] and intricate response mechanisms evolved in managing oxidative stresses [87]. Exposure to toxic chemicals must be considered to be one of the major classes of environmental stresses, and as such it is important to consider the strategies of the stress response against toxic chemicals that have been evolved in the given species within the context of the survival of that species. What one can notice is the common denominator among those defensive strategies that there are multiple and redundant pathways to ensure the success of each type of stress response (i.e., redundancy of stress signaling). This can be also seen in the case of signaling of the ligand-activated AHR, which is now accepted to rely on more than one pathway to ensure that its messages are properly delivered. That is where the nongenomic signaling comes as one of the means to assist the above mission of the AHR. Inflammation is only one of those defensive stress response mechanisms (Fig. 13.1). Indeed, the occurrence of inflammation as the result of toxic actions of xenobiotics has been already pointed out [88]. It must be added that the main point of selecting inflammation as the major theme of this chapter has been the availability of precedents indicating the intimate involvement of nongenomic signaling in mediating inflammatory responses in many organisms [4–6]. One of the main reasons for the adoption of nongenomic response mechanisms may be the necessity of activating the inflammatory responses as rapidly as possible. Another reason may be the necessity that inflammatory responses must be closely coordinated with changes in other major pathways, such as activation of glucocorticoid signaling, triggering the receptor signaling for growth factors, activating many pathways involved in decision-making process for apoptosis, and inducing changes in nutritional metabolic activities of cells. Thus, mechanisms of protein kinase-based signal transduction that can simultaneously affect the functions of many substrate proteins at once though rapid changes in phosphorylation activities as the means of transducing the initial inflammatory signaling become the convenient vehicle for this type of stress response mechanism, which appears to be rooted in the very early stages of evolutionary selection for survival of organisms [89].

While much more work will definitely be needed to firmly establish the nongenomic pathway as one of the major routes of toxic signaling of the ligand-activated AHR, it is also becoming clear that new perspectives on the biological meaning of ligand-induced signaling of the AHR are badly needed as we cannot just continue the path of relying solely on the classical pathway beyond its natural limitations. As explained above, one of the major functions of the ligand-activated AHR is likely activation of a host of "stress responses" that are intended to act as prosurvival mechanisms. In this regard, AHR plays just as important roles as the Toll-like receptors, although the latter receptors have received far more research attentions in the past from scientists from a variety of disciplines than the AHR. Recent findings on the action of TCDD to cause profound changes not only in the innate immunity but also in the adaptive immunity [90, 91] underscore the above notion, since the changes in adaptive immunity caused by AHR could also result in certain immune dysfunctions such as allergy and autoimmunity just as the ligand-activated Toll-like receptors do. Certainly many pieces of key evidence are still missing to allow us to critically compare these two major types of receptors, particularly on the role of AHR. Nevertheless, it is important to start addressing such a critical question on the role of AHR, and in this regard the above description, despite its sometimes risk taking approaches, offers a fresh viewpoint, which might help stimulate the interests of other scientists, and hence the main justification of proposing this new paradigm in this field of toxicology.

ACKNOWLEDGMENTS

The author would like to thank sincerely the contributions of the past and present members of this laboratory for their diligent efforts to clarify many aspects of nongenomic signaling of TCDD. Without their hard works dedicated to the overall goal of understanding this novel route of action of TCDD over the past 25 years, it would not have been possible to complete this chapter. The majority of the experimental works presented in this chapter has been supported by Research Grants R01-ES05233 and P01-ES05707 from the National Institute of Environmental Health Sciences, Research Triangle Park, NC, USA.

REFERENCES

1. McMillan, B. J. and Bradfield, C. A. (2007). The aryl hydrocarbon receptor sans xenobiotics: endogenous function in genetic model systems. *Molecular Pharmacology*, 72, 487–498.
2. Delescluse, C., Lemaire, G., de Sousa, G., and Rahmani, R. (2000). Is CYP1A1 induction always related to AHR signaling pathway? *Toxicology*, 153, 73–82.
3. Weiss, U. (2008). Inflammation. *Nature*, 454, 427.
4. Stellato, C. (2004). Post-transcriptional and nongenomic effects of glucocorticoids. *Proceedings of American Thoracic Society*, 1, 255–263.
5. Burgermeister, E. and Seger, R. (2007). MAPK kinases as nucleo-cytoplasmic shuttles for PPARγ. *Cell Cycle*, 6, 1539–1548.
6. Stice, J. P. and Knowlton, A. A. (2008). Estrogen, NF-κB, and the heat shock response. *Molecular Medicine*, 14, 517–527.
7. Cancela, L., Nemere, I., and Norman, A. W. (1988). 1α,25 (OH)$_2$ vitamin D3: a steroid hormone capable of producing pleiotropic receptor-mediated biological responses by both genomic and nongenomic mechanisms. *Journal of Steroid Biochemistry*, 30, 33–9.
8. Vasudevan, N. and Pfaff, D. W. (2007). Membrane-initiated actions of estrogens in neuroendocrinology: emerging principles. *Endocrine Reviews*, 28, 1–19.
9. Lange, C. A. (2004). Making sense of cross-talk between steroid hormone receptors and intracellular signaling pathways: who will have the last word? *Molecular Endocrinology*, 18, 269–278.
10. Freeman, M. R., Cinar, B., and Lu, M. L. (2005). Membrane rafts as potential sites of nongenomic hormonal signaling in prostate cancer. *Trends in Endocrinology and Metabolism*, 16, 273–279.
11. Carlson, D. B. and Perdew, G. H. (2002). A dynamic role for the Ah receptor in cell signaling? Insights from a diverse group of Ah receptor interacting proteins. *Journal of Biochemical and Molecular Toxicology*, 16, 317–325.
12. Boutwell, R. K., Takigawa, M., Verma, A. K., and Ashendel, C. L. (1983). Observations on the mechanism of skin tumor promotion by phorbol esters. *Princess Takamatsu Symposium*, 14, 177–193.
13. Hecker, E. (1985). Cell membrane associated protein kinase C as receptor of diterpene ester co-carcinogens of the tumor promoter type and the phenotypic expression of tumors. *Arzneimittelforschung*, 35, 1890–1903.
14. Bell, R. M., Hannun, Y. A., and Loomis, C. R. (1986). Mechanism of regulation of protein kinase C by lipid second messengers. *Symposium on Fundamental Cancer Research*, 39, 145–156.
15. Das, U. N. (1991). Arachidonic acid as a mediator of some of the actions of phorbolmyristate acetate, a tumor promoter and inducer of differentiation. *Prostaglandins Leukotrienes and Essential Fatty Acids*, 42, 241–244.
16. Haffner, S. M. (2000). Clinical relevance of the oxidative stress concept. *Metabolism*, 49, 30–34.
17. Kundu, J. K., Shin, Y. K., and Surh, Y. J. (2006). Resveratrol modulates phorbol ester-induced pro-inflammatory signal transduction pathways in mouse skin *in vivo*: NF-κB and AP-1 as prime targets. *Biochemical Pharmacology*, 72, 1506–1515.
18. Garg, R., Ramchandani, A. G., and Maru, G. B. (2008). Curcumin decreases 12-*O*-tetradecanoylphorbol-13-acetate-induced protein kinase C translocation to modulate downstream targets in mouse skin. *Carcinogenesis*, 29, 1249–1257.
19. Cossins, A., Fraser, J., Hughes, M., and Gracey, A. (2006). Post-genomic approaches to understanding the mechanisms of environmentally induced phenotypic plasticity. *Journal of Experimental Biology*, 209, 2328–2336.
20. Osterloh, A. and Breloer, M. (2008). Heat shock proteins: linking danger and pathogen recognition. *Medical Microbiology and Immunology*, 197, 1–8.
21. Barton, G. M. (2008). A calculated response: control of inflammation by the innate immune system. *Journal of Clinical Investigation*, 118, 413–420.
22. Mazina, O., Park, S., Sano, H., Wong, P., and Matsumura, F. (2004). Studies on the mechanism of rapid activation of protein tyrosine phosphorylation activities, particularly c-Src kinase, by TCDD in MCF10A. *Journal of Biochemical and Molecular Toxicology*, 18, 313–321.
23. Park, S., Mazina, O., Kitagawa, A., Wong, P., and Matsumura, F. (2004). TCDD causes suppression of growth and differentiation of MCF10A, human mammary epithelial cells by interfering with their insulin receptor signaling through c-Src kinase and ERK activation. *Journal of Biochemical and Molecular Toxicology*, 18, 322–331.
24. Park, S. and Matsumura, F. (2006). Characterization of anti-apoptotic action of TCDD as a defensive cellular stress response reaction against the cell damaging action of ultra-violet irradiation in an immortalized normal human mammary epithelial cell line, MCF10A. *Toxicology*, 217, 139–146.
25. Dong, B. and Matsumura, F. (2008). Roles of cytosolic phospholipase A2 and Src kinase in the early action of 2,3,7,8-tetrachlorodibenzo-*p*-dioxin through a nongenomic pathway in MCF10A cells. *Molecular Pharmacology*, 74, 255–263.
26. Dong, B and Matsumura, F. (2009). The conversion of rapid TCCD nongenomic signals to persistent inflammatory effects via select protein kinases in MCF10A cells. *Molecular Endocrinology*, 23, 549–558.
27. Puga, A., Nebert, D. W., and Carrier, F. (1992). Dioxin induces expression of c-fos and c-jun proto-oncogenes and a large increase in transcription factor AP-1. *DNA and Cell Biology*, 11, 269–281.
28. Matsumura, F., Brewster, D. W., Madhukar, B. V., and Bombick, D. W. (1984). Alteration of rat hepatic plasma membrane functions by 2,3,7,8-tetrachlorodibenzo-*p*-dioxin (TCDD). *Archives of Environmental Contamination and Toxicology*, 13, 509–515.
29. Medzhitov, R. (2008). Origin and physiological roles of inflammation. *Nature*, 454, 428–435.
30. Sciullo, E. M., Vogel, C. F., Li, W., and Matsumura, F. (2008). Initial and extended inflammatory messages of the nongenomic

signaling pathway of the TCDD-activated Ah receptor in U937 macrophages. *Archives of Biochemistry and Biophysics, 480,* 143–155.

31. Li, W. and Matsumura, F. (2008). Significance of the nongenomic, inflammatory pathway in mediating the toxic action of TCDD to induce rapid and long-term cellular responses in 3T3-L1 adipocytes. *Biochemistry, 47,* 13997–14008.

32. Li, W., Vogel, C. F., Fujiyoshi, P., and Matsumura, F. (2008). Development of a human adipocyte model derived from human mesenchymal stem cells (hMSC) as a tool for toxicological studies on the action of TCDD. *Biological Chemistry, 389,* 169–177.

33. Dong, B., Nishimura, N., Vogel, C. F., Tohyama, C., and Matsumura, F. (2010). TCDD-induced cyclooxygenase-2 expression is mediated by the nongenomic pathway in mouse MMDD1 macula densa cells and kidneys. *Biochemical Pharmacology, 79,* 487–497.

34. Bombick, D. W., Madhukar, B. V., Brewster, D. W., and Matsumura, F. (1985). TCDD (2,3,7,8-tetrachlorodibenzo-*p*-dioxin) causes increases in protein kinases particularly protein kinase C in the hepatic plasma membrane of the rat and the guinea pig. *Biochemical and Biophysical Research Communications, 127,* 296–302.

35. Bombick, D. W. and Matsumura, F. (1987). 2,3,7,8-Tetrachlorodibenzo-*p*-dioxin causes elevation of the levels of the protein tyrosine kinase pp60c-src. *Journal of Biochemical Toxicology, 2,* 141–154.

36. Madhukar, B. V., Ebner, K., Matsumura, F., Bombick, D. W., Brewster D. W., and Kawamoto, T. (1988). 2,3,7,8-Tetrachlorodibenzo-*p*-dioxin causes an increase in protein kinases associated with epidermal growth factor receptor in the hepatic plasma membrane. *Journal of Biochemical Toxicology, 3,* 261–277.

37. Sewall, C. H., Lucier, G. W., Tritscher, A. M., and Clark, G. C. (1993). TCDD-mediated changes in hepatic epidermal growth factor receptor may be a critical event in the hepatocarcinogenic action of TCDD. *Carcinogenesis, 14,* 1885–1893.

38. Kohn, M. C. and Portier, C. J. (1994). A model of effects of TCDD on expression of rat liver proteins. *Progress in Clinical Biology Research, 387,* 211–222.

39. Abbott, B. D. and Birnbaum, L. S. (1990). TCDD-induced altered expression of growth factors may have a role in producing cleft palate and enhancing the incidence of clefts after coadministration of retinoic acid and TCDD. *Toxicology and Applied Pharmacology, 106,* 418–432.

40. Abbott, B. D., Buckalew, A. R., and Leffler, K. E. (2005). Effects of epidermal growth factor (EGF), transforming growth factor-alpha (TGFα), and 2,3,7,8-tetrachlorodibenzo-*p*-dioxin on fusion of embryonic palates in serum-free organ culture using wild-type, EGF knockout, and TGFα knockout mouse strains. *Birth Defects Research, Part A, 73,* 447–454.

41. Matsumura, F. (1994). How important is the protein phosphorylation pathway in the toxic expression of dioxin-type chemicals? *Biochemical Pharmacology, 48,* 215–224.

42. Kitamura, N., Wong, P., and Matsumura, F. (2006). Mechanistic investigation on the cause for reduced toxicity of TCDD in wa-1 homozygous TGFα mutant strain of mice as compared its matching wild-type counterpart, C57BL/6J mice. *Journal of Biochemical and Molecular Toxicology, 20,* 151–158.

43. Köhle, C., Gschaidmeier, H., Lauth, D., Topell, S., Zitzer, H., and Bock, K. W. (1999). 2,3,7,8-Tetrachlorodibenzo-*p*-dioxin (TCDD)-mediated membrane translocation of c-Src protein kinase in liver WB-F344 cells. *Archives of Toxicology, 73,* 152–158.

44. Haarmann-Stemmann, T., Bothe, H., and Abel, J. (2009). Growth factors, cytokines and their receptors as downstream targets of arylhydrocarbon receptor (AHR) signaling pathways. *Biochemical Pharmacology, 77,* 508–520.

45. Bombick, D. W., Jankun, J., Tullis, K., and Matsumura, F. (1988). 2,3,7,8-Tetrachlorodibenzo-*p*-dioxin causes increases in expression of c-erb-A and levels of protein-tyrosine kinases in selected tissues of responsive mouse strains. *Proceedings of the National Academy of Sciences of the United States of America, 85,* 4128–4132.

46. Bagchi, D., Bagchi, M., Tang, L., and Stohs, S. J. (1997). Comparative *in vitro* and *in vivo* protein kinase C activation by selected pesticides and transition metal salts. *Toxicology Letters, 91,* 31–37.

47. Weber, T. J., Chapkin, R. S., Davidson, L. A., and Ramos, K. S. (1996). Modulation of protein kinase C-related signal transduction by 2,3,7,8-tetrachlorodibenzo-*p*-dioxin exhibits cell cycle dependence. *Archives of Biochemistry and Biophysics, 328,* 227–232.

48. Williams, S. R., Son, D. S., and Terranova, P. F. (2004). Protein kinase C delta is activated in mouse ovarian surface epithelial cancer cells by 2,3,7,8-tetrachlorodibenzo-*p*-dioxin (TCDD). *Toxicology, 195,* 1–17.

49. Kim, S. Y., Lee, H. G., Choi, E. J., Park, K. Y., and Yang, J. H. (2007). TCDD alters PKC signaling pathways in developing neuronal cells in culture. *Chemosphere, 67,* S421–S427.

50. Wölfle, D., Schmutte, C., and Marquardt, H. (1993). Effects of 2,3,7,8-tetrachlorodibenzo-*p*-dioxin on protein kinase C and inositol phosphate metabolism in primary cultures of rat hepatocytes. *Carcinogenesis, 14,* 2283–2287.

51. Rich, T. C. and Karpen, J. W. (2002). Review article: cyclic AMP sensors in living cells: what signals can they actually measure? *Annals of Biomedical Engineering, 30,* 1088–1099.

52. Matsumura, F. (2009). The significance of the nongenomic pathway in mediating inflammatory signaling of the dioxin-activated Ah receptor to cause toxic effects. *Biochemical Pharmacology, 77,* 608–626.

53. Vogel, C. F., Sciullo, E., Park, S., Liedtke, C., Trautwein, C., and Matsumura, F. (2004). Dioxin increases C/EBPβ transcription by activating cAMP/protein kinase A. *Journal of Biological Chemistry, 279,* 8886–8894.

54. McKenzie, K. E., Bandyopadhyay, G. K., Imagawa, W., Sun, K., and Nandi, S. (1994). Omega-3 and omega-6 fatty acids and PGE2 stimulate the growth of normal but not tumor mouse mammary epithelial cells: evidence for alterations in the signaling pathways in tumor cells. *Prostaglandins Leukotrienes and Essential Fatty Acids, 51,* 437–443.

55. Oesch-Bartlomowicz, B. and Oesch, F. (2009). Role of cAMP in mediating AHR signaling. *Biochemical Pharmacology*, 77, 627–641.

56. de Oliveira, S. K., Hoffmeister, M., Gambaryan, S., Müller-Esterl, W., Guimaraes, J. A., and Smolenski, A. P. (2007). Phosphodiesterase 2A forms a complex with the co-chaperone XAP2 and regulates nuclear translocation of the aryl hydrocarbon receptor. *Journal of Biological Chemistry*, 282, 13656–13663.

57. Filhol, O. and Cochet, C. (2000). Protein kinase CK2 in health and disease: cellular functions of protein kinase CK2: a dynamic affair. *Cell and Molecular Life Science*, 66, 1830–1839.

58. Enan, E. and Matsumura, F. (1995). Regulation by 2,3,7,8-tetrachlorodibenzo-*p*-dioxin (TCDD) of the DNA binding activity of transcriptional factors via nuclear protein phosphorylation in guinea pig adipose tissue. *Biochemical Pharmacology*, 50, 1199–1206.

59. Ashida, H., Nagy, S., and Matsumura, F. (2000). 2,3,7,8-Tetrachlorodibenzo-*p*-dioxin (TCDD)-induced changes in activities of nuclear protein kinases and phosphatases affecting DNA binding activity of c-Myc and AP-1 in the livers of guinea pigs. *Biochemical Pharmacology*, 59, 741–751.

60. Singh, N. N. and Ramji, D. P. (2008). Protein kinase CK2, and important regulator of the inflammatory response? *Journal of Molecular Medicine*, 86, 887–897.

61. Adenuga, D. and Rahman, I. (2010). Protein kinase CK2-mediated phosphorylation of HDAC2 regulates co-repressor formation, deacetylase activity and acetylation of HDAC2 by cigarette smoke and aldehydes. *Archives of Biochemistry and Biophysics*, 498, 62–73.

62. Vogel, C. F., Nishimura, N., Sciullo, E., Wong, P., Li, W., and Matsumura, F. (2007). Modulation of the chemokines KC and MCP-1 by 2,3,7,8-tetrachlorodibenzo-*p*-dioxin (TCDD) in mice. *Archives of Biochemistry and Biophysics*, 461, 169–175.

63. Liu, P. C. and Matsumura, F. (2007). TCDD suppresses insulin-responsive glucose transporter (GLUT-4) gene expression through C/EBP nuclear transcription factors in 3T3-L1 adipocytes. *Journal of Biochemical and Molecular Toxicology*, 20, 79–87.

64. Vogel, C. F., Sciullo, E., Li, W., Wong, P., Lazennec, G., and Matsumura, F. (2007). RelB, a new partner of aryl hydrocarbon receptor-mediated transcription. *Molecular Endocrinology*, 21, 2941–2955.

65. Ohtake, F., Fujii-Kuriyama, Y., and Kato, S. (2009). AHR acts as an E3 ubiquitin ligase to modulate steroid receptor functions. *Biochemical Pharmacology*, 77, 474–484.

66. Gomez-Duran, A., Carvajal-Gonzalez, J. M., Mulero-Navarro, S., Santiago-Josefat, B., Puga, A., and Fernandez-Salguero, P. M. (2007). Fitting a xenobiotic receptor into cell homeostasis: how the dioxin receptor interacts with TGFβ signaling. *Biochemical Pharmacology*, 77, 700–712.

67. Vogel, C. F. and Matsumura, F. (2009). A new cross-talk between the aryl hydrocarbon receptor and RelB, a member of the NF-κB family. *Biochemical Pharmacology*, 77, 734–745.

68. Vogel, C. F., Goth, S. R., Dong, B., Pessah, I. N., and Matsumura, F. (2008). Aryl hydrocarbon receptor signaling mediates expression of indoleamine 2,3-dioxygenase. *Biochemical and Biophysical Research Communication*, 375, 331–335.

69. Quintana, F. J., Basso, A. S., Iglesias, A. H., Korn, T., Farez, M. F., Bettelli, E., Caccamo, M., Oukka, M., and Weiner, H. L. (2008). Control of T_{reg} and T_H17 cell differentiation by the aryl hydrocarbon receptor. *Nature*, 453, 65–71.

70. Dunlap, D. Y., Ikeda, I, Nagashima, H., Vogel, C. F., and Matsumura, F. (2002). Effects of src-deficiency on the expression of *in vivo* toxicity of TCDD in a strain of c-src knockout mice procured through six generations of backcrossings to C57BL/6 mice. *Toxicology*, 172, 125–141.

71. Vogel, C. F. and Matsumura, F. (2003). Interaction of 2,3,7,8-tetrachlorodibenzo-*p*-dioxin (TCDD) with induced adipocyte differentiation in mouse embryonic fibroblasts (MEFs) involves tyrosine kinase c-Src. *Biochemical Pharmacology*, 66, 1231–1244.

72. Couture, L. A., Abbott, B. D., and Birnbaum L. S. (1990). A critical review of the developmental toxicity and teratogenicity of 2,3,7,8-tetrachlorodibenzo-*p*-dioxin: recent advances toward understanding the mechanism. *Teratology*, 42, 619–627.

73. Moore, J. A., Gupta, B. N., Zinkl, J. G., and Vos, J. G. (1973). Postnatal effects of maternal exposure to 2,3,7,8-tetrachlorodibenzo-*p*-dioxin (TCDD). *Environmental Health Perspectives*, 5, 81–85.

74. Couture-Haws, L., Harris, M. W., McDonald, M. M., Lockhart, A. C., and Birnbaum, L. S. (1991). Hydronephrosis in mice exposed to TCDD-contaminated breast milk: identification of the peak period of sensitivity and assessment of potential recovery. *Toxicology and Applied Pharmacology*, 107, 413–428.

75. Nishimura, N, Matsumura, F., Vogel, C. F., Nishimura, H., Yonemoto, J., Yoshioka, W., and Tohyama, C. (2008). Critical role of cyclooxygenase-2 activation in pathogenesis of hydronephrosis caused by lactational exposure of mice to dioxin. *Toxicology and Applied Pharmacology*, 231, 374–383.

76. Seibert, K., Masferrer, J. L., Needleman, P., and Salvemini, D. (1996). Pharmacological manipulation of cyclo-oxygenase-2 in the inflamed hydronephrotic kidney. *British Journal of Pharmacology*, 117, 1016–1020.

77. Yang, F. and Bleich, D. (2004). Transcriptional regulation of cyclooxygenase-2 gene in pancreatic beta-cells. *Journal of Biological Chemistry*, 279, 35403–35411.

78. Kinehara, M., Fukuda, I., Yoshida, K., and Ashida, H. (2009). Aryl hydrocarbon receptor-mediated induction of the cytosolic phospholipase A(2)alpha gene by 2,3,7,8-tetrachlorodibenzo-*p*-dioxin in mouse hepatoma Hepa-1c1c7 cells. *Journal of Bioscience and Bioengineering*, 108, 277–281.

79. Harris, S. L. and Levine, A. J. (2005). The p53 pathway: positive and negative feedback loops. *Oncogene*, 24, 2899–2908.

80. Shtil, A. A. and Azare, J. (2005). Redundancy of biological regulation as the basis of emergence of multidrug resistance. *International Reviews on Cytology*, 246, 1–29.

81. Kung, T., Murphy, K. A., and White, L. A. (2009). The aryl hydrocarbon receptor (AHR) pathway as a regulatory pathway for cell adhesion and matrix metabolism. *Biochemical Pharmacology, 77,* 536–546.
82. Vezina, C. M., Lin, T. M., and Peterson, R. E. (2009). AHR signaling in prostate growth, morphogenesis, and disease. *Biochemical Pharmacology, 77,* 566–576.
83. Hanneman, W. H., Legare, M. E., Barhoumi, R., Burghardt, R. C., Safe, S., and Tiffany-Castiglioni, E. (1996). Stimulation of calcium uptake in cultured rat hippocampal neurons by 2,3,7,8-tetrachlorodibenzo-*p*-dioxin. *Toxicology, 112,* 19–28.
84. Ray, S. and Swanson, H. I. (2009). Activation of the aryl hydrocarbon receptor by TCDD inhibits senescence: a tumor promoting event? *Biochemical Pharmacology, 77,* 681–688.
85. Taylor, M. J., Lucier, G. W., Mahler, J. F., Thompson, M., Lockhart, A. C., and Clark, G. C. (1992). Inhibition of acute TCDD toxicity by treatment with anti-tumor necrosis factor antibody or dexamethasone. *Toxicology and Applied Pharmacology, 117,* 126–132.
86. Takeda, K. (2005). Evolution and integration of innate immune recognition systems: the Toll-like receptors. *Journal of Endotoxin Research, 11,* 51–55.
87. Cullinan, S. B. and Diehl, J. A. (2006). Coordination of ER and oxidative stress signaling: the PERK/Nrf2 signaling pathway. *International Journal of Biochemistry and Cell Biology, 38,* 317–332.
88. Ganey, P. E. and Roth, R. A. (2001). Concurrent inflammation as a determinant of susceptibility to toxicity from xenobiotic agents. *Toxicology, 169,* 195–208.
89. Roux, P. P. and Blenis, J. (2004). ERK and p38 MAPK-activated protein kinases: a family of protein kinases with diverse biological functions. *Microbiology and Molecular Biology Reviews, 68,* 320–344.
90. Esser, C. (2009). The immune phenotype of AHR null mouse mutants: not a simple mirror of xenobiotic receptor overactivation. *Biochemical Pharmacology, 77,* 597–607.
91. Kirkvliet N. I. (2009). AHR-mediated immunomodulation: the role of altered transcription. *Biochemical Pharmacology, 77,* 746–760.

14

INTERSPECIES HETEROGENEITY IN THE HEPATIC TRANSCRIPTOMIC RESPONSE TO AHR ACTIVATION BY DIOXIN

Paul C. Boutros

14.1 MOTIVATION AND OVERVIEW

As described in earlier chapters, the aryl hydrocarbon receptor (AHR) is a major regulator of both toxicological responses and basal physiology. It is evolutionarily ancient, with ancestors known in all animal taxa. The evolution of the AHR and its associated signaling pathways is described in Chapter 27, while its function in invertebrates is outlined in Chapter 28. Given this strong conservation of the AHR primary protein sequence, many have hypothesized that AHR *function* has also been conserved over evolutionary time. This hypothesis is bolstered by the toxicity of the dioxin class of compounds to a broad range of species. Given that the transcription factor activity of the AHR appears essential for its mediation of toxic responses, it appears likely that any conservation of AHR function will be caused by a conservation of AHR target genes across species. Several groups have sought to directly test this hypothesis by identifying and comparing AHR target genes across different mammalian species. This chapter outlines those studies and aims to address two key questions. First, which AHR target genes are conserved and which are not? Second, what does the pattern of conservation tell us about the function of the AHR? First, the techniques and approaches used to answer these questions are overviewed. Next, the work of several groups on diverse model organisms is outlined. These studies come to the surprising conclusion that AHR target genes are poorly conserved: the vast majority of responses to AHR ligands are species specific. After noting some technical limitations with current studies, I speculate on the potential consequences of these findings for our broad understanding of AHR biology.

14.2 WHY FOCUS ON THE LIVER?

In this chapter, we focus on hepatic differences in AHR target genes, as assessed at the transcriptional level. This largely reflects the current literature—most studies have focused on hepatic changes in mRNA levels. But why? There are two basic reasons: the central role of the liver in AHR function and in TCDD-mediated toxicities, and the fact that the liver is a near-ideal organ for microarray studies.

14.2.1 Known Hepatic TCDD Toxicities

The liver is a central organ for TCDD-mediated toxicities. As outlined elsewhere in this book, exposure to TCDD induces a range of responses in the liver. These include extensive necrosis (particularly in rabbit), steatosis (abnormal lipid retention), hepatocellular hypertrophy (enlargement of liver cells), the formation of multinucleated hepatocytes (a potential precursor to tumors), and immune cell infiltration [1]. Changes in mouse liver can even be observed within the first 24 h following TCDD exposure [2] and are therefore perhaps the very earliest morphological (rather than molecular) changes following TCDD exposure.

A recent step in a long line of elegant transgenic animal studies from Dr. Chris Bradfield's lab has highlighted the critical role of the liver in TCDD-mediated toxicities. The Bradfield lab has been successively generating transgenic mouse models highlighting key aspects of TCDD biology. After the initial demonstrations that $Ahr^{-/-}$ mice were refractory to TCDD toxicities [3, 4], successive transgenic mouse models have been used to demonstrate the essential

---

The AH Receptor in Biology and Toxicology, First Edition. Edited by Raimo Pohjanvirta.
© 2012 John Wiley & Sons, Inc. Published 2012 by John Wiley & Sons, Inc.

roles of DNA binding [5], nuclear translocation [6], and heterodimerization [7]. Furthermore, functional AHR protein in hepatocytes is required for most of liver toxicities, including hepatomegaly, inflammation, lipid accumulation, ALT induction, and most of the TCDD-induced upregulation of *Cyp1a1*, *Cyp1a2*, and *Cyp1b1* [8]. In their most recent work along these lines, Nukaya et al. demonstrate that the presence of functional ARNT protein (and hence presumably the presence of functional AHR:ARNT dimers) in hepatocytes is required for major hepatic toxicities. For example, the TCDD-induced upregulation of *Cyp1a1*, *Cyp1a2*, *Cyp1b1*, and *Ahrr* were all ablated in transgenic mice with hepatocyte-specific knockout of ARNT. Dioxin-induced liver damage was absent in these mice—liver weights were unaltered, focal inflammation was absent, and steatosis was not observed [9]. These data strongly suggest that the extensive liver damage caused by dioxin exposure in mice originates from hepatocyte-based transcriptional changes.

Furthermore, one study has also shown that TCDD induces more extensive transcriptional alterations in the liver (297 genes) than in the kidney (17 genes) 19 h after TCDD exposure. Interestingly, this intertissue difference was not as pronounced when untreated $Ahr^{-/-}$ and $Ahr^{+/+}$ mice were compared to one another (471 genes altered in liver versus 379 genes altered in kidney) [10]. Recent unpublished data from our group indicates that both the hypothalamus and the white adipose tissue also display substantially fewer transcriptomic alterations in response to TCDD than does the liver, at least in rats (Boutros and Pohjanvirta, unpublished data). These data suggest that the physiological centrality of the liver in dioxin toxicity is also mirrored by a priority in the number and magnitude of dioxin-responsive transcriptional alterations, even at relatively early time points.

14.2.2 Easy "Array-Ability"

The vast majority of AHR- and TCDD-related microarray studies have focused on the liver [2, 11–23], although there have been important exceptions [10, 24, 25]. In particular, most of the earlier array studies focused specifically on liver or cell lines derived from normal or malignant liver cells [11, 26, 27]. While this is in part because of the central role of the liver in TCDD toxicities described above, it also has some technical origins.

Some of the major challenges in a microarray experiment revolve around cellular heterogeneity and RNA availability. In both these respects, the liver is a near-ideal organ for study. It is a large organ that is relatively easy to dissect, and it lacks the large number of RNases present in tissues such as the pancreas—the presence of significant nuclease concentrations can substantially impinge on the ability to extract intact, undegraded RNA from a tissue. Similarly, the liver is a relatively homogeneous tissue at the cellular level. Hepatocytes are by far the dominant cell type, and the different lobes of the liver show only modest differences. Thus, there is minimal intratissue heterogeneity to confound microarray studies.

14.2.3 Caveats on Liver Studies

Although the liver is an important tissue for dioxin toxicities in most species, there are some noteworthy differences. Even between mice and rats there are some divergences in the specific hepatotoxicities that arise from dioxin exposure. These are generally differences in degree rather than in nature: for example, fat accumulation, inflammatory cell infiltration, and hydropic degeneration are more pronounced in mice than in rats. Rats tend to be more sensitive to effects on liver (and body) weight gain [28]. Similarly, there are some cellular differences, with a greater extent of apoptosis and necrosis in mice. The only species-specific hepatotoxicity of which the author is aware of involves giant, multinucleated hepatocytes, which appear in rats but not mice [2, 14, 29–32].

It should also be noted, however, that the microarray analysis of liver is not definitively related to toxicity. For example, toxicities may arise from nonhepatic events and therefore do not depend on liver mRNA levels. Indeed, there is a report that hematopoietic cells contribute to TCDD-induced hepatic lesions [29]. Alternatively, while the liver is predominantly comprised of hepatocytes, there are several other cell types present at lower frequencies. These changes could lie out of the dynamic range of current microarrays. In particular, the Kupffer cells have been implicated in hepatotoxicities induced by a variety of toxic and carcinogenic agents [33]. The stellate cells (also called Ito cells) may also be involved, as they play a central role in retinoid storage and metabolism [34], which are strongly linked to TCDD toxicities [28].

14.3 TECHNIQUES AND APPROACHES

14.3.1 Microarrays and Transcriptomic Data Analysis

To date, all interspecies comparisons of liver have employed functional genomics approaches and, in particular, microarrays. This is not because proteomic approaches are not viable [15], but rather because of their increased complexity in sample preparation, experimental execution, and data analysis. In contrast, a transcriptomic study can be readily performed by virtually any molecular biology laboratory by exploiting widespread and generally affordable core facilities.

Almost all the key studies in this area have used microarray approaches. Microarrays are a well-established way of performing transcriptome-wide profiling [35–37], but there are a variety of other options. For example, high-throughput real-time RT-PCR has been used to study hundreds of genes in hundreds of samples [38, 39]. More recently, so-called medium-throughput platforms such as NanoString [40] and OpenArray provide the promise of increasingly rapid medium-throughput profiling of mRNA abundances. On the upper end, microarrays are gradually being replaced by direct RNA sequencing, which provides greater coverage and information content at a rapidly declining price [41]. Given these significant technological advances, it is likely that the conclusions derived from microarray studies will soon be revisited in greater detail.

Nevertheless, current microarray approaches can assess the mRNA abundances in a relative manner for ~20,000 genes in a parallel experiment. It is important to note that essentially all microarrays provide relative measurements—the absolute signal intensities are almost meaningless because they are confounded with technical factors such as differences in hybridization efficiency, fluorophore incorporation, and cross-hybridization, although attempts have been made [42, 43]. Thus, the results described here have all been derived by comparing a single gene across multiple conditions, rather than by comparing one gene to others.

14.3.2 Ortholog Identification

Once microarray experiments have been used to identify a set of dioxin-responsive genes in two or more species, it is necessary to compare these lists. This is a surprisingly difficult challenge for two reasons. First, it is challenging because the definition of a gene is in serious flux, thanks to the work of large consortia, such as the ENCODE project [44, 45]. Without clear definitions of a gene, it becomes very challenging (and indeed controversial) to determine how to define the evolutionary relationships among regions of sequences across different species [46–49]. The most common approach in use involves making unambiguous mappings of microarray probes to gene IDs [50–52], and then to map these gene IDs to the HomoloGene database of homologs [53]. HomoloGene mostly uses simple sequence alignment approaches to identify evolutionarily related genes, and so this approach is generally thought to be conservative. Nevertheless, it is clear that simple sequence similarity analysis cannot ensure that a gene harbors similar functions in two separate species [48].

14.3.3 Strains and Species

When comparing a biological phenomenon across multiple species, it is helpful to select species that have sufficient physiological and molecular similarities that convergences and divergences in response can be clearly recognized. The ortholog identification problem is one aspect of this challenge. It is clearly easy to identify orthologous gene pairs among species that display significant synteny, such as mouse and human [54].

For questions revolving around the effects of dioxins on humans, it might conceivably be advantageous to perform studies on humans and other primates. This is, of course, highly unethical! Instead, model systems are widely used. The selection of appropriate model systems is complex, but involves considerations such as experimental tractability, generation time, and similarity to humans. For these reasons, mice and rats are widely used in interspecies studies. For comparisons with humans, immortalized cell culture model systems or primary hepatocytes can be used. It is well established that the use of single cell-type cultures (primary or immortalized) will not fully represent the *in vivo* response of a complex organ. Indeed, persistent *in vivo* to *in vitro* differences have been associated with the use of culture models [55].

Even when using model species such as mice and rats, there are important genetic selections to be made. Each of these common rodent model organisms is represented by dozens of genetically distinct strains. These strains can differ dramatically in their sensitivity to dioxins [56, 57] and in their transcriptomic profiles [58–60]. As a result, strain differences can greatly impact interspecies comparisons. Indeed, it has been reported that in studies of up to six rat strains, the majority of dioxin-responsive genes in rat were strain specific and were altered in only a single strain [17, 22].

14.4 RAT VERSUS MOUSE

The most extensive interspecies comparisons to date have been made between rats and mice. There have been two separate transcriptomic studies [2, 21] along with the development of a very important transgenic animal model [61]. A summary of these studies is given in Table 14.1.

14.4.1 Initial Surprise: Boverhof et al.

The first study of interspecies transcriptomic alterations was due to the work of Dr. Timothy Zacharewski's group at Michigan State University [2]. They chose to focus on immature, ovariectomized female animals. This immediately creates a potential source of differences from the other studies described later, which primarily focused on male animals or culture-based model systems. They used C57BL/6 mice and Sprague Dawley rats and employed an older cDNA microarray platform that represented only 3022 unique rat genes (along with 7885 unique mouse genes). To supplement their rat data set with more comprehensive genome-wide coverage, they downloaded and reanalyzed a publicly available data set of male Sprague Dawley expression profiles from an oligonucleotide platform [14]. Together, these three data sets gave a list of 3087 orthologous gene pairs.

The authors found that there were approximately equivalent numbers of genes altered by TCDD exposure in mouse (238) and rat (201). Of these, only 33 genes were differen-

TABLE 14.1 Experimental Design of Rat-Mouse Comparison Studies

	Boverhof	Boutros
Mouse strain	C57BL/6	C57BL/6
Mouse gender	Female/ovariectomized	Male
Mouse age	Immature	15 weeks
Mouse exposure time(s)	2, 4, 8, 12, 18, 24, 72, 168 h	19
Mouse exposure dose(s)	0.001, 0.01, 0.1, 1, 10, 100, 300 μg/kg	1000 μg/kg
Rat strain	Sprague Dawley	Long-Evans
Rat gender	Female/ovariectomized and male	Male
Rat age	Immature	12 weeks
Rat exposure times	2, 4, 8, 12, 18, 24, 72, 168 h	19 h
Rat Exposure Doses	0.001, 0.01, 0.1, 1, 10, 30, 100 μg/kg	100 μg/kg
Orthologous genes	3087	8125
Technology	cDNA + Affymetrix	Affymetrix
Rat genes changed	201	200
Mouse genes changed	238	278

tially expressed in both strains when the authors employed a rigorous statistical analysis. The authors did not report any dose–response-based differences, despite a relatively extensive dose–response profile of seven doses ranging from 0.001 to 300 μg/kg TCDD. To try to systematize these data, the authors performed a detailed pathway analysis. They found that genes altered in both species were enriched for xenobiotic metabolizing enzymes and genes involved in lipid and amino acid metabolism (e.g., *Cyp1a1* and *Nqo1*). Lipid metabolizing genes were also enriched in the set of genes differentially expressed only in mice, and again in the set of genes differentially expressed only in rats. In contrast, growth factors were enriched in rats alone, while immune-related genes were altered only in mice [2]. Because of the temporal nature of the analysis, it is unclear whether these alterations reflect primary differences induced by the AHR or are manifestations of the few differences in toxicological responses between mice and rats (outlined in Section 14.2.3). The authors concluded by performing an analysis of AHR binding sites (aryl hydrocarbon response elements (AHREs)) using techniques they had described earlier [62]. They found evidence that about 85% (94/111) of the genes tested showed evidence of AHREs. Furthermore, almost half of all genes (53/111) showed positionally conserved AHREs, which are particularly strong evidence of functional associations [63].

Taken together, then, Boverhof and coworkers showed that the mouse and rat respond in dramatically different ways to TCDD. Strengths of this work include the extensive dose–response and time course analysis, detailed histological and clinical chemistry assessments of the species-specific toxicities, and a comprehensive bioinformatics pipeline. However, the study was limited to only 3087 homologous gene pairs, used a relatively older (and error-prone) cDNA microarray technology, and for mixed male and female rats in the same analysis.

14.4.2 Validation of the Surprise: Boutros et al.

The results from the Boverhof study were surprising, so in 2008, while working in the lab of Dr. Allan Okey and in collaboration with Dr. Raimo Pohjanvirta, I set out to validate them in an independent group of animals [21]. There were slight differences in the model systems used in the two studies (Table 14.1). While both studies used C57BL/6 mice, our work used Long-Evans rats rather than the Sprague Dawley rats of the Boverhof study. Moreover, while the Boverhof work used an extensive time course analysis, we elected to focus on a single time point, prior to the appearance of overt toxicities: 19 h. Furthermore, we selected a single dose that was an approximately constant multiple of the LD_{50} for each species. This approach was intended to generate as great a similarity as possible between mouse and rat experiments. We employed a newer oligonucleotide microarray platform that evaluated 8125 ortholog pairs—about 2.6 times more than in the Boverhof study. We reasoned that the startling interspecies discordance might have resulted from surveying only a limited and biased fraction of the transcriptome.

In many ways, our analysis fully replicated the results of the earlier study. Both studies found approximately equal numbers of genes altered by TCDD in rat (201 versus 200), in mice (238 versus 278), and in both species (33 genes in each). Furthermore, the fold changes of the genes differentially expressed in our study were extremely closely correlated between the two studies in both mice (Spearman's $\rho = 0.83$; $p = 4.4 \times 10^{-8}$) and rats (Spearman's $\rho = 0.88$; $p = 1.1 \times 10^{-9}$). Both studies showed extensive presence of AHREs in dioxin-responsive genes and identified enrichment of xenobiotic metabolizing enzymes in species-independent genes and of lipid metabolizing genes in species-dependent ones.

Outside this core of major similarities, there were also two interesting differences between the two studies. First, while the fold changes of the two studies were extremely well correlated, the specific gene calls were not. The Boverhof study identified 33 rat–mouse common genes. In total, 28 of these were present on the arrays used in our "validation" study. In our study, 12 of these were not changed in either species, 6 were changed in mice only, 2 in rats only, and 8 in both species. Thus, out of the 56 predictions (28 for each species), only 24 were validated. This is a major discrepancy and suggests that even the use of newer array platforms can lead to a very highly elevated false-negative rate. Second, several of the pathway results were not replicated between the two studies. For example, we saw no enrichment in lipid metabolizing enzymes in the species-independent cohort.

Finally, our study added several new aspects to attempt to understand the interspecies divergence. First, we looked at the genomic localization of differentially expressed genes to see if there was clustering of these to specific regions. This did not occur in mouse or rat. Second, we looked at the upstream regulatory regions of each differentially expressed gene and performed a detailed database search of known transcription factor binding sites (TFBSs). While ~60 TFBSs were enriched in the promoters of one of the gene lists, we showed that in general these transcription factors reflected architectural differences between mice and rats rather than TCDD-associated effects. For example, the p53 binding motif was enriched in the upstream regulatory regions of TCDD-responsive mouse genes, but not in TCDD-responsive rat genes. Third, we performed an unbiased sequence analysis of the promoter regions to try to identify novel sequences associated with dioxin exposure. We identified two such sequences but were unable to match them to any known transcription factor family.

Overall, the Boutros study validated the broad conclusions of the Boverhof work [2, 21]. One concern raised, however, was that the microarray technologies used have high false-negative rates that bias us to consistently underestimate the concordance of mouse and rat responses.

14.4.3 Ratonized Mice

The *Ahr* itself differs substantially between rats and mice. In particular, while the two genes produce proteins with similar affinity for TCDD [64], they do differ substantially in sequence in their C-terminus. This is the region of the protein responsible for transactivation and indeed has been implicated in the heightened dioxin resistance of Han/Wistar (*Kuopio*) rats [65]. However, beyond this single locus, the rat and mouse genomes differ substantially in myriad ways [66]. Therefore, it is unclear how much of the differential response between mice and rats is a function of alterations in the primary structure of the AHR protein, and how much is a function of differences in promoter architecture between the two species, as indicated by the differences in p53 binding sites noted above [21].

To address these key issues, Dr. Raimo Pohjanvirta generated an entirely new model system by generating transgenic mice harboring the wild-type rat *Ahr*. Although full transcriptomic studies of these animals have not yet been published, these animals have already provided some useful insights into species differences in dioxin sensitivity. Mice bearing the wild-type rat AHR (rWT mice) showed comparable mRNA induction to wild-type C57BL/6 mice for Cyp1a1 (males: liver, lung, and testis; females: liver and ovary), Cyp1a2 (male and female liver), and Cyp1b1 (male and female liver). Ratonized mice showed somewhat earlier onset of TCDD toxicities, with some animals succumbing as early as 4 days after a 500 µg/kg TCDD (compared to 14 days for the first wild-type animal). These data suggest that the rat AHR is fully competent within the mouse system and indicate that differences in AHR itself may not be the primary explanation for the species differences in the specific transcripts induced by TCDD.

14.5 HUMANS VERSUS RODENTS

For obvious ethical reasons, there have been far fewer studies of humans than of rodent models. Several alternative model systems can be used in place of *in vivo* studies. First, Dr. Jay Silkworth's lab investigated the response of primary human hepatocytes to TCDD and compared it to that of rats. Second, Dr. Gary Perdew's group investigated a transgenic mouse model bearing the human AHR—a "humanized" counterpart to the "ratonized" mouse described above (Section 14.4.3). A summary of these studies is given in Table 14.2.

14.5.1 Rats: Carlson

Primary hepatocytes can be prepared from human tissue that cannot be used for transplants. They are generally isolated with a collagenase digestion and then kept in flasks with a mixture of varied supplements (e.g., MEM, L-glutamine, insulin, nonessential amino acids, etc.). Carlson took just such an approach to study the response of human liver to TCDD. Their human liver was derived from three separate donors: two Caucasians (a 41-year-old male and a 46-year-old female) and a 46-year-old African-American male [67]. As a comparison to their human data, Carlson and coworkers focused on rat hepatocytes. To try to minimize the confounding effects of culturing, the rat cells were also used as primary hepatocyte cultures. In this case, six female Sprague Dawley rats were used. However, the authors merged these six samples into two pools of three samples each. Thus, the experiment involved only three human samples and two pooled rat samples. The use of only three samples to reflect the significant genetic diversity of human populations immediately places severe limits on the generalizability of this study. Furthermore, only a single sample from each exposure group was used, further reducing the statistical power and robustness of this study. Although direct polymorphisms in the AHR are rare and not likely to be functional [57, 68], it remains likely that there are significant modifier alleles (i.e., *trans* effects) that alter interindividual responses to TCDD and other AHR ligands. Indeed, it is well known that increased sample size is associated with improved statistical power and can be used to reduce the rate of both false positives and false negatives [69].

Primary cultures were treated for 48 h with a range of TCDD concentrations from 10^{-14} to $10^{-6.5}$ M (seven

TABLE 14.2 Experimental Design of Human-Rodent Comparison Studies

	Carlson	Flaveny
Rodent source	Hepatocytes derived from rats	Hepatocytes derived from $Ahr^{b/b}$ mice
Rodent strain	Sprague Dawley	C57BL/6J
Rodent gender	Female	Not given
Rodent age	Not given	Not given
Rodent exposure time(s)	48 h	6 h
Rodent exposure dose(s)	Seven doses from 10^{-14} to $10^{-6.5}$ M	10 nM
Human source	Hepatocytes derived from three human donors	Hepatocytes derived from $Ahr^{human/human}$ mice
Human gender	Male, female, male	Not given
Human age	41, 56, 46 years	Not given
Human exposure time(s)	48 h	6 h
Human exposure dose(s)	Seven doses from 10^{-14} to $10^{-6.5}$ M	10 nM
Orthologous genes	4158–4190	~34,000
Technology	Affymetrix	Affymetrix
Rat genes changed	97	
Genes changed in WT		2707
Human genes changed	57	
Genes changed in *Ahr*-humanized hepatocytes		1965

separate doses used). It is unclear how well these *in vitro* doses match with the doses used in various *in vivo* systems, but the doses do clearly induce maximal CYP1A1 induction in both mice and rats, lending confidence to this selection. The authors also used a very sophisticated modeling procedure to estimate the dose–response model for each gene. Separate models were fit for each gene in each species, and ProbeSets that did not respond in at least two cell cultures/pools were removed (i.e., each gene considered responded in 2/2 rat pools or at least 2/3 human samples). After a fold-change filter (twofold threshold) was applied, human samples in which a gene was entirely responsive were removed from the modeling for that gene (only). A modified version of the Hill equation was applied, and a variety of linear and nonlinear dose–response models were tested to maximize the chance that an appropriate model could be fit for each gene. This sophisticated approach greatly increases the effective power of the study by ensuring application of optimal techniques to noise-reduced data [67].

After this complex experimental procedure, the authors identified 47 genes that exhibited dose-responsive TCDD dependency in humans and 75 with similar behavior in rats. Strikingly, only four genes were common between the two species, and these were the well-characterized TCDD-responsive genes *CYP1A1*, *CYP1B1*, *CYP1A2*, and *NQO1*. The authors performed Gene Ontology functional analysis and identified only a single differential pathway, which is likely attributable to random chance. So, despite some technical caveats, this study clearly demonstrates minimal overlap in transcriptomic response to TCDD between rat and human hepatocytes.

14.5.2 Mice: Flaveny

Just as Dr. Pohjanvirta sought to clarify the comparison of rat and mouse responses to TCDD by generating a transgenic animal, so too did Dr. Gary Perdew seek to improve our understanding of the human AHR. His group generated a mouse harboring the human AHR, derived hepatocyte cultures from them, and exposed these cultures to TCDD. The authors used relatively similar culturing conditions to those employed by Carlson and coworkers [67, 70] and demonstrated maximal induction of common AHR-responsive genes such as *Cyp1a1*, *Cyp1a2*, and *Cyp1b1*.

They found a surprisingly large number of TCDD-responsive genes 48 h after exposure. In part, this may be due to confounding of their results with circadian effects—vehicle control samples were treated only for 6 h. Nevertheless, the authors observed 2852 differentially expressed genes in murine hepatocytes harboring a mouse *Ahr* and 1965 in murine hepatocytes harboring a human *AHR*. The contrast to the *in vivo* Boverhof and Boutros studies noted above is striking: those studies identified 238 and 278 dioxin-regulated genes, respectively. In fact, the sheer number of dioxin-responsive genes in this study exceeds that found in anything described in the literature, outside of a few late-stage studies of animals where overt toxicities have become manifest [20].

Flaveny and coworkers then proceeded to perform a pathway analysis to try to understand the nature of the genes identified as species specific. Interestingly, they found that the murine AHR regulated more genes involved in metabolism and membrane transport, whereas the human isoform regulated genes involved in proliferation and immune response. These differences, again, do not concord well with those described in earlier studies. It is intriguing that the two primary hepatocyte studies show both the fewest [67] and the most [70] changes in gene expression of all the studies evaluated.

14.6 UNANSWERED QUESTIONS

14.6.1 Importance of Strain

These conflicting data raise a number of key questions. First, how important are strain differences? Some work [17, 22] suggests that strain differences can be as large as, or larger than, differences between species. To truly understand AHR-induced gene batteries in rodents, it may be necessary to profile a relatively large number of strains to explore their differential response. With the advent of affordable next-generation sequencing, this may also bring opportunities to comprehensively evaluate the genomes of multiple rat and mouse strains. It may be possible to correlate these genome sequences with the lists of induced genes in different strains to identify the genetic differences that underlie these transcriptional divergences. However, given the vast differences between rat, mouse, and human differentially expressed genes, it is possible that strain simply does not matter. Rather, perhaps the key genes involved in dioxin-related toxicities can be identified from the interspecies analysis, independent of the strain used. Systematic analysis of multiple strains using common protocols and time points may help to address this issue.

14.6.2 Are Common Genes Functional?

All these comparative studies are premised on three assumptions. First, they assume that the liver is a critical organ for the toxicological effects induced by dioxins. Second, they assume that these toxicological effects are sufficiently similar in multiple species to be caused by a single mechanism. Third, they presuppose that this unified mechanism is likely to occur at the transcriptomic level via regulation of individual mRNAs.

While each of these assumptions is reasonable at face value, there is no guarantee that they are valid. For example, while there is extensive evidence (reviewed in Section 14.2.1) that the liver is a critical tissue for mediating TCDD toxicities [9], this is not a given. A conditional hepatocyte-specific *Ahr*-null mouse model has been generated by the Bradfield lab and used to demonstrate that several liver toxicities are dependent on hepatic *Ahr*, while thymic involution is not [8]. This model is one definitive way of probing the importance of the liver for various toxicological end points. Similarly, the assumption that there is a singular manifestation of dioxin-induced toxicity across species is certainly reasonable, but there are some species-specific differences to consider (see Section 14.2.3). It is unclear whether these differences are sufficiently large as to indicate a completely separate mechanistic basis.

However, the biggest problems of using comparative transcriptomics to identify genes associated with dioxin toxicities arise from assuming that, because the AHR acts as a transcription factor, the key genes involved in AHR-mediated toxicities must necessarily have altered mRNA levels. To the contrary, there are several alternative hypotheses. First, the AHR protein has been implicated, in some controversial work, as an E3 ubiquitin ligase that targets other nuclear proteins [71, 72]. If this work can be confirmed, that would indicate that the function of the AHR may involve regulation of protein stability and localization, which would not necessarily be reflected at the mRNA level. Second, the critical genes may not themselves be significantly changed at the mRNA level—rather, they may be altered posttranscriptionally, for example, by alternative splicing [73]. Third, the same toxic end point could be reached through multiple independent mechanisms. For example, in one organism it might require dysregulation of three particular transporters, whereas in another an entirely different set of five transporters would be involved. This is another way of saying that similarity of phenotype does not necessitate similarity of mechanism.

One approach to assessing these common interspecies dioxin-inducible genes would involve validation of them in alternative model systems. For example, the genes identified as dioxin responsive in mice and rats could then be evaluated in the unique, dioxin-resistant Han/Wistar (*Kuopio*) rat model system. A gene that was induced by dioxins in sensitive mice and rats, but not in resistant rats, would be a particularly strong candidate to mechanistically regulate toxicities. Similarly, these genes could be evaluated at the protein level—genes altered only in mRNA abundance but not in protein abundance would be poor candidates to regulate toxic outcomes. Finally, dose–response and time course studies could be performed. One would predict that toxicity-associated genes would exhibit dose–response and time course profiles that would approximately mirror (or slightly precede) those of the toxic outcomes they regulate. Ultimately, the decisive experiment might involve testing the toxicity of transgenic animals in which key candidate genes have been genetically ablated.

14.6.3 How Many of These Effects Are Technical in Origin?

As alluded to several times above, it is unclear just how many of the alterations in mRNA abundance described in the various studies are real and how many are technical artifacts. The three order of magnitude range in the number of dioxin-responsive genes (from tens to thousands [67, 70]) across the various studies gives one pause. Are the studies showing the largest numbers of genes particularly prone to false positives? Are those with particularly few changes prone to false negatives? The comparison of the Boverhof and Boutros studies suggests that there is a significant false-negative rate that may in part account for the paucity of species-independent responses to dioxins.

Similarly, one must question the comparability of the various time points and doses used in these studies. It is particularly curious that the two primary hepatocyte studies show the widest variation in the number of perturbed genes [67, 70], suggesting a potential technical difference. Comprehensive studies that use a standardized range of doses and time points may improve predictability. Furthermore, by focusing on early time points, future studies may be able to avoid the very significant confounding effects of secondary toxicities [20].

An alternative experimental approach might involve studying the physical association of the AHR with chromatin using chromatin immunoprecipitation (ChIP) studies. Although ChIP studies are laborious and challenging to analyze [74], they would focus the analysis on AHR-regulated changes in mRNA abundance, thereby providing a more specific analysis. There have been multiple successful reports of AHR ChIP studies in the recent literature, suggesting that this would be a viable approach [75–77].

14.7 CONCLUSIONS

The strong evolutionary conservation of the AHR primary protein sequence suggests that the function of this gene has also been conserved over evolutionary time. This hypothesis is bolstered by the toxicity of the dioxin class of compounds to a broad range of species. Because the transcription factor activity of the AHR appears essential to mediation of toxic responses, several groups have sought to identify genes associated with dioxin toxicities by identifying and comparing AHR target genes across mammalian species. These studies have identified a small core of AHR target genes that are conserved, but these include genes generally not believed to be associated with toxicities (i.e., *Nqo1*, *Cyp1a1*, *Cyp1b1*, and *Cyp1a2*). Instead, the vast majority of AHR target genes appear to be species specific. While this may, in part, be due to technical challenges in the various studies, these data have now been reproduced by four separate groups. This suggests that the AHR is highly sensitive to genomic context and that the small number of species-independent genes may truly regulate many key dioxin toxicities. Alternatively, it may be that the liver is not the most appropriate organ for this type of study and that indeed interspecies comparative toxicogenomics could be used to identify the organ most closely associated with the initiation of toxicity. Additional work in this area is clearly indicated and has the potential to yield exciting new hypotheses into the mechanisms of dioxin toxicities.

REFERENCES

1. Pohjanvirta, R. and Tuomisto, J. (1994). Short-term toxicity of 2,3,7,8-tetrachlorodibenzo-*p*-dioxin in laboratory animals: effects, mechanisms, and animal models. *Pharmacological Reviews*, 46, 483–549.
2. Boverhof, D. R., Burgoon, L. D., Tashiro, C., Sharratt, B., Chittim, B., Harkema, J. R., Mendrick, D. L., and Zacharewski, T. R. (2006). Comparative toxicogenomic analysis of the hepatotoxic effects of TCDD in Sprague Dawley rats and C57BL/6 mice. *Toxicological Sciences*, 94, 398–416.
3. Schmidt, J. V., Su, G. H., Reddy, J. K., Simon, M. C., and Bradfield, C. A. (1996). Characterization of a murine Ahr null allele: involvement of the Ah receptor in hepatic growth and development. *Proceedings of the National Academy of Sciences of the United States of America*, 93, 6731–6736.
4. Fernandez-Salguero, P., Pineau, T., Hilbert, D. M., McPhail, T., Lee, S. S., Kimura, S., Nebert, D. W., Rudikoff, S., Ward, J. M., and Gonzalez, F. J. (1995). Immune system impairment and hepatic fibrosis in mice lacking the dioxin-binding Ah receptor. *Science*, 268, 722–726.
5. Bunger, M. K., Glover, E., Moran, S. M., Walisser, J. A., Lahvis, G. P., Hsu, E. L., and Bradfield, C. A. (2008). Abnormal liver development and resistance to 2,3,7,8-tetrachlorodibenzo-*p*-dioxin toxicity in mice carrying a mutation in the DNA-binding domain of the aryl hydrocarbon receptor. *Toxicological Sciences*, 106, 83–92.
6. Bunger, M. K., Moran, S. M., Glover, E., Thomae, T. L., Lahvis, G. P., Lin, B. C., and Bradfield, C. A. (2003). Resistance to 2,3,7,8-tetrachlorodibenzo-*p*-dioxin toxicity and abnormal liver development in mice carrying a mutation in the nuclear localization sequence of the aryl hydrocarbon receptor. *Journal of Biological Chemistry*, 278, 17767–17774.
7. Walisser, J. A., Bunger, M. K., Glover, E., Harstad, E. B., and Bradfield, C. A. (2004). Patent ductus venosus and dioxin resistance in mice harboring a hypomorphic Arnt allele. *Journal of Biological Chemistry*, 279, 16326–16331.
8. Walisser, J. A., Glover, E., Pande, K., Liss, A. L., and Bradfield, C. A. (2005). Aryl hydrocarbon receptor-dependent liver development and hepatotoxicity are mediated by different cell types. *Proceedings of the National Academy of Sciences of the United States of America*, 102, 17858–17863.
9. Nukaya, M., Walisser, J. A., Moran, S. M., Kennedy, G. D., and Bradfield, C. A. (2010). Aryl hydrocarbon receptor nuclear translocator in hepatocytes is required for aryl hydrocarbon receptor-mediated adaptive and toxic responses in liver. *Toxicological Sciences*, 118, 554–563.
10. Boutros, P. C., Bielefeld, K. A., Pohjanvirta, R., and Harper, P. A. (2009). Dioxin-dependent and dioxin-independent gene batteries: comparison of liver and kidney in AHR-null mice. *Toxicological Sciences*, 112, 245–256.
11. Frueh, F. W., Hayashibara, K. C., Brown, P. O., and Whitlock, J. P., Jr. (2001). Use of cDNA microarrays to analyze dioxin-induced changes in human liver gene expression. *Toxicology Letters*, 122, 189–203.
12. Vezina, C. M., Walker, N. J., and Olson, J. R. (2004). Subchronic exposure to TCDD, PeCDF, PCB126, and PCB153:

effect on hepatic gene expression. *Environmental Health Perspectives, 112*, 1636–1644.

13. Boverhof, D. R., Burgoon, L. D., Tashiro, C., Chittim, B., Harkema, J. R., Jump, D. B., and Zacharewski, T. R. (2005). Temporal and dose-dependent hepatic gene expression patterns in mice provide new insights into TCDD-mediated hepatotoxicity. *Toxicological Sciences, 85*, 1048–1063.

14. Fletcher, N., Wahlstrom, D., Lundberg, R., Nilsson, C. B., Nilsson, K. C., Stockling, K., Hellmold, H., and Hakansson, H. (2005). 2,3,7,8-Tetrachlorodibenzo-p-dioxin (TCDD) alters the mRNA expression of critical genes associated with cholesterol metabolism, bile acid biosynthesis, and bile transport in rat liver: a microarray study. *Toxicology and Applied Pharmacology, 207*, 1–24.

15. Pastorelli, R., Carpi, D., Campagna, R., Airoldi, L., Pohjanvirta, R., Viluksela, M., Hakansson, H., Boutros, P. C., Moffat, I. D., Okey, A. B., and Fanelli, R. (2006). Differential expression profiling of the hepatic proteome in a rat model of dioxin resistance: correlation with genomic and transcriptomic analyses. *Molecular & Cellular Proteomics, 5*, 882–894.

16. Hayes, K. R., Zastrow, G. M., Nukaya, M., Pande, K., Glover, E., Maufort, J. P., Liss, A. L., Liu, Y., Moran, S. M., Vollrath, A. L., and Bradfield, C. A. (2007). Hepatic transcriptional networks induced by exposure to 2,3,7,8-tetrachlorodibenzo-p-dioxin. *Chemical Research in Toxicology, 20*, 1573–1581.

17. Franc, M. A., Moffat, I. D., Boutros, P. C., Tuomisto, J. T., Tuomisto, J., Pohjanvirta, R., and Okey, A. B. (2008). Patterns of dioxin-altered mRNA expression in livers of dioxin-sensitive versus dioxin-resistant rats. *Archives of Toxicology, 82*, 809–830.

18. N'Jai, A., Boverhof, D. R., Dere, E., Burgoon, L. D., Tan, Y. S., Rowlands, J. C., Budinsky, R. A., Stebbins, K. E., and Zacharewski, T. R. (2008). Comparative temporal toxicogenomic analysis of TCDD- and TCDF-mediated hepatic effects in immature female C57BL/6 mice. *Toxicological Sciences, 103*, 285–297.

19. Ovando, B. J., Ellison, C. A., Vezina, C. M., and Olson, J. R. (2010). Toxicogenomic analysis of exposure to TCDD, PCB126 and PCB153: identification of genomic biomarkers of exposure to AhR ligands. *BMC Genomics, 11*, 583.

20. Boutros, P. C., Yao, C. Q., Watson, J. D., Wu, A. H., Moffat, I. D., Prokopec, S. D., Smith, A. B., Okey, A. B., and Pohjanvirta, R. (2011). Hepatic transcriptomic responses to TCDD in dioxin-sensitive and dioxin-resistant rats during the onset of toxicity. *Toxicology and Applied Pharmacology, 251*, 119–129.

21. Boutros, P. C., Yan, R., Moffat, I. D., Pohjanvirta, R., and Okey, A. B. (2008). Transcriptomic responses to 2,3,7,8-tetrachlorodibenzo-p-dioxin (TCDD) in liver: comparison of rat and mouse. *BMC Genomics, 9*, 419.

22. Moffat, I. D., Boutros, P. C., Chen, H., Okey, A. B., and Pohjanvirta, R. (2010). Aryl hydrocarbon receptor (AHR)-regulated transcriptomic changes in rats sensitive or resistant to major dioxin toxicities. *BMC Genomics, 11*, 263.

23. Tijet, N., Boutros, P. C., Moffat, I. D., Okey, A. B., Tuomisto, J., and Pohjanvirta, R. (2006). Aryl hydrocarbon receptor regulates distinct dioxin-dependent and dioxin-independent gene batteries. *Molecular Pharmacology, 69*, 140–153.

24. Karyala, S., Guo, J., Sartor, M., Medvedovic, M., Kann, S., Puga, A., Ryan, P., and Tomlinson, C. R. (2004). Different global gene expression profiles in benzo[a]pyrene- and dioxin-treated vascular smooth muscle cells of AHR-knockout and wild-type mice. *Cardiovascular Toxicology, 4*, 47–73.

25. Guo, J., Sartor, M., Karyala, S., Medvedovic, M., Kann, S., Puga, A., Ryan, P., and Tomlinson, C. R. (2004). Expression of genes in the TGF-β signaling pathway is significantly deregulated in smooth muscle cells from aorta of aryl hydrocarbon receptor knockout mice. *Toxicology and Applied Pharmacology, 194*, 79–89.

26. Puga, A., Maier, A., and Medvedovic, M. (2000). The transcriptional signature of dioxin in human hepatoma HepG2 cells. *Biochemical Pharmacology, 60*, 1129–1142.

27. Ishida, S., Jinno, H., Tanaka-Kagawa, T., Ando, M., Ohno, Y., Ozawa, S., and Sawada, J. (2002). Characterization of human CYP1A1/1A2 induction by DNA microarray and alpha-naphthoflavone. *Biochemical and Biophysical Research Communications, 296*, 172–177.

28. Fletcher, N., Hanberg, A., and Hakansson, H. (2001). Hepatic vitamin A depletion is a sensitive marker of 2,3,7,8-tetrachlorodibenzo-p-dioxin (TCDD) exposure in four rodent species. *Toxicological Sciences, 62*, 166–175.

29. Thurmond, T. S., Silverstone, A. E., Baggs, R. B., Quimby, F. W., Staples, J. E., and Gasiewicz, T. A. (1999). A chimeric aryl hydrocarbon receptor knockout mouse model indicates that aryl hydrocarbon receptor activation in hematopoietic cells contributes to the hepatic lesions induced by 2,3,7,8-tetrachlorodibenzo-p-dioxin. *Toxicology and Applied Pharmacology, 158*, 33–40.

30. Chang, H., Wang, Y. J., Chang, L. W., and Lin, P. (2005). A histochemical and pathological study on the interrelationship between TCDD-induced AhR expression, AhR activation, and hepatotoxicity in mice. *Journal of Toxicology and Environmental Health A, 68*, 1567–1579.

31. Pande, K., Moran, S. M., and Bradfield, C. A. (2005). Aspects of dioxin toxicity are mediated by interleukin 1-like cytokines. *Molecular Pharmacology, 67*, 1393–1398.

32. Pohjanvirta, R., Kulju, T., Morselt, A. F., Tuominen, R., Juvonen, R., Rozman, K., Mannisto, P., Collan, Y., Sainio, E. L., and Tuomisto, J. (1989). Target tissue morphology and serum biochemistry following 2,3,7,8-tetrachlorodibenzo-p-dioxin (TCDD) exposure in a TCDD-susceptible and a TCDD-resistant rat strain. *Fundamental and Applied Toxicology, 12*, 698–712.

33. Roberts, R. A., Ganey, P. E., Ju, C., Kamendulis, L. M., Rusyn, I., and Klaunig, J. E. (2007). Role of the Kupffer cell in mediating hepatic toxicity and carcinogenesis. *Toxicological Sciences, 96*, 2–15.

34. Senoo, H. (2004). Structure and function of hepatic stellate cells. *Medical Electron Microscopy, 37*, 3–15.

35. Schena, M., Shalon, D., Davis, R. W., and Brown, P. O. (1995). Quantitative monitoring of gene expression patterns with a complementary DNA microarray. *Science, 270*, 467–470.

36. Golub, T. R., Slonim, D. K., Tamayo, P., Huard, C., Gaasenbeek, M., Mesirov, J. P., Coller, H., Loh, M. L., Downing, J. R., Caligiuri, M. A., et al. (1999). Molecular classification of cancer: class discovery and class prediction by gene expression monitoring. *Science*, 286, 531–537.

37. Diehn, M., Eisen, M. B., Botstein, D., and Brown, P. O. (2000). Large-scale identification of secreted and membrane-associated gene products using DNA microarrays. *Nature Genetics*, 25, 58–62.

38. Lau, S. K., Boutros, P. C., Pintilie, M., Blackhall, F. H., Zhu, C. Q., Strumpf, D., Johnston, M. R., Darling, G., Keshavjee, S., Waddell, T. K., et al. (2007). Three-gene prognostic classifier for early-stage non small-cell lung cancer. *Journal of Clinical Oncology*, 25, 5562–5569.

39. Shi, L., Reid, L. H., Jones, W. D., Shippy, R., Warrington, J. A., Baker, S. C., Collins, P. J., de Longueville, F., Kawasaki, E. S., Lee, K. Y., et al. (2006). The MicroArray Quality Control (MAQC) project shows inter- and intraplatform reproducibility of gene expression measurements. *Nature Biotechnology*, 24, 1151–1161.

40. Geiss, G. K., Bumgarner, R. E., Birditt, B., Dahl, T., Dowidar, N., Dunaway, D. L., Fell, H. P., Ferree, S., George, R. D., Grogan, T., et al. (2008). Direct multiplexed measurement of gene expression with color-coded probe pairs. *Nature Biotechnology*, 26, 317–325.

41. Ozsolak, F., Platt, A. R., Jones, D. R., Reifenberger, J. G., Sass, L. E., McInerney, P., Thompson, J. F., Bowers, J., Jarosz, M., and Milos, P. M. (2009). Direct RNA sequencing. *Nature*, 461, 814–818.

42. Hekstra, D., Taussig, A. R., Magnasco, M., and Naef, F. (2003). Absolute mRNA concentrations from sequence-specific calibration of oligonucleotide arrays. *Nucleic Acids Research*, 31, 1962–1968.

43. Held, G. A., Grinstein, G., and Tu, Y. (2003). Modeling of DNA microarray data by using physical properties of hybridization. *Proceedings of the National Academy of Sciences of the United States of America*, 100, 7575–7580.

44. Birney, E., Stamatoyannopoulos, J. A., Dutta, A., Guigo, R., Gingeras, T. R., Margulies, E. H., Weng, Z., Snyder, M., Dermitzakis, E. T., Thurman, R. E., et al. (2007). Identification and analysis of functional elements in 1% of the human genome by the ENCODE pilot project. *Nature*, 447, 799–816.

45. Gerstein, M. B., Bruce, C., Rozowsky, J. S., Zheng, D., Du, J., Korbel, J. O., Emanuelsson, O., Zhang, Z. D., Weissman, S., and Snyder, M. (2007). What is a gene, post-ENCODE? History and updated definition. *Genome Research*, 17, 669–681.

46. Koonin, E. V. (2001). An apology for orthologs—or brave new memes. *Genome Biology*, 2, COMMENT1005.

47. Korf, I., Flicek, P., Duan, D., and Brent, M. R. (2001). Integrating genomic homology into gene structure prediction. *Bioinformatics*, 17 (Suppl. 1), S140–S148.

48. Ponting, C. P. (2001). Issues in predicting protein function from sequence. *Briefings in Bioinformatics*, 2, 19–29.

49. Baldauf, S. L. (2003). Phylogeny for the faint of heart: a tutorial. *Trends in Genetics*, 19, 345–351.

50. Dai, M., Wang, P., Boyd, A. D., Kostov, G., Athey, B., Jones, E. G., Bunney, W. E., Myers, R. M., Speed, T. P., Akil, H., et al. (2005). Evolving gene/transcript definitions significantly alter the interpretation of GeneChip data. *Nucleic Acids Research*, 33, e175.

51. Mecham, B. H., Klus, G. T., Strovel, J., Augustus, M., Byrne, D., Bozso, P., Wetmore, D. Z., Mariani, T. J., Kohane, I. S., and Szallasi, Z. (2004). Sequence-matched probes produce increased cross-platform consistency and more reproducible biological results in microarray-based gene expression measurements. *Nucleic Acids Research*, 32, e74.

52. Mecham, B. H., Wetmore, D. Z., Szallasi, Z., Sadovsky, Y., Kohane, I., and Mariani, T. J. (2004). Increased measurement accuracy for sequence-verified microarray probes. *Physiological Genomics*, 18, 308–315.

53. Wheeler, D. L., Church, D. M., Lash, A. E., Leipe, D. D., Madden, T. L., Pontius, J. U., Schuler, G. D., Schriml, L. M., Tatusova, T. A., Wagner, L., and Rapp, B. A. (2002). Database resources of the National Center for Biotechnology Information: 2002 update. *Nucleic Acids Research*, 30, 13–16.

54. Kent, W. J., Baertsch, R., Hinrichs, A., Miller, W., and Haussler, D. (2003). Evolution's cauldron: duplication, deletion, and rearrangement in the mouse and human genomes. *Proceedings of the National Academy of Sciences of the United States of America*, 100, 11484–11489.

55. Dere, E., Boverhof, D. R., Burgoon, L. D., and Zacharewski, T. R. (2006). In vivo–in vitro toxicogenomic comparison of TCDD-elicited gene expression in Hepa1c1c7 mouse hepatoma cells and C57BL/6 hepatic tissue. *BMC Genomics*, 7, 80.

56. Pohjanvirta, R. and Tuomisto, J. (1987). Han/Wistar rats are exceptionally resistant to TCDD. II. *Archives of Toxicology*, (Suppl. 11), 344–347.

57. Okey, A. B., Franc, M. A., Moffat, I. D., Tijet, N., Boutros, P. C., Korkalainen, M., Tuomisto, J., and Pohjanvirta, R. (2005). Toxicological implications of polymorphisms in receptors for xenobiotic chemicals: the case of the aryl hydrocarbon receptor. *Toxicology and Applied Pharmacology*, 207, 43–51.

58. Daniels, G. M. and Buck, K. J. (2002). Expression profiling identifies strain-specific changes associated with ethanol withdrawal in mice. *Genes, Brain and Behavior*, 1, 35–45.

59. Pritchard, C., Coil, D., Hawley, S., Hsu, L., and Nelson, P. S. (2006). The contributions of normal variation and genetic background to mammalian gene expression. *Genome Biology*, 7, R26.

60. Pritchard, C. C., Hsu, L., Delrow, J., and Nelson, P. S. (2001). Project normal: defining normal variance in mouse gene expression. *Proceedings of the National Academy of Sciences of the United States of America*, 98, 13266–13271.

61. Pohjanvirta, R. (2009). Transgenic mouse lines expressing rat AH receptor variants—a new animal model for research on AH receptor function and dioxin toxicity mechanisms. *Toxicology and Applied Pharmacology*, 236, 166–182.

62. Sun, Y. V., Boverhof, D. R., Burgoon, L. D., Fielden, M. R., and Zacharewski, T. R. (2004). Comparative analysis of dioxin response elements in human, mouse and rat genomic sequences. *Nucleic Acids Research*, 32, 4512–4523.

63. Wasserman, W. W., Palumbo, M., Thompson, W., Fickett, J. W., and Lawrence, C. E. (2000). Human-mouse genome comparisons to locate regulatory sites. *Nature Genetics, 26*, 225–228.
64. Denison, M. S., Vella, L. M., and Okey, A. B. (1986). Structure and function of the Ah receptor for 2,3,7,8-tetrachlorodibenzo-*p*-dioxin. Species difference in molecular properties of the receptors from mouse and rat hepatic cytosols. *Journal of Biological Chemistry, 261*, 3987–3995.
65. Pohjanvirta, R., Wong, J. M., Li, W., Harper, P. A., Tuomisto, J., and Okey, A. B. (1998). Point mutation in intron sequence causes altered carboxyl-terminal structure in the aryl hydrocarbon receptor of the most 2,3,7,8-tetrachlorodibenzo-*p*-dioxin-resistant rat strain. *Molecular Pharmacology, 54*, 86–93.
66. Gibbs, R. A., Weinstock, G. M., Metzker, M. L., Muzny, D. M., Sodergren, E. J., Scherer, S., Scott, G., Steffen, D., Worley, K. C., Burch, P. E., et al. (2004). Genome sequence of the Brown Norway rat yields insights into mammalian evolution. *Nature, 428*, 493–521.
67. Carlson, E. A., McCulloch, C., Koganti, A., Goodwin, S. B., Sutter, T. R., and Silkworth, J. B. (2009). Divergent transcriptomic responses to aryl hydrocarbon receptor agonists between rat and human primary hepatocytes. *Toxicological Sciences, 112*, 257–272.
68. Okey, A. B., Boutros, P. C., and Harper, P. A. (2005). Polymorphisms of human nuclear receptors that control expression of drug-metabolizing enzymes. *Pharmacogenetics and Genomics, 15*, 371–379.
69. Begun, A. (2008). Power estimation of the *t* test for detecting differential gene expression. *Functional & Integrative Genomics, 8*, 109–113.
70. Flaveny, C. A., Murray, I. A., and Perdew, G. H. (2010). Differential gene regulation by the human and mouse aryl hydrocarbon receptor. *Toxicological Sciences, 114*, 217–225.
71. Ohtake, F., Baba, A., Takada, I., Okada, M., Iwasaki, K., Miki, H., Takahashi, S., Kouzmenko, A., Nohara, K., Chiba, T., et al. (2007). Dioxin receptor is a ligand-dependent E3 ubiquitin ligase. *Nature, 446*, 562–566.
72. Ohtake, F., Takeyama, K., Matsumoto, T., Kitagawa, H., Yamamoto, Y., Nohara, K., Tohyama, C., Krust, A., Mimura, J., Chambon, P., et al. (2003). Modulation of oestrogen receptor signalling by association with the activated dioxin receptor. *Nature, 423*, 545–550.
73. Falahatpisheh, M. H. and Ramos, K. S. (2003). Ligand-activated Ahr signaling leads to disruption of nephrogenesis and altered Wilms' tumor suppressor mRNA splicing. *Oncogene, 22*, 2160–2171.
74. Ponzielli, R., Boutros, P. C., Katz, S., Stojanova, A., Hanley, A. P., Khosravi, F., Bros, C., Jurisica, I., and Penn, L. Z. (2008). Optimization of experimental design parameters for high-throughput chromatin immunoprecipitation studies. *Nucleic Acids Research, 36*, e144.
75. Celius, T., Roblin, S., Harper, P. A., Matthews, J., Boutros, P. C., Pohjanvirta, R., and Okey, A. B. (2008). Aryl hydrocarbon receptor-dependent induction of flavin-containing monooxygenase mRNAs in mouse liver. *Drug Metabolism and Disposition, 36*, 2499–2505.
76. Sartor, M. A., Schnekenburger, M., Marlowe, J. L., Reichard, J. F., Wang, Y., Fan, Y., Ma, C., Karyala, S., Halbleib, D., Liu, X., et al. (2009). Genomewide analysis of aryl hydrocarbon receptor binding targets reveals an extensive array of gene clusters that control morphogenetic and developmental programs. *Environmental Health Perspectives, 117*, 1139–1146.
77. Ahmed, S., Valen, E., Sandelin, A., and Matthews, J. (2009). Dioxin increases the interaction between aryl hydrocarbon receptor and estrogen receptor alpha at human promoters. *Toxicological Sciences, 111*, 254–266.

15

DIOXIN-ACTIVATED AHR: TOXIC RESPONSES AND THE INDUCTION OF OXIDATIVE STRESS

Sidney J. Stohs and Ezdihar A. Hassoun

15.1 INTRODUCTION

2,3,7,8-Tetrachlorodibenzo-*p*-dioxin (TCDD) is one of the most potent toxicants known to man and is prototypical of many halogenated aromatic hydrocarbons that occur as by-products of the chemical industry. They exist as environmental contaminants and may also occur in nature. Much has been learned over the past 30 years regarding the binding of a ligand, such as TCDD to the aryl hydrocarbon receptor (AHR), subsequent gene regulation and expression, influence on other signal transduction mechanisms, and other mechanistic aspects, as well as the ultimate adverse health consequences of this cascade of events including, but is not limited to, teratogenicity, carcinogenicity, and immunological and developmental effects.

An event that appears to be common to all tissues exhibiting toxicological consequences as a result of exposure to TCDD and related congeners is binding to the AHR followed by the production of reactive oxygen species (ROS) and subsequent development of oxidative stress, leading to oxidative DNA damage, extensive lipid peroxidation, damage to membranes and organelles, loss of tissue function, and programmed cell death. Multiple and diverse enzymatic pathways, membranes, organelles, and cells are involved in the generation of oxidative stress, and numerous studies demonstrating this pleiotropic effect in diverse tissues and organs will be reviewed.

The first study that led to the belief of the involvement of free radicals and ROS in response to TCDD was the report by Stohs et al. [1], demonstrating production of lipid peroxidation (determined by the production of thiobarbituric acid reactive substances (TBARS) and conjugated dienes) in the microsomes of hepatic tissues of rats exposed to TCDD. The involvement of ROS was greatly expanded upon in subsequent studies emanating from the authors' laboratory and led to the involvement of other laboratories in exploring and assessing the role of oxidative stress in the toxicity of dioxin-activated AHR. The role of oxidative stress in the toxicity of TCDD was initially reviewed by Stohs [2] and subsequently by Dalton et al. [3] and Reichard et al. [4].

15.2 ROLE OF THE AHR IN PRODUCTION OF OXIDATIVE STRESS

Multiple lines of experimental evidence demonstrate and support the AHR as a mediator of oxidative stress. One of the first studies to clearly implicate the AHR in oxidative stress involved a comparison of selected oxidative effects of TCDD in guinea pigs (highly sensitive), Syrian golden hamsters (highly resistant), and rats (intermediate sensitivity) [5]. Excellent correlations were shown to exist between toxic sensitivity to TCDD, activation of aryl hydrocarbon hydroxylase, inhibition of glutathione peroxidase activity, and hepatic lipid peroxidation. In another study, hepatic lipid peroxidation induced by TCDD was shown to occur at a low dose (500 ng/kg) in C57BL/6 mice that are known to carry a high-affinity AHR allele, compared with a higher dose (5 μg/kg) required to produce a similar effect in the DBA/2 mice that are known to possess a low-affinity allele [6].

In a study involving congenic C57BL/6 mice that were responsive or resistant to TCDD and differed at the AHR

The AH Receptor in Biology and Toxicology, First Edition. Edited by Raimo Pohjanvirta.
© 2012 John Wiley & Sons, Inc. Published 2012 by John Wiley & Sons, Inc.

locus, TCDD was shown to produce a dose-dependent increase in superoxide anion production by the peritoneal lavage cells of responsive mice, but produced a significant increase in this ROS in the cells of resistant DBA/2 mice at a dose that was approximately 10-fold higher than the maximal dose used for the responsive mice [7]. These findings agree well with the relative lethality of TCDD in these two mouse strains [8].

Hassoun and Stohs [9] examined the ability of TCDD, endrin, and lindane to induce oxidative stress in fetal and placental tissues of pregnant C57BL/6 and DBA/2 mice 48 h after single oral doses on day 12 of gestation. Production of superoxide anion, DNA single-strand breaks, and lipid peroxidation was determined. The results of the study suggested that the effects were mediated at least, in part, by the AHR. TCDD was also identified as a more potent inducer of oxidative stress than endrin or lindane. Under these same conditions, higher levels of the lipid metabolic products malondialdehyde, formaldehyde, acetaldehyde, and acetone were produced in maternal sera and amniotic fluids of the TCDD-responsive mice compared to the resistant mice [10]. These results in addition to providing evidence for the involvement of oxidative stress in the fetotoxicity of TCDD also support the possible role of the AHR in producing oxidative stress in fetal and placental tissues.

Sugihara et al. [11] showed that induction by TCDD as well as 3-methylcholanthrene (3-MC) of xanthine oxidase/xanthine dehydrogenase activity, a source of ROS, was mediated by the AHR. Both TCDD and 3-MC induced this enzyme system in responsive AHR(+/+) mice, while no induction by TCDD or 3-MC occurred in null mice AHR(−/−). Furthermore, TCDD-induced lipid peroxidation was demonstrated in the responsive, but not the nonresponsive, mice. The investigators concluded that induction of xanthine oxidase/xanthine dehydrogenase may contribute to oxidative stress and the various toxicities of dioxins.

Inactivation of aconitase activity, which is a measure of oxidative stress, occurred in response to TCDD in C57BL/6 mice, but not in resistant DBA/2 mice [4, 12]. The studies also showed that iron administration potentiated TCDD-induced toxicity, including hepatic porphyria in C57BL/6 mice, via an oxidative process, but did not do so in the DBA/2 mice. These results further underscore the role of the AHR in TCDD-induced oxidative stress.

The use of species of animals with varying sensitivities to TCDD, together with the use of responsive and nonresponsive mouse strains and the application of AHR knockout (null) mice in the aforementioned studies, clearly demonstrates the role of the AHR in the production of ROS and oxidative stress by TCDD and related congeners. To further support this contention, additional discussion regarding the kinds of ROS produced and the cellular and subcellular sites of production is provided in the following sections.

15.3 OXIDATIVE STRESS AND TOXICITY

Is there a relationship between oxidative stress produced in response to the interaction of TCDD and its congeners with the AHR and the subsequent development of toxicological effects? Several of the first studies to demonstrate the possible involvement of ROS and free radicals tested the abilities of antioxidants to suppress TCDD-induced toxicity and lethality in rats. Stohs et al. [13–16] treated rats with butylated hydroxyanisole (BHA), vitamin E, and retinol acetate during and after administration of a lethal dose of TCDD. Daily treatment of female rats with BHA protected 100% of the animals against TCDD lethality [14]. Retinol acetate (vitamin A) and vitamin E resulted in 30% and 10% survival of a lethal dose of TCDD after 25 days [13], but did not provide long-term protection [15]. The protective effects were associated with reduced microsomal lipid peroxidation, increased glutathione peroxidase activity, and decreased aryl hydrocarbon hydroxylase activity.

Subsequent studies [14] showed that TCDD produced a significant decrease in hepatic glutathione content, and when BHA was no longer given daily to the animals, all animals died. Death was presumably due to the short half-life of BHA, relative to the long half-life of TCDD. Additional experimentation confirmed that BHA provided partial protection against TCDD-induced lipid peroxidation, inhibition of glutathione peroxidase activity, and losses in liver, thymus, and body weights [16].

Because of the inhibition of selenium-dependent glutathione peroxidase by TCDD [5], the effect of dietary selenium on various parameters, including survival, were assessed [17, 18]. Optimum dietary selenium (0.1 ppm) provided partial protection against the toxic and lethal effects of TCDD, and good correlations were shown to exist between survival of the animals and selenium-dependent glutathione peroxidase and aryl hydrocarbon hydroxylase activities, as well as lipid peroxidation.

The modulation of TCDD-induced fetotoxicity and oxidative stress by vitamin E succinate and ellagic acid in embryonic and placental tissues of C57BL/6J mice was evaluated by Hassoun et al. [19]. Both antioxidants significantly decreased TCDD-induced fetal growth retardation, fetal deaths, and placental weight reduction, as well as production of superoxide anion, lipid peroxidation, and DNA single-strand breaks in fetal and placental tissues. Ellagic acid provided better protection than vitamin E succinate against TCDD-induced fetal growth retardation and lipid peroxidation in the two tissues.

In a subchronic exposure study [20], vitamin E succinate and ellagic acid were given every other day for 90 days to rats in conjunction with a TCDD daily dose of 46 ng/kg. TCDD administration resulted in significant increases in the production of superoxide anion, lipid peroxidation, and DNA single-strand breaks in the cerebral cortex and hippocampus of the brain, but not in the cerebellum or brain stem. Furthermore, coadministration of vitamin E succinate or ellagic acid provided significant protection against these TCDD-induced effects. In this case, vitamin E succinate provided better protection compared to ellagic acid. The results indicated the involvement of oxidative stress in TCDD-induced neurotoxicity following subchronic exposure, as well as the protective role of these antioxidants against possible production of neurotoxic effects by TCDD.

In another study that involved acute administration of TCDD to C57BL/6J mice, the protective effects of vitamin A and vitamin E succinate against the induction of various toxic responses by TCDD were assessed [21]. Both vitamins were shown to significantly reduce TCDD-induced total body and thymus weight losses and hepatomegaly, as well as production of superoxide anion and DNA single-strand breaks in peritoneal lavage cells. These results further indicate a role for oxidative stress in the acute toxicity of TCDD. Vitamin E was also shown to protect against TCDD-induced testicular damage and weight loss in testis, epididymis, seminal vesicles, and ventral prostate of rats, as well as against the associated changes in various biomarkers of oxidative stress, including production of hydrogen peroxide and lipid peroxidation, and decreases in the activities of the antioxidant enzymes glutathione peroxidase, glutathione reductase, catalase, and superoxide dismutase [22].

The chemoprotective effects of resveratrol against TCDD-induced toxicity were tested both *in vitro* and *in vivo* [23, 24]. Resveratrol was shown to inhibit TCDD-induced expression of CYP1A1 and CYP1B2 in cultured human mammary epithelial cells [23]. It also inhibited catechol estrogen-mediated oxidative DNA damage by blocking the formation of catechol estrogens and scavenging reactive ROS [23]. Furthermore, resveratrol attenuated TCDD-induced hepatomegaly, thymic atrophy, and hepatic lipid peroxidation after both oral and subcutaneous administration to mice with the oral route shown to provide greater protection based on its more favorable pharmacokinetics [24].

Daily administration of a tea melanin to mice was shown to protect against TCDD-induced lipid peroxidation, inhibition of glutathione peroxidase, loss of body weight, hepatomegaly, and alterations in reduced and oxidized glutathione, as well as expression of CYP1A1 gene [25]. Curcumin, another powerful antioxidant, was shown to attenuate nuclear levels of AHR and AHR nuclear translocator (ARNT)-mediated CYP1A1 and CYP1B1 induction by TCDD in mammalian cells [26]. The studies suggested that this mode of action may be responsible for the ability of curcumin to prevent TCDD-induced malignant transformation of mammalian cells.

An interesting approach to attenuating the oxidative stress-related effects of TCDD has been the use of photobiomodulation [27]. Chicken eggs were injected with TCDD and exposed daily to red light at a wavelength of 670 nm, and upon hatching, the kidneys were collected and assayed.

TCDD exposure suppressed the activities of the antioxidant enzymes glutathione peroxidase, glutathione reductase, superoxide dismutase, catalase, and glutathione S-transferase, and it reduced glutathione and ATP levels and increased lipid peroxidation relative to controls. However, daily phototherapy was shown to reverse the TCDD-induced changes in the various biomarkers of oxidative stress.

Preincubation of the insulin-secreting beta cell line INS-1E with the antioxidant dehydroascorbic acid was found to protect partially against TCDD-induced mitochondrial depolarization and inhibition of insulin secretion [28]. In a study by Shertzer [29] in mice, the antioxidant 4b,5,9b,10-tetrahydroindeno[1,2-b]indole was shown to attenuate the TCDD-induced changes in mitochondrial respiration and membrane fluidity, as well as a large increase in mitochondrial oxygen consumption. TCDD exposure also increased hepatic lipid peroxidation and mitochondrial production of hydrogen peroxide and superoxide anion.

Pohjanvirta et al. [30] suggested that lipid peroxidation produced by TCDD may represent, at least in part, a consequence of weight loss and the wasting syndrome rather than the cause thereof. These studies involved the use of pair-fed, TCDD-responsive (Long-Evans), and resistant (Han/Wistar) rats. BHA was shown to provide partial protection against TCDD-induced lethality in responsive rats. Wahba et al. [31] also used pair-fed (Sprague Dawley) rats and concluded that although some similarities existed between TCDD-treated and pair-fed rats, the overall biochemical changes observed following TCDD treatment could not be attributed to weight loss associated with hypophagia. Greater hepatic lipid peroxidation, DNA single-strand breaks, and aryl hydrocarbon hydroxylase activity, as well as greater inhibition of glutathione peroxidase activity, occurred in response to TCDD as opposed to pair-feeding. TCDD administration also produced greater thymic involution. The reason for the differences in these two studies is not clear [30, 31]. A role for ARH in the wasting syndrome was demonstrated by Uno et al. [32] using CYP1A1 (−/−) mice. They concluded that CYP1A1 contributes to TCDD-induced toxicity and lethality. The role of CYP1A1 in the production of ROS is described in the next section.

In summary, a large number of studies have demonstrated the protective effects of a wide range of antioxidants against both the TCDD-induced production of ROS and the toxic effects of the compound. These results lend strong support to the presumption that TCDD toxic effects are at least, in part, due to an altered redox state in various tissues, leading to oxidative stress and oxidative tissue damage.

15.4 SOURCES AND TYPES OF REACTIVE OXYGEN SPECIES

Studies on dioxin-induced ROS production indicate involvement of multiple sources for the production of multiple types of the species in multiple tissues. Various *in vivo* and *in vitro* studies were initially conducted by Stohs et al. [2, 33, 34] and subsequently by others [4, 35, 36] to determine the types of ROS produced in response to TCDD. The results of these studies led to the conclusion that superoxide anion, hydrogen peroxide, and hydroxyl radical were all produced in response to TCDD and other related environmental toxicants. Further studies indicated the essential role of iron in the TCDD-induced production of porphyria [37], lipid peroxidation [38, 39], and DNA damage [40, 41], as well as the role of the AHR in TCDD-induced alteration in iron and heme metabolism [1, 2, 42].

More than one enzyme system appears to be involved in the production of ROS in response to TCDD. Initial studies focused on the role of the microsomal enzymes, including the cytochrome P450 system [5, 13–16], in TCDD-induced lipid peroxidation, reduction in antioxidant enzymes, and modulation of reduced glutathione levels. The dioxin-inducible AHR battery of genes contains at least six genes in mice that play important roles in the toxic effects of TCDD and its congeners, including the production of oxidative stress [43]. At least two cytochrome P450 genes are involved [43–50] and are positively regulated by TCDD and other ligands that bind to the AHR. In general, CYP1A1 [45] and CYP1B1 [46], but not CYP1A2 [48], appeared to play critical roles in TCDD-induced oxidative stress. Furthermore, differential dose and time relationships exist in response to TCDD with respect to these cytochromes [47]. Shertzer et al. have shown that CYP1A2 protects against ROS production in the mouse liver microsomes [50]. CYP1A2 appeared to act as an electron sink, preventing the degeneration of hydrogen peroxide and the oxidation of lipids associated with microsomal membranes. Uncoupling-mediated generation of ROS in microsomes by TCDD and related compounds may also contribute to oxidative stress [51].

Mitochondria also constitute a major site for the production of ROS in response to dioxins. Studies by Wahba et al. [41] were the first to demonstrate the potential of mitochondria to generate ROS in response to TCDD, by incubating hepatic mitochondria or microsomes from TCDD-treated rats with nuclei from untreated animals [41]. While the microsomes and mitochondria of the TCDD-treated rats resulted in increases in nuclear DNA single-strand breaks, the microsomes and mitochondria of the control animals showed no effect on DNA strand breaks. Subsequent studies by these and other investigators directly demonstrated enhanced production of ROS by hepatic mitochondria in rats treated with TCDD [33, 35, 52]. Production of the ROS superoxide anion and hydrogen peroxide is believed to involve disruption of the mitochondrial electron transport system [35] and the involvement of iron or iron–sulfur proteins [33].

Studies in knockout AHR($-/-$) and wild-type mice indicated that mitochondrial oxidative stress is AHR dependent [53]. TCDD-induced thiol (glutathione) changes in cytosol and mitochondria were shown to be dependent on AHR. The study also indicated the existence of relationships between mitochondrial glutathione uptake, hydrogen peroxide production, and mitochondrial and nuclear DNA damage, with greater DNA damage occurring in mitochondria compared to nuclei.

TCDD-induced modulation of various respiratory chain complexes in mitochondria has been examined. Forgacs et al. [54] demonstrated the dose- and time-dependent relationships of dioxin with the responsiveness of gene-encoded proteins associated with electron transport complex I (NADH dehydrogenase), III (cytochrome *c* reductase), IV (cytochrome *c* oxidase), and V (ATP synthase) using high-throughput PCR analysis and concluded that TCDD alters the expression of genes associated with mitochondrial electron transport chain function.

Senft et al. [55] observed elevation of succinate-stimulated mitochondrial hydrogen peroxide production for up to 8 weeks, with significantly lower levels of hepatic ATP in mice treated with TCDD. Aly and Domenech [56] reported significant decreases in the mitochondrial membrane potential and respiratory chain complexes II and IV induced by TCDD in isolated hepatocytes, which were associated with inhibition of antioxidant enzymes, increases in production of superoxide anion, hydrogen peroxide, and lipid peroxidation. The reason for the differences in effects on several respiratory complex components is not clear but may be related to the different experimental systems and techniques employed in the various studies.

Chen et al. [57] have eloquently examined the effects of TCDD–induced mitochondrial dysfunction and apoptosis in human trophoblast-like JAR cells. TCDD was shown to produce increased oxidative damage and dysfunction in the mitochondria of treated cells, associated with time-dependent increases in lipid peroxides and 8-hydroxy-2'-deoxyguanosine, a biomarker of DNA damage. Reductions in mitochondrial copy number and ATP content and increases in mitochondrial DNA deletions, apoptosis, cytochrome *c* release, p53 accumulation, Bax overexpression, and sequential caspase 3 activation were also observed. The authors concluded that the apoptotic effects of TCDD may be due to oxidative damage and mitochondrial dysfunction.

In addition to the production of ROS via the electron transport chain, mitochondrial cytochromes under AHR control and not under AHR control were shown to increase in response to TCDD and may also be involved in ROS production [58]. The studies also suggested that ROS production in response to TCDD in mice may involve not only electron transport complex components in mitochondria but also cytochromes associated with the mitochondria, as well as the microsomes.

A third source of ROS in response to TCDD and its congeners involves the activation of macrophages with subsequent tissue infiltration and the resultant production of hydrogen peroxide and superoxide anion. Stohs [2] initially indicated the possible contribution of this source, based on a review of the dioxin-induced edema with macrophage infiltration in various species, including rats, mice, monkeys, chicks, and humans. Exposure of macrophages and other phagocytic cells to appropriate stimuli results in a respiratory burst that involves the NADPH-dependent one-electron reduction of oxygen to superoxide anion [59]. The enzyme NADPH oxidase is involved in this process and is normally dormant until activated by a large number of stimuli, including TCDD. TCDD and other toxicants were shown to induce superoxide anion when incubated with rat peritoneal macrophages [34]. This investigation also showed that TCDD was able to enhance peritoneal membrane apparent microviscosity, indicating a significant decrease in membrane viscosity.

Studies by Alsharif et al. [7, 60, 61] clearly demonstrated significant and dose- and time-dependent production of superoxide anion by peritoneal lavage (primarily macrophage) cells of rats after oral administration of TCDD. They also showed that activation of oxygen to superoxide anion preceded the formation of DNA single-strand breaks in these cells [61]. Based on the use of TCDD-responsive (C56BL/6J) and TCDD nonresponsive (DBA/2) mice, TCDD effects on ROS production by the macrophages were shown to be mediated by the AHR [7]. Subsequent studies on the role of AHR in ROS production by macrophages have demonstrated significant induction by the AHR ligand 3-MC of the expression of AHR mRNA during human monocyte differentiation [62, 63]. Furthermore, 3-MC induced the expression of the CYP1A1 mRNA. The investigators concluded that AHR may play a role in the function of macrophages, associated with the activation of environmental carcinogens. In addition, these results support the involvement of monocytes in the production of ROS in response to dioxins and indicate the presence of AHR in human monocytes and macrophages.

On the basis of the studies by Clark et al. [64] that implicated tumor necrosis factor alpha (TNF-α) in the toxicity of TCDD in responsive mice, Alsharif et al. [65] postulated that TNF-α may act as an amplifying loop in TCDD-induced oxidative stress. The studies involved administration of TNF-α antibody to mice previously treated with TCDD and showed significant decreases in peritoneal lavage cell activation, hepatic DNA single-strand breaks, and mitochondrial and microsomal lipid peroxidation. Subsequent studies by Moos et al. [66] confirmed the earlier results by demonstrating increased production and secretion of TNF-α by peritoneal cells of mice treated with TCDD. Cheon et al. [67] extended these observations by showing that TCDD-induced production of TNF-α in differentiated human macrophages is AHR dependent and involves activation of epidermal growth factor receptor (EGFR).

Furthermore, extracellular signal-regulated kinase (ERK) is a downstream effector of EGFR activation. Thus, TCDD may induce sequential activation of AHR, EGFR, and ERK, leading to the increased production of TNF-α.

Based on the above studies, it is concluded that TCDD induces ROS production and the resultant oxidative stress via at least three major systems, namely, microsomal and the associated cytochromes, the mitochondrial electron transport chain, and a macrophage-related oxidative burst. While the microsomal system has been most extensively studied, the mitochondrial and macrophage pathways for ROS production by TCDD have received less attention. The relative importance of these three systems is not known. However, one system may prove to be more important in one tissue or organ than in another. Thus, all three systems may make major contributions to the massive and widespread production of oxidative stress in response to TCDD and related congeners, which may help explain the subsequent occurrence of the extensive tissue damage.

15.5 OTHER INDICATORS OF OXIDATIVE STRESS

As noted above, dioxin-induced alterations in iron distribution and availability are believed to contribute to oxidative stress. Sweeney et al. [37] showed that iron deficiency prevented TCDD-induced liver toxicity and porphyria in mice. Addition of iron to microsomes from untreated animals enhanced lipid peroxidation more than adding iron to microsomes from TCDD-treated rats [38, 68], suggesting greater availability of catalytically active iron in microsomes of TCDD-treated animals. Addition of the iron chelator desferrioxamine (DFX) to microsomes from control and TCDD-treated rats decreased lipid peroxidation to the same basal levels, although the level of lipid peroxidation was sevenfold higher in the microsomes of treated animals in the absence of DFX [38]. Thus, greater levels of iron may have been available to catalyze lipid peroxidation in microsomes from TCDD-treated rats. Pretreatment of rats with DFX prior to and during treatment with TCDD modulated subcellular distribution of iron and decreased microsomal lipid peroxidation [69].

The role of dietary iron on various indicators of TCDD-induced oxidative stress has been studied. TCDD failed to induce hepatic lipid peroxidation in female rats fed an iron-deficient diet, providing further evidence for the role of iron in TCDD-induced lipid peroxidation and oxidative stress [38, 39]. Subsequent studies demonstrated that TCDD produced differential effects with respect to the distribution of iron [70, 71], as well as copper and magnesium, in subcellular hepatic fractions of male and female rats [70]. The hepatic iron content of female rats was found to be twofold higher than male rats and agreed with the observed levels of lipid peroxidation. Also, TCDD administration increased iron content of the mitochondria while decreased that of the microsomes in both groups of rats [70, 71]. Hence, the mitochondria may play a greater role than microsomes in ROS production in rats.

Studies by Smith et al. [12] showed that iron administration potentiated TCDD-induced hepatic porphyria and hepatocellular damage and elevated various hepatic enzymes. Differences were also shown between some iron metabolism proteins in TCDD-responsive and TCDD-resistant (DBA/2) mice, and perturbations in iron regulatory proteins were proposed. Subsequent experiments by Davies et al. [42] on C57BL/6J responsive and C57BL/6J knockout AHR(−/−) mice indicated that AHR is a key factor in TCDD-induced porphyria in mice and therefore in the perturbations in iron metabolism. Knowing that these toxic effects are correlated with the production of oxidative stress, it can be concluded that iron plays a significant role in TCDD-induced oxidative stress.

Alsharif et al. [72] examined the effects of TCDD on the fluidity of rat liver membranes. Dose- and time-dependent increases in TCDD-induced lipid peroxidation of mitochondrial, microsomal, and plasma membranes were observed. Concomitantly, significant decreases in membrane fluidity as determined by fluorescence polarization and anisotropy parameter values occurred in response to TCDD. Excellent inverse correlations existed between lipid peroxidation and membrane fluidity, indicating that oxidative stress may produce membrane lipid peroxidation, which is reflected in altered membrane fluidity and presumably altered membrane function. Shertzer [29] has also demonstrated TCDD-induced changes in the membrane fluidity of mouse hepatic mitochondria.

A relationship clearly exists between oxidative stress and altered calcium homeostasis [2, 73–78]. An increase in calcium associated with rat hepatic mitochondria, microsomes, and cytosol occurs in response to TCDD [73], which may be due, at least in part, to enhanced oxidative membrane damage. A number of studies have subsequently reported an increase in intracellular calcium in response to TCDD effects associated with heart [74], neurons [75, 76], pancreatic cells [77], and mouse hepatoma cells [78].

Sul et al. [76] have demonstrated associations between DNA damage, lipid peroxidation, tau phosphorylation and increased intracellular calcium, in response to TCDD, in neuroblastoma cells, and they concluded that TCDD exposure induces neurotoxicity via these complex and interrelated mechanisms. The role of calcium to activate phospholipase, endonucleases, and proteases, as well as the contribution of these mechanisms to TCDD-induced cellular toxicity and death, is well known [79, 80]. Increases in intracellular calcium were found to parallel increases in DNA damage and lipid peroxidation, but inversely correlated with nonprotein sulfhydryl and NADPH content [81]. The results of these studies provide strong evidence regarding the contribution of calcium influx and altered calcium homeostasis to the cascade of events associated with the toxic manifestations of TCDD and related congeners.

Further studies to demonstrate that TCDD induces oxidative stress have involved the concurrent determination of malondialdehyde, formaldehyde, acetaldehyde, and acetone in the urine of rats after exposure to TCDD [82]. Hydrazones of the lipid metabolites in urine were prepared with 2,4-dinitorphenylhydrazine and separated by high pressure liquid chromatography (HPLC). Formaldehyde was found to be excreted in greatest amounts, followed by acetone. The studies also compared the effects of TCDD on the excretion of these same metabolites with those of other toxic compounds, such as paraquat, endrin, and carbon tetrachloride [82].

Subsequent studies by Bagchi et al. [83] assessed the time-dependent effects of TCDD on serum and urine levels of formaldehyde, acetaldehyde, acetone, and malondialdehyde in *ad libitum* and pair-fed rats using HPLC. Up to a threefold increase in acetone levels with lesser amounts of other lipid oxidation products were observed in the sera of the TCDD-treated rats. When serum levels of malondialdehyde were determined by HPLC and compared with results obtained using the thiobarbituric acid (TBA) method, similar time courses following TCDD administration were demonstrated, but higher values were observed for the less specific colorimetric TBA method. The findings confirmed the utility of the less complicated TBA method and identified HPLC as an accurate system for possible investigation of altered lipid metabolism in disease states and following exposure to environmental toxicants [83].

In addition to the production of DNA single-strand breaks [19, 20, 40, 41, 61, 69, 76, 81, 84], 8-hydroxy-deoxyguanosine (8-OHDG) and 8-oxo-deoxyguanosine (8-oxo-DG) were also identified as biomarkers of oxidative DNA damage in response to TCDD and its congeners. Treatment of mice with 5 µg TCDD per day for three consecutive days resulted in 20-fold increases in urinary 8-OHDG that persisted for at least 8 weeks [85]. Similar results were shown in marine mammals exposed to coplanar polychlorinated biphenyls (PCBs), which are dioxin-like congeners [86].

Wyde et al. [87] noted the formation of 8-oxo-DG adducts in the livers of TCDD-treated female rats, concluding that this adduct formation occurred most likely in response to an oxidative imbalance. Greater hepatic levels of 8-oxo-DG were shown in intact compared to ovariectomized TCDD-treated rats [88]. Most recently, Chen et al. [57] demonstrated increased 8-OHDG and decreased mitochondrial DNA copy number production by TCDD in JAR, which is a human trophoblast-like cell line. Taken together, these results provide convincing evidence of oxidative DNA damage production in response to TCDD and its congeners.

15.5.1 Oxidative Stress and Hepatotoxicity

The liver is the primary target organ for TCDD toxicity [89–93] and has been the tissue most frequently employed in mechanistic studies [2, 3, 94–97] regarding the toxic manifestations of TCDD and its congeners. Similarly, initial studies concerning the role of ROS and oxidative stress focused primarily on the liver [1, 2, 5, 13–18, 33, 38–41, 52, 97]. These studies clearly demonstrated the roles of lipid peroxidation, DNA damage, production of ROS, altered membrane fluidity, and modulation of glutathione and glutathione metabolizing enzymes in the hepatotoxicty of TCDD and related chemicals [2–4, 97].

Most initial studies with dioxins involved the use of high, overtly toxic doses. However, a number of studies to determine whether environmentally relevant exposure to dioxins results in the induction of oxidative stress and oxidative tissue damage were subsequently conducted. Hassoun et al. [98] assessed the abilities of TCDD and selected congeners to induce oxidative stress in the livers of rats after subchronic (13 weeks) exposure. TCDD doses ranging from 10 to 100 ng/kg per day induced significant increases in hepatic superoxide anion, lipid peroxidation, and DNA. The studies also involved two TCDD congeners, namely, 2,3,4,7,8-pentachlorodibenzofuran (PeCDF) and 3,3′,4,4′,5-pentachlorobiphenyl (PCB126), and found them to be much less potent than TCDD. Those studies were further extended to investigate similar effects in mice exposed acutely and subchronically to TCDD [99]. Subchronic administration of 150 ng TCDD/kg/day resulted in significant increases in hepatic superoxide anion and lipid peroxidation, and higher tissue concentrations of TCDD were required to elicit an acute response as opposed to a subchronic response.

In a subsequent chronic (30 week) study, Hassoun et al. [84] investigated the abilities of TCDD, PeCDF, and PCB126, as well as toxic equivalent mixtures (TEQs), to induce oxidative stress in the liver. The three chemicals and the TEQs produced dose-dependent increases in the biomarkers of oxidative stress, namely, superoxide anion production, lipid peroxidation, and DNA single-strand breaks. The doses required to induce oxidative damage were approximately half of those required to produce similar effects after subchronic exposure and over 50-fold less than that used for the acute toxicity studies.

In another low-dose study, Kern et al. [100] injected rats daily with 30 ng TCDD/kg/day for 8 weeks and assessed hepatic and adipose tissue oxidative stress-related enzymes. Significant decreases in glutathione peroxidase and catalase activities were observed that were exacerbated by high-fat–sucrose diet. Thus, diet may play an important role in the toxicity of TCDD at low doses.

The role of the tumor suppressor gene p53 in the hepatotoxicity of TCDD and other xenobiotics was examined by Bagchi et al. [101]. Hepatic lipid peroxidation, superoxide anion production, and DNA fragmentation were all found to be greater in response to TCDD in p53-deficient mice compared to control (C57BL/6NTac) mice. Similar results were obtained for endrin, naphthalene, and chromium(VI). Thus, these results provide evidence for the involvement of p53 tumor suppressor gene not only in TCDD toxicity but also in the toxicity of structurally diverse chemicals known to induce oxidative stress.

15.5.2 Oxidative Stress and Human Toxicity

Several reviews, including various chapters in this book, have summarized the adverse health effects associated with exposure to dioxins [92, 97, 102–107]. In summary, those effects include, but are not limited to, chloracne, various cancers, thyroid dysfunction, reproductive and developmental defects, diabetes, compromised immune functions, impaired neurodevelopment, liver damage, and wasting syndrome. These and other adverse effects have been demonstrated in animal studies and nearly all have been linked to oxidative stress. Persistent alterations in the AHR pathway in response to human exposure to TCDD have been demonstrated and summarized by Steenland et al. [103]. The diverse, nontissue-specific effects of TCDD in humans, which are supported by the much more extensive animal studies, are consistent with the involvement of oxidative stress-mediated responses.

15.5.3 Oxidative Stress and Carcinogenesis

Long-term treatment of rodents with dioxins results in the development of tumors in liver, lung, skin, thyroid, oral cavity, and other sites [96, 108], and development of liver tumors, primarily in female rats exposed to dioxins, was used as basis for qualitative risk assessment [96]. The roles of ROS and oxidative stress in human carcinogenesis have been a primary focus of research that has been summarized in various reviews [106, 107, 109–112], and the application of oxidative stress biomarkers to the assessment of cancer risk by dioxins has been reviewed by Yoshida and Ogawa [97].

Unregulated or prolonged production of ROS was shown to result in DNA damage, as well as the modification of intracellular signaling and communication, gene expression, membrane structure and function, calcium homeostasis, tumor markers, and other cellular functions [109–113]. ROS formation via the uncoupling of human cytochrome P450 1B1 was found to have a possible contribution to the carcinogencity of dioxin-like polychlorinated biphenyls [107]. Mitochondrial generation of ROS associated with modulation of phosphorylation, activation, oxidation, and DNA binding of a wide range of transcription factors that result in alterations in target gene expression has been reviewed by Verschnoor et al. [112]. TCDD was also found to activate mitochondria, resulting in stress signaling and increased invasiveness of C2C12 cells, an action suggested to contribute to TCDD-induced tumor progression [113].

In summary, the above studies provide evidence for the contribution of oxidative stress in chemical carcinogenesis, and TCDD and its congeners produce significant amounts of ROS, suggesting that as a plausible mechanism for dioxin-induced carcinogenesis. As previously discussed, general toxicity of TCDD and its congeners is associated with multiple sources of ROS. With respect to dioxin-induced carcinogenesis, primary emphasis has focused on the microsomal cytochromes and the mitochondrial electron transport system.

15.5.4 Oxidative Stress and Testicular Toxicity

The TCDD testicular effects, including atrophy, morphological changes, and decreased spermatogenesis in various species of animals, have been reviewed by Al-Bayati et al. [114]. These authors were the first to present results demonstrating a role for oxidative stress in this toxicity. The results indicated the association of production of increases in testicular lipid peroxidation and iron contents and decreases in testicular superoxide dismutase and glutathione peroxidase with decreases in testes weights of rats exposed to TCDD [114].

Latchoumycandane and Mathur [22] demonstrated decreases in the weights of testes, epididymis, seminal vesicles, and ventral prostate that were associated with significant declines in the activities of the antioxidant enzymes superoxide dismutase, glutathione peroxidase, glutathione reductase, and catalase in rats exposed to TCDD. They also demonstrated partial protective effects for vitamin E against TCDD-induced modulations of those biomarkers upon coadministration of the two compounds to rats. Later studies by the same group of investigators demonstrated dose-dependent depletion of antioxidant defenses in epididymal sperm of rats exposed to TCDD [115, 116]. They further demonstrated dose-dependent decreases in mitochondrial and microsomal antioxidant enzyme activities of testes associated with concomitant increases in testicular hydrogen peroxide and lipid peroxidation production in rats exposed to TCDD [117, 118]. TCDD administration to mice was found to result in significantly lower levels of mRNA for antioxidant enzymes in testis when compared to controls [119]. The studies further demonstrated increased TGF-β1 activity, activation of the receptor-activated protein Smad2, and increased transcription factor regulator mitogen-activated protein kinase (MAPK), along with the major transcription factors c-jun and ATF-3.

Lactational exposure of mouse dams to 1 μg TCDD/kg/day for 4 days after parturition resulted in significant decreases in all antioxidant enzyme activities of the male reproductive system, as well as in testosterone levels. These effects were associated with upregulation and expression of p53 and Bax [120]. Thus, oxidative stress may be a major mediator of TCDD-induced adverse events in the male reproductive system, and p53 may play a role.

15.5.5 Oxidative Stress and Neurotoxicity

The first indication for the production of oxidative damage by TCDD in brain was reported by Al-Bayati et al. [121], showing that acute exposure of rats to TCDD resulted in increased whole brain lipid peroxidation 6 days postadministration. An examination of the biochemical effects of acute exposure to TCDD and related compounds on the CNS revealed induction of cytochrome P450-related enzyme activities with no major changes in catecholaminogenic neurotransmitter [122]. These may suggest a possible association between cytochrome P450-related enzymes and the induction of oxidative stress in brain by TCDD.

A series of more detailed studies on the induction of oxidative stress in brain tissues have been conducted by Hassoun et al. [20, 84, 98, 123–126]. Oral administration of up to 150 ng TCDD/kg/day to female mice for 13 weeks resulted in dose-dependent increases in whole brain superoxide anion, lipid peroxidation, and DNA single-strand breaks [123]. The subchronic exposure of female rats to low doses of TCDD and two of its congeners, PeCDF and PCB126, produced dose-dependent increases in brain superoxide anion, lipid peroxidation, and DNA single-strand breaks [98]. A maximal response for TCDD was found to be produced by doses ranging between 7 and 34 ng/kg/day, but higher doses were required by the other two congeners to produce responses similar to those of TCDD [98].

Hassoun et al. [84] have also examined the ability of TCDD and its congeners PeCDF and PCB126, as well as TEQ mixtures of these chemicals, to induce oxidative stress in the rat brain after chronic (30 weeks) exposure, where dose-dependent increases in the production of superoxide anion, lipid peroxidation, and DNA single-strand breaks were produced by all the tested compounds and mixtures. When rat brains were dissected into cerebral cortex, hippocampus, cerebellum, and brain stem following subchronic exposure to TCDD, dose-dependent increases in the production of superoxide anion and lipid peroxidation were observed in cerebral cortex and hippocampus but not cerebellum and brain stem, indicating that these two regions of the brain may be more vulnerable than the other regions [124].

Subsequent studies, as previously noted [20], demonstrated significant reductions in superoxide anion, lipid peroxidation, and DNA single-strand breaks production in cerebral cortex and hippocampus of the rat brain when vitamin E succinate or ellagic acid were coadminstered with TCDD subchronically. The effects of subchronic administration of vitamin E succinate or ellagic acid with TCDD on antioxidant enzyme activities and reduced glutathione levels of various regions of the rat brain were also examined [125]. The results demonstrated significant effects of each antioxidant on certain enzymes in certain brain regions and suggested involvement of different pathways in the protective actions of the two antioxidants [125].

The relationships between superoxide anion production and the levels of different biogenic amines were determined in brain regions of rats exposed to TCDD subchronically [126]. Strong correlations were observed between superoxide anion and norepinephrine (NE) in all brain regions and also between superoxide anion and 5-hydroxytryptamine (5-HT)/5-hydroxy-indole 3-acetic acid (HIAA) in the cerebral cortex and hippocampus, suggesting the contribution of biogenic amines and their metabolites, especially NE and 5-HT/HIAA, to superoxide anion overproduction in those regions.

The role of p53 tumor suppressor gene in TCDD-induced oxidative stress in brain as well as liver tissues has also been examined [101]. Treatment of p53-deficient mice with TCDD resulted in significantly increases in superoxide anion production, lipid peroxidation, and DNA fragmentation in brain tissues of those mice compared to responsive control mice. The results of induction of various oxidative stress biomarkers in brain were found to be similar to those in the liver of the same p53-deficient and TCDD-responsive mice [101].

Kim and Yang [127] have examined the neurotoxic effects of TCDD in cerebellar granule cells. TCDD-induced dose-dependent increases in protein kinase C (PKC) activity, ROS production, and intracellular calcium concentration effects were shown to be AHR dependent. These results and a growing body of evidence indicate a role for oxidative stress in the neurotoxicity of dioxins, and these effects can occur at exposure levels near those present for humans in the environment.

15.5.6 Oxidative Stress and Embryotoxicity

The relationship between TCDD-induced fetotoxicity and induction of oxidative stress in responsive mice was first examined by Hassoun et al. [128]. Significant increases in fetal and placental DNA single-strand breaks and lipid peroxidation were observed 2 days after administering 30 µg TCDD/kg to pregnant animals on day 12 of gestation. In addition, significant increases in the lipid metabolites malondialdehyde, formaldehyde, acetaldehyde, and acetone were observed in the amniotic fluids of TCDD-treated mice compared to the control. Subsequent studies by the same investigators have demonstrated the induction by TCDD of significantly greater oxidative stress in embryonic and placental tissues [9], as well as greater levels of lipid metabolites in the amniotic fluids and sera [10] of pregnant responsive C57BL/6J mice, compared to the nonresponsive DBA/2J mice [9, 10]. Coadministration of vitamin E succinate or ellagic acid with TCDD to pregnant mice attenuated TCDD-induced fetal growth retardation and death and placental weight reduction but not TCDD-induced teratogenic effects. In addition, induction of the biomarkers of oxidative stress, including superoxide anion production, lipid peroxidation, and DNA single-strand breaks by TCDD in fetal and placental tissues, was significantly reduced by those antioxidants. Taken together, the results of these studies strongly affirm a relationship between TCDD-induced oxidative stress and fetotoxicity, but not teratogenicity.

While Willey et al. [129] have demonstrated production of teratogenic effects in embryos of lacZ pregnant mice when treated with TCDD, Mimura et al. [130] have demonstrated loss of TCDD teratogenic responses in embryos of pregnant mice lacking the AHR. Thus, the studies demonstrated a correlation between the TCDD-elicited embryotoxicity and the AHR activation. However, no association with ROS production was made in either study.

TCDD developmental toxicity was also examined in chickens, where TCDD injection into fertilized eggs was shown to produce dose-dependent suppression of the activities of the hepatic antioxidant enzymes glutathione peroxidase, glutathione reductase, and superoxide dismutase [131]. The studies concluded that exposure of chick embryos to low doses of TCDD decreases the ability to scavenge ROS, which may contribute to oxidative stress and subsequent embryotoxicity.

15.5.7 Oxidative Stress and Cardiotoxicity

TCDD and PCBs were found to be risk factors for cardiovascular diseases [132, 133]. The first study suggesting the involvement of oxidative stress in TCDD-induced cardiotoxicity was that of Albro et al. [134], where lipofuscin, a pigment considered to be a by-product of lipid peroxidation, was found to accumulate in the hearts of rats treated with TCDD. Hermansky et al. [135] conducted a more in-depth study on the biochemical and functional effects of TCDD on the hearts of female rats. Six days after treatment with TCDD, a small but significant increase in lipid peroxidation with a concomitant decrease in superoxide dismutase activity occurred in the heart. No histopathological changes were observed at this time point, but blood pressure and resting heart rate were significantly decreased. Furthermore, serum thyroxine level in the TCDD-treated animals was 66% less than that of the controls.

Kopf et al. [136] have demonstrated changes in blood pressure, heart weight, left ventricular hypertrophy, and serum triglycerides, associated with increased superoxide anion production in mice treated with TCDD. The cardiovascular embryotoxicity of TCDD has been reviewed by Goldstone and Stegeman [137], with oxidative stress and growth factor modulation being implicated.

The review by Humblet et al. [138] discusses possible causation of dioxin-induced cardiotoxicity based on animal studies and included oxidative stress, inflammation, perturbation of calcium homeostasis, signaling pathways, and mitochondrial dysfunction. The data suggest the existence of a strong association between dioxin exposure and human mortality from ischemic heart disease, but a lesser association with cardiovascular disease. The involvement of oxidative stress as a key contributor to diabetic cardiomyopathy has been reviewed by Khullar et al. [139].

15.5.8 Oxidative Stress and TCDD Toxicity in Other Organs/Tissues

The ability of TCDD to induce oxidative stress in kidneys was demonstrated by Al-Bayati et al. [121]. Microsomes from kidneys of rats treated with acute doses of TCDD exhibited significant increase in lipid peroxidation. Subchronic (13 weeks) administration of TCDD to mice at very low doses resulted in significant decreases in reduced glutathione levels in kidneys [96].

Lu et al. [140] examined the nephrotoxic effects of PCBs (Aroclor 1254) and TCDD, alone and in combination, in rats after daily exposure for 12 days. All treatments induced nephrotoxicity based on significant increases in serum creatinine and blood urea nitrogen and histopathological changes that were associated with oxidative stress and induction of renal CYP1A1 expression. Most effects were found to be greater in the combined exposure group compared to the TCDD- or PCBs-treated groups.

TCDD was also found to induce biomarkers of oxidative stress in other animal tissues, such as the adrenals [141] and thymus [121]. Furthermore, the immunotoxicity of several xenobiotics including dioxins was found to involve a multifaceted cascade of events, including alterations in mitochondrial electron transport, ROS, antioxidants and antioxidant enzymes, cell signaling, and receptors [142].

Human exposure to TCDD was found to be associated with hyperinsulinemia and insulin resistance [143]. Although no studies to demonstrate a direct relationship between production of this effect and oxidative stress are available, TCDD was shown to increase intracellular calcium in the pancreas, resulting in continuous insulin release by the beta cells [74]. Since the relationship between oxidative stress and altered calcium homeostasis is well documented [73–78], it is likely that oxidative stress is involved in the diabetogenic effect of TCDD.

15.6 CONCLUSIONS

A growing body of information indicates that the diverse toxicological effects of polyhalogenated and polycyclic hydrocarbons, including the dioxins, involve a common cascade of events following activation of the AHR. This series of events entails production of ROS and oxidative stress in conjunction with modulation of intracellular redox status, inhibition of antioxidant enzymes, oxidative tissue damage, including lipid peroxidation, DNA and membrane (organelle and cell) damage, and programmed cell death. In addition, altered iron and calcium homeostasis, activation of protein kinase C, enhanced release of TNF-α, induction of various cell signaling mechanisms, stimulation of oncogene expression, and inhibition of tumor suppressor genes also occur. These effects lead to a plethora of tissue damaging events throughout the body, and as a consequence, diverse toxicological effects of dioxins and related chemicals are produced.

REFERENCES

1. Stohs, S. J., Hassan, M. Q., and Murray, W. J. (1983). Lipid peroxidation as a possible cause of TCDD toxicity. *Biochemical and Biophysical Research Communications*, 111, 854–859.
2. Stohs, S. J. (1990). Oxidative stress induced by 2,3,7,8-tetrachlorodibenzo-p-dioxin (TCDD). *Free Radicals in Biology and Medicine*, 9, 79–90.
3. Dalton, T. P, Puga, A., and Shertzer, H.G. (2002). Induction of cellular oxidative stress by aryl hydrocarbon receptor activation. *Chemico-Biological Interactions*, 141, 77–95.
4. Reichard, J. F, Dalton T. P., Shertzer, H. G., and Puga, A. (2006). Induction of oxidative stress responses by dioxin and other ligands of the aryl hydrocarbon receptor. *Dose Response*, 3, 306–331.
5. Hassan, M. Q., Stohs, S. J., and Murray, W. J. (1983). Comparative ability of TCDD to induce lipid peroxidation in rats, guinea pigs and Syrian golden hamsters. *Bulletin of Environmental Contamination and Toxicology*, 31, 649–657.
6. Mohammadpour, H., Murray, W. J., and Stohs, S. J. (1988). 2,3,7,8-Tetracholorodibenzo-p-dioxin (TCDD)-induced lipid peroxidation in genetically responsive and non-responsive mice. *Archives of Environmental Contamination and Toxicology*, 17, 645–650.
7. Alsharif, N. Z., Lawson, T., and Stohs, S. J. (1994). Oxidative stress induced by 2,3,7,8-tetrachlorodibenzo-p-dioxin is mediated by the aryl hydrocarbon (Ah) receptor complex. *Toxicology*, 92, 39–51.
8. Poland, A. and Glover, E. (1975). Genetic expression of aryl hydrocarbon hydroxylase by 2,3,7,8-tetrachlorodibenzo-p-dioxin: evidence for a receptor mutation in genetically nonresponsive mice. *Molecular Pharmacology*, 11, 389–398.
9. Hassoun, E. A and Stohs, S. J. (1996). TCDD, endrin and lindane induced oxidative stress in fetal and placental tissues of C57BL/6J and DBA/2J mice. *Comparative Biochemistry and Physiology, Part C*, 115, 11–18.
10. Hassoun, E. A., Bagchi, D., and Stohs, S. J. (1996). TCDD, endrin and lindane induced increases in lipid metabolites in maternal sera and amniotic fluids of pregnant C57BL/6J and DBA/2J mice. *Research Communications in Molecular Pathology and Pharmacology*, 94, 157–169.

11. Sugihara, K., Kitamura, S., Yamada, T., Ohta, S., Yamashita, K., Yasuda, M., and Fujii-Kuriyama, Y. (2001). Aryl hydrocarbon receptor (AHR)-mediated induction of xanthine oxidase/xanthine dehydrogenase activity by 2,3,7,8-tetrachlorodibenzo-p-dioxin. *Biochemical and Biophysical Research Communications*, 281, 1093–1099.

12. Smith, A. G., Clothier, B., Robinson, S., Scullion, M. J., Carthew, P., Edwards, R., Luo, J., Lim, C. K., and Toledano, M. (1998). Interaction between iron metabolism and 2,3,7,8-tetrachlorodibenzo-p-dioxin in mice with variants of the Ahr gene: a hepatic oxidative mechanism. *Molecular Pharmacology*, 53, 52–61.

13. Stohs, S. J, Hassan, M. Q., and Murray, W. J. (1984). Effects of BHA, D-alpha-tocopherol and retinol acetate on TCDD-mediated changes in lipid peroxidation, glutathione peroxidase activity and survival. *Xenobiotica*, 14, 533–537.

14. Hassan, M. Q., Stohs, S. J., and Murray, W. J. (1985). Inhibition of TCDD-induced lipid peroxidation, glutathione peroxidase activity and toxicity by BHA and glutathione. *Bulletin of Environmental Contamination and Toxicology*, 34, 787–796.

15. Hassan, M. Q., Stohs, S. J., and Murray, W. J. (1985). Effects of vitamins E and A on 2,3,7,8-tetrachlorodibenzo-p-dioxin (TCDD)-induced lipid peroxidation and other biochemical changes in the rat. *Archives of Environmental Contamination and Toxicology*, 14, 437–442.

16. Hassan, M. Q., Mohhammadpour, H., Hermansky, S. J., Murray, W. J., and Stohs, S. J. (1987). Comparative effects of BHA and ascorbic acid on the toxicity of 2,3,7,8-tetrachlorodibenzo-p-dioxin (TCDD) in rats. *General Pharmacology*, 18, 547–550.

17. Stohs, S. J., Hassan, M. Q., and Murray, W. J. (1984). Induction of lipid peroxidation and inhibition of glutathione peroxidase by TCDD. *Banbury Report*, 18, 241–253.

18. Hassan, M. Q., Stohs, S. J., Murray, W. J., and Birt, D. F. (1985). Dietary selenium, glutathione peroxidase activity, and toxicity of 2,3,7,8-tetrachlorodibenzo-p-dioxin. *Journal of Toxicology and Environmental Health*, 15, 405–415.

19. Hassoun, E. A., Walter, A. C., Alsharif, N. Z., and Stohs, S. J. (1997). Modulation of TCDD-induced fetotoxicity and oxidative stress in embryonic and placental tissues of C57/6J mice by vitamin E succinate and ellagic acid. *Toxicology*, 124, 27–37.

20. Hassoun, E. A., Vodhanel, J., and Abushaban, A. (2004). The modulatory effects of ellagic acid and vitamin E succinate on TCDD-induced oxidative stress in different brain regions of rats after subchronic exposure. *Journal of Biochemical and Molecular Toxicology*, 18, 196–203.

21. Alsharif, N. Z. and Hassoun, E. A. (2004). Protective effects of vitamin A and vitamin E succinate against 2,3,7,8-tetrachlorodibenzo-p-dioxin (TCDD)-induced body wasting, hepatomegaly, thymic atrophy, production of reactive oxygen species and DNA damage in C57BL/6J mice. *Basic and Clinical Pharmacology and Toxicology*, 95, 131–138.

22. Latchoumycandane, C. and Mathur, P. P. (2002). Effects of vitamin E on reactive oxygen species-mediated 2,3,7,8-tetrachlorodibenzo-p-dioxin toxicity in rat testis. *Journal of Applied Toxicology*, 22, 345–351.

23. Chen, A. H., Hurh, Y. J., Na, J. H., Chun, Y. J., Kim, D. H., Kang, K. S., Cho, M. H., and Suhr, Y. J. (2004). Resveratrol inhibits TCDD-induced expression of CYP1A1 and CYP1B1 and catechol estrogen-mediated oxidative DNA damage in cultured human mammary epithelial cells. *Carcinogenesis*, 25, 2005–2013.

24. Ishida, T., Takeda, T., Koga, T., Yahata, M., Ike, A., Kuramoto, C., Taketoh, J., Hashiguchi, I., Akamine, A., Ishii, Y., and Yamada, H. (2009). Attenuation of 2,3,7,8-tetrachlorodibenzo-p-dioxin toxicity by resveratrol: a comparative study with different routes of administration. *Biological & Pharmaceutical Bulletin*, 32, 876–881.

25. Hung, Y. C., Huang, G. S., Sava, V. M., Blagodarsky, V. A., and Hong, M. Y. (2006). Protective effects of tea melanin against 2,3,7,8-tetrachlorodibenzo-p-dioxin-induced toxicity: antioxidant activity and aryl hydrocarbon receptor suppressive effect. *Biological & Pharmaceutical Bulletin*, 29, 2284–2291.

26. Choi, H., Chun, Y. S., Shin, Y. J., Ye, S. K., Kim, M. S., and Park, J. W. (2008). Curcumin attenuates cytochrome P450 induction in response to 2,3,7,8-tetrachlorodibenzo-p-dioxin by ROS-dependently degrading AhR and ARNT. *Cancer Science*, 99, 2518–2524.

27. Lim, J., Sanders, R. A., Yeager, R. L., Millsap, D. S., Watkins, J. B., Eells, J. T., and Henshel, D. S. (2008). Attenuation of TCDD-induced oxidative stress by 670 nm photobiomodulation in developmental chicken kidney. *Journal of Biochemical and Molecular Toxicology*, 22, 230–239.

28. Martino, L., Novelli, M., Masini, M., Chimenti, D., Piaggi, S., Masiello, P., and De Tata, V. (2009). Dehydroascorbate protection against dioxin-induced toxicity in the beta-cell line INS-1E. *Toxicology Letters*, 189, 27–34.

29. Shertzer, H. G. (2010). Protective effects of the antioxidant 4b,5,9b,10-tetrahydroindeno[1,2-b]indole against TCDD toxicity in C57BL/6J mice. *International Journal of Toxicology*, 29, 40–48.

30. Pohjanvirta, R., Sankari, S., Kulju, T., Naukkarinen, A., Ylinen, M., and Tuomisto, J. (1990). Studies on the role of lipid peroxidation in the acute toxicity of TCDD in rats. *Pharmacology & Toxicology*, 66, 399–408.

31. Wahba, Z. Z., Murray, W. J., Hassan, M. Q., and Stohs, S. J. (1989). Comparative effects of pair-feeding and 2,3,7,8-tetrachlorodibenzo-p-dioxin (TCDD) on various biochemical parameters in female rats. *Toxicology*, 59, 311–323.

32. Uno, S., Dalton, T. P., Sinclair, P. R., Gorman, N., Wang, B., Smith, A. G., Miller, M. L., Shertzer, H. G., and Nebert, D. W. (2004). Cyp1a1(−/−) male mice: protection against high-dose TCDD-induced lethality and wasting syndrome, and resistance to intrahepatocyte lipid accumulation and uroporphyria. *Toxicology and Applied Pharmacology*, 196, 410–421.

33. Stohs, S. J., Al-Bayati, Z. F., Hassan, M. Q., Murray, W. J., and Mohammadpour, H. A. (1986). Glutathione peroxidase and reactive oxygen species in TCDD-induced lipid peroxidation.

Advances in Experimental Medicine and Biology, 197, 357–365.

34. Bagchi, M. and Stohs, S. J. (1993). *In vitro* induction of reactive oxygen species by 2,3,7,8-tetrachlorodibenzo-*p*-dioxin, endrin and lindane in rat peritoneal macrophages, and hepatic mitochondria and microsomes. *Free Radicals in Biology and Medicine, 14,* 11–18.

35. Senft, A. P., Dalton, T. P., Nebert, D. W., Genter, M. B., Puga, A., Hutchinson, R. J., Kerzee, J. K., Uno, S., and Shertzer, H. G. (2002). Mitochondrial reactive oxygen production is dependent on the aromatic hydrocarbon receptor. *Free Radicals in Biology and Medicine, 33,* 1268–1278.

36. Shertzer, H. G., Clay, C. D., Genter, M. B., Chames, M. C., Schneider, S. N., Oakley, G. G., Nebert, D. W., and Dalton, T. P. (2004). Uncoupling-mediated generation of reactive oxygen by halogenated aromatic hydrocarbons in mouse liver microsomes. *Free Radicals in Biology and Medicine, 36,* 618–631.

37. Sweeney, G. D., Jones, K. G., Cole, F. M., Basford, D., and Krestynski, F. (1979). Iron deficiency prevents liver toxicity of 2,3,7,8-tetrachlorodibenzo-*p*-dioxin. *Science, 204,* 332–335.

38. Al-Bayati, Z.A.F. and Stohs, S. J. (1987). The role of iron in 2,3,7,8-tetrachlorodibenzo-*p*-dioxin (TCDD)-induced lipid peroxidation in rat liver microsomes. *Toxicology Letters, 38,* 115–121.

39. Al-Turk, W. A., Shara, M. A., Mohammadpour, H., and Stohs, S. J. (1988). Dietary iron and 2,3,7,8-tetrachlorodibenzo-*p*-dioxin induced alterations in hepatic lipid peroxidation, glutathione content and body weight. *Drug and Chemical Toxicology, 11,* 55–70.

40. Wahba, Z. Z., Lawson, T. A., and Stohs, S. J. (1988). Induction of hepatic DNA single strand breaks in rats by 2,3,7,8-tetrachlorodibenzo-*p*-dioxin (TCDD). *Cancer Letters, 39,* 381–386.

41. Wahba, Z. Z., Lawson, T. A., Murray, W. A., and Stohs, S. J. (1989). Factors influencing the induction of DNA single strand breaks in rats by 2,3,7,8-tetrachlorodibenzo-*p*-dioxin (TCDD). *Toxicology, 58,* 57–69.

42. Davies, R., Clothier, B., Robinson, S. W., Edwards, R. E., Greaves, P., Luo, J., Gant, T. W., Chernova, T., and Smith, A. G. (2008). Essential role of the AH receptor in the dysfunction of heme metabolism induced by 2,3,7,8-tetrachlorodibenzo-*p*-dioxin. *Chemical Research in Toxicology, 21,* 330–340.

43. Nebert, D. W., Petersen, D. D., and Fornace, A. J. Jr., (1990). Cellular responses to oxidative stress: the [Ah] gene battery as a paradigm. *Environmental Health Perspectives, 88,* 13–25.

44. Nebert, D. W. and Duffy, J. J. (1997). How knockout mouse lines will be used to study the role of drug-metabolizing enzymes and their receptors during reproduction and development, and in environmental toxicity, cancer, and oxidative stress. *Biochemical Pharmacology, 53,* 249–254.

45. Morel, Y. and Barouki, R. (1998). Down-regulation of cytochrome P4501A1 gene promoter by oxidative stress: critical contribution of nuclear factor 1. *Journal of Biological Chemistry, 273,* 26969–26976.

46. Abel, J., Li, W., Dohr, O., Vogel, C., and Donat, S. (1996). Dose–response relationship of cytochrome P4501b1 mRNA induction and 2,3,7,8-tetrachlorodibenzo-*p*-dioxin in livers of C57BL/6J and DBA/2J mice. *Archives of Toxicology, 70,* 510–513.

47. Santostefano, M. J., Ross, D. G., Savas, U., Jefcoate, C. R., and Birnbaum, L. S. (1997). Differential time-course and dose-response relationships of TCDD-induced CYP1B1, CYP1A1, and CYP1A2 proteins in rats. *Biochemical and Biophysical Research Communications, 233,* 20–24.

48. Slezek, B. P., Diliberto, J. J., and Birnbaum, L. S. (1999). 2,3,7,8-Tetrachlorodibenzo-*p*-dioxin-mediated oxidative stress in CYP1A2 knockout (CYP1A2−/−) mice. *Biochemical and Biophysical Research Communications, 264,* 376–379.

49. Matsumura, F. (2003). On the significance of the role of cellular stress response reactions in the toxic actions of dioxin. *Biochemical Pharmacology, 66,* 527–540.

50. Shertzer, H. G., Clay, C. D., Genter, M. B., Schneider, S. N., Nebert, D. W., and Dalton, T. P. (2004). CYP1A2 protects against reactive oxygen production in mouse liver microsomes. *Free Radicals in Biology and Medicine, 36,* 605–617.

51. Shertzer, H. G., Clay, C. D., Genter, M. B., Chames, M. C., Schneider, S. N., Oakley, G. G., Nebert, D. W., and Dalton, T. P. (2004). Uncoupling-mediated generation of reactive oxygen by halogenated aromatic hydrocarbons in mouse liver microsomes. *Free Radicals in Biology and Medicine, 36,* 618–631.

52. Stohs, S. J., Alsharif, N. Z., Shara, M. A., Al-Bayati, Z.A.F., and Wahba, Z. Z. (1990). Evidence for the induction of an oxidative stress in rat hepatic mitochondria by 2,3,7,8-tetrachlorodibenzo-*p*-dioxin (TCDD). In *Biologically Reactive Intermediates IV* (Wittmer, C. M.,et al., Eds) Plenum Press, New York, pp. 827–831.

53. Shen, D., Dalton, T. P., Nebert, D. W., and Shertzer, H. G. (2005). Glutathione redox state regulates mitochondrial reactive oxygen production. *Journal of Biological Chemistry, 280,* 25305–25312.

54. Forgacs, A. L., Burgoon, L. D., Lynn, S. G., Lapres, J. J., and Zacharewski, T. (2010). Effects of TCDD on the expression of nuclear encoded mitochondrial genes. *Toxicology and Applied Pharmacology 246,* 58–65.

55. Senft, A. P., Dalton, T. P., Nebert, D. W., Genter, M. B., Hutchinson, R. J., and Shertzer, H. G. (2002). Dioxin increases reactive oxygen production in mouse liver mitochondria. *Toxicology and Applied Pharmacology, 178,* 15–21.

56. Aly, H. A. and Domenech, O. (2009). Cytotoxicity and mitochondrial dysfunction of 2,3,7,8-tetracholordibenzo-*p*-dioxin (TCDD) in isolated rat hepatocytes. *Toxicology Letters, 191,* 79–87.

57. Chen, S. C., Liao, T. L., Wei, Y. H., Tzeng, C. R., and Kao, S. H. (2010). Endocrine disruptor, dioxin (TCDD)-induced mitochondrial dysfunction and apoptosis in human trophoblast-like JAR cells. *Molecular Human Reproduction, 16,* 361–372.

58. Genter, M. B., Clay, C. D., Dalton, T. P., Nebert, D. W., and Shertzer, H. G. (2006). Comparison of mouse hepatic

mitochondrial versus microsomal cytochromes P450 following TCDD treatment. *Biochemical and Biophysical Research Communications*, 342, 1375–1381.

59. Witz, G. and Czerniecki, B. J. (1989). Tumor promoters differ in their ability to stimulate superoxide anion radical production by murine peritoneal exudates cells following *in vivo* administration. *Carcinogenesis*, 10, 807–811.

60. Alsharif, N. Z. and Stohs, S. J. (1992). The activation of rat peritoneal macrophages by 2,3,7,8-tetrachlorodibenzo-*p*-dioxin. *Chemosphere*, 25, 899–904.

61. Alsharif, N. Z., Schlueter, W. J., and Stohs, S. J. (1994). Stimulation of NADPH-dependent reactive oxygen species formation and DNA damage by 2,3,7,8-tetrachlorodibenzo-*p*-dioxin in rat peritoneal lavage cells. *Archives of Environmental Contamination and Toxicology*, 26, 392–397.

62. Hayashi, S., Okabe-Kado, J., Honma, Y., and Kawajiri, K. (1995). Expression of Ah receptor (TCDD receptor) during human monocytic differentiation. *Carcinogenesis*, 16, 1403–1409.

63. Komura, K., Hayashi, S., Makino, I., Poellinger, L., and Tanaka, H. (2001). Aryl hydrocarbon receptor/dioxin receptor in human monocytes and macrophages. *Molecular and Cellular Biochemistry*, 226, 107–118.

64. Clark, G. C., Taylor, M. J., Tritscher, A. M., and Lucier, G. W. (1991). Tumor necrosis factor involvement in 2,3,7,8-tetrachlorodibenzo-*p*-dioxin-mediated endotoxin hypersensitivity in C57BL/6J mice congenic at the Ah locus. *Toxicology and Applied Pharmacology*, 111, 422–431.

65. Alsharif, N. Z., Hassoun, E., Bagchi, M., Lawson, T., and Stohs, S. J. (1994). The effects of anti-TNF-alpha antibody and dexamethasone on TCDD-induced oxidative stress in mice. *Pharmacology*, 48, 127–136.

66. Moos, A. B., Oughton, J. A., and Kerkvliet, N. I. (1997). The effects of 2,3,7,8-tetrachlorodibenzo-*p*-dioxin (TCDD) on tumor necrosis factor (TNF) production by peritoneal cells. *Toxicology Letters*, 90, 145–153.

67. Cheon, H., Woo, Y. S., Lee, J. Y., Kim, H. S., Kim, H. J., Cho, S., Won, N. H., and Sohn, J. (2007). Signaling pathway for 2,3,7,8-tetrachlorodibenzo-*p*-dioxin-induced TNF-alpha production in differentiated THP-1 human macrophages. *Experimental and Molecular Medicine*, 39, 524–534.

68. Albro, P. W., Corbett, J. T., and Schroeder, J. L. (1986). Effects of 2,3,7,8-tetrachlorodibenzo-*p*-dioxin on lipid peroxidation in microsomal systems in vitro. *Chemico-Biological Interactions*, 57, 301–313.

69. Wahba, Z. Z., Murray, W. J., and Stohs, S. J. (1990). Desferrioxamine induced alterations in hepatic iron distribution, DNA damage and lipid peroxidation in control and 2,3,7,8-tetrachlorodibenzo-*p*-dioxin treated rats. *Journal of Applied Toxicology*, 10, 119–124.

70. Wahba, Z. Z., Al-Bayati, Z.A.F., and Stohs, S. J. (1988). Effect of 2,3,7,8-tetrachlorodibenzo-*p*-dioxin (TCDD) on hepatic distribution of iron, copper, zinc and magnesium in rats. *Journal of Biochemical Toxicology*, 3, 195–214.

71. Al-Bayati, Z.A.F., Stohs, S. J., and Al-Turk, W. A. (1987). TCDD, dietary iron and hepatic iron distribution in female rats. *Bulletin of Environmental Contamination and Toxicology*, 38, 300–307.

72. Alsharif, N. Z., Grandjean, C. J., Murray, W. J., and Stohs, S. J. (1990). 2,3,7,8-Tetrachlorodibenzo-*p*-dioxin (TCDD)-induced decrease in fluidity of rat liver membranes. *Xenobiotica*, 20, 979–988.

73. Al-Bayati, Z.A.F., Murray, W. J., Pankaskie, M. C., and Stohs, S. J. (1988). 2,3,7,8-Tetrachlorodibenzo-*p*-dioxin (TCDD) induced perturbation of calcium distribution in rats. *Research Communications in Chemical Pathology and Pharmacology*, 60, 47–56.

74. Canga, L., Levi, R., and Rifkind, A. B. (1988). Heart as a target organ in 2,3,7,8-tetrachlorodibenzo-*p*-dioxin toxicity: decreased beta-adrenergic responsiveness and evidence of increased intracellular calcium. *Proceedings of the National Academy of Sciences of the United States of America*, 85, 905–909.

75. Hanneman, W. H., Legare, M. E., Barthoumi, R., Burghardt, R. C., Safe, S., and Tinnany-Castiglioni, E. (1996). Stimulation of calcium uptake in cultured rat hippocampal neurons by 2,3,7,8-tetrachlorodibenzo-*p*-dioxin. *Toxicology*, 112, 19–28.

76. Sul, D., Kim, H. S., Cho, E. K., Lee, M., Kim, Y. H., Jung, W. W., Hwang, K. W., and Park, S. Y. (2009). 2,3,7,8-TCDD neurotoxicity in neuroblastoma cells is caused by increased oxidative stress, intracellular calcium levels, and tau phosphorylation. *Toxicology*, 255, 65–71.

77. Kim, Y. H., Shim, Y. J., Shin, Y. J., Sul, D., Lee, E., and Min, B. H. (2009). 2,3,7,8-Tetrachlorodibenzo-*p*-dioxin (TCDD) induces calcium flux through T-type calcium channels and enhances lysosomal exocytosis and insulin secretion in INS-1 cells. *International Journal of Toxicology*, 28, 151–161.

78. Puga, A., Hoffer, A., Zhou, S., Bohm, J. M., Leikauf, G. D., and Shertzer, H. G. (1997). Sustained increase in intracellular free calcium and activation of cyclooxygenase-2 expression in mouse hepatoma cells treated with dioxin. *Biochemical Pharmacology*, 54, 1287–1296.

79. Al-Bayati, Z.A.F. and Stohs, S. J. (1991). The possible role of phospholipase A2 in hepatic microsomal lipid peroxidation induced by 2,3,7,8-tetrachlorodibenzo-*p*-dioxin in rats. *Archives of Environmental Contamination and Toxicology*, 20, 361–365.

80. Orrenius, S., McConkey, D. J., Bellomo, G., and Nicotera, P. (1989). Role of Ca^{2+} in toxic cell killing. *Trends in Pharmacological Sciences*, 10, 281–285.

81. Stohs, S. J., Shara, M. A., Alsharif, N. Z., Wahba, Z. Z., and Al-Bayati, Z.A.F. (1990). 2,3,7,8-Tetrachlorodibenzo-*p*-dioxin (TCDD)-induced oxidative stress in female rats. *Toxicology and Applied Pharmacology*, 106, 126–135.

82. Shara, M. A., Dickson, P. H., Bagchi, D., and Stohs, S. J. (1992). Excretion of formaldehyde, malondialdehyde, acetaldehyde and acetone in the urine of rats in response to 2,3,7,8-tetrachlorodibenzo-*p*-dioxin, paraquat, endrin and carbon tetrachloride. *Journal of Chromatography*, 576, 221–233.

83. Bagchi, D., Shara, M. A., Hassoun, E. A., and Stohs, S. J. (1993). Time-dependent effects of 2,3,7,8-tetrachlorodibenzo-*p*-dioxin (TCDD) on serum and urine levels of

malondialdehyde, formaldehyde, acetaldehyde and acetone in rats. *Toxicology and Applied Pharmacology*, 123, 83–88.

84. Hassoun, E. A., Wang, H., Abushaban, A., and Stohs, S. J. (2002). Induction of oxidative stress in the tissues of rats after chronic exposure to TCDD, 2,3,4,7,8-pentachlorodibenzofuran and 3,3',4,4', 5-pentachlorobiphenyl. *Journal of Toxicology and Environmental Health*, 65, 825–842.

85. Shertzer, H. G., Nebert, D. W., Puga, A., Ary, M., Sonntag, D., Dixon, K., Robinson, L. J., Cianciolo, E., and Dalton, T. P. (1998). Dioxin causes a sustained oxidative stress response in the mouse. *Biochemical and Biophysical Research Communications*, 253, 44–48.

86. Li, C. S., Wu, K. Y., Chang-Chien, G. P., and Chou, C. C. (2005). Analysis of oxidative DNA damage 8-hydroxy-2'-deoxyguanosine as a biomarker of exposures to persistent pollutants for marine mammals. *Environmental Science & Technology*, 39, 2455–2460.

87. Wyde, M. E., Wong, V. A., Kim, A. H., Lucier, G. W., and Walker, N. J. (2001). Induction of hepatic 8-oxo-deoxyguanosine by 2,3,7,8-tetrachlorodibenzo-p-dioxin in Sprague-Dawley rats is female specific and estrogen-dependent. *Chemical Research in Toxicology*, 14, 849–855.

88. Tritscher, A. M., Seacat, A. M., Yager, J. D., Groopman, J. D., Miller, B. D., Sutter, T. R., and Lucier, G. W. (1996). Increased oxidative DNA damage in livers of 2,3,7,8-tetrachlorodibenzo-p-dioxin treated intact but not ovariectomized rats. *Cancer Letters*, 98, 219–225.

89. Fowler, B. A., Lucier, G. W., Brown, H. W., and McDaniel, O. S. (1973). Ultrastructural changes in rat liver cells following a single oral dose of TCDD. *Environmental Health Perspectives*, 5, 141–148.

90. Lucier, G. W., McDaniel, O. S., Hook, G. E., Fowler, B. A., Sonawane, B. R., and Faeder, E. (1973). TCDD-induced changes in rat liver microsomal enzymes. *Environmental Health Perspectives*, 5, 199–209.

91. Birnbaum, L. S. and Tuomisto, J. (2000). Non-carcinogenic effects of TCDD in animals. *Food Additives & Contaminants*, 17, 275–288.

92. Mandal, P. K. (2005). Dioxin: a review of its environmental effects and its aryl hydrocarbon receptor biology. *Journal of Comparative Physiology B*, 175, 221–230.

93. Lee, J. H., Wada, T., Febbraio, M., He, J., Matsubara, T., Lee, M. J., Gonzalez, F. J., and Xie, W. (2010). A novel role for the dioxin receptor in fatty acid metabolism and hepatic steatosis. *Gastroenterology*, 139, 653–663.

94. Kohn, M. C., Lucier, G. W., Clark, G. C., Sewell, C., Tritscher, A. M., and Portier, C. J. (1993). A mechanistic model of effects of dioxin on gene expression in the rat liver. *Toxicology and Applied Pharmacology*, 120, 138–154.

95. Maruyama, W. and Aoki, Y. (2006). Estimated cancer risk of dioxins to humans using a bioassay and physiologically based pharmacokinetic model. *Toxicology and Applied Pharmacology*, 214, 188–198.

96. Knerr, S. and Schrenk, D. (2006). Carcinogenicity of 2,3,7,8-tetrachlorodibenzo-p-dioxin in experimental models. *Molecular Nutrition & Food Research*, 50, 897–907.

97. Yoshida, R. and Ogawa, Y. (2000). Oxidative stress induced by 2,3,7,8-tetrachlorodibenzo-p-dioxin: an application of oxidative stress markers to cancer risk assessment of dioxins. *Industrial Health*, 38, 5–14.

98. Hassoun, E. A., Li, F., Abushaban, A., Stohs, S. J. (2000). The relative abilities of TCDD and its congeners to induce oxidative stress in the hepatic and brain tissues of rats after subchronic exposure. *Toxicology*, 145, 103–113.

99. Slezek, B. P., Hatch, G. E., DeVito, M. J., Diliberto, J. J., Slade, R., Crissman, K., Hassoun, E., and Birnbaum, L. S. (2000). Oxidative stress in female B6C3F1 mice following acute and subchronic exposure to 2,3,7,8-tetrachlorodibenzo-p-dioxin (TCDD). *Toxicological Sciences*, 54, 390–398.

100. Kern, P. A., Fishman, R. B., Song, W., Brown, A. D., and Fonseca, V. (2002). The effect of 2,3,7,8-tertachlorodibenzo-p-dioxin (TCDD) on oxidative enzymes in adipocytes and liver. *Toxicology*, 171, 117–125.

101. Bagchi, D., Balmoori, J., Ye, X., Williams, C. B., and Stohs, S. J. (2000). Role of p53 tumor suppressor gene in the toxicity of TCDD, endrin, naphthalene, and chromium (VI) in liver and brain tissues of mice. *Free Radicals in Biology and Medicine*, 28, 895–903.

102. Mukerjee, D. (1998). Health impact of polychlorinated dibenzo-p-dioxins: a critical review. *Journal of the Air & Waste Management Association*, 48, 157–165.

103. Steenland, K., Betazzi, P., Baccarelli, A., and Kogevinas, M. (2004). Dioxin revisited: developments since 1997 IARC classification of dioxin as a human carcinogen. *Environmental Health Perspectives*, 112, 1265–1268.

104. Foster, W. G., Maharaj-Briceno, S., and Cyr, D. G. (2010). Dioxin-induced changes in epididymal sperm count and spermatogenesis. *Environmental Health Perspectives*, 118, 458–464.

105. Michalek, J. E. and Pavuk, M. (2008). Diabetes and cancer in veterans of Operation Ranch Hand after adjustment for calendar period, days of spraying, and time spent in Southeast Asia. *Journal of Occupational and Environmental Medicine*, 50, 330–340.

106. Lin, P. H., Lin, C. H., Huang, C. C., Chuang, M. C., and Lin, P. (2007). 2,3,7,8-Tetrachlorodibenzo-p-dioxin (TCDD) induces oxidative stress, DNA strand breaks, and poly(ADP-ribose) polymerase-1 activation in human breast carcinoma cell lines. *Toxicology Letters*, 172, 146–158.

107. Green, R. M., Hodges, N. J., Chipman, J. K., O'Donovan, M. R., and Graham, M. (2008). Reactive oxygen species from the uncoupling of human cytochrome P450 1B1 may contribute to the carcinogenicity of dioxin-like polychlorinated biphenyls. *Mutagenesis*, 23, 457–463.

108. Mann, P. C. (1997). Selected lesions of dioxin in laboratory rodents. *Toxicological Pathology*, 25, 72–79.

109. Klaunig, J. E., Kamendulis, L. M., and Hocevar, B. A. (2010). Oxidative stress and oxidative damage in carcinogenesis. *Toxicological Pathology*, 38, 96–109.

110. Klaunig, J. E. and Kamendulis, L. M. (2004). The role of oxidative stress in carcinogenesis. *Annual Review of Pharmacology and Toxicology*, 44, 239–267.

111. Liou, G. Y. and Storz, P. (2010). Reactive oxygen species in cancer. *Free Radical Research, 44*, 479–496.

112. Verschnoor, M. L., Wilson, L. A., and Singh, G. (2010). Mechanisms associated with mitochondrial-generated reactive oxygen species in cancer. *Canadian Journal of Physiology and Pharmacology, 88*, 204–219.

113. Biswas, G., Srinivasan, S., Anandatheerthavarada, K., and Avadhani, N. G. (2008). Dioxin-mediated tumor progression through activation of mitochondria-to-nucleus stress signaling. *Proceedings of the National Academy of Sciences of the United States of America, 105*, 186–196.

114. Al-Bayati, Z.A.F., Wahba, Z. Z., and Stohs, S. J. (1988). 2,3,7,8-Tetrachlorodibenzo-p-dioxin (TCDD)-induced alterations in lipid peroxidation, enzymes and divalent cations in rat testis. *Xenobiotica, 18*, 1281–1289.

115. Latchoumycandane, C., Chitra, C., and Mathur, P. (2002). Induction of oxidative stress in rat epididymal sperm after exposure to 2,3,7,8-tetrachlorodibenzo-p-dioxin. *Archives of Toxicology, 76*, 113–118.

116. Latchoumycandane, C., Chitra, K. C., and Mathur, P. P. (2003). 2,3,7,8-Tetrachlorodibenzo-p-dioxin (TCDD) induces oxidative stress in the epididymis and epididymal sperm of adult rats. *Archives of Toxicology, 77*, 280–284.

117. Latchoumcandane, C., Chitra, K. C., and Mathur, P. P. (2002). The effect of 2,3,7,8-tetrachlorodibenzo-p-dioxin on the antioxidant system in mitochondrial and microsomal fractions of rat testis. *Toxicology, 171*, 127–135.

118. Dhanabalan, S. and Mathur, P. P. (2009). Low dose of 2,3,7,8-tetrachlorodibenzo-p-dioxin induces testicular oxidative stress in adult rats under the influence of corticosterone. *Experimental and Toxicologic Pathology, 61*, 415–423.

119. Jin, M. H., Hong, C. H., Lee, H. Y., Kang, H. J., and Han, S. W. (2008). Enhanced TGF-β1 is involved in 2,3,7,8-tetrachlorodibenzo-p-dioxin (TCDD) induced oxidative stress in C57BL/6 mouse testis. *Toxicology Letters, 178*, 202–209.

120. Jin, M. H., Hong, C. H., Lee, H. Y., Kang, H. J., and Han, S. W. (2010). Toxic effects of lactational exposure to 2,3,7,8-tetrachlorodibenzo-p-dioxin (TCDD) on development of male reproductive system: involvement of antioxidants, oxidants, and p53. *Environmental Toxicology, 25*, 1–8.

121. Al-Bayati, Z.A.F., Murray, W. J., and Stohs, S. J. (1987). 2,3,7,8-Tetracholordibenzo-p-dioxin-induced lipid peroxidation in hepatic and extrahepatic tissues of male and female rats. *Archives of Environmental Contamination and Toxicology, 16*, 159–166.

122. Unkila, M., Pojanvirta, R., and Tuomisto, J. (1995). Biochemical effects of 2,3,7,8-tetrachlorodibenzo-p-dioxin (TCDD) and related compounds on the central nervous system. *International Journal of Biochemistry & Cell Biology, 27*, 443–455.

123. Hassoun, E. A., Wilt, S. C., Devito, M. J., Van Birgelen, A., Alsharif, N. Z., Birnbaum, L. S., and Stohs, S. J. (1998). Induction of oxidative stress in brain tissues of mice after subchronic exposure to 2,3,7,8-tetrachlorodibenzo-p-dioxin. *Toxicological Sciences, 42*, 23–27.

124. Hassoun, E. A., Ghafri, M., and Abushaban, A. (2003). The role of antioxidant enzymes in TCDD-induced oxidative stress in various brain regions of rats after subchronic exposure. *Free Radicals in Biology and Medicine, 35*, 1028–1036.

125. Hassoun, E. A., Vodhanel, J., Holden, B., and Abushaban, A. (2006). The effects of ellagic acid and vitamin E succinate on antioxidant enzyme activities and glutathione levels in different brain regions of rats after subchronic exposure to TCDD. *Journal of Toxicology and Environmental Health, Part A, 69*, 381–393.

126. Byers, J., Masters, K., Sarver, J. G., and Hassoun, E. A. (2006). Association between the levels of biogenic amines and superoxide anion production in brain regions of rats after subchronic exposure to TCDD. *Toxicology, 228*, 291–198.

127. Kim, S. Y. and Yang, J. H. (2005). Neurotoxic effects of 2,3,7,8-tetrachlorodibenzo-p-dioxin in cerebellar granule cells. *Experimental and Molecular Medicine, 28*, 58–64.

128. Hassoun, E. A., Bagchi, D., and Stohs, S. J. (1995). Evidence of 2,3,7,8-tetrachlorodibenzo-p-dioxin (TCDD)-induced tissue damage in fetal and placental tissues and changes in amniotic fluid lipid metabolites of pregnant CF1 mice. *Toxicology Letters, 76*, 245–250.

129. Willey, J. J., Stripp, B. R., Baggs, R. B., and Gasiewicz, T. A. (1998). Aryl hydrocarbon receptor activation in genital tubercule, palate, and other embryonic tissues of 2,3,7,8-tetrachlorodibenzo-p-dioxin-responsive lacZ mice. *Toxicology and Applied Pharmacology, 151*, 33–44.

130. Mimura, J., Yamashita, K., Nakamura, K., Morita, M., Takagi, T. N., Nakao, K., Ema, M., Sogawa, K., Yasuda, M., Katsuki, M., and Fujii-Kuriyama, Y. (1997). Loss of teratogenic response to 2,3,7,8-tetrachlorodibenzo-p-dioxin (TCDD) in mice lacking the Ah (dioxin) receptor. *Genes to Cells, 2*, 645–654.

131. Lim, J., DeWitt, J. C., Sanders, R. A., Watkins, J. B., 3rd, and Henshel, D. S. (2007). Suppression of endogenous antioxidant enzymes by 2,3,7,8-tertrachlorodibenzo-p-dioxin-induced oxidative stress in chicken liver during development. *Archives of Environmental Contamination and Toxicology, 52*, 590–595.

132. Dalton, T. P., Kerzee, J. K., Wang, D., Miller, M., Dieter, M. Z., Lorenz, J. N., Shertzer, H. G., Nebert, D. W., and Puga, A. (2001). Dioxin exposure is an environmental risk factor for ischemic heart disease. *Cardiovascular Toxicology, 1*, 285–298.

133. Lind, P. M., Orberg, J., Edlund, U. B., Sjblom, L., and Lind, L. (2004). The dioxin-like pollutant PCB126 (3,3′4,4′5-pentachlorobiphenyl) affects risk factors for cardiovascular disease. *Toxicology Letters, 150*, 293–299.

134. Albro, P. W., Corbett, J. T., Harris, M., and Lawson, L. D. (1978). Effects of 2,3,7,8-tetrachlorodibenzo-p-dioxin on lipid profiles in tissues of Fisher rat. *Chemico-Biological Interactions, 23*, 315–330.

135. Hermansky, S. J., Holcslaw, T. L., Murray, W. J., Markin, R. S., and Stohs, S. J. (1988). Biochemical and functional effects of 2,3,7,8-tetrachlorodibenzo-p-dioxin (TCDD) on the heart of female rats. *Toxicology and Applied Pharmacology, 95*, 175–184.

136. Kopf, P. G., Huwe, J. K., and Walker, M. K. (2008). Hypertension, cardiac hypertrophy and impaired vascular relaxation

induced by 2,3,7,8-tetrachlorodibenzo-*p*-dioxin are associated with increased superoxide. *Cardiovascular Toxicology*, 8, 181–193.

137. Goldstone, H. M. and Stegeman, J. J. (2006). Molecular mechanisms of 2,3,7,8-tetrachlorodibenzo-*p*-dioxin cardiovascular embryotoxicity. *Drug Metabolism Reviews*, 38, 261–298.

138. Humblet, O., Birnbaum, L., Rimm, E., Mittleman, M. A., and Hauser, R. (2008). Dioxins and cardiovascular disease mortality. *Environmental Health Perspectives*, 116, 1443–1448.

139. Khullar, M., Al-Shudiefat, A. A., Ludke, A., Binepal, G., and Singal, P. K. (2010). Oxidative stress: a key contributor to diabetic cardiomyopathy. *Canadian Journal of Physiology and Pharmacology*, 88, 233–240.

140. Lu, C. F., Wang, Y. M., Peng, S. Q., Zou, L. B., Tan, D. H., Liu, G., Fu, Z., Wang, Q. X., and Zhao, J. (2009). Combined effects of repeated administration of 2,3,7,8-tetrachlorodibenzo-*p*-dioxin and polychlorinated biphenyls on kidneys of male rats. *Archives of Environmental Contamination and Toxicology*, 57, 767–776.

141. Bestervelt, L. L., Piper, D. W., Pitt, J. A., and Piper, W. N. (1994). Lipid peroxidation in the adrenal glands of male rats exposed to 2,3,7,8-tetrachlorodibenzo-*p*-dioxin (TCDD). *Toxicology Letters*, 70, 139–145.

142. Kovacic, P. and Somanthan, R. (2008). Integrated approach to immunotoxicity;electron transfer, reactive oxygen species, antioxidants, cell signaling, and receptors. *Journal of Receptors and Signal Transduction Research*, 28, 323–346.

143. Cranmer, M., Louie, S., Kennedy, R. H., Kern, P. A., and Fonseca, V. A. (2000). Exposure to 2,3,7,8-tetrachlorodibenzo-*p*-dioxin (TCDD) is associated with hyperinsulinemia and insulin resistance. *Toxicological Sciences*, 56, 431–436.

16

DIOXIN ACTIVATED AHR AND CANCER IN LABORATORY ANIMALS

DIETER SCHRENK AND MARTIN CHOPRA

16.1 CARCINOGENICITY FINDINGS

PCDD/F and among those especially TCDD cause tumors in rodent models (rats, mice, and hamsters). Target organs for TCDD carcinogenicity are the liver, thyroid, lung, skin, oral cavity, and other sites. Since TCDD causes cancer in various organs in different species, it has been termed a multiple-site, multiple-species carcinogen. Though TCDD itself is not genotoxic, it causes tumors in the absence of initiating agents after long-term treatment. Following the pretreatment of animals with genotoxic agents, TCDD enhances tumor formation in the skin, lung, and liver [1, 2], thus acting as a tumor promoter.

TCDD was classified as a human carcinogen by the International Agency for Research on Cancer (IARC) in 1997 [3], mainly based on limited evidence of carcinogenicity to humans (derived from follow-up of workers who had been heavily exposed in industrial accidents), and sufficient evidence of carcinogenicity in experimental animals, taking into account that the evolutionarily conserved AHR is involved in carcinogenicity of TCDD in laboratory animals and mechanistically functions the same way in humans [3, 4].

Different hypotheses have been presented trying to explain the carcinogenicity of TCDD. It was proposed that the mode of action consists of both regenerative repair and tumor promotion, while today it is widely acknowledged that inhibition of apoptosis of cancer precursor cells plays a central role in tumor promotion by TCDD [5, 6].

16.1.1 Liver Carcinogenicity

TCDD is one of the most potent liver carcinogens in rodents. In a 2-year study, Kociba et al. [7] found an increased formation of hepatic adenoma and carcinoma mostly in female rats, while male rats showed carcinogenicity only at the highest dose level. A very similar dose–response relationship was found based on data by Mayes et al. [8] in a treatment experiment with various Aroclors (technical PCB mixtures) when the TEQ levels instead of the total PCB doses were used as dose metric [9].

Besides being carcinogenic, TCDD acts as a tumor-promoting agent in rodent liver; that is, it strongly enhances the yield of liver tumors in animals that have been pretreated with a genotoxic, initiating carcinogen such as diethylnitrosamine [10]. Attempts to investigate the role of the AHR in the carcinogenicity/tumor-promoting action of TCDD led to some indirect evidence. Tumor promotion studies with less potent AHR agonists among the group of 2,3,7,8-substituted PCDD congeners revealed a lower potency according to the potency as AHR agonists [11]. However, no data have been published on the carcinogenicity of TCDD in AHR-deficient animals. Viluksela et al. [12] studied the liver tumor-promoting activity of TCDD in the sensitive Long-Evans (L-E) and the resistant Han/Wistar (H/W) rats differing >1000-fold in their sensitivity to the acute lethality of TCDD. Female rats were partially hepatectomized, initiated with nitrosodiethylamine (diethyl-nitrosamine, DEN), and treated with TCDD for 20 weeks. H/W rats were exceptionally resistant to the induction of

---

The AH Receptor in Biology and Toxicology, First Edition. Edited by Raimo Pohjanvirta.
© 2012 John Wiley & Sons, Inc. Published 2012 by John Wiley & Sons, Inc.

preneoplastic liver foci by TCDD, the resistance being associated with an altered transactivation domain of the AHR. Interestingly, induction of foci was related to hepatotoxicity but not to CYP1A1 activity in the liver.

16.1.2 Skin Carcinogenicity

After local initiation with a genotoxic carcinogen, TCDD treatment of the skin resulted in enhanced tumor formation in mice [13]. The tumor-promoting potency of TCDD in these experiments was much higher in C57BL/6 than in DBA/2 mice, which is in accordance with the presence of a high- or low-affinity AHR. Furthermore, systemic application of TCDD can also result in dermal tumors, for example, in Syrian golden hamsters [14]. Hébert et al. [15] applied various PCDD/Fs with different affinities to the AHR, or TEF values, respectively. Their findings support a crucial role of the AHR in skin tumor promotion by AHR agonists.

16.1.3 Cancer in Other Organs

Other tissues showing increased tumor rates in experimental animals after chronic treatment with TCDD and other persistent AHR agonists are the hard palate, nasal turbinate, tongue, oral mucosa, lung, thyroid, uterus, integumentary system (fibrosarcoma), and hematopoietic system. Details on the species, gender, way and duration of application, and dosage regimen are given in Table 16.1.

16.2 MECHANISMS OF DIOXIN-MEDIATED CARCINOGENICITY

16.2.1 DNA Alterations

Findings on the preferential liver carcinogenicity of TCDD in female versus male Sprague Dawley (SD) rats [7] inspired investigations on gender-specific differences in hepatic effects. Notably, a more pronounced pattern of noncarcinogenic alterations in the livers of female rats was found both with TCDD [7] and with Aroclors [8]. In the latter experiments, however, the role of the nondioxin-like (ndl) constituents cannot be ruled out, although recent studies with pure nondioxin-like PCBs are not in favor of a marked hepatotoxicity of this group of compounds either in male or in female rats [16].

Obviously, the marked hepatotoxic effects of TCDD in female SD rats are accompanied by a relevant degree of oxidative stress although the sources of reactive oxygen species (ROS) generation under these conditions are not entirely clear. Wyde et al. [17] found a significant increase in 8-oxo-dG, a hallmark of oxidative DNA damage, in female but not in male SD rats after TCDD treatment. Ovariectomy abolished oxidative DNA damage, whereas supplementation with estradiol reconstituted it in part.

16.2.2 Changes in Expression of Genes Relevant in Carcinogenesis

In the rat uterus, TCDD inhibited the 17β-estradiol-induced and constitutive expression of the cellular protooncogene c-fos [18]. In mice, TCDD treatment resulted in an increase of c-ras expression in the hepatic plasma membranes [19]. In mouse hepatoma cells *in vitro*, TCDD led to increased c-fos and c-jun expression and to a large increase in the transcription factor AP-1 [20], later found to be triggered by AHR-dependent and -independent pathways [21]. Later, Weiss et al. [22] reported that induction of c-jun depends on activation of p38 mitogen-activated protein kinase (MAPK) by an AHR-dependent mechanism.

In the MDA-MB-468 breast cancer cell line, being ER-negative and AHR-positive, TCDD caused a rapid and sustained induction of transforming growth factor alpha (TGFα) gene expression and secreted protein, which was suggested to be responsible for the antimitogenic effect of TCDD in those cells [23]. In 1998, Enan and Matsumura [24] reported that c-src protein kinase is associated with the AHR in various cell types and contributes to the stimulation of tyrosine protein kinase activity seen after TCDD treatment in those cells or tissues. Changes in protein kinase activities were also found to be responsible for TCDD-dependent increased AP-1 and decreased c-Myc binding to their respective response element DNAs in guinea pig hepatocytes [25]. Reiners et al. [26] found that stable transfection of the Ha-ras oncogene in MCF-10A cells suppressed AHR function. In human choriocarcinoma (BeWo) cells, TCDD induced telomerase activity mediated through AHR signaling and ER-independent c-Myc signaling [27]. Jun NH$_2$-terminal kinase (JNK) was identified as a trigger of epithelial cell plasticity under the influence of AHR activation [28]. In rat liver oval cells in culture, TCDD treatment led to induction of the transcription factor JunD, contributing to deregulation of the cell cycle [29]. In human liver L02 cells, exposure to TCDD increased c-Myc protein levels and the phosphorylation of Akt and GSK3 beta [30].

16.2.3 Inhibition of Apoptosis

Several research groups have studied the effects of TCDD on (mostly chemically induced) apoptosis in different cellular systems. Apoptosis is considered to be essential in maintaining cellular homeostasis, especially in eliminating genetically aberrant cells. Evasion of apoptosis is widely accepted to be an important mechanism in cancer development [31, 32]. Since TCDD is both a carcinogen in long-term carcinogenesis assays and a potent liver tumor promoter in rodents [2], it is tempting to hypothesize that an inhibition of spontaneous and chemically induced apoptosis in the rodent liver is the underlying mechanism in the carcinogenicity of TCDD.

TABLE 16.1 Long-Term Carcinogenicity Studies with AHR Agonists in Animal Models

Animal Model	Treatment	Target Organ	Tumorigenicity (Lowest Effective Daily Dose)	References
Rat, male	TCDD in the diet at 22, 210, 2200 ppt for 2 years	Hard palate, nasal turbinates, tongue	Squamous cell carcinoma (0.1 μg/kg bw)	[7]
Rat, female		Liver, liver, hard palate, nasal turbinates, tongue, lung	Hyperplastic nodules (0.01 μg/kg bw) Hepatocellular carcinoma (0.1 μg/kg bw) Squamous cell carcinoma (0.1 μg/kg bw)	
Rat, male	TCDD orally, by gavage at 0.01, 0.05, 0.5 μg/kg bw per day for 104 weeks	Thyroid, liver	Follicular cell adenoma (dose related) Neoplastic nodules (dose related)	[84]
Rat, female		Thyroid, liver	Follicular cell adenoma (0.5 μg/kg bw) Neoplastic nodules (0.5 μg/kg bw)	
Rat, female	TCDD orally, by gavage at 3, 10, 22, 46, 100 ng/kg bw per day, 5 days a week, for 104 weeks	Liver, lung, oral mucosa, uterus	Cholangiocarcinoma (22 ng/kg bw) Hepatocellular adenoma (100 ng/kg bw) Cystic keratinizing epithelioma (100 ng/kg bw) Gingival squamous cell carcinoma (100 ng/kg bw) Squamous cell carcinoma (46 ng/kg bw)	[85]
Rat, female	TCDD, 2,3,4,7,8-pentachlorodibenzofuran, and PCB 126 at various dose levels by gavage, 5 days per week for up to 105 weeks.	Liver, lung	Hepatocellular adenoma (various dose combinations) Cholangiocarcinoma (various dose combinations) Cystic keratinizing epithelioma (various dose combinations)	[86]
Mouse, male	TCDD 0.007, 0.7, 7.0 μg/kg bw for 1 year, orally once a week; observed over lifetime	Liver	Hepatocellular adenoma and carcinoma (0.7 μg/kg bw)	[87]
Mouse, male	TCDD 0.01, 0.05, 0.5 μg/kg bw for 104 weeks, orally twice a week	Liver	Hepatocellular carcinoma (dose related)	[85]

TABLE 16.1 (*Continued*)

Animal Model	Treatment	Target Organ	Tumorigenicity (Lowest Effective Daily Dose)	References
Mouse, female	TCDD 0.04, 0.2, 2.0 µg/kg bw for 104 weeks, orally twice a week	Liver, thyroid hematopoietic system, skin	Hepatocellular carcinoma (dose related) Follicle cell adenoma (dose related) Lymphoma (dose related) Subcutaneous fibrosarcoma (dose related)	[85]
Mouse, female	TCDD 0.005 µg per animal, skin three times per week for 104 weeks	Integumentary system	Fibrosarcoma (0.001 µg per animal)	[88]
Mouse, male	TCDD 2.5, 5.0 µg/kg bw for 52 weeks, orally once a week, followed until 104 weeks	Liver	Hepatocellular carcinoma (2.5 µg/kg b.w.)	[89]
Mouse, female	TCDD 2.5, 5.0 mg/kg bw for 52 weeks, orally once a week, followed until 104 weeks	Liver	Hepatocellular carcinoma (2.5 µg/kg b.w.)	[89]
Mouse, male	TCDD 1, 30, 60 mg/kg bw i.p. once a week for 5 weeks; observed until 78 weeks of age	Thymus, liver	Lymphoma (dose related) Hepatocellular adenoma and carcinoma (dose related)	[89]
Mouse, female	TCDD 1, 30, 60 µg/kg bw i.p. once a week for 5 weeks; observed until 78 weeks of age	Thymus, liver	Lymphoma (dose related) Hepatocellular adenoma and carcinoma (dose related)	[89]
Hamster, male	TCDD 50, 100 µg/kg bw i.p. or s.c. six times at 4-week intervals; observed for 1 year	Skin	Squamous cell carcinoma (100 µg/kg bw)	[14]

In fact, Stinchcombe et al. [33] found a suppression of apoptosis in preneoplastic GSTP-positive foci following initiation with DEN and tumor promotion with TCDD. The occurrence of apoptotic bodies was elevated in preneoplastic foci compared to normal tissue in initiated control animals. In contrast to proliferation, apoptosis was reduced almost to background levels in initiated animals treated with TCDD. These data for the first time suggested that tumor promotion by TCDD might be mediated by an inhibition of cell death rather than by the induction of cell proliferation [33]. The inhibitory effect of TCDD on DEN-induced apoptosis *in vivo* was confirmed in both female H/W and L-E rats by terminal deoxynucleotidyl transferase-mediated X-dUTP nick-end labeling (TUNEL) [34].

Later, it was shown in several studies that TCDD inhibited apoptosis in primary hepatocytes isolated from male Wistar rats. TCDD attenuated both UVC light- and 2-acetylaminofluorene (2-AAF)-induced apoptosis in this *in vitro* test system [35–37]. The antiapoptotic effect of TCDD in these cells was shown to be dependent on both AHR activation and protein biosynthesis [37, 38]. Further studies confirmed the inhibiting effects of TCDD on apoptosis in continuous human hepatoma-derived cell lines *in vitro* [37, 39].

AHR knockout mice were reported to exhibit abnormally small livers compared to wild-type animals. In one of the three independently generated AHR-deficient mouse lines, this is accompanied by an enhanced expression of TGFβ and an increase in the rate of liver cell apoptosis, indicating a role for the AHR in normal liver development that might be based on the regulation of apoptosis [40]. This and the fact that TCDD also causes liver malignancies in mice [2] might indicate a role for the AHR in apoptosis regulation in mice. It is unclear whether the fact that the apoptosis-inhibiting effect of TCDD could not be observed in liver cells derived from mice [41, 42] is due to experimental reasons or to species differences between mice and rats or humans.

Which mechanisms mediate apoptosis inhibition by TCDD and related substances? In 1996, Wörner and Schrenk observed a reduction in the amount of immunoprecipitated p53 following high doses of UVC irradiation in primary hepatocytes from male Wistar rats treated with TCDD *in vitro* [35]. A later study found the p53-modulating effect of TCDD being accompanied by a concentration-dependent hyperphosphorylation of p53. The phosphorylation site was not assessed in this study, but it was shown that c-src was involved in the hyperphosphorylation of p53 by TCDD. Removal of this kinase from liver cell homogenates abolished p53 phosphorylation by TCDD [36]. C-src can associate with the AHR [24] and was proposed to influence p53 signaling [43], although the exact mechanism of crosstalk between these two proteins is unclear. No inhibitory effect of TCDD on the expression of known p53 target genes following low-dose UVC irradiation (10 and 50 J/m^2) (e.g., *apaf-1*, *bax*) could be observed in primary rat hepatocytes despite a clear antiapoptotic effect of TCDD on nuclear apoptosis at these UV doses [37].

In HepG2 human hepatoma cells, TCDD caused a reduction in both total p53 and p53 phosphorylated at ser15 following etoposide treatment. The observed effects were abolished by transfection of the cells with siRNA against *anterior gradient 2* [39]. Anterior gradient 2 counteracts p53 activation by reducing its phosphorylation and silencing its transactivating activity without affecting cell cycle control [44].

The only *in vivo* study addressing the effects of TCDD on p53 characteristics was conducted by Pääjärvi and collaborators. In this study, female rats were treated with DEN to induce liver cell apoptosis. TCDD treatment inhibited the rate of TUNEL-positive cells and this effect was accompanied by a decrease in both total p53 and p53 phosphorylated at ser15. The suppressive effect of TCDD on DEN-induced p53 was proposed to be mediated by the induction of Mdm2 ubiquitin ligase, which is known to propagate p53 degradation [34]. In a later study, the same group tested several PCBs in HepG2 cells and could show that 6 out of 20 tested nondioxin-like PCBs caused ser166 Mdm2 phosphorylation and tyr204 ERK phosphorylation and attenuated the p53 response following etoposide and leptomycin B treatment. The same was seen with TCDD and the dioxin-like PCB 126, indicating that several polyhalogenated aromatic hydrocarbons inhibit p53 characteristics following genotoxic treatment in these cells [45].

TCDD is able to inhibit chemically or UV irradiation-induced apoptosis in liver cells derived from different species both *in vivo* and *in vitro*. An inhibition of apoptosis in the liver of animals following genotoxic treatment by TCDD could be the underlying mechanism of its potent liver tumor-promoting potential. The inhibition of spontaneous apoptosis of genetically aberrant liver cells might furthermore be the reason for TCDD causing liver tumors in long-term carcinogenicity studies in the absence of genotoxic tumor initiators. Although an apoptosis-inhibiting effect could be shown in different human hepatoma cell lines (HepG2 and Huh-7), there are no studies testing whether this also holds true for normal human liver cells (e.g., primary human hepatocytes). Furthermore, there is no convincing evidence from epidemiological studies indicating that TCDD exposure increases the risk for liver cancer in humans [3, 46, 47].

TCDD inhibits not only apoptosis in liver cell systems, but also epidermal growth factor (EGF) withdrawal-induced apoptosis in normal human mammary epithelial cells MCF-10A, as shown by Davis et al. [48]. Papers published by the group of Burchiel proposed that TCDD counteracts EGF withdrawal-induced apoptosis in MCF-10A cells by inducing TGFα that causes Akt and ERK phosphorylation. First, the authors could show that EGF receptor activation is a prerequisite in this mechanism. Addition of EGF or TGFα to cells incubated in growth factor-deficient medium restored viability and inhibited apoptosis, and employing an EGFR inhibitor abolished apoptosis inhibition by TCDD [49, 50]. Inhibition of apoptosis by TCDD in MCF-10A cells could also be observed following apoptosis induction by a wide variety of different apoptogenic stimuli [51].

Park and Matsumura employed a number of chemical inhibitors to characterize the inhibitory action of TCDD on UVC-induced apoptosis in MCF-10A cells. Treatment of the cells with a TGFα antibody, inhibitors of c-Src kinase, and inhibitors of ERK reversed the inhibition of apoptosis by TCDD [51]. Since TGFα and LPS exhibited the same antiapoptotic effect as TCDD, the authors concluded that TCDD generally acts in an antiapoptotic manner as a kind of general stress response in these cells.

A possible mechanism attenuating the p53 response following cellular stress by TCDD was elucidated in the human breast adenocarcinoma cell line MCF-7; that is, TCDD caused a decrease in the expression of p53 and its target gene Dusp5 following incubation under mild hypoxic conditions. This was proposed to be mediated by an induction of the human Mdm2 homolog Hdm2. Downregulation of Dusp5 by TCDD or siRNA transfection resulted in enhanced phosphorylation of ERK [52].

An inhibition of physiological mammary gland apoptosis by TCDD could have implications in the development of breast cancer. In animal studies, however, TCDD administration did not cause mammary gland malignancies [2]. On the other hand, epidemiological data in humans revealed a slightly increased risk for breast cancer [46, 47].

Another interesting study looking at the effects of TCDD on apoptosis induced by genotoxic damage was conducted in the mouse myoblastoma cell line C2C12. Here, apoptosis induced by staurosporine was attenuated by TCDD as assessed by TUNEL staining. This effect was reversed by cotreatment with FK506, an inhibitor of calcineurin. The authors of this study concluded that by directly targeting mitochondrial transcription and the induction of mitochondrial stress signaling, TCDD activated a calcineurin A-sensitive pathway leading to NF-κB activity. These effects were stated to be AHR independent, since C2C12 cells do not express AHR. Furthermore, employing the AHR antagonist α-naphthoflavone or ARNT siRNA did not reverse the effects by TCDD on the above-described pathway [53]. This is a rather surprising result, since all other studies done to elucidate apoptosis inhibition by TCDD identified AHR activation as an essential step. Nevertheless, it is highly interesting that TCDD is able to directly affect mitochondria. The underlying mode of action is not known.

Except for the study by Biswas et al. (see above), all other authors found direct or indirect evidence for AHR activation being a prerequisite for the inhibition of apoptosis by TCDD. The effects could be reversed by AHR siRNA knockdown [39], AHR knockout [34], or chemically antagonizing the AHR [34, 37, 38, 50, 51]. Chopra et al. found protein translation to be needed for the inhibition of UVC-induced apoptosis in primary rat hepatocytes by TCDD. This would implicate that classic genomic AHR signaling is sufficient to mediate an inhibition of apoptosis [38]. Albeit in recent years nongenomic AHR signaling was discussed [54], the induction of AHR target genes appears to be a vital step in counteracting apoptosis. This is further emphasized by the fact that in several studies the induction of certain AHR target genes was shown to be crucial for apoptosis inhibition: *anterior gradient 2* [39] or *tgf-α* [49]. Induction of other genes also played a role in the p53-modulating effect of TCDD in some studies [34, 53].

A number of studies analyzed the influence of cellular AHR status on apoptosis in different test systems in the absence of exogenous inducers. AHR knockout mice exhibit abnormally small livers and an enhanced rate of hepatic apoptosis [40]. The same holds true for embryo fibroblasts from these animals that showed delayed growth characteristics and augmented apoptosis. It was suggested that the AHR in the absence of exogenous ligands controls the cell cycle via a pathway involving TGFβ and the mitotic kinases Cdc2 and Plk [55]. Interestingly, both *cdc2* and *plk* expressions are negatively regulated by p53 [56, 57]. Furthermore, this implies that an endogenous role of the AHR is to govern p53 characteristics and, thereby, might function in regulating cell cycle control and apoptosis.

In a study from the group of Puga, embryonic fibroblasts from AHR knockout mice carrying a TET-OFF-regulated vector expressing the AHR were used to investigate the mechanisms that lead to apoptosis induction by the loss of AHR. The downregulation of AHR expression was associated with the induction of bax- and caspase-dependent apoptosis. Furthermore, treatment of AHR-expressing cells with TCDD protected them from etoposide-induced cytotoxicity, whereas in the absence of the receptor, this was not the case. Transfection of these cells with siRNA downregulating the proapoptotic transcription factor E2F1 inhibited apoptosis induction by AHR loss. In cells expressing both AHR and E2F1, a direct interaction of these proteins could be shown. It was proposed that the AHR is able to inhibit the proapoptotic function of E2F1 by direct interference and antagonizes E2F1 functions such as the expression of its target genes apaf1 and p73 [58]. E2F1 is also a regulator of p53 function [59–61], and it would be highly interesting to further analyze the impact of this protein on apoptosis inhibition by TCDD-activated AHR *in vivo*.

Wong et al. generated MCF-10AT1 cells that overexpressed AHR after incubating them with estradiol for 20 passages. These cells showed enhanced proliferation in comparison to normal cells, again indicating a role for the AHR in cell homeostasis. Treatment of these cells with 3-methoxy-nitroflavone, an AHR antagonist, inhibited proliferation. Next, the effects of AHR status on apoptosis were assessed. It could be shown that estradiol-selected cells were more resistant to apoptosis induced by UVC irradiation, H_2O_2, staurosporine, or TGF-β1 than control cells. Antagonizing the AHR reversed the protective effect [62]. All these effects were observed in the absence of exogenous AHR ligands. It is not clear whether they were mediated by unknown endogenous ligands or ligand-independent mechanisms. Wong and coworkers also tested whether the change in AHR status affected inhibition of UVC-induced apoptosis by TCDD, finding that TCDD inhibited UVC-induced apoptosis in normal cells only; that is, it had no further suppressive effect in AHR-overexpressing cells [62]. No further experiments were conducted in this study to elucidate the mechanism by which the AHR inhibited apoptosis in this system.

Resistance toward apoptosis is a well-accepted hallmark of cancer [32]. Thus, it appears quite possible that the inhibition of apoptosis, spontaneous or chemically induced, by TCDD in different target tissues might play a role in the tumorigenicity of this compound. So far, several signaling pathways have been proposed to mediate these effects. Generally it can be said that the effects are mediated by the AHR and p53 appears to be a key regulator. Furthermore, Akt and ERK were shown to transduce the antiapoptotic effects. But still, the exact mechanism of how an activation of the AHR inhibits apoptosis is unclear.

16.2.4 Suppression of Intercellular Communication

Effects of AHR ligands on intercellular communication (IC) have been investigated mainly under the aspect of tumor promotion. First studies in the C3H/10T1/2 mouse fibroblast cell line [63] and in the V79 Chinese hamster fibroblast cell line [64] with TCDD were negative. However, in hepatoma cell lines such as Hepa 1c1c7, AHR-related suppression of IC could be found with TCDD and DL-PCBs [65]. Coincubation of rat hepatocytes with TCDD and α-naphthoflavone abolished downregulation of gap junctional IC (GJIC) by TCDD. Similarly, cotreatment with a cAMP analog prevented downregulation of GJIC by TCDD. These results suggest that this effect is mediated, at least in part, through the AHR, possibly due to changes in levels of the connexin 32 gap junction mRNA [66]. Similar results were reported with TCDD and the AHR agonist indolo[3,2-b]carbazole (ICZ) in the rat liver epithelial cell line WB-F344 [67]. In MCF-7 cells, Gakhar et al. [68] found that TCDD decreases GJIC by phosphorylating connexin 43 (Cx43) via PKCα signaling and affects the localization of Cx43.

16.2.5 Stimulation of Proliferation of Preneoplastic Cells/Stem Cells

Lucier et al. [69] reported an increased proliferation in DEN-initiated/TCDD-promoted female rat liver that was diminished after ovariectomy. Apparently, preneoplastic hepatocytes mostly contributed to this effect. In preneoplastic rat liver, Buchmann et al. [70] found no increase in proliferation of normal hepatocytes and a slight increase in preneoplastic foci after treatment of the animals with TCDD or heptachlorodibenzo-p-dioxin. Bauman et al. [71] found that TCDD caused an overall inhibition of hepatocyte proliferation after partial hepatectomy in female rats. A periportal pattern of cell proliferation was observed in the TCDD-partial hepatectomy group compared to the panlobular pattern of cell proliferation in the control-partial hepatectomy group. A mitoinhibitory action of TCDD was also found in hepatocytes in culture [72], while Schrenk et al. [73] had reported a comitogenic effect of TCDD and EGF on rat hepatocyte DNA synthesis *in vitro*.

16.2.6 "Indirect" Genotoxicity Mediated by AHR Activation

The liver carcinogen TCDD leads to the development of hepatoma and hepatocellular carcinoma after chronic application to rats. Besides being a potent liver tumor promoter in rodents [10], TCDD is also carcinogenic without pretreatment with a genotoxic carcinogen [7]. Since TCDD is not genotoxic in classical test batteries *in vitro* or *in vivo*, it has been speculated that alternative mechanisms including epigenetic alterations and secondary oxidative DNA damage may play a role in the liver carcinogenicity of TCDD. This hypothesis was supported by several findings. First, TCDD is a potent liver carcinogen in female Sprague Dawley rats but to a much lower extent in males [7]. Second, chronic (30 weeks) TCDD treatment was reported to lead to oxidative DNA damage in rat liver, mainly in females. After ovariectomy, the levels of 8-oxo-guanin, a hallmark of oxidative DNA damage, were much lower, that is, in the range found in males. Substitution with estradiol (E_2) restored the original effect in ovariectomized females [17]. Since the source of ROS generated upon TCDD treatment in intact female rats is unknown, several possibilities have been discussed [2]. One is induction of cytochromes P450 (CYPs) by TCDD treatment via activation of the AHR. The massive amount of certain induced CYPs in the hepatocellular endoplasmic reticulum may thus lead to a relevant leakage of ROS into the nucleus. Another option is the metabolic activation of E_2, leading to reactive intermediates. It has been described that CYPs catalyze the formation of 2-OH and 4-OH-E_2, the latter being partially transformed into an electrophilic quinone [74]. In fact, incubations of MCF-7 cells with 3,4-estrone quinone (3,4-EQ) resulted in the formation of covalently bound DNA adducts [75].

16.3 AHR AND PRO- VERSUS ANTICARCINOGENICITY

In breast cancer cell lines including a putative mammary cancer stem cell line, AHR ligand activation inhibited invasiveness and anchorage-independent growth. These effects were correlated with the ability of exogenous AHR agonists to promote differentiation [76].

The synthesis of selective AHR modulators, which lack the AHR-mediated toxic responses induced by TCDD, has led to compounds that are capable of inhibiting mammary tumor growth in rodents. The compounds bind the AHR and induce a pattern of AHR–estrogen receptor crosstalk similar to TCDD [77]. The authors suggest that such modulators may be an important new class of drugs for clinical treatment of breast cancer.

In transgenic AHR-deficient mice, given a single i.p injection of DEN, hepatic tumor formation, inflammatory cytokine release, and DNA damage were significantly higher than in wild-type animals [78]. The authors suggest that AHR in its basal state in the absence of a xenobiotic ligand functions as a tumor suppressor gene and that its silencing may be associated with cancer progression.

In transgenic AHR-deficient mouse prostate (TRAMP) mice, prostate tumor incidence was greater than that in wild-type TRAMP mice. The authors suggest that the AHR inhibits prostate carcinogenesis in TRAMP mice by interfering with neuroendocrine differentiation, since AHR-deficient mice also exhibit altered smaller dorsolateral and anterior prostates [79].

In contrast, in human gliomas, TGFβ signaling, positively correlated with constitutive AHR activity, is considered as an important mediator of the malignant phenotype. In human gliomas and glioblastomas, expression and nuclear staining of the AHR were found. In malignant glioma cells, treatment with the AHR antagonist CH-223191 reduced clonogenic survival and invasiveness [80].

In mice, expression of a constitutively active AHR (CA-AHR transgenic mice) induced tumors in the glandular part of the stomach [81]. Subsequently, it was found that osteopontin was negatively regulated by the AHR and that downregulation of its expression correlated with development of stomach tumors in CA-AHR transgenic mice [82]. Moennikes et al. [83] found significantly more liver tumors in male CA-AHR transgenic mice treated with a single dose of DEN at 6 weeks of age than in AHR wild-type mice. A microarray-based gene expression profiling analysis revealed downregulation in the liver of CA-AHR transgenic mice of a cluster of genes encoding heat shock proteins, including GRP78/BiP, Herp1, Hsp90, DnaJ (Hsp40) homolog B1, and Hsp105.

16.4 FINDINGS IN HUMANS

TCDD was evaluated as carcinogenic to humans (group 1 carcinogen) by the International Agency for Research on Cancer in 1997 [3]. This classification was based on the following supporting evidence:

- TCDD is a multisite carcinogen in experimental animals that has been shown by several lines of evidence to act through a mechanism involving the AHR
- AHR receptor is highly conserved in an evolutionary sense and functions the same way in humans as in experimental animals
- Tissue concentrations are similar both in heavily exposed human populations in which an overall increased cancer risk was observed and in rats exposed to carcinogenic regimens in bioassays

Other PCDD/F are not classifiable as to their carcinogenicity to humans (group 3) due to inadequate data.

In the IARC 1997 evaluation, the strongest evidence for the carcinogenicity of TCDD from epidemiological studies of high-exposure cohorts was for all cancers combined, rather than for any specific site. The relative risk for all cancers combined in the most highly exposed and longer latency subcohorts was 1.4. More recent data showed this relative risk to be in the range of 1.4–2.0. It should be borne in mind that the general population is exposed to levels far below those experienced by these cohorts [3].

The most recent data on the Seveso cohort did not reveal an increase in the cancer risk. An excess of lymphatic and hematopoietic tissue neoplasms was observed, however. An increased risk of breast cancer was detected in females after 15 years since the accident. Interestingly, no cancer cases were observed among subjects diagnosed with chloracne early after the accident [47].

16.5 CONCLUSIONS

TCDD exerts a variety of toxic effects that are believed to be mostly, if not exclusively, mediated by activation of the AHR. Chronic treatment of rodents with TCDD can cause liver cancer. Although oxidative DNA damage was shown to occur in female TCDD-treated rats in an estrogen-dependent manner, conventional *in vivo* and *in vitro* genotoxicity tests with TCDD are negative. Among other toxicities, TCDD is the most potent liver and skin tumor promoter in rodent models. An inhibition of hepatocyte apoptosis was proposed to be a causative event in liver tumor promotion by TCDD, allowing genetically aberrant cells to evade apoptosis execution. Toward this end, modulation of p53 characteristics has been shown to be a consistent effect of TCDD in genotoxin-challenged liver cells. The exact mechanisms of suppression of apoptosis, however, are unknown. In addition to suppression of apoptosis, AHR agonists were reported to inhibit gap junctional intercellular communication, especially in liver-derived cells. Furthermore, depending on the cell type, TCDD induces the expression or levels of certain proto-oncogenes, most notably of c-jun and c-fos, and can activate c-src, which was found to be associated with the receptor. In contrast to liver cells, activated AHR, especially by certain ligands, is able to prevent proliferation of estrogen-dependent mammary tumor cells. In the absence of exogenous ligand, the AHR was shown to inhibit tumor development and growth in certain cell types and tissues, most notably the mammary gland and the prostate.

REFERENCES

1. Holder, J. W. and Menzel, H. M. (1989). Analysis of 2,3,7,8-TCDD tumor promotion activity and its relationship to cancer. *Chemosphere, 19*, 861–868.
2. Knerr, S. and Schrenk, D. (2006). Carcinogenicity of 2,3,7,8-tetrachlorodibenzo-*p*-dioxin in experimental models. *Molecular Nutrition and Food Research, 50*, 897–907.
3. International Agency for Research on Cancer (1997). *Monographs on the Evaluation of Carcinogenic Risks to Humans.* Vol 69: Polychlorinated Dibenzo-para-Dioxins and Polychlorinated Dibenzofurans. International Agency for Research on Cancer, Lyon.
4. McGregor, D. B., Partensky, C., Wilbourn, J., and Rice, J. M. (1998). An IARC evaluation of polychlorinated dibenzo-*p*-dioxins and polychlorinated dibenzofurans as risk factors in human carcinogenesis. *Environmental Health Perspectives, 106* (Suppl 2), 755–760.
5. Hattis, D., Chu, M., Rahmioglu, N., Goble, R., Verma, P., Hartman, K., and Kozlak, M. (2009). A preliminary operational classification system for nonmutagenic modes of action for carcinogens. *Critical Reviews in Toxicology, 39*, 97–138.
6. Simon, T., Aylward, L. L., Kirman, C. R., Rowlands, J. C., and Budinsky, R. A. (2009). Estimates of cancer potency of 2,3,7,8-tetrachlorodibenzo(p)dioxin using linear and nonlinear dose–response modeling and toxicokinetics. *Toxicological Sciences, 112*, 490–506.
7. Kociba, R. J., Keyes, D. G., Beyer, J. E., Carreon, R. M., Wade, C. E., Dittenber, D. A., Kalnins, R. P., Frauson, L. E., Park, C. N., Barnard, S. D., Hummel, R. A., and Humiston, C. G. (1978). Results of a two-year chronic toxicity and oncogenicity study of 2,3,7,8-tetrachlorodibenzo-*p*-dioxin in rats. *Toxicology and Applied Pharmacology, 46*, 279–303.
8. Mayes, B. A., McConnell, E. E., Neal, B. H., Brunner, M. J., Hamilton, S. B., Sullivan, T. M., Peters, A. C., Ryan, M. J., Toft, J. D., Singer, A. W., Brown, J. F. Jr., Menton, R. G., and Moore, J. A. (1998). Comparative carcinogenicity in Sprague-Dawley rats of the polychlorinated biphenyl mixtures Aroclors 1016, 1242, 1254, and 1260. *Toxicological Sciences, 41*, 62–76.
9. EFSA (2005). Opinion of the Scientific Panel on contaminants in the food chain [CONTAM] related to the presence of non dioxin-like polychlorinated biphenyls (PCB) in feed and food. *EFSA Journal, 284*, 1–137.
10. Pitot, H. C., Goldsworthy, T., Campbell, H. A., and Poland, A. (1980). Quantitative evaluation of the promotion by 2,3,7,8-tetrachlorodibenzo-*p*-dioxin of hepatocarcinogenesis from diethylnitrosamine. *Cancer Research, 40*, 3616–3620.
11. Schrenk, D., Buchmann, A., Dietz, K., Lipp, H. P., Brunner, H., Sirma, H., Münzel, P., Hagenmaier, H., Gebhardt, R., and Bock, K. W. (1994). Promotion of preneoplastic foci in rat liver with 2,3,7,8-tetrachlorodibenzo-*p*-dioxin, 1, 2,3,4,6,7,8-heptachlorodibenzo-*p*-dioxin and a defined mixture of 49 polychlorinated dibenzo-*p*-dioxins. *Carcinogenesis, 15*, 509–515.
12. Viluksela, M., Bager, Y., Tuomisto, J. T., Scheu, G., Unkila, M., Pohjanvirta, R., Flodström, S., Kosma, V. M., Mäki-Paakkanen, J., Vartiainen, T., Klimm, C., Schramm, K. W., Wärngård, L., and Tuomisto, J. (2000). Liver tumor-promoting activity of 2,3,7,8-tetrachlorodibenzo-*p*-dioxin (TCDD) in TCDD-sensitive and TCDD-resistant rat strains. *Cancer Research, 60*, 6911–6920.
13. Poland, A., Palen, D., and Glover, E. (1982). Tumour promotion by TCDD in skin of HRS/J hairless mice. *Nature, 300*, 271–273.
14. Rao, M. S., Subbarao, V., Prasad, J. D., and Scarpelli, D. G. (1988). Carcinogenicity of 2,3,7,8-tetrachlorodibenzo-*p*-dioxin in the Syrian golden hamster. *Carcinogenesis, 9*, 1677–1679.
15. Hébert, C. D., Harris, M. W., Elwell, M. R., and Birnbaum, L. S. (1990). Relative toxicity and tumor-promoting ability of 2,3,7,8-tetrachlorodibenzo-*p*-dioxin (TCDD), 2, 3,4,7,8-pentachlorodibenzofuran (PCDF), and 1,2,3,4,7,8-hexachlorodibenzofuran (HCDF) in hairless mice. *Toxicology and Applied Pharmacology, 102*, 362–377.
16. NTP (2006). NTP toxicology and carcinogenesis studies of a binary mixture of 3,3′,4,4′,5-pentachlorobiphenyl (PCB 126) (CAS No. 57465-28-8) and 2,2′,4,4′,5,5′-hexachlorobiphenyl (PCB 153) (CAS No. 35065-27-1) in female Harlan Sprague-Dawley rats (gavage studies). *National Toxicology Program Technical Report Series, 530*, 1–258.
17. Wyde, M. E., Wong, V. A., Kim, A. H., Lucier, G. W., and Walker, N. J. (2001). Induction of hepatic 8-oxo-deoxyguanosine adducts by 2,3,7,8-tetrachlorodibenzo-*p*-dioxin in Sprague-Dawley rats is female-specific and estrogen-dependent. *Chemical Research in Toxicology, 14*, 849–855.
18. Astroff, B., Eldridge, B., and Safe, S. (1991). Inhibition of the 17β-estradiol-induced and constitutive expression of the cellular protooncogene c-fos by 2,3,7,8-tetrachlorodibenzo-*p*-dioxin (TCDD) in the female rat uterus. *Toxicology Letters, 56*, 305–315.
19. Tullis, K., Olsen, H., Bombick, D. W., Matsumura, F., and Jankun, J. (1992). TCDD causes stimulation of c-ras expression in the hepatic plasma membranes *in vivo* and *in vitro*. *Journal of Biochemical Toxicology, 7*, 107–116.
20. Puga, A., Nebert, D. W., and Carrier, F. (1992). Dioxin induces expression of c-fos and c-jun proto-oncogenes and a large increase in transcription factor AP-1. *DNA and Cell Biology, 11*, 269–281.
21. Hoffer, A., Chang, C. Y., and Puga, A. (1996). Dioxin induces transcription of fos and jun genes by Ah receptor-dependent and -independent pathways. *Toxicology and Applied Pharmacology, 141*, 238–247.
22. Weiss, C., Faust, D., Dürk, H., Kolluri, S. K., Pelzer, A., Schneider, S., Dietrich, C., Oesch, F., and Göttlicher, M. (2005). TCDD induces c-jun expression via a novel Ah (dioxin) receptor-mediated p38-MAPK-dependent pathway. *Oncogene, 24*, 4975–4983.

23. Wang, W. L., Porter, W., Burghardt, R., and Safe, S. H. (1997). Mechanism of inhibition of MDA-MB-468 breast cancer cell growth by 2,3,7,8-tetrachlorodibenzo-*p*-dioxin. *Carcinogenesis*, *18*, 925–933.

24. Enan, E. and Matsumura, F. (1996). Identification of c-Src as the integral component of the cytosolic Ah receptor complex, transducing the signal of 2,3,7,8-tetrachlorodibenzo-*p*-dioxin (TCDD) through the protein phosphorylation pathway. *Biochemical Pharmacology*, *52*, 599–612.

25. Ashida, H., Nagy, S., and Matsumura, F. (2000). 2,3,7,8-Tetrachlorodibenzo-*p*-dioxin (TCDD)-induced changes in activities of nuclear protein kinases and phosphatases affecting DNA binding activity of c-Myc and AP-1 in the livers of guinea pigs. *Biochemical Pharmacology*, *59*, 741–751.

26. Reiners, J. J., Jr., Jones, C. L., Hong, N., Clift, R. E., and Elferink, C. (1997). Downregulation of aryl hydrocarbon receptor function and cytochrome P450 1A1 induction by expression of Ha-ras oncogenes. *Molecular Carcinogenesis*, *19*, 91–100.

27. Sarkar, P., Shiizaki, K., Yonemoto, J., and Sone, H. (2006). Activation of telomerase in BeWo cells by estrogen and 2,3,7,8-tetrachlorodibenzo-*p*-dioxin in co-operation with c-Myc. *International Journal of Oncology*, *28*, 43–51.

28. Diry, M., Tomkiewicz, C., Koehle, C., Coumoul, X., Bock, K. W., Barouki, R., and Transy, C. (2006). Activation of the dioxin/aryl hydrocarbon receptor (AhR) modulates cell plasticity through a JNK-dependent mechanism. *Oncogene*, *25*, 5570–5574.

29. Weiss, C., Faust, D., Schreck, I., Ruff, A., Farwerck, T., Melenberg, A., Schneider, S., Oesch-Bartlomowicz, B., Zatloukalová, J., Vondráček, J., Oesch, F., and Dietrich, C. (2008). TCDD deregulates contact inhibition in rat liver oval cells via Ah receptor, JunD and cyclin A. *Oncogene*, *27*, 2198–2207.

30. An., J., Yang, D. Y, Xu, Q. Z., Zhang, S. M., Huo, Y. Y., Shang, Z. F., Wang, Y., Wu, D. C., and Zhou, P. K. (2008). DNA-dependent protein kinase catalytic subunit modulates the stability of c-Myc oncoprotein. *Molecular Cancer*, *7*, 32.

31. Wyllie, A. H. (1997). Apoptosis and carcinogenesis. *European Journal of Cell Biology*, *73*, 189–197.

32. Hanahan, D. and Weinberg, R. A. (2000). The hallmarks of cancer. *Cell*, *100*, 57–70.

33. Stinchcombe, S., Buchmann, A., Bock, K. W., and Schwarz, M. (1995). Inhibition of apoptosis during 2,3,7,8-tetrachlorodibenzo-*p*-dioxin-mediated tumour promotion in rat liver. *Carcinogenesis*, *16*, 1271–1275.

34. Pääjärvi, G., Viluksela, M., Pohjanvirta, R., Stenius, U., and Högberg, J. (2005). TCDD activates mdm2 and attenuates the p53 response to DNA damaging agents. *Carcinogenesis*, *26*, 201–208.

35. Wörner, W. and Schrenk, D. (1996). Influence of liver tumor promoters on apoptosis in rat hepatocytes induced by 2-acetylaminofluorene, ultraviolet light, or transforming growth factor β1. *Cancer Research*, *56*, 1272–1278.

36. Wörner, W. and Schrenk, D. (1998). 2,3,7,8-Tetrachlorodibenzo-*p*-dioxin suppresses apoptosis and leads to hyperphosphorylation of p53 in rat hepatocytes. *Environmental Toxicology and Pharmacology*, *6*, 239–247.

37. Chopra, M., Dharmarajan, A. M., Meiss, G., and Schrenk, D. (2009). Inhibition of UV-C light-induced apoptosis in liver cells by 2,3,7,8-tetrachlorodibenzo-*p*-dioxin. *Toxicological Sciences*, *111*, 49–63.

38. Chopra, M., Gährs, M., Haben, M., Michels, C., and Schrenk, D. (2010). Inhibition of apoptosis by 2,3,7,8-tetrachlorodibenzo-*p*-dioxin depends on protein biosynthesis. *Cell Biology and Toxicology*, *26*, 391–401.

39. Ambolet-Camoit, A., Bui, L. C., Pierre, S., Chevallier, A., Marchand, A., Coumoul, X., Garlatti, M., Andreau, K., Barouki, R., and Aggerbeck, M. (2010). 2,3,7,8-Tetrachlorodibenzo-*p*-dioxin counteracts the p53 response to a genotoxicant by up-regulating expression of the metastasis marker AGR2 in the hepatocarcinoma cell line HepG2. *Toxicological Sciences*, *115*, 501–512.

40. Gonzalez, F. J. and Fernandez-Salguero, P. (1998). The aryl hydrocarbon receptor: studies using AHR-null mice. *Drug Metabolism and Disposition*, *26*, 1194–1198.

41. Christensen, J. G., Gonzales, A. J., Cattley, R. C., and Goldsworthy, T. L. (1998). Regulation of apoptosis in mouse hepatocytes and alteration of apoptosis by nongenotoxic carcinogens. *Cell Growth and Differentiation*, *9*, 815–825.

42. Schreck, I., Chudziak, D., Schneider, S., Seidel, A., Platt, K. L., Oesch, F., and Weiss, C. (2009). Influence of aryl hydrocarbon-(Ah) receptor and genotoxins on DNA repair expression and cell survival of mouse hepatoma cells. *Toxicology*, *259*, 91–96.

43. Ping-Yuan, L., Hung-Jen, L., Meng-Jiun, L., Feng-Ling, Y., Hsue-Yin, H., Jeng-Woei, L., and Wen-Ling, S. (2006). Avian reovirus activates a novel proapoptotic signal linking src to p53. *Apoptosis*, *11*, 2179–2193.

44. Pohler, E., Craig, A. L., Cotton, J., Lawrie, L., Dillon, J. F., Ross, P., Kernohan, N., and Hupp, T. R. (2004). The Barrett's antigen anterior gradient-2 silences the p53 transcriptional response to DNA damage. *Molecular and Cellular Proteomics*, *3*, 534–547.

45. Al-Anati, L., Högberg, J., and Stenius, U. (2009). Non-dioxin-like-PCB phosphorylate mdm2 at ser166 and attenuate the p53 response in HepG2 cells. *Chemico-Biological Interactions*, *182*, 191–198.

46. Kogevinas, M., Becher, H., Benn, T., Bertazzi, P. A., Boffetta, P., Bueno-de-Mesquita, H. B., Coggon, D., Colin, D., Flesch-Janys, D., Fingerhut, M., Green, L., Kauppinen, T., Littorin, M., Lynge, E., Mathews, J. D., Neuberger, M., Pearce, N., and Saracci, R. (1997). Cancer mortality in workers exposed to phenoxy herbicides, chlorophenols, and dioxins. *American Journal of Epidemiology*, *145*, 1061–1075.

47. Pesatori, A. C., Consonni, D., Rubagotti, M., Grillo, P., and Bertazzi, P. A. (2009). Cancer incidence in the population exposed to dioxin after the "Seveso accident": twenty years of follow-up. *Environmental Health*, *8*, 39.

48. Davis, J. W., II, Melendez, K., Salas, V. M., Lauer, F. T., and Burchiel, S. W. (2000). 2,3,7,8-Tetrachlorodibenzo-*p*-dioxin (TCDD) inhibits growth factor withdrawal-induced apoptosis in the human mammary epithelial cell line, MCF-10A. *Carcinogenesis*, *21*, 881–886.

49. Davis, J. W., II, Lauer, F. T., Burdick, A. D., Hudson, L. G., and Burchiel, S. W. (2001). Prevention of apoptosis by 2,3,7,8-tetrachlorodibenzo-p-dioxin (TCDD) in the MCF-10A cell line: correlation with increased transforming growth factor α production. *Cancer Research*, *61*, 3314–3320.

50. Davis, J. W., Jr., Burdick, A. D., Lauer, F. T., and Burchiel, S. W. (2003). The aryl hydrocarbon receptor antagonist, 3'methoxy-4'nitroflavone, attenuates 2,3,7,8-tetrachlorodibenzo-p-dioxin-dependent regulation of growth factor signaling and apoptosis in the MCF-10A cell line. *Toxicology and Applied Pharmacology*, *188*, 42–49.

51. Park, S. and Matsumura, F. (2006). Characterization of anti-apoptotic action of TCDD as a defensive cellular stress response reaction against the cell damaging action of ultraviolet irradiation in an immortalized normal human mammary epithelial cell line, MCF10A. *Toxicology*, *217*, 139–146.

52. Seifert, A., Taubert, H., Hombach-Klonisch, S., Fischer, B., and Navarrete Santos, A. (2009). TCDD mediates inhibition of p53 and activation of ERα signaling in MCF-7 cells at moderate hypoxic conditions. *International Journal of Oncology*, *35*, 417–424.

53. Biswas, G., Srinivasan, S., Anandatheerthavarada, H. K., and Avadhani, N. G. (2008). Dioxin-mediated tumor progression through activation of mitochondria-to-nucleus stress signaling. *Proceedings of the National Academy of Sciences of the United States of America*, *105*, 186–191.

54. Matsumura, F. (2009). The significance of the nongenomic pathway in mediating inflammatory signaling of the dioxin-activated Ah receptor to cause toxic effects. *Biochemical Pharmacology*, *77*, 608–626.

55. Elizondo, G., Fernandez-Salguero, P., Sheikh, M. S., Kim, G.-Y., Fornace, A. J., Lee, K. S., and Gonzalez, F. J. (2000). Altered cell cycle control at the G_2/M phases in aryl hydrocarbon receptor-null embryo fibroblast. *Molecular Pharmacology*, *57*, 1056–1063.

56. Yun, J., Chae, H.-D., Choy, H. E., Chung, J., Yoo, H.-S., Han, M.-H., and Shin, D. Y. (1999). p53 negatively regulates cdc2 transcription via the CCAAT-binding NF-Y transcription factor. *Journal of Biological Chemistry*, *274*, 29677–29682.

57. Kho, P. S., Wang, Z., Zhuang, L., Li, Y., Chew, J.-L., Ng, H.-H., Liu, E. T., and Yu, Q. (2004). p53-regulated transcriptional program associated with genotoxic stress-induced apoptosis. *Journal of Biological Chemistry*, *279*, 21183–21192.

58. Marlowe, J. L., Fan, Y., Chang, X., Peng, L., Knudsen, E. S., Xia, Y., and Puga, A. (2008). The aryl hydrocarbon receptor binds to E2F1 and inhibits E2F1-induced apoptosis. *Molecular Biology of the Cell*, *19*, 3263–3271.

59. Powers, J. T., Hong, S., Mayhew, C. N., Rogers, P. M., Knudsen, E. S., and Johnson, D. G. (2004). E2F1 uses the ATM signaling pathway to induce p53 and Chk2 phosphorylation and apoptosis. *Molecular Cancer Research*, *2*, 203–214.

60. Fogal, V., Hsieh, J.-K., Royer, C., Zhong, S., and Lu, X. (2005). Cell cycle-dependent nuclear retention of p53 by E2F1 requires phosphorylation of p53 at ser315. *EMBO Journal*, *24*, 2768–2782.

61. Polager, S. and Ginsberg, D. (2009). p53 and E2f: partners in life and death. *Nature Reviews Cancer*, *9*, 738–748.

62. Wong, P. S., Li, W., Vogel, C. F., and Matsumura, F. (2009). Characterization of MCF mammary epithelial cells overexpressing the arylhydrocarbon receptor (AhR). *BMC Cancer*, *9*, 234.

63. Boreiko, C. J., Abernethy, D. J., Sanchez, J. H., and Dorman, B. H. (1986). Effect of mouse skin tumor promoters upon [^{3}H] uridine exchange and focus formation in cultures of C3H/10T1/2 mouse fibroblasts. *Carcinogenesis*, *7*, 1095–1099.

64. Lincoln, D. W., 2nd, Kampcik, S. J., and Gierthy, J. F. (1987). 2,3,7,8-Tetrachlorodibenzo-p-dioxin (TCDD) does not inhibit intercellular communication in Chinese hamster V79 cells. *Carcinogenesis*, *8*, 1817–1820.

65. De Haan, L. H., Simons, J. W., Bos, A. T., Aarts, J. M., Denison, M. S., and Brouwer, A. (1994). Inhibition of intercellular communication by 2,3,7,8-tetrachlorodibenzo-p-dioxin and dioxin-like PCBs in mouse hepatoma cells (Hepa1c1c7): involvement of the Ah receptor. *Toxicology and Applied Pharmacology*, *129*, 283–293.

66. Baker, T. K., Kwiatkowski, A. P., Madhukar, B. V., and Klaunig, J. E. (1995). Inhibition of gap junctional intercellular communication by 2,3,7,8-tetrachlorodibenzo-p-dioxin (TCDD) in rat hepatocytes. *Carcinogenesis*, *16*, 2321–2326.

67. Herrmann, S., Seidelin, M., Bisgaard, H. C., and Vang, O. (2002). Indolo[3,2-b]carbazole inhibits gap junctional intercellular communication in rat primary hepatocytes and acts as a potential tumor promoter. *Carcinogenesis*, *23*, 1861–1868.

68. Gakhar, G., Schrempp, D., and Nguyen, T. A. (2009). Regulation of gap junctional intercellular communication by TCDD in HMEC and MCF-7 breast cancer cells. *Toxicology and Applied Pharmacology*, *235*, 171–181.

69. Lucier, G. W., Tritscher, A., Goldsworthy, T., Foley, J., Clark, G., Goldstein, J., and Maronpot, R. (1991). Ovarian hormones enhance 2,3,7,8-tetrachlorodibenzo-p-dioxin-mediated increases in cell proliferation and preneoplastic foci in a two-stage model for rat hepatocarcinogenesis. *Cancer Research*, *51*, 1391–1397.

70. Buchmann, A., Stinchcombe, S., Körner, W., Hagenmaier, H., and Bock, K. W. (1994). Effects of 2,3,7,8-tetrachloro- and 1,2,3,4,6,7,8-heptachlorodibenzo-p-dioxin on the proliferation of preneoplastic liver cells in the rat. *Carcinogenesis*, *15*, 1143–1150.

71. Bauman, J. W., Goldsworthy, T. L., Dunn, C. S., and Fox, T. R. (1995). Inhibitory effects of 2,3,7,8-tetrachlorodibenzo-p-dioxin on rat hepatocyte proliferation induced by 2/3 partial hepatectomy. *Cell Proliferation*, *28*, 437–451.

72. Bock, K. W., Gschaidmeier, H., Bock-Hennig, B. S., and Eriksson, L. C. (2000). Density-dependent growth of normal and nodular hepatocytes. *Toxicology*, *144*, 51–56.

73. Schrenk, D., Karger, A., Lipp, H. P., and Bock, K. W. (1992). 2,3,7,8-Tetrachlorodibenzo-p-dioxin and ethinylestradiol as co-mitogens in cultured rat hepatocytes. *Carcinogenesis*, *13*, 453–456.

74. Jefcoate, C. R., Liehr, J. G., Santen, R. J., Sutter, T. R., Yager, J. D., Yue, W., Santner, S. J., Tekmal, R., Demers, L., Pauley, R.,

Naftolin, F., Mor, G., and Berstein, L. (2000). Tissue-specific synthesis and oxidative metabolism of estrogens. *Journal of the National Cancer Institute Monographs*, 27, 95–112.

75. Roy, D. and Abul-Hajj, Y. J. (1997). Estrogen-nucleic acid adducts: guanine is major site for interaction between 3,4-estrone quinone and COIII gene. *Carcinogenesis*, 18, 1247–1249.

76. Hall, J. M., Barhoover, M. A., Kazmin, D., McDonnell, D. P., Greenlee, W. F., and Thomas, R. S. (2010). Activation of the aryl-hydrocarbon receptor inhibits invasive and metastatic features of human breast cancer cells and promotes breast cancer cell differentiation. *Molecular Endocrinology*, 24, 359–369.

77. Khan, S., Barhoumi, R., Burghardt, R., Liu, S., Kim, K., and Safe, S. (2006). Molecular mechanism of inhibitory aryl hydrocarbon receptor-estrogen receptor/Sp1 cross talk in breast cancer cells. *Molecular Endocrinology*, 20, 2199–2214.

78. Fan, Y., Boivin, G. P., Knudesen, E. S., Nebert, D. W., Xia, Y., and Puga, A. (2010). The aryl hydrocarbon receptor functions as a tumor suppressor of liver carcinogenesis. *Cancer Research*, 70, 212–220.

79. Fritz, W. A., Lin, T. M., Cardiff, R. D., and Peterson, R. E. (2007). The aryl hydrocarbon receptor inhibits prostate carcinogenesis in TRAMP mice. *Carcinogenesis*, 28, 497–505.

80. Gramatzki, D., Pantazis, G., Schittenhelm, J., Tabatabai, G., Köhle, C., Wick, W., Schwarz, M., Weller, M., and Tritschler, I. (2009). Aryl hydrocyrbon receptor inhibition downregulates the TGF-β/Smad pathway in human glioblastoma cells. *Oncogene*, 28, 2593–2605.

81. Andersson, P, McGuire, J., Rubio, C., Gradin, K., Whitelaw, M. L., Petterson, S., Hanberg, A., and Poewllinger, L. (2002). A constitutively active dioxin/aryl hydrocarbon receptor induces stomach tumors. *Proceedings of the National Academy of Sciences of the United States of America*, 99, 9990–9995.

82. Kuznetsov, N. V., Andersson, P., Gradin, K., Stein, P., Dieckmann, A., Pettersson, S., Hanberg, A., and Poellinger, L. (2005). The dioxin/arly hydrocarbon receptor mediates downregulation of osteopontin gene expression in a mouse model of gastric tumorigenesis. *Oncogene*, 24, 3216–3222.

83. Moennikes, O., Loeppen, S., Buchmann, A., Andersson, P., Ittrich, C., Poellinger, L., and Schwarz, M. (2004). A constitutively active dioxin/aryl hydrocarbon receptor promotes hepatocarcinogenesis in mice. *Cancer Research*, 64, 4707–4710.

84. NTP (1982). United States National Toxicology Program. Technical Report Series No. 209, DHEW Publication No. (NIH) 82-1765, Research Triangle Park, NC.

85. NTP (2005). Technical Report Series No. 521, Draft Abstract, Research Triangle Park, NC.

86. NTP (2006). United States National Toxicology Program. Toxicology and carcinogenesis studies of a mixture of 2,3,7,8-tetrachlorodibenzo-*p*-dioxin (TCDD) (CAS No. 1746-01-6) 2,3,4,7,8-pentachlorodibenzofuran (PeCDF) (CAS No. 57117-31-4), and 3,3',4,4',5-pentachlorobiphenyl (PCB 126) (CAS No. 57465-28-8) in female Harlan Sprague-Dawley rats (gavage studies). *National Toxicology Program Technical Report Series*, 526, 1–180.

87. Tóth, K., Somfai-Relle, S., Sugár, J., and Bence, J. (1979). Carcinogenicity testing of herbicide 2,4,5-trichlorophenoxyethanol containing dioxin and of pure dioxin in Swiss mice. *Nature*, 278, 548–549.

88. NTP (1982). United States National Toxicology Program. Technical Report Series No. 201, DHEW Publication No. (NIH) 82-1757, Research Triangle Park, NC.

89. Della Porta, G., Dragani, T. A., and Sozzi, G. (1987). Carcinogenic effects of infantile and long-term 2,3,7,8-tetrachlorodibenzo-*p*-dioxin treatment in the mouse. *Tumori*, 73, 99–107.

17

TERATOGENIC IMPACT OF DIOXIN ACTIVATED AHR IN LABORATORY ANIMALS*

BARBARA D. ABBOTT

17.1 INTRODUCTION: AHR-MEDIATED DEVELOPMENTAL TOXICITY

The reproductive and developmental toxicity of ligands for the aryl hydrocarbon receptor (AHR) is well documented for many species, including birds, fish, and mammals, and there is evidence for developmental toxicity in humans. The reproductive and developmental toxicity of 2,3,7,8-tetrachlorodibenzo-p-dioxin (TCDD) has been extensively studied in laboratory animals and in this book; Chapter 22 examines effects on development of fish and birds, and in Chapters 29–31 the developmental effects on liver, heart, and ovarian development and function are discussed. This chapter examines the induction of cleft palate (CP) and hydronephrosis (HN) following TCDD exposure, presents evidence of the dependence of these responses on AHR-mediated signaling, and reviews the current understanding of growth factor signaling involved in the mechanisms by which TCDD produces these responses.

In laboratory studies, the adult hamster, rat, and guinea pig exhibit markedly different sensitivity to the toxic effects of TCDD, with an almost 5000-fold difference in acute lethal potency. In contrast, in the developing embryo the difference in sensitivity between species is only around 10-fold for induction of embryo/fetal lethality [1]. Based on these observations, the developing embryo/fetus is uniquely sensitive to TCDD and there is little difference between species in the vulnerability of the embryo. Developmental toxicities (e.g., mortality, growth inhibition, effects on liver weight, neurobehavioral effects, endocrine disruption, and malformations) of TCDD have been studied extensively in laboratory animals [2, 3]. Among the most studied teratogenic effects of TCDD are isolated clefts of the secondary palate and hydronephrosis. TCDD and other halogenated aromatic hydrocarbons with a strong affinity for the AHR produce these defects in mice after exposure *in utero* [4–6]. These responses occur at doses that increase the maternal liver weight but generally produce no other toxicity in the pregnant or lactating adult female.

17.2 HYDRONEPHROSIS: SENSITIVE PERIOD, ETIOLOGY, AND RECOVERY

HN is produced at lower doses of TCDD than CP and is a very sensitive indicator of gestational exposure. HN can be produced at doses that do not induce CP. The sensitive period for TCDD induction of HN extends from early gestation into the postnatal period [7]. Exposure of mouse pups to TCDD on postnatal day (PND) 1 or 4 produces HN with peak for postnatal induction on PND1; after monitoring up to PND26, there was no recovery of the kidneys from this structural defect [8]. In a cross-foster study, HN was induced by exposure *in utero* or lactationally, with the greatest severity in pups exposed both *in utero* and lactationally, and there was no recovery among those pups after monitoring up to PND67 [9]. The etiology of the defect may vary depending on whether the exposure occurred *in utero* or after birth. Exposure of C57BL/6N mice to TCDD on gestation day (GD) 10 to 12 μg TCDD/kg dam body weight produced a

---

* This chapter has been reviewed by the National Health and Environmental Effects Research Laboratory, U.S. EPA. The use of trade names is for identification only and does not constitute endorsement by the U.S. EPA.

The AH Receptor in Biology and Toxicology, First Edition. Edited by Raimo Pohjanvirta.
© 2012 John Wiley & Sons, Inc. Published 2012 by John Wiley & Sons, Inc.

high incidence and severity of HN in fetuses examined on GD16 or 17. The lumen of the ureters were narrow and obstructed and, as late as GD15, regions of the ureter were completely occluded by ureteric epithelial cells [10]. Extracellular matrix (ECM) composition of the tubules and glomeruli of the kidney were also affected and thinner in fetuses with this exposure regimen [11]. HN induced by exposure early in gestation was a consequence of altered regulation of proliferation and differentiation of the ureteric epithelial cells lining the lumen and this occurred prior to production of urine by the kidney. Later in gestation (generally around GD14 in the mouse), the fetal kidney begins producing urine, and occlusion of the ureteric lumen results in increased pressure in the kidney from accumulating urine, resulting in thinning and damage to the blastema. By GD16 occlusion was at least partially ameliorated, apparently by death of cells in the central area of the occlusion, allowing some urine to flow to the bladder, although hydroureter and HN persisted and remained severe on GD17. Occlusion of the ureteric lumen was less likely to occur if the exposure to TCDD was initiated after birth. Other mechanisms have been suggested for production of HN following lactational exposure. A "functional" HN was proposed in which polyuria could overwhelm the peristaltic transfer of urine to the bladder. In that study, C57BL/6J dams were dosed on PND1 with 10 μg TCDD/kg and pups were exposed lactationally. Ureteric obstruction was not detected, but altered gene expression in the kidney suggested effects on kidney transporters that could contribute to polyuria-mediated HN [12].

17.3 CLEFT PALATE: SENSITIVE PERIOD AND ETIOLOGY

The period of sensitivity for induction of CP in the mouse lies between GD6 and GD12, with the peak at GD12 [4, 13]. This period of sensitivity reflects the developmental events in formation of the secondary palate. The formation of the secondary palate involves production of palatine shelves that extend above the tongue and fuse to form the roof of the oral cavity. The initial critical events include formation of the first visceral arch, migration of neural crest cells into the arch, and formation of the maxillary and mandibular processes from this arch. This occurs early in embryonic life, between GD9 and 11 in the mouse. Subsequent events necessary for formation of the secondary palate occur in the mouse between GD11 and 14 (with slight variations in different strains of mice) and include outgrowth of the palatal shelves from the maxillary arch, elevation of the shelves above the tongue, expansion and contact of the opposing shelves, and adhesion and fusion of the shelves [14]. TCDD has the potential to interfere with all the critical processes involved in palatogenesis. The etiology of CP induction can depend on both the dose and the timing of exposure. Exposure to very high levels may affect the size of the palatal shelves such that the opposing shelves never contact and thus cannot fuse. However, exposure to TCDD on GD10 or 12 can be at a level that produces CP in all the exposed fetuses, but still allows formation of palatal shelves of normal size that elevate and make contact along the medial edges, but fail to fuse. Pratt et al. [15] exposed C57BL/6J mice to a single subcutaneous dose of 100 μg TCDD/kg on GD11 and using frozen sections found the palatal shelves on GD14 to be in contact above the tongue but not fused and on GD17 all fetuses had CP. The conclusion was that TCDD interfered with adhesion or fusion of the opposing shelves. Subsequent studies confirmed this etiology using lower exposures and oral administration. Oral dosing of pregnant C57BL6N mice on GD10 or 12 with 24 μg TCDD/kg resulted in production of palatal shelves that came into contact above the tongue, but failed to fuse due to effects on the medial edge epithelium (MEE) [13]. The effects on proliferation and differentiation of the MEE appear critical to the failure of the palates to fuse. These effects are linked to alterations in expression and signaling through growth factor pathways, including the epidermal growth factor receptor (EGFR) and transforming growth factor beta (TGFβ) pathways.

17.4 REQUIREMENT FOR AHR EXPRESSION IN TERATOGENIC RESPONSES

AHR and aryl hydrocarbon receptor nuclear translocator (ARNT) are expressed throughout development and their mRNA and proteins have been localized from GD10–16 in the mouse to specific tissues [16–19]. AHR and ARNT are expressed in spatiotemporal patterns in the developing palate, kidney, and ureter of the mouse [20–22]. Expression patterns similar to those of the mouse palate were reported for human fetal palatal shelves on GD42–59 [23, 24]. Demonstrating the presence of AHR and ARNT in the target tissues was an important step toward establishing a role for AHR signaling in the teratogenic responses. Rapid distribution of TCDD to the target tissues was also documented within 30 min of dosing. After oral dosing of the pregnant mouse on GD12, TCDD was distributed to placenta, fetal liver, and palate [15, 25]. Similar distribution patterns were shown in TCDD-exposed pregnant rats [26]. Direct activation of AHR to a transcriptionally functional form was demonstrated in the embryo using a dioxin-responsive *lacZ* transgenic mouse model. Activation of AHR after exposure to TCDD was detected in GD14 mice in the genital tubercle (strongly reactive), craniofacial tissues, and other regions [27].

Expression of AHR, distribution of TCDD to the target tissue, and activation of the receptor in the embryo certainly suggest a role for AHR in mediating the teratogenic response to TCDD. However, the strongest evidence for the role of AHR in response to TCDD was demonstrated in AHR knockout (KO) mice. Several laboratories generated AHR KO mice and all the models share some characteristic features, including resistance to TCDD toxicity, reduced reproductive capacity, and altered vascular development, particularly patent ductus venosus [18, 28, 29]. Chapters 13 and 29–31 provide in-depth information on the phenotype of these mice, evaluate consequences to specific organ systems during development in the absence of AHR expression, and discuss the requirement for AHR in major toxic outcomes of TCDD exposure. Here, the focus will be limited to the use of AHR KO mice in determining the importance of AHR in TCDD-induced teratogenicity. In a study using AHR KO mice, oral exposure on GD12.5 to 40 μg TCDD/kg did not produce CP or HN in the AHR KO fetuses, but in heterozygous fetuses HN but not CP was present, suggesting that the mechanisms may differ for induction of these defects [18]. A study using AHR KO mice from a different laboratory also found that oral exposure to 25 μg TCDD/kg on GD10 did not produce a significant increase in CP or HN in the AHR KO fetuses but was effective in producing these defects in wild-type fetuses [30]. Together these developmental studies demonstrate the requirement for AHR to mediate the teratogenic effects of TCDD.

The maternal AHR KO genotype is reported to affect the sensitivity of the fetus to teratogenic effects, but this is likely to be a consequence of more circulating TCDD reaching the embryos, as the investigators found lower levels of TCDD in the livers of KO dams compared to the wild type [31]. Similarly, in a Cyp1a2 KO dam, less TCDD will be bound and sequestered in the maternal liver in the absence of Cyp1a2, resulting in an increased exposure of the embryos and high sensitivity to CP, HN, and embryo lethality [32].

Additional insight can be gained from transgenic models in which the nuclear localization of AHR is blocked [33], as well as in another model in which AHR translocates to the nucleus and associates with ARNT, but fails to bind to dioxin response elements (DREs) upstream of regulated genes [34]. In the model in which AHR did not translocate to the nucleus, AHR retained the ability to associate with XAP2 and bind hsp90 and ligand. After exposure of GD12.5 palates to 3.3 nM TCDD in organ culture for 4 days, none of the heterozygous or wild-type palates fused, but all of the mutant palates in which AHR did not translocate to the nucleus fused, showing that translocation of AHR to the nucleus was required [33]. In the model in which binding to the DRE is prevented, pregnant mice were injected intraperitoneally (IP) with 128 μg TCDD/kg on GD10 and none of the 52 exposed fetuses had CP or HN, showing that DNA binding of the AHR complex was required [34].

AHR antagonists, α-naphthoflavone, and resveratrol were tested for their ability to block the toxicity of TCDD. Both agents were successful in reducing the incidence of CP and HN, as well as modulating the severity of HN and hydroureter in the fetuses exposed to both TCDD and antagonist [35, 36].

Collectively, these studies provide convincing evidence that AHR expression, translocation to the nucleus, and binding to DRE are required for mediation of TCDD induction of CP and HN.

17.5 HUMAN PALATAL AHR EXPRESSION AND RESPONSE IN ORGAN CULTURE AND THE HUMANIZED MOUSE MODEL

Human AHR and ARNT have been sequenced [37, 38] and polymorphisms of AHR and related proteins and potential health consequences are explained in detail in Chapter 31. The relationship of AHR structure and species-specific variation in sensitivity is discussed in Chapter 14. Adverse health outcomes related to TCDD exposure in humans are presented in Chapter 23. There is limited information regarding expression and activation of AHR and ARNT in human embryos. However, studies have characterized the expression of AHR, ARNT, and growth factors in human embryonic palates and evaluated responses to TCDD exposure in organ culture [24, 39–41]. There are similarities in the patterns of AHR and ARNT expression in mouse and human palates. In human medial epithelial cells, specific patterns were identified for AHR and ARNT [41]. In GD40–59 human palates, AHR mRNA and protein were detected in epithelial and mesenchymal cells. In epithelial cells, AHR protein localization was nuclear and cytoplasmic, whereas it was only nuclear in mesenchymal cells. In a double-staining immunohistochemical analysis, AHR and ARNT were both expressed in palatal epithelial cells, but mesenchymal cells could express either protein alone or both proteins. AHR palatal expression declined with increasing gestational age (Table 17.1) and exposure to 1×10^{-8} M TCDD in organ culture increased AHR protein expression (Table 17.2) [24]. This differs somewhat from the response in the mouse palate in which AHR protein and mRNA decreased after exposure to TCDD (Table 17.2, oral dosing of 24 μg TCDD/kg on GD10) [20]. In human palates, ARNT expression was mainly nuclear in epithelial and mesenchymal cells, did not change with age, and was also increased by TCDD exposure in culture (Tables 17.1 and 17.2). In human palates, exposure to TCDD in organ culture resulted in time- and concentration-dependent induction of Cyp1a1 mRNA [40].

Palate organ culture models were also used to evaluate the potential for human palatal shelves to respond to TCDD with effects on MEE that might interfere with fusion. In a Trowell-like model, palatal shelves were placed on supported membranes above medium, whereas in a submerged model, palates were floating freely in the medium in flasks on a rocker [39]. In both models, GD45–54 left and right palatal shelves were cultured separately, one in control medium and one in TCDD-containing medium, and exposed to various

TABLE 17.1 Summary of Human and Mouse Embryonic Gene Expression in Palatogenesis

	Mouse GD12–16 (Control Palates)	Human GD40–59 (Not Cultured)
AHR	Increase in mRNA expression GD12–14, epithelial protein expression high GD14	Decrease with age in epithelium and mesenchyme
ARNT	Increase in mRNA expression GD12–14	No change with age, expression is nuclear in epithelium and mesenchyme
EGF receptor	Decrease in medial epithelial cells prior to fusion	Increased in mesenchyme and expressed in medial epithelium through fusion events
TGFα	Decreases with age, expression in epithelium and mesenchyme similar	Expressed through palatogenesis, increase with age
EGF	Increases with age, expression in epithelium > mesenchyme	No change with age, expression in epithelium > mesenchyme
TGFβ1	No change with age, epithelium and mesenchyme similar	No change with age for epithelium, increase in mesenchyme with age, mesenchyme > epithelium
TGFβ2	High in early stages, becomes regional, decreases in mesenchyme	Expression varied between individuals, no trend detected
TGFβ3	Higher in medial epithelium at contact and fusion	Not higher in medial cells, epithelium > mesenchyme.

Data from Refs 13, 20, 21, 72–74.

concentrations of TCDD for 2–5 days, depending on the gestational age at the start of culture [39]. At a concentration of 1×10^{-8} M TCDD, MEE proliferation and differentiation were changed and expression of EGFR, EGF, TGFα, and TGFβ were altered [39]. The effects on MEE prevented removal of peridermal cells and resulted in a proliferating, multilayered squamous epithelium. These effects interfere with palatal fusion, and similar outcomes were observed for mouse palates exposed to TCDD *in vivo* or *in vitro*. Thus, human palates are capable of responding to TCDD; however, the concentration of TCDD required was approximately 200 times higher than that needed to produce these responses in mouse palates. In addition, the expression of AHR was determined to be about 350 times lower in human palate compared to mouse palate [23]. Although only limited information is available regarding TCDD levels in the human fetus, Schecter et al. reported approximately 1.4 ppt in human embryos on a lipid adjusted basis [42]. After exposure to 1×10^{-8} M TCDD in culture, the exposure that produced morphological effects, the human palates contained 5.3×10^6 pg TCDD/g tissue (ppt), a level considerably higher than any other reported for human embryos.

An overall summary of the studies in human and mouse palates is presented in Tables 17.1 and 17.2. It is apparent that the mouse and human have somewhat different expression patterns for AHR, ARNT, and growth factor genes that regulate proliferation and differentiation of the MEE. Also, the responses to TCDD differ somewhat between mouse and human palates. Lower expression of AHR and a requirement for higher concentrations of TCDD to produce effects in human MEE suggest that the human palates are less responsive to TCDD than the mouse and that disruption of palatogenesis is unlikely to occur at the levels reported in the human fetus.

The AHR humanized mouse model provides an additional means for comparison of the sensitivity of mouse and human to the teratogenic effects of TCDD. In the humanized AHR (hAHR) mouse, the mouse AHR was replaced by human AHR, and thus human AHR protein was expressed in a C57BL/6J mouse [43]. Wild-type C57BL/6J mice are of the

TABLE 17.2 Summary of Human and Mouse Palatal Medial Epithelial Cell Response to TCDD

	AHR	ARNT	EGFR	EGF	TGFα	TGFβ1	TGFβ2	TGFβ3
Human GD45–54 palates cultured with 1×10^{-8} M TCDD for 2–5 days								
Protein	⇑	⇑	⇑	⇑	⇑	⇓	⇑	⇑
mRNA	⇓	⇑	⇔	⇑	⇑	⇑	⇔	⇓
Mouse *in vivo*: dosed orally on GD10 at 24 µg TCDD/kg; palates collected GD14								
Protein	⇓		⇑ EGFR	⇔	⇓	⇑	⇑	
mRNA	⇓							
Mouse *in vivo*: dosed orally on GD12 at 24 µg TCDD/kg; palates collected GD14								
Protein			⇑ EGFR and ⇑ EGF binding	⇔	⇔	⇓	⇔	
mRNA	⇔	⇔						

Relative to control: ⇑ = increase, ⇓ = decrease, ⇔ = unchanged; blank = data not available.
Data from Refs 13, 20, 21, 72–74.

Ahr$^{b-1/b-1}$ genotype and are highly sensitive to TCDD toxicity, whereas the DBA/2 mice with the Ahr$^{d/d}$ allele are less sensitive to TCDD. Pregnant wild-type C57BL/6J, wild-type DBA/2 mice, and hAHR mice were dosed IP with 40 µg TCDD/kg on GD12.5. Based on induction of genes regulated by AHR, such as Cyp1a1 and Cyp1a2, the hAHR mice were functionally less responsive to TCDD than DBA/2 or C57BL/6J mice. *In utero* exposure to TCDD induced CP in C57BL/6J and DBA/2 fetuses (100%, 30% incidence, respectively), but hAHR fetuses did not develop CP. Although TCDD-exposed hAHR fetuses developed HN at an incidence comparable to that of TCDD-exposed C57BL/6J and DBA/2 mice, the severity of the defect was significantly lower in hAHR fetuses [43]. These responses were independent of maternal genotype. Thus, human AHR protein expressed in the hAHR mouse retains specific functionality distinct from that of mouse AHR. Based on these outcomes, and results showing that hAHR and DBA/2 AHR have similar dissociation constants for TCDD, the authors concluded that factors other that ligand affinity are involved in determining responsiveness [44, 45].

The humanized mouse outcomes also support the conclusion from the palatal organ culture model that human fetal palates are relatively insensitive to the effects of TCDD. An investigation of the effects of dioxin in defoliants sprayed in Vietnam during 1954–1973 revealed that 20 years after the war, there was no difference between incidences of cleft lip/palate in Ben Tre Province and that in Japanese populations [46]. However, the development of HN, but not CP, in hAHR mice also suggests that these defects diverge in their etiology and the sensitivity of the human fetus to effects of TCDD on urinary tract health remains unresolved.

17.6 AHR AND SIGNALING PATHWAYS: MECHANISMS FOR TERATOGENIC RESPONSES

AHR activation triggers cell signaling and also influences synthesis and remodeling of extracellular matrix. Thus, AHR influences regulation of processes essential to development, including progression of the cell cycle, proliferation, differentiation, apoptosis, and cellular migration [47–51]. There is evidence for crosstalk between AHR and growth factor-mediated signaling pathways, including EGFR and members of the TGFβ family. EGFR is a ligand-activated signal transduction pathway, and ligands include EGF, TGFα, amphiregulin, epiregulin, betacellulin, and heparin binding EGF. Activation of the EGFR pathway initiates signaling via mitogen-activated protein kinase (MAPK) pathways. Members of the MAPK pathway, extracellular signal-regulated kinase (ERK), and jun N-terminal kinase (JNK) are modulators of ARNT and AHR activity and increased or decreased activity appears to be tissue specific [52, 53]. TCDD activates ERK and JNK pathway signaling and stimulation of MAPKs appears to be necessary for induction of AHR regulation of gene transcription. Ma and Babish [54] proposed a model of TCDD-mediated dysregulation in which the AHR-ligand complex activates MAPK kinase-mediated signal transduction. The topic of AHR and signal transduction pathway interactions is covered in depth in Chapter 8. Chapter 32 examines the role of AHR in regulation of cell cycle control and proliferation, and Chapter 33 discusses the involvement of AHR in cell adhesion and migration. In this chapter, the interactions between AHR and growth factor pathways will be discussed relative to the induction of AHR-mediated teratogenic outcomes.

17.6.1 TGFβ

The TGFβ family includes numerous ligands and receptors and signaling through this pathway influences almost every cellular process, including proliferation, differentiation, apoptosis, ECM remodeling, and epithelial–mesenchymal transformation [48]. TGFβ1, 2, and 3 are expressed in the developing palate and expression is altered by TCDD exposure. The specific effect depends on the gestational stage at which exposure is initiated (Table 17.2). AHR does not appear to directly regulate expression of these growth factors, as DREs have not been identified in EGF, TGFα, or TGFβ gene regulatory regions. At least in the case of TGFβ1, stabilization of mRNA after TCDD exposure appears to account for the increased level of mRNA in fibroblasts [55]. In the palate, the role of TGFβ1 and 2 is likely regulation of mesenchymal cell proliferation, but TGFβ3 is critical to the adhesion and fusion of the opposing shelves. TGFβ3 is expressed in the MEE with peak levels just prior to adherence and fusion [56]. Inhibition of TGFβ3 activity can prevent fusion of palatal shelves *in vitro* [57]. All TGFβ3 KO mice exhibit CP and this is attributed to failure of the opposing shelves to adhere and fuse. In TGFβ3 KO fetuses, the MEE remains in the midline seam and epithelial basement membrane is not degraded [58, 59]. In palatal organ culture, addition of TGFβ3 to the medium allows adhesion and fusion of TGFβ3 KO palatal shelves [60, 61]. Just prior to fusion, MEE cells develop abundant filopodial-like protrusions that appear to be coated with proteoglycan filaments and it is postulated that interaction of these structures is critical to adherence and fusion [60, 61]. The TGFβ3 KO palates have a reduced presence of filopodia, but exposure in culture to TGFB3 restores development of the structures and induces expression of chondroitin sulfate proteoglycans on the MEE surface. Interestingly, exposure to TCDD in culture also reduces development of the filopodia. Addition of TGFβ3 reverses this effect and prevents the dioxin-induced block in palatal fusion [61]. However, the lack of filopodia in MEE exposed to TCDD is likely to be secondary to effects of TCDD that lead to a differentiated, stratified, squamous epithelium with ongoing proliferation and inappropriate expression of EGFR and its ligands [13]. Clearly, TGFβ3 plays a key role in preparation of the MEE for adhesion and fusion, and exogenous TGFβ3 *in vitro* can reverse effects of TCDD that interfere with fusion, possibly overriding effects of TCDD on EGFR that lead to MEE proliferation and differentiation. However, it is not clear whether exposure of mouse embryos to TCDD reduces palatal MEE expression of TGFβ3. A review of the literature revealed no published data regarding effects of TCDD on TGFβ3 expression in the mouse palatal MEE. In HepG2 cells, TCDD did not interfere with TGFβ3 signaling, as monitored by effects on SMAD2 phosphorylation [61]. In human palates exposed to TCDD in culture, TGFβ3 protein levels increased and mRNA was reduced or unchanged [24].

AHR-mediated effects on TGFβ1, 2, and 3 expression and function in the palate may alter differentiation of MEE and, through effects on underlying mesenchyme, could influence the epithelial–mesenchymal interactions important in the preparation for adhesion and fusion. However, it is likely that these effects work in concert with the influence of activated AHR on EGFR signaling, as there is evidence for a significant role of that pathway in the induction of CP by TCDD. It is of interest that in the mouse, sensitivity to TCDD-induced cleft palate is modified by a locus on chromosome 3 and genes in close association to this locus include EGF, as well as fibroblast growth factor, which influences the expression of TGFβ3 [62].

17.6.2 EGFR Signal Transduction Pathway: Cleft Palate

In forming the secondary palate, just prior to contact of opposing shelves, expression of EGFR decreases in the medial epithelial cells and these cells stop proliferating [13]. The basement membrane adjacent to the epithelial cells in the medial seam degrades and epithelial cells are removed by one or more processes, including migration of the cells to the surface, transformation from an epithelial to a mesenchymal phenotype, and cell death. In the mouse embryo, after exposure to TCDD, the palatal shelves elevate and come into contact, but the medial epithelial cells continue to express high levels of EGFR and proliferate to form a stratified, squamous, oral-like epithelium [63]. The excessive proliferation of the medial epithelial cells and the ongoing expression of EGFR are accompanied by effects on EGF and TGFα expression (Table 17.2). After oral dosing of pregnant mice on GD12 with 24 µg TCDD/kg, EGF protein decreases in oral epithelium and mesenchymal cells, but remains unchanged in MEE. TGFα protein expression is not different from controls in embryos exposed on GD12, but exposure on GD10 to the same dose decreases expression of TGFα (Table 17.2).

As cell signaling through the EGFR pathway regulates epithelial proliferation and differentiation, and TCDD affects expression of the receptor and its ligands, it is possible that the induction of the aberrant MEE phenotype depends on the activation of EGFR signaling. This hypothesis was examined further in EGF and TGFα KO mice. The EGF KO mice exposed to 24 µg TCDD/kg body weight did not develop CP although that dose produced a significant induction of CP in wild-type mice. Even at 50 µg TCDD/kg, the EGF KO mice do not show a significant increase in CP, although doses of 100 µg or higher produce CP. The TGFα KO mice (on a C57BL/6J background) develop CP at doses and incidences comparable to the wild type. These outcomes suggest that TGFα expression is not required for the induction of CP by TCDD and that expression of EGF is a major factor in mediating the induction of CP. It is important to note that the TGFα knockout mice still express EGF, and if the response to TCDD depends on expression of EGF, then these mice would be expected to respond as they do. The role of EGFR signaling in CP was also examined in a study using mice heterozygous for EGFR knockout [64]. The dams were dosed orally on GD10 at levels ranging from 1.5 to 106 µg TCDD/kg and litters containing wild-type, heterozygous, and EGFR KO fetuses were evaluated for CP on GD18. The interpretation of this study was weakened by the limited number of fetuses in the EGFR KO groups. Although the authors reported a significant trend with dose, it seemed to be driven by the 100% incidence of CP in the highest exposure group, (only two fetuses were in the 106 µg/kg dose group and both had CP; two control EGFR KO fetuses also had CP). That outcome was similar to that observed in the EGF KO fetuses where very high exposures also induced CP.

The requirement for EGF in mediating TCDD-induced CP was also examined using palate organ culture that allowed the availability of growth factor to be manipulated [65]. Wild-type and EGF KO GD12 midfacial tissues were suspended in medium and cultured for 4 days. In that culture model, palatal shelves grew, elevated, and fused during culture. In unsupplemented medium, the EGF KO palates fused in the presence of TCDD; however, in medium supplemented with 2 ng EGF/mL medium, the EGF KO palates failed to fuse when exposed to TCDD [65].

Further evidence for a role of EGF in palatogenesis was provided by *in vitro* experiments in which exogenous EGF stimulated continued proliferation of the MEE and palates failed to fuse [66, 67]. More recently, it was shown that EGF stimulation of MEE DNA synthesis and cell proliferation in palate organ culture was dependent on EGF signaling via ERK1/2 [68]. However, EGF must be at higher than endogenous levels to interfere with palatogenesis, and peak activation of ERK1/2 was achieved by adding 50 ng EGF/mL medium (compared to the 2 ng EGF/mL medium used to restore palatal responses to TCDD in the EGF KO palates).

Although evidence supports an important role for the EGFR pathway in production of CP after activation of AHR, the mechanism is likely to be complex. EGFR has other ligands whose expression and response to TCDD have not been characterized in the palate. Epiregulin, a potent mitogen and ligand for EGFR, has a DRE in its promoter region and was directly upregulated by activated AHR [69]. It is possible that in the EGF KO mouse, epiregulin or one of the other EGFR ligands is responsible for the induction of CP in the EGF KO embryo exposed to the high TCDD doses.

17.6.3 EGFR Signal Transduction Pathway: Hydronephrosis

Studies in EGF KO mice suggested that EGFR ligands other than EGF or TGFα were involved in mediation of HN, as both EGF and TGFα KO fetuses developed HN after TCDD exposure. Although the EGF KO mice were resistant to TCDD-induced CP, HN was induced at high incidence and severity in the EGF KO fetuses at doses of 1 μg TCDD/kg or higher and the response seemed exacerbated in the EGF KO fetuses compared to wild type [70, 71]. The requirement for EGFR in mediating HN was also examined in mice heterozygous for EGFR knockout [64]. The dams were dosed orally on GD10 at levels ranging from 1.5 to 106 μg TCDD/kg and litters containing wild-type, heterozygous, and EGFR KO fetuses were evaluated for HN on GD18. There were limited numbers of fetuses in the EGFR KO groups, but TCDD induced HN in EGFR KO fetuses in a dose-related fashion. The data were somewhat more convincing than that reported for CP (discussed previously), as 7 of 11 fetuses at 30.9 μg TCDD/kg and 7 out of 7 at 55.6 μg TCDD/kg had HN. The divergent outcomes for CP and HN in the EGF KO mice and the induction of HN in EGFR KO mice suggested that cellular pathways other than EGFR were involved and that the mechanisms differed between the urinary tract and the palate.

17.7 SUMMARY

AHR and ARNT are expressed in mouse and human palatal shelves and in the urinary tract of the mouse fetus. AHR expression, translocation to the nucleus, binding to DRE, and activation are required for mediation of TCDD induction of CP and HN. Although the human palate requires a higher exposure than the mouse, cellular responses to TCDD exposure are similar in the mouse and human palates, with induction of proliferation of medial epithelial cells and differentiation to a stratified, squamous epithelium that blocks fusion of the opposing shelves. AHR activation by TCDD affects regulation of EGFR, its ligands, and TGFβ family members in palate and urinary tract. EGF appears to play a major role in the induction of CP, but not HN, following AHR activation. The mechanisms for induction of CP and HN appear to differ; however, both appear to involve effects on proliferation and differentiation of epithelial cells. The etiologies for AHR-mediated CP and HN are undoubtedly complex and likely to involve interactions between signaling pathways and effects on multiple, tightly regulated cellular processes.

REFERENCES

1. Kransler, K. M., McGarrigle, B. P., and Olson, J. R. (2007). Comparative developmental toxicity of 2,3,7,8-tetrachlorodibenzo-p-dioxin in the hamster, rat and guinea pig. *Toxicology, 229*, 214–225.

2. Brouwer, A., Ahlborg, U. G., Van den Berg, M., Birnbaum, L. S., Boersma, E. R., Bosveld, B., Denison, M. S., Gray, L. E., Hagmar, L., Holene, E., et al. (1995). Functional aspects of developmental toxicity of polyhalogenated aromatic hydrocarbons in experimental animals and human infants. *European Journal of Pharmacology, 293*, 1–40.

3. Couture, L. A., Abbott, B. D., and Birnbaum, L. S. (1990). A critical review of the developmental toxicity and teratogenicity of 2,3,7,8-tetrachlorodibenzo-p-dioxin: recent advances toward understanding the mechanism. *Teratology, 42*, 619–627.

4. Birnbaum, L. S., Weber, H., Harris, M. W., Lamb, J. C., IV, and McKinney, J. D. (1985). Toxic interaction of specific polychlorinated biphenyls and 2,3,7,8-tetrachlorodibenzo-p-dioxin: increased incidence of cleft palate in mice. *Toxicology and Applied Pharmacology, 77*, 292–302.

5. Courtney, K. D. and Moore, J. A. (1971). Teratology studies with 2,4,5-trichlorophenoxyacetic acid and 2,3,7,8-tetrachlorodibenzo-p-dioxin. *Toxicology and Applied Pharmacology, 20*, 396–403.

6. Neubert, D., Zens, P., Rothenwallner, A., and Merker, H. J. (1973). A survey of the embryotoxic effects of TCDD in mammalian species. *Environmental Health Perspectives, 5*, 67–79.

7. Couture, L. A., Harris, M. W., and Birnbaum, L. S. (1990). Characterization of the peak period of sensitivity for the induction of hydronephrosis in C57BL/6N mice following exposure to 2,3,7, 8-tetrachlorodibenzo-p-dioxin. *Fundamental and Applied Toxicology, 15*, 142–150.

8. Couture-Haws, L., Harris, M. W., McDonald, M. M., Lockhart, A. C., and Birnbaum, L. S. (1991). Hydronephrosis in mice exposed to TCDD-contaminated breast milk: identification of the peak period of sensitivity and assessment of potential recovery. *Toxicology and Applied Pharmacology, 107*, 413–428.

9. Couture-Haws, L., Harris, M. W., Lockhart, A. C., and Birnbaum, L. S. (1991). Evaluation of the persistence of hydronephrosis induced in mice following *in utero* and/or lactational exposure to 2,3,7,8-tetrachlorodibenzo-p-dioxin. *Toxicology and Applied Pharmacology, 107*, 402–412.

10. Abbott, B. D., Birnbaum, L. S., and Pratt, R. M. (1987). TCDD-induced hyperplasia of the ureteral epithelium produces hydronephrosis in murine fetuses. *Teratology, 35*, 329–334.

11. Abbott, B. D., Morgan, K. S., Birnbaum, L. S., and Pratt, R. M. (1987). TCDD alters the extracellular matrix and basal lamina of the fetal mouse kidney. *Teratology, 35*, 335–344.

12. Nishimura, N., Matsumura, F., Vogel, C. F., Nishimura, H., Yonemoto, J., Yoshioka, W., and Tohyama, C. (2008). Critical role of cyclooxygenase-2 activation in pathogenesis of hydronephrosis caused by lactational exposure of mice to dioxin. *Toxicology and Applied Pharmacology, 231,* 374–383.

13. Abbott, B. D. and Birnbaum, L. S. (1989). TCDD alters medial epithelial cell differentiation during palatogenesis. *Toxicology and Applied Pharmacology, 99,* 276–286.

14. Meng, L., Bian, Z., Torensma, R., and Von den Hoff, J. W. (2009). Biological mechanisms in palatogenesis and cleft palate. *Journal of Dental Research, 88,* 22–33.

15. Pratt, R. M., Dencker, L., and Diewert, V. M. (1984). 2,3,7,8-Tetrachlorodibenzo-*p*-dioxin-induced cleft palate in the mouse: evidence for alterations in palatal shelf fusion. *Teratogenesis, Carcinogenesis, and Mutagenesis, 4,* 427–436.

16. Abbott, B. D., Birnbaum, L. S., and Perdew, G. H. (1995). Developmental expression of two members of a new class of transcription factors: I. Expression of aryl hydrocarbon receptor in the C57BL/6N mouse embryo. *Developmental Dynamics, 204,* 133–143.

17. Abbott, B. D. and Probst, M. R. (1995). Developmental expression of two members of a new class of transcription factors: II. Expression of aryl hydrocarbon receptor nuclear translocator in the C57BL/6N mouse embryo. *Developmental Dynamics, 204,* 144–155.

18. Mimura, J., et al. (1997). Loss of teratogenic response to 2,3,7,8-tetrachlorodibenzo-*p*-dioxin (TCDD) in mice lacking the Ah (dioxin) receptor. *Genes to Cells, 2,* 645–654.

19. Jain, S., Maltepe, E., Lu, M. M., Simon, C., and Bradfield, C. A. (1998). Expression of ARNT, ARNT2, HIF1 alpha, HIF2 alpha and Ah receptor mRNAs in the developing mouse. *Mechanisms of Development, 73,* 117–123.

20. Abbott, B. D., Perdew, G. H., and Birnbaum, L. S. (1994). Ah receptor in embryonic mouse palate and effects of TCDD on receptor expression. *Toxicology and Applied Pharmacology, 126,* 16–25.

21. Abbott, B. D., Schmid, J. E., Brown, J. G., Wood, C. R., White, R. D., Buckalew, A. R., and Held, G. A. (1999). RT-PCR quantification of AHR, ARNT, GR, and CYP1A1 mRNA in craniofacial tissues of embryonic mice exposed to 2,3,7,8-tetrachlorodibenzo-*p*-dioxin and hydrocortisone. *Toxicological Sciences, 47,* 76–85.

22. Bryant, P. L., Clark, G. C., Probst, M. R., and Abbott, B. D. (1997). Effects of TCDD on Ah receptor, ARNT, EGF, and TGF-alpha expression in embryonic mouse urinary tract. *Teratology, 55,* 326–337.

23. Abbott, B. D., Buckalew, A. R., Diliberto, J. J., Wood, C. R., Held, G., Pitt, J. A., and Schmid, J. E. (1999). AhR, ARNT, and CYP1A1 mRNA quantitation in cultured human embryonic palates exposed to TCDD and comparison with mouse palate *in vivo* and in culture. *Toxicological Sciences, 47,* 62–75.

24. Abbott, B. D., Probst, M. R., Perdew, G. H., and Buckalew, A. R. (1998). AH receptor, ARNT, glucocorticoid receptor, EGF receptor, EGF, TGF alpha, TGF beta 1, TGF beta 2, and TGF beta 3 expression in human embryonic palate, and effects of 2,3,7,8-tetrachlorodibenzo-*p*-dioxin (TCDD). *Teratology, 58,* 30–43.

25. Abbott, B. D., Birnbaum, L. S., and Diliberto, J. J. (1996). Rapid distribution of 2,3,7,8-tetrachlorodibenzo-*p*-dioxin (TCDD) to embryonic tissues in C57BL/6N mice and correlation with palatal uptake *in vitro*. *Toxicology and Applied Pharmacology, 141,* 256–263.

26. Hurst, C. H., Abbott, B. D., DeVito, M. J., and Birnbaum, L. S. (1998). 2,3,7,8-Tetrachlorodibenzo-*p*-dioxin in pregnant Long Evans rats: disposition to maternal and embryo/fetal tissues. *Toxicological Sciences, 45,* 129–136.

27. Willey, J. J., Stripp, B. R., Baggs, R. B., and Gasiewicz, T. A. (1998). Aryl hydrocarbon receptor activation in genital tubercle, palate, and other embryonic tissues in 2,3,7,8-tetrachlorodibenzo-*p*-dioxin-responsive lacZ mice. *Toxicology and Applied Pharmacology, 151,* 33–44.

28. Fernandez-Salguero, P. M., Ward, J. M., Sundberg, J. P., and Gonzalez, F. J. (1997). Lesions of aryl-hydrocarbon receptor-deficient mice. *Veterinary Pathology, 34,* 605–614.

29. Schmidt, J. V., Su, G. H., Reddy, J. K., Simon, M. C., and Bradfield, C. A. (1996). Characterization of a murine Ahr null allele: involvement of the Ah receptor in hepatic growth and development. *Proceedings of the National Academy of Sciences of the United States of America, 93,* 6731–6736.

30. Peters, J. M., Narotsky, M. G., Elizondo, G., Fernandez-Salguero, P. M., Gonzalez, F. J., and Abbott, B. D. (1999). Amelioration of TCDD-induced teratogenesis in aryl hydrocarbon receptor (AhR)-null mice. *Toxicological Sciences, 47,* 86–92.

31. Thomae, T. L., Glover, E., and Bradfield, C. A. (2004). A maternal Ahr null genotype sensitizes embryos to chemical teratogenesis. *Journal of Biological Chemistry, 279,* 30189–30194.

32. Dragin, N., Dalton, T. P., Miller, M. L., Shertzer, H. G., and Nebert, D. W. (2006). For dioxin-induced birth defects, mouse or human CYP1A2 in maternal liver protects whereas mouse CYP1A1 and CYP1B1 are inconsequential. *Journal of Biological Chemistry, 281,* 18591–18600.

33. Bunger, M. K., Moran, S. M., Glover, E., Thomae, T. L., Lahvis, G. P., Lin, B. C., and Bradfield, C. A. (2003). Resistance to 2,3,7,8-tetrachlorodibenzo-*p*-dioxin toxicity and abnormal liver development in mice carrying a mutation in the nuclear localization sequence of the aryl hydrocarbon receptor. *Journal of Biological Chemistry, 278,* 17767–17774.

34. Bunger, M. K., Glover, E., Moran, S. M., Walisser, J. A., Lahvis, G. P., Hsu, E. L., and Bradfield, C. A. (2008). Abnormal liver development and resistance to 2,3,7,8-tetrachlorodibenzo-*p*-dioxin toxicity in mice carrying a mutation in the DNA-binding domain of the aryl hydrocarbon receptor. *Toxicological Sciences, 106,* 83–92.

35. Jang, J. Y., et al. (2008). Antiteratogenic effect of resveratrol in mice exposed *in utero* to 2,3,7,8-tetrachlorodibenzo-*p*-dioxin. *European Journal of Pharmacology, 591,* 280–283.

36. Jang, J. Y., et al. (2007). Antiteratogenic effects of alpha-naphthoflavone on 2,3,7,8-tetrachlorodibenzo-*p*-dioxin (TCDD) exposed mice *in utero*. *Reproductive Toxicology, 24,* 303–309.

37. Dolwick, K. M., Schmidt, J. V., Carver, L. A., Swanson, H. I., and Bradfield, C. A. (1993). Cloning and expression of a human Ah receptor cDNA. *Molecular Pharmacology*, 44, 911–917.

38. Hoffman, E. C., Reyes, H., Chu, F. F., Sander, F., Conley, L. H., Brooks, B. A., and Hankinson, O. (1991). Cloning of a factor required for activity of the Ah (dioxin) receptor. *Science*, 252, 954–958.

39. Abbott, B. D. and Birnbaum, L. S. (1991). TCDD exposure of human embryonic palatal shelves in organ culture alters the differentiation of medial epithelial cells. *Teratology*, 43, 119–132.

40. Abbott, B. D., Held, G. A., Wood, C. R., Buckalew, A. R., Brown, J. G., and Schmid, J. (1999). AhR, ARNT, and CYP1A1 mRNA quantitation in cultured human embryonic palates exposed to TCDD and comparison with mouse palate *in vivo* and in culture. *Toxicological Sciences*, 47, 62–75.

41. Abbott, B. D., Probst, M. R., and Perdew, G. H. (1994). Immunohistochemical double-staining for Ah receptor and ARNT in human embryonic palatal shelves. *Teratology*, 50, 361–366.

42. Schecter, A., Startin, J., Wright, C., Papke, O., Ball, M., and Lis, A. (1996). Concentrations of polychlorinated dibenzo-p-dioxins and dibenzofurans in human placental and fetal tissues from the U. S. in placentas from Yu-Cheng exposed mothers. *Chemosphere*, 32, 551–557.

43. Moriguchi, T., et al. (2003). Distinct response to dioxin in an arylhydrocarbon receptor (AHR)-humanized mouse. *Proceedings of the National Academy of Sciences of the United States of America*, 100, 5652–5657.

44. Ema, M., Ohe, N., Suzuki, M., Mimura, J., Sogawa, K., Ikawa, S., and Fujii-Kuriyama, Y. (1994). Dioxin binding activities of polymorphic forms of mouse and human arylhydrocarbon receptors. *Journal of Biological Chemistry*, 269, 27337–27343.

45. Micka, J., Milatovich, A., Menon, A., Grabowski, G. A., Puga, A., and Nebert, D. W. (1997). Human Ah receptor (AHR) gene: localization to 7p15 and suggestive correlation of polymorphism with CYP1A1 inducibility. *Pharmacogenetics*, 7, 95–101.

46. Natsume, N., Kawai, T., and Le, H. (1998). In Vietnam, many congenital anomalies are believed to result from the scattering of defoliants, including dioxin. *The Cleft Palate-Craniofacial Journal*, 35, 183.

47. Gomez-Duran, A., Carvajal-Gonzalez, J. M., Mulero-Navarro, S., Santiago-Josefat, B., Puga, A., and Fernandez-Salguero, P. M. (2009). Fitting a xenobiotic receptor into cell homeostasis: how the dioxin receptor interacts with TGFβ signaling. *Biochemical Pharmacology*, 77, 700–712.

48. Haarmann-Stemmann, T., Bothe, H., and Abel, J. (2009). Growth factors, cytokines and their receptors as downstream targets of arylhydrocarbon receptor (AhR) signaling pathways. *Biochemical Pharmacology*, 77, 508–520.

49. Kung, T., Murphy, K. A., and White, L. A. (2009). The aryl hydrocarbon receptor (AhR) pathway as a regulatory pathway for cell adhesion and matrix metabolism. *Biochemical Pharmacology*, 77, 536–546.

50. Puga, A., Ma, C., and Marlowe, J. L. (2009). The aryl hydrocarbon receptor cross-talks with multiple signal transduction pathways. *Biochemical Pharmacology*, 77, 713–722.

51. Ma, C., Marlowe, J. L., and Puga, A. (2009). The aryl hydrocarbon receptor at the crossroads of multiple signaling pathways. *EXS*, 99, 231–257.

52. Chen, S., Operana, T., Bonzo, J., Nguyen, N., and Tukey, R. H. (2005). ERK kinase inhibition stabilizes the aryl hydrocarbon receptor: implications for transcriptional activation and protein degradation. *Journal of Biological Chemistry*, 280, 4350–4359.

53. Tan, Z., Huang, M., Puga, A., and Xia, Y. (2004). A critical role for MAP kinases in the control of Ah receptor complex activity. *Toxicological Sciences*, 82, 80–87.

54. Ma, X. and Babish, J. G. (1999). Activation of signal transduction pathways by dioxins. In *Molecular Biology of the Toxic Response* (Puga, A. and Wallace, K. B.,Eds). Taylor & Francis, Philadelphia, pp. 493–516.

55. Chang, X., Fan, Y., Karyala, S., Schwemberger, S., Tomlinson, C. R., Sartor, M. A., and Puga, A. (2007). Ligand-independent regulation of transforming growth factor β1 expression and cell cycle progression by the aryl hydrocarbon receptor. *Molecular and Cellular Biology*, 27, 6127–6139.

56. Fitzpatrick, D. R., Denhez, F., Kondaiah, P., and Akhurst, R. J. (1990). Differential expression of TGF beta isoforms in murine palatogenesis. *Development*, 109, 585–595.

57. Brunet, C. L., Sharpe, P. M., and Ferguson, M. W. (1995). Inhibition of TGF-β3 (but not TGF-β1 or TGF-β2) activity prevents normal mouse embryonic palate fusion. *International Journal of Developmental Biology*, 39, 345–355.

58. Kaartinen, V., Cui, X. M., Heisterkamp, N., Groffen, J., and Shuler, C. F. (1997). Transforming growth factor-β3 regulates transdifferentiation of medial edge epithelium during palatal fusion and associated degradation of the basement membrane. *Developmental Dynamics*, 209, 255–260.

59. Proetzel, G., et al. (1995). Transforming growth factor-β3 is required for secondary palate fusion. *Nature Genetics*, 11, 409–414.

60. Taya, Y., O'Kane, S., and Ferguson, M. W. (1999). Pathogenesis of cleft palate in TGF-β3 knockout mice. *Development*, 126, 3869–3879.

61. Thomae, T. L., Stevens, E. A., and Bradfield, C. A. (2005). Transforming growth factor-β3 restores fusion in palatal shelves exposed to 2,3,7,8-tetrachlorodibenzo-p-dioxin. *Journal of Biological Chemistry*, 280, 12742–12746.

62. Thomae, T. L., Stevens, E. A., Liss, A. L., Drinkwater, N. R., and Bradfield, C. A. (2006). The teratogenic sensitivity to 2,3,7,8-tetrachlorodibenzo-p-dioxin is modified by a locus on mouse chromosome 3. *Molecular Pharmacology*, 69, 770–775.

63. Abbott, B. D. and Birnbaum, L. S. (1998). Dioxins and teratogenesis. In *Molecular Biology of the Toxic Response* (Puga, A. and Wallace, K.,Eds). Taylor & Francis, Washington, DC, pp. 439–447.

64. Miettinen, H. M., Huuskonen, H., Partanen, A. M., Miettinen, P., Tuomisto, J. T., Pohjanvirta, R., and Tuomisto, J. (2004). Effects of epidermal growth factor receptor deficiency and

2,3,7,8-tetrachlorodibenzo-*p*-dioxin on fetal development in mice. *Toxicology Letters*, 150, 285–291.

65. Abbott, B. D., Buckalew, A. R., and Leffler, K. E. (2005). Effects of epidermal growth factor (EGF), transforming growth factor-α (TGFα), and 2,3,7,8-tetrachlorodibenzo-*p*-dioxin on fusion of embryonic palates in serum-free organ culture using wild-type, EGF knockout, and TGFα knockout mouse strains. *Birth Defects Research, Part A*, 73, 447–454.

66. Hassell, J. R. (1975). The development of rat palatal shelves *in vitro*. An ultrastructural analysis of the inhibition of epithelial cell death and palate fusion by the epidermal growth factor. *Developmental Biology*, 45, 90–102.

67. Tyler, M. S. and Pratt, R. M. (1980). Effect of epidermal growth factor on secondary palatal epithelium *in vitro*: tissue isolation and recombination studies. *Journal of Embryology and Experimental Morphology*, 58, 93–106.

68. Yamamoto, T., Cui, X. M., and Shuler, C. F. (2003). Role of ERK1/2 signaling during EGF-induced inhibition of palatal fusion. *Developmental Biology*, 260, 512–521.

69. Patel, R. D., Kim, D. J., Peters, J. M., and Perdew, G. H. (2006). The aryl hydrocarbon receptor directly regulates expression of the potent mitogen epiregulin. *Toxicological Sciences*, 89, 75–82.

70. Abbott, B. D., Buckalew, A. R., DeVito, M. J., Ross, D., Bryant, P. L., and Schmid, J. E. (2003). EGF and TGF-α expression influence the developmental toxicity of TCDD: dose response and AhR phenotype in EGF, TGF-α, and EGF + TGF-α knockout mice. *Toxicological Sciences*, 71, 84–95.

71. Bryant, P. L., Schmid, J. E., Fenton, S. E., Buckalew, A. R., and Abbott, B. D. (2001). Teratogenicity of 2,3,7,8-tetrachlorodibenzo-*p*-dioxin (TCDD) in mice lacking the expression of EGF and/or TGF-α. *Toxicological Sciences*, 62, 103–114.

72. Abbott, B. D. and Birnbaum, L. S. (1990). Effects of TCDD on embryonic ureteric epithelial EGF receptor expression and cell proliferation. *Teratology*, 41, 71–84.

73. Abbott, B. D., Harris, M. W., and Birnbaum, L. S. (1992). Comparisons of the effects of TCDD and hydrocortisone on growth factor expression provide insight into their interaction in the embryonic mouse palate. *Teratology*, 45, 35–53.

74. Abbott, B. D., Perdew, G. H., Buckalew, A. R., and Birnbaum, L. S. (1994). Interactive regulation of Ah and glucocorticoid receptors in the synergistic induction of cleft palate by 2,3,7,8-tetrachlorodibenzo-*p*-dioxin and hydrocortisone. *Toxicology and Applied Pharmacology*, 128, 138–150.

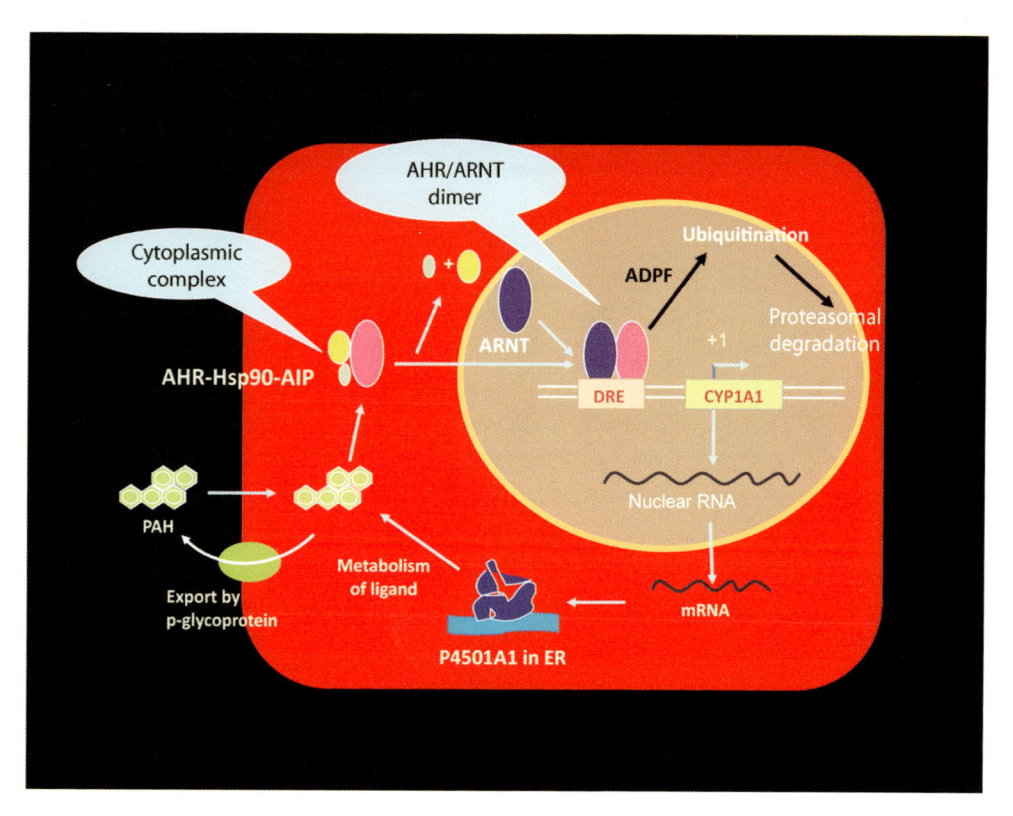

FIGURE 2.2 The "classical" pathway of AHR signaling for CYP1A1 induction.

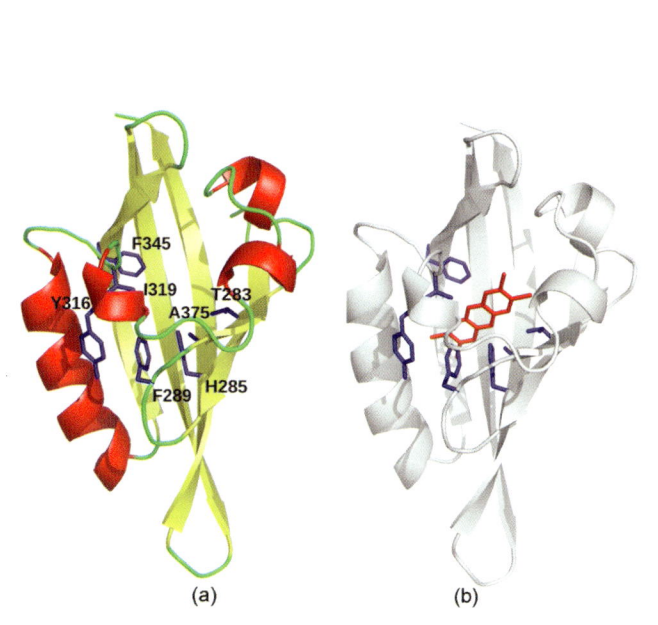

FIGURE 4.2 Homology model of the mouse AHR ligand binding domains and docking of TCDD in the binding cavity. (a) A cartoon representation of the modeled mAHR ligand binding domain with the "TCDD binding fingerprint" of conserved residues within the AHR ligand binding cavity necessary for optimal high-affinity TCDD binding [71]. (b) The lowest energy docked conformation of TCDD (blue sticks) within the binding cavity relative to the "binding fingerprint" residues.

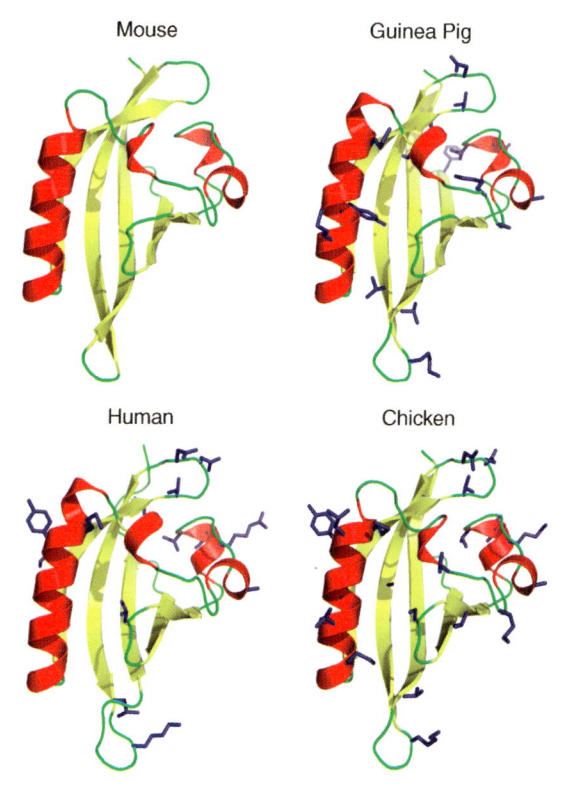

FIGURE 4.3 Cartoon representation of the modeled structures of the mouse, guinea pig, human, and chicken AHR ligand binding domains. Residues that are different from those in the mouse AHR ligand binding domain are indicated as blue sticks.

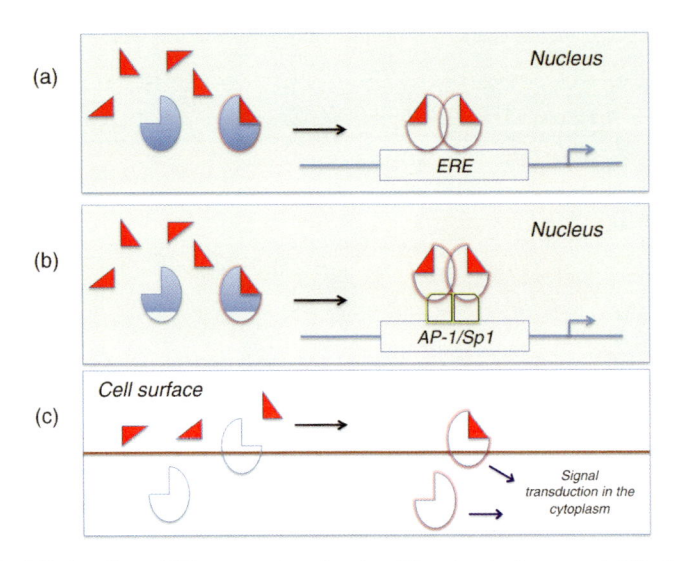

FIGURE 9.2 ER signaling: different modes of action. The classical, *genomic* ER signaling occurs through direct binding of ligand-bound ER dimers to ERE sequences in the regulatory regions of estrogen-responsive genes (a). Another type of *genomic* activity occurs through indirect association with promoters by protein–protein interactions with other transcription factors, such as AP-1 or Sp1 (b). The second mode of action, referred to as *nongenomic* activity, has been shown to involve interactions with cytoplasmic signal transduction proteins, such as kinases (c).

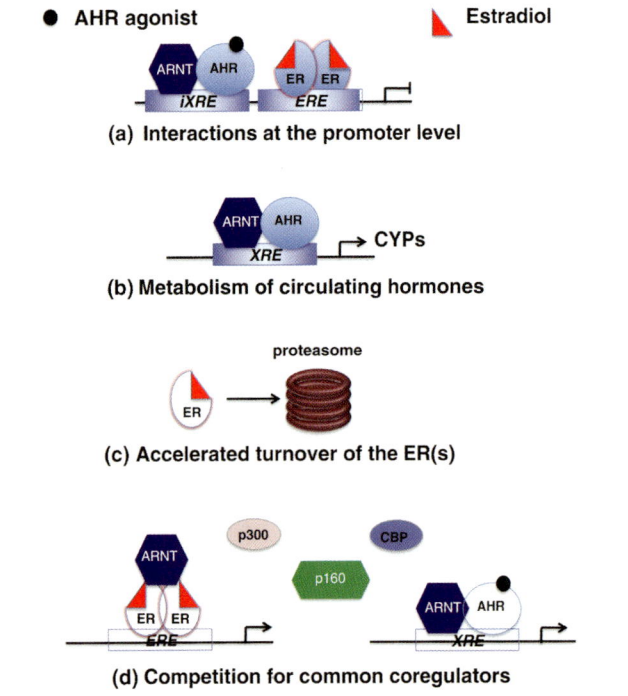

FIGURE 9.3 ER–AHR crosstalk mechanisms. (a) ER transactivation is repressed by the binding of liganded AHR to XREs, which blocks the ER binding or formation of the ERα/Sp1 or ERα/AP-1 complex. (b) AHR ligands can activate the CYP1A1 and 1A2 enzymatic activities. Given that circulating estrogens are metabolized in the liver, predominantly by 1A1 and 1A2, exposure to AHR agonists could thus lead to altered hormonal balance. (c) TCDD has been shown to enhance degradation of ERα by targeting it to the proteasome, thus lowering the cellular levels of ER. (d) Simultaneous induction of ER and AHR leads to competition for common regulatory factors or coactivators, thus inhibiting full transcriptional activity of the receptors.

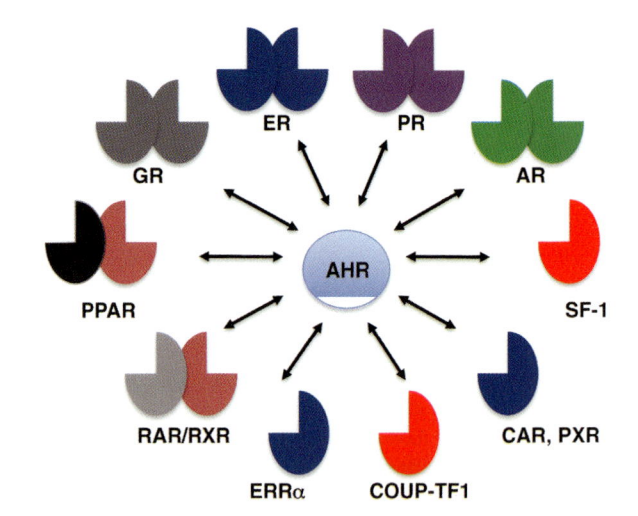

FIGURE 9.4 Overview of AHR–NR crosstalk. There are many indications of the AHR pathway converging with various NR signaling pathways.

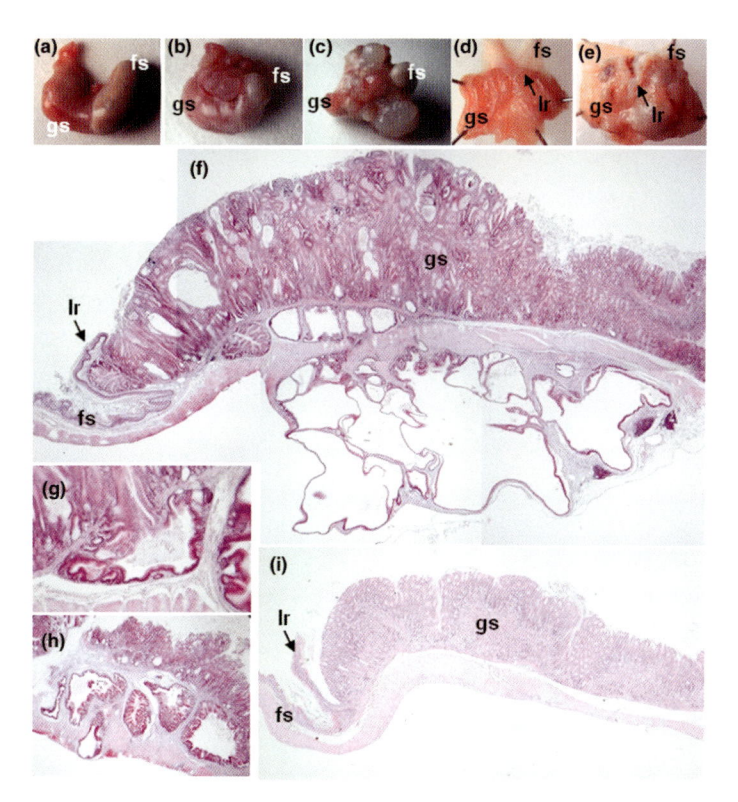

FIGURE 26.1 Stomach tumors in the CA-AHR mice. External views of stomachs from (a) wild-type mouse, (b) homozygous 6-month-old CA-AHR male (A3 line) with cysts in the inner curvature, and (c) homozygous 15-month-old CA-AHR male placed on major curvature with large cysts originating from minor curvature with expansion into dorsal and ventral directions. Inside view of stomachs from (d) wild-type mouse and (e) 6-month-old male of Y8 strain. Note the thickened limiting ridge. (f and g) HE-stained stomach (the same as in part (b)). (f) Section of the stomach showing unaffected forestomach (fs), thickened limiting ridge (lr), and glandular stomach (gs) part with aberrant structures and filled with cysts throughout all layers. (g) Close-up of cystic structure penetrating through the muscularis mucosa. (h) Section from a different part showing muscle layers with numerous cysts lined with glandular epithelium. (i) HE-stained stomach from a wild-type mouse. Abbreviations: fs, forestomach; gs, glandular stomach; lr, limiting ridge.

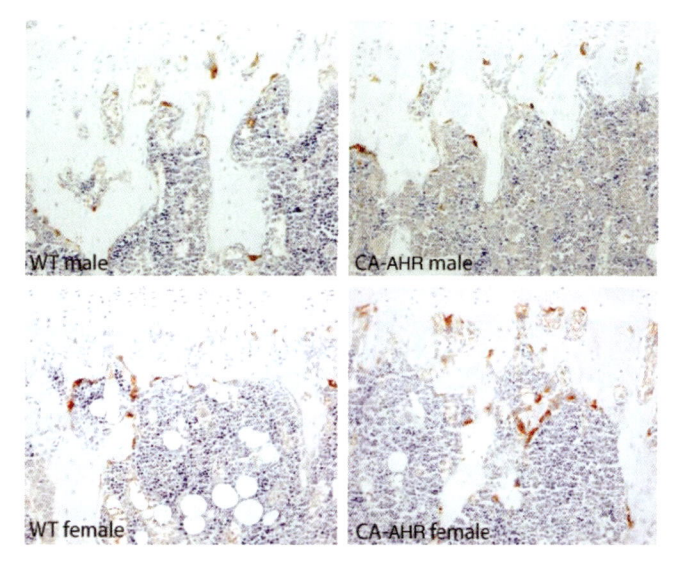

FIGURE 26.2 Immunohistochemistry of cathepsin K-expressing osteoclasts in proximal tibia of wild-type (WT) and transgenic male and female mice.

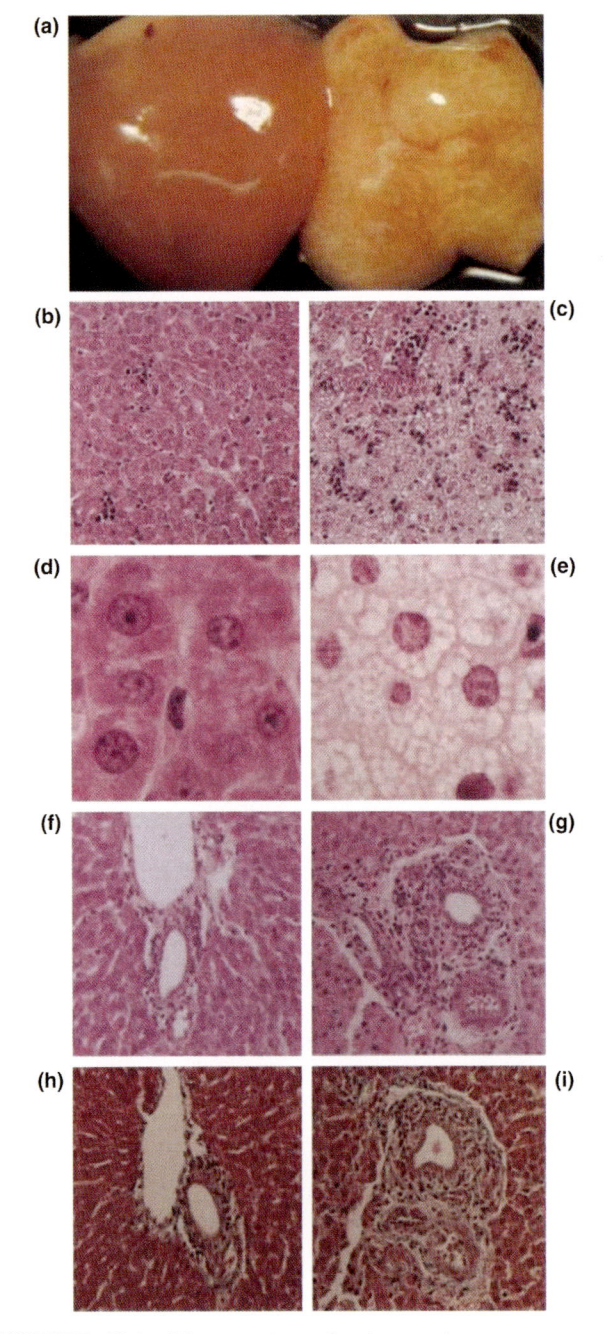

FIGURE 29.1 Macroscopic and microscopic appearance of mouse livers. (a) Macroscopic appearance of livers from 1-week-old wild-type (all at left) and AHR-null (all at right) mice. (b and c) Hematoxylin–eosin-stained sections of livers from 1-week-old wild-type and AHR-null mice demonstrating extensive microvesicular fatty metamorphosis and extramedullary hematopoiesis in the AHR-null animals. (d and e) Higher power view of 1-week-old wild-type and AHR-null livers. (f and g) Hematoxylin–eosin-stained liver sections from 2-week-old wild-type and AHR-null mice demonstrating hypercellularity and fibrosis of the portal tract in the AHR-null animals. (h and i) Trichrome attaining for connective tissue highlighting the portal fibrosis in AHR-null mouse liver compared to wild types.

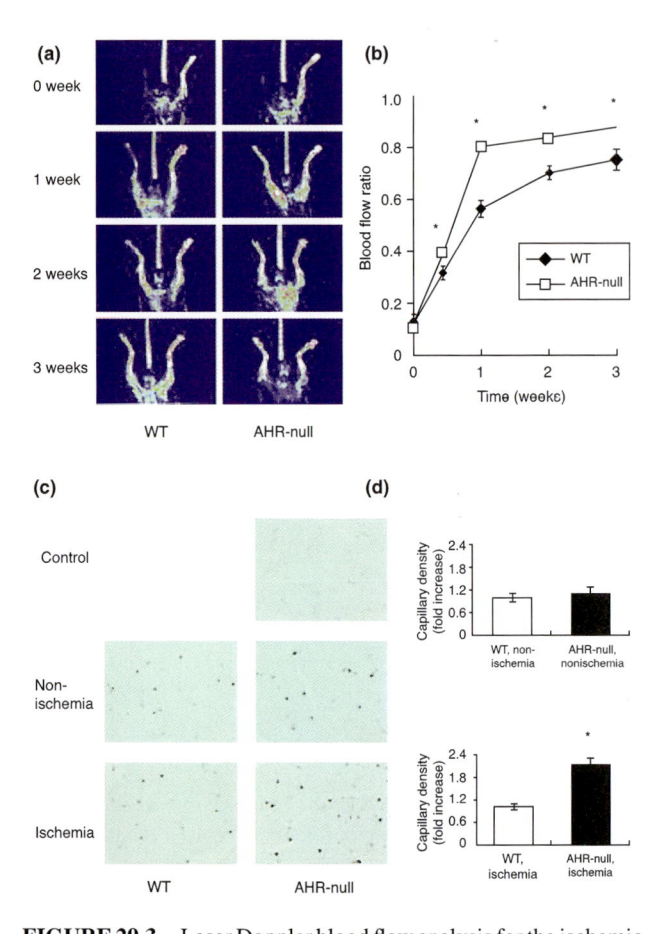

FIGURE 29.3 Laser Doppler blood flow analysis for the ischemic hindlimb of AHR-null or wild-type mice. (a) Laser Doppler perfusion imaging of blood flow in the ischemic hindlimb measured immediately (time 0), 1, 2, and 3 weeks after surgery. Blood flow is color coded, with normal perfusion indicated by red and a marked reduction in blood flow indicated by blue. (b) Quantitation of blood flow expressed as the ratio of blood flow in the ischemic (left) hindlimb to that in the normal (right) hindlimb. Data are mean ± SEM of values from eight animals per group. $^*P < 0.05$ versus corresponding value for wild-type mice. Capillary density in skeletal muscle of the ischemic or nonischemic hindlimb of AHR-null or wild-type mice. (c) Immunostaining of ischemic or nonischemic tissue with antibodies to CD31 (brown) at 3 weeks after arterial ligation. The upper panel (control) represents ischemic tissue from an AHR-null mouse stained with secondary antibodies only. Scale bar = 100 μm. (d) Quantitation of capillary density in ischemic or nonischemic tissue at 3 weeks after surgery. Data are mean ± SEM of values from eight animals per group. $^*P < 0.05$ versus wild-type mice.

+/+   −/−

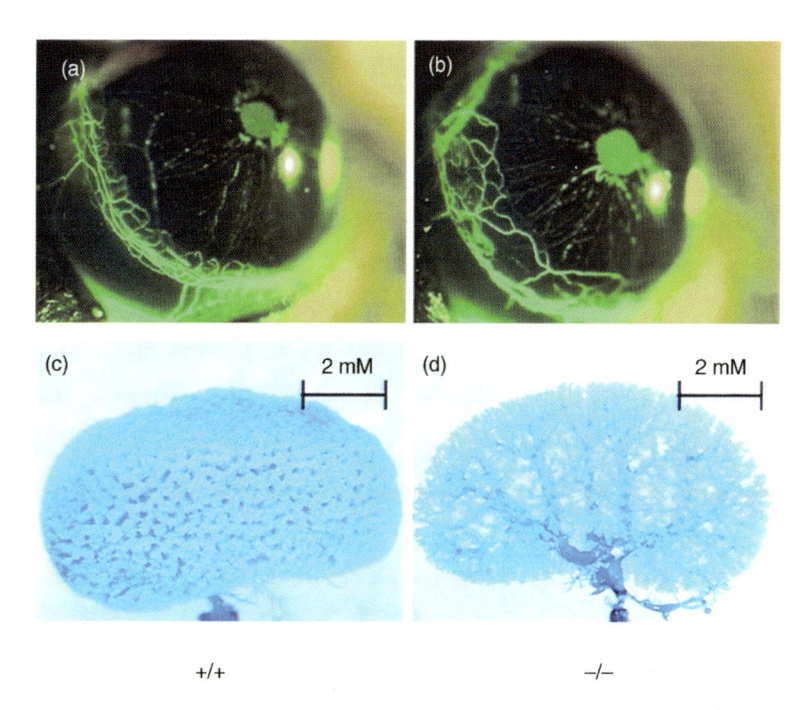

FIGURE 30.2 Altered vascular development in the eyes and kidneys of $Ahr^{-/-}$ mice. Limbal vessels structures (a and b) and corrosion casting of the kidneys (c and d) of $Ahr^{+/+}$ and $Ahr^{-/-}$ mice ($n \geq 4$ per genotype) [34].

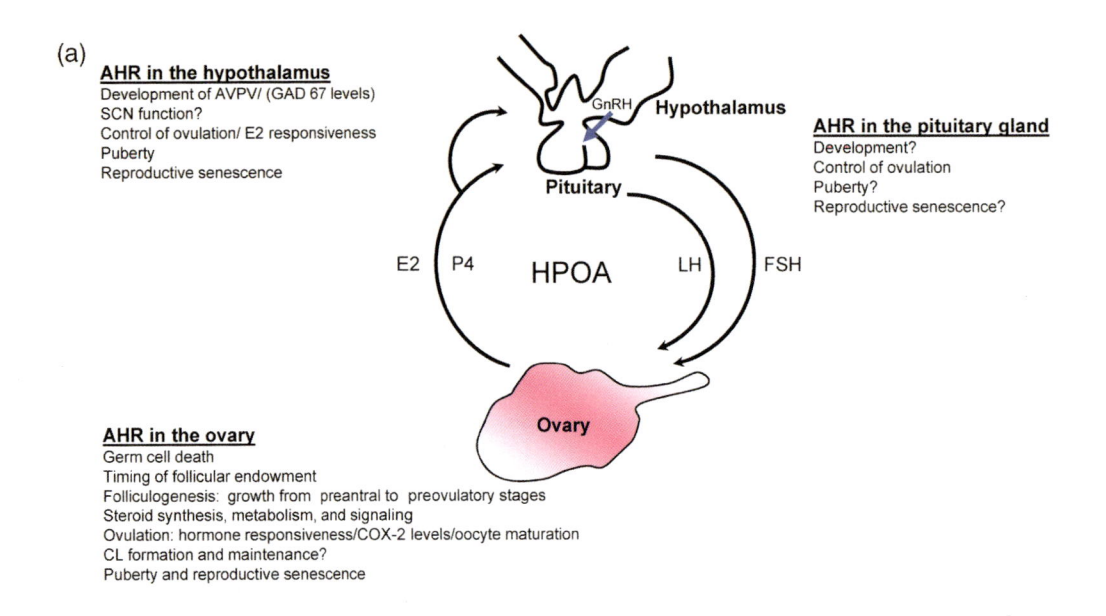

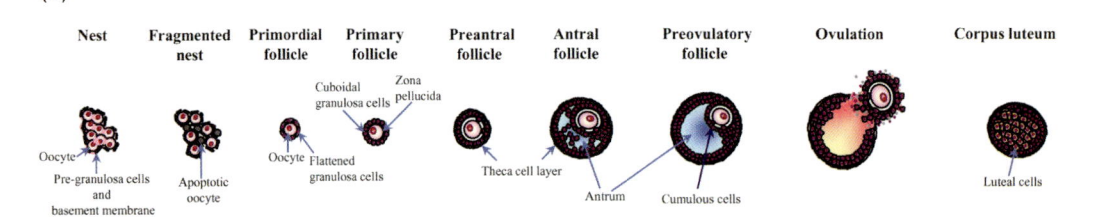

(a)

AHR in the hypothalamus
Development of AVPV/ (GAD 67 levels)
SCN function?
Control of ovulation/ E2 responsiveness
Puberty
Reproductive senescence

AHR in the pituitary gland
Development?
Control of ovulation
Puberty?
Reproductive senescence?

GnRH Hypothalamus
Pituitary
E2 P4 **HPOA** LH FSH
Ovary

AHR in the ovary
Germ cell death
Timing of follicular endowment
Folliculogenesis: growth from preantral to preovulatory stages
Steroid synthesis, metabolism, and signaling
Ovulation: hormone responsiveness/COX-2 levels/oocyte maturation
CL formation and maintenance?
Puberty and reproductive senescence

(b)

Nest | Fragmented nest | Primordial follicle | Primary follicle | Preantral follicle | Antral follicle | Preovulatory follicle | Ovulation | Corpus luteum

Oocyte — Pre-granulosa cells and basement membrane
Apoptotic oocyte
Oocyte — Flattened granulosa cells
Cuboidal granulosa cells — Zona pellucida
Theca cell layer
Antrum — Cumulous cells
Luteal cells

FIGURE 31.1 (a) Involvement of the AHR in controlling the hypothalamus–pituitary–ovarian axis. The female reproductive system is controlled by a feedback system conducted by the HPOA. Hypothalamic GnRH stimulates the anterior pituitary to secrete LH and FSH. The FSH binds to receptors on ovarian follicles and stimulates follicle growth. LH binds to receptors on ovarian cells to stimulate ovulation. The ovarian steroids E2 and progesterone (P4) produced in response to pituitary hormones can feed back to the hypothalamus and anterior pituitary and suppress GnRH, FSH, and LH release via negative feedback. The various roles that the AHR may play in control of the HPOA are listed. (b) The major stages of mammalian folliculogenesis. In mice, primordial follicle formation occurs postnatally and involves oocyte nest breakdown and oocyte death via apoptosis. Primordial follicles develop into primary follicles once the granulosa cells acquire a cuboidal shape. Primary follicles develop into preantral follicles when two or more layers of granulosa cells surround the oocyte and an additional layer of somatic cells called the theca forms outside the basement membrane. Preantral follicle development does not require stimulation by the pituitary gonadotropins. At puberty, the pituitary secreted FSH supports further granulosa cell proliferation and formation of an antrum. Ovulation of the dominant follicles in mice occurs in response to a rise in pituitary produced LH. After ovulation, the remaining somatic cells differentiate into luteal cells to form the corpus luteum.

AHR in pregnancy
Proliferation in morulae and blastocyst embryos
Implantation
Uterine decidualization
Fetal–maternal interactions
Placental development
Vascular remodeling of the placenta during late gestation

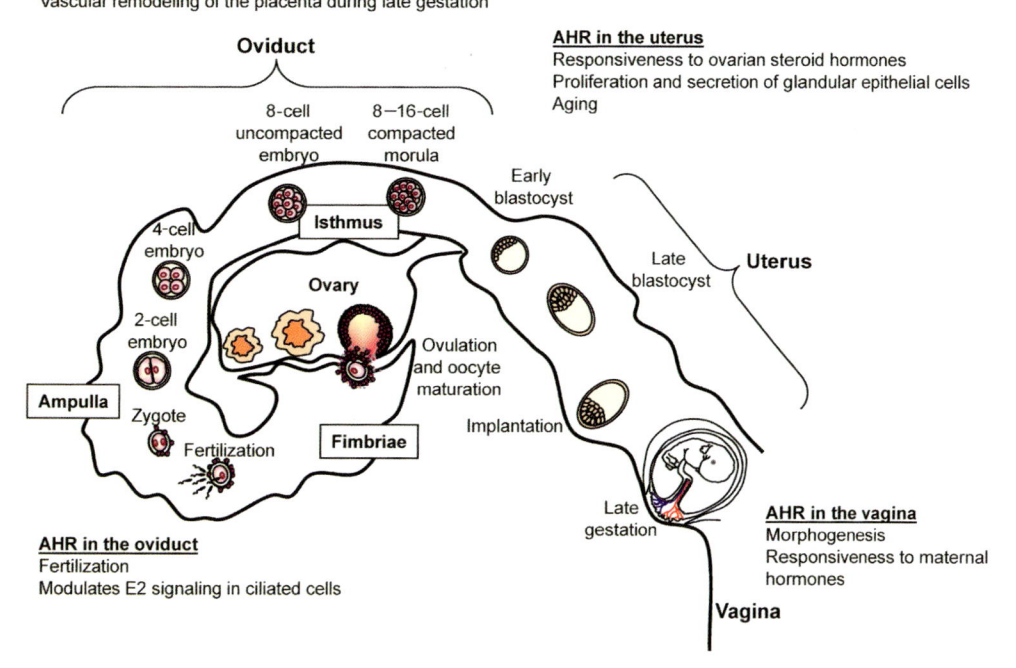

AHR in the uterus
Responsiveness to ovarian steroid hormones
Proliferation and secretion of glandular epithelial cells
Aging

AHR in the vagina
Morphogenesis
Responsiveness to maternal hormones

AHR in the oviduct
Fertilization
Modulates E2 signaling in ciliated cells

FIGURE 31.2 Involvement of the AHR in supporting pregnancy. Following ovulation, fertilization occurs in the ampulla of the oviduct and the embryo undergoes several rounds of mitotic cell division, forming the morula. The propulsion of the embryo through the ampulla and isthmus of the oviduct to the uterotubal junction involves muscle contractions, ciliate activity, and flow of secretions. The morula stage embryo enters the uterine lumen and transforms into a blastocyst. The blastocyst further develops and differentiates before it attaches to the uterine lining to initiate the process of implantation. During late gestation, vascular remodeling of the placenta provides the developing fetus with the required nutrients and oxygen. The various roles that the AHR may play in controlling pregnancy and in development and functioning of the supporting female organs are listed.

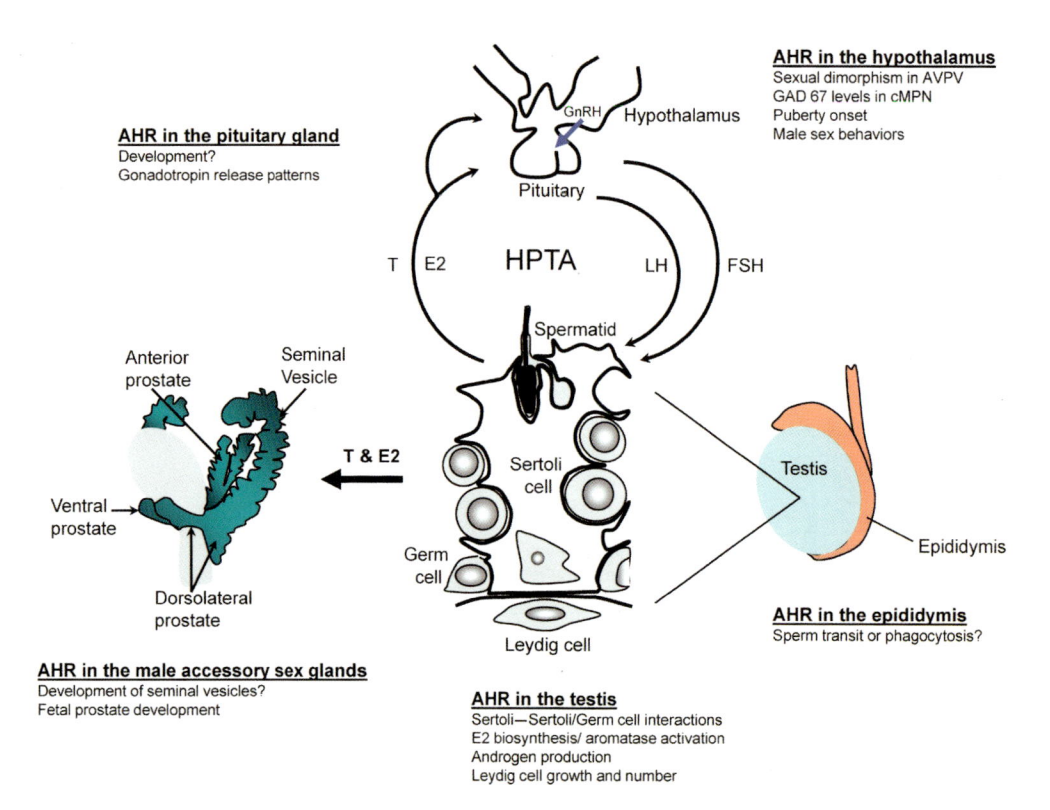

FIGURE 31.3 Involvement of the AHR in controlling the hypothalamus–pituitary–testis axis. During adulthood, the primary role of the testis is spermatogenesis. The process of spermatogenesis is mainly under the control of an endocrine feedback loop regulated by the HPTA. Hypothalamic GnRH stimulates the anterior pituitary to secrete LH and FSH into the bloodstream. The LH targets Leydig cells in the testis to produce primarily testosterone (T) that can be aromatized to E2. FSH stimulates Sertoli cells in the testis to support spermatogenesis. While T and E2 have local effects on spermatogenesis, they also have negative feedback effects on both hypothalamic GnRH secretion and pituitary LH and FSH secretion. The Sertoli cells support the developing germ cells and control release of mature spermatids. The Leydig cells support spermatogenesis by producing steroids. The steroids produced by the testis are necessary for the proper development and functioning of the sexually dimorphic brain regions and male accessory sex glands (i.e., epididymis, prostate, and seminal vesicles). The various roles that the AHR may play in controlling the HPTA and the male accessory sex glands are listed.

18

THE DEVELOPMENTAL TOXICITY OF DIOXIN TO THE DEVELOPING MALE REPRODUCTIVE SYSTEM IN THE RAT: RELEVANCE OF THE AHR FOR RISK ASSESSMENT

DAVID R. BELL

18.1 INTRODUCTION

2,3,7,8-Tetrachlorodibenzo-*p*-dioxin (TCDD) is the iconic ligand and agonist of the aryl hydrocarbon receptor (AHR) [1, 2]. It is one of many halogenated dioxins or furans that can agonize the AHR, and is part of a much broader array of compounds that can activate the AHR [3, 4]. The significance of this lies in the facts that the potent biological effects of these agents are mediated by binding to, and agonism of [5], the AHR, and that some of these compounds are relatively potent toxicants as a result of their agonism of the AHR. The toxicity affects a variety of physiological systems, affecting development, carcinogenesis, and a variety of organs including skin, liver, and thymus, and TCDD has an LD50 of $\sim$1 µg/kg body weight in the guinea pig.

The relatively potent toxicity of these compounds, coupled with the fact that they are present throughout the environment, led to a problem. It was necessary to develop regulatory limits for these compounds, so that human health was protected from the adverse effects that these compounds can cause. Moreover, the fact that multiple dioxin and furan compounds activate the same receptor led to risk assessments based upon the idea that all of these compounds act through a common mechanism, activation of the AHR [6, 7] (see the chapter on TEFs). Thus, the AHR, and activation of the AHR by ligands, is enshrined in risk assessments undertaken by several regulatory bodies [8–10].

While it is apparent that dioxins and related compounds have acute toxic effects in human, notably chloracne [11], these effects would be expected to be among the less potent of TCDD's toxic effects. However, the more potent toxic effects would be expected to result from chronic exposure, and it is much more difficult to establish a relationship between chronic exposure and health effects in human populations. Even for the end point of carcinogenicity in occupationally exposed humans, where exposure was up to three magnitudes above background levels of exposure, it has been difficult to establish whether dioxin causes cancer in humans [12–18]. Given the uncertainty in estimating the risks of TCDD in human populations, it is not possible to base a risk assessment on the human data. Consequently, several risk assessments are based upon the most sensitive effects seen in animal species. The most sensitive end point for TCDD-induced, and AHR-mediated, toxicity was identified by the UK Committee on Toxicity as effects on the male reproductive system of the rat, subsequent to developmental exposure to TCDD. Risk assessments were undertaken by several regulatory bodies around 2001 [8–10], and this end point was used as the basis for setting a tolerable level of exposure for human populations. The nature of this effect in animals, and the study of this effect, is the subject of this chapter.

18.2 INITIAL STUDIES ON AHR-MEDIATED DEVELOPMENTAL REPRODUCTIVE TOXICITY OF TCDD

In 1992, Mably et al. published a series of papers detailing the consequences of a single dose of TCDD to a pregnant

---

The author writes in a personal capacity, and the opinions in this work do not necessarily reflect the opinion of the European Chemicals Agency.

The AH Receptor in Biology and Toxicology, First Edition. Edited by Raimo Pohjanvirta.
© 2012 John Wiley & Sons, Inc. Published 2012 by John Wiley & Sons, Inc.

Holtzman rat on gestational day (GD) 15, notably the effects on the reproductive system of the male offspring [19–21]. These papers provided evidence of an exceptionally potent effect of TCDD on male reproductive end points (lowest observed adverse effect level (LOAEL) < 64 ng TCDD/kg body weight) and showed that the developmental toxicity of TCDD was much more potent than many other effects. For example, the LOAEL for induction of hepatic ethoxyresorufin O-deethylase activity (normally considered a sensitive marker) was 100 ng TCDD/kg [22], or a LOAEL of 10 μg TCDD/kg on various male reproductive end points in adult rats [23]. Moreover, the fact that developmental exposure to TCDD causes a defect in the male reproductive system, which is seen only in the adult male rat after exposure during development, exposes a key weakness in conventional (OECD 414) developmental toxicity tests. The importance of this finding is reflected in the fact that there are over 640 citations of the three papers by Mably et al. [19–21], as of April 2010.

It is important to examine some key methodological issues and findings in Mably et al.'s work. Groups of approximately nine pregnant Holtzman rats were dosed on GD 15 with TCDD (0, 64, 160, 400, and 1000 ng TCDD/kg), and the animals allowed to litter and nurse pups. The offspring were examined at various times, notably postnatal days (PNDs) 32, 49, 63, and 120. There were multiple significant effects of TCDD exposure on the offspring, with decreases in live birth index, cauda epididymal sperm count, testicular sperm production, testis weight, cauda epididymis weight, seminal vesicle weight, ventral prostate weight, anogenital distance, body weight gain, a delay in the time of testis descent, and multiple alterations in sexual behavior. There are several important aspects of this study. First, the fact that so many potent effects were seen provides a corroboration that any individual effect does not arise simply from a chance finding. Second, it is known that antiandrogens exert their effects by binding to androgen receptor, and thereby produce a whole series of mechanistically interlinked consequences on male sexual and accessory organs [24]; thus, the concordance between effects on sperm levels and accessory sexual organs provides evidence that is consistent with a common mode of causation, and corroborates the results. Third, the results reveal an extraordinary potency of TCDD in this experimental system; not only did TCDD induce statistically significant effects after a single maternal dose of 64 ng TCDD/kg, but the authors were also able to undertake power calculations, and stated, "From the dose–response curve for cauda epididymal sperm number (Fig. 4 [sic-actually Figure 5]) the LOAEL can be estimated to be substantially lower than 0.064 μg TCDD/kg (the lowest dose tested)" [19]. The potency and the dose responsiveness of some of these effects are shown for rats killed on PND 63 in Fig. 18.1a. Cauda epididymal sperm levels, daily sperm production, and the weight of the right epididymis, ventral prostate, and seminal vesicles are all significantly decreased in a dose-dependent way, across multiple doses.

18.2.1 Effects of TCDD Are Mediated via the AHR

The effects of TCDD in this system could reasonably be assumed to be mediated via the AHR. Not only is it the case that all potent effects of TCDD tested are mediated through the AHR, but it is also possible to study this effect in mice. In comparison to the rat, the mouse shows greatly reduced sensitivity to TCDD [28], requiring a dose of ~15 μg TCDD/kg to produce an effect on the ventral prostate, that is, a dose approximately 200-fold higher than that used for the rat. This species difference in sensitivity cannot be explained by the well-known strain differences in the AHR that are present in mice [1], since the results were obtained in the C57BL/6 mouse, a strain that is known to be sensitive to TCDD as a result of the sequence of its AHR gene [29]. The effects of TCDD have been well characterized in the C57BL/6 mouse [30], and the decrease in prostate weight after developmental exposure to TCDD has been shown to be absent in the AHR-null mouse [31]. Hence, one can show that the effects of TCDD on the prostrate in mouse are mediated by the AHR. It is an interesting question as to whether the AHR dependency of a dose of 15 μg TCDD/kg to decrease prostate weight in mouse is sufficient evidence to be able to conclude that the effects putatively seen in the rat at a dose of 64 ng TCDD/kg are also mediated by the AHR.

18.2.2 Dose Dependency of Effects of TCDD

The difference in susceptibility to developmental dose of TCDD between rats and mice illustrates effectively that the dose of TCDD that is given to pregnant animals is a fundamentally important variable. In the rat, a dose of greater than ~800 ng TCDD/kg on GD 15 to the pregnant rat results in a significant increase in perinatal lethality, typically on the order of ~10–30% of offspring [21, 25, 32–36]. Although there are reports that gestational exposure to 800 ng TCDD/kg does not cause a statistically significant increase in perinatal lethality [37], many such studies lack the statistical power (i.e., have insufficient litters) to detect an effect killing 10% of offspring. The perinatal lethality of high doses (>800 ng TCDD/kg on GD 15) is accompanied by a variety of effects on growth retardation [38], and Nishimura et al. have shown that there is significant perturbation of a number of essential systems, including thyroid and kidney function [39–41]. The severity and potentially confounding nature of such severe effects on reproductive parameters [42] suggest that it is not necessarily sensible to extrapolate from effects on reproductive parameters seen at doses in excess of 800 ng TCDD/kg to effects seen at doses of 200 ng TCDD/kg and below, where there is no measurable lethality or growth retardation. This chapter will therefore focus on dose levels

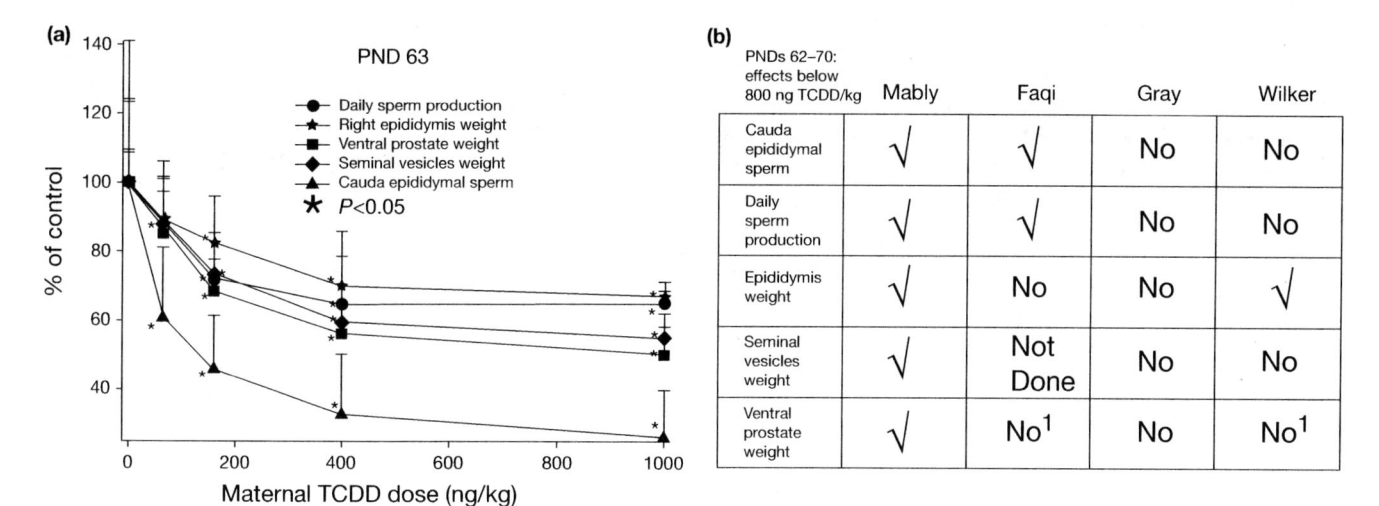

FIGURE 18.1 Male reproductive effects of TCDD after developmental administration. (a) Some of the effects noted by Mably et al. [19–21]. Holtzman rats were dosed on GD 15 with the indicated dose of TCDD (ng/kg), and daily sperm production (circle), right epididymis weight (star), ventral prostate weight (square), seminal vesicle weight (diamond), and cauda epididymal sperm number (triangle) were measured in the F1 males on PND 63. Results are normalized to control (set as 100%), and are presented as mean ± SD. The "*" symbol indicates $P < 0.05$; note that at 64 ng/kg, right epididymis weight, seminal vesicles weight, and ventral prostate weight are not significantly different from control. (b) Comparison of effects of TCDD on F1 males at PNDs 62–70, for an approximate maternal TCDD dose of approximately 500 ng/kg and below. Data from Refs 19, 21, 25, 26, and 27 are compared on the basis of stated statistically significant results. The "√" symbol means that there was a statistically significant effect. "No" means no statistically significant effect, "Not Done" means that the measurement is not reported, and the superscript "1" refers to measurements on whole prostate, rather than ventral prostate weight.

less than 1 μg TCDD/kg, and these are of most importance for risk assessment.

18.3 REPRODUCIBILITY OF THE EFFECTS OF TCDD ON THE DEVELOPING MALE REPRODUCTIVE SYSTEM

These developmental effects of <1 μg TCDD/kg on the male reproductive system were studied in three other laboratories prior to 2000 [25–27], using either an acute dose of TCDD on GD 15 [25, 27] or a chronic dosing regime [26], which is approximated to an equivalent acute dose on the basis of the hepatic TCDD concentration on GD 21 [43] (David R. Bell, unpublished data). These studies revealed markedly different outcomes to the diverse spectrum of effects seen by Mably et al. [19–21]. For example, the results of replication at PNDs 62–70 are shown in Fig. 18.1b for doses of TCDD less than 800 ng/kg. After a maternal dose of <800 ng TCDD/kg, the offspring at PNDs 62–70 had statistically significant changes in 2/5 [26], 0/5 [25], and 1/5 [27] of the five findings in Fig. 18.1b reported by Mably et al. However, two of these three studies reported statistically significant reductions in epididymal sperm levels at doses below 800 ng TCDD/kg [25, 26]. While the magnitude of the reduction in epididymal sperm levels and the overall potency and spectrum of effects seen were much less than those observed by Mably et al. [19–21], the consistency between the three studies was sufficient for reductions in sperm levels to be regarded as the most potent adverse effect of TCDD, and it was used as the basis for some risk assessments [8–10].

Given the importance of this scientific area to the risk assessments, seven additional studies have been published examining the effects of developmental exposure to TCDD on the male reproductive system of offspring [32, 33, 37, 44–47] (see the paper by Rebourcet et al. [48] for an additional discussion).[1] The consistency of the outcomes from all of these studies is examined. This chapter concentrates on effects seen at maternal doses of less than

---

[1] Subsequent to the analyses presented in the text, Rebourcet et al. published a further study of the developmental reproductive toxicity of TCDD in rats [48]. At the top dose of 0.2 μg TCDD/kg body weight, there was no statistically significant effect on epididymal sperm reserves or daily sperm production, when using the litter as the basis of analysis. There were no detectable effects on testis or epididymal histology. There was substantial induction of CYP1A1 RNA in the livers of offspring at PNDs 5 and 28. Two reviews of the developmental reproductive toxicity of TCDD have been published [49, 50].

800 ng TCDD/kg (or equivalent), since these are the doses that are of interest for risk assessment, and on effects on PNDs 62–70 and PND 120+, since sperm measurements at earlier time points have been deprecated as unreliable [51, 52]. Effects on epididymal sperm levels and on prostate weight are further considered, because of the magnitude and potency of these effects in the original papers of Mably et al.

18.3.1 Effects on Prostate Weight

Figure 18.2a shows the data from studies examining the effects at PNDs 62–70 on prostate weight. With the exception of the original report [21], there are no significant effects of TCDD on prostate weight at doses below 1 μg/kg at PNDs 62–70. There are significant effects on prostate weight at doses of ~1 μg/kg, when TCDD is frankly toxic. The fact that eight studies showed no effect on prostate weight at PNDs 62–70 at doses below 1 μg TCDD/kg calls into question why the low-dose effects of TCDD on prostate in the rat were not reproducible. The studies have sufficient statistical power to detect the effect reported by Mably et al.; for example, Bell et al. [33] report a 90% power for detecting a 10% decrease in ventral prostate weight, compared with the ~40% decrease in ventral prostate weight reported by Mably et al.

In F1 rats at PND 120+, Mably et al. reported no statistically significant effect of maternal TCDD treatment on ventral prostate weight below 1 μg TCDD/kg (Fig. 18.2b). A decrease in ventral prostate weight was seen after an acute dose of 200 ng TCDD/kg [45], or a chronic dose approximately equivalent to an acute dose of 400 ng TCDD/kg [44]; notably, neither of these two studies accounted for litter as a source of variation in the statistical analysis, and the failure to account for litter effects can lead to spuriously inflated estimations of significance [53]. Bell et al. [33] reported a statistically significant increase in prostate weight after chronic dosing of 8 ng TCDD/kg/day (approximately equivalent to an acute dose of 200 ng TCDD/kg). Thus, after developmental exposure to <1 μg TCDD/kg, statistically significant reductions in prostate weight have been reported in only two studies, both of which used an inappropriate statistical technique.

18.3.2 Effects on Epididymal Sperm Levels

Figure 18.2c and d shows the effect of maternal exposure to TCDD on epididymal sperm levels in F1 rats, at PNDs 62–70 and 120+, respectively. Faqi et al. [26] reported a decrease in epididymal sperm at ~50 ng TCDD/kg (the chronic dose regime used was approximated to an acute dose on the basis of liver TCDD concentration; David R. Bell, unpublished data), Gray et al. [25] saw effects at 200 ng TCDD/kg in adult (but not PND 63) rats, and Simanainen et al. [47] saw a reduction after 300 ng TCDD/kg in line C rats, but not at lower doses (Fig. 18.2c and d). The magnitude of these reductions in epididymal sperm count was less than that observed by Mably et al. [19], and was frequently within 30% of the values obtained from control rats. Bell et al. [32] observed a statistically significant increase in epididymal sperm counts of 30–38% at the top two doses of TCDD, noted that these values were not accompanied by an effect on testicular sperm production and were within the range of historical control epididymal sperm counts, and concluded that the statistical significance of these results arose from random variation. Statistically significant reductions in epididymal sperm numbers have been observed in only 4 out of 11 studies at doses below 1 μg TCDD/kg, and in only 3 studies at doses below 300 ng TCDD/kg.

18.3.3 Effects on Delays in Development

Table 18.1 shows measurements of developmental delay in the offspring of TCDD-exposed dams, and TCDD-induced delay in balanopreputial separation (BPS, a marker of male puberty) has been demonstrated in each study where it has been measured. After a single dose of TCDD on GD 15, Gray et al. [25] reported a delay after a maternal dose of 200 ng TCDD/kg, Yonemoto et al. [37] reported a significant delay after a maternal dose of 200 ng TCDD/kg, and Bell et al. [32] found no significant effect at 50 or 200 ng TCDD/kg, but only at the highest dose of 1000 ng TCDD/kg; however, in the last experiment, the data at 200 ng TCDD/kg were just above the threshold for statistical significance. After subchronic dosing of TCDD, Faqi et al. [26] reported a significant delay of unspecified magnitude in their medium- and high-dose groups, and Bell et al. [33] showed that all three dose groups (2.4, 8, and 46 ng TCDD/kg/day) caused a significant delay in BPS. Although maternal TCDD administration reduces body weight gain in offspring, the body weight of males at PND 21 or 42 did not correlate with the delay in BPS, thus excluding decreased weight gain as a cause of delayed BPS [33]. Thus, delay in BPS is consistently found to be an adverse effect in the offspring of animals dosed with TCDD, and may be the most sensitive adverse effect. The direct comparison of acute and chronic dosing of TCDD in the same strain of rats using the same methodology [32, 33] provides evidence that chronic dosing of TCDD has more potent effects in offspring.

18.3.4 Statistical Analysis and Overall Consistency of Reported Effects

The original observations by Mably et al. [21] showed that administration to dams of 160 or 400 ng TCDD/kg reduced ventral prostate weight in F1 males, and these effects were highly significant at PND 63 but not in adult rats. In repeat studies, only 2 (out of 10 studies) show a statistically significant decrease in prostate weight after maternal administration of <1 μg TCDD/kg; both these studies show an effect at a time point when Mably et al. saw no effect, and

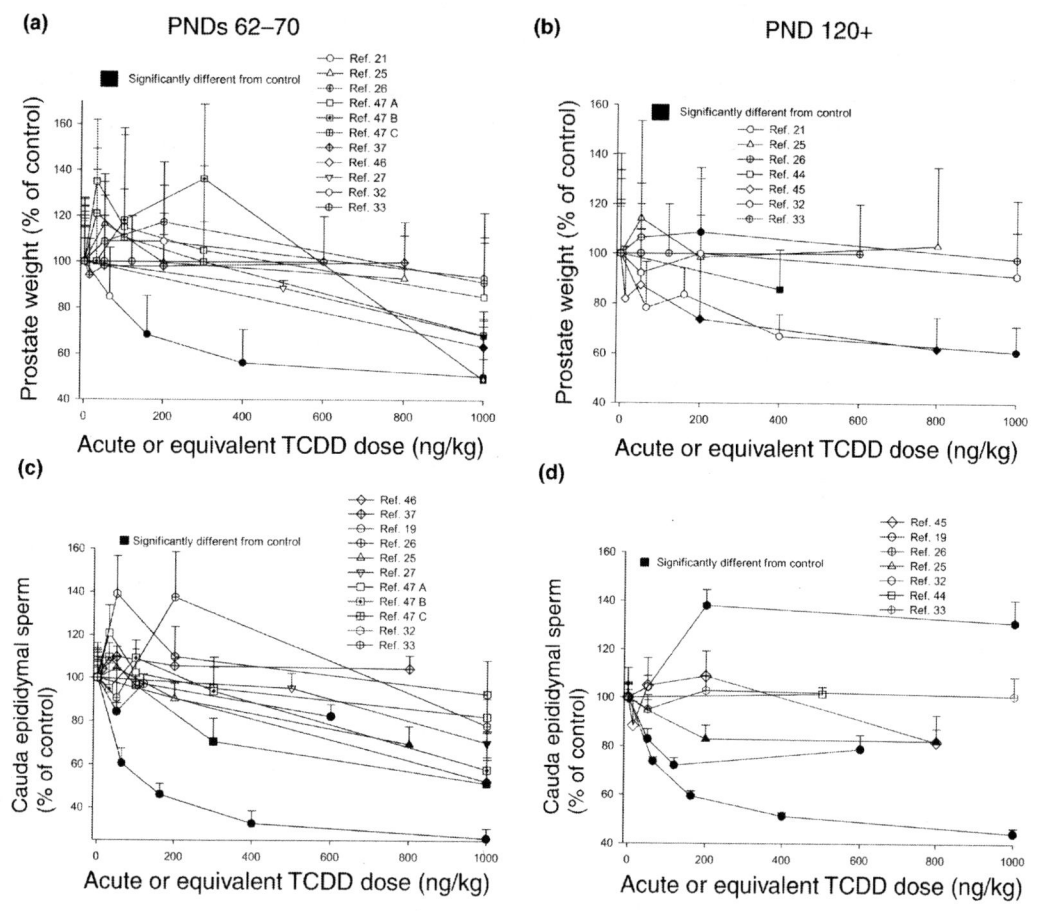

FIGURE 18.2 Developmental exposure to TCDD and reproductive endpoints in male F1. (a) Effects on prostate weight at PNDs 62–70. All experiments are normalized to control values of 100%, and results are shown as mean ± SD. Statistical significance at $P < 0.05$ is shown by filled symbols. Data are from Ref. 21 (circle), Ref. 25 (triangle), Ref. 26 (circle with cross), Ref. 47 (line A: square; line B: square with dot; line C: square with cross), Ref. 37 (diamond with cross), Ref. 46 (diamond), Ref. 27 (inverted triangle), Ref. 32 (hexagon), and Ref. 33 (hexagon with cross). Results are ventral prostate weight, except for Refs 32, 27, and 26, which are prostate weight. Doses greater than 1 μg TCDD/kg are not shown. Studies with chronic dosing of TCDD are shown by the equivalent acute doses, based on tissue concentrations of TCDD [43]. (b) Effects on prostate weight at PND 120+. All experiments are normalized to control values of 100%, and results are shown as mean ± SD. Statistical significance at $P < 0.05$ is shown by filled symbols. Data are from Ref. 21 (circle), Ref. 25 (triangle), Ref. 26 (circle with cross), Ref. 44 (square), Ref. 45 (diamond), Ref. 32 (hexagon), and Ref. 33 (hexagon with cross). Results are ventral prostate weight, except for Refs 32 and 26, which are prostate weight. (c) Effects on epididymal sperm count at PNDs 62–70. All experiments are normalized to control values of 100%, and results are shown as mean ± SEM. Statistical significance at $P < 0.05$ is shown by filled symbols. Data are from Ref. 19 (circle), Ref. 25 (triangle), Ref. 26 (circle with cross), Ref. 47 (line A: square; line B: square with dot; line C: square with cross), Ref. 37 (diamond with cross), Ref. 46 (diamond), Ref. 27 (inverted triangle), Ref. 32 (hexagon), and Ref. 33 (hexagon with cross). Doses greater than 1 μg TCDD/kg are not shown. Studies with chronic dosing of TCDD are shown by the equivalent acute doses, based on tissue concentrations of TCDD [43]. (d) Effects on epididymal sperm counts at PND 120+. All experiments are normalized to control values of 100%, and results are shown as mean ± SEM. Statistical significance at $P < 0.05$ is shown by filled symbols. Data are from Ref. 19 (circle), Ref. 25 (triangle), Ref. 26 (circle with cross), Ref. 44 (square), Ref. 45 (diamond), Ref. 32 (hexagon), and Ref. 33 (hexagon with cross).

these studies [44, 45] do not account for litter differences as a source of variation in statistical analyses; failure to account for interlitter variation may result in spuriously inflated estimations of significance [53]. Eight of the 11 studies show no significant decrease in prostate weight from control values after maternal doses of <1 μg TCDD/kg, so the original

TABLE 18.1 Effect of Maternal Dose of TCDD on Developmental Delay

Study	Rat Strain	Dosing Regime	Dose Level	Developmental End Point
[26]	Wistar	Loading and weekly maintenance dose, subcutaneous dose	0.3 μg/kg loading dose, followed by 0.06 μg/kg/week	"Age at preputial separation was slightly delayed," no statistical analysis shown ($n = 17–22$) [0.06 μg/kg loading dose, followed by 0.012 μg/kg/week]
[44]	Holtzman	Loading and weekly maintenance dose, by gavage	0.4 μg/kg loading dose, followed by 0.08 μg/kg/week	N.D. [not applicable]
[33]	Wistar(Han)	Chronic dietary dosing for >12 weeks	0.046 μg/kg/day	Yes, delay in BPS of 1.8, 1.9, and 4.4 days ($n = 18–25$) [0.0024 μg/kg/day]
[19–21]	Holtzman	Single dose on GD 15, oral dose	1 μg/kg	Yes, testis descent and eye opening. BPS N.D. ($n = 9$) [0.16 μg/kg testis descent]
[34]	Holtzman	Single oral dose on GD 15; also *in utero* and/or lactational	1 μg/kg	Yes, 2.4- or 3.4-day delay in BPS ($n = 9–11$) [1 μg/kg]
[54]	Long–Evans hooded	Single dose on GD 15 or GD 8, by gavage	1 μg/kg	Yes, 3.6-day delay in BPS ($n = 6–8$) [1 μg/kg]
[35]	Holtzman	Single dose on GD 15, oral dose	1 μg/kg	Yes, 2.1-day delay in BPS ($n = 30–32$) [1 μg/kg]
[27]	Sprague Dawley	Single dose on GD 15, by gavage	2 μg/kg	N.D. [not applicable]
[36]	Holtzman	Single dose on GD 15, oral dose	1 μg/kg	Yes, 2-day delay in BPS ($n = 34–39$) [1 μg/kg]
[25]	Long–Evans hooded	Single dose on GD 15, by gavage	0.8 μg/kg	Yes, delay in eye opening, delay in BPS of 1.5 and 3.1 days ($n = 10–12$) [0.05 and 0.2 μg/kg, respectively]
[45]	Holtzman	Single dose on GD 15, per os	0.8 μg/kg	N.D. [not applicable]
[46]	Sprague Dawley	Single dose on GD 15 or GD 18, oral dose	1 μg/kg	N.D. [not applicable]
[47]	Wistar(Han) × Long–Evans crosses	Single dose on GD 15, by gavage	1 μg/kg	N.D. [not applicable]
[37]	Long–Evans	Single dose on GD 15, oral dose	0.8 μg/kg	Yes, delay in BPS ($n = 9–12$) [0.2 μg/kg]
[32]	Wistar(Han)	Single dose on GD 15, by gavage	1 μg/kg	Yes, delay in BPS of 2.8 days ($n = 15–21$) [1 μg/kg]

a The strain and TCDD dosing regime are indicated for each study. The top dose level of TCDD is given per kg body weight. Measures of developmental delay are shown and effect size indicated, together with the number of litters (n) and the lowest dose level at which the effect was detected, in square brackets. N.D. = not done.

report of marked decreases in ventral prostate weight in F1 rats after a maternal dose of <1 μg TCDD/kg [21] is not reproducible in the majority of laboratories.

For analysis of epididymal sperm counts, 4 studies out of 11 show a reduction after maternal administration of <1 μg TCDD/kg [19, 25, 26, 47]. The reductions in epididymal sperm count at maternal doses of <1 μg TCDD/kg are less than 30% from control values, except for Ref. 19. Ashby et al. [55] have shown the importance of careful interpretation of minimal changes in testicular sperm levels by using historical control data, and epididymal sperm counts doubtless require similar consideration of historical control data. It must be perilous to interpret reductions of 30% or less in epididymal sperm levels as being treatment-related reductions, especially when the relationship with dose is ambiguous [25, 26]. The fact that 7 of the 11 studies (Fig. 18.2) find no significant decrease after maternal dosing of <1 μg TCDD/kg on F1 epididymal sperm counts calls into question whether this effect is reproducible. The studies that found an effect of <1 μg TCDD/kg of TCDD on prostate weight did not find an effect on epididymal sperm levels, and vice versa, except the studies of Mably et al. [19, 21]. The demonstration of adverse effects on sperm counts and on accessory sexual organs provides a consistency that is causally related to the mechanism of action, for example, through antiandrogenic signaling [56]. Figure 18.2 shows that this corroboration through effects on multiple end points is not present for epididymal sperm count and prostate weights, and is similarly evanescent for all the other end points shown in Fig. 18.1a (data not shown).

18.4 STRAIN DIFFERENCES IN RESPONSE TO TCDD

The toxicology of TCDD has been instrumental in revealing profound species and strain differences in toxicology, from the marked differences between guinea pig and mouse in acute toxicity through to the differences in susceptibility of different mouse strains [1]. In the case of mouse strains, it has been shown that much (but not all) of the differences between mouse strains can be attributed to strain differences in the AHR [26]. However, the understanding of the contribution of specific genetic factors to the susceptibility of rats to toxicity from TCDD remains incomplete.

The induction of CYP1A1 RNA has been investigated in eight rat strains [57] at two dose levels of TCDD. Although the authors felt able to characterize rat strains into "low" and "high" responders on the basis of induction of CYP1A1 RNA, the use of end point (rather than real-time) PCR for analysis, coupled with incomplete characterization of the dose–response curves, renders this characterization questionable. Any conclusion that the postulated increase in responsiveness to CYP1A1 induction was associated with increased responsiveness for measures of toxicity would be without factual basis. Strain differences in toxicity in the rat have been examined. The Wistar(Han) (Kuopio) rat is resistant to specific toxic effects of TCDD [58, 59], and one allele for resistance to the acute lethality of TCDD in a Long–Evans × Wistar(Han) cross was subsequently shown to cosegregate with an AHR allele [60, 61], suggesting that the AHR allele was responsible for the phenotype. The polymorphism in the Wistar(Han) (Kuopio) AHR has been characterized [62, 63]. However, it was discovered that the original rat strains used were an inbred Long–Evans line and a small colony of outbred Wistar(Han) rats (derived from Zentralinstitut fur Versuchstierzucht GmbH, Hannover, FRG). The inbred Long–Evans rats had significantly different susceptibility to the acute lethality of TCDD compared with outbred Long–Evans rats, and the small Finnish colony of Wistar (Han) rats had different susceptibility to the acute lethality of TCDD, compared with a different batch of Wistar(Han) rats [64], and hence the Finnish rats were named as the inbred L-E (Turku/AB) and outbred H/W (Kuopio) strains. It is therefore not clear to what extent the characteristics of the Kuopio strain of Wistar(Han) rats are representative for other Wistar(Han), or Wistar, strain rats.

Much less is known about the role of genetic factors in controlling sensitivity to TCDD for developmental toxicity. It has been shown that the inbred Long–Evans and H/W (Kuopio) rats display differential effects upon TCDD exposure [65], with different spectra of teratogenic effects at doses of 5 or 10 μg TCDD/kg body weight. The genetic components that underlie these differential effects are unclear; although the Long–Evans and H/W (Kuopio) rats were crossed and the offspring typed for genetic status and acute lethality in adults, the segregation of susceptibility to developmental toxicity in the resulting animals was not examined, and it is unknown whether the susceptibility to acute lethality cosegregates with patterns of developmental toxicity to TCDD. Simanainen et al. found that line C rats were more sensitive to TCDD, compared with line A and B rats [47], specifically with slight reductions in daily sperm production and epididymal sperm reserves at 0.3 and 1 μg TCDD/kg body weight. The importance of thoroughly characterizing the genetics of the rat strain was shown by Bell et al. [43] and Jiang et al. [66], who characterized the developmental toxicity of TCDD in the CRL:WI(Han) rat, which is a Wistar (Han) rat closely related to the H/W (Kuopio) strain. The CRL:WI(Han) rat showed ~20% offspring lethality after doses of 1 μg/kg on GD 15 [32], or 46 ng/kg/day [33]—this is in striking contrast to the resistance of the adult Wistar(Han) rat to lethality from doses as high at 10,000 μg/kg [58, 59]. Direct genotyping of rats failed to yield any credible correlation with the AHR allele [63] with multiple measures of developmental toxicity in two independent studies in the CRL:WI(Han) rat [66].

Studies of developmental reproductive toxicity in the rat have featured a variety of rat strains, including the Holtzman, Sprague Dawley, Wistar, Wistar(Han), and Long–Evans rats. However, there is little evidence that the strain is an important determinant of toxicity. For example, the statistical power was a key determinant of being able to detect lethality in offspring, but offspring lethality was detected in Long–Evans, Holtzman, and Wistar(Han) rats after a maternal dose of ~1 μg TCDD/kg [38]. Likewise, Table 18.1 shows that developmental delays were consistently detected in a variety of rat strains. It is difficult to explain the effects of TCDD on developmental reproductive toxicity by appealing to strain differences, since Holtzman rats [19, 44, 45], Long–Evans rats [25, 37], and Wistar/Wistar(Han) rats [26, 32, 33] have all been used with markedly divergent results.

18.5 CONCLUSIONS

The failure to replicate the magnitude or variety of responses caused by maternal doses of <1 μg TCDD/kg in the original work of Mably et al. [19–21], since the risk assessments in 2001, is now a matter of record. While there are reports of statistically significant adverse effects in offspring after maternal administration of <1 μg TCDD/kg, these reports are in the minority for prostate weight and epididymal sperm counts, and in the case of epididymal sperm counts are frequently within the range of historical variation seen in other laboratories [32, 46]. It is unclear why it has not been possible to replicate these findings, despite extensive efforts, and there are numerous potential explanations that have not been tested, including the effect of diet, strain drift, rat housing, and so on. High maternal doses of TCDD cause multiple severe adverse effects, and mechanistic studies at doses >0.8 μg TCDD/kg are likely to be confounded by indirect effects. Maternal pharmacokinetics of TCDD vary considerably between acute and chronic dosing [38, 43], and these two differing dosing regimes have been shown to impact the potency of TCDD at inducing adverse effects [32, 33, 38, 43]. Understanding how and when TCDD operates to cause adverse effects in F1 animals after low dose maternal exposure is a key research need, with consequences for current risk evaluations of dioxins and dioxin-like compounds.

REFERENCES

1. Poland, A. and Knutson, J. C. (1982). 2,3,7,8-Tetrachlorodibenzo-*para*-dioxin and related halogenated aromatic hydrocarbons: examination of the mechanism of toxicity. *Annual Review of Pharmacology and Toxicology*, 22, 517–554.
2. Bradshaw, T. D. and Bell, D. R. (2009). Relevance of the aryl hydrocarbon receptor (AhR) for clinical toxicology. *Clinical Toxicology*, 47, 632–642.
3. Denison, M. S. and Nagy, S. R. (2003). Activation of the aryl hydrocarbon receptor by structurally diverse exogenous and endogenous chemicals. *Annual Review of Pharmacology and Toxicology*, 43, 309–334.
4. Denison, M. S., Pandini, A., Nagy, S. R., Baldwin E. P., and Bonati, L. (2002). Ligand binding and activation of the Ah receptor. *Chemico-Biological Interactions*, 141, 3–24.
5. Bazzi, R., Bradshaw, T. D., Rowlands, J. C., Stevens, M. F. G., and Bell, D. R. (2009). 2-(4-Amino-3-methylphenyl)-5-fluorobenzothiazole is a ligand and shows species-specific partial agonism of the aryl hydrocarbon receptor. *Toxicology and Applied Pharmacology*, 237, 102–110.
6. Van den Berg, M., Birnbaum, L., Bosveld, A. T. C., Brunstrom, B., Cook, P., Feeley, M., Giesy, J. P., Hanberg, A., Hasegawa, R., Kennedy, S. W., Kubiak, T., Larsen, J. C., van Leeuwen, F. X. R., Liem, A. K. D., Nolt, C., Peterson, R. E., Poellinger, L., Safe, S., Schrenk, D., Tillitt, D., Tysklind, M., Younes, M., Waern, F., and Zacharewski, T. (1998). Toxic equivalency factors (TEFs) for PCBs, PCDDs, PCDFs for humans and wildlife. *Environmental Health Perspectives*, 106, 775–792.
7. Van den Berg, M., Birnbaum, L. S., Denison, M., De Vito, M., Farland, W., Feeley, M., Fiedler, H., Håkansson, H., Hanberg, A., Haws, L., Rose, M., Safe, S., Schrenk, D., Tohyama, C., Tritscher, A., Tuomisto, J., Tysklind, M., Walker, N., and Peterson, R. E. (2006). The 2005 World Health Organization reevaluation of human and mammalian toxic equivalency factors for dioxins and dioxin-like compounds. *Toxicological Sciences*, 93, 223–241.
8. Committee on Toxicity of Chemicals in Food, Consumer Products and the Environment (2001). Statement on the tolerable daily intake for dioxins and dioxin-like polychlorinated biphenyls. COT/2001/07.
9. SCF (2001). Opinion of the Scientific Committee on Food on the risk assessment of dioxins and dioxin-like PCBs in food. Update based on new scientific information available since the adoption of the SCF opinion of 22nd November 2000. Adopted on May 30, 2001.
10. JECFA (2001). 57th Meeting of the Joint FAO/WHO Expert Committee on Food Additives, Rome, June 5–14, 2001. Summary and Conclusions, I. WHO Technical Report Series 909, pp. 121–149.
11. Geusau, A., Abraham, K. Geissler, K., Sator, M. O., Stingl, G., and Tschachler, E. (2001). Severe 2,3,7,8-tetrachlorodibenzo-*p*-dioxin (TCDD) intoxication: clinical and laboratory effects. *Environmental Health Perspectives*, 109, 865–869.
12. Collins, J. J., Bodner, K., Aylward, L. L., Wilken, M., and Bodnar, C. M. (2009). Mortality rates among trichlorophenol workers with exposure to 2,3,7,8-tetrachlorodibenzo-*p*-dioxin. *American Journal of Epidemiology*, 170, 501–506.
13. Onozuka, D., Yoshimura, T., Kaneko, S., and Furue, M. (2009). Mortality after exposure to polychlorinated biphenyls and polychlorinated dibenzofurans: a 40-year follow-up study of Yusho patients. *American Journal of Epidemiology*, 169, 86–95.
14. Steenland, K., Bertazzi, P., Baccarelli, A., and Kogevinas, M. (2004). Dioxin revisited: developments since the 1997 IARC classification of dioxin as a human carcinogen. *Environmental Health Perspectives*, 112, 1265–1268.
15. Stayner, L., Steenland, K., Dosemeci, M., and Hertz-Picciotto, I. (2003). Attenuation of exposure–response curves in occupational cohort studies at high exposure levels. *Scandinavian Journal of Work Environment & Health*, 29, 317–324.
16. Starr, T. B. (2003). Significant issues raised by meta-analyses of cancer mortality and dioxin exposure. *Environmental Health Perspectives*, 111, 1443–1447.
17. Bodner, K. M., Collins, J. J., Bloemen, L. J., and Carson, M. L. (2003). Cancer risk for chemical workers exposed to 2,3,7,8-tetrachlorodibenzo-*p*-dioxin. *Occupational and Environmental Medicine*, 60, 672–675.
18. Steenland, K., Deddens, J., and Piacitelli, L. (2001). Risk assessment for 2,3,7,8-tetrachlorodibenzo-*p*-dioxin (TCDD) based on an epidemiologic study. *American Journal of Epidemiology*, 154, 451–458.
19. Mably, T. A., Bjerke, D. L., Moore, R. W., Gendronfitzpatrick, A., and Peterson, R. E. (1992). In utero and lactational exposure of male rats to 2,3,7,8-tetrachlorodibenzo-*para*-dioxin. 3. Effects on spermatogenesis and reproductive capability. *Toxicology and Applied Pharmacology*, 114, 118–126.
20. Mably, T. A., Moore, R. W., Goy, R. W., and Peterson, R. E. (1992). In utero and lactational exposure of male rats to 2,3,7,8-tetrachlorodibenzo-*para*-dioxin. 2. Effects on sexual behavior and the regulation of luteinizing hormone secretion in adulthood. *Toxicology and Applied Pharmacology*, 114, 108–117.
21. Mably, T. A., Moore, R. W., and Peterson, R. E. (1992). *In utero* and lactational exposure of male rats to 2,3,7,8-tetrachlorodibenzo-*para*-dioxin. 1. Effects on androgenic status. *Toxicology and Applied Pharmacology*, 114, 97–107.
22. Vandenheuvel, J. P., Clark, G. C., Kohn, M. C., Tritscher, A. M., Greenlee, W. F., Lucier, G. W., and Bell, D. A. (1994). Dioxin-responsive genes: examination of dose–response relationships

using quantitative reverse transcriptase-polymerase chain reaction. *Cancer Research*, 54, 62–68.

23. Simanainen, U., Adamsson, A., Tuomisto, J. T., Miettinen, H. M., Toppari, J. Tuomisto, J., and Viluksela, M. (2004). Adult 2,3,7,8-tetrachlorodibenzo-*p*-dioxin (TCDD) exposure and effects on male reproductive organs in three differentially TCDD-susceptible rat lines. *Toxicological Sciences*, 81, 401–407.

24. Mylchreest, E., Cattley, R. C., and Foster, P. M. D. (1998). Male reproductive tract malformations in rats following gestational and lactational exposure to di(*n*-butyl) phthalate: an antiandrogenic mechanism? *Toxicological Sciences*, 43, 47–60.

25. Gray, L. E., Ostby, J. S., and Kelce, W. R. (1997). A dose–response analysis of the reproductive effects of a single gestational dose of 2,3,7,8-tetrachlorodibenzo-*p*-dioxin in male Long–Evans hooded rat offspring. *Toxicology and Applied Pharmacology*, 146, 11–20.

26. Faqi, A. S., Dalsenter, P. R., Merker, H. J., and Chahoud, I. (1998). Reproductive toxicity and tissue concentrations of low doses of 2,3,7,8-tetrachlorodibenzo-*p*-dioxin in male offspring rats exposed throughout pregnancy and lactation. *Toxicology and Applied Pharmacology*, 150, 383–392.

27. Wilker, C., Johnson, L., and Safe, S. (1996). Effects of developmental exposure to indole-3-carbinol or 2,3,7,8-tetrachlorodibenzo-*p*-dioxin on reproductive potential of male rat offspring. *Toxicology and Applied Pharmacology*, 141, 68–75.

28. Theobald, H. M. and Peterson, R. E. (1997). *In utero* and lactational exposure to 2,3,7,8-tetrachlorodibenzo-rho-diox in: effects on development of the male and female reproductive system of the mouse. *Toxicology and Applied Pharmacology*, 145, 124–135.

29. Poland, A., Palen, D., and Glover, E. (1994). Analysis of the 4 alleles of the murine aryl-hydrocarbon receptor. *Molecular Pharmacology*, 46, 915–921.

30. Ko, K., Theobald, H. M., and Peterson, R. E. (2002). *In utero* and lactational exposure to 2,3,7,8-tetrachlorodibenzo-*p*-dioxin in the C57BL/6J mouse prostate: lobe-specific effects on branching morphogenesis. *Toxicological Sciences*, 70, 227–237.

31. Lin, T. M., Ko, K., Moore, R. W., Simanainen, U., Oberley, T. D., and Peterson, R. E. (2002). Effects of aryl hydrocarbon receptor null mutation and *in utero* and lactational 2,3,7,8-tetrachlorodibenzo-*p*-dioxin exposure on prostate and seminal vesicle development in C57BL/6 mice. *Toxicological Sciences*, 68, 479–487.

32. Bell, D. R., Clode, S., Fan, M. Q., Fernandes, A., Foster, P. M. D., Jiang, T., Loizou, G., MacNicoll, A., Miller, B. G., Rose, M., Tran, L., and White, S. (2007). Toxicity of 2,3,7,8-tetrachlorodibenzo-*p*-dioxin in the developing male Wistar(Han) rat. I. No decrease in epididymal sperm count after a single acute dose. *Toxicological Sciences*, 99, 214–223.

33. Bell, D. R., Clode, S., Fan, M. Q., Fernandes, A., Foster, P. M. D., Jiang, T., Loizou, G., MacNicoll, A., Miller, B. G., Rose, M., Tran, L., and White, S. (2007). Toxicity of 2,3,7,8-tetrachlorodibenzo-*p*-dioxin in the developing male Wistar(Han) rat. II. Chronic dosing causes developmental delay. *Toxicological Sciences*, 99, 224–233.

34. Bjerke, D. L. and Peterson, R. E. (1994). Reproductive toxicity of 2,3,7,8-tetrachlorodibenzo-*p*-dioxin in male rats: different effects of *in utero* versus lactational exposure. *Toxicology and Applied Pharmacology*, 127, 241–249.

35. Roman, B. L., Sommer, R. J., Shinomiya, K., and Peterson, R. E. (1995). *In utero* and lactational exposure of the male rat to 2,3,7,8-tetrachlorodibenzo-*p*-dioxin: impaired prostate growth and development without inhibited androgen production. *Toxicology and Applied Pharmacology*, 134, 241–250.

36. Sommer, R. J., Ippolito, D. L., and Peterson, R. E. (1996). *In utero* and lactational exposure of the male Holtzman rat to 2,3,7,8-tetrachlorodibenzo-*p*-dioxin: decreased epididymal and ejaculated sperm numbers without alterations in sperm transit rate. *Toxicology And Applied Pharmacology*, 140, 146–153.

37. Yonemoto, J., Ichiki, T. Takei, T., and Tohyama, C. (2005). Maternal exposure to 2,3,7,8-tetrachlorodibenzo-*p*-dioxin and the body burden in offspring of Long–Evans rats. *Environmental Health and Preventive Medicine*, 10, 21–32.

38. Bell, D. R., Clode, S., Fan, M., Fernandes, A., Foster, P. M. D., Jiang, T., Loizou, G., MacNicoll, A., Miller, B. G., Rose, M., Tran, L., and White, S. (2010). Interpretation of studies on the developmental reproductive toxicology of 2,3,7,8-tetrachlorodibenzo-*p*-dioxin in male offspring. *Food and Chemical Toxicology*, 48, 1439–1447.

39. Nishimura, N., Yonemoto, J., Miyabara, Y., Sato, M., and Tohyama, C. (2003). Rat thyroid hyperplasia induced by gestational and lactational exposure to 2,3,7,8-tetrachlorodibenzo-*p*-dioxin. *Endocrinology*, 144, 2075–2083.

40. Nishimura, N., Yonemoto, J., Nishimura, H., Ikushiro, S., and Tohyama, C. (2005). Disruption of thyroid hormone homeostasis at weaning of Holtzman rats by lactational but not *in utero* exposure to 2,3,7,8-tetrachlorodibenzo-*p*-dioxin. *Toxicological Sciences*, 85, 607–614.

41. Nishimura, N., Yonemoto, J., Nishimura, H., and Tohyama, C. (2006). Localization of cytochrome P450 1A1 in a specific region of hydronephrotic kidney of rat neonates lactationally exposed to 2,3,7,8-tetrachlorodibenzo-*p*-dioxin. *Toxicology*, 227, 117–126.

42. Carney, E. W., Zablotny, C. L., Marty, M. S., Crissman, J. W., Peterson, P., Woolhiser, M., and Holsapple, M. (2004). The effects of feed restriction during *in utero* and postnatal development in rats. *Toxicological Sciences*, 82, 237–249.

43. Bell, D. R., Clode, S., Fan, M. Q., Fernandes, A., Foster, P. M. D., Jiang, T., Loizou, G., MacNicoll, A., Miller, B. G., Rose, M., Tran, L., and White, S. (2007). Relationships between tissue levels of 2,3,7,8-tetrachlorodibenzo-*p*-dioxin (TCDD), mRNAs, and toxicity in the developing male Wistar(Han) rat. *Toxicological Sciences*, 99, 591–604.

44. Ikeda, M., Tamura, M., Yamashita, J., Suzuki, C., and Tomita, T. (2005). Repeated *in utero* and lactational 2,3,7,8-tetrachlorodibenzo-*p*-dioxin exposure affects male gonads in offspring, leading to sex ratio changes in F-2 progeny. *Toxicology and Applied Pharmacology*, 206, 351–355.

45. Ohsako, S., Miyabara, Y., Nishimura, N., Kurosawa, S., Sakaue, M., Ishimura, R., Sato, M., Takeda, K., Aoki, Y., Sone, H., Tohyama, C., and Yonemoto, J. (2001). Maternal exposure to a low dose of 2,3,7,8-tetrachlorodibenzo-p-dioxin (TCDD) suppressed the development of reproductive organs of male rats: dose-dependent increase of mRNA levels of 5 alpha-reductase type 2 in contrast to decrease of androgen receptor in the pubertal ventral prostate. *Toxicological Sciences*, 60, 132–143.

46. Ohsako, S., Miyabara, Y., Sakaue, M., Ishimura, R., Kakeyama, M., Izumi, H., Yonemoto, J., and Tohyama, C. (2002). Developmental stage-specific effects of perinatal 2,3,7,8-tetrachlorodibenzo-p-dioxin exposure on reproductive organs of male rat offspring. *Toxicological Sciences*, 66, 283–292.

47. Simanainen, U., Haavisto, T., Tuomisto, J. T., Paranko, J., Toppari, J., Tuomisto, J., Peterson, R. E., and Viluksela, M. (2004). Pattern of male reproductive system effects after *in utero* and lactational 2,3,7,8-tetrachlorodibenzo-p-dioxin (TCDD) exposure in three differentially TCDD-sensitive rat lines. *Toxicological Sciences*, 80, 101–108.

48. Rebourcet, D., Odet, F., Verot, A., Combe, E., Meugnier, E., Pesenti, S., Leduque, P., Dechaud, H., Magre, S., and Le Magueresse-Battistoni, B. (2010). The effects of an *in utero* exposure to 2,3,7,8-tetrachloro-dibenzo-p-dioxin on male reproductive function: identification of Ccl5 as a potential marker. *International Journal of Andrology*, 33, 413–424.

49. Foster, W. G., Maharaj-Briceno, S., and Cyr, D. G. (2010). Dioxin-induced changes in epididymal sperm count and spermatogenesis. *Environmental Health Perspectives*, 118, 458–464.

50. Bell, D. R., Clode, S., Fan, M. Q., Fernandes, A., Foster, P. M. D., Jiang, T., Loizou, G., MacNicoll, A., Miller, B. G., Rose, M., Tran, L., and White, S. (2010). Interpretation of studies on the developmental reproductive toxicology of 2,3,7,8-tetrachloro-dibenzo-p-dioxin in male offspring. *Food and Chemical Toxicology*, 48, 1439–1447.

51. Creasy, D. M. (2003). Evaluation of testicular toxicology: a synopsis and discussion of the recommendations proposed by the society of toxicologic pathology. *Birth Defects Research, Part B*, 68, 408–415.

52. Lanning, L. L., Creasy, D. M., Chapin, R. E., Mann, P. C., Barlow, N. J., Regan, K. S., and Goodman, D. G. (2002). Recommended approaches for the evaluation of testicular and epididymal toxicity. *Toxicologic Pathology*, 30, 507–520.

53. Weil, C. S. (1970). Selection of valid number of sampling units and a consideration of their combination in toxicological studies involving reproduction, teratogenesis or carcinogenesis. *Food and Cosmetics Toxicology*, 8, 177–182.

54. Gray, L. E., Kelce, W. R., Monosson, E., Ostby, J. S., and Birnbaum, L. S. (1995). Exposure to TCDD during development permanently alters reproductive function in male Long–Evans rats and hamsters: reduced ejaculated and epididymal sperm numbers and sex accessory gland weights in offspring with normal androgenic status. *Toxicology and Applied Pharmacology*, 131, 108–118.

55. Ashby, J., Tinwell, H., Lefevre, P. A., Joiner, R., and Haseman, J. (2003). The effect on sperm production in adult Sprague-Dawley rats exposed by gavage to bisphenol a between postnatal days 91–97. *Toxicological Sciences*, 74, 129–138.

56. Howdeshell, K. L., Rider, C. V., Wilson, V. S., and Gray, L. E. (2008). Mechanisms of action of phthalate esters, individually and in combination, to induce abnormal reproductive development in male laboratory rats. *Environmental Research*, 108, 168–176.

57. Jana, N. R., Sarkar, S., Yonemoto, J., Tohyama, C., and Sone, H. (1998). Strain differences in cytochrome P4501A1 gene expression caused by 2,3,7,8-tetrachlorodibenzo-p-dioxin in the rat liver: role of the aryl hydrocarbon receptor and its nuclear translocator. *Biochemical and Biophysical Research Communications*, 248, 554–558.

58. Pohjanvirta, R. and Tuomisto, J. (1994). Short-term toxicity of 2,3,7,8-tetrachlorodibenzo-p-dioxin in laboratory animals: effects, mechanisms, and animal models. *Pharmacological Reviews*, 46, 483–549.

59. Pohjanvirta, R., Tuomisto, J., Vartiainen, T., and Rozman, K. (1987). Han/Wistar rats are exceptionally resistant to TCDD. I. *Pharmacology & Toxicology*, 60, 145–150.

60. Pohjanvirta, R. (1990). TCDD resistance is inherited as an autosomal dominant trait in the rat. *Toxicology Letters*, 50, 49–56.

61. Tuomisto, J. T., Viluksela, M., Pohjanvirta, R., and Tuomisto, J. (1999). The AH receptor and a novel gene determine acute toxic responses to TCDD: segregation of the resistant alleles to different rat lines. *Toxicology and Applied Pharmacology*, 155, 71–81.

62. Moffat, I. D., Roblin, S., Harper, P. A., Okey, A. B., and Pohjanvirta, R. (2007). Aryl hydrocarbon receptor splice variants in the dioxin-resistant rat: tissue expression and transactivational activity. *Molecular Pharmacology*, 72, 956–966.

63. Pohjanvirta, R., Wong, J. M. Y., Li, W., Harper, P. A., Tuomisto, J., and Okey, A. B. (1998). Point mutation in intron sequence causes altered carboxyl-terminal structure in the aryl hydrocarbon receptor of the most 2,3,7,8-tetrachlorodibenzo-p-dioxin-resistant rat strain. *Molecular Pharmacology*, 54, 86–93.

64. Pohjanvirta, R. and Tuomisto, J. (1990). Letter to the editor. *Toxicology and Applied Pharmacology*, 105, 508–509.

65. Huuskonen, H., Unkila, M., Pohjanvirta, R., and Tuomisto, J. (1994). Developmental toxicity of 2,3,7,8-tetrachlorodibenzo-p-dioxin (TCDD) in the most TCDD-resistant and TCDD-susceptible rat strains. *Toxicology and Applied Pharmacology*, 124, 174–180.

66. Jiang, T., Bell, D. R., Clode, S., Fan, M. Q., Fernandes, A., Foster, P. M. D., Loizou, G., MacNicoll, A., Miller, B. G., Rose, M., Tran, L., and White, S. (2009). A truncation in the aryl hydrocarbon receptor of the CRL:WI(Han) rat does not affect the developmental toxicity of TCDD. *Toxicological Sciences*, 107, 512–521.

19

TCDD, AHR, AND IMMUNE REGULATION

Nancy I. Kerkvliet

19.1 INTRODUCTION

The immunotoxicity of TCDD has been studied for over three decades, and the literature has been extensively reviewed elsewhere [1–4]. This chapter will incorporate the older literature as necessary for background context (without specific references) but will focus in greater detail on the most recent studies that address the underlying immunological mechanisms that appear to drive the immunological effects of TCDD. Although studies using TCDD have provided significant insight into the potential endogenous role of the AHR in immunity, this topic is addressed separately in Chapter 34, along with a discussion of potential endogenous ligands for AHR. The reader is also referred to Chapter 34 for an overview of the immune system and a review of the cells involved in the generation of immune responses.

19.2 OVERVIEW OF TCDD's IMMUNOMODULATORY EFFECTS

The long history of interest in the immune effects of TCDD stems from the fact that the adaptive immune response is exceedingly sensitive to suppression by this small molecule that was never intentionally manufactured. Rather, it was formed as a contaminant during the manufacture of other products, many of which were used as pesticides, such as the Agent Orange defoliant used during the Vietnam War in the late 1960s. Because of this history, TCDD has been referred to as "the most toxic man-made environmental chemical." However, TCDD is not "toxic" in the classic sense of a poison. Rather, a lethal dose of TCDD causes a slow wasting disease that leads to death in about 3 weeks in rats and mice. At 10–100-fold lower doses, TCDD causes numerous changes in the physiology of several organs, with the immune system being one of the most sensitive.

In mice, a single dose of TCDD in the low µg/kg range suppresses both antibody-mediated and cell-mediated immune responses to a wide range of model antigens, including sheep red blood cells (SRBCs) and allogeneic tumor cells, as well as foreign proteins such as keyhole limpet hemocyanin (KLH) and chicken ovalbumin (OVA). Likewise, animals exposed to TCDD show increased severity of symptoms and/or increased incidence of disease-induced mortality when challenged with a variety of infectious agents, including *Salmonella*, *Listeria*, influenza virus, Moloney sarcoma virus, and herpes type 2 virus, indicating that the immune system is compromised in a biologically significant manner by TCDD. The immunosuppressive effects of TCDD are also seen in the context of immune-mediated diseases with suppressed allergic responses to a variety of allergens [5] as well as suppression of immune responses in animal models of autoimmune diseases such as type 1 diabetes (T1D) [6] and multiple sclerosis [7]. The suppressive effects of TCDD on adaptive immune responses are dose dependent and occur at doses that do not produce overt toxicity. The suppression of adaptive immunity is not induced via elevation of the adrenal-derived stress hormones [8, 9], and even though TCDD causes thymic atrophy at high doses, suppression of the immune response occurs at lower doses and is thymic independent, as adult thymectomized mice remain sensitive to immune suppression by TCDD [10].

19.2.1 Role of the AHR in TCDD's Immune Modulation

TCDD acts through activation of the AHR to suppress immune responses. Mice that do not express a functional

The AH Receptor in Biology and Toxicology, First Edition. Edited by Raimo Pohjanvirta.
© 2012 John Wiley & Sons, Inc. Published 2012 by John Wiley & Sons, Inc.

AHR are refractory to the immunosuppressive effects of TCDD [11]. Since most, if not all, of the cells involved in immune responses express AHR, there are many potential targets for direct effects of TCDD. In addition, although a basal level of AHR is seen in many types of cells, AHR expression increases upon activation by a variety of stimuli, including mitogens, antigens, and various cytokines [12–14]. This induced expression implicates a role for AHR signaling in regulating immune responses. However, to the limited extent that it has been studied, the absence of a functional AHR does not appear to affect the ability of mice to generate a normal adaptive immune response [11, 15, 16]. On the other hand, AHR KO mice tend to produce more inflammatory cytokines and show a tendency toward hyperinflammatory responses and hypersensitivity to certain diseases [15, 17–19]. Taken together, these data support an endogenous role for the AHR in downregulating immune function and implicate the existence of endogenous AHR ligands. This topic is covered in detail in Chapter 34.

19.2.2 Effects of TCDD on Innate Immunity

In comparison to the overwhelming evidence for suppression of adaptive immune responses by TCDD, innate immune responses appear to be more resistant to suppression and, in some cases, show enhanced responsiveness following TCDD treatment [2]. Some of the earliest functional studies showed that TCDD did not alter the phagocytic or microbicidal properties of macrophages nor suppress the cytotoxic function of natural killer cells. On the other hand, TCDD exposure was associated with increased numbers of neutrophils and macrophages at the site of antigen challenge, but even though Mac-1$^+$ peritoneal cells expressed increased level of MHC Class II, no effect on antigen presentation was found [20]. Likewise, TCDD did not appear to affect antigen presentation by the adherent cell component of Mishell–Dutton cultures [21]. TCDD-treated mice and rats show increased mortality following endotoxin exposure, and this hypersensitivity has been linked to reduced hepatic clearance [22] and increased production of inflammatory mediators such as TNF and IL-1 [23]. Vogel et al. [24] found that the transcripts of keratinocyte chemoattractant (KC) and monocyte chemoattractant protein 1 (MCP-1) were increased in liver and spleen of mice exposed to TCDD, along with increased expression of the macrophage marker F4/80, suggesting that TCDD exposure increases tissue infiltration by inflammatory macrophages. However, it is important to note that inflammatory cytokine production is not universally increased by TCDD exposure. Inflammatory cytokines produced as a direct consequence of an adaptive immune response are reduced in TCDD-treated animals as would be expected when the adaptive response is suppressed [25].

The role of innate immune responses in the lung as a mechanism for enhanced virus-induced mortality in TCDD-treated mice has been studied extensively in the Lawrence laboratory [26]. TCDD has been shown to increase mortality of influenza-infected mice in association with increased numbers of neutrophils and macrophages in the lung and depletion of neutrophils (Gr-1$^+$ cells) reduces influenza mortality [27]. A fourfold increase in pulmonary IFN-γ level was also found in TCDD-treated mice in association with elevated inducible nitric oxide synthase (iNOS) while TNF-α, IL-1, and IFN-α/β levels were not altered [28]. Bone marrow chimera studies suggested that the AHR signals driving the TCDD response emanate from nonhematopoietic cells [29, 30]. The mechanisms underlying the increase in neutrophils seen in the lungs of TCDD-treated mice have not been elucidated despite extensive characterization of factors that regulate neutrophil responses [27, 28].

Increased numbers of Gr-1$^+$ cells were also found in TCDD-treated mice in association with a failure to reject an allogeneic tumor graft, leading to testing of the hypothesis that TCDD induced Gr-1$^+$ Mac-1$^+$ myeloid suppressor cells (MSCs) [31]. Consistent with MSC activity, CD11b$^+$ Gr-1$^+$ cells isolated from TCDD- but not vehicle-treated mice suppressed the development of CTL activity when added to mixed lymphocyte–tumor cell cultures [32]. Also consistent with MSC activity, this suppressive effect *in vitro* required cell-to-cell contact. However, *in vivo* depletion of CD11b$^+$ Gr-1$^+$ cells failed to affect TCDD-induced suppression of the allo-CTL response, arguing against an immunoregulatory role for the cells *in vivo* [32]. Immunohistochemical analysis of the spleen suggested that TCDD was increasing the production of myeloid cells in the red pulp, perhaps as a compensatory response to the growing tumor allograft.

Another important component of the innate immune response is the dendritic cell (DC). Recent studies have focused on the role of DCs in TCDD immunotoxicity because of their important role in antigen presentation to T cells and initiation of the adaptive immune response [33]. Several studies have found changes in the number and function of DCs in TCDD-treated mice that appear to implicate these cells in suppression of adaptive immunity. Vorderstrasse and Kerkvliet [34, 35] characterized the temporal and dose-related effects of TCDD on the expression of costimulatory molecules on splenic DCs, and the ability of splenic DCs from TCDD-treated mice to activate T cells *in vitro*. Unexpectedly, TCDD increased rather than decreased the expression of several accessory molecules on DCs, including MHC Class II, ICAM-1, CD40, and CD24. DCs from TCDD-treated mice also produced more IL-12 and stimulated a higher T-cell proliferative response in a mixed lymphocyte reaction. However, TCDD did not affect the ability of DCs to phagocytose latex beads, to present KLH to KLH-specific T cells, or to process and present OVA to OVA peptide-specific transgenic T cells either *in vivo* or *in vitro*. Bankoti et al. [36] validated many of these effects of TCDD on inflammatory bone marrow-derived DCs (BMDCs), including increased expression of MHC Class II and other costimulatory molecules on different subsets of DCs. They also reported that TCDD increased LPS- and/or CpG-induced IL-6 and TNF-α production but decreased IL-12p70 and NO production. IL-10 secretion was not affected by TCDD. Despite these changes, TCDD did not alter the ability of BMDCs to activate antigen-specific T cells.

In addition to alterations in DC activation, TCDD appears to influence DC survival. Treatment with TCDD resulted in a significant reduction in the number of DCs recovered from spleens of TCDD-treated mice within 1 week after TCDD treatment [34]. Bankoti et al. [37] validated and extended these finding to show that TCDD selectively depleted the CD11c^{hi}CD8α$^-$33D1$^+$ subset but not the regulatory CD11c^{hi} CD8α$^+$ DEC205$^+$ subset of splenic DCs. Contrary to results from in vitro studies [38, 39], the loss of DCs was independent of Fas but associated with an increase in expression of CCR7, a chemokine receptor that plays a role in DC migration. Interestingly, a recent study reported fewer DCs in the draining lymph nodes of TCDD-treated, influenza-infected mice, suggesting that TCDD may inhibit the migration of DCs following their activation [40].

Since TCDD does not appear to suppress the activation of DCs, an alternative possibility is that TCDD depletes the stimulatory DCs and selectively spares and/or activates DCs associated with tolerance induction. This hypothesis is supported by the phenotypic data of Bankoti et al. [36]. Tolerogenic DCs stimulate the induction of unresponsiveness in T cells, in part by the induction of regulatory T (T_{reg}) cells (reviewed in Ref. 41). Tolerogenic DCs have been characterized by various parameters depending on the tissue and stimulatory conditions. Induction of the tryptophan metabolizing enzyme indoleamine-2,3-dioxygenase (IDO), production of various suppressive cytokines such as IL-10 and TGFβ, and expression of a variety of negative costimulatory ligands have been associated with tolerogenic DCs. Vogel et al. [42] reported that TCDD induces IDO1 and IDO2 expression in a human DC cell line. In addition, the expression of DC-STAMP and IL-8 was significantly elevated by TCDD treatment, whereas upregulation of DC-CK1 was blocked during DC differentiation in the presence of TCDD. Bankoti et al. [36] also reported increased expression of IDO transcripts in DCs from TCDD-treated mice.

19.2.3 Lymphocytes as Direct AHR-Expressing Targets of TCDD

19.2.3.1 B Lymphocytes The direct AHR-dependent effects of TCDD on LPS-induced B-cell differentiation leading to depressed IgM secretion and reduced numbers of antibody-forming cells have been studied for many years by Kaminski and colleagues [43]. Using comparative studies of CH12.LX (AHR-expressing) and BCL-1 (AHR-deficient) B-cell lines, they have identified several key genes that are regulated by AHR including the Ig heavy-chain 3′alpha enhancer [44, 45], suppressor of cytokine signaling 2 (SOCS2) [46], paired box 5 isoform a (Pax5a) [47, 48], B lymphocyte-induced maturation protein-1 (Blimp-1), and AP-1 [49]. Subsequent in vivo studies in LPS-treated mice demonstrated that TCDD caused a dose-dependent suppression of Igmu chain, Igkappa chain, IgJ chain, XBP-1, and Blimp-1 in CD19$^+$ B cells [50]. TCDD also suppressed the normal increases in Blimp-1 protein expression necessary to allow B-cell differentiation. Recently, Bhattacharya et al. [51] used computational systems biology modeling to identify two feedback loops that link changes in expression of Blimp-1, Bcl-6, and Pax5 via a bistable switch to plasma cell differentiation. The authors hypothesize that TCDD may raise the threshold dose of antigen (e.g., LPS) required to trigger the bistable switch, thereby suppressing the B-cell to plasma cell differentiation process. Further studies have revealed 1893 regions in the DNA of LPS-activated B cells that show increased AHR binding after TCDD treatment [52]. These data, when paired with data from gene expression microarrays, revealed multiple nodes of the B-cell differentiation network that were altered by TCDD that may influence B-cell function.

19.2.3.2 T Lymphocytes Like B cells, direct AHR-dependent effects of TCDD on T cells have been identified. Many of these findings emanate from early studies in the author's laboratory using the P815 mastocytoma allograft model that showed TCDD potently suppressed the generation of allospecific CTL and alloantibody responses [25]. The mechanism of suppression by TCDD was traced to an early event in the activation of CD4$^+$ T cells that are required to generate CTL and alloantibody responses. CD4$^+$ T cells were subsequently identified as the critical AHR-expressing target for suppression of the allo-CTL response using an acute graft versus host (GVH) model, where activation of the donor T cells could be tracked using flow cytometry [16]. In this model, donor T cells from C57Bl/6 (B6) H-2^{bb} mice are injected intravenously into an F1 cross between B6 and DBA/2 (D2) mice (B6D2F1). The F1 host (H-2^{bd}) does not recognize the donor B6 T cells as foreign, whereas the donor B6 T cells respond to the Class 1 and Class II alloantigens expressed on the cells of the F1 host to generate a CD4$^+$ T cell-dependent antihost CTL response. The allospecific CTL response generated by the donor T cells in the F1 host was suppressed by treatment of the host with TCDD, but only if the donor T cells expressed AHR [48]. AHR expression in both the CD4$^+$ and CD8$^+$ subsets of donor T cells played a role in the suppression of the CTL response by TCDD, but the CD4$^+$ T cells clearly played the major role.

The AHR-dependent effects of TCDD on the activation and differentiation of the donor T cells were evaluated by comparing the responses of donor CD4$^+$ and CD8$^+$ T cells from AHR$^{+/+}$ and AHR$^{-/-}$ mice in vehicle- and TCDD-treated host mice. TCDD did not alter proliferation of the donor T cells but significantly altered expression of markers of T-cell differentiation on both subsets [53, 54]. Major changes included increased expression of CD25 and CTLA-4 and decreased expression of CD62L. These changes were not seen if the donor T cells did not express AHR. The high level of CD25 and CTLA-4 expressed suggested that AHR activation by TCDD could be promoting the differentiation of T$_{reg}$ cells. In fact, both CD4$^+$ and CD8$^+$ donor T cells from TCDD-treated mice were capable of suppressing the proliferative response on naïve T cells in vitro [53, 54], an attribute associated with T$_{reg}$ function. However, despite the similarities in phenotype and function between CD4$^+$ and CD8$^+$ T cells from TCDD-treated hosts, activation of AHR within CD4$^+$ T cells was found to drive the phenotypic changes in both T-cell subsets [54]. If AHR expression was limited to the CD8$^+$ cells, TCDD effects were restricted to downregulation of CD62L in the CD8$^+$ T cells themselves. Interestingly, a constitutively active AHR in T cells also results in decreased CD62L expression but only a modest effect on CD25 and no suppression of the CTL response [55]. The possibility that AHR activation in CD4$^+$ T cells induces CD8$^+$ T cells to become T$_{reg}$ cells instead of CTL effectors is a novel mechanism for suppressing the development of CTL activity.

On the other hand, a different mechanism appears to underlie the AHR-dependent suppression of the CTL response during influenza virus infection. The time course for suppression of the antiviral CTL response by TCDD is nearly identical to the allo-CTL response [56, 57] even though a primary CTL response to influenza virus does not depend on CD4$^+$ T cells. Furthermore, the antiviral CTL response in CD4-deficient mice is still suppressed by TCDD (B. P. Lawrence, personal communication). The CTL response to influenza is also suppressed even if the CD8 T cells do not express the AHR, suggesting that a nonlymphocyte target is mediating the suppression [58]. Recent studies show that TCDD significantly impairs the ability of DCs from influenza-infected mice to activate influenza-specific CD8 T cells in vitro, reducing their proliferation and differentiation into CTL [40]. As previously mentioned, TCDD may also affect the trafficking of DCs from the lung to the regional lymph node [40], which could also contribute to suppression of the CTL response.

A DC-dependent effect of TCDD on adaptive immunity has been postulated for a long time, yet other studies have failed to find a direct link. In the GVH model, for example, the host DC is the primary allostimulating cell. However, reducing the AHR responsiveness of the F1 host cells by using B6 AHR$^{-/-}$ mice as one parental strain had no influence on the ability of TCDD to suppress the CTL response, suggesting that host DCs were not primary targets [55]. Likewise, Shepherd et al. [59] reported that activating DCs with agonistic anti-CD40 antibody failed to overcome TCDD's suppression of the antibody response to OVA despite increased IL-12 production and increased expression of costimulatory molecule CD86. It is not yet clear if the DCs from influenza-infected mice are uniquely sensitive to TCDD (perhaps due to infection by the virus) or if the involvement and sensitivity of CD4$^+$ T cells to TCDD in other models of adaptive immunity have overshadowed the role of the DCs in TCDD's suppression of adaptive immune responses.

19.2.4 Is AHR Driving the Differentiation of CD4$^+$ T Cells to T$_{reg}$ Cells?

Regulatory T cells represent the primary mechanism by which the body protects itself against misguided or excessive immune responses that may produce harmful immune pathology (reviewed in Refs 60 and 61). Naturally occurring CD4$^+$ T$_{reg}$ cells that develop in the thymus express high levels of CD25 and the transcription factor Foxp3, while antigen-induced adaptive CD4$^+$ T$_{reg}$ cells in the periphery (T$_R$1 and T$_H$3 subsets) are more heterogeneous and may or may not express CD25 or Foxp3. However, Foxp3 is known to confer suppressive activity on CD25$^-$ cells and is thus considered to be a definitive marker for T$_{reg}$ cells. T$_{reg}$ cells use many different mechanisms to suppress inflammation and control immune responses. T$_{reg}$ cells may target DCs or directly inhibit developing effector T and B cells. Both cytokine-mediated and contact-dependent suppressive mechanisms have been described that include increased IL-10, TGFβ, CTLA-4, and programmed death (PD)-1, as well as tryptophan depletion and enhanced adenosine metabolism.

Evidence for induction of T$_{reg}$ cells by TCDD was first reported by Funatake et al. [53] who reported high levels of CD25 and CTLA4 expression on donor CD4$^+$ T cells from TCDD-treated mice during an acute GVHR, along with the ability to suppress the proliferation of naïve activated T cells in vitro. Depletion of existing natural CD4$^+$CD25$^+$ T$_{reg}$ cells from the donor population prior to injection into F1 hosts did not influence the percentage of CD4$^+$CD25$^+$ cells, suggesting that TCDD was not expanding the natural CD4$^+$CD25$^+$ T$_{reg}$ population. The absence of Foxp3 expression in the donor T cells further suggested that TCDD was inducing adaptive T$_{reg}$ cells. The CD4$^+$CD25$^+$ cells from TCDD-treated mice produced increased amounts of IL-10 and expressed higher levels of mRNA for several genes that have been associated with T$_{reg}$ cells, including TGFβ3, Blimp-1, and granzyme B, as well as genes associated with the IL-12-Rb2 signaling pathway [62]. Since the development of the T$_{reg}$ phenotype by TCDD was entirely dependent on the expression of AHR in the donor CD4$^+$ cells, these results suggest that TCDD-activated AHR regulates many of the same genes that Foxp3 regulates in CD4$^+$ T cells.

On the other hand, there are studies that indicate AHR influences Foxp3 expression. AHR response elements that bind AHR are present in the Foxp3 promoter, and binding is increased in the presence of TCDD [7]. Some studies have shown that TCDD increases Foxp3 expression in $CD4^+$ T cells *in vitro*; however, the induction of Foxp3 expression was seen only with high concentrations of TCDD (100–160 nM) [7]. Lower concentrations of TCDD that fully activate known AHR-regulated genes (1–10 nM) do not alter transcript levels of Foxp3 in $CD4^+$ T cells cultured under a variety of conditions (Rohlman and Kerkvliet, unpublished data). The frequency of $Foxp3^+$ cells is not altered in mice lacking a functional AHR indicating that AHR is not required for Foxp3 expression. And yet, an increased frequency of $Foxp3^+CD4^+$ T cells has been associated with TCDD treatment in several animal models of autoimmune disease, including mouse models of multiple sclerosis (EAE) [7], type 1 diabetes in NOD mice [6], autoimmune colitis [63], and autoimmune uveitis [64]. It is not known whether this increase in $Foxp3^+$ T cells represents the induction of adaptive T_{reg} cells or the preservation/expansion of the natural T_{reg} population under autostimulating conditions.

The induction of adaptive AHR-dependent T_{reg} cells by TCDD would provide a holistic mechanism to explain the breadth of immune responses that TCDD is capable of suppressing. However, the signaling pathways leading to adaptive T_{reg} induction have not been elucidated. In the GVH model, the AHR in the T cells is clearly driving the differentiation. In other models, such as those in which increases in $Foxp3^+$ T cells are observed, the AHR in non-T cells (e.g., DCs) might be the primary initiator of altered T-cell differentiation. This potential divergence in phenotype and mechanism is supported by the findings of recent studies using human cells. Gandhi et al. [65] showed that different polarizing conditions *in vitro* influenced the induction of human T_{reg} cells by TCDD. In the absence of TGFβ, TCDD-induced T_{reg} cells were $Foxp3^-$, produced IL-10, and controlled responder T-cell proliferation through granzyme B, a phenotype consistent with the mouse T_{reg} phenotype in the GVH model [62]. However, if TGFβ was added to the culture, the T_{reg} cells induced by TCDD were $Foxp3^+$ and $CD39^+$, and suppression of responder T-cell proliferation was dependent on CD39, a cell surface enzyme that metabolizes ATP [66]. Foxp3 expression appeared to be regulated by AHR-induced Smad1 expression [65]. TCDD also increased the expression of Aiolos, a transcription factor that interacted with Foxp3 to silence IL-2 expression.

In another recent study, Apetoh et al. [67] found that AHR expression was highly upregulated in mouse $CD4^+$ T cells when IL-27 was used to induce the differentiation of T_r1-type regulatory T cells. When TCDD was added to the cultures, the percentage of IL-10 producing $Foxp3^-$ cells was increased by twofold, and secreted IL-10 was increased by threefold. IL-21, an autocrine growth factor that increases IL-10 expression, was also increased by TCDD. Further studies showed that AHR physically interacted with c-Maf to transactivate both IL-10 and IL-21 promoters. Interestingly, AHR did not require exogenous ligand to transactivate these genes suggesting a role for endogenous AHR ligands in inducing T_r1 cells.

Taken together, these findings support a model in which specific cytokines present during antigen-induced activation of $CD4^+$ T cells lead to increased expression of AHR and thus increased efficacy of TCDD to induce or enhance the differentiation of diverse types of regulatory T cells in both human and murine systems. The ability of TCDD to induce T_{reg} cells appears to supersede the generation of effector T cells, including T_H1, T_H2, and T_H17 cells. Further, this suggests that selective AHR ligands that mimic TCDD can be developed for clinical use in the treatment of immune-mediated diseases.

REFERENCES

1. Lawrence, B. P. and Kerkvliet, N. I. (2007). Immune modulation by TCDD and related polyhalogenated aromatic hydrocarbons. In *Immunotoxicology and Immunopharmacology*, (Luebke, R., House, R., and Kimber, I., Eds). CRC Press, Boca Raton, FL, pp. 239–258.

2. Kerkvliet, N. I. (2003). Immunotoxicology of dioxins and related chemicals. In *Dioxins and Health*, 2nd edition (Schecter, A. and Gasiewicz, T. A., Eds) Wiley, Hoboken, NJ, pp. 299–328.

3. Kerkvliet, N. I. (2009). AHR-mediated immunomodulation: the role of altered gene transcription. *Biochemical Pharmacology*, 77, 746–760.

4. Marshall, N. B. and Kerkvliet, N. I. (2010). Dioxin and immune regulation: emerging role of aryl hydrocarbon receptor in the generation of regulatory T cells. *Annals of the New York Academy of Sciences*, 1183, 25–37.

5. Luebke, R. W., Copeland, C. B., Daniels, M., Lambert, A. L., and Gilmour, M. I. (2001). Suppression of allergic immune responses to house dust mite (HDM) in rats exposed to 2,3,7,8-TCDD. *Toxicological Sciences*, 62, 71–79.

6. Kerkvliet, N. I., Steppan, L. B., Vorachek, W., Oda, S., Farrer, D., et al. (2009). Activation of aryl hydrocarbon receptor by TCDD prevents diabetes in NOD mice and increases $Foxp3^+$ T cells in pancreatic lymph nodes. *Immunotherapy*, 1, 539–547.

7. Quintana, F. J., Basso, A. S., Iglesias, A. H., Korn, T., Farez, M. F., et al. (2008) Control of T_{reg} and T_H17 cell differentiation by the aryl hydrocarbon receptor. *Nature*, 453, 65–71.

8. De Krey, G. K., Baecher-Steppan, L., Deyo, J. A., Smith, B., and Kerkvliet, N. I. (1993). Polychlorinated biphenyl-induced immune suppression: castration, but not adrenalectomy or RU 38486 treatment, partially restores the suppressed cytotoxic T lymphocyte response to alloantigen. *Journal of Pharmacology and Experimental Therapeutics*, 267, 308–315.

9. De Krey, G. K. and Kerkvliet, N. I. (1995). Suppression of cytotoxic T lymphocyte activity by 2,3,7,8-tetrachlorodibenzo-

p-dioxin occurs *in vivo*, but not *in vitro*, and is independent of corticosterone elevation. *Toxicology*, 97, 105–112.

10. Kerkvliet, N. I. and Brauner, J. A. (1987). Mechanisms of 1,2,3,4,6,7,8-heptachlorodibenzo-*p*-dioxin (HpCDD)-induced humoral immune suppression: evidence of primary defect in T-cell regulation. *Toxicology and Applied Pharmacology*, 87, 18–31.

11. Vorderstrasse, B. A., Steppan, L. B., Silverstone, A. E., and Kerkvliet, N. I. (2001). Aryl hydrocarbon receptor-deficient mice generate normal immune responses to model antigens and are resistant to TCDD-induced immune suppression. *Toxicology and Applied Pharmacology*, 171, 157–164.

12. Crawford, R. B., Holsapple, M. P., and Kaminski, N. E. (1997). Leukocyte activation induces aryl hydrocarbon receptor up-regulation, DNA binding, and increased Cyp1a1 expression in the absence of exogenous ligand. *Molecular Pharmacology*, 52, 921–927.

13. Veldhoen, M., Hirota, K., Westendorf, A. M., Buer, J., Dumoutier, L., et al. (2008). The aryl hydrocarbon receptor links T_H17-cell-mediated autoimmunity to environmental toxins. *Nature*, 453, 106–109.

14. Kimura, A., Naka, T., Nohara, K., Fujii-Kuriyama, Y., and Kishimoto, T. (2008). Aryl hydrocarbon receptor regulates Stat1 activation and participates in the development of T_h17 cells. *Proceedings of the National Academy of Sciences of the United States of America*, 105, 9721–9726.

15. Rodriguez-Sosa, M., Elizondo, G., Lopez-Duran, R. M., Rivera, I., Gonzalez, F. J., et al. (2005). Over-production of IFN-gamma and IL-12 in AhR-null mice. *FEBS Letters*, 579, 6403–6410.

16. Kerkvliet, N. I., Shepherd, D. M., and Baecher-Steppan, L. (2002). T lymphocytes are direct, aryl hydrocarbon receptor (AhR)-dependent targets of 2,3,7,8-tetrachlorodibenzo-*p*-dioxin (TCDD): AhR expression in both $CD4^+$ and $CD8^+$ T cells is necessary for full suppression of a cytotoxic T lymphocyte response by TCDD. *Toxicology and Applied Pharmacology*, 185, 146–152.

17. Sekine, H., Mimura, J., Oshima, M., Okawa, H., Kanno, J., et al. (2009) Hypersensitivity of aryl hydrocarbon receptor-deficient mice to lipopolysaccharide-induced septic shock. *Molecular and Cellular Biology*, 29, 6391–6400.

18. Thatcher, T. H., Maggirwar, S. B., Baglole, C. J., Lakatos, H. F., Gasiewicz, T. A., et al. (2007). Aryl hydrocarbon receptor-deficient mice develop heightened inflammatory responses to cigarette smoke and endotoxin associated with rapid loss of the nuclear factor-kappaB component RelB. *American Journal of Pathology*, 170, 855–864.

19. Sanchez, Y., de Dios Rosado, J., Vega, L., Elizondo, G., Estrada-Muñiz, E., et al. (2010). The unexpected role for the aryl hydrocarbon receptor on susceptibility to experimental toxoplasmosis. *Journal of Biomedicine and Biotechnology*, 2010, 1–16.

20. Kerkvliet, N. I. and Oughton, J. A. (1993). Acute inflammatory response to sheep red blood cell challenge in mice treated with 2,3,7,8-tetrachlorodibenzo-*p*-dioxin (TCDD): phenotypic and functional analysis of peritoneal exudate cells. *Toxicology and Applied Pharmacology*, 119, 248–257.

21. Dooley, R. K. and Holsapple, M. P. (1988). Elucidation of cellular targets responsible for tetrachlorodibenzo-*p*-dioxin (TCDD)-induced suppression of antibody responses. I. The role of the B lymphocyte. *Immunopharmacology*, 16, 167–180.

22. Rosenthal, G. J., Lebetkin, E., Thigpen, J. E., Wilson, R., Tucker, A. N., et al. (1989). Characteristics of 2,3,7,8-tetrachlorodibenzo-*p*-dioxin induced endotoxin hypersensitivity: association with hepatotoxicity. *Toxicology*, 56, 239–251.

23. Clark, G. C., Taylor, M. J., Tritscher, A. M., and Lucier, G. W. (1991). Tumor necrosis factor involvement in 2,3,7,8-tetrachlorodibenzo-*p*-dioxin-mediated endotoxin hypersensitivity in C57BL/6J mice congenic at the Ah locus. *Toxicology and Applied Pharmacology*, 111, 422–431.

24. Vogel, C. F., Nishimura, N., Sciullo, E., Wong, P., Li, W., et al. (2007). Modulation of the chemokines KC and MCP-1 by 2,3,7,8-tetrachlorodibenzo-*p*-dioxin (TCDD) in mice. *Archives of Biochemistry and Biophysics*, 461, 169–175.

25. Kerkvliet, N. I., Baecher-Steppan, L., Shepherd, D. M., Oughton, J. A., Vorderstrasse, B. A., et al. (1996). Inhibition of TC-1 cytokine production, effector cytotoxic T lymphocyte development and alloantibody production by 2,3,7,8-tetrachlorodibenzo-*p*-dioxin. *Journal of Immunology*, 157, 2310–2319.

26. Head, J. L. and Lawrence, B. P. (2009). The aryl hydrocarbon receptor is a modulator of anti-viral immunity. *Biochemical Pharmacology*, 77, 642–653.

27. Teske, S., Bohn, A. A., Regal, J. F., Neumiller, J. J., and Lawrence, B. P. (2005). Activation of the aryl hydrocarbon receptor increases pulmonary neutrophilia and diminishes host resistance to influenza A virus. *American Journal of Physiology: Lung Cellular and Molecular Physiology*, 289, L111–L124.

28. Neff-LaFord, H. D., Vorderstrasse, B. A., and Lawrence, B. P. (2003). Fewer CTL, not enhanced NK cells, are sufficient for viral clearance from the lungs of immunocompromised mice. *Cellular Immunology*, 226, 54–64.

29. Neff-LaFord, H., Teske, S., Bushnell, T. P., and Lawrence, B. P. (2007). Aryl hydrocarbon receptor activation during influenza virus infection unveils a novel pathway of IFN-gamma production by phagocytic cells. *Journal of Immunology*, 179, 247–255.

30. Teske, S., Bohn, A. A., Hogaboam, J. P., and Lawrence, B. P. (2008). Aryl hydrocarbon receptor targets pathways extrinsic to bone marrow cells to enhance neutrophil recruitment during influenza virus infection. *Toxicological Sciences*, 102, 89–99.

31. Gabrilovich, D. I. and Nagaraj, S. (2009). Myeloid-derived suppressor cells as regulators of the immune system. *Nature Reviews Immunology*, 9, 162–174.

32. Choi, J. Y., Oughton, J. A., and Kerkvliet, N. I. (2003). Functional alterations in $CD11b^+Gr-1^+$ cells in mice injected with allogeneic tumor cells and treated with 2,3,7,8-tetrachlorodibenzo-*p*-dioxin. *International Immunopharmacology*, 3, 553–570.

33. Banchereau, J. and Steinman, R. M. (1998). Dendritic cells and the control of immunity. *Nature*, 392, 245–252.

34. Vorderstrasse, B. A. and Kerkvliet, N. I. (2001). 2,3,7,8-Tetrachlorodibenzo-*p*-dioxin affects the number and function of

murine splenic dendritic cells and their expression of accessory molecules. *Toxicology and Applied Pharmacology, 171,* 117–125.

35. Vorderstrasse, B. A., Dearstyne, E. A., and Kerkvliet, N. I. (2003). Influence of 2,3,7,8-tetrachlorodibenzo-*p*-dioxin on the antigen-presenting activity of dendritic cells. *Toxicological Sciences, 72,* 103–112.

36. Bankoti, J., Rase, B., Simones, T., and Shepherd, D. M. (2010). Functional and phenotypic effects of AhR activation in inflammatory dendritic cells. *Toxicology and Applied Pharmacology, 246,* 18–28.

37. Bankoti, J., Burnett, A., Navarro, S., Miller, A. K., Rase, B., et al. (2010) Effects of TCDD on the fate of naive dendritic cells. *Toxicological Sciences, 115,* 422–434.

38. Ruby, C. E., Funatake, C. J., and Kerkvliet, N. I. (2005). 2,3,7,8 Tetrachlorodibenzo-*p*-dioxin (TCDD) directly enhances the maturation and apoptosis of dendritic cells *in vitro*. *Journal of Immunotoxicology, 1,* 159–166.

39. Ruby, C. E., Leid, M., and Kerkvliet, N. I. (2002). 2,3,7,8-Tetrachlorodibenzo-*p*-dioxin suppresses tumor necrosis factor-alpha and anti-CD40-induced activation of NF-kappaB/Rel in dendritic cells: p50 homodimer activation is not affected. *Molecular Pharmacology, 62,* 722–728.

40. Jin, G. B., Moore, A. J., Head, J. L., Neumiller, J. J., and Lawrence, B. P. (2010). Aryl hydrocarbon receptor activation reduces dendritic cell function during influenza virus infection. *Toxicological Sciences, 116,* 514–522.

41. Pulendran, B., Tang, H., and Manicassamy, S. (2010). Programming dendritic cells to induce T_H2 and tolerogenic responses. *Nature Immunology, 11,* 647–655.

42. Vogel, C. F., Goth, S. R., Dong, B., Pessah, I. N., and Matsumura, F. (2008). Aryl hydrocarbon receptor signaling mediates expression of indoleamine 2,3-dioxygenase. *Biochemical Biophysical Research Communications, 375,* 331–335.

43. Sulentic, C. E. and Kaminski, N. E. (2011). The long winding road toward understanding the molecular mechanisms for B cell suppression by 2,3,7,8-tetrachlorodibenzo-*p*-dioxin. *Toxicological Sciences, 120 (Suppl. 1),* S171–S191.

44. Sulentic, C. E., Zhang, W., Na, Y. J., and Kaminski, N. E. (2004). 2,3,7,8-Tetrachlorodibenzo-*p*-dioxin, an exogenous modulator of the 3′alpha immunoglobulin heavy chain enhancer in the CH12. LX mouse cell line. *Journal of Pharmacology and Experimental Therapeutics, 309,* 71–78.

45. Sulentic, C. E., Holsapple, M. P., and Kaminski, N. E. (2000). Putative link between transcriptional regulation of IgM expression by 2,3,7,8-tetrachlorodibenzo-*p*-dioxin and the aryl hydrocarbon receptor/dioxin-responsive enhancer signaling pathway. *Journal of Pharmacology and Experimental Therapeutics, 295,* 705–716.

46. Boverhof, D. R., Tam, E., Harney, A. S., Crawford, R. B., Kaminski, N. E., et al. (2004). 2,3,7,8-Tetrachlorodibenzo-*p*-dioxin induces suppressor of cytokine signaling 2 in murine B cells. *Molecular Pharmacology, 66,* 1662–1670.

47. Yoo, B. S., Boverhof, D. R., Shnaider, D., Crawford, R. B., Zacharewski, T. R., et al. (2004). 2,3,7,8-Tetrachlorodibenzo-*p*-dioxin (TCDD) alters the regulation of Pax5 in lipopolysaccharide-activated B cells. *Toxicological Sciences, 77,* 272–279.

48. Schneider, D., Manzan, M. A., Crawford, R. B., Chen, W., and Kaminski, N. E. (2008). 2,3,7,8-Tetrachlorodibenzo-*p*-dioxin-mediated impairment of B cell differentiation involves dysregulation of paired box 5 (Pax5) isoform, Pax5a. *Journal of Pharmacology and Experimental Therapeutics, 326,* 463–474.

49. Schneider, D., Manzan, M. A., Yoo, B. S., Crawford, R. B., and Kaminski, N. (2009). Involvement of Blimp-1 and AP-1 dysregulation in the 2,3,7,8-tetrachlorodibenzo-*p*-dioxin-mediated suppression of the IgM response by B cells. *Toxicological Sciences, 108,* 377–388.

50. North, C. M., Crawford, R. B., Lu, H., and Kaminski, N. E. (2009). Simultaneous *in vivo* time course and dose response evaluation for TCDD-induced impairment of the LPS-stimulated primary IgM response. *Toxicological Sciences, 112,* 123–132.

51. Bhattacharya, S., Conolly, R. B., Kaminski, N. E., Thomas, R. S., Andersen, M. E., et al. (2010). A bistable switch underlying B-cell differentiation and its disruption by the environmental contaminant 2,3,7,8-tetrachlorodibenzo-*p*-dioxin. *Toxicological Sciences, 115,* 51–65.

52. De Abrew, K. N., Kaminski, N. E., and Thomas, R. S. (2010). An integrated genomic analysis of aryl hydrocarbon receptor-mediated inhibition of B-cell differentiation. *Toxicological Sciences, 118,* 454–469.

53. Funatake, C. J., Marshall, N. B., Steppan, L. B., Mourich, D. V., and Kerkvliet, N. I. (2005). Cutting edge: activation of the aryl hydrocarbon receptor by 2,3,7,8-tetrachlorodibenzo-*p*-dioxin generates a population of $CD4^+ CD25^+$ cells with characteristics of regulatory T cells. *Journal of Immunology, 175,* 4184–4188.

54. Funatake, C. J., Marshall, N. B., and Kerkvliet, N. I. (2008). 2,3,7,8-Tetrachlorodibenzo-*p*-dioxin alters the differentiation of alloreactive $CD8^+$ T cells toward a regulatory T cell phenotype by a mechanism that is dependent on aryl hydrocarbon receptor in $CD4^+$ T cells. *Journal of Immunotoxicology, 5,* 81–91.

55. Funatake, C. J., Ao, K., Suzuki, T., Murai, H., Yamamoto, M., et al. (2009). Expression of constitutively-active aryl hydrocarbon receptor in T-cells enhances the down-regulation of CD62L, but does not alter expression of CD25 or suppress the allogeneic CTL response. *Journal of Immunotoxicology, 6,* 194–203.

56. Warren, T. K., Mitchell, K. A., and Lawrence, B. P. (2000). Exposure to 2,3,7,8-tetrachlorodibenzo-*p*-dioxin (TCDD) suppresses the humoral and cell-mediated immune responses to influenza A virus without affecting cytolytic activity in the lung. *Toxicological Sciences, 56,* 114–123.

57. Lawrence, B. P., Warren, T. K., and Luong, H. (2000). Fewer T lymphocytes and decreased pulmonary influenza virus burden in mice exposed to 2,3,7,8-tetrachlorodibenzo-*p*-dioxin (TCDD). *Journal of Toxicology and Environmental Health A, 61,* 39–53.

58. Lawrence, B. P., Roberts, A. D., Neumiller, J. J., Cundiff, J. A., and Woodland, D. L. (2006). Aryl hydrocarbon receptor acti-

vation impairs the priming but not the recall of influenza virus-specific $CD8^+$ T cells in the lung. *Journal of Immunology, 177*, 5819–5828.

59. Shepherd, D. M., Steppan, L. B., Hedstrom, O. R., and Kerkvliet, N. I. (2001). Anti-CD40 treatment of 2,3,7,8-tetrachlorodibenzo-*p*-dioxin (TCDD)-exposed C57Bl/6 mice induces activation of antigen presenting cells yet fails to overcome TCDD-induced suppression of allograft immunity. *Toxicology and Applied Pharmacology, 170*, 10–22.

60. Palomares, O., Yaman, G., Azkur, A. K., Akkoc, T., Akdis, M., et al. (2010). Role of T_{reg} in immune regulation of allergic diseases. *European Journal of Immunology, 40*, 1232–1240.

61. Sakaguchi, S., Wing, K., and Yamaguchi, T. (2009). Dynamics of peripheral tolerance and immune regulation mediated by T_{reg}. *European Journal of Immunology, 39*, 2331–2336.

62. Marshall, N. B., Vorachek, W. R., Steppan, L. B., Mourich, D. V., and Kerkvliet, N. I. (2008). Functional characterization and gene expression analysis of $CD4^+$ $CD25^+$ regulatory T cells generated in mice treated with 2,3,7,8-tetrachlorodibenzo-*p*-dioxin. *Journal of Immunology, 181*, 2382–2391.

63. Zhang, L., Ma, J., Takeuchi, M., Usui, Y., Hattori, T., et al. (2009). Activation of aryl hydrocarbon receptor suppresses experimental autoimmune uveoretinitis by inducing differentiation of regulatory T cells. *Investigative Opthalmology and Visual Science, 51*, 2109–2117.

64. Takamura, T., Harama, D., Matsuoka, S., Shimokawa, N., Nakamura, Y., et al. (2010). Activation of the aryl hydrocarbon receptor pathway may ameliorate dextran sodium sulfate-induced colitis in mice. *Immunology and Cell Biology, 88*, 685–689.

65. Gandhi, R., Kumar, D., Burns, E. J., Nadeau, M., Dake, B., et al. (2010). Activation of the aryl hydrocarbon receptor induces human type 1 regulatory T cell-like and $Foxp3^+$ regulatory T cells. *Nature Immunology, 11*, 846–853.

66. Borsellino, G., Kleinewietfeld, M., Di Mitri, D., Sternjak, A., Diamantini, A., et al. (2007). Expression of ectonucleotidase CD39 by $Foxp3^+$ T_{reg} cells: hydrolysis of extracellular ATP and immune suppression. *Blood, 110*, 1225–1232.

67. Apetoh, L., Quintana, F. J., Pot, C., Joller, N., Xiao, S., et al. (2010). The aryl hydrocarbon receptor interacts with c-Maf to promote the differentiation of type 1 regulatory T cells induced by IL-27. *Nature Immunology, 11*, 854–861.

20

EFFECTS OF DIOXINS ON TEETH AND BONE: THE ROLE OF AHR

Matti Viluksela, Hanna M. Miettinen, and Merja Korkalainen

20.1 INTRODUCTION

Calcified tissues are not considered typical targets of toxic effects of chemicals, and therefore, bones and especially teeth are seldom in the focus of toxicity studies designed to elucidate the toxicological profile of chemicals. However, the dependence of development and maintenance of these tissues on a complex interaction of different cell types, hormones, growth factors, and nutritional factors makes them potentially sensitive targets of toxic effects of AHR ligands and other chemicals that are able to modify these functions.

The toxicological profile of dioxins and the basic features of the AHR signaling pathway had already been clarified quite extensively prior to the first findings on their effects on teeth and bone. Subsequently, more focused studies revealed these effects in dioxin-exposed human populations, laboratory animals, and *in vitro* models. Experimental studies have scrutinized the role of the AHR in mediating these effects, and the fact that the same type of effects have been observed in human populations at low exposure levels indicates that they belong to the most sensitive end points of dioxin toxicity.

Tooth development is initiated during fetal organogenesis and continues to early adulthood. Early phases of tooth development are characterized by complex reciprocal and sequential interactions between epithelial and mesenchymal cells derived from oral epithelium (dental lamina), ectomesenchymal neural crest, and foregut endoderm [1]. Several hundred genes are expressed during tooth development, and the current knowledge is presented in the Gene Expression in Tooth database (http://bite-it.helsinki.fi). The successive stages of tooth development according to their morphological features are initiation, bud, cap, and bell stages. Tooth morphogenesis takes place at the bud stage, and cell differentiation takes place at cap and bell stages, after which matrix formation and mineralization take place. Unlike bone, teeth do not undergo remodeling after mineralization. Therefore, developmental defects induced by chemicals or other factors are permanent and can be observed later in life even long time after exposure. This feature makes teeth highly useful indicators of toxicity.

Bones originate mainly from mesodermal mesenchyme, and bone-forming osteoblasts are differentiated from mesenchymal stem cells. Two mechanisms and their specific signaling pathways are involved in bone formation [2]. In endochondral ossification, a cartilage model of bone is formed and subsequently replaced with bone as it gradually grows. Cranial and jugular bones are derived from ectomesenchymal neural crest and they undergo direct intramembranous ossification, which takes place without cartilageous intermediate phase. Most skeletal elements are ossified at birth, but the growth of bones continues until the cartilageous growth plates are ossified. Physical integrity of bones is maintained by continuous remodeling, which involves resorption of existing bone by osteoclasts that originate from hematopoietic stem cells, and formation of new bone matrix and its subsequent mineralization by osteoblasts. Both bone modeling during development and remodeling are potentially sensitive targets to chemical insults, but remodeling may repair chemically induced damage.

The AH Receptor in Biology and Toxicology, First Edition. Edited by Raimo Pohjanvirta.
© 2012 John Wiley & Sons, Inc. Published 2012 by John Wiley & Sons, Inc.

20.2 EFFECTS OF DIOXINS ON TEETH

The first indications about the influence of dioxin-like compounds on teeth emerged almost simultaneously from highly exposed human populations and high-dose experimental studies, and in both cases developing teeth were involved. Hara [3] studied the health status of PCB-exposed Japanese capacitor workers and their children, and reported about mottled enamel and carious teeth in children of exposed female workers. At the same time, McNulty [4] exposed juvenile rhesus macaques to the commercial PCB mixture Aroclor 1242 for 40 days and observed squamous metaplasia in several specialized epithelial structures including enamel-secreting ameloblasts of unerupted teeth. This alteration was associated with severely deformed jaws due to cystic periodontal lesions around unerupted teeth. Also, epidermal pearls and cysts were observed throughout the jawbone. Mass poisoning after accidental contamination of rice oil with dioxin-like impurities in Japan in 1968 (Yusho) and in Taiwan in 1979 (Yu-Cheng) resulted in toxic effects in teeth of children [5–7]. These included natal teeth, tooth chipping, caries in primary dentition, abnormal roots, delayed eruption, and a high frequency of missing teeth. In accordance with these findings, childhood exposure to TCDD during the Seveso accident in 1976 was associated with permanent developmental dental aberrations, most notably hypodontia (missing permanent teeth due to failure in their development) [8] that could be observed 25 years after the accident.

The first evidence about high sensitivity of developing teeth to dioxins came from the pioneering studies of Alaluusua et al. [9, 10], who showed that exposure of children to PCDD/Fs via mother's milk at the Northern European background levels was correlated with the incidence and severity of hypomineralization defects of the first permanent molars. In utero/lactational exposure studies in rats indicated that disturbed molar development of the offspring is one of the most sensitive end points of TCDD toxicity and provided biological plausibility for the findings in children [11, 12]. Dose-dependent changes were smaller molar size, delayed eruption, complete block of development of the third molars, increased susceptibility to caries, and some changes in enamel mineral composition. In mouse offspring, decreased size and altered shape of mandibles [13–15], increased fluctuating asymmetry of molars [16], and missing third molars [17] were reported after in utero/lactational exposure to low doses of TCDD. Also, in vitro studies have shown that TCDD interferes with the morphogenesis and dental matrix mineralization of embryonic mouse teeth [18–20].

In addition to rodents, fish, minks, and primates are also sensitive to TCDD-induced developmental dental defects. TCDD-treated rainbow trout sac fry exhibited craniofacial malformations and absence of teeth [21]. Juvenile minks had proliferation of the squamous epithelium in periodontal ligament, loose and displaced incisors, prominent canine teeth, and maxillary and mandibular osteolysis after they were fed TCDD-containing diet for 36 days [22]. Incomplete calcification, accelerated eruption of teeth, and malshaped and missing teeth were reported in rhesus monkey offspring exposed to low doses of TCDD during gestation and lactation [23].

Exposure of adult rats to TCDD was shown to affect only the continuously erupting incisors, and the effects were observed at higher dose levels than those seen after in utero/lactational exposure [24–27]. The incisor defects included pulpal perforation, odontoblast and pulpal cell death, and the consequent arrest of dentin formation. Also, precocious squamous metaplasia was observed in the postsecretory enamel organ.

20.3 EFFECTS OF DIOXINS ON BONE

Clinical examination of pre- and perinatal babies exposed in Yusho and Yu-Cheng accidents produced the first reports on bone effects of dioxin-like compounds [7, 28]. Spotty calcification of skulls, large fontanels, and clinodactyly were observed. More recent epidemiological studies provide only minimal support to an association between a lower exposure to dioxin-like compounds and bone-related effects. Wives of Swedish fishermen from the coast of the organochlorine-contaminated Baltic Sea eating more than one meal per month of fatty fish from the Baltic Sea had a slightly increased incidence of osteoporotic fractures compared to wives of fishermen from the less contaminated Swedish west coast, who ate at most one such meal per month [29]. However, in a further study with individual exposure assessment using serum PCB 153 and p,p'-DDE as indicators of organochlorine exposure, no association with bone mineral density (BMD) or bone metabolism markers was found [30]. In another Swedish study, exposure to dioxin-like PCB 118 showed a slight negative association with BMD [31].

Further evidence of toxic effects of dioxin-like compounds on bone relies largely on experimental findings. Cleft palate is a well-characterized skeletal developmental defect of TCDD observed in teratology studies at relatively high dose levels [32, 33]. Dose–response studies with TCDD using peripheral quantitative computed tomography (pQCT) and biomechanical testing showed alterations in bone geometry, BMD, and mechanical strength in long bones of rats both after long-term exposure of adult animals [34, 35] and in offspring after in utero/lactational exposure [36]. The characteristic findings were decreased cross-sectional and medullary areas, decreased trabecular area, and decreased breaking force and stiffness of tibia. BMD showed only minor and variable alterations; increases in cortical BMD were observed after adult exposure [35] and decreases after in utero/lactational exposure [36]. Most of the bone effects observed after in utero/lactational exposure were reversed at the age of 1 year. In mice, lactational exposure to TCDD resulted in decreased cortical thickness, decreased BMD, and increased osteoid surface as a result of lowered osteoblastic bone formation [37].

Bone effects induced by dioxin-like compounds have also been observed in fish [21, 38–40], birds [41, 42], and primates. Offspring of rhesus monkeys exposed to low doses of TCDD *in utero*/lactationally showed decreased mechanical strength of femur and some minor geometric alterations [43]. It is noteworthy that these observations were made as late as at the age of 7 years. Several *in vitro* studies have been essential in clarifying mechanistic information on effects of dioxin-like compounds on bone formation [44–47] and resorption [45, 48]. It was also shown that differentiating osteoblasts and osteoclasts are highly sensitive to TCDD, because concentrations as low as 100 fM cause significant alterations in differentiation [45].

20.4 AHR SIGNALING PATHWAY IN TEETH AND BONE

The first step in addressing the plausibility for AHR-mediated effects on teeth and bone has been to demonstrate the presence of AHR and ARNT during the critical window of sensitivity. Further evidence of the functionality of the pathway is derived from studies on the effects of TCDD on well-defined AHR-mediated end points, such as CYP1A1 induction, studies on the effects of AHR antagonists and non-dioxin AHR ligands, and studies with transgenic animal models, that is, AHR-null mice and mice with a constitutively active AHR (CA-AHR). The outcome of these studies is summarized in Table 20.1.

Studies on the mechanism of TCDD-induced cleft palate in mice were the first to reveal the expression of the AHR and ARNT in developing palate. Using *in situ* hybridization and immunohistochemistry, Abbott et al. [49] demonstrated the expression of AHR mRNA and protein in epithelial and mesenchymal cells of palatal shelves in gestational day (GD) 14 mouse embryos. In the mesenchyme, high levels of AHR were found in the regions of bone formation. Interestingly, contrary to the other mesenchymal regions, the AHR expression showed a strong nuclear localization. TCDD treatment was also shown to downregulate the expression of AHR at the transcriptional level. Further studies confirmed AHR, ARNT, and CYP1A1 mRNA expression in cultured human and mouse palates [50, 60] and in mouse craniofacial tissues *in vivo* [51]. In addition, TCDD exposure resulted in a time- and concentration-related expression of CYP1A1 confirming that the AHR pathway is functional in developing palate. Human embryonic palatal tissue expressed about 350 times less AHR and the CYP1A1 response was about 1500 times lower compared to the palatal tissue of C57BL/6N mice. Therefore, it can be speculated that humans are resistant to dioxin-induced cleft palate.

Abbott et al. also studied the overall spatial and temporal patterns of AHR and ARNT mRNA and protein expression in C57BL/6N mouse embryos on GDs 10–16 [61, 62]. Regions of bone formation exhibited a high level of AHR expression on GD 12, which coincides with the first morphological signs of the development of the skeletal system, and continued to the end of the observation period until GD 16. In general, the expression was associated with regions of rapid proliferation and differentiation, and the expression showed a strong nuclear pattern. Also, in tooth buds AHR was expressed simultaneously with the first signs of their development on GD 13. The expression of ARNT in developing bones and tooth buds followed the same temporal and spatial patterns as the expression of AHR and with strong nuclear localization [62].

In a more detailed study on developing mouse teeth covering the tooth development from GD 12 until postnatal day (PND) 12, the expression patterns of AHR and ARNT were shown to be related to the stage of tooth development [63]. The expression of both AHR and ARNT was detected in early tooth germ epithelium, but not in undifferentiated dental mesenchyme and epithelium from the bud stage to the bell stage. However, as odontoblasts and ameloblasts differentiate, the expression is upregulated. An intense expression was detected in newly differentiated secretory odontoblasts and ameloblasts, and the intense expression continued throughout the secretory phase. The authors concluded that the TCDD-induced arrest of tooth development at the early bud stage could be mediated by the AHR and speculated that during the developmental stages of dental morphogenesis when the expression of AHR and ARNT is weak or absent, other pathways such as the tyrosine kinase/*c-src* pathway [64] and epidermal growth factor (EGFR) [19] could potentially mediate dioxin effects.

In line with the findings of Sahlberg and coworkers, intense AHR expression was detected in rat molar ameloblasts and odontoblasts on PNDs 9 and 22 [65]. Immunostaining was more intense in ameloblasts and at an earlier time point. Also, CYP1A1 was detected in both cell types. Lactational exposure to high maternal doses of TCDD on PND 1 did not affect the AHR or CYP1A1 expression on PND 9, but resulted in a dose-dependent and almost complete loss of expression of both proteins from both cell types on PND 22. However, some AHR and CYP1A1 expression was still detected in the papillary layer of the outer enamel epithelium. The authors suggested that the loss of CYP1A1 is secondary to strongly repressed AHR expression and reflects impaired function of secretory dental cells.

Expression of AHR protein as well as AHR and ARNT mRNA was shown in rat calvarial osteoblast-like cells and mouse calvarial clonal preosteoblastic MC3T3-E1 cells [52], as well as in primary bone cell cultures isolated from newborn rats [48]. Immunostaining of rat bone cells, histological sections from $AHR^{+/+}$ and $AHR^{-/-}$ mouse tibia,

TABLE 20.1 Evidence for Functionality of AHR Pathway in Developing Teeth and Bone and Its Role in Mediating Dioxin Effects

Test System	Effect	References
Embryonic mouse palate *in vivo*	TCDD-induced AHR downregulation	[49]
Human and mouse embryonic palates *in vitro*, embryonic mouse craniofacial tissues *in vivo*	Time- and concentration-dependent induction of CYP1A1 by TCDD	[50, 51]
Embryonic chicken periosteal osteogenesis model and rat stromal cell bone nodule formation model *in vitro*	Resveratrol fully or partially antagonized TCDD-induced decrease in bone formation, expression of markers of osteogenesis, and mineralization	[47]
Zebrafish embryos	TCDD activated AHR-regulated reporter gene system in the eye, nose, and vertebra of developing embryos and induced craniofacial and vertebral dysmorphogenesis	[40]
Medaka fish embryos	AHR antagonist α-naphthoflavone prevented TCDD-induced vertebral malformations	[39]
Rat calvarial osteoblast-like cells and mouse calvarial preosteoblastic cell line MC3T3-E1 *in vitro*, mouse embryos *in vivo*	AHR agonist 3-methylcholathrene induced largely similar changes to those induced by TCDD	[52, 53]
Mouse RAW264.7 cell line differentiating into osteoclasts *in vitro*	AHR agonist benzo[*a*]pyrene showed inhibitory effects on osteoclast differentiation similar to TCDD. Resveratrol prevented these effects	[54, 55]
Rat osteosarcoma cell line UMR-106 *in vitro*	TCDD-induced CYP1A1 induction and osteopontin downregulation	[56]
Neonatal rat calvarial osteoblasts *in vitro*	TCDD-induced translocation of AHR into the nucleus, binding to DRE, and induction of CYP1A1 and Cox-2. AHR antagonist MNF prevented these effects	[46]
Differentiating rat and mouse bone marrow stromal cells *in vitro*	TCDD-induced increase in AHRR expression. TCDD-induced decrease in RUNX2, alkaline phosphatase, and osteocalcin expression did not take place in cells from $AHR^{-/-}$ mice	[45]
Adult L-E and H/W rats *in vivo*	Relative potency of tetra-, penta-, hexa-, and heptachlorinated dioxins in inducing incisor pulpal perforations followed the rank order of the TEF_{WHO} values	[26]
Ahr^b/Ahr^b and Ahr^d/Ahr^d mice *in vivo*	Male mice with high binding affinity AHR (Ahr^b/Ahr^b) were more sensitive than male mice with low binding affinity AHR (Ahr^d/Ahr^d) to alterations in molar and mandibular shape induced by *in utero*/lactational exposure to TCDD	[14]
$AHR^{+/+}$ and $AHR^{-/-}$ mice *in vivo*	No major bone changes in TCDD-treated $AHR^{-/-}$ mice	[57]
Mice with constitutively active AHR *in vivo*	Bone phenotype of female CA-AHR mice has some similarities to bones of TCDD-treated animals	[58]
Zebrafish embryos	AHR antagonist α-naphthoflavone suppressed the expression of *FoxQ1b* and *Cyp1A*. A morpholino antisense oligonucleotide designed against zebrafish *Ahr2* blocked TCDD-induced translation of FoxQ1b mRNA and prevented TCDD-induced craniofacial defects	[59]

and pure human osteoclast cultures revealed a strong AHR expression in both bone-forming osteoblasts and bone-resorbing osteoclasts, except in bones of $AHR^{-/-}$ mice [48]. Expression was very prominent in osteoclasts, which suggests that they are potential targets of dioxin-induced bone toxicity. Expression of AHR and ARNT was also detected in bone marrow stem cells induced to differentiate to osteoblasts [45].

Overall, evidence from *in vivo* and *in vitro* studies on coordinated and stage-specific expression pattern of AHR and ARNT and on a functional AHR signaling pathway implies that AHR-mediated effects on tooth and bone development are plausible. In addition, the characteristic expression pattern peaking during differentiation suggests that these proteins play a role in the normal tooth and bone development. Their role, however, is poorly understood.

20.5 PHYSIOLOGICAL ROLE OF AHR IN BONE: STUDIES WITH TRANSGENIC MODELS

AHR is not essential for embryonic development, and $AHR^{-/-}$ mice are viable without major skeletal or dental abnormalities [66–68]. This may be due to gene redundancy; that is, other genes may compensate for the loss of AHR. Nevertheless, $Ahr^{-/-}$ genotype has been associated with some changes in the bone phenotype, which are suggestive of reduced osteogenesis. These include a lower incidence of large interfrontal bones in unexposed $AHR^{-/-}$ mice [68]. Furthermore, in bone marrow cells from $AHR^{-/-}$ mice, the differentiation of osteoblasts induced to form mineralized multilayered nodules *in vitro*; the formation of nodules was significantly reduced by 70% compared to cells from $AHR^{+/+}$ mice [46]. Similarly, the expression of the differentiation markers alkaline phosphatase and osteocalcin in differentiating osteoblasts derived from bone marrow stem cells of $AHR^{-/-}$ mice was reduced [45].

Comparison of long bones of $AHR^{-/-}$ and $AHR^{+/+}$ mice revealed gender-specific differences in the bone phenotype [57]. Male $AHR^{-/-}$ mice had smaller cross-sectional area, trabecular area, and periosteal and endosteal circumferences in tibial metaphysis, while female $AHR^{-/-}$ mice had higher trabecular BMD in metaphysis and lower cortical BMD, thickness, and area in diaphysis. Data from a tibial microarchitecture study using high-resolution X-ray microtomography (μCT) indicated a slightly lower trabecular thickness and lower mechanical strength of tibia compared to $AHR^{+/+}$ mice.

Another view on the physiological role of AHR in development and maintenance of bone phenotype is obtained from studies in transgenic mice expressing CA-AHR [58]. This model was suggested to mimic low-dose long-term exposure to AHR ligands. Although the expression of CYP1A1 mRNA was equal in femurs of both genders of CA-AHR mice (but hardly detectable in wild-type mice), the bone phenotype of these mice was gender specific as female mice showed several alterations in bone geometry, density, and biomechanical properties, but in male mice the only difference was increased trabecular BMD. In female mice, trabecular area and bone mineral content were increased in metaphysis, while endosteal and periosteal circumferences and cross-sectional area were increased and cortical and total BMD decreased in mid-diaphysis. Based on analysis of bone markers in serum, markers of osteoblast and osteoclast differentiation, and a stereological analysis, osteoclast size and bone-resorbing activity were increased and the activity of bone-forming osteoblasts apparently decreased compared to wild-type females. Overall, in this transgenic model continuous AHR activation is associated with female-specific increase in bone loss and more ductile bones with wider diameter.

In spite of some similarities, the observed alterations in bones of female CA-AHR mice were not identical to those in TCDD-treated animals. Similarities include decreased cortical BMD in these mice, in TCDD-treated wild-type ($AHR^{+/+}$) mice [57], and *in utero*/lactationally exposed rats [36], and a decreased osteoblast activity in CA-AHR females and in differentiating rat and mouse osteoblasts [45]. On the other hand, long bones of CA-AHR females were more ductile and less brittle as indicated by increased displacement and decreased stiffness, while the bones of TCDD-treated rats had reduced bending breaking force [35, 36]. Also, trabecular area and bone size (cross-sectional area and circumferences) were increased in CA-AHR females, but decreased in TCDD-treated rats [34–36], and no significant differences were found in responses between male and female offspring after *in utero*/lactational TCDD treatment [36]. These data suggest that AHR-related mechanisms alone are not sufficient to account for all bone effects of TCDD.

Findings from transgenic models suggest that the AHR has a physiological role in osteoblast differentiation and bone modeling, which is also reflected in bone phenotype in terms of geometry, mineral density, and biomechanical properties. Gender-specific bone phenotype due to transgenic modification of AHR strongly suggests that crosstalk between AHR and estrogen receptors is involved.

20.6 ROLE OF AHR IN EFFECTS OF DIOXINS ON TEETH

In experimental studies with adult rats at relatively high doses of TCDD, pulpal perforation, odontoblast and pulpal cell death, and a consequent arrest of dentin formation of continuously erupting incisors were observed [24, 25]. Also, precocious squamous metaplasia was observed in the post-secretory enamel organ. Therefore, in rodent incisors the targets of dioxin exposure are ectomesenchymal odontoblasts and, to a lesser extent, epithelial enamel organ, and the altered function of these cells results in impaired matrix formation [25]. Pulpal perforations are observable already 8 days after exposure, and the rank order of relative potencies among tetra-, penta-, hexa-, and heptachlorinated dioxins is in agreement with the TEF_{WHO} values [26, 27] pointing to a pivotal role of the AHR in mediating these effects. There were no sensitivity differences for this defect between Long–Evans (L-E; *Turku/AB*) rats that express the wild-type rat AHR and TCDD-resistant Han/Wistar (H/W; *Kuopio*) rats. H/W rats have a truncated AHR transactivation domain (TCDD resistance allele Ahr^{hw}) as a result of a point mutation leading to an insertion/deletion-type alteration at the 3′-end of the coding region of AHR [69, 70] (see Chapter 12).

Developing molars proved to be much more sensitive to TCDD than incisors of adult rats. Defects in incisors were observed at a total dose of 17 μg/kg TCDD [25], but molar defects in the offspring already at a single maternal dose of 30 ng/kg given on GD 15 [11, 12]. Thus, disturbed molar development after *in utero*/lactational exposure is one of the most sensitive end points of TCDD-induced toxicity. Alterations in molars included delayed eruption, smaller size, a complete block of development of the third molars, and increased susceptibility to caries. The frequency of missing third molars was higher, the earlier during gestation the exposure was initiated. The highest frequency was observed when TCDD was administered on GD 11, 2 days prior to the initiation of molar development [71]. Based on time course studies, the critical window of sensitivity for the agenesis of the third molar is during the early morphogenesis, from tooth initiation to early bud stage, which is from GD 20 to PND 0 in rats. The period of high sensitivity ends at birth, and on PND 1 a very high dose of TCDD is required to arrest the molar development [72]. Mechanistically, the critical window of sensitivity involves the period when the dental epithelium governs tooth development by inducing the mesenchyme [73]. After the bud stage, when the critical window ends, the signaling proceeds in the opposite direction, that is, from dental mesenchyme to epithelium. *In situ* hybridization and immunohistochemical studies have shown that AHR and ARNT are expressed during the bud stage in dental epithelium, but not in mesenchyme, and neither between the bud and bell stages [63]. Accordingly, exposure of embryonic mouse teeth to TCDD *in vitro* resulted in increased and advanced apoptosis in dental epithelial cells, but not in mesenchymal cells [20]. Based on these findings, the primary target is the dental epithelium, where TCDD is likely to interfere with epithelial–mesenchymal interactions, and involvement of AHR signaling is plausible. The developing epithelium is affected by TCDD in several other organs as well, such as prostate [74], seminal vesicle [75], mammary gland [76], and salivary gland [77], and the common feature is high sensitivity.

Because the AHR is not known to regulate tooth development, it is likely that the activation of AHR initiates a cascade of events interfering with downstream factors essential for tooth development [78]. EGFR is one of the suggested targets that contributes to the developmental effects of TCDD [19, 77, 79]. When embryonic mouse teeth were exposed to TCDD *in vitro*, depolarization of odontoblasts and ameloblasts, failures in mineralization of dentin matrix and enamel matrix deposition, and complete disruption of cuspal morphology were observed in teeth from wild-type mice [19]. However, teeth from EGFR$^{-/-}$ mice showed only slightly abnormal cuspal morphology. Similarly, studies in mice deficient in the EGFR ligands EGF or transforming growth factor α (TGFα), or both EGF and TGFα, highlighted the importance of EGFR signaling in responsiveness to TCDD-induced cleft palate and hydronephrosis, although EGF and TGFα are not absolutely required for these effects [79] (see Chapter 17).

In order to find out if impaired mineralization in cultured embryonic mouse molar explants treated with 1 μM TCDD is associated with altered gene expression, Kiukkonen et al. [18] studied the expression of dentin sialophosphoprotein (*Dspp*), *Bono1*, and matrix metalloproteinase-20 (*MMP-20*). The protein products of these genes are linked to matrix mineralization. *In situ* hybridization did not reveal differences in localization or intensity of *Bono1* or *MMP-20*, but the expression of *Dspp* was reduced or prevented in secretory odontoblasts and decreased in presecretory ameloblasts. The protein product of *Dspp* is a precursor protein that is post-translationally cleaved to dentin phosphoprotein (DPP), which has a role in initiating mineralization, and dentin sialoprotein (DSP). In accordance with these findings, *Dspp* knockout mice exhibit defective dentin mineralization, which is similar to human dentinogenesis imperfecta type III [80]. It is therefore plausible that reduced expression of *Dspp* plays a role in TCDD-induced mineralization defects.

Analysis of total RNA from embryonic mouse teeth exposed to TCDD *in vitro* for 24 h on Affymetrix microarrays identified 31 genes with altered expression at least by a fold factor of 2 [81]. These were genes active in cellular metabolism, intracellular organelles, or membranes, and as expected, also included *Cyp1a1* (15.9-fold) and *Cyp1b1* (5.96-fold). However, only modest changes were observed in genes implicated in molar tooth development according to the Gene Expression in Tooth database (http://bite-it.helsinki.fi). The most highly upregulated genes were *Follistatin* (*Fst*; 1.61-fold) and *Runx2* (1.96-fold by qPCR), and the most downregulated genes were *Transforming growth factor beta 1* (*TGFβ1*; 1.72-fold by qPCR), *Retinoid X receptor beta* (*Rxrb*; 1.56-fold), and *Cyclin-dependent kinase inhibitor 1a* (*p21*; 3.13-fold). The observed modifications in gene expression may contribute to the TCDD-induced alterations in tooth morphogenesis.

Fluctuating asymmetry or random variation between left and right sides in bilaterally symmetrical characters is a measure of developmental instability that is potentially increased by environmental stress. Fluctuating asymmetry of molar shape and size was studied in the F_2 generation of intercross between C57BL/6J and AKR/J mice that differ in their sensitivity to TCDD [16]. *In utero*/lactational TCDD exposure resulted in a small, but significant increase in fluctuating asymmetry of molar shape. In spite of an attempt to identify TCDD-responsive genes affecting molar size, shape, or asymmetry, no such genes interacting directly with TCDD could be located.

The role of the AHR in mediating the effects of TCDD on molar and mandibular development was studied by

Keller et al. [14] using congenic mice with a common C57BL/6 genetic background but differing only at the *Ahr* locus, that is, Ahr^b/Ahr^b genotype with high binding affinity AHR and Ahr^d/Ahr^d genotype with low binding affinity AHR. Pregnant females were given a single dose of TCDD at 0, 0.01, 0.1, or 1 μg/kg on GD 13. Molar and mandible size was not affected in either of the genotypes. Molar and mandibular shape was altered significantly in Ahr^b/Ahr^b but not in Ahr^d/Ahr^d male mice at the highest dose level. In females, molar shape was affected similarly in both strains at the highest dose level only, but mandibular shape was not affected in either strain. In spite of the complication caused by gender interactions, the study showed that Ahr^b/Ahr^b male mice are more sensitive than Ahr^d/Ahr^d male mice, and therefore, *Ahr* genotype is involved in TCDD-induced molar and mandibular effects. Low responsiveness of the mice with the C57BL/6 genetic background was not expected, but a subsequent study [17] showed that C57BL/6J mice are more resistant to TCDD-induced molar defects than other mouse strains with the Ahr^b/Ahr^b genotype (see below).

Keller et al. compared the sensitivity of five inbred mouse strains to alterations in the presence of third molars [17], size, shape, and asymmetry of molars [82], and mandible size and shape [15] induced by *in utero*/lactational TCDD exposure. All of the studied strains have the same high-affinity ligand binding Ahr^b allele. Several Ahr^b allelic variants have been identified, and of the studied strains C57BL/6J and C57BL/10J have the b-1 variant (95 kDa), while CBA/J, C3H/HeJ, and BALB/cByJ strains have the b-2 variant (104 kDa) [83]. Missing third molars were observed only in CBA/J and C3H/HeJ mice, but not in C57BL/6J, BALB/cByJ, or C57BL/10J mice. Similarly, CBA/J and C3H/HeJ mice were the most sensitive to altered molar shape (but not size or asymmetry), while C57BL/6J mice were the least sensitive. Thus, it seems plausible that in mice the b-2 allele is required for the molar defects, and an interaction with another gene product missing from BALB/cByJ is also necessary. TCDD also altered mandible size and shape, but the relative sensitivity was not identical to that for missing third molars [15]. Strain differences in mandible shape were observed only in male mice, and C3H/HeJ and C57BL/6J mice were the most sensitive, while BALB/cByJ and CBA/J mice were the least affected. Molar size was altered only in C3H/HeJ male mice. These finding suggest that other genes beyond the *Ahr* locus are also mediating and/or modifying the effects of TCDD on molar and mandibular development in mice.

20.7 ROLE OF AHR IN EFFECTS OF DIOXIN-LIKE COMPOUNDS ON BONE

The current understanding on the function of AHR signaling pathway and its role in the mediation of dioxin-induced effects on bone development is largely based on studies utilizing differentiating bone cells *in vitro*, zebrafish embryos, and transgenic animal models. Employing chicken periosteal osteogenesis and rat stromal cell bone nodule formation models, Singh et al. [47] reported about inhibition of osteogenesis due to impaired differentiation of osteoblasts. Expression of the bone-associated proteins type I collagen, osteopontin, bone sialoprotein, and alkaline phosphatase was decreased and mineralization reduced. All these effects were inhibited by the pure AHR antagonist resveratrol (3,5,4′-trihydroxystilbene). In the well-differentiated rat osteosarcoma cell line UMR-106, TCDD treatment resulted in induction of CYP1A1 and downregulation of osteopontin mRNA [56]. In osteoblasts from neonatal rat calvaria induced to differentiate to form bone nodules, the expression of AHR mRNA and protein was induced during differentiation, but returned back to control levels during matrix maturation, when the expression of osteocalcin is induced [46]. Ligand-dependent activation of AHR proved to be functional in these cells, because exposure to TCDD resulted in translocation of AHR into the nucleus and gel shift assay indicated binding of the activated receptor to the DRE. Furthermore, TCDD treatment concentration-dependently increased the expression of CYP1A1 and cyclooxygenase-2 (Cox-2) proteins and reduced the activity of the osteoblast differentiation marker alkaline phosphatase, thereby modulating the differentiation of osteoblasts. The fact that these effects were inhibited by the AHR antagonist 3′-methoxy-4′-nitroflavone (MNF) confirmed that they were mediated by AHR activation. AHR and ARNT expression was not affected by TCDD. It was also reported that exposure of osteoblasts to lead acetate at 0.1 μM upregulates the expression of AHR mRNA and protein.

AHR signaling pathway was also shown to be functional in differentiating osteoblasts derived from rat and mouse bone marrow stem cells [45]. In these cells, TCDD treatment concentration-dependently increased the expression of AHR repressor (AHRR). TCDD did not deplete the AHR, but resulted in a slight concentration-dependent decrease in ARNT expression after exposure for 10 days. Effects of TCDD on osteoblast differentiation were studied by monitoring the stage-specific differentiation markers RUNX2 (proliferation), alkaline phosphatase (matrix maturation), and osteocalcin (mineralization). TCDD treatment concentration-dependently decreased the expression of mRNA of all the differentiation markers and the activity of alkaline phosphatase. All these changes were AHR dependent, because TCDD-treated cells from $AHR^{-/-}$ mice did not differ from unexposed control cells. Impaired differentiation of these osteoblasts was also associated with significant alterations in expression of 18 individual proteins involved in cytoskeleton organization and biogenesis, actin filament-based processes, and protein transport and folding [84]. These changes were reflected in increased cell adhesion and actin stress fibers in cytoskeleton as observed by confocal microscopy. Furthermore, decreased expression of calcium binding proteins may explain the observed decrease in calcium deposition. Using a MetaCore pathway mapping tool, most of the altered proteins converged into a network of interactions, the two principal regulatory proteins of which are *c-fos* and *c-Myc*. In addition, the transcription factors p53 and c-Jun were identified, and it is suggested that these factors are involved in TCDD-induced inhibition of osteoblast differentiation.

Studies on bone-resorbing osteoclasts showed that these cells exhibit a very strong expression of AHR, but bone resorption by mature osteoclasts is not affected by TCDD [48]. In order to study the effects of TCDD on osteoclast differentiation, Korkalainen et al. [45] exposed to it bone marrow-derived hematopoietic stem cells induced to differentiate to osteoclasts. TCDD significantly decreased the number of cells with characteristic markers of osteoclasts, that is, multinucleated cells positive for tartrate-resistant acid phosphatase (TRACP) isoenzyme 5b and formation of filamentous actin ring structures. Furthermore, resorption activity of these cells was decreased in terms of the area of resorption pits formed on bovine bone slices. It is noteworthy that both osteoblast and osteoclast differentiation were affected already at extremely low (100 fM) TCDD concentrations.

Mattingly et al. [40] showed that the AHR signaling pathway is fully operational during the development of craniofacial tissues and bones of fish. Increased levels of AHR mRNA were expressed from 4 h after fertilization throughout the development. Using a transgenic zebrafish model expressing AHR-regulated green fluorescent protein reporter construct, Mattingly et al. demonstrated TCDD-induced reporter gene expression in the eye, nose, and along the vertebra of developing embryos, and a few days later the exposed fish had gross dysmorphogenesis in craniofacial and vertebral development exactly at the sites of reporter gene expression.

In embryos of medaka fish, TCDD-induced bone defects proved to be the most sensitive end points as they were observed already at picomolar concentrations [39]. The AHR antagonist α-naphthoflavone (α-NF) concentration-dependently prevented the formation of TCDD-induced bone defects, that is, curved vertebral column, deformed neural and hemal spines, and decreased mineralization. Treatment of the embryos with a slightly higher concentration of α-NF alone also caused degeneration of the posterior end of the spinal cord, but not spine defects. Similarly, exposure of adult fish to α-NF alone via diet for 2 months resulted in the loss of all posterior fins. The authors concluded that constitutive activation of AHR is required for normal bone development and maintenance of posterior fins in fish.

Two recent studies utilized microarray technology for identifying transcriptional responses to a low TCDD concentration in zebrafish embryos [59, 85]. Xiong et al. [85] focused on immediate responses on gene expression in the jaw and exposed zebrafish embryos to TCDD (1 ng/mL) at 96 h postfertilization (hpf). TCDD altered the expression of 193 genes by at least twofold in the jaw, and of these genes 24 are known to participate in cartilage or bone development. The most strongly affected gene was *sox9b*, which is a transcription factor essential for chondrogenesis, and in 12 h the expression of this gene was reduced by 15-fold. Reduced *sox9b* expression contributes to the TCDD-induced craniofacial defects, because heterozygous *sox9b*(+/−) mutants carrying only one copy of the *sox9b* gene were more sensitive to TCDD-induced jaw malformations than wild-type embryos. Also, morpholino knockdown of *sox9b* expression alone resulted in a TCDD-like jaw phenotype. Finally, injection of sox9b mRNA into zebrafish embryo at 1–4-cell stage prior to TCDD treatment prevented malformations in 14% of *sox9b*-injected embryos. Altogether these results demonstrate that reduced *sox9b* expression subsequent to TCDD treatment is both necessary and sufficient to induce craniofacial defects in developing zebrafish.

In another microarray study, zebrafish embryos were exposed to TCDD (1 nM) at 1–24 or 6–7 hpf, and RNA from whole embryo was studied [59]. Comparison of control and TCDD-treated embryos uncovered 65 genes with at least a twofold difference in expression levels. Of these genes, *FoxQ1b* belonging to the evolutionarily conserved family of forkhead box transcription factors proved to be upregulated in TCDD-treated embryos by 7- and 10-fold at 24 and 48 hpf, respectively. Induction of *FoxQ1b* was more rapid than that of *Cyp 1A*, and the expression increased over the course of the experiment and was still present at 144 hpf. *In situ* hybridization indicated that at 24 hpf the expression was not discretely localized, but at 48 hpf a strong expression was found specifically in structures of lower jaw, in particular Meckel's cartilage and associated evolutionarily conserved structures. TCDD-induced structural abnormalities were present in the same structures.

FoxQ1b gene is likely to be regulated by AHR, because six putative AHR response elements were identified in the promoter region of this gene, and α-NF suppressed the expression of both *FoxQ1b* and *Cyp1A* [59]. Furthermore, a morpholino antisense oligonucleotide designed against zebrafish *Ahr2* blocked TCDD-induced translation of FoxQ1b mRNA and prevented the characteristic craniofacial defects. The role of misregulation of this gene in observed craniofacial defects is emphasized by the fact that knocking out of mouse *Foxq1* is also associated with cranial abnormalities [86]. These studies showed that activation of the AHR by TCDD upregulates the expression of *FoxQ1b*, and that this misregulation is likely to contribute to the craniofacial defects in zebrafish.

In mammals, the role of AHR in dioxin-induced bone effects *in vivo* has been studied by comparing responses in rats with wild-type AHR to those in rat strains or lines with truncated AHR transactivation domain (Ahr^{hw}) [69, 70] both after adult [34, 35] and *in utero*/lactational exposure [36], as well as in $AHR^{+/+}$ and $AHR^{-/-}$ mice after adult exposure [57]. Alterations such as decreases in bone length, cross-sectional area and circumference, trabecular area, and mechanical strength were reported in adult L-E and H/W rats after TCDD exposure for 20 weeks [34, 35]. Comparison of dose responses indicated a significantly lower sensitivity in TCDD-resistant H/W rats with the altered Ahr^{hw} allele. The Ahr^{hw} allele seems to protect rats from bone effects of *in utero*/lactational TCDD exposure, because decreases in bone length, cross-sectional area, and BMD and reduced bending breaking force and stiffness of tibia, femur, and femoral neck were seen only in line C rats with the wild-type AHR, but not in line A rats with the altered receptor.

$AHR^{+/+}$ and $AHR^{-/-}$ mice were exposed weekly to a high dose of TCDD (total dose 200 μg/kg) for 10 weeks and the bone effects were compared using pQCT, μCT, and three-point bending test [57]. In TCDD-treated $AHR^{+/+}$ mice, the volume and BMD of trabecular bone of tibiae were increased due to a higher number of trabeculae and lower trabecular separation. In addition, the trabeculae were more isotropically oriented, more interconnected, and more planar in shape. The bones were weaker as there were decreases in the maximal bending breaking force, yield bending breaking force, deformation, and maximal energy absorption. No major changes were found in bones of $AHR^{-/-}$ mice.

The pattern of effects of dioxin-like compounds on bone cells as characterized in a number of studies includes impaired differentiation and reduced functionality in terms of matrix formation and mineralization, and decreased resorbing activity. Alterations in BMD observed *in vivo* depend on the bone type and timing of exposure, but nearly all studies report a decreased cross-sectional area and mechanical strength. The role of AHR in mediating these effects of dioxin-like compounds seems to be crucial. In this regard, it is interesting that two other AHR ligands, 3-methylcholathrene and benzo[a]pyrene, have produced largely similar effects in osteoblasts and osteoclast-like cells *in vitro* and in mice *in vivo* [52, 53], as well as in RAW264.7 cells differentiating into osteoclasts [54, 55].

20.8 CONCLUSIONS

There is convincing evidence that teeth and bone belong to the target organs of dioxin toxicity, and the development of these tissues is particularly sensitive and therefore significant from the risk assessment point of view. AHR plays a significant role in mediating effects of dioxins on differentiation of tooth- and bone-specific cell types, epithelial–mesenchymal interactions during development, secretory and other functional activity of these cells, apoptosis, and morphology of teeth and bone. However, other factors beyond AHR are also likely to be involved in mediating or modifying effects of dioxins on teeth and bone. Several potentially significant target genes of dioxins have also been identified in developing teeth, craniofacial tissues, and bone cells.

REFERENCES

1. Berkovitz, B. W. B., Holland, G. R., and Moxham, B. J. (2002). *Oral Anatomy, Histology and Embryology*, 3rd ed. Mosby, Edinburgh, pp. 378.

2. Olsen, B. R. (2006). Bone embryology. In *Primer on the Metabolic Bone Diseases and Disorders of Mineral Metabolism. Section I. Morphogenesis, Structure, and Cell Biology of Bone* (Lian, J. B. and Goldring, S. R.,Eds) The American Society for Bone and Mineral Research, pp. 2–6.

3. Hara, I. (1985). Health status and PCBs in blood of workers exposed to PCBs and of their children. *Environmental Health Perspectives*, 59, 85–90.

4. McNulty, W. P. (1985). Toxicity and fetotoxicity of TCDD, TCDF and PCB isomers in rhesus macaques (Macaca mulatta). *Environmental Health Perspectives*, 60, 77–88.

5. Kashimoto, T. and Miyata, H. (1987). Differences between Yusho and other kinds of poisoning involving only PCBs. In *PCBs in the Environment* (Ward, J. S.,Ed.) CRC Press, Boca Raton, FL, pp. 1–26.

6. Lan, S. J., Yen, Y. Y., Ko, Y. C., and Chen, E. R. (1989). Growth and development of permanent teeth germ of transplacental Yu-Cheng babies in Taiwan. *Bulletin of Environmental Contamination and Toxicology*, 42, 931–934.

7. Rogan, W. J., Gladen, B. C., Hung, K.-L., Koong, S.-L., Shih, L.-Y., Taylor, J. S., Wu, Y.-C., Yang, D., Ragan, N. B., and Hsu, C.-C. (1988). Congenital poisoning by polychlorinated biphenyls and their contaminants in Taiwan. *Science*, 241, 334–336.

8. Alaluusua, S., Calderara, P., Gerthoux, P. M., Lukinmaa, P.-L., Kovero, O., Needham, L., Patterson, D. G. J., Tuomisto, J., and Mocarelli, P. (2004). Developmental dental aberrations after the dioxin accident in Seveso. *Environmental Health Perspectives*, 112, 1313–1318.

9. Alaluusua, S., Lukinmaa, P.-L., Torppa, J., Tuomisto, J., and Vartiainen, T. (1999). Developing teeth as biomarker of dioxin exposure. *Lancet*, 353, 206.

10. Alaluusua, S., Lukinmaa, P.-L., Vartiainen, T., Partanen, M., Torppa, J., and Tuomisto, J. (1996). Polychlorinated dibenzo-p-dioxins and dibenzofurans via mother's milk may cause developmental defects in the child's teeth. *Environmental Toxicology and Pharmacology*, 1, 193–197.

11. Kattainen, H., Tuukkanen, J., Simanainen, U., Tuomisto, J. T., Kovero, O., Lukinmaa, P.-L., Alaluusua, S., Tuomisto, J., and Viluksela, M. (2001). *In utero*/lactational TCDD exposure impairs molar tooth development in rats. *Toxicology and Applied Pharmacology*, 174, 216–224.

12. Miettinen, H. M., Sorvari, R., Alaluusua, S., Murtomaa, M., Tuukkanen, J., and Viluksela, M. (2006). The effect of perinatal TCDD exposure on caries susceptibility in rats. *Toxicological Sciences*, 91, 568–575.

13. Allen, D. E. and Leamy, L. J. (2001). 2,3,7,8-Tetrachlorodibenzo-*p*-dioxin affects size and shape, but not asymmetry, of mandibles in mice. *Ecotoxicology*, 10, 167–176.
14. Keller, J. M., Huang, J. C., Huet-Hudson, Y., and Leamy, L. J. (2007). The effects of 2,3,7,8-tetrachlorodibenzo-*p*-dioxin on molar and mandible traits in congenic mice: a test of the role of the Ahr locus. *Toxicology*, 242, 52–62.
15. Keller, J. M., Zelditch, M. L., Huet, Y. M., and Leamy, L. J. (2008). Genetic differences in sensitivity to alterations of mandible structure caused by the teratogen 2,3,7,8-tetrachlorodibenzo-*p*-dioxin. *Toxicological Pathology*, 36, 1006–1013.
16. Keller, J. M., Allen, D. E., Davis, C. R., and Leamy, L. J. (2007). 2,3,7,8-Tetrachlorodibenzo-*p*-dioxin affects fluctuating asymmetry of molar shape in mice, and an epistatic interaction of two genes for molar size. *Heredity*, 98, 259–267.
17. Keller, J. M., Huet-Hudson, Y. M., and Leamy, L. J. (2007). Qualitative effects of dioxin on molars vary among inbred mouse strains. *Archives of Oral Biology*, 52, 450–454.
18. Kiukkonen, A., Sahlberg, C., Lukinmaa, P. L., Alaluusua, S., Peltonen, E., and Partanen, A. M. (2006). 2,3,7,8-Tetrachlorodibenzo-*p*-dioxin specifically reduces mRNA for the mineralization-related dentin sialophosphoprotein in cultured mouse embryonic molar teeth. *Toxicology and Applied Pharmacology*, 216, 399–406.
19. Partanen, A. M., Alaluusua, S., Miettinen, P. J., Thesleff, I., Tuomisto, J., Pohjanvirta, R., and Lukinmaa, P.-L. (1998). Epidermal growth factor receptor as a mediator of developmental toxicity of dioxin in mouse embryonic teeth. *Laboratory Investigation*, 78, 1473–1481.
20. Partanen, A. M., Kiukkonen, A., Sahlberg, C., Alaluusua, S., Thesleff, I., Pohjanvirta, R., and Lukinmaa, P.-L. (2004). Developmental toxicity of dioxin to mouse embryonic teeth *in vitro*: arrest of tooth morphogenesis involves stimulation of apoptotic program in the dental epithelium. *Toxicology and Applied Pharmacology*, 194, 24–33.
21. Hornung, M. W., Spitsbergen, J. M., Peterson, R. E., and Jan, T. S. (1999). 2,3,7,8-Tetrachlorodibenzo-*p*-dioxin alters cardiovascular and craniofacial development and function in sac fry of rainbow trout (Oncorhynchus mykiss). *Toxicological Sciences*, 47, 40–51.
22. Render, J. A., Bursian, S. J., Rosenstein, D. S., and Aulerich, R. J. (2001). Squamous epithelial proliferation in the jaws of mink fed diets containing 3,3′,4,4′,5-pentachlorobiphenyl (PCB 126) or 2,3,7,8-tetrachlorodibenzo-*p*-dioxin (TCDD). *Veterinary and Human Toxicology*, 43, 22–26.
23. Yasuda, I., Yasuda, M., Sumida, H., Tsusaki, H., Arima, A., Ihara, T., Kubota, S., Asaoka, K., Tsuga, K., and Akagawa, Y. (2005). In utero and lactational exposure to 2,3,7,8-tetrachlorodibenzo-*p*-dioxin (TCDD) affects tooth development in rhesus monkeys. *Reproductive Toxicology*, 20, 21–30.
24. Alaluusua, S., Lukinmaa, P.-L., Pohjanvirta, R., Unkila, M., and Tuomisto, J. (1993). Exposure to 2,3,7,8-tetrachloro-para-dioxin leads to defective dentin formation and pulpal perforation in rat incisor tooth. *Toxicology*, 81, 1–13.
25. Kiukkonen, A., Viluksela, M., Sahlberg, C., Alaluusua, S., Tuomisto, J. T., Tuomisto, J., and Lukinmaa, P.-L. (2002). Response of the incisor tooth to 2,3,7,8-tetrachlorodibenzo-*p*-dioxin in a dioxin-resistant and a dioxin-sensitive rat strain. *Toxicological Sciences*, 69, 482–489.
26. Simanainen, U., Tuomisto, J. T., Tuomisto, J., and Viluksela, M. (2002). Structure–activity relationships and dose responses of polychlorinated dibenzo-*p*-dioxins (PCDDs) for short-term effects in TCDD-resistant and TCDD-sensitive rat strains. *Toxicology and Applied Pharmacology*, 181, 38–47.
27. Simanainen, U., Tuomisto, J. T., Tuomisto, J., and Viluksela, M. (2003). Dose–response analysis of short-term effects of 2,3,7,8-tetrachlorodibenzo-*p*-dioxin in three differentially susceptible rat lines. *Toxicology and Applied Pharmacology*, 187, 128–136.
28. Yamashita, F. and Hayashi, M. (1985). Fetal PCB syndrome: clinical features, intrauterine growth retardation and possible alteration in calcium metabolism. *Environmental Health Perspectives*, 59, 41–45.
29. Wallin, E., Rylander, L., and Hagmar, L. (2004). Exposure to persistent organochlorine compounds through fish consumption and the incidence of osteoporotic fractures. *Scandinavian Journal of Work, Environment & Health*, 30, 30–35.
30. Wallin, E., Rylander, L., Jonssson, B. A., Lundh, T., Isaksson, A., and Hagmar, L. (2005). Exposure to CB-153 and *p,p′*-DDE and bone mineral density and bone metabolism markers in middle-aged and elderly men and women. *Osteoporosis International*, 16, 2085–2094.
31. Hodgson, S., Thomas, L., Fattore, E., Lind, P. M., Alfven, T., Hellstrom, L., Hakansson, H., Carubelli, G., Fanelli, R., and Jarup, L. (2008). Bone mineral density changes in relation to environmental PCB exposure. *Environmental Health Perspectives*, 116, 1162–1166.
32. Birnbaum, L. S. (1995). Developmental effects of dioxins and related endocrine disrupting chemicals. *Toxicology Letters*, 82–83 734–750.
33. Courtney, D. K. and Moore, J. A. (1971). Teratology studies with 2,4,5-trichlorophenoxyacetic acid and 2,3,7,8-tetrachlorodibenzo-*p*-dioxin. *Toxicology and Applied Pharmacology*, 20, 396–403.
34. Herlin, M., Kalantari, F., Stern, N., Sand, S., Larsson, S., Viluksela, M., Tuomisto, J. T., Tuomisto, J., Tuukkanen, J., Jämsä, T., Lind, P. M., and Håkansson, H. (2010). Quantitative characterization of changes in bone geometry, mineral density and biomechanical properties in two rat strains with different Ah-receptor structures after long-term exposure to 2,3,7,8-tetrachlorodibenzo-*p*-dioxin. *Toxicology*, 273, 1–11.
35. Jämsä, T., Viluksela, M., Tuomisto, J. T., Tuomisto, J., and Tuukkanen, J. (2001). Effects of 2,3,7,8-tetrachlorodibenzo-*p*-dioxin on bone in two rat strains with different aryl hydrocarbon receptor structures. *Journal of Bone and Mineral Research*, 16, 1812–1820.
36. Miettinen, H. M., Pulkkinen, P., Jämsä, T., Koistinen, J., Simanainen, U., Tuomisto, J., Tuukkanen, J., and Viluksela, M. (2005). Effects of *in utero* and lactational TCDD exposure on bone development in differentially sensitive rat lines. *Toxicological Sciences*, 85, 1003–1012.
37. Nishimura, N., Nishimura, H., Ito, T., Miyata, C., Izumi, K., Fujimaki, H., and Matsumura, F. (2009). Dioxin-induced

up-regulation of the active form of vitamin D is the main cause for its inhibitory action on osteoblast activities, leading to developmental bone toxicity. *Toxicology and Applied Pharmacology, 236*, 301–309.

38. Carvalho, P. S., Noltie, D. B., and Tillitt, D. E. (2004). Intrastrain dioxin sensitivity and morphometric effects in swim-up rainbow trout (*Oncorhynchus mykiss*). *Comparative Biochemistry and Physiology, 137*, 133–142.

39. Kawamura, T. and Yamashita, I. (2002). Aryl hydrocarbon receptor is required for prevention of blood clotting and for the development of vasculature and bone in the embryos of medaka fish, Oryzias latipes. *Zoological Science, 19*, 303–319.

40. Mattingly, C. J., McLachlan, J. A., and Toscano, W. A. J. (2001). Green fluorescent protein (GFP) as a marker of aryl hydrocarbon receptor (AhR) function in developing zebrafish (*Danio rerio*). *Environ Health Perspectives, 109*, 845–849.

41. Hoffman, D. J., Melancon, M. J., Klein, P. N., Rice, C. P., Eisemann, J. D., Hines, R. K., Spann, J. W., and Pendleton, G. W. (1996). Developmental toxicity of PCB 126 (3,3′,4,4′,5-pentachlorobiphenyl) in nestling American kestrels (Falco sparverius). *Fundamental and Applied Toxicology, 34*, 188–200.

42. Summer, C. L., Giesy, J. P., Bursian, S. J., Render, J. A., Kubiak, T. J., Jones, P. D., Verbrugge, D. A., and Aulerich, R. J. (1996). Effects induced by feeding organochlorine-contaminated carp from Saginaw Bay, Lake Huron, to laying White Leghorn hens. II. Embryotoxic and teratogenic effects. *Journal of Toxicology and Environmental Health, 49*, 409–438.

43. Hermsen, S. A., Larsson, S., Arima, A., Muneoka, A., Ihara, T., Sumida, H., Fukusato, T., Kubota, S., Yasuda, M., and Lind, P. M. (2008). In utero and lactational exposure to 2,3,7,8-tetrachlorodibenzo-p-dioxin (TCDD) affects bone tissue in rhesus monkeys. *Toxicology, 253*, 147–152.

44. Gierthy, J. F., Silkworth, J. B., Tassinari, M., Stein, G. S., and Lian, J. B. (1994). 2,3,7,8-Tetrachlorodibenzo-p-dioxin inhibits differentiation of normal diploid rat osteoblasts in vitro. *Journal of Cellular Biochemistry, 54*, 231–238.

45. Korkalainen, M., Kallio, E., Olkku, A., Nelo, K., Ilvesaro, J., Tuukkanen, J., Mahonen, A., and Viluksela, M. (2009). Dioxins interfere with differentiation of osteoblasts and osteoclasts. *Bone, 44*, 1134–1142.

46. Ryan, E. P., Holz, J. D., Mulcahey, M., Sheu, T. J., Gasiewicz, T. A., and Puzas, J. E. (2007). Environmental toxicants may modulate osteoblast differentiation by a mechanism involving the aryl hydrocarbon receptor. *Journal of Bone and Mineral Research, 22*, 1571–1580.

47. Singh, S. U., Casper, R. F., Fritz, P. C., Sukhu, B., Ganss, B., Girard, B. J., Savouret, J. F., and Tenenbaum, H. C. (2000). Inhibition of dioxin effects on bone formation *in vitro* by a newly described aryl hydrocarbon receptor antagonist, resveratrol. *Journal of Endocrinology, 167*, 183–195.

48. Ilvesaro, J., Pohjanvirta, R., Tuomisto, J., Viluksela, M., and Tuukkanen, J. (2005). Bone resorption by aryl hydrocarbon receptor-expressing osteoclasts is not disturbed by TCDD in short-term cultures. *Life Sciences, 77*, 1351–1366.

49. Abbott, B. D., Perdew, G. H., and Birnbaum, L. S. (1994). Ah receptor in embryonic mouse palate and effects of TCDD on receptor expression. *Toxicology and Applied Pharmacology, 126*, 16–25.

50. Abbott, B. D., Held, G. A., Wood, C. R., Buckalew, A. R., Brown, J. G., and Schmid, J. (1999). AhR, ARNT, and CYP1A1 mRNA quantitation in cultured human embryonic palates exposed to TCDD and comparison with mouse palate *in vivo* and in culture. *Toxicological Sciences, 47*, 62–75.

51. Abbott, B. D., Schmid, J. E., Brown, J. G., Wood, C. R., White, R. D., Buckalew, A. R., and Held, G. A. (1999). RT-PCR quantification of AHR, ARNT, GR, and CYP1A1 mRNA in craniofacial tissues of embryonic mice exposed to 2,3,7,8-tetrachlorodibenzo-p-dioxin and hydrocortisone. *Toxicological Sciences, 47*, 76–85.

52. Naruse, M., Ishihara, Y., Miyagawa-Tomita, S., Koyama, A., and Hagiwara, H. (2002). 3-Methylchloanthrene, which binds to the arylhydrocarbon receptor, inhibits proliferation and differentiation of osteoblasts *in vitro* and ossification *in vivo*. *Endocrinology, 143*, 3575–3581.

53. Naruse, M., Otsuka, E., Naruse, M., Ishihara, Y., Miyagawa-Tomita, S., and Hagiwara, H. (2004). Inhibition of osteoclast formation by 3-methylcholanthrene, a ligand for arylhydrocarbon receptor: suppression of osteoclast differentiation factor in osteogenic cells. *Biochemical Pharmacology, 67*, 119–127.

54. Voronov, I., Heersche, J. N., Casper, R. F., Tenenbaum, H. C., and Manolson, M. F. (2005). Inhibition of osteoclast differentiation by polycyclic aryl hydrocarbons is dependent on cell density and RANKL concentration. *Biochemical Pharmacology, 70*, 300–307.

55. Voronov, I., Li, K., Tenenbaum, H. C., and Manolson, M. F. (2008). Benzo[a]pyrene inhibits osteoclastogenesis by affecting RANKL-induced activation of NF-kappaB. *Biochemical Pharmacology, 75*, 2034–2044.

56. Wejheden, C., Brunnberg, S., Hanberg, A., and Lind, P. M. (2006). Osteopontin: a rapid and sensitive response to dioxin exposure in the osteoblastic cell line UMR-106. *Biochemical and Biophysical Research Communications, 341*, 116–120.

57. Herlin, M., Aula, A., Miettinen, H. M., Korkalainen, M., Finnilä, M., Tuukkanen, J., Håkansson, H., and Viluksela, M., (2009). The role of AhR in dioxin-induced modulation of bone microarchitecture and mechanical strength. *Toxicology Letters, 189S*, S197.

58. Wejheden, C., Brunnberg, S., Larsson, S., Lind, P. M., Andersson, G., and Hanberg, A. (2010). Transgenic mice with a constitutively active aryl hydrocarbon receptor display a gender-specific bone phenotype. *Toxicological Sciences, 114*, 48–58.

59. Planchart, A. and Mattingly, C. J. (2010). 2,3,7,8-Tetrachlorodibenzo-p-dioxin upregulates FoxQ1b in zebrafish jaw primordium. *Chemical Research in Toxicology, 23*, 480–487.

60. Abbott, B. D., Probst, M. R., and Perdew, G. H. (1994). Immunohistochemical double-staining for Ah receptor and ARNT in human embryonic palatal shelves. *Teratology, 50*, 361–366.

61. Abbott, B. D., Birnbaum, L. S., and Perdew, G. H. (1995). Developmental expression of two members of a new class of transcription factors. I. Expression of aryl hydrocarbon receptor in the C57Bl/6N mouse embryo. *Developmental Dynamics*, 204, 133–143.
62. Abbott, B. D. and Probst, M. R. (1995). Developmental expression of two members of a new class of transcription factors. II. Expression of aryl hydrocarbon receptor nuclear translocator in the C57BL/6N mouse embryo. *Developmental Dynamics*, 204, 144–155.
63. Sahlberg, C., Pohjanvirta, R., Gao, Y., Alaluusua, S., Tuomisto, J., and Lukinmaa, P.-L. (2002). Expression of the mediator of dioxin toxicity, aryl hydrocarbon receptor (AHR) and the AHR nuclear translocator (ARNT), is developmentally regulated in mouse teeth. *The International Journal of Developmental Biology*, 46, 295–300.
64. Tiffee, J. C., Xing, L., Nilsson, S., and Boyce, B. F. (1999). Dental abnormalities associated with failure of tooth eruption in src knockout and op/op mice. *Calcified Tissue International*, 65, 53–58.
65. Gao, Y., Sahlberg, C., Kiukkonen, A., Alaluusua, S., Pohjanvirta, R., Tuomisto, J., and Lukinmaa, P.-L. (2004). Lactational exposure of Han/Wistar rats to 2,3,7,8-tetrachlorodibenzo-p-dioxin interferes with enamel maturation and retards dentin mineralization. *Journal of Dental Research*, 83, 139–144.
66. Fernandez-Salguero, P., Pineau, T., Hilbert, D. M., McPhail, T., Lee, S. S. T., Kimura, S., Nebert, D. W., Rudikoff, S., Ward, J. M., and Gonzalez, F. J. (1995). Immune system impairment and hepatic fibrosis in mice lacking the dioxin-binding Ah receptor. *Science*, 268, 722–726.
67. Mimura, J., Yamashita, K., Nakamura, K., Morita, M., Takagi, T. N., Nakao, K., Ema, M., Sogawa, K., Yasuda, M., Katsuki, M., and Fuiji-Kuriyama, Y. (1997). Loss of teratogenic response to by 2,3,7,8-tetrachlorodibenzo-p-dioxin (TCDD) in mice lacking the Ah (dioxin) receptor. *Genes to Cell*, 2, 645–654.
68. Peters, J. M., Narotsky, M. G., Elizondo, G., Fernandez-Salguero, P. M., Gonzalez, F. J., and Abbot, B. D. (1999). Amelioration of TCDD-induced teratogenesis in aryl hydrocarbon receptor (AhR)-null mice. *Toxicological Sciences*, 47, 86–92.
69. Pohjanvirta, R., Wong, J. M. Y., Li, W., Harper, P. A., Tuomisto, J., and Okey, A. B. (1998). Point mutation in intron sequence causes altered C-terminal structure in the AH receptor of the most TCDD-resistant rat strain. *Molecular Pharmacology*, 54, 86–93.
70. Tuomisto, J. T., Viluksela, M., Pohjanvirta, R., and Tuomisto, J. (1999). The Ah receptor and a novel gene determine acute toxic responses to TCDD: segregation of the resistant alleles to different rat lines. *Toxicology and Applied Pharmacology*, 155, 71–81.
71. Miettinen, H. M., Alaluusua, S., Tuomisto, J., and Viluksela, M. (2002). Effect of in utero and lactational TCDD exposure on rat molar development: the role of exposure time. *Toxicology and Applied Pharmacology*, 184, 57–66.
72. Lukinmaa, P.-L., Sahlberg, C., Leppäniemi, A., Partanen, A.-M., Pohjanvirta, R., Tuomisto, J., and Alaluusua, S. T. (2001). Arrest of rat molar tooth development by lactational dioxin exposure. *Toxicology and Applied Pharmacology*, 173, 38–47.
73. Thesleff, I. and Nieminen, P. (2001). Tooth induction. In *Encyclopedia of Life Sciences*. Nature Publishing Group, pp. 1–8.
74. Theobald, H. M., Roman, B. L., Lin, T. M., Ohtani, S., Chen, S. W., and Peterson, R. E. (2000). 2,3,7,8-Tetrachlorodibenzo-p-dioxin inhibits luminal cell differentiation and androgen responsiveness of the ventral prostate without inhibiting prostatic 5α-dihydrotestosterone formation or testicular androgen production in rat offspring. *Toxicological Sciences*, 58, 324–338.
75. Hamm, J. T., Sparrow, B. R., Wolf, D., and Birnbaum, L. S. (2000). In utero and lactational exposure to 2,3,7,8-tetrachlorodibenzo-p-dioxin alters postnatal development of seminal vesicle epithelium. *Toxicological Sciences*, 54, 424–430.
76. Fenton, S. E., Hamm, J. T., Birnbaum, L. S., and Youngblood, G. L. (2002). Persistent abnormalities in the rat mammary gland following gestational and lactational exposure to 2,3,7,8-tetrachlorodibenzo-p-dioxin (TCDD). *Toxicological Sciences*, 67, 63–74.
77. Kiukkonen, A., Sahlberg, C., Partanen, A.-M., Alaluusua, S., Pohjanvirta, R., Tuomisto, J., and Lukinmaa, P.-L. (2006). Interference by 2,3,7,8-tetrachlorodibenzo-p-dioxin with cultured mouse submandibular gland branching morphogenesis involves reduced epidermal growth factor receptor signaling. *Toxicology and Applied Pharmacology*, 212, 200–211.
78. Alaluusua, S. and Lukinmaa, P. L. (2006). Developmental dental toxicity of dioxin and related compounds: a review. *International Dental Journal*, 56, 323–331.
79. Abbott, B. D., Buckalew, A. R., DeVito, M. J., Ross, D., Bryant, P. L., and Schmid, J. E. (2003). EGF and TGF-alpha expression influence the developmental toxicity of TCDD: dose response and AhR phenotype in EGF, TGF-alpha, and EGF + TGF-alpha knockout mice. *Toxicological Sciences*, 71, 84–95.
80. Sreenath, T., Thyagarajan, T., Hall, B., Longenecker, G., D'Souza, R., Hong, S., Wright, J. T., MacDougall, M., Sauk, J., and Kulkarni, A. B. (2003). Dentin sialophosphoprotein knockout mouse teeth display widened predentin zone and develop defective dentin mineralization similar to human dentinogenesis imperfecta type III. *Journal of Biological Chemistry*, 278, 24874–24880.
81. Sahlberg, C., Peltonen, E., Lukinmaa, P. L., and Alaluusua, S. (2007). Dioxin alters gene expression in mouse embryonic tooth explants. *Journal of Dental Research*, 86, 600–605.
82. Keller, J. M., Huet-Hudson, Y., and Leamy, L. J. (2008). Effects of 2,3,7,8-tetrachlorodibenzo-p-dioxin on molar development among non-resistant inbred strains of mice: a geometric morphometric analysis. *Growth Development and Aging*, 71, 3–16.
83. Poland, A., Glover, E., and Taylor, B. A. (1987). The murine Ah locus: a new allele and mapping to chromosome 12. *Molecular Pharmacology*, 32, 471–478.
84. Carpi, D., Korkalainen, M., Airoldi, L., Fanelli, R., Håkansson, H., Muhonen, V., Tuukkanen, J., Viluksela, M., and Pastorelli, R. (2009). Dioxin-sensitive proteins in differentiating osteoblasts: effects on bone formation in vitro. *Toxicological Sciences*, 108, 330–343.

85. Xiong, K. M., Peterson, R. E., and Heideman, W. (2008). Aryl hydrocarbon receptor-mediated down-regulation of sox9b causes jaw malformation in zebrafish embryos. *Molecular Pharmacology, 74,* 1544–1553.

86. Goering, W., Adham, I. M., Pasche, B., Manner, J., Ochs, M., Engel, W., and Zoll, B. (2008). Impairment of gastric acid secretion and increase of embryonic lethality in Foxq1-deficient mice. *Cytogenetic and Genome Research, 121,* 88–95.

21

IMPACTS OF DIOXIN-ACTIVATED AHR SIGNALING IN FISH AND BIRDS

Michael T. Simonich and Robert L. Tanguay

21.1 INTRODUCTION

The archetype dioxin 2,3,7,8-TCDD elicits a wide array of toxic and teratogenic effects in fish and birds, almost all of which depend on signaling via the aryl hydrocarbon receptor (AHR) [1–4]. During early development of freshwater and saltwater fish, TCDD exposure elicits the same hallmark defects including mortality, arrested development and associated ischemia, yolk sac edema, pericardial edema, anemia, hemorrhage, uninflated swimbladder, severe heart and vasculature defects, and craniofacial malformation [5–7]. Far fewer bird species have been assessed, but craniofacial malformation, cardiovascular defects, hepatotoxicity, and edema are generally observed. In fish and birds, as in mammals, the AHR functions as a ligand-activated transcription factor in heterodimeric complex with the AHR nuclear translocator (ARNT). Both AHR and ARNT are basic helix–loop–helix transcription factors [6, 8]. The AHR–ARNT complex is translocated to the nucleus where it binds xenobiotic response elements (XREs) resulting in the transactivation of dioxin-dependent genes and ensuing toxicity [5].

Interspecies variance in TCDD sensitivity is well established and bird species display a far greater range of sensitivity to dioxin than do fish species. The domestic chicken, for instance, is 10–1000-fold more sensitive to developmental dioxin exposure than fish-eating species, for example, the common tern and the common cormorant [9–11]. These avian interspecies differences belie very strong sequence conservation among avian AHRs as well as the presence of all the canonical AHR pathway members known in mammals and fish. As yet undefined AHR structural differences and other signaling factors/targets are determinants of the sensitivity differences.

Recent evidence also suggests that the AHR is required for normal developmental and physiological signaling events distinct from xenobiotic-elicited signaling [2, 12]. For instance, AHR-null mice exhibit hepatic bile duct fibrosis and impaired spleen and lymph node development, aberrant liver vascularization, cardiac hypertrophy and hypertension, and aberrant prostate and seminal vesicle development [13–18].

The most studied AHR-responsive genes include *CYP1A* and *CYP1B1* [3, 4, 19]. Considerable effort has defined other interacting/modulating factors including cytoplasmic AIP (XAP2, ARA9), p23, HSP90, nuclear factors including coactivators, and the estrogen receptor alpha [1, 20–23]. The majority of research efforts have been aimed at understanding the first steps in toxic and adaptive responses to AHR ligands, namely, AHR activation, nuclear translocation, dimerization, factor recruitment, and direct transcriptional responses. Relatively few *in vivo* studies have been aimed at identifying the role of downstream regulated genes in toxicity, but such studies will be critical to mechanistically resolve dioxin toxicity.

21.2 AHR PATHWAY MEMBERS IN FISH

Piscine AHR signal transduction has only been intensively studied in zebrafish due to the extraordinary utility of that model [6]. Zebrafish possess three *AHR* genes, *AHR1a* and *1b* and *AHR2*, while mammals have only one [24, 25]. *AHR1* and *AHR2* have also been cloned in fathead minnow and four AHR genes (*AHR1b-1*, *AHR1b-2*, *AHR2a*, and *AHR2b*) were

---

The AH Receptor in Biology and Toxicology, First Edition. Edited by Raimo Pohjanvirta.
© 2012 John Wiley & Sons, Inc. Published 2012 by John Wiley & Sons, Inc.

identified in the medaka genome [26, 27]. Full-length AHR2s have been described in zebrafish, medaka, Atlantic tomcod, rainbow trout, and red sea bream [7, 26, 28–30]. Full-length cDNAs for zebrafish *AHR2* (*zfAHR2*) and *ARNT2* (*zfARNT2*) have been cloned, and their translation products have been characterized [30, 31]. zfAHR2 and zfARNT2b form a functional heterodimer *in vitro* that specifically recognizes dioxin response elements (DREs) in gel shift experiments and induces DRE-driven transcription in response to TCDD exposure in COS-7 cells [30, 31]. In addition to the induction of several *CYP* genes, dioxin also induces the expression of the zebrafish AHR repressor (zfAHRR), which has been shown to have functional DREs in its proximal promoter [32]. The AHRR is closely related to the *AHR* and is capable of dimerizing with ARNT to form a negative feedback loop with AHR, repressing AHR signaling by competition for binding to DREs, as well as by novel mechanisms that are independent of DRE binding by AHRR [33–35]. Two AHRR isoforms, AHRRa and AHRRb, have been characterized in zebrafish embryos [34]. AHR2 regulates the TCDD-induced expression of both AHRRa and AHRRb. This is evident from AHR2 knockdown via morpholino oligonucleotide (MO) severely inhibiting the TCDD-dependent induction of both AHRRa and AHRRb. No data exist yet for a role for either AHR1a or AHR1b in regulating AHRRa or AHRRb expression in response to dioxin or related agonists.

Using individual AHRR knockdowns, Jenny et al. showed that, in the absence of TCDD exposure, AHRRa morphants recapitulated TCDD embryo toxicity, whereas AHRRb morphants did not, suggesting that AHRRa and AHRRb have distinct functions in the zebrafish embryo [34]. If the toxic end points elicited by TCDD or related xenobiotics result from hyperactivation of the AHR, as is widely suspected, then the recapitulation of TCDD embryo toxicity by AHRRa knockdown in untreated embryos suggests that AHR is both constitutively active during development and tightly regulated by AHRRa [34].

21.3 AHR PATHWAY MEMBERS IN BIRDS

The domestic chicken has been an important model for studying AHR signaling [9, 20, 22, 23]. Full-length AHR cDNAs have been cloned and described in the domestic chicken and common tern, both being orthologous to the mammalian AHR1 [9]. Full-length cDNAs for distinct AHR1 and AHR2 isoforms have also been cloned and described from common cormorant and black-footed albatross [11]. ARNT1 has been identified from domestic chicken and ARNT1 and ARNT2 have been identified from common cormorant [36, 37]. *In vitro* expressed avian AHR1 and AHR2 isoforms specifically bind TCDD, and both isoforms are transcriptionally active in *in vitro* reporter transactivation assays, but AHR2 induction is markedly lower than AHR1 in cormorant and albatross [9, 11]. Hepatic mRNA expression of AHR2 is also much lower in these species, suggesting that AHR1 is the putative dominant isoform in birds [11]. Avian AHR1 dimerizes preferentially with avian ARNT1.

Downstream, the known transactivation targets of TCDD signaling are few. Avian *CYP1A4* and *CYP1A5*, the orthologs of mammalian *CYP1A1* and *CYP1A2*, respectively, are transcriptionally induced in the chicken by TCDD treatment [38]. Importantly, the canonical basis of this induction, that is, the presence of DREs in the promoter/enhancer regions of avian *CYP1A*s and their transactivation by an avian AHR/ARNT complex, appears to be present and operational in unrelated bird species such as cormorant and chicken [39]. Other factors, such as AHR repressors, remain to be detected in birds. Also, the functional requirement for avian *CYP1A* induction in early developmental stage TCDD toxicity is unknown.

21.4 TCDD-ELICITED DEVELOPMENTAL TOXICITY IN FISH

Zebrafish larvae exposed to waterborne TCDD as early as 6 h postfertilization (hpf) display a variety of deformities including pericardial edema, yolk sac edema, craniofacial malformations, and mortality [40–43]. The craniofacial structures and cardiovascular system are the most severely developmentally impacted by TCDD in fish. These same early-stage toxicities are also observed in lake trout, brook trout, rainbow trout, fathead minnow, channel catfish, lake herring, medaka, white sucker, northern pike, killifish, and red sea bream [7, 44–50]. The earliest signs of TCDD toxicity observed in the cardiovascular system of zebrafish larvae were pericardial edema and reduced blood flow to the trunk at 72–77 hpf. Yolk sac edema and reduced blood flow to gills and head were more apparent by 96 hpf. It is noteworthy to mention that TCDD does not impact the initial development of the vasculature, but affects the maintenance of peripheral vascular beds after they form [41]. TCDD causes cardiac malformation and heart failure in zebrafish embryos through dramatic downregulation of cell cycle progression genes, effectively stopping heart growth [51]. Cardiac valve development is also affected by TCDD. Zebrafish embryos exposed to TCDD fail to form functional heart valves in the developing heart, though TCDD does not prevent the initial specification of the valve locations [52].

21.5 IMPACTS OF SUBLETHAL TCCD EXPOSURE

Most of our knowledge of the impacts of dioxin-activated AHR in fish (see above) comes from TCDD doses that were ultimately lethal or near lethal. A recent study indicated that sublethal transient dose regimens during early zebrafish development can induce latent TCDD toxicity in adults, as well as in their offspring whose only possible exposure to TCDD was as a gamete [53]. Both the survival of F1 offspring and their subsequent reproductive capacity were significantly reduced by transient TCDD exposure of the mother during early development. This striking finding may be rooted in putative epigenetic changes from exposure to endocrine disrupting chemicals during development [53–60]. Evidence from other models suggests that early developmental exposure to endocrine disruptors can induce permanent functional changes that do not manifest until later in life [54, 57–59, 61–64].

Potent activation of AHR signaling by the TCDD ligand is thus seen to perturb an integrated and complex array of developmental processes. The underlying mechanisms of these perturbations are mostly unknown, awaiting discovery and definition of the roles of specific AHR-responsive genes in mediating these developmental responses. This has been challenging because of the integrated complexity of developmental processes, but mechanism discovery in the zebrafish model has been fruitful.

21.6 CANONICAL TARGETS OF DIOXIN-ACTIVATED AHR: A SMALL PART OF THE PICTURE

The prototypical downstream AHR target genes are *CYP1A1*, *CYP1A2*, *CYP1B1*, and *NQO1* [3, 4, 19, 65, 66]. Recently, the *CYP1C* family, newly discovered in fish, was also shown to be an important AHR target in the eye and heart [67]. The significance of the CYP targets is still unclear. Increased expression of CYP1A by TCDD has been proposed to mediate toxic responses in zebrafish. However, antisense repression of *CYP1A* failed to prevent TCDD toxicity in developing zebrafish [68]. AHR2-dependent transcription of *CYP1B* is also induced by TCDD in developing embryos, but like *CYP1A*, antisense repression of *CYP1B* also failed to prevent TCDD developmental toxicity [69]. Collectively, at least for early-stage TCDD exposure, the induction of CYP1A and CYP1B is a parallel transcriptional response. In other fish species, the *CYP* gene responses to TCDD, though much less intensively characterized, are similar to zebrafish. In birds, TCDD response characterization for *CYP* genes has been even more species limited, but of those birds examined, the response of avian *CYP* orthologs appears to be conserved. Analyses of CYP function in other fish and birds via antisense repression have not yet been reported.

In adult zebrafish, acute TCDD exposure altered not only *CYP1A*, but also *CYP1B1*, *CYP1C1*, and *CYP1C2* expression [70]. Importantly, the expression of cyclooxygenase (COX)-1 and COX-2b in the mesenteric artery was also altered as a result of AHR activity [70]. COX-2a has also recently been shown to be necessary and sufficient for mediating some TCDD developmental toxicity in zebrafish [71]. *COX* genes have key roles in cardiovascular development and function and their induction by TCDD provides an important insight into the different tissues and range of targets affected by TCDD-activated AHR.

21.7 TOXICOGENOMIC APPROACH TO AHR TARGETS

Toxicogenomic approaches have significantly advanced our understanding of the transcriptional changes following AHR activation in zebrafish. For instance, a study of heart-specific transcriptional responses to TCDD, aimed at understanding the molecular mechanisms leading to cardiovascular dysfunction, identified changes in genes important in xenobiotic metabolism, cell proliferation, heart contractility, and heart development [51, 72]. A study of transcriptional responses to TCDD in the developing zebrafish jaw uncovered sox9b, a chondrogenic transcription factor, as the most significantly repressed transcript [73]. MO knockdown of sox9b during jaw development in the absence of TCDD recapitulated the TCDD malformation. Conversely, injection of sox9b mRNA into 1–4-cell stage embryos rescued the malformation from subsequent TCDD exposure. Thus, sox9b repression is necessary and sufficient for producing the TCDD jaw malformation [73]. Another study of TCDD-induced jaw transcriptional changes uncovered FoxQ1b, a forkhead box family transcription factor, as strongly upregulated by TCDD in the developing zebrafish jaw [74]. A functional role for FoxQ1b in TCCD-mediated jaw abnormality has not been determined, but it is likely another critical target of dioxin-activated AHR signaling. Hundreds of other transcriptional changes elicited by dioxin-induced AHR activation in zebrafish await evaluation of their role in mediating specific toxic responses to TCDD [5, 72–75].

21.8 THE FIN REGENERATION MODEL: A UNIQUE PLATFORM FOR TCDD TOXICOGENOMICS

Zebrafish caudal fin regeneration after amputation is a tightly regulated orchestration of multiple signaling pathways. For two recent reviews of this model, see Refs 6 and 76. Our laboratory has previously demonstrated that TCDD inhibits zebrafish fin regeneration at both the adult and the larval stage [77, 78]. Morphological analysis demonstrated that TCDD affects several components involved in cellular differentiation and extracellular matrix (ECM) composition in adult tissue regenerates [79]. Gene expression studies performed in adult regenerating fin tissue after exposure to TCDD revealed a cluster of genes important for xenobiotic metabolism, cellular differentiation, and extracellular matrix composition [79]. Genomic analysis conducted in larval regenerating fin tissue after AHR activation also uncovered genes important for xenobiotic metabolism, cellular differentiation, signal transduction, and extracellular matrix composition, similar to the results of the adult zebrafish study [79, 80]. The largest category of transcripts misregulated by TCDD was related to ECM composition and metabolism. TCDD exposure resulted in repression of 34 of the 41 transcripts in this category suggesting that TCDD impairs the maturation of the ECM. This could occur by directly altering the expression of ECM genes or by impacting the genes responsible for controlling cell differentiation and matrix maturation [79]. R-Spondin1, a secreted protein capable of promoting Wnt/β-catenin signaling, was the most highly induced transcript, while *sox9b*, a chondrogenic transcription factor, was the gene most repressed by TCDD, similar to the *sox9b* response in jaw [73, 79–81].

It is noteworthy that in addition to *R-Spondin1*, a cluster of Wnt target genes were also altered by AHR activation in the regenerates of TCDD-exposed zebrafish [79, 80]. A mechanistic possibility is that R-Spondin1, as a Wnt ligand, is a TCDD-dependent upstream modulator of AHR-dependent signaling, thereby inhibiting tissue regeneration. In agreement with this mechanism, TCDD-elevated expression of *R-Spondin1* was sufficient to inhibit regeneration, and R-Spondin1 morphants, in the presence of TCDD, regenerated their fin tissue. This suggests that induction of the Wnt ligand during epimorphic regeneration inhibits normal regenerative signaling [80]. A parallel mechanism of TCDD inhibition of regeneration may be posttranslational regulation of β-catenin via sox9b. It has been proposed that sox9 competes with Tcf/Lef for binding to β-catenin, leading to increased β-catenin degradation [82]. In fact, TCDD-elicited AHR signaling, with its subsequent downregulation of *sox9b* transcription, leads to the overaccumulation of β-catenin and the upregulation of many β-catenin-responsive genes in the regenerating tissues of TCDD-exposed adult and larval zebrafish [79, 80]. A recent report in mice showed that AHR has a critical role in suppression of intestinal carcinogenesis by mediating a previously unknown ligand-dependent E3 ubiquitin ligase degradation of β-catenin [83]. The ligands in this case were both naturally generated dietary metabolites and the exogenous ligands 3-methylcholanthrene and β-naphthoflavone. This would seem to be at odds with the TCDD-activated AHR activity in zebrafish that leads to overaccumulation of β-catenin. We do not yet know if a similar ubiquitin ligase response to the AHR ligand types used in the mouse study is operational in zebrafish. If so, it may be that TCDD-activated AHR signaling exceeds the response to other ligands.

TCDD-elicited AHR signaling results in the strong induction of R-Spondin1 and the reciprocally strong repression of sox9b in larval and adult zebrafish. R-Spondin is a Wnt ligand, and sox9b is a negative regulator of β-catenin accumulation. Collectively, we now know that knockdown of AHR signaling via zfAHR2 or zfARNT1, or of Wnt signaling via R-Spondin1 and sox9b modulation, restores regeneration competency, otherwise lost upon TCDD exposure [77]. TCDD appears to inhibit fin regeneration by AHR-signaled misregulation of Wnt/β-catenin signaling. Importantly, the regenerative and cardiovascular end points of TCDD exposure are mechanistically distinct. Even though both end points are AHR2/ARNT1 dependent, R-Spondin1 morphants fully regenerate their tissues in the presence of TCDD, but still manifest the full cardiovascular effects of TCDD [80]. Thus, the uncoupling of two major AHR-dependent responses to TCDD is beginning to define tissue-specific toxicity mechanisms. Still elusive is the mechanism of regulation. Does AHR-signaled induction of R-Spondin1 simply repress *sox9b* transcription, or is an additional posttranscriptional or protein level regulation operative? Further studies will help us to understand the interaction between AHR and Wnt signaling pathways.

21.9 CONCLUSIONS

Identification of new AHR roles, downstream target genes, and AHR pathway crosstalk with processes such as tissue regeneration firmly establish the necessity of AHR signaling in development and homeostasis. The impacts of dioxin-activated AHR signaling seen in zebrafish are astonishingly complex, but with the advent of toxicogenomic approaches in zebrafish and other models, we are finally beginning to resolve the mechanism of TCDD developmental toxicity. The zebrafish has proven particularly well suited to toxicogenomic studies for several reasons, not the least of which is that it is a fish that, along with fish-eating animals, experiences the most severe chronic dioxin exposures in nature. Research on dioxin-activated AHR impacts in fish-eating birds has primarily consisted of TCDD sensitivity correlations with *CYP* gene inductions. There are inherent distinct advantages and disadvantages for all research models; however, with the recent advances in comparative genomics, we now see remarkable similarities at the molecular level across taxa. Integrative approaches where the advantages of individual research models are exploited will speed the rate at which current information gaps are filled. The use of zebrafish and other models will continue to help define the genes downstream of AHR activation that are responsible for tissue-specific responses.

REFERENCES

1. Carlson, D. B. and Perdew, G. H. (2002). A dynamic role for the Ah receptor in cell signaling? Insights from a diverse group of Ah receptor interacting proteins. *Journal of Biochemical and Molecular Toxicology*, 16, 317–325.

2. Mitchell, K. A. and Elferink, C. J. (2009). Timing is everything: consequences of transient and sustained AhR activity. *Biochemical Pharmacology*, 77, 947–956.
3. Schmidt, J. V. and Bradfield, C. A. (1996). Ah receptor signaling pathways. *Annual Review of Cell and Developmental Biology*, 12, 55–89.
4. Sutter, T. R. and Greenlee, W. F. (1992). Classification of members of the Ah gene battery. *Chemosphere*, 25, 223–226.
5. Carney, S. A., Prasch, A. L., Heideman, W., and Peterson, R. E. (2006). Understanding dioxin developmental toxicity using the zebrafish model. *Birth Defects Research, Part A*, 76, 7–18.
6. Mathew, L. K., Simonich, M. T., and Tanguay, R. L. (2009). AHR-dependent misregulation of Wnt signaling disrupts tissue regeneration. *Biochemical Pharmacology*, 77, 498–507.
7. Yamauchi, M., Kim, E. Y., Iwata, H., Shima, Y., and Tanabe, S. (2006). Toxic effects of 2,3,7,8-tetrachlorodibenzo-p-dioxin (TCDD) in developing red seabream (*Pagrus major*) embryo: an association of morphological deformities with AHR1, AHR2 and CYP1A expressions. *Aquatic Toxicology*, 80, 166–179.
8. Huang, Z. J., Edery, I., and Rosbash, M. (1993). PAS is a dimerization domain common to *Drosophila* period and several transcription factors. *Nature*, 364, 259–262.
9. Karchner, S. I., Franks, D. G., Kennedy, S. W., Hahn, and M. E. (2006). The molecular basis for differential dioxin sensitivity in birds: role of the aryl hydrocarbon receptor. *Proceedings of the National Academy of Sciences of the United States of America*, 103, 6252–6257.
10. Kennedy, S. W., Lorenzen, A., Jones, S. P., Hahn, M. E., and Stegeman, J. J. (1996). Cytochrome P4501A induction in avian hepatocyte cultures: a promising approach for predicting the sensitivity of avian species to toxic effects of halogenated aromatic hydrocarbons. *Toxicology and Applied Pharmacology*, 141, 214–230.
11. Yasui, T., Kim, E. Y., Iwata, H., Franks, D. G., Karchner, S. I., Hahn, M. E., and Tanabe, S. (2007). Functional characterization and evolutionary history of two aryl hydrocarbon receptor isoforms (AhR1 and AhR2) from avian species. *Toxicological Sciences*, 99, 101–117.
12. Puga, A., Ma, C., and Marlowe, J. L. (2009). The aryl hydrocarbon receptor cross-talks with multiple signal transduction pathways. *Biochemical Pharmacology*, 77, 713–722.
13. Fernandez-Salguero, P., et al. (1995). Immune system impairment and hepatic fibrosis in mice lacking the dioxin-binding Ah receptor [see comments]. *Science*, 268, 722–726.
14. Fernandez-Salguero, P. M., Ward, J. M., Sundberg, J. P., and Gonzalez, F. J. (1997). Lesions of aryl-hydrocarbon receptor-deficient mice. *Veterinary Pathology*, 34, 605–614.
15. Lahvis, G. P., Lindell, S. L., Thomas, R. S., McCuskey, R. S., Murphy, C., Glover, E., Bentz, M., Southard, J., and Bradfield, C. A. (2000). Portosystemic shunting and persistent fetal vascular structures in aryl hydrocarbon receptor-deficient mice. *Proceedings of the National Academy of Sciences of the United States of America*, 97, 10442–10447.
16. Lin, T. M., Ko, K., Moore, R. W., Simanainen, U., Oberley, T. D., and Peterson, R. E. (2002). Effects of aryl hydrocarbon receptor null mutation and *in utero* and lactational 2,3,7,8-tetrachlorodibenzo-p-dioxin exposure on prostate and seminal vesicle development in C57BL/6 mice. *Toxicological Sciences*, 68, 479–487.
17. Schmidt, J. V., Su, G. H., Reddy, J. K., Simon, M. C., and Bradfield, C. A. (1996). Characterization of a murine Ahr null allele: involvement of the Ah receptor in hepatic growth and development. *Proceedings of the National Academy of Sciences of the United States of America*, 93, 6731–6736.
18. Thackaberry, E. A., Gabaldon, D. M., Walker, M. K., and Smith, S. M. (2002). Aryl hydrocarbon receptor null mice develop cardiac hypertrophy and increased hypoxia-inducible factor-1alpha in the absence of cardiac hypoxia. *Cardiovascular Toxicology*, 2, 263–274.
19. Hankinson, O. (1995). The aryl hydrocarbon receptor complex. *Annual Review of Pharmacology and Toxicology*, 35, 307–340.
20. Beischlag, T. V. and Perdew, G. H. (2005). ER alpha–AHR–ARNT protein–protein interactions mediate estradiol-dependent transrepression of dioxin-inducible gene transcription. *Journal of Biological Chemistry*, 280, 21607–21611.
21. Hankinson, O. (2005). Role of coactivators in transcriptional activation by the aryl hydrocarbon receptor. *Archives of Biochemistry and Biophysics*, 433, 379–386.
22. Kumar, M. B., Tarpey, R. W., and Perdew, G. H. (1999). Differential recruitment of coactivator RIP140 by Ah and estrogen receptors. Absence of a role for LXXLL motifs. *Journal of Biological Chemistry*, 274, 22155–22164.
23. Wang, W. D., Wang, Y., Wen, H. J., Buhler, D. R., and Hu, C. H. (2004). Phenylthiourea as a weak activator of aryl hydrocarbon receptor inhibiting 2,3,7,8-tetrachlorodibenzo-p-dioxin-induced CYP1A1 transcription in zebrafish embryo. *Biochemical Pharmacology*, 68, 63–71.
24. Hahn, M. (2002). Aryl hydrocarbon receptors: diversity and evolution. *Chemico-Biological Interactions*, 141, 131.
25. Hahn, M. E., Karchner, S. I., Shapiro, M. A., and Perera, S. A. (1997). Molecular evolution of two vertebrate aryl hydrocarbon (dioxin) receptors (AHR1 and AHR2) and the PAS family. *Proceedings of the National Academy of Sciences of the United States of America*, 94, 13743–13748.
26. Hanno, K., Oda, S., and Mitani, H. (2010). Effects of dioxin isomers on induction of AhRs and CYP1A1 in early developmental stage embryos of medaka (*Oryzias latipes*). *Chemosphere*, 78, 830–839.
27. Karchner, S. I., Powell, W. H., and Hahn, M. E. (1999). Identification and functional characterization of two divergent aryl hydrocarbon receptors (AHR1 and AhR2) in teleost *Fundulus heteroclitus*. *Journal of Biological Chemistry*, 274, 33814–33824.
28. Abnet, C. C., Tanguay, R. L., Hahn, M. E., Heideman, W., and Peterson, R. E. (1999). Two forms of aryl hydrocarbon receptor type 2 in rainbow trout (*Oncorhynchus mykiss*). Evidence for differential expression and enhancer specificity. *Journal of Biological Chemistry*, 274, 15159–15166.
29. Roy, N. K. and Wirgin, I. (1997). Characterization of the aromatic hydrocarbon receptor gene and its expression in Atlantic tomcod. *Archives of Biochemistry and Biophysics*, 344, 373–386.

30. Tanguay, R. L., Abnet, C. C., Heideman, W., and Peterson, R. E. (1999). Cloning and characterization of the zebrafish (*Danio rerio*) aryl hydrocarbon receptor. *Biochimica et Biophysica Acta, 1444*, 35–48.

31. Tanguay, R. L., Andreasen, E., Heideman, W., and Peterson, R. E. (2000). Identification and expression of alternatively spliced aryl hydrocarbon nuclear translocator 2 (ARNT2) cDNAs from zebrafish with distinct functions. *Biochimica et Biophysica Acta, 1494*, 117–128.

32. Karchner, S. I., Franks, D. G., Powell, W. H., and Hahn, M. E. (2002). Regulatory interactions among three members of the vertebrate aryl hydrocarbon receptor family: AHR repressor, AHR1, and AHR2. *Journal of Biological Chemistry, 277*, 6949–6959.

33. Evans, B. R., Karchner, S. I., Allan, L. L., Pollenz, R. S., Tanguay, R. L., Jenny, M. J., Sherr, D. H., and Hahn, M. E. (2008). Repression of aryl hydrocarbon receptor (AHR) signaling by AHR repressor: role of DNA binding and competition for AHR nuclear translocator. *Molecular Pharmacology, 73*, 387–398.

34. Jenny, M. J., Karchner, S. I., Franks, D. G., Woodin, B. R., Stegeman, J. J., and Hahn, M. E. (2009). Distinct roles of two zebrafish AHR repressors (AHRRa and AHRRb) in embryonic development and regulating the response to 2,3,7,8-tetrachlorodibenzo-*p*-dioxin. *Toxicological Sciences, 110*, 426–441.

35. Mimura, J., Ema, M., Sogawa, K., and Fujii-Kuriyama, Y. (1999). Identification of a novel mechanism of regulation of Ah (dioxin) receptor function. *Genes & Development, 13*, 20–25.

36. Catron, T., Mendiola, M. A., Smith, S. M., Born, J., and Walker, M. K. (2001). Hypoxia regulates avian cardiac Arnt and HIF-1alpha mRNA expression. *Biochemical and Biophysical Research Communications, 282*, 602–607.

37. Lee, J. S., Kim, E. Y., Iwata, H., and Tanabe, S. (2007). Molecular characterization and tissue distribution of aryl hydrocarbon receptor nuclear translocator isoforms, ARNT1 and ARNT2, and identification of novel splice variants in common cormorant (*Phalacrocorax carbo*). *Comparative Biochemistry and Physiology, Part C, 145*, 379–393.

38. Mahajan, S. S. and Rifkind, A. B. (1999). Transcriptional activation of avian CYP1A4 and CYP1A5 by 2,3,7,8-tetrachlorodibenzo-*p*-dioxin: differences in gene expression and regulation compared to mammalian CYP1A1 and CYP1A2. *Toxicology and Applied Pharmacology, 155*, 96–106.

39. Lee, J. S., Kim, E. Y., and Iwata, H. (2009). Dioxin activation of CYP1A5 promoter/enhancer regions from two avian species, common cormorant (*Phalacrocorax carbo*) and chicken (*Gallus gallus*): association with aryl hydrocarbon receptor 1 and 2 isoforms. *Toxicology and Applied Pharmacology, 234*, 1–13.

40. Andreasen, E. A., Spitsbergen, J. M., Tanguay, R. L., Stegeman, J. J., Heideman, W., and Peterson, R. E. (2002). Tissue-specific expression of AHR2, ARNT2, and CYP1A in zebrafish embryos and larvae: effects of developmental stage and 2,3,7,8-tetrachlorodibenzo-*p*-dioxin exposure. *Toxicological Sciences, 68*, 403–419.

41. Henry, T. R., Spitsbergen, J. M., Hornung, M. W., Abnet, C. C., and Peterson, R. E. (1997). Early life stage toxicity of 2,3,7,8-tetrachlorodibenzo-*p*-dioxin in zebrafish (*Danio rerio*). *Toxicology and Applied Pharmacology, 142*, 56–68.

42. Teraoka, H., Dong, W., Ogawa, S., Tsukiyama, S., Okuhara, Y., Niiyama, M., Ueno, N., Peterson, R. E., and Hiraga, T. (2002). 2,3,7,8-Tetrachlorodibenzo-*p*-dioxin toxicity in the zebrafish embryo: altered regional blood flow and impaired lower jaw development. *Toxicological Sciences, 65*, 192–199.

43. Teraoka, H., Dong, W., Okuhara, Y., Iwasa, H., Shindo, A., Hill, A. J., Kawakami, A., and Hiraga, T. (2006). Impairment of lower jaw growth in developing zebrafish exposed to 2,3,7,8-tetrachlorodibenzo-*p*-dioxin and reduced hedgehog expression. *Aquatic Toxicology, 78*, 103–113.

44. Cantrell, S. M., Lutz, L. H., Tillitt, D. E., and Hannink, M. (1996). Embryotoxicity of 2,3,7,8-tetrachlorodibenzo-*p*-dioxin (TCDD): the embryonic vasculature is a physiological target for TCDD-induced DNA damage and apoptotic cell death in medaka (*Orizias latipes*). *Toxicology and Applied Pharmacology, 141*, 23–34.

45. Elonen, G. E., Sphear, R. L., Holcombe, G. W., and Johnson, R. D. (1998). Comparative toxicity of 2,3,7,8-tetrachlorodibenzo-*p*-dioxin to seven freshwater species during early life-stage development. *Environmental Toxicology and Chemistry, 17*, 472–483.

46. Hornung, M. W., Spitsbergen, J. M., and Peterson, R. E. (1999). 2,3,7,8-Tetrachlorodibenzo-*p*-dioxin alters cardiovascular and craniofacial development and function in sac fry of rainbow trout (*Oncorhynchus mykiss*). *Toxicological Sciences, 47*, 40–51.

47. Kawamura, T. and Yamashita, I. (2002). Aryl hydrocarbon receptor is required for prevention of blood clotting and for the development of vasculature and bone in the embryos of medaka fish, *Oryzias latipes*. *Zoological Science, 19*, 309–319.

48. Toomey, B. H., Bello, S., Hahn, M. E., Cantrell, S., Wright, P., Tillitt, D. E., and Di Giulio, R. T. (2001). 2,3,7,8-Tetrachlorodibenzo-*p*-dioxin induces apoptotic cell death and cytochrome P4501A expression in developing *Fundulus heteroclitus* embryos. *Aquatic Toxicology, 53*, 127–138.

49. Walker, M. K. and Peterson, R. E. (1991). Potencies of polychlorinated dibenzo-*p*-dioxin, dibenzofuran and biphenyl congeners, relative to 2,3,7,8-tetrachlorodibenzo-*p*-dioxin for producing early life stage mortality in rainbow trout (*Oncorhynchus mykiss*). *Aquatic Toxicology, 21*, 219–238.

50. Walker, M. K., Spitsbergen, J. M., Olson, J. R., and Peterson, R. E. (1991). Tetrachlorodibenzo-*p*-dioxin (TCDD) toxicity during early life stage development of lake trout (*Salvelinus namaycush*). *Canadian Journal of Fisheries and Aquatic Sciences, 48*, 875–883.

51. Chen, J., Carney, S. A., Peterson, R. E., and Heideman, W. (2008). Comparative genomics identifies genes mediating cardiotoxicity in the embryonic zebrafish heart. *Physiological Genomics, 33*, 148–158.

52. Mehta, V., Peterson, R. E., and Heideman, W. (2008). 2,3,7,8-Tetrachlorodibenzo-*p*-dioxin exposure prevents cardiac valve formation in developing zebrafish. *Toxicological Sciences, 104*, 303–311.

53. King Heiden, T. C., Spitsbergen, J., Heideman, W., and Peterson, R. E. (2009). Persistent adverse effects on health and reproduction caused by exposure of zebrafish to 2,3,7,8-tetrachlorodibenzo-p-dioxin during early development and gonad differentiation. *Toxicological Sciences*, *109*, 75–87.

54. Anway, M. D., Leathers, C., and Skinner, M. K. (2006). Endocrine disruptor vinclozolin induced epigenetic transgenerational adult-onset disease. *Endocrinology*, *147*, 5515–5523.

55. Anway, M. D., Memon, M. A., Uzumcu, M., and Skinner, M. K. (2006). Transgenerational effect of the endocrine disruptor vinclozolin on male spermatogenesis. *Journal of Andrology*, *27*, 868–879.

56. Anway, M. D. and Skinner, M. K. (2006). Epigenetic transgenerational actions of endocrine disruptors. *Endocrinology*, *147*, S43–S49.

57. Chang, H. S., Anway, M. D., Rekow, S. S., and Skinner, M. K. (2006). Transgenerational epigenetic imprinting of the male germline by endocrine disruptor exposure during gonadal sex determination. *Endocrinology*, *147*, 5524–5541.

58. Heindel, J. J. (2005). The fetal basis of adult disease: role of environmental exposures. Introduction. *Birth Defects Research, Part A*, *73*, 131–132.

59. Heindel, J. J. (2007). Role of exposure to environmental chemicals in the developmental basis of disease and dysfunction. *Reproductive Toxicology*, *23*, 257–259.

60. Jirtle, R. L. and Skinner, M. K. (2007). Environmental epigenomics and disease susceptibility. *Nature Reviews Genetics*, *8*, 253–262.

61. Jefferson, W. N., Padilla-Banks, E., and Newbold, R. R. (2007). Disruption of the developing female reproductive system by phytoestrogens: genistein as an example. *Molecular Nutrition & Food Research*, *51*, 832–844.

62. Jefferson, W. N., Padilla-Banks, E., and Newbold, R. R. (2007). Disruption of the female reproductive system by the phytoestrogen genistein. *Reproductive Toxicology*, *23*, 308–316.

63. Nayyar, T., Bruner-Tran, K. L., Piestrzeniewicz-Ulanska, D., and Osteen, K. G. (2007). Developmental exposure of mice to TCDD elicits a similar uterine phenotype in adult animals as observed in women with endometriosis. *Reproductive Toxicology*, *23*, 326–336.

64. Uzumcu, M. and Zachow, R. (2007). Developmental exposure to environmental endocrine disruptors: consequences within the ovary and on female reproductive function. *Reproductive Toxicology*, *23*, 337–352.

65. Nebert, D. W., Dalton, T. P., Okey, A. B., and Gonzalez, F. J. (2004). Role of aryl hydrocarbon receptor-mediated induction of the CYP1 enzymes in environmental toxicity and cancer. *Journal of Biological Chemistry*, *279*, 23847–23850.

66. Tijet, N., Boutros, P. C., Moffat, I. D., Okey, A. B., Tuomisto, J., and Pohjanvirta, R. (2006). Aryl hydrocarbon receptor regulates distinct dioxin-dependent and dioxin-independent gene batteries. *Molecular Pharmacology*, *69*, 140–153.

67. Jonsson, M. E., Orrego, R., Woodin, B. R., Goldstone, J. V., and Stegeman, J. J. (2007). Basal and 3,3′,4,4′,5-pentachlorobiphenyl-induced expression of cytochrome P450 1A, 1B and 1C genes in zebrafish. *Toxicology and Applied Pharmacology*, *221*, 29–41.

68. Carney, S. A., Peterson, R. E., and Heideman, W. (2004). 2,3,7,8-Tetrachlorodibenzo-p-dioxin activation of the aryl hydrocarbon receptor/aryl hydrocarbon receptor nuclear translocator pathway causes developmental toxicity through a CYP1A-independent mechanism in zebrafish. *Molecular Pharmacology*, *66*, 512–521.

69. Yin, H. C., Tseng, H. P., Chung, H. Y., Ko, C. Y., Tzou, W. S., Buhler, D. R., and Hu, C. H. (2008). Influence of TCDD on zebrafish CYP1B1 transcription during development. *Toxicological Sciences*, *103*, 158–168.

70. Bugiak, B. and Weber, L. P. (2009). Hepatic and vascular mRNA expression in adult zebrafish (*Danio rerio*) following exposure to benzo[*a*]pyrene and 2,3,7,8-tetrachlorodibenzo-p-dioxin. *Aquatic Toxicology*, *95*, 299–306.

71. Teraoka, H., Kubota, A., Dong, W., Kawai, Y., Yamazaki, K., Mori, C., Harada, Y., Peterson, R. E., and Hiraga, T. (2009). Role of the cyclooxygenase 2-thromboxane pathway in 2,3,7,8-tetrachlorodibenzo-p-dioxin-induced decrease in mesencephalic vein blood flow in the zebrafish embryo. *Toxicology and Applied Pharmacology*, *234*, 33–40.

72. Carney, S. A., Chen, J., Burns, C. G., Xiong, K. M., Peterson, R. E., and Heideman, W. (2006). Aryl hydrocarbon receptor activation produces heart-specific transcriptional and toxic responses in developing zebrafish. *Molecular Pharmacology*, *70*, 549–561.

73. Xiong, K. M., Peterson, R. E., and Heideman, W. (2008). Aryl hydrocarbon receptor-mediated down-regulation of sox9b causes jaw malformation in zebrafish embryos. *Molecular Pharmacology*, *74*, 1544–1553.

74. Planchart, A. and Mattingly, C. J. (2010). 2,3,7,8-Tetrachlorodibenzo-p-dioxin upregulates FoxQ1b in zebrafish jaw primordium. *Chemical Research in Toxicology*, *23*, 480–487.

75. Handley-Goldstone, H. M., Grow, M. W., and Stegeman, J. J. (2005). Cardiovascular gene expression profiles of dioxin exposure in zebrafish embryos. *Toxicological Sciences*, *85*, 683–693.

76. Tal, T. L., Franzosa, J. A., and Tanguay, R. L. (2009). Molecular signaling networks that choreograph epimorphic fin regeneration in zebrafish: a mini-review. *Gerontology*, *56*, 231–240.

77. Mathew, L. K., Andreasen, E. A., and Tanguay, R. L. (2006). Aryl hydrocarbon receptor activation inhibits regenerative growth. *Molecular Pharmacology*, *69*, 257–265.

78. Zodrow, J. M. and Tanguay, R. L. (2003). 2,3,7,8-Tetrachlorodibenzo-p-dioxin inhibits zebrafish caudal fin regeneration. *Toxicological Sciences*, *76*, 151–161.

79. Andreasen, E. A., Mathew, L. K., and Tanguay, R. L. (2006). Regenerative growth is impacted by TCDD: gene expression analysis reveals extracellular matrix modulation. *Toxicological Sciences*, *92*, 254–269.

80. Mathew, L. K., Sengupta, S. S., Ladu, J., Andreasen, E. A., and Tanguay, R. L. (2008). Crosstalk between AHR and Wnt signaling through R-Spondin1 impairs tissue regeneration in zebrafish. *FASEB Journal, 22,* 3087–3096.

81. Kim, K. A., Zhao, J., Andarmani, S., Kakitani, M., Oshima, T., Binnerts, M. E., Abo, A., Tomizuka, K., and Funk, W. D. (2006). R-Spondin proteins: a novel link to beta-catenin activation. *Cell Cycle, 5,* 23–26.

82. Akiyama, H., et al. (2004). Interactions between Sox9 and beta-catenin control chondrocyte differentiation. *Genes & Development, 18,* 1072–1087.

83. Kawajiri, K., et al. (2009). Aryl hydrocarbon receptor suppresses intestinal carcinogenesis in Apc$^{Min/+}$ mice with natural ligands. *Proceedings of the National Academy of Sciences of the United States of America, 106,* 13481–13486.

22

ADVERSE HEALTH OUTCOMES CAUSED BY DIOXIN-ACTIVATED AHR IN HUMANS*

SALLY S. WHITE, SUZANNE E. FENTON, AND LINDA S. BIRNBAUM

22.1 INTRODUCTION

Humans, like other vertebrates, possess and express the aryl hydrocarbon receptor (AHR), and signaling through this pathway is believed to be the means by which dioxin exerts human toxicity, as it does in laboratory animal species. For the purposes of streamlining this discussion, the term "dioxin" or "dioxins" will be used as an umbrella term to refer to the 7 polychlorinated dibenzo-p-dioxins (PCDDs; including 2,3,7,8-tetrachlorodibenzo-p-dioxin, TCDD), 10 polychlorinated dibenzofurans (PCDFs), and 12 polychlorinated biphenyls (PCBs) that exhibit dioxin-like activity with respect to AHR activation and toxicity. This definition derives from meetings organized by the World Health Organization, convened in order to reach and reevaluate a consensus concerning the toxic equivalency factors used to evaluate health risks of compounds with the same molecular mechanism of action as TCDD [1, 2]. In addition, as noted at these meetings, some other halogenated dioxins and furans also have the capacity to bind AHR, and thus may be described as dioxin-like. Here we discuss specific epidemiological evidence describing the adverse health effects of dioxin exposure in humans.

In vivo studies, as discussed elsewhere in this book, have examined the AHR-mediated toxicity of dioxin in a variety of laboratory species and diverse end points of interest. AHR knockout mice, for example, have demonstrated clearly the requirement for the AHR signaling pathway in the canonical profile for dioxin toxicity (Chapter 12). Furthermore, functional polymorphisms that result in multiple AHR isoforms appear to contribute substantially to the degree of the toxic outcome (Chapter 12). Studies of mice expressing human AHR also exhibit dioxin-induced toxicity, though with somewhat altered toxic response profiles than their wild-type controls [3], further demonstrating that AHR mediates dioxin toxicity irrespective of the species from which the receptor derives.

It is important to note that, as articulated in the EPA's Reanalysis of Key Issues Related to Dioxin Toxicity and Response to National Academy of Sciences Comments [4], if the potential carcinogenicity of dioxins is mediated through AHR, a key event in this "... mode of action is binding to and activating AHR; however, downstream events leading to tumor formation are uncertain and may be tissue specific."

Humans receive dioxin exposure through multiple routes. The primary route of exposure for nonoccupationally exposed populations is through the ingestion of contaminated food. For example, in areas with high environmental contamination with dioxin, such as the Great Lakes region of the Midwestern United States where dioxins have entered the local food webs, humans may be exposed through consumption of local fish or other wildlife. In general, dietary intake of dioxin is highest among those with diets high in meat and dairy products, because of the tendency of dioxin to concentrate in fat and bioaccumulate up trophic levels of food webs. Dietary exposures largely occur as a result of "recycling" of existing dioxin in the ambient environment. The primary sources of new dioxins to the environment,

---

*This chapter is the work of National Institutes of Health (NIH) employees. However, the statements, opinions, and conclusions contained herein represent those of the authors and not the NIH or the U.S. government. The research described in this chapter has been reviewed by the National Institute of Environmental Health Sciences and approved for publication. Approval does not signify that the contents necessarily reflect the views of the Agency, nor does the mention of trade names or commercial products constitute endorsement or recommendation for use.

The AH Receptor in Biology and Toxicology, First Edition. Edited by Raimo Pohjanvirta.
© 2012 John Wiley & Sons, Inc. Published 2012 by John Wiley & Sons, Inc.

however, in the second decade of the twenty-first century are forest fires and open waste incineration.

Fortunately, unusually high dioxin exposures have been limited to a relatively small number of people, resulting from occupational exposures or accidental environmental releases of dioxin. Much of what is known today about the human health effects of dioxin exposure derives from epidemiological studies of these high-exposure populations. These studies have provided the primary evidence to drive both the EPA's and the International Agency for Research on Cancer (IARC)'s decision to characterize dioxins as either "carcinogenic to humans" or "likely to be carcinogenic to humans," and as a "Group 1 human carcinogen," respectively [4, 5]. In addition, from these and more recent studies, it has become clear that newborn children are exposed to dioxin both transplacentally [6] and through breast milk [7], making it very important to understand the exposure levels of the various age groups involved in these studies.

22.2 EPIDEMIOLOGICAL STUDY COHORTS

While humans globally are exposed to dioxin, a few epidemiological cohorts have been well studied to understand the adverse consequences of exposure to human health. Here, those cohorts are introduced and described.

22.2.1 NIOSH Cohort

The National Institute of Occupational Safety and Health (NIOSH) initiated the study of a cohort of occupationally TCDD-exposed American workers in 1978. This population—at Dow Chemical and at Monsanto, across 12 plants—worked with 2,4,5-trichlorophenol (2,4,5-TCP), which was used to make the herbicide 2,4,5-trichlorophenoxyacetic acid (2,4,5-T) and was frequently contaminated with TCDD [8, 9]. This represented the largest occupational cohort of TCDD-exposed workers, with 5172 participants and spanning over two decades. Studies on this cohort included both serum TCDD measurements and back-extrapolated estimates of serum TCDD concentrations, and have focused primarily upon cancer incidence and mortality [10–12].

22.2.2 BASF Cohort

BASF—originally named for the Baden Aniline and Soda Factory—is a German chemical company in existence since the mid-nineteenth century that experienced two major accidents at its manufacturing plants [13, 14]. In 1953, an autoclaving accident during the production of 2,4,5-TCP occurred at a plant in Ludwigshafen, Germany, causing a release of TCDD that exposed 74 employees. Thirty-five years later, in 1988 another accident in a BASF plant led to the release of TCDD, this time the result of a process involving the extrusion blending of thermoplastic polyesters with brominated flame retardants, the latter being dioxin precursors (particularly, precursors of brominated dioxins or furans, which may exhibit AHR activity). Peak exposures occurred during accidents, as well as during the subsequent cleanup. Unlike the NIOSH cohort, the BASF cohort addressed a population with acute, high-dose exposure to dioxin.

22.2.3 Hamburg Cohort

The Hamburg cohort followed 1600 herbicide workers all exposed at one plant producing phenoxy herbicides and chlorophenols in Hamburg, Germany, between 1950 and 1984 [15]. As with the NIOSH cohort, the Hamburg cohort represents workers with relatively constant, continuous exposures over the duration of their work experiences, in contrast to the BASF cohort and other populations having high exposures following accidental releases. However, although serum TCDD measurements were investigated, it is known that workers at the Hamburg plant also received extensive exposure to mixed dioxins and dioxin-like compounds, as a result of the varied chemical processes being conducted there [16].

22.2.4 Seveso Cohort

An explosion of a 2,4,5-TCP reactor in the ICMESA chemical company in Meda, Italy, in 1976 resulted in the accidental release of 2,4,5-T, TCP, TCDD, and other agents to the local environment [17, 18]. The dioxin release in this event traveled as far as 6 km from the site of emission, covering the densely populated area of Seveso, Italy. Within days, general illness was observed throughout the Seveso community. The deaths of livestock were also observed, and the emergency slaughter of remaining exposed livestock was conducted in order to prevent TCDD from entering the food chain. The cohort of Seveso was managed slightly differently from prior cohorts, with "zones" created based upon proximity to (1) the geographic epicenter of the contamination and (2) concentrations of pollutant in the soil, the latter being a function of wind direction, not only proximity. The region closest to the release was denoted Zone A, and included a population of 736, most of whom left the area following the accident. The next region was denoted Zone B, which exhibited lower soil contamination from Zone A, and included a population of 4737. The region furthest from the release was denoted Zone R, and represented a reference population of 31,800 individuals minimally exposed to dioxin by the accident. While the mean and median blood levels vary between the three zones, there is a wide range of exposure within each and there are some people in Zone R who have higher levels than some people in Zone A. Another unusual hallmark of the Seveso cohort and the studies that derived from it was that unlike the occupational cohorts that included mostly or entirely men, it allowed for risk characterization for health effects in women, children, and a second generation of children born from parents who were potentially exposed during this accidental emission. A final unique characteristic of this cohort was the fact that serum was drawn and stored from nearly all individuals in these zones, regardless of age soon after the accident, and at regular intervals thereafter, for use in future research [19]. Like the BASF cohort, the Seveso cohort addresses acute, high-dose accidental exposures that are followed by low-level exposures from the ambient environment.

22.2.5 Ranch Hand Cohort

Operation Ranch Hand was the name of a U.S. Air Force program designed to apply defoliant during the Vietnam War between 1962 and 1971, in order to reduce enemy ground cover. The Ranch Hand cohort included only U.S. Air Force personnel, and specifically only those who applied herbicides by aerial spray. One of the principal herbicides applied was known as Agent Orange, and contained a mixture of 2,4,5-T and 2,4-D, but was regrettably contaminated with dioxin [20, 21]. The studies of this cohort were designed to follow the more than 2000 participants longitudinally over 20 years, and examine their health status, reproductive outcomes, and mortality.

22.2.6 Times Beach Cohort

The Times Beach cohort followed approximately 300 former community members of Times Beach, Missouri. This now defunct town was evacuated and condemned after, in 1972, its dirt roads were sprayed with dioxin-contaminated waste oil, in order to control dust. The presence of dioxin was discovered only after many horses in the area died, and concerned owners contacted the CDC to investigate. The application of this contaminated oil resulted in the exposure of the entire town's population to dioxin, the subsequent evacuation of all residents, and the ultimate destruction and incineration of the town's structures and topsoil [22, 23]. This population was not longitudinally evaluated for health effects, as the move away from the region resulted in many participants being scattered throughout the United States.

22.3 CANCER OUTCOMES IN EPIDEMIOLOGICAL COHORTS

Assessments of the effect of exposure on the incidence of all cancers combined, as well as overall cancer mortality, are frequently addressed in studies of these and other epidemiological cohorts. It should be noted that not all studies evaluated the same end points, and cohort composition sometimes also precluded evaluation of certain end points.

22.3.1 All Cancers Combined

Within the NIOSH cohort, workers were exposed 15–37 years prior and 1052 deaths occurred among the 5172 participants. Mortality from all cancers combined was elevated in early studies (265 deaths; standardized mortality ratio (SMR): 1.2; 95% confidence interval (CI): 1.0–1.3) [9], and more recent analyses supported this, as well as a dose–response relationship between cancer risk and cumulative serum TCDD [10]. A subcohort within this population included workers with 1 or more years of work exposure to TCDD, in addition to a minimum 20-year latency period (1520 workers). Among this group of workers, mortality from all cancers combined was more notably elevated than among the cohort as a whole (114 events; SMR: 1.5; 95% CI: 1.2–1.8) [9].

Within the relatively small BASF cohort of workers exposed as a result of the 1953 accident, a clear increase in cancer mortality with increasing cumulative TCDD dose was evident (estimated and back-calculated to the time of exposure) [13, 24], where those in the category of >1 μg TCDD/kg body weight after 20 or more years since the first exposure exhibited significantly increased mortality from all cancers combined (13 deaths; SMR: 2.0; 95% CI: 1.1–3.4).

Studies on the Hamburg cohort [16, 25] observed an increase in all cancer mortality in association with TCDD-like exposures. The initial SMRs were calculated, using as a reference the national mortality statistics for West Germany and deaths in a cohort of male gas workers; total cancer SMRs were 1.24 (95% CI: 1.0–1.52) and 1.39 (95% CI: 1.10–1.75), respectively, among men. Within these studies, observations across exposure groups with 20 years or more of occupational exposure supported a dose–response relationship, with respect to total cancer mortality (SMR = 1.87 comparing all of Germany and 1.82 comparing gas workers). Among men who began employment before 1955, the SMRs were 1.61 and 1.87, respectively. The group of men with the highest suspected TCDD exposures had significant SMRs of 1.42 and 1.78, respectively. Only 7% of the women were in the high-exposure cohort, compared with 40% of the men, and no increased cancer mortality risk was found among these women [16].

Ecological studies of the Seveso cohort found no increase in mortality from all cancers combined, but the incidence of all cancers combined 15–20 years after exposure was elevated in men (only). This observation illustrates that substantial latency plays a key role in recognition of cancer as a consequence of dioxin exposure. This cohort is still being followed and it is hypothesized that in time the significance of this overall effect will change. This male and female cohort provides valuable information on the association between exposure to TCDD and cancer since the accident had resulted in exposure to primarily TCDD and not a mixture of dioxins. In addition, the apparent follow-up and documentation of this study exceeds expectations, in that over 99% of the cohort can be traced.

Findings from one study on the Ranch Hand cohort did not observe an increase in cancer mortality, but did identify an increased risk of cancer at any site among those in the high-dioxin category (initial body burden measure of greater than 118.5 ppt), who spent no more than 2 years in Southeast Asia (RR: 2.02) [26]. A later, follow-up study found that in this high-dioxin exposure category, for all-site cancer, the relative risk was 0.9 (95% CI: 0.6–1.4). However, in this same study, when isolating those cohort participants who sprayed for more than 30 days, and during the period when Agent Orange was most heavily used, a clear, significant increase in all-site cancer was identified (RR: 2.2; 95% CI: 1.1–4.4) [21]. No increase in cancer mortality was seen at this later date [27]. None of these studies observed associations between participation in Operation Ranch Hand and incidence of any site-specific cancer.

Because the Times Beach cohort was both small and scattered, in conjunction with the long latency to frank cancer presentation following carcinogen exposure, limited cancer assessments have been made.

22.3.2 Respiratory and Upper Respiratory Cancers

Lung cancer occurrence was not elevated among the full NIOSH cohort of TCDD-exposed workers. However, within the subcohort described above, the SMR for lung cancer mortality was 1.4 (40 deaths; 95% CI: 1.0–1.9) [9]. In the BASF cohort, present at the 1953 accident, respiratory cancer and respiratory cancer mortality increased with increasing exposure [13]. Within the high-dose group of >1 μg TCDD/kg body weight after 20 or more years since the first exposure, there were eight cases of such cancers (standardized incidence ratio (SIR): 2.0; 95% CI: 0.9–3.9) and six deaths from these (SMR: 3.1; 95% CI: 1.1–6.7). In the Hamburg exposure cohort, Flesch-Janys et al. [25] observed a significant increase in respiratory cancers (SMR: 1.51; 95% CI: 1.07–2.08). No increase in mortality from respiratory cancers has been observed within the Seveso cohort, though the relatively young age of cohort participants and latency to cancer-caused mortality should be taken into consideration [17]. No clear trend to increased incidence of respiratory cancer was observed in the Ranch Hand cohort [26], though the latency to frank respiratory cancer has been an interest of the Institute of Medicine (IOM) [20] in understanding potential risk for the cohort.

22.3.3 Lymphoma

No increase in frequency of non-Hodgkin lymphoma was observed in the NIOSH cohort, including within the higher cumulative exposure subcohort described [9]. Within the Hamburg cohort, lymphatic and hematopoietic cancer increases were observed, with a SMR of 2.16 (95% CI: 1.11–3.77) [25], with a significant increase in lymphosarcoma (SMR: 3.73; 95% CI: 1.20–8.71). In Seveso, an increase in lymphatic and hematopoietic cancers was observed within Zones A and B, and described as primarily non-Hodgkin. An increase was also observed in lymphatic and hematopoietic cancers in Seveso children aged 0–19 years (nine cases; RR: 1.6; 95% CI: 0.7–3.4), though Hodgkin's was more prevalent than non-Hodgkin (three cases; RR: 2.0; 95% CI: 0.5–7.6) [28]. No assessment of exposure effect on non-Hodgkin lymphoma occurrence appears to have been made for the BASF cohort, potentially due to the smaller cohort size. The IOM [20], in assessing Veteran's health with respect to Agent Orange, declared that there existed "sufficient evidence of an association" "between exposure to herbicides" and both Hodgkin's and non-Hodgkin lymphoma, as well as chronic lymphocytic leukemia.

22.3.4 Soft Tissue Sarcoma

Soft tissue sarcomas are fairly rare tumors, and thus large cohorts would be needed to assess their incidence in a population. Soft tissue sarcoma mortality was marginally increased within the NIOSH cohort, based upon four deaths (SMR: 3.4; 95% CI: 0.9–8.7) [9]. Furthermore, among the previously described subcohort, soft tissue sarcoma was the only substantially elevated cancer (three events; SMR: 9.2; 95% CI: 1.9–27) [9]. Although soft tissue sarcomas were not increased in either of the highly exposed zones in Seveso, an increase in frequency was observed within the reference population of Zone R (which had less exposure on average than Zones A or B). Studies from neither the BASF cohort nor the Hamburg cohort (or others) appear to have addressed soft tissue sarcoma, again likely due to the small population size available for comparison, and the rarity of the tumor.

22.3.5 Other Cancers

A number of other cancers have at times been reported to be affected within these cohorts, though not all studies have reported on all cancers. Within the BASF cohort, digestive system cancer incidence rates and mortality (SMR: 1.46; 95% CI: 1.13–1.89) from gastrointestinal cancers were both significantly elevated in association with dioxin exposure, without taking into account other significant factors, such as age, cigarette smoking, and BMI, all found to further increase some cancers [13]. The Hamburg cohort exhibited significant excess in rectal cancer compared to controls (SMR: 2.3; 95% CI: 1.05–4.37) and breast cancer mortality was raised among female workers (7% of the highly exposed population; SMR 2.15) [25].

In the Seveso cohort, increases in occurrence of a number of other cancers were found to be associated with dioxin exposure. Warner et al. [18] found that median exposure was slightly higher among breast cancer cases from the Seveso area; however, the small case number and the exclusion of those breast cancer patients who died before 1996 reduce confidence in the meaning of these results. Most importantly, insufficient time may have elapsed for the development of breast cancer assuming a latency of 20–40 years. Most interesting from this study was the observation of the clear association between individual serum TCDD concentration and breast cancer incidence, specifically the significantly increased hazard ratio (2.1; 95% CI: 1.0–4.6) for breast cancer associated with a 10-fold increase in serum TCDD concentration. No increase in mortality from breast cancer has thus far been associated with exposure in that cohort. Within Zones A and B, increases in the occurrence of multiple myeloma, biliary tract cancer, vaginal cancer, and gastrointestinal cancer were all observed. In the analysis of 0–19-year-olds exposed at Seveso, both ovarian (two cases) and thyroid cancer (two cases) incidences were elevated above those predicted [28].

22.4 NONCANCER END POINTS

Other noncancerous adverse health effects are also associated with dioxin exposure. These outcomes were tracked less consistently across the epidemiological cohorts, because most studies were initially aimed only to identify effects on cancer and mortality. Thus, where cohorts are still in existence (or existing data remain uninvestigated), certain of these end points are still being addressed and reported today. Some observations of certain accepted effects, such as chloracne, have derived from very small populations, case studies, or anecdotal reports.

22.4.1 Chloracne

Chloracne is a frequently observed adverse health outcome of high-dose dioxin exposure in humans, particularly in those occupationally exposed. While the presence of chloracne provides definitive proof of exposure to dioxins, the absence of chloracne does not preclude prior exposure.

Within the NIOSH cohort, chloracne was observed in workers with chronic daily exposure, and this chronic exposure appeared to be a primary determinant [29]. In exposure accidents, chloracne was consistently observed in at least some of those exposed. The dermatological effects of chloracne also persisted over many years among those affected individuals in the NIOSH cohort with prior chronic exposure. Following the 1953 BASF accident, 7 of the exposed were hospitalized for chloracne, and 66 of the 74 exposed were affected with chloracne (TCDD was not identified as causative for chloracne until 1957) [30]. Cohort members exposed during the 1988 accident, however, did not exhibit chloracne, whatsoever [31]. Members of the Hamburg cohort were observed to have developed chloracne as early as 1954. However, data on chloracne were inconsistently collected, and thus no conclusions can be reached. Chloracne was also observed in the Seveso cohort, but largely resolved after cessation of exposure. All of the Seveso cases, except one, resolved within 10 years of exposure [32]. Interestingly, chloracne appeared only in those aged 16 years or younger [33]. Chloracne was not observed in either the Ranch Hand cohort or the Times Beach cohort; however, this may be attributable to the prolonged time between the exposures and the assessments, as data from Seveso suggest (over 10 years) [32].

22.4.2 Metabolic Effects

22.4.2.1 Diabetes Several studies of these cohorts exposed to high levels of dioxins have suggested that one of the plausible diseases associated with dioxins is type 2 diabetes. In the BASF cohort, mean fasting glucose levels were slightly elevated and associated with contemporaneous TCDD, but not with respect to back-calculated TCDD levels [14]. In the Ranch Hand cohort [34], glucose abnormalities (RR: 1.4; 95% CI: 1.1–1.8), diabetes prevalence (RR: 1.5; 95% CI: 1.2–2.0), and the use of oral medications to control diabetes (RR: 2.3; 95% CI: 1.3–3.9) were increased, whereas the time until diabetes was diagnosed decreased with dioxin exposure (shorter latency). Serum insulin abnormalities (RR: 3.4; 95% CI: 1.9–6.1) increased with dioxin exposure in nondiabetics. This association is supported in subjects exposed to high levels of dioxins in Seveso [35] and Korean Vietnam veterans exposed to Agent Orange [36]. In Seveso, increased mortality from diabetes was observed in Zones B and R (Zone B > Zone R) after 15 years, but not in the highest exposure zone, Zone A. This may have reflected the much smaller number of persons in Zone A, with only two cases observed in Zone A, given its much smaller population. These results indicate an adverse relation between dioxin exposure and diabetes mellitus, glucose metabolism, and insulin production.

22.4.2.2 Lipids Within studies of plant workers, including the NIOSH, BASF, and Hamburg cohorts, no effects on cholesterol or triglyceride levels were observed [14, 37]. Unlike the BASF cohort, however, the NIOHS cohort exhibited a slight increase in triglyceride levels with increasing serum TCDD, though the influence of factors such as BMI and smoking on triglyceride levels could not be determined [14, 37].

The accidents at Seveso and Times Beach also did not appear to result in altered cholesterol or triglyceride levels among exposed individuals [22, 32, 38]. Within the Ranch Hand cohort, however, high exposure to TCDD was correlated with both increased total cholesterol (with serum TCDD concentrations greater than 33.3 pg/g) and increased serum triglyceride levels (with serum TCDD concentrations greater than 15 pg/g) [39].

22.4.3 Thyroid

Examination of thyroid function within these cohorts has been limited. Calvert et al. [40] found no effect of exposure on thyroid-stimulating hormone (TSH) or thyroxine (T4) within the NIOSH cohort. Within the BASF cohort, a morbidity study spanning 1953 through 1989, of workers employed during the year following the 1953 accident, reported that TSH, T4, and thyroxine binding globulin (TGB) were all in the normal range [14]. However, T4 and TGB levels were positively associated with both current and back-calculated serum TCDD concentrations in workers. In the Ranch Hand cohort, TSH means were elevated for the high-exposure group (serum TCDD above 94 ppt at first examination) at two of the five examinations, and were increasing with increasing exposure at four of the five [41].

In the Seveso cohort, thyroid effects were observed in the children of cohort members, who had been exposed more than 18 years prior to the births of these children. Neonatal basal TSH means increased with maternal dioxin exposure as determined by zone, and were positively correlated with concurrent maternal plasma TCDD concentrations [42].

22.4.4 Cardiovascular Health

There have been myriad efforts to examine the effect of dioxin exposure on the risk of developing ischemic heart disease (IHD). The National Academy of Sciences has concluded that dioxin exposure appears to be associated with IHD mortality. The IOM's Veterans and Agent Orange: Update 2008, however, considered the evidence for this to be limited or suggestive, concluding instead that dioxin may contribute to some upstream risk factors for developing IHD [20]. Furthermore, the association is modest, and most studies informing this conclusion could not be adjusted for confounders, such as smoking or body mass index. In a recent review, Humblet et al. [43] concluded that dioxin exposure is associated with mortality from both ischemic heart disease and all cardiovascular diseases. Two studies of the Ranch Hand cohort found no increase in risk of death from any circulatory disease [27, 44]. Studies on the Seveso cohort, however, suggest that there may be links between dioxin exposure and circulatory disease incidence, though the influence of stress resulting from the accident may have contributed to this increase [17].

22.4.5 Immunological Effects

Higher circulating levels of multiple immunoglobulins and complement proteins were associated with TCDD in the BASF cohort. Increases in circulating IgA may have resulted from liver carcinomas (no statistically significant increase in incidence of this cancer, thus not described earlier).

Circulating IgA was also elevated in association with TCDD in the Ranch Hand cohort. Few immunological measures were made in the Seveso cohort, including circulating immunoglobulin levels, complement protein levels, and lymphocyte counts and activities. Those that were measured appeared only slightly affected in exposed children, compared to those in unexposed.

22.4.6 Reproductive Effects

Males within the NIOSH cohort exhibited potentially non-monotonic effects on FSH, LH, and testosterone. In the Ranch Hand cohort, no effect was observed on sperm count, LH, FSH, testosterone, or testicular end points. Other cohorts addressed few reproductive end points.

Extensive studies of reproductive outcomes within the Seveso cohort have been undertaken, compared to the other cohorts. Females within this cohort are still being evaluated in the Seveso Women's Health Study and exhibited significantly lower hazard ratios for uterine fibroids with increased serum TCDD (inverse correlation), suggesting the possible antiestrogenicity of TCDD in the uterus, unlike in the breast [45]. However, the estrogenicity of TCDD appears to be potentially not only tissue dependent, but also dependent upon developmental stage of the individual at the time of the exposure. A doubling in risk for endometriosis was observed in women exhibiting a serum TCDD concentration of 100 ppt or more, but this was neither significant nor illustrative of a clear dose response [46]. A strong association was also observed between individual serum TCDD concentration and breast cancer (see above). An association of earlier menopause onset with serum TCDD concentration was also observed, although no associations were observed between serum TCDD and ovulation, progesterone or estradiol, or age at menarche [47]. In a study of highly exposed Belgian adolescents, a doubling of their serum dioxin concentration increased the odds of delayed breast development by 2.3 ($p = 0.02$) [48].

In Seveso, time to pregnancy—based upon the duration spent attempting to conceive—was associated with dioxin exposure, where this time was increased by 25% for every 10-fold increase in serum TCDD concentration (odds ratio (OR): 0.75; 95% CI: 0.60–0.95) [49]. Furthermore, infertility—defined here as 12+ months attempting—was also associated with dioxin, and every 10-fold increase in serum TCDD concentration generated a doubling of infertility risk (OR: 1.9; 95% CI: 1.1–3.2).

22.5 DEVELOPMENTAL EFFECTS

Unfortunately, there are substantial gender limitations of occupational studies, due to bias toward males in the workplace. For this reason, many of the affected developmental end points described here have been observed in the Seveso cohort, in which people of all ages (including infants) were highly exposed, compared to a mostly adult population in the occupational exposure cohorts.

In the Seveso cohort, sex ratios of offspring were altered [50], wherein excess females were born between 1977 and 1996, with an increasing probability of female offspring associated with increased paternal serum TCDD concentration. This effect was even more prominent among fathers who were exposed before the age of 19 years, who had significantly more daughters than sons, exhibiting a sex ratio of 0.38 (95% CI: 0.30–0.47).

Males in Seveso who were exposed early in childhood (mean age 6.2 years) exhibited reduced sperm count and total motile sperm, as well as decreased estradiol, at age of 22–31 years [51]. However, in this same study, males from this cohort who were peripubertal (mean age 13.2 years) at the time of exposure exhibited the inverse trend of increased sperm count and total motile sperm, and decreased estradiol at age of 32–39 years. Both exposure windows yielded increases in follicle-stimulating hormone. Males who were exposed as adults (mean age 21.5 years) exhibited no such effects of exposure at age of 40–47 years. A more recent study examined males born in the 6 years immediately following the accidental exposure of Seveso. In this study of young men (mean current age 22.5 years), Mocarelli et al. [52] observed that men who had been breast-fed between 1977 and 1984 exhibited the same trends in adult sperm quality as men exposed in their early childhood at the time of the accident (reported in Ref. 51). This trend was not observed in bottle-fed men, and indicates that lactational exposure to dioxin may permanently alter sperm quality. Furthermore, it reinforces the significance of timing of dioxin exposure, with respect to life stage, and the expected health outcomes resulting from exposure.

As described above, thyroid effects were also observed in offspring [42], and b-TSH in neonates (Zone A > Zone B > Zone R) was positively correlated with maternal serum TCDD in women of reproductive age in 1994–2005 who had been exposed as children. Birth weight and spontaneous abortion rates were also addressed in this cohort [53], but no association with maternal exposure was observed. However, an interesting point of note on this cohort is that the youngest individuals in Seveso had the highest exposures, and thus they may not yet have had sufficient temporal opportunity to bear children. More studies may well illuminate the effects of early-life exposure on a second generation.

As described previously, various cancer incidences among Seveso children (aged 0–19 years) were associated with having lived in the zones of higher exposure [28]. This may be potentially looked upon as a developmental effect of exposure, and differences between the types of cancers observed in these children as compared to adults across the cohorts suggest further the differential response of the developing human to dioxin, even with respect to its carcinogenic profile.

In Chapaevsk—a Russian town highly contaminated with dioxins from historic industrial activity—longitudinal studies of peripubertal boys suggest that dioxin exposures can diminish normal growth patterns occurring at this life stage [54]. Assessing serum dioxin concentrations, the boys in the highest exposure quintile as compared to the lowest quintile had significantly lower body mass index measurements. Serum PCB concentrations in this cohort also appeared to alter normal growth, and were correlated with reduced height.

The IOM [20], in their update on Veteran's Health and Agent Orange, discussed the potential for effects of Veterans' Agent Orange exposure on the health of their children and grandchildren, but found there was insufficient or inadequate evidence suggesting increased risk for either altered reproductive function or childhood cancers. They acknowledged the increasing understanding of epigenetic mechanisms underlying the transgenerational transmission of adverse health effects, and its contribution to the elevated plausibility of transmission of effects to the children of Veterans, but articulated that more research would need to be conducted, both epidemiological and experimental research, into the potential epigenetically mediated transgenerational effects of dioxin exposure.

22.6 CONCLUSIONS

The evaluations completed within the cohort studies have clearly identified some of the predominant cancers associated with chronic, high-dose dioxin exposures, such as soft tissue sarcoma. Observed more consistently across cohorts, however, has been the increased risk of mortality due to all cancers combined. These findings, in conjunction with critical animal studies, have led dioxin to be classified as "carcinogenic to humans" or "likely carcinogenic to humans" by U.S. EPA and as a "Group 1 human carcinogen" by IARC [4, 5]. Both agencies have clearly articulated that all epidemiological evidence concerning carcinogenicity points to an increase in all cancers combined, as opposed to a site- or tissue-specific carcinogenic profile for dioxin. The indiscriminate nature of the carcinogenicity of this prototypic AHR agonist is consistent with what we know of the ubiquity of AHR, with respect to both its temporal and spatial distribution in the individual and its broad roles in endogenous activities.

However, much has been garnered from the more recent Seveso studies, in which young children were exposed to a burst of dioxin *in utero*, during infancy, childhood, and/or puberty. The numerous reports of altered developmental programming, from reproductive to metabolic end points, have also earned dioxin the title of endocrine disruptor [51].

There are clearly critical periods of life that leave individuals more or less susceptible to the effects of dioxin, and the ongoing studies may reveal yet another list of long-term health effects from this pervasive and persistent pollutant, as well as further illuminate the complex and varied endogenous roles for AHR over the course of development.

REFERENCES

1. Van den Berg, M., Birnbaum, L., Bosveld, A. T., Brunström, B., Cook, P., Feeley, M., Giesy, J. P., Hanberg, A., Hasegawa, R., Kennedy, S. W., Kubiak, T., Larsen, J. C., van Leeuwen, F. X., Liem, A. K., Nolt, C., Peterson, R. E., Poellinger, L., Safe, S., Schrenk, D., Tillitt, D., Tysklind, M., Younes, M., Waern, F., and Zacharewski T. (1998). Toxic equivalency factors (TEFs) for PCBs, PCDDs, PCDFs for humans and wildlife. *Environmental Health Perspectives*, *106*, 775–792.

2. Van den Berg, M., Birnbaum, L. S., Denison, M., De Vito, M., Farland, W., Feeley, M., Fiedler, H., Hakansson, H., Hanberg, A., Haws, L., Rose, M., Safe, S., Schrenk, D., Tohyama, C., Tritscher, A., Tuomisto, J., Tysklind, M., Walker, N., and Peterson, R. E. (2006). The 2005 World Health Organization reevaluation of human and mammalian toxic equivalency factors for dioxins and dioxin-like compounds. *Toxicological Sciences*, *93*, 223–241.

3. Moriguchi, T., Motohashi, H., Hosoya, T., Nakajima, O., Takahashi, S., Ohsako, S., Aoki, Y., Nishimura, N., Tohyama, C., Fujii-Kuriyama, Y., and Yamamoto, M. (2003). Distinct response to dioxin in an arylhydrocarbon receptor (AHR)-humanized mouse. *Proceedings of the National Academy of Sciences of the United States of America*, *100*, 5652–5657.

4. U.S., Environmental Protection Agency, (U.S., EPA) (2010). EPA's Reanalysis of Key Issues Related to Dioxin Toxicity and Response to NAS Comments. EPA/600/R-10/038A. http://cfpub.epa.gov/ncea/iris_drafts/recordisplay.cfm?deid=222203.

5. International Agency for Research on Cancer (IARC), (1997). *IARC Monographs on the Evaluation of Carcinogenic Risks to Humans*, Vol. 69. International Agency for Research on Cancer, Lyon, France.

6. Pedersen, M., Halldorsson, T. I., Mathiesen, L., Mose, T., Brouwer, A., Hedegaard, M., Loft, S., Kleinjans, J. C., Besselink, H., and Knudsen, L. E. (2010). Dioxin-like exposures and effects on estrogenic and androgenic exposures and micronuclei frequency in mother–newborn pairs. *Environment International*, *36*, 344–351.

7. LaKind, J. S. (2007). Recent global trends and physiologic origins of dioxins and furans in human milk. *Journal of Exposure Science and Environmental Epidemiology*, *17*, 510–524.

8. Fingerhut, M. A., Halperin, W. E., Marlow, D. A., Piacitelli, L. A., Honchar, P. A., Sweeney, M. H., Greife, A. L., Dill, P. A., Steenland, K., and Suruda, A. J. (1991). Cancer mortality in workers exposed to 2,3,7,8-tetrachlorodibenzo-p-dioxin. *New England Journal of Medicine*, *324*, 212–218.

9. Fingerhut, M. A., Halperin, W. E., Marlow, D. A., Piaciteli, L. A., Honchar, P. A., Sweeney, M. H., Greife, A. L., Dil, P. A., Steenland, K., and Suruda, A. J. (1991). Mortality among U.S. workers employed in the production of chemicals contaminated with 2,3,7,8-tetrachlorodibenzo-p-dioxin (TCDD). NIOSH Final Report PB91125971. National Technical Information Service, Springfield, VA.

10. Steenland, K., Deddens, J., and Piacitelli, L. (2001). Risk assessment for 2,3,7,8-tetrachlorodibenzo-p-dioxin (TCDD) based on an epidemiologic study. *American Journal of Epidemiology*, *154*, 451–458.

11. Cheng, H., Aylward, L., Beall, C., Starr, T. B., Brunet, R. C., Carrier, G., and Delzell, E. (2006). TCDD exposure–response analysis and risk assessment. *Risk Analysis*, *26*, 1059–1071.

12. Collins, J. J., Bodner, K., Aylward, L. L., Wilken, M., and Bodnar, C. M. (2009). Mortality rates among trichlorophenol workers with exposure to 2,3,7,8-tetrachlorodibenzo-p-dioxin. *American Journal of Epidemiology*, *170*, 501–506.

13. Ott, M. G. and Zober, A. (1996). Cause specific mortality and cancer incidence among employees exposed to 2,3,7,8-TCDD after a 1953 reactor accident. *Journal of Occupational and Environmental Medicine*, *53*, 606–612.

14. Ott, M. G., Zober, A., and Germann, C. (1994). Laboratory results for selected target organs in 138 individuals occupationally exposed to TCDD. *Chemosphere*, *29*, 2423–2437.

15. Becher, H., Steindorf, K., and Flesch-Janys, D. (1998). Quantitative cancer risk assessment for dioxins using an occupational cohort. *Environmental Health Perspectives*, *106*, 663–670.

16. Manz, A., Berger, J., Dwyer, J. H., Flesch-Janys, D., Nagel, S., and Waltsgott, H. (1991). Cancer mortality among workers in chemical plant contaminated with dioxin. *Lancet*, *338*, 959–964.

17. Bertazzi, P. A., Consonni, D., Bachetti, S., Rubagotti, M., Baccarelli, A., Zochetti, C., and Pesatori, A. C. (2001). Health effects of dioxin exposure: a 20-year mortality study. *American Journal of Epidemiology*, *153*, 1031–1044.

18. Warner, M., Eskenazi, B., Mocarelli, P., Gerthoux, P. M., Samuels, S., Needham, L., Patterson, D., and Brambilla, P. (2002). Serum dioxin concentrations and breast cancer risk in the Seveso women's health study. *Environmental Health Perspectives*, *110*, 625–628.

19. Mocarelli, P., Patterson, D. J., Marocchi, A., and Needham, L. (1990). Pilot study (phase II) for determining polychlorinated dibenzo-p-dioxin (PCDD) and polychlorinated dibenzofuran (PCDF) levels in serum of Seveso, Italy, residents collected at the time of exposure: future plans. *Chemosphere*, *20*, 967–974.

20. Institute of Medicine (IOM) Committee to Review the Health Effects in Vietnam Veterans of Exposure to Herbicides (Seventh Biennial Update), (2009). Veterans and Agent Orange: Update 2008,pp. 708. http://www.nap.edu/catalog/12662.html.

21. Michalek, J. E. and Pavuk, M. (2008). Diabetes and cancer in veterans of Operation Ranch Hand after adjustment for calendar period, days of spraying, and time spent in Southeast Asia. *Journal of Occupational and Environmental Medicine*, *50*, 330–340.

22. Hoffman, R. E., Stehr-Green, P. A., Webb, K. B., Evans, R. G., Knutsen, A. P., Schramm, W. F., Staake, J. L., Gibson, B. B., and Steinberg, K. K. (1986). Health effects of long-term exposure to 2,3,7,8-tetrachlorodibenzo-p-dioxin. *Journal of the American Medical Association, 255,* 2031–2038.

23. Evans, R. G., Webb, K. B., Knutsen, A. P., Roodman, S. T., Roberts, D. W., Bagby, J. R., Garrett, W. A., Jr., and Andrews, J. S., Jr. (1988). A medical follow-up of the health effects of long-term exposure to 2,3,7,8-tetrachlorodibenzo-p-dioxin. *Archives of Environmental Health, 43,* 273–278.

24. Zober, A., Messerer, P., and Huber, P. (1990). Thirty-four year mortality follow-up of BASF employees exposed to 2,3,7,8-TCDD after the 1953 accident. *International Archives of Occupational and Environmental Health, 62,* 139–157.

25. Flesch-Janys, D., Steindorf, K., Gurn, P., and Becher H. (1998). Estimation of the cumulated exposure to polychlorinated dibenzo-p-dioxins/furans and standardized mortality ratio analysis of cancer mortality by dose in an occupationally exposed cohort. *Environmental Health Perspectives, 106* (Suppl. 2), 655–62.

26. Akhtar, F. Z., Garabrant, D. H., Ketchum, N. S., and Michalek, J. E. (2004). Cancer in US Air Force veterans of the Vietnam War. *Journal of Occupational and Environmental Medicine, 46,* 123–136.

27. Ketchum, N. S., Michalek, J. E., and Pavuk, M. (2007). Mortality, length of life, and physical examination attendance in participants of the Air Force Health Study. *Military Medicine, 172,* 53–57.

28. Pesatori, A. C., Consonni, D., Tironi, A., Zocchetti, C., Fini, A., and Bertazzi, P. A. (1993). Cancer in a young population in a dioxin-contaminated area. *International Journal of Epidemiology, 22,* 1010–1013.

29. Steenland, K. and Deddens, J. (2003). Dioxin: exposure–response analyses and risk assessment. *Industrial Health, 41,* 175–180.

30. Zober, A., Ott, M. G., and Messerer, P. (1994). Morbidity follow up study of BASF employees exposed to 2,3,7,8-tetrachlorodibenzo-p-dioxin (TCDD) after a 1953 chemical reactor incident. *Journal of Occupational and Environmental Medicine, 51,* 479–486.

31. Zober, A., Schilling, D., Ott, M. G., Schauwecker, P., Riemann, J. F., and Messerer, P. (1998). Helicobacter pylori infection: prevalence and clinical relevance in a large company. *Journal of Occupational and Environmental Medicine, 40,* 586–594.

32. Assennato, G., Cervino, D., Emmett, E. A., Longo, G., and Merlo, F. (1989). Follow-up of subjects who developed chloracne following TCDD exposure at Seveso. *American Journal of Industrial Medicine, 16,* 119–225.

33. Mocarelli, P., Needham, L. L., Marocchi, A., Patterson, D. G., Jr., Brambilla, P., Gerthoux, P. M., Meazza, L., and Carreri, V. (1991). Serum concentrations of 2,3,7,8-tetrachlorodibenzo-p-dioxin and test results from selected residents of Seveso, Italy. *Journal of Toxicology and Environmental Health, 32,* 357–366.

34. Henriksen, G. L., Ketchum, N. S., Michalek, J. E., and Swaby, J. A. (1997). Serum dioxin and diabetes mellitus in veterans of Operation Ranch Hand. *Epidemiology, 8,* 252–258.

35. Pesatori, A. C., Consonni, D., Bachetti, S., Zocchetti, C., Bonzini, M., Baccarelli, A., and Bertazzi, P. A. (2003). Short- and long-term morbidity and mortality in the population exposed to dioxin after the "Seveso accident". *Industrial Health, 41,* 127–138.

36. Kim, J. S., Lim, H. S., Cho, S. I., Cheong, H. K., and Lim, M. K. (2003). Impact of Agent Orange exposure among Korean Vietnam veterans. *Industrial Health, 41,* 149–157.

37. Calvert, G. M., Willie, K. K., Sweeney, M. H., Fingerhut, M. A., and Halperin, W. E. (1996). Evaluation of serum lipid concentrations among U.S. workers exposed to 2,3,7,8-tetrachlorodibenzo-p-dioxin. *Archives of Environmental Health, 51,* 100–107.

38. Mocarelli, P., Marocchi, A., Brambilla, P., Gerthoux, P., Young, D. S., and Mantel, N. (1986). Clinical laboratory manifestations of exposure to dioxin in children. A six-year study of the effects of an environmental disaster near Seveso, Italy. *Journal of the American Medical Association, 256,* 2687–2695.

39. Wolfe, W. H., Michalek, J. E., Miner, J. C., Rahe, A., Silva, J., Thomas, W. F., Grubbs, W. D., Lustik, M. B., Karrison, T. G., Roegner, R. H., et al. (1990). Health status of Air Force veterans occupationally exposed to herbicides in Vietnam. I. Physical health. *Journal of the American Medical Association, 264,* 1824–1831.

40. Calvert, G. M., Sweeney, M. H., Deddens, J., and Wall, D. K. (1999). An evaluation of diabetes mellitus, serum glucose, and thyroid function among U.S. workers exposed to 2,3,7,8-tetrachlorodibenzo-p-dioxon. *Occupational and Environmental Medicine, 56,* 270–276.

41. Pavuk, M., Schecter, A. J., Akhtar, F. Z., and Michalek, J. E. (2003). Serum 2,3,7,8-tetrachlorodibenzo-p-dioxin (TCDD) levels and thyroid function in Air Force veterans of the Vietnam War. *Annals of Epidemiology, 13,* 335–343.

42. Baccarelli, A., Giacomini, S. M., Corbetta, C., Landi, M. T., Bonzini, M., Consonni, D., Grillo, P., Patterson, D. G., Pesatori, A. C., and Bertazzi, P. A. (2008). Neonatal thyroid function in Seveso 25 years after maternal exposure to dioxin. *PLoS Medicine, 5,* e161.

43. Humblet, O., Birnbaum, L., Rimm, E., Mittleman, M. A., and Hauser, R. (2008). Dioxins and cardiovascular disease mortality. *Environmental Health Perspectives, 116,* 1443–1448.

44. Michalek, J. E., Wolfe, W. H., and Miner, J. C. (1990). Health status of Air Force veterans occupationally exposed to herbicides in Vietnam. II. Mortality. *Journal of the American Medical Association, 264,* 1832–1836.

45. Eskenazi, B., Warner, M., Samuels, S., Young, J., Gerthoux, P. M., Needham, L., Patterson, D., Olive, D., Gavoni, N., Vercellini, P., and Mocarelli, P. (2007). Serum dioxin concentrations and risk of uterine leiomyoma in the Seveso Women's Health Study. *American Journal of Epidemiology, 166,* 79–87.

46. Eskenazi, B., Mocarelli, P., Warner, M., Samuels, S., Vercellini, P., Olive, D., Needham, L. L., Patterson, D. G., Jr., Brambilla, P., Gavoni, N., Casalini, S., Panazza, S., Turner, W., and Gerthoux, P. M. (2002). Serum dioxin concentrations and endometriosis: a

cohort study in Seveso, Italy. *Environmental Health Perspectives*, *110*, 629–634.

47. Warner, M., Samuels, S., Mocarelli, P., Gerthoux, P. M., Needham, L., Patterson, D. G., Jr., and Eskenazi, B. (2004). Serum dioxin concentrations and age at menarche. *Environmental Health Perspectives*, *112*, 1289–1292.
48. Den Hond, E., Roels, H. A., Hoppenbrouwers, K., Nawrot, T., Thijs, L., Vandermeulen, C., Winneke, G., Vanderschueren, D., and Staessen, J. A. (2002). Sexual maturation in relation to polychlorinated aromatic hydrocarbons: Sharpe and Skakkebaek's hypothesis revisited. *Environmental Health Perspectives*, *110*, 771–776.
49. Eskenazi, B., Warner, M., Marks, A. R., Samuels, S., Needham, L., Brambilla, P., and Mocarelli, P. (2010). Serum dioxin concentrations and time to pregnancy. *Epidemiology*, *21*, 224–231.
50. Mocarelli, P., Gerthoux, P. M., Ferrari, E., Patterson, D. G., Jr., Kieszak, S. M., Brambilla, P., Vincoli, N., Signorini, S., Tramacere, P., Carreri, V., Sampson, E. J., Turner, W. E., and Needham, L. L. (2000). Paternal concentrations of dioxin and sex ratio of offspring. *Lancet*, *355*, 1858–1863.
51. Mocarelli, P., Gerthoux, P. M., Patterson, D. G., Jr., Milani, S., Limonta, G., Bertona, M., Signorini, S., Tramacere, P., Colombo, L., Crespi, C., Brambilla, P., Sarto, C., Carreri, V., Sampson, E. J., Turner, W. E., and Needham, L. L. (2008). Dioxin exposure, from infancy through puberty, produces endocrine disruption and affects human semen quality. *Environmental Health Perspectives*, *116*, 70–77.
52. Mocarelli, P., Gerthoux, P. M., Needham, L. L., Patterson, D. G., Jr., Limonta, G., Falbo, R., Signorini, S., Bertona, M., Crespi, C., Sarto, C., Scott, P. K., Turner, W. E., and Brambilla, P. (2011). Perinatal exposure to low doses of dioxin can permanently impair human semen quality. *Environmental Health Perspectives*, *119*, 713–718.
53. Eskenazi, B., Mocarelli, P., Warner, M., Chee, W. Y., Gerthoux, P. M., Samuels, S., Needham, L. L., and Patterson, D. G., Jr. (2003). Maternal serum dioxin levels and birth outcomes in women of Seveso, Italy. *Environmental Health Perspectives*, *111*, 947–953.
54. Burns, J. S., Williams, P. L., Sergeyev, O., Korrick, S., Lee, M. M., Revich, B., Altshul, L., Del Prato, J. T., Humblet, O., Patterson, D. G., Jr., Turner, W. E., Needham, L. L., Starovoytov, M., and Hauser, R. (2011). Serum dioxins and polychlorinated biphenyls are associated with growth among Russian boys. *Pediatrics*, *127*, e59–e68.

23

THE TOXIC EQUIVALENCY PRINCIPLE AND ITS APPLICATION IN DIOXIN RISK ASSESSMENT

Jouko Tuomisto

23.1 INTRODUCTION

Dioxin-like compounds are typically present as complex mixtures both at source level and in human and animal organisms. Because the potencies vary by over several orders of magnitude, simply adding up the amounts or concentrations does not give useful information on total toxicity, even if the mechanism of action of different compounds were the same. Therefore, relative toxicity indexes were created originally for risk management purposes, but they have subsequently been used also for scientific purposes.

In short, dioxin-like compounds demonstrating essentially AHR-based mechanism of action have each been given a congener-specific toxicity equivalence factor (TEF) to compare their toxicities with the toxicity of 2,3,7,8-tetrachlorodibenzo-p-dioxin (TCDD), the most toxic congener. The latest TEF assessment uses TEF values from 0.00003 to 1.0 [1]. These TEF values are used simply by multiplying the amount or concentration of a certain congener by its TEF. This gives the amount or concentration toxicologically equivalent to TCDD. Such an "impact-corrected" value can be used as if the congener were TCDD, and all congeners in a mixture can now be summed up to give a total toxicity equivalence or TCDD equivalence (TEQ) of the mixture (TEQ $= \sum_n [\text{TEF}_i \cdot A_i]$, where A_i is the quantity of each individual congener). This ideally should provide the toxic amount expressed as TCDD, and any data available on the most thoroughly studied TCDD may be assumed to apply for a mixture expressed as TEQ and vice versa. It is important to note, however, that the TEFs are not scientifically calculated numbers, they are consensus values based on relative potencies (REPs) of the compounds but also involving a fair amount of scientific judgment by experts.

23.2 HISTORY OF TEF

Toxicity equivalence concept was developed during the mid-1980s to meet the requirements of managing toxicity and risks of mixtures of dioxin-like compounds [2]. The TEF methodology was first proposed by the Ontario Ministry of the Environment in 1984. First, only polychlorinated dibenzo-p-dioxins and dibenzofurans were included [3]. In 1987, the U.S. Environmental Protection Agency adopted an interim TEF approach [4]. Other countries, among others the Nordic Countries, adopted slightly different TEF concepts [2, 5]. An international effort was made to harmonize the different TEF schemes proposed in different countries [6]. This was denoted as International TEF or I-TEF and it was adopted also by U.S. EPA [7, 8]. Soon TEFs were proposed also for dioxin-like PCBs [9]. Preparations of the first WHO-ECEH/IPCS consultation were then started [10], and all relevant mammalian experimental data including PCBs were compiled as a database. The criteria for including a PCB compound were (1) the congener is structurally similar to the PCDDs and PCDFs, (2) the congener binds to the AHR, (3) the congener elicits dioxin-like biochemical and toxic responses, and (4) the congener is persistent and accumulates in the food chain [10].

After compiling a PCB database and including PCBs, the database was recommended to be expanded to include information on PCDDs and PCDFs. This formed the basis for the next reassessment in 1997. This assessment included sepa-

The AH Receptor in Biology and Toxicology, First Edition. Edited by Raimo Pohjanvirta.
© 2012 John Wiley & Sons, Inc. Published 2012 by John Wiley & Sons, Inc.

TABLE 23.1 WHO 1998 TEFs for Humans, Mammals, Fish, and Birds

	TEF		
Congener	Humans/Mammals	Fish	Birds
TCDD	1	1	1
1,2,3,7,8-PentaCDD	1	1	1
1,2,3,4,7,8-HexaCDD	0.1	0.5	0.05
1,2,3,4,7,8-HexaCDD	0.1	0.01	0.01
1,2,3,6,7,8-HexaCDD	0.1	0.01	0.1
1,2,3,7,8,9-HexaCDD	0.1	0.01	0.1
1,2,3,4,6,7,8-HeptaCDD	0.01	0.001	<0.001
OctaCDD	0.0001	<0.0001	0.0001
2,3,7,8-TetraCDF	0.1	0.05	1
1,2,3,7,8-PentaCDF	0.05	0.05	0.1
2,3,4,7,8-PentaCDF	0.5	0.5	1
1,2,3,4,7,8-HexaCDF	0.1	0.1	0.1
1,2,3,6,7,8-HexaCDF	0.1	0.1	0.1
1,2,3,7,8,9-HexaCDF	0.1	0.1	0.1
2,3,4,6,7,8-HexaCDF	0.1	0.1	0.1
1,2,3,4,7,8,9-HeptaCDF	0.01	0.01	0.01
1,2,3,4,7,8,9-HeptaCDF	0.01	0.01	0.01
OctaCDF	0.0001	<0.0001	0.0001
3,4,4',5-TetraCB (81)	0.0001	0.0005	0.1
3,3',4,4'-TetraCB (77)	0.0001	0.0001	0.05
3,3',4,4',5-PentaCB (126)	0.1	0.005	0.1
3,3',4,4',5,5'-HexaCB (169)	0.01	0.00005	0.001
2,3,3',4,4'-PentaCB (105)	0.0001	<0.000005	0.0001
2,3,4,4',5-PentaCB (114)	0.0005	<0.000005	0.0001
2,3',4,4',5-PentaCB (118)	0.0001	<0.000005	0.00001
2',3,4,4',5-PentaCB (123)	0.0001	<0.000005	0.00001
2,3,3',4,4',5-HexaCB (156)	0.0005	<0.000005	0.0001
2,3,3',4,4',5'-HexaCB (157)	0.0005	<0.000005	0.0001
2,3',4,4',5,5'-HexaCB (167)	0.00001	<0.000005	0.00001
2,3,3',4,4',5,5'-HeptaCB (189)	0.0001	<0.000005	0.00001

rate TEFs for fish and birds (Table 23.1) [11]. WHO 1998 TEFs are assigned values, not calculated values, and the expert panel recommended TEFs for 29 compounds including the 17 laterally substituted PCDDs/PCDFs, 4 non-*ortho* PCBs, and 8 mono-*ortho* PCBs [11].

23.3 PREREQUISITES OF THE TEF CONCEPT

The validity criteria used in several assessments [1, 10, 11] for inclusion in the TEF concept include that the compound must

- show a structural relationship to the PCDDs and PCDFs;
- bind to the AHR;
- elicit AHR-mediated biochemical and toxic responses;
- be persistent and accumulate in the food chain.

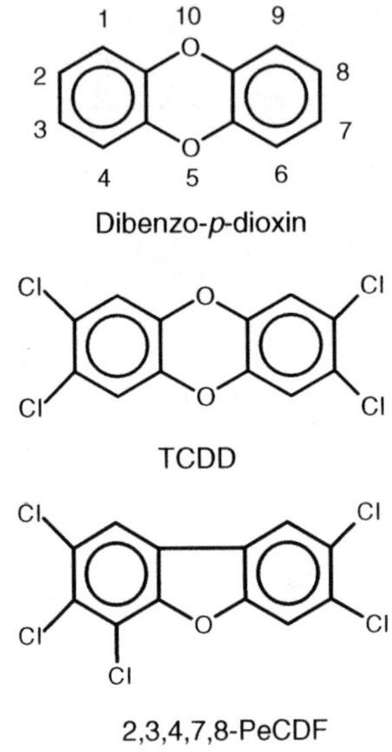

FIGURE 23.1 Structures of dibenzo-*p*-dioxin, TCDD, and 2,3,4,7,8-PeCDF.

These requirements mean that there must be a chlorine atom present in PCDDs and PCDFs in positions 2, 3, 7, and 8, the so-called "lateral chlorines" (Fig. 23.1). Other compounds show no substantial binding to the AHR, and they are more easily metabolized. Thus, 7 dibenzo-*p*-dioxins and 10 dibenzofurans have been given a TEF value. Each chlorine additional to 2, 3, 7, and 8 usually decreases the potency by several fold.

For PCBs, a basic requirement in addition to the lateral chlorines is that the molecule is able to assume a planar (flat) conformation. The biphenyl molecule is able to rotate along its interring C–C axis, but the flat conformation is not energetically favored, and each substitute at *ortho* (2 or 6) positions will further hinder the molecule from assuming a planar conformation (Fig. 23.2). Therefore, only non-*ortho* PCBs show noteworthy dioxin-like activity, and mono-*ortho* PCBs are borderline or very weakly active.

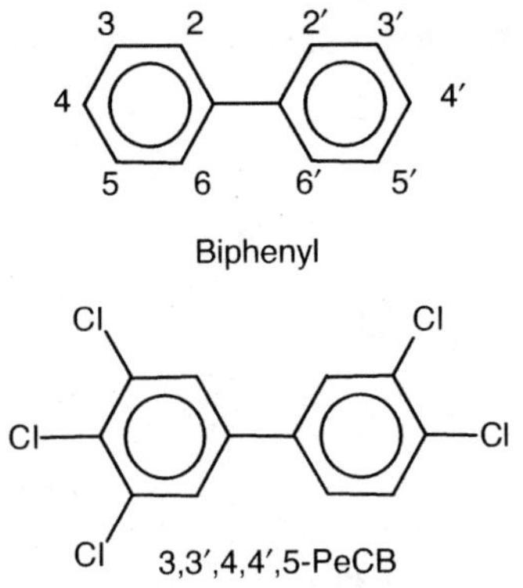

FIGURE 23.2 Structures of biphenyl and 3,3',4,4',5-PeCB (PCB 126).

Because TEF values are for the most part derived from toxicity studies in animals, they refer by and large to oral intake. If *in vivo* studies are missing, *in vitro* studies have been used for some congeners.

The two most important prerequisites of being able to use a common metric for a group of compounds are a common mechanism of action and an additive effect. All compounds given a TEF value act by activating the aryl hydrocarbon receptor. The receptor theory presumes that a compound binding to a receptor may activate the receptor and be an agonist, or block the receptor from active compounds and be an antagonist, or be a partial agonist or agonist–antagonist with some activity but at the same time blocking the receptor from more active compounds.

Hence, there are two intrinsic properties of each compound binding to a receptor, efficacy and potency. Efficacy or intrinsic activity means the ability to activate the receptor, and it often shows as the maximal response the compound can elicit. Potency usually associates with affinity to the receptor, and the higher the potency the lower concentrations suffice to occupy the receptor. Therefore, potency, in essence, defines the doses causing the toxic effects among similarly intrinsically active compounds. Different elimination may naturally modify this correlation.

There are numerous examples of groups of compounds binding to the same receptors but eliciting different responses; for example, among opioids morphine is an agonist, fentanyl is a similar agonist with higher potency, naloxone is an antagonist, and pentazocine is a partial agonist relieving pain to some extent like morphine but at the same time inhibiting the effects of more effective morphine. Therefore, a possibility of (partial) antagonism has to be considered when dealing with mixtures of similarly acting compounds. Potentiating synergism (supra-additive effect) is not very likely among similarly acting compounds, because it typically requires sites of action at two different points but regulating the same end point (e.g., inhibition of two different enzymes) [12]. To some extent, the validity of dioxin TEFs has been analyzed in studies comparing the effects of TCDD to those of supposedly equivalent mixtures (see Section 23.6).

Setting TEF values on the basis of animal experiments assumes that potency differences compared with TCDD are the same in the studied species and in humans. This may or may not be the case. In general, the extrapolation is probably reasonable, but a few exceptions have been found. Han/Wistar Kuopio rats are more sensitive to higher chlorinated dioxins than to TCDD [13]. PCDDs, especially higher chlorinated, have been suggested to be less potent in human cell models suggesting needs to improve the predictive value of TEF [14].

Comparisons of ligand binding and gene expression of various ligands have revealed that human AHR has a higher relative affinity for indirubin than mouse AHR while the latter has a higher affinity for TCDD [15]. Moreover, among sets of genes induced by mouse or human AHRs, only a minority is common to both [16]; the same applies between mice and rats [17]. This clearly gives possibilities for different responses in different species.

Another kind of difficulty arises from the fact that we are not dealing with acute single doses when assessing the risks of dioxin-like compounds. In fact, persistence has been used as a validity criterion for inclusion in the TEF concept. Many dioxins have half-lives of years in humans, but they may vary from about a year to about 20 years [18]. This means very different tendencies for different congeners to cumulate over the lifetime, and hence to achieve different levels in the body after a similar daily intake. Since TEF is a single number, one attempts to take into consideration also different elimination rates of congeners, for example, by giving higher priority to long-term studies over short-term studies. This is a very rough and arbitrary way of dealing with this problem, because usually only animal studies are available as a basis, and there is no guarantee that differences in half-lives of some weeks in rats would compare in a reasonable manner to the differences in half-lives of many years in humans of the same congeners. It also treats different congeners differently, because there are long-term studies on some but not all of them. Thus, some TEFs are valid for acute intake, and some for long-term intake, typically in rats.

A related problem with persistence is the fact that the spectrum of congeners at the source, for example, the stack of a factory, is usually quite different from the spectrum of congeners in humans or animals. This means that the efficiency of different congeners to be transported from their site of origin to their targets is different. Therefore, the TEF assessed at the target end is not very useful when regulating the emissions or concentrations in sediment or soil. Usually, the higher chlorinated congeners predominate at the source, but typically lower chlorinated species dominate TEQs at the level of target organism. Some possible ways to deal with these problems will be suggested later in this chapter.

All these problems exemplify the difficulty of expressing with a single number several dimensions that are not necessarily parallel. However, the TEF concept has been the best we have, and it has served reasonably well in the risk management of dioxin-like compounds.

23.4 WHO 2006 TEF ASSESSMENT AND MAJOR CHANGES TO THE PREVIOUS ASSESSMENTS

The latest assessment of TEF values was performed by World Health Organization International Programme on Chemical Safety expert meeting in Geneva in 2005 [1]. The evaluation was performed according to the same principles as the previous WHO assessments (see above), but a few systematic changes were made (Table 23.2). Previously, increments of 0.01, 0.05, 0.1, and so on were used, but in the new assessment

TABLE 23.2 WHO 2006 TEF Values

Congener	TEF
Chlorinated dibenzo-*p*-dioxins	
TCDD	1
1,2,3,7,8-PentaCDD	1
1,2,3,4,7,8-HexaCDD	0.1
1,2,3,6,7,8-HexaCDD	0.1
1,2,3,7,8,9-HexaCDD	0.1
1,2,3,4,6,7,8-HeptaCDD	0.01
OctaCDD	0.0003
Chlorinated dibenzofurans	
2,3,7,8-TetraCDF	0.1
1,2,3,7,8-PentaCDF	0.03
2,3,4,7,8-PentaCDF	0.3
1,2,3,4,7,8-HexaCDF	0.1
1,2,3,6,7,8-HexaCDF	0.1
1,2,3,7,8,9-HexaCDF	0.1
2,3,4,6,7,8-HexaCDF	0.1
1,2,3,4,6,7,8-HeptaCDF	0.01
1,2,3,4,7,8,9-HeptaCDF	0.01
OctaCDF	0.0003
Non-*ortho*-substituted PCBs	
3,3′,4,4′-TetraCB (77)	0.0001
3,4,4′,5-TetraCB (81)	0.0003
3,3′,4,4′,5-PentaCB (126)	0.1
3,3′,4,4′,5,5′-HexaCB (169)	0.03
Mono-*ortho*-substituted PCBs	
2,3,3′,4,4′-PentaCB (105)	0.00003
2,3,4,4′,5-PentaCB (114)	0.00003
2,3′,4,4′,5-PentaCB (118)	0.00003
2′,3,4,4′,5-PentaCB (123)	0.00003
2,3,3′,4,4′,5-HexaCB (156)	0.00003
2,3,3′,4,4′,5′-HexaCB (157)	0.00003
2,3′,4,4′,5,5′-HexaCB (167)	0.00003
2,3,3′,4,4′,5,5′-HeptaCB (189)	0.00003

half order of magnitude increments on a logarithmic scale were used, that is, 0.01, 0.03, 0.1, 0.3, and so on. This is in line with the receptor concept predicting responses as a function of log dose, and it also makes it easier to describe with statistical methods the uncertainty of TEFs. As a default, all TEF values were assumed to vary in uncertainty by at least one order of magnitude. Consequently, a TEF of 0.1 infers a degree of uncertainty bounded by 0.03 and 0.3. Thus, the TEF is a central value with a degree of uncertainty assumed to be at least ±half a log, which is one order of magnitude [1]. It is very important to emphasize this uncertainty to the users of TEFs. In a complex mixture, the errors due to uncertainties of different congeners may to some extent cancel out, so the uncertainty of the total TEQ may be expected to be somewhat less.

Another systematic change was the use of a single value of 0.00003 for all mono-*ortho* PCBs. The data on mono-*ortho* PCBs is variable for most congeners, and many studies are plagued by the presence of more active impurities in the studied preparations [19]. It was considered not feasible to give different TEF values for all different mono-*ortho* PCBs on the basis of the present data, and they were given a single value 0.00003 as a group.

23.5 HOW THE ASSESSMENT WAS DONE

The first step of the TEF assessment in 2005 was collection of all relevant studies and assessing first the relative potencies of various dioxin-like compounds [20]. This resulted in a database built on the previous database collected in 1993 for the first WHO TEF assessment [10]. A list for the requirements of "ideal" REP studies was determined, and it included the requirement of a full dose–response curve both for the congener to be studied and for the reference TCDD, the same route of administration to animals of the same species, strain, sex, and age under the same conditions, and that the maximal response should be the same for both the congener and TCDD, and the dose–response curves parallel. Then the REP should be calculated at the level of ED_{50}. If the full dose–response curve is not available, lowest observed effect doses or benchmark doses could be used, but at the cost of more uncertainty in the REP [1].

The database [20] revealed clearly that there are significant gaps of knowledge even among the most studied congeners such as 2,3,4,7,8-pentachlorodibenzofuran (2,3,4,7,8-PeCDF) and PCB 126. *In vivo* studies were considered to have the highest priority, because they combine both toxicokinetic and toxicodynamic aspects. *In vitro* studies were considered to contribute to establishing the AHR-mediated mechanism of action and explain possible differences in species sensitivity. In most studies, TCDD had been used as a reference compound, but some PCBs had been compared with PCB 126 (3,3′,4,4′,5-PeCB). This was found acceptable for rat but not necessarily mouse studies [1].

Another typical finding was that the REP values derived from different studies varied sometimes by several orders of magnitude [20]. This makes the conclusions somewhat challenging, and the user of TEF/TEQ must understand the uncertainties especially in the individual TEF values, but to some extent also in the total TEQ of a mixture. However, usually in the case of those congeners contributing most to a TEQ in biological samples (e.g., 2,3,4,7,8-PeCDF, PCB 126), the variation in REP values based on *in vivo* studies was less than or around one order of magnitude.

The starting point to TEF assessment was to compare the 1998 TEF with the distribution of unweighted REP values, preferring *in vivo* data. The 75th percentile of the *in vivo* REP distribution in the available peer-reviewed studies was used as an initial decision point to review the 1998 TEF value for that congener. This leads to a deliberate exaggeration of toxicity that may be criticized, since we are now dealing with relative potencies, not absolute potencies, and it could be argued that the 50th percentile should be the starting point. In other words, increasing the estimate of toxicity of the compound to be assessed will automatically decrease the estimate of toxicity of TCDD as compared with the said compound. If this is done systematically for all compounds in a mixture, the relative impact of TCDD versus other congeners to total toxicity of the mixture will be perplexed. However, this was only the starting point, and at the end of the day most TEFs of the important congeners, perhaps with the exception of 1,2,3,7,8-PeCDD and 1,2,3,6,7,8-HxCDD, were between 50th and 75th percentiles.

Still there is more likelihood of systematic error toward too high than too low TEF [1], and consequently also the TEQ of a mixture is more likely assessed too high than too low. This may not be so much of a problem in risk management, but if TEQs are used in research to investigate the effects of a mixture, the investigators should be aware of this. For instance, in epidemiological studies using TEQ, any comparison with TCDD studies such as studies on Seveso or Vienna TCDD accidents [21, 22] may result in some discrepancy. The error (for some reason called conservative approach) is likely to be somewhat less in using WHO 2005 TEF values than in using previous TEF values, because WHO 2005 TEQs are 10–25% lower than WHO 1998 TEQs in most matrixes including human samples [1, 23].

Some researchers have found clearly larger discrepancies. Gray et al. [24] calculated that, for pharmacokinetic reasons, a TEQ mixture of TCDD, 2,3,4,7,8-PeCDF, and PCB 126 was three- to fivefold less potent as carcinogen than predicted by 1998 TEFs. The biggest discrepancy found was a 10-fold overestimate of the TEF of PeCDF. The data were from National Toxicology Program 2-year rat study [25]; in the original study, the TEF of PeCDF was also found to be too high, and the true TEF value for carcinogenesis was estimated between 0.16 and 0.34. This is supported by Budinsky et al. [26].

There is still a question of mono-*ortho* PCBs even if their TEF values were lowered in the latest assessment [1]. Purification of test compounds containing more active impurities is still expected to lower their TEF values [19]. It is noteworthy that the fish TEF values for mono-*ortho* PCBs are very low, and their assessment is mainly based on residue analysis rather than experimental dosing with test compounds possibly contaminated with more active non-*ortho* PCBs or furans [11].

23.6 STUDIES ON ADDITIVITY OF DIOXIN-LIKE TOXICITY

As stated above, additivity of effects is a prerequisite of rational use of the TEF concept. Rozman's group assessed the acute toxicity of four dioxin congeners, and their mixture with each contributing one-fourth of the toxicity. The results supported strict additivity [27]. Other end points such as immature rat ovulation model resulted in similar conclusions with three dioxins [28] as well as with 2,3,4,7,8-PeCDF, 3,3',4,4',5-PeCB, and 2,2',5,5'-TeCB [29]. Also, subchronic studies of 13 weeks on the four dioxins supported additivity as to lethality and other signs of frank toxicity [30] and biochemical parameters [31].

Fattore et al. [32] studied vitamin A reduction after a long-term exposure to several PCDDs and PCDFs and their mixtures. The study indicated a somewhat less-than-expected effect by the mixture in male rats, but the result was interpreted to essentially confirm the additive mechanism. In a long-term study, accumulation of different congeners depends on their respective half-lives, and including congeners with shorter half-life than that of TCDD may lead to a lessened effect.

Several PCDDs, PCDFs, and PCBs were studied in pregnant rats for subchronic toxicity [33] and for reproductive outcomes [34], and there were signs of antagonism, although the results were interpreted to reasonably predict the toxicity of a mixture. In fact, none of the above studies formally calculated additivity, but it may be accepted that no frank antagonism was seen in these studies among the most potent dioxin-like compounds.

Toyoshiba et al. [35] and Walker et al. [25] evaluated the TEF approach in a 2-year rat cancer bioassay including TCDD, 2,3,4,7,8-PeCDF, 3,3',4,4',5-PeCB, and the mixture of the three compounds. Here a thorough mathematical approach was followed. Enzyme induction data at various time points questioned if either additivity or the WHO 1998 TEF values are true for the studied furan or PCB congener [35]. On the other hand, additivity seemed to hold for carcinogenicity, and the dose response of the mixture could be predicted from a combination of the individual compounds [25]. However, the potency of PeCDF seemed to be overestimated in the previous TEF values as described above.

Real world may be somewhat different. If there are significant antagonists, they should be sought among less effective dioxin-like compounds that might bind to the receptor but not elicit dioxin-like responses and thus have no TEF values. As yet, relatively weak antagonists have been found, for example, among PCBs (e.g., PCB 153) [36–39]. However, Aroclor 1254 and TCDD induced the known AHR battery of genes like CYP genes in a TEQ dose-dependent manner [40].

Additivity has also been studied *in vitro*. Measuring dioxin-like activity by CALUX assay suggested an antagonism of TCDD responses by several technical PCBs [41] and PCB 126 [42]. Inhibition of the induction of CYP1A1 has been showed by nonplanar PCB compounds [43, 44], and the crucial step seems to be the formation of AHR–ligand–DRE (dioxin-responsive element) complex [44–46].

Relatively high concentrations of polybrominated diphenyl ethers were found to antagonize CYP1A1 induction caused by TCDD [47].

It seems that several compounds not eliciting dioxin-like responses (e.g., nonplanar PCBs, PBDEs) may bind to AHR and thereby prevent the formation of a proper dioxin–AHR complex and its binding to the DRE. The significance of this phenomenon at environmental concentrations is unknown. Among the most potent dioxin-like compounds (dioxins, furans, non-*ortho* PCBs), additivity seems to be satisfactory to justify the use of the TEF concept.

The above studies deal with additivity of known mixtures. It is another issue to estimate additivity and especially antagonism in real-world mixtures possibly containing uncharacterized compounds and natural AHR ligands. In one study, POPs were extracted from Baltic herring and subjected to toxicological studies [48]. Expected biochemical effects were seen, but there was no TCDD control group for calculating the expected effect on the TEQ basis of analyzed compounds. Studies on reconstituted mixtures of chemicals, for example, detected in breast milk suggest additive effects for some end points but antagonism for others; synergism was not detected [12]. Reconstituted mixtures, of course, do not contain possible undetected chemicals.

It is quite possible that there are unknown antagonists in natural mixtures decreasing the effects of known dioxin-like chemicals, and on the other side a number of agonists that might be more important than the known dioxin-like chemicals [49, 50].

23.7 HOW RISK MANAGERS SHOULD RESPOND TO CHANGED TEF VALUES

Some risk managers have been asking whether or not limit values should be changed because the TEQ in many products is lower when using the WHO 2005 TEF values than when using the WHO 1998 TEF values. It should be pointed out that TEF values are only an attempt to compare the toxicity of a congener as accurately as possible with that of TCDD. If more recent studies indicate that the toxicity of a congener was previously estimated too high in comparison to TCDD, introduction of a new TEF value only gives a better new estimate [1]. Because most toxicity studies were performed using TCDD as the test compound, a decrease of TEQ by 10–30 % [1, 23, 51] only gives a better estimate of the true risk of toxicity of the mixture and is no reason to change previous limit values. As indicated above, even the new TEF values are more likely an overestimate than an underestimate of toxicity as compared with that of TCDD.

23.8 IS INCLUSION OF ADDITIONAL COMPOUNDS NEEDED TO MAKE TEQ MORE COMPREHENSIVE?

In addition to chlorinated dioxins and furans and dioxin-like PCBs, several groups of persistent compounds are AHR ligands. These include polybrominated or mixed halogenated dioxins and furans, and polyhalogenated naphthalenes. At least so far their concentrations seem to be lower, although not very well known due to the high number of compounds. Therefore, more studies would be needed and by increasing information their inclusion in the TEF scheme should be considered to increase the accuracy of total TEQ [1]. At the present state of knowledge, this has not been considered feasible.

23.9 USE OF TEQs TO DESCRIBE THE RESULTS OF DIOXIN BIOASSAYS (e.g., CALUX)

Several reporter gene bioassay methods have been developed to analyze dioxin-like activity in biological or environmental samples [52–54]. In the commonly used methods, firefly luciferase gene is linked to the AHR in genetically modified hepatoma-derived cell lines. Thus, AHR activation gives a light response that can be detected and measured. These cells respond to all compounds activating the AHR be they synthetic or natural. This is both their strength and their weakness. Often the result is expressed as TEQ to describe total dioxin-like activity in the sample.

The term used for bioassay results is highly important to avoid confusion. One of the important reasons for using bioassays is the cost of high-resolution MS analysis with a cumbersome purification process. However, if the bioassays detect *in vitro*, in addition to the chlorinated dioxin-like compounds, a number of compounds that are perhaps not so relevant *in vivo*, the use of the term TEQ is clearly misleading. To improve the precision of the analysis, most laboratories now include a chemical purification of samples to be assayed, but then the cost also increases and even doing so it is not certain that the two analyses measure the same thing. Different cell preparations seem to give different results [55]. Hence, the term induction equivalent (IEQ) has also been used rather than TEQ [56]. This would be more honest, and a separate term might require some further defining to characterize the degree of purification of samples. Under ideal conditions after rigorous purification, good correlations have been established between chemically and biologically assessed concentrations of dioxin-like compounds [57].

23.10 AHR ACTIVITY OF NONCHLORINATED NATURAL COMPOUNDS

A number of chemicals, including many natural compounds, bind to the AHR and some of them activate it at very low concentrations, sometimes comparable to those of dioxins (reviewed in Refs 49 and 58). Some of the natural compounds are exogenous and the intake is mainly via food [59], and some are endogenously formed in the body [60]. Surprisingly, little *in vivo* information is available after about 20 years of *in vitro* experimentation. This clearly hampers the risk assessment of these compounds and also has important and unforeseen consequences in the risk assessment of low levels of dioxin-like compounds. Natural compounds are widely distributed in vegetables, fruits, teas, and herbal dietary products [49]. This issue will be dealt with more thoroughly in Chapter 24.

The natural ligands of the AHR (sometimes called NAH-RAs, natural AHR agonists) have not been given TEF values, because they do not fulfill the requirements presented in the beginning of this chapter, notably that of persistence. In fact, there are several separate issues here. Should these compounds be given TEF values? Are these the long-searched natural ligands of the AHR? Is it likely that they potentiate or antagonize the toxicity of dioxin-like compounds? Do they somehow change the TEF/TEQ concept as a whole? The last question was discussed during the latest TEF reevaluation, but it was concluded that while these compounds may impact the magnitude and overall toxic effects produced by a defined amount of TEQ, they do not impact the determination of individual REP or TEF values for dioxin-like chemicals [1].

These compounds may be relatively short-lived due to fast metabolism, but they seem to be continuously present in the diet and activate the AHR. Many are advocated to prevent cancer [61–63] or inhibit cancer growth [64, 65]. Some are already being tested in clinical trials [66]; actually these compounds have multiple effects in oncogenic signaling pathways [62, 67, 68]. Does this challenge the relevance of the present very low background concentrations of chlorinated dioxins in our diet? Does this mean that the regulatory advice on limiting fish consumption is superfluous? These questions are outside the scope of this chapter (see Chapter 24), but the TEF question is fully relevant, and will be briefly dealt with below.

While these compounds are active *in vitro*—and some of them are very potent AHR agonists—they seem to be rapidly metabolized in mammals. In the very few animal or human studies available, they have failed to produce frank toxicity *in vivo* [69–71]. While low, bioavailability is not zero, and some of the most sensitive responses such as enzyme induction can be seen *in vivo* [69, 72] as well as some dioxin-like reproductive effects on male offspring [73]. Natural AHR ligands may also be formed in human body [60]. For comparison, it should be noted that dioxin-like persistent chemicals at present TEQ intake levels have not been proven to cause any effects *in vivo* either.

Recently, the contribution of natural compounds to bioassay-determined dioxin-like activity was assessed in blood samples from 10 volunteers [56]. The activities were analyzed both during baseline diet and after dietary interventions. It was demonstrated that compared with chemical analysis a direct bioassay resulted in several orders of magnitude higher levels, and this was still augmented by high-vegetable diet or indole-3-carbinol supplements. This seems to mean that even if the compounds are rapidly metabolized, their huge excess in food guarantees a continuous and relatively high impact in the body as compared with chlorinated dioxin-like ligands.

This will clearly require more serious scrutiny *in vivo* both in animals and in man. If it can be shown that at the present background concentrations the persistent dioxin-like compounds are in fact a minor constituent of chemicals activating the AHR, this will change our whole concept of dioxin toxicity at low intake levels. The first task is to make it possible to compare the concentrations in a meaningful way. TEF in the present form is not ideal tool for this, because both bioavailability and elimination of these compounds are very different from those of persistent compounds. An internal TEF meaning TEF based on toxicity as a function of the level of a compound at the site of action would be a step to make better comparisons possible (see below).

This brings another question: What is actually the crucial site of action of dioxin toxicity? Two examples of rather effective barriers toward polar compounds in the body are blood–brain barrier and blood–testis barrier. Also, placenta is rather restrictive in passing polar compounds to the fetus. Therefore, any study defining the TEF on the basis of either intake or even blood or liver concentration does not necessarily reflect the true situation at the site of toxic action. This is a true uncertainty even in the present TEF assessment; it cannot be taken as granted that different congeners of dioxins or PCBs enter the fetus at the same ratio they are present in food. This question becomes even more important, when the natural polar AHR agonists are included in the assessment.

23.11 PROBLEMS RELATED TO DIFFERENT KINETICS OF COMPOUNDS AND THE USE OF TEF AT VARIOUS LEVELS OF ENVIRONMENTAL DISTRIBUTION

TEFs are based primarily on potency, but because *in vivo* and long-term data are preferred [1, 10] in establishing TEFs ("intake TEF"), kinetic factors (absorption, distribution, elimination) are also involved. Long-term studies in animals only partially help and the use is based on unproven assumptions. Half-lives of different congeners vary; for example, reference half-life of PCB 126 in humans is 1.6 years and of TCDD is 7.2 years [18]. This means a 4.5-fold higher steady-state body burden for TCDD at the same daily intake level. Therefore, TCDD would be higher a risk after lifetime exposure if all other conditions are equal, and this should show in the TEF of PCB 126. Moreover, the preferred use of long-term animal studies assumes that the differences in accumulation are similar in the experimental animal and in humans. This cannot be taken as granted.

It should be noted that while the mammalian TEF values are based on intake, fish and bird TEFs are based on residue analysis, that is, tissue concentration [11]. Therefore, they are not directly comparable.

More accurate risk assessment could be performed, if TEFs were based on internal concentrations or body burdens. The possibility of internal TEF was discussed in the latest TEF assessment [1], but it was not found feasible as yet because of scarce data. Aiming at internal TEF would allow defining the kinetic factors separately. Then it would not matter, if they should be different in the experimental animal and in humans. If natural polar compounds are included in the TEF concept, these questions become absolutely crucial, and might even require organ-specific characterization. Because of the obvious effects on enzyme induction, much of dioxin work has concentrated around the liver as the target, but the crucial toxic effects do not necessarily depend on liver concentrations but those in the fetus, skin, endocrine organs, or even the brain.

Another source of uncertainty is the use of TEQs for various matrices such as emissions, sediments, and contaminated soil. It ignores the fact that all congeners are not transferred to the same extent to human beings. Need for future research on the transport was emphasized already during the 1997 assessment [11], but not much data are available even now. The latest TEF assessment panel expressed its concern on this, because the present TEQ methodology is primarily meant for estimating exposure via dietary intake [1]. Therefore, it was recommended that congener-specific equations be used throughout the model, because the fate and transport properties differ widely between congeners.

A congener-specific intake fraction (iF) might help to involve different transport of different congeners from sources to humans. iF denotes the fraction of emission that is inhaled or ingested by human population. It is used especially in air pollution management as a robust and easily understandable tool [74–76].

There is very little research on this issue [77]. Changes in congener spectra from one matrix level to another were tested in high consumers of Baltic herring [78]. Congener concentrations were compared in air emissions, in fish, and in fishermen. Accumulation of each congener into the next matrix was compared with that of TCDD (=1). Concentrations of most congeners were lower in fishers than in emissions, relative to TCDD, down to a factor of 1/60. This was sorted out stepwise from emission to fish and from fish to fisher. Transport from emission to herring was lower for congeners (especially higher chlorinated PCDD/Fs) than for TCDD. This is in line with the results on feeding salmon with various PCDD and PCDF congeners [79], and with the intake fractions determined from the source to fish [77]. On the other hand, levels of higher chlorinated dioxins were higher in fishers relative to TCDD than in fish, and furans were lower than dioxins [78]. This might suggest longer half-lives or better absorption of higher dioxins and shorter half-lives or less effective absorption of furans.

The likelihood of different congeners to reach people thus varies by orders of magnitude, and kinetic differences modify accumulation. Therefore, relevance could be improved by dividing intake fraction, kinetic factors, and TEF into separate entities. An internal TEF would be here more appropriate than intake TEF. This gives the formula MS-TEQ = $\sum_n [iF_i \cdot B_i \cdot iTEF_i \cdot A_i]$, where MS-TEQ is matrix-specific TEQ, iF_i is relative intake fraction specific to the matrix and congener compared with TCDD, B_i is toxicokinetic factor (different for short-term risk and long-term risk, also relative to TCDD), $iTEF_i$ is internal TEF, and A_i is the amount of the congener in the matrix. This segregates intake, kinetics, and effect into separate entities and TEF describes only the last step directly relevant to relative human toxicity. As yet, there seems to be too little data to do this systematically, but research here should be encouraged.

23.12 USE OF TEQs IN SCIENTIFIC STUDIES

Although TEQs were developed for risk management purposes, they have been widely used in exposure studies to illustrate the total exposure (reviewed in Refs 23, 80, and 81) and in a few epidemiological studies. A noteworthy set of studies is the follow-up of TEQs in breast milk originally organized as an international intercalibration exercise of analytical laboratories, later organized by the World Health Organization [82–84]. By and large these studies showed large differences between exposure levels in different countries, and then a dramatic decrease of the levels, in some countries down to about 10% of the likely peak levels in 1970s [85, 86].

Surprisingly, exhaustive congener-specific analysis on the Yu-Cheng incident in Taiwan was only reported 15 years after the incident [87]. The PCDD/F TEQ levels in Yu-Cheng victims were still 46 times higher at 15 years than those in the general population in Taiwan, and PCDFs contributed most to TEQ (44% versus PCDDs 12%).

Because dioxin-like compounds are always present in the environment as complex mixtures, it is usually not rational to seek for associations of diseases with single compounds, and TEQ may be a reasonable proxy of the total exposure. Surprisingly, few epidemiological studies have been published, possibly due to the high cost of a congener-specific dioxin analysis. Some examples are given below to illustrate the usefulness and limitations of using TEQ.

In a relatively large case–control study, the association of soft tissue sarcoma with dioxins and individual TEQs were studied [88]. There was no increase in sarcoma risk when dioxin TEQ increased; on the contrary, there was a decreasing trend. Concentrations of individual congeners correlated with each other, and when the four congeners contributing most to TEQ were individually analyzed, the result with each of them was in line with that on the TEQ. A possibility of

J-shaped dose–response curve was discussed with a decreased cancer risk at moderate concentrations of dioxin-like compounds [89].

It is interesting to note here the postulated cancer-preventing properties of natural AHR ligands [61, 63, 90]. It is completely plausible that one physiological function of the AHR is to respond to foreign and endogenous compounds that may be toxic and carcinogenic, and activate a metabolic machinery to eliminate them [49, 68, 90]. The interpretations will be complicated by the myriad of intertwining effects and crosstalk between various cell signaling systems [90]. At any rate, certain level of AHR activation seems beneficial and physiological, but excessive activation inducing or inhibiting many genes inappropriately causes toxicity [91, 92].

Lymphomas are another type of malignancies connected with dioxins, chlorophenols, and phenoxy acids, and there are many studies using indirect exposure methods such as patient recall or work history [93]. Seveso studies after TCDD accident indicate some increase in hematopoietic malignancies, especially lymphoid leukemia and lymphomas [94]. An American case–control study revealed an association with several organochlorines, including an association with TEQs [95]. In a Canadian case–control study, PCDD/Fs and non-*ortho* PCBs were not studied, but an association was found with total PCBs and mono-*ortho* congeners 118 and 156 [96]. Without further information, it is hard to know if AHR-based mechanism is involved and if the total TEQs correlate with the mono-*ortho* compounds as a proxy.

Mortality in Finnish fisher families with high TEQs (twice those in controls) was compared with that in the control population. A lowered total mortality was found, but also lowered mortality from neoplasms in males [97]. There are many confounding factors possibly explaining a lowered total mortality, particularly low cardiac mortality due to beneficial omega-3 fatty acids in fish. However, it is noteworthy that cancer mortality was also low. In lymphomas, a nonsignificant increase was seen in fishermen but a decrease in wives.

Thyroid hormone deficiency with consequent neurobehavioral developmental effects is a possible effect of dioxins on the basis of experimental studies, but there is uncertainty on its relevance at the present intake levels and previous studies are ambiguous [98]. In a recent German cohort study, no impact of TEQ levels on thyroid function or neurodevelopment was seen [99]. In Swedish families, thyroid hormone levels in babies were compared with POP levels in breast milk [100]. In this study, a simple regression analysis revealed weak associations between TEQ levels and TSH levels at 3 weeks of age, and no associations with thyroid hormones T_3 and T_4. After analyzing several confounding factors, the association disappeared, and the authors emphasize the importance of controlling confounding, which was not properly done in several previous studies.

Another endocrinological problem associated with dioxins is insulin resistance and type 2 diabetes [101, 102]. An association between serum TEQs and a modest increase in insulin resistance is possible and mechanistically plausible. The difficulty of low-level studies is to exclude the possibility of high-energy diet and use of animal fats leading to both diabetes and higher POP levels [101]. Also, birth weight has been correlated with maternal TEQ levels, and a weak inverse association was found, but this disappeared when birth order was taken into account [103]. Birth order is an important confounder, because dioxin levels decrease during each lactation period, and average birth weight increases in later pregnancies.

There are several epidemiological studies on the risk of endometriosis associated with dioxin concentrations expressed as TEQ. In a minireview, Heiler et al. [104] concluded that a majority of 12 studies did not observe a significant association with organochlorines (TEQ was measured only in some of the studies), but there was some evidence that deep endometric nodules may be associated with higher level of dioxin-like compounds and marker PCBs. Of two more recent case–control studies, one [105] showed similar low levels of TEQs in both patients and controls, and the other (using CALUX bioassay) found marginally higher concentrations in the patients [106]. Therefore, the issue remains open as yet. Also, male infertility has been associated with dioxins, but a recent study measuring TEQs did not find any association [107]. In a Danish–Finnish collaborative study, no convincing evidence was found associating cryptorchidism with TEQ concentrations in mothers' placenta or breast milk [108].

23.13 CONCLUSIONS

TEF concept was an innovative and important way of assessing and managing the risks of dioxin-like compounds as a group. It has been widely used in analytical studies to describe the intake of dioxin-like compounds, and in risk management mainly based on such analyses. Meaningful animal studies are relatively fragmentary, but by and large they confirm additivity and the usefulness of the concept as long as we are dealing with PCDDs, PCDFs, and non-*ortho* PCBs. Proper epidemiological studies using TEQs as a measure of exposure are surprisingly few. Therefore, the risks of dioxins are in essence derived from animal experiments using TCDD, and extrapolated to humans by using TEQs. The available epidemiological studies are not able to confirm whether the hypothesis justifying the present TEQ use in this extrapolation is correct or not. However, it is the best we have at the moment.

A real challenge to the TEF concept and dioxin risk assessment as a whole is inclusion of all compounds activating the AHR. The brominated compounds and other dioxin-like synthetic compounds so far not given TEF values seem to be a minor part and may not cause major problems.

However, natural AHR agonists may revolutionize dioxin risk assessment, if they are given TEF or IEF values and included in the assessment. If they can be shown to cause a continuous baseline activation of the AHR that is much higher than that caused by the persistent chemicals, the risks (or even benefits) of low concentrations have to be reconsidered.

REFERENCES

1. Van den Berg, M., Birnbaum, L. S., Denison, M., De Vito, M., Farland, W., Feeley, M., Fiedler, H., Hakansson, H., Hanberg, A., Haws, L., Rose, M., Safe, S., Schrenk, D., Tohyama, C., Tritscher, A., Tuomisto, J., Tysklind, M., Walker, N., and Peterson, R. E. (2006). The 2005 World Health Organization reevaluation of human and mammalian toxic equivalency factors for dioxins and dioxin-like compounds. *Toxicological Sciences*, 93, 223–241.
2. Ahlborg, U. G., Brouwer, A., Fingerhut, M. A., Jacobson, J. L., Jacobson, S. W., Kennedy, S. W., Kettrup, A. A. F., Koeman, J. H., Poiger, H., Rappe, C., Safe, S. H., Seegal, R. F., Tuomisto, J., and Van den Berg, M. (1992). Impact of polychlorinated dibenzo-p-dioxins, dibenzofurans, and biphenyls on human and environmental health, with special emphasis on application of the toxic equivalency factor concept. *European Journal of Pharmacology*, 228, 179–199.
3. Ontario Ministry of the Environment (1985). Scientific Criteria Document for standard development. Polychlorinated dibenzo-p-dioxins (PCDDs) and polychlorinated dibenzofurans (PCDFs). Report No. 4184, Ministry of Environment.
4. U.S. EPA (1987). Interim procedures for estimating risks associated with exposures to mixtures of chlorinated dibenzo-p-dioxins and -dibenzofurans (CDDs/CCFs). EPA/625/3-87/012.
5. Ahlborg, U. G. (1988). Nordisk dioxinriskbedömning (Nordic risk assessment of dioxins). National Institute of Environmental Medicine.
6. NATO/CCMS (1988). International toxicity equivalency factor (I-TEF) method of risk assessment for complex mixtures of dioxins and related compounds. Report No. 176, NATO/CCMS (North Atlantic Treaty Organization, Committee on the Challenges of Modern Society).
7. Barnes, D., Kutz, F., and Bottimore, D. (1989). Interim procedures for estimating risks associated with exposure to mixtures of chlorinated dibenzo-p-dioxins and -dibenzofurans and 1989 update. EPA/625/3-89/016, Risk Assessment Forum, U.S. Environmental Protection Agency.
8. Barnes, D. G. (1991). Toxicity equivalents and EPA's risk assessment of 2,3,7,8-TCDD. *Science of the Total Environment*, 104, 73–86.
9. Safe, S. (1990). Polychlorinated-biphenyls (PCBs), dibenzo-para-dioxins (PCDDs), dibenzofurans (PCDFs), and related compounds: environmental and mechanistic considerations which support the development of toxic equivalency factors (TEFs). *Critical Reviews in Toxicology*, 21, 51–88.
10. Ahlborg, U. G., Becking, G. C., Birnbaum, L. S., Brouwer, A., Derks, H. J. G. M., Feeley, M., Golor, G., Hanberg, A., Larsen, J. C., Liem, A. K. D., Safe, S. H., Schlatter, C., Waern, F., Younes, M., and Yrjanheikki, E. (1994). Toxic equivalency factors for dioxin-like PCBs: report on a WHO-ECEH and IPCS consultation, December 1993. *Chemosphere*, 28, 1049–1067.
11. Van den Berg, M., Birnbaum, L., Bosveld, A. T. C., Brunstrom, B., Cook, P., Feeley, M., Giesy, J. P., Hanberg, A., Hasegawa, R., Kennedy, S. W., Kubiak, T., Larsen, J. C., van Leeuwen, F. X. R., Liem, A. K. D., Nolt, C., Peterson, R. E., Poellinger, L., Safe, S., Schrenk, D., Tillitt, D., Tysklind, M., Younes, M., Waern, F., and Zacharewski, T. (1998). Toxic equivalency factors (TEFs) for PCBs, PCDDs, PCDFs for humans and wildlife. *Environmental Health Perspectives*, 106, 775–792.
12. Desaulniers, D., Leingartner, K., Musicki, B., Yagminas, A., Xiao, G. H., Cole, J., Marro, L., Charbonneau, M., and Tsangt, B. K. (2003). Effects of postnatal exposure to mixtures of non-ortho-PCBs, PCDDs, and PCDFs in prepubertal female rats. *Toxicological Sciences*, 75, 468–480.
13. Pohjanvirta, R., Unkila, M., Linden, J., Tuomisto, J. T., and Tuomisto, J. (1995). Toxic equivalency factors do not predict the acute toxicities of dioxins in rats. *European Journal of Pharmacology*, 293, 341–353.
14. Sutter, C. H., Rahman, M., and Sutter, T. R. (2006). Uncertainties related to the assignment of a toxic equivalency factor for 1,2,3,4,6,7,8,9-octachlorodibenzo-p-dioxin. *Regulatory Toxicology and Pharmacology*, 44, 219–225.
15. Flaveny, C. A., Murray, I. A., Chiaro, C. R., and Perdew, G. H. (2009). Ligand selectivity and gene regulation by the human aryl hydrocarbon receptor in transgenic mice. *Molecular Pharmacology*, 75, 1412–1420.
16. Flaveny, C. A., Murray, I. A., and Perdew, G. H. (2010). Differential gene regulation by the human and mouse aryl hydrocarbon receptor. *Toxicological Sciences*, 114, 217–225.
17. Boutros, P. C., Yan, R., Moffat, I. D., Pohjanvirta, R., and Okey, A. B. (2008). Transcriptomic responses to 2,3,7,8-tetrachlorodibenzo-p-dioxin (TCDD) in liver: comparison of rat and mouse. *BMC Genomics*, 9, 419.
18. Milbrath, M. O., Wenger, Y., Chang, C., Emond, C., Garabrant, D., Gillespie, B. W., and Jolliet, O. (2009). Apparent half-lives of dioxins, furans, and polychlorinated biphenyls as a function of age, body fat, smoking status, and breast-feeding. *Environmental Health Perspectives*, 117, 417–425.
19. Peters, A. K., Leonards, P. E., Zhao, B., Bergman, A., Denison, M. S., and Van den Berg, M. (2006). Determination of *in vitro* relative potency (REP) values for mono-*ortho* polychlorinated biphenyls after purification with active charcoal. *Toxicology Letters*, 165, 230–241.
20. Haws, L. C., Su, S. H., Harris, M., DeVito, M. J., Walker, N. J., Farland, W. H., Finley, B., and Birnbaum, L. S. (2006). Development of a refined database of mammalian relative potency estimates for dioxin-like compounds. *Toxicological Sciences*, 89, 4–30.
21. Eskenazi, B., Mocarelli, P., Warner, M., Samuels, S., Vercellini, P., Olive, D., Needham, L. L., Patterson, D. G., Brambilla,

P., Gavoni, N., Casalini, S., Panazza, S., Turner, W., and Gerthoux, P. M. (2002). Serum dioxin concentrations and endometriosis: a cohort study in Seveso, Italy. *Environmental Health Perspectives, 110,* 629–634.

22. Geusau, A., Abraham, K., Geissler, K., Sator, M. O., Stingl, G., and Tschachler, E. (2001). Severe 2,3,7,8-tetrachlorodibenzo-*p*-dioxin (TCDD) intoxication: clinical and laboratory effects. *Environmental Health Perspectives, 109,* 865–869.

23. Scott, L. L. F., Unice, K. M., Scott, P., Nguyen, L. M., Haws, L. C., Harris, M., and Paustenbach, D. (2008). Evaluation of PCDD/F and dioxin-like PCB serum concentration data from the 2001–2002 National Health and Nutrition Examination Survey of the United States population. *Journal of Exposure Science and Environmental Epidemiology, 18,* 524–532.

24. Gray, M. N., Aylward, L. L., and Keenan, R. E. (2006). Relative cancer potencies of selected dioxin-like compounds on a body-burden basis: comparison to current toxic equivalency factors (TEFs). *Journal of Toxicology and Environmental Health, Part A, 69,* 907–917.

25. Walker, N. J., Crockett, P. W., Nyska, A., Brix, A. E., Jokinen, M. P., Sells, D. M., Hailey, J. R., Easterling, M., Haseman, J. K., Yin, M., Wyde, M. E., Bucher, J. R., and Portier, C. J. (2005). Dose-additive carcinogenicity of a defined mixture of "dioxin-like compounds". *Environmental Health Perspectives, 113,* 43–48.

26. Budinsky, R. A., Paustenbach, D., Fontaine, D., Landenberger, B., and Starr, T. B. (2006). Recommended relative potency factors for 2,3,4,7,8-pentachlorodibenzofuran: the impact of different dose metrics. *Toxicological Sciences, 91,* 275–285.

27. Stahl, B. U., Kettrup, A., and Rozman, K. (1992). Comparative toxicity of 4 chlorinated dibenzo-*p*-dioxins (CDDs) and their mixture. 1. Acute toxicity and toxic equivalency factors (TEFs). *Archives of Toxicology, 66,* 471–477.

28. Gao, X., Son, D. S., Terranova, P. F., and Rozman, K. K. (1999). Toxic equivalency factors of polychlorinated dibenzo-*p*-dioxins in an ovulation model: validation of the toxic equivalency concept for one aspect of endocrine disruption. *Toxicology and Applied Pharmacology, 157,* 107–116.

29. Gao, X., Terranova, P. F., and Rozman, K. K. (2000). Effects of polychlorinated dibenzofurans, biphenyls, and their mixture with dibenzo-*p*-dioxins on ovulation in the gonadotropin-primed immature rat: support for the toxic equivalency concept. *Toxicology and Applied Pharmacology, 163,* 115–124.

30. Viluksela, M., Stahl, B. U., Birnbaum, L. S., Schramm, K. W., Kettrup, A., and Rozman, K. K. (1998). Subchronic/chronic toxicity of a mixture of four chlorinated dibenzo-*p*-dioxins in rats. I. Design, general observations, hematology, and liver concentrations. *Toxicology and Applied Pharmacology, 151,* 57–69.

31. Viluksela, M., Stahl, B. U., Birnbaum, L. S., and Rozman, K. K. (1998). Subchronic/chronic toxicity of a mixture of four chlorinated dibenzo-*p*-dioxins in rats. II. Biochemical effects. *Toxicology and Applied Pharmacology, 151,* 70–78.

32. Fattore, E., Trossvik, C., and Hakansson, H. (2000). Relative potency values derived from hepatic vitamin A reduction in male and female Sprague-Dawley rats following subchronic dietary exposure to individual polychlorinated dibenzo-*p*-dioxin and dibenzofuran congeners and a mixture thereof. *Toxicology and Applied Pharmacology, 165,* 184–194.

33. Van Birgelen, A. P. J. M., Van der Kolk, J., Fase, K. M., Bol, I., Poiger, H., Brouwer, A., and Van den Berg, M. (1994). Toxic potency of 3,3′,4,4′,5-pentachlorobiphenyl relative to and in combination with 2,3,7,8-tetrachlorodibenzo-*p*-dioxin in a subchronic feeding study in the rat. *Toxicology and Applied Pharmacology, 127,* 209–221.

34. Hamm, J. T., Chen, C. Y., and Birnbaum, L. S. (2003). A mixture of dioxins, furans, and non-*ortho* PCBs based upon consensus toxic equivalency factors produces dioxin-like reproductive effects. *Toxicological Sciences, 74,* 182–191.

35. Toyoshiba, H., Walker, N. J., Bailer, A. J., and Portier, C. J. (2004). Evaluation of toxic equivalency factors for induction of cytochromes P450 CYP1A1 and CYP1A2 enzyme activity by dioxin-like compounds. *Toxicology and Applied Pharmacology, 194,* 156–168.

36. Morrissey, R. E., Harris, M. W., Diliberto, J. J., and Birnbaum, L. S. (1992). Limited PCB antagonism of TCDD-induced malformations in mice. *Toxicology Letters, 60,* 19–25.

37. Battershill, J. M. (1994). Review of the safety assessment of polychlorinated-biphenyls (PCBs) with particular reference to reproductive toxicity. *Human & Experimental Toxicology, 13,* 581–597.

38. Safe, S. (1997). Limitations of the toxic equivalency factor approach for risk assessment of TCDD and related compounds. *Teratogenesis, Carcinogenesis, and Mutagenesis, 17,* 285–304.

39. Haag-Gronlund, M., Johansson, N., Fransson-Steen, R., Hakansson, H., Scheu, G., and Warngard, L. (1998). Interactive effects of three structurally different polychlorinated biphenyls in a rat liver tumor promotion bioassay. *Toxicology and Applied Pharmacology, 152,* 153–165.

40. Silkworth, J. B., Carlson, E. A., McCulloch, C., Illouz, K., Goodwin, S., and Sutter, T. R. (2008). Toxicogenomic analysis of gender, chemical, and dose effects in livers of TCDD- or Aroclor 1254-exposed rats using a multifactor linear model. *Toxicological Sciences, 102,* 291–309.

41. Schroijen, C., Windal, I., Goeyens, L., and Baeyens, W. (2004). Study of the interference problems of dioxin-like chemicals with the bio-analytical method CALUX. *Talanta, 63,* 1261–1268.

42. Sanctorum, H., Elskens, M., and Baeyens, W. (2007). Bioassay (CALUX) measurements of 2,3,7,8-TCDD and PCB 126: interference effects. *Talanta, 73,* 185–188.

43. Aarts, J. M. M. J. G., Denison, M. S., Cox, M. A., Schalk, M. A. C., Garrison, P. M., Tullis, K., Dehaan, L. H. J., and Brouwer, A. (1995). Species-specific antagonism of AH receptor action by 2,2′,5,5′-tetrachlorobiphenyl and 2,2′,3,3′,4,4′-hexachlorobiphenyl. *European Journal of Pharmacology, 293,* 463–474.

44. Suh, J. H., Kang, J. S., Yang, K. H., and Kaminski, N. E. (2003). Antagonism of aryl hydrocarbon receptor-dependent induction of CYP1A1 and inhibition of IgM expression by di-*ortho*-substituted polychlorinated biphenyls. *Toxicology and Applied Pharmacology, 187,* 11–21.

45. Petrulis, J. R. and Bunce, N. J. (2000). Competitive behavior in the interactive toxicology of halogenated aromatic compounds. *Journal of Biochemical and Molecular Toxicology*, *14*, 73–81.
46. Chen, G. S. and Bunce, N. J. (2004). Interaction between halogenated aromatic compounds in the Ah receptor signal transduction pathway. *Environmental Toxicology*, *19*, 480–489.
47. Peters, A. K., Sanderson, J. T., Bergman, A., and van den Berg, M. (2006). Antagonism of TCDD-induced ethoxyresorufin-O-deethylation activity by polybrominated diphenyl ethers (PBDEs) in primary cynomolgus monkey (*Macaca fascicularis*) hepatocytes. *Toxicology Letters*, *164*, 123–132.
48. Stern, N., Oberg, M., Casabona, H., Trossvik, C., Manzoor, E., Johansson, N., Lind, M., Orberg, J., Feinstein, R., Johansson, A., Chu, I., Poon, R., Yagminas, A., Brouwer, A., Jones, B., and Hakansson, H. (2002). Subchronic toxicity of Baltic herring oil and its fractions in the rat. II. Clinical observations and toxicological parameters. *Pharmacology & Toxicology*, *91*, 232–244.
49. Denison, M. S. and Nagy, S. R. (2003). Activation of the aryl hydrocarbon receptor by structurally diverse exogenous and endogenous chemicals. *Annual Review of Pharmacology and Toxicology*, *43*, 309–334.
50. Howard, G. J., Schlezinger, J. J., Hahn, M. E., and Webster, T. F. (2010). Generalized concentration addition predicts joint effects of aryl hydrocarbon receptor agonists with partial agonists and competitive antagonists. *Environmental Health Perspectives*, *118*, 666–672.
51. Wittsiepe, J., Fürst, P., and Wilhelm, M. (2007). The 2005 World Health Organization re-evaluation of TEFs for dioxins and dioxin-like compounds: what are the consequences for German human background levels? *International Journal of Hygiene and Environmental Health*, *210*, 335–339.
52. Murk, A. J., Legler, J., Denison, M. S., Giesy, J. P., van de Guchte, C., and Brouwer, A. (1996). Chemical-activated luciferase gene expression (CALUX): a novel *in vitro* bioassay for Ah receptor active compounds in sediments and pore water. *Fundamental and Applied Toxicology*, *33*, 149–160.
53. Windal, I., Denison, M. S., Birnbaum, L. S., Van Wouwe, N., Baeyens, W., and Goeyens, L. (2005). Chemically activated luciferase gene expression (CALUX) cell bioassay analysis for the estimation of dioxin-like activity: critical parameters of the CALUX procedure that impact assay results. *Environmental Science & Technology*, *39*, 7357–7364.
54. Takigami, H., Suzuki, G., and Sakai, S. (2008). Application of bioassays for the detection of dioxins and dioxin-like compounds in wastes and the environment. In *Interdisciplinary Studies on Environmental Chemistry: Biological Responses to Chemical Pollution* (Murakami, Y., Nakayama, K., Kitamura, S.-I., Iwata, H., and Tanabe, S., Eds). Terrapub, pp. 87–94.
55. Van der Heiden, E., Bechoux, N., Muller, M., Sergent, T., Schneider, Y., Larondelle, Y., Maghuin-Rogister, G., and Scippo, M. (2009). Food flavonoid aryl hydrocarbon receptor-mediated agonistic/antagonistic/synergic activities in human and rat reporter gene assays. *Analytica Chimica Acta*, *637*, 337–345.
56. Connor, K. T., Harris, M. A., Edwards, M. R., Budinsky, R. A., Clark, G. C., Chu, A. C., Finley, B. L., and Rowlands, J. C. (2008). AH receptor agonist activity in human blood measured with a cell-based bioassay: evidence for naturally occurring AH receptor ligands *in vivo*. *Journal of Exposure Science and Environmental Epidemiology*, *18*, 369–380.
57. Denison, M. S., Zhao, B., Baston, D. S., Clark, G. C., Murata, H., and Han, D. (2004). Recombinant cell bioassay systems for the detection and relative quantitation of halogenated dioxins and related chemicals. *Talanta*, *63*, 1123–1133.
58. Mandlekar, S., Hong, J., and Kong, A. T. (2006). Modulation of metabolic enzymes by dietary phytochemicals: a review of mechanisms underlying beneficial versus unfavorable effects. *Current Drug Metabolism*, *7*, 661–675.
59. De Waard, W. J., Aarts, J. M. M. J. G., Peijnenburg, A. C. M., De Kok, T. M. C. M., Van Schooten, F.-J., and Hoogenboom, L. A. P. (2008). Ah receptor agonist activity in frequently consumed food items. *Food Additives and Contaminants*, *25*, 779–787.
60. Wincent, E., Amini, N., Luecke, S., Glatt, H., Bergman, J., Crescenzi, C., Rannug, A., and Rannug, U. (2009). The suggested physiologic aryl hydrocarbon receptor activator and cytochrome P4501 substrate 6-formylindolo[3,2-*b*]carbazole is present in humans. *Journal of Biological Chemistry*, *284*, 2690–2696.
61. Kim, Y. S. and Milner, J. A. (2005). Targets for indole-3-carbinol in cancer prevention. *Journal of Nutritional Biochemistry*, *16*, 65–73.
62. Hayes, J. D., Kelleher, M. O., and Eggleston, I. M. (2008). The cancer chemopreventive actions of phytochemicals derived from glucosinolates. *European Journal of Nutrition*, *47*, 73–88.
63. Kawajiri, K., Kobayashi, Y., Ohtake, F., Ikuta, T., Matsushima, Y., Mimura, J., Pettersson, S., Pollenz, R. S., Sakaki, T., Hirokawa, T., Akiyama, T., Kurosumi, M., Poellinger, L., Kato, S., and Fujii-Kuriyama, Y. (2009). Aryl hydrocarbon receptor suppresses intestinal carcinogenesis in Apc$^{Min/+}$ mice with natural ligands. *Proceedings of the National Academy of Sciences of the United States of America*, *106*, 13481–13486.
64. Safe, S. and McDougal, A. (2002). Mechanism of action and development of selective aryl hydrocarbon receptor modulators for treatment of hormone-dependent cancers (review). *International Journal of Oncology*, *20*, 1123–1128.
65. Ahmad, A., Sakr, W. A., and Rahman, K. M. W. (2010). Anticancer properties of indole compounds: mechanism of apoptosis induction and role in chemotherapy. *Current Drug Targets*, *11*, 652–666.
66. Naik, R., Nixon, S., Lopes, A., Godfrey, K., Hatem, M. H., and Monaghan, J. M. (2006). A randomized phase II trial of indole-3-carbinol in the treatment of vulvar intraepithelial neoplasia. *International Journal of Gynecological Cancer*, *16*, 786–790.
67. Higdon, J. V., Delage, B., Williams, D. E., and Dashwood, R. H. (2007). Cruciferous vegetables and human cancer risk: epidemiologic evidence and mechanistic basis. *Pharmacological Research*, *55*, 224–236.

68. Weng, J., Tsai, C., Kulp, S. K., and Chen, C. (2008). Indole-3-carbinol as a chemopreventive and anti-cancer agent. *Cancer Letters, 262*, 153–163.

69. Pohjanvirta, R., Korkalainen, M., McGuire, J., Simanainen, U., Juvonen, R., Tuomisto, J. T., Unkila, M., Viluksela, M., Bergman, J., Poellinger, L., and Tuomisto, J. (2002). Comparison of acute toxicities of indolo[3,2-*b*]carbazole (ICZ) and 2,3,7,8-tetrachlorodibenzo-*p*-dioxin (TCDD) in TCDD-sensitive rats. *Food and Chemical Toxicology, 40*, 1023–1032.

70. Leibelt, D. A., Hedstrom, O. R., Fischer, K. A., Pereira, C. B., and Williams, D. E. (2003). Evaluation of chronic dietary exposure to indole-3-carbinol and absorption-enhanced 3,3′-diindolylmethane in Sprague-Dawley rats. *Toxicological Sciences, 74*, 10–21.

71. de Waard, P. W. J., Peijnenburg, A. A. C. M., Baykus, H., Aarts, J. M. M. J. G., Hoogenboom, R. L. A. P., van Schooten, F. J., and de Kok, T. M. C. M. (2008). A human intervention study with foods containing natural Ah-receptor agonists does not significantly show AhR-mediated effects as measured in blood cells and urine. *Chemico-Biological Interactions, 176*, 19–29.

72. Crowell, J. A., Page, J. G., Levine, B. S., Tomlinson, M. J., and Hebert, C. D. (2006). Indole-3-carbinol, but not its major digestive product 3,3′-diindolylmethane, induces reversible hepatocyte hypertrophy and cytochromes P450. *Toxicology and Applied Pharmacology, 211*, 115–123.

73. Wilker, C., Johnson, L., and Safe, S. (1996). Effects of developmental exposure to indole-3-carbinol or 2,3,7,8-tetrachlorodibenzo-*p*-dioxin on reproductive potential of male rat offspring. *Toxicology and Applied Pharmacology, 141*, 68–75.

74. Bennett, D. H., McKone, T. E., Evans, J. S., Nazaroff, W. W., Margni, M. D., Jolliet, O., and Smith, K. R. (2002). Defining intake fraction. *Environmental Science & Technology, 36*, 206A–211A.

75. Tainio, M., Sofiev, M., Hujo, M., Tuomisto, J. T., Loh, M., Jantunen, M. J., Karppinen, A., Kangas, L., Karvosenoja, N., Kupiainen, K., Porvari, P., and Kukkonen, J. (2009). Evaluation of the European population intake fractions for European and Finnish anthropogenic primary fine particulate matter emissions. *Atmospheric Environment, 43*, 3052–3059.

76. Ries, F. J., Marshall, J. D., and Brauer, M. (2009). Intake fraction of urban wood smoke. *Environmental Science & Technology, 43*, 4701–4706.

77. Hirai, Y., Sakai, S., Watanabe, N., and Takatsuki, H. (2004). Congener-specific intake fractions for PCDDs/DFs and Co-PCBs: modeling and validation. *Chemosphere, 54*, 1383–1400.

78. Tuomisto, J., Leino, O., Kiviranta, H., and Tuomisto, J. T. (2006). Use of intake fraction to improve dioxin risk assessment. *Toxicology Letters, 164*, S148–S149.

79. Berntssen, M. H. G., Giskegjerde, T. A., Rosenlund, G., Torstensen, B. E., and Lundebye, A. (2007). Predicting World Health Organization toxic equivalency factor dioxin and dioxin-like polychlorinated biphenyl levels in farmed Atlantic salmon (*Salmo salar*) based on known levels in feed. *Environmental Toxicology and Chemistry, 26*, 13–23.

80. Srogi, K. (2008). Levels and congener distributions of PCDDs, PCDFs and dioxin-like PCBs in environmental and human samples: a review. *Environmental Chemistry Letters, 6*, 1–28.

81. Patterson, D. G., Jr., Turner, W. E., Caudill, S. P., and Needham, L. L. (2008). Total TEQ reference range (PCDDs, PCDFs, cPCBs, mono-PCBs) for the US population 2001–2002. *Chemosphere, 73*, S261–S277.

82. Brouwer, A., Ahlborg, U. G., van Leeuwen, F. X. R., and Feeley, M. M. (1998). Report of the WHO working group on the assessment of health risks for human infants from exposure to PCDDs, PCDFs and PCBs. *Chemosphere, 37*, 1627–1643.

83. van Leeuwen, F. X. R., Feeley, M., Schrenk, D., Larsen, J. C., Farland, W., and Younes, M. (2000). Dioxins: WHO's tolerable daily intake (TDI) revisited. *Chemosphere, 40*, 1095–1101.

84. Colles, A., Koppen, G., Hanot, V., Nelen, V., Dewolf, M., Noel, E., Malisch, R., Kotz, A., Kypke, K., Biot, P., Vinkx, C., and Schoeters, G. (2008). Fourth WHO-coordinated survey of human milk for persistent organic pollutants (POPs): Belgian results. *Chemosphere, 73*, 907–914.

85. Noren, K. and Meironyte, D. (2000). Certain organochlorine and organobromine contaminants in Swedish human milk in perspective of past 20–30 years. *Chemosphere, 40*, 1111–1123.

86. Tuomisto, J. (2007). Safe food: crucial for child development. In *Children's Health and the Environment in Europe: A Baseline Assessment* (Dalbokova, D., Krzyzanowski, M., and Lloyd, S., Eds). WHO Regional Office for Europe, Copenhagen, pp. 88–97.

87. Hsu, J. F., Guo, Y. L., Yang, S. Y., and Liao, P. C. (2005). Congener profiles of PCBs and PCDD/Fs in Yucheng victims fifteen years after exposure to toxic rice-bran oils and their implications for epidemiologic studies. *Chemosphere, 61*, 1231–1243.

88. Tuomisto, J. T., Pekkanen, J., Kiviranta, H., Tukiainen, E., Vartiainen, T., and Tuomisto, J. (2004). Soft-tissue sarcoma and dioxin: a case–control study. *International Journal of Cancer, 108*, 893–900.

89. Tuomisto, J., Pekkanen, J., Kiviranta, H., Tukiainen, E., Vartiainen, T., Viluksela, M., and Tuomisto, J. T. (2005). Dioxin cancer risk: example of hormesis? *Dose Response, 3*, 332–341.

90. Barouki, R., Coumoul, X., and Fernandez-Salguero, P. M. (2007). The aryl hydrocarbon receptor, more than a xenobiotic-interacting protein. *FEBS Letters, 581*, 3608–3615.

91. Tijet, N., Boutros, P. C., Moffat, I. D., Okey, A. B., Tuomisto, J., and Pohjanvirta, R. (2006). Aryl hydrocarbon receptor regulates distinct dioxin-dependent and dioxin-independent gene batteries. *Molecular Pharmacology, 69*, 140–153.

92. Okey, A. B. (2007). An aryl hydrocarbon receptor odyssey to the shores of toxicology: the Deichmann Lecture, International Congress of Toxicology – XI. *Toxicological Sciences, 98*, 5–38.

93. Hardell, L., Eriksson, M., and Degerman, A. (1994). Exposure to phenoxyacetic acids, chlorophenols, or organic solvents in relation to histopathology, stage, and anatomical localization of non-Hodgkin's lymphoma. *Cancer Research, 54*, 2386–2389.

94. Pesatori, A. C., Consonni, D., Rubagotti, M., Grillo, P., and Bertazzi, P. A. (2009). Cancer incidence in the population exposed to dioxin after the "Seveso accident": twenty years of follow-up. *Environmental Health*, 8, 39.
95. De Roos, A. J., Hartge, P., Lubin, J. H., Colt, J. S., Davis, S., Cerhan, J. R., Severson, R. K., Cozen, W., Patterson, D. G., Needham, L. L., and Rothman, N. (2005). Persistent organochlorine chemicals in plasma and risk of non-Hodgkin's lymphoma. *Cancer Research*, 65, 11214–11226.
96. Spinelli, J. J., Ng, C. H., Weber, J., Connors, J. M., Gascoyne, R. D., Lai, A. S., Brooks-Wilson, A. R., Le, N. D., Berry, B. R., and Gallagher, R. P. (2007). Organochlorines and risk of non-Hodgkin lymphoma. *International Journal of Cancer*, 121, 2767–2775.
97. Turunen, A. W., Verkasalo, P. K., Kiviranta, H., Pukkala, E., Jula, A., Männisto, S., Räsänen, R., Marniemi, J., and Vartiainen, T. (2008). Mortality in a cohort with high fish consumption. *International Journal of Epidemiology*, 37, 1008–1017.
98. Giacomini, S. M., Hou, L. F., Bertazzi, P. A., and Baccarelli, A. (2006). Dioxin effects on neonatal and infant thyroid function: routes of perinatal exposure, mechanisms of action and evidence from epidemiology studies. *International Archives of Occupational and Environmental Health*, 79, 396–404.
99. Wilhelm, M., Wittsiepe, J., Lemm, F., Ranft, U., Krämer, U., Fürst, P., Röseler, S., Greshake, M., Imöhl, M., Eberwein, G., Rauchfuss, K., Kraft, M., and Winneke, G. (2008). The Duisburg birth cohort study: influence of the prenatal exposure to PCDD/Fs and dioxin-like PCBs on thyroid hormone status in newborns and neurodevelopment of infants until the age of 24 months. *Mutation Research*, 659, 83–92.
100. Darnerud, P. O., Lignell, S., Glynn, A., Aune, M., Tornkvist, A., and Stridsberg, M. (2010). POP levels in breast milk and maternal serum and thyroid hormone levels in mother–child pairs from Uppsala, Sweden. *Environment International*, 36, 180–187.
101. Everett, C. J., Frithsen, I. L., Diaz, V. A., Koopman, R. J., Simpson, W. M., Jr., and Mainous, A. G. (2007). Association of a polychlorinated dibenzo-*p*-dioxin, a polychlorinated biphenyl, and DDT with diabetes in the 1999–2002 National Health and Nutrition Examination Survey. *Environmental Research*, 103, 413–418.
102. Chang, J., Chen, H., Su, H., Liao, P., Guo, H., and Lee, C. (2010). Dioxin exposure and insulin resistance in Taiwanese living near a highly contaminated area. *Epidemiology*, 21, 56–61.
103. Vartiainen, T., Jaakkola, J. J. K., Saarikoski, S., and Tuomisto, J. (1998). Birth weight and sex of children and the correlation to the body burden of PCDDs/PCDFs and PCBs of the mother. *Environmental Health Perspectives*, 106, 61–66.
104. Heiler, J., Donnez, J., and Lison, D. (2008). Organochlorines and endometriosis: a mini-review. *Chemosphere*, 71, 203–210.
105. Niskar, A. S., Needham, L. L., Rubin, C., Turner, W. E., Martin, C. A., Patterson, D. G., Jr., Hasty, L., Wong, L., and Marcus, M. (2009). Serum dioxins, polychlorinated biphenyls, and endometriosis: a case–control study in Atlanta. *Chemosphere*, 74, 944–949.
106. Simsa, P., Mihalyi, A., Schoeters, G., Koppen, G., Kyama, C. M., Den Hond, E. M., Fueloep, V., and D'Hooghe, T. M. (2010). Increased exposure to dioxin-like compounds is associated with endometriosis in a case–control study in women. *Reproductive Biomedicine Online*, 20, 681–688.
107. Cok, I., Donmez, M. K., Satiroglu, M. H., Aydinuraz, B., Henkelmann, B., Shen, H., Kotalik, J., and Schramm, K. (2008). Concentrations of polychlorinated dibenzo-*p*-dioxins (PCDDs), polychlorinated dibenzofurans (PCDFs), and dioxin-like PCBs in adipose tissue of infertile men. *Archives of Environmental Contamination and Toxicology*, 55, 143–152.
108. Krysiak-Baltyn, K., Toppari, J., Skakkebaek, N. E., Jensen, T. S, Virtanen, H. E., Schramm, K.-W., Shen, H., Vartiainen, T., Kiviranta, H., Taboureau, O., Audouze, K., Brunak, S., and Main, K.-M.: Association between chemical pattern in breast milk and congenital cryptorchidism: modeling of complex human exposures (submitted).
109. Virtanen, H. E., Koskenniemi, J. J., Sundqvist, E., Main, K. M., Kiviranta, H., Tuomisto, J. T., Tuomisto, J., Viluksela, M., Vartiainen, T., Skakkebaek, N. E., and Toppari J.: Associations between congenital cryptorchidism in newborn boys and levels of dioxins and PCBs in placenta (submitted).

24

AHR-ACTIVE COMPOUNDS IN THE HUMAN DIET

STEPHEN SAFE, GAYATHRI CHADALAPAKA, AND INDIRA JUTOORU

24.1 INTRODUCTION

The aryl hydrocarbon receptor (AHR) is a ligand-activated transcription factor that forms a heterodimeric complex with the AHR nuclear translocator (Arnt) protein. This ligand-bound complex interacts with cis-acting dioxin-responsive elements (DREs) in target gene promoters or with other nuclear factors to modulate gene expression [1, 2]. The AHR was initially discovered and characterized by Poland et al. using the environmental toxicant 2,3,7,8-tetrachlorodibenzo-p-dioxin (TCDD) and structurally related halogenated aromatics as model receptor ligands [3]. Radiolabeled TCDD and other radiolabeled analogs bind the AHR and there was a rank order correlation between structure–binding and structure–activity relationships for TCDD and several isosteric halogenated aromatic polychlorinated dibenzo-p-dioxins (PCDDs), polychlorinated dibenzofurans (PCDFs), and polychlorinated biphenyls (PCBs) [4, 5]. Figure 24.1 illustrates some of the most avid halogenated aromatic AHR ligands; these compounds are toxic and cause the classical AHR-mediated toxicities that include body weight loss, thymic atrophy and immunotoxicities, porphyria, and reproductive/developmental problems [1, 2]. In addition, AHR ligands also induce highly characteristic biochemical responses including the induction of *CYP1A1* gene expression and CYP1A1-dependent activities (ethoxyresorufin-O-deethylase (EROD) activity). As indicated in other chapters, the structure–activity relationships among halogenated aromatic pollutants and industrial by-products led to the development of the toxic equivalency factor (TEF) approach for estimating the TCDD equivalents (TEQs) for mixtures of halogenated aromatics that have been identified on extracts of environmental samples, fish, wildlife, and humans [6, 7]. The TEF approach has been widely used for risk assessment and management of complex mixtures of environmental contaminants and TEQ-based regulations have been used to decrease emissions of these compounds. TEQs for dietary intakes of halogenated aromatics have been extensively published; however, as indicated below, the potential adverse human health impacts of halogenated aromatics (TEQs) are a complex and debatable concept that will be discussed in more detail at the end of this chapter.

24.2 THE CONCEPT OF AHR ANTAGONISTS AND SELECTIVE AHR MODULATORS

The nuclear receptor (NR) family of ligand-activated transcription factors has 48 members including the steroid hormone receptors that have a ligand-dependent activation pathway similar to that of the AHR [8, 9]. Steroid hormone receptors are important drug targets and both agonists and antagonists have been developed for treating various hormone-related diseases [10]. For example, tamoxifen was identified as an estrogen receptor (ER) antagonist in the human breast and this antiestrogenic drug has been used by millions of women worldwide to treat early-stage hormone-responsive breast cancer [10–13]. It was also observed in many of the early studies with tamoxifen that this compound exhibited species- and tissue-specific ER antagonist and agonist activities. The ER agonist activity of tamoxifen in humans is associated with the increased risk of endometrial cancer in women who have had prolonged use of tamoxifen for treatment of breast cancer [14].

Based on the experience with NRs, research in this laboratory initially focused on development of AHR antagonists that could block the toxic responses induced by AHR

The AH Receptor in Biology and Toxicology, First Edition. Edited by Raimo Pohjanvirta.
© 2012 John Wiley & Sons, Inc. Published 2012 by John Wiley & Sons, Inc.

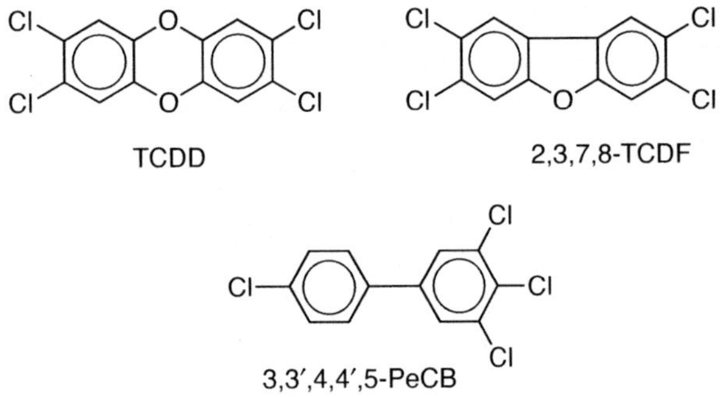

FIGURE 24.1 Structures of TCDD, 2,3,7,8-tetrachlorodibenzofuran (2,3,7,8-TCDF), and 3,3′,4,4′,5-pentachlorobiphenyl (3,3′,4,4′,5-PeCB).

agonists such as TCDD and related halogenated aromatics. Since the most toxic AHR agonists among the PCDDs and PCDFs contain four lateral (2, 3, 7, and 8) chloro substituents (Fig. 24.1), it was hypothesized that compounds with only three or two lateral chlorines plus the addition of a "metabolizable" group (e.g., methyl) should interact with and block the AHR with minimal induction of AHR agonist activity. Several alternate-substituted (1, 3, 6, 8 or 2, 4, 6, 8) PCDFs containing a methyl group were identified as potential AHR antagonists and the prototypical compound used was 6-methyl-1,3,8-trichlorodibenzofuran (MCDF) (Fig. 24.2). Results of several studies showed that MCDF inhibited induction of CYP1A1 by TCDD in cell culture and *in vivo*, and using a radiolabeled analog of MCDF it was shown that MCDF binds the AHR to form a transcriptionally inactive nuclear AHR complex. Subsequent studies show that MCDF also inhibits TCDD-induced immunotoxicity, porphyria, teratogenicity, and other toxic responses demonstrating classical AHR antagonist activity [15–21].

Another well-characterized AHR-mediated response has been the inhibition of 17β-estradiol (E2)-induced effects in breast and endometrial cancer cells, the rodent uterus, and rodent mammary tumors [22–25]. The mechanisms of ligand (TCDD)-activated inhibitory AHR–ERα crosstalk are complex [26], and initial studies using MCDF showed that this compound did not block TCDD-induced antiestrogenic activity in the rodent uterus but instead MCDF acted as an AHR agonist [27–30]. Subsequent studies have demonstrated that MCDF resembles TCDD as an antiestrogen in multiple tissues/species and this compound was a potent inhibitor of mammary tumor growth in carcinogen-induced Sprague Dawley rats [31, 32]. The development of this compound as a potential drug for treatment of breast cancer represented one of the first examples of a selective AHR modulator (SAHRM) that exhibits tissue/species-specific AHR antagonist and agonist activities [33, 34]. α-Naphthoflavone (ANF) was also identified as an AHR antagonist particularly for inhibition of ligand-induced CYP1A1 [35]. However, in some cell lines it has been shown that low concentrations of ANF ($\leq 1.0\,\mu M$) exhibit AHR antagonist activity whereas $10\,\mu M$ ANF in MCF-7 breast cancer cells was a full AHR agonist for induction of CYP1A1 [36]. 3-Methoxy-4′-nitroflavone (MNF) was initially identified during a screen of synthetic flavonoids for AHR agonist and antagonist activities and in MCF-7 human breast cancer cells MNF inhibited TCDD-induced CYP1A1 expression/activity and blocked nuclear accumulation of the bound AHR complex [37]. Other reports show that several methoxyflavone derivatives also exhibit AHR antagonist activity; however, there is evidence that MNF can be both an AHR antagonist and agonist [38]. A more recent study has characterized 2-methyl-2H-pyrazole-3-carboxylic acid (2-methyl-4-o-tolylazo-phenyl)-amide (CH-223191) as a "pure" AHR antagonist [39]. CH-223191 inhibited TCDD-induced CYP1A1 expression in both *in vitro* and *in vivo* models and also inhibited TCDD-induced hepatotoxicity and the wasting syndrome in rats. The potential AHR agonist activity of this compound (or lack thereof) remains to be determined.

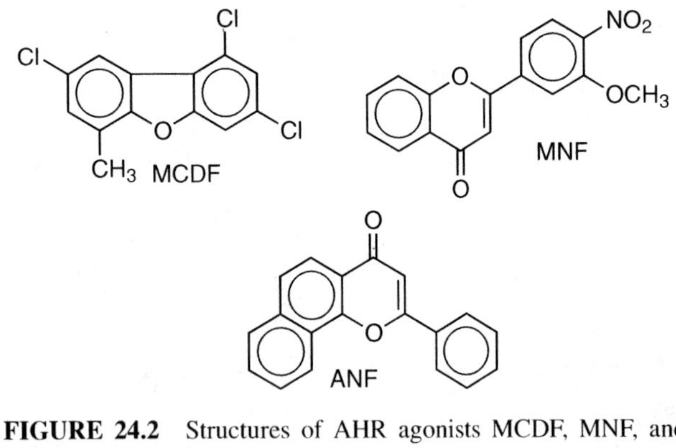

It is apparent that most compounds initially identified as AHR antagonists also exhibit tissue/species-specific AHR agonist activities and thereby represent examples of SAHRMs [33, 34]. These observations are consistent with results obtained for tamoxifen that is a selective ER modulator (SERM) and ligands that bind other steroid hormone receptors and NRs have been characterized as selective receptor modulators (SRMs) [10–12, 40–43]. The extensive development of SRMs for different receptors is based on their tissue-specific receptor agonist and antagonist activities and their many applications as therapeutic agents. The effects of any receptor ligand including compounds that bind and activate the AHR depend on several factors, namely (i) the unique ligand-induced conformational change in the receptor, (ii) interaction of the bound receptor complex with *cis*-elements, (iii) changes in *cis*-element sequences and modifications of promoter DNA, (iv) recruitment of coactivator, corepressor, and other coregulatory proteins, and (v) other factors that affect gene expression [40] (Fig. 24.3). Changes in one or more of these parameters dictate the tissue-specific agonist or antagonist activity of a receptor ligand.

FIGURE 24.2 Structures of AHR agonists MCDF, MNF, and ANF.

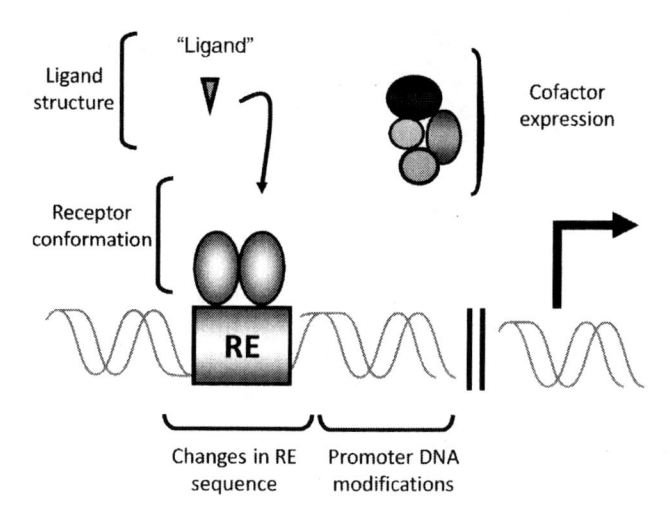

FIGURE 24.3 Factors that affect the tissue-specific receptor agonists and antagonist activity of a ligand (adapted from Ref. 40).

This concept of a SRM or SAHRM is essential for understanding the activity of diet-derived AHR ligands and their potential adverse or health promoting effects.

24.3 NATURALLY OCCURRING AHR AGONISTS AND ANTAGONISTS IN THE DIET

Initial studies on the AHR primarily focused on TCDD, other halogenated aromatic pollutants (e.g., Fig. 24.1), and also polycyclic aromatic hydrocarbons such as 3-methylcholanthrene (3-MC) and benzo[a]pyrene and their interactions with the AHR. However, research over the past 20–25 years has identified an increasingly large number of compounds that bind and/or activate or deactivate the AHR and these include endogenous biochemicals, industrial compounds, pesticides, pharmaceutical agents, and natural products that are part of the human diet (Figs. 24.4 and 24.5) [44–46]. Most of these AHR ligands bind with moderate or low affinity to the AHR compared to TCDD and for those compounds that are AHR agonists their potencies as inducers of CYP1A1-dependent activities are usually >100 to 1000 times lower than TCDD. In the following discussion, the AHR agonist and antagonist activities of dietary compounds will be discussed; however, it should be noted that the assignment of activities is provisional due to the limited number of assays that have been used in many of these studies. Most reports on AHR agonist activities of various chemicals have examined the effects of individual compounds on induction of CYP1A1-dependent responses that also include activation of AHR-responsive CYP1A1 promoter–reporter or DREx–reporter ($x = 2$ to 5) constructs transiently or stably transfected into various cancer cell lines. The reporter genes are usually luciferase or green fluorescent protein and these assays can also be used to detect AHR agonists or antagonists by

FIGURE 24.4 Phytochemicals that exhibit AHR agonist and/or antagonist activities.

FIGURE 24.5 AHR-active drugs and other compounds that are also associated with human exposures.

determining inhibition of TCDD-induced responses. In addition, other assays such as ligand-induced AHR transformation or inhibition of TCDD-induced transformation of the AHR complex followed by analysis and quantitation of the transformed complexes in gel mobility shift assays can be used. The designation of any dietary-derived AHR agonist/antagonist as a SAHRM requires multiple tissue- and species-specific end points; however, as indicated below even among the dietary compounds there is some evidence that they exhibit SAHRM-like activity.

24.3.1 I3C, DIM, and Related Compounds

Consumption of cruciferous vegetables has been associated with health benefits including decreased risks for some cancers and this activity has been linked to several phytochemicals including isothiocyanates, indole-3-carbinol (I3C), and diindolylmethane (DIM). Bjeldanes et al. first reported that I3C, DIM, and related I3C condensation products competitively bound the AHR and their relative binding affinities, compared to TCDD [10], were 2.6×10^{-7} and 7.8×10^{-5} for I3C and DIM, respectively [47]. Based on the relatively weak AHR binding affinities for I3C and DIM, these compounds were investigated as partial AHR antagonists in T47D breast cancer cells treated with TCDD. The results showed a concentration-dependent inhibition of TCDD-induced CYP1A1-dependent responses, and both I3C and DIM also inhibited formation of a radiolabeled nuclear AHR complex in T47D cells treated with [^{3}H]-TCDD [48]. These results were complemented by more recent studies in MCF-7 cells that show that DIM (up to 20 µM) inhibited AHR–XRE binding and induction of COX-2 mRNA levels in cells treated with TCDD [49]. Like ANF, DIM exhibits concentration-dependent AHR antagonist/agonist activities: at concentrations <30 µM antagonist activity is observed; however, as the concentrations increase (50 and 100 µM), DIM becomes a full AHR agonist [50]. Several other dimerization products of I3C and other tryptophan-derived nitrogen-containing heteroaromatics also exhibit AHR agonist activities [47]. These results demonstrate that I3C and DIM are both AHR agonists and antagonists and can be classified as SAHRMs.

24.3.2 Flavonoids and Polyphenolics

ANF, BNF, and various methoxyflavones have been characterized as AHR antagonists and agonists; however, the naturally occurring bioflavonoids that widely occur in fruits, vegetables, and other plants constitute a major dietary source of AHR ligands. Flavonoids have been characterized as both AHR antagonists and agonists and there has been considerable variability in the results due to the different assays used [51–62]. Ashida et al. [51] have extensively used inhibition of TCDD-induced transformation of rat hepatic cytosol followed by gel mobility shift assays as a screening tool for examining AHR-mediated effects of flavonoids. Their results showed that flavones (chrysin, baicailein, apigenin, luteolin, tangeretin, and luteolin glucoside), flavonols (galangin, kaempferol, fisetin, morin, quercetin, myrcetin, tamarixetin, isorahmietin, and quercetin glycosides), and flavonones (naringenin, eriodictoyl, hesperitin, and naringin) all inhibited TCDD-induced transformation of rat hepatic cytosol with IC$_{50}$ values ranging from 0.22 µM (galangin) to 28 µM (luteolin glucoside). Moreover, among these compounds only chrysin and quercetin induce transformation of rat hepatic cytosol. A competitive AHR binding (displacement) assay was also used to screen flavonoids and other phytochemicals [52]. There was a good correspondence between the results of the transformation and binding assays; however, the quercetin glycosides and naringin (also a glycoside) were inactive in the binding assays.

Several studies have more thoroughly examined the AHR agonist and antagonist activities of individual bioflavonoids. For example, diosmin, diosmetin, and quercetin induced CYP1A1 mRNA and protein levels in MCF-7 cells, whereas kaempferol (in MCF-7 cells) and galangin (in Hepa-1 mouse cancer cells) exhibit prototypical AHR antagonist activities [56, 57]. Studies in this laboratory [62] showed that among 13 bioflavonoids only chrysin and baicalein were active in Hepa-1, HepG2, and MCF-7 cells; genistein, daidzein, and apigenin were also active in Hepa-1 but not the other cell lines. Cell context-dependent differences were also observed for four of these compounds as AHR antagonists: luteolin was an AHR antagonist in both HepG2 and MCF-7 cells; myricetin, kaempferol, and quercetin were AHR antagonist in MCF-7 (but not HepG2) cells. A recent study has also demonstrated that the AHR agonist/antagonist activities of quercetin, chrysin, and genistein in three different cell lines (T47D, H4IIE, HepG2) depend on time, concentration, species, and cell context [53].

Paraoxonase-1 (*Pon-1*) is an Ah-responsive gene that has multiple activities including cardioprotection, and in HuH7 human hepatoma cells 3-MC induced *Pon-1* gene expression whereas TCDD was a relatively weak agonist [63]. Surprisingly, quercetin and other polyphenolics also induced *Pon-1* and this was an unusual example of AHR-dependent induction in which ligand potencies were different from the "normal" potencies (e.g., CYP1A1 induction), where TCDD is the most active compound. The authors showed that the ligand potency differences for induction of *Pon-1* versus *CYP1A1* were associated with their promoter DRE sequences in which the consensus pentanucleotide core DRE sequence in the *Pon-1* promoter differed by one base from the *CYP1A1* DRE (GCGTG) that has been identified in several Ah-responsive gene promoters. These results demonstrate the promoter-dependent effects (Fig. 24.3) in determining the activity of AHR agonists and also confirm that bioflavonoids are SAHRMs.

Curcumin, dibenzoylmethane (DBM), and green tea polyphenolics such as epigallocatechin gallate (EGCG) and epigallocatechin (EGC) are potent antioxidants that have also been characterized as AHR antagonists [64–66]. Similar results have been obtained for black theaflavins and other pigments/components of tea leaves [67–69]. The precise mechanism of AHR inhibition by these compounds is complex; however, at least one pathway involves binding to heat shock protein 90 that induces an AHR conformation that does not bind consensus DREs [70, 71].

Curcumin is a component of turmeric species (spices) and has been extensively characterized as an anticancer agent that acts through multiple pathways. Curcumin is also an AHR agonist in MCF-7 and oral squamous cell carcinoma cells and induces *CYP1A1* gene expression and transforms the AHR [72, 73]. However, there is also evidence that curcumin exhibits AHR antagonist activities and these effects may be related to cell context, concentration, and other pathways affected by curcumin [72–75]. Dibenzoylmethane is a constituent of licorice that also exhibits partial AHR agonist and antagonist activities [76]. Thus, it is evident that the polyphenolic phytochemical compounds with known chemoprotective and anticancer activity also exhibit AHR agonist and antagonist activities and are SAHRMs. The role of their AHR-mediated responses in the health benefits of these compounds is unclear; however, it is likely that this activity will be a contributing factor.

24.3.3 Resveratrol

trans-3,4′,5-Trihydroxystilbene or resveratrol is another polyphenolic phytochemical found in red grapes, red wine, and other fruits that exhibits antioxidant, anti-inflammatory, and anticancer activities and there is an extensive literature on the putative health benefits of this compound. Several studies report that resveratrol is an AHR ligand and an AHR antagonist; however, the effects on the AHR pathway are highly variable among studies. For example, in human HepG2 liver cancer cells, resveratrol inhibited TCDD-induced *CYP1A1* gene expression and blocked formation of nuclear AHR binding activity but did not competitively displace [^{3}H]-TCDD from the AHR *in vitro* [77]. Similar results for resveratrol were observed by the same group using benzo[*a*]pyrene as an AHR agonist [78]. In contrast, another report showed that resveratrol directly bound the AHR, and in T47D human breast cancer cells resveratrol induced formation of a nuclear receptor complex that bound DNA but did not induce transcription (of *CYP1A1*) [79]. We also observed that resveratrol inhibited CYP1A1 via a posttranscriptional pathway [80]. A recent study used the ChIP assay to show that resveratrol inhibited TCDD-induced recruitment of the AHR complex and pol II to the *CYP1A1* and *CYP1B1* gene promoters in multiple cell lines [81]. Despite these differences in the AHR agonist activities of resveratrol, there is evidence that this compound inhibits the classical TCDD-induced teratogenic responses such as cleft palate, renal pelvic dilation, and uteric dilatations in mice [82]. Resveratrol also induced AHR agonist activity in experimental autoimmune encephalomyelitis (EAE)-induced mice used as a model for studying multiple sclerosis. Resveratrol induced AHR-dependent apoptosis in activated T cells in this model and the protective effects of resveratrol are similar to the AHR-dependent protective effects reported for TCDD in the EAE mouse model [83, 84].

24.3.4 Plant Extracts

Plant herbal food extracts have also been investigated for their AHR agonist and antagonist activities. Herbal extracts from ginseng, white oak bark, ginko balboa, licorice, fo-ti, and black cohosh stimulated AHR transformation/DNA binding and induced DRE-dependent gene expression. Among 21 food extracts, only corn and jalapeno pepper extract induced AHR transformation and gene expression [84]. Another extensive assay on 39 food extracts focused on their inhibition of AHR activation (Ah immunoassay and DRE-dependent luciferase) and several extracts exhibited AHR antagonist activity [85]. Other plant/herbal extracts also exhibit AHR antagonist activities [86–90] suggesting that these extracts resemble their constitutive components, some of which have been discussed in this chapter, namely, that they exhibit AHR antagonist and agonist activities and thereby resemble SAHRMs.

24.4 HUMAN EXPOSURES TO OTHER AHR-ACTIVE COMPOUNDS

24.4.1 Exposures to Indigoids, Drugs, and Endogenous Compounds

The presentation in this chapter has highlighted human exposure to TCDD and related compounds and the vast array of natural AHR-active phytochemicals in foods. In addition, there are several other sources of human exposure to AHR-active compounds that include pharmaceutical agents and endogenous biochemicals. For example, indirubin is a component of Chinese medicines and indirubin, indigo, and related compounds have been widely used in dyes. These compounds have been detected in human urine and have been characterized as AHR agonists [91, 92]. In addition, there is increasing evidence that many drugs are also AHR-active and these include NSAIDs such as sulindac and salicylamide, oltipraz, omeprazole, benzothiazole derivatives, and retinoids; these compounds exhibit both AHR antagonist and agonist activities [93–98]. A recent study demonstrated that 4-hydroxytamoxifen, the active metabolite of tamoxifen, was an AHR agonist, and other SERMs have also been identified as SAHRMs [99]. The results observed for tamoxifen and its metabolite 4-hydroxytamoxifen were particularly important since tamoxifen is one of the most successful and widely used anticancer drugs and has been prescribed for millions of breast cancer patients. Hu et al. [100] identified 596 drugs for which there was some evidence for induction of CYP1A1; however, evaluation of a subset (147) of these agents showed that 59% of the compounds that induced CYP1A1 *in vivo* did not bind or activate the AHR *in vitro*. Moreover, further evaluation of a subset of these drugs that exhibited AHR agonist activity showed that they did not induce "dioxin-like" toxicity.

24.4.2 Endogenous Biochemicals

Endogenous biochemicals that are found in body tissues and serum and as urinary metabolites are another important source of exposure to AHR-active compounds. Some of these biochemicals include the heme degradation products bilirubin and biliverdin (agonists), 7-ketocholesterol (antagonist), kynurenic acid (agonist), prostaglandins (agonists), leukotriene A4 (agonists), 12(R)-hydroxy-5,8,10,14-eicosatetraenoic acid (12(R)-HETE) (agonist), and the tryptophan photoproduct 6-formylindolo[3,2-b]-carbazole (FICZ) (agonist) [101–107]. Although both TCDD and FICZ typically induce CYP1A1 in several assay systems, FICZ and TCDD exhibit different AHR-dependent activities in the EAE mouse model where TCDD suppressed and FICZ enhanced EAE through differential modulation of T_{reg} cells [83, 108] demonstrating that FICZ is also a SAHRM.

24.4.3 Other Persistent Environmental Contaminants

The development of TEFs and their use for determination of mixture TEQs has been useful for estimating TCDD equivalents for complex mixtures of halogenated aromatic compounds; however, the potential adverse impacts on animal and human health associated with TEQs have been questioned [109, 110]. Human exposures to halogenated aromatic hydrocarbons (HAHs) such as TCDD and those congeners that have been assigned TEF values [6, 7] are invariably accompanied by other AHR-inactive HAHs that include many PCBs such as 2,2′,4,4′,5,5′-hexachlorobiphenyl (PCB 153). Several studies have investigated interactions between AHR-active HAHs such as TCDD or 3,3′,4,4′,5-PeCB and AHR-inactive HAHs including PCB 153 and there are numerous examples of antagonist interactions [111–127]. Commercial PCB mixtures including Aroclor 1254 and PCB 153, a dominant PCB congener in environmental samples, inhibit several TCDD-induced responses including CYP1A1, immunotoxicity, and cleft palate [111–116] and the antagonistic effects of Aroclor 1254 are observed at Aroclor 1254/TCDD ratios that are similar to ratios of total PCBs/(PCDD + PCDF) (TEQ) in serum and blood samples from a Japanese cohort [128]. Antagonism of TCDD-induced responses by polybrominated diphenyl ethers (PBDEs) has also been reported [129, 130]; however, the precise mechanism of HAHs as inhibitors of TCDD-induced responses is unclear. Nevertheless, the presence of these compounds in environmental and human samples complicates and compromises the TEF/TEQ approach for risk assessment of HAHs.

24.4.4 Polynuclear Aromatic Hydrocarbons

Polynuclear aromatic hydrocarbons (PAHs) such as benzo[a] pyrene were among the first chemicals identified as AHR agonists [131]; these compounds typically induce CYP1A1. However, their induction of HAH-mediated toxic responses is minimal due to their rapid metabolism. Nevertheless, there is considerable human exposure to PAHs from the air and cooked or barbecued meat.

24.5 OVERALL IMPACTS OF DIETARY INTAKES OF AHR-ACTIVE CHEMICALS

The major human health concerns regarding AHR-active compounds in the diet have focused on PCDDs, PCBs, and PCDF, and among these congeners the estimated intake is generally <5 pg/kg body weight per day in most countries [132, 133]. However, the amounts can be significantly increased for individuals consuming high levels of fish or other products contaminated with higher levels of HAHs. As previously indicated, the intake of HAH TEQs is less than the estimated 735,000,000 pg of I3C associated with consumption of cruciferous vegetables or the $(1.2–5) \times 10^6$ pg of PAHs in cooked foods [110]. Moreover, up to 1 g of flavonoids may be consumed each day [134]. A recent study reported HAH TEQ blood levels of approximately 20 pg/g lipid and this value can be compared to levels of the endogenous AHR antagonist 7-ketocholesterol where plasma concentrations range from 20 to 200 μM or 8,000,000 to 80,000,000 pg/mL [135]. Serum levels of flavonoids can be in the low μM concentrations and it was suggested that the levels of these compounds may protect against dioxin-like TEQs [51]. In most studies, the potential interactions between different structural classes of AHR-active compounds have been estimated based on *in vitro* biotransformation (of the AHR) or CYP1A1 induction assays. However, this approach is complicated by the fact that many AHR-active compounds in the diet are SAHRMs and may exhibit tissue-specific AHR agonist or antagonist activities. Moreover, there is some evidence from *in vitro* studies that there are biological differences even among TCDD and related halogenated aromatics [136]. The TEF/TEQ approach for risk assessment of dioxin-like compounds also assumes that the contribution of each congener is additive. It should be evident from the results presented in this chapter that human health applications of the TEQs for halogenated aromatics are problematic due to the relatively high exposure to other AHR agonists and antagonists in the diet.

ACKNOWLEDGMENTS

The financial support of the Syd Kyle Chair endowment and the National Institutes of Health (CA142697) is greatly acknowledged.

REFERENCES

1. Gu, Y. Z., Hogenesch, J. B., and Bradfield, C. A. (2000). The PAS superfamily: sensors of environmental and developmental signals. *Annual Review of Pharmacology and Toxicology*, 40, 519–561.
2. Wilson, C. L. and Safe, S. (1998). Mechanisms of ligand-induced aryl hydrocarbon receptor-mediated biochemical and toxic responses. *Toxicologic Pathology*, 26, 657–671.
3. Poland, A., Glover, E., and Kende, A. S. (1976). Stereospecific, high affinity binding of 2,3,7,8-tetrachlorodibenzo-p-dioxin by hepatic cytosol: evidence that the binding species is receptor for induction of aryl hydrocarbon hydroxylase. *Journal of Biological Chemistry*, 251, 4936–4946.
4. Safe, S. (1990). Polychlorinated biphenyls (PCBs), dibenzo-p-dioxins (PCDDs), dibenzofurans (PCDFs) and related compounds: environmental and mechanistic considerations which support the development of toxic equivalency factors (TEFs). *CRC Critical Reviews in Toxicology*, 26, 371–399.
5. Safe, S. (1986). Comparative toxicology and mechanism of action of polychlorinated dibenzo-p-dioxins and dibenzofurans. *Annual Review of Pharmacology and Toxicology*, 26, 371–399.
6. Van den Berg, M., Birnbaum, L., Bosveld, A. T. C., et al. (1998). Toxic equivalency factors (TEFs) for PCBs, PCDDs, PCDFs for humans and wildlife. *Environmental Health Perspectives*, 106, 775–792.
7. Van den Berg, M., Birnbaum, L. S., Denison, M., et al. (2006). The 2005 World Health Organization reevaluation of human and mammalian toxic equivalency factors for dioxins and dioxin-like compounds. *Toxicological Sciences*, 93, 223–241.
8. Olefsky, J. M. (2001). Nuclear receptor minireview series. *Journal of Biological Chemistry*, 276, 36863–36864.
9. O'Malley, B. W. (2005). A life-long search for the molecular pathways of steroid hormone action. *Molecular Endocrinology*, 19, 1402–1411.
10. Jordan, V. C. (2009). A century of deciphering the control mechanisms of sex steroid action in breast and prostate cancer: the origins of targeted therapy and chemoprevention. *Cancer Research*, 68, 1243–1254.
11. Jordan, V. C. (2008). The 38th David A. Karnofsky lecture: the paradoxical actions of estrogen in breast cancer—survival or death? *Journal of Clinical Oncology*, 26, 3073–3082.
12. Jordan, V. C. and O'Malley, B. W. (2007). Selective estrogen-receptor modulators and antihormonal resistance in breast cancer. *Journal of Clinical Oncology*, 25, 5815–5824.
13. Jordan, V. C. (2003). Tamoxifen: a most unlikely pioneering medicine. *Nature Reviews Drug Discovery*, 3, 205–213.
14. Bernstein, L., Deapen, D., Cerhan, J. R., Schwartz, S. M., Liff, J., McGann-Maloney, E., Perlman, J. A., and Ford, L. (1999). Tamoxifen therapy for breast cancer and endometrial cancer risk. *Journal of National Cancer Institute*, 91, 1654–1662.
15. Astroff, B. and Safe, S. (1989). 6-Substituted-1,3,8-trichlorodibenzofurans as 2,3,7,8-tetrachlorodibenzo-p-dioxin antagonists in the rat: structure–activity relationships. *Toxicology*, 33, 231–236.
16. Astroff, B., Zacharewski, T., Safe, S., Arlotto, M. P., Parkinson, A., Thomas, P., and Levin, W. (1988). 6-Methyl-1,3,8-trichlorodibenzofuran as a 2,3,7,8-tetrachlorodibenzo-p-dioxin antagonist: inhibition of the induction of rat cytochrome P-450 isozymes and related monooxygenase activities. *Molecular Pharmacology*, 33, 231–236.
17. Harris, M., Zacharewski, T., Astroff, B., and Safe, S. (1989). Partial antagonism of 2,3,7,8-tetrachlorodibenzo-p-dioxin-mediated induction of aryl hydrocarbon hydroxylase by 6-methyl-1,3,8-trichlorodibenzofuran: mechanistic studies. *Molecular Pharmacology*, 35, 729–735.
18. Bannister, R., Biegel, L., Davis, D., Astroff, B., and Safe, S. (1989). 6-Methyl-1,3,8-trichlorodibenzofuran (MCDF) as a 2,3,7,8-tetrachlorodibenzo-p-dioxin antagonist in C57BL/6 mice. *Toxicology*, 54, 139–150.
19. Yao, C. and Safe, S. (1989). 2,3,7,8-Tetrachlorodibenzo-p-dioxin-induced porphyria in genetically inbred mice: partial antagonism and mechanistic studies. *Toxicology and Applied Pharmacology*, 100, 208–216.
20. Piskorska-Pliszczynska, J., Astroff, B., Zacharewski, T., Harris, M., Rosengren, R., Morrison, V., Safe, L., and Safe, S. (1991). Mechanism of action of 2,3,7,8-tetrachlorodibenzo-p-dioxin antagonists: characterization of 6-[^{125}I]methyl-8-iodo-1,3-dichlorodibenzofuran–Ah receptor complexes. *Archives of Biochemistry and Biophysics*, 284, 193–200.
21. Santostefano, M., Piskorska-Pliszczynska, J., Morrison, V., and Safe, S. (1992). Effects of ligand structure on the *in vitro* transformation of the rat cytosolic aryl hydrocarbon receptor. *Archives of Biochemistry and Biophysics*, 297, 73–79.
22. Kociba, R. J., Keyes, D. G., Beger, J., et al. (1978). Results of a 2-year chronic toxicity and oncogenicity study of 2,3,7,8-tetrachlorodibenzo-p-dioxin (TCDD) in rats. *Toxicology and Applied Pharmacology*, 46, 279–303.
23. Umbreit, T. H., Hesse, E. J., Macdonald, G. J., and Gallo, M. A. (1988). Effects of TCDD–estradiol interactions in three strains of mice. *Toxicology Letters*, 40, 1–9.
24. Romkes, M. and Safe, S. (1988). Comparative activities of 2,3,7,8-tetrachlorodibenzo-p-dioxin and progesterone as antiestrogens in the female rat uterus. *Toxicology and Applied Pharmacology*, 92, 368–380.
25. Romkes, M., Piskorska-Pliszczynska, J., and Safe, S. (1987). Effects of 2,3,7,8-tetrachlorodibenzo-p-dioxin on hepatic and

26. Safe, S. and Wormke, M. (2003). Inhibitory aryl hydrocarbon receptor–estrogen receptor alpha cross-talk and mechanisms of action. *Chemical Research in Toxicology*, 16, 807–816.
27. Astroff, B. and Safe, S. (1991). 6-Alkyl-1,3,8-trichlorodibenzofurans as antiestrogens in female Sprague-Dawley rats. *Toxicology*, 69, 187–197.
28. Zacharewski, T., Harris, M., Biegel, L., Morrison, V., Merchant, M., and Safe, S. (1992). 6-Methyl-1,3,8-trichlorodibenzofuran (MCDF) as an antiestrogen in human and rodent cancer cell lines: evidence for the role of the Ah receptor. *Toxicology and Applied Pharmacology*, 113, 311–318.
29. Sun, G. and Safe, S. (1997). Antiestrogenic activities of alternate-substituted polychlorinated dibenzofurans in MCF-7 human breast cancer cells. *Cancer Chemotherapy and Pharmacology*, 40, 239–244.
30. Dickerson, R., Keller, L. H., and Safe, S. (1995). Alkyl polychlorinated dibenzofurans and related compounds as antiestrogens in the female rat uterus: structure–activity studies. *Toxicology and Applied Pharmacology*, 135, 287–298.
31. McDougal, A., Wilson, C., and Safe, S. (1997). Inhibition of 7,12-dimethylbenz[a]anthracene-induced rat mammary tumor growth by aryl hydrocarbon receptor agonists. *Cancer Letters*, 120, 53–63.
32. McDougal, A., Wormke, M., Calvin, J., and Safe, S. (2001). Tamoxifen-induced antitumorigenic/antiestrogenic action synergized by a selective Ah receptor modulator. *Cancer Research*, 61, 3901–3907.
33. Safe, S., Qin, C., and McDougal, A. (1999). Development of selective aryl hydrocarbon receptor modulators (SARMs) for treatment of breast cancer. *Expert Opinion on Investigational Drugs*, 8, 1385–1396.
34. Safe, S. (1992). MCDF. *Drugs of the Future*, 17, 564–565.
35. Gasiewicz, T. A. and Rucci, G. (1991). α-Naphthoflavone acts as an antagonist of 2,3,7,8-tetrachloro-dibenzo-p-dioxin by forming an inactive complex with the Ah receptor. *Molecular Pharmacology*, 40, 607–612.
36. Merchant, M., Krishnan, V., and Safe, S. (1993). Mechanism of action of alpha-naphthoflavone as an Ah receptor antagonist in MCF-7 human breast cancer cells. *Toxicology and Applied Pharmacology*, 120, 179–185.
37. Lu, Y. F., Santostefano, M., Cunningham, B. D., Threadgill, M. D., and Safe, S. (1996). Substituted flavones as aryl hydrocarbon (Ah) receptor agonists and antagonists. *Biochemical Pharmacology*, 51, 1077–1087.
38. Zhou, J. and Gasiewicz, T. A. (2003). 3′-Methoxy-4′-nitroflavone, a reported aryl hydrocarbon receptor antagonist, enhances CYP1A1 transcription by a dioxin responsive element-dependent mechanism. *Archives of Biochemistry and Biophysics*, 416, 68–80.
39. Kim, S. H., Henry, E. C., Kim, D. K., Kim, Y. H., Shin, K. J., Han, M. S., Lee, T. G., Kang, J. K., Gasiewicz, T. A., Ryu, S. H., and Suh, P. G. (2006). Novel compound 2-methyl-2H-pyrazole-3-carboxylic acid (2-methyl-4-o-tolylazo-phenyl)-amide (CH-223191) prevents 2,3,7,8-TCDD-induced toxicity by antagonizing the aryl hydrocarbon receptor. *Molecular Pharmacology*, 69, 1871–1878.
40. Katzenellenbogen, J. A., O'Malley, B. W., and Katzenellenbogen, B. A. (1996). Tripartite steroid hormone receptor pharmacology interaction with multiple effector sites as a basis for the cell- and promoter-specific action of these hormones. *Molecular Endocrinology*, 10, 119–131.
41. Smith, C. L. and O'Malley, B. W. (2004). Coregulator function: a key to understanding tissue specificity of selected receptor modulators. *Endocrine Reviews*, 25, 45–71.
42. Klinge, C.-M. (2000). Estrogen receptor interactions with co-activators and corepressors. *Steroids*, 65, 227–251.
43. Krishnan, V., Heath, H., and Bryant, H. U. (2000). Mechanism of action of estrogens and selective estrogen receptor modulators. *Vitamins & Hormones*, 60, 123–147.
44. Seidel, S. D., Li, V., Winter, G. M., Rogers, W. J., Martinez, E. I., and Denison, M. S. (2000). Ah receptor-based chemical screening bioassays application and limitations for the detection of Ah receptor of Ah receptor agonists. *Toxicological Sciences*, 55, 107–115.
45. Gasiewicz, T. A., Kende, A. S., Rucci, G., Whitney, B., and Wiley, J. J (1996). Analysis of structural requirements for Ah receptor antagonist activity: ellipticines, flavones, and related compounds. *Biochemical Pharmacology*, 52, 1787–1803.
46. Denison, M. S., Seidel, S. D., Rogers, W. J., et al. (1998). Natural and synthetic ligands for the Ah receptor. *Molecular Biology Approaches to Toxicology*, 3–33.
47. Bjeldanes, L. F., Kim, J. Y., Grose, K. R., Bartholomew, J. C., and Bradfield, C. A. (1991). Aromatic hydrocarbon responsiveness-receptor agonists generated from indole-3-carbinol *in vitro* and *in vivo*: comparisons with 2,3,7,8-tetrachlorodibenzo-p-dioxin. *Proceedings of the National Academy of Sciences of the United States of America*, 88, 9543–9547.
48. Chen, I., Safe, S., and Bjeldanes, L. (1996). Indole-3-carbinol and diindolylmethane as aryl hydrocarbon (Ah) receptor agonists and antagonists in T47D human breast cancer cells. *Biochemical Pharmacology*, 51, 1069–1076.
49. Degner, S. C., Papoutsis, A. J., Selmin, O., and Romagnolo, F. (2009). Targeting of aryl hydrocarbon receptor-mediated activation of cyclooxygenase-2 expression by the indole-3-carbinol metabolite 3,3′-diindolylmethane in breast cancer cells. *Journal of Nutrition*, 139, 26–32.
50. Chen, I., McDougal, A., Wang, F., and Safe, S. (1998). Aryl hydrocarbon receptor-mediated antiestrogenic and antitumorigenic activity of diindolylmethane. *Carcinogenesis*, 19, 1631–1639.
51. Ashida, H., Fukuda, I., Yamashita, T., and Kanazawa, K. (2000). Flavones and flavonols at dietary levels inhibit a transformation of aryl hydrocarbon receptor induced by dioxin. *FEBS Letters*, 476, 213–217.
52. Amakura, Y., Tsutsumi, T., Sasaki, K., Yoshida, T., and Maitani, T. (2003). Screening of the inhibitory effect of vegetable constituents on the aryl hydrocarbon receptor-mediated activity induced by 2,3,7,8-tetrachlorodibenzo-p-dioxin. *Biological & Pharmaceutical Bulletin*, 26, 1754–1760.

53. Van der Heiden, E., Bechoux, N., Muller, M., Sergent, T., Schneider, Y. J., Larondelle, Y., Maghuin-Rogister, G., and Scippo, M. L. (2009). Food flavonoid aryl hydrocarbon receptor-mediated agonistic/antagonistic/synergic activities in human and rat reporter gene assays. *Analytica Chimica Acta*, *637*, 337–345.

54. Fukuda, I., Mukai, R., Kawase, M., Yoshida, K., and Ashida, H. (2007). Interaction between the aryl hydrocarbon receptor and its antagonists, flavonoids. *Biochemical and Biophysical Research Communications*, *359*, 822–827.

55. Quadri, S. A., Qadri, A. N., Hahn, M. E., Mann, K. K., and Sherr, D. H. (2000). The bioflavonoid galangin blocks aryl hydrocarbon receptor activation and polycyclic aromatic hydrocarbon-induced pre-B cell apoptosis. *Molecular Pharmacology*, *58*, 515–525.

56. Ciolino, H. P., Wang, T. T., and Yeh, G. C. (1998). Diosmin and diosmetin are agonists of the aryl hydrocarbon receptor that differentially affect cytochrome P450 1A1 activity. *Cancer Research*, *58*, 2754–2760.

57. Ciolino, H. P., Daschner, P. J., and Yeh, G. C. (1999). Dietary flavonols quercetin and kaempferol are ligands of the aryl hydrocarbon receptor that affect CYP1A1 transcription differentially. *Biochemical Journal*, *340*, 715–722.

58. Allen, S. W., Mueller, L., Williams, S. N., Quattrochi, L. C., and Raucy, J. (2001). The use of a high-volume screening procedure to assess the effects of dietary flavonoids on human CYP1A1 expression. *Drug Metabolism and Disposition*, *29*, 1074–1079.

59. Han, E. H., Kim, J. Y., and Jeong, H. G. (2006). Effect of biochanin A on the aryl hydrocarbon receptor and cytochrome P450 1A1 in MCF-7 human breast carcinoma cells. *Archives of Pharmacal Research*, *29*, 570–576.

60. Puppala, D., Gairola, C. G., and Swanson, H. I. (2007). Identification of kaempferol as an inhibitor of cigarette smoke-induced activation of the aryl hydrocarbon receptor and cell transformation. *Carcinogenesis*, *28*, 639–647.

61. Mukai, R., Satsu, H., Shimizu, M., and Ashida, H. (2009). Inhibition of P-glycoprotein enhances the suppressive effect of kaempferol on transformation of the aryl hydrocarbon receptor. *Bioscience, Biotechnology, and Biochemistry*, *73*, 1635–1639.

62. Zhang, S., Qin, C., and Safe, S. H. (2003). Flavonoids as aryl hydrocarbon receptor agonists/antagonists: effects of structure and cell context. *Environmental Health Perspectives*, *111*, 1877–1882.

63. Gouédard, C., Barouki, R., and Morel, Y. (2004). Dietary polyphenols increase paraoxonase 1 gene expression by an aryl hydrocarbon receptor-dependent mechanism. *Molecular and Cellular Biology*, *24*, 5209–5222.

64. Williams, S. N., Shih, H., Guenette, D. K., Brackney, W., Denison, M. S., Pickwell, G. V., and Quattrochi, L. C. (2000). Comparative studies on the effects of green tea extracts and individual tea catechins on human CYP1A gene expression. *Chemico-Biological Interactions*, *128*, 211–229.

65. Chan, H. Y., Wang, H., Tsang, D. S., Chen, Z. Y., and Leung, L. K. (2003). Screening of chemopreventive tea polyphenols against PAH genotoxicity in breast cancer cells by a XRE-luciferase reporter construct. *Nutrition and Cancer*, *46*, 93–100.

66. Palermo, C. M., Hernando, J. I., Dertinger, S. D., Kende, A. S., and Gasiewicz, T. A. (2003). Identification of potential aryl hydrocarbon receptor antagonists in green tea. *Chemical Research in Toxicology*, *16*, 865–872.

67. Fukuda, I., Sakane, I., Yabushita, Y., Sawamura, S., Kanazawa, K., and Ashida, H. (2005). Black tea theaflavins suppress dioxin-induced transformation of the aryl hydrocarbon receptor. *Bioscience, Biotechnology, and Biochemistry*, *69*, 883–890.

68. Hung, Y. C., Huang, G. S., Sava, V. M., Blagodarsky, V. A., and Hong, M. Y. (2006). Protective effects of tea melanin against 2,3,7,8-tetrachlorodibenzo-p-dioxin-induced toxicity: antioxidant activity and aryl hydrocarbon receptor suppressive effect. *Biological & Pharmaceutical Bulletin*, *29*, 2284–2291.

69. Fukuda, I., Sakane, I., Yabushita, Y., Kodoi, R., Nishiumi, S., Kakuda, T., Sawamura, S., Kanazawa, K., and Ashida, H. (2004). Pigments in green tea leaves (*Camellia sinensis*) suppress transformation of the aryl hydrocarbon receptor induced by dioxin. *Journal of Agricultural and Food Chemistry*, *52*, 2499–2506.

70. Palermo, C. M., Westlake, C. A., and Gasiewicz, T. A. (2005). Epigallocatechin gallate inhibits aryl hydrocarbon receptor gene transcription through an indirect mechanism involving binding to a 90 kDa heat shock protein. *Biochemistry*, *44*, 5041–5052.

71. Hughes, D., Guttenplan, J. B., Marcus, C. B., Subbaramaiah, K., and Dannenberg, A. J. (2008). Heat shock protein 90 inhibitors suppress aryl hydrocarbon receptor-mediated activation of CYP1A1 and CYP1B1 transcription and DNA adduct formation. *Cancer Prevention Research*, *1*, 485–493.

72. Ciolino, H. P., Daschner, P. J., Wang, T. T., and Yeh, G. C. (1998). Effect of curcumin on the aryl hydrocarbon receptor and cytochrome P450 1A1 in MCF-7 human breast carcinoma cells. *Biochemical Pharmacology*, *56*, 197–206.

73. Rinaldi, A. L., Morse, M. A., Fields, H. W., Rothas, D. A., Pei, P., Rodrigo, K. A., Renner, R. J., and Mallery, S. R. (2002). Curcumin activates the aryl hydrocarbon receptor yet significantly inhibits (—)-benzo(*a*)pyrene-7*R*-trans-7,8-dihydrodiol bioactivation in oral squamous cell carcinoma cells and oral mucosa. *Cancer Research*, *62*, 5451–5456.

74. Nishiumi, S., Yoshida, K., and Ashida, H. (2007). Curcumin suppresses the transformation of an aryl hydrocarbon receptor through its phosphorylation. *Archives of Biochemistry and Biophysics*, *466*, 267–273.

75. Goergens, A., Frericks, M., and Esser, C. (2009). The arylhydrocarbon receptor is only marginally involved in the antileukemic effects of its ligand curcumin. *Anticancer Research*, *29*, 4657–4664.

76. MacDonald, C. J., Ciolino, H. P., and Yeh, G. C. (2001). Dibenzoylmethane modulates aryl hydrocarbon receptor function and expression of cytochromes P50 1A1, 1A2, and 1B1. *Cancer Research*, *61*, 3919–3924.

77. Ciolino, H. P., Daschner, P. J., and Yeh, G. C. (1998). Resveratrol inhibits transcription of CYP1A1 in vitro by preventing activation of the aryl hydrocarbon receptor. *Cancer Research*, 58, 5707–5712.

78. Ciolino, H. P. and Yeh, G. C. (1999). Inhibition of aryl hydrocarbon-induced cytochrome P-450 1A1 enzyme activity and CYP1A1 expression by resveratrol. *Molecular Pharmacology*, 56, 760–767.

79. Casper, R. F., Quesne, M., Rogers, I. M., Shirota, T., Jolivet, A., Milgrom, E., and Savouret, J. F. (1999). Resveratrol has antagonist activity on the aryl hydrocarbon receptor: implications for prevention of dioxin toxicity. *Molecular Pharmacology*, 56, 784–790.

80. Lee, J. E. and Safe, S. (2001). Involvement of a post-transcriptional mechanism in the inhibition of CYP1A1 expression by resveratrol in breast cancer cells. *Biochem. Pharmacol.*, 15, 1113–1124.

81. Beedanagari, S. R., Bebenek, I., Bui, P., and Hankinson, O. (2009). Resveratrol inhibits dioxin-induced expression of human CYP1A1 and CYP1B1 by inhibiting recruitment of the aryl hydrocarbon receptor complex and RNA polymerase II to the regulatory regions of the corresponding genes. *Toxicological Sciences*, 110, 61–67.

82. Jang, J. Y., Park, D., Shin, S., Jeon, J. H., Choi, B. I., Joo, S. S., Hwang, S. Y., Nahm, S. S., and Kim, Y. B. (2008). Antiteratogenic effect of resveratrol in mice exposed in utero to 2,3,7,8-tetrachlorodibenzo-p-dioxin. *European Journal of Pharmacology*, 591, 280–283.

83. Quintana, F. J., Basso, A. S., Iglesias, A. H., Korn, T., Farez, M. F., Bettelli, E., Caccamo, M., Oukka, M., and Weiner, H. L. (2008). Control of T_{reg} and T_H17 cell differentiation by the aryl hydrocarbon receptor. *Nature*, 453, 65–71.

84. Jeuken, A., Keser, B. J., Khan, E., Brouwer, A., Koeman, J., and Denison, M. S. (2003). Activation of the Ah receptor by extracts of dietary herbal supplements, vegetables, and fruits. *Journal of Agricultural and Food Chemistry*, 51, 5478–5487.

85. Amakura, Y., Tsutsumi, T., Nakamura, M., Kitagawa, H., Fujino, J., Sasaki, K., Yoshida, T., and Toyoda, M. (2002). Preliminary screening of the inhibitory effect of food extracts on activation of the aryl hydrocarbon receptor induced by 2,3,7,8-tetrachlorodibenzo-p-dioxin. *Biological & Pharmaceutical Bulletin*, 25, 272–274.

86. van Ede, K., Li, A., Antunes-Fernandes, E., Mulder, P., Peijnenburg, A., and Hoogenboom, R. (2008). Bioassay directed identification of natural aryl hydrocarbon-receptor agonists in marmalade. *Analytica Chimica Acta*, 617, 238–245.

87. Collins, N. H., Lessey, E. C., DuSell, C. D., McDonnell, D. P., Fowler, L., Palomino, W. A., Illera, M. J., Yu, X., Mo, B., Houwing, A. M., and Lessey, B. A. (2009). Characterization of antiestrogenic activity of the Chinese herb, *Prunella vulgaris*, using in vitro and in vivo (mouse xenograft) models. *Biology of Reproduction*, 80, 375–383.

88. Nishiumi, S., Yabushita, Y., Fukuda, I., Mukai, R., Yoshida, K., and Ashida, H. (2006). Molokhia (*Corchorus olitorius* L.) extract suppresses transformation of the aryl hydrocarbon receptor induced by dioxins. *Food and Chemical Toxicology*, 44, 250–260.

89. Park, Y. K., Fukuda, I., Ashida, H., Nishiumi, S., Yoshida, K., Daugsch, A., Sato, H. H., and Pastore, G. M. (2005). Suppressive effects of ethanolic extracts from propolis and its main botanical origin on dioxin toxicity. *Journal of Agricultural and Food Chemistry*, 53, 10306–10309.

90. Mukai, R., Fukuda, I., Nishiumi, S., Natsume, M., Osakabe, N., Yoshida, K., and Ashida, H. (2008). Cacao polyphenol extract suppresses transformation of an aryl hydrocarbon receptor in C57BL/6 mice. *Journal of Agricultural and Food Chemistry*, 56, 10399–10405.

91. Adachi, J., Mori, Y., Matsui, S., Takigami, H., Fujino, J., Kitagawa, H., Miller, C. A.,3rd, Kato, T., Saeki, K., and Matsuda, T. (2001). Indirubin and indigo are potent aryl hydrocarbon receptor ligands present in human urine. *Journal of Biological Chemistry*, 276, 31475–31478.

92. Guengerich, F. P., Martin, M. V., McCormick, W. A., Nguyen, L. P., Glover, E., and Bradfield, C. A. (2004). Aryl hydrocarbon receptor response to indigoids in vitro and in vivo. *Archives of Biochemistry and Biophysics*, 423, 309–316.

93. Ciolino, H. P., MacDonald, C. J., Memon, O. S., Bass, S. E., and Yeh, G. C. (2006). Sulindac regulates the aryl hydrocarbon receptor-mediated expression of Phase 1 metabolic enzymes in vivo and in vitro. *Carcinogenesis*, 27, 1586–1592.

94. Macdonald, C. J., Ciolino, H. P., and Yeh, G. C. (2004). The drug salicylamide is an antagonist of the aryl hydrocarbon receptor that inhibits signal transduction induced by 2,3,7,8-tetrachlorodibenzo-p-dioxin. *Cancer Research*, 64, 429–434.

95. Cho, J. I. and Kim, S. G. (2003). Oltipraz inhibits 3-methylcholanthrene induction of CYP1A1 by CCAAT/enhancer-binding protein activation. *Journal of Biological Chemistry*, 278, 44103–44112.

96. Soprano, D. R., Gambone, C. J., Sheikh, S. N., Gabriel, J. L., Chandraratna, R. A. S., Sprano, D. J., and Kochhar, D. M. (2001). The synthetic retinoid AGN 193109 but not retinoic acid elevates CYP1A1 levels in mouse embryos and Hepa-1c1c7 cells. *Toxicology and Applied Pharmacology*, 174, 153–159.

97. Dzeletovic, N., McGuire, J., Daujat, M., Tholander, J., Emal, M., Fujii-Kuriyama, Y., Bergman, J., Maurel, P., and Poellinger, L. (1997). Regulation of dioxin receptor function by omeprazole. *Journal of Biological Chemistry*, 272, 12705–12713.

98. Loaiza-Pérez, A., Trapani, V., Hose, C., Singh, S. S., Trepel, J. B., Stevens, M. F., Bradshaw, T. D., and Sausville, E. A. (2002). Aryl hydrocarbon receptor mediates sensitivity of MCF-7 breast cancer cells to antitumor agent 2-(4-amino-3-methylphenyl) benzothiazole. *Molecular Pharmacology*, 61, 13–19.

99. DuSell, C. D., Nelson, E. R., Wittmann, B. M., Fretz, J. A., Kazmin, D., Thomas, R. S., Pike, J. W., and McDonnell, D. P. (2010). Regulation of aryl hydrocarbon receptor function by selective estrogen receptor modulators. *Molecular Endocrinology*, 24, 33–46.

100. Hu, W., Sorrentino, C., Denison, M. S., Kolaja, K., and Fielden, M. R. (2007). Induction of CYP1A1 is a nonspecific biomarker of aryl hydrocarbon receptor activation: results of

large scale screening of pharmaceuticals and toxicants *in vivo* and *in vitro*. *Molecular Pharmacology*, *71*, 1475–1486.

101. Sinal, C. J. and Bend, J. R. (1997). Aryl hydrocarbon receptor-dependent induction of Cyp1a1 by bilirubin in mouse hepatoma Hepa 1c1c7 cells. *Molecular Pharmacology*, *52*, 590–599.

102. Phelan, D., Winter, G. M., Rogers, W. J., Lam, J. C., and Denison, M. S. (1998). Activation of the Ah receptor signal transduction pathway by bilirubin and biliverdin. *Archives of Biochemistry and Biophysics*, *357*, 155–163.

103. Seidel, S. D., Winters, G. M., Rogers, W. J., Ziccardi, M. H., Li, V., Keser, B., and Denison, M. S. (2001). Activation of the Ah receptor signaling pathway by prostaglandins. *Journal of Biochemical and Molecular Toxicology*, *15*, 187–196.

104. Chiaro, C. R., Patel, R. D., and Perdew, G. H. (2008). 12(R)-Hydroxy-5(Z),8(Z),10(E),14(Z)-eicosatetraenoic acid [12(R)-HETE], an arachidonic acid derivative, is an activator of the aryl hydrocarbon receptor. *Molecular Pharmacology*, *74*, 1649–1656.

105. Chiaro, C. R., Morales, J. L., Prabhu, K. S., and Perdew, G. H. (2008). Leukotriene A4 metabolites are endogenous ligands for the Ah receptor. *Biochemistry*, *47*, 8445–8455.

106. DiNatale, B. C., Murray, I. A., Schroeder, J. C., Flaveny, C. A., Lahoti, T. S., Laurenzana, E. M., Omiecinski, C. J., and Perdew, G. H. (2010). Kynurenic acid is a potent endogenous aryl hydrocarbon receptor ligand that synergistically induces interleukin-6 in the presence of inflammatory signaling. *Toxicological Sciences*, *115*, 89–97.

107. Wincent, E., Amini, N., Luecke, S., Glatt, H., Bergman, J., Crescenzi, C., Rannug, A., and Rannug, U. (2009). The suggested physiologic aryl hydrocarbon receptor activator and cytochrome P4501 substrate 6-formylindolo[3,2-b]carbazole is present in humans. *Journal of Biological Chemistry*, *284*, 2690–2696.

108. Veldhoen, M., Hirota, K., Westendorf, A. M., Buer, J., Dumoutier, L., Renauld, J. C., and Stockinger, B. (2008). The aryl hydrocarbon receptor links T_H17-cell-mediated autoimmunity to environmental toxins. *Nature*, *453*, 106–169.

109. Safe, S. H. (1998). Development, validation and problems with the TEF approach for risk assessment of dioxins and related compounds. *Journal of Animal Science*, *76*, 134–141.

110. Safe, S. (1998). Limitations of the toxic equivalency factor approach for risk assessment of TCDD and related compounds. *Teratogenesis, Carcinogenesis, and Mutagenesis*, *17*, 285–304.

111. Bannister, R., Davis, D., Zacharewski, T., Tizard, I., and Safe, S. (1987). Aroclor 1254 as a 2,3,7,8-tetrachlorodibenzo-*p*-dioxin antagonist: effects on enzyme activity and immunotoxicity. *Toxicology*, *46*, 29–42.

112. Haake, J. M., Safe, S., Mayura, K., and Phillips, T. D. (1987). Aroclor 1254 as an antagonist of the teratogenicity of 2,3,7,8-tetrachlorodibenzo-*p*-dioxin. *Toxicology Letters*, *38*, 299–306.

113. Biegel, L., Harris, M., Davis, D., Rosengren, R., Safe, L., and Safe, S. (1989). 2, 2′,4,4′,5,5′-Hexachlorobiphenyl as a 2,3,7,8-tetrachlorodibenzo-*p*-dioxin antagonist in C57BL/6J mice. *Toxicology and Applied Pharmacology*, *97*, 561–571.

114. Davis, D. and Safe, S. (1989). Dose–response immunotoxicities of commercial polychlorinated biphenyls (PCBs) and their interaction with 2,3,7,8-tetrachlorodibenzo-*p*-dioxin. *Toxicology Letters*, *48*, 35–43.

115. Biegel, L., Howie, L., and Safe, S. (1989). Polychlorinated biphenyl (PCB) congeners as 2,3,7,8-TCDD antagonists: teratogenicity studies. *Chemosphere*, *19*, 955–958.

116. Davis, D. and Safe, S. (1990). Immunosuppressive activities of polychlorinated biphenyls in C57BL/6 mice: structure–activity relationships as Ah receptor agonists and partial antagonists. *Toxicology*, *63*, 97–111.

117. Morrissey, R. E., Harris, M. W., Diliberto, J. J., and Birnbaum, L. S. (1992). Limited PCB antagonism of TCDD-induced malformations in mice. *Toxicology Letters*, *60*, 19–25.

118. Harper, N., Connor, K., Steinberg, M., and Safe, S. (1995). Immunosuppressive activity of polychlorinated biphenyl mixtures and congeners: non-additive (antagonistic) interactions. *Fundamental and Applied Toxicology*, *27*, 131–139.

119. Zhao, F., Mayura, K., Kocurek, N., Edwards, J. F., Kubena, L. F., Safe, S. H., and Phillips, T. D. (1997). Inhibition of 3,3′,4,4′,5-pentachlorobiphenyl-induced chicken embryotoxicity by 2,2′,4,4′,5,5′-hexachlorobiphenyl. *Fundamental and Applied Toxicology*, *35*, 1–8.

120. Zhao, F., Mayura, K., Harper, N., Safe, S. H., and Phillips, T. D. (1997). Inhibition of 3,3′,4,4′,5-pentachlorobiphenyl-induced fetal cleft palate and immunotoxicity in C57BL/6 mice by 2,2′,4,4′,5,5′-hexachlorobiphenyl. *Chemosphere*, *34*, 1605–1613.

121. Tysklind, M., Boxveld, A. T. C., Andersson, P., Verhallen, E., Sinnige, T., Seinen, W., Rappe, C., and Van den Berg, M. (1995). Inhibition of ethoxyresorufin-O-deethylase (EROD) activity in mixtures of 2,3,7,8-tetrachlorodibenzo-*p*-dioxin and polychlorinated biphenyls. *Environmental Science & Pollution Research*, *4*, 211–216.

122. Davis, D. and Safe, S. (1988). Immunosuppressive activities of polychlorinated dibenzofuran congeners: quantitative structure–activity relationships and interactive effects. *Toxicology and Applied Pharmacology*, *94*, 141–149.

123. Harris, G. E., Metcalfe, T. L., Metcalfe, C. D., and Huestis, S. Y. (1995). Embryotoxicity of extracts from Lake Ontario rainbow trout (*Oncorhynchus mykiss*) to Japanese medaka (*Oryzias latipes*). *Environmental Toxicology & Chemistry*, *13*, 1393–1403.

124. Keys, B., Piskorska-Pliszczynska, J., and Safe, S. (1986). Polychlorinated dibenzofurans as 2,3,7,8-TCDD antagonists: *in vitro* inhibition of monooxygenase enzyme induction. *Toxicology Letters*, *31*, 151–158.

125. Bosveld, A. T. C., Verhallen, E., Seinen, W., and Van den Berg, M. (1995). Mixture interactions in the *in vitro* CYP1A1 induction bioassay using chicken embryo hepatocytes. *Organohalogen Compounds*, *25*, 309–312.

126. Aarts, J. M., Denison, M. S., Cox, M. A., Schalk, M. A., Garrison, P. M., Tullis, K., de Haan, L. H., and Brouwer, A. (1995). Species-specific antagonism of Ah receptor action by 2,2′,5,5′-tetrachloro- and 2,2′,3,3′,4,4′-hexachlorobiphenyl. *European Journal of Pharmacology*, *293*, 463–474.

127. Suh, J., Kang, J. S., Yang, K. H., and Kaminski, N. E. (2003). Antagonism of aryl hydrocarbon receptor-dependent induction of CYP1A1 and inhibition of IgM expression by di-*ortho*-substituted polychlorinated biphenyls. *Toxicology and Applied Pharmacology*, 187, 11–21.

128. Yoshioka, E., Yuasa, M., Kishi, R., Iida, T., and Furue, M. (2010). Relationship between the concentrations of polychlorinated dibenzo-*p*-dioxins, polychlorinated dibenzofurans, and polychlorinated biphenyls in maternal blood and those in breast milk. *Chemosphere*, 78, 185–192.

129. Chen, G. and Bunce, N. J. (2003). Polybrominated diphenyl ethers as Ah receptor agonists and antagonists. *Toxicological Sciences*, 76, 310–320.

130. Peters, A. K., Sanderson, J. T., Bergman, A., and Van den Berg, M. (2006). Antagonism of TCDD-induced ethoxyresorufin-O-deethylation activity by polybrominated diphenyl ethers (PBDEs) in primary cynmolgus monkey (*Macaca fascicularis*) hepatocytes. *Toxicology Letters*, 164, 123–132.

131. Piskorska-Pliszczynska, J., Keys, B., Safe, S., and Newman, M. S. (1986). The cytosolic receptor binding affinities and AHH induction potencies of 29 polynuclear aromatic hydrocarbons. *Toxicology Letters*, 34, 67–74.

132. Bilau, M., Matthys, C., Baeyens, W., Bruckers, L., De Backer, G., Den Hond, E., Keune, H., Koppen, G., Nelen, V., Schoeters, P., Van Larebeke, N, Willems, J. L., and De Henauw, S. (2008). Dietary exposure to dioxin-like compounds in three age groups: results from the Flemish environment and health study. *Chemosphere*, 70, 584–592.

133. Fromme, H., Albrecht, M., Boehmer, S., Buchner, K., Mayer, R., Liebl, B., Wittsiepe, J., and Bolte, G. (2009). Intake and body burden of dioxin-like compounds in Germany: the INES study. *Chemosphere*, 76, 1457–1463.

134. Verdeal, K. and Ryan, D. S. (1979). Naturally-occurring estrogens in plant foodstuffs: a review. *Journal of Food Protection*, 42, 577–583.

135. Dzeletovic, S., Breuer, O., Lund, E., and Diczfalusy, U. (1995). Determination of cholesterol oxidation products in human plasma by isotope dilution-mass spectrometry. *Analytical Biochemistry*, 225, 73–80.

136. Zhang, S., Rowlands, C., and Safe, S. (2008). Ligand-dependent interactions of the Ah receptor with coactivators in a mammalian two-hybrid assay. *Toxicology and Applied Pharmacology*, 227, 196–206.

25

MODULATION OF AHR FUNCTION BY HEAVY METALS AND DISEASE STATES

ANWAR ANWAR-MOHAMED AND AYMAN O. S. EL-KADI

25.1 INTRODUCTION

The aryl hydrocarbon receptor (AHR) is a cytosolic transcription factor that mediates many toxic and carcinogenic effects in animals and humans. The AHR is constitutively present in the cytosol as an inactive complex attached to two molecules of heat shock protein 90 (HSP90), the 23 kDa heat shock protein (p23), and a 43 kDa protein known as AHR inhibitory protein (AIP) or hepatitis B virus X-associated protein 2 (XAP2) [1]. HSP90 mainly serves as a chaperone with dual functions of preventing nuclear translocation and premature binding with DNA binding partners such as the aryl hydrocarbon nuclear translocator (ARNT) and keeping the AHR in a configuration that favors ligand binding [2]. The AIP or XAP2 enhances the sensitivity and magnitude of ligand-induced signaling through the AHR; moreover, it binds to both AHR and HSP90 stabilizing the AIP–HSP90–AHR complex [3]. The hydrophobic AHR inducers enter the cell by diffusion and bind to the AHR ligand binding domain. Upon ligand binding, the AHR–ligand complex dissociates from the cytoplasmic complex and translocates to the nucleus where it associates with ARNT [1]. The whole complex then acts as a transcription factor that binds to a specific DNA recognition sequence, termed the xenobiotic-responsive element (XRE), located in the promoter region of a number of AHR-regulated genes.

It is generally accepted that the activation of AHR in vertebrates causes the toxic and carcinogenic effects of a wide array of environmental contaminants such as 2,3,7,8-tetrachlorodibenzo-p-dioxin (TCDD), polychlorinated biphenyls (PCBs), and polycyclic and/or halogenated aromatic hydrocarbons (PAHs or HAHs) [4]. As a consequence of its activation, the AHR induces the transcription of many phase I and phase II genes. Among these genes are those encoding four phase I enzymes (cytochrome P450 1A1 (CYP1A1), CYP1A2, CYP1B1, and CYP2S1) and four phase II enzymes (NAD(P)H:quinone oxidoreductase-1 (NQO1), glutathione S-transferase A1 (GSTA1), cytosolic aldehyde dehydrogenase-3, and UDP-glucuronosyltransferase 1A6 (UGT1A6)) [1].

The AHR is a member of the basic helix–loop–helix (bHLH)/Per-ARNT-Sim (PAS) family of transcription proteins that have been shown to be involved in regulation of development, circadian rhythms, neurogenesis, metabolism, and stress response to hypoxia [5]. Moreover, recent studies using AHR knockout mice have shown additional functions to the AHR beyond xenobiotic metabolism. As such, ablation of AHR gene in mice leads to cardiovascular diseases, hepatic fibrosis, reduced liver size, teratogenicity, immune dysfunctions, nephropathy, and neurotoxicity suggesting the presence of biological functions that is likely contributing to the overall toxic response as a consequence of its activation [6].

In this chapter, we will present current knowledge about the effects of heavy metals, in particular, arsenic (As^{3+}), mercury (Hg^{2+}), lead (Pb^{2+}), cadmium (Cd^{2+}), chromium (Cr^{6+}), copper (Cu^{2+}), and vanadium (V^{5+}), on the AHR activity biomarkers, and the mutual interaction between different disease states and AHR.

The AH Receptor in Biology and Toxicology, First Edition. Edited by Raimo Pohjanvirta.
© 2012 John Wiley & Sons, Inc. Published 2012 by John Wiley & Sons, Inc.

25.2 ACTIVATION OF THE AHR

25.2.1 Ligand-Dependent Activation of the AHR

In vitro studies showed that the binding site for HSP90 within the AHR is overlapping the ligand binding site [7] and masking the AHR nuclear localization sequence (NLS) [8]. Because the three-dimensional structure of the AHR is not determined yet, quantitative structure–activity relationship studies are commonly used to gain insight into the nature of the ligand–receptor interactions. Theoretically, there are two hypotheses for AHR interaction with its ligands [9]. First, electrostatic interaction, in which effective interaction of the ligand with the receptor depends on the molecular electrostatic potential around the ligand [9]. For example, all dioxin compounds that were able to activate the AHR share a unique molecular charge distribution pattern, which was dramatically changed with the chlorination pattern [9]. The second hypothesis is based on molecular polarizability and the distance between the receptor and the ligand [9]. In this regard, it has been shown that the AHR pocket can bind planar ligands with maximum dimensions of $14 \text{Å} \times 12 \text{Å} \times 5 \text{Å}$, which mainly depends upon the ligand's electronic and thermodynamic features [10].

Although ARNT and the AHR of all species are about 20% identical in amino acid sequence, ARNT does not have any ligand binding capacity and therefore appears to be free from any repressive effect by HSP90 [11]. Some evidence suggests that ARNT promotes dissociation of the AHR–HSP90 complex and targets the AHR to its nuclear site of action [12].

25.2.2 Ligand-Independent Regulation of the AHR

There are several lines of evidence to support the presence of a ligand-independent regulation of the AHR. Using an antagonist similar to 3′-methoxy-4′-aminoflavone (MNF) that can block the induction of CYP1A1 by TCDD, it was shown that omeprazole can induce CYP1A1 despite the presence of the antagonist [13]. This specific mechanism has been shown to be tyrosine kinase dependent as it was blocked by herbimycin A, while it is not independent of the AHR, as the nuclear accumulation of the DNA binding of the AHR was observed [13]. Interestingly, the induction of CYP1A1 by TCDD was not affected by herbimycin A, suggesting two distinct pathways for AHR activation [14]. In light of the extensive evidence that pharmacological inhibitors mediate potential nonspecific effects, these studies must be interpreted with caution. α-Naphthoflavone is a partial agonist/antagonist depending on the concentration tested [15]. At low concentrations (<10 μM), α-naphthoflavone acts as an antagonist. However, at higher concentrations (>10 μM), α-naphthoflavone acts as an agonist with a reduced affinity to AHR [16]. This agonism was also prevented by herbimycin A [17]. A more direct evidence for the presence of ligand-independent regulation of AHR can be demonstrated by the constitutive expression of CYP1A1 in a number of human lung cancer cell lines despite the absence of an exogenous AHR ligand [18]. Importantly, one must be cautious in supporting such hypothesis as there might be an indirect generation of AHR ligands. To illustrate this line of thought, malassezin (2-(1*H*-indol-3-ylmethyl)-1*H*-indol-3-carbaldehyde) that is extracted from a strain of yeast is a nonclassical AHR ligand agonist by itself, but treatment with catalytic HCl liberates indolo[3,2-*b*]carbazole (ICZ) that itself is a classic AHR ligand agonist [19].

25.3 CROSSTALKS OF THE AHR SIGNALING PATHWAY

A number of divergent points of crosstalk of the AHR signaling pathway with other signal transduction pathways will be discussed in this section. These crosstalks can be classified into three major categories: (1) coactivators and corepressors involved in AHR regulation, (2) phosphorylation cascades involved in AHR regulation, and (3) crosstalks of AHR with other nuclear receptors.

25.3.1 Coactivators and Corepressors Involved in AHR Regulation

Data on AHR and other signal transduction pathways support a role for coactivators in crosstalk between signaling molecules. It has been well documented that the chromatin core is comprised of a pair of histone proteins. Coactivators play an essential role in remodeling chromatin structure and relieving the transcription repressive effects of nucleosomes [20]. Coactivators that increase histone acetylation of the chromatin result in transcriptional activation, whereas corepressors increase histone deacetylase activity, causing transcriptional repression. The AHR signaling pathway is modulated by several nuclear coactivators such as the CREB binding protein (CBP), p300, steroid receptor coactivators 1 and 2 (SRC-1 and -2), receptor interacting protein 140 (RIP140), estrogen receptor-associated protein of 140 kDa (ERAP140), silencing mediator for retinoic acid and thyroid hormone receptors (SMRT), and ATPase-dependent chromatin remodeling factors such as Brahma-related gene 1 (BRG-1) [21].

Functional interactions of AHR and estrogen receptor (ER) with ERAP140 and SMRT suggest a possible competition for limited pools of these coregulators, or a mutual inhibitory crosstalk between AHR and ER [22]. Two independent studies have shown that the corepressor SMRT directly interacts with AHR. However, due to the conflicting findings between the two studies, which might be due to cell-specific effects, it has become difficult to generalize an effect of SMRT on the AHR-dependent gene expression.

The coactivator RIP140 interacts directly with AHR in different cell types [23]. The recruitment of this coactivator by AHR or other transcription factors leads to enhanced XRE-driven luciferase reporter activity [23]. Studies using AHR-null embryonic fibroblasts showed that when these cells are transfected with AHR vector, the resultant AHR protein was bound to p300/CBP and was required for p300 DNA synthesis [24]. In 293T cells, AHR binds to different SRC-1 family proteins. Similarly, overexpression of different SRC-1 proteins increased the ligand-dependent XRE-driven luciferase reporter activity in Hepa 1c1c7 cells [25]. Importantly, competition for CBP did not seem to mediate the AHR crosstalks with ER and NF-κB [26]. However, the crosstalk between AHR and NF-κB is thought to be through mutual interactions with SRC-1, p300, and CBP coactivators leading to mutual inhibitory effect between the two transcription factors [27].

25.3.2 Phosphorylation Cascades Involved in AHR Regulation

Because AHR is a "phosphoprotein," its phosphorylation status plays an important role in modulating its activity [28]. Protein kinases (PKs), mitogen-activated protein kinases (MAPKs), and tyrosine kinases (TKs) have been implicated in AHR signaling [29]. Thus, studies investigating the phosphorylation of AHR may provide novel mechanism(s) for ligand-independent activation of AHR.

25.3.2.1 PKCs The first evidence supporting a positive interaction between AHR and PKC was demonstrated by studies on mice in which the inhibition of PKC antagonized ligand-activated AHR/ARNT DNA binding and subsequently decreased *Cyp1* gene expression [30]. Similarly, several laboratories have independently demonstrated that PKC activation increased XRE-driven reporter activity and blocking PKC activity with 12-O-tetradecanoylphorbol-13-acetate (TPA) decreased AHR-mediated induction of CYP1A1 and CYP1A2 mRNA and CYP1A1 activity [31]. Intriguingly, some early reports on PKC indicated that phosphorylation of one of the AHR core complex proteins AHR, ARNT, HSP90, p23, or XAP2 was necessary for AHR-mediated induction of CYP1A1 mRNA and AHR/ARNT binding to DNA. Moreover, inhibition of PKC by TPA or staurosporine in mice Hepa-1 cells, human keratinocytes, and MCF-7 cells decreased the TCDD-mediated induction of CYP1A1 mRNA and catalytic activity via inhibiting the AHR/ARNT DNA binding [31].

AHR contains a NLS composed of amino acid residues 13–16 and 37–39, and a nuclear export signal (NES) [32]. In contrast to the previous studies, it was shown that phosphorylation of NLS at Ser12 and Ser36 by PKC inhibited the nuclear accumulation of AHR [32]. In addition, when these Ser residues were substituted with Ala, AHR nuclear translocation was not affected by PKC-mediated phosphorylation, whereas replacement with Asp retained the mutant AHR in the cytoplasm [32]. Thus, despite the conflicting reports about the role of PKC in AHR regulation, it is apparent that AHR activation is tightly regulated by PKC that might be cell and species specific.

25.3.2.2 MAPKs MAPKs are serine threonine kinases involved in inflammatory responses, apoptosis, cell growth, and further mitogenic and developmental events. The three families of MAPKs are extracellular signal-regulated kinases (ERK1/2), c-Jun N-terminal/stress-activated protein kinases (JNK/SAPK), and the p38s; they are all important intracellular signal transduction mediators [33]. MAPK activities are controlled by MAPK kinase kinase–MAPK kinase (MAPKKK–MAPKK) signaling cascades, in which MAPKs are activated by MAPKK-dependent phosphorylation, and MAPKKs are activated by MAPKKK-dependent phosphorylation [33]. As a general rule, ERK1 and ERK2 are involved in regulating mitogenic and developmental events, and the four p38 isoforms are involved in inflammatory responses, apoptosis, and cell cycle regulation [34]. The three JNK isoforms are mainly involved in cellular signaling, immune system, stress-induced apoptosis, carcinogenesis, and diabetes [34].

Three different AHR ligands, TCDD, benzo[*a*]pyrene (B[*a*]P), and B[a]P diol epoxide, activate JNK in mouse Hepa-1 cells, human lung carcinoma A549 cells, AHR-deficient CV-1 cells, and AHR-positive and AHR-deficient mouse embryonic fibroblasts, suggesting that TCDD-mediated activation of MAPK is independent of AHR [35]. Conversely, TCDD-stimulated MAPKs appear to be important for the induction of CYP1A1 [35]. For example, TCDD-induced modulation of epithelial morphology causes the activation of JNK. These TCDD-mediated effects can be mimicked by constitutive expression of AHR [36]. Furthermore, ablation of JNK2 and ERK decreases TCDD-mediated induction of CYP1A1 mRNA in mouse thymus and testis [35]. It has been further noted that the induction of CYP1A1 and CYP1B1 mRNA and protein levels in response to UV radiation in human keratinocytes was partially due to JNK and p38 activation [37]. The interaction between AHR and ERK appears to be critically linked as ERK inhibitors were shown to prolong TCDD-induced AHR degradation [38]. Interestingly, in Hepa 1c1c7 cells it was shown that overexpression of ERK1 promoted AHR degradation, implying an important role of ERK in AHR proteolysis [38]. Furthermore, phosphorylation of Ser68 in AHR NES by p38 activated AHR export from the nucleus, prior to its degradation [39]. Moreover, constitutively active MEK1, which is a MAPKK upstream of ERK1/2, increased TCDD-mediated induction of CYP1A1 mRNA via AHR [40]. Based on the previous studies investigating the crosstalk between MAPKs and AHR, it can be concluded that AHR ligands contribute to the upregulation of several MAPKs that will consequently exert a positive effect on AHR nuclear accumulation and nuclear export.

25.3.2.3 TKs

Bombick et al. were the first to report an AHR-dependent modulation of tyrosine kinase [41]. In addition, several reports suggested a requirement for tyrosine phosphorylation in AHR transactivation potential [41]. Direct interactions of the AHR complex with pp60^{src}, a tyrosine kinase, were observed in mouse hepatic cytosol using a cell-free system [42]. Tyrosine phosphorylation has also been suggested as a requirement for AHR/ARNT complex DNA binding. For example, phosphorylation sites in two tyrosine domains of the C-terminus of the AHR are required for the formation of the functional AHR/ARNT heterodimer [43]. Furthermore, phosphorylation in a single tyrosine domain of the N-terminus is essential for proper recognition of the AHR for PKC-dependent phosphorylation, for binding of the AHR to its cognate DNA sequence, and for full transcriptional activity [44].

25.3.3 Crosstalks of AHR with Other Nuclear Receptors

For the last couple of decades, extensive studies have been made to investigate the possible crosstalks between different nuclear receptors (NRs) and the AHR. Of interest, there have been several attempts to explain these crosstalks. Generally, there are several theories explaining these crosstalks, that is, competitive binding of different NRs to a DNA binding site, selective dimerization with other NRs prior to the DNA binding step, and finally binding of different ligands that would probably affect the recruitment of a wide array of coactivators and/or corepressors. Therefore, in the following section we will focus on seven NR crosstalks with AHR, namely, nuclear factor erythroid 2-related factor-2 (Nrf2), ER, glucocorticoid receptor (GR), retinoid activated receptors and retinoid X receptors (RARs and RXRs), NF-κB, activator protein-1 (AP-1), and hypoxia-inducible factor 1α (HIF-1α).

25.3.3.1 Nrf2

The XRE was identified to be the DNA motif that upregulates a battery of genes including phase I and phase II drug metabolizing enzymes. Similarly, the antioxidant-responsive element (ARE) was identified to be the DNA motif that upregulates specific phase II genes such as *NQO1*, *GSTA1*, *UGT1A6*, *ALDH3*, and heme oxygenase (*HO-1*) through the Nrf2/ARE signaling pathway. Nrf2 is a redox-sensitive member of the cap 'n' collar basic leucine zipper (CNC bZip) family of transcription factors [45]. In response to oxidative stress, Nrf2 dissociates from its cytoplasmic tethering polypeptide, Kelch-like ECH associating protein 1 (Keap1), translocates into the nucleus, dimerizes with a musculoaponeurotic fibrosarcoma (MAF) protein, and thereafter binds to and activates the ARE [46]. The proximity of the CYP1A1 promoter (XRE) and ARE suggested a crosstalk and functional overlap between the two signaling pathways [47, 48].

Conglomerates of studies have shown that bifunctional inducers, which activate both XRE and ARE signaling pathways, require a direct crosstalk between the XRE- and ARE-mediated pathways for the induction of several phase II genes [48]. Of interest, it has been reported that the induction of *NQO1* by selective ARE inducers requires the presence of the AHR, suggesting a more direct crosstalk between the XRE- and ARE-mediated pathways [47]. Furthermore, it has been suggested that mouse Nrf2 is under the control of AHR as AHR ligands increased Nrf2 mRNA transcripts [47]. Conversely, it was demonstrated that the expression of AHR, and subsequently CYP1A1, in addition to CYP2B1 partially depends on Nrf2 in Hepa 1c1c7 cells, implying that Nrf2 modulates AHR, constitutive androstane receptor (CAR), and their downstream targets [49]. Reduction of AHR mRNA levels in Nrf2 knockout mice compared to wild-type mice provides further support to this hypothesis [49]. Moreover, AHR mRNA levels were increased in Keap1 knockout mice, inferring a direct effect of Nrf2 in regulating AHR [49].

25.3.3.2 ER

Ohtake et al. have demonstrated that the estrogenic action of AHR agonists could be exerted through a direct interaction between the AHR/ARNT complex and the unliganded ER in the absence of 17β-estradiol [50]. The use of AHR and ER knockout mice provided further support to this hypothesis. As such, 3-methylcholanthrene (3MC) was unable to activate the estrogen-responsive genes, namely, *c-Fos* and vascular endothelial growth factor (*VEGF*), in both AHR and ER knockout mice [50]. Incongruously, Hoivik et al. found no effect of estrogen on the CYP1A1 induction in both mouse hepatoma Hepa 1c1c7 and human breast cancer MCF-7 cells [51].

To date, no estrogen-responsive elements (EREs) in the *CYP1A1* gene have been identified. However, mutual inhibitory effect between the binding of ERα and AHR to their corresponding response elements has been previously reported and is a matter of debate [52]. For example, estradiol had no effect on the AHR/XRE binding, while on the other hand ERα-mediated suppression of induced CYP1A1 was successfully reversed by both ER antagonist and coexpression of nuclear factor-1 (NF-1), a transcription factor that interacts with both AHR and ERα. Therefore, these results suggested a direct crosstalk between AHR and ERα through competing on a common transcription factor NF-1 [53]. The competition between AHR and ERα is not limited only to NF-1 as it was shown that both nuclear receptors also compete for several other coactivators such as RIP140, ERAP140, and SMRT [22, 23]. Conversely, AHR ligands were shown to downregulate ER-dependent gene expression in human MCF-7 cells and in rodent estrogen-responsive tissues [54]. Importantly, TCDD was shown to inhibit the interaction of ERα with its ligand and its response element [55]. Thus, these results suggest a potential competitive crosstalk between AHR and ERα for common coactivators [22].

The contradictory effects of estradiol on the AHR-regulated genes could be attributed to species-specific effects, the concentration of estradiol tested, and the tissue origin of the cell line utilized. These factors will determine the degree and the direction of response upon exposure to estradiols, while the cell line-specific effects will be in fact related to the changing levels of certain transcription factors or coactivators among different cell lines from the same species.

25.3.3.3 GR Previous reports have demonstrated that the inducibility of CYP1A1 by different PAHs and HAHs, which are known to be potent AHR ligands, is potentiated by the action of GR [56]. The *CYP1A1* gene first intron contains three GREs [56], while exon 1 is a noncoding region and the initiation codon of *CYP1A1* gene expression is located within exon 2 [56]. Thus, binding of ligand-activated GR to the GRE sequences in *CYP1A1* first intron will interact with the initiation complex (AHR/ARNT) on the XRE, and consequently enhance the level of induction of CYP1A1 mRNA by AHR ligands [56]. Importantly, however, the GR will not be able to initiate the transcription process alone in the absence of the initiation complex.

In contrast to CYP1A1, *CYP1B1* gene expression was suppressed by dexamethasone in fibroblasts via GR-dependent mechanism. An explanation offered for this awkward response is that dexamethasone effect was mediated by a 256 bp DNA fragment carrying the XRE response element but not the GRE [57]. Thus, one may speculate that dexamethasone might act differentially to downregulate *CYP1B1* gene expression. In addition, the modulation of both *CYP1A1* and *CYP1B1* probably involves protein–protein interactions between the GR and other transcription factors such as AHR or competition for a common coactivator [58].

25.3.3.4 RARs and RXRs Retinoic acid (RA) has been identified as the most potent vitamin A metabolite that regulates a wide array of physiological processes including growth, differentiation, cell proliferation, and morphogenesis [59]. The physiological effects of RA are mediated by nuclear proteins RAR-α,β,γ and RXR-α,β,γ [60]. It is believed that RXRs are the master regulators among other RA receptors because they dimerize either with themselves to form homodimers or with most of the nuclear transcription factors forming heterodimers [61]. These receptor complexes then interact with the DNA *cis*-acting RA response element (RARE) to modulate the transcription of target genes [62].

The effect of RA on the regulation of *CYP1A1* is contradictory. For example, studies carried on keratinocytes showed that RA was able to downregulate or upregulate CYP1A1 gene expression [61]. Vecchini et al. demonstrated that in keratinocytes the *CYP1A1* gene promoter contains an unusual RARE element [61]. In contrast, other studies on hepatocytes showed that RA had a minimal effect on CYP1A1 or CYP1A2 mRNAs, while selective ligands for RARs and RXRs caused a pronounced decrease in hepatic CYP1A2 expression *in vivo* [63].

25.3.3.5 NF-κB NF-κB is a family of transcription factors that plays a critical role in regulating gene expression [64]. The NF-κB family is composed of six known proteins, NF-κB1, NF-κB2, RelA, RelB, c-Rel, and v-Rel, that can form homodimers as well as heterodimers with each other to bind to enhancer sequences [65]. Coimmunoprecipitation assays in Hepa 1c1c7 [66] and human breast cancer [67] cell extracts demonstrated physical and functional interactions between AHR and RelA subunit of NF-κB. These studies suggested that activation of one signaling pathway could significantly downregulate the other. This has been demonstrated experimentally in Hepa 1c1c7 cells in which activation of NF-κB suppressed the expression of *Cyp1a1* at the transcription level [27]. Although it is not clear whether such interaction occurs at the cytoplasmic or nuclear levels, several studies suggested that the interaction of NF-κB and AHR primarily occurs in the cytoplasm since ARNT was not found to dimerize with RelA in the absence of a ligand [66]. It has been reported that unactivated AHR and NF-κB in the cytoplasm are kept away by being sequestered by their inhibitory proteins, HSP90 and inhibitory κB protein (IκB), respectively. However, once activated by TCDD and tumor necrosis factor-α (TNF-α), AHR and NF-κB, respectively, would then interact [66]. Although the details of cytoplasmic interactions of RelA and AHR are still undetermined, transient transfection of Hepa 1c1c7 cells with AHR did not alter IκB levels, suggesting that the repressive effects are not mediated through the induction of IκB [66]. On the other hand, the observation of a competition between RelA and AHR for binding to transcriptional coactivators and corepressors strongly suggests a nuclear crosstalk between AHR and NF-κB. This was supported by Tian and coworkers who demonstrated that activation of NF-κB by TNF-α suppressed *Cyp1a1* gene expression through abolishing histone acetylation, which is an initial step for gene expression, resulting in inactivation of the *Cyp1a1* promoter [27]. Furthermore, the suppressive effect of β-naphthoflavone β-NF, a potent AHR ligand, on κB enhancer-driven luciferase reporter gene was reversed by the AHR antagonist, α-naphthoflavone [66].

Another postulated mechanism for the suppression of AHR by NF-κB activation is through activation of aryl hydrocarbon receptor repressor (AHRR). In this regard, NF-κB binding site (κB) was found in the promoter region of AHRR; therefore, activation of NF-κB will result in induction of AHRR expression that heterodimerizes with

ARNT and subsequently suppresses AHR activation and the expression of its regulated genes [68].

25.3.3.6 AP-1 AP-1 is a heterodimeric complex of leucine zipper proteins, c-Jun and c-Fos, which are involved in a wide range of physiological and pathological conditions, such as cell proliferation, apoptosis, cell cycle control, tumor promotion, and carcinogenesis [64]. Upon activation by a large number of stimuli, including proinflammatory cytokines, oxidative stress, and tumor promoters, AP-1 binds to TPA-responsive elements (TREs) within the promoter regions of several target genes [64]. AP-1 activity has been shown to be regulated by MAPK signaling pathways such as JNK, ERK1/2, and p38 [64]. Once activated, JNK translocates to the nucleus where it phosphorylates c-Jun to potentiate its transcriptional activity, which results in the induction of *c-Jun* and other AP-1 target gene transcription [64].

A well-established link between the AP-1 signaling pathway and the expression of AHR-regulated genes was demonstrated previously [69, 70]. The role of AP-1 in the modulation of *CYP1A1* and *CYP1A2* gene expression is controversial. Several previous studies have shown that TCDD inhibited LPS-induced DNA binding and transcriptional activity of AP-1 in murine lymphoma WT CH12.LX, but not in AHR-deficient BCL-1 cells [69]. Furthermore, the observation that AHR antagonists attenuated TCDD-induced inhibition of AP-1 binding in CH12.LX cells [69] strongly suggests a coordination between AHR and AP-1 signaling pathways. In contrast, treatment of Hepa 1c1c7 cells with TCDD or B[*a*]P caused an increase in c-Fos and c-Jun mRNA levels, which was associated with an increase in the DNA binding activity of AP-1, suggesting that AP-1 activation requires a functional AHR–XRE complex [71]. In addition, it has been shown that induction of CYP1A2 activity in HepG2 cells in response to 3MC is mediated through activation of AP-1 DNA binding [72].

25.3.3.7 HIF-1α The dimerization partner of AHR, ARNT, is sometimes called HIF-1β [73]. In addition to binding with AHR, ARNT dimerizes with HIF-1α to form HIF-1 complex [74]. HIF-1α is also a member of the bHLH/PAS family of transcription factors. Upon the formation of HIF-1 and the subsequent binding to the hypoxia response element (HRE), the induction of transcription of hypoxia-related genes such as VEGF and platelet-derived growth factor (PDGF) is initiated [75]. Because of the potential competition between AHR and HIF-1α over ARNT, certain studies have supported the notion that the limiting cellular factor ARNT would influence the intensity of activation between the two pathways [76]. In this sense, reciprocal crosstalk between hypoxia and TCDD signal transduction pathways has been demonstrated to occur *in vitro* and *in vivo* [76]. Hypoxia on one hand would downregulate the expression of AHR-regulated genes despite the presence of dioxin, while on the other hand increased oxygen supplementation would reverse these effects [76].

25.4 THE EFFECTS OF HEAVY METALS ON AHR ACTIVITY BIOMARKERS

There are several approaches that can be used to measure AHR activity. Assays employing biomarkers often provide a powerful tool for determining biological effects and their underlying mechanism(s) [77]. In general, these assays can complement or even replace the use of analytical chemistry-based assays [77]. Increasingly, bioassays employing cultured cells or cellular extracts are being developed and used to detect AHR activity. Examples of these assays would include AHR binding, *CYP1A1* gene expression, AHR/ARNT/XRE binding, CYP1A activity, and finally XRE-driven luciferase reporter activity. Therefore, in the current section we are going to shed light on the effects of heavy metals on these important biomarkers.

25.4.1 Arsenic (As^{3+})

Early reports have demonstrated that As^{3+} inhibited the β-NF-mediated induction of the CYP1A1-dependent 7-ethoxyresorufin-O-deethylation (EROD) activity in the liver and kidney but not lung of guinea pig [78]. In contrast, As^{3+} potentiated the β-NF-mediated induction of CYP1A1 catalytic activity in the lungs while decreasing β-NF-mediated induction of CYP1A1 activity in the kidneys and liver [78]. In Wistar rats, As^{3+} decreased total hepatic CYP450 content and monooxygenase activities of several CYP450s including CYP1A1 [79]. Similarly, studies on primary cultures of chick and rat hepatocytes showed that As^{3+} decreased total CYP450 and 3MC-mediated induction of CYP1A1 activity in chick hepatocytes and CYP1A1 mRNA, protein, and catalytic activity in rat hepatocytes [80, 81]. The effect of As^{3+} was also tested in primary human hepatocytes in which As^{3+} decreased PAH-mediated induction of CYP1A2 but not CYP1A1 at mRNA levels, while it decreased protein and catalytic activity levels of both isozymes in these cells [82]. In mouse Hepa 1c1c7 cells, we have shown that As^{3+}, in the presence of several AHR ligands, inhibited CYP1A1 catalytic activity while potentiating its mRNA and protein levels [83].

Although the effect of As^{3+} on CYP1A1 activity does not always parallel its effect on the expression on CYP1A1 mRNA that reflects AHR activity, almost all studies have reported a decrease in CYP1A1 catalytic activity in hepatic and extrahepatic tissues and cells in response to As^{3+}. Thus, multiple, but common, underlying pathways may be involved. As such, As^{3+}-dependent decrease in CYP1A1 catalytic activity was accompanied by a decrease, an increase, or no change in its mRNA levels. In addition to AHR-regulated CYP450s, As^{3+} also decreases the catalytic activity levels of other CYP450s that are not regulated by AHR [84]. Thus, it is apparent that As^{3+} may have a direct effect on the function of the CYP450 protein, independent of its effect on transcriptional regulation. As such, it has been well documented that As^{3+} interacts with critical cysteine residues of many intercellular proteins, thus altering their functions [85]. For example, As^{3+} has been shown to prevent the activation of NF-κB via interacting with cysteine 179 in the activation loop of the IκB kinase catalytic subunit, and subsequently inhibiting the dissociation of IκB from NF-κB, which is a necessity for NF-κB activation [86]. On the other hand, CYP450 activity critically depends on the binding of iron heme to sulfur atom of a conserved cysteine residue in the apoprotein [87]. Therefore, As^{3+} might inhibit CYP1A1 catalytic activity by competing with heme for binding to the critical cysteine residue in the apoprotein [88]. Another possibility is that metal-induced reactive oxygen species (ROS) may oxidize thiol groups in cysteine residues, directly or indirectly through the formation of reactive nitrogen species, of the CYP1A1 protein causing loss of protein function [89]. Furthermore, ROS may also interact with the heme Fe^{2+} leading to heme destruction and enzyme inactivation [90].

HO-1, a 32 kDa enzyme, catalyzes the oxidative conversion of heme into biliverdin and subsequently bilirubin that serves an important role in protecting cells from oxidative damage caused by free radicals [91]. HO-1 regulation occurs through the redox-sensitive Nrf2/ARE signaling pathway. Of interest, HO-1 anchors to the endoplasmic reticulum membrane via a stretch of hydrophobic residues at its C-terminus [92]. Thus, it is expected to interact with CYP450s that are also endoplasmic reticulum-bound enzymes.

As^{3+}, via accumulating in the mitochondria and altering cellular respiration, has been shown to stimulate the production of superoxide ($O_2^{\bullet-}$) and hydrogen peroxide (H_2O_2). Using Hepa 1c1c7 cells, we have shown that As^{3+} alone increased HO-1 mRNA that coincided with increased cellular glutathione levels, either to compensate the oxidative stress production by As^{3+} or as a direct response to oxidative stress [93]. Of interest, As^{3+} decreased the TCDD-mediated induction of CYP1A1 catalytic activity, and this inhibition was further potentiated when glutathione was depleted.

Almost all studies on the As^{3+}-mediated effect on CYP1A1 catalytic activity suspected a role of HO-1 in the degradation of its heme group creating a hollow functionless protein [21]. HO-1 activity was elevated in primary cultures of chick hepatocytes treated with As^{3+} [80]. We have previously demonstrated that HO-1 mRNA was elevated and total cellular heme content was decreased in Hepa 1c1c7 cells treated with As^{3+}. We therefore speculated that the induction of HO-1 may contribute to the inhibition of CYP1A1 activity by As^{3+}, despite the presence of other interplaying mechanisms. In the primary cultures of chick and rat hepatocytes, the effect of As^{3+} on CYP1A1 catalytic activity was not reversed by mesoporphyrin, an inhibitor for HO-1 [80]. Moreover, the addition of heme increased HO-1 activity to similar levels in the absence and presence of As^{3+}, while the decrease in CYP1A1 activity was only observed in the presence of As^{3+} [80]. Thus, it was concluded that the elevated levels of HO-1 alone may not be responsible for As^{3+}-mediated effects on CYP1A1 catalytic activity.

Recently, however, we have shown that As^{3+} decreased the TCDD-mediated induction of CYP1A1 catalytic activity that coincided with increased HO-1 mRNA and decreased total cellular heme content. Upon using tin mesoporphyrin (SnMP) as HO-1 competitive inhibitor, supplementation with external heme, treatment with hemoglobin as carbon monoxide scavenger, or transfecting HepG2 cells with siRNA for HO-1, there was a partial restoration of the inhibition of TCDD-mediated induction of CYP1A1 catalytic activity [94]. Thus, these results imply that HO-1 is partially involved in the As^{3+}-mediated downregulation of CYP1A1 at the catalytic activity level [94].

In human breast cancer (T-47D) cells, B[a]P-mediated induction of CYP1A1 mRNA was not affected by As^{3+} treatment although B[a]P-mediated induction of CYP1A1 protein and catalytic activity levels was decreased upon exposing the cells to As^{3+} [95]. This decrease in CYP1A1 activity coincided with an increase in HO-1 mRNA. In the absence of a decrease in CYP1A1 mRNA levels, As^{3+} may have decreased the CYP1A1 protein half-life through an increase in its protein degradation. In our studies, however, we have shown that As^{3+} had no effect on CYP1A1 protein stability [96].

The uncertainties in the mechanisms involved in the modulation of CYP1A1 activity and protein level by As^{3+} are also accompanied by many questions regarding the effect of this metal on mRNA levels. As^{3+} failed to attenuate the B[a]P-mediated induction of CYP1A1 mRNA, yet decreased its catalytic activity [97, 98]. In agreement with these studies, pretreatment of human lung adenocarcinoma cells (CL3) with As^{3+} did not affect the B[a]P-mediated induction of CYP1A1 mRNA [99]. Further discrepancies arose with the emergence of data on the effect of As^{3+} on XRE-driven luciferase reporter gene in Hepa-1 cells. Maier et al. have shown that despite having no effect on TCDD-induced CYP1A1 mRNA, As^{3+} decreased luciferase activity in cells transfected with the XRE-driven luciferase reporter gene [100]. The discrepancy between the effect of As^{3+} on CYP1A1 mRNA and its effect on the XRE-driven luciferase reporter gene was not explained.

In human HepG2 cells, it was previously demonstrated that As^{3+} alone did not affect CYP1A1 mRNA but reduced the benzo[k]fluoranthene-induced levels upon treatment with As^{3+} [101]. However, similar to earlier reports, As^{3+} alone decreased the XRE-driven luciferase reporter activity, and inhibition of TCDD-mediated induction of CYP1A1 mRNA and XRE-dependent luciferase activity was also observed in human hepatoma Huh7 cells [102]. Furthermore, it was demonstrated that the actions of As^{3+} on blocking CYP1A1 induction by TCDD were primarily through altering CYP1A1 transcription, possibly through inhibiting the recruitment of polymerase II to the CYP1A1 promoter, which is independent of the regulatory mechanisms initiated by As^{3+}-induced cell arrest [103]. Using HepG2 cells, we have also shown that As^{3+} decreased the TCDD-mediated induction of CYP1A1 mRNA, protein, and catalytic activity levels in a dose-dependent manner. In addition, As^{3+} decreased the XRE-driven luciferase reporter activity [94].

Later studies contradicted previously reported data as As^{3+} was able to induce AHR nuclear translocation in Hepa-1 cells with the same efficiency as TCDD [104]. The increase in nuclear accumulation of the AHR resulted in an increased expression of CYP1A1 mRNA when the cells were treated with As^{3+} alone and potentiation of B[a]P-mediated induction of CYP1A1 mRNA [104]. Thus, these results imply the presence of species-specific differences in the effect of As^{3+} on CYP1A1 mRNA expression.

We have previously shown that As^{3+} alone was able to increase CYP1A1 mRNA while potentiating the TCDD-, 3-MC, BNF, and B[a]P-mediated induction of CYP1A1 mRNA in Hepa 1c1c7 cells [83]. In addition, the increase in mRNA levels was translated to an increase in CYP1A1 protein levels.

Mechanistically, As^{3+} alone increased CYP1A1 mRNA in Hepa 1c1c7 cells in a time- and dose-dependent manner, and this inducibility was completely abolished after the addition of the RNA polymerase inhibitor actinomycin D, implying a requirement of *de novo* RNA synthesis [105]. As^{3+} was able also to increase the XRE-dependent luciferase reporter activity despite not increasing the AHR nuclear accumulation.

To confirm the role of the AHR in the regulation of CYP1A1 by As^{3+}, cycloheximide, a protein synthesis inhibitor that inhibits a labile protein required for the proteolysis of the AHR [106], caused superinduction of the CYP1A1 mRNA [107]. Similarly, Z-Leu-Leu-Leu-CHO (MG-132), a 26-proteasome inhibitor that stabilizes the AHR protein, caused further induction of the CYP1A1 mRNA in response to As^{3+} treatment. Besides its transcriptional effect, As^{3+} also increased the stability of CYP1A1 mRNA transcripts induced by TCDD. Thus, we concluded that As^{3+} increases CYP1A1 transcriptionally, through an AHR-dependent mechanism, and stabilizes CYP1A1 mRNA transcripts.

The mechanisms by which As^{3+} induces CYP1A1 mRNA transcription through the AHR are still at large. However, heavy metals may alter some cellular metabolic pathways leading to the enhanced production of endogenous AHR ligands such as biliverdin and bilirubin [10]. However, despite the fact that elevated pulmonary CYP1A1 expression was associated with an increase in total plasma bilirubin concentrations, the administration of bilirubin to the lung through intratracheal injection did not cause any increase in CYP1A1 mRNA [108, 109]. Moreover, As^{3+}-induced oxidative stress may enhance the release of arachidonic acid from glycerol phospholipids, which has also been shown to be an endogenous AHR ligand [10, 110]. As^{3+} might also bind to vicinal thiols in Hsp90, disrupting the interaction between Hsp90 and AHR, causing the NLS to be exposed and subsequently causing AHR nuclear accumulation [104].

Another postulated mechanism for the As^{3+}-mediated induction of CYP1A1 implicates posttranscriptional modifications of histones. As^{3+} induces phosphorylation [111, 112] and acetylation [111] of histone H3 through ERK, p38, and the Akt1 pathways. In addition, histone deacetylation has been shown to be involved in the inducibility of CYP1A1 mRNA transcription in human and mouse hepatoma cells [113] and HeLa cells [114]. Thus, As^{3+} might induce CYP1A1 through modifying the phosphorylation and acetylation of these histones.

Although many possible mechanisms might be involved in the As^{3+}-mediated induction of CYP1A1 mRNA in mouse hepatoma cells, much less information is available to explain the controversial effects of this metal on CYP1A1 expression in other species. The summarized effects of As^{3+} on AHR activity biomarkers are shown in Table 25.1.

25.4.2 Mercury (Hg^{2+})

Although organic Hg^{2+} is far more toxic than inorganic Hg^{2+}, almost all studies concerning the effect of this metal on the AHR signaling pathway are utilizing the inorganic form. The studies by Vakharia et al. were among the first to report the effect of Hg^{2+} on CYP1A1 [82]. It was shown that Hg^{2+} potently inhibited CYP1A2 activity in freshly isolated human hepatocytes, yet its potency was less than that of As^{3+} in inhibiting the AHR ligand-mediated induction of CYP1A1 protein and catalytic activity levels. However, in HepG2 cells Hg^{2+} failed to decrease the B[a]P-, benzo[b]fluoranthene-, benzo[a]anthracene-, and benzo[k]fluoranthene-mediated induction of CYP1A1 catalytic activity and benzo[k]fluoranthene-mediated induction of CYP1A1 mRNA, yet it did inhibit dibenzo[a,h]anthracene-mediated induction of CYP1A1 catalytic activity [98]. Interestingly, Hg^{2+} potentiated the benzo[k]fluoranthene-mediated induction of CYP1A1 protein. Thus, these results imply the presence of AHR ligand-specific effects of Hg^{2+} on CYP1A1 expression.

In shark rectal gland (SRG) cells, Hg^{2+} increased CYP1A1 protein levels while its activity was not measured [115]. In contrast, Hg^{2+} significantly decreased the β-naphthoflavone-mediated induction of CYP1A1 activity in hepatic sea bass (*Dicentrarchus labrax* L.) microsomes and was partially reversed by glutathione, implicating a

TABLE 25.1 The Effect of As^{3+} on AHR Activity Biomarkers

	Species (Tissue/Cell line)	Changes	References
In vivo	Rats (liver, lung)	↓ Total hepatic CYP450 and CYP1A1 activities ↓ Hepatic CYP1A1 activity ↑ Pulmonary CYP1A1 mRNA, protein, and activity	[79, 115]
	Guinea pig (kidneys, liver, lung)	↓ CYP1A1 EROD activity ↓ β-NF-mediated induction of CYP1A1 activity in kidneys and liver ↑ β-NF-mediated induction of CYP1A1 activity in the lung	[78]
In vitro	Human (hepatocytes)	↓ PAH-mediated induction of CYP1A1 protein and activity	[82]
	Human adenocarcinoma (T-47D cells)	↔ B[a]P-mediated induction of CYP1A1 mRNA ↓ B[a]P-mediated induction of CYP1A1 protein and activity	[95, 97, 98]
	Human non-small-cell lung carcinoma (Cl3 cells)	↔ B[a]P-mediated induction of CYP1A1 mRNA	[99]
	Human hepatoma (HepG2 cells)	↔ CYP1A1 mRNA ↓ BKF-mediated induction of CYP1A1 mRNA ↓ TCDD-induced XRE-dependent luciferase activity ↔ CYP1A1 mRNA half-life	[101, 103]
	Human hepatoma (Huh7 cells)	↓ TCDD-mediated induction of CYP1A1 mRNA ↓ XRE-dependent luciferase activity	[102]
	Rat (hepatocytes)	↓ 3-MC-mediated induction of CYP1A1 mRNA, protein, and catalytic activity	[80]
	Mouse hepatoma (Hepa-1 cells)	↔ B[a]P-mediated induction of CYP1A1 activity ↔ TCDD-mediated induction of CYP1A1 mRNA ↓ TCDD-induced XRE-dependent luciferase activity	[104]
	Mouse hepatoma (Hepa 1c1c7 cells)	↑ AHR translocation ↑ CYP1A1 mRNA ↑↑ AHR ligand-mediated induction of CYP1A1 mRNA and protein ↓ AHR ligand-mediated induction of CYP1A1 activity ↑ XRE-dependent luciferase activity ↑ CYP1A1 mRNA half-life ↔ CYP1A1 protein stability	[83, 93, 103–105]
	Chick (hepatocytes)	↓ Total CYP450 ↓ 3-MC-mediated induction of CYP1A1 activity	[81]

ROS-dependent mechanism in the regulation of CYP1A1 by Hg^{2+} [116].

We have also shown that Hg^{2+} alone was able to increase CYP1A1 mRNA without affecting its protein or catalytic activity levels in Hepa 1c1c7 cells. Moreover, using multiple AHR ligands, we have shown that Hg^{2+} potentiated the AHR ligand-mediated induction of CYP1A1 mRNA and protein levels while decreasing its activity levels. An increase in CYP1A1 protein half-life accounted for the potentiation in CYP1A1 protein levels. However, the decrease in activity levels could only be attributed to a decrease in total cellular heme content [117].

Mechanistically, Hg^{2+} alone increased CYP1A1 mRNA in a time- and concentration-dependent manner while potentiating the TCDD-mediated induction of CYP1A1 mRNA in a time- and TCDD concentration-dependent manner [118]. Transcriptionally, actinomycin D completely blocked the Hg^{2+}-mediated induction of CYP1A1 mRNA, implying requirement of *de novo* RNA synthesis. In the presence of cycloheximide and/or MG-132, Hg^{2+}-mediated effect on CYP1A1 mRNA was further potentiated suggesting the presence of a transcriptional AHR-dependent mechanism. In addition, Hg^{2+} transformed AHR in guinea pig hepatic cytosol (*in vitro*) and induced nuclear accumulation of transformed AHR in Hepa 1c1c7 cells (*in vivo*) [118]. In contrast to the effect of As^{3+}, Hg^{2+} had no effect on CYP1A1 mRNA transcripts' half-life. Thus, it is apparent that Hg^{2+} increases CYP1A1 mRNA solely through a transcriptional mechanism that depends on AHR function.

The cell-specific response to Hg^{2+} was clearly demonstrated, as Hg^{2+} decreased the TCDD-mediated induction of CYP1A1 protein and catalytic activity in HepG2 cells [119]. Perdew and coworkers have demonstrated that there are distinctive differences in the properties of unliganded human and mouse AHR including cellular localization, nuclear translocation, and the effects of XAP on both processes [120, 121]. Studies by Suzuki and Nohara provided further insight into these postulated mechanisms [122]. As such, factors that could be responsible for the species-specific characteristics of AHR functions could be summarized in three major components: the nuclear translocation, transcription initiation via remodeling of chromatin, and finally proteasomal degradation of AHR. It was shown that, in Hepa 1c1c7 cells, TCDD treatment causes the recruitment of coactivator CBP but not p300 to the CYP1A1 promoter region [122]. On the other hand, in HepG2, TCDD treatment caused the recruitment of p300 but not CBP to the CYP1A1 promoter region [122]. Thus, we may speculate that TCDD, although a potent AHR ligand and CYP1A1 inducer, might be still functioning differentially across species. Another important difference between Hepa 1c1c7 and HepG2 that might participate in the species-specific responses is the sensitivity of AHR to proteasomal degradation upon exposure to TCDD, with the HepG2 AHR being less sensitive than the Hepa 1c1c7 AHR.

In order to dissect the mechanisms by which Hg^{2+} induces AHR translocation and subsequently *Cyp1a1* transcription, the role of the redox-sensitive transcription factors, NF-κB and AP-1, was tested [123]. Both transcription factors are activated by changes in the redox status of cells and we have shown the induction of such environment in Hepa 1c1c7 cells by Hg^{2+}. In addition, previous reports have shown a mutual inhibitory interaction between the AHR and NF-κB and attenuation of AP-1 DNA binding by TCDD in an AHR-dependent manner [66, 124]. Other studies have shown that these transcription factors share a number of coactivators, providing an explanation to this phenomenon as described earlier [125].

We previously demonstrated that Hg^{2+} increased NF-κB and AP-1 binding to κB-RE and TRE as evidenced by electrophoretic mobility shift assay (EMSA). In addition, the NF-κB activator phorbol 12-myristate 13-acetate (PMA) decreased Hg^{2+}-mediated induction of HO-1 mRNA and CYP1A1 mRNA and activity. In contrast, the NF-κB inhibitor pyrrolidine dithiocarbamate (PDTC) potentiated the Hg^{2+}-mediated induction of HO-1 mRNA and CYP1A1 mRNA and activity in the mouse hepatoma Hepa 1c1c7 cells, implying a crucial role for NF-κB in the regulation of CYP1A1 by Hg^{2+}.

Inhibitors of AP-1 upstream its signaling pathway, such as the JNK inhibitor, SB600125, and the P38 (MAPK) inhibitor, SB203580, had similar effects to those of the NF-κB activator, PMA. On the other hand, ERK inhibitor, U0126, had similar effects to those of NF-κB inhibitor, PDTC. Altogether, these results imply that AP-1 and MAPKs might also be key players in the regulation of CYP1A1 by Hg^{2+}. Furthermore, these results imply the requirement of NF-κB and/or MAPK [35] and subsequently AP-1 in the regulation of CYP1A1.

25.4.3 Lead (Pb^{2+})

In HepG2 cells and primary human hepatocytes, Pb^{2+} decreased CYP1A1 protein and catalytic activity levels in the absence and presence of different PAHs without significantly affecting CYP1A1 mRNA levels [82, 98]. In agreement with these studies, we found that TCDD-mediated induction of CYP1A1 protein and catalytic activity levels in HepG2 cells was reduced in the presence of Pb^{2+} compared to that of TCDD alone [123].

Interestingly, Pb^{2+} alone was able to increase CYP1A1 mRNA but not its protein levels. Upon coadministration of different AHR ligands with Pb^{2+}, the AHR ligand-mediated induction of CYP1A1 activity was reduced in a Pb^{2+} concentration- and AHR ligand-dependent manner [117]. In contrast, Pb^{2+} potentiated the TCDD-mediated induction of CYP1A1 mRNA in a dose- and time-dependent manner [118]. Actinomycin D completely abolished the Pb^{2+}-mediated induction of CYP1A1 mRNA, implying a requirement for *de novo* RNA synthesis. The AHR also appears to have a predominant transcriptional activity in this induction as cotreatment of Hepa 1c1c7 cells with Pb^{2+} in the presence of cycloheximide and/or MG-132 potentiated the Pb^{2+}-mediated induction of CYP1A1 mRNA. Pb^{2+} also induced AHR protein transformation *in vitro* in untreated guinea pig hepatic cytosol and nuclear accumulation of transformed AHR *in vivo* in Hepa 1c1c7 cells. Similar to the effect of Hg^{2+}, Pb^{2+} did not alter CYP1A1 mRNA half-life, implying that Pb^{2+}-mediated induction of CYP1A1 mRNA was not due to increasing the transcripts' stability.

In an attempt to investigate the role of the redox-sensitive transcription factors, NF-κB and AP-1, in the modulation of *Cyp1a1* gene expression by Pb^{2+}, we first confirmed the induction of oxidative stress by Pb^{2+} [123]. We found that Pb^{2+} was able to increase ROS production and decrease total cellular glutathione content in Hepa 1c1c7 cells. The pro-oxidant buthionine-(*S,R*)-sulfoximine potentiated Pb^{2+}-mediated induction of HO-1 and CYP1A1 mRNA levels, while further suppressing the TCDD-mediated induction of CYP1A1 catalytic activity by Pb^{2+}. Importantly, Pb^{2+} increased the binding of NF-κB and AP-1 to their corresponding response elements. In addition, the NF-κB activator, PMA, decreased the Pb^{2+}-mediated induction of both

HO-1 and CYP1A1 mRNAs, while further potentiating the suppressive effect of Pb^{2+} on the TCDD-mediated induction of CYP1A1 catalytic activity. In contrast, the NF-κB inhibitor, PDTC, potentiated the Pb^{2+}-mediated induction of HO-1 mRNA but inhibited the Pb^{2+}-mediated induction of CYP1A1 mRNA. Finally, the Pb^{2+}-mediated inhibition of the TCDD-mediated induction of CYP1A1 catalytic activity was reversed in the presence of PDTC.

The controversy between the cell-, species-, and AHR ligand-specific effects of Pb^{2+} is worth studying in order to dissect the exact mechanisms. Moreover, the use of pharmacological inhibitors might not be the ultimate correct approach in mechanistic studies as these inhibitors might act to interfere with the AHR signaling pathway aside from their meant effect. Thus, the best approaches identified to date are either the use of deficient cell lines or in vivo models, or utilizing siRNAs to knock down certain genes in order to confirm their roles in the signaling pathway of interest [126].

25.4.4 Cadmium (Cd^{2+})

Numerous studies have shown that Cd^{2+} exposure affects total hepatic CYP450 and monooxygenase activities in different mammalian systems [127–130]. For example, Cd^{2+} significantly increases CYP1A1, but not CYP1A2 catalytic activity in HepG2 cells [131, 132]. Incongruously, CYP1A1 was not detected in the livers or kidneys of smokers or nonsmokers exposed to Cd^{2+} [132]. In the presence of inducers, however, Cd^{2+} selectively inhibited dibenzo[a,h]anthracene-mediated induction of CYP1A1 activity, but had no effect on its mRNA levels in HepG2 cells [98]. Similarly, in primary human hepatocytes, Cd^{2+} inhibited the PAH-mediated induction of CYP1A1 protein and catalytic activity without affecting its mRNA levels [82]. Of interest, Cd^{2+} was able to induce XRE-dependent luciferase reporter activity in HepG2 cells [131]. Surprisingly, in human fibroblast cells (GM0637 and GM4429), Cd^{2+} increased CYP1A1 mRNA and catalytic activity and potentiated the AHR-mediated toxicity of B[a]P [133].

In male Sprague Dawley rats, Cd^{2+} increased hepatic CYP1A1 catalytic activity without affecting total hepatic CYP450 [134], while it decreased total intestinal CYP450 and CYP1A1 activities [135]. In mouse Hepa-1 cells, Cd^{2+} had no effects on CYP1A1 mRNA in the absence and presence of TCDD, and was unable to alter XRE-dependent luciferase reporter activity [100]. In Antarctic fish (Trematomus bernacchii) livers treated with B[a]P and black sea bream, rainbow trout, and top minnow hepatic cell lines treated with 3MC, Cd^{2+} completely suppressed B[a]P- and 3MC-mediated induction of CYP1A1 mRNA, protein, and catalytic activity levels [136].

In Hepa 1c1c7 cells, we have demonstrated that Cd^{2+} alone was able to induce CYP1A1 mRNA while potentiating AHR ligand-mediated induction of CYP1A1 mRNA and protein in a Cd^{2+} concentration- and AHR ligand-dependent manner [96]. At the transcriptional level, Cd^{2+} induced CYP1A1 mRNA in a time- and concentration-dependent manner. The transcription inhibitor, actinomycin D, completely blocked Cd^{2+}-mediated induction of CYP1A1 mRNA implying a requirement of de novo RNA synthesis. Cd^{2+} was also shown to be a potent inducer of AHR nuclear accumulation with a subsequent increase in the XRE-dependent luciferase reporter activity [96]. Cd^{2+} did not affect CYP1A1 mRNA stability, indicating a predominant transcriptional mechanism for the increase in CYP1A1 mRNA. In determining the role of oxidative stress in the modulation of CYP1A1 by Cd^{2+}, we observed that pretreatment of Hepa 1c1c7 cells with the pro-oxidant buthionine-(S,R)-sulfoximine potentiated the Cd^{2+}-mediated induction of CYP1A1 and HO-1 mRNA levels in the presence and absence of TCDD. N-Acetylcysteine, on the other hand, protected against Cd^{2+}-induced modulation of CYP1A1 mRNA [93].

On the contrary, although Cd^{2+} increased CYP1A1 protein stability, it decreased AHR ligand-mediated induction of CYP1A1 catalytic activity, which coincided with a potent increase in HO-1 mRNA and a decrease in glutathione content [83]. This inhibition occurred simultaneously with a decrease in total cellular heme content. Interestingly, buthionine-(S,R)-sulfoximine also potentiated, while N-acetylcysteine protected against, the suppressive effect of Cd^{2+} on the TCDD-mediated induction of CYP1A1 activity, further confirming the role of ROS in the inhibition of protein function [93].

25.4.5 Chromium (Cr^{6+})

In contrast to other metals, very little information is available about the effect of Cr^{6+} on AHR. In different species of fish, Cr^{6+} decreased hepatic CYP1A1 catalytic activity [116, 137]. The addition of glutathione potentiated this inhibition, possibly by enhancing the reduction of Cr^{6+} to Cr^{5+}, Cr^{4+}, and Cr^{3+}, and the subsequent production of ROS that has been shown to alter the AHR signaling pathway as discussed earlier [116].

In more extensive work in Hepa-1 cells, it has been shown that Cr^{6+} decreases TCDD-mediated induction of CYP1A1 mRNA, protein, and activity in a concentration-dependent manner [100]. Cr^{6+} also inhibited the XRE-dependent luciferase reporter activity but failed to attenuate TCDD-induced nuclear accumulation of the AHR protein. Interestingly, although Cr^{6+} generates oxidative stress rapidly [138], it inhibited B[a]P-induced mutagenesis when it was coadministered with the AHR ligand, but not when administered before or after B[a]P, suggesting a direct interaction between Cr^{6+} and B[a]P [139]. Thus, the inhibition of CYP1A1 expression may play a role in the reduction of B[a]P metabolism and subsequent DNA damage. Furthermore, the use of buthionine-(S,R)-sulfoximine or N-acetylcysteine did not affect the inhibition of XRE-dependent luciferase reporter activity by Cr^{6+}, suggesting involvement of an alternative mechanism in the Cr^{6+}-mediated inhibition of XRE activation.

Mechanistically, using Hepa 1c1c7 cells, we demonstrated that Cr^{6+} induced constitutive CYP1A1 mRNA levels in a concentration- and time-dependent manner [96]. The pro-oxidant buthionine-(S,R)-sulfoximine caused further potentiation to the Cr^{6+}-mediated induction of CYP1A1 mRNA, while N-acetylcysteine protected against this induction. Thus, we concluded that ROS production by Cr^{6+} is a key player in CYP1A1 mRNA induction, which occurs primarily through an AHR-dependent transcriptional mechanism.

Despite its induction of Cyp1a1 transcription, Cr^{6+} inhibited inducible CYP1A1 mRNA expression. The inhibition of inducible but not constitutive gene expression by Cr^{6+} could be attributable to its effect on histone deacetylase activity. Recently, Cr^{6+} was shown to inhibit B[a]P-mediated release of histone deacetylase-1 (HDAC-1) from chromatin, preventing the recruitment of the coactivator p300 to the CYP1A1 transcriptional domain, hence blocking the induction of gene transcription [140].

Similar to other metals, Cr^{6+} decreased AHR ligand-mediated induction of CYP1A1 activity while causing an increase in CYP1A1 protein stability [96]. In addition, the ROS-mediated mechanism was also involved since buthionine-(S,R)-sulfoximine and N-acetylcysteine potentiate, or protected against, the decrease in CYP1A1 activity by Cr^{6+}.

25.4.6 Copper (Cu^{2+})

In the Antarctic fish (*T. bernacchii*) liver, Cu^{2+} significantly inhibited B[a]P-mediated induction of hepatic CYP1A protein and catalytic activity, but no effects were observed at the transcriptional level [136]. On the other hand, in striped bass fish (*Morone saxatilis*) the transcription of *CYP1A1* was not affected by any tested Cu^{2+} concentration in any tissue [141]. In HepG2 cells, Cu^{2+} was able to increase CYP1A1 mRNA in a time- and concentration-dependent manner [142] and potentiated the TCDD-mediated induction of CYP1A1 protein and catalytic activity levels [119]. Previous studies by Kim et al. [143] have reported that as a result of impaired NADPH-P450 reductase, Cu^{2+} inhibited the hepatic CYP450-catalyzed reactions in mice and rats. In addition, it has been reported that Cu^{2+} deficiency increased CYP1A1 activity in rat small intestine [144], implying a suppressive effect of Cu^{2+} on CYP1A1.

In Hepa 1c1c7 cells, Cu^{2+} at low doses induced constitutive CYP1A1 mRNA [117]. The transcription- and AHR-dependent requirements for Cu^{2+}-mediated effect on CYP1A1 mRNA were confirmed by cotreatment of Hepa 1c1c7 cells with actinomycin D or cycloheximide and/or MG-132. Intriguingly, Cu^{2+} successfully transformed untreated guinea pig hepatic cytosol AHR protein while it failed to induce its nuclear accumulation in Hepa 1c1c7 cells as evident by cytosolic and nuclear EMSAs, respectively. Using different ligands, Cu^{2+} decreased the AHR ligand-mediated increase in CYP1A1 mRNA in a concentration- and ligand-dependent manner. Moreover, it was able to decrease the TCDD-mediated induction of CYP1A1 mRNA in a time- and TCDD dose-dependent manner while not affecting its stability.

It was shown that Cu^{2+} increased the production of ROS that coincided with an increase in HO-1 mRNA and a decrease in cellular glutathione content; thus, the role of the redox-sensitive transcription factors NF-κB and AP-1 in the regulation of *Cyp1a1* gene by Cu^{2+} was also investigated [123]. Similar to what was seen with other metals, Cu^{2+} induced the binding of NF-κB and AP-1 to their corresponding responsive elements. Furthermore, the NF-κB activator PMA decreased the Cu^{2+}-mediated induction of CYP1A1 mRNA while the NF-κB inhibitor PDTC caused a significant increase in Cu^{2+}-mediated induction of CYP1A1 mRNA. Thus, it is apparent that NF-κB plays a role in the regulation of CYP1A1 expression by Cu^{2+}. However, the role of AP-1 is less clear than NF-κB as JNK inhibition decreased, but ERK inhibition further potentiated, the Cu^{2+}-mediated induction of CYP1A1 mRNA. Meanwhile, the inhibition of p38 had no effect. Thus, individual MAPKs, and not AP-1, have a role in the modulation of CYP1A1 expression by Cu^{2+}.

It is worth mentioning that Cu^{2+} decreased AHR ligand-mediated induction of CYP1A1 protein and activity in a Cu^{2+} concentration- and AHR ligand-dependent manner [117]. Yet, at the posttranslational level, Cu^{2+} increased the CYP1A1 protein stability. Thus, the decrease in CYP1A1 protein and subsequent decrease in activity is primarily caused by a decrease in AHR ligand-induced mRNA levels. The pro-oxidant buthionine-(S,R)-sulfoximine did not affect HO-1 or CYP1A1 mRNA levels while it potentiated the suppressive effect of Cu^{2+} on the TCDD-mediated induction of CYP1A1 catalytic activity, implicating a minor, but direct role for ROS in the inhibition of CYP1A1 activity by Cu^{2+}. In addition, the inhibition of NF-κB had no effect on the TCDD-mediated induction of CYP1A1 activity, implying that NF-κB does not affect Cu^{2+}-mediated modulation of inducible CYP1A1 expression. Similar to their effect on the CYP1A1 mRNA levels, JNK inhibition potentiated, while ERK inhibition reversed, the suppressive effect of Cu^{2+} on the TCDD-mediated induction of CYP1A1 activity. Thus, inhibition of CYP1A1 mRNA and activity is mediated primarily through a JNK-dependent mechanism. The effects of heavy metals on AHR activity biomarkers are summarized in Table 25.2.

25.4.7 Vanadium (V^{5+})

Compared to other metal species, much less information is available on the effect of V^{5+} on the AHR. We have recently

TABLE 25.2 The Effect of Heavy Metals on AHR Activity Biomarkers

Metal	Species (Tissue/Cell line)	Changes	References
Hg^{2+}	Sea bass (liver)	↓ β-NF-mediated induction of CYP1A1 activity	[116]
	Human (hepatocytes)	↓ AHR ligand-mediated induction of CYP1A1 protein and activity	[82]
	Human hepatoma (HepG2 cells)	↔ AHR ligand-mediated induction of CYP1A1 mRNA and activity	[98, 119]
		↓ TCDD-mediated induction of CYP1A1 protein and activity	
	Mouse hepatoma (Hepa 1c1c7 cells)	↑ AHR transformation and translocation	[117, 118]
		↑↑ AHR ligand-mediated induction of CYP1A1 mRNA and protein	
		↓ AHR ligand-mediated induction of CYP1A1 activity	
Pb^{2+}	Shark rectal gland (SRG cells)	↑ CYP1A1 protein levels	[115]
	Human (hepatocytes)	↔ CYP1A1 mRNA	[82]
		↓ PAH-mediated increase in CYP1A1 protein and activity	
		↓ CYP1A1 activity	
	Human hepatoma (HepG2 cells)	↔ CYP1A1 mRNA	[98, 119]
		↓ TCDD-mediated induction of CYP1A1 protein and activity	
		↓ CYP1A1 activity	
		↓ PAH-mediated increase in CYP1A1 activity	
	Mouse hepatoma (Hepa 1c1c7 cells)	↑ AHR transformation and translocation	[117, 118]
		↑ AHR ligand-mediated induction of CYP1A1 mRNA and protein	
		↓ AHR ligand-mediated induction of CYP1A1 activity	
		↔ CYP1A1 mRNA half-life	
		↑ CYP1A1 protein half-life	
Cd^{2+}	Rat (liver, intestine, and kidneys)	↑ CYP1A1 hepatic and renal protein	[134, 135]
		↓ CYP1A1 hepatic and renal activity	
		↑ Hepatic CYP1A activity	
		↓ Intestinal CYP1A1 activity	
	Antarctic fish (liver)	↓ B[a]P-mediated induction of CYP1A1 mRNA, protein, and activity	[136]
	Human (hepatocytes)	↔ PAH-mediated induction of CYP1A1 mRNA	[82]
		↓ PAH-mediated induction of CYP1A1 protein and activity	
	Human hepatoma (HepG2 cells)	↑ XRE expression	[98, 131]
		↔ DMBA-mediated induction of CYP1A1 mRNA	
		↑ CYP1A1 activity	
		↓ DMBA-mediated induction of CYP1A1 activity	
	Human skin fibroblasts (GM0637 and GM4429 cells)	↑ CYP1A1 mRNA and activity	[133]
	Mouse hepatoma (Hepa-1 cells)	↔ XRE-dependent luciferase activity	[100]
		↔ TCDD-mediated induction of CYP1A1 mRNA	
	Mouse hepatoma (Hepa 1c1c7 cells)	↑ AHR translocation	[83, 105]
		↑ XRE-dependent luciferase activity	
		↑ CYP1A1 mRNA	
		↑↑ AHR ligand-mediated induction of CYP1A1 mRNA and protein	
		↓ AHR ligand-mediated induction of CYP1A1 activity	
		↔ CYP1A1 mRNA stability	
		↑ CYP1A1 protein stability	
	Sea bream, rainbow trout (hepatocytes)	↓ 3-MC-mediated induction of CYP1A1 mRNA, protein, and activity	

TABLE 25.2 (Continued)

Metal	Species (Tissue/Cell line)	Changes	References
Cr^{6+}	Mouse hepatoma (Hepa-1 cells)	↔ AHR translocation ↓ XRE-dependent luciferase activity ↓ TCDD-mediated induction of CYP1A1 mRNA, protein, and activity	[100]
	Mouse hepatoma (Hepa 1c1c7 cells)	↑ AHR translocation ↑ XRE-dependent luciferase activity ↑ CYP1A1 mRNA ↑↑ AHR ligand-mediated induction of CYP1A1 mRNA and protein ↓ AHR ligand-mediated induction of CYP1A1 activity ↔ CYP1A1 mRNA half-life ↑ CYP1A1 protein half-life	[83, 96]
Cu^{2+}	Antarctic fish (liver)	↔ B[a]P-mediated induction of CYP1A mRNA ↓ B[a]P-mediated induction of CYP1A protein and activity	[136]
	Human hepatoma (HepG2 cells)	↑ CYP1A1 mRNA ↑↑ TCDD-mediated induction of CYP1A1 protein and catalytic activity	[119, 142]
	Mouse hepatoma (Hepa 1c1c7 cells)	↑ AHR transformation ↔ AHR translocation ↑ CYP1A1 mRNA ↓ AHR ligand-mediated induction of CYP1A1 mRNA, protein, and activity ↔ CYP1A1 mRNA half-life ↑ CYP1A1 protein half-life	[117]

demonstrated that V^{5+} was able to decrease the TCDD-mediated induction of CYP1A1 mRNA, protein, catalytic activity levels and XRE-dependent luciferase reporter activity in both Hepa 1c1c7 and HepG2 cells [145, 146]. This inhibition in XRE function coincided with a similar inhibition in the nuclear accumulation of the transformed AHR in Hepa 1c1c7 cells. Furthermore, the inhibition of AHR nuclear accumulation was not due to a decrease in the AHR protein expression levels. Our results also showed that V^{5+} did not significantly alter the HO-1 mRNA level or total cellular heme content. Thus, V^{5+} modulates CYP1A1 function solely by inhibiting transcription in an AHR-dependent manner. Although V^{5+} failed to induce HO-1 [147], V^{5+} was found to induce the expression of HIF-1α and subsequently VEGF through the phosphatidylinositol 3-kinase/Akt pathway and ROS [148]. Since HIF-1α and AHR share ARNT for hypoxia- and AHR-mediated signaling pathways, respectively, we might suggest that hypoxia inhibits TCDD-mediated induction of CYP1A1 expression due to the competition between HIF-1α and the AHR for ARNT [149]. Furthermore, the observed effect of V^{5+} on CYP1A1 might be due to two other possible mechanisms. First, V^{5+} activates the p38 MAPK and subsequently induces the transcription of NF-κB that would inhibit the AHR activity [150, 151]. Second, it has been well documented that cytosolic AHR requires phosphorylation prior to ligand binding; thus, it might have been that V^{5+} through affecting MAPK inhibited the phosphorylation of AHR with a subsequent inhibition to its ATP-dependent nuclear translocation [152]. The postulated effect of heavy metals on CYP1A1 expression is summarized in Fig. 25.1.

25.5 DISEASE STATES AND AHR

Although the AHR expression levels are altered in different disease states, the cause–effect relationship between AHR on one side and the disease states on the other side remains a matter of debate. Therefore, in this section we are going to illustrate the recent findings with regard to the mutual interaction between AHR and different disease states, namely, carcinogenesis, reproduction and teratogenicity, cardiovascular diseases, immune diseases, and neural diseases.

25.5.1 Carcinogenesis

The carcinogenicity of AHR ligands including TCDD has been recently found to involve the AHR [153]. Animal studies (multisite carcinogenicity) and increased overall

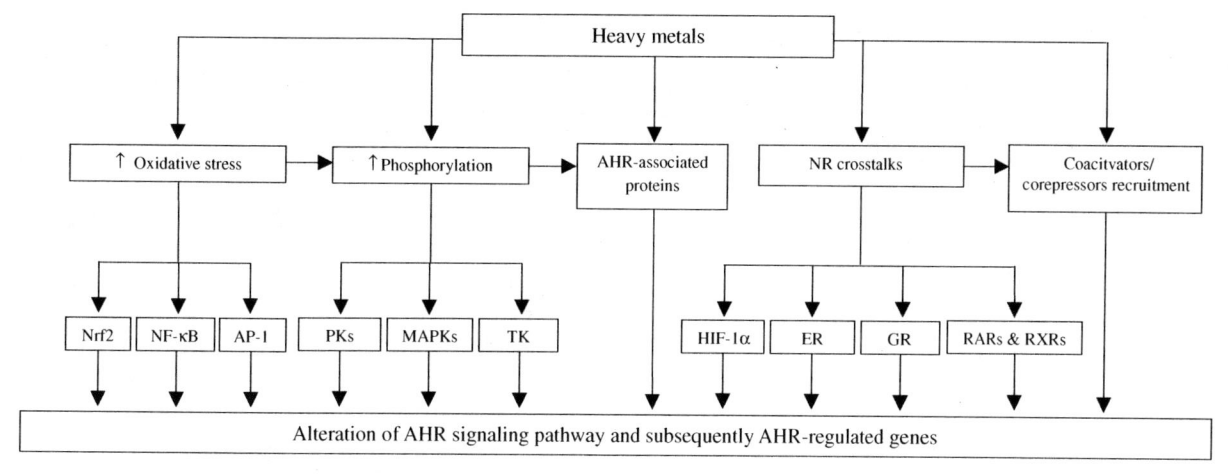

FIGURE 25.1 Postulated mechanisms for the effect of heavy metals on AHR activity biomarkers.

cancer mortality from industrial cohorts in several countries (Netherlands, Germany, and United States) lend support to this hypothesis [4]. Epidemiological data from the industrial cohorts with high exposure to AHR ligands showed that the pattern of increased types of cancer (lung, gastrointestinal, soft tissue sarcoma, non-Hodgkin lymphoma) varied considerably between these cohorts [4]. For instance, the standard mortality ratios for non-Hodgkin lymphoma were 4.6 and 3.8 in German and Dutch studies [154, 155], while the value in the United States was 0.9 [156]. Because most industrial cohorts were exposed to TCDD as a trace cocontaminant along with other contaminants such as chlorinated phenols, phenoxy herbicides, PCBs, PAHs, and metals, the increased incidence of cancer in these human cohorts could not be only TCDD dependent but rather AHR dependent. Importantly, association between AHR ligands including PAHs or HAHs and breast cancer incidences was reported [157], particularly in genetically susceptible subsets of women [158]. The risk of breast cancer recurrence was associated with adipose tissue PCB concentration. However, several recent prospective and retrospective studies failed to confirm association between HAH exposure and increased breast cancer risk [158].

TCDD-induced tumors have been observed in different rodent models that were species, age, and sex variable. With the fact that TCDD is not genotoxic by itself, it is likely that these carcinogenic effects are associated with the action of TCDD as a tumor promoter rather than tumor initiator [159]. Major target organs for the TCDD-induced tumorigenesis are the liver, thyroid, oral cavity, and lung in rats and the liver, thymus, and skin in mice [160]. In a 2-year feeding study, TCDD induced hepatocellular carcinogenesis in female but not male Sprague Dawley rats that was estrogen dependent [161]. On the contrary and in the same study, it was shown that the antitumorigenic effects of TCDD were due to blocking ER action [161]. This inhibitory effect of TCDD resulted in inhibition of multiple age-dependent spontaneous tumors, including pituitary adenomas, pituitary adenocarcinomas, interfollicular adenoma of the thyroid, benign uterine tumors, and benign neoplasma of the mammary gland [162].

The AHR appears to have constitutive expression and activity in established cell lines such as the adult T-cell leukemia (ATL) and in some primary ATL cell cultures, suggesting the existence of oncogenic activity associated with AHR expression in this tumor type [6]. In contrast, it was shown that acute lymphoblastic leukemia (ALL) REH and chronic myeloid leukemia (CML) K562 cell lines did not express AHR and had higher levels of AHR promoter methylation [6]. Interestingly, the AHR was methylated in 33% of the 21 human primary ALL tumors [6], suggesting that AHR silencing in this tumor type might be reflecting an AHR-mediated tumor suppression activity.

By far the most frequently studied set of genes induced by the AHR are those encoding phase I CYP450 enzymes, CYP1A1, CYP1A2, and CYP1B1, which metabolize at least some environmental AHR ligands into mutagenic intermediates [163]. The role of CYP1 enzymes in mediating PAH toxicity can be demonstrated from epidemiological studies showing a correlation between the high CYP1 inducibility phenotype in cigarette smokers and cancers of the lung, larynx, and oral cavity [164]. Xenobiotics have been shown to be activated by the CYP1 family into genotoxic and carcinogenic metabolites. Carcinogenic PAHs are activated by several drug metabolizing enzymes to highly mutagenic metabolites responsible for their toxicity [165]. The major activation pathway for PAHs, however, is through the formation of diol epoxides by members of the CYP450 and epoxide hydrolase enzymes [165]. B[a]P, the prototype carcinogenic PAH, is metabolized by CYP1A1 to B[a]P-7,8-oxide that is subsequently hydrolyzed by epoxide hydrolase to (+)- and (−)-B[a]P-7,8-diol. CYP1A1 is then again responsible for the conversion of these metabolic intermediates into the toxic bay region metabolites, (−)-B[a]P-7,8-diol-9,10-oxide-1, (+)-B[a]P-7,8-diol-9,10-oxide-2,

(+)-B[a]P-7,8-diol-9,10-oxide-1, and (−)-B[a]P-7,8-diol-9,10-oxide-2 [165]. All four epoxides are highly mutagenic, yet (+)-B[a]P-7,8-diol-9,10-oxide-2 was identified as the most tumorgenic of the metabolites [166]. The toxicity of other bay region epoxide metabolites has similarly been reported for other PAHs, including benzo[a]anthracene-3,4-diol, benzo[b]fluoranthene-9,10-diol, and benzo[c]phenanthrene-3,4-diol [165].

Other members of the CYP1 family are also involved in xenobiotic activation. CYP1B1 metabolizes PAHs and aryl and heterocyclic amines to carcinogenic metabolites [167]. In fact, Cyp1b1-null mice were not able to metabolize 7,12-DMBA and were resistant to the formation lymphomas and other malignancies caused by 7,12-DMBA [168]. Furthermore, CYP1B1 is the principal precipitant of PAH-mediated immunotoxicity [39]. Interestingly, in vitro studies demonstrated a 10-fold higher activity for CYP1B1 than CYP1A1 in converting B[a]P to B[a]P-7,8-diol in a reconstituted human enzyme system [169]. However, similar rates of activation were reported for CYP1A1 and CYP1B1 in the activation of various PAH diols, including (−)-B[a]P-7,8-diol, 7,12-DMBA-3,4-diol, and B[a]A-3,4-diol [170, 171]. On the other hand, CYP1A2 demonstrated different enzyme kinetics as it had slower rates in the activation of these diols [165].

Although there has been no clear link between exposure to heavy metals such as As^{3+} and V^{5+} and their mediated effects on AHR activity to carcinogenesis, there have been few attempts to explain some of these anticancer effects. As such, As^{3+} (Trisenox™) has been successfully used for the treatment of acute promyelocytic leukemia and multiple myeloma [172, 173]. In addition, it is also being considered for treatment against solid tumors [173]. Similarly, V^{5+} compounds exert protective effects against chemical-induced carcinogenesis in animals, by modifying various xenobiotic enzymes, possibly the AHR-regulated CYP1 family. Thus, V^{5+} inhibits carcinogen-derived active metabolites' generation through this mechanism. Moreover, recent studies have suggested V^{5+} as an effective nonplatinum metal antitumor agent [174]. As^{3+} and V^{5+} are believed to mediate their anticancer effect mainly through three different mechanisms: first by inactivation of the carcinogen-generating metabolizing enzymes such as CYP1A1, second through affecting cell proliferation, and finally, through inducing cellular oxidative stress [175].

25.5.2 Reproduction and Teratogenicity

Previous studies have demonstrated that lack of AHR gene expression in AHR-deficient mice along with the unscheduled activation of AHR by HAHs results in adverse effects in female reproduction organs leading to impaired reproduction [176]. For example, female AHR-deficient mice exhibit defects in multiple reproductive aspects, such as conception, litter size, and pup survival rate [177]. These defects are further reflected in the decreased number of ovulated oocytes and disordered estrous cycle [176]. Both estrous cycle and oocyte development in the ovary are controlled by the complementary actions of the anterior pituitary hormones, luteinizing hormone (LH) and follicle-stimulating hormone (FSH), and ovarian steroid hormones such as androgens, estrogens, and progestins [178]. In normal estrous cycle, these hormonal signals induce the maturation of granulosa cells surrounding the oocyte triggering follicle rupture that would subsequently cause the release of the oocyte into the fallopian tube [179]. In this regard, the number of mature follicles and corpora lutea (CL) were significantly reduced in the superovulation process in AHR-deficient mice compared to wild-type mice [180]. Although the serum LH and FSH levels were not changed between AHR-deficient and wild-type mice, at all stages of the estrous cycle, estradiol synthesis was lower in the AHR-deficient mice and was shown to be responsible for these reproduction defects [180].

During the past couple of decades, efforts have been made to better understand the relationship between estradiol synthesis-mediated reproduction defects and AHR deficiency. It is well documented that estrogen is synthesized from cholesterol in the ovary [181], and it is therefore speculated that all the enzymatic machinery involved in estradiol synthesis is present in the ovary. Specifically, FSH secreted from the anterior pituitary binds to FSH receptor (FSHR) and stimulates CYP450 aromatase (CYP19) gene expression [182], a rate-limiting enzyme in estrogen synthesis that converts testosterone to estradiol [182]. In contrast, other steroidogenic enzymes such as CYP11A, CYP450 17α-hydroxylase (CYP17), 3β-hydroxysteroid dehydrogenase (3β-SHD), and 17β-hydroxysteroid dehydrogenase (17β-SHD) are constitutively expressed [183]. In this regard, it has been demonstrated that the expression of these constitutive enzymes with the exception of CYP19 was not changed between AHR-deficient and wild-type mice [184]. Interestingly, in AHR-deficient mice CYP19 was not induced, compared to wild-type mice, in the proestrous stage leading to significantly lower ovarian estradiol concentrations [177]. Furthermore, the reproductive defects in CYP19- or ER-deficient mice were similar to those observed in AHR-deficient mice confirming its role in regulating CYP19 and subsequently controlling the levels of estradiol [176].

In addition to regulating estrous cycle, AHR has been shown to play an important role during the preimplantation period. As such, AHR is constitutively expressed in the uterine vasculature and developing tissues between the embryo and mother [185]. In addition to reduced fertility, the AHR-deficient female mice are unable to raise their pups to the weaning age [186], and these pups in return have lower survival rates than those raised by wild-type female mice, an impairment that might be more related to mammary glands.

Activating the AHR beyond physiological limits by means of exposure to potent inducers such as TCDD affects developing embryos of several species. Thus, it has become apparent that TCDD-induced teratogenesis requires AHR, and the first step in this cascade is the ligand activation of AHR. In developing mice, prenatal exposure to AHR ligands such as TCDD results in teratogenic effects at doses below those that can cause maternal or embryo/fetal toxicity [187]. Teratogenic effects mediated by AHR include thymic involution, cleft palate, hydronephrosis, aberrant cardiac development, and inhibition of ventral, dorsolateral, and anterior prostatic bud development [188]. The thymic involution and hydronephrosis were also observed in mice harboring mutant AHR that is constitutively active in the absence of a ligand [189]. These results highlight the importance of proper AHR regulation during development.

The AHR nonresponsive mouse strain DBA/2 expresses an AHR encoded by the Ahr^d allele that binds TCDD with low affinity [190]. In contrast, the C57BL/6 strain expresses an AHR that is encoded by AHR^{b1} allele that binds TCDD with high affinity [188]. Comparing the fetuses of both strains revealed that DBA/2 fetuses exposed to TCDD developed cleft palate and hydronephrosis to a much lower extent than C57BL/6 fetuses [188]. Similarly, AHR-deficient fetuses displayed minimal incidences of cleft palate and hydronephrosis when compared to AHR wild-type fetuses [188]. Therefore, the existence of AHR and the affinity of AHR to TCDD are undeniably important in determining the teratogenic outcome.

Despite the prominent role of AHR in mediating teratogenicity, the embryonic activity of most CYP450s is low to negligible, and their role in xenobiotic bioactivation is still at large [191]. However, it was previously reported that embryo cultures exposed to B[a]P and 2-acetylaminofluorene have elevated levels of both CYP1A1 and CYP1A2 mediating the bioactivation of these xenobiotics [192]. In humans, CYP3A5 and CYP2C19 are the most abundant forms throughout pregnancy, and therefore they might play a role in the teratogen bioactivation [191]. Human CYP450-mediated bioactivation has been implicated yet not proven in fetal hydantoin syndrome caused by in utero exposure to phenytoin [193]. The activities of some CYP450s increase in the fetus later and could contribute to the teratogenic outcome. In contrast, there are some developmentally expressed CYP450s such as CYP1B1/2 in rodents and CYP3A7 in humans that have high activity during embryonic development but decline rapidly at birth. These enzymes might play an important role in the bioactivation of teratogens such as B[a]P and aflatoxin B1 that is highly metabolized by CYP3A7 [193].

25.5.3 Cardiovascular Diseases

The AHR protein expression has been most commonly studied in the liver where its concentration ranges from 20 to 300 fmol/mg of cytosolic protein [194, 195]. In contrast, very little is known about the expression and function of the AHR in the heart. Studies investigating this phenotype have demonstrated that defects in the resolution of fetal vascular structure were responsible for this phenotype [177]. In the embryonic vasculature, the flow of blood partially bypasses the liver sinusoids via a shunt known as ductus venosus (DV) [196], which directly connects the inferior vena cava and the portal vein. In normally developing embryos, the DV closes shortly after birth, forcing oxygen- and nutrient-rich blood to migrate through liver sinusoids [177]. In the absence of AHR, the DV remains open throughout adulthood, and as a consequence this will lead to reduced postnatal liver growth due to deprivation of nutrients and oxygen [177]. In addition to the DV phenotype, AHR-deficient mice have been shown to suffer from cardiac hypertrophy, hypertension, and elevated levels of the potent vasoconstrictors endothelin-1 and angiotensin II [197]. Interestingly, mounting evidence indicates that vascular AHR signaling is mechanistically linked to shear stress that can be mimicked by blood flow through vascular system [198]. As such, several independent reports have demonstrated that AHR is highly activated by cellular exposure to fluid shear. Interestingly, recent evidence indicated that fluid shear stress activates the AHR through a direct effect on serum low-density lipoprotein (LDL) function and structure; however, AHR activation through modified LDL does not seem to be solely through shear-induced modifications [199].

Early studies on the expression of AHR in different organ systems of healthy human subjects showed that AHR mRNA was expressed in several organs in the following order (from higher to lower): placenta > lung > heart > kidney > liver [2]. The intensity and localization of AHR was analyzed in the left ventricle of healthy persons as well as ischemic and dilative cardiomyopathy subjects [200]. In this study, it was shown that AHR protein levels were twofold higher in dilative and ischemic cardiomyopathy than in healthy subjects, suggesting a possible cause–effect relationship between AHR overexpression and cardiovascular diseases.

Studies conducted on embryonic chicken heart showed that both AHR and ARNT are constitutively expressed during the cardiogenesis process in the myocytes of aorta, atrium, ventricle, and atrioventricular canal, but not in the endothelial cells lining the heart valves and septa [201, 202]. Surprisingly, although both chicken embryonic heart and liver express the same level of the AHR protein, TCDD was able to induce the hepatic but not cardiac *CYP1A1* gene [202]. This discrepancy suggests that the TCDD-mediated signaling cascades of cardiac AHR are different from those in the liver, and implies the presence of a tissue-specific transcriptional activation of the AHR [202].

ARNT serves as a common dimer partner for both AHR and HIF-1α. A target gene downstream the signaling cascade of HIF-1α is the VEGF [203]. Licht et al. showed that the expression of HIF-1α in the cardiovascular system is essential for embryonic cardiogenesis and blood vessel development, in which inhibition of HIF-1α activity resulted in a thin ventricular wall, disarranged endocardium, and impaired remodeling of the embryonic mouse vascular system [204]. Of particular interest, C57BL/6J mouse embryo exhibited a high level of HIF-1α expression in the myocardial wall and atrioventricular canal of the developing heart [205].

Although little is known about the expression of CYP450s in the cardiovascular system, the expression of cardiac AHR-regulated CYP450s remains a matter of debate. Studies on AHR-deficient and wild-type mice demonstrated very low to negligible levels of cardiac constitutive expression of CYP1A1 and CYP1A2 mRNA and catalytic activities [206]. In contrast, CYP1B1 mRNA is predominantly expressed at a significant level in the heart of both strains representing a total of 13% of total cardiac CYP450s [206]. This was also observed in adult, but not fetal human heart [206]. At the inducible level, both cardiac *CYP1A1* and *CYP1B1* genes were significantly induced by different AHR ligands such as B[*a*]P, PCB 126, β-naphthoflavone, and 3MC in different mammalian species [207]. Furthermore, cardiac CYP1A1-dependent EROD activity was induced by fourfold in PCB 126-treated wild-type mice, while this induction was completely abolished in the AHR-deficient mice [207]. Similarly, Thackaberry et al. showed that TCDD was able to induce *Cyp1a1* and *Cyp1b1* gene expression by 13- and 5-fold, respectively, while it did not alter *Cyp1a2* gene expression in wild-type mouse fetuses [208]. In contrast to rodents, studies on human and chicken embryos demonstrated persistent expression of CYP1A1 in the cardiovascular tissue. For example, embryonic chicken heart has shown a strong constitutive expression of CYP1A1 protein [209].

The importance of these AHR-regulated CYP450s lies behind their prominent role in the degradation, synthesis, and/or metabolism of cardioactive endogenous substances such as arachidonic acid, prostaglandins, and thyroid hormones that have been shown to be effectively contributing to the pathogenesis of cardiovascular diseases [210]. Although the exact mechanism of this effect is not fully understood, it has been experimentally demonstrated that the induction of CYP1A1 mediated the metabolism of arachidonic acid to hydroxyeicosatrienoic acid (HETE), epoxyeicosatrienoic acid (EET), and PGE2 metabolites in hearts of both rat and chicken embryos [211]. Furthermore, CYP1B1 can metabolize arachidonic acid to both mid-chain HETEs and EETs, suggesting that these endogenous metabolites are involved in AHR-mediated cardiotoxicity [212]. Bearing in mind that HETE is a potent vasoconstrictor, and EET is a vasodilator, the exposure to different AHR ligands will be detrimental in the cardiovascular pathophysiology. In this regard, we have recently demonstrated that 3MC and B[*a*]P induced cardiac hypertrophy in male Sprague Dawley rats partially due to increased HETE production by the AHR-regulated CYP450s, CYP1A1 and CYP1B1 [213].

The cardiotoxic effects of As^{3+} were reported earlier in patients exposed to pesticides, metallurgy, textiles, glass, and paint [214]. Due to the fact that As^{3+} is now being used for the treatment of refractory acute promyelocytic leukemia, its cardiotoxic effects have also been well documented [214]. As^{3+} has known cardiographic abnormalities that include low flat T-wave, sinus tachycardia, prolonged QT intervals, atrioventricular blocks, multifocal ventricular tachycardia, and ventricular fibrillation [215]. The most common change is prolongation of the QT interval, seen in 38% ($n = 99$) of patients treated with As^{3+} for advanced malignancies [216]. A^{3+}-induced myocarditis has also been reported in the literature [217]. In contrast, Cu^{2+} supplementation has been shown to prevent or ameliorate cardiomyopathy [218]. As such, it has been observed that dietary Cu^{2+} supplementation (20 mg Cu^{2+}/kg diet) can reverse an experimentally induced cardiac hypertrophy and improve heart contractile function in mice with established heart hypertrophy and dysfunction induced by ascending aortic constriction [219]. Conversely, it was also shown that ascending aortic constriction caused a decrease in cardiac Cu^{2+} levels, which can be corrected by dietary Cu^{2+} supplementation [220]. Altogether, despite the reported cardiac effects of both As^{3+} and Cu^{3+}, there have been no attempts to correlate these effects to the alteration in AHR-regulated CYP450s.

25.5.4 Immune Diseases/Inflammation

Immunotoxicology occurs via adverse interference by xenobiotics with the immune system, such as allergic reactions, drug-induced autoimmunity, or immunosupression by TCDD [221]. In this regard, TCDD has been shown to be immunosuppressive in laboratory animals. This immunosuppressive effect has been shown to be limited not only to TCDD, but also to other AHR ligands such as PAHs and HAHs [222]. In general, the immunosuppressive effects of these AHR ligands are decreased host resistance to infectious disease, and suppressed humoral and cell-mediated immune responses [222]. In addition to its immunosuppressive effect, TCDD promotes inflammatory responses via upregulating the production of inflammatory cytokines such as interleukin-1 (IL-1) and tumor necrosis factor-α (TNF-α) [223].

In an effort to distinguish AHR-mediated TCDD toxicity from those resulting from alternative pathways, the AHR-deficient mice were compared with wild-type mice regarding the role of AHR in immunosupression [224]. AHR-deficient mice are defective in T-cell differentiation and are more susceptible to bacterial infection. Most of the cells that participate in immune responses either constitutively or inducibly express AHR at a certain level throughout the responsive process. Importantly, many of the genes involved in immune response such as IL-1Rs, IL-6, IL-18, and all nine Toll-like receptors (TLRs) have XRE sequences in their promoters, although it is not clear whether or not AHR is directly involved in controlling their expression [39].

In contrast to the previously mentioned effect, studies carried out on AHR-deficient mice have shown that treatment with LPS rendered these mice more susceptible to septic shock [225]. In addition, these mice were also hypersensitive to LPS-induced lethal shock compared to their littermate wild-type mice [225]. Upon treatment with LPS, the serum levels of IL-6, TNF-α, and IL-1β were higher in AHR-deficient mice compared to wild-type mice [225]. In agreement with these *in vivo* studies, it was shown that isolated LPS-treated peritoneal macrophages from AHR-deficient mice had much higher IL-6 secretion than those from wild-type mice [225]. In an effort to confirm the role of AHR in the release of IL-6 from these macrophages, AHR-deficient mice macrophages transfected with AHR expression plasmid showed suppression of IL-6 expression upon LPS treatment [226]. Conversely, it was also demonstrated that LPS treatment enhances AHR expression in peritoneal macrophages [226]. An explanation offered to explain AHR suppressive effect is that AHR by forming a complex with Stat1 and NF-κB leads to inhibition of the promoter activity of the *IL-6* gene [226]. With regard to IL-1β secretion, it was shown that AHR activated the expression of *Pai-2*, an inhibitor of caspase-1 activation that functions in the process of IL-1β secretion [227]. Treatment of wild-type macrophages with AHR ligands markedly enhanced *Pai-2* expression, and similarly transduction of *Pai-2* expression in AHR-deficient macrophages restored IL-1β secretion [228]. Activated AHR together with NF-κB but not ARNT directly enhanced *Pai-2* gene expression by binding to its promoter [225, 228]. Using human macrophage cell line U937, it was shown that TCDD induced the expression of IL-8, Bcl (B-cell chemoattractant), and Baff (B-cell activating factor) [229, 230]. These increases in gene expression were shown to happen with a concomitant binding of the AHR–RelB complex to κB binding site in the promoters of these genes [229]. Interestingly, this process was ARNT independent suggesting the presence of a new transcriptional mechanism that involves TCDD-activated AHR.

It has been previously reported that AHR ligands stimulate naïve T cells (Th0) to differentiate into helper T cells that either suppress or accelerate the immune responses by modulating effector T-cell proliferation and cytokine secretion [39]. Interestingly, it was also shown that the profound suppression of acute graft versus host (GVH) response in TCDD-treated mice was associated with the generation of donor-derived $CD4^+$ $CD25^+$ T_{reg} cells that was AHR dependent [231]. In addition, the promoter of the *FoxP3* gene, a transcription factor that plays a master role in T_{reg} cell differentiation, contains functional XRE sequences that are responsible for inducing this gene in the presence of activated AHR [225]. Furthermore, chromatin immunoprecipitation (ChIP) assay revealed that AHR is recruited to the *Foxp3* promoter [225].

Previous studies have demonstrated that the AHR-regulated CYP450s such as CYP1A1 and CYP1A2 were resistant to the downregulation induced by interferon unlike other CYP450s such as CYP3A2, CYP2C11, CYP2C12, and CYP2E1 that were significantly downregulated in response to interferon [232]. In primary rat and human hepatocyte cells, the media obtained from activated Kupffer cells in addition to IL-1 decreased the total P450 content of these cells [233]. Subsequently, recombinant cytokines including IL-1α, IL-1β, IL-6, TNF, and tumor growth factor were shown to suppress the expression of CYP1A1 and CYP1A2 in rodents *in vivo* and in HepG2 cells *in vitro* [234]. Importantly, the suppressive effects of TNF-α and IL-1β are at least in part due to a direct NF-κB action [66]. In this regard, a few studies have suggested that in addition to the transcriptional inhibition of CYP1A1 and CYP1A2 by these cytokines, these cytokines might also increase the turnover rate of these transcripts [235].

It has been clearly demonstrated that TCDD causes chronic and sustained oxidative stress in animals, likely due to the production of reactive oxygenated metabolites by CYP1 enzymes [236]. The production of reactive oxygen species occurs also during inflammation and/or infections [237]. Thus, ROS produced *in vivo* and *in vitro* have been implicated in the cascade of events leading to the loss of CYP450s during inflammation [238]. As such, it has been shown that hydrogen peroxide and oxidative stress cause loss of hepatic CYP1A1 and CYP1A2 mRNAs in isolated rat hepatocytes [239].

Metallothionein serves as one of the points of intersection for many of the immune and autoimmune disease states [218]. Although this small stress response protein is not a member of the heat shock protein family, it serves many roles in both normal and stressed cells, acting as a reservoir of essential heavy metals such as Cu^{2+}, as a scavenger for heavy metal toxicants such as Hg^{2+}, Cd^{2+}, and their associated free radicals, and as a regulator of transcription factor activity [240]. With the striking fact that AHR wild-type mice have relatively lower expression of metallothioneins compared to AHR-deficient mice [241], it might be a possibility that in the presence of AHR and the consequent reduction of functional metallothionein, animals exposed to either Cd^{2+} or Hg^{2+} would have an elevation in these heavy metals-mediated immunotoxicity [218].

25.5.5 Neuronal Diseases

During development, exposure to TCDD and related compounds affects a wide array of brain functions even at low doses [242]. An earlier report showed that exposure of rats to 10–100 nM stimulated Ca^{2+} uptake in hippocampal neurons within 40 s [243]. Moreover, Ca^{2+} uptake was antagonized by the calcium channel blocker nifedipine [243]. In the nervous system of zebrafish, TCDD-induced effects were shown to be exerted through ligand-activated AHR together with ARNT [242]. In addition to zebrafish, AHR and its partner ARNT were identified in the olfactory bulb, cerebral and cerebellar cortices, and hippocampus of TCDD-exposed rats [244]. Moreover, it has been suggested that TCDD could have significant endocrine-disruptive effects on the gonadal and thyroid hormone axes, as well as neural-disrupting action on neural transmission and neural network formation [245].

According to the literature, Han–Wistar rats are among the most resistant of all mammals to TCDD toxicity [242]. The reason behind this resistance is a mutation in the AHR gene that alters and truncates the C-terminal transactivation domain of the AHR protein [242]. Of interest, Han–Wistar rats show normal acute induction of the CYP1 family genes. With the fact that TCDD-mediated toxicity in rats develops over weeks and sometimes over months of exposure time [246], it is reasonable to conclude that the TCDD-induced neural toxicity necessitates acute and chronic AHR activity to mediate these effects.

Exposure to environmental toxins such as PAHs and HAHs in genetically predisposed individuals might have an important role in etiopathogenesis of neurodegenerative disorders [247]. For example, Parkinson's disease has been associated with mutations in the *CYP2D6* and *CYP1A1* genes [248]. In addition, the hepatic CYP1A2 has been shown to be involved in the metabolism of 1-methyl-4-phenyl-1,2,3,6-tetrahydropyridie to the neurotoxic metabolite 1-methyl-4-phenylpyridinium ion that produces Parkinson's like syndrome [249]. In addition to CYP1A1 and CYP1A2, CYP1B1 is highly expressed in the human blood–brain barrier, brain microvessels, brain glial cells, and dura matter [250, 251]. Therefore, due to its high level of expression, CYP1B1 might be a contributor to neurodegenerative disorders.

Pb^{2+} is a ubiquitous environmental pollutant in industrial environment that poses a serious threat to human health [252]. It has been demonstrated that heavy metals may play a role in the pathogenesis of diseases of nervous system, such as multiple sclerosis, amylotrophic sclerosis, and Alzheimer's [253]. For example, Waterman et al. have shown that Pb^{2+} exposure leads to the production of autoantibodies against neural proteins, including myelin basic protein and glial fibrillary acidic protein, concluding that Pb^{2+} can aggravate neurological disease by increasing the immunogenicity of nervous system proteins [254]. However, the role of AHR-regulated CYP450s in this regulation is undeniably at large.

REFERENCES

1. Nebert, D. W. and Duffy, J. J. (1997). How knockout mouse lines will be used to study the role of drug-metabolizing enzymes and their receptors during reproduction and development, and in environmental toxicity, cancer, and oxidative stress. *Biochemical Pharmacology*, 53, 249–254.
2. Dolwick, K. M., Swanson, H. I., and Bradfield, C. A. (1993). *In vitro* analysis of Ah receptor domains involved in ligand-activated DNA recognition. *Proceedings of the National Academy of Sciences of the United States of America*, 90, 8566–8570.
3. Kazlauskas, A., Poellinger, L., and Pongratz, I. (2000). The immunophilin-like protein XAP2 regulates ubiquitination and subcellular localization of the dioxin receptor. *Journal of Biological Chemistry*, 275, 41317–41324.
4. Safe, S. (2001). Molecular biology of the Ah receptor and its role in carcinogenesis. *Toxicology Letters*, 120, 1–7.
5. Crews, S. T. and Fan, C. M. (1999). Remembrance of things PAS: regulation of development by bHLH–PAS proteins. *Current Opinion in Genetics & Development*, 9, 580–587.
6. Barouki, R., Coumoul, X., and Fernandez-Salguero, P. M. (2007). The aryl hydrocarbon receptor, more than a xenobiotic-interacting protein. *FEBS Letters*, 581, 3608–3615.
7. Whitelaw, M. L., Gottlicher, M., Gustafsson, J. A., and Poellinger, L. (1993). Definition of a novel ligand binding domain of a nuclear bHLH receptor: co-localization of ligand and hsp90 binding activities within the regulable inactivation domain of the dioxin receptor. *EMBO Journal*, 12, 4169–4179.
8. Ikuta, T., Eguchi, H., Tachibana, T., Yoneda, Y., and Kawajiri, K. (1998). Nuclear localization and export signals of the human aryl hydrocarbon receptor. *Journal of Biological Chemistry*, 273, 2895–2904.
9. Mhin, B. J., Lee, J. E., and Choi, W. (2002). Understanding the congener-specific toxicity in polychlorinated dibenzo-*p*-dioxins: chlorination pattern and molecular quadrupole moment. *Journal of the American Chemical Society*, 124, 144–148.
10. Denison, M. S. and Nagy, S. R. (2003). Activation of the aryl hydrocarbon receptor by structurally diverse exogenous and endogenous chemicals. *Annual Review of Pharmacology and Toxicology*, 43, 309–334.
11. Hankinson, O. (1995). The aryl hydrocarbon receptor complex. *Annual Review of Pharmacology and Toxicology*, 35, 307–340.
12. Pollenz, R. S. (2002). The mechanism of AH receptor protein down-regulation (degradation) and its impact on AH receptor-mediated gene regulation. *Chemico-Biological Interactions*, 141, 41–61.
13. Ma, Q. and Whitlock, J. P., Jr. (1996). The aromatic hydrocarbon receptor modulates the Hepa 1c1c7 cell cycle and differentiated state independently of dioxin. *Molecular and Cellular Biology*, 16, 2144–2150.
14. Kasai, S. and Kikuchi, H. (2010). The inhibitory mechanisms of the tyrosine kinase inhibitors herbimycin A, genistein, and tyrphostin B48 with regard to the function of the aryl hydrocarbon receptor in Caco-2 cells. *Bioscience, Biotechnology, and Biochemistry*, 74, 36–43.
15. Timme-Laragy, A. R., Cockman, C. J., Matson, C. W., and Di Giulio, R. T. (2007). Synergistic induction of AHR regulated genes in developmental toxicity from co-exposure to two model PAHs in zebrafish. *Aquatic Toxicology*, 85, 241–250.

16. Carver, L. A., Jackiw, V., and Bradfield, C. A. (1994). The 90-kDa heat shock protein is essential for Ah receptor signaling in a yeast expression system. *Journal of Biological Chemistry*, *269*, 30109–30112.
17. Kikuchi, H., Hossain, A., Yoshida, H., and Kobayashi, S. (1998). Induction of cytochrome P-450 1A1 by omeprazole in human HepG2 cells is protein tyrosine kinase-dependent and is not inhibited by alpha-naphthoflavone. *Archives of Biochemistry and Biophysics*, *358*, 351–358.
18. McLemore, T. L., Adelberg, S., Czerwinski, M., Hubbard, W. C., Yu, S. J., Storeng, R., Wood, T. G., Hines, R. N., and Boyd, M. R. (1989). Altered regulation of the cytochrome P4501A1 gene: novel inducer-independent gene expression in pulmonary carcinoma cell lines. *Journal of the National Cancer Institute*, *81*, 1787–1794.
19. Wille, G., Mayser, P., Thoma, W., Monsees, T., Baumgart, A., Schmitz, H. J., Schrenk, D., Polborn, K., Steglich, W., and Malassezin, A. (2001). A novel agonist of the arylhydrocarbon receptor from the yeast *Malassezia furfur*. *Bioorganic & Medicinal Chemistry*, *9*, 955–960.
20. Hankinson, O. (2005). Role of coactivators in transcriptional activation by the aryl hydrocarbon receptor. *Archives of Biochemistry and Biophysics*, *433*, 379–386.
21. Anwar-Mohamed, A., Elbekai, R. H., and El-Kadi, A. O. (2009). Regulation of CYP1A1 by heavy metals and consequences for drug metabolism. *Expert Opinion on Drug Metabolism & Toxicology*, *5*, 501–521.
22. Nguyen, T. A., Hoivik, D., Lee, J. E., and Safe, S. (1999). Interactions of nuclear receptor coactivator/corepressor proteins with the aryl hydrocarbon receptor complex. *Archives of Biochemistry and Biophysics*, *367*, 250–257.
23. Kumar, M. B., Tarpey, R. W., and Perdew, G. H. (1999). Differential recruitment of coactivator RIP140 by Ah and estrogen receptors. Absence of a role for LXXLL motifs. Journal of Biological Chemistry, *274*, 22155–22164.
24. Tohkin, M., Fukuhara, M., Elizondo, G., Tomita, S., and Gonzalez, F. J. (2000). Aryl hydrocarbon receptor is required for p300-mediated induction of DNA synthesis by adenovirus E1A. *Molecular Pharmacology*, *58*, 845–851.
25. Beischlag, T. V., Wang, S., Rose, D. W., Torchia, J., Reisz-Porszasz, S., Muhammad, K., Nelson, W. E., Probst, M. R., Rosenfeld, M. G., and Hankinson, O. (2002). Recruitment of the NCoA/SRC-1/p160 family of transcriptional coactivators by the aryl hydrocarbon receptor/aryl hydrocarbon receptor nuclear translocator complex. *Molecular and Cellular Biology*, *22*, 4319–4333.
26. Harnish, D. C., Scicchitano, M. S., Adelman, S. J., Lyttle, C. R., and Karathanasis, S. K. (2000). The role of CBP in estrogen receptor cross-talk with nuclear factor-kappaB in HepG2 cells. *Endocrinology*, *141*, 3403–3411.
27. Ke, S., Rabson, A. B., Germino, J. F., Gallo, M. A., and Tian, Y. (2001). Mechanism of suppression of cytochrome P-450 1A1 expression by tumor necrosis factor-alpha and lipopolysaccharide. *Journal of Biological Chemistry*, *276*, 39638–39644.
28. Chen, Y. H. and Tukey, R. H. (1996). Protein kinase C modulates regulation of the CYP1A1 gene by the aryl hydrocarbon receptor. *Journal of Biological Chemistry*, *271*, 26261–26266.
29. Long, W. P., Pray-Grant, M., Tsai, J. C., and Perdew, G. H. (1998). Protein kinase C activity is required for aryl hydrocarbon receptor pathway-mediated signal transduction. *Molecular Pharmacology*, *53*, 691–700.
30. Carrier, F., Owens, R. A., Nebert, D. W., and Puga, A. (1992). Dioxin-dependent activation of murine Cyp1a-1 gene transcription requires protein kinase C-dependent phosphorylation. *Molecular and Cellular Biology*, *12*, 1856–1863.
31. Okino, S. T., Pendurthi, U. R., and Tukey, R. H. (1992). Phorbol esters inhibit the dioxin receptor-mediated transcriptional activation of the mouse Cyp1a-1 and Cyp1a-2 genes by 2,3,7,8-tetrachlorodibenzo-*p*-dioxin. *Journal of Biological Chemistry*, *267*, 6991–6998.
32. Ikuta, T., Kobayashi, Y., and Kawajiri, K. (2004). Phosphorylation of nuclear localization signal inhibits the ligand-dependent nuclear import of aryl hydrocarbon receptor. *Biochemical and Biophysical Research Communications*, *317*, 545–550.
33. Cobb, M. H. and Goldsmith, E. J. (2000). Dimerization in MAP-kinase signaling. *Trends Biochemical Sciences*, *25*, 7–9.
34. Weston, C. R. and Davis, R. J. (2007). The JNK signal transduction pathway. *Current Opinion in Cell Biology*, *19*, 142–149.
35. Tan, Z., Chang, X., Puga, A., and Xia, Y. (2002). Activation of mitogen-activated protein kinases (MAPKs) by aromatic hydrocarbons: role in the regulation of aryl hydrocarbon receptor (AHR) function. *Biochemical Pharmacology*, *64*, 771–780.
36. Diry, M., Tomkiewicz, C., Koehle, C., Coumoul, X., Bock, K. W., Barouki, R., and Transy, C. (2006). Activation of the dioxin/aryl hydrocarbon receptor (AhR) modulates cell plasticity through a JNK-dependent mechanism. *Oncogene*, *25*, 5570–5574.
37. Shimizu, H., Banno, Y., Sumi, N., Naganawa, T., Kitajima, Y., and Nozawa, Y. (1999). Activation of p38 mitogen-activated protein kinase and caspases in UVB-induced apoptosis of human keratinocyte HaCaT cells. *Journal of Investigative Dermatology*, *112*, 769–774.
38. Chen, S., Operana, T., Bonzo, J., Nguyen, N., and Tukey, R. H. (2005). ERK kinase inhibition stabilizes the aryl hydrocarbon receptor: implications for transcriptional activation and protein degradation. *Journal of Biological Chemistry*, *280*, 4350–4359.
39. Fujii-Kuriyama, Y. and Mimura, J. (2005). Molecular mechanisms of AhR functions in the regulation of cytochrome P450 genes. *Biochemical and Biophysical Research Communications*, *338*, 311–317.
40. Andrieux, L., Langouet, S., Fautrel, A., Ezan, F., Krauser, J. A., Savouret, J. F., Guengerich, F. P., Baffet, G., and Guillouzo, A. (2004). Aryl hydrocarbon receptor activation and cytochrome P450 1A induction by the mitogen-activated protein kinase inhibitor U0126 in hepatocytes. *Molecular Pharmacology*, *65*, 934–943.
41. Bombick, D. W. and Matsumura, F. (1987). TCDD (2,3,7,8-tetrachlorodibenzo-*p*-dioxin) causes an increase in protein

tyrosine kinase activities at an early stage of poisoning *in vivo* in rat hepatocyte membranes. *Life Sciences, 41,* 429–436.

42. Blankenship, A. and Matsumura, F. (1997). 2, 3,7,8-Tetrachlorodibenzo-*p*-dioxin-induced activation of a protein tyrosine kinase, pp60^{src,} in murine hepatic cytosol using a cell-free system. *Molecular Pharmacology, 52,* 667–675.

43. Mahon, M. J. and Gasiewicz, T. A. (1995). Ah receptor phosphorylation: localization of phosphorylation sites to the C-terminal half of the protein. *Archives of Biochemistry and Biophysics, 318,* 166–174.

44. Minsavage, G. D., Park, S. K., and Gasiewicz, T. A. (2004). The aryl hydrocarbon receptor (AhR) tyrosine 9, a residue that is essential for AhR DNA binding activity, is not a phosphoresidue but augments AhR phosphorylation. *Journal of Biological Chemistry, 279,* 20582–20593.

45. Itoh, K., Chiba, T., Takahashi, S., Ishii, T., Igarashi, K., Katoh, Y., Oyake, T., Hayashi, N., Satoh, K., Hatayama, I., Yamamoto, M., and Nabeshima, Y. (1997). An Nrf2/small Maf heterodimer mediates the induction of phase II detoxifying enzyme genes through antioxidant response elements. *Biochemical and Biophysical Research Communications, 236,* 313–322.

46. Ma, Q., Kinneer, K., Bi, Y., Chan, J. Y., and Kan, Y. W. (2004). Induction of murine NAD(P)H:quinone oxidoreductase by 2,3,7,8-tetrachlorodibenzo-*p*-dioxin requires the CNC (cap 'n' collar) basic leucine zipper transcription factor Nrf2 (nuclear factor erythroid 2-related factor 2): cross-interaction between AhR (aryl hydrocarbon receptor) and Nrf2 signal transduction. *The Biochemical Journal, 377,* 205–213.

47. Miao, W., Hu, L., Scrivens, P. J., and Batist, G. (2005). Transcriptional regulation of NF-E2 p45-related factor (NRF2) expression by the aryl hydrocarbon receptor-xenobiotic response element signaling pathway: direct cross-talk between phase I and II drug-metabolizing enzymes. *Journal of Biological Chemistry, 280,* 20340–20348.

48. Kohle, C. and Bock, K. W. (2006). Activation of coupled Ah receptor and Nrf2 gene batteries by dietary phytochemicals in relation to chemoprevention. *Biochemical Pharmacology, 72,* 795–805.

49. Shin, S., Wakabayashi, N., Misra, V., Biswal, S., Lee, G. H., Agoston, E. S., Yamamoto, M., and Kensler, T. W. (2007). NRF2 modulates aryl hydrocarbon receptor signaling: influence on adipogenesis. *Molecular and Cellular Biology, 27,* 7188–7197.

50. Ohtake, F., Takeyama, K., Matsumoto, T., Kitagawa, H., Yamamoto, Y., Nohara, K., Tohyama, C., Krust, A., Mimura, J., Chambon, P., Yanagisawa, J., Fujii-Kuriyama, Y., and Kato, S. (2003). Modulation of oestrogen receptor signalling by association with the activated dioxin receptor. *Nature, 423,* 545–550.

51. Hoivik, D., Willett, K., Wilson, C., and Safe, S. (1997). Estrogen does not inhibit 2,3,7,8-tetrachlorodibenzo-*p*-dioxin-mediated effects in MCF-7 and Hepa 1c1c7 cells. *Journal of Biological Chemistry, 272,* 30270–30274.

52. Kharat, I. and Saatcioglu, F. (1996). Antiestrogenic effects of 2,3,7,8-tetrachlorodibenzo-*p*-dioxin are mediated by direct transcriptional interference with the liganded estrogen receptor. Cross-talk between aryl hydrocarbon- and estrogen-mediated signaling. *Journal of Biological Chemistry, 271,* 10533–10537.

53. Ricci, M. S., Toscano, D. G., Mattingly, C. J., and Toscano, W. A., Jr. (1999). Estrogen receptor reduces CYP1A1 induction in cultured human endometrial cells. *Journal of Biological Chemistry, 274,* 3430–3438.

54. Safe, S., Wang, F., Porter, W., Duan, R., and McDougal, A. (1998). Ah receptor agonists as endocrine disruptors: antiestrogenic activity and mechanisms. *Toxicology Letters, 102–103,* 343–347.

55. Gierthy, J. F., Spink, B. C., Figge, H. L., Pentecost, B. T., and Spink, D. C. (1996). Effects of 2,3,7,8-tetrachlorodibenzo-*p*-dioxin, 12-O-tetradecanoylphorbol-13-acetate and 17beta-estradiol on estrogen receptor regulation in MCF-7 human breast cancer cells. *Journal of Cellular Biochemistry, 60,* 173–184.

56. Monostory, K., Kohalmy, K., Prough, R. A., Kobori, L., and Vereczkey, L. (2005). The effect of synthetic glucocorticoid, dexamethasone on CYP1A1 inducibility in adult rat and human hepatocytes. *FEBS Letters, 579,* 229–235.

57. Brake, P. B., Zhang, L., and Jefcoate, C. R. (1998). Aryl hydrocarbon receptor regulation of cytochrome P4501B1 in rat mammary fibroblasts: evidence for transcriptional repression by glucocorticoids. *Molecular Pharmacology, 54,* 825–833.

58. Konig, H., Ponta, H., Rahmsdorf, H. J., and Herrlich, P. (1992). Interference between pathway-specific transcription factors: glucocorticoids antagonize phorbol ester-induced AP-1 activity without altering AP-1 site occupation *in vivo. EMBO Journal, 11,* 2241–2246.

59. Gambone, C. J., Hutcheson, J. M., Gabriel, J. L., Beard, R. L., Chandraratna, R. A., Soprano, K. J., and Soprano, D. R. (2002). Unique property of some synthetic retinoids: activation of the aryl hydrocarbon receptor pathway. *Molecular Pharmacology, 61,* 334–342.

60. Chambon, P. (1996). A decade of molecular biology of retinoic acid receptors. *FASEB Journal, 10,* 940–954.

61. Vecchini, F., Lenoir-Viale, M. C., Cathelineau, C., Magdalou, J., Bernard, B. A., and Shroot, B. (1994). Presence of a retinoid responsive element in the promoter region of the human cytochrome P4501A1 gene. *Biochemical and Biophysical Research Communications, 201,* 1205–1212.

62. Mader, S., Leroy, P., Chen, J. Y., and Chambon, P. (1993). Multiple parameters control the selectivity of nuclear receptors for their response elements. Selectivity and promiscuity in response element recognition by retinoic acid receptors and retinoid X receptors. *Journal of Biological Chemistry, 268,* 591–600.

63. Howell, S. R., Shirley, M. A., and Ulm, E. H. (1998). Effects of retinoid treatment of rats on hepatic microsomal metabolism and cytochromes P450. Correlation between retinoic acid receptor/retinoid X receptor selectivity and effects on metabolic enzymes. *Drug Metabolism and Disposition, 26,* 234–239.

64. Shen, G., Jeong, W. S., Hu, R., and Kong, A. N. (2005). Regulation of Nrf2, NF-kappaB, and AP-1 signaling pathways by chemopreventive agents. *Antioxidants & Redox Signaling*, 7, 1648–1663.

65. Tian, Y., Rabson, A. B., and Gallo, M. A. (2002). Ah receptor and NF-kappaB interactions: mechanisms and physiological implications. *Chemico-Biological Interactions*, 141, 97–115.

66. Tian, Y., Ke, S., Denison, M. S., Rabson, A. B., and Gallo, M. A. (1999). Ah receptor and NF-kappaB interactions, a potential mechanism for dioxin toxicity. *Journal of Biological Chemistry*, 274, 510–515.

67. Kim, D. W., Gazourian, L., Quadri, S. A., Romieu-Mourez, R., Sherr, D. H., and Sonenshein, G. E. (2000). The RelA NF-kappaB subunit and the aryl hydrocarbon receptor (AhR) cooperate to transactivate the c-myc promoter in mammary cells. *Oncogene*, 19, 5498–5506.

68. Baba, T., Mimura, J., Gradin, K., Kuroiwa, A., Watanabe, T., Matsuda, Y., Inazawa, J., Sogawa, K., and Fujii-Kuriyama, Y. (2001). Structure and expression of the Ah receptor repressor gene. *Journal of Biological Chemistry*, 276, 33101–33110.

69. Suh, J., Jeon, Y. J., Kim, H. M., Kang, J. S., Kaminski, N. E., and Yang, K. H. (2002). Aryl hydrocarbon receptor-dependent inhibition of AP-1 activity by 2,3,7,8-tetrachlorodibenzo-p-dioxin in activated B cells. *Toxicology and Applied Pharmacology*, 181, 116–123.

70. Yao, K. S., Hageboutros, A., Ford, P., and O'Dwyer, P. J. (1997). Involvement of activator protein-1 and nuclear factor-kappaB transcription factors in the control of the DT-diaphorase expression induced by mitomycin C treatment. *Molecular Pharmacology*, 51, 422–430.

71. Hoffer, A., Chang, C. Y., and Puga, A. (1996). Dioxin induces transcription of fos and jun genes by Ah receptor-dependent and -independent pathways. *Toxicology and Applied Pharmacology*, 141, 238–247.

72. Quattrochi, L. C., Shih, H., and Pickwell, G. V. (1998). Induction of the human CYP1A2 enhancer by phorbol ester. *Archives of Biochemistry and Biophysics*, 350, 41–48.

73. Eguchi, H., Ikuta, T., Tachibana, T., Yoneda, Y., and Kawajiri, K. (1997). A nuclear localization signal of human aryl hydrocarbon receptor nuclear translocator/hypoxia-inducible factor 1beta is a novel bipartite type recognized by the two components of nuclear pore-targeting complex. *Journal of Biological Chemistry*, 272, 17640–17647.

74. Salceda, S., Beck, I., and Caro, J. (1996). Absolute requirement of aryl hydrocarbon receptor nuclear translocator protein for gene activation by hypoxia. *Archives of Biochemistry and Biophysics*, 334, 389–394.

75. Laughner, E., Taghavi, P., Chiles, K., Mahon, P. C., and Semenza, G. L. (2001). HER2 (neu) signaling increases the rate of hypoxia-inducible factor 1alpha (HIF-1alpha) synthesis: novel mechanism for HIF-1-mediated vascular endothelial growth factor expression. *Molecular and Cellular Biology*, 21, 3995–4004.

76. Chan, W. K., Yao, G., Gu, Y. Z., and Bradfield, C. A. (1999). Cross-talk between the aryl hydrocarbon receptor and hypoxia inducible factor signaling pathways. Demonstration of competition and compensation. *Journal of Biological Chemistry*, 274, 12115–12123.

77. Hahn, M. E. (2002). Biomarkers and bioassays for detecting dioxin-like compounds in the marine environment. *The Science of the Total Environment*, 289, 49–69.

78. Falkner, K. C., McCallum, G. P., Cherian, M. G., and Bend, J. R. (1993). Effects of acute sodium arsenite administration on the pulmonary chemical metabolizing enzymes, cytochrome P-450 monooxygenase, NAD(P)H:quinone acceptor oxidoreductase and glutathione S-transferase in guinea pig: comparison with effects in liver and kidney. *Chemico-Biological Interactions*, 86, 51–68.

79. Siller, F. R., Quintanilla-Vega, B., Cebrian, M. E., and Albores, A. (1997). Effects of arsenite pretreatment on the acute toxicity of parathion. *Toxicology*, 116, 59–65.

80. Jacobs, J. M., Nichols, C. E., Andrew, A. S., Marek, D. E., Wood, S. G., Sinclair, P. R., Wrighton, S. A., Kostrubsky, V. E., and Sinclair, J. F. (1999). Effect of arsenite on induction of CYP1A, CYP2B, and CYP3A in primary cultures of rat hepatocytes. *Toxicology and Applied Pharmacology*, 157, 51–59.

81. Jacobs, J., Roussel, R., Roberts, M., Marek, D., Wood, S., Walton, H., Dwyer, B., Sinclair, P., and Sinclair, J. (1998). Effect of arsenite on induction of CYP1A and CYP2H in primary cultures of chick hepatocytes. *Toxicology and Applied Pharmacology*, 150, 376–382.

82. Vakharia, D. D., Liu, N., Pause, R., Fasco, M., Bessette, E., Zhang, Q. Y., and Kaminsky, L. S. (2001). Effect of metals on polycyclic aromatic hydrocarbon induction of CYP1A1 and CYP1A2 in human hepatocyte cultures. *Toxicology and Applied Pharmacology*, 170, 93–103.

83. Elbekai, R. H. and El-Kadi, A. O. (2004). Modulation of aryl hydrocarbon receptor-regulated gene expression by arsenite, cadmium, and chromium. *Toxicology*, 202, 249–269.

84. Albores, A., Cebrian, M. E., Bach, P. H., Connelly, J. C., Hinton, R. H., and Bridges, J. W. (1989). Sodium arsenite induced alterations in bilirubin excretion and heme metabolism. *Journal of Biochemical Toxicology*, 4, 73–78.

85. Del Razo, L. M., Quintanilla-Vega, B., Brambila-Colombres, E., Calderon-Aranda, E. S., Manno, M., and Albores, A. (2001). Stress proteins induced by arsenic. *Toxicology and Applied Pharmacology*, 177, 132–148.

86. Kapahi, P., Takahashi, T., Natoli, G., Adams, S. R., Chen, Y., Tsien, R. Y., and Karin, M. (2000). Inhibition of NF-kappa B activation by arsenite through reaction with a critical cysteine in the activation loop of Ikappa B kinase. *Journal of Biological Chemistry*, 275, 36062–36066.

87. Gonzalez, F. J. (1988). The molecular biology of cytochrome P450s. *Pharmacological Reviews*, 40, 243–288.

88. Vernhet, L., Allain, N., Le Vee, M., Morel, F., Guillouzo, A., and Fardel, O. (2003). Blockage of multidrug resistance-associated proteins potentiates the inhibitory effects of arsenic trioxide on CYP1A1 induction by polycyclic aromatic hydrocarbons. *Journal of Pharmacology and Experimental Therapeutics*, 304, 145–155.

89. Bogdan, C. (2001). Nitric oxide and the regulation of gene expression. *Trends in Cell Biology*, 11, 66–75.

90. Lowe, G. M., Hulley, C. E., Rhodes, E. S., Young, A. J., and Bilton, R. F. (1998). Free radical stimulation of tyrosine kinase and phosphatase activity in human peripheral blood mononuclear cells. *Biochemical and Biophysical Research Communications*, 245, 17–22.

91. Marilena, G. (1997). New physiological importance of two classic residual products: carbon monoxide and bilirubin. *Biochemical and Molecular Medicine*, 61, 136–142.

92. Schuller, D. J., Wilks, A., Ortiz de Montellano, P., and Poulos, T. L. (1998). Crystallization of recombinant human heme oxygenase-1. *Protein Science*, 7, 1836–1838.

93. Elbekai, R. H. and El-Kadi, A. O. (2005). The role of oxidative stress in the modulation of aryl hydrocarbon receptor-regulated genes by As^{3+}, Cd^{2+}, and Cr^{6+}. *Free Radical Biology & Medicine*, 39, 1499–1511.

94. Anwar-Mohamed, A. and El-Kadi, A. O. (2010). Arsenite down-regulates cytochrome P450 1A1 at the transcriptional and posttranslational levels in human HepG2 cells. *Free Radical Biology & Medicine*, 48, 1399–1409.

95. Spink, D. C., Katz, B. H., Hussain, M. M., Spink, B. C., Wu, S. J., Liu, N., Pause, R., and Kaminsky, L. S. (2002). Induction of CYP1A1 and CYP1B1 in T-47D human breast cancer cells by benzo[a]pyrene is diminished by arsenite. *Drug Metabolism and Disposition*, 30, 262–269.

96. Elbekai, R. H. and El-Kadi, A. O. (2007). Transcriptional activation and posttranscriptional modification of Cyp1a1 by arsenite, cadmium, and chromium. *Toxicology Letters*, 172, 106–119.

97. Wu, S. J., Spink, D. C., Spink, B. C., and Kaminsky, L. S. (2003). Quantitation of CYP1A1 and 1B1 mRNA in polycyclic aromatic hydrocarbon-treated human T-47D and HepG2 cells by a modified bDNA assay using fluorescence detection. *Analytical Biochemistry*, 312, 162–166.

98. Vakharia, D. D., Liu, N., Pause, R., Fasco, M., Bessette, E., Zhang, Q. Y., and Kaminsky, L. S. (2001). Polycyclic aromatic hydrocarbon/metal mixtures: effect on PAH induction of CYP1A1 in human HEPG2 cells. *Drug Metabolism and Disposition*, 29, 999–1006.

99. Ho, I. C. and Lee, T. C. (2002). Arsenite pretreatment attenuates benzo[a]pyrene cytotoxicity in a human lung adenocarcinoma cell line by decreasing cyclooxygenase-2 levels. *Journal of Toxicology and Environmental Health*, 65, 245–263.

100. Maier, A., Dalton, T. P., and Puga, A. (2000). Disruption of dioxin-inducible phase I and phase II gene expression patterns by cadmium, chromium, and arsenic. *Molecular Carcinogenesis*, 28, 225–235.

101. Bessette, E. E., Fasco, M. J., Pentecost, B. T., and Kaminsky, L. S. (2005). Mechanisms of arsenite-mediated decreases in benzo[k]fluoranthene-induced human cytochrome P4501A1 levels in HepG2 cells. *Drug Metabolism and Disposition*, 33, 312–320.

102. Chao, H. R., Tsou, T. C., Li, L. A., Tsai, F. Y., Wang, Y. F., Tsai, C. H., Chang, E. E., Miao, Z. F., Wu, C. H., and Lee, W. J. (2006). Arsenic inhibits induction of cytochrome P450 1A1 by 2,3,7,8-tetrachlorodibenzo-p-dioxin in human hepatoma cells. *Journal of Hazardous Materials*, 137, 716–722.

103. Bonzo, J. A., Chen, S., Galijatovic, A., and Tukey, R. H. (2005). Arsenite inhibition of CYP1A1 induction by 2,3,7,8-tetrachlorodibenzo-p-dioxin is independent of cell cycle arrest. *Molecular Pharmacology*, 67, 1247–1256.

104. Kann, S., Huang, M. Y., Estes, C., Reichard, J. F., Sartor, M. A., Xia, Y., and Puga, A. (2005). Arsenite-induced aryl hydrocarbon receptor nuclear translocation results in additive induction of phase I genes and synergistic induction of phase II genes. *Molecular Pharmacology*, 68, 336–346.

105. Elbekai, R. H. and El-Kadi, A. O. (2008). Arsenite and cadmium, but not chromium, induce NAD(P)H:quinone oxidoreductase 1 through transcriptional mechanisms, in spite of post-transcriptional modifications. *Toxicology In Vitro*, 22, 1184–1190.

106. Joiakim, A., Mathieu, P. A., Elliott, A. A., and Reiners, J. J., Jr. (2004). Superinduction of CYP1A1 in MCF10A cultures by cycloheximide, anisomycin, and puromycin: a process independent of effects on protein translation and unrelated to suppression of aryl hydrocarbon receptor proteolysis by the proteasome. *Molecular Pharmacology*, 66, 936–947.

107. Ma, Q., Renzelli, A. J., Baldwin, K. T., and Antonini, J. M. (2000). Superinduction of CYP1A1 gene expression. Regulation of 2,3,7,8-tetrachlorodibenzo-p-dioxin-induced degradation of Ah receptor by cycloheximide. *Journal of Biological Chemistry*, 275, 12676–12683.

108. Seubert, J. M., Sinal, C. J., and Bend, J. R. (2002). Acute sodium arsenite administration induces pulmonary CYP1A1 mRNA, protein and activity in the rat. *Journal of Biochemical and Molecular Toxicology*, 16, 84–95.

109. Seubert, J. M., Webb, C. D., and Bend, J. R. (2002). Acute sodium arsenite treatment induces Cyp2a5 but not Cyp1a1 in the C57Bl/6 mouse in a tissue (kidney) selective manner. *Journal of Biochemical and Molecular Toxicology*, 16, 96–106.

110. Schaldach, C. M., Riby, J., and Bjeldanes, L. F. (1999). Lipoxin A4: a new class of ligand for the Ah receptor. *Biochemistry*, 38, 7594–7600.

111. Li, J., Gorospe, M., Hutter, D., Barnes, J., Keyse, S. M., and Liu, Y. (2001). Transcriptional induction of MKP-1 in response to stress is associated with histone H3 phosphorylation–acetylation. *Molecular and Cellular Biology*, 21, 8213–8224.

112. He, Z., Ma, W. Y., Liu, G., Zhang, Y., Bode, A. M., and Dong, Z. (2003). Arsenite-induced phosphorylation of histone H3 at serine 10 is mediated by Akt1, extracellular signal-regulated kinase 2, and p90 ribosomal S6 kinase 2 but not mitogen- and stress-activated protein kinase 1. *Journal of Biological Chemistry*, 278, 10588–10593.

113. Shibazaki, M., Takeuchi, T., Ahmed, S., and Kikuchi, H. (2004). Suppression by p38 MAP kinase inhibitors (pyridinyl imidazole compounds) of Ah receptor target gene activation by 2,3,7,8-tetrachlorodibenzo-p-dioxin and the possible mechanism. *Journal of Biological Chemistry*, 279, 3869–3876.

114. Nakajima, M., Iwanari, M., and Yokoi, T. (2003). Effects of histone deacetylation and DNA methylation on the constitutive

and TCDD-inducible expressions of the human CYP1 family in MCF-7 and HeLa cells. *Toxicology Letters, 144,* 247–256.

115. Ke, Q., Yang, Y., Ratner, M., Zeind, J., Jiang, C., Forrest, J. N., Jr., and Xiao, Y. F. (2002). Intracellular accumulation of mercury enhances P450 CYP1A1 expression and Cl⁻ currents in cultured shark rectal gland cells. *Life Sciences, 70,* 2547–2566.

116. Oliveira, M., Santos, M. A., and Pacheco, M. (2004). Glutathione protects heavy metal-induced inhibition of hepatic microsomal ethoxyresorufin O-deethylase activity in *Dicentrarchus labrax* L. *Ecotoxicology and Environmental Safety, 58,* 379–385.

117. Korashy, H. M. and El-Kadi, A. O. (2004). Differential effects of mercury, lead and copper on the constitutive and inducible expression of aryl hydrocarbon receptor (AHR)-regulated genes in cultured hepatoma Hepa 1c1c7 cells. *Toxicology, 201,* 153–172.

118. Korashy, H. M. and El-Kadi, A. O. (2005). Regulatory mechanisms modulating the expression of cytochrome P450 1A1 gene by heavy metals. *Toxicological Sciences, 88,* 39–51.

119. Korashy, H. M. and El-Kadi, A. O. (2008). Modulation of TCDD-mediated induction of cytochrome P450 1A1 by mercury, lead, and copper in human HepG2 cell line. *Toxicology In Vitro, 22,* 154–158.

120. Ramadoss, P., Petrulis, J. R., Hollingshead, B. D., Kusnadi, A., and Perdew, G. H. (2004). Divergent roles of hepatitis B virus X-associated protein 2 (XAP2) in human versus mouse Ah receptor complexes. *Biochemistry, 43,* 700–709.

121. Ramadoss, P. and Perdew, G. H. (2005). The transactivation domain of the Ah receptor is a key determinant of cellular localization and ligand-independent nucleocytoplasmic shuttling properties. *Biochemistry, 44,* 11148–11159.

122. Suzuki, T. and Nohara, K. (2007). Regulatory factors involved in species-specific modulation of arylhydrocarbon receptor (AhR)-dependent gene expression in humans and mice. *Journal of Biochemistry, 142,* 443–452.

123. Korashy, H. M. and El-Kadi, A. O. (2008). The role of redox-sensitive transcription factors NF-kappaB and AP-1 in the modulation of the Cyp1a1 gene by mercury, lead, and copper. *Free Radical Biology & Medicine, 44,* 795–806.

124. Ruby, C. E., Leid, M., and Kerkvliet, N. I. (2002). 2,3,7,8-Tetrachlorodibenzo-*p*-dioxin suppresses tumor necrosis factor-alpha and anti-CD40-induced activation of NF-kappaB/Rel in dendritic cells: p50 homodimer activation is not affected. *Molecular Pharmacology, 62,* 722–728.

125. Tian, Y., Ke, S., Chen, M., and Sheng, T. (2003). Interactions between the aryl hydrocarbon receptor and P-TEFb. Sequential recruitment of transcription factors and differential phosphorylation of C-terminal domain of RNA polymerase II at cyp1a1 promoter. *Journal of Biological Chemistry, 278,* 44041–44048.

126. Ichim, T. E., Li, M., Qian, H., Popov, I. A., Rycerz, K., Zheng, X., White, D., Zhong, R., and Min, W. P. (2004). RNA interference: a potent tool for gene-specific therapeutics. *American Journal of Transplantation, 4,* 1227–1236.

127. Wagstaff, D. D. (1973). Stimulation of liver detoxication enzymes by dietary cadmium acetate. *Bulletin of Environmental Contamination and Toxicology, 10,* 328–332.

128. Eaton, D. L., Stacey, N. H., Wong, K. L., and Klaassen, C. D. (1980). Dose–response effects of various metal ions on rat liver metallothionein, glutathione, heme oxygenase, and cytochrome P-450. *Toxicology and Applied Pharmacology, 55,* 393–402.

129. Anjum, F., Raman, A., Shakoori, A. R., and Gorrod, J. W. (1992). An assessment of cadmium toxicity on cytochrome P-450 and flavin monooxygenase-mediated metabolic pathways of dimethylaniline in male rabbits. *Journal of Environmental Pathology, Toxicology and Oncology, 11,* 191–195.

130. Alexidis, A. N., Rekka, E. A., and Kourounakis, P. N. (1994). Influence of mercury and cadmium intoxication on hepatic microsomal CYP2E and CYP3A subfamilies. *Research Communications in Molecular Pathology and Pharmacology, 85,* 67–72.

131. Dehn, P. F., White, C. M., Conners, D. E., Shipkey, G., and Cumbo, T. A. (2004). Characterization of the human hepatocellular carcinoma (hepg2) cell line as an *in vitro* model for cadmium toxicity studies. *In Vitro Cellular & Developmental Biology, 40,* 172–182.

132. Baker, J. R., Satarug, S., Reilly, P. E., Edwards, R. J., Ariyoshi, N., Kamataki, T., Moore, M. R., and Williams, D. J. (2001). Relationships between non-occupational cadmium exposure and expression of nine cytochrome P450 forms in human liver and kidney cortex samples. *Biochemical Pharmacology, 62,* 713–721.

133. States, J. C., Quan, T., Hines, R. N., Novak, R. F., and Runge-Morris, M. (1993). Expression of human cytochrome P450 1A1 in DNA repair deficient and proficient human fibroblasts stably transformed with an inducible expression vector. *Carcinogenesis, 14,* 1643–1649.

134. Iscan, M., Ada, A. O., Coban, T., Kapucuoglu, N., Aydin, A., and Isimer, A. (2002). Combined effects of cadmium and nickel on testicular xenobiotic metabolizing enzymes in rats. *Biological Trace Element Research, 89,* 177–190.

135. Rosenberg, D. W. and Kappas, A. (1991). Induction of heme oxygenase in the small intestinal epithelium: a response to oral cadmium exposure. *Toxicology, 67,* 199–210.

136. Benedetti, M., Martuccio, G., Fattorini, D., Canapa, A., Barucca, M., Nigro, M., and Regoli, F. (2007). Oxidative and modulatory effects of trace metals on metabolism of polycyclic aromatic hydrocarbons in the Antarctic fish *Trematomus bernacchii*. *Aquatic Toxicology, 85,* 167–175.

137. Ahmad, I., Maria, V. L., Oliveira, M., Pacheco, M., and Santos, M. A. (2006). Oxidative stress and genotoxic effects in gill and kidney of *Anguilla anguilla* L. exposed to chromium with or without pre-exposure to beta-naphthoflavone. *Mutation Research, 608,* 16–28.

138. Chen, F., Ye, J., Zhang, X., Rojanasakul, Y., and Shi, X. (1997). One-electron reduction of chromium(VI) by alpha-lipoic acid and related hydroxyl radical generation, dG hydroxylation and nuclear transcription factor-kappaB activation. *Archives of Biochemistry and Biophysics, 338,* 165–172.

139. Tesfai, Y., Davis, D., and Reinhold, D. (1998). Chromium can reduce the mutagenic effects of benzo[a]pyrene diolepoxide in normal human fibroblasts via an oxidative stress mechanism. *Mutation Research*, 416, 159–168.

140. Wei, Y. D., Tepperman, K., Huang, M. Y., Sartor, M. A., and Puga, A. (2004). Chromium inhibits transcription from polycyclic aromatic hydrocarbon-inducible promoters by blocking the release of histone deacetylase and preventing the binding of p300 to chromatin. *Journal of Biological Chemistry*, 279, 4110–4119.

141. Geist, J., Werner, I., Eder, K. J., and Leutenegger, C. M. (2007). Comparisons of tissue-specific transcription of stress response genes with whole animal endpoints of adverse effect in striped bass (*Morone saxatilis*) following treatment with copper and esfenvalerate. *Aquatic Toxicology*, 85, 28–39.

142. Song, M. O. and Freedman, J. H. (2005). Expression of copper-responsive genes in HepG2 cells. *Molecular and Cellular Biochemistry*, 279, 141–147.

143. Kim, J. S., Ahn, T., Yim, S. K., and Yun, C. H. (2002). Differential effect of copper(II) on the cytochrome P450 enzymes and NADPH-cytochrome P450 reductase: inhibition of cytochrome P450-catalyzed reactions by copper(II) ion. *Biochemistry*, 41, 9438–9447.

144. Johnson, W. T. and Smith, T. B. (1994). Copper deficiency increases cytochrome P450-dependent 7-ethoxyresorufin-O-deethylase activity in rat small intestine. *Proceedings of the Society for Experimental Biology and Medicine*, 207, 302–308.

145. Anwar-Mohamed, A. and El-Kadi, A. O. (2008). Downregulation of the carcinogen-metabolizing enzyme cytochrome P450 1a1 by vanadium. *Drug Metabolism and Disposition*, 36, 1819–1827.

146. Abdelhamid, G., Anwar-Mohamed, A., Badary, O. A., Moustafa, A. A., and El-Kadi, A. O. (2010). Transcriptional and posttranscriptional regulation of CYP1A1 by vanadium in human hepatoma HepG2 cells. *Cell Biology and Toxicology*, 26, 421–434.

147. Anwar-Mohamed, A. and El-Kadi, A. O. (2009). Downregulation of the detoxifying enzyme NAD(P)H:quinone oxidoreductase 1 by vanadium in Hepa 1c1c7 cells. *Toxicology and Applied Pharmacology*, 236, 261–269.

148. Gao, N., Ding, M., Zheng, J. Z., Zhang, Z., Leonard, S. S., Liu, K. J., Shi, X., and Jiang, B. H. (2002). Vanadate-induced expression of hypoxia-inducible factor 1 alpha and vascular endothelial growth factor through phosphatidylinositol 3-kinase/Akt pathway and reactive oxygen species. *Journal of Biological Chemistry*, 277, 31963–31971.

149. Kim, J. E. and Sheen, Y. Y. (2000). Inhibition of 2,3,7,8-tetrachlorodibenzo-*p*-dioxin (TCDD)-stimulated Cyp1a1 promoter activity by hypoxic agents. *Biochemical Pharmacology*, 59, 1549–1556.

150. Jaspers, I., Samet, J. M., Erzurum, S., and Reed, W. (2000). Vanadium-induced kappaB-dependent transcription depends upon peroxide-induced activation of the p38 mitogen-activated protein kinase. *American Journal of Respiratory Cell and Molecular Biology*, 23, 95–102.

151. Chen, F., Demers, L. M., Vallyathan, V., Ding, M., Lu, Y., Castranova, V., and Shi, X. (1999). Vanadate induction of NF-kappaB involves IkappaB kinase beta and SAPK/ERK kinase 1 in macrophages. *Journal of Biological Chemistry*, 274, 20307–20312.

152. Wang, X. and Safe, S. (1994). Development of an *in vitro* model for investigating the formation of the nuclear Ah receptor complex in mouse Hepa 1c1c7 cells. *Archives of Biochemistry and Biophysics*, 315, 285–292.

153. Moennikes, O., Loeppen, S., Buchmann, A., Andersson, P., Ittrich, C., Poellinger, L., and Schwarz, M. (2004). A constitutively active dioxin/aryl hydrocarbon receptor promotes hepatocarcinogenesis in mice. *Cancer Research*, 64, 4707–4710.

154. Kogevinas, M., Kauppinen, T., Winkelmann, R., Becher, H., Bertazzi, P. A., Bueno-de-Mesquita, H. B., Coggon, D., Green, L., Johnson, E., Littorin, M., et al. (1995). Soft tissue sarcoma and non-Hodgkin's lymphoma in workers exposed to phenoxy herbicides, chlorophenols, and dioxins: two nested case–control studies. *Epidemiology*, 6, 396–402.

155. Hooiveld, M., Heederik, D. J., Kogevinas, M., Boffetta, P., Needham, L. L., Patterson, D. G., Jr., and Bueno-de-Mesquita, H. B. (1998). Second follow-up of a Dutch cohort occupationally exposed to phenoxy herbicides, chlorophenols, and contaminants. *American Journal of Epidemiology*, 147, 891–901.

156. Fingerhut, M. A., Sweeney, M. H., Halperin, W. E., and Schnorr, T. M. (1991). The epidemiology of populations exposed to dioxin. *IARC Scientific Publications*, 31–50.

157. Mukherjee, S., Koner, B. C., Ray, S., and Ray, A. (2006). Environmental contaminants in pathogenesis of breast cancer. *Indian Journal of Experimental Biology*, 44, 597–617.

158. Schlezinger, J. J., Liu, D., Farago, M., Seldin, D. C., Belguise, K., Sonenshein, G. E., and Sherr, D. H. (2006). A role for the aryl hydrocarbon receptor in mammary gland tumorigenesis. *Biological Chemistry*, 387, 1175–1187.

159. Tritscher, A. M., Goldstein, J. A., Portier, C. J., McCoy, Z., Clark, G. C., and Lucier, G. W. (1992). Dose–response relationships for chronic exposure to 2,3,7,8-tetrachlorodibenzo-*p*-dioxin in a rat tumor promotion model: quantification and immunolocalization of CYP1A1 and CYP1A2 in the liver. *Cancer Research*, 52, 3436–3442.

160. Huff, J., Lucier, G., and Tritscher, A. (1994). Carcinogenicity of TCDD: experimental, mechanistic, and epidemiologic evidence. *Annual Review of Pharmacology and Toxicology*, 34, 343–372.

161. Kociba, R. J. and Schwetz, B. A. (1982). Toxicity of 2,3,7,8-tetrachlorodibenzo-*p*-dioxin (TCDD). *Drug Metabolism Reviews*, 13, 387–406.

162. Mann, P. C. (1997). Selected lesions of dioxin in laboratory rodents. *Toxicologic Pathology*, 25, 72–79.

163. Castell, J. V., Donato, M. T., and Gomez-Lechon, M. J. (2005). Metabolism and bioactivation of toxicants in the lung. The *in vitro* cellular approach. *Experimental and Toxicologic Pathology*, 57, (Suppl. 1), 189–204.

164. Senft, A. P., Dalton, T. P., Nebert, D. W., Genter, M. B., Puga, A., Hutchinson, R. J., Kerzee, J. K., Uno, S., and Shertzer, H. G.

(2002). Mitochondrial reactive oxygen production is dependent on the aromatic hydrocarbon receptor. *Free Radical Biology & Medicine*, 33, 1268–1278.
165. Shimada, T. (2006). Xenobiotic-metabolizing enzymes involved in activation and detoxification of carcinogenic polycyclic aromatic hydrocarbons. *Drug Metabolism and Pharmacokinetics*, 21, 257–276.
166. Kapitulnik, J., Wislocki, P. G., Levin, W., Yagi, H., Jerina, D. M., and Conney, A. H. (1978). Tumorigenicity studies with diol-epoxides of benzo(a)pyrene which indicate that (+/−)-trans-7beta,8alpha-dihydroxy-9alpha,10alpha-epoxy-7,8,9,10-tetrahydrobenzo(a)pyrene is an ultimate carcinogen in newborn mice. *Cancer Research*, 38, 354–358.
167. Shimada, T., Hayes, C. L., Yamazaki, H., Amin, S., Hecht, S. S., Guengerich, F. P., and Sutter, T. R. (1996). Activation of chemically diverse procarcinogens by human cytochrome P-450 1B1. *Cancer Research*, 56, 2979–2984.
168. Buters, J. T., Sakai, S., Richter, T., Pineau, T., Alexander, D. L., Savas, U., Doehmer, J., Ward, J. M., Jefcoate, C. R., and Gonzalez, F. J. (1999). Cytochrome P450 CYP1B1 determines susceptibility to 7,12-dimethylbenz[a]anthracene-induced lymphomas. *Proceedings of the National Academy of Sciences of the United States of America*, 96, 1977–1982.
169. Shimada, T., Gillam, E. M., Oda, Y., Tsumura, F., Sutter, T. R., Guengerich, F. P., and Inoue, K. (1999). Metabolism of benzo[a]pyrene to trans-7,8-dihydroxy-7,8-dihydrobenzo[a]pyrene by recombinant human cytochrome P450 1B1 and purified liver epoxide hydrolase. *Chemical Research in Toxicology*, 12, 623–629.
170. Amin, S., Desai, D., Dai, W., Harvey, R. G., and Hecht, S. S. (1995). Tumorigenicity in newborn mice of fjord region and other sterically hindered diol epoxides of benzo[g]chrysene, dibenzo[a,l]pyrene (dibenzo[def,p]chrysene), 4H-cyclopenta[def]chrysene and fluoranthene. *Carcinogenesis*, 16, 2813–2817.
171. Conney, A. H. (1982). Induction of microsomal enzymes by foreign chemicals and carcinogenesis by polycyclic aromatic hydrocarbons: G. H. A. Clowes Memorial Lecture. *Cancer Research*, 42, 4875–4917.
172. Hussein, M. A. (2001). Arsenic trioxide: a new immunomodulatory agent in the management of multiple myeloma. *Medical Oncology*, 18, 239–242.
173. Murgo, A. J. (2001). Clinical trials of arsenic trioxide in hematologic and solid tumors: overview of the National Cancer Institute Cooperative Research and Development Studies. *Oncologist*, 6, (Suppl. 2), 22–28.
174. Kostova, I. (2009). Titanium and vanadium complexes as anticancer agents. *Anticancer Agents in Medicinal Chemistry*, 9, 827–842.
175. Evangelou, A. M. (2002). Vanadium in cancer treatment. *Critical Reviews in Oncology/Hematology*, 42, 249–265.
176. Baba, T., Mimura, J., Nakamura, N., Harada, N., Yamamoto, M., Morohashi, K., and Fujii-Kuriyama, Y. (2005). Intrinsic function of the aryl hydrocarbon (dioxin) receptor as a key factor in female reproduction. *Molecular and Cellular Biology*, 25, 10040–10051.
177. McMillan, B. J. and Bradfield, C. A. (2007). The aryl hydrocarbon receptor sans xenobiotics: endogenous function in genetic model systems. *Molecular Pharmacology*, 72, 487–498.
178. Gong, J. G., Campbell, B. K., Bramley, T. A., Gutierrez, C. G., Peters, A. R., and Webb, R. (1996). Suppression in the secretion of follicle-stimulating hormone and luteinizing hormone, and ovarian follicle development in heifers continuously infused with a gonadotropin-releasing hormone agonist. *Biology of Reproduction*, 55, 68–74.
179. Tornell, J., Billig, H., and Hillensjo, T. (1991). Regulation of oocyte maturation by changes in ovarian levels of cyclic nucleotides. *Human Reproduction*, 6, 411–422.
180. Barnett, K. R., Tomic, D., Gupta, R. K., Miller, K. P., Meachum, S., Paulose, T., and Flaws, J. A. (2007). The aryl hydrocarbon receptor affects mouse ovarian follicle growth via mechanisms involving estradiol regulation and responsiveness. *Biology of Reproduction*, 76, 1062–1070.
181. Gruber, C. J., Tschugguel, W., Schneeberger, C., and Huber, J. C. (2002). Production and actions of estrogens. *New England Journal of Medicine*, 346, 340–352.
182. Gonzalez-Robayna, I. J., Falender, A. E., Ochsner, S., Firestone, G. L., and Richards, J. S. (2000). Follicle-stimulating hormone (FSH) stimulates phosphorylation and activation of protein kinase B (PKB/Akt) and serum and glucocorticoid-induced kinase (Sgk): evidence for A kinase-independent signaling by FSH in granulosa cells. *Molecular Endocrinology*, 14, 1283–1300.
183. Hatano, O., Takayama, K., Imai, T., Waterman, M. R., Takakusu, A., Omura, T., and Morohashi, K. (1994). Sex-dependent expression of a transcription factor, Ad4BP, regulating steroidogenic P-450 genes in the gonads during prenatal and postnatal rat development. *Development*, 120, 2787–2797.
184. Hollingshead, B. D., Patel, R. D., and Perdew, G. H. (2006). Endogenous hepatic expression of the hepatitis B virus X-associated protein 2 is adequate for maximal association with aryl hydrocarbon receptor-90-kDa heat shock protein complexes. *Molecular Pharmacology*, 70, 2096–2107.
185. Bremer, S., Brittebo, E., Dencker, L., Knudsen, L. E., Mathisien, L., Olovsson, M., Pazos, P., Pellizzer, C., Paulesu, L. R., Schaefer, W., Schwarz, M., Staud, F., Stavreus-Evers, A., and Vahankangas, K. (2007). In vitro tests for detecting chemicals affecting the embryo implantation process. The report and recommendations of ECVAM Workshop 62: a strategic workshop of the EU ReProTect project. *Alternatives to Laboratory Animals*, 35, 421–439.
186. Abbott, B. D., Schmid, J. E., Pitt, J. A., Buckalew, A. R., Wood, C. R., Held, G. A., and Diliberto, J. J. (1999). Adverse reproductive outcomes in the transgenic Ah receptor-deficient mouse. *Toxicology and Applied Pharmacology*, 155, 62–70.
187. Li, B., Liu, H. Y., Dai, L. J., Lu, J. C., Yang, Z. M., and Huang, L. (2006). The early embryo loss caused by 2,3,7,8-tetrachlorodibenzo-p-dioxin may be related to the accumulation of this compound in the uterus. *Reproductive Toxicology*, 21, 301–306.

188. Wells, P. G., Lee, C. J., McCallum, G. P., Perstin, J., and Harper, P. A. (2010). Receptor- and reactive intermediate-mediated mechanisms of teratogenesis. *Handbook of Experimental Pharmacology*, 196, 131–162.
189. Choi, S. S., Miller, M. A., and Harper, P. A. (2006). In utero exposure to 2,3,7,8-tetrachlorodibenzo-p-dioxin induces amphiregulin gene expression in the developing mouse ureter. *Toxicological Sciences*, 94, 163–174.
190. Reichard, J. F., Dalton, T. P., Shertzer, H. G., and Puga, A. (2005). Induction of oxidative stress responses by dioxin and other ligands of the aryl hydrocarbon receptor. *Dose Response*, 3, 306–331.
191. Hines, R. N. (2008). The ontogeny of drug metabolism enzymes and implications for adverse drug events. *Pharmacology & Therapeutics*, 118, 250–267.
192. Juchau, M. R., Lee, Q. P., and Fantel, A. G. (1992). Xenobiotic biotransformation/bioactivation in organogenesis-stage conceptual tissues: implications for embryotoxicity and teratogenesis. *Drug Metabolism Reviews*, 24, 195–238.
193. Winn, L. M. and Wells, P. G. (1995). Phenytoin-initiated DNA oxidation in murine embryo culture, and embryo protection by the antioxidative enzymes superoxide dismutase and catalase: evidence for reactive oxygen species-mediated DNA oxidation in the molecular mechanism of phenytoin teratogenicity. *Molecular Pharmacology*, 48, 112–120.
194. Hahn, M. E. (1998). The aryl hydrocarbon receptor: a comparative perspective. *Comparative Biochemistry and Physiology, Part C*, 121, 23–53.
195. Swanson, H. I. and Bradfield, C. A. (1993). The AH-receptor: genetics, structure and function. *Pharmacogenetics*, 3, 213–230.
196. Venkat-Raman, N., Murphy, K. W., Ghaus, K., Teoh, T. G., Higham, J. M., and Carvalho, J. S. (2001). Congenital absence of portal vein in the fetus: a case report. *Ultrasound in Obstetrics and Gynecology*, 17, 71–75.
197. Lund, A. K., Goens, M. B., Kanagy, N. L., and Walker, M. K. (2003). Cardiac hypertrophy in aryl hydrocarbon receptor null mice is correlated with elevated angiotensin II, endothelin-1, and mean arterial blood pressure. *Toxicology and Applied Pharmacology*, 193, 177–187.
198. Korashy, H. M. and El-Kadi, A. O. (2006). The role of aryl hydrocarbon receptor in the pathogenesis of cardiovascular diseases. *Drug Metabolism Reviews*, 38, 411–450.
199. McMillan, B. J. and Bradfield, C. A. (2007). The aryl hydrocarbon receptor is activated by modified low-density lipoprotein. *Proceedings of the National Academy of Sciences of the United States of America*, 104, 1412–1417.
200. Mehrabi, M. R., Steiner, G. E., Dellinger, C., Kofler, A., Schaufler, K., Tamaddon, F., Plesch, K., Ekmekcioglu, C., Maurer, G., Glogar, H. D., and Thalhammer, T. (2002). The arylhydrocarbon receptor (AhR), but not the AhR-nuclear translocator (ARNT), is increased in hearts of patients with cardiomyopathy. *Virchows Archives*, 441, 481–489.
201. Catron, T., Mendiola, M. A., Smith, S. M., Born, J., and Walker, M. K. (2001). Hypoxia regulates avian cardiac Arnt and HIF-1alpha mRNA expression. *Biochemical and Biophysical Research Communications*, 282, 602–607.
202. Kanzawa, N., Kondo, M., Okushima, T., Yamaguchi, M., Temmei, Y., Honda, M., and Tsuchiya, T. (2004). Biochemical and molecular biological analysis of different responses to 2,3,7,8-tetrachlorodibenzo-p-dioxin in chick embryo heart and liver. *Archives of Biochemistry and Biophysics*, 427, 58–67.
203. Semenza, G. L., Agani, F., Iyer, N., Kotch, L., Laughner, E., Leung, S., and Yu, A. (1999). Regulation of cardiovascular development and physiology by hypoxia-inducible factor 1. *Annals of the New York Academy of Sciences*, 874, 262–268.
204. Licht, A. H., Muller-Holtkamp, F., Flamme, I., and Breier, G. (2006). Inhibition of hypoxia-inducible factor activity in endothelial cells disrupts embryonic cardiovascular development. *Blood*, 107, 584–590.
205. Jain, S., Maltepe, E., Lu, M. M., Simon, C., and Bradfield, C. A. (1998). Expression of ARNT, ARNT2, HIF1 alpha, HIF2 alpha and Ah receptor mRNAs in the developing mouse. *Mechanisms of Development*, 73, 117–123.
206. Choudhary, D., Jansson, I., Stoilov, I., Sarfarazi, M., and Schenkman, J. B. (2005). Expression patterns of mouse and human CYP orthologs (families 1-4) during development and in different adult tissues. *Archives of Biochemistry and Biophysics*, 436, 50–61.
207. Granberg, A. L., Brunstrom, B., and Brandt, I. (2000). Cytochrome P450-dependent binding of 7,12-dimethylbenz[a]anthracene (DMBA) and benzo[a]pyrene (B[a]P) in murine heart, lung, and liver endothelial cells. *Archives of Toxicology*, 74, 593–601.
208. Thackaberry, E. A., Jiang, Z., Johnson, C. D., Ramos, K. S., and Walker, M. K. (2005). Toxicogenomic profile of 2,3,7,8-tetrachlorodibenzo-p-dioxin in the murine fetal heart: modulation of cell cycle and extracellular matrix genes. *Toxicological Sciences*, 88, 231–241.
209. Walker, M. K., Pollenz, R. S., and Smith, S. M. (1997). Expression of the aryl hydrocarbon receptor (AhR) and AhR nuclear translocator during chick cardiogenesis is consistent with 2,3,7,8-tetrachlorodibenzo-p-dioxin-induced heart defects. *Toxicology and Applied Pharmacology*, 143, 407–419.
210. Rifkind, A. B., Gannon, M., and Gross, S. S. (1990). Arachidonic acid metabolism by dioxin-induced cytochrome P-450: a new hypothesis on the role of P-450 in dioxin toxicity. *Biochemical and Biophysical Research Communications*, 172, 1180–1188.
211. Annas, A., Brunstrom, B., Brandt, I., and Brittebo, E. B. (1998). Induction of ethoxyresorufin O-deethylase (EROD) and endothelial activation of the heterocyclic amine Trp-P-1 in bird embryo hearts. *Archives of Toxicology*, 72, 402–410.
212. Choudhary, D., Jansson, I., Stoilov, I., Sarfarazi, M., and Schenkman, J. B. (2004). Metabolism of retinoids and arachidonic acid by human and mouse cytochrome P450 1b1. *Drug Metabolism and Disposition*, 32, 840–847.
213. Aboutabl, M. E., Zordoky, B. N., and El-Kadi, A. O. (2009). 3-Methylcholanthrene and benzo(a)pyrene modulate cardiac cytochrome P450 gene expression and arachidonic acid metabolism in male Sprague Dawley rats. *British Journal of Pharmacology*, 158, 1808–1819.

214. Butany, J., Ahn, E., and Luk, A. (2009). Drug-related cardiac pathology. *Journal of Clinical Pathology, 62*, 1074–1084.
215. Ohnishi, K., Yoshida, H., Shigeno, K., Nakamura, S., Fujisawa, S., Naito, K., Shinjo, K., Fujita, Y., Matsui, H., Takeshita, A., Sugiyama, S., Satoh, H., Terada, H., and Ohno, R. (2000). Prolongation of the QT interval and ventricular tachycardia in patients treated with arsenic trioxide for acute promyelocytic leukemia. *Annals of Internal Medicine, 133*, 881–885.
216. Barbey, J. T., Pezzullo, J. C., and Soignet, S. L. (2003). Effect of arsenic trioxide on QT interval in patients with advanced malignancies. *Journal of Clinical Oncology, 21*, 3609–3615.
217. Hall, J. C. and Harruff, R. (1989). Fatal cardiac arrhythmia in a patient with interstitial myocarditis related to chronic arsenic poisoning. *Southern Medical Journal 82*, 1557–1560.
218. Lynes, M. A., Kang, Y. J., Sensi, S. L., Perdrizet, G. A., and Hightower, L. E. (2007). Heavy metal ions in normal physiology, toxic stress, and cytoprotection. *Annals of the New York Academy of Sciences, 1113*, 159–172.
219. Jiang, Y., Reynolds, C., Xiao, C., Feng, W., Zhou, Z., Rodriguez, W., Tyagi, S. C., Eaton, J. W., Saari, J. T., and Kang, Y. J. (2007). Dietary copper supplementation reverses hypertrophic cardiomyopathy induced by chronic pressure overload in mice. *Journal of Experimental Medicine, 204*, 657–666.
220. Uriu-Adams, J. Y., Scherr, R. E., Lanoue, L., and Keen, C. L. (2010). Influence of copper on early development: prenatal and postnatal considerations. *Biofactors, 36*, 136–152.
221. Johnson, C. W., Williams, W. C., Copeland, C. B., DeVito, M. J., and Smialowicz, R. J. (2000). Sensitivity of the SRBC PFC assay versus ELISA for detection of immunosuppression by TCDD and TCDD-like congeners. *Toxicology, 156*, 1–11.
222. Silkworth, J. B., Lipinskas, T., and Stoner, C. R. (1995). Immunosuppressive potential of several polycyclic aromatic hydrocarbons (PAHs) found at a Superfund site: new model used to evaluate additive interactions between benzo[*a*]pyrene and TCDD. *Toxicology, 105*, 375–386.
223. Fan, F., Yan, B., Wood, G., Viluksela, M., and Rozman, K. K. (1997). Cytokines (IL-1beta and TNFalpha) in relation to biochemical and immunological effects of 2,3,7,8-tetrachlorodibenzo-*p*-dioxin (TCDD) in rats. *Toxicology, 116*, 9–16.
224. Kerkvliet, N. I., Shepherd, D. M., and Baecher-Steppan, L. (2002). T lymphocytes are direct, aryl hydrocarbon receptor (AhR)-dependent targets of 2,3,7,8-tetrachlorodibenzo-*p*-dioxin (TCDD): AhR expression in both CD4[+] and CD8[+] T cells is necessary for full suppression of a cytotoxic T lymphocyte response by TCDD. *Toxicology and Applied Pharmacology, 185*, 146–152.
225. Sekine, H., Mimura, J., Oshima, M., Okawa, H., Kanno, J., Igarashi, K., Gonzalez, F. J., Ikuta, T., Kawajiri, K., and Fujii-Kuriyama, Y. (2009). Hypersensitivity of aryl hydrocarbon receptor-deficient mice to lipopolysaccharide-induced septic shock. *Molecular and Cellular Biology, 29*, 6391–6400.
226. Kimura, A., Naka, T., Nakahama, T., Chinen, I., Masuda, K., Nohara, K., Fujii-Kuriyama, Y., and Kishimoto, T. (2009). Aryl hydrocarbon receptor in combination with Stat1 regulates LPS-induced inflammatory responses. *Journal of Experimental Medicine, 206*, 2027–2035.
227. Mimura, J. and Fujii-Kuriyama, Y. (2003). Functional role of AhR in the expression of toxic effects by TCDD. *Biochimica et Biophysica Acta, 1619*, 263–268.
228. Matsumura, F. (2009). The significance of the nongenomic pathway in mediating inflammatory signaling of the dioxin-activated Ah receptor to cause toxic effects. *Biochemical Pharmacology, 77*, 608–626.
229. Vogel, C. F. and Matsumura, F. (2009). A new cross-talk between the aryl hydrocarbon receptor and RelB, a member of the NF-kappaB family. *Biochemical Pharmacology, 77*, 734–745.
230. Vogel, C. F., Sciullo, E., and Matsumura, F. (2007). Involvement of RelB in aryl hydrocarbon receptor-mediated induction of chemokines. *Biochemical and Biophysical Research Communications, 363*, 722–726.
231. Funatake, C. J., Marshall, N. B., Steppan, L. B., Mourich, D. V., and Kerkvliet, N. I. (2005). Cutting edge: activation of the aryl hydrocarbon receptor by 2,3,7,8-tetrachlorodibenzo-*p*-dioxin generates a population of CD4[+] CD25[+] cells with characteristics of regulatory T cells. *Journal of Immunology, 175*, 4184–4188.
232. Cribb, A. E., Delaporte, E., Kim, S. G., Novak, R. F., and Renton, K. W. (1994). Regulation of cytochrome P-4501A and cytochrome P-4502E induction in the rat during the production of interferon alpha/beta. *Journal of Pharmacology and Experimental Therapeutics, 268*, 487–494.
233. Muntane-Relat, J., Ourlin, J. C., Domergue, J., and Maurel, P. (1995). Differential effects of cytokines on the inducible expression of CYP1A1, CYP1A2, and CYP3A4 in human hepatocytes in primary culture. *Hepatology, 22*, 1143–1153.
234. Vrzal, R., Ulrichova, J., and Dvorak, Z. (2004). Aromatic hydrocarbon receptor status in the metabolism of xenobiotics under normal and pathophysiological conditions. *Biomedical Papers of the Medical Faculty of the University of Palacky Olomouc, Czech Republic, 148*, 3–10.
235. Delaporte, E. and Renton, K. W. (1997). Cytochrome P4501A1 and cytochrome P4501A2 are downregulated at both transcriptional and post-transcriptional levels by conditions resulting in interferon-alpha/beta induction. *Life Sciences, 60*, 787–796.
236. Dalton, T. P., Puga, A., and Shertzer, H. G. (2002). Induction of cellular oxidative stress by aryl hydrocarbon receptor activation. *Chemico-Biological Interactions, 141*, 77–95.
237. Coussens, L. M. and Werb, Z. (2002). Inflammation and cancer. *Nature, 420*, 860–867.
238. Symons, A. M. and King, L. J. (2003). Inflammation, reactive oxygen species and cytochrome P450. *Inflammopharmacology, 11*, 75–86.
239. Renton, K. W. (2001). Alteration of drug biotransformation and elimination during infection and inflammation. *Pharmacological Therapy, 92*, 147–163.
240. Vasak, M. (2005). Advances in metallothionein structure and functions. *Journal of Trace Elements in Medicine and Biology, 19*, 13–17.

241. Tijet, N., Boutros, P. C., Moffat, I. D., Okey, A. B., Tuomisto, J., and Pohjanvirta, R. (2006). Aryl hydrocarbon receptor regulates distinct dioxin-dependent and dioxin-independent gene batteries. *Molecular Pharmacology, 69,* 140–153.

242. Mandal, P. K. (2005). Dioxin: a review of its environmental effects and its aryl hydrocarbon receptor biology. *Journal of Comparative Physiology B, 175,* 221–230.

243. Xie, A., Walker, N. J., and Wang, D. (2006). Dioxin (2,3,7,8-tetrachlorodibenzo-*p*-dioxin) enhances triggered afterdepolarizations in rat ventricular myocytes. *Cardiovascular Toxicology, 6,* 99–110.

244. Huang, P., Ceccatelli, S., Hoegberg, P., Sten Shi, T. J., Hakansson, H., and Rannug, A. (2003). TCDD-induced expression of Ah receptor responsive genes in the pituitary and brain of cellular retinol-binding protein (CRBP-I) knockout mice. *Toxicology and Applied Pharmacology, 192,* 262–274.

245. Brouwer, A., Longnecker, M. P., Birnbaum, L. S., Cogliano, J., Kostyniak, P., Moore, J., Schantz, S., and Winneke, G. (1999). Characterization of potential endocrine-related health effects at low-dose levels of exposure to PCBs. *Environmental Health Perspectives, 107,* (Suppl. 4), 639–649.

246. Walker, N. J., Miller, B. D., Kohn, M. C., Lucier, G. W., and Tritscher, A. M. (1998). Differences in kinetics of induction and reversibility of TCDD-induced changes in cell proliferation and CYP1A1 expression in female Sprague-Dawley rat liver. *Carcinogenesis, 19,* 1427–1435.

247. Bains, J. S. and Shaw, C. A. (1997). Neurodegenerative disorders in humans: the role of glutathione in oxidative stress-mediated neuronal death. *Brain Research, Brain Research Reviews, 25,* 335–358.

248. Tan, E. K., Khajavi, M., Thornby, J. I., Nagamitsu, S., Jankovic, J., and Ashizawa, T. (2000). Variability and validity of polymorphism association studies in Parkinson's disease. *Neurology, 55,* 533–538.

249. Coleman, T., Ellis, S. W., Martin, I. J., Lennard, M. S., and Tucker, G. T. (1996). 1-Methyl-4-phenyl-1,2,3,6-tetrahydropyridine (MPTP) is N-demethylated by cytochromes P450 2D6, 1A2 and 3A4: implications for susceptibility to Parkinson's disease. *Journal of Pharmacology and Experimental Therapeutics, 277,* 685–690.

250. Dauchy, S., Dutheil, F., Weaver, R. J., Chassoux, F., Daumas-Duport, C., Couraud, P. O., Scherrmann, J. M., De Waziers, I., and Decleves, X. (2008). ABC transporters, cytochromes P450 and their main transcription factors: expression at the human blood–brain barrier. *Journal of Neurochemistry, 107,* 1518–1528.

251. Dutheil, F., Dauchy, S., Diry, M., Sazdovitch, V., Cloarec, O., Mellottee, L., Bieche, I., Ingelman-Sundberg, M., Flinois, J. P., de Waziers, I., Beaune, P., Decleves, X., Duyckaerts, C., and Loriot, M. A. (2009). Xenobiotic-metabolizing enzymes and transporters in the normal human brain: regional and cellular mapping as a basis for putative roles in cerebral function. *Drug Metabolism and Disposition, 37,* 1528–1538.

252. Rahbari, M. and Goharrizi, A. S. (2009). Adsorption of lead (II) from water by carbon nanotubes: equilibrium, kinetics, and thermodynamics. *Water Environment Research, 81,* 598–607.

253. Mishra, K. P. (2009). Lead exposure and its impact on immune system: a review. *Toxicology In Vitro, 23,* 969–972.

254. Waterman, S. J., el-Fawal, H. A., and Snyder, C. A. (1994). Lead alters the immunogenicity of two neural proteins: a potential mechanism for the progression of lead-induced neurotoxicity. *Environmental Health Perspectives, 102,* 1052–1056.

26

TRANSGENIC MICE WITH A CONSTITUTIVELY ACTIVE AHR: A MODEL FOR HUMAN EXPOSURE TO DIOXIN AND OTHER AHR LIGANDS

PATRIK ANDERSSON, SARA BRUNNBERG, CAROLINA WEJHEDEN, LORENZ POELLINGER, AND ANNIKA HANBERG

26.1 BACKGROUND

Exposure to dioxin elicits a plethora of biologic responses ranging from biochemical alterations, such as enzyme induction, through overtly toxic responses including tumor promotion and a lethal wasting syndrome. However, the most sensitive adverse effects observed in multiple species appear to be on fetal development, including effects on the developing reproductive, immune, and nervous systems [1–3]. The mechanisms behind these effects are not yet understood, but it is generally assumed that the toxicity processes are initiated by the dioxin or dioxin-like compound binding to and activating the AHR and thereby altering gene expression.

Humans are exposed to a variety of different dioxin-like compounds. This class of toxicologically important environmental contaminants includes polychlorinated dibenzo-*p*-dioxins (PCDDs), polychlorinated dibenzofurans (PCDFs), and polychlorinated biphenyls (PCBs). Each subgroup contains a large number of congeners displaying widely varying toxic potencies. Based on the assumed common mechanism of action (via binding to the AHR) and that these compounds differ only in potency (quantitatively), a toxic equivalency factor (TEF) system has been developed to handle the assessment of all these compounds [4]. The TEF system is currently widely used to assess risks with dioxins and dioxin-like PCBs to humans. Due to high lipophilicity and stability, these compounds tend to bioaccumulate in tissue lipid and in the food chain. Food is therefore the major source for human exposure to dioxins and PCBs, especially fatty foods of animal origin, such as fish, meat, egg, and dairy products.

The risk assessments of dioxins and dioxin-like PCBs performed by WHO and EU led to major concerns [1–3]. The tolerable daily intake for humans, 2 pg TEQ/kg body weight, has been estimated to be within the range of human exposures occurring in the general population. To further improve the risk assessment of dioxin-like compounds as well as the present TEF scheme, more knowledge on the mechanism(s) and mode(s) of action of dioxin-like compounds is needed.

We have developed a transgenic mouse model expressing a constitutively active Ah receptor (CA-AHR), that is, mimicking exposure to dioxin-like pollutants acting through this receptor. The CA-AHR mouse model can be used to easily study which effects are caused by the activated AHR, that is, for which effects the TEF concept is valid. In addition, the CA-AHR model also resembles human exposure to dioxin-like compounds in that the activation of the AHR is continuous from early development and onward, a situation difficult to mimic in regular toxicity studies.

26.2 THE CA-AHR MICE

A transgenic mouse model expressing a constitutively activated AHR has been established in our laboratories [5]. The receptor has been made constitutively active by deleting the minimal ligand binding domain within the receptor without disrupting the DNA binding function of the protein [6]. This receptor has been expressed in mice that show modest levels of expression of the transgene in numerous tissues [5]. The constitutively active receptor is acting as if TCDD, or another ligand, were continuously present.

The AH Receptor in Biology and Toxicology, First Edition. Edited by Raimo Pohjanvirta.
© 2012 John Wiley & Sons, Inc. Published 2012 by John Wiley & Sons, Inc.

In order to fully characterize the CA-AHR model, a thorough histopathological study of most organs in both sexes and at four different ages (3, 6, 9, and 12 months) has been performed, as well as molecular and cellular studies.

26.2.1 A Constitutively Active AHR *In Vivo*: the CA-AHR Construct

The C-terminal part of the AHR ligand binding domain (residues 288–421) encompasses the full PAS B region. Deletion of these amino acids gives a mutant that not only fails to interact with Hsp90 and shows constitutive nuclear localization, but also binds ARNT and exhibits potent transactivation in a constitutive manner in the absence of ligand [6]. Conditional expression of the CA-AHR (by the tet-off system) in the human breast cancer cell line MCF-7 results in increased expression of CYP1A1 as well as inhibition of estrogen-stimulated proliferation of these cells [7].

The immune system has been described as one of the most sensitive organ systems for dioxin toxicity, and we therefore aimed to preferentially express the CA-AHR in lymphatic tissues. Expression of the coding sequence for CA-AHR was regulated by an expression vector consisting of a strong promoter (SRα, a modified viral SV40 promoter) and the Eµ enhancer from the immunoglobulin heavy-chain promoter.

The CA-AHR mice are of the conventional transgenic type; that is, the expression construct was randomly integrated into the genome. Thus, the endogenous, ligand-dependent AHR is still expressed and functional in these mice. The mice were generated on a mixed genetic background (C57BL6/J and CBA), but despite having different origins, the remaining wild-type AHRs in the C57Bl/6 and CBA strains are both of the highly responsive type [8].

Analysis of mice of a mixed genetic background can be problematic. In parallel to most of the phenotype characterization described in this chapter, two transgenic lines of mice were backcrossed for more than 10 generations with the inbred strain C57Bl/6. However, most data presented here were generated on mice with a mixed genetic background. There were no apparent qualitative differences seen between mice of mixed or homogenous genetic background.

26.2.2 Expression Pattern of CA-AHR

As expected, CA-AHR was expressed in lymphatic organs, for example, thymus, spleen, lymph nodes, and in enriched T and B cells and bone marrow. Considering the expression construct used, it was surprising to find that the CA-AHR was also expressed and transcriptionally active in all organs that were analyzed, for example, liver, lung, muscle, skin, brain, heart, kidney, testis, uterus, and the entire gastrointestinal tract [5].

26.2.3 Constitutive Transcriptional Activity of the CA-AHR *In Vivo*

Constitutive transcriptional activity of the CA-AHR was demonstrated by analyzing expression of the AHR target gene CYP1A1. In the absence of any exogenous AHR ligands, increased expression of CYP1A1 was shown in many tissues by RT-PCR and RNA blotting, and in stomach sections by both *in situ* hybridization and immunohistochemistry [5].

Homozygous CA-AHR animals showed higher CYP1A1 mRNA levels than heterozygous mice. The induction of other established target genes such as CYP1A2 and AHRR [5] and NMO1 [9] provided additional evidence for the transcriptional activity of the CA-AHR *in vivo*. However, the expression levels of CA-AHR and the target gene CYP1A1 did not show a direct correlation. For instance, liver, muscle, skin, and heart showed considerably higher CYP1A1 expression compared to thymus and spleen even though the CA-AHR was expressed at a higher level in the latter two organs [5]. This observation indicates that there are other, tissue-specific factors that influence expression of the CYP1A1 gene.

When the transcriptional activity of the CA-AHR was compared to wild-type animals exposed to different doses of TCDD, the CYP1A1 induction levels in homozygous CA-AHR were similar to those in wild-type animals treated with 3 and 0.3 µg TCDD/kg in thymus and liver, respectively [5], indicating that the activity of the CA-AHR in these tissues corresponds to exposure to a relatively low dose of TCDD. The CA-AHR activity in the glandular part of the stomach corresponded to a single TCDD dose of 10 µg/kg or higher [5].

The random genomic integration of a transgenic construct can result in disruption of vital genes, which could result in a significant phenotype that is unrelated to the added gene of interest. To control for this, mice from three independent CA-AHR lines (called Y8, A3, and Y4) were analyzed. Homozygous mice of the Y4 line were difficult to obtain and the reason for this is unknown. Expression of the CA-AHR and CYP1A1 did not seem to differ markedly between heterozygous mice of these three lines and mice from all three lines developed the stomach tumors described below.

26.2.4 Sex Ratios and Body Weight

Homozygous CA-AHR animals were fertile. Mating heterozygous Y8 and A3 animals yielded homozygous, heterozygous, and negative progeny in approximately the expected Mendelian 1:2:1 ratio, indicating that expression of CA-AHR with this construct did not result in fetal mortality. An increased frequency of females has been reported from several human populations exposed to TCDD or other AHR ligands [10, 11]. However, homozygous CA-AHR pups did not show any significant alterations in the sex ratio compared to wild-type pups. In fact, in CA-AHR mice of the Y8 and A3 founder lines, the frequency of males was 50% and 57%, respectively, compared to 55% in wild-type mice.

Exposure of experimental animals to high doses of TCDD results in a lethal wasting syndrome that is characterized by decreased food intake and dramatic weight loss [12]. Compared to wild-type animals, the body weight of male homozygous CA-AHR animals of the A3 and Y8 founder lines was lower, beginning at 27 and 37 weeks of age, respectively. Although observed long before the A3 animals become moribund, the decreased body weight gain is probably associated with the increased mortality discussed below.

26.2.5 Increased Mortality

A true mortality study was not performed due to ethical reasons, but CA-AHR mice showed a shortened life span, where homozygous male mice from both the Y8 and the A3 founder lines were more affected than female mice and animals from the A3 line showed the highest mortality when these two founder lines were compared [5]. These deaths were not preceded by any long-term apparent clinical signs of illness. When body weights of older A3 animals were recorded by daily intervals, a sudden drop in body weight (by 5% over 24 h) could occasionally be noted. Within 24 h of this body weight loss, those animals became moribund and were sacrificed. Thus, the process leading to death appears to be very rapid. Although the increased mortality correlates well with the development of stomach tumors, the ultimate cause of this premature mortality is still unknown.

26.2.6 Stomach Tumors

During necropsy of some of the first moribund mice, large cysts were found on the stomachs (Fig. 26.1b and c). These findings were surprising and triggered a more systematic characterization. The first macroscopically visible lesions appeared around 3 months of age and were seen as small cysts filled with clear content found in the minor curvature of the stomach. In older animals, the cysts were more numerous and varied in color from clear, white, and beige to brown and black (Fig. 26.1c). In animals around 12 months of age, the cystic stomachs were sometimes found to adhere to surrounding organs, such as liver, pancreas, spleen, and fat. On the inside of the stomach, the so-called limiting ridge separates the squamous epithelium of the forestomach from the glandular part of the stomach. This structure was substantially enlarged in the CA-AHR animals (Fig. 26.1e). Bleeding was sometimes observed in the mucosa [5].

26.2.6.1 Histopathological Characterization of the Stomach Lesions Over time, several pathologists experienced in work with experimental animals and/or specialists in gastrointestinal pathology have contributed with histopathological assessments of the stomach lesions found in the CA-AHR animals. The descriptions have ranged from "not tumors," "congenital malformation," to "gastric adenocarcinoma." The lesions show a rare combination of characteristics not fitting with established criteria for the established terminology. For reasons discussed at the end of this section, the term "tumor" was chosen for glandular structures penetrating into the submucosa and beyond.

The earliest microscopical alterations in the CA-AHR glandular stomach (at 9 weeks of age) were cystic dilatations of basally located glands. Starting at 10 weeks, isolated cystic glands penetrated through the submucosa, the muscularis propria, and reaching the subserosa. These tumors were found to be actively proliferating as demonstrated with staining for the proliferation marker PCNA. Van Gieson stain excluded herniation as the cause of this penetration. Ten weeks and older CA-AHR animals also showed hyperplasia of the foveolar cells and expanding glandular structures within the limiting ridge structure [5, 13].

In older animals, tumors were found to penetrate into the muscularis propria at several locations, resulting in a substantially thickened stomach wall with multiple cysts of varying size. This could sometimes give a quite bizarre overall appearance (Fig. 26.1f). The tumors were composed by highly differentiated gastric cells, such as foveolar and parietal cells as well as cells of cardiopyloric type, but metaplastic epithelium (squamous metaplasia) was also observed. In addition, many tumors were surrounded by large amounts of connective tissue, vessels, and well-organized lymphatic foci and fatty tissue [5, 13]. Lesions with these characteristics are classified in human pathology as hamartomatous tumors. Although tumor-laden stomachs were found to adhere to adjacent organs [5], penetration into these organs or metastases to other sites was not found.

Epithelial cells express specific types of mucins depending on their location in the gastrointestinal tract. This can change during the progression of cells toward a more tumor-like cellular phenotype. The mucin content was characterized by different histological stains, for example, PAS, Alcian Blue pH 2.5, and High Iron Diamide, demonstrating altered mucin expression in several cell types, for example, acid mucins that are normally not found in stomach. This indicated intestinal metaplasia of these cells [5, 13].

A general assessment of the overall severity of the stomach tumors demonstrated a difference between heterozygous and homozygous animals as well as between males and females [5]. These differences were confirmed in a more quantitative characterization where several parameters were analyzed on hematoxylin–eosin (HE)-stained stomach sections [13], and they are in good agreement with the differences observed with regard to mortality [5]. Apart from isolated dilatations of glands and occasional findings of goblet cells in older animals [13], none of the lesions described above was found in wild-type animals.

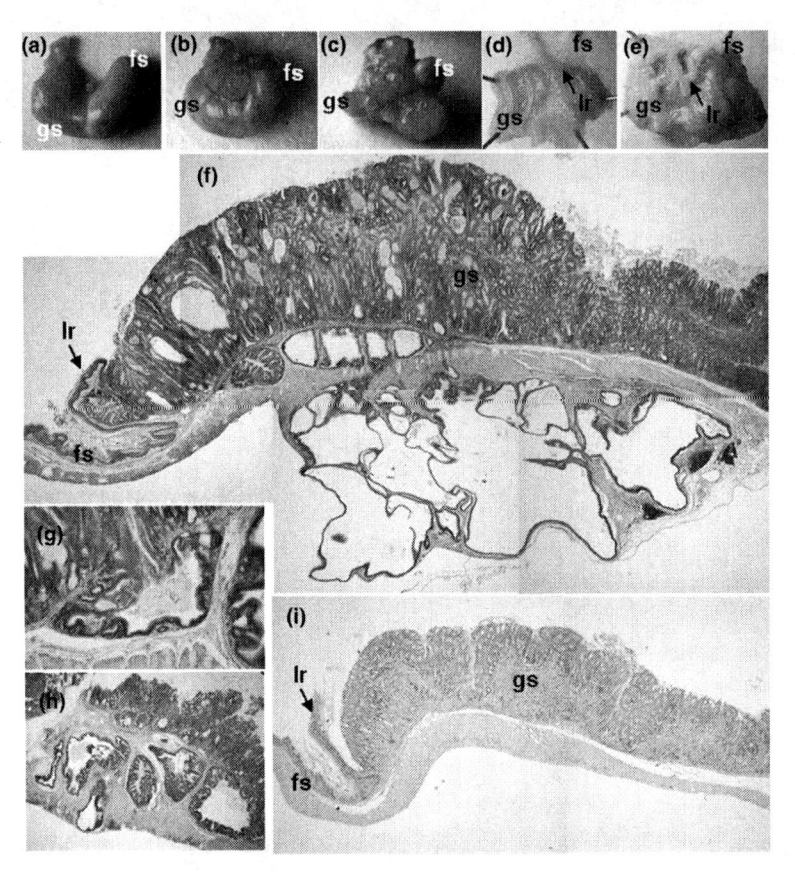

FIGURE 26.1 Stomach tumors in the CA-AHR mice. External views of stomachs from (a) wild-type mouse, (b) homozygous 6-month-old CA-AHR male (A3 line) with cysts in the inner curvature, and (c) homozygous 15-month-old CA-AHR male placed on major curvature with large cysts originating from minor curvature with expansion into dorsal and ventral directions. Inside view of stomachs from (d) wild-type mouse and (e) 6-month-old male of Y8 strain. Note the thickened limiting ridge. (f and g) HE-stained stomach (the same as in part (b)). (f) Section of the stomach showing unaffected forestomach (fs), thickened limiting ridge (lr), and glandular stomach (gs) part with aberrant structures and filled with cysts throughout all layers. (g) Close-up of cystic structure penetrating through the muscularis mucosa. (h) Section from a different part showing muscle layers with numerous cysts lined with glandular epithelium. (i) HE-stained stomach from a wild-type mouse. Abbreviations: fs, forestomach; gs, glandular stomach; lr, limiting ridge. (See the color version of this figure in Color Plates section.)

26.2.6.2 Decrease in Parietal Cells and Altered Proliferation Zone

A distinct decrease of the parietal cell region in CA-AHR mice was observed, which was supported by decreased mRNA expression of the β-subunit of the parietal cell marker H^+/K^+ ATPase [13]. Quantification of the combined parietal/chief cell region in HE-stained sections demonstrated a decrease of this region in 16 weeks or older males of the A3 founder line and in 52 weeks old males and females of the Y8 founder line [13].

Proliferating cells are normally found in the narrow isthmus region of the stomach. In CA-AHR stomachs, the proliferation marker PCNA could be found throughout the parietal/chief cell region. Exposure of wild-type animals to TCDD resulted in expansion of the narrow isthmus zone similar to the CA-AHR mice [5]. This suggests that not only the CA-AHR, but also the ligand-activated AHR can affect the proliferative status of cells in the gastric mucosa.

26.2.6.3 Benign or Malignant Lesions?

Considering common definitions of tumors and cancer, the lesions observed in the glandular stomach of the CA-AHR mice display some really paradoxical characteristics. The cells that constitute the penetrating glands are highly differentiated and may even at a small scale be organized like gastric mucosa of zymogenic type or of antral/pyloral type [13]. The columnar cells retained the capability of secreting neutral mucins [13], and no signs belonging to commonly accepted criteria for dysplasia or malignancy on a cellular level were found. Despite this lack of classical markers of neoplasia, "uncontrollably" proliferating cells from the gastric mucosa penetrated into the submucosa, muscularis propria, and reached the subserosa. Squamous and intestinal metaplasia was observed and stomachs adhered to adjacent organs and increased mortality was observed in the CA-AHR animals. Thus, highly differentiated cells (indicating nontumor) show uncontrolled and progressive proliferation (indicating neoplasia/tumor) with properties of invasion that most likely contribute to the increased mortality (indicating cancer/malignancy). Until further properties of these cells have been evaluated (such as the potential for ectopic growth in immunodeficient nude mice), the neutral term "tumor" for describing cells that penetrated into the submucosa and beyond has conservatively been used.

26.2.6.4 Stomach as a Target Organ for AHR Ligands
At the time of the initial characterization of the CA-AHR mice, the stomach was rarely mentioned as a target organ for AHR-mediated toxicity. However, in the review by Poland and Knutson from 1982 [14], several examples of hyperplastic lesions in the stomach of monkeys and cows were described. A more careful analysis of the literature between 1973 and 1990 identified numerous examples where stomach lesions strikingly similar to the CA-AHR mice had been observed after exposure to various AHR agonists. These included exposure to commercial mixtures of PCBs (Aroclors) in rhesus monkeys [15–18] and in rats [19, 20] or exposure of the dioxin-like 3,4,3′,4′-TCB (PCB 77) in rhesus monkeys [21–23] and marmoset monkeys [24]. Rhesus monkeys [25] and rats [26] exposed to commercial mixtures of polybrominated biphenyls or exposure of rhesus monkeys [27, 28] or hairless mice [29] to TCDD-like dibenzofurans also produced similar stomach lesions.

Although the severity of the lesions differs, a pattern can be found in these previously published studies that is strikingly similar to the findings in the CA-AHR mice: a marked hyperplasia of the surface mucous cells and a remarkable decrease in the number of parietal and chief cells; in most studies, cystic dilatations were seen at the base of the gastric glands and with glandular structures penetrating into the submucosa, often associated with foci of lymphatic cells. Intestinal metaplasia, as assessed with Alcian Blue staining is also observed. Many of the other investigators also note that only the zymogenic region of the stomach is affected; the antrum/pyloric region or other parts of the gastrointestinal tract do not show any alterations. Using [^{3}H]-thymidine incorporation as a marker for proliferation, rhesus monkeys treated with PCB showed labeling of almost the entire mucosa below the most apical cell layers [22], just like the findings seen in the CA-AHR mice.

26.2.6.5 Dysregulated Gene Expression in the Glandular Stomach: Osteopontin The stomach tumors observed in the CA-AHR mice did not seem to have any counterpart in the general pathology literature, and we chose to search for dysregulated genes with an unbiased approach, the so-called suppressive subtraction hybridization. To detect relatively early changes in gene expression, glandular stomachs from 90-day-old male CA-AHR mice were compared to those of wild-type males. A total of 37 clones representing upregulated and 16 clones representing genes downregulated in the CA-AHR stomachs were found [9].

One of the affected genes was osteopontin. Northern blot analysis of RNA demonstrated decreased osteopontin expression in mice of both the Y8 and A3 founder lines being 8 weeks and older [9]. Moreover, osteopontin expression seemed to be decreased also in other organs of the CA-AHR mice, such as brain, bone, and kidney [9, 30]. Cultured wild-type hepatoma cells exposed to TCDD showed a rapid, but transient decrease in osteopontin expression. In contrast, no effect on osteopontin expression was seen in mutant, dioxin-resistant hepatoma cells [31] that express either substantially reduced AHR levels (the C12 mutant) or a functionally defective ARNT protein [9]. This suggests that the decrease in osteopontin expression observed in the CA-AHR mice and in TCDD exposed cells depends on both the AHR and ARNT.

The role, if any, of osteopontin in the development of the CA-AHR stomach lesions is not clear. Several studies have reported an association between *increased* expression of osteopontin and malignant and metastatic progression of several different human tumors, including cancers of the breast, lung, bladder, kidney, colon [32], and stomach [33]. Our findings apparently contradict the positive association observed by others. However, using osteopontin-deficient mice in a chemically induced skin squamous cell carcinoma assay showed that these mice demonstrated accelerated tumor growth and progression, with a greater number of metastases [34]. Thus, the role of osteopontin in tumor development appears to be complex and warrants further investigation.

26.2.7 Effects in the Immune System

26.2.7.1 Thymus In analogy with observations in TCDD-exposed experimental animals [12, 14, 35], the weight of thymus in the CA-AHR mice was decreased [5]. Thymus atrophy is one of the most characteristic effects of TCDD exposure. CA-AHR mice showed a lower thymus weight at 6 months (females) and 9 months of age (both sexes). Histopathological examination identified no morphological changes in the thymus of CA-AHR mice. CYP1A1 staining was detected in endothelial cells lining the blood vessels of CA-AHR and TCDD-treated wild-type mice, whereas no corresponding staining occurred in untreated wild-type mice [36]. In addition, individual cells within the medulla stained positive for CYP1A1 in CA-AHR and TCDD-treated mice. This staining was much more intense in TCDD-treated mice (10 μg/kg, 3 days earlier) as compared to the transgenic CA-AHR mice and did not occur in untreated wild-type mice.

26.2.7.2 Spleen No effect on the relative spleen weights of CA-AHR mice was observed, which is in agreement with data reported by Nohara et al. from a T cell-specific CA-AHR model (see below) [37]. Histopathological analysis showed that both hematopoiesis and hemosiderin deposition were increased in 12-month-old CA-AHR males [36]. In female CA-AHR mice, hemosiderin deposition, but not hematopoiesis, was increased. Expression of CYP1A1 protein was detected in transgenic, TCDD-treated and control wild-type mice. The pattern of expression was similar in all mice with CYP1A1 staining in the red pulp and no staining in the white pulp. The staining in the red pulp from CA-AHR and TCDD-treated wild-type mice was slightly more intense than in the control wild-type mice. The expression of CYP1A1 mRNA was confirmed by RT-PCR analysis. Constitutive expression of CYP1A1 in the spleen of wild-type mice has previously been observed [38].

26.2.7.3 Effects Observed on T-Cell Populations Experimental animals exposed to TCDD show atrophy of the thymus [12, 14, 35]. Moreover, prenatal exposure to TCDD results in altered distribution of thymic $CD4^+$ and $CD8^+$ populations [39]. Decreased thymus weight is seen in the CA-AHR mice [5]. FACS analysis of thymocytes in adult CA-AHR animals did not show any altered distribution of single-positive $CD4^+$ or $CD8^+$ populations. However, although not statistically significant, the $CD4^+$ population appeared to be decreased and the $CD8^+$ population appeared to be increased in newborn CA-AHR animals. In contrast, peripheral T cells from the axillary and inguinal lymph nodes showed indications of an increase in population of $CD4^+$ and a decreased size of the $CD8^+$ T-cell population. No such differences were observed in spleen or mesenterial lymph nodes [40].

A similar transgenic mouse expressing a constitutively active mutant of AHR under the regulation of the T cell-specific CD2 promoter has been generated [37]. The transgene is expressed in $CD4^+$ and $CD8^+$ T cells with induced CYP1A1 mRNA in the thymus and spleen. A reduced thymus weight was seen in female mice, but the immune response of these mice only partially mimicked the immune response of wild-type mice exposed to TCDD [37, 41].

26.2.7.4 Increase in Mature Bone Marrow-Derived B Cells More consistent effects were observed in the B-cell compartment of CA-AHR mice. Consistent with previous studies of TCDD-exposed mice [42], FACS analysis showed an increased fraction of mature B cells in the bone marrow of CA-AHR mice. Moreover, the fraction of mature B cells recognized by high levels of B220 staining and low level of staining for the monoclonal antibody 493 was also increased in the spleen of young CA-AHR mice [40]. The reason for the observed enlargement of the mature B cell population is unknown.

26.2.7.5 Decreased Peritoneal Population of B1 Cells In contrast to conventional B cells (also called B2 cells) that develop in the bone marrow, a population of B cells that express CD5 (B1 cells) originates from fetal liver and omentum, forming a distinct peripheral B-cell population that is mainly present in the peritoneal and pleural cavities [43]. This population is considered to be self-renewing.

In the CA-AHR mice, a subpopulation of the peritoneal B1 cells (B1a), expressing the surface marker CD5, was substantially reduced in both the Y8 and A3 founder lines. According to the classical definition of B1a cells [44], these cells express low levels of B220. The $CD5^+ B220^{lo}$ population was reduced in the CA-AHR mice, demonstrating a selective loss of classical B1a cells. In contrast, the CD5-expressing B cells that remained in CA-AHR mice expressed substantially higher levels of B220 than those in wild-type mice [40]. The lack or reduction of the B1 population in the peritoneum has been reported in many different mouse models, most of which have deficiencies in molecules that modulate B-cell receptor signalling. It would be interesting to study if the CA-AHR or ligand activation of AHR affects BCR signalling.

An alternative mechanism could perhaps be provided by the observed reduction in expression of osteopontin. This protein is also called Eta-1 for Early T lymphocyte activation-1. It is normally expressed in activated T cells and macrophages and plays an important role in cell-mediated (type-1) immunity [45, 46]. Mice lacking osteopontin show decreased resistance to viral (*Herpes simplex*) and bacterial infections (*Listeria monocytogenes*) [45]. In contrast, mice overexpressing osteopontin demonstrate an increased population of peritoneal B1 cells [47]. It is tempting to speculate that the decrease in osteopontin mRNA expression levels observed in CA-AHR mice may contribute to the decreased population of B1 cells.

26.2.8 General Pathological and Molecular Characterization

26.2.8.1 Liver Liver enlargement is a highly reproducible effect in many rodent species after dioxin exposure and is reported to correlate with hepatocellular hypertrophy and enzyme induction [12]. In CA-AHR mice, the liver is enlarged in both male [5] and female mice [36]. This effect is associated with induced expression of CYP1A1 and CYP1A2. Histologically, hepatic hematopoiesis was slightly increased in CA-AHR mice, especially in females. TCDD is a well-established carcinogen and tumor promoter [48]. Consistently, hepatocellular adenoma has been observed in 2 out of 11 CA-AHR mice [36]. This liver tumorigenic potential of CA-AHR was also demonstrated in a tumor promotion study performed with male CA-AHR mice initiated with N-nitrosodiethylamine (DEN) [49]. Compared to TCDD-exposed mice, the degree of hepatic CYP1A1 mRNA induction is quite modest in relation to the relatively high transcriptional activity of CA-AHR in other tissues [5]. A possible explanation may be that in exposure studies, TCDD is preferably distributed to the liver where it would reach a proportionally higher concentration compared to most extrahepatic tissues, and this does not occur in CA-AHR mice. In addition, relatively few hepatocytes in CA-AHR mice express high levels of CYP1A1. Thus, their CYP1A1 mRNA is being diluted in mRNA analysis of the whole tissue. However, the hepatocellular tumors observed in CA-AHR mice support the significance of expressing the active CA-AHR in hepatocytes.

The hepatic phase II enzyme UDP-glucuronosyltransferase 1A6 (UGT1A6) is another AHR target gene [12]. In line with this, the mRNA expression level of UGT1A6 was induced in the liver of CA-AHR mice. Glucuronidation of, for example, thyroxine (T4) is mediated by UGT1A6 and results in a complex that can be excreted via bile and feces. Therefore, increased levels of UGT often result in decreased serum levels of free T4, which might be related to an adverse effect on neurobehavioral development observed in experimental animals [12, 50, 51]. Analysis of free thyroxine (FT4) in serum of adult male and female CA-AHR mice showed, however, no statistical difference between wild-type and CA-AHR mice. The average concentrations of FT4 in serum were 9.81 ± 2.96 pmol/L ($n=5$) in wild-type female mice and 11.3 ± 2.44 pmol/L ($n=10$) in CA-AHR female mice. In males, the serum levels were 9.54 ± 0.74 pmol/L ($n=5$) in wild-type mice and 8.23 ± 2.22 pmol/L ($n=6$) in CA-AHR mice. However, only adult mice have yet been studied and they could have compensated for reduced levels that could have occurred earlier in life.

26.2.8.2 Lung
No effects on lung weight or any histopathological abnormalities were observed at any age in CA-AHR mice. Intense CYP1A1 staining of epithelial cells lining the bronchi and the bronchioles was observed in both CA-AHR and TCDD-exposed wild-type mice [36]. Both nonciliated Clara and ciliated cells were stained, and this staining occurred also in unexposed wild-type mice. Strong CYP1A1 expression was also observed in the alveolar compartment of CA-AHR and TCDD-exposed wild-type mice, but not in control wild-type mice.

The localization of CYP1A1 observed in the lungs of both CA-AHR and TCDD-treated mice agrees well with previous reports of 3-methylcholanthrene (3-MC)-treated mice [52–54]. The constitutive expression of CYP1A1 in bronchi and bronchiolar epithelium in untreated wild-type mice is consistent with the earlier observed constitutive CYP1A1 mRNA levels and high levels of expression of endogenous AHR mRNA levels [5]. Constitutive levels of CYP1A1 in the lung have previously been reported [52] and interpreted to reflect exposure to potential endogenous AHR ligands [55].

26.2.8.3 Kidney
In 6–12-month-old male CA-AHR mice, the relative weight of the kidneys was increased significantly by 19–75%, as compared to the age-matched wild-type animal [36]. No similar effect was observed in females. Increased kidney weight has previously been reported in female rats exposed to TCDD [56]. Histopathological examination of the kidney showed a moderate tubular vacuolization in CA-AHR males but not in female CA-AHR mice or in wild-type mice [36]. Tubular vacuoles were observed already at 3 months of age in male CA-AHR mice, and the frequency and severity increased with age. The increased tubular vacuolization observed in the CA-AHR males could possibly give a reasonable explanation to the increased kidney weight. Hydronephrosis is a teratogenic effect following TCDD exposure to pregnant mice [57], which is AHR dependent [58]. However, no sign of hydronephrosis was seen in the CA-AHR mouse, probably due to the low activity of the CA-AHR in the kidney. The CYP1A1 staining of the kidney showed a similar pattern in CA-AHR and TCDD-treated mice [36]. Cells expressing CYP1A1 could be detected in all regions of the kidney (the cortex, the medulla, and the pelvis), while no staining occurred in untreated wild-type mice. There was a strong staining of cells in the glomerulus and in the linings of blood vessels in the cortex and medulla of CA-AHR and TCDD-exposed mice. In addition, occasional tubuli cells were also positively stained. CYP1A1 staining patterns in the kidneys of CA-AHR and TCDD-treated mice are consistent with those reported for 3-MC-exposed mice [52, 53].

26.2.8.4 Heart
Male CA-AHR mice showed a consistent increase in heart-to-body weight ratio. This effect was evident already at relatively early age (58 days) and could be seen in all older age groups (up to 12 months of age). No morphological changes were, however, observed [36]. Exposure studies have shown an increased heart-to-body weight ratio in mice exposed to TCDD *in utero* [59, 60]. Somewhat contradictory, also AHRKO mice show an increase in the relative weight of the heart together with morphological changes, such as fibrosis and smooth muscle cell hyperplasia of coronary arteries and arterioles [59, 61–63]. Thus, it appears that the developing heart is a sensitive target for changes in AHR activity. Interestingly, the increased heart-to-body weight ratio was seen only in male CA-AHR. Such a sex-specific effect has not been discussed in previous papers showing cardiac hypertrophy after TCDD exposure or AHR deficiency [60, 62].

26.2.8.5 Male Reproductive Organs
Exposure to TCDD *in utero* and during lactation has been reported to cause

diverse changes in the reproductive system of male rodents [64]. In 3-month-old male CA-AHR mice, the weights of both ventral prostate and testis were decreased (Brunnberg et al., submitted), which is in agreement with results from previous dioxin exposure studies [59, 65]. The weight decrease of the ventral prostate of CA-AHR mice was relatively modest (33%), which may indicate a relatively low activity of the CA-AHR. Published results on testis weight after *in utero* exposure to TCDD in rodents have been somewhat conflicting [65–67]. This may be due to different experimental protocols since it is likely that there is a critical gestational age window for exposure to affect the testis weight. In the present mouse model, the CA-AHR is most likely expressed and active from early fetal life and onward, covering such sensitive window(s). More than 70% of the testicular weight is constituted by the seminiferous tubules and therefore a decrease in testicular sperm count may cause a decrease in testis weight [68]. Consequently, the observed decrease in epididymal sperm count of 45% in CA-AHR mice seems to mimic the effects reported from *in utero* exposure of dioxins (Brunnberg et al., submitted).

26.2.8.6 Female Reproductive Organs
The CA-AHR transgene has been stably incorporated into the estrogen-dependent human breast cancer cell line MCF-7. Notably, a decrease in estrogen-dependent growth in these cells was observed when the CA-AHR was expressed, in perfect agreement with previous studies on dioxin-exposed cell cultures [69]. This result shows that AHR, when activated, has the potential to interfere with normal hormonal functions.

The genetically activated AHR also affected the female reproductive tract. In adult CA-AHR females, the weight of uterus was reduced when compared to wild-type mice (Brunnberg et al., submitted). This result agrees with the general hypothesis that an activated AHR acts in an antiestrogenic manner [70]. When immature CA-AHR mice were exposed to E2, no antagonizing effect of CA-AHR could be seen on the uterine weight. The lack of effect in this short-term study may reflect that previous studies with the CA-AHR mouse have shown that the activity of the mutant AHR corresponds to a relatively low dose of dioxin [5]. If this is true also for the uterine activity, the activity of CA-AHR may be too low to counteract the activity of E2 with regard to the uterus weight increase by exogenous E2. Such a dose dependency has in fact previously been shown in a study with TCDD [71]. In contrast to these results, immature CA-AHR mice, not coadministered with E2, showed a growth *promoting* effect of the CA-AHR on the uterus to a similar extent as after E2 treatment of wild-type mice at the same age (Brunnberg et al., submitted). Interestingly, also the AHR agonist 3-MC increased uterine wet weight in ovariectomized mice as well as several other estrogen-responsive genes and this effect was suppressed in the presence of estradiol [72, 73]. Accordingly, 3-MC-activated AHR and CA-AHR were shown to act as transcriptional coregulators for the unliganded ERalpha [72, 73]. In addition, TCDD has also shown weak estrogen-like activity on uterine gene expression in the absence of estradiol [74, 75].

26.2.8.7 Bone Tissue
The importance of a functional AHR in dioxin-mediated bone toxicity has been demonstrated by studying rats with different AHR structures [76]. The rat strain with a dysfunctional transactivation domain (H/W; see Chapter 12) displayed resistance to dioxin exposure, while the bone tissue in rats with a functional AHR (L E; see Chapter 12) was considerably more affected. Bones from female L-E rats displayed altered growth, modeling, and physical properties when exposed to TCDD. The H/W strain displayed some alterations, albeit at much higher doses of TCDD compared to L-E rats [76]. In CA-AHR mice, the most striking effect in bone tissue was a gender difference. Females showed effects on several bone-related parameters, while males were almost unaffected [30].

CA-AHR females displayed changes at both trabecular and cortical parts of tibia. The trabecular area and trabecular bone mineral content (BMC) were increased in CA-AHR females, while the trabecular bone mineral density (BMD) was unchanged [30]. However, an increased trabecular area and content might be recorded even though there is increased resorption. This can be due to endocortical resorption that lowers the volumetric cortical BMD at the inner surface so that it falls below the threshold value for cortical bone and into the range for trabecular bone [77].

Alterations in the cortical part of bones from female CA-AHR mice were also reflected in the physical properties of the bones, where the stiffness was decreased and the displacement was increased suggesting softer bones [30]. The stiffness of the bones from female rats exposed to TCDD was also reduced at a concentration of 17 µg TCDD/kg [76]. Altered physical properties of the bones can be due to altered content, altered dimensions, or altered quality of the bone tissue. All three of these factors likely contribute to the phenotype observed in female CA-AHR: the volumetric cortical BMD (content) decreased, and the bones became slightly wider (dimensions) and softer (quality). Hence, a constitutively active AHR interferes with female bone tissue homeostasis in multiple ways.

Analysis of the serum markers CTX and TRAP 5b suggests that the mean osteoclast activity was increased in the female CA-AHR mice. CTX is a marker for bone resorption and measures collagen degradation, while TRAP 5b is a marker for the number of osteoclasts. The osteoclast volume density was increased in the female transgenic mice. Osteoclast size is positively related to osteoclast activity; that is, larger osteoclasts resorb more actively [78, 80]. Hence, in the transgenic female mice the bone resorption seems to be elevated, resulting in a decreased bone mass. Interestingly,

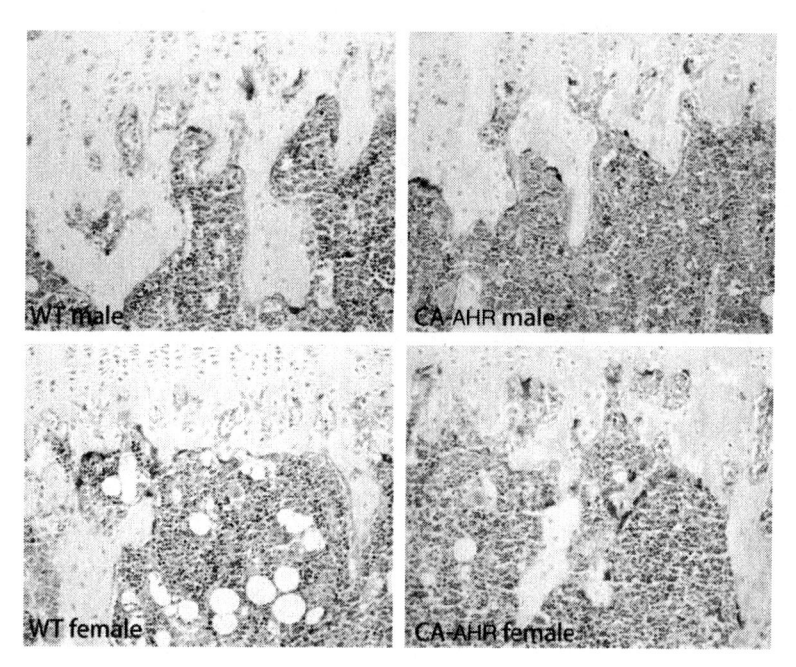

FIGURE 26.2 Immunohistochemistry of cathepsin K-expressing osteoclasts in proximal tibia of wild-type (WT) and transgenic male and female mice. (See the color version of this figure in Color Plates section.)

the mRNA expression levels of several osteoclast genes involved in the function of osteoclasts (e.g., TRAP, cathepsin K, and MMP-9) were decreased in female CA-AHR. The expression levels of the early osteoclast differentiation marker were, however, normal. Moreover, the RANKL/OPG ratio was increased indicating a slightly elevated drive for osteoclast formation. One explanation may be that increased osteoclast differentiation and elevated bone resorption can activate regulatory feedback mechanisms to decrease the activity of the mature osteoclasts in an attempt to normalize the resorption.

The serum level of the bone formation marker PINP was increased, while ALP, another bone formation marker, was decreased in bone extracts from female mice [30]. The mRNA expression levels of collagen type I and other osteoblast markers, for example, Runx2, were decreased in bones from the female CA-AHR mice, suggesting AHR-mediated disturbances also in bone formation. Inhibited osteoblast differentiation by dioxins has previously been demonstrated [81–87], although the mechanisms still remain unknown. Runx2 is the master gene for osteoblast differentiation and this factor displayed a decreased mRNA expression in CA-AHR females. Hence, it is possible that Runx2 is one of the key mediators in dioxin-induced inhibition of osteoblast differentiation.

The analyses revealed that the resorptive activity of the osteoclasts was increased. In sections from tibia (stained for cathepsin K as a marker for mature and active osteoclasts), the relative osteoclast volume was increased in the trabecular part of the metaphysis of female transgenic mice, while no difference was seen in male CA-AHR mice (Fig. 26.2) [30].

A constitutively active AHR results in a gender-specific bone phenotype, where the females are more severely affected than the males. Interestingly, two epidemiological studies in Sweden have implied a similar gender-specific response in humans in cohorts of Swedish fishermen and their wives from the Swedish east coast (exposed to higher levels of dioxins in fish) and west coast (exposed to lower levels of dioxins in fish). The first study showed that women, but not men, on the east coast had a significantly increased risk for vertebral fractures [88]. The second study with these cohorts showed that the east coast wives, with a higher intake of fatty fish, had an increased fracture incidence compared to the west coast wives consuming less fatty fish. No such association was detected for the men [89].

26.3 CONCLUSIONS

A transgenic mouse model expressing a constitutively activated AHR has been established in our laboratories. The receptor is acting as if TCDD, or another agonist, were continuously present. We have shown that classical effects of dioxins (e.g., liver enlargement and thymus atrophy) appear in the CA-AHR mouse and that the functional activity of the mutated AHR corresponds to a relatively low dose of TCDD and that the CA-AHR is expressed in numerous tissues. Also, the uterus, sperm count, bone tissue, and

B lymphocytes have been shown to be affected in the CA-AHR mice. Moreover, these mice develop invasive tumors of the glandular stomach at early age that correlates with increased mortality. We have so far identified osteopontin and estrogen as important and sensitive biomarkers for effects mediated by AHR. Our results further suggest that effects caused by an activated AHR can differ between males and females, as well as between different life stages.

Our results further strengthen the validity of the TEF concept; that is, the toxic effects caused by dioxins and other AHR ligands are results of activation of the AHR. Since the CA-AHR is continuously active, in most organs at a relatively low level and from early development, this model resembles the human exposure scenario and provides an excellent tool for further studies on the mechanisms behind the most sensitive effects of AHR ligands, without needing to disturb experimental animals by dosing during sensitive periods.

REFERENCES

1. EU/SCF (2000). Opinion of the Scientific Committee on Food on the risk assessment of dioxins and dioxin-like PCBs in food. SCF/CS/CNTM/DIOXIN/8 Final. European Commission, Brussels, Belgium.
2. EU/SCF (2001). Opinion of the Scientific Committee on Food on the risk assessment of dioxins and dioxin-like PCBs in food. CS/CNTM/DIOXIN/20 Final. European Commission, Brussels, Belgium.
3. WHO-ECEH/IPCS (2000). Assessment of the health risk of dioxins: re-evaluation of the tolerable daily intake (TDI). *Food Additives and Contaminants*, 17, 223–369.
4. Van den Berg, M., Birnbaum, L. S., Denison, M., De Vito, M., Farland, W., Feeley, M., Fiedler, H., Hakansson, H., Hanberg, A., Haws, L., Rose, M., Safe, S., Schrenk, D., Tohyama, C., Tritscher, A., Tuomisto, J., Tysklind, M., Walker, N., and Peterson, R. E. (2006). The 2005 World Health Organization reevaluation of human and Mammalian toxic equivalency factors for dioxins and dioxin-like compounds. *Toxicological Sciences*, 93, 223–241.
5. Andersson, P., McGuire, J., Rubio, C., Gradin, K., Whitelaw, M. L., Pettersson, S., Hanberg, A., and Poellinger, L. (2002). A constitutively active dioxin/aryl hydrocarbon receptor induces stomach tumors. *Proceedings of the National Academy of Sciences of the United States of America*, 99, 9990–9995.
6. McGuire, J., Okamoto, K., Whitelaw, M. L., Tanaka, H., and Poellinger, L. (2001). Definition of a dioxin receptor mutant that is a constitutive activator of transcription: delineation of overlapping repression and ligand binding functions within the PAS domain. *Journal of Biological Chemistry*, 276, 41841–41849.
7. Kohle, C., Hassepass, I., Bock-Hennig, B. S., Walter Bock, K., Poellinger, L., and McGuire, J. (2002). Conditional expression of a constitutively active aryl hydrocarbon receptor in MCF-7 human breast cancer cells. *Archives of Biochemistry and Biophysics*, 402, 172–179.
8. Poland, A. and Glover, E. (1990). Characterization and strain distribution pattern of the murine Ah receptor specified by the Ahd and Ahb-3 alleles. *Molecular Pharmacology*, 38, 306–312.
9. Kuznetsov, N. V., Andersson, P., Gradin, K., Stein, P., Dieckmann, A., Pettersson, S., Hanberg, A., and Poellinger, L. (2005). The dioxin/aryl hydrocarbon receptor mediates down-regulation of osteopontin gene expression in a mouse model of gastric tumourigenesis. *Oncogene*, 24, 3216–3222.
10. Mocarelli, P., Gerthoux, P. M., Ferrari, E., Patterson, D. G., Kieszak, S. M., Brambilla, P., Vincoli, N., Signorini, S., Tramacere, P., Carreri, V., Sampson, E. J., Turner, W. E., and Needham, L. L. (2000). Paternal concentrations of dioxin and sex ratio of offspring. *Lancet*, 355, 1858–1863.
11. Ryan, J. J., Amirova, Z., and Carrier, G. (2002). Sex ratios of children of Russian pesticide producers exposed to dioxin. *Environmental Health Perspectives*, 110, A699–A701.
12. Pohjanvirta, R. and Tuomisto, J. (1994). Short-term toxicity of 2,3,7,8-tetrachlorodibenzo-*p*-dioxin in laboratory animals: effects, mechanisms, and animal models. *Pharmacological Reviews*, 46, 483–549.
13. Andersson, P., Rubio, C., Poellinger, L., and Hanberg, A. (2005). Gastric hamartomatous tumours in a transgenic mouse model expressing an activated dioxin/Ah receptor. *Anticancer Research*, 25, 903–911.
14. Poland, A. and Knutson, J. C. (1982). 2,3,7,8-Tetrachlorodibenzo-*p*-dioxin and related halogenated aromatic hydrocarbons: examination of the mechanism of toxicity. *Annual Reviews in Pharmacology and Toxicology*, 22, 517–554.
15. McConnell, E. E., Hass, J. R., Altman, N., and Moore, J. A. (1979). A spontaneous outbreak of polycholorinated biphenyl (PCB) toxicity in rhesus monkeys (*Macaca mulatta*): toxicopathology. *Lab Animal Science*, 29, 666–673.
16. Allen, J. R., Carstens, L. A., and Barsotti, D. A. (1974). Residual effects of short-term, low-level exposure of nonhuman primates to polychlorinated biphenyls. *Toxicology and Applied Pharmacology*, 30, 440–451.
17. Allen, J. R. and Norback, D. H. (1973). Polychlorinated biphenyl- and triphenyl-induced gastric mucosal hyperplasia in primates. *Science*, 179, 498–499.
18. Becker, G. M., McNulty, W. P., and Bell, M. (1979). Polychlorinated biphenyl-induced morphologic changes in the gastric mucosa of the rhesus monkey. *Laboratory Investigation*, 40, 373–383.
19. Morgan, R. W., Ward, J. M., and Hartman, P. E. (1981). Aroclor 1254-induced intestinal metaplasia and adenocarcinoma in the glandular stomach of F344 rats. *Cancer Research*, 41, 5052–5059.
20. Ward, J. M. (1985). Proliferative lesions of the glandular stomach and liver in F344 rats fed diets containing Aroclor 1254. *Environmental Health Perspectives*, 60, 89–95.
21. Silverman, S., Rosenquist, C. J., and McNulty, W. P. (1979). Radiographic study of gastric hyperplasia induced by polychlorinated biphenyls in the rhesus monkey. *Investigations in Radiology*, 14, 65–69.

22. Becker, G. M. and McNulty, W. P. (1984). Gastric epithelial cell proliferation in monkeys fed 3,4,3′,4′-tetrachlorobiphenyl. *Journal of Pathology*, 143, 267–274.

23. McNulty, W. P., Becker, G. M., and Cory, H. T. (1980). Chronic toxicity of 3,4,3′,4′- and 2,5,2′,5′-tetrachlorobiphenyls in rhesus macaques. *Toxicology and Applied Pharmacology*, 56, 182–190.

24. van den Berg, K. J., Zurcher, C., Brouwer, A., and van Bekkum, D. W. (1988). Chronic toxicity of 3,4,3′,4′-tetrachlorobiphenyl in the marmoset monkey (*Callithrix jacchus*). *Toxicology*, 48, 209–224.

25. Lambrecht, L. K., Barsotti, D. A., and Allen, J. R. (1978). Responses of nonhuman primates to a polybrominated biphenyl mixture. *Environmental Health Perspectives*, 23, 139–145.

26. Gupta, B. N., McConnell, E. E., Moore, J. A., and Haseman, J. K. (1983). Effects of a polybrominated biphenyl mixture in the rat and mouse. II. Lifetime study. *Toxicology and Applied Pharmacology*, 68, 19–35.

27. Brewster, D. W., Elwell, M. R., and Birnbaum, L. S. (1988). Toxicity and disposition of 2,3,4,7,8-pentachlorodibenzofuran (4PeCDF) in the rhesus monkey (*Macaca mulatta*). *Toxicology and Applied Pharmacology*, 93, 231–246.

28. McNulty, W. P., Pomerantz, I., and Farrell, T. (1981). Chronic toxicity of 2,3,7,8-tetrachlorodibenzofuran for rhesus macaques. *Food & Cosmetics Toxicology*, 19, 57–65.

29. Hebert, C. D., Harris, M. W., Elwell, M. R., and Birnbaum, L. S. (1990). Relative toxicity and tumor-promoting ability of 2,3,7,8-tetrachlorodibenzo-*p*-dioxin (TCDD), 2,3,4,7,8-pentachlorodibenzofuran (PCDF), and 1,2,3,4,7,8-hexachlorodibenzofuran (HCDF) in hairless mice. *Toxicology and Applied Pharmacology*, 102, 362–377.

30. Wejheden, C., Brunnberg, S., Larsson, S., Lind, P. M., Andersson, G., and Hanberg, A. (2010). Transgenic mice with a constitutively active aryl hydrocarbon receptor display a gender-specific bone phenotype. *Toxicological Sciences*, 114, 48–58.

31. Hankinson, O. (1994). A genetic analysis of processes regulating cytochrome P4501A1 expression. *Advances in Enzyme Regulation*, 34, 159–171.

32. Agrawal, D., Chen, T., Irby, R., Quackenbush, J., Chambers, A. F., Szabo, M., Cantor, A., Coppola, D., and Yeatman, T. J. (2002). Osteopontin identified as lead marker of colon cancer progression, using pooled sample expression profiling. *Journal of National Cancer Institute*, 94, 513–521.

33. Ue, T., Yokozaki, H., Kitadai, Y., Yamamoto, S., Yasui, W., Ishikawa, T., and Tahara, E. (1998). Co-expression of osteopontin and CD44v9 in gastric cancer. *International Journal of Cancer*, 79, 127–132.

34. Crawford, H. C., Matrisian, L. M., and Liaw, L. (1998). Distinct roles of osteopontin in host defense activity and tumor survival during squamous cell carcinoma progression *in vivo*. *Cancer Research*, 58, 5206–5215.

35. Fletcher, N., Hanberg, A., and Hakansson, H. (2001). Hepatic vitamin a depletion is a sensitive marker of 2,3,7,8-tetrachlorodibenzo-*p*-dioxin (TCDD) exposure in four rodent species. *Toxicological Sciences*, 62, 166–175.

36. Brunnberg, S., Andersson, P., Lindstam, M., Paulson, I., Poellinger, L., and Hanberg, A. (2006). The constitutively active Ah receptor (CA-Ahr) mouse as a potential model for dioxin exposure: effects in vital organs. *Toxicology*, 224, 191–201.

37. Nohara, K., Pan, X., Tsukumo, S., Hida, A., Ito, T., Nagai, H., Inouye, K., Motohashi, H., Yamamoto, M., Fujii-Kuriyama, Y., and Tohyama, C. (2005). Constitutively active aryl hydrocarbon receptor expressed specifically in T-lineage cells causes thymus involution and suppresses the immunization-induced increase in splenocytes. *Journal of Immunology*, 174, 2770–2777.

38. Kimura, S., Gonzalez, F. J., and Nebert, D. W. (1986). Tissue-specific expression of the mouse dioxin-inducible P(1)450 and P(3)450 genes: differential transcriptional activation and mRNA stability in liver and extrahepatic tissues. *Molecular and Cellular Biology*, 6, 1471–1477.

39. Gehrs, B. C., Riddle, M. M., Williams, W. C., and Smialowicz, R. J. (1997). Alterations in the developing immune system of the F344 rat after perinatal exposure to 2,3,7,8-tetrachlorodibenzo-*p*-dioxin. II. Effects on the pup and the adult. *Toxicology*, 122, 229–240.

40. Andersson, P., Ridderstad, A., McGuire, J., Pettersson, S., Poellinger, L., and Hanberg, A. (2003). A constitutively active aryl hydrocarbon receptor causes loss of peritoneal B1 cells. *Biochemical and Biophysical Research Communications*, 302, 336–341.

41. Nohara, K., Suzuki, T., Ao, K., Murai, H., Miyamoto, Y., Inouye, K., Pan, X., Motohashi, H., Fujii-Kuriyama, Y., Yamamoto, M., and Tohyama, C. (2009). Constitutively active aryl hydrocarbon receptor expressed in T cells increases immunization-induced IFN-gamma production in mice but does not suppress T(h)2-cytokine production or antibody production. *International Immunology*, 21, 769–777.

42. Thurmond, T. S. and Gasiewicz, T. A. (2000). A single dose of 2,3,7,8-tetrachlorodibenzo-*p*-dioxin produces a time- and dose-dependent alteration in the murine bone marrow B-lymphocyte maturation profile. *Toxicological Sciences*, 58, 88–95.

43. Kantor, A. B. and Herzenberg, L. A. (1993). Origin of murine B cell lineages. *Annual Review of Immunology*, 11, 501–538.

44. Herzenberg, L. A., Stall, A. M., Lalor, P. A., Sidman, C., Moore, W. A., and Parks, D. R. (1986). The Ly-1 B cell lineage. *Immunological Reviews*, 93, 81–102.

45. Ashkar, S., Weber, G. F., Panoutsakopoulou, V., Sanchirico, M. E., Jansson, M., Zawaideh, S., Rittling, S. R., Denhardt, D. T., Glimcher, M. J., and Cantor, H. (2000). Eta-1 (osteopontin): an early component of type-1 (cell-mediated) immunity. *Science*, 287, 860–864.

46. Denhardt, D. T., Noda, M., O'Regan, A. W., Pavlin, D., and Berman, J. S. (2001). Osteopontin as a means to cope with environmental insults: regulation of inflammation, tissue remodeling, and cell survival. *Journal of Clinical Investigation*, 107, 1055–1061.

47. Iizuka, J., Katagiri, Y., Tada, N., Murakami, M., Ikeda, T., Sato, M., Hirokawa, K., Okada, S., Hatano, M., Tokuhisa, T., and Uede, T. (1998). Introduction of an osteopontin gene confers the

increase in B1 cell population and the production of anti-DNA autoantibodies. *Laboratory Investigations*, 78, 1523–1533.

48. IARC (1997). IARC Working Group on the Evaluation of Carcinogenic Risks to Humans: polychlorinated dibenzo-*para*-dioxins and polychlorinated dibenzofurans. Lyon, France, 4–11 February 1997. *IARC Monographs Evaluating Carcinogenic Risks to Humans*, 69, 1–631.

49. Moennikes, O., Loeppen, S., Buchmann, A., Andersson, P., Ittrich, C., Poellinger, L., and Schwarz, M. (2004). A constitutively active dioxin/aryl hydrocarbon receptor promotes hepatocarcinogenesis in mice. *Cancer Research*, 64, 4707–4710.

50. Brouwer, A., Morse, D. C., Lans, M. C., Schuur, A. G., Murk, A. J., Klasson-Wehler, E., Bergman, A., and Visser, T. J. (1998). Interactions of persistent environmental organohalogens with the thyroid hormone system: mechanisms and possible consequences for animal and human health. *Toxicology and Industrial Health*, 14, 59–84.

51. Schuur, A. G., Legger, F. F., van Meeteren, M. E., Moonen, M. J., van Leeuwen-Bol, I., Bergman, A., Visser, T. J., and Brouwer, A. (1998). *In vitro* inhibition of thyroid hormone sulfation by hydroxylated metabolites of halogenated aromatic hydrocarbons. *Chemical Research in Toxicology*, 11, 1075–1081.

52. Anderson, L. M., Ward, J. M., Park, S. S., Jones, A. B., Junker, J. L., Gelboin, H. V., and Rice, J. M. (1987). Immunohistochemical determination of inducibility phenotype with a monoclonal antibody to a methylcholanthrene-inducible isozyme of cytochrome P-450. *Cancer Research*, 47, 6079–6085.

53. Dey, A., Jones, J. E., and Nebert, D. W. (1999). Tissue- and cell type-specific expression of cytochrome P450 1A1 and cytochrome P450 1A2 mRNA in the mouse localized *in situ* hybridization. *Biochemical Pharmacology*, 58, 525–537.

54. Forkert, P. G., Vessey, M. L., Park, S. S., Gelboin, H. V., and Cole, S. P. (1989). Cytochromes P-450 in murine lung. An immunohistochemical study with monoclonal antibodies. *Drug Metabolism and Disposition*, 17, 551–555.

55. Song, J., Clagett-Dame, M., Peterson, R. E., Hahn, M. E., Westler, W. M., Sicinski, R. R., and DeLuca, H. F. (2002). A ligand for the aryl hydrocarbon receptor isolated from lung. *Proceedings of the National Academy of Sciences of the United States of America*, 99, 14694–14699.

56. Van Birgelen, A. P., Van der Kolk, J., Fase, K. M., Bol, I., Poiger, H., Brouwer, A., and Van den Berg, M. (1995). Subchronic dose–response study of 2,3,7,8-tetrachlorodibenzo-*p*-dioxin in female Sprague-Dawley rats. *Toxicology and Applied Pharmacology*, 132, 1–13.

57. Moore, J. A., Gupta, B. N., Zinkl, J. G., and Vos, J. G. (1973). Postnatal effects of maternal exposure to 2,3,7,8-tetrachlorodibenzo-*p*-dioxin (TCDD). *Environmental Health Perspectives*, 5, 81–85.

58. Mimura, J., Yamashita, K., Nakamura, K., Morita, M., Takagi, T. N., Nakao, K., Ema, M., Sogawa, K., Yasuda, M., Katsuki, M., and Fujii-Kuriyama, Y. (1997). Loss of teratogenic response to 2,3,7,8-tetrachlorodibenzo-*p*-dioxin (TCDD) in mice lacking the Ah (dioxin) receptor. *Genes to Cells*, 2, 645–654.

59. Lin, T. M., Ko, K., Moore, R. W., Buchanan, D. L., Cooke, P. S., and Peterson, R. E. (2001). Role of the aryl hydrocarbon receptor in the development of control and 2,3,7,8-tetrachlorodibenzo-*p*-dioxin-exposed male mice. *Journal of Toxicology and Environmental Health A*, 64, 327–342.

60. Thackaberry, E. A., Nunez, B. A., Ivnitski-Steele, I. D., Friggins, M., and Walker, M. K. (2005). Effect of 2,3,7,8-tetrachlorodibenzo-*p*-dioxin on murine heart development: alteration in fetal and postnatal cardiac growth, and postnatal cardiac chronotropy. *Toxicological Sciences*, 88, 242–249.

61. Fernandez-Salguero, P. M., Ward, J. M., Sundberg, J. P., and Gonzalez, F. J. (1997). Lesions of aryl-hydrocarbon receptor-deficient mice. *Veterinary Pathology*, 34, 605–614.

62. Lund, A. K., Goens, M. B., Nunez, B. A., and Walker, M. K. (2006). Characterizing the role of endothelin-1 in the progression of cardiac hypertrophy in aryl hydrocarbon receptor (AhR) null mice. *Toxicology and Applied Pharmacology*, 212, 127–135.

63. Lund, A. K., Peterson, S. L., Timmins, G. S., and Walker, M. K. (2005). Endothelin-1-mediated increase in reactive oxygen species and NADPH oxidase activity in hearts of aryl hydrocarbon receptor (AhR) null mice. *Toxicological Sciences*, 88, 265–273.

64. Vezina, C. M., Lin, T. M., and Peterson, R. E. (2009). AHR signaling in prostate growth, morphogenesis, and disease. *Biochemical Pharmacology*, 77, 566–576.

65. Ohsako, S., Miyabara, Y., Sakaue, M., Ishimura, R., Kakeyama, M., Izumi, H., Yonemoto, J., and Tohyama, C. (2002). Developmental stage-specific effects of perinatal 2,3,7,8-tetrachlorodibenzo-*p*-dioxin exposure on reproductive organs of male rat offspring. *Toxicological Sciences*, 66, 283–292.

66. Adamsson, A., Simanainen, U., Viluksela, M., Paranko, J., and Toppari, J. (2009). The effects of 2,3,7,8-tetrachlorodibenzo-*p*-dioxin on foetal male rat steroidogenesis. *International Journal of Andrology*, 32, 575–585.

67. Faqi, A. S., Dalsenter, P. R., Merker, H. J., and Chahoud, I. (1998). Reproductive toxicity and tissue concentrations of low doses of 2,3,7,8-tetrachlorodibenzo-*p*-dioxin in male offspring rats exposed throughout pregnancy and lactation. *Toxicology and Applied Pharmacology*, 150, 383–392.

68. Jahn, A. I. and Gunzel, P. K. (1997). The value of spermatology in male reproductive toxicology: do spermatologic examinations in fertility studies provide new and additional information relevant for safety assessment? *Reproductive Toxicology*, 11, 171–178.

69. Köhle, C., Hassepass, I., Bock-Hennig, B. S., Walter Bock, K., Poellinger, L., and McGuire, J. (2002). Conditional expression of a constitutively active aryl hydrocarbon receptor in MCF-7 human breast cancer cells. *Archives of Biochemistry and Biophysics*, 402, 172–179.

70. Safe, S. and Wormke, M. (2003). Inhibitory aryl hydrocarbon receptor–estrogen receptor alpha cross-talk and mechanisms of action. *Chemical Research in Toxicology*, 16, 807–816.

71. Wyde, M. E., Seely, J., Lucier, G. W., and Walker, N. J. (2000). Toxicity of chronic exposure to 2,3,7,8-tetrachlorodibenzo-*p*-dioxin in diethylnitrosamine-initiated ovariectomized rats implanted with subcutaneous 17 beta-estradiol pellets. *Toxicological Sciences, 54*, 493–499.

72. Ohtake, F., Baba, A., Fujii-Kuriyama, Y., and Kato, S. (2008). Intrinsic AhR function underlies cross-talk of dioxins with sex hormone signalings. *Biochemical and Biophysical Research Communications, 370*, 541–546.

73. Ohtake, F., Takeyama, K., Matsumoto, T., Kitagawa, H., Yamamoto, Y., Nohara, K., Tohyama, C., Krust, A., Mimura, J., Chambon, P., Yanagisawa, J., Fujii-Kuriyama, Y., and Kato, S. (2003). Modulation of oestrogen receptor signalling by association with the activated dioxin receptor. *Nature, 423*, 545–550.

74. Boverhof, D. R., Kwekel, J. C., Humes, D. G., Burgoon, L. D., and Zacharewski, T. R. (2006). Dioxin induces an estrogen-like, estrogen receptor-dependent gene expression response in the murine uterus. *Molecular Pharmacology, 69*, 1599–1606.

75. Watanabe, H., Suzuki, A., Goto, M., Ohsako, S., Tohyama, C., Handa, H., and Iguchi, T. (2004). Comparative uterine gene expression analysis after dioxin and estradiol administration. *Journal of Molecular Endocrinology, 33*, 763–771.

76. Jämsa, T., Viluksela, M., Tuomisto, J. T., Tuomisto, J., and Tuukkanen, J. (2001). Effects of 2,3,7,8-tetrachlorodibenzo-*p*-dioxin on bone in two rat strains with different aryl hydrocarbon receptor structures. *Journal of Bone Mineralization Research, 16*, 1812–1820.

77. Gasser, J. A. (2003). Bone measurements by peripheral quantitative computed tomography in rodents. In *Methods in Molecular Medicine: Bone Research Protocols*, Vol. 80 (Helfrich, M. H. and Ralston, S. H., Eds). Humana Press, Totowa, NJ, pp. 323–341.

78. Lees, R. L. and Heersche, J. N. (1999). Macrophage colony stimulating factor increases bone resorption in dispersed osteoclast cultures by increasing osteoclast size. *Journal of Bone Mineralization Research, 14*, 937–945.

79. Lees, R. L., Sabharwal, V. K., and Heersche, J. N. (2001). Resorptive state and cell size influence intracellular pH regulation in rabbit osteoclasts cultured on collagen-hydroxyapatite films. *Bone, 28*, 187–194.

80. Piper, K., Boyde, A., and Jones, S. J. (1992). The relationship between the number of nuclei of an osteoclast and its resorptive capability *in vitro*. *Anatomy & Embryology, 186*, 291–299.

81. Gierthy, J. F., Silkworth, J. B., Tassinari, M., Stein, G. S., and Lian, J. B. (1994). 2,3,7,8-Tetrachlorodibenzo-*p*-dioxin inhibits differentiation of normal diploid rat osteoblasts *in vitro*. *Journal of Cellular Biochemistry, 54*, 231–238.

82. Korkalainen, M., Kallio, E., Olkku, A., Nelo, K., Ilvesaro, J., Tuukkanen, J., Mahonen, A., and Viluksela, M. (2009). Dioxins interfere with differentiation of osteoblasts and osteoclasts. *Bone, 44*, 1134–1142.

83. Naruse, M., Ishihara, Y., Miyagawa-Tomita, S., Koyama, A., and Hagiwara, H. (2002). 3-Methylcholanthrene, which binds to the arylhydrocarbon receptor, inhibits proliferation and differentiation of osteoblasts *in vitro* and ossification *in vivo*. *Endocrinology, 143*, 3575–3581.

84. Singh, S. U., Casper, R. F., Fritz, P. C., Sukhu, B., Ganss, B., Girard, B., Jr., Savouret, J. F., and Tenenbaum, H. C. (2000). Inhibition of dioxin effects on bone formation *in vitro* by a newly described aryl hydrocarbon receptor antagonist, resveratrol. *Journal of Endocrinology, 167*, 183–195.

85. Guo, L., Zhao, Y. Y., Zhao, Y. Y., Sun, Z. J., Liu, H., and Zhang, S. L. (2007). Toxic effects of TCDD on osteogenesis through altering IGFBP-6 gene expression in osteoblasts. *Biological and Pharmaceutical Bulletin, 30*, 2018–2026.

86. Ryan, E. P., Holz, J. D., Mulcahey, M., Sheu, T. J., Gasiewicz, T. A., and Puzas, J. E. (2007). Environmental toxicants may modulate osteoblast differentiation by a mechanism involving the aryl hydrocarbon receptor. *Journal of Bone Mineralization Research, 22*, 1571–1580.

87. Carpi, D., Korkalainen, M., Airoldi, L., Fanelli, R., Hakansson, H., Muhonen, V., Tuukkanen, J., Viluksela, M., and Pastorelli, R. (2009). Dioxin-sensitive proteins in differentiating osteoblasts: effects on bone formation *in vitro*. *Toxicological Sciences, 108*, 330–343.

88. Alveblom, A. K., Rylander, L., Johnell, O., and Hagmar, L. (2003). Incidence of hospitalized osteoporotic fractures in cohorts with high dietary intake of persistent organochlorine compounds. *International Archives of Occupational and Environmental Health, 76*, 246–248.

89. Wallin, E., Rylander, L., and Hagmar, L. (2004). Exposure to persistent organochlorine compounds through fish consumption and the incidence of osteoporotic fractures. *Scandinavian Journal of Work and Environmental Health, 30*, 30–35.

PART IV

AHR AS A PHYSIOLOGICAL REGULATOR

27

STRUCTURAL AND FUNCTIONAL DIVERSIFICATION OF AHRs DURING METAZOAN EVOLUTION

Mark E. Hahn and Sibel I. Karchner

The aryl hydrocarbon receptor (AHR) was discovered through its dual roles in regulating the inducible expression of xenobiotic metabolizing enzymes such as cytochrome P450 1A (CYP1A) and in mediating the toxicity of 2,3,7,8-tetrachlorodibenzo-*p*-dioxin (TCDD) in mammals [1, 2]. Because of this, for many years the AHR was known only to toxicologists. Early in the history of AHR research, however, it was recognized by toxicologists that study of TCDD and the AHR would likely lead to novel insights about cell biology and physiology, including AHR roles beyond those of relevance in the response to chemicals [3–5]. Those predictions are being realized, aided in part by research on the comparative biology and evolutionary history of the AHR and AHR-related genes in diverse groups of animals.

A variety of species and cell culture model systems have been used to study the AHR. The species can be categorized with respect to the kinds of insight they provide (Table 27.1). *Biomedical model species* typically are established models that are used to obtain fundamental understanding of biological processes involving AHRs and the molecular mechanisms underlying AHR function; studies in these species have special relevance for human health. *Evolutionary models* occupy key phylogenetic positions; studies of AHR in these species help us understand the molecular evolution of AHR genes and proteins, and inform inferences about ancestral and derived functions. *Environmental models* provide information about the properties of AHRs in species likely to be targets of contaminants that act through AHR-dependent mechanisms; studies in these species provide insight into AHR characteristics that may be useful in predicting sensitivity to these compounds, and thus of value in ecological risk assessment. All of these species contribute to a comparative perspective on AHR structure and function.

In this chapter, we summarize the structural and functional diversity and evolutionary history of AHRs, focusing on what is known about AHRs and their functions in chordate animals—especially vertebrates. Other chapters in this book explore in greater depth the role of invertebrate AHR homologs in development (Chapter 28) and the toxicological importance of AHRs in fish and birds (Chapter 21).

27.1 ORIGINS AND EVOLUTION OF THE AHR IN (Eu)METAZOANS

The AHR is an ancient protein. This can be inferred from studies showing that AHR homologs exist in most major groups of animals, including deuterostomes (chordates, hemichordates, echinoderms) and the two major clades of protostome invertebrates (ecdysozoans and lophotrochozoans). Together, protostomes and deuterostomes comprise the clade of bilaterian metazoans, whose most recent common ancestor lived approximately 570 million years ago (MYA) [6, 7]. Moreover, AHR homologs have also been identified in sequenced genomes of cnidarians such as the sea anenome *Nematostella* [8] and in a placozoan, *Trichoplax* [9]. Bilaterians, cnidarians, and placozoans form the clade *Eumetazoa*, whose last common ancestor lived approximately 600 MYA [7]. The presence of AHR homologs in modern representatives of all these groups (Fig. 27.1) provides strong evidence that the original eumetazoan animal

---

The AH Receptor in Biology and Toxicology, First Edition. Edited by Raimo Pohjanvirta.
© 2012 John Wiley & Sons, Inc. Published 2012 by John Wiley & Sons, Inc.

TABLE 27.1 Animal Models for AHR Research

	Biomedical Models	Evolutionary Models	Environmental Targets and Models
Mammals	Human, mouse	Opossum, platypus	Whales, seals, mink, polar bear
Birds	Chicken	Chicken	Terns, albatross, cormorant
Reptiles		*Chrysemys*	*Chrysemys* (turtle)
Amphibians	*Xenopus* spp. (frog)	*Xenopus* spp.	
Bony fishes	Zebrafish, medaka	*Takifugu* (fugu), *Tetraodon*	Killifish (*Fundulus*), rainbow trout, Atlantic salmon
Cartilaginous fishes		Spiny dogfish	Dogfish, skate, sharks
Jawless fishes		Lamprey, hagfish	
Invertebrates (deuterostomes)	Amphioxus, *Strongylocentrotus* (sea urchin)	*Ciona*, amphioxus, *Strongylocentrotus*	
Invertebrates (protostomes)	*Drosophila*, *C. elegans*	*Drosophila*, *C. elegans*	*Mya*, *Mytilus*, *Dreissena*
Invertebrates (basal metazoans)		*Nematostella*, *Trichoplax*	

See the text for explanations.

possessed an AHR homolog. AHR homologs have not yet been found in modern representatives of more ancient animals, such as sponges [10]. It is not clear whether the AHR gene arose after the divergence of the eumetazoan and sponge lineages, or was present in their common ancestor but was subsequently lost in the sponge lineage. Genes encoding bHLH–PAS proteins were likely present in the first metazoans [11].

What physiological functions were served by the ancestral AHR? Although we cannot resurrect the long-extinct original eumetazoan, we can make inferences about its AHR by studying the functions of AHRs found in modern eumetazoan taxa. Most of the invertebrate AHRs have not yet been characterized. However, AHR homologs from the nematode *Caenorhabditis elegans* (AHR-1) and the arthropod *Drosophila melanogaster* (spineless (ss)) have been studied in detail, as described in several reports [5, 12–15] and reviewed elsewhere in this book (Chapter 28). These studies, some involving loss-of-function experiments, have revealed important roles for AHR-1 and ss as developmental regulatory proteins controlling the differentiation of neurons, appendages (legs, antennae), and photoreceptors [12, 16–20]. Similarly, the mammalian AHR also has important roles in development (Chapter 29). The fact that developmental roles are shared by AHRs in distantly related taxa such as nematodes, arthropods, and chordates (see below) supports the idea that the ancestral AHR first evolved as a developmental regulatory protein [21].

The evolutionary emergence of chordates and especially the highly successful vertebrate chordates (jawless fish, cartilaginous fish, bony fish, and tetrapods such as amphibians, reptiles, birds, and mammals) was accompanied by diversification of AHR genes as well as other genes in the bHLH–PAS family. This diversification was enabled by a series of whole-genome duplications (WGDs), two of which occurred early in vertebrate evolution [22–24]. As a result, the bHLH–PAS family in vertebrates is characterized by gene subfamilies containing multiple paralogs.[1] Thus, for each invertebrate bHLH–PAS gene—including AHR—there exist typically two to four orthologous[1] genes in vertebrates [15, 21, 27]. In some taxa, additional gene and genome duplications and gene losses have further shaped AHR diversity.

27.2 AHR DIVERSITY IN VERTEBRATES AND OTHER CHORDATES

27.2.1 Historical Context: Discovery of AHR Multiplicity

The initial cloning and genomic studies of AHR in mammals identified a single gene [28–30]. The first hint that there might be additional AHR diversity (i.e., multiple genes) in some vertebrates came from studies that identified duplicate AHRs (AHR1 and AHR2) in fish, including the teleost *Fundulus heteroclitus* (Atlantic killifish) and a cartilaginous fish, the smooth dogfish *Mustelus canis* [15, 31]. The origin of these duplicate AHR genes was uncertain, but the presence of orthologous duplicates in both bony and cartilaginous fish suggested that they arose early in vertebrate evolution, prior to the divergence of the mammalian and fish lineages [32]. This predicted that the ancestral vertebrate had two (at least) AHRs, and thus that AHR multiplicity might be widespread in modern vertebrates, including possibly in mammals.

The situation became more interesting when an AHR-related gene, designated AHR repressor (AHRR), was discovered in mice [33] and other mammals [34–36].

---

[1] As defined by Fitch [25], paralogs are homologs in one species that are the products of gene duplication. Orthologs are homologs arising from speciation, that is, they exist in different species. Where there has been duplication of an ortholog in one species, the genes are co-orthologs in relation to an ortholog in another species [26]. Orthology and paralogy refer to evolutionary relationships; orthologs need not share the same function.

An ortholog of mouse AHRR was identified in fish and shown to be distinct from AHR2, revealing a vertebrate AHR subfamily with at least three members (AHR1, AHR2, and AHRR) [37].

27.2.2 AHR Homologs in Basal Chordates and Other Nonvertebrate Deuterostome Taxa

AHR homologs are known to occur in a variety of invertebrate deuterostomes (echinoderms and invertebrate chordates) (Fig. 27.1), but the functions of these proteins are not well known. However, because these taxa span the evolutionary transition from invertebrates to vertebrates, studies in these groups may be especially valuable for informing our understanding of AHR evolution. For example, the sea urchin (*Strongylocentrotus purpuratus*), an echinoderm that is widely used as a model in developmental biology, has two *AHR* genes, one of which is more closely related to AHR homologs in *D. melanogaster* and *C. elegans*, and another that is more closely related to mammalian AHRs [38, 39] (Karchner and Hahn, unpublished results). AHR homologs have also been identified in a hemichordate, the acorn worm *Saccoglossus kowalevskii* [40, 41], the urochordate *Ciona intestinalis* [42], and the cephalochordate amphioxus (*Branchiostoma floridae*) [43].

27.2.3 AHR Diversity: Orthologs and Paralogs in Vertebrate Groups

Over the past decade, information about AHR diversity in vertebrates has expanded rapidly as cloning methods have become more accessible and sequenced genomes have provided new insight [27]. Inferring the homologous relationships among these AHRs (e.g., assigning orthology) and the timing of the duplication events that led to this diversity is challenging [44]. Evidence comes primarily from phylogenetic analyses coupled with comparative genomic studies that identify shared syntenic relationships (arrangement of AHR genes on chromosomes and occurrence of AHR genes in association with other genes on the same chromosomes in different species). Here we summarize briefly our current understanding of AHR multiplicity in the major vertebrate taxa.

27.2.3.1 Jawless Fishes
The jawless fishes, the earliest diverging living vertebrates, are of great interest because of the potential insight they may provide concerning the functional properties of the ancestral vertebrate AHR. A single AHR is known from a lamprey (*Petromyzon marinus*), but attempts to identify an AHR or evidence for AHR-dependent signaling in the other major lineage of jawless fishes, hagfish, have not yet succeeded [15, 45–47].

27.2.3.2 Cartilaginous Fishes
The cartilaginous fishes (sharks, skates, and rays) and bony fishes diverged early in vertebrate evolution (~450 MYA) [48]. Although early cloning studies had identified AHR1 and AHR2 genes in some cartilaginous fish [15, 49], additional AHR multiplicity has been discovered recently. Studies in the spiny dogfish shark (*Squalus acanthias*) and other sharks and a skate have revealed a novel AHR form (AHR3) that has not been found in any other group of animals [27, 50, 51]. The three shark AHRs are distinct in their molecular functional properties [50].

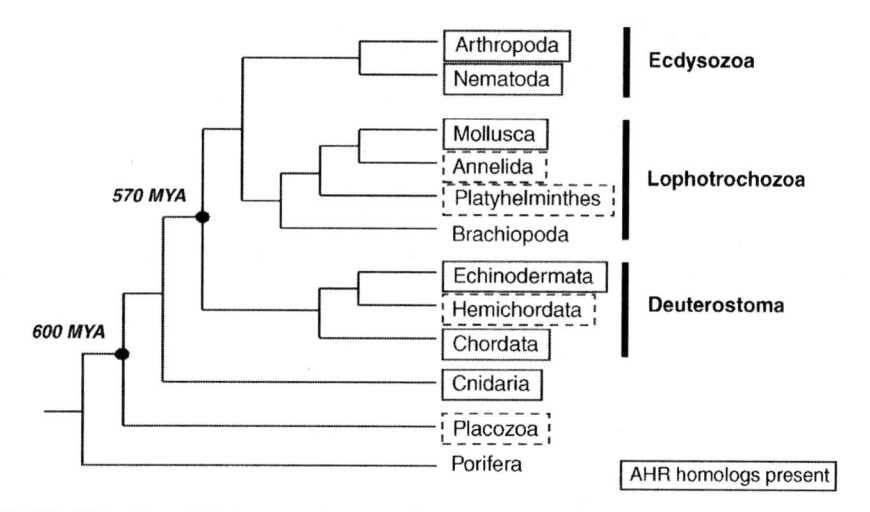

FIGURE 27.1 The AHR is an ancient eumetazoan protein. The tree shows the relationships of selected metazoan (animal) taxa. Ecdysozoa and Lophotrochozoa together comprise the protostomes. Protostomes and deuterostomes are bilaterian animals. Solid boxes represent groups containing species from which AHR homologs have been verified by cloning. Dashed boxes occur around groups with evidence of AHR homologs from sequenced genomes. See the text for additional information.

27.2.3.3 Bony Fish Comprising approximately 50% of all vertebrate species, this large group is a rich source of interesting information concerning AHR diversity and provides a variety of research models that are proving useful in understanding AHR evolution and function. Nearly all species of bony fish investigated so far possess multiple AHR genes—as many as six in the case of some salmonids [52, 53]. The number of AHR genes varies by species, but all of the genes fall into the AHR1 clade or AHR2 clade. In species such as the Atlantic tomcod [54] and Atlantic killifish [32], where only one or two AHR genes have been described, sequencing of their genome or transcriptome may reveal additional AHR diversity. Among species with sequenced genomes, there are three AHRs in zebrafish, four in medaka (*Oryzias latipes*) and *Tetraodon nigroviridis*, and five in fugu (*Takifugu rubripes*) [21, 27, 55–60].

27.2.3.4 Amphibians AHR diversity has not been studied extensively in amphibians. However, insensitivity of some amphibians to TCDD [61–64] makes this an interesting group in which to investigate AHR function. Two AHR1 paralogs [65, 66] and an AHRR [67] have been identified in *Xenopus laevis*, which has a well-known tetraploid ancestry. There are not yet any reports of an amphibian AHR2, but given the lack of attention paid to amphibian AHRs, it is premature to draw conclusions about AHR diversity in this group.

27.2.3.5 Reptiles and Birds Reptiles and birds form a monophyletic group of vertebrates, and so might be expected to share similar features with regard to AHR diversity and signaling. Only a few reptilian species have been examined for AHR diversity, and there is no functional information, but both AHR1 and AHR2 forms have been identified in turtles [45, 68, 69]. AHRs have been more intensively studied in birds, in part because the chick embryo was instrumental early on as a model system for elucidating AHR-dependent changes in gene expression [70–72] and because birds—especially fish-eating birds—have been among the wildlife species most impacted by dioxin-like compounds that act through the AHR [73]. In addition to possessing AHR1 and AHR2 forms [74–79], some birds possess a second AHR1-like gene [21, 27, 77].

27.2.3.6 Mammals The first AHR was identified in mice [28, 29] and mammalian AHR forms have been the subject of intensive research, as detailed throughout this book. For the purposes of this chapter, the key features of mammalian AHR are that (1) it occurs as a single gene in most eutherian mammals that have been examined, and (2) it not only is required for TCDD toxicity [80, 81] but also has important roles in normal development [82–85]. Although most commonly studied mammals, including humans and laboratory rodents, have a single AHR (AHR1), we have found AHR2 genes (unpublished results) in prototherian mammals (e.g., platypus [86]), metatherian mammals (e.g., opossum [87]), and some eutherians (e.g., cow), consistent with our earlier suggestion [32] that AHR2 was present in an ancestral mammal and has been lost in some mammalian lineages.

27.2.4 Sources of AHR Diversity

From an evolutionary perspective, there is interest in understanding how the current patterns of AHR diversity came about—what processes and pressures have shaped AHR evolution in different taxa? One influential process is genome duplication. Evidence accumulating over the past 15 years now strongly supports the idea that two whole-genome duplications (designated 1R and 2R) occurred early in the evolution of the vertebrates [22, 23, 88]. As illustrated by *Hox* gene clusters [22, 89], many genes that exist singly in genomes of invertebrates are found as multiples (often four) in vertebrates. We have proposed that AHRR originated as a duplicate of AHR in one of these WGDs [37], and that in general the diversity of bHLH–PAS proteins in vertebrate animals as compared to invertebrates can be traced back to the 1R and 2R WGDs [15, 21, 27, 90].

An additional WGD that helped shape AHR diversity occurred at the base of the ray-finned fish lineage, approximately 350 MYA (3R [91–95]). This fish-specific WGD is thought to be the origin of many of the fish AHR paralogs, including those found in zebrafish (Ahr1a and Ahr1b), fugu (AHR1a and AHR1b; AHR2a and AHR2b), medaka, and other species [21, 27, 57].

Despite the importance of WGDs in establishing AHR diversity, individual gene duplication and loss events have also contributed to existing patterns. Most notably, AHR1 and AHR2—originally proposed to be the result of a WGD [15, 32]—are now thought to have originated as an isolated, tandem gene duplication [27, 57]. Lineage-specific AHR gene losses have also helped shape AHR diversity [27].

27.2.5 Genomic Organization of AHR Loci

Inferring the relationships and evolutionary history of AHR homologs found in different species can be challenging. Phylogenetic analysis is a valuable tool in such assessments [15], especially when informed by the latest understanding of the evolutionary relationships among the species from which the AHR sequences are obtained. Phylogenetic analysis has limitations, however, especially when the entire complement of AHR genes in each species is not known. Comparative genomics—mapping the genomic locations and identifying shared synteny of AHR and other genes in different genomes—is a complementary approach that can contribute to establishing orthologous and paralogous relationships [89, 96, 97]. The identification of AHR1 and AHR2 in tandem in several genomes from fish, birds, and mammals [15, 27, 57, 60, 77] provided evidence that these two AHR forms did not arise from a WGD, but rather from an isolated, tandem duplication. Phylogenetic analyses demonstrated that this duplication nevertheless occurred early in vertebrate evolution, and that AHR2 forms found in cartilaginous and bony fish, in birds and reptiles, and in some mammals are orthologous [15, 27].

27.2.6 Polymorphic Variants of AHR

Another source of AHR diversity is mutation leading to intraspecific diversity. Numerous polymorphic variants occur in mice [98, 99], rats [100], and humans [101, 102]. AHR variants have also been described in two species of fish, Atlantic tomcod (*Microgadus tomcod*) and Atlantic killifish (*F. heteroclitus*) [54, 103], in which selection for low-functioning variants may be responsible for adaptive evolution in response to pollution [104, 105].

27.3 EVOLUTION OF AHR LIGAND BINDING AND OTHER FUNCTIONAL PROPERTIES

Although AHR homologs exist in a wide variety of taxa, there are important differences in the functional characteristics of these proteins. Among the most striking are differences in the ability to engage in high-affinity interactions with ligands such as TCDD. Variations in ligand binding properties occur across broad taxonomic groups (phylum-level differences) as well as among and within species and among paralogs. In addition, even for those AHRs that are able to bind planar ligands, there are species-specific differences in structure–binding relationships—the relative ability of different ligands to interact with each AHR.

27.3.1 Binding of TCDD, the Prototypical AHR Ligand

Some of the earliest studies of AHRs in rodents demonstrated a strong correlation between the binding affinity for TCDD and *in vivo* sensitivity to TCDD effects (gene induction and toxicity) [1, 106, 107] as well as a strong correlation between the relative binding affinity of various halogenated aromatic hydrocarbons and their biological potency [1, 72, 108]. Because ligand binding affinity plays such an important role in determining the effects of AHR ligands, there is great interest in understanding how AHR orthologs and paralogs in diverse species vary in this property.

One of the most dramatic differences in ligand binding properties is seen when comparing AHR homologs in vertebrate and invertebrate species. In contrast to vertebrate AHRs—most of which bind TCDD and other compounds with high affinity—invertebrate AHRs do not display high-affinity binding of TCDD, TCDD analogs, or PAH-like compounds such as β-naphthoflavone [14, 109]. Because AHR-1, ss, and other invertebrate AHRs lack the ability to bind halogenated dioxins and other typical ligands of vertebrate AHRs [14, 21, 109] and appear to be constitutively active [13, 110], it has been suggested that they act in a ligand-independent manner. Although a role for endogenous ligands cannot be ruled out, it is clear that the ligand binding properties of vertebrate and invertebrate AHRs are quite distinct. Based on this, we suggested several years ago that the ability of the AHR to engage in high-affinity interactions with planar aromatic ligands was a vertebrate innovation [21]. Further discussion of the differences between invertebrate and vertebrate AHRs can be found in earlier reviews [5, 21, 111] and elsewhere in this book (Chapter 28).

Although the distinction between invertebrate and most vertebrate AHRs is striking, one can find dramatic differences in ligand binding properties also among vertebrate AHRs. Some AHRs appear to be unable to bind TCDD and related ligands; examples include zebrafish Ahr1a (one of three zebrafish AHRs) [56, 57] and fugu AHR2C (one of five fugu AHRs) [60]. More commonly, one sees quantitative differences in the binding affinity of AHRs for TCDD, as measured by equilibrium dissociation constants (K_d) derived from saturation binding experiments. The most well-known example is that of the so-called "responsive" and "nonresponsive" strains of mouse, which possess AHRs that differ by ~10-fold in binding affinity for TCDD [1, 107], a property linked to a single amino acid difference [98]. A similar variation in affinity for TCDD, linked to the same amino acid difference, is seen in comparisons of "responsive" mice and humans [112, 113]. These studies highlight the importance of certain amino acid residues identified by comparisons among species and strains, but other amino acids in the ligand binding domain and elsewhere in the AHR also influence ligand binding and other functional properties. Some of the important residues are being revealed by structural modeling combined with site-directed mutagenesis [114, 115].

Studies in wildlife are also revealing variation in AHR ligand binding affinity that appears to explain the differential sensitivity of some species or populations. Some amphibians, including frogs, are known for their insensitivity to TCDD [61–64]. Consistent with this, the two AHRs identified in *X. laevis* exhibit very low binding affinity for TCDD [66]. Similarly, the insensitivity of the common tern (*Sterna hirundo*) to effects of halogenated aromatic hydrocarbons including TCDD appears to be explained by the structure of its AHR ligand binding domain, and in particular by two amino acid differences as compared to the dioxin-sensitive chicken (*Gallus gallus*) [75]. Additional studies involving dozens of other bird species show that the identity of amino acids present at these same two positions can predict the species' sensitivity to dioxin-like compounds [79, 116].

The AHR can also serve as a target for selection in the presence of pollutants, and in at least one case this has resulted in changes in the ligand binding characteristics of AHRs in the impacted population. The population of Atlantic tomcod (*Microgadus tomcod*) inhabiting the PCB-contaminated Hudson River (New York) has evolved resistance to the effects of these compounds [117]. A recent study demonstrated that fish at this site possess an AHR variant with reduced binding affinity for TCDD and impaired ability to activate transcription [104]. PCB-resistant populations of the killifish *F. heteroclitus* in several locations have also undergone selection for AHR variants, but in those cases the functional changes appear to be independent of ligand binding [103, 105, 118].

The selection for AHR variants with altered functional characteristics at polluted sites demonstrates that rapid adaptive evolution of AHR can occur in response to changing environmental conditions. The absence of ligand binding capabilities of some AHR paralogs in zebrafish and fugu (see above) suggests that the presence of AHR duplicates may relax evolutionary constraints on AHR function. Such observations might provide insight into the origin of AHRR, an AHR homolog that acts in a negative feedback loop to repress AHR-dependent signaling [33, 119, 120]. AHRRs have been identified in mammals, birds, amphibians, and bony fish [33, 36, 37, 67, 121, 122], showing that the origin of this gene predates the divergence of tetrapod and bony fish lineages. Unlike most AHRs, AHRRs do not bind TCDD [37] and they appear to act exclusively as transcriptional repressors. We have suggested that the AHRR evolved from an AHR duplicate that appeared as a result of one of the two whole-genome duplications occurring early in vertebrate evolution [27]. Released from constraints imposed by the need to retain AHR functions, it would have been free to accumulate changes that eliminated its ability to bind ligands and activate transcription.

27.3.2 Ligand Structure–Binding Relationships

The variations in ability of AHRs from different species and populations to bind the prototypical AHR ligand TCDD are well known, at least at a phenomenological level. Species differences in structure–binding relationships are less well understood, but recent findings suggest that such differences may be widespread and important. Research in several laboratories has established that species differences in binding affinity for TCDD do not necessarily predict relative affinities for other ligands. For example, although the human AHR has lower affinity for TCDD as compared to AHRs from "responsive strains" of mice, it has a higher relative affinity for some other ligands, such as flavonoids [123] or indole derivatives [124]. The ability of a compound to act as an AHR antagonist also varies by species [125–127] and some antagonists are ligand selective [128, 129]. Additional complexity is provided by the identification of "selective AHR modulators" (SAHRMs) that can activate a subset of AHR-regulated genes [130–132].

Species-specific structure–binding relationships also occur in nonmammalian species. Examples include the relatively low potency of mono-*ortho* PCBs in fish as compared to mammals [133–136] and the dramatic differences in the relative potencies of PCB and PCDF congeners among bird species [137–139]. Together, these findings illustrate the heterogeneity in functional properties of AHRs and AHR–ligand interactions among species and underscore the complexity that impedes a comprehensive understanding of AHR function.

27.4 EVOLUTION OF AHR ROLES IN DEVELOPMENT AND PHYSIOLOGY

27.4.1 Mammalian AHR Has Multiple Functions

Although originally identified and studied because of its importance in toxicology, the mammalian AHR is now known to have multiple functions that appear distinct from its involvement in regulating the expression of xenobiotic metabolizing enzymes and in mediating the toxicity of dioxin-like compounds. The nontoxicological roles of AHR, including those involving the cardiovascular, reproductive, and immune systems, are described elsewhere in this book and in several recent reviews [5, 111, 140–145]. An underlying theme in these endogenous or physiological functions is one of AHR playing an important role in development and differentiation (Table 27.2).

The emerging roles of AHR in developmental processes in mammals may provide insight into the original function of this pleiotropic transcription factor. The developmental roles of mammalian AHRs appear to parallel the established functions of invertebrate AHR homologs in controlling the development and differentiation of neurons, appendages (legs, antennae), and photoreceptors [12, 16–20]. Although the specific processes in which AHR is involved are not identical in flies, worms, and mice, it seems highly likely that there could be evolutionarily conserved molecular interactions and genes that underlie some of these processes and are shared targets of vertebrate and invertebrate AHRs. Alternatively, the AHR might have been recruited to newly emerging developmental signaling pathways in early vertebrates.

27.4.2 Evidence for Subfunction Partitioning Among AHR Paralogs: The Value of Nonmammalian Models

Identifying all functions of a pleiotropic protein like AHR is challenging. Studies in mammals have exploited species and strain differences in AHR properties and the technical

TABLE 27.2 Phylogenetic Perspective on AHR Functions

		AHR Roles			
Taxon	Number of AHR Genes[a]	Physiological/ Developmental	Adaptive: Ligand Binding	Adaptive: Regulation of XMEs[b]	Toxic
Placozoans	1	?	?	?	?
Cnidarians	1	?	No[c]	?	?
Nematodes	1	Yes	No	No	No
Arthropods	1	Yes	No	No[d]	No
Urochordates	1	?	No	?	?
Jawless fish	1	?	Low affinity	No	?
Cartilaginous fish	4	?	Yes	Yes	?
Bony fish	4	Yes	Yes	Yes	Yes
Birds	3	?	Yes	Yes	Yes
Prototherian/metatherian mammals	2	?	Yes	?	?
Eutherian mammals	1	Yes	Yes	Yes	Yes

Because developmental roles are widely shared by AHRs (including all taxa in which they have been studied), the ancestral role of AHR most likely was as a developmental regulator. Roles of AHRs in adaptation and toxicity, present only in vertebrates, are viewed as derived.
[a] Typical number; some variability occurs among species.
[b] XME = xenobiotic metabolizing enzymes (e.g., CYP1A1).
[c] A. M. Reitzel et al., manuscript in preparation.
[d] But see Ref. 146.

advantages of various mammalian model systems to gain insight into its physiological and toxicological roles. In the same way, nonmammalian models have unique features that might be exploited to enhance our understanding of AHR's many facets. A key advantage of some nonmammalian models is the presence of multiple AHR paralogs. Theoretical and empirical studies show how duplicated genes often diverge in function, sometimes partitioning the multiple functions of their progenitor gene (subfunctionalization or subfunction partitioning) [93, 147, 148]. Subfunctionalization can occur through partitioning of expression domains or through complementary changes in protein function.

Are the multiple roles of the vertebrate AHR partitioned among the multiple fish AHRs? Studies in zebrafish and killifish have identified AHR2 as the predominant AHR form regulating the response to TCDD and non-*ortho*-PCBs in embryos [149–155]. In contrast, the functions of fish AHR1 forms are not well understood [156]. We have suggested that AHR1 may be involved in some of the physiological AHR functions in fish [57], but a definitive assessment of this question will require more complete loss-of-function experiments.

In birds, the relative roles of AHR1 and AHR2 may be reversed. Although only a few studies have compared properties of avian AHR1 and AHR2, those studies suggest that AHR2 is less active than AHR1 in the presence of TCDD [77, 157]. These results are consistent with other data showing a strong relationship between the ligand binding affinity of AHR1 and sensitivity of the species to TCDD [75, 79, 116].

The AHR-related gene AHRR serves as a good example of the potential for novel insights from studying duplicated genes. Two AHRR genes (*ahrra* and *ahrrb*) have been identified in zebrafish, and evidence suggests that they are co-orthologs of the single mammalian AHRR [121]. Studies using paralog-specific knockdown of AHRRa or AHRRb expression by injection of morpholino-modified antisense oligonucleotides in embryos showed that their functions were distinct. Knockdown of AHRRb resulted in enhanced induction of CYP1A and other AHR-regulated CYP1 forms by TCDD, consistent with a role for this protein in dampening the response to TCDD through negative feedback [158]. In contrast, knockdown of AHRRa led to embryonic phenotypes that resembled those seen in TCDD-treated, noninjected embryos, even when the AHRRa morphants were not exposed to TCDD [158]. Those results were interpreted as supporting a role for this AHRR in controlling the constitutive activity of one of the zebrafish AHRs during development. Thus, the results of this experiment point to two possible functions of mammalian AHRR—one involving regulation of AHR activated by exogenous signals and the other involving repression of the constitutive activity of AHR during a critical phase of development.

27.5 CONCLUSIONS

Despite the recent progress made in elucidating molecular mechanisms of AHR action, novel physiological functions involving AHR-dependent signaling, and the phylogenetic distribution and comparative properties of AHRs in a variety of species, our understanding of AHR biology remains rudimentary. AHR diversity—especially that found in non-mammalian species, in which multiple AHR forms often occur—provides a rich source of potential insight regarding the many functions of AHR. The ongoing explosion of genomic information and emerging techniques for genetic manipulation of nontraditional models [159–161] suggest that the next decade will see major advances in our understanding of the physiological and toxicological roles of this fascinating but enigmatic protein.

ACKNOWLEDGMENTS

We gratefully acknowledge the U.S. National Institutes of Health for support in preparation of this chapter and long-term funding that has allowed us to conduct research on the comparative biology of AHRs: grants R01ES006272 and P42ES007381 (Superfund Basic Research Program at Boston University). We also acknowledge valuable support from the WHOI Sea Grant program with funding from the National Oceanic and Atmospheric Administration. We also thank all of our colleagues at WHOI and elsewhere who have contributed to our research on the comparative biology of the AHR over many years.

REFERENCES

1. Poland, A., Glover, E., and Kende, A. S. (1976). Stereospecific, high-affinity binding of 2,3,7,8-tetrachlorodibenzo-*p*-dioxin by hepatic cytosol. *Journal of Biological Chemistry*, *251*, 4936–4946.
2. Okey, A. B. (2007). An aryl hydrocarbon receptor odyssey to the shores of toxicology: the Deichmann Lecture, International Congress of Toxicology – XI. *Toxicological Sciences*, *98*, 5–38.
3. Poland, A. and Kende, A. (1975). 2,3,7,8-Tetrachlorodibenzo-*p*-dioxin: environmental contaminant and molecular probe. *Federation Proceedings*, *35*, 2404–2411.
4. Nebert, D. W. and Karp, C. L. (2008). Endogenous functions of the aryl hydrocarbon receptor (AHR): intersection of cytochrome P450 1 (CYP1)-metabolized eicosanoids and AHR biology. *Journal of Biological Chemistry*, *283*, 36061–36065.
5. McMillan, B. J. and Bradfield, C. A. (2007). The aryl hydrocarbon receptor sans xenobiotics: endogenous function in genetic model systems. *Molecular Pharmacology*, *72*, 487–498.
6. Peterson, K. J., Lyons, J. B., Nowak, K. S., Takacs, C. M., Wargo, M. J., and McPeek, M. A. (2004). Estimating metazoan divergence times with a molecular clock. *Proceedings of the National Academy of Sciences of the United States of America*, *101*, 6536–6541.
7. Peterson, K. J. and Butterfield, N. J. (2005). Origin of the Eumetazoa: testing ecological predictions of molecular clocks against the Proterozoic fossil record. *Proceedings of the National Academy of Sciences of the United States of America*, *102*, 9547–9552.
8. Putnam, N. H., Srivastava, M., Hellsten, U., Dirks, B., Chapman, J., Salamov, A., Terry, A., Shapiro, H., Lindquist, E., Kapitonov, V. V., Jurka, J., Genikhovich, G., Grigoriev, I. V., Lucas, S. M., Steele, R. E., Finnerty, J. R., Technau, U., Martindale, M. Q., and Rokhsar, D. S. (2007). Sea anemone genome reveals ancestral eumetazoan gene repertoire and genomic organization. *Science*, *317*, 86–94.
9. Srivastava, M., Begovic, E., Chapman, J., Putnam, N. H., Hellsten, U., Kawashima, T., Kuo, A., Mitros, T., Salamov, A., Carpenter, M. L., Signorovitch, A. Y., Moreno, M. A., Kamm, K., Grimwood, J., Schmutz, J., Shapiro, H., Grigoriev, I. V., Buss, L. W., Schierwater, B., Dellaporta, S. L., and Rokhsar, D. S. (2008). The *Trichoplax* genome and the nature of placozoans. *Nature*, *454*, 955–960.
10. Simionato, E., Ledent, V., Richards, G., Thomas-Chollier, M., Kerner, P., Coornaert, D., Degnan, B. M., and Vervoort, M. (2007). Origin and diversification of the basic helix–loop–helix gene family in metazoans: insights from comparative genomics. *BMC Evolutionary Biology*, *7*, 33.
11. Sebe-Pedros, A., de Mendoza, A., Lang, B. F., Degnan, B. M., and Ruiz-Trillo, I. (2011). Unexpected repertoire of metazoan transcription factors in the unicellular holozoan *Capsaspora owczarzaki*. *Molecular Biology and Evolution*, *28*, 1241–1254.
12. Duncan, D. M., Burgess, E. A., and Duncan, I. (1998). Control of distal antennal identity and tarsal development in *Drosophila* by spineless-aristapedia, a homolog of the mammalian dioxin receptor. *Genes & Development*, *12*, 1290–1303.
13. Emmons, R. B., Duncan, D., Estes, P. A., Kiefel, P., Mosher, J. T., Sonnenfeld, M., Ward, M. P., Duncan, I., and Crews, S. T. (1999). The spineless-aristapedia and tango bHLH–PAS proteins interact to control antennal and tarsal development in *Drosophila*. *Development*, *126*, 3937–3945.
14. Powell-Coffman, J. A., Bradfield, C. A., and Wood, W. B. (1998). *Caenorhabditis elegans* orthologs of the aryl hydrocarbon receptor and its heterodimerization partner the aryl hydrocarbon receptor nuclear translocator. *Proceedings of the National Academy of Sciences of the United States of America*, *95*, 2844–2849.
15. Hahn, M. E., Karchner, S. I., Shapiro, M. A., and Perera, S. A. (1997). Molecular evolution of two vertebrate aryl hydrocarbon (dioxin) receptors (AHR1 and AHR2) and the PAS family. *Proceedings of the National Academy of Sciences of the United States of America*, *94*, 13743–13748.
16. Wernet, M. F., Mazzoni, E. O., Celik, A., Duncan, D. M., Duncan, I., and Desplan, C. (2006). Stochastic spineless expression creates the retinal mosaic for colour vision. *Nature*, *440*, 174–180.
17. Kim, M. D., Jan, L. Y., and Jan, Y. N. (2006). The bHLH–PAS protein Spineless is necessary for the diversification of dendrite morphology of *Drosophila* dendritic arborization neurons. *Genes & Development*, *20*, 2806–2819.
18. Qin, H. and Powell-Coffman, J. A. (2004). The *Caenorhabditis elegans* aryl hydrocarbon receptor, AHR-1, regulates neuronal development. *Developmental Biology*, *270*, 64–75.

19. Huang, X., Powell-Coffman, J. A., and Jin, Y. (2004). The AHR-1 aryl hydrocarbon receptor and its co-factor the AHA-1 aryl hydrocarbon receptor nuclear translocator specify GABAergic neuron cell fate in C. elegans. *Development, 131*, 819–828.

20. Qin, H., Zhai, Z., and Powell-Coffman, J. A. (2006). The *Caenorhabditis elegans* AHR-1 transcription complex controls expression of soluble guanylate cyclase genes in the URX neurons and regulates aggregation behavior. *Developmental Biology, 298*, 606–615.

21. Hahn, M. E. (2002). Aryl hydrocarbon receptors: diversity and evolution. *Chemico-Biological Interactions, 141*, 131–160.

22. Hoegg, S. and Meyer, A. (2005). Hox clusters as models for vertebrate genome evolution. *Trends in Genetics, 21*, 421–424.

23. Blomme, T., Vandepoele, K., De Bodt, S., Simillion, C., Maere, S., and Van de Peer, Y. (2006). The gain and loss of genes during 600 million years of vertebrate evolution. *Genome Biology, 7*, R43.

24. Dehal, P. and Boore, J. L. (2005). Two rounds of whole genome duplication in the ancestral vertebrate. *PLoS Biology, 3*, e314.

25. Fitch, W. M. (1970). Distinguishing homologous from analogous proteins. *Systematic Zoology, 19*, 99–113.

26. Meyer, A. and Mindell, D. P. (2001). Homology evolving. *Trends in Ecology and Evolution, 16*, 434–440.

27. Hahn, M. E., Karchner, S. I., Evans, B. R., Franks, D. G., Merson, R. R., and Lapseritis, J. M. (2006). Unexpected diversity of aryl hydrocarbon receptors in non-mammalian vertebrates: insights from comparative genomics. *Journal of Experimental Zoology, 305A*, 693–706.

28. Burbach, K. M., Poland, A., and Bradfield, C. A. (1992). Cloning of the Ah receptor cDNA reveals a distinctive ligand-activated transcription factor. *Proceedings of the National Academy of Sciences of the United States of America, 89*, 8185–8189.

29. Ema, M., Sogawa, K., Watanabe, N., Chujoh, Y., Matsushita, N., Gotoh, O., Funae, Y., and Fujii-Kuriyama, Y. (1992). cDNA cloning and structure of mouse putative Ah receptor. *Biochemical and Biophysical Research Communications, 184*, 246–253.

30. Schmidt, J. V., Carver, L. A., and Bradfield, C. A. (1993). Molecular characterization of the murine *Ahr* gene. Organization, promoter analysis, and chromosomal assignment. *Journal of Biological Chemistry, 268*, 22203–22209.

31. Hahn, M. E. and Karchner, S. I. (1995). Evolutionary conservation of the vertebrate Ah (dioxin) receptor: amplification and sequencing of the PAS domain of a teleost Ah receptor cDNA. *Biochemical Journal, 310*, 383–387.

32. Karchner, S. I., Powell, W. H., and Hahn, M. E. (1999). Identification and functional characterization of two highly divergent aryl hydrocarbon receptors (AHR1 and AHR2) in the teleost *Fundulus heteroclitus*. Evidence for a novel subfamily of ligand-binding basic helix–loop–helix–Per–ARNT–Sim (bHLH–PAS) factors. *Journal of Biological Chemistry, 274*, 33814–33824.

33. Mimura, J., Ema, M., Sogawa, K., and Fujii-Kuriyama, Y. (1999). Identification of a novel mechanism of regulation of Ah (dioxin) receptor function. *Genes & Development, 13*, 20–25.

34. Nagase, T., Ishikawa, K., Kikuno, R., Hirosawa, M., Nomura, N., and Ohara, O. (1999). Prediction of the coding sequences of unidentified human genes. XV. The complete sequences of 100 new cDNA clones from brain which code for large proteins *in vitro*. *DNA Research, 6*, 337–345.

35. Baba, T., Mimura, J., Gradin, K., Kuroiwa, A., Watanabe, T., Matsuda, Y., Inazawa, J., Sogawa, K., and Fujii-Kuriyama, Y. (2001). Structure and expression of the Ah receptor repressor gene. *Journal of Biological Chemistry, 276*, 33101–33110.

36. Korkalainen, M., Tuomisto, J., and Pohjanvirta, R. (2004). Primary structure and inducibility by 2,3,7,8-tetrachlorodibenzo-*p*-dioxin (TCDD) of aryl hydrocarbon receptor repressor in a TCDD-sensitive and a TCDD-resistant rat strain. *Biochemical and Biophysical Research Communications, 315*, 123–131.

37. Karchner, S. I., Franks, D. G., Powell, W. H., and Hahn, M. E. (2002). Regulatory interactions among three members of the vertebrate aryl hydrocarbon receptor family: AHR repressor, AHR1, and AHR2. *Journal of Biological Chemistry, 277*, 6949–6959.

38. Sodergren, E., Weinstock, G. M., Davidson, E. H., Cameron, R. A., Gibbs, R. A., Angerer, R. C., Angerer, L. M., Arnone, M. I., Burgess, D. R., Burke, R. D., Coffman, J. A., Dean, M., Elphick, M. R., Ettensohn, C. A., Foltz, K. R., Hamdoun, A., Hynes, R. O., Klein, W. H., Marzluff, W., McClay, D. R., Morris, R. L., Mushegian, A., Rast, J. P., Smith, L. C., Thorndyke, M. C., Vacquier, V. D., Wessel, G. M., Wray, G., Zhang, L., Elsik, C. G., Ermolaeva, O., Hlavina, W., Hofmann, G., Kitts, P., Landrum, M. J., Mackey, A. J., Maglott, D., Panopoulou, G., Poustka, A. J., Pruitt, K., Sapojnikov, V., Song, X., Souvorov, A., Solovyev, V., Wei, Z., Whittaker, C. A., Worley, K., Durbin, K. J., Shen, Y., Fedrigo, O., Garfield, D., Haygood, R., Primus, A., Satija, R., Severson, T., Gonzalez-Garay, M. L., Jackson, A. R., Milosavljevic, A., Tong, M., Killian, C. E., Livingston, B. T., Wilt, F. H., Adams, N., Belle, R., Carbonneau, S., Cheung, R., Cormier, P., Cosson, B., Croce, J., Fernandez-Guerra, A., Geneviere, A. M., Goel, M., Kelkar, H., Morales, J., Mulner-Lorillon, O., Robertson, A. J., Goldstone, J. V., Cole, B., Epel, D., Gold, B., Hahn, M. E., Howard-Ashby, M., Scally, M., Stegeman, J. J., Allgood, E. L., Cool, J., Judkins, K. M., McCafferty, S. S., Musante, A. M., Obar, R. A., Rawson, A. P., Rossetti, B. J., Gibbons, I. R., Hoffman, M. P., Leone, A., Istrail, S., Materna, S. C., Samanta, M. P., Stolc, V., Tongprasit, W., Tu, Q., Bergeron, K. F., Brandhorst, B. P., Whittle, J., Berney, K., Bottjer, D. J., Calestani, C., Peterson, K., Chow, E., Yuan, Q. A., Elhaik, E., Graur, D., Reese, J. T., Bosdet, I., Heesun, S., Marra, M. A., Schein, J., Anderson, M. K., Brockton, V., Buckley, K. M., Cohen, A. H., Fugmann, S. D., Hibino, T., Loza-Coll, M., Majeske, A. J., Messier, C., Nair, S. V., Pancer, Z., Terwilliger, D. P., Agca, C., Arboleda, E., Chen, N., Churcher, A. M., Hallbook, F., Humphrey, G. W., Idris, M. M., Kiyama, T., Liang, S., Mellott, D., Mu, X., Murray, G.,

Olinski, R. P., Raible, F., Rowe, M., Taylor, J. S., Tessmar-Raible, K., Wang, D., Wilson, K. H., Yaguchi, S., Gaasterland, T., Galindo, B. E., Gunaratne, H. J., Juliano, C., Kinukawa, M., Moy, G. W., Neill, A. T., Nomura, M., Raisch, M., Reade, A., Roux, M. M., Song, J. L., Su, Y. H., Townley, I. K., Voronina, E., Wong, J. L., Amore, G., Branno, M., Brown, E. R., Cavalieri, V., Duboc, V., Duloquin, L., Flytzanis, C., Gache, C., Lapraz, F., Lepage, T., Locascio, A., Martinez, P., Matassi, G., Matranga, V., Range, R., Rizzo, F., Rottinger, E., Beane, W., Bradham, C., Byrum, C., Glenn, T., Hussain, S., Manning, F. G., Miranda, E., Thomason, R., Walton, K., Wikramanayke, A., Wu, S. Y., Xu, R., Brown, C. T., Chen, L., Gray, R. F., Lee, P. Y., Nam, J., Oliveri, P., Smith, J., Muzny, D., Bell, S., Chacko, J., Cree, A., Curry, S., Davis, C., Dinh, H., Dugan-Rocha, S., Fowler, J., Gill, R., Hamilton, C., Hernandez, J., Hines, S., Hume, J., Jackson, L., Jolivet, A., Kovar, C., Lee, S., Lewis, L., Miner, G., Morgan, M., Nazareth, L. V., Okwuonu, G., Parker, D., Pu, L. L., Thorn, R., and Wright, R. (2006). The genome of the sea urchin *Strongylocentrotus purpuratus*. *Science*, *314*, 941–952.

39. Goldstone, J. V., Hamdoun, A., Cole, B. J., Howard-Ashby, M., Nebert, D. W., Scally, M., Dean, M., Epel, D., Hahn, M. E., and Stegeman, J. J. (2006). The chemical defensome: environmental sensing and response genes in the *Strongylocentrotus purpuratus* genome. *Developmental Biology*, *300*, 366–384.

40. Freeman, R. M., Jr., Wu, M., Cordonnier-Pratt, M. M., Pratt, L. H., Gruber, C. E., Smith, M., Lander, E. S., Stange-Thomann, N., Lowe, C. J., Gerhart, J., and Kirschner, M. (2008). cDNA sequences for transcription factors and signaling proteins of the hemichordate Saccoglossus kowalevskii: efficacy of the expressed sequence tag (EST) approach for evolutionary and developmental studies of a new organism. *Biological Bulletin*, *214*, 284–302.

41. Lowe, C. J. (2008). Molecular genetic insights into deuterostome evolution from the direct-developing hemichordate *Saccoglossus kowalevskii*. *Philosophical Transactions of the Royal Society B*, *363*, 1569–1578.

42. Dehal, P., Satou, Y., Campbell, R. K., Chapman, J., Degnan, B., De Tomaso, A., Davidson, B., Di Gregorio, A., Gelpke, M., Goodstein, D. M., Harafuji, N., Hastings, K. E., Ho, I., Hotta, K., Huang, W., Kawashima, T., Lemaire, P., Martinez, D., Meinertzhagen, I. A., Necula, S., Nonaka, M., Putnam, N., Rash, S., Saiga, H., Satake, M., Terry, A., Yamada, L., Wang, H. G., Awazu, S., Azumi, K., Boore, J., Branno, M., Chin-Bow, S., DeSantis, R., Doyle, S., Francino, P., Keys, D. N., Haga, S., Hayashi, H., Hino, K., Imai, K. S., Inaba, K., Kano, S., Kobayashi, K., Kobayashi, M., Lee, B. I., Makabe, K. W., Manohar, C., Matassi, G., Medina, M., Mochizuki, Y., Mount, S., Morishita, T., Miura, S., Nakayama, A., Nishizaka, S., Nomoto, H., Ohta, F., Oishi, K., Rigoutsos, I., Sano, M., Sasaki, A., Sasakura, Y., Shoguchi, E., Shin-i, T., Spagnuolo, A., Stainier, D., Suzuki, M. M., Tassy, O., Takatori, N., Tokuoka, M., Yagi, K., Yoshizaki, F., Wada, S., Zhang, C., Hyatt, P. D., Larimer, F., Detter, C., Doggett, N., Glavina, T., Hawkins, T., Richardson, P., Lucas, S., Kohara, Y., Levine, M., Satoh, N., and Rokhsar, D. S. (2002). The draft genome of *Ciona intestinalis*: insights into chordate and vertebrate origins. *Science*, *298*, 2157–2167.

43. Putnam, N. H., Butts, T., Ferrier, D. E., Furlong, R. F., Hellsten, U., Kawashima, T., Robinson-Rechavi, M., Shoguchi, E., Terry, A., Yu, J. K., Benito-Gutierrez, E. L., Dubchak, I., Garcia-Fernandez, J., Gibson-Brown, J. J., Grigoriev, I. V., Horton, A. C., de Jong, P. J., Jurka, J., Kapitonov, V. V., Kohara, Y., Kuroki, Y., Lindquist, E., Lucas, S., Osoegawa, K., Pennacchio, L. A., Salamov, A. A., Satou, Y., Sauka-Spengler, T., Schmutz, J., Shin, I. T., Toyoda, A., Bronner-Fraser, M., Fujiyama, A., Holland, L. Z., Holland, P. W., Satoh, N., and Rokhsar, D. S. (2008). The amphioxus genome and the evolution of the chordate karyotype. *Nature*, *453*, 1064–1071.

44. Kuraku, S. (2010). Palaeophylogenomics of the vertebrate ancestor—impact of hidden paralogy on hagfish and lamprey gene phylogeny. *Integrative and Comparative Biology*, *50*, 124–129.

45. Hahn, M. E., Poland, A., Glover, E., and Stegeman, J. J. (1994). Photoaffinity labeling of the Ah receptor: phylogenetic survey of diverse vertebrate and invertebrate species. *Archives of Biochemistry and Biophysics*, *310*, 218–228.

46. Hahn, M. E., Woodin, B. R., Stegeman, J. J., and Tillitt, D. E. (1998). Aryl hydrocarbon receptor function in early vertebrates: inducibility of cytochrome P4501A in agnathan and elasmobranch fish. *Comparative Biochemistry and Physiology*, *120C*, 67–75.

47. Hahn, M. E., Sakai, J. A., Greninger, D., Franks, D. G., Merson, R. R., and Karchner, S. I. (2004). Structural and functional characterization of the aryl hydrocarbon receptor in an early diverging vertebrate, the lamprey *Petromyzon marinus*. *Marine Environmental Research*, *58*, 137–138.

48. Venkatesh, B., Kirkness, E. F., Loh, Y. H., Halpern, A. L., Lee, A. P., Johnson, J., Dandona, N., Viswanathan, L. D., Tay, A., Venter, J. C., Strausberg, R. L., and Brenner, S. (2007). Survey sequencing and comparative analysis of the elephant shark (*Callorhinchus milii*) genome. *PLoS Biology*, *5*, e101.

49. Betka, M., Welenc, A., Franks, D. G., Hahn, M. E., and Callard, G. V. (2000). Characterization of two aryl hydrocarbon receptor (AhR) mRNA forms in *Squalus acanthias* and stage-specific expression during spermatogenesis. *Bulletin of the Mount Desert Island Biological Laboratory*, *39*, 110–112.

50. Merson, R. R., Hersey, S. P., Zalobowski, T. W., Albanese, A. R., Franks, D. G., and Hahn, M. E. (2009). Aryl hydrocarbon receptors (AHR) of sharks: structural and functional divergence among AHR paralogs. *Toxicological Sciences (The Toxicologist Supplement)*, *108*, 15 (Abstract #81).

51. Merson, R. R., Mattingly, C. J., and Planchart, A. J. (2009). Tandem duplication of aryl hydrocarbon receptor (AHR) genes in the genome of the spiny dogfish shark (*Squalus acanthias*). *Bulletin of the Mount Desert Island Biological Laboratory*, *48*, 43–44.

52. Hansson, M. C., Wittzell, H., Persson, K., and von Schantz, T. (2003). Characterization of two distinct aryl hydrocarbon receptor (AhR2) genes in Atlantic salmon (*Salmo salar*) and evidence for multiple AhR2 gene lineages in salmonid fish. *Gene*, *303*, 197–206.

53. Hansson, M. C., Wittzell, H., Persson, K., and von Schantz, T. (2004). Unprecedented genomic diversity of AhR1 and AhR2 genes in Atlantic salmon (*Salmo salar* L.). *Aquatic Toxicology*, 68, 219–232.

54. Roy, N. K. and Wirgin, I. (1997). Characterization of the aromatic hydrocarbon receptor gene and its expression in Atlantic tomcod. *Archives of Biochemistry and Biophysics*, 344, 373–386.

55. Tanguay, R. L., Abnet, C. C., Heideman, W., and Peterson, R. E. (1999). Cloning and characterization of the zebrafish (*Danio rerio*) aryl hydrocarbon receptor. *Biochimica et Biophysica Acta*, 1444, 35–48.

56. Andreasen, E. A., Hahn, M. E., Heideman, W., Peterson, R. E., and Tanguay, R. L. (2002). The zebrafish (*Danio rerio*) aryl hydrocarbon receptor type 1 (zfAHR1) is a novel vertebrate receptor. *Molecular Pharmacology*, 62, 234–249.

57. Karchner, S. I., Franks, D. G., and Hahn, M. E. (2005). AHR1B, a new functional aryl hydrocarbon receptor in zebrafish: tandem arrangement of *ahr1b* and *ahr2* genes. *Biochemical Journal*, 392, 153–161.

58. Kawamura, T. and Yamashita, I. (2002). Aryl hydrocarbon receptor is required for prevention of blood clotting and for the development of vasculature and bone in the embryos of medaka fish, *Oryzias latipes*. *Zoological Science*, 19, 309–319.

59. Hanno, K., Oda, S., and Mitani, H. (2010). Effects of dioxin isomers on induction of AhRs and CYP1A1 in early developmental stage embryos of medaka (*Oryzias latipes*). *Chemosphere*, 78, 830–839.

60. Karchner, S. I. and Hahn, M. E. (2004). Pufferfish (*Fugu rubripes*) aryl hydrocarbon receptors: unusually high diversity in a compact genome. *Marine Environmental Research*, 58, 139–140.

61. Jung, R. E. and Walker, M. K. (1997). Effects of 2,3,7,8-tetrachlorodibenzo-*p*-dioxin (TCDD) on development of anuran amphibians. *Environmental Toxicology and Chemistry*, 16, 230–240.

62. Beatty, P. W., Holscher, M. A., and Neal, R. A. (1976). Toxicity of 2,3,7,8-tetrachloridibenzo-*p*-dioxin in larval and adult forms of *Rana catesbeiana*. *Bulletin of Environmental Contamination and Toxicology*, 16, 578–581.

63. Laub, L. B., Jones, B. D., and Powell, W. H. (2009). Responsiveness of a *Xenopus laevis* cell line to the aryl hydrocarbon receptor ligands 6-formylindolo[3,2-*b*]carbazole (FICZ) and 2,3,7,8-tetrachlorodibenzo-*p*-dioxin (TCDD). *Chemico-Biological Interactions*, 183, 202–211.

64. Jonsson, M. E., Berg, C., Goldstone, J. V., and Stegeman, J. J. (2011). New CYP1 genes in the frog *Xenopus* (*Silurana*) *tropicalis*: induction patterns and effects of AHR agonists during development. *Toxicology and Applied Pharmacology*, 250, 170–183.

65. Ohi, H., Fujita, Y., Miyao, M., Saguchi, K., Murayama, N., and Higuchi, S. (2003). Molecular cloning and expression analysis of the aryl hydrocarbon receptor of *Xenopus laevis*. *Biochemical and Biophysical Communications*, 307, 595–599.

66. Lavine, J. A., Rowatt, A. J., Klimova, T., Whitington, A. J., Dengler, E., Beck, C., and Powell, W. H. (2005). Aryl hydrocarbon receptors in the frog *Xenopus laevis*: two AhR1 paralogs exhibit low affinity for 2,3,7,8-tetrachlorodibenzo-*p*-dioxin (TCDD). *Toxicological Sciences*, 88, 60–72.

67. Zimmermann, A. L., King, E. A., Dengler, E., Scogin, S. R., and Powell, W. H. (2008). An aryl hydrocarbon receptor repressor from *Xenopus laevis*: function, expression, and role in dioxin responsiveness during frog development. *Toxicological Sciences*, 104, 124–134.

68. Barley, A. J., Spinks, P. Q., Thomson, R. C., and Shaffer, H. B. (2010). Fourteen nuclear genes provide phylogenetic resolution for difficult nodes in the turtle tree of life. *Molecular Phylogenetics and Evolution*, 55, 1189–1194.

69. Marquez, E. C. (2010). Cloning of cDNAs for estrogen receptor alpha, aromatase, and aryl hydrocarbon receptors and environmental effects on gene expression in the painted turtle (*Chrysemys picta*) and the red-eared slider (*Pseudemys scripta*). Ph.D. thesis, Boston University, Boston, MA.

70. Poland, A. and Glover, E. (1973). Chlorinated dibenzo-*p*-dioxins: potent inducers of delta-aminolevulinic acid synthetase and aryl hydrocarbon hydroxylase. II. A study of the structure–activity relationship. *Molecular Pharmacology*, 9, 736–747.

71. Poland, A., and Glover, E. (1973). 2,3,7,8-Tetrachlorodibenzo-*p*-dioxin: potent inducer of δ-aminolevulinic acid synthetase. *Science*, 179, 476–477.

72. Poland, A. and Glover, E. (1977). Chlorinated biphenyl induction of aryl hydrocarbon hydroxylase activity: a study of the structure–activity relationship. *Molecular Pharmacology*, 13, 924–938.

73. Giesy, J. P., Ludwig, J. P., and Tillitt, D. E. (1994). Deformities in birds of the Great Lakes region: assigning causality. *Environmental Science and Technology*, 28, 128A–135A.

74. Karchner, S. I., Kennedy, S. W., Trudeau, S., and Hahn, M. E. (2000). Towards a molecular understanding of species differences in dioxin sensitivity: initial characterization of Ah receptor cDNAs in birds and an amphibian. *Marine Environmental Research*, 50, 51–56.

75. Karchner, S. I., Franks, D. G., Kennedy, S. W., and Hahn, M. E. (2006). The molecular basis for differential dioxin sensitivity in birds: role of the aryl hydrocarbon receptor. *Proceedings of the National Academy of Sciences of the United States of America*, 103, 6252–6257.

76. Yasui, T., Kim, E. Y., Iwata, H., and Tanabe, S. (2004). Identification of aryl hydrocarbon receptor 2 in aquatic birds;cDNA cloning of AHR1 and AHR2 and characteristics of their amino acid sequences. *Marine Environmental Research*, 58, 113–118.

77. Yasui, T., Kim, E. Y., Iwata, H., Franks, D. G., Karchner, S. I., Hahn, M. E., and Tanabe, S. (2007). Functional characterization and evolutionary history of two aryl hydrocarbon receptor isoforms (AhR1 and AhR2) from avian species. *Toxicological Sciences*, 99, 101–117.

78. Walker, M. K., Heid, S. E., Smith, S. M., and Swanson, H. I. (2000). Molecular characterization and developmental

expression of the aryl hydrocarbon receptor from the chick embryo. *Comparative Biochemistry and Physiology*, *126C*, 305–319.

79. Head, J. A., Hahn, M. E., and Kennedy, S. W. (2008). Key amino acids in the aryl hydrocarbon receptor predict dioxin sensitivity in avian species. *Environmental Science & Technology*, *42*, 7535–7541.

80. Fernandez-Salguero, P., Hilbert, D. M., Rudikoff, S., Ward, J. M., and Gonzalez, F. J. (1996). Aryl-hydrocarbon receptor-deficient mice are resistant to 2,3,7,8-tetrachlorodibenzo-p-dioxin-induced toxicity. *Toxicology and Applied Pharmacology*, *140*, 173–179.

81. Mimura, J., Yamashita, K., Nakamura, K., Morita, M., Takagi, T., Nakao, K., Ema, M., Sogawa, K., Yasuda, M., Katsuki, M., and Fujii-Kuriyama, Y. (1997). Loss of teratogenic response to 2,3,7,8-tetrachlorodibenzo-p-dioxin (TCDD) in mice lacking the Ah (dioxin) receptor. *Genes to Cells*, *2*, 645–654.

82. Schmidt, J. V., Su, G. H.-T., Reddy, J. K., Simon, M. C., and Bradfield, C. A. (1996). Characterization of a murine Ahr null allele: involvement of the Ah receptor in hepatic growth and development. *Proceedings of the National Academy of Sciences of the United States of America*, *93*, 6731–6736.

83. Lahvis, G. P., Lindell, S. L., Thomas, R. S., McCuskey, R. S., Murphy, C., Glover, E., Bentz, M., Southard, J., and Bradfield, C. A. (2000). Portosystemic shunting and persistent fetal vascular structures in aryl hydrocarbon receptor-deficient mice. *Proceedings of the National Academy of Sciences of the United States of America*, *97*, 10442–10447.

84. Robles, R., Morita, Y., Mann, K. K., Perez, G. I., Yang, S., Matikainen, T., Sherr, D. H., and Tilly, J. L. (2000). The aryl hydrocarbon receptor, a basic helix–loop–helix transcription factor of the PAS gene family, is required for normal ovarian germ cell dynamics in the mouse. *Endocrinology*, *141*, 450–453.

85. Benedict, J. C., Lin, T. M., Loeffler, I. K., Peterson, R. E., and Flaws, J. A. (2000). Physiological role of the aryl hydrocarbon receptor in mouse ovary development. *Toxicological Sciences*, *56*, 382–388.

86. Warren, W. C., Hillier, L. W., Marshall Graves, J. A., Birney, E., Ponting, C. P., Grutzner, F., Belov, K., Miller, W., Clarke, L., Chinwalla, A. T., Yang, S. P., Heger, A., Locke, D. P., Miethke, P., Waters, P. D., Veyrunes, F., Fulton, L., Fulton, B., Graves, T., Wallis, J., Puente, X. S., Lopez-Otin, C., Ordonez, G. R., Eichler, E. E., Chen, L., Cheng, Z., Deakin, J. E., Alsop, A., Thompson, K., Kirby, P., Papenfuss, A. T., Wakefield, M. J., Olender, T., Lancet, D., Huttley, G. A., Smit, A. F., Pask, A., Temple-Smith, P., Batzer, M. A., Walker, J. A., Konkel, M. K., Harris, R. S., Whittington, C. M., Wong, E. S., Gemmell, N. J., Buschiazzo, E., Vargas Jentzsch, I. M., Merkel, A., Schmitz, J., Zemann, A., Churakov, G., Kriegs, J. O., Brosius, J., Murchison, E. P., Sachidanandam, R., Smith, C., Hannon, G. J., Tsend-Ayush, E., McMillan, D., Attenborough, R., Rens, W., Ferguson-Smith, M., Lefevre, C. M., Sharp, J. A., Nicholas, K. R., Ray, D. A., Kube, M., Reinhardt, R., Pringle, T. H., Taylor, J., Jones, R. C., Nixon, B., Dacheux, J. L., Niwa, H., Sekita, Y., Huang, X., Stark, A., Kheradpour, P., Kellis, M., Flicek, P., Chen, Y., Webber, C., Hardison, R., Nelson, J., Hallsworth-Pepin, K., Delehaunty, K., Markovic, C., Minx, P., Feng, Y., Kremitzki, C., Mitreva, M., Glasscock, J., Wylie, T., Wohldmann, P., Thiru, P., Nhan, M. N., Pohl, C. S., Smith, S. M., Hou, S., Nefedov, M., de Jong, P. J., Renfree, M. B., Mardis, E. R., and Wilson, R. K. (2008). Genome analysis of the platypus reveals unique signatures of evolution. *Nature*, *453*, 175–183.

87. Mikkelsen, T. S., Wakefield, M. J., Aken, B., Amemiya, C. T., Chang, J. L., Duke, S., Garber, M., Gentles, A. J., Goodstadt, L., Heger, A., Jurka, J., Kamal, M., Mauceli, E., Searle, S. M., Sharpe, T., Baker, M. L., Batzer, M. A., Benos, P. V., Belov, K., Clamp, M., Cook, A., Cuff, J., Das, R., Davidow, L., Deakin, J. E., Fazzari, M. J., Glass, J. L., Grabherr, M., Greally, J. M., Gu, W., Hore, T. A., Huttley, G. A., Kleber, M., Jirtle, R. L., Koina, E., Lee, J. T., Mahony, S., Marra, M. A., Miller, R. D., Nicholls, R. D., Oda, M., Papenfuss, A. T., Parra, Z. E., Pollock, D. D., Ray, D. A., Schein, J. E., Speed, T. P., Thompson, K., VandeBerg, J. L., Wade, C. M., Walker, J. A., Waters, P. D., Webber, C., Weidman, J. R., Xie, X., Zody, M. C., Graves, J. A., Ponting, C. P., Breen, M., Samollow, P. B., Lander, E. S., and Lindblad-Toh, K. (2007). Genome of the marsupial *Monodelphis domestica* reveals innovation in non-coding sequences. *Nature*, *447*, 167–177.

88. Kuraku, S., Meyer, A., and Kuratani, S. (2009). Timing of genome duplications relative to the origin of the vertebrates: did cyclostomes diverge before or after? *Molecular Biology and Evolution*, *26*, 47–59.

89. Sundstrom, G., Larsson, T. A., and Larhammar, D. (2008). Phylogenetic and chromosomal analyses of multiple gene families syntenic with vertebrate Hox clusters. *BMC Evolutionary Biology*, *8*, 254.

90. Hahn, M. E., Merson, R. R., and Karchner, S. I. (2005). Xenobiotic receptors in fishes: structural and functional diversity and evolutionary insights. In *Biochemistry and Molecular Biology of Fishes*. Vol. 6 *Environmental Toxicology* (Moon, T. W. and Mommsen, T. P., Eds). Elsevier, The Netherlands.

91. Amores, A., Force, A., Yan, Y.-L., Joly, L., Amemiya, C., Fritz, A., Ho, R. K., Langeland, J., Prince, V., Wang, Y.-L., Westerfield, M., Ekker, M., and Postlethwait, J. H. (1998). Zebrafish hox clusters and vertebrate genome evolution. *Science*, *282*, 1711–1714.

92. Meyer, A. and Van de Peer, Y. (2005). From 2R to 3R: evidence for a fish-specific genome duplication (FSGD). *Bioessays*, *27*, 937–945.

93. Postlethwait, J., Amores, A., Cresko, W., Singer, A., and Yan, Y. L. (2004). Subfunction partitioning, the teleost radiation and the annotation of the human genome. *Trends in Genetics*, *20*, 481–490.

94. Vandepoele, K., De Vos, W., Taylor, J. S., Meyer, A., and Van de Peer, Y. (2004). Major events in the genome evolution of vertebrates: paranome age and size differ considerably between ray-finned fishes and land vertebrates. *Proceedings of the National Academy of Sciences of the United States of America*, *101*, 1638–1643.

95. Christoffels, A., Koh, E. G., Chia, J. M., Brenner, S., Aparicio, S., and Venkatesh, B. (2004). Fugu genome analysis provides evidence for a whole-genome duplication early during the

evolution of ray-finned fishes. *Molecular Biology and Evolution*, 21, 1146–1151.

96. Postlethwait, J. H. (2007). The zebrafish genome in context: ohnologs gone missing. *Journal of Experimental Zoology, Part B*, 308, 563–577.

97. Postlethwait, J. H. (2006). The zebrafish genome: a review and msx gene case study. In *Vertebrate Genomes*, Vol. 2 Karger, Basel.

98. Poland, A., Palen, D., and Glover, E. (1994). Analysis of the four alleles of the murine aryl hydrocarbon receptor. *Molecular Pharmacology*, 46, 915–921.

99. Thomas, R. S., Penn, S. G., Holden, K., Bradfield, C. A., and Rank, D. R. (2002). Sequence variation and phylogenetic history of the mouse Ahr gene. *Pharmacogenetics*, 12, 151–163.

100. Pohjanvirta, R., Wong, J. M. Y., Li, W., Harper, P. A., Tuomisto, J., and Okey, A. B. (1998). Point mutation in intron sequence causes altered carboxyl-terminal structure in the aryl hydrocarbon receptor of the most 2,3,7,8-tetrachlorodibenzo-p-dioxin-resistant rat strain. *Molecular Pharmacology*, 54, 86–93.

101. Harper, P. A., Wong, J. M. Y., Lam, M. S. M., and Okey, A. B. (2002). Polymorphisms in the human AH receptor. *Chemico-Biological Interactions*, 141, 161–187.

102. Okey, A. B., Franc, M. A., Moffat, I. D., Tijet, N., Boutros, P. C., Korkalainen, M., Tuomisto, J., and Pohjanvirta, R. (2005). Toxicological implications of polymorphisms in receptors for xenobiotic chemicals: the case of the aryl hydrocarbon receptor. *Toxicology and Applied Pharmacology*, 207, 43–51.

103. Hahn, M. E., Karchner, S. I., Franks, D. G., and Merson, R. R. (2004). Aryl hydrocarbon receptor polymorphisms and dioxin resistance in Atlantic killifish (*Fundulus heteroclitus*). *Pharmacogenetics*, 14, 131–143.

104. Wirgin, I., Roy, N. K., Loftus, M., Chambers, R. C., Franks, D. G., and Hahn, M. E. (2011). Mechanistic basis of resistance to PCBs in Atlantic tomcod from the Hudson River. *Science*, 331, 1322–1325.

105. Reitzel, A. M., Karchner, S. I., Franks, D. G., Evans, B. R., Nacci, D. E., Champlin, D., Vieira, V., and Hahn, M. E. (2011). Genetic diversity in aryl hydrocarbon receptor (AHR) loci in PCB-sensitive and PCB-resistant populations of Atlantic killifish (*Fundulus heteroclitus*). manuscript in preparation.

106. Poland, A. and Glover, E. (1980). 2,3,7,8-Tetrachlorodibenzo-p-dioxin: segregation of toxicity with the Ah locus. *Molecular Pharmacology*, 17, 86–94.

107. Okey, A. B., Vella, L. M., and Harper, P. A. (1989). Detection and characterization of a low affinity form of cytosolic Ah receptor in livers of mice nonresponsive to induction of cytochrome P_1-450 by 3-methylcholanthrene. *Molecular Pharmacology*, 35, 823–830.

108. Safe, S. (1990). Polychlorinated biphenyls (PCBs), dibenzo-p-dioxins (PCDDs), dibenzofurans (PCDFs), and related compounds: environmental and mechanistic considerations which support the development of toxic equivalency factors (TEFs). *CRC Critical Reviews in Toxicology*, 21, 51–88.

109. Butler, R. B., Kelley, M. L., Powell, W. H., Hahn, M. E., and Van Beneden, R. J. (2001). An aryl hydrocarbon receptor homologue from the soft-shell clam, *Mya arenaria*: evidence that invertebrate AHR homologues lack TCDD and BNF binding. *Gene*, 278, 223–234.

110. Céspedes, M. A., Galindo, M. I., and Couso, J. P. (2010). Dioxin toxicity *in vivo* results from an increase in the dioxin-independent transcriptional activity of the aryl hydrocarbon receptor. *PLoS ONE*, 5, e15382.

111. Nguyen, L. P. and Bradfield, C. A. (2008). The search for endogenous activators of the aryl hydrocarbon receptor. *Chemical Research in Toxicology*, 21, 102–116.

112. Ema, M., Ohe, N., Suzuki, M., Mimura, J., Sogawa, K., Ikawa, S., and Fujii-Kuriyama, Y. (1994). Dioxin binding activities of polymorphic forms of mouse and human aryl hydrocarbon receptors. *Journal of Biological Chemistry*, 269, 27337–27343.

113. Ramadoss, P. and Perdew, G. H. (2004). Use of 2-azido-3-[^{125}I]iodo-7,8-dibromodibenzo-p-dioxin as a probe to determine the relative ligand affinity of human versus mouse aryl hydrocarbon receptor in cultured cells. *Molecular Pharmacology*, 66, 129–136.

114. Pandini, A., Denison, M. S., Song, Y., Soshilov, A. A., and Bonati, L. (2007). Structural and functional characterization of the aryl hydrocarbon receptor ligand binding domain by homology modeling and mutational analysis. *Biochemistry*, 46, 696–708.

115. Pandini, A., Soshilov, A. A., Song, Y., Zhao, J., Bonati, L., and Denison, M. S. (2009). Detection of the TCDD binding-fingerprint within the Ah receptor ligand binding domain by structurally driven mutagenesis and functional analysis. *Biochemistry*, 48, 5972–5983.

116. Farmahin, R., Wu, D., Bursian, S. J., Crump, D., Giesy, J. P., Hahn, M. E., Jones, S. P., Karchner, S. I., Mundy, L. J., Zwiernik, M. J., and Kennedy, S. W. (2010). The ligand binding domain—the key to the classification of avian sensitivity to dioxin-like compounds. *Toxicology Letters*, 196S, S116–S117.

117. Wirgin, I. and Waldman, J. R. (2004). Resistance to contaminants in North American fish populations. *Mutation Research*, 552, 73–100.

118. Hahn, M. E., Karchner, S. I., Franks, D. G., Evans, B. R., Nacci, D., Champlin, D., and Cohen, S. (2005). Mechanism of PCB- and dioxin-resistance in fish in the Hudson River Estuary: role of receptor polymorphisms. Final report, Hudson River Foundation, New York.

119. Haarmann-Stemmann, T. and Abel, J. (2006). The arylhydrocarbon receptor repressor (AhRR): structure, expression, and function. *Biological Chemistry*, 387, 1195–1199.

120. Hahn, M. E., Allan, L. L., and Sherr, D. H. (2009). Regulation of constitutive and inducible AHR signaling: complex interactions involving the AHR repressor. *Biochemical Pharmacology*, 77, 485–497.

121. Evans, B. R., Karchner, S. I., Franks, D. G., and Hahn, M. E. (2005). Duplicate aryl hydrocarbon receptor repressor genes (ahrr1 and ahrr2) in the zebrafish *Danio rerio*: structure,

function, evolution, and AHR-dependent regulation *in vivo*. *Archives of Biochemistry and Biophysics*, 441, 151–167.

122. Lee, J. S., Kim, E. Y., Nomaru, K., and Iwata, H. (2011). Molecular and functional characterization of aryl hydrocarbon receptor repressor from the chicken (*Gallus gallus*): interspecies similarities and differences. *Toxicological Sciences*, 119, 319–334.

123. Flaveny, C. A., Murray, I. A., Chiaro, C. R., and Perdew, G. H. (2009). Ligand selectivity and gene regulation by the human aryl hydrocarbon receptor in transgenic mice. *Molecular Pharmacology*, 75, 1412–1420.

124. DiNatale, B. C., Murray, I. A., Schroeder, J. C., Flaveny, C. A., Lahoti, T. S., Laurenzana, E. M., Omiecinski, C. J., and Perdew, G. H. (2010). Kynurenic acid is a potent endogenous aryl hydrocarbon receptor ligand that synergistically induces interleukin-6 in the presence of inflammatory signaling. *Toxicological Sciences*, 115, 89–97.

125. Boitano, A. E., Wang, J., Romeo, R., Bouchez, L. C., Parker, A. E., Sutton, S. E., Walker, J. R., Flaveny, C. A., Perdew, G. H., Denison, M. S., Schultz, P. G., and Cooke, M. P. (2010). Aryl hydrocarbon receptor antagonists promote the expansion of human hematopoietic stem cells. *Science*, 329, 1345–1348.

126. Henry, E. C. and Gasiewicz, T. A. (2008). Molecular determinants of species-specific agonist and antagonist activity of a substituted flavone towards the aryl hydrocarbon receptor. *Archives of Biochemistry and Biophysics*, 472, 77–88.

127. Zhou, J. G., Henry, E. C., Palermo, C. M., Dertinger, S. D., and Gasiewicz, T. A. (2003). Species-specific transcriptional activity of synthetic flavonoids in guinea pig and mouse cells as a result of differential activation of the aryl hydrocarbon receptor to interact with dioxin-responsive elements. *Molecular Pharmacology*, 63, 915–924.

128. Whelan, F., Hao, N., Furness, S. G., Whitelaw, M. L., and Chapman-Smith, A. (2010). Amino acid substitutions in the aryl hydrocarbon receptor ligand binding domain reveal YH439 as an atypical AhR activator. *Molecular Pharmacology*, 77, 1037–1046.

129. Zhao, B., Degroot, D. E., Hayashi, A., He, G., and Denison, M. S. (2010). CH223191 is a ligand-selective antagonist of the Ah (dioxin) receptor. *Toxicological Sciences*, 117, 393–403.

130. Zhang, S., Rowlands, C., and Safe, S. (2008). Ligand-dependent interactions of the Ah receptor with coactivators in a mammalian two-hybrid assay. *Toxicology and Applied Pharmacology*, 227, 196–206.

131. Murray, I. A., Krishnegowda, G., DiNatale, B. C., Flaveny, C., Chiaro, C., Lin, J. M., Sharma, A. K., Amin, S., and Perdew, G. H. (2010). Development of a selective modulator of aryl hydrocarbon (Ah) receptor activity that exhibits anti-inflammatory properties. *Chemical Research in Toxicology*, 23, 955–966.

132. Murray, I. A., Morales, J. L., Flaveny, C. A., Dinatale, B. C., Chiaro, C., Gowdahalli, K., Amin, S., and Perdew, G. H. (2010). Evidence for ligand-mediated selective modulation of aryl hydrocarbon receptor activity. *Molecular Pharmacology*, 77, 247–254.

133. Gooch, J. W., Elskus, A. A., Kloepper-Sams, P. J., Hahn, M. E., and Stegeman, J. J. (1989). Effects of *ortho* and non-*ortho* substituted polychlorinated biphenyl congeners on the hepatic monooxygenase system in scup (*Stenotomus chrysops*). *Toxicology and Applied Pharmacology*, 98, 422–433.

134. Walker, M. K. and Peterson, R. E. (1991). Potencies of polychlorinated dibenzo-*p*-dioxin, dibenzofuran, and biphenyl congeners, relative to 2,3,7,8-tetrachlorodibenzo-*p*-dioxin, for producing early life stage mortality in rainbow trout (*Oncorhynchus mykiss*). *Aquatic Toxicology*, 21, 219–238.

135. Abnet, C. C., Tanguay, R. L., Heideman, W., and Peterson, R. E. (1999). Transactivation activity of human, zebrafish, and rainbow trout aryl hydrocarbon receptors expressed in COS-7 cells: greater insight into species differences in toxic potency of polychlorinated dibenzo-*p*-dioxin, dibenzofuran, and biphenyl congeners. *Toxicology and Applied Pharmacology*, 159, 41–51.

136. Hestermann, E. V., Stegeman, J. J., and Hahn, M. E. (2000). Relative contributions of affinity and intrinsic efficacy to aryl hydrocarbon receptor ligand potency. *Toxicology and Applied Pharmacology*, 168, 160–172.

137. Kennedy, S. W., Lorenzen, A., Jones, S. P., Hahn, M. E., and Stegeman, J. J. (1996). Cytochrome P4501A induction in avian hepatocyte cultures: a promising approach for predicting the sensitivity of avian species to toxic effects of halogenated aromatic hydrocarbons. *Toxicology and Applied Pharmacology*, 141, 214–230.

138. Herve, J. C., Crump, D. L., McLaren, K. K., Giesy, J. P., Zwiernik, M. J., Bursian, S. J., and Kennedy, S. W. (2010). 2,3,4,7,8-Pentachlorodibenzofuran is a more potent cytochrome P4501A inducer than 2,3,7,8-tetrachlorodibenzo-*p*-dioxin in herring gull hepatocyte cultures. *Environmental Toxicology and Chemistry*, 29, 2088–2095.

139. Herve, J. C., Crump, D., Jones, S. P., Mundy, L. J., Giesy, J. P., Zwiernik, M. J., Bursian, S. J., Jones, P. D., Wiseman, S. B., Wan, Y., and Kennedy, S. W. (2010). Cytochrome P4501A induction by 2,3,7,8-tetrachlorodibenzo-*p*-dioxin and two chlorinated dibenzofurans in primary hepatocyte cultures of three avian species. *Toxicological Sciences*, 113, 380–391.

140. Vezina, C. M., Lin, T. M., and Peterson, R. E. (2009). AHR signaling in prostate growth, morphogenesis, and disease. *Biochemical Pharmacology*, 77, 566–576.

141. Singh, K. P., Casado, F. L., Opanashuk, L. A., and Gasiewicz, T. A. (2009). The aryl hydrocarbon receptor has a normal function in the regulation of hematopoietic and other stem/progenitor cell populations. *Biochemical Pharmacology*, 77, 577–587.

142. Hernandez-Ochoa, I., Karman, B. N., and Flaws, J. A. (2009). The role of the aryl hydrocarbon receptor in the female reproductive system. *Biochemical Pharmacology*, 77, 547–559.

143. Esser, C. (2009). The immune phenotype of AhR null mouse mutants: not a simple mirror of xenobiotic receptor overactivation. *Biochemical Pharmacology*, 77, 597–607.

144. Stevens, E. A., Mezrich, J. D., and Bradfield, C. A. (2009). The aryl hydrocarbon receptor: a perspective on potential roles in the immune system. *Immunology*, 127, 299–311.

145. McIntosh, B. E., Hogenesch, J. B., and Bradfield, C. A. (2010). Mammalian Per–Arnt–Sim proteins in environmental adaptation. *Annual Reviews in Physiology*, 72, 625–645.

146. Brown, R. P., McDonnell, C. M., Berenbaum, M. R., and Schuler, M. A. (2005). Regulation of an insect cytochrome P450 monooxygenase gene (CYP6B1) by aryl hydrocarbon and xanthotoxin response cascades. *Gene*, 358, 39–52.

147. Force, A., Lynch, M., Pickett, F. B., Amores, A., Yan, Y.-L., and Postlethwait, J. H. (1999). Preservation of duplicate genes by complementary, degenerative mutations. *Genetics*, 151, 1531–1545.

148. Lynch, M., and Force, A. (2000). The probability of duplicate gene preservation by subfunctionalization. *Genetics*, 154, 459–473.

149. Prasch, A. L., Teraoka, H., Carney, S. A., Dong, W., Hiraga, T., Stegeman, J. J., Heideman, W., and Peterson, R. E. (2003). Aryl hydrocarbon receptor 2 mediates 2,3,7,8-tetrachlorodibenzo-p-dioxin developmental toxicity in zebrafish. *Toxicological Sciences*, 76, 138–150.

150. Dong, W., Teraoka, H., Tsujimoto, Y., Stegeman, J. J., and Hiraga, T. (2004). Role of aryl hydrocarbon receptor in mesencephalic circulation failure and apoptosis in zebrafish embryos exposed to 2,3,7,8-tetrachlorodibenzo-p-dioxin. *Toxicological Sciences*, 77, 109–116.

151. Antkiewicz, D. S., Peterson, R. E., and Heideman, W. (2006). Blocking expression of AHR2 and ARNT1 in zebrafish larvae protects against cardiac toxicity of 2,3,7,8-tetrachlorodibenzo-p-dioxin. *Toxicological Sciences*, 94, 175–182.

152. Billiard, S. M., Timme-Laragy, A. R., Wassenberg, D. M., Cockman, C., and Di Giulio, R. T. (2006). The role of the aryl hydrocarbon receptor pathway in mediating synergistic developmental toxicity of polycyclic aromatic hydrocarbons to zebrafish. *Toxicological Sciences*, 92, 526–536.

153. Mathew, L. K., Andreasen, E. A., and Tanguay, R. L. (2006). Aryl hydrocarbon receptor activation inhibits regenerative growth. *Molecular Pharmacology*, 69, 257–265.

154. Clark, B. W., Matson, C. W., Jung, D., and Di Giulio, R. T. (2010). AHR2 mediates cardiac teratogenesis of polycyclic aromatic hydrocarbons and PCB-126 in Atlantic killifish (*Fundulus heteroclitus*). *Aquatic Toxicology*, 99, 232–240.

155. Jönsson, M. E., Jenny, M. J., Woodin, B. R., Hahn, M. E., and Stegeman, J. J. (2007). Role of AHR2 in the expression of novel cytochrome P450 1 family genes, cell cycle genes, and morphological defects in developing zebra fish exposed to 3,3′,4,4′,5-pentachlorobiphenyl or 2,3,7,8-tetrachlorodibenzo-p-dioxin. *Toxicological Sciences*, 100, 180–193.

156. Incardona, J. P., Day, H. L., Collier, T. K., and Scholz, N. L. (2006). Developmental toxicity of 4-ring polycyclic aromatic hydrocarbons in zebrafish is differentially dependent on AH receptor isoforms and hepatic cytochrome P4501A metabolism. *Toxicology and Applied Pharmacology*, 217, 308–321.

157. Lee, J. S., Kim, E. Y., and Iwata, H. (2009). Dioxin activation of CYP1A5 promoter/enhancer regions from two avian species, common cormorant (*Phalacrocorax carbo*) and chicken (*Gallus gallus*): association with aryl hydrocarbon receptor 1 and 2 isoforms. *Toxicology and Applied Pharmacology*, 234, 1–13.

158. Jenny, M. J., Karchner, S. I., Franks, D. G., Woodin, B. R., Stegeman, J. J., and Hahn, M. E. (2009). Distinct roles of two zebrafish AHR repressors (AHRRa and AHRRb) in embryonic development and regulating the response to 2,3,7,8-tetrachlorodibenzo-p-dioxin. *Toxicological Sciences*, 110, 426–441.

159. Doyon, Y., McCammon, J. M., Miller, J. C., Faraji, F., Ngo, C., Katibah, G. E., Amora, R., Hocking, T. D., Zhang, L., Rebar, E. J., Gregory, P. D., Urnov, F. D., and Amacher, S. L. (2008). Heritable targeted gene disruption in zebrafish using designed zinc-finger nucleases. *Nature Biotechnology*, 26, 702–708.

160. Meng, X., Noyes, M. B., Zhu, L. J., Lawson, N. D., and Wolfe, S. A. (2008). Targeted gene inactivation in zebrafish using engineered zinc-finger nucleases. *Nature Biotechnology*, 26, 695–701.

161. Foley, J. E., Yeh, J. R., Maeder, M. L., Reyon, D., Sander, J. D., Peterson, R. T., and Joung, J. K. (2009). Rapid mutation of endogenous zebrafish genes using zinc finger nucleases made by Oligomerized Pool ENgineering (OPEN). *PLoS ONE*, 4, e4348.

28

INVERTEBRATE AHR HOMOLOGS: ANCESTRAL FUNCTIONS IN SENSORY SYSTEMS

Jo Anne Powell-Coffman and Hongtao Qin

28.1 INTRODUCTION

28.1.1 Mammalian AHR Is the Dioxin Receptor

The mammalian aryl hydrocarbon receptor (AHR) was initially characterized as the gene that mediated the toxic effects of dioxins and related pollutants [1]. Halogenated aromatic hydrocarbons, including polychlorinated biphenyls (PCBs) and dioxins, are widespread environmental contaminants. Since dioxins are not easily metabolized, they have enduring health effects [2]. 2,3,7,8-Tetrachlorodibenzo-p-dioxin (TCDD) binds to AHR, resulting in persistent and ectopic activation of the AHR transcription complex. TCDD exposure has been associated with chloracne, cancer, and a broad range of developmental defects [3].

The AHR gene encodes a DNA binding protein containing basic helix–loop–helix domains and PAS motifs (the acronym PAS refers to the Per–Arnt–Sim domain) (reviewed in Ref. 4). AHR binds to ARNT, another member of the bHLH–PAS family, to form a heterodimeric DNA binding complex [5–7]. The mammalian AHR–ARNT complex binds to the consensus sequence 5′-(T/G)NGCGTG, which has been termed the dioxin response element or the xenobiotic response element (XRE) [8–10].

In the absence of an activating ligand, human AHR is bound to a cytoplasmic complex of chaperones, including HSP90 and XAP2/AIP. The chaperone proteins have been shown to be important for folding AHR and for sequestering the transcription factor in the cytoplasm [11–15]. Ligand binding induces conformational changes in AHR that enable it to translocate to the nucleus. Some endogenous compounds have been shown to activate AHR, but the presumptive ligands that control the physiological functions of AHR during normal development are still unknown (reviewed in Refs 16 and 17). The AHR–ARNT transcriptional complex has been shown to regulate genes involved in detoxification and stress response, most notably cytochrome P450 genes, as well as genes that regulate signal transduction and development [18–23]. Mice carrying deletions in the Ahr gene exhibit vascular defects that result in decreased liver size [24–27]. Collectively, these data suggest that mammalian AHR has at least two general functions. First, AHR promotes adaptive responses to planar aromatic hydrocarbons and related compounds. Second, AHR has important functions during normal development. These developmental roles are likely modulated by endogenous AHR ligands and other interactions (reviewed in Ref. 28).

28.1.2 Comparing and Contrasting Vertebrate and Invertebrate AHR Homologs

The presence of AHR homologs in both invertebrates and vertebrates demonstrates that the ancestral AHR gene arose over 500 million years ago [29]. Invertebrate AHR homologs were discovered first in nematodes, insects, and mollusks [30–32]. The invertebrate AHR proteins have been shown to share important molecular properties with their mammalian cognates, and they can interact with ARNT to form DNA binding complexes that recognize XRE sequences [30, 32, 33]. However, while human AHR binds specifically to radioactively labeled dioxins or β-naphthoflavone, the AHR proteins from *Caenorhabditis elegans*, *Drosophila melanogaster*, or *Mya arenaria* do not bind these compounds [30, 32]. The TCDD binding properties of AHR first evolved in early jawed vertebrates [29]. It is possible that

---

The AH Receptor in Biology and Toxicology, First Edition. Edited by Raimo Pohjanvirta.
© 2012 John Wiley & Sons, Inc. Published 2012 by John Wiley & Sons, Inc.

invertebrate AHR proteins do not bind small-molecule ligands. Alternatively, invertebrate AHR proteins may be modulated by an array of endogenous compounds or cofactors that have not been identified yet.

This chapter describes the functions of AHR homologs in *D. melanogaster* and *C. elegans*. Genetic studies in these model organisms have shown that the invertebrate homologs of AHR have key roles in the development of cells and structures that sense the environment.

28.2 CAENORHABDITIS ELEGANS AHR-1

28.2.1 The *C. elegans* Aryl Hydrocarbon Receptor Complex: AHR-1 and AHA-1

The free-living nematode *C. elegans* is a powerful genetic model system for the discovery and characterization of cellular and molecular processes that are common to metazoans. An adult hermaphrodite is only ~1 mm long and consists of 959 somatic cells, including 302 neurons. The cell lineage is essentially invariant, and this detailed description of wild-type development provides an exceptionally strong foundation for deciphering mutant phenotypes. Individual cells can be identified by their relative position and morphology and by cell type-specific markers [34–36].

The *C. elegans* orthologs of AHR and ARNT are encoded by the *ahr-1* (aryl hydrocarbon receptor-related) and *aha-1* (*ahr-1* associated) genes. The AHR-1 and AHA-1 proteins interact to bind DNA fragments containing xenobiotic response elements (5′-KNGCGTG) with sequence specificity *in vitro* [32]. Like mammalian AHR proteins, *C. elegans* AHR-1 can bind to HSP90 in rabbit reticulocyte lysates [32]. *C. elegans* AHR-1 does not bind the XAP2 chaperone [37].

Complementary strategies have been employed to test the hypothesis that *C. elegans* AHR-1 binds TCDD or other known ligands of human AHR. Biochemical studies have established that *C. elegans* AHR-1 does not bind radiolabeled TCDD, 2-azido-3-iodo-7,8-dibromodibenzo-*p*-dioxin, or β-naphthoflavone [30, 32]. Yeast expression assays have been employed to search for ligands. When the amino-terminal bHLH domains of *C. elegans* AHR-1 or murine AHR were replaced with the DNA binding and dimerization domains of LexA, then the fusion proteins did not activate expression of reporter genes. When β-naphthoflavone was added to these yeast strains, the fusion protein containing murine AHR sequences induced the expression of the reporter, but the fusion protein containing *C. elegans* AHR-1 did not [32, 38]. In additional unpublished studies, we assayed a broad spectrum of ligands known to activate mammalian AHR and did not find any that activated *C. elegans* AHR-1 in the yeast assay system. In transgenic *C. elegans*, AHR-1 fused to green fluorescent protein (GFP) is localized to the nuclei of certain neurons and blast cells, suggesting that exogenous ligands are not required for nuclear translocation of AHR-1 [39].

AHA-1, the *C. elegans* ortholog of ARNT, is the only known partner for AHR-1 [32]. AHA-1 can form complexes with other transcription factors in the bHLH–PAS family. For example, AHA-1 dimerizes with HIF-1 to form the hypoxia-inducible factor DNA binding complex [40]. AHA-1 is expressed in most, if not all, cells [40]. Deletion mutations in the *aha-1* gene result in larval lethality. Viability can be rescued by expressing *aha-1* in a subset of pharyngeal cells. Using this strategy, the phenotypes of *ahr-1* mutants have been compared to those of animals that lack *aha-1* expression in most cells. All *ahr-1* functions that have been investigated also require *aha-1* function [41].

28.2.2 Overview of the *C. elegans* AHR-1 Expression Pattern

AHR-1 expression is predominantly neuronal. Analyses of P*ahr-1*:GFP reporter constructs in which GFP expression was directed by *ahr-1* regulatory sequences revealed a dynamic expression pattern during embryonic and larval development. Upon hatching, *C. elegans* larvae have 550 nuclei, and some of their cells will continue to divide during larval development to contribute to a more mature nervous system. P*ahr-1*:GFP is expressed in 28 neurons, several blast cells, and 2 phasmid socket cells during the early larval stages. The neurons that express P*ahr-1*:GFP include ALNR/ALNL, AQR/PQR, AVM/PVM, BDUR/BDUL, PLMR/PLML, PLNR/PLNL, PHCL/PHCR, PVWL/PVWR, RMEL/RMER, SDQR/SDQL, and URXR/URXL. These neurons belong to multiple neuronal subtypes, including sensory neurons, interneurons, and GABAergic neurons [39, 41]. The P*ahr-1*:GFP reporter is expressed in the descendents of the QR and QL neuroblasts, and, as described further below, *ahr-1* is required for appropriate migration and differentiation of these cells. The T.pa, T.ppa, and T.ppp blast cells in the tail also express P*ahr-1*:GFP, as do all of their descendents, including the PHso1 and PHso2 phasmid socket cells. P*ahr-1*:GFP is also expressed in the MI and I3 neurons in the pharynx and the G2 and W blast cells [39].

28.2.3 Loss-of-Function Mutations in *C. elegans* *ahr-1* Cause Defects in the Development of Specific Neurons and Neuroblasts

Animals carrying mutations that abolish *ahr-1* function exhibit defects in neuronal development. The aberrations in *ahr-1*-deficient mutants include changes in gene expression indicative of cell fate changes and altered cell or axonal migration and morphology. These defects do not compromise viability or fecundity in standard lab culture conditions, but they show that *ahr-1* plays important roles in the development or the function of neurons belonging to multiple lineages and neuronal circuits [39, 41, 42].

In *ahr-1*-deficient animals, certain GABAergic neurons adopt inappropriate cell fates. The four RME neurons produce the γ-amino butyric acid (GABA) neurotransmitter and innervate head muscles to control foraging behaviors. Based on axon morphology, cell lineage, and gene expression, RME neurons can be classified into two subgroups, RMEL/RMER (left and right) and RMED/RMEV (dorsal and ventral). The *ahr-1*:GFP reporter is expressed in RMEL and RMER, but not in RMED or RMEV. Huang et al. [41] discovered that RMEL/RMER neurons adopted cell fates similar to those of RMED/RMEV in *ahr-1*-deficient animals. This change was characterized by changes in gene expression and by additional axonal processes. Further, they demonstrated that ectopic expression of *ahr-1* in RMED and RMEV was sufficient to cause the neurons to adopt morphologies similar to those of RMER and RMEL. These experiments suggest that AHR-1 functions as a binary switch to specify the fates of the RME neurons (RMER/RMEL subtype versus the RMED/RMEV subtype) [41].

The function of *ahr-1* is also essential for the appropriate migration and development of Q neuroblasts. Upon hatching, *C. elegans* first-stage larvae have two Q neuroblasts, QL on the left side and QR on the right side. These cells migrate in opposite directions and give rise to progeny cells, all of which express *Pahr-1*:GFP. Descendents of QR include three neurons: AVM, AQR, and SDQR. In *ahr-1*-deficient mutants, AVM and SDQR are mispositioned along the anterior–posterior axis. In addition, ~12% of AVM axons have defects in axonal morphology, such as axon branching or additional processes [39].

C. elegans ahr-1 also has a role in the development of the PLM neurons, which are located in the tail and transduce touch stimuli. In approximately one-third of *ahr-1*-deficient mutants, at least one of the PLM cell bodies is displaced anteriorly. In rare cases, an additional PLM cell is present. This suggests that *ahr-1* inhibits another cell from adopting a PLM-like cell fate [39].

28.2.4 *C. elegans* AHR-1 Expression and Oxygen-Sensitive Behaviors

C. elegans AHR-1 has key roles in hyperoxia avoidance and feeding behaviors. The wild-type strain of *C. elegans* most commonly used in laboratory studies (Bristol N2) does not aggregate on lawns of bacterial food. However, other strains carrying loss-of-function mutations in the *npr-1* neuropeptide receptor form clusters on lawns of *E. coli* food [43–45]. Interestingly, this behavior is regulated by environmental oxygen levels [44, 45]. In the absence of food, wild-type *C. elegans* aerotaxes from 21% oxygen (room air) to lower oxygen levels (5–12%). Aggregation on bacterial lawns is thought to create less hyperoxic microenviroments [44]. Interestingly, these behaviors also influence the organism's susceptibility to certain pathogens [46, 47]. If *C. elegans* are provided with food at 1% oxygen, then this experience modifies the behavior such that the animals prefer lower oxygen levels [48]. Social feeding behavior is controlled by complex neural circuits that integrate information regarding environmental oxygen levels and nutritional cues [49–51].

AHR-1 acts acutely in two neurons, URXR and URXL, to influence aggregation behavior. These neurons contact the pseudocoelomic fluid and are integral to the neural networks that govern the behavior [52, 53]. Transgenes that drive *ahr-1* expression in URXR and URXL can largely rescue aggregation behavior in *ahr-1*, *npr-1* double mutants. URX neurons born in *ahr-1*-deficient embryos are not permanently disabled. Induction of *ahr-1* expression after the URX neurons are established can rescue social feeding behavior. Key functions of the AHR-1 transcription complex in these neurons include driving the expression of *npr-1* and promoting the expression of soluble guanylate cyclases, including *gcy-32*, *gcy-34*, and *gcy-35* [42]. The GCY-35 protein has been shown to bind oxygen, and animals that lack a functional *gcy-35* gene do not exhibit normal hyperoxia avoidance behaviors [44, 45]. Thus, the AHR-1 transcription complex drives expression of oxygen-sensing guanylyl cyclases in URX neurons. Upon binding oxygen, the GCY-35 complex produces cGMP to promote social feeding behavior and hyperoxia avoidance.

28.3 DROSOPHILA MELANOGASTER SPINELESS

28.3.1 The *Drosophila* AHR Complex: Spineless and Tango

In the fruit fly *D. melanogaster*, the homologs of mammalian AHR and ARNT are encoded by the *spineless* and *tango* genes, respectively [31, 54]. *Drosophila* Spineless protein is similar in structure to the mammalian dioxin receptor, and all five splice junctions in the bHLH and PAS domains are conserved, providing strong support for the conclusion that *Drosophila* and mammalian AHR homologs arose from a common ancestral gene [31]. In binding studies, *Drosophila* Spineless does not bind radiolabeled TCDD or β-naphthoflavone [30]. Sequence alignments also suggest that some of the key residues that are required for mammal AHR to bind TCDD are not conserved in Spineless [30, 31]. In mouse Hepa 1c1c7 or *Drosophila* S2 cells, studies of chimeric *AHR* proteins have shown that substitution of the murine ligand binding domain with analogous sequences from Spineless results in a constitutively active transcription factor [55]. While it is still possible that Spineless localization or activity is modulated by an unidentified endogenous ligand in some contexts, the protein does not bind well-known agonists of mammalian AHR and appears to be constitutively active in certain cells.

Multiple lines of evidence demonstrate that the Spineless and Tango proteins interact and form a complex that binds DNA with sequence specificity. The two proteins bind in yeast two-hybrid assays [33]. In transient transfection assays, Spineless and Tango induce expression of a reporter containing an XRE in the 5′ regulatory domain [33], and the two proteins cooperate to bind XRE sequences in electrophoretic gel shift assays [56]. Ectopic expression of Spineless is sufficient to cause nuclear localization of Tango, providing further evidence that the proteins interact *in vivo* [33]. Like mammalian ARNT or *C. elegans* AHA-1, *Drosophila* Tango has multiple bHLH–PAS dimerization partners, and its functions are essential [33, 54, 57]. Although *spineless* appears to have roles in the peripheral nervous system that are independent of *tango* [58], the two genes clearly work in concert to direct many other cell fate decisions. Loss-of-function alleles of *tango* act as dominant enhancers, intensifying the bristle phenotypes seen in animals heterozygous for a *spineless* deficiency. In genetic mosaics, *tango*-deficient somatic clones have similar defects in antenna, leg, and bristle development as seen in *spineless*-deficient animals [33].

28.3.2 Overview of the *spineless* Expression Pattern

spineless transcripts are expressed in many tissues during embryonic, larval, and pupal development. In embryos, *spineless* mRNA is strongly expressed in embryonic stage 8 in a crescent just anterior to the cephalic furrow. As embryogenesis proceeds, spineless is expressed in the developing peripheral nervous system and in the forming eye–antennal discs [31]. Expression of Spineless protein in sensory neurons has been confirmed by antibody staining [58]. As development proceeds, spineless is expressed in imaginal discs that ultimately give rise to antennae, eyes, legs, and wings. In second instar larvae, spineless is expressed in the antennal disc and in the presumptive tarsal region of the leg disc. Later in larval development, spineless is expressed in wing discs and in developing bristle cells [31]. *spineless* is also expressed in the fly retina [59].

28.3.3 Discovery of *spineless*: Roles in Antenna and Leg Development

Over 85 years ago, Bridges and Morgan identified the first *spineless* allele as a mutation that shortened almost all sensory bristles [60]. Bristles have stereotyped positions and morphologies, and they are closely associated with sensory neurons. Additional genetic screens identified null mutations in *spineless* that cause transformation of the distal part of the antenna to the distal part of the second leg and deletion of the most tarsal region of the leg [31, 61]. Ectopic expression of spineless in the imaginal discs resulted in the formation of antennal structures in novel locations, demonstrating that *spineless* has an instructive role in antennal specification [31]. This function appears to be conserved in other insects. The ortholog of *spineless* in the red flour beetle *Tribolium castaneum* is required for adult antennal specification [62–64].

28.3.4 Stochastic Variations in *spineless* Expression Underpin Photoreceptor Cell Fate Differences in the Eye

The compound eye of adult *Drosophila* consists of about 800 hexagonal ommatidia, and each one is an assembly of eight photoreceptor (PR) and accessory cells. Each PR expresses a single rhodopsin that senses and transduces a specific range of light wavelengths. The inner R7 PRs are present in two subtypes. Approximately 70% of the R7 PR cells express the opsin Rh4. The remaining 30% of the R7 cells express Rh3. Importantly, the two R7 subtypes induce differing fates in the underlying R8 cells. Thus, the *Drosophila* eye is a functional mosaic of ommatidia, defined by different R7/R8 subtypes, and this pattern enables color vision [59, 65–67].

Random variations of spineless expression in R7 cells are thought to generate the two types of R7 photoreceptor cells. Several lines of evidence support this model. First, *spineless* mRNA is transiently expressed in 60–80% of R7 cells shortly before the rhodopsins are expressed (mid-pupation). Second, *spineless* is required for the R7 (Rh4) subtype. In *spineless* mutants, 100% of R7 cells express Rh3. Finally, expression of *spineless* throughout the eye causes all the photoreceptors to adopt the R7 (Rh4) fate. Indeed, expression of *spineless* later in development, at the time of rhodopsin expression, can still drive Rh4 expression in R7 cells. This stochastic expression of *spineless* is thought to be controlled at the transcriptional level, as the *spineless* mRNA expression pattern in R7 cells can be recapitulated by a reporter containing a 1.6 kb enhancer from the *spineless* gene [59].

28.3.5 Roles for *spineless* in the Morphology of Sensory Neurons

There are four classes of dendritic arborization (da) sensory neurons in *Drosophila*, and each is characterized by a stereotyped dendritic pattern [68]. In animals carrying loss-of-function mutations in *spineless*, the cell bodies and axons of da neurons appear normal, but the morphologies of the dendrites are altered [58]. The effects of the *spineless* mutation depend on the class of neuron: relatively simple dendritic processes (in class I and class II da neurons) overgrow, while complex dendritic trees (in class III or class IV neurons) are reduced. Thus, spineless function provides greater diversity in the dendritic architecture of da neurons [58]. Spineless protein is expressed at similar levels throughout the peripheral nervous system [31, 58], suggesting that the effects of Spineless on dendritic morphology are determined by developmental context.

Remarkably, mutations in *tango* do not recapitulate the dendritic phenotypes caused by *spineless* mutations [58]. This suggests that the Spineless protein acts independently of Tango to influence the morphologies of da neurons. Further studies of the *Drosophila* peripheral nervous system may reveal novel mechanisms for Spineless function.

28.4 CONCLUDING REMARKS

A theme that has emerged from these studies is that invertebrate AHR homologs fine-tune gene expression and cell fates in sensory cells. In *C. elegans*, the AHR-1 transcription complex drives the expression of soluble guanylate cyclases in the URX neurons, thereby facilitating oxygen-sensitive behaviors [42]. In *D. melanogaster*, Spineless promotes the expression of a specific rhodopsin gene in the R7 cells of the retina [59]. In these two examples, late expression of the AHR homologs after the cells are formed is sufficient to restore expression of downstream genes. In other contexts, loss-of-function mutations in invertebrate AHR homologs can cause dramatic transformations in cells or structures with sensory functions [31, 41].

Future studies will continue to investigate the question of whether invertebrate AHR proteins are modulated by small-molecule ligands. Recent data show that invertebrate AHR proteins are constitutively active in some contexts [55]. PAS domain-containing proteins in prokaryotes bind a wide range of cofactors, including chromophores and hemes (reviewed in Refs 4 and 69). Some PAS domains in higher eukaryotes have also been shown to bind heme, and structural studies suggest that individual PAS domains may bind other endogenous small ligands [70–72]. This leaves open the intriguing possibility that invertebrate AHR proteins are modulated by interactions with undiscovered cofactors or small molecules in some contexts, thereby influencing the ways in which organisms sense and respond to environmental cues.

REFERENCES

1. Burbach, K. M., Poland, A., and Bradfield, C. A. (1992). Cloning of the Ah-receptor cDNA reveals a distinctive ligand-activated transcription factor. *Proceedings of the National Academy of Sciences of the United States of America, 89*, 8185–8189.
2. Pirkle, J. L., Wolfe, W. H., Patterson, D. G., Needham, L. L., Michalek, J. E., Miner, J. C., Peterson, M. R., and Phillips, D. L. (1989). Estimates of the half-life of 2,3,7,8-tetrachlorodibenzo-*p*-dioxin in Vietnam veterans of Operation Ranch Hand. *Journal of Toxicology and Environmental Health, 27*, 165–171.
3. White, S. S. and Birnbaum, L. S. (2009). An overview of the effects of dioxins and dioxin-like compounds on vertebrates, as documented in human and ecological epidemiology. *Journal of Environmental Science and Health, Part C, 27*, 197–211.
4. Gu, Y. Z., Hogenesch, J. B., and Bradfield, C. A. (2000). The PAS superfamily: sensors of environmental and developmental signals. *Annual Review of Pharmacology and Toxicology, 40*, 519–561.
5. Denison, M. S., Fisher, J. M., and Whitlock, J. P. (1989). Protein–DNA interactions at recognition sites for the dioxin–Ah receptor complex. *Journal of Biological Chemistry, 264*, 16478–16482.
6. Henry, E. C., Rucci, G., and Gasiewicz, T. A. (1994). Purification to homogeneity of the heteromeric DNA-binding form of the aryl hydrocarbon receptor from rat liver. *Molecular Pharmacology, 46*, 1022–1027.
7. Hoffman, E. C., Reyes, H., Chu, F. F., Sander, F., Conley, L. H., Brooks, B. A., and Hankinson, O. (1991). Cloning of a factor required for activity of the Ah (dioxin) receptor. *Science, 252*, 954–958.
8. Yao, E. F. and Denison, M. S. (1992). DNA sequence determinants for binding of transformed Ah receptor to a dioxin-responsive enhancer. *Biochemistry, 31*, 5060–5067.
9. Bacsi, S. G., Reisz-Porszasz, S., and Hankinson, O. (1995). Orientation of the heterodimeric aryl hydrocarbon (dioxin) receptor complex on its asymmetric DNA recognition sequence. *Molecular Pharmacology, 47*, 432–438.
10. Swanson, H. I., Chan, W. K., and Bradfield, C. A. (1995). DNA binding specificities and pairing rules of the Ah receptor, ARNT, and SIM proteins. *Journal of Biological Chemistry, 270*, 26292–26302.
11. Ma, Q. and Whitlock, J. P. (1997). A novel cytoplasmic protein that interacts with the Ah receptor, contains tetratricopeptide repeat motifs, and augments the transcriptional response to 2,3,7,8-tetrachlorodibenzo-*p*-dioxin. *Journal of Biological Chemistry, 272*, 8878–8884.
12. Carver, L. A. and Bradfield, C. A. (1997). Ligand-dependent interaction of the aryl hydrocarbon receptor with a novel immunophilin homolog *in vivo*. *Journal of Biological Chemistry, 272*, 11452–11456.
13. Meyer, B. K., Pray-Grant, M. G., Vanden Heuvel, J. P., and Perdew, G. H. (1998). Hepatitis B virus X-associated protein 2 is a subunit of the unliganded aryl hydrocarbon receptor core complex and exhibits transcriptional enhancer activity. *Molecular and Cellular Biology, 18*, 978–988.
14. Petrulis, J. R., Kusnadi, A., Ramadoss, P., Hollingshead, B., and Perdew, G. H. (2003). The hsp90 co-chaperone XAP2 alters importin beta recognition of the bipartite nuclear localization signal of the Ah receptor and represses transcriptional activity. *Journal of Biological Chemistry, 278*, 2677–2685.
15. Lin, B. C., Nguyen, L. P., Walisser, J. A., and Bradfield, C. A. (2008). A hypomorphic allele of aryl hydrocarbon receptor-associated protein-9 produces a phenocopy of the AHR-null mouse. *Molecular Pharmacology, 74*, 1367–1371.
16. Nguyen, L. P. and Bradfield, C. A. (2008). The search for endogenous activators of the aryl hydrocarbon receptor. *Chemical Research in Toxicology, 21*, 102–116.
17. Furness, S. G. and Whelan, F. (2009). The pleiotropy of dioxin toxicity: xenobiotic misappropriation of the aryl hydrocarbon

receptor's alternative physiological roles. *Pharmacology and Therapeutics*, 124, 336–353.

18. Whitlock, J. P., Jr., (1987). The regulation of gene expression by 2,3,7,8-tetrachlorodibenzo-*p*-dioxin. *Pharmacological Reviews*, 39, 147–161.

19. Nebert, D. W., Dalton, T. P., Okey, A. B., and Gonzalez, F. J. (2004). Role of aryl hydrocarbon receptor-mediated induction of the CYP1 enzymes in environmental toxicity and cancer. *Journal of Biological Chemistry*, 279, 23847–23850.

20. Puga, A., Maier, A., and Medvedovic, M. (2000). The transcriptional signature of dioxin in human hepatoma HepG2 cells. *Biochemical Pharmacology*, 60, 1129–1142.

21. Fletcher, N., Wahlstrom, D., Lundberg, R., Nilsson, C. B., Nilsson, K. C., Stockling, K., Hellmold, H., and Håkansson, H. (2005). 2,3,7,8-Tetrachlorodibenzo-*p*-dioxin (TCDD) alters the mRNA expression of critical genes associated with cholesterol metabolism, bile acid biosynthesis, and bile transport in rat liver: a microarray study. *Toxicology and Applied Pharmacology*, 207, 1–24.

22. Hayes, K. R., Zastrow, G. M., Nukaya, M., Pande, K., Glover, E., Maufort, J. P., Liss, A. L., Liu, Y., Moran, S. M., Vollrath, A. L., and Bradfield, C. A. (2007). Hepatic transcriptional networks induced by exposure to 2,3,7,8-tetrachlorodibenzo-*p*-dioxin. *Chemical Research in Toxicology*, 20, 1573–1581.

23. Flaveny, C. A., Murray, I. A., and Perdew, G. H. (2010). Differential gene regulation by the human and mouse aryl hydrocarbon receptor. *Toxicological Sciences*, 114, 217–225.

24. Lahvis, G. P., Lindell, S. L., Thomas, R. S., McCuskey, R. S., Murphy, C., Glover, E., Bentz, M., Southard, J., and Bradfield, C. A. (2000). Portosystemic shunting and persistent fetal vascular structures in aryl hydrocarbon receptor-deficient mice. *Proceedings of the National Academy of Sciences of the United States of America*, 97, 10442–10447.

25. Lahvis, G. P., Pyzalski, R. W., Glover, E., Pitot, H. C., McElwee, M. K., and Bradfield, C. A. (2005). The aryl hydrocarbon receptor is required for developmental closure of the ductus venosus in the neonatal mouse. *Molecular Pharmacology*, 67, 714–720.

26. Fernandez-Salguero, P., Pineau, T., Hilbert, D. M., McPhail, T., Lee, S. S., Kimura, S., Nebert, D. W., Rudikoff, S., Ward, J. M., and Gonzalez, F. J. (1995). Immune system impairment and hepatic fibrosis in mice lacking the dioxin-binding Ah receptor. *Science*, 268, 722–726.

27. Schmidt, J. V., Su, G. H., Reddy, J. K., Simon, M. C., and Bradfield, C. A. (1996). Characterization of a murine Ahr null allele: involvement of the Ah receptor in hepatic growth and development. *Proceedings of the National Academy of Sciences of the United States of America*, 93, 6731–6736.

28. McMillan, B. J. and Bradfield, C. A. (2007). The aryl hydrocarbon receptor sans xenobiotics: endogenous function in genetic model systems. *Molecular Pharmacology*, 72, 487–498.

29. Hahn, M. E. (2002). Aryl hydrocarbon receptors: diversity and evolution. *Chemico-Biological Interactions*, 141, 131–160.

30. Butler, R. A., Kelley, M. L., Powell, W. H., Hahn, M. E., and Van Beneden, R. J. (2001). An aryl hydrocarbon receptor (AHR) homologue from the soft-shell clam, *Mya arenaria*: evidence that invertebrate AHR homologues lack 2,3,7,8-tetrachlorodibenzo-*p*-dioxin and beta-naphthoflavone binding. *Gene*, 278, 223–234.

31. Duncan, D. M., Burgess, E. A., and Duncan, I. (1998). Control of distal antennal identity and tarsal development in *Drosophila* by spineless-aristapedia, a homolog of the mammalian dioxin receptor. *Genes & Development*, 12, 1290–1303.

32. Powell-Coffman, J. A., Bradfield, C. A., and Wood, W. B. (1998). *Caenorhabditis elegans* orthologs of the aryl hydrocarbon receptor and its heterodimerization partner the aryl hydrocarbon receptor nuclear translocator. *Proceedings of the National Academy of Sciences of the United States of America*, 95, 2844–2849.

33. Emmons, R. B., Duncan, D., Estes, P. A., Kiefel, P., Mosher, J. T., Sonnenfeld, M., Ward, M. P., Duncan, I., and Crews, S. T. (1999). The spineless-aristapedia and tango bHLH–PAS proteins interact to control antennal and tarsal development in *Drosophila*. *Development*, 126, 3937–3945.

34. Sulston, J. E., Schierenberg, E., White, J. G., and Thomson, J. N. (1983). The embryonic cell lineage of the nematode *Caenorhabditis elegans*. *Developmental Biology*, 100, 64–119.

35. White, J. G., Southgate, E., Thomson, J. N., and Brenner, S. (1986). The structure of the nervous system in the nematode *C. elegans*. *Philosophical Transactions of the Royal Society B*, 314, 1–340.

36. Hall, D. H. and Altun, Z. F. (2008). *C. elegans Atlas*. Cold Spring Harbor Laboratory Press, New York.

37. Bell, D. R. and Poland, A. (2000). Binding of aryl hydrocarbon receptor (AhR) to AhR-interacting protein. The role of hsp90. *Journal of Biological Chemistry*, 275, 36407–36414.

38. Carver, L. A., Jackiw, V., and Bradfield, C. A. (1994). The 90-kDa heat shock protein is essential for Ah receptor signaling in a yeast expression system. *Journal of Biological Chemistry*, 269, 30109–30112.

39. Qin, H. and Powell-Coffman, J. A. (2004). The *Caenorhabditis elegans* aryl hydrocarbon receptor, AHR-1, regulates neuronal development. *Developmental Biology*, 270, 64–75.

40. Jiang, H., Guo, R., and Powell-Coffman, J. A. (2001). The *Caenorhabditis elegans hif-1* gene encodes a bHLH–PAS protein that is required for adaptation to hypoxia. *Proceedings of the National Academy of Sciences of the United States of America*, 98, 7916–7921.

41. Huang, X., Powell-Coffman, J. A., and Jin, Y. (2004). The AHR-1 aryl hydrocarbon receptor and its co-factor the AHA-1 aryl hydrocarbon receptor nuclear translocator specify GABAergic neuron cell fate in *C. elegans*. *Development*, 131, 819–828.

42. Qin, H., Zhai, Z., and Powell-Coffman, J. A. (2006). The *Caenorhabditis elegans* AHR-1 transcription complex controls expression of soluble guanylate cyclase genes in the URX neurons and regulates aggregation behavior. *Developmental Biology*, 298, 606–615.

43. de Bono, M. and Bargmann, C. I. (1998). Natural variation in a neuropeptide Y receptor homolog modifies social behavior and food response in *C. elegans*. *Cell*, 94, 679–689.

44. Gray, J. M., Karow, D. S., Lu, H., Chang, A. J., Chang, J. S., Ellis, R. E., Marletta, M. A., and Bargmann, C. I. (2004). Oxygen sensation and social feeding mediated by a *C. elegans* guanylate cyclase homologue. *Nature, 430*, 317–322.

45. Cheung, B. H., Arellano-Carbajal, F., Rybicki, I., and de Bono, M. (2004). Soluble guanylate cyclases act in neurons exposed to the body fluid to promote *C. elegans* aggregation behavior. *Current Biology, 14*, 1105–1111.

46. Styer, K. L., Singh, V., Macosko, E., Steele, S. E., Bargmann, C. I., and Aballay, A. (2008). Innate immunity in *Caenorhabditis elegans* is regulated by neurons expressing NPR-1/GPCR. *Science, 322*, 460–464.

47. Reddy, K. C., Andersen, E. C., Kruglyak, L., and Kim, D. H. (2009). A polymorphism in *npr-1* is a behavioral determinant of pathogen susceptibility in *C. elegans. Science, 323*, 382–384.

48. Cheung, B. H., Cohen, M., Rogers, C., Albayram, O., and de Bono, M. (2005). Experience-dependent modulation of *C. elegans* behavior by ambient oxygen. *Current Biology, 15*, 905–917.

49. Rogers, C., Persson, A., Cheung, B., and de Bono, M. (2006). Behavioral motifs and neural pathways coordinating O_2 responses and aggregation in *C. elegans. Current Biology, 16*, 649–659.

50. Chang, A. J., Chronis, N., Karow, D. S., Marletta, M. A., and Bargmann, C. I. (2006). A distributed chemosensory circuit for oxygen preference in *C. elegans. PLoS Biology, 4*, e274.

51. Zimmer, M., Gray, J. M., Pokala, N., Chang, A. J., Karow, D. S., Marletta, M. A., Hudson, M. L., Morton, D. B., Chronis, N., and Bargmann, C. I. (2009). Neurons detect increases and decreases in oxygen levels using distinct guanylate cyclases. *Neuron, 61*, 865–879.

52. Coates, J. C. and de Bono, M. (2002). Antagonistic pathways in neurons exposed to body fluid regulate social feeding in *Caenorhabditis elegans. Nature, 419*, 925–929.

53. de Bono, M., Tobin, D. M., Davis, M. W., Avery, L., and Bargmann, C. I. (2002). Social feeding in *Caenorhabditis elegans* is induced by neurons that detect aversive stimuli. *Nature, 419*, 899–903.

54. Sonnenfeld, M., Ward, M., Nystrom, G., Mosher, J., Stahl, S., and Crews, S. (1997). The *Drosophila* tango gene encodes a bHLH–PAS protein that is orthologous to mammalian Arnt and controls CNS midline and tracheal development. *Development, 124*, 4571–4582.

55. Kudo, K., Takeuchi, T., Murakami, Y., Ebina, M., and Kikuchi, H. (2009). Characterization of the region of the aryl hydrocarbon receptor required for ligand dependency of transactivation using chimeric receptor between *Drosophila* and *Mus musculus. Biochimica et Biophysica Acta, 1789*, 477–486.

56. Kozu, S., Tajiri, R., Tsuji, T., Michiue, T., Saigo, K., and Kojima, T. (2006). Temporal regulation of late expression of Bar homeobox genes during *Drosophila* leg development by Spineless, a homolog of the mammalian dioxin receptor. *Developmental Biology, 294*, 497–508.

57. Ward, M. P., Mosher, J. T., and Crews, S. T. (1998). Regulation of bHLH–PAS protein subcellular localization during *Drosophila* embryogenesis. *Development, 125*, 1599–1608.

58. Kim, M. D., Jan, L. Y., and Jan, Y. N. (2006). The bHLH–PAS protein Spineless is necessary for the diversification of dendrite morphology of *Drosophila* dendritic arborization neurons. *Genes & Development, 20*, 2806–2819.

59. Wernet, M. F., Mazzoni, E. O., Celik, A., Duncan, D. M., Duncan, I., and Desplan, C. (2006). Stochastic spineless expression creates the retinal mosaic for colour vision. *Nature, 440*, 174–180.

60. Bridges, C. B. and Morgan, T. H. (1923) The third-chromosome group of mutant characters of *Drosophila melanogaster*. Publication No. 327, Carnegie Institute of Washington, Washington, DC.

61. Struhl, G. (1982). Spineless-aristapedia: a homeotic gene that does not control the development of specific compartments in *Drosophila. Genetics, 102*, 737–749.

62. Shippy, T. D., Yeager, S. J., and Denell, R. E. (2009). The *Tribolium* spineless ortholog specifies both larval and adult antennal identity. *Development Genes and Evolution, 219*, 45–51.

63. Toegel, J. P., Wimmer, E. A., and Prpic, N. M. (2009). Loss of spineless function transforms the *Tribolium* antenna into a thoracic leg with pretarsal, tibiotarsal, and femoral identity. *Development Genes and Evolution, 219*, 53–58.

64. Angelini, D. R., Kikuchi, M., and Jockusch, E. L. (2009). Genetic patterning in the adult capitate antenna of the beetle *Tribolium castaneum. Developmental Biology, 327*, 240–251.

65. Feiler, R., Bjornson, R., Kirschfeld, K., Mismer, D., Rubin, G. M., Smith, D. P., Socolich, M., and Zuker, C. S. (1992). Ectopic expression of ultraviolet-rhodopsins in the blue photoreceptor cells of *Drosophila*: visual physiology and photochemistry of transgenic animals. *Journal of Neuroscience, 12*, 3862–3868.

66. Papatsenko, D., Sheng, G., and Desplan, C. (1997). A new rhodopsin in R8 photoreceptors of *Drosophila*: evidence for coordinate expression with Rh3 in R7 cells. *Development, 124*, 1665–1673.

67. Chou, W. H., Hall, K. J., Wilson, D. B., Wideman, C. L., Townson, S. M., Chadwell, L. V., and Britt, S. G. (1996). Identification of a novel *Drosophila* opsin reveals specific patterning of the R7 and R8 photoreceptor cells. *Neuron, 17*, 1101–1115.

68. Grueber, W. B., Jan, L. Y., and Jan, Y. N. (2002). Tiling of the *Drosophila* epidermis by multidendritic sensory neurons. *Development, 129*, 2867–2878.

69. Taylor, B. L. and Zhulin, I. B. (1999). PAS domains: internal sensors of oxygen, redox potential, and light. *Microbiology and Molecular Biology Reviews, 63*, 479–506.

70. Kaasik, K. and Lee, C. C. (2004). Reciprocal regulation of haem biosynthesis and the circadian clock in mammals. *Nature, 430*, 467–471.

71. Dioum, E. M., Rutter, J., Tuckerman, J. R., Gonzalez, G., Gilles-Gonzalez, M. A., and McKnight, S. L. (2002). NPAS2: a gas-responsive transcription factor. *Science, 298*, 2385–2387.

72. Scheuermann, T. H., Tomchick, D. R., Machius, M., Guo, Y., Bruick, R. K., and Gardner, K. H. (2009). Artificial ligand binding within the HIF2alpha PAS-B domain of the HIF2 transcription factor. *Proceedings of the National Academy of Sciences of the United States of America, 106*, 450–455.

29

ROLE OF AHR IN THE DEVELOPMENT OF THE LIVER AND BLOOD VESSELS

Sahoko Ichihara

29.1 INTRODUCTION

The aryl hydrocarbon receptor (AHR) is a ligand-activated transcription factor with multiple functions in adaptive metabolism, development, and dioxin toxicity in a variety of organs and cell systems. Phenotypes observed in AHR-null mice suggest organ-dependent physiological functions, and three AHR-deficient mice have been generated thus far to elucidate the signaling pathway mediating such physiological functions of AHR. One of the most consistent phenotypes among the AHR-null mouse models is a small liver. This phenotype results from a persistent ductus venosus and thereby portocaval shunt in these mice. Recently, it was also found that ischemia-induced angiogenesis is enhanced in AHR-null mice. This chapter will focus on the role of AHR in developing liver and blood vessels, and highlight possible mechanisms linking phenotypes and AHR-modulated genotypes. The role of AHR in angiogenesis and vascular development is also discussed in this chapter.

29.2 LIVER PHENOTYPE IN THE AHR-NULL MICE

AHR is a member of the basic helix–loop–helix family of transcription factors and is considered to mediate pleiotropic biological responses such as teratogenesis, tumor promotion, epithelial hyperplasia, and the induction of drug metabolizing enzymes to environmental contaminants such as 2,3,7,8-tetrachlorodibenzo-p-dioxin (TCDD) [1]. AHR is constitutively expressed in many mammalian tissues, with the highest levels of mRNA detected in liver, kidney, lung, heart, thymus, and placenta [2, 3]. To investigate the physiological and teratogenic role of AHR, three different laboratories independently generated gene-targeting mice with a disruption to the murine *AHR* gene introduced by homologous recombination [4–6]. While these mice manifested somewhat different abnormal phenotypes, one of the most consistent features across all three models is a liver phenotype. Schmidt et al. [4] demonstrated that the livers in their AHR-null animals were 25% smaller than AHR (+/+) littermates (wild-type). This difference in relative liver weight persisted at 4 weeks of age. The livers of heterozygous animals were not different in size from wild-type mice at any age. The livers of these AHR-null mice at 1 week of age were also pale and mottled in appearance with a spongy texture (Fig. 29.1a). This phenotype was not apparent in newborn animals and in most cases had completely disappeared by 2 weeks of age. Histological analysis of 1-week-old AHR-null livers showed extensive microvesicular fatty metamorphosis of hepatocytes and prolonged extramedullary hematopoiesis compared to wild-type mice (Fig. 29.1b–g). The fatty metamorphosis was patchy in all AHR-null livers, and was not consistently pericentral or periportal. By 3 weeks of age, all evidence of fatty metamorphosis had disappeared and only small pockets of hematopoietic cells remained in the livers of AHR-null mice. In addition, the majority of AHR-null mouse livers revealed mild to moderate hypercellularity with thickening and fibrosis in the portal regions by 2 weeks of age (Fig. 29.1h and i) [4].

In another AHR-null mouse model, the liver phenotype also differed both quantitatively and qualitatively from wild-type animals [5]. Livers from these AHR-null mice at 4 weeks

---

The AH Receptor in Biology and Toxicology, First Edition. Edited by Raimo Pohjanvirta.
© 2012 John Wiley & Sons, Inc. Published 2012 by John Wiley & Sons, Inc.

of age constituted $2.9 \pm 0.3\%$ of total body mass, compared to $6.1 \pm 0.4\%$ in wild-type or heterozygous animals. Histological examination revealed normal general structure of the hepatic lobules, and these animals developed pronounced fibrosis in the portal tract. Some mice also showed mild to moderate inflammatory changes in the bile ducts, and this phenomenon was already apparent at 3 weeks of age. Smooth muscle actin immunostaining of liver sections from these AHR-null mice revealed increased numbers of small arteries and arterioles in the portal areas [7]. Hepatic tumors were found in some of the AHR-null mice at 11–13 months of age, including hepatocellular adenoma and hepatocellular carcinoma. *Helicobacter hepaticus* hepatitis was not found in any of these mice. Liver tumors were never found in control mice of the same age.

These observations from the AHR-null mice populations together suggested that AHR plays an important role in liver development and a fundamental role in cell and organ physiology and homeostasis.

29.3 THE ROLE OF AHR IN VASCULAR DEVELOPMENT

To obtain a physiological explanation for the AHR-null phenotype observed in mice, Lahvis et al. [8] looked for changes in the liver vasculature of AHR-null mice given that relative liver weight and hepatocyte size decrease in response to starvation or the introduction of portosystemic shunts [9, 10]. Colloidal carbon uptake and microsphere perfusion studies indicated that 56% of portal blood flow bypasses the liver sinusoids in AHR-null mice [8]. To determine the nature of the portosystemic shunt, Lahvis et al. [8] also collected serial angiograms. In wild-type mice, contrast medium flowed into the portal vein, and then immediately into the portal branches of the liver (Fig. 29.2a). After filling the major branching veins, contrast entered the suprahepatic inferior vena cava (IVC) (Fig. 29.2c) and then flowed retrogradely, filling the infrahepatic IVC (Fig. 29.2d). However, contrast medium in both male and female AHR-null mice flowed from the portal vein directly to the IVC. Contrast filled the IVC first (Fig. 29.2h), and only slowly filled the individual branches of the liver (Fig. 29.2i–l). These angiographic results suggested that shunting is consistent with patent ductus venosus in adult AHR-null animals [8]. The same authors also evaluated hepatic vascular development in AHR-null mice neonates using 3D reconstructions from serial histological sections [11]. They found portosystemic shunting in the livers of these animals that would lead to smaller hepatocyte size, providing that the AHR is necessary for resolution of the ductus venosus, a shunt that normally closes within 2 days after birth in mice, and that AHR contributes to the resolution of a number of vascular structures.

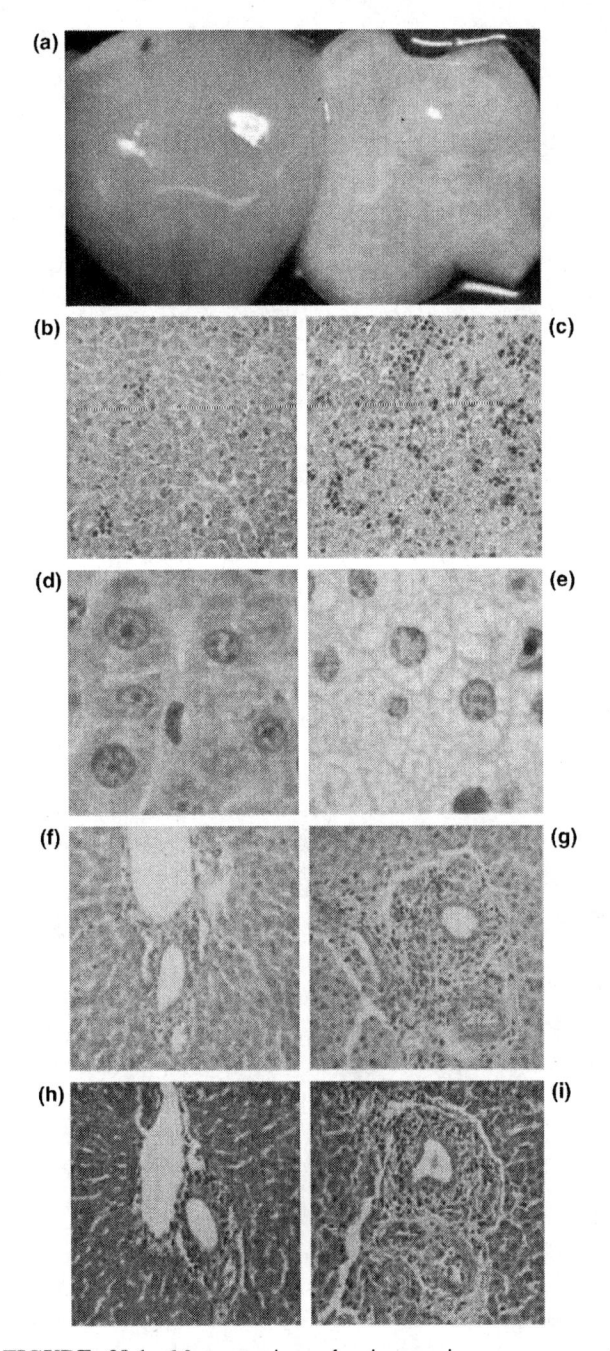

FIGURE 29.1 Macroscopic and microscopic appearance of mouse livers. (a) Macroscopic appearance of livers from 1-week-old wild-type (all at left) and AHR-null (all at right) mice. (b and c) Hematoxylin–eosin-stained sections of livers from 1-week-old wild-type and AHR-null mice demonstrating extensive microvesicular fatty metamorphosis and extramedullary hematopoiesis in the AHR-null animals. (d and e) Higher power view of 1-week-old wild-type and AHR-null livers. (f and g) Hematoxylin–eosin-stained liver sections from 2-week-old wild-type and AHR-null mice demonstrating hypercellularity and fibrosis of the portal tract in the AHR-null animals. (h and i) Trichrome attaining for connective tissue highlighting the portal fibrosis in AHR-null mouse liver compared to wild types. (See the color version of this figure in Color Plates section.)

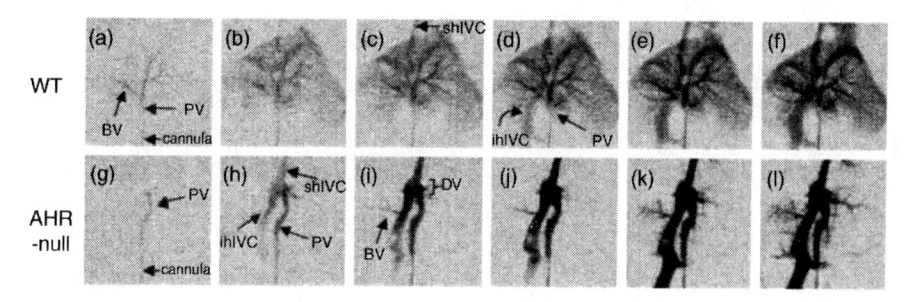

FIGURE 29.2 Angiography indicated patent ductus venosus. Continuous X-ray images were taken at approximately 10 s intervals after contrast was injected into the portal veins of wild-type (a–f) and AHR-null (g–l) mice. Representative serial radiographs are presented from left to right, showing the portal vein (PV), infrahepatic inferior vena cava (ihIVC), suprahepatic inferior vena cava (shIVC), ductus venosus (DV), and branching vessels (BV).

AHR-null mice as early as embryonic day 15.5 also show necrotic lesions in the fetal liver peripheries, with an increasing incidence up to postnatal day 1 and resolution by 2 weeks postpartum [12]. Further examination of the adult AHR-null animals revealed that the smaller livers were predominantly due to decreases in the size of both the left and right lobes, corresponding to regions of decreased perfusion and hepatic necrosis observed in the fetal liver tissue. These data support the hypothesis that normal ductus venosus closure is related to a generalized postpartum decrease in the diameter of major vessels of the liver. Inefficiency of the neonatal vasoconstriction may explain the failure of ductus venosus closure in AHR-null mice. Fetal vascular structures were also observed in other organs of AHR-null mice, including remnants of neonatal architecture in the eyes [8]. Like the ductus venosus and sinusoidal anastomoses, the hyaloid artery is a component of the neonatal vasculature that resolves early in development [13]. Intravital microscopy demonstrated an immature sinusoidal architecture in the liver and persistent hyaloid arteries in the eyes of adult AHR-null mice [8]. The vascular architecture of kidneys in AHR-null mice is also altered as shown by retrograde perfusion of the kidney with resin from the renal vein and out the renal artery [8].

Given that increased levels of mature transforming growth factor-β (TGF-β) were correlated with the induction of increased fibrosis [14, 15] and decreased rates of cell proliferation [16] in rodent liver, the relationship between TGF-β expression and apoptosis in AHR-null mice was analyzed. Zaher et al. [17] demonstrated marked increases in active TGF-β1 and TGF-β3 proteins, and elevated numbers of hepatocytes undergoing apoptosis, in livers from AHR-null mice compared with wild-type littermates. Furthermore, primary cultures of hepatocytes from AHR-null mice exhibited an elevated secretion of active TGF-β into the conditioned media compared with cultures from wild-type hepatocytes [17]. Apoptosis was also more pronounced in primary cultures of hepatocytes from the AHR-null mice, and these cells also showed an early onset of chromatin condensation, nuclear fragmentation, and DNA laddering together with very low levels of DNA synthesis. Moreover, AHR-null mice showed a threefold increase in the concentration of retinyl palmitate and a threefold accumulation in retinoic acid and retinol in liver compared with controls [18]. In fact, the AHR-null livers showed a marked decrease in their ability to metabolize retinoic acid compared with heterozygous and wild-type mice. This increase in retinoid accumulation and tissue transglutaminase II activity could account for activation of the latent TGF-β complex, resulting in altered deposition of collagen, abnormal cell cycle control, and increased fibrosis in the livers of AHR-null mice. Andreola et al. [19] demonstrated that livers from AHR-null mice fed a vitamin A-deficient diet had depleted retinoids, reduced TGF-βs and downstream signaling molecules, and decreased collagen deposition, consistent with the absence of liver fibrosis. These findings indicated that the phenotypic abnormalities in AHR-null mice could be mediated in part by abnormal levels of active TGF-β and altered cell cycle control.

29.4 THE PHENOTYPE OF LIVER IN SEVERAL GENE-TARGETING MICE RELATED TO AHR

Several gene-targeting mice related to AHR were created to confirm the proposed physiological functions of AHR. Bunger et al. [20] generated mice carrying a mutation in the AHR nuclear localization/dioxin response element (DRE) binding domain (designated as Ahr^{nls}) by homologous recombination. Relative liver weights in $Ahr^{nls/nls}$ animals were 27% smaller than those in wild-type animals, similar to those reported previously in AHR-null mice. Time lapse angiography showed that contrast medium in the $Ahr^{nls/nls}$ mice flowed directly from the portal vein into the IVC. The shunt between the portal vein and the IVC was clearly visible as a short segment running perpendicular to both the portal vein and IVC within the liver. This vascular pattern is consistent with patent ductus venosus in AHR-null mice. Walisser et al. [21, 22] generated a hypomorphic or low-expressing allele of the Ahr or $Arnt$ loci (designated Ahr^{fxneo}

or $Arnt^{fxneo}$). Similar to the AHR-null mice, injection of contrast agent into the portal vein of the $Ahr^{fxneo/fxneo}$ mice revealed a direct shunt consistent with a patent ductus venosus; contrast agent injected into the portal vein flowed directly between the portal vein and the IVC [21]. The contrast agent was also found to enter the liver via the portal vein and immediately flowed through the ductus venosus into the IVC [22]. Aberrations in liver development observed in these mice were identical to those observed in mice harboring a null allele at the Ahr locus. Moreover, the trypan blue perfusion and autoradiography experiments demonstrated that two distinct mutations in the AHR signaling pathway (Ahr^{fxneo} and $Arnt^{fxneo}$) generate a qualitatively similar pattern of portocaval shunting to AHR-null mice. These findings define the temporal regulation of receptor activation during normal ontogeny and provide evidence to support the idea that receptor activation and AHR–ARNT heterodimerization are essential for normal vascular development.

Walisser et al. [23] also generated mice harboring a conditional allele of Ahr, designated as Ahr^{fx}. Conditional inactivation of the Ahr^{fx} allele was then accomplished by Cre-mediated deletion of exon 2 (designated as $Ahr^{fx/fx}$-Cre^{Tek}), which contains the region encoding the basic helix–loop–helix domain essential for DNA binding [24]. Liver angiography confirmed that the ductus venosus remained open in the majority of $Ahr^{fx/fx}Cre^{Tek}$ mice, whereas the $Ahr^{fx/fx}$ and $Ahr^{fx/fx}Cre^{Alb}$ mice showed normal liver perfusion and ductus venosus closure. Around 80% of the $Ahr^{fx/fx}Cre^{Tek}$ mice displayed a patent ductus venosus, a similar frequency to that seen in the Ahr and $Arnt$ mutants [22, 23]. A number of models proposed that AHR takes part in important signaling events that are independent of DRE binding or even ARNT dimerization [25, 26]. AHR may be involved in cellular signal transduction mechanisms independent of interactions with ARNT or DREs. Bunger et al. [27] generated a mouse model expressing AHR protein capable of ligand binding, interactions with chaperone proteins, functional heterodimerization with ARNT, and nuclear translocation, but is unable to bind DREs (designated as $Ahr^{dbd/dbd}$). The $Ahr^{dbd/dbd}$ mice had livers 25% smaller than wild-type littermate controls, similar to AHR-null mice. Histopathological analysis of livers taken from these $Ahr^{dbd/dbd}$ mice at postnatal days (PNDs) 7, 14, and 21 revealed a transient microvesicular steatosis around PND 7, which resolved by PND 14; this phenotype was identical to livers from age-matched AHR-null mice [27]. In the $Ahr^{dbd/dbd}$ mice, contrast agent also flowed directly from the portal vein to the IVC. The ductus venosus in the $Ahr^{dbd/dbd}$ mice was also clearly visible as a short segment running perpendicular to both the portal vein and the IVC. All $Ahr^{dbd/dbd}$ mice scored positive for this structure by trypan blue perfusion and showed a vascular pattern consistent with the frequency of patent ductus venosus seen in AHR-null mice. The $Ahr^{dbd/dbd}$ mice exhibit a patent ductus venosus, suggesting that DNA binding is necessary for AHR-mediated liver developmental signaling.

29.5 THE GENE EXPRESSION PROFILES ON THE AHR SIGNALING PATHWAY IN LIVERS EXPOSED TO DIOXIN

TCDD is a severely toxic environmental pollutant that causes various biological effects in mammals. Several studies have analyzed the gene expression profiles of both hepatocyte cell lines treated with TCDD using cDNA microarray [28–30] and gene expression profiling and *in vivo* material using serial analysis of gene expression (SAGE) [31] to identify the specific gene expression profiles induced by xenobiotics and to delineate the molecular mechanisms of toxicity. Frueh et al. [28] verified the regulation of proto-oncogene cot and human enhancer of filamentation-1, genes involved in cellular proliferation, as well as metallothionein and plasminogen activator inhibitor, genes involved in cellular signaling and regeneration, in human liver HepG2 cells exposed to TCDD. Other studies demonstrated that TCDD also altered the expression of a large array of genes involved in apoptosis, cytokine production, angiogenesis [29], cholesterol metabolism, bile acid biosynthesis, and bile transport [30], in addition to genes encoding drug metabolizing enzymes and stress responses, cytoskeleton-related proteins, signal transduction, and plasma proteins [31]. Yoon et al. [32] identified the genes involved in hepatotoxicity and hepatocarcinogenesis caused by a single treatment of TCDD at the dose of 100 mg/kg body weight in wild-type and AHR-null mice. Most of the genes were associated with chemotaxis, inflammation, carcinogenesis, acute-phase responses, immune responses, cell metabolism, cell proliferation, signal transduction, and tumor suppression. Other groups performed comparative toxicogenomic analysis of the hepatotoxic effects of TCDD between rats and mice [33, 34]. These data indicated that species-specific toxicity might be mediated by differences in gene expression, which could explain the wide range of species sensitivities and will have important implications in risk assessment strategies. Boutros et al. [35, 36] defined and compared transcriptional responses to dioxin exposure in the liver and kidney of wild-type and AHR-null adult mice, and found many genes affected by the Ahr genotype in both tissues. These data demonstrated the basal role of AHR in liver and kidney, and support a role for AHR in development and normal physiology [36].

29.6 THE ROLE OF AHR IN ANGIOGENESIS

Angiogenesis, the development of new blood vessels from preexisting vessels, is a tightly controlled physiological process that is also associated with pathological conditions such as tumor growth, diabetic retinopathy, and ischemic disease. Both hypoxia and inflammation contribute to the regulation of new vessel growth under ischemic conditions [37]. Vascular endothelial growth factor (VEGF) is a key angiogenic factor produced by ischemic tissues and growing tumors [38], and upregulation of VEGF expression at the transcriptional level is thought to induce progressive development of the collateral circulation [39]. Activation of the VEGF gene is mediated via binding of hypoxia-inducible factor-1α (HIF-1α) to the hypoxia response elements (HREs) present in the promoter region of the gene [40]. HIF-1α belongs to the Per–Arnt–Sim (PAS) family of basic helix–loop–helix transcription factors and dimerizes with another basic helix–loop–helix transcription factor, HIF-1β [41]. The AHR also contains a basic helix–loop–helix motif and functions together with the AHR nuclear translocator (ARNT), which is identical to HIF-1β [42]. Ichihara et al. [43] demonstrated that ischemia-induced angiogenesis was markedly enhanced in AHR-null mice compared to wild-type animals, using a surgical model of ischemia (Fig. 29.3a and b). When capillary density was evaluated as a measure of vascularity at the level of the microcirculation, collateral capillary density in the ischemic hindlimb was significantly greater for AHR-null mice than for wild-type animals (Fig. 29.3c and d). Ischemia induced upregulation of HIF-1α and ARNT expressions in AHR-null mice, as well as that of target gene for the transcription factors, VEGF. DNA binding activity of the HIF-1α–ARNT complex and association of HIF-1α and ARNT with the VEGF gene promoter was also increased by ischemia to a greater extent in AHR-null mice than in wild-type animals [43]. In addition, Thackaberry et al. [44] demonstrated cardiac hypertrophy at 5 months of age in AHR-null mice and increased expression of HIF-1α protein and VEGF mRNA in the heart. Other reports also demonstrated that AHR negatively affects angiogenesis in cancer using AHR-null C57BL/6J transgenic adenocarcinoma of the mouse prostate (TRAMP) mice [45, 46]. The malignant prostate tumors rarely developed in wild-type TRAMP mice, but the incidence was greatly increased in the heterozygous and AHR-null TRAMP mice [45]. The expression of molecular markers of neuroendocrine differentiation, including chromogranin A and neurophlin-1, was elevated in prostate tumors compared to tumor-free ventral prostates, suggesting that AHR inhibits prostate carcinogenesis in the C57BL/6J TRAMP mice by interfering with neuroendocrine differentiation. Vanadate, an inducer of VEGF through the HIF-1α–ARNT pathway, also induced VEGF protein in a dose-dependent fashion in heterozygous and AHR-null TRAMP cultures, but not in wild-type cultures [46]. These results suggested that AHR, when present, would interact with ARNT, thus minimizing the ability of ARNT to interact with stabilized HIF-1 to induce VEGF production. However, when the AHR is absent, ARNT would be free to interact with stabilized HIF-1, resulting in VEGF production and a greater likelihood of tumor growth.

The effects of benzo[a]pyrene (B[a]P) in cigarette smoke and of other polycyclic and halogenated aromatic hydrocarbons in the environment are mediated through AHR [47]. To determine the effect of smoking on angiogenesis, Ichihara et al. [48] treated mice with B[a]P. At a dose of 125 mg/kg per week, oral exposure to B[a]P significantly inhibited the increase in blood flow induced by hypoxia in wild-type mice. This observation is consistent with previous results [49] showing that mice exposed to cigarette smoke show markedly impaired angiogenesis in response to surgically induced hindlimb ischemia. The amounts of interleukin-6 and of VEGF mRNA in the ischemic hindlimb of wild-type mice were reduced by exposure to B[a]P. These various effects of B[a]P are markedly attenuated in AHR-null mice, and this could be attributed to maintenance of interleukin-6 expression and consequent promotion of angiogenesis through upregulation of VEGF. Furthermore, TCDD or 3-methylcholanthrene decreased the VEGF expression under several experimental conditions, such as in the case of coronary endothelial tube formation in chicken embryos [50, 51] and human umbilical vein endothelial cells *in vitro* [52]. These further experiments suggest that the enhancement of angiogenesis with the associated enhancement of ischemia-induced VEGF expression in AHR-null mice or the AHR-dependent reduction of VEGF expression by TCDD and other AHR ligands may result in a tilted balance toward the AHR/ARNT pathway instead of the HIF-1α/ARNT pathway. On the other hand, the essential chaperones or transcriptional coactivators with which the AHR–ARNT heterodimer interacts, including the 90 kDa heat shock protein (Hsp90), the cAMP response element binding protein (CREB) binding protein (CBP), and steroid receptor coactivator-1 (Src-1), also modulate gene transcription mediated by HIF-1α–ARNT [53]. Increased availability of those cofactors common to both the HIF-1α pathway and the disrupted AHR signaling cascade might contribute to the increased transactivation activity of HIF-1α–ARNT apparent in AHR-null mice. Given that the hypoxia and dioxin response pathways compete for limiting cellular factors, the dimerization of ARNT with HIF-1α and consequent activation of genes involved in the regulation of adaptation to low oxygen tension are likely facilitated in AHR-null mice. However, Pollenz et al. [54] showed that activation of one pathway did not inhibit signaling by the other pathway as a result of competitive dimerization of ARNT with AHR or HIF-1α *in vitro*. The AHR pathway also interferes with other ARNT-independent pathways, such as the progesterone- and estrogen-responsive pathways [55], suggesting that competition for other cofactors might also account for this crosstalk.

29.7 THE ROLE OF AHR IN THE PLACENTA AND ENDOTHELIAL CELLS

The maintenance of the placental vasculature is essential for sustaining normal fetal growth. Ishimura et al. [56] demonstrated that fetal death was accompanied by placental

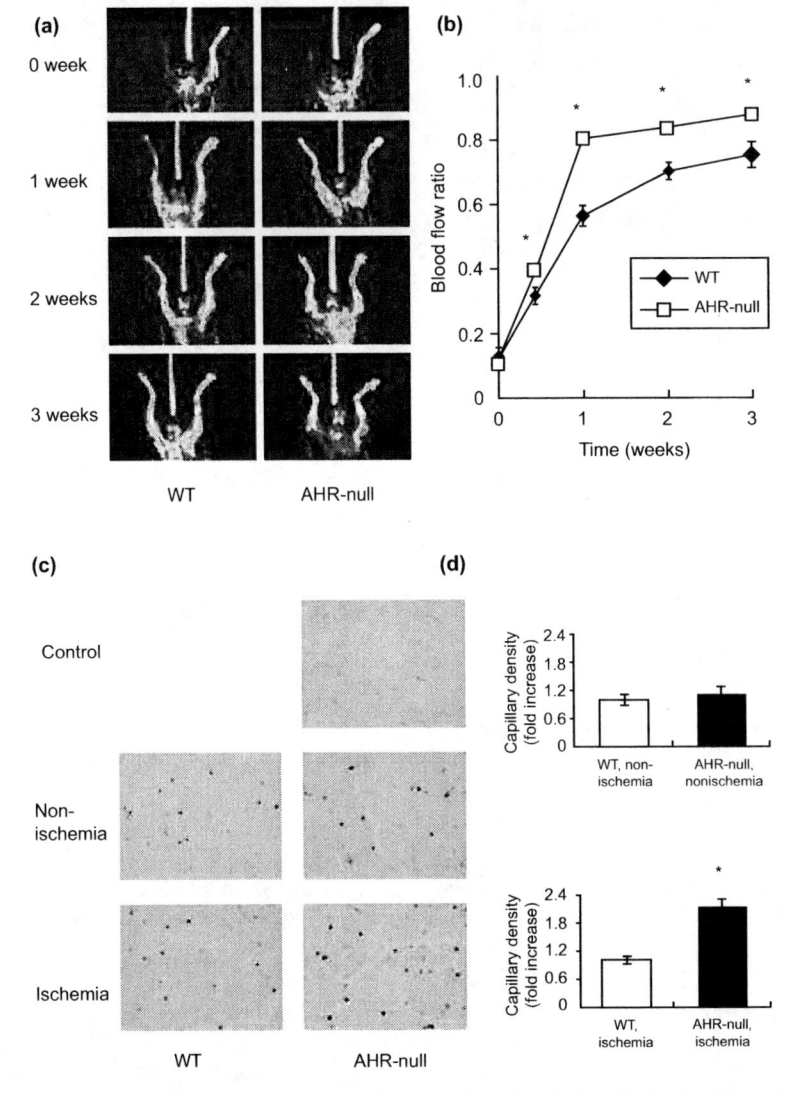

FIGURE 29.3 Laser Doppler blood flow analysis for the ischemic hindlimb of AHR-null or wild-type mice. (a) Laser Doppler perfusion imaging of blood flow in the ischemic hindlimb measured immediately (time 0), 1, 2, and 3 weeks after surgery. Blood flow is color coded, with normal perfusion indicated by white and a marked reduction in blood flow indicated by gray. (b) Quantitation of blood flow expressed as the ratio of blood flow in the ischemic (left) hindlimb to that in the normal (right) hindlimb. Data are mean ± SEM of values from eight animals per group. *$P < 0.05$ versus corresponding value for wild-type mice. Capillary density in skeletal muscle of the ischemic or nonischemic hindlimb of AHR-null or wild-type mice. (c) Immunostaining of ischemic or nonischemic tissue with antibodies to CD31 at 3 weeks after arterial ligation. The upper panel (control) represents ischemic tissue from an AHR-null mouse stained with secondary antibodies only. Scale bar = 100 μm. (d) Quantitation of capillary density in ischemic or nonischemic tissue at 3 weeks after surgery. Data are mean ± SEM of values from eight animals per group. *$P < 0.05$ versus wild-type mice. (See the color version of this figure in Color Plates section.)

hypoxia upon exposure to TCDD. In order to evaluate the effects of low-dose TCDD on placental function, they analyzed two comprehensive methods, representational difference analysis and DNA microarray technology [57]. Then, they showed that activation of the interferon signaling pathway and induction of antiangiogenesis factors by TCDD might have a role in causing the inhibition of neovascularization, resulting in the hypoxia state of placenta and increased incidence of fetal death. They also showed that lack of dilatation of both maternal blood sinusoids and fetal capillaries, existence of large size trophoblast cells, and the downregulated Tie2 mRNA level among the VEGF/VEGFR and Ang/Tie2 systems were observed in TCDD-exposed placentas, suggesting that vascular remodeling was suppressed by TCDD exposure [58]. Furthermore, TCDD suppressed the apoptosis of trophoblast cells with a concomitant increase in the incidence of fetal death under hypoxia condition. There results indicated that crosstalk between the HIF-mediated pathway and AHR-mediated pathway is considered to play an important role in the vascular remodeling.

By generating a targeted disruption of the *Arnt* locus in the mouse, Maltepe et al. [59] showed that *Arnt*-null embryonic stem cells fail to activate genes that normally respond to low oxygen tension. *Arnt*-null embryonic stem cells also failed to respond to a decrease in glucose concentration, indicating that ARNT is crucial in the response to hypoxia and to hypoglycemia. Several studies implicate AHR would regulate endothelial function. Dabir et al. [60] demonstrated that AHR was expressed in aortic endothelial cells (ECs), activated, and bound to the promoter in response to high-glucose stimulation of ECs. The constitutively active form of AHR induced activation of the thrombospondin-1 gene promoter. In response to high-glucose stimulation, AHR was found in complex with Egr-1 and activator protein-2, which are two other nuclear transcription factors activated by glucose in ECs that have not been previously detected in complex with AHR. The activity of the DNA binding complex was regulated by glucose through the activation of hexosamine pathway and intracellular glycosylation. These observations suggested that the AHR was activated by high glucose that links AHR to the physiological regulation of gene expression by glucose and the pathological effects of hyperglycemia in the vasculature.

29.8 CONCLUSIONS

AHR has been studied as a receptor for environmental contaminants and as a mediator of chemical toxicity. However, additional roles for AHR in normal vascular development were recently identified. In fetal AHR-null mice, the ductus venosus remains patent, resulting in significant portocaval shunting and liver atrophy. Moreover, abnormal hepatic circulation, characterized by anastomotic sinusoidal vessels, leads to decreased perfusion and necrosis of the liver periphery in this strain. Given that the VEGF system is upregulated by HIF-1α/ARNT under hypoxic conditions during vasculogenesis, leading to activation of the Ang/Tie2 system in vascular remodeling, it seems that ischemia-induced angiogenesis is markedly enhanced in AHR-null mice. The vascular remodeling in the placenta was suppressed by TCDD exposure. The inhibitory effect of the AHR-mediated signaling pathway on HIF-1α signaling is thus due to downregulation of active HIF-1α. The findings discussed indicate that AHR plays a fundamental role in liver development and in ischemia-induced angiogenesis.

REFERENCES

1. Swanson, H. I. and Bradfield, C. A. (1993). The AH-receptor: genetics, structure and function. *Pharmacogenetics*, 3, 213–230.
2. Dolwick, K. M., Schmidt, J. V., Carver, L. A., Swanson, H. I., and Bradfield, C. A. (1993). Cloning and expression of a human Ah receptor cDNA. *Molecular Pharmacology*, 44, 911–917.
3. Hayashi, S., Watanabe, J., Nakachi, K., Eguchi, H., Gotoh, O., and Kawajiri, K. (1994). Interindividual difference in expression of human Ah receptor and related P450 genes. *Carcinogenesis*, 15, 801–806.
4. Schmidt, J. V., Su, G. H., Reddy, J. K., Simon, M. C., and Bradfield, C. A. (1996). Characterization of a murine Ahr null allele: involvement of the Ah receptor in hepatic growth and development. *Proceedings of the National Academy of Sciences of the United States of America*, 93, 6731–6736.
5. Fernandez-Salguero, P. M., Pineau, T., Hilbert, D. M., McPhail, T., Lee, S. S., Kimura, S., Nebert, D. W., Rudikoff, S., Ward, J. M., and Gonzalez, F. J. (1995). Immune system impairment and hepatic fibrosis in mice lacking the dioxin-binding Ah receptor. *Science*, 268, 722–726.
6. Mimura, J., Yamashita, K., Nakamura, K., Morita, M., Takagi, T. N., Nakao, K., Ema, M., Sogawa, K., Yasuda, M., Katsuki, M., and Fujii-Kuriyama, Y. (1997). Loss of teratogenic response to 2,3,7,8-tetrachlorodibenzo-*p*-dioxin (TCDD) in mice lacking the Ah (dioxin) receptor. *Genes to Cells*, 2, 645–654.
7. Fernandez-Salguero, P. M., Ward, J. M., Sundberg, J. P., and Gonzalez, F. J. (1997). Lesions of aryl-hydrocarbon receptor-deficient mice. *Veterinary Pathology*, 34, 605–614.
8. Lahvis, G. P., Lindell, S. L., Thomas, R. S., McCuskey, R. S., Murphy, C., Glover, E., Bentz, M., Southard, J., and Bradfield, C. A. (2000). Portosystemic shunting and persistent fetal vascular structures in aryl hydrocarbon receptor-deficient mice. *Proceedings of the National Academy of Sciences of the United States of America*, 97, 10442–10447.
9. Dubuisson, L., Bioulac-Sage, P., Bedin, C., and Balabaud, C. (1984). Hepatocyte ultrastructure in the rat after long term portacaval anastomosis: a morphometric study. *Journal of Submicroscopical Cytology*, 16, 283–287.
10. Schröder, R., Müller, O., and Bircher, J. (1985). The portacaval and splenocaval shunt in the normal rat. A morphometric and functional reevaluation. *Journal of Hepatology*, 1, 107–123.
11. Lahvis, G. P., Pyzalski, R. W., Glover, E., Pitot, H. C., McElwee, M. K., and Bradfield, C. A. (2005). The aryl hydrocarbon receptor is required for developmental closure of the ductus venosus in the neonatal mouse. *Molecular Pharmacology*, 67, 714–720.
12. Harstad, E. B., Guite, C. A., Thomae, T. L., and Bradfield, C. A. (2006). Liver deformation in *Ahr*-null mice: evidence for aberrant hepatic perfusion in early development. *Molecular Pharmacology*, 69, 1534–1541.
13. Ko, M. K., Chi, J. G., and Chang, B. L. (1985). Hyaloid vascular pattern in the human fetus. *Journal of Pediatric Ophthalmology & Strabismus*, 22, 188–193.

14. Oberhammer, F. A., Pavelka, M., Sharma, S., Tiefenbacher, R., Purchio, A. F., Bursch, W., and Schulte-Hermann, R. (1992). Induction of apoptosis in cultured hepatocytes and in regressing liver by transforming growth factor β1. *Proceedings of the National Academy of Sciences of the United States of America*, 89, 5408–5412.

15. Bernasconi, P., Torchiana, E., Confalonieri, P., Brugnoni, R., Barresi, R., Mora, M., Correlio, F., Morandi, L., and Mantegazza, R. (1995). Expression of transforming growth factor-β1 in dystrophic patient muscle correlates with fibrosis. *Journal of Clinical Investigation*, 96, 1137–1144.

16. Sanderson, N., Factor, V., Nagy, P., Kopp, J., Kondaiah, P., Wakefield, L., Roberts, A. B., Sporn, M. B., and Thorgeirsson, S. S. (1995). Hepatic expression of mature transforming growth factor β1 in transgenic mice results in multiple tissue lesions. *Proceedings of the National Academy of Sciences of the United States of America*, 92, 2572–2576.

17. Zaher, H., Fernandez-Salguero, P. M., Letterio, J., Sheikh, M. S., Fornace, A. J., Jr., Roberts, A. B., and Gonzalez, F. J. (1998). The involvement of aryl hydrocarbon receptor in the activation of transforming growth factor-β and apoptosis. *Molecular Pharmacology*, 54, 313–321.

18. Andreola, F., Fernandez-Salguero, P. M., Chiantore, M. V., Petkovich, M. P., Gonzalez, F. J., and De Luca, L. M. (1997). Aryl hydrocarbon receptor knockout mice (AHR$^{-/-}$) exhibit liver retinoid accumulation and reduced retinoic acid metabolism. *Cancer Research*, 57, 2835–2838.

19. Andreola, F., Calvisi, D. F., Elizondo, G., Jakowlew, S. B., Mariano, J., Gonzalez, F. J., and De Luca, L. M. (2004). Reversal of liver fibrosis in aryl hydrocarbon receptor null mice by dietary vitamin A depletion. *Hepatology*, 39, 157–166.

20. Bunger, M. K., Moran, S. M., Glover, E., Thomae, T. L., Lahvis, G. P., Lin, B. C., and Bradfield, C. A. (2003). Resistance to 2,3,7,8-tetrachlorodibenzo-p-dioxin toxicity and abnormal liver development in mice carrying a mutation in the nuclear localization sequence of the aryl hydrocarbon receptor. *Journal of Biological Chemistry*, 278, 17767–17774.

21. Walisser, J. A., Bunger, M. K., Glover, E., and Bradfield, C. A. (2004). Gestational exposure of *Ahr* and *Arnt* hypomorphs to dioxin rescues vascular development. *Proceedings of the National Academy of Sciences of the United States of America*, 101, 16677–16682.

22. Walisser, J. A., Bunger, M. K., Glover, E., Harstad, E. B., and Bradfield, C. A. (2004). Patent ductus venosus and dioxin resistance in mice harboring a hypomorphic *Arnt* allele. *Journal of Biological Chemistry*, 279, 16326–16331.

23. Walissern, J. A., Glover, E., Pande, K., Liss, A. L., and Bradfield, C. A. (2005). Aryl hydrocarbon receptor-dependent liver development and hepatotoxicity are mediated by different cell types. *Proceedings of the National Academy of Sciences of the United States of America*, 102, 17858–17863.

24. Fukunaga, B. N., Probst, M. R., Reisz-Porszasz, S., and Hankinson, O. (1995). Identification of functional domains of the aryl hydrocarbon receptor. *Journal of Biological Chemistry*, 270, 29270–29278.

25. Ge, N. L. and Elferink, C. J. (1998). A direct interaction between the aryl hydrocarbon receptor and retinoblastoma protein. Linking dioxin signaling to the cell cycle. *Journal of Biological Chemistry*, 273, 22708–22713.

26. Puga, A., Barnes, S. J., Dalton, T. P., Chang, C., Knudsen, E. S., and Maier, M. A. (2000). Aromatic hydrocarbon receptor interaction with the retinoblastoma protein potentiates repression of E2F-dependent transcription and cell cycle arrest. *Journal of Biological Chemistry*, 275, 2943–2950.

27. Bunger, M. K., Glover, E., Moran, S. M., Walisser, J. A., Lahvis, G. P., Hsu, E. L., and Bradfield, C. A. (2008). Abnormal liver development and resistance to 2,3,7,8-tetrachlorodibenzo-p-dioxin toxicity in mice carrying a mutation in the DNA-binding domain of the aryl hydrocarbon receptor. *Toxicological Sciences*, 106, 83–92.

28. Frueh, F. W., Hayashibara, K. C., Brown, P. O., and Whitlock, J. P., Jr (2001). Use of cDNA microarrays to analyze dioxin-induced changes in human liver gene expression. *Toxicology Letters*, 6, 189–203.

29. Zeytun, A., McKallip, R. J., Fisher, M., Camacho, I., Nagarkatti, M., and Nagarkatti, P. S. (2002). Analysis of 2,3,7,8-tetrachlorodibenzo-p-dioxin-induced gene expression profile *in vivo* using pathway-specific cDNA arrays. *Toxicology*, 178, 241–260.

30. Fletcher, N., Wahlstrom, D., Lundberg, R., Nilsson, C. B., Nilsson, K. C., Stockling, K., Hellmold, H., and Håkansson, H. (2005). 2,3,7,8-Tetrachlorodibenzo-p-dioxin (TCDD) alters the mRNA expression of critical genes associated with cholesterol metabolism, bile acid biosynthesis, and bile transport in rat liver: a microarray study. *Toxicology and Applied Pharmacology*, 207, 1–24.

31. Kurachi, M., Hashimoto, S., Obata, A., Nagai, S., Nagahata, T., Inadera, H., Sone, H., Tohyama, C., Kaneko, S., Kobayashi, K., and Matsushima, K. (2002). Identification of 2,3,7,8-tetrachlorodibenzo-p-dioxin-responsive genes in mouse liver by serial analysis of gene expression. *Biochemical and Biophysical Research Communications*, 292, 368–377.

32. Yoon, C. Y., Park, M., Kim, B. H., Park, J. Y., Park, M. S., Jeong, Y. K., Kwon, H., Jung, H. K., Kang, H., Lee, Y. S., and Lee, B. J. (2006). Gene expression profile by 2,3,7,8-tetrachlorodibenzo-p-dioxin in the liver of wild-type (AhR$^{+/+}$) and aryl hydrocarbon receptor-deficient (AhR$^{-/-}$) mice. *Journal of Veterinary Medical Science*, 68, 663–668.

33. Boverhof, D. R., Burgoon, L. D., Tashiro, C., Sharratt, B., Chittim, B., Harkema, J. R., Mendrick, D. L., and Zacharewski, T. R. (2006). Comparative toxicogenomic analysis of the hepatotoxic effects of TCDD in Sprague Dawley rats and C57BL/6 mice. *Toxicological Sciences*, 94, 398–416.

34. Boutros, P. C., Yan, R., Moffat, I. D., Pohjanvirta, R., and Okey, A. B. (2008). Transcriptomic responses to 2,3,7,8-tetrachlorodibenzo-p-dioxin (TCDD) in liver: comparison of rat and mouse. *BMC Genomics*, 9, 419–435.

35. Tijet, N., Boutros, P. C., Moffat, I. D., Okey, A. B., Tuomisto, J., and Pohjanvirta, R. (2006). Aryl hydrocarbon receptor regulates distinct dioxin-dependent and dioxin-independent gene batteries. *Molecular Pharmacology*, 69, 140–153.

36. Boutros, P. C., Bielefeld, K. A., Pohjanvirta, R., and Harper, P. A. (2009). Dioxin-dependent and dioxin-independent gene batteries: comparison of liver and kidney in AHR-null mice. *Toxicological Sciences, 112,* 245–256.
37. Carmeliet, P. (2000). Mechanisms of angiogenesis and arteriogenesis. *Nature Medicine, 6,* 389–395.
38. Ferrara, N., Houck, K., Jakeman, L., and Leung, D. W. (1992). Molecular and biological properties of the vascular endothelial growth factor family of proteins. *Endocrine Reviews, 13,* 18–32.
39. Takeshita, S., Zheng, L. P., Brogi, E., Kearney, M., Pu, L.-Q., Bunting, S., Ferrara, N., Symes, J. F., and Isner, J. M. (1994). Therapeutic angiogenesis: a single intraarterial bolus of vascular endothelial growth factor augments revascularization in a rabbit ischemic hind limb model. *Journal of Clinical Investigation, 93,* 662–670.
40. Forsythe, J. A., Jiang, B. H., Iyer, N. V., Agani, F., Leung, S. W., Koos, R. D., and Semenza, G. L. (1996). Activation of vascular endothelial growth factor gene transcription by hypoxia-inducible factor 1. *Molecular and Cellular Biology, 16,* 4604–4613.
41. Wang, G. L., Jiang, B.-H., Rue, E. A., and Semenza, G. L. (1995). Hypoxia-inducible factor 1 is a basic-helix–loop–helix–PAS heterodimer regulated by cellular O_2 tension. *Proceedings of the National Academy of Sciences of the United States of America, 92,* 5510–5514.
42. Reyes, H., Reisz-Porszansz, S., and Hankinson, O. (1992). Identification of the Ah receptor nuclear translocator protein (Arnt) as a component of the DNA binding form of the Ah receptor. *Science, 256,* 1193–1195.
43. Ichihara, S., Yamada Y, Ichihara G, Nakajima T, and Murahara T. (2007). A role of the aryl hydrocarbon receptor in regulation of ischemia-induced angiogenesis. *Arteriosclerosis, Thrombosis, and Vascular Biology, 27,* 1297–1304.
44. Thackaberry, E. A., Smith, S. M., Gabaldno, D. M., and Walker, M. K. (2002). Aryl hydrocarbon receptor null mice develop cardiac hypertrophy and increased hypoxia-inducible factor-1α in the absence of cardiac hypoxia. *Cardiovascular Toxicology, 2,* 263–273.
45. Fritz, W. A., Lin, T. M., Cardiff, R. D., and Peterson, R. E. (2007). The aryl hydrocarbon receptor inhibits prostate carcinogenesis in TRAMP mice. *Carcinogenesis, 28,* 497–505.
46. Fritz, W. A., Lin, T. M., and Peterson, R. E. (2008). The aryl hydrocarbon receptor (AhR) inhibits vanadate-induced vascular endothelial growth factor (VEGF) production in TRAMP prostates. *Carcinogenesis, 29,* 1077–1082.
47. Schmidt, J. V. and Bradfield, C. A. (1996). Ah receptor signaling pathways. *Annual Review of Cell and Developmental Biology, 12,* 55–89.
48. Ichihara, S., Yamada, Y., Gonzalez, F. J., Nakajima, T., Murohara, T., and Ichihara, G. (2009). Inhibition of ischemia-induced angiogenesis by benzo[*a*]pyrene in a manner dependent on the aryl hydrocarbon receptor. *Biochemical and Biophysical Research Communications, 381,* 44–49.
49. Michaud, S.-É., Ménard, C., Guy, L.-G., Gennaro, G., and Rivard, A. (2003). Inhibition of hypoxia-induced angiogenesis by cigarette smoke exposure: impairment of the HIF-1α/VEGF pathway. *FASEB Journal, 17,* 1150–1152.
50. Ivnitski-Steele, I. D., Sanchez, A., and Walker, M. K. (2004). 2,3,7,8-Tetrachlorodibenzo-*p*-dioxin reduces myocardial hypoxia and vascular endothelial growth factor expression during chick embryo development. *Birth Defects Research, Part A, 70,* 51–58.
51. Ivnitski-Steele, I. D., Friggens, M., Chavez, M., and Walker, M. K. (2005). 2,3,7,8-Tetrachlorodibenzo-*p*-dioxin (TCDD) inhibition of coronary vasculogenesis is mediated, in part, by reduced responsiveness to endogenous angiogenic stimuli, including vascular endothelial growth factor A (VEGF-A). *Birth Defects Research, Part A, 73,* 440–446.
52. Juan, S. H., Lee, J. L., Ho, P. Y., Lee, Y. H., and Lee, W. S. (2006). Antiproliferative and antiangiogenic effects of 3-methylcholanthrene, an aryl-hydrocarbon receptor agonist, in human umbilical vascular endothelial cells. *European Journal of Pharmacology, 530,* 1–8.
53. Hankinson, O. (2005). Role of coactivators in transcriptional activation by the aryl hydrocarbon receptor. *Archives of Biochemistry and Biophysics, 433,* 379–386.
54. Pollenz, R. S., Davarinos, N. A., and Shearer, T. P. (1999). Analysis of aryl hydrocarbon receptor-mediated signaling during physiological hypoxia reveals lack of competition for the aryl hydrocarbon nuclear translocator transcription. *Molecular Pharmacology, 56,* 1127–1137.
55. Wormke, M., Stoner, M., Saville, B., and Safe, S. (2000). Crosstalk between estrogen receptor α and the aryl hydrocarbon receptor in breast cancer cells involves unidirectional activation of proteasomes. *FEBS Letters, 478,* 109–112.
56. Ishimura, R., Ohsako, S., Kawakami, T., Sakaue, M., Aoki, Y., and Tohyama, C. (2002). Altered protein profile and possible hypoxia in the placenta of 2,3,7,8-tetrachlorodibenzo-*p*-dioxin-exposed rats. *Toxicology and Applied Pharmacology, 185,* 197–206.
57. Mizutani, T., Yoshino, M., Satake, T., Nakagawa, M., Ishimura, R., Tohyama, C., Kokame, K., Kangawa, K., and Miyamoto, K. (2004). Identification of 2,3,7,8-tetrachlorodibenzo-*p*-dioxin (TCDD)-inducible and -suppressive genes in the rat placenta: induction of interferon-regulated genes with possible inhibitory roles for angiogenesis in the placenta. *Endocrine Journal, 51,* 69–77.
58. Ishimura, R., Kawakami, T., Ohsako, S., Nohara, K., and Tohyama, C. (2006). Suppressive effect of 2,3,7,8-tetrachlorodibenzo-*p*-dioxin on vascular remodeling that takes place in the normal labyrinth zone of rat placenta during late gestation. *Toxicological Sciences, 91,* 265–274.
59. Maltepe, E., Schmidt, J. V., Baunoch, D., Bradfield, C. A., and Simon, M. C. (1997). Abnormal angiogenesis and response to glucose and oxygen deprivation in mice lacking the protein ARNT. *Nature, 386,* 403–407.
60. Dabir, P., Marinic, T. E., Krukovets, I., and Stenina, O. I. (2008). Aryl hydrocarbon receptor is activated by glucose and regulates the thrombospondin-1 gene promoter in endothelial cells. *Circulation Research, 102,* 1558–1565.

30

INVOLVEMENT OF THE AHR IN CARDIAC FUNCTION AND REGULATION OF BLOOD PRESSURE

JASON A. SCOTT AND MARY K. WALKER

30.1 INTRODUCTION

The role of AHR in the cardiovascular system is relatively understudied in comparison to its role in other systems and in response to xenobiotics. It has been long known that the activation of AHR by xenobiotics produces cardiovascular toxicity in a variety of species (reviewed in Ref. 1 and described elsewhere in this book); however, it was not until relatively recently that a direct role of AHR in cardiovascular development and homeostasis had been described. Genetic mouse models have provided a glimpse into the cardiovascular phenotypes produced in the absence of AHR, both global and endothelial-specific, and with the loss of a single AHR allele. Together with flow-mediated responses in AHR signaling *in vitro*, AHR appears to play a role in mediating vascular development, cardiac and vascular endothelial cell homeostasis, and blood pressure regulation. However, due to the abnormalities in vascular development and the sensitized response to hypoxia in adulthood in AHR knockout models, the definitive roles of the AHR in the cardiovascular system have not been completely elucidated.

30.2 BACKGROUND: CARDIOVASCULAR PHYSIOLOGY

The mechanisms and pathways involved in regulating cardiovascular function are highly complex and are beyond the scope of this chapter. However, we have briefly outlined various physiological pathways relevant to the role of AHR in the cardiovascular system to provide adequate background for those readers who are less familiar with cardiovascular physiology and its regulation.

30.2.1 Sympathetic Nervous System

The sympathetic nervous system (SNS) is a key mediator of total peripheral resistance in the vasculature and is essential in the control of blood pressure. Arterial baroreceptors, chemoreceptors, and cardiopulmonary receptors provide a constant sensory input of peripheral blood pressure and chemistry through communication with integration centers in the CNS, including the medullary cardiovascular centers in the medulla oblongata (reviewed in Ref. 2). The sensory information is then passed to the peripheral vasculature via an efferent pathway through ganglia. The efferent SNS neural fibers innervate arterioles and cause vasoconstriction via the release of norepinephrine. Norepinephrine interacts with α_1-adrenoceptors on vascular smooth muscle cells resulting in a phospholipase C/inositol triphosphate (IP$_3$)-mediated increase in intracellular Ca^{2+} and vasoconstriction. An increase in SNS activity also increases cardiac contractility and heart rate, which increase cardiac output, and activates the renin–angiotensin system (RAS) in an effort to increase blood pressure. In addition, angiotensin (Ang) II, an active component of the RAS (see below), can increase SNS via Ang II receptor binding sites in medullary cardiovascular centers [3]. An overactive SNS or altered afferent inputs have been associated with hypertension and heart failure in humans [2].

30.2.2 Endothelin-1

Endothelin-1 (ET-1) is a potent vasoconstrictor and cardiac hypertrophic peptide that has been linked to the pathogenesis

The AH Receptor in Biology and Toxicology, First Edition. Edited by Raimo Pohjanvirta.
© 2012 John Wiley & Sons, Inc. Published 2012 by John Wiley & Sons, Inc.

of cardiovascular dysfunction and hypertension. ET-1 is predominantly produced in the vascular endothelium, but is also induced in cardiac tissue, lung, fibroblasts, and the nervous system. It is induced by a variety of stimuli including Ang II, inflammatory cytokines, hypoxia, norepinephrine, and shear stress [4]. The vasoconstrictive properties of ET-1 are mediated by the activation of ET_A and ET_B receptors on vascular smooth muscle cells. ET_B receptors are also located on the vascular endothelium and function to counteract vasoconstriction by releasing vasodilators (nitric oxide (NO) and cyclooxygenase metabolites) and removing ET-1 from circulation (reviewed in Ref. 5). An elevation of circulating and/or endothelium-derived ET-1 has been linked to several forms of cardiovascular disease, including essential and pulmonary hypertension, chronic heart failure, atherosclerosis, and renal vascular dysfunction in chronic kidney disease [5].

30.2.3 Renin–Angiotensin System

The RAS is a complex axis that plays an integral role in the regulation of basal blood pressure and the pathogenesis of hypertension. Upon activation of the RAS, renin is released from granular cells in the kidney, which converts circulating angiotensinogen to Ang I. In the classic pathway, angiotensin converting enzyme (ACE), released from pulmonary endothelial cells, then converts Ang I to Ang II, a potent vasoconstrictive peptide that acts on the SNS, kidneys, pituitary gland, and arterioles to increase blood pressure, cardiac output, and water retention. In recent years, however, the complexity of the RAS has truly emerged. Localized tissue RAS has been identified in numerous organs, including heart, brain, adipose tissue, and vessels, and evidence for an intracellular RAS has been reported. Further, the discovery of ACE2, neutral endopeptidase, and chymase, which catalyze the formation of other vasoactive Ang isoforms, has led to the identification of key biologically active mediators of the RAS, including Ang 1–12 and Ang 1–7. Finally, recent discovery of (pro)renin receptor has added an additional layer of complexity to the once simplistic classic view of the RAS [6]. The RAS is activated by reduced blood pressure, blood volume, plasma sodium, and kidney perfusion; however, circulating levels of Ang II can be influenced by other factors, including ET-1. It has been suggested that ET-1 can induce an ET_A-mediated increase in plasma renin activity and ACE activity [7–9]. The mechanism of Ang II-mediated vasoconstriction is predominantly mediated by activation of IP_3 and an increase in intracellular Ca^{2+} concentration, but is also influenced by NAD(P)H oxidase (Nox)-derived reactive oxygen species (ROS) [10]. Ang II is also known to increase expression of vasoconstrictive vascular adrenergic receptors (α_{1D}-adrenoceptors), which have been implicated in hypertension [11, 12]. Furthermore, Ang II can have growth promoting effects on cardiac myocytes and fibroblasts.

30.2.4 Vascular Nitric Oxide and Endothelial NO Synthase

In vascular endothelial cells, constitutively expressed endothelial NO synthase (eNOS) produces basal levels of NO, an endothelial-derived relaxing factor, which plays a key role in regulating blood pressure (reviewed in Ref. 13). eNOS is a calcium/calmodulin-dependent enzyme with a heme center that uses (6R)-5,6,7,8-tetrahydrobiopterin (BH_4) as a cofactor to catalyze NO production from L-arginine. NO diffuses from the endothelium to vascular smooth muscle cells and binds to soluble guanylate cyclase, which catalyzes the formation of cyclic guanine monophosphate (cGMP). cGMP activates cGMP-dependent protein kinases, which results in vasodilation via a decrease in intracellular Ca^{2+} concentrations and myosin light chain activity, and/or phosphorylation of myosin light chain phosphatase. Altered NO signaling results in loss of endothelial-derived vasorelaxation, termed "endothelial dysfunction," which has been implicated in the development of hypertension and atherosclerosis (reviewed in Ref. 13). In addition to being a vasorelaxant, NO inhibits vascular smooth muscle proliferation [14].

30.2.5 Shear Stress

Shear stress (SS) is the frictional force imposed on the vascular endothelium as blood flows through vessels. SS is an important hemodynamic stress in regulating vascular tone and homeostasis. Laminar SS (10–30 dyn/cm^2) occurs in linear regions of the vasculature and elicits a response in vascular endothelial cells that is generally considered to be antiatherogenic, antithrombotic, antiproliferative, and anti-inflammatory (reviewed in Ref. 15). Turbulent blood flow near branching or partially occluded areas creates oscillatory low shear forces, which are linked to endothelial dysfunction and development of atherosclerosis. The vasoprotective effect of laminar SS is mediated, in part, by an upregulation and activation of eNOS and a subsequent increase in NO production (reviewed in Ref. 16), and a decrease in the production of ET-1 and ACE [17, 18].

30.3 LOCALIZATION OF AHR IN THE CARDIOVASCULAR SYSTEM

Due to the low levels of constitutively expressed AHR in cardiovascular tissue, the localization of AHR has been inferred indirectly from the inducible expression of AHR-responsive genes (e.g., *CYP1A1*). Using radiolabeled substrates for CYP1A1 after induction with AHR ligands, CYP1A1 appears to be present in several blood vessels of the heart, kidneys, and liver *in vivo* [19, 20]. CYP1A1 is induced in coronary arteries, capillaries, and veins of heart; portal veins of the liver; and afferent and efferent arteries, and glomerular and peritubular capillaries in kidneys [19, 20]. Notably, in these studies, CYP1A1 is located exclusively in the endothelium in these vessels. Similarly, inducible CYP1A1 protein is confined to the endothelium of mouse aorta and mesenteric arterioles [21], and is detected in primary endothelial cells of cerebral arteries [22]. However, CYP1A1 protein has also been reported in human coronary smooth muscle cells *in vitro* [23], and CYP1A1 and CYP1B1 are inducible in murine vascular smooth muscle cells [24]. Constitutive CYP1A1 mRNA expression is also detected in the left ventricles of rats and humans [25, 26].

Thus, the indirect evidence of AHR localization suggests that AHR may be ubiquitously expressed in the cardiovascular system; however, there is overwhelming evidence that AHR is predominantly located in the vascular endothelium. Given the critical role of endothelial tissue in cardiovascular function, it is not surprising that a role for AHR in cardiovascular homeostasis and the pathogenesis of cardiovascular disease has been suggested.

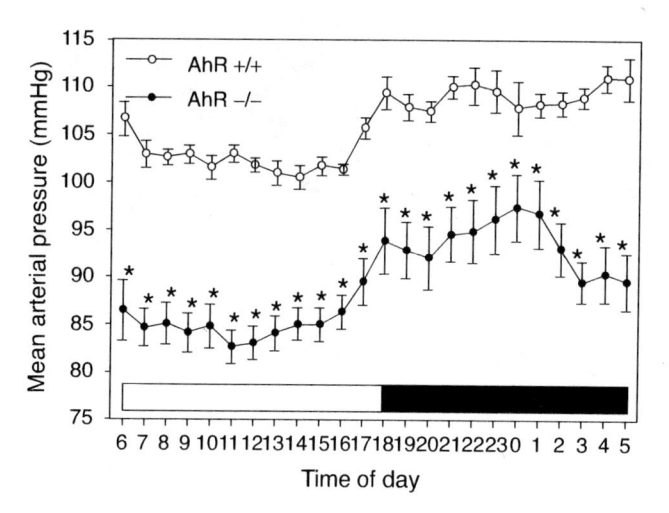

FIGURE 30.1 Mean arterial pressure of 3–4-month-old $Ahr^{+/+}$ and $Ahr^{-/-}$ mice over a 24 h period ($n = 8–10$ per genotype). Data represent mean ± SEM and were analyzed by repeated-measures, two-way ANOVA. $^*p < 0.05$, compared to $Ahr^{+/+}$. Reprinted from [36] with permission from Elsevier.

30.4 ROLE OF AHR IN CARDIOVASCULAR SYSTEM: INSIGHTS FROM GENETIC MOUSE MODELS

30.4.1 Cardiovascular Phenotype of AHR Knockout (−/−) Mice (Normoxic State)

The most pronounced cardiovascular effects described in mice in which AHR has been genetically deleted ($Ahr^{-/-}$) include cardiomegaly with left ventricular hypertrophy; kidney hypertrophy; vascular defects in the liver, kidneys, and eyes; and a significant decrease in blood pressure [27–36]. The cardiovascular phenotype of $Ahr^{-/-}$ mice is paralleled by an increase in circulating and tissue levels of a potent vasoactive and cell growth promoting peptide, ET-1, and by a decrease in activity and responsiveness of the RAS [36, 37].

$Ahr^{-/-}$ mice are hypotensive when measured at 3 months of age (Fig. 30.1) [36]. Mean arterial pressure (MAP) and both systolic and diastolic pressures are significantly lower during both light and dark photoperiods in $Ahr^{-/-}$ mice with no difference in heart rate from $Ahr^{+/+}$ mice [36]. The basal vascular tone observed in $Ahr^{-/-}$ mice is mediated, in part, by ET-1 but not Ang II. ET_A receptor antagonism with PD155080 significantly reduced MAP in $Ahr^{-/-}$ mice (versus basal levels); however, they are completely refractory to ACE inhibition with captopril [36]. Furthermore, the change in MAP after SNS inhibition using prazosin (selective smooth muscle α-adrenergic receptor blocker) or hexamethonium (ganglionic blocker) is not significantly different between $Ahr^{+/+}$ and $Ahr^{-/-}$ mice [36], suggesting similar SNS control over basal vascular tone. This observation is confirmed by comparable norepinephrine levels in urine. An increase in aortic NO and eNOS levels in $Ahr^{-/-}$ mice suggests that there may be a significant contribution of NO in the pathogenesis of the observed hypotension. However, pharmacological blockade of NOS fails to increase blood pressure in $Ahr^{-/-}$ mice, suggesting that increased NO does not play a role in the observed hypotension.

The vascular defects observed in $Ahr^{-/-}$ mice include the presence of remnants of neonatal vascular structures and an aberrant vascular architecture in the kidneys [34, 35]. Portosystemic shunting is observed in $Ahr^{-/-}$ mice, which results from a failed closure of the ductus venosus (DV). The DV is a fetal vessel that shunts blood away from the liver into the inferior vena cava, serving to increase blood flow to the developing brain. In wild-type ($Ahr^{+/+}$) mice, the DV disappears between 24 and 48 h of age; however, the DV persists in $Ahr^{-/-}$ mice (see Fig. 29.2) [35]. A smaller liver results as a consequence of reduced perfusion during development. Lahvis et al. [35] suggest that portosystemic shunting in $Ahr^{-/-}$ mice may result from a failure to constrict major blood vessels (portal and umbilical veins), as they observe a constriction of portal and umbilical veins prior to DV closure in $Ahr^{+/+}$ mice, which fails to occur in $Ahr^{-/-}$ mice. In addition, the hyaloid artery, a fetal vessel that supplies the developing eye and resolves postpartum, persists in most (64%) 12-week-old $Ahr^{-/-}$ mice (Fig. 30.2a and b) [34]. Corrosion casting revealed an altered vascular pattern in the kidney of $Ahr^{-/-}$ mice compared to $Ahr^{+/+}$ kidneys (Fig. 30.2b and c) [34]. It is important to note that although $Ahr^{-/-}$ mice exhibit aberrant kidney vasculature and an increase in kidney weight, measurements indicative of kidney function (e.g., blood urea nitrogen, plasma creatinine, plasma electrolytes, and urine osmolality) suggest that no renal filtration deficits exist [36] (unpublished data).

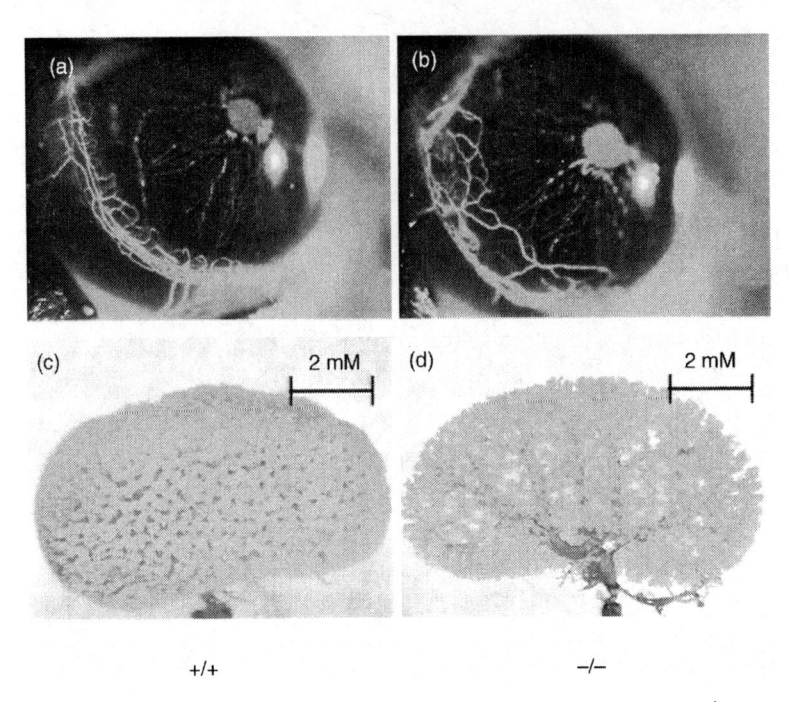

FIGURE 30.2 Altered vascular development in the eyes and kidneys of $Ahr^{-/-}$ mice. Limbal vessels structures (a and b) and corrosion casting of the kidneys (c and d) of $Ahr^{+/+}$ and $Ahr^{-/-}$ mice ($n \geq 4$ per genotype) [34]. Copyright 2000 National Academy of Sciences, USA. (See the color version of this figure in Color Plates section.)

Cardiomegaly in $Ahr^{-/-}$ mice is associated with cardiac hypertrophy, a twofold increase in cardiomyocyte size, and a decrease in cardiac output and stroke volume, without an increase in signature hypertrophic markers (e.g., α-skeletal actin, atrial natriuretic factor (ANF)) and with minimal fibrosis [27, 28]. Thus, the observed cardiac hypertrophy is not secondary to pressure or volume overload. There is some discrepancy in the earliest time point with measurable changes in heart weight. Vasquez et al. [28] identified an increase in heart weight at 4 weeks of age; however, Thackaberry et al. [30] reported a maternal genotype-specific increase in ventricle thickness and heart size in $Ahr^{-/-}$ neonates, and even in some $Ahr^{-/-}$ embryos at days 14.5 and 17.5 of gestation. Both $Ahr^{-/-}$ and heterozygote neonates born to $Ahr^{-/-}$ females have increased heart weights and cardiac hypertrophy when compared to neonates born to $Ahr^{+/+}$ females [30]. In addition, $Ahr^{-/-}$ mice exhibit an increase in hypertrophic markers (ANF and β-myosin heavy chain (β-MHC)) and ventricular wall thickness at days 14.5 and 17.5 of gestation. It is noteworthy that these embryonic and neonatal pathologies were measured at modest altitude prior to the identification of altered blood pressure regulation mediated by hypoxemia [31–33, 37] (discussed below), and that in contrast to Thackaberry et al. [30], all other studies have generated $Ahr^{-/-}$ genotypes by mating $Ahr^{-/-}$ males and heterozygous females or heterozygote littermates. Thus, it is uncertain if altitude and hypoxia played a role in the embryonic and neonatal phenotypes of $Ahr^{-/-}$ mice; however, the results suggest that the cardiovascular pathologies observed with the loss of AHR in mice may be significantly influenced by the maternal genotype.

It is unclear how much the developmental pathologies observed in $Ahr^{-/-}$ mice directly contribute to altered cardiovascular function and blood pressure homeostasis in adult mice. The role of AHR in blood pressure regulation is further complicated by altitude- and hypoxemia-induced differences in blood pressure.

30.4.2 AHR KO Phenotype at Modest Altitude

The first assessments of blood pressure in $Ahr^{-/-}$ mice reveal two opposing phenotypes depending on altitude. At sea level (SUNY Upstate Medical University, Syracuse, NY; altitude: 124 m), AHR KOs are normotensive at 5 months of age, but hypotensive at 8 months [28]. Conversely, at modest altitude (University of New Mexico, Albuquerque, NM; altitude: 1632 m) a significant increase in MAP in $Ahr^{-/-}$ mice is first evident at 2 months of age (+10 mmHg versus $Ahr^{+/+}$), with a further increase in MAP at 5 months (+21 mmHg) [31]. The increase in blood pressure at modest altitude was confirmed by similar observations by Villalobos-Molina et al. [38] (Mexico City, Mexico; altitude: 2240 m). Despite the elevated blood pressure, $Ahr^{-/-}$ mice at modest altitude displayed gross phenotypic similarities to those at sea level [27, 28], including portosystemic shunting, kidney hypertrophy, and cardiomegaly with left ventricular (LV) hypertrophy (Fig. 30.3) [29, 31].

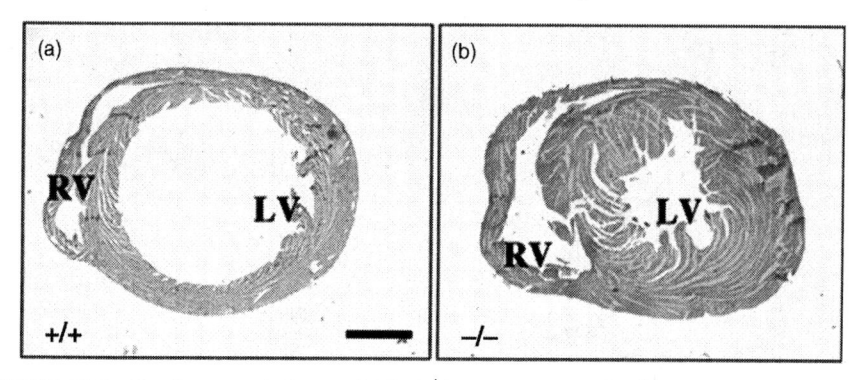

FIGURE 30.3 Cardiac morphology of $Ahr^{+/+}$ ($n=6$) and $Ahr^{-/-}$ mice ($n=4$). RV, right ventricle; LV, left ventricle. Scale bar = 1 mm. With kind permission from Springer Science + Business Media: [29] figure 2.

Similar to $Ahr^{-/-}$ mice at sea level, the earliest observable change in phenotype in $Ahr^{-/-}$ mice at modest altitude is an increase in heart weight, which occurs as early as 6 weeks [33]. By 2 months of age, $Ahr^{-/-}$ mice are hypertensive and have a significant increase in concentrations of plasma and tissue (cardiac, kidney, and lung) ET-1, plasma Ang II, expression of cardiac hypertrophic markers (β-MHC, ANF, and β-myosin light chain 2V (β-MLC2V)), cardiac and renal fibrosis, and perivascular collagen deposition [31, 33]. A time-dependent increase in MAP, ET-1, and Ang II is evident at 5 and 4 months of age for the cardiac hypertrophic markers β-MHC and ANF, respectively [31, 33]. In addition, a significant increase in cardiac and systemic oxidative stress is observed in 3-month-old $Ahr^{-/-}$ mice. Lund et al. [32] observe a 10-fold increase in superoxide anion ($O_2^{\bullet-}$) levels in cardiac tissue, which is mediated by Nox enzymes. This increase in $O_2^{\bullet-}$ is paralleled by an increase in mRNA expression of membrane-bound and regulatory Nox subunits ($gp91^{phox}$, $p47^{phox}$, and $p67^{phox}$) and a concomitant increase in thiobarbituric acid reactive substances (TBARS), a measurement of oxidative damage. Furthermore, plasma 8-isoprostane concentration, an index of systemic oxidative stress, is significantly elevated in 3-month-old $Ahr^{-/-}$ mice [32]. Echocardiography further confirms an aberrant cardiac morphology in 3- and 4-month-old $Ahr^{-/-}$ mice, as they displayed a decrease in dimension, mass, posterior wall thickness, and percent fractional shortening of the LV [31, 33]. In addition, increases in expression of markers of left ventricular hypertrophy (osteopontin and collagen I) in cardiac tissue [33] and in activity of aortic $α_{1D}$-adrenoceptor activity [38] are observed in 4-month-old $Ahr^{-/-}$ mice.

The observed phenotype at modest altitude is largely mediated by the increase in ET-1 and Ang II. Independently, inhibition of ET_A receptor signaling or ACE using BQ-123 (100 nmol/kg/day) and captopril (4 mg/kg/day), respectively, significantly reduced blood pressure, heart weight, LV hypertrophy, cardiac hypertrophic markers, and fibrosis in $Ahr^{-/-}$ mice; however, blood pressure is not fully attenuated to $Ahr^{+/+}$ levels [31, 33]. Furthermore, ET_A inhibition significantly reduces cardiac ($O_2^{\bullet-}$, TBARS) and systemic (plasma 8-isoprostane) oxidative stress, but only fully attenuates Nox subunit expression [32]. ACE inhibition also reduces $α_{1D}$-adrenoceptor expression and activity in $Ahr^{-/-}$ mice [38]. Thus, ET-1 and Ang II significantly influence the hypertensive phenotype observed in $Ahr^{-/-}$ mice at modest altitude. Moreover, BQ-123 reduces circulating levels of Ang II and captopril decreases plasma concentrations of ET-1, albeit at a high concentration (400 mg/kg/day) [31], suggesting that the two peptides likely exacerbate each other's pathological effects.

The hypertension observed in $Ahr^{-/-}$ mice at modest altitude is mediated by hypoxia. The $Ahr^{-/-}$ mice at modest altitude are hypertensive, hypoxemic, hypercapnic, and acidotic compared to $Ahr^{+/+}$ mice, with a slight, but insignificant, decrease in hematocrit and hemoglobin [37]. $Ahr^{-/-}$ mice also express cardiac markers of hypoxia, including an increase in hypoxia-inducible factor-1α (HIF-1α) protein, and mRNA expression of a HIF-1α-responsive neovascularization gene, vascular endothelial growth factor (*VEGF*) [29]. When $Ahr^{-/-}$ mice are transferred from a modest altitude to sea level (Michigan State University, East Lansing, MI; altitude: 255 m), they display a decrease in MAP and plasma ET-1, although ET-1 levels remained significantly higher than in $Ahr^{+/+}$ mice (Fig. 30.4) [37]. Furthermore, these changes are mimicked by exposing $Ahr^{-/-}$ mice residing at modest altitude to sea level O_2 concentrations (PIO_2: 150 mmHg) [37]. An 11-day exposure to sea level O_2 steadily normalized arterial PO_2 and PCO_2, with a concomitant decrease in MAP (Fig. 30.5) and plasma ET-1 to levels observed in $Ahr^{-/-}$ mice at sea level; however, ET-1 remains higher than in $Ahr^{+/+}$ mice [37]. In this study, no differences are observed in heart rate among all genotypes and treatments, and no changes are observed in all parameters measured in $Ahr^{+/+}$ mice either transferred to sea level or exposed to sea level O_2. Although the observed changes in the $Ahr^{-/-}$ mice when exposed to sea level O_2 are an expected response when moved from a hypoxic to a normoxic state, the lack of signs typical of chronic hypoxia and

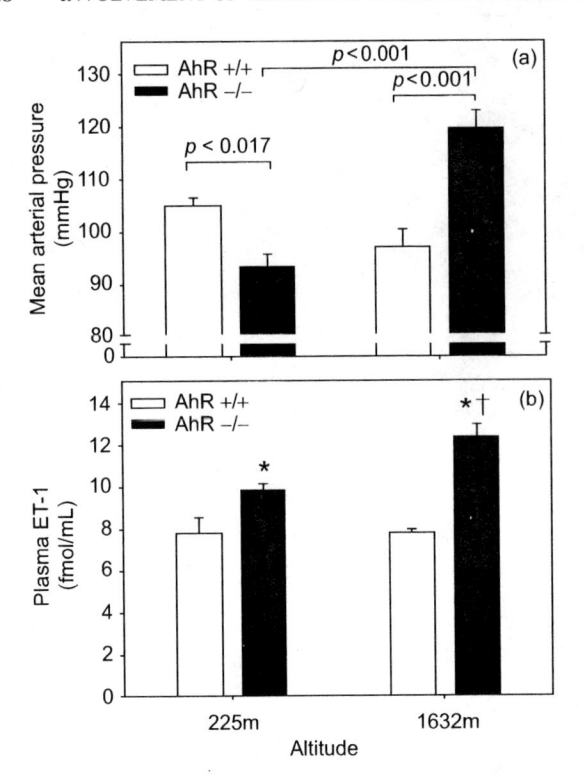

FIGURE 30.4 (a) Mean arterial pressure and (b) plasma ET-1 concentrations in 4-month-old $Ahr^{+/+}$ and $Ahr^{-/-}$ mice residing at sea level (225 m; Michigan State University) and at modest altitude (1632 m; University of New Mexico). Two-way ANOVA demonstrated that significant differences are detected based on location, and location by genotype for both end points ($p < 0.009$). (b) *Significant increase compared to $Ahr^{+/+}$ ($p < 0.014$); †significant increase ($p < 0.05$) compared to $Ahr^{-/-}$ at 225 m. Reprinted from [37] with permission.

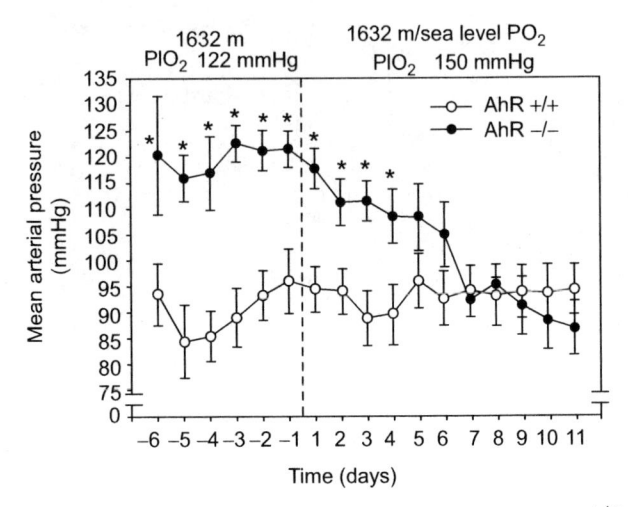

FIGURE 30.5 Mean arterial pressure in 4-month-old $Ahr^{+/+}$ and $Ahr^{-/-}$ mice residing at modest altitude (1632 m) and after exposure to sea level PIO_2 of 150 mmHg for 11 days. Repeated-measures, two-way ANOVA demonstrated significant differences with genotype ($p < 0.02$), time ($p < 0.001$), and genotype–time interaction ($p < 0.001$). *Significant increase ($p < 0.05$) compared to $Ahr^{+/+}$. Reprinted from [37] with permission.

sustained hypertension, including tachycardia, pulmonary hypertension, right ventricular hypertrophy, and an increase in erythropoiesis (hematocrit), is puzzling. Nevertheless, the lack of these signs may suggest that the $Ahr^{-/-}$ mice have an altered adaptive response to hypoxia, considering $Ahr^{+/+}$ mice are also obtained from sea level conditions and brought to a modest altitude.

Interestingly, after multiple generations at modest altitude, $Ahr^{-/-}$ mice are now hypotensive [36]. Thus, it appears that the $Ahr^{-/-}$ mice have adapted to the altitude; however, the physiological mechanism driving this adaptation remains uncertain. It may, in part, be due to an altered RAS. Hypotensive $Ahr^{-/-}$ mice at modest altitude have similar levels of plasma renin activity and ACE activity, rate-limiting steps in the RAS, compared to $Ahr^{+/+}$ mice, and show a reduced response to ACE inhibition and Ang II stimulation [36]. Given the correlation between Ang II levels and α_{1D}-adrenoceptor expression and activity [38], the reduced RAS activation in hypotensive $Ahr^{-/-}$ mice may be paralleled by a decrease in vascular α_{1D}-adrenoceptors, and a concomitant decrease in SNS control of vascular tone to $Ahr^{+/+}$ levels. It remains to be determined whether an altered RAS is common to all hypotensive $Ahr^{-/-}$ mice, or is an adaptative response to chronic hypoxia at modest altitude.

The striking observation when comparing the previously hypertensive $Ahr^{-/-}$ mice to the induced, adapted, or sea level $Ahr^{-/-}$ mice is that, apart from blood pressure, the phenotypes are remarkably similar.

30.4.3 Cardiovascular Consequence of a Loss of a Single AHR Allele

Ahr heterozygous ($+/-$) mice are superficially similar in phenotype to $Ahr^{+/+}$ mice. The loss of a single AHR allele does not result in the portosystemic shunting, nor an increase in kidney, liver, or heart weight observed in $Ahr^{-/-}$ mice [27, 36]. Furthermore, $Ahr^{+/-}$ mice are normotensive at 3–4 months of age. Despite these similarities, $Ahr^{+/-}$ mice do exhibit a gene dose effect of AHR function and are markedly different in the degree to which various physiological pathways contribute to basal vascular tone compared to both $Ahr^{-/-}$ and $Ahr^{+/+}$ mice [36]. It appears that $Ahr^{+/-}$ mice maintain normal basal tone with an increase in RAS activation and ET-1 signaling (Fig. 30.6) [36]. When compared to $Ahr^{+/+}$, $Ahr^{+/-}$ mice have significant increases in the rate-limiting steps of Ang II formation and the downstream indices of RAS activation, including increases in plasma renin activity, ACE, sodium retention, and urine electrolyte concentration. ACE inhibition with captopril effectively reduces the MAP of all AHR genotypes, although $Ahr^{+/-}$ mice are affected to the greatest degree [36]. Similarly, ACE inhibition in combination with ET_A antagonism using

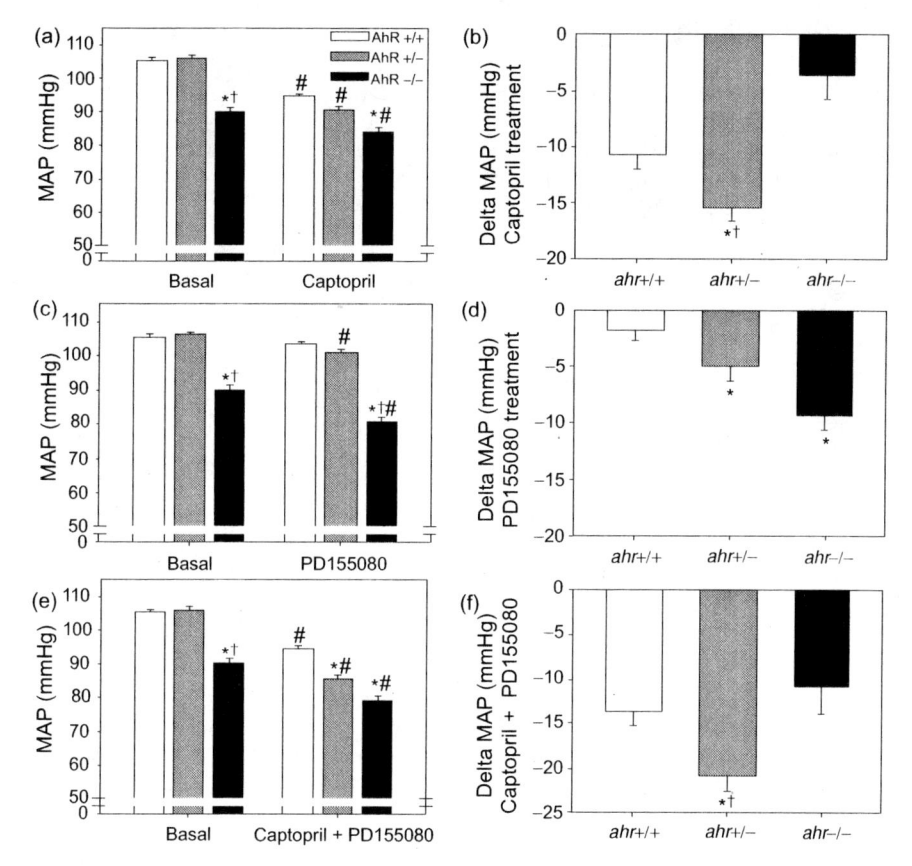

FIGURE 30.6 Mean arterial pressure (left column) and change in MAP (right column) following exposure to ACE (captopril, 4 mg/kg in drinking water for 3 days) or ET_A receptor (PD155080, 50 mg/kg/day for 3 days) antagonists, alone or in combination. Data represent mean ± SEM and were analyzed by repeated-measures two-way ANOVA (panels (a), (c), and (e)) or one-way ANOVA, with post-hoc Holm–Sidak comparisons. Panels (a), (c), and (e): $^*p < 0.05$ compared to $Ahr^{+/+}$ and $^†p < 0.05$ compared to $Ahr^{+/-}$ within the same treatment group; $^#p < 0.05$ compared to untreated, basal MAP of the same genotype. Panels (b), (d), and (f): $^*p < 0.05$ compared to $Ahr^{+/+}$ and $^†p < 0.05$ compared to $Ahr^{-/-}$. Reprinted from [36] with permission from Elsevier.

PD155080 reduces the MAP of $Ahr^{+/-}$ mice to the greatest degree when compared to $Ahr^{-/-}$ and $Ahr^{+/+}$ mice. ET_A antagonism alone significantly reduces MAP in $Ahr^{+/-}$ and $Ahr^{-/-}$ mice to a similar degree, with no effect in $Ahr^{+/+}$ mice. There are no differences in the degree of SNS control of basal tone among all genotypes, as evidenced by similar changes in MAP after prazosin (1 μg/g, i.p.) and hexamethonium (30 μg/g, i.p.) administration [36]. Similar to $Ahr^{-/-}$ mice, $Ahr^{+/-}$ mice exhibit a greater vasoconstriction to phenylephrine (PE) when NOS is inhibited, suggesting an increase in basal NO activity relative to $Ahr^{+/+}$ mice; however, this increase in $Ahr^{+/-}$ mice is not associated with an increase in eNOS protein expression for a refractory response to vasoconstriction after ACE inhibition, as is observed in $Ahr^{-/-}$ mice [36].

There has been a report of an altered cardiac phenotype in $Ahr^{+/-}$ mice, although the changes were detected at modest altitude during the time hypertension was observed in $Ahr^{-/-}$ mice. In this study, $Ahr^{+/-}$ mice appear as a phenotypic intermediate between $Ahr^{-/-}$ and $Ahr^{+/+}$ mice with respect to the development of cardiac hypertrophy [29]. $Ahr^{+/-}$ develop an increase in LV thickness at 5 months of age with a concomitant increase in an early hypertrophic marker (ANF [39]), although a significant increase in heart weight was not observed until 7 months. In contrast, $Ahr^{-/-}$ mice have an increase in heart weight by 5 months and express hypertrophic markers typically expressed later in the progression of cardiac hypertrophy (β-MHC and β-MLC2V [40]). These cardiac phenotypes are indicative of pressure overload or concentric hypertrophy, which is typically observed in hypertensive states [41]. Thus, these observations may suggest that similar to $Ahr^{-/-}$ mice [31, 33, 37], $Ahr^{+/-}$ mice are hypertensive, or were developing hypertension, but this is purely speculative. However, the cardiac effects in $Ahr^{+/-}$ mice are independent of increases in markers typical of hypoxemia, HIF-1α and VEGF mRNA expression [29].

30.4.4 Endothelial-Specific KO of AHR

Activation of AHR by xenobiotics, potential endogenous compounds, and vascular hemodynamic forces has demonstrated an important role of AHR in vascular endothelium. AHR-responsive genes (e.g., *CYP1A1*) are highly induced almost exclusively in the vascular endothelium following exposure to halogenated and polycyclic aromatic hydrocarbons, as well as dietary polyphenols [19–21, 42, 43]. In addition, CYP1A1 and CYP1B1 are highly induced following shear stress in aortic endothelial cells [44–46] (discussed below), and AHR has been shown to be activated by potentially relevant endogenous compounds in endothelial cells, including oxidized low-density lipoprotein and glucose [47, 48].

Transgenic mice with an endothelial-specific deletion of AHR were generated using *Cre/lox* technology. In this model, exon 2 of *Ahr* is flanked by lox P sites (termed "floxed" and annotated as $Ahr^{fx/fx}$) containing two 13 bp palindromic sequences bridged by an 8 bp spacer. Lox P sites are recognition sequences for Cre recombinase (or Cre), a bacteriophage topoisomerase that facilitates the site-specific excision of floxed sequences. Endothelial-specific expression of Cre is accomplished by the insertion of a construct that includes promoter/enhancer elements derived from *TEK tyrosine kinase* (*TEK*; also known as *Tie2*), a gene expressed exclusively in endothelial cells, to drive Cre expression (annotated as Cre^{Tek}). The specific mutations and breeding used to generate these transgenic models are described elsewhere [49, 50]. It is important to note that a deletion of *Ahr* has also been observed in hematopoietic tissues, including bone marrow, spleen, and thymus, of $Ahr^{fx/fx}Cre^{Tek}$ mice [50].

Preliminary studies of $Ahr^{fx/fx}Cre^{Tek}$ mice reveal a role of endothelial AHR in regulating the closure of the DV. Portosystemic shunting is observed in $Ahr^{fx/fx}Cre^{Tek}$ mice [50], suggesting that the absence of AHR in vascular endothelium is mediating at least one, if not more, of the pathological phenotypes of $Ahr^{-/-}$ mice. Walisser et al. [50] argue that the loss of AHR in the endothelial cells of the hepatic sinusoids, more so than in the DV, may be mediating the failure to close the DV considering that the sinusoids retain a fetal, anastomotic pattern in $Ahr^{-/-}$ mice. A reduction in liver weight, a consequence of a patent DV and observed in $Ahr^{-/-}$ mice [34, 35], is observed in $Ahr^{fx/fx}Cre^{Tek}$ mice [51]. An increase in kidney weight is also observed in $Ahr^{fx/fx}Cre^{Tek}$ mice, similar to $Ahr^{-/-}$ mice [51].

$Ahr^{fx/fx}Cre^{Tek}$ mice are modestly hypotensive (approximately −10 mmHg), suggesting that blood pressure regulation is partially dependent on AHR activation in the vascular endothelium (Fig. 30.7) [51]. A decrease in MAP is observed over an entire 24 h period in $Ahr^{fx/fx}Cre^{Tek}$ mice; however, the decrease in MAP during the dark (active) photoperiod tends to be greater. During the recording of blood pressure, there is no difference in overall activity and heart rate between $Ahr^{+/+}$ and $Ahr^{fx/fx}Cre^{Tek}$ mice. There is a significantly greater role of SNS in mediating basal vascular tone when AHR is absent in the vascular endothelium, as both prazosin and hexamethonium reduce MAP to a larger degree in $Ahr^{fx/fx}Cre^{Tek}$ mice than $Ahr^{+/+}$ mice [51]. An intraperitoneal injection of Ang II (30 µg/kg) results in a similar increase in MAP between $Ahr^{fx/fx}Cre^{Tek}$ and $Ahr^{+/+}$ mice over the first few minutes; however, $Ahr^{fx/fx}Cre^{Tek}$ mice have a significantly reduced capacity to maintain vasoconstriction (Fig. 30.8). By 20 min post-Ang II administration, $Ahr^{+/+}$ mice still exhibit a ∼30% increase in MAP over baseline, while $Ahr^{fx/fx}Cre^{Tek}$ mice exhibit a ∼10% increase [51]. There are no significant differences in response to NO inhibition with LNNA (L-N^G-nitroarginine) both *in vivo* and in *ex vivo* aortic reactivity studies observed in $Ahr^{fx/fx}Cre^{Tek}$ mice. Furthermore, aortas from $Ahr^{fx/fx}Cre^{Tek}$ mice show no differences in PE- or potassium chloride-induced vasoconstriction [51]. Further studies are needed to confirm these vasoreactivity studies in resistance arterioles, and to identify the specific mechanisms that maintain basal vascular tone and mediate the observed hypotension in $Ahr^{fx/fx}Cre^{Tek}$ mice.

Given the difficulty with conclusively defining a role of AHR in the cardiovascular system because of the vascular developmental effects observed in both $Ahr^{-/-}$ and $Ahr^{fx/fx}Cre^{Tek}$ mice, current studies are being conducted to generate an inducible endothelial cell AHR knockout mouse. $Ahr^{fx/fx}$ mice will be mated to *Tg*(Tek-Cre/ESR1)1Arnd mice, which express a fusion protein with a mutated estrogen receptor binding domain that inactivates Cre recombinase in the absence of a ligand of the binding domain [52]. Tamoxifen, an estrogen receptor antagonist, will then be administered at a chosen time point to interact with the binding domain of the fusion protein, activating Cre recombinase and subsequently excising of the floxed *Ahr* solely in the

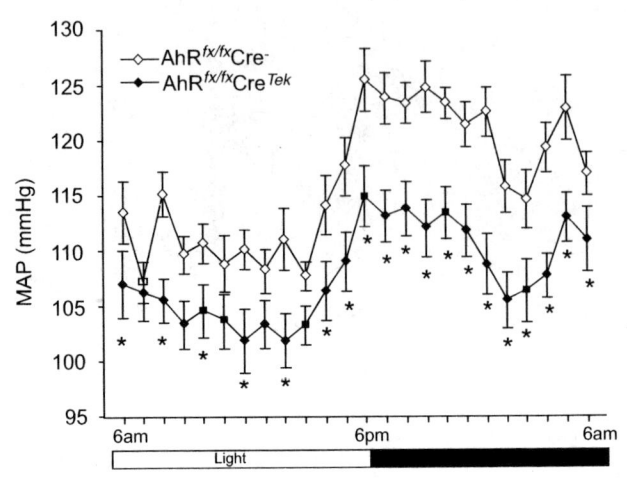

FIGURE 30.7 Mean arterial pressure of wild-type ($Ahr^{fx/fx}Cre^-$) and endothelial cell AHR knockout ($Ahr^{fx/fx}Cre^{Tek}$) mice over a 24 h period (n = 8–10 per genotype). *$p < 0.05$ compared to $Ahr^{fx/fx}Cre^-$ mice at the same time point. Reprinted from [51] with permission from Elsevier.

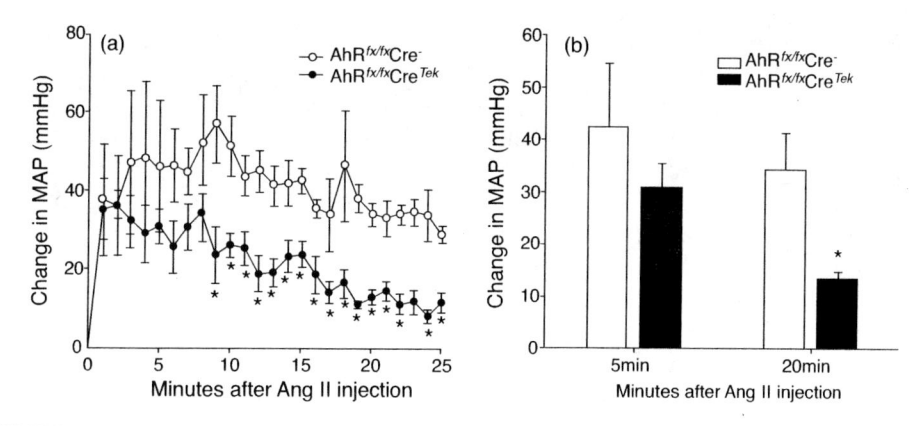

FIGURE 30.8 Change in mean arterial pressure after Ang II (30 μg/kg, i.p.) injection in wild-type ($Ahr^{fx/fx}Cre^-$) and endothelial cell AHR knockout ($Ahr^{fx/fx}Cre^{Tek}$) mice ($n = 4$ per genotype). *$p < 0.05$ compared to $Ahr^{fx/fx}Cre^-$ mice at the same time point. Reprinted from [51] with permission from Elsevier.

endothelial cells. Thus, in these mice, *Ahr* will be normally expressed during embryo, fetal, and postnatal development, allowing for the normal vascularization to occur, after which *Ahr* will be excised from endothelial cells in adult mice to determine its role in vascular tone and blood pressure regulation independent of its role during vascular development.

Nevertheless, the current mouse models support a role for AHR activation in endothelial cells in the maintenance of vascular tone, which is consistent with *in vitro* data suggesting a potential regulatory role of AHR and its inducible genes in vascular homeostasis [44–46] (discussed below).

30.5 ROLE OF AHR IN SS AND POTENTIAL SS-INDUCED ENDOGENOUS AHR LIGANDS

There is overwhelming *in vitro* evidence that AHR is induced and activated in vascular endothelial cells following laminar shear stress. Physiologically relevant levels of laminar SS (10–30 dyn/cm^2) results in a rapid induction of CYP1A1 and CYP1B1 mRNA and protein expression and activity [44–46, 53, 54] mediated by AHR [45]. In addition, AHR (mRNA) itself is induced by laminar SS (15 dyn/cm^2) as early 1 h, which is sustained throughout the 24 h study [45]. Induction of AHR mRNA expression is mediated by specific mitogen-activated protein kinases, as inhibiting c-Jun N-terminal kinase (JNK) and p38 with SP600125 and SB203580, respectively, significantly reduces expression of both AHR and CYP1A1 mRNA. In contrast, inhibition of extracellular signal regulating kinase (ERK) has no affect on AHR and CYP1A1 mRNA levels.

The response of CYP1A1 and CYP1B1 expression is both time and magnitude dependent. In human umbilical vein endothelial cells (HUVECs), Han et al. [45] observe a roughly 25-fold increase in 1A1 mRNA and protein 2 h after the onset of a 15 dyn/cm^2 laminar SS, which plateaus at 4 h and is sustained through 24 h. CYP1A1 activity, as measured by ethoxyresorufin-O-deethylase (EROD) activity, follows a similar trend, although a significant increase following SS is observed as early as 1 h and is maximal by 2 h [45]. CYP1A1 protein expression increases nearly threefold at 5 dyn/cm^2 and nearly sixfold at 15 dyn/cm^2, compared to static conditions. Similar trends in CYP1A1 mRNA, protein, and activity are observed in bovine aortic endothelial cells (BAECs) [45]. At 25 dyn/cm^2, Conway et al. [46] observe greatly elevated mRNA expression of both CYP1A1 and CYP1B1 (roughly 30- and 430-fold, respectively) in HUVECs, beginning at 24 h after the onset of SS and remaining elevated through 72 h. A P450 Glo assay, a measurement of the combined activity of CYP1A1 and CYP1B1, detects a 7.2-fold increase in CYP1A1/1B1 activity in stressed cells when compared to static conditions [46]. In this study, a significant increase in CYP1A1 mRNA and protein expression is not observed until the magnitude of SS reaches 15 dyn/cm^2, and until 25 dyn/cm^2 for CYP1B1. Similar mRNA expression patterns of CYP1A1 and CYP1B1 in both HUVECs and human aortic endothelial cells (HAECs) exposed to 25 dyn/cm^2 are reported by Eskin et al. [44]. Furthermore, SS-induced activation of AHR is significantly weaker or absent during nonlaminar flow. Physiologically relevant turbulent SS (1.5 dyn/cm^2) induces mRNA expression of AHR and CYP1A1 in HUVECs; however, the levels measured are significantly (twofold) lower than those observed during laminar SS (6 dyn/cm^2) [45]. Reversing SS from 15 to 1 dyn/cm^2 fully attenuates the expression of CYP1A1 and CYP1B1 mRNA [46]. Thus, SS-induced AHR activation is likely only physiologically relevant in arteries (versus veins) and in regions exposed to laminar SS.

Interestingly, treatment of HUVECs with β-naphthoflavone (β-NF; 6 μg/mL), a potent AHR ligand, reveals that the relative inducibility and expression of AHR-mediated genes in the vascular endothelium are largely stimulus dependent. While SS consistently induces remarkably higher levels of CYP1B1 mRNA than CYP1A1 in all endothelial cell types [44–46], β-NF induces 4.5-fold higher levels of CYP1A1 than CYP1B1 in HUVECs [44].

In vivo studies identifying the localization of AHR and CYP1A1 support observed responses to laminar SS *in vitro*. In mice, AHR is localized to the endothelium of the thoracic aorta and greater curvature of aorta, areas experiencing laminar SS, but is absent in the lesser curvature of the aorta, an area of turbulent SS [46]. CYP1A1 mRNA expression in these aortic sections confirms this observation. In addition, immunohistochemical staining of CYP1A1 in cross sections of human coronary arteries, which also experience laminar SS, is only observed in vascular endothelium [46]. In contrast, CYP1B1 is observed in the endothelium, media, and adventitia of human coronary arteries, suggesting other AHR-independent regulatory mechanisms for CYP1B1, which has been reported [24].

Together, the time-, magnitude-, and flow-dependent increase in AHR activation, and localization of AHR in areas of laminar SS support the hypothesis that a potent AHR ligand is produced during SS and that AHR likely serves as an atheroprotective pathway. These conclusions are further supported by *in vitro* evidence of an AHR-mediated reduction in endothelial cell proliferation and the presence of an AHR ligand in SS-conditioned media.

Laminar SS induces an arrest in endothelial cell proliferation, likely to restrict vascular wall permeability and to maintain vascular homeostasis [55]. Han et al. [45] observe a reversed suppression of endothelial cell proliferation in HUVECs after a knockdown of AHR using small interfering RNA (siRNA). AHR siRNA also prevents the dephosphorylation of phosphorylated retinoblastoma (pRb) protein and the induction of p21^{Cip1}, a cyclin-dependent kinase inhibitor [45], which supports an AHR/pRb-mediated antiproliferative action [56] upon AHR activation during laminar SS-exposed vascular endothelium.

Furthermore, CYP1A1 siRNA effectively reduced laminar SS-induced expression of thrombospondin-1, an antiangiogenic and antiproliferative glycoprotein, suggesting the influence of CYP1A1 metabolite(s) in cell cycle arrest [46].

There have been several reports of an active AHR ligand produced during SS; however, they remain largely unidentified. SS-conditioned media significantly induce CYP1A1 and/or CYP1B1 mRNA expression in various naïve cells [44, 47]. McMillan and Bradfield [47] identified a modified low-density lipoprotein (LDL) produced during SS (12 dyn/cm^2) that activates AHR and induces CYP1A1 mRNA via the interaction with xenobiotic-responsive elements (XRE) in hepatoma (H1L6.1c3) cells. Furthermore, a NaOCl-induced oxidized LDL is a potent activator of AHR. Notably, laminar shear stress induces a significant increase in ROS production [57], which could, in theory, activate AHR ligands and subsequently enhance the induction of antioxidant genes mediated by Nrf2 (nuclear factor (erythroid-derived 2)-like 2) [58, 59]. McMillan and Bradfield [47] also noted that LDL may indirectly activate AHR through a downstream product of arachidonic acid (AA) release during SS. Certainly, several AA metabolites, including CYP1A1- and CYP1B1-derived metabolites, have been shown to induce AHR [60, 61]; however, there are conflicting reports regarding the release of AA during SS (unpublished observations reported in Ref. 47). Nonetheless, the direct role of AA metabolites in mediating an AHR response in the cardiovascular system remains uncertain.

30.6 INFLUENCE OF EXOGENOUS LIGANDS ON AHR FUNCTION IN THE CARDIOVASCULAR SYSTEM

It is important to acknowledge that AHR has a dual role in the vascular endothelium (as well as other tissues). In addition to its obvious role in cardiovascular development and blood pressure regulation, AHR also serves as a detoxifying pathway. It is well known that AHR and often CYP1 enzymes mediate the toxicity of various xenobiotics, including polycyclic and halogenated aromatic hydrocarbons (PAHs and HAHs), with the vascular endothelium as a primary target [1, 21, 62–64] (described elsewhere in this book). PAHs and HAHs produce numerous developmental and homeostatic cardiovascular pathologies in a variety of species (reviewed in Ref. 1), and the degree of toxicity has been linked to AHR and CYP1A1 polymorphisms in humans and mice [65–69]. The cardiovascular pathologies and toxic mechanisms are beyond the scope of this chapter, but are reviewed elsewhere [1].

Notably, human epidemiology studies have linked HAH exposure to hypertension, further suggesting a role for AHR in the vasculature and blood pressure regulation. Army and Air Force Vietnam veterans exposed to the HAH AHR agonist, TCDD, as a contaminant in the defoliant Agent Orange exhibited a significantly higher incidence of hypertension [70], while a recent National Health and Nutritional Examination Survey (NHANES) data have linked HAH exposure and hypertension in the general public in the United States. Thus, we are continuously exposed to dietary AHR ligands, whether potentially beneficial (e.g., dietary flavonoids [71]) or toxic (e.g., HAHs), and how this alters, or even aids, the role of endogenous ligand-mediated effects in the cardiovascular system remains to be elucidated.

30.7 CONCLUSIONS

The AHR plays a clear role in vascular development, closure of the ductus venosus, regulation of cardiomyocyte proliferation, and a vasoprotective role during laminar shear stress. Moreover, the endothelial-specific AHR knockout model demonstrates that the influence of AHR in cardiovascular physiology is largely through its action in endothelial cells. However, the vascular abnormalities observed in AHR genetic mouse models, together with the altered response to hypoxia at modest altitudes, make it very difficult to define the specific role of AHR in cardiac function and blood pressure regulation. Despite these complications, some common observations identify potential roles of AHR in cardiovascular function. In particular, evidence suggests that AHR may influence ET-1 and RAS signaling, two pathways with significant contributions to blood pressure regulation, and cardiac and vascular hypertrophy/remodeling.

The data obtained from $Ahr^{-/-}$ and $Ahr^{+/-}$ mice collectively suggest that AHR may negatively regulate ET-1 signaling, as evidenced by (1) a consistent, gene dose-dependent elevation of ET-1 or ET-1 signaling with the loss of an Ahr allele that is independent of altitude-induced hypoxia (Figs. 30.5 and 30.7) [31–33, 36, 37]; and (2) the lack of attenuated ET-1 or ET-1 signaling with a decrease in factors known to induce ET-1 (e.g., hypoxia, Ang II, and shear stress). ET-1 remains elevated during normoxic and hypotensive (i.e., low shear stress) states in $Ahr^{-/-}$ mice [36, 37] and a decrease in RAS activation (i.e., Ang II) is not paralleled by a decrease in ET-1 signaling when $Ahr^{-/-}$ mice become hypotensive [36]. It is noteworthy that inflammatory cytokines may contribute to the elevated increase in ET-1, as they are known to induce ET-1 [72, 73] and be suppressed by AHR activation [74].

The influence of AHR on the RAS is less clearly defined; however, all AHR mouse models show an altered RAS. Normoxic, hypotensive $Ahr^{-/-}$ mice are refractory to ACE inhibition (Fig. 30.6) [36], while hypoxic, hypertensive $Ahr^{-/-}$ mice at modest altitude had an increase in plasma Ang II and a significant reduction in MAP and cardiac abnormalities after ACE inhibition [31–33]. In $Ahr^{+/-}$ mice, RAS activation was significantly increased to a sufficient degree to maintain a normotensive state [36]. Finally, endothelial-specific AHR knockouts ($Ahr^{fx/fx}Cre^{Tek}$) had no changes in all rate-limiting measurements of the RAS (plasma renin activity, ACE), but had a decreased ability to maintain vasoconstriction after Ang II infusion when compared to wild-type controls (Fig. 30.8) [51]. Nonetheless, more direct evidence is needed to confirm the mechanisms by which AHR influences both RAS and ET-1 signaling in the cardiovascular system.

The *in vitro* studies of AHR function during hemodynamic stress not only provide strong evidence that AHR plays an antiatherogenic and antiproliferative role in the vascular endothelium during laminar shear stress, but also provide some of the first convincing evidence of the production and action of a endogenous ligand in the cardiovascular system. Whether the endogenous ligand *in vivo* is a modified or oxidized LDL remains to be determined.

This area of research is still in its infancy and is complicated by experimental conditions and developmental abnormalities; however, it is undeniable that AHR has a role in cardiovascular development and homeostasis in mammals. It is anticipated that a more definable and conclusive role of AHR in blood pressure regulation will be elucidated with the current use of inducible AHR knockout mice, which will bypass the developmental abnormalities associated with the loss of AHR.

REFERENCES

1. Korashy, H. M. and El-Kadi, A. O. S. (2006). The role of aryl hydrocarbon receptor in the pathogenesis of cardiovascular diseases. *Drug Metabolism Reviews*, 38, 411–450.

2. Malpas, S. C. (2010). Sympathetic nervous system overactivity and its role in the development of cardiovascular disease. *Physiological Reviews*, 90, 513–557.

3. Allen, A. M., MacGregor, D. P., McKinley, M. J., and Mendelsohn, F. A. (1999). Angiotensin II receptors in the human brain. *Regulatory Peptides*, 79, 1–7.

4. Kedzierski, R. M. and Yanagisawa, M. (2001). Endothelin system: the double-edged sword in health and disease. *Annual Review of Pharmacology and Toxicology*, 41, 851–876.

5. Schneider, M. P., Boesen, E. I., and Pollock, D. M. (2007). Contrasting actions of endothelin ET_A and ET_B receptors in cardiovascular disease. *Annual Review of Pharmacology and Toxicology*, 47, 731–759.

6. Ferrario, C. M. (2010). New physiological concepts of the renin–angiotensin system from the investigation of precursors and products of angiotensin I metabolism. *Hypertension*, 55, 445–452.

7. Kawaguchi, H., Sawa, H., and Yasuda, H. (1990). Endothelin stimulates angiotensin I to angiotensin II conversion in cultured pulmonary artery endothelial cells. *Journal of Molecular and Cellular Cardiology*, 22, 839–842.

8. Campia, U., Cardillo, C., and Panza, J. A. (2004). Ethnic differences in the vasoconstrictor activity of endogenous endothelin-1 in hypertensive patients. *Circulation*, 109, 3191–3195.

9. Xia, Y. and Karmazyn, M. (2004). Obligatory role for endogenous endothelin in mediating hypertrophic effects of phenylephrine and angiotensin II in neonatal rat ventricular myocytes: evidence for two distinct mechanisms for endothelin regulation. *Journal of Pharmacology and Experimental Therapeutics*, 310, 43–51.

10. Sowers, J. R. (2002). Hypertension, angiotensin II, and oxidative stress. *New England Journal of Medicine*, 346, 1999–2001.

11. Villalobos-Molina, R. and Ibarra, M. (1999). Vascular α_{1D}-adrenoceptors: are they related to hypertension? *Archives of Medical Research*, 30, 347–352.

12. Gisbert, R., Ziani, K., Miquel, R., Noguera, M. A., Ivorra, M. D., Anselmi, E., and Docon, P. (2002). Pathological role of a constitutively active population of α_{1D}-adrenoceptors in arteries of spontaneously hypertensive rats. *British Journal of Pharmacology*, 135, 206–216.

13. Liu, V. W. T. and Huang, P. L. (2008). Cardiovascular roles of nitric oxide: a review of insights from nitric oxide synthase gene disrupted mice. *Cardiovascular Research*, 77, 19–29.
14. Nakati, T., Nakayam, M., and Kato, R. (1990). Inhibition by nitric oxide and nitric oxide-producing vasodilators of DNA synthesis in vascular smooth muscle cells. *European Journal of Pharmacology*, 189, 347–353.
15. Traub, O. and Berk, B. C. (1998). Laminar shear stress: mechanisms by which endothelial cells transduce an atheroprotective force. *Atherosclerosis, Thrombosis, and Vascular Biology*, 18, 677–685.
16. Gimbrone, M. A., Jr., Topper, J. N., Nagel, T., Anderson, K. R., and Garcia-Cardeña, G. (2000). Endothelial dysfunction, hemodynamic forces, and atherogenesis. *Annals of the New York Academy of Sciences*, 902, 230–239.
17. Vanhoutte, P. M. (1989). Endothelium and control of vascular function: state of the art lecture. *Hypertension*, 13, 658–667.
18. Rieder, M. J., Carmona, R., Krieger, J. E., Pritchard, K. A., Jr., and Greene, A. S. (1997). Suppression of angiotensin-converting enzyme expression and activity by shear stress. *Circulation Research*, 80, 312–319.
19. Brittebo, E. B. (1994). Metabolic activation of the food mutagen Trp-P-1 in endothelial cells of heart and kidney in cytochrome P450-induced mice. *Carcinogenesis*, 4, 667–672.
20. Granberg, A. L., Brunstrom, B., and Brandt, I. (2000). Cytochrome P450-dependent binding of 7,12-dimethylbenz[a]anthracene (DMBA) and benzo[a]pyrene (B[a]P) in murine heart, lung, and liver endothelial cells. *Archives of Toxicology*, 10, 593–601.
21. Kopf, P. G., Scott, J. A., Agbor, L. N., Boberg, J. R., Elased, K. M., Huwe, J. K., and Walker, M. K. (2010). Cytochrome P4501A1 is required for vascular dysfunction and hypertension induced by 2,3,7,8-tetrachlorodibenzo-*p*-dioxin. *Toxicological Sciences*, 117, 537–546.
22. Filbrandt, C. R., Wu, Z., Zlokovic, B., Opanashuk, L., and Gasiewicz, T. A. (2004). Presence and functional activity of the aryl hydrocarbon receptor in isolated murine cerebral vascular endothelial cells and astrocytes. *Neurotoxicology*, 25, 605–616.
23. Dubey, R. K., Jackson, E. K., Gillespie, D. G., Zacharia, L. C., and Imthurn, B. (2004). Catecholamines block the antimitogenic effect of estradiol on human coronary artery smooth muscle cells. *Journal of Clinical Endocrinology & Metabolism*, 8, 3922–3931.
24. Kerzee, J. K. and Ramos, K. S. (2001). Constitutive and inducible expression of *Cyp1a1* and *Cyp1b1* in vascular smooth muscle cells. *Circulation Research*, 89, 573–582.
25. Thum, T. and Borlak, J. (2000). Cytochrome P450 monooxygenase gene expression and protein activity in cultures of adult cardiomyocytes of the rat. *British Journal of Pharmacology*, 8, 1745–1752.
26. Thum, T. and Borlak, J. (2002). Testosterone, cytochrome P450, and cardiac hypertrophy. *FASEB Journal*, 16, 1537–1549.
27. Fernandez-Salguero, P. M., Ward, J. M., Sunberg, J. P., and Gonzalez, F. J. (1997). Lesion of aryl-hydrocarbon receptor-deficient mice. *Veterinary Pathology*, 34, 605–614.
28. Vasquez, A., Atallah-Yunes, N., Smith, F. C., You, X., Chase, S. E., Silverstone, A. E., and Vikstrom, K. L. (2003). A role for the aryl hydrocarbon receptor in cardiac physiology and function as demonstrated by AhR knockout mice. *Cardiovascular Toxicology*, 3, 153–163.
29. Thackaberry, E. A., Gabaldon, D. M., Walker, M. K., and Smith, S. M. (2002). Aryl hydrocarbon receptor mice develop cardiac hypertrophy and increased hypoxia-inducible factor 1-α in the absence of cardiac hypoxia. *Cardiovascular Toxicology*, 2, 263–273.
30. Thackaberry, E. A., Bedrick, E. J., Goens, M. B., Danielson, L., Lund, A. K., Gabaldon, D., Smith, S. M., and Walker, M. K. (2003). Insulin regulation in AhR-null mice: embryonic cardiac enlargement, neonatal macrosomia, and altered insulin regulation and response in pregnant and aging AhR-null females. *Toxicological Sciences*, 76, 407–417.
31. Lund, A. K., Goens, M. B., Kanagy, N. L., and Walker, M. K. (2003). Cardiac hypertrophy in aryl hydrocarbon receptor null mice is correlated with elevated angiotensin II, endothelin-1, and mean arterial blood pressure. *Toxicology and Applied Pharmacology*, 193, 177–187.
32. Lund, A. K., Peterson, S. L., Timmins, G. S., and Walker, M. K. (2005). Endothelin-1 mediated increase in reactive oxygen species and NAD(P)H oxidase activity in hearts of aryl hydrocarbon receptor (AhR) null mice. *Toxicological Sciences*, 88, 265–273.
33. Lund, A. K., Goens, M. B., Nuñez, B. A., and Walker, M. K. (2006). Characterizing the role of endothelin-1 in the progression of cardiac hypertrophy in aryl hydrocarbon receptor (AhR) null mice. *Toxicology and Applied Pharmacology*, 212, 127–135.
34. Lahvis, G. P., Lindell, S. L., Thomas, R. S., McCuskey, R. S., Murphy, C., Glover, E., Bentz, M., Southard, J., and Bradfield, C. A. (2000). Portosystemic shunting and persistent fetal vascular structures in aryl hydrocarbon receptor-deficient mice. *Proceedings of the National Academy of Sciences of the United States of America*, 97, 10442–10447.
35. Lahvis, G. P., Pyzalski, R. W., Glover, E., Pitot, H. C., McElwee, M. K., and Bradfield, C. A. (2005). The aryl hydrocarbon receptor is required for closure of the ductus venosus in the neonatal mouse. *Molecular Pharmacology*, 67, 714–720.
36. Zhang, N., Agbor, L. N., Scott, J. A., Zalobowski, T., Elased, K. M., Trujillo, N. A., Skelton Duke, M., Wolf, V., Walsh, M. T., Born, J. L., Felton, L. A., Wang, J., Wang, W., Kanagy, N. L., and Walker, M. K. (2010). An activated renin–angiotensin system maintains normal blood pressure in aryl hydrocarbon receptor heterozygous mice, but not in null mice. *Biochemical Pharmacology*, 15, 197–204.
37. Lund, A. K., Agbor, L. N., Zhang, N., Baker, A., Zhao, H., Fink, G. D., Kanagy, N. L., and Walker, M. K. (2008). Loss of the aryl hydrocarbon receptor induces hypoxemia, endothelin-1, and systemic hypertension at modest altitude. *Hypertension*, 51, 803–809.
38. Villalobos-Molina, R., Vázquez-Cuevas, F. G., López-Guerrero, J. J., Figueroa-García, M. C., Gallardo-Ortíz, I. A., Ibarra, M., Rodríguez-Sosa, M., Gonzalez, F. J., and Elizondo, G. (2008). Vascular α_{1D}-adrenoceptors are overexpressed in aorta

of aryl hydrocarbon receptor null mouse: role of angiotensin II. *Autonomic & Autacoid Pharmacology*, 28, 61–67.

39. Mercadier, J. J., Samuel, J. L., Michel, J. B., Zongazo, M. A., de la Bastie, D., Lompre, A. M., Wisnewsky, C., Rappaport, L., Levy, B., and Schwartz, K. (1998). Atrial natriuretic factor gene expression in rat ventrical during experimental hypertension. *American Journal of Physiology*, 257, H979–H987.

40. Lee, H. R., Henderson, S. A., Reynolds, R., Dunnmon, P., Yuan, D., and Chien, K. R. (1988). Alpha 1-adrenergic stimulation of cardiac gene transcription in neonatal rat myocardial cells. *Journal of Biological Chemistry*, 263, 7352–7358.

41. Frohlich, E. D., Apstein, C., and Chobanian, A. V. (1992). The heart in hypertension. *New England Journal of Medicine*, 327, 998–1008.

42. Gouédard, C., Barouki, R., and Morel, Y. (2004). Dietary polyphenols increase paraoxonase 1 gene expression by an aryl hydrocarbon receptor-dependent mechanism. *Molecular and Cellular Biology*, 24, 5209–5222.

43. Kopf, P. G. and Walker, M. K. (2010). 2,3,7,8-Tetrachlorodibenzo-*p*-dioxin increases reactive oxygen species production in human endothelial cells via induction of cytochrome P4501A1. *Toxicology and Applied Pharmacology*, 245, 91–99.

44. Eskin, S. G., Turner, N. A., and McIntire, L. V. (2004). Endothelial cell cytochrome P450 1A1 and 1B1: up-regulation by shear stress. *Endothelium*, 11, 1–10.

45. Han, Z., Miwa, Y., Obikane, H., Mitsumata, M., Takahashi-Yanaga, F., Morimoto, S., and Sasaguri, T. (2008). Aryl hydrocarbon receptor mediates laminar shear stress-induced CYP1A1 activation and cell cycle arrest in vascular endothelial cells. *Cardiovascular Research*, 77, 809–818.

46. Conway, D. E., Sakurai, Y., Weiss, D., Vega, J. D., Taylor, W. R., Jo, H., Eskin, S., Marcus, C. B., and McIntire, L. V. (2009). Expression of CYP1A1 and CYP1B1 in human endothelial cells: regulation by fluid shear stress. *Cardiovascular Research*, 81, 669–677.

47. McMillan, B. J. and Bradfield, C. A. (2007). The aryl hydrocarbon receptor is activated by modified low-density lipoprotein. *Proceedings of the National Academy of Sciences of the United States of America*, 104, 1412–1417.

48. Dabir, P., Marinic, T. E., Krukovets, I., and Stenina, O. I. (2008). Aryl hydrocarbon receptor is activated by glucose and regulates thrombospondin-1 gene promoter in endothelial cells. *Circulation Research*, 102, 1558–1565.

49. Walisser, J. A., Bunger, M. K., Glover, E., Harstad, E. B., and Bradfield, C. A. (2004). Patent ductus venosus and dioxin resistance in mice harboring a hypomorphic *Arnt* allele. *Journal of Biological Chemistry*, 279, 16326–16331.

50. Walisser, J. A., Glover, E., Pande, K., Liss, A. L., and Bradfield, C. A. (2005). Aryl hydrocarbon receptor-dependent liver development and hepatotoxicity are mediated by different cell types. *Proceedings of the National Academy of Sciences of the United States of America*, 102, 17858–17863.

51. Agbor, L. N., Elased, K. M., and Walker, M. K. (20011). Endothelial cell-specific aryl hydrocarbon receptor knockout mice exhibit hypotension mediated, in part, by an attenuated angiotensin II responsiveness. *Biochemical Pharmacology*, 82, 514–523.

52. Forde, A., Constein, R., Gröne, H.-J., Hämmerling, G., and Arnold, B. (2002). Temporal Cre-mediated recombination exclusively in endothelial cells using Tie2 regulatory elements. *Genesis*, 33, 191–197.

53. Garcia-Cardeña, G., Comander, J., Anderson, K. R., Blackman, B. R., and Gimbrone, M. A., Jr., (2001). Biomechanical activation of vascular endothelium as a determinant of its functional phenotype. *Proceedings of the National Academy of Sciences of the United States of America*, 98, 4478–4485.

54. Dekker, R. J., van Soest, S., Fontijn, R. D., Salamanca, S., deGroot, P. G., VanBravel, E., Pannekoek, H., and Horrevoets, A. J. G. (2002). Prolonged fluid shear stress induces a distinct set of endothelial cell genes, most specifically lung Kruppel-like factor (KLF2). *Blood*, 100, 1689–1698.

55. Akimoto, S., Mitsumata, M., Sagaguri, T., and Yoshida, Y. (2000). Laminar shear stress inhibits vascular endothelial cell proliferation by inducing cyclin-dependent kinase inhibitor $p21^{Sdi1/Cip1/Waf1}$. *Circulation Research*, 86, 185–190.

56. Ge, N. L. and Elferink, C. J. (1998). A direct interaction between the aryl hydrocarbon receptor and retinoblastoma protein. Linking dioxin signaling to the cell cycle. *Journal of Biological Chemistry*, 273, 22708–22713.

57. Warabi, E. Takabe, W., Minami, T., Inoue, K., Itoh, K., Yamamoto, M., Ishii, T., Kodama, T., and Noguchi, N. (2007). Shear stress stabilizes NF-E2-related factor 2 and induces antioxidant genes in endothelial cells: role of reactive oxygen/nitrogen species. *Free Radical Biology & Medicine*, 42, 260–269.

58. Ma, Q., Kinneer, K., Bi, Y., Chan, J. Y., and Kan, Y. W. (2004). Induction of murine NAD(P)H:quinone oxidoreductase by 2,3,7,8-tetrachlorodibenzo-*p*-dioxin requires the CNC (cap 'n' collar) basic leucine zipper transcription factor Nrf2 (nuclear factor erythroid 2-related factor 2): cross-interaction between AhR (aryl hydrocarbon receptor) and Nrf2 signal transduction. *Biochemical Journal*, 377, 205–213.

59. Marchand, A., Barouki, R., and Garlatti, M. (2004). Regulation of NAD(P)H:quinone oxidoreductase 1 gene expression by CYP1A1 activity. *Molecular Pharmacology*, 65, 1029–1037.

60. Chiaro, C. R., Morales, J. L., Prabhu, K. S., and Perdew, G. H. (2008). Leukotriene A4 metabolites are endogenous ligands for the Ah receptor. *Biochemistry*, 47, 8445–8455.

61. Chiaro, C. R., Patel, R. D., and Perdew, G. H. (2008). 12(*R*)-Hydroxy-5(*Z*),8(*Z*),10(*E*),14(*Z*)-eicosatetraenoic acid [12(*R*)-HETE], an arachidonic acid derivative, is an activator or the aryl hydrocarbon receptor. *Molecular Pharmacology*, 74, 1649–1656.

62. Fernandez-Salguero, P. M., Hilbert, D. M., Rudikoff, S., Ward, J. M., and Gonzalez, F. J. (1996). Aryl-hydrocarbon receptor-deficient mice are resistant to 2,3,7,8-tetrachlorodibenzo-*p*-dioxin-induced toxicity. *Toxicology and Applied Pharmacology*, 140, 173–179.

63. Peters, J. M., Narotsky, M. G., Elizondo, G., Fernandez-Salguero, P. M., Gonzalez, F. J., and Abbott, B. D. (1999).

Amelioration of TCDD-induced teratogenesis in aryl hydrocarbon receptor (AhR)-null mice. *Toxicological Sciences*, 47, 86–92.

64. Uno, S., Dalton, T. P., Sinclair, P. R., Gorman, N., Wang, B., Smith, A. G., Miller, M. L., Shertzer, H. G., and Nebert, D. W. (2004). *Cyp1a1(-/-)* male mice: protection against high-dose TCDD-induced lethality and wasting syndrome, and resistance to intrahepatocyte lipid accumulation and uroporphyria. *Toxicology and Applied Pharmacology*, 196, 410–421.

65. Okey, A. B., Vella, L. M., and Harper, P. A. (1989). Detection and characterization of a low affinity form of cystolic Ah receptor in livers of mice nonresponsive to induction of cytochrome P1-450 by 3-methylcholanthrene. *Molecular Pharmacology*, 35, 823–830.

66. Chang, C.-Y., Smith, D. R., Prasad, V. S., Sidman, C. L., Nebert, D. W., and Puga, A. (1993). Ten nucleotide differences, five of which cause amino acid changes, are associated with the Ah receptor locus polymorphism of C57BL/6 and DBA/2 mice. *Pharmacogenetics*, 3, 312–321.

67. Poland, A., Palen, D., and Glover, E. (1994). Analysis of the four alleles of the murine aryl hydrocarbon receptor. *Molecular Pharmacology*, 46, 915–921.

68. Wang, X. L., Greco, M., Sim, A. S., Duarte, N., Wang, J., and Wilcken, D. E. L. (2002). Effect of CYP1A1 *Msp*I polymorphism on cigarette smoking related coronary artery disease and diabetes. *Atherosclerosis*, 162, 391–397.

69. Wang, X. L., Raveendran, M., and Wang, J. (2003). Genetic influence on cigarette-induced cardiovascular disease. *Progress in Cardiovascular Diseases*, 45, 361–382.

70. Kang, H. K., Dalager, N. A., Needham, L. L., Patterson, D. G., Lees, P. S. J., Yates, K., and Matanoski, G. M. (2006). Health status of Army Chemical Corps Vietnam veterans who sprayed defoliant in Vietnam. *American Journal of Industrial Medicine*, 49, 875–884.

71. Tutel'yan, V. A., Gapparov, M. M., Telegin, L. Yu., Devichenskii, V. M., and Pevnitskii, L. A. (2003). Flavonoids and resveratrol as regulators of Ah-receptor activity: protection from dioxin toxicity. *Bulletin of Experimental Biology and Medicine*, 136, 533–539.

72. Woods, M., Mitchell, J. A., Wood, E. G., Barker, S., Walcot, N. R., Rees, G. M., and Warner, T. D. (1999). Endothelin-1 is induced by cytokines in human vascular smooth muscle cells: evidence for intracellular endothelin-converting enzyme. *Molecular Pharmacology*, 55, 902–909.

73. Woods, M., Wood, E. G., Mitchell, J. A., and Warner, T. D. (2000). Signal transduction pathways involved in cytokine stimulation of endothelin-1 release from human vascular smooth muscle cells. *Journal of Cardiovascular Pharmacology*, 36, S407–S409.

74. Patel, R. D., Murray, I. A., Flaveny, C. A., Kusnadi, A., and Perdew, G. H. (2009). Ah receptor represses acute-phase response gene expression without binding to its cognate receptor response element. *Laboratory Investigation*, 89, 695–707.

31

INVOLVEMENT OF THE AHR IN DEVELOPMENT AND FUNCTIONING OF THE FEMALE AND MALE REPRODUCTIVE SYSTEMS

BETHANY N. KARMAN, ISABEL HERNÁNDEZ-OCHOA, AYELET ZIV-GAL, AND JODI A. FLAWS

31.1 INTRODUCTION

Proper functioning of the female and male reproductive systems begins during early embryonic development, continues from puberty through adulthood, and ends at the onset of reproductive senescence. The major processes regulating female reproductive function include sex determination, sexual differentiation, folliculogenesis, puberty, cyclicity, sexual behavior, ovulation, fertilization, implantation, pregnancy, and parturition. Similar to the female, some major processes regulating male reproductive function also include sex determination, sexual differentiation, puberty, and sexual behavior. Reproductive processes that differ from the female, but are essential for male reproductive function, are androgen production, spermatogenesis, and sperm transfer. These complex processes in female and male reproduction rely on development of a proper genetic program in each tissue and maintenance of highly coordinated webs of communication between various cell types, within each tissue and between the reproductive tissues themselves.

This chapter describes emerging evidence that the aryl hydrocarbon receptor (AHR) is an important biological factor involved in the development, regulation, and functioning of some of the female and male reproductive processes. A brief description of the development and functioning of the female and male reproductive systems is followed by a review of the literature supporting a pleiotropic role for the AHR in these processes. Both activation of the AHR pathway by ligands, deficiency of the AHR in AHR "knockout" (AHRKO) mouse models, and single nucleotide genetic polymorphisms (SNPs) in members of the AHR pathway lead to adverse phenotypes in the reproductive system and result in impaired reproductive function. Thus, the following sections focus on the literature utilizing these models from a viewpoint of understanding the endogenous function of the AHR in both the female and male reproductive systems.

31.2 THE AHR REGULATES FEMALE REPRODUCTIVE FUNCTIONS

Functioning of the female reproductive system is absolutely dependent on a tightly regulated feedback system conducted by the hypothalamus–pituitary–ovarian axis (HPOA) (Fig. 31.1a) [1]. During adulthood, the hypothalamus, also known as the "central relay station," collects and integrates signals from the ovary and the environment, and relays them to the anterior pituitary gland. Specifically, the hypothalamus synthesizes and releases gonadotropin-releasing hormone (GnRH) from neurons in the medial preoptic, anterior hypothalamic areas, and arcuate nucleus through the hypophyseal portal system to the anterior pituitary gland. GnRH stimulates the anterior pituitary to synthesize and release follicle-stimulating hormone (FSH) and luteinizing hormone (LH) from the gonadotrophs. FSH binds to receptors on ovarian follicles and stimulates follicle growth and estrogen production. LH also binds to receptors on ovarian cells and can stimulate ovulation. Ovarian steroids produced in response to pituitary hormones can feedback to the hypothalamus and anterior pituitary and suppress GnRH, FSH, and LH release via negative feedback [2]. The ovarian produced sex steroid hormones (i.e., estrogens, progestins, and androgens) are also

---

The AH Receptor in Biology and Toxicology, First Edition. Edited by Raimo Pohjanvirta.
© 2012 John Wiley & Sons, Inc. Published 2012 by John Wiley & Sons, Inc.

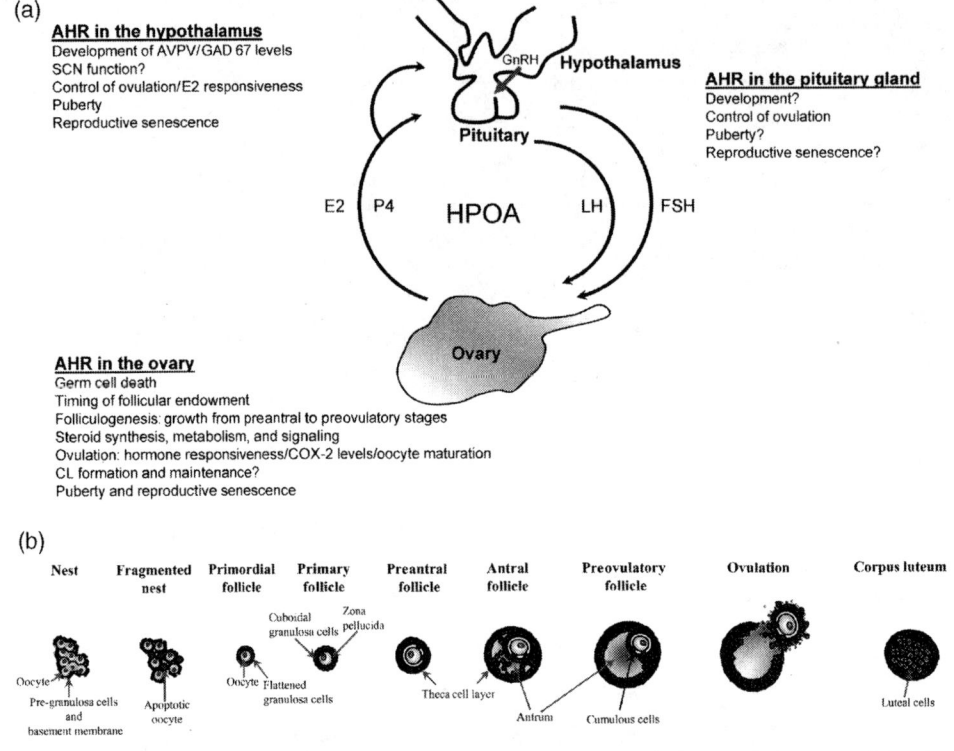

FIGURE 31.1 (a) Involvement of the AHR in controlling the hypothalamus–pituitary–ovarian axis. The female reproductive system is controlled by a feedback system conducted by the HPOA. Hypothalamic GnRH stimulates the anterior pituitary to secrete LH and FSH. The FSH binds to receptors on ovarian follicles and stimulates follicle growth. LH binds to receptors on ovarian cells to stimulate ovulation. The ovarian steroids E2 and progesterone (P4) produced in response to pituitary hormones can feed back to the hypothalamus and anterior pituitary and suppress GnRH, FSH, and LH release via negative feedback. The various roles that the AHR may play in control of the HPOA are listed. (b) The major stages of mammalian folliculogenesis. In mice, primordial follicle formation occurs postnatally and involves oocyte nest breakdown and oocyte death via apoptosis. Primordial follicles develop into primary follicles once the granulosa cells acquire a cuboidal shape. Primary follicles develop into preantral follicles when two or more layers of granulosa cells surround the oocyte and an additional layer of somatic cells called the theca forms outside the basement membrane. Preantral follicle development does not require stimulation by the pituitary gonadotropins. At puberty, the pituitary secreted FSH supports further granulosa cell proliferation and formation of an antrum. Ovulation of the dominant follicles in mice occurs in response to a rise in pituitary produced LH. After ovulation, the remaining somatic cells differentiate into luteal cells to form the corpus luteum. (See the color version of this figure in Color Plates section.)

responsible for regulating the supporting organs of the female reproductive system such as the oviduct, uterus, and vagina that are vital for successful fertilization, implantation, and supporting pregnancy [3].

Various studies have identified the AHR at all sites in the HPOA. The AHR has been localized in the female rodent hypothalamus by *in situ* hybridization, immunohistochemistry, and reverse transcription-polymerase chain reactions (RT-PCR) [4–6]. Specifically, the AHR is present in female sexually dimorphic hypothalamic regions in the preoptic area (POA) and suprachiasmatic nucleus (SCN), areas known to regulate puberty onset, ovulation, and timing of reproductive senescence [5, 7–16]. In addition, studies have reported its presence in the brains of other vertebrates such as humans and fish [17, 18]. Localization of the AHR in the female pituitary has been limited to a handful of studies conducted in the rat [4, 5, 19, 20]. However, many studies have localized the AHR in the ovary of a variety of vertebrate species. Specifically, studies using various techniques such as immunohistochemistry, radiolabeled AHR ligand binding assays, RT-PCR, and Western blots have localized the AHR protein and transcripts to various ovarian follicle cell types (oocytes, granulosa cells, and theca cells) in fish, nonhuman primates, humans, pigs, rats, mice, and rabbits [18, 21–27].

Given that the AHR is expressed in all sites of the HPOA and it is conserved in the ovary across multiple species, it is likely that the receptor functions in regulating the female reproductive system. Besides its presence in the HPOA, a growing amount of evidence indicates that the AHR has biological roles in the female reproductive system from fetal life through reproductive senescence. A brief review of the basic functions of the female reproductive system followed by an overview of some of the studies supporting biological roles for the AHR in these functions is provided below. For simplicity, the development of the reproductive organs will be described in the mouse or the rat.

31.2.1 The AHR Regulates Development of Female Neuroendocrine Functions

31.2.1.1 The AHR Regulates Female Hypothalamic Development and Function The female hypothalamus begins to develop during fetal life. During late gestation in the rodent, sex-specific nuclei differentiate in the female hypothalamus, which are critical for ovulation and female sexual behaviors [28]. The neurons of the sexually dimorphic regions of the hypothalamic preoptic area are formed around embryonic day (ED) 16.5 in the rat [28]. Formation of the POA involves mechanisms that control cell migration, tissue patterning, cell-to-cell communication, and cell fate [28]. During late gestation and early neonatal life in the female rodent, specific morphological differences in the nuclei of the female POA emerge. One area of the POA that is essential for ovulation called the anteroventral periventricular nucleus (AVPV) has 2.5 times more gamma-aminobutyric acid (GABA)/glutamatergic (Glu) neurons in females than in males. This area is also packed with estrogen receptor alpha (ESR1)- and beta (ESR2)-positive neurons that have more physical contacts with GnRH neurons in females than in males [28, 29]. Neurons in the AVPV are responsible for most of the estradiol (E2)-sensitive input to the GnRH neurons. The AVPV is a structure that is critical for sex-specific and E2-dependent LH surge release and ovulation [5].

Several studies indicate that the AHR is involved in development and functioning of the female hypothalamus. AHR and the aryl hydrocarbon receptor nuclear translocator (ARNT) protein and mRNA expression first appear in the neuroepithelium of the developing mouse brain around EDs 10–15, a time before specific brain nuclei differentiate in the rodent brain [30, 31]. Pravettoni et al. reported that the levels of AHR mRNA increase in the developing female rat hypothalamus during the early stages of sexual differentiation between EDs 16 and 19, while levels of ARNT mRNA are constant [4]. They also reported that AHR and ARNT mRNAs are present in neurons, glia, and astrocytes isolated from the female fetal hypothalamus [4].

Other studies investigating the AHR pathways in neonatal rat brains found that the *Ahr* gene is expressed in virtually all GABA/Glu neurons in the AVPV that contain estrogen receptors (ESRs) [5]. These are the neurons responsible for puberty onset and ovulation in the female. Further studies demonstrated that gestational exposure to the potent AHR ligand, 2,3,7,8-tetrachlorodibenzo-*p*-dioxin (TCDD) (1 μg/kg maternal body mass; single oral dose), abolishes the sex differences in the level of glutamic acid decarboxylase (GAD 67) mRNA expression in GABA neurons of the female AVPV [5]. GAD 67 is the enzyme required for GABA synthesis and proper functioning of the GABAergic neurons. Thus, the data suggest that the AHR could act as a transcription factor for GAD 67 since the GAD 67 promoter contains multiple canonical aryl hydrocarbon receptor response element (AHRE) sequences [5]. Thomsen et al. also reported that hairy and enhancer of split homolog-1 (HES-1), another gene that plays a role in GABAergic neuronal development in mammals, is also a target of the AHR [29, 32]. Importantly, female rats exposed *in utero* to TCDD have permanent changes in adult reproductive function such as early puberty, acyclicity, and premature reproductive senescence [33, 34]. These findings suggest a role for the AHR in development of the sexually dimorphic brain regions responsible for regulating puberty onset, ovulation, and timing of reproductive senescence.

31.2.1.2 The AHR May Regulate Anterior Pituitary Gland Development Very little is known about a role for the AHR during pituitary gland development. The anterior endocrine portion of the pituitary develops from a diverticulum of ectodermal cells from the roof of the oral cavity called Rathke's pouch, which is fully formed by ED 12.5 in the mouse [35]. The gonadotrophs, cells that synthesize FSH and LH in the anterior pituitary gland, develop in parallel with the female sexually dimorphic brain regions around ED 16 in the mouse [35]. Similar to brain development, the proper development of the anterior pituitary gland also involves mechanisms that control cell migration, tissue patterning, cell-to-cell communication, and cell fate [35]. Interestingly, potential DNA binding sites within the 5′-flanking region of the AHR gene have been identified for two important transcription factors for pituitary development, LIM homeobox protein 3 (LHX3) and POU domain, class 1, transcription factor 1 (POU1F1) [36]. This suggests a transcriptional mechanism for developmental regulation of the AHR in the pituitary. Also of interest, germline mutations in the aryl hydrocarbon receptor interacting protein (AIP) have been associated with familial pituitary cancers, further suggesting a developmental role for AHR pathways in the pituitary [37]. Further support for this statement comes from the observation that overall systemic levels of the AHR appear higher in fetuses and neonates than in adults [29]. Thus, it is very likely that this receptor plays crucial roles during organogenesis and differentiation of the pituitary, which could predetermine functions later in life, such as puberty onset, ovulation, and timing of reproductive senescence.

31.2.2 The AHR Regulates Development of the Embryonic and Neonatal Ovary

The ovary develops from a bipotential gonad during embryonic life. This process requires the induction of the female molecular pathway [38, 39]. Recent work in the field of sexual determination has uncovered factors that induce a cascade of genes that drive differentiation of the ovary, while inhibiting male determining factors. Some of the newly identified members of the ovary determination cascade include R-spondin homolog 1 (RSPO1), wingless-related MMTV integration site 4 (WNT4), and β-catenin (CTNNB1) [39, 40].

Primordial germ cells (PGCs) colonize the genital ridge around ED 10.5 in the mouse [39, 41–43]. The colonizing germ cells rapidly proliferate and remain connected by cellular bridges due to incomplete cytokinesis, forming clusters (Fig. 31.1b) [41, 44]. Somatic cells of the developing ovary also proliferate extensively and some of the cells envelop the germ cell clusters in a single layer with an outer basement membrane, forming germ cell nests [45, 46]. The germ cells become mitotically arrested and enter meiosis around ED 12.5. At this time, the germ cells become known as oocytes [43]. The timing of meiotic entry in the mouse ovary is governed by the ability of the XX germ cells to receive a retinoic acid signal originating from the mesonephros, which leads to induction of *stimulated by retinoic acid gene 8* (*Stra8*), in the developing germline [47–51].

The oocytes progress through the early stages of meiotic prophase I, until they become arrested in the diplotene stage around ED 17.5 [43]. Concurrently, many of the germ cells die via programmed cell death, in a process termed attrition [51]. The process of attrition occurs during each phase of oogenesis in both mitotic and postmitotic germ cells [51, 52]. However, most of the germ cells are lost postmitotically between ED 13.5 and postnatal day (PND) 5 [41, 52, 53]. A massive wave of germ cell death occurs from the time of birth until PND 5. This wave of germ cell death is thought to aid in germ cell nest breakdown and primordial follicle assembly. During this period, there is an approximately 70% reduction in the number of oocytes [54]. The individual surviving oocytes that break away from the nests are then assembled into primordial follicles consisting of a single oocyte, a monolayer of pre-granulosa cells, and a basement membrane encompassing the oocyte. The process of primordial follicle assembly is generally complete by PND 8 in the mouse [41]. The primordial follicles formed during this period represent a finite pool available to the female for the entire reproductive life span [41, 55].

Several studies support the hypothesis that the AHR is involved in cell death/survival signaling in the developing ovary and that the AHR controls both oocyte development and ultimately follicle pool formation. In one study, ED 13.5 AHRKO and wild-type (WT) ovaries were cultured for 72 h in the absence of hormonal support with the purpose of inducing apoptosis [26]. Interestingly, the AHRKO fetal ovaries had significantly higher numbers of nonapoptotic oocytes compared to the WT fetal ovaries after identical culture conditions [26]. This suggests that the AHR could be involved in regulating programmed cell death in embryonic oocytes still contained within nests. Furthermore, two independent groups performed primordial follicle counts and both groups reported that the neonatal AHRKO ovaries contain approximately twice as many primordial follicles during PNDs 2–4 when compared to the WT neonatal ovaries of the same age [26, 56]. This could be the result of excessive numbers of oocytes that quickly assemble into primordial follicles shortly after birth or production of primordial follicles that are resistant to cell death signals during early neonatal life in the AHRKO ovaries.

Other evidence supporting a cell death/survival role for the AHR in the ovary comes from studies demonstrating that the AHR ligand, dimethylbenz[*a*]anthracene (DMBA; 1 μM), and one of its metabolites, 9,10-dimethylbenz[*a*]anthracene-3,4-dihydrodiol (DMBA-DHD; 0.1–1 μM), induce apoptosis in primordial follicles from WT cultured neonatal mouse ovaries, whereas no apoptosis occurs in AHRKO neonatal ovaries [57]. Finally, two studies established that activation of the AHR by DMBA, but not by TCDD, accelerates germ cell death in the mouse ovary both embryonically and neonatally by inducing the proapototic factor BCL2-associated X protein (BAX) through functional AHREs in the Bax promoter [57, 58]. Interestingly, substitution of an existing guanine or cytosine to an adenine in the Bax promoter three bases downstream of a core AHRE sequence renders the gene inducible by TCDD [57]. This suggests ligand-dependent discrimination of AHREs in the Bax promoter and that other cell death/survival factors involved in ovarian development could also be under transcriptional control of the AHR depending on ligand binding. This suggestion is supported by studies that have shown that exposure to polychlorinated biphenyls (PCBs) (1–100 μg per chicken egg) promotes germ cell survival in chicken neonatal ovaries, while TCDD (1 μg/kg; single oral dose) does not cause germ cell loss in rats [59, 60]. These differences in biological effects elicited by AHR ligands could also be due to differences between species, doses, and ligand–AHR affinity.

Collectively, these existing data support pleiotropic roles for the AHR in fetal and neonatal ovarian development possibly by regulating cell death factors that contribute to the survival or death of ovarian germ cells and/or somatic cells. This regulation could affect the process of follicular endowment by controlling the ratio of somatic to germ cells and in turn the amount of specific growth factors in the fetal and neonatal ovary. The size and reproductive capacity of the primordial follicle pool involve a dynamic interplay between programmed cell death and proliferation of somatic and germ cells both embryonically and neonatally [61]. Future studies are warranted to measure cell survival/death factors in AHRKO and WT ovaries in the absence or presence of known AHR ligands. Additional studies quantifying the ratio of somatic cells and germ cells and the amount of cell death in these ovaries will aid in delineating between a cell death or survival role for the AHR during fetal and neonatal life. Further, it may be useful in the future to develop germ cell- or granulosa cell-specific knockout animal models to distinguish whether the impaired fertility in the global AHRKO animal is mainly due to a germ cell or somatic cell defect, or both. In addition, it may be crucial to examine involvement of the AHR in other processes important to early ovarian development, such as the initiation and progression of meiosis I in the developing female germline during fetal life using AHRKO and WT ovaries in the presence or absence of AHR ligands. During the first meiotic prophase, the crucial steps of chromosomal condensation, synapsis, homologous recombination, and chromatin decondensation occur [62]. Any errors in these steps can lead to germ cell death or defective oocytes, altering the process of follicular endowment [63].

31.2.3 The AHR Regulates Folliculogenesis

The follicle is the main functional unit of the adult ovary. Proper functioning of the ovary during adult life relies on recruitment of cohorts of primordial follicles into the growing pool. There are two broad categories of follicle recruitment in the ovary: initial activation of primordial follicles from the resting pool, which occurs throughout the female life span until reproductive senescence, and cyclic recruitment of a limited number of small follicles for ovulation from the growing cohort, which occurs only after puberty [39, 64]. A follicle must go through several developmental stages in a process called folliculogenesis before it is capable of ovulation. Folliculogenesis is a complex physiological process that relies on autocrine, paracrine, juxtacrine, and endocrine factors [39].

The major stages of folliculogenesis can be classified into several categories (Fig. 31.1b). The names commonly used to describe these follicle stages are primordial, primary, preantral, antral, and preovulatory [65]. During transition from the primordial to primary follicle, the morphological appearance of the granulosa cells surrounding the oocyte changes from squamous to cuboidal, and the oocyte increases in size [41, 66]. Primary follicles can then grow into preantral follicles. During this process, the granulosa cells proliferate, producing multiple layers of granulosa cells surrounding the oocyte. In addition, the oocyte secretes a ring of proteins that form the zona pellucida. Stromal cells begin to condense around the outer basement membrane to form additional layers of spindle-shaped cells called the theca cells [3]. The theca cells develop into two distinct layers, an inner highly vascular theca interna and a fibrous theca externa. Intercellular communication between the different follicle cell types is partly maintained through gap junctions [67, 68]. Once follicles develop a theca layer, they are capable of producing E2. The theca cells are responsible for producing androgens.

The androgens then diffuse to nearby granulosa cells where they are converted by cytochrome P450, family 19, subfamily A, polypeptide 1 (aromatase) to E2 [39].

As the follicle continues to grow, fluid begins to accumulate between the proliferating granulosa cells, ultimately forming an antral space. Once the follicle obtains an antral space, it is called an antral follicle and has entered the gonadotropin-dependent growth phase. Starting from this phase, follicle differentiation and growth is partly under the control of the pituitary gonadotropin hormones LH and FSH. LH stimulates the synthesis of androgens by binding to LH receptors (LHCGR) on the thecal cells and FSH stimulates the conversion of androgens to E2 by binding to follicle-stimulating hormone receptors (FSHRs) on the granulosa cells. The E2 produced by the follicles promotes granulosa cell proliferation and follicle growth [69]. As the volume of the follicular fluid increases, the antral space expands, subsequently increasing the size of the follicle. During the later stages of antral follicle growth, the oocyte becomes suspended in the antral fluid and is surrounded by a layer of granulosa cells called the cumulus oophorus. Only follicles at this stage of development are called preovulatory and are capable of ovulation [39].

During ovulation, the follicle ruptures and releases the oocyte for fertilization. The remaining somatic cells form the corpus luteum (CL) (Fig. 31.1b). The cells of the CL become specialized and their primary role is to produce progesterone, which is critical for maintenance of early pregnancy. In the absence of pregnancy hormones, the CL regresses in a luteolytic process leading to a drop in progesterone levels, a process necessary for maintaining reproductive cyclicity [70].

In the mouse, ovulation occurs approximately every 4–5 days [71]. Only a few follicles grow from the primordial to the preovulatory stage and release an oocyte during ovulation. Most follicles (>95%) reach the preantral/antral stage and undergo atresia via programmed cell death [41, 72]. Female infertility can be the result of problems in follicle pool maintenance, follicle recruitment, follicle growth, follicle atresia, steroidogenesis, and ovulation [39, 73, 74].

Several studies show that the AHR regulates ovarian follicle growth. The first line of evidence comes from studies suggesting that the AHR regulates the number of preantral and antral follicles present in the ovary. First, histological evaluation of ovaries from AHRKO and WT mice at various ages revealed that PND 53 AHRKO ovaries contain significantly reduced numbers of preantral and antral follicles, whereas the numbers of primordial and primary follicles are similar in both genotypes [56]. Second, rats exposed *in utero* on gestational days (GDs) 7–13 to ligands for the AHR such as PCBs (2.5 mg/kg) have reduced numbers of preantral and antral follicles compared to unexposed rats [75]. These findings suggest that the AHR may be an important regulator of follicle growth from the preantral to antral stages, but that it is not important for maintenance of the primordial follicle pool or growth of follicles from the primordial to primary stages [56].

In recent years, it has become clear that the AHR plays a pivotal role in ovarian follicular growth. One study found that cultured antral follicles isolated from AHRKO mice ovaries have slower growth compared to antral follicles isolated from WT ovaries, as demonstrated by a smaller follicle diameter without a change in percentage of antrum/follicle volume [76]. Upon closer examination, immunohistochemical analysis for a marker of cell proliferation, proliferating cell nuclear antigen (PCNA), revealed decreased staining in the granulosa cells of preantral and antral follicles from AHRKO ovaries compared to WT ovaries at various ages [76]. Concurrently, mRNA and protein levels of two cell cycle progression factors, cyclin D2 (CCND2) and cyclin-dependent kinase 4 (CDK4), were decreased in isolated AHRKO antral follicles compared to isolated WT antral follicles [76]. This suggests that the AHR pathway could interact with cell cycle pathways to modulate granulosa cell proliferation in the ovary. Interestingly, one later study found that the AHR forms a protein complex with CDK4 and cyclin D1 (CCND1) in the absence of exogenous ligands and facilitates cell cycle progression in human breast cancer cells [77]. This finding could suggest a potential mechanism for control of these cell cycle factors by the AHR in the ovary as well. Furthermore, Bussmann et al. demonstrated that β-naphthoflavone, a ligand and agonist of the AHR, increases the proliferation of cultured rat granulosa cells by amplifying the mitogenic actions of FSH and E2 [78]. Collectively, these studies suggest that the AHR regulates follicle growth by promoting granulosa cell proliferation, by either directly or indirectly interacting with cell cycle pathways.

Another process that affects the number of follicles in the ovary is the process of follicular atresia [41, 72]. Atresia is a normal degenerative process that is thought to begin with apoptosis of granulosa cells and is responsible for eliminating more than 95% of the ovarian follicle reserve by the time of reproductive senescence [79, 80]. To evaluate whether the AHR plays a role in atresia of follicles in the adult ovary, Benedict et al. compared the percent of atretic preantral and antral follicles in AHRKO and WT ovaries at various ages, and found that AHRKO and WT ovaries contain a similar percentage of atretic follicles [81]. They also found similar amounts of DNA fragmentation, a major indicator of apoptosis, in isolated antral follicles from equine chorionic gonadotropin primed immature AHRKO and WT mice [81]. Similarly, PND 22 rats gestationally exposed to TCDD (1 μg/kg maternal body mass; single oral dose) have reduced numbers of preantral and antral follicles with no increase in apoptotic cell death in their ovaries compared to rats exposed to vehicle alone [82]. Collectively, these studies suggest that the AHR is a regulator of preantral and antral follicle growth, but may not play a role in regulating atresia in the adult ovary.

31.2.3.1 The AHR Modulates Steroid Synthesis, Metabolism, and Signaling in the Ovary

A growing amount of evidence indicates that the AHR is a regulator of steroid biosynthesis, steroid responsiveness, and possibly steroid metabolism in the ovary. The steroidogenic pathway relies on both the ability of the theca cells to produce androgens and the ability of the granulosa cells to aromatize the androgens, primarily androstenedione to E2. Binding of LH to the LHCGRs in theca cells stimulates a cascade of steroidogenic enzymes such as cytochrome P450 cholesterol side chain cleavage (P450scc), 3β-hydroxysteroid dehydrogenase (3β-HSD), and cytochrome P450 17α-hydroxylase (P450c17) to convert cholesterol to androstenedione in theca cells [83–85]. Binding of FSH from the anterior pituitary to FSHRs in granulosa cells stimulates the expression of aromatase, which is responsible for converting the theca produced aromatizable androgen, androstenedione, to E2 [39, 86]. Signaling of the steroid hormones mainly occurs through autocrine signals between their receptors within the ovary and endocrine signaling through their receptors in the brain and other organs [1]. In the ovary, E2 stimulates the growth of antral follicles to the preovulatory stage by binding to ESR1 and ESR2 [85].

Several studies utilizing AHRKO mouse models indicate that the AHR regulates follicle growth by modulating the levels of E2 through the steroidogenic pathway and controlling hormone receptors. Specifically, both circulating and follicle produced E2 levels are significantly reduced in AHRKO animals compared to WT animals [76]. These low levels of E2 coincide with reduced follicle growth of cultured antral follicles isolated from AHRKO mice compared to cultured follicles isolated from WT mice [76]. Interestingly, Barnett et al. were able to restore AHRKO follicle growth to WT levels after exogenous E2 treatments, suggesting that the low E2 levels in AHRKO mice are what contribute to the reduced follicle growth [76].

Similarly, Baba et al. reported significantly reduced intraovarian E2 levels after gonadotropin treatment in AHRKO ovaries compared to WT ovaries [87]. Interestingly, they were able to demonstrate that the reduced E2 levels in the AHRKO mice are partly due to the lack of activation of aromatase expression and activity. Specifically, they found that the AHR mRNA is constitutively expressed in the WT ovary after gonadotropin treatment and that the AHR is necessary for proper expression of the *aromatase* gene. They established that the AHR directly activates the *aromatase* gene through interaction with AHREs in the ovary-specific promoter region of the *aromatase* gene and through cooperation with an orphan nuclear receptor called nuclear receptor subfamily 5, group A, member 1 (Nr5a1). In addition, they suggested that the aryl hydrocarbon receptor repressor

(AHRR) may be responsible for regulating AHR activity in the ovary, since AHRR mRNA levels were increased 6 and 7 h after hCG injection. These data support a novel mechanism that involves AHRR control of the AHR and aromatase activation by the AHR to control E2 synthesis in the ovary, giving one explanation for why the AHRKO mice have lower aromatase and E2 levels in their ovaries compared to WT mice [87].

Another mechanism responsible for regulating E2 levels in the ovary is ligand–receptor interactions. Appropriate binding of FSH, LH, and E2 to their corresponding receptors is required for timely expression of the steroidogenic enzymes and steroid production. Interestingly, levels of ESR1, ESR2, FSHR, and LHCGR mRNA were measured and found to be significantly reduced in antral follicles isolated from AHRKO mice compared to WT mice [76, 88]. This suggests that in addition to controlling E2 levels by modulating *aromatase* expression, the AHR may also regulate *Esr1*, *Esr2*, *Fshr*, and *Lhcgr* in antral follicles.

In addition to the studies utilizing the AHRKO mice, numerous studies using various AHR ligands support involvement of the AHR in regulating E2 levels. Specifically, prepubertal female rats (PNDs 14–21) exposed in utero and lactationally to TCDD (1 μg/kg maternal body mass; single oral dose) have a similar phenotype to the AHRKO mice in that they have reduced serum E2 levels compared to unexposed mice [89, 90]. Furthermore, treatment with TCDD ranging from 3.1 pM to 10 nM reduces E2 production in theca and granulosa cell coculture systems [91–94].

Similar to the downregulation of the *aromatase* gene in the absence of the AHR, TCDD treatment of cultured granulosa cells results in a downregulation of aromatase as well as other steroidogenic enzymes in the cascade. For example, one study found that TCDD exposure reduces aromatase and P450scc mRNAs in cultured rat granulosa cells [91]. Furthermore, another study demonstrated that TCDD inhibits E2 synthesis by decreasing the supply of androgens in human luteinized granulosa cells through decreasing P450c17 activity and protein levels [94]. In contrast, the reduced E2 levels in the AHRKO mice do not seem to be the result of a reduced ability of the follicles to produce androgens, since intraovarian testosterone levels are similar between AHRKO and WT gonadotropin-stimulated mice [87]. Further, Baba et al. found that another AHR ligand, DMBA (50 mg/kg; single i.p. dose), induces *aromatase* gene expression in WT mouse ovaries, leading to a rise in E2 regardless of the stage of the estrous cycle. These differences in results could be due to endogenous actions of the AHR versus induction of the AHR pathway by AHR ligands, as well as differences between species, doses, and ligand–AHR affinity.

Similar to the AHRKO mice, ovaries from TCDD-exposed mice (5 μg/kg; single i.p. dose) express lower levels of ESR1 compared to ovaries from vehicle-treated control mice [95, 96]. Further, granulosa cells exposed to TCDD (1 pM to 10 nM range) express lower levels of FSHR and LHCGR compared to untreated controls [97, 98]. Collectively, these studies suggest that the AHR can interact with the E2 signaling pathway at multiple sites in the ovary to activate or repress estrogen-responsive genes such as *Esr1* and *Esr2*, as well as to modulate the levels and activity of steroidogenic enzymes such as aromatase and P450c17.

In addition to helping control the E2 biosynthesis and signaling pathways, the AHR may also regulate E2 metabolism in the ovary. AHR activation has been shown to increase metabolism of E2 by inducing the cytochrome P450 monooxygenases, CYP1A1 (cytochrome P450, family 1, subfamily A, polypeptide 1) and CYP1B1 (cytochrome P450, family 1, subfamily B, polypeptide 1) [99–101]. CYP1A1 and CYP1B1 are responsible for converting E2 to catechol estrogens, 2-hydroxyestradiol and 4-hydroxyestradiol, respectively [101]. One study demonstrated that these metabolites can inhibit hormonally stimulated rat granulosa cell proliferation in culture [78]. This suggests a role for estradiol metabolites in regulation of granulosa cell proliferation. Interestingly, in the absence of exogenous AHR ligands, the levels of CYP1B1 mRNA are decreased on the morning of estrus and increased on the evening of proestrus in the rat ovary [102]. Similarly, another study found that CYP1A1 activity was present in the rat ovary, and was highest during the proestrous phase of the ovarian cycle [103]. In addition, CYP1A1 mRNA was constitutively expressed in mouse oocytes before maturation [104]. It is possible that the AHR affects follicle growth by modulating the levels and activity of the cytochrome P450 monooxygenases, ultimately controlling the levels of E2 and its metabolites in the ovary. More studies are needed to understand if the AHR controls E2 metabolizing enzymes in the ovary ultimately affecting follicle growth.

Although it has not been directly tested, it is also possible that the AHR could be affecting follicle growth and E2 levels through the insulin signaling pathway. This suggestion is supported by a study that revealed that 7-month-old AHRKO females have decreased systemic glucose uptake and reduced fasting plasma insulin levels compared to WT animals [105]. Since insulin has been shown to promote follicle growth *in vitro*, another hypothesis for the decreased follicle growth in the AHRKO ovaries could be due to altered insulin signaling in these animals [69, 106]. Several studies have reported that insulin can also stimulate aromatase mRNA expression and activity in human granulosa cells in culture, while also increasing E2 levels, suggesting that the enhanced follicle growth in insulin-treated follicles is partly due to increased conversion of androgens to E2 [107, 108] Interestingly, one study also demonstrated that granulosa cells from 3-day estradiol-treated immature rats showed a concentration-dependent increase in cyclin D2 mRNA expression in response to insulin [106]. Collectively, these studies support a proliferative role for insulin in the ovary through a mechanism that enhances E2 synthesis and may also involve the AHR.

31.2.4 The AHR Regulates Ovulation

Ovulation is a complex process that requires a tightly coordinated feedback system in the HPOA (Fig. 31.1a). In preparation for ovulation, the ovarian follicles selected for ovulation become vascularized and respond preferentially to LH and FSH secreted by the gonadotrophs in the anterior pituitary [109]. This stimulates the maturing preovulatory follicles to produce increasing amounts of E2. The release of LH and FSH from the anterior pituitary is driven by pulsatile secretion of GnRH from neurons in the AVPV of the hypothalamus [110]. Once the E2 levels produced by the cohort of preovulatory follicles reach a threshold level, a positive feedback effect on GnRH-secreting neurons is triggered and followed by a subsequent ovulatory surge of LH and FSH [110]. The presence of follicle produced progesterone is also necessary to induce the ovulatory release of gonadotropins from the anterior pituitary [111]. During ovulation, LH stimulates granulosa cells in the preovulatory follicles to increase production of progesterone, synthesize progesterone receptors (PRs), and decrease E2 secretion [111]. LH also activates an enzyme called cyclooxygenase-2 (COX-2), which increases production of substances called prostaglandins [112]. Prostaglandins are responsible for initiating vasodilation, immunomodulation, inflammation, and promotion of proteolytic enzymes and smooth muscle contraction, which are all processes necessary for follicle rupture and ovulation [109]. Concurrently, oocyte maturation occurs when meiosis resumes and arrests again in diplotene of meiosis II until fertilization (Fig. 31.2) [39].

Multiple studies indicate that the AHR plays crucial roles in ovulation. Specifically, both animals exposed to various AHR ligands and the AHRKO mice exhibit significantly reduced numbers of ovulated ova or even blocked ovulation in response to gonadotropin treatment compared to controls [87, 88, 113–120]. Interestingly, these studies provide evidence that the AHR could affect ovulation at all levels of the HPOA. Several studies utilizing AHR ligands support a role for the AHR in regulating ovulation at the level of the hypothalamus. For example, exposure to TCDD and other

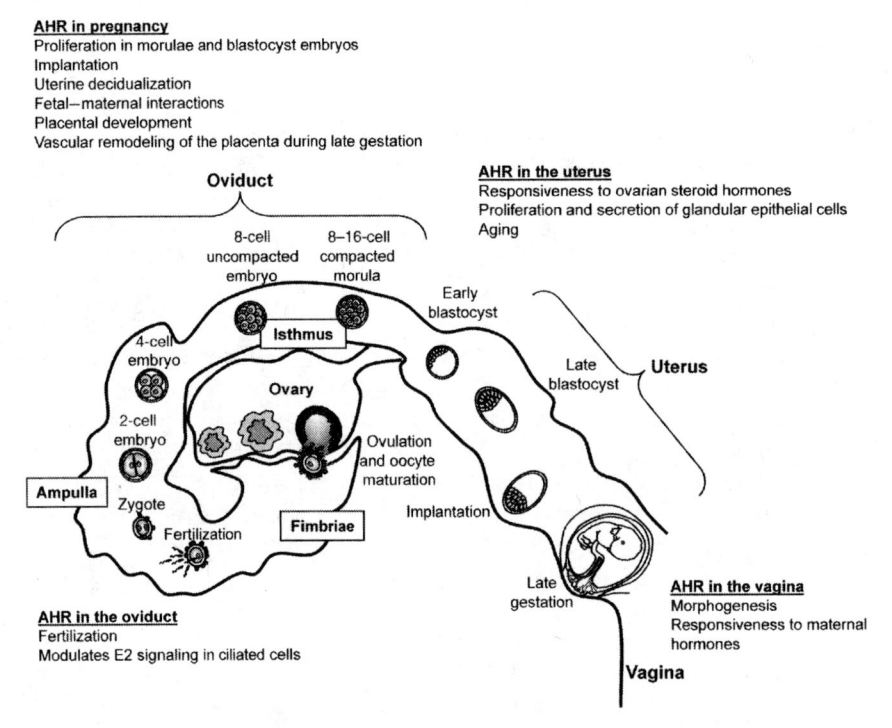

FIGURE 31.2 Involvement of the AHR in supporting pregnancy. Following ovulation, fertilization occurs in the ampulla of the oviduct and the embryo undergoes several rounds of mitotic cell division, forming the morula. The propulsion of the embryo through the ampulla and isthmus of the oviduct to the uterotubal junction involves muscle contractions, ciliate activity, and flow of secretions. The morula stage embryo enters the uterine lumen and transforms into a blastocyst. The blastocyst further develops and differentiates before it attaches to the uterine lining to initiate the process of implantation. During late gestation, vascular remodeling of the placenta provides the developing fetus with the required nutrients and oxygen. The various roles that the AHR may play in controlling pregnancy and in development and functioning of the supporting female organs are listed. (See the color version of this figure in Color Plates section.)

related AHR ligands in gonadotropin primed immature female rats results in an attenuated and altered pattern of preovulatory LH and FSH secretion without a decrease in circulating E2 levels compared to the controls [121]. Administration of a 10-fold higher than physiological level of E2 to TCDD-exposed animals completely reversed the TCDD effects on gonadotropin secretion and ovulation [122]. This led to the conclusion that the attenuation of gonadotropin levels after TCDD treatment was not due to reduced circulating E2, but rather due to a reduced responsiveness of the hypothalamus and pituitary to the positive feedback of estrogens. The same investigators were able to use tamoxifen, an estrogen receptor antagonist, to reverse the ability of E2 to restore ovulation in TCDD-treated rats, suggesting a mechanism that involves AHR- and E2-mediated pathways to control ovulation [122].

Interestingly, though it has not been directly tested, the AHR could also be regulating ovulation through control of the central pacemaker in the SCN. The AHR and members of the AHR pathway are present in the SCN, and female ovariectomized AHRKO mice display different activity rhythms compared to the WT mice [7, 123]. Since the SCN is responsible for collecting environmental cues and signals from the peripheral organs to coordinate the timing of ovulation, a dysregulation of the AHR in the SCN could alter the timing of receptivity in the female hypothalamus and block ovulation [14, 124].

Some studies also suggest that the AHR may control ovulation at the level of the pituitary. One study reported that AHR mRNA was present in adult female rat pituitaries and levels increased after ovariectomy (OVX) [19]. E2 administration attenuated the AHR mRNA levels in the OVX animals, suggesting crosstalk between the E2 and AHR pathways in the pituitary [19]. In addition, rat pituitary halves cultured in the presence of GnRH and TCDD (0.1–100 nM) responded with a dose response release of LH into the media, which was not attenuated by a GnRH antagonist [125]. This is evidence that the AHR pathway could be functioning in the pituitary to regulate gonadotropin release, but does not require GnRH receptivity.

In addition to a role for the AHR during ovulation in the hypothalamus and pituitary gland, convincing evidence shows that it regulates ovulation at the level of the ovary. First, it is possible that the AHR itself is activated in ovarian follicles by gonadotropins. In support of this, one study reported that AHR mRNA expression was induced in granulosa cells isolated from macaque monkeys during controlled ovarian stimulation cycles after administration of an ovulatory bolus of human chorionic gonadotropin (hCG) [126]. Furthermore, several studies support the hypothesis that the reduced ovulation rate in the AHRKO mice is mostly due to an ovarian defect rather than due to defects in the hypothalamus or pituitary. Specifically, no differences were found in the levels of circulating LH and FSH between the AHRKO and WT mice during all stages of the estrous cycle [88]. In addition, the numbers of functional gonadotrophs in the anterior pituitary were similar between AHRKO and WT mice [87], and circulating levels of LH were similar in both AHRKO and WT mice after administration of a GnRH agonist, emphasizing a biological role for the AHR in the ovary rather than in the hypothalamus and pituitary gland [87]. This was surprising since as mentioned before, the AHRKO animals have significantly lower circulating levels of E2, and reduced E2 levels in other animal models have resulted in a hypothalamus–pituitary axis that is hypersensitive to the feedback of ovarian steroids with increased output of pituitary produced FSH and LH [127–130].

While E2 treatment restores LH and FSH surges in TCDD-exposed female rats, exogenous GnRH administration leads to a supranormal LH and FSH surge and only partially restores ovulation [131]. A failure of the supranormal gonadotropin surge to completely restore ovulation in rats receiving TCDD and GnRH suggests direct effects of AHR ligands on the ovary, decreasing its responsiveness to gonadotropins. This further suggests a role for the AHR in facilitating responsiveness of ovarian follicles to the hormones required for ovulation. In support of this, Baba et al. found that intraovarian E2 concentrations from superovulated mice were significantly lower in AHRKO mice compared to WT mice, suggesting a reduced responsiveness of antral follicles to gonadotropins [87]. Furthermore, Barnett et al. reported that cycling AHRKO mice had reduced transcript levels for FSHR and LHCGR in isolated antral follicles compared to those from WT cycling mice [88]. Interestingly, they also demonstrated that the AHR may directly regulate the *Fshr* gene through a functional AHRE in the *Fshr* promoter, suggesting a direct mechanism for control of this receptor by the AHR [88]. Further, they found that the AHRKO mice had a reduced ability to ovulate in response to exogenous gonadotropins compared to WT mice, demonstrated by a significant reduction in the number of ova in AHRKO mice compared to the WT mice after treatment [88]. Similarly, Baba et al. demonstrated only a partial rescue in the number of ovulated ova in superovulated AHRKO mice administered increasing amounts of exogenous E2, compared to WT mice, indicating a reduced responsiveness to E2 in AHRKO mice compared to WT mice [87]. This reduced responsiveness to E2 could be at any of the HPOA sites, not only at the level of the preovulatory follicle. Collectively, these studies support a role for the AHR in regulating ovulation through controlling follicular responsiveness to hormones.

Another event that is controlled by the ovulatory LH surge is oocyte maturation, which requires the resumption of meiosis and progression to the diplotene stage of meiosis II where the oocyte arrests again until fertilization [39]. Interestingly, experimental evidence supports the involvement of the AHR in oocyte maturation. Specifically, both AHR and

CYP1A1 levels increase after *in vitro* maturation of bovine oocytes in the absence of AHR ligands [23, 132]. Further, in the presence of AHR ligands (PCBs; 0.1–10 μg/mL), oocyte maturation is disrupted *in vitro* [133].

Finally, during the ovulation process, LH is responsible for activating the COX-2 enzyme in the preovulatory follicles, a factor required for activation of the cascade of events leading to follicle rupture [109]. Studies show that the AHR is involved in this process. Specifically, animal studies have shown that in rats exposed to TCDD (32 μg/kg; single oral dose), ovulation is blocked and levels of ovarian COX-2 are reduced [120]. Furthermore, several studies report that COX-2 is regulated by an AHR-dependent pathway in nonovarian cell types [134, 135].

Collectively, these studies support a role for the AHR in the process of ovulation at all the sites of the HPOA, possibly by regulating development during sexual differentiation, modulating responsiveness to hormones, and/or regulating oocyte maturation, follicle rupture, and ultimately oocyte release. Though these studies suggest mechanisms involving crosstalk between the E2 and the AHR pathways, further investigation is needed to directly test this and other mechanisms of action. It appears that both animals exposed to TCDD and AHRKO animals exhibit a reduced responsiveness to E2 at all sites of the HPOA. To better understand this potential crosstalk, studies could compare levels and distribution of ESRs in the hypothalamus and pituitaries of AHRKO, WT, and animals treated with AHR ligands. Further, other factors essential for ovulation, such as the PR, androgen receptor (AR), and factors necessary for follicle rupture could be compared in antral follicles from AHRKO, WT, and animals exposed to various AHR ligands [69, 109, 136].

31.2.5 The AHR Regulates the Corpus Luteum

The AHR may be involved in regulating the corpus luteum. After ovulation, differentiation of the luteinized granulosa cells to luteal cells, production of progesterone by the luteal cells, and the demise of the CL are all essential functions that are important for maintaining cyclicity, promoting implantation, and supporting early pregnancy [136, 137]. Proper functioning of the CL involves extensive vascularization, hypertrophy of the luteinized granulosa cells, and demise of the luteal cells and vasculature in the absence of pregnancy hormones in some species [138].

Barnett et al. examined serial sections of the AHRKO and WT ovaries and found that AHRKO ovaries had significantly fewer CLs when compared to the WT ovaries at PND 53. They also found that immature gonadotropin-stimulated AHRKO mice had fewer CLs in their ovaries compared to immature gonadotropin-stimulated WT mice [88]. Similarly, Baba et al. reported a reduction in the number of CLs to a level almost histologically undetectable by PND 63 in the AHRKO ovaries [87]. Notably, ovarian weight can change relative to reproductive status and stage of the mouse estrous cycle, and the CLs make up a large part of the adult ovarian mass and are an indication of ovulation [70]. Baba et al. reported a significant reduction in ovarian wet weight in adult AHRKO mice when compared to their WT littermates [87]. Similarly, Benedict et al. found that the gonadotropin-stimulated immature AHRKO ovaries had a significantly reduced weight when compared to the stimulated immature WT ovaries [81, 88]. Interestingly, TCDD treatment *in vivo* also results in smaller ovaries with fewer CLs than vehicle treated [121].

It is possible that the reduced ovarian weight and reduction in the number of CLs in the AHRKO ovaries are not only due to an ovulation defect, but are also due to a failure in CL formation and function. Since the AHR pathway has been shown to crosstalk with factors in the angiogenesis pathway and angiogenesis is crucial for proper functioning of the CL, it is possible that the AHR is a regulator of angiogenesis in the CL [138, 139]. This possibility could be examined in future studies.

31.2.6 The AHR Regulates the Oviduct, Fertilization, and Early Embryo Development

Fertilization and early embryo development involves a complex interplay between the oviduct and the gametes before and after fertilization (Fig. 31.2). The oviduct is an important female reproductive organ because it enables reception, transport, and maturation of male and female gametes, their fusion, and supports early embryo development. The oviduct is a thin muscular tube that includes the fimbriae, the ampulla, the isthmus, and the uterotubal junction. It is mainly comprised of a thin muscular layer lined by an inner epithelial layer made of mostly ciliated and secretory cells. These cells are sensitive to circulating ovarian-derived steroid hormones (mostly progesterone and estradiol). In response to steroid hormones, the cells in the oviduct become ciliated and produce secretions important for fertilization and early embryo development. Muscle contractions, activity of cilia, and tubular secretions in the oviduct are required for movement of both gametes and blastocysts [140, 141].

Although it has not been extensively studied, some evidence suggests that the AHR may be an important factor for normal oviduct function, fertilization, and early embryo development (Fig. 31.2). First, AHR protein is located in the cytoplasm of both secretory and ciliated cells of the rabbit ampulla and isthmus [27]. The AHR has also been detected in the nucleus of stromal cells in the rabbit oviduct during the preimplantation period [27]. Its presence in the nucleus is suggestive of recruitment to transcription sites [142], and indicates active AHR signaling in stromal cells of the oviduct. In addition, the AHR has been detected in the human oviduct [22]. Interestingly, Hombach-Klonisch et al. demonstrated that activation of the AHR by TCDD in a telomerase-immortalized porcine oviduct epithelial cell line resulted in downregulation of the ESR1 [143]. Since proliferation, differentiation, and motility of ciliated cells in the oviduct are coordinated by regulation of ESR1 and PR, this downregulation could suggest a role for the AHR in modulating these receptors or in their signaling pathways [140, 144, 145]. Furthermore, the main site of fertilization, the ampullar region of the oviduct, expresses high levels of ESR1 protein and mRNA in the epithelial cells during the ovarian midcycle [146]. Thus, the AHR could play a role in fertilization by controlling the morphological changes and proliferative activity of the ciliated cells by modulating the estrogen signaling pathway in the oviduct. This possibility is supported by studies suggesting that the AHR modulates the ESR1 signaling pathway in other tissues [147, 148].

Interestingly, in addition to its possible role in regulating the oviduct environment, the AHR plays a role in regulating the early embryo itself. One study showed that CYP1A1 mRNA is increased by more than 100-fold in the mouse ovum just 12 h after fertilization, an indication of AHR pathway activation [104]. In addition, AHR transcripts and protein were detected in morulae and blastocyst stage preimplantation mouse embryos [149]. Further, the percentage of blastocyst formation was reduced when mouse blastocysts were cultured with AHR antisense RNA [149]. Maternal TCDD exposure leads to disrupted morphogenesis at the compaction stage (8–16 cell) of preimplantation embryos, which is suggestive for a role for the AHR at these stages [150]. Although this is suggestive of a role for the AHR in early embryogenesis and possibly in preparation for implantation, more studies are necessary to directly test these possibilities.

31.2.7 The AHR Regulates the Uterus, Implantation, and Placentation

The primary function of the uterus is to respond to the shifting levels of steroid hormones during the ovarian cycle in preparation to receive the conceptus and nourish it through term. The adult uterus is mainly comprised of two layers, a thick outer layer called the myometrium and a thinner internal layer called the endometrium. The myometrium consists of smooth muscle, whereas the endometrium consists of a stromal matrix over which lies a columnar luminal epithelium with glandular epithelium extensions that penetrate the stroma [3, 151].

Cyclic changes in the uterus depend on the steroids produced by the ovary and can be divided into two major stages, the proliferative and secretory phases. A rise in estrogen during the follicular phase in the ovary causes a thickening of the endometrium due to proliferation of both the stromal and epithelial cells. This is called the proliferative phase. After ovulation, the proliferative phase is followed by the secretory phase or preimplantation period. During this period, the uterus becomes distended due to synthesis and secretions of glycoproteins, sugars, and amino acids by the endometrial glands in response to a rise in progesterone. This process of remodeling and differentiation prepares the uterus for implantation of the embryonic blastocyst. During implantation, the embryonic blastocyst establishes contact with the luminal epithelium and maternal blood vessels in a process called decidualization [3, 151].

Numerous studies provide evidence that the AHR is important for uterine functions including proliferation, secretion, and establishment and maintenance of the fetal–maternal interactions required for pregnancy (Fig. 31.2). First, studies have identified the presence of AHR as well as its dimerization partner ARNT in the uterus [152–155]. Specifically, one study found AHR and ARNT mRNA and protein expression in both the epithelium and stroma of the mouse uterus [152]. In addition, AHR and ARNT mRNA and protein are expressed in human endometrium and myometrium from both reproductive-age and postmenopausal woman [153–155].

Second, the AHR–ARNT pathway can be activated in the uterus of both rats and humans in the presence of AHR ligands. For example, one study reported induction of CYP1A1 and CYP1B1 in the uterus of hypophysectomized rats after treatment with the AHR agonist β-naphthoflavone (70 mg/kg; single i.p. dose) [156]. Further, other studies have shown activation of the AHR–ARNT pathway by induction of CYP1A1 and CYP1B1 mRNA in human endometrial explant cultures after TCDD exposure (1 pM to 10 nM) [154, 157]. This coincides with a downregulation of AHR mRNA in human cultured endometrial cells [158]. AHR expression in various uterine cell types and activation of the AHR–ARNT pathway in mammalian uteri are supportive of a role for the AHR in uterine physiology.

Another line of evidence that the AHR is involved in uterine function comes from studies focused on the presence and localization of the AHR in human uteri. Küchenhoff et al. found that the expression of both AHR mRNA and protein in the uterus depends on the stage of the menstrual cycle in premenopausal cycling woman [153]. Specifically, some patients expressed the receptor during the proliferative phase of the menstrual cycle, while more than 70% expressed the receptor around the time of ovulation [153]. Due to an induction of this receptor during the proliferative phase and around the time of ovulation, it is likely that the AHR may crosstalk with the estrogen signaling pathways since ESRs are upregulated during these stages [159]. In addition, the cytoplasmic location of the AHR suggests that it is not transcriptionally active, but may be acting through a nongenomic pathway in these glandular cells, as has been demonstrated in a breast cancer cell line [147, 148]. A further study, however, observed both nuclear and cytoplasmic AHR protein expression in cultured endometrial stromal cells isolated from healthy women during the proliferative phase [155]. This observation suggests that in fact the AHR may translocate to the nucleus, where it acts as a transcription factor driving the expression of responsive genes in endometrial stromal cells.

Interestingly, Küchenhoff et al. also found that the presence of the AHR in the endometrium of premenopausal woman had a negative correlation with increasing age, possibly indicating a loss of function or a change in the endocrine status with increasing age [153]. Another study conducted by Khorram et al. also examined the expression of the AHR in human uterine endothelial tissue and found constitutive mRNA expression during both the proliferative and secretory phases without a decrease in expression with increasing age of the women [22]. Interestingly, they also found that the endometrium isolated from postmenopausal women taking hormone replacement therapy had increased levels of AHR mRNA compared to the endometrium from postmenopausal women not taking hormone replacement therapy, which suggests that the AHR can be regulated by exogenous hormones in the uterus [22]. Notably, the differences seen in the pattern of AHR expression between the Küchenhoff et al. and Khorram et al. studies could simply be due to small sample size, differences in the methods used to detect the AHR, differences in the classification of uterine cycle stage, and differences in the health status of the women participating in the studies.

In support of the localization and cyclic changes of the AHR in glandular epithelial cells of the uterine endometrium, studies have shown that the AHR may actually play a role in the proliferation and secretory functions of these cells as well. Specifically, Mueller et al. reported an upregulation of protein biosynthesis and secretion of glycodelin from secretory phase primary human endometrial epithelial cells after TCDD treatment in culture (10 nM) [160]. These studies further demonstrated that this upregulation may be due to transcriptional regulation of the *glycodelin* gene in an AHRE-dependent manner. Since glycodelin is essential for embryo implantation, this may suggest a mechanism by which the AHR regulates implantation in the uterus [160]. Normally, E2 acts through the ESR1 to induce uterine epithelial proliferation as well as secretory protein production [152]. Thus, another study investigating the estrogenic properties of TCDD and the AHR pathway in the uterus found that mRNA abundance for an epithelial secretory protein, lactoferrin, and proliferation in the luminal epithelium of AHRKO and WT mice are stimulated in response to E2 treatment [152]. However, TCDD (5 μg/kg; single i.p. dose) exposure in conjunction with E2 treatment results in partial inhibition of epithelial proliferation and mRNA expression of lactoferrin in WT mice, but not in the AHRKO mice [152]. This is supportive of a regulatory role for the AHR in both uterine epithelial cell proliferation and secretory function of the glandular cells in the uterus. Ultimately, this suggests interaction between the E2 signaling pathways and the AHR pathways, since in the absence of the AHR signaling pathway, E2 and TCDD were unable to inhibit the functioning of these cells.

It is also possible that similar to the *glycodelin* gene and in response to E2, the AHR may directly inhibit the expression of *lactoferrin* and other genes in the uterus through AHREs. In support of this, one recent study has shown that AHR expression in the uterus is directly regulated by E2 treatment in the OVX mouse uterus (0.5 and 4 μg/kg; s.c. once daily for three consecutive days) [161]. Interestingly, after E2 treatment, AHR mRNA levels along with the mRNA levels for other members of the AHR pathway were significantly downregulated, while the AHR protein levels were increased in the mouse uterus [161]. This is supportive of the hypothesis that the E2 signaling pathway is involved in changes seen in the AHR expression patterns during the human menstrual cycle. The inverse relationship between mRNA and protein levels in this study could be due to a regulation of AHR protein stability as well as transcription.

Some of the most convincing evidence that the AHR may have a biological function in the uterus comes from studies examining the reproductive phenotype of the AHRKO mice. AHRKO mice show a decrease in litter size after three to four pregnancies [162] and a higher abortion rate [163] compared to WT mice. They also have a reduction in the number of implantation sites and number of pups (live and dead) found at birth [163]. As the animals age, they have hypertrophic uteri with vascular mineralization and thrombosis [162].

In addition to its involvement in regulating the proliferative and secretory responses to E2 in the uterus, the AHR may also play an important role in supporting early pregnancy. First, AHR expression patterns in the rabbit uterus have been shown to be influenced by pregnancy state and blastocyst implantation, indicating control by both endogenous maternal steroid hormones and fetal–maternal interactions [164, 165]. Specifically, during the preimplantation period in the pregnant rabbit, the AHR was found in both the cytoplasm and the nucleus of the endometrial epithelial cells, while in nonpregnant uteri it was restricted to a small cytoplasmic area apical to the nucleus [164]. Shortly prior to the anticipated time of implantation, stromal cells and the epithelium of the uterine glands, but no longer the luminal epithelial cells, express the AHR protein [164]. Upon attachment of the gestational day 7 blastocyst, both AHR and ARNT expressions were observed in the mesometrial obplacental region, while later in pregnancy they were more or less entirely restricted to the stromal decidual cells [165]. Similarly, Kitajima et al. found that before implantation in the mouse uterus, AHR expression was localized in blood vessels of the stroma and in smooth muscle cells with no nuclear staining [166]. During uterine invasion of the embryo and during decidualization, the AHR became localized in the nucleus of the luminal epithelial cells and in the differentiated stroma cells [166]. Finally, as the pregnancy became established, AHR expression was prominent in the vasculature of the decidualizing zone as well as in the vasculature and spongiotrophoblasts of the developing placenta [166].

Interestingly, several studies have demonstrated that TCDD exposure during the period of implantation decreases the number of surviving implanted embryos [166, 167]. Specifically, Li et al. demonstrated that exposure to TCDD during implantation inhibits uterine decidualization and leads to a reduction in maternal serum progesterone levels, further suggesting that the AHR in the uterus may be regulated by maternal hormones [167]. Thus, changes in the pattern of AHR expression in the uterus and embryo before and during implantation are controlled by maternal hormones and are possibly needed to maintain pregnancy. In summary, these results are highly suggestive of functional roles for the AHR during the establishment of fetal–maternal interactions as well as a role in embryogenesis and placental development.

Besides involvement of the AHR during early pregnancy, a growing amount of evidence indicates that the AHR is necessary for vascular remodeling of the placenta, a process essential to provide the requirements of oxygen and nutrients to the developing fetus during the later stages of pregnancy [139]. Specifically, one study demonstrated that maternal exposure to known AHR ligands results in abnormalities in the placental vasculature and consequently fetal intrauterine growth restriction (IUGR). Specifically, conceptuses were isolated from GD 15.5 pregnant AHR heterozygous (C57BL/6-Ahr^{tm1Bra}) mice exposed before conception to polycyclic aromatic hydrocarbons (PAHs). The fetoplacental vasculature was evaluated and quantified by microcomputed tomography. The evaluation revealed that the arterial surface area and volume of the fetal arterial vasculature of the placenta were significantly reduced in the WT conceptuses, leading to IUGR after maternal PAH exposure, while the AHRKO conceptuses escaped these changes in vasculature and IUGR [168].

One proposed mechanism for how the AHR controls the process of vascular remodeling is a crosstalk between the AHR-mediated pathway and the hypoxia-inducible factor (HIF)-mediated pathway. Vascular development is stimulated under hypoxic conditions, and is dependent on transcription factors known as HIFs [139, 169]. Under hypoxic conditions, such as during pregnancy, one of the HIFs, HIF-1α, forms a heterodimer with ARNT and acts as a transcription factor for genes involved in the adaptation to hypoxia, such as vascular endothelial growth factor (VEGF). Since ARNT is a common transcription factor that shares a role in both the hypoxia pathway and the AHR pathway, it has been proposed that the crosstalk between the two pathways is due to competition for ARNT [139]. It is also possible that the two transcription factors compete for common transcriptional coactivators, such as CBP/p300. Collectively, these studies demonstrate that the AHR may play a role in later pregnancy and inappropriate activation of the AHR pathway by ligands could lead to impaired vascularization of the placenta. Though these studies highly suggest a biological role for the AHR in the process of placental vascularization during late gestation, this suggestion has not been directly tested and further investigation is warranted.

31.2.8 The AHR Regulates Development and Functioning of the Vagina

The vagina is an important female reproductive organ because it is the initial site of sperm deposition and plays a critical role in supporting the pelvic organs. It is also important for delivering a pregnancy. The vagina is an elastic canal composed of four layers: (1) an epithelial layer composed of noncornified stratified squamous cells; (2) a subepithelial layer of dense connective tissue; (3) a layer of smooth muscle; and (4) a layer of loose connective tissue. The vagina is a dynamic structure that distends sufficiently for parturition and displays cyclic changes in response to ovarian steroids [3, 141, 170].

Several lines of evidence indicate that the AHR may play a role in regulating development of the vagina (Fig. 31.2). Studies have found malformations in the external genitalia, such as the presence of a vaginal thread that leads to delayed vaginal opening, in rats exposed to TCDD during gestation (1 μg/kg maternal body mass; single oral dose) [60, 171–173] as well as in hamsters [174]. Interestingly, it was discovered that exposure to TCDD during embryonic life interferes with vaginal development by impairing regression of the Wolffian ducts and by preventing fusion of the Müllerian ducts [172, 173], events required for normal development of the vagina and other organs such as the uterus and cervix [3]. Together, these findings suggest that the AHR may modulate hormone signaling and control factors that are involved in cell movement and proliferation during the critical period of vaginal development [172].

Involvement of the AHR in functioning of the adult vagina has not been widely studied. One published study investigated AHR protein expression patterns in pregnant and nonpregnant rabbit vaginas [27]. Specifically, AHR localization changed from cytoplasmic in the nonpregnant rabbit to nuclear staining in the basal layer of the vaginal epithelium on day 6 of pregnancy [27]. These results suggest that the AHR is responsive to maternal hormones and may be involved in regulating structural changes in the vagina necessary for supporting pregnancy. Collectively, these studies are supportive of a functional role for the AHR in development and functioning of the vagina. More studies, however, are needed to confirm whether this is the case. For example, studies could compare development of the vagina during sexual differentiation or during the different stages of pregnancy in the AHRKO and WT mice in the presence or absence of ligands.

31.2.9 The AHR Regulates Puberty Onset, Cyclicity, Fertility, and Senescence

Due to the increasing amount of evidence that the AHR is a factor that regulates the development and functioning of the HPOA and supporting female organs, it is not surprising that there is experimental evidence supporting the idea that the AHR regulates puberty onset, cyclicity, fertility status, and timing of reproductive senescence. These reproductive processes are absolutely dependent on the normal development and functioning of the HPOA and the supporting female organs [3]. Specifically, gestational and/or lactational exposure to various AHR ligands results in delayed puberty compared to controls, as measured by the timing of vaginal opening in rats [33, 34, 175–177]. While ligand studies support a role for the AHR in timing of puberty onset, studies examining female AHRKO mice did not observe differences in the timing of vaginal opening compared to the WT mice. Thus, at this time, it is unclear whether the AHR pathway is directly involved in regulating puberty onset.

Further, due to the reports that in the absence of the AHR pathway and in the presence of AHR ligands, circulating E2 levels are reduced and ovulation is impaired, it is not surprising that there are studies that support a role for the AHR in regulating estrous/menstrual cyclicity. Specifically, TCDD exposure (10 μg/kg; single oral dose) causes irregular estrous cycles in rats, which are characterized by the loss of proestrous and estrous phases [114] or by persistent estrus or diestrus at earlier age compared to unexposed rats [177, 178]. In addition, a follow-up study conducted in women accidentally exposed to high concentrations of TCDD after an industrial accident revealed alterations in menstrual cyclicity [179]. Specifically, there was a significant association between irregular menstrual cycles in women who were premenarchal at the time of the accidental exposure and high serum concentrations of TCDD (10–100 ppt) [179].

Interestingly, while studies using TCDD support a role for the AHR in regulating estrous/menstrual cyclicity, studies using AHRKO mouse models reveal contrasting results. Two studies independently measured estrous cyclicity in AHRKO and WT mice by vaginal lavage over a 20-day period. The study conducted by Baba et al. found significantly disordered estrous cycles in AHRKO mice compared to WT mice [87], whereas Barnett et al. observed no difference between AHRKO and WT mice regarding the amount of time spent in each stage of the cycle [88]. The observed differences between the two studies could be due to the age of the animals used, genetic background effects, and differences in environmental factors [180].

Numerous studies suggest that the AHR regulates female fertility [87, 162, 163, 181, 182]. The earliest indication that the AHR is a regulator of fertility came from studies using various mouse strains with different responses to exogenous ligands [183]. The mouse strains were developed from progenitors possessing a high-affinity AHR receptor (AHR-responsive strain), as well as from progenitors possessing a low-affinity AHR receptor [181]. These studies showed that mice possessing a high-affinity AHR receptor had greater fertility, fitness, and longevity compared to mice possessing the low-affinity receptor. Specifically, female mice having the high-affinity receptor exhibited enhanced breeding efficiency and gave birth to more pups, suggesting that AHR activity plays a role in successful fertility [181]. Studies done much later using the various AHRKO mice support this observation. Specifically, Abbott et al. found that the AHRKO females had a higher abortion rate, a reduction in the number of implantation sites, and a reduction in the number of pups found at birth compared to the WT females [163]. The AHRKO females also had difficulty surviving pregnancy and lactation [163]. Fernandez-Salguero et al. observed that the AHRKO females had a decrease in litter size after three to four pregnancies [162]. Similarly, Baba et al. reported that the AHRKO females had fewer and smaller litters when compared to the WT females [87]. Furthermore, studies have observed reduced conception rates and increased abortion rates in humans and animal models after exposure to various exogenous AHR ligands [139, 148, 166–168, 184]. Thus, the AHR is an important factor for female fertility by regulating development and functioning of the reproductive system as described in detail in this chapter.

Reproductive senescence is a complex biological process and is defined as loss of reproductive function with age. In females, reproductive senescence is characterized by low E2 production due to depletion of the ovarian follicle reserve and high FSH levels [185, 186]. Further, studies have shown that senescence is also marked by age-related changes in the ability of E2 to coordinate the neuroendocrine events that lead to regular preovulatory GnRH surges [187]. These events contribute to the onset of irregular estrous cycles and eventually acyclicity.

Interestingly, both humans and animals exposed to various AHR ligands experience an earlier reproductive senescence. One study found that there was a nonmonotonic dose-related association with increasing risk for earlier menopause and serum levels of TCDD (up to 100 ppt) among a cohort of woman who were accidentally exposed to high levels of TCDD during their premenopausal period [188].

Animal studies also support the hypothesis that the AHR may play a role in regulating the timing of reproductive senescence. Some studies observed that a single prenatal exposure to TCDD (0.5–1 μg/kg maternal body mass; single oral dose) increased the incidence of premature constant estrus in rats [33, 34]. Follow-up studies demonstrated that prepubertal exposure to TCDD also leads to a reduction in circulating E2 levels, and an increase in serum FSH levels, without depletion of the follicular reserves in mature adult female rats [177]. These data suggest that activation of the

AHR by TCDD could be directly affecting ovarian function, which then leads to endocrine disruption and early senescence. Further investigation by Valdez et al. found differential regulation of global gene expression in the ovaries from older animals that were chronically exposed to TCDD during development and adult life compared to unexposed animals [189]. They suggested that the early onset of acyclicity in the TCDD-exposed rats may be due to altered E2 biosynthesis in the ovary because one of the downregulated genes in the ovaries from exposed animals was *CYP17A1*, an important substrate in the estrogen biosynthesis pathway [189].

Another line of evidence that the AHR could be a regulating reproductive senescence at the level of the ovary comes from the studies conducted using the AHRKO mice. As mentioned earlier in the chapter, AHR deficiency does not lead to increased FSH levels like in the TCDD-exposed animals experiencing early constant estrus [88]. However, much like the TCDD-exposed animals in premature constant estrus, the AHRKO mice produce less E2 without a reduction in follicle reserves [76, 190]. It would be interesting to examine estrous cyclicity, circulating FSH levels, and levels of ovarian steroidogenic enzymes in older AHRKO mice to determine if they experience an earlier reproductive senescence compared to the WT mice. Collectively, these studies support an ovarian role for the AHR in regulating the timing of senescence.

31.3 THE AHR REGULATES MALE REPRODUCTIVE FUNCTIONS

The male reproductive system is mainly comprised of the hypothalamus–pituitary–testis axis (HPTA) (Fig. 31.3). The primary role of the testis during the perinatal period is androgen production and germ cell maintenance. The androgen production drives the development of the male reproductive organs and sexually dimorphic brain regions that are responsible for male-specific gonadotropin release patterns and sex behavior [3]. During adulthood, the primary role of the testis is spermatogenesis. The process of spermatogenesis

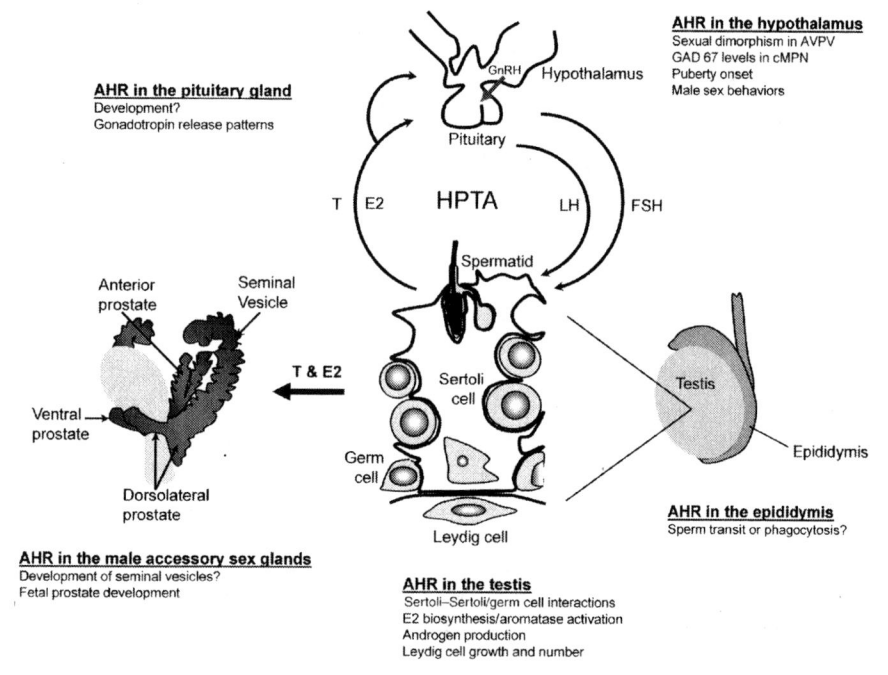

FIGURE 31.3 Involvement of the AHR in controlling the hypothalamus–pituitary–testis axis. During adulthood, the primary role of the testis is spermatogenesis. The process of spermatogenesis is mainly under the control of an endocrine feedback loop regulated by the HPTA. Hypothalamic GnRH stimulates the anterior pituitary to secrete LH and FSH into the bloodstream. The LH targets Leydig cells in the testis to produce primarily testosterone (T) that can be aromatized to E2. FSH stimulates Sertoli cells in the testis to support spermatogenesis. While T and E2 have local effects on spermatogenesis, they also have negative feedback effects on both hypothalamic GnRH secretion and pituitary LH and FSH secretion. The Sertoli cells support the developing germ cells and control release of mature spermatids. The Leydig cells support spermatogenesis by producing steroids. The steroids produced by the testis are necessary for the proper development and functioning of the sexually dimorphic brain regions and male accessory sex glands (i.e., epididymis, prostate, and seminal vesicles). The various roles that the AHR may play in controlling the HPTA and the male accessory sex glands are listed. (See the color version of this figure in Color Plates section.)

is mainly under the control of an endocrine feedback loop regulated by the HPTA (Fig. 31.3). Hypothalamic GnRH stimulates the anterior pituitary to secrete LH and FSH into the bloodstream. The LH targets Leydig cells in the testis to produce primarily testosterone that can be aromatized to E2. FSH stimulates Sertoli cells in the testis to support spermatogenesis. While testosterone and E2 have local effects on spermatogenesis, they also have negative feedback effects on both hypothalamic GnRH secretion and pituitary LH and FSH secretion [191].

The steroids produced by the adult testis and their signaling pathways are absolutely necessary for the process of spermatogenesis and proper functioning of the male reproductive tract [192–195]. The male reproductive tract is mainly comprised of the testis, epididymis, and the accessory sex glands (Fig. 31.3) [3]. The epididymis and accessory sex glands are important for sperm maturation, concentration, motility, and transfer [196, 197]. Various studies have identified the AHR at all sites in the HPTA [4, 5, 198–205]. Further, a growing amount of evidence indicates that the AHR plays a role in development of the male sexually dimorphic brain regions, male sex determination, testis function, epididymal function, and development and functioning of the accessory sex glands (Fig. 31.3). The involvement of the AHR in these processes is described below. For simplicity, the development of the reproductive organs is described in the mouse or the rat.

31.3.1 The AHR Regulates Male Neuroendocrine Development and Function

Some very convincing evidence indicates that the AHR may play a physiological role in development of the regions of the brain responsible for controlling male-specific sexual behavior and gonadotropin release patterns. First, as in the female, AHR and ARNT are expressed in the fetal and neonatal male rat hypothalamus, and localized in both neurons and glia [4]. Interestingly, males express much higher AHR levels in the hypothalamus than females and the expression increases during the period of sexual differentiation between GD 16 and PND 5 [4]. In addition, many members of the AHR pathway such as AHR, ARNT, ARNT2, and AHRR are expressed in the adult male rat hypothalamus [6, 198, 206].

Interestingly, males exposed to TCDD during the critical period of sexual dimorphic brain development have significantly higher numbers of GABA/Glu neurons in the AVPV compared to controls [29]. The number of neurons in the AVPV depends on the presence of T, which is aromatized to E2 during male development [207, 208]. As mentioned earlier, E2 signals the GnRH neurons through GABA/Glu neurons in the AVPV. These GABA/Glu neurons also coexpress ESR1, ESR2, and AHR, suggesting a role for the AHR pathway in these neurons [29]. It is unclear which mechanism TCDD uses to control the numbers of neurons in the AVPV, but it does not seem to involve AHR/ESR crosstalk, since *in utero* TCDD exposure does not affect ESR-dependent sex differences in the level of progestin receptors in the male brain [5]. One mechanism that has been proposed is transcriptional regulation of genes in the cell death pathway by the AHR, such as BAX [29]. BAX has been shown to play a role in the development of sexually dimorphic brain regions and the AHR has been shown to regulate cell death through the BAX promoter in other tissues [57, 58, 207].

31.3.1.1 The AHR Regulates Male Puberty Onset and Gonadotropin Release Patterns
In addition to altering the number of GABA/Glu neurons in the AVPV, exposure to TCDD during gestation results in males with delayed puberty that exhibit a feminine pattern of gonadotropin release consisting of an LH surge in response to exogenous E2 and progesterone during adult life [209, 210]. Specifically, time to separation of the prepuce from the glans penis (an index of pubertal development) is delayed in males exposed to TCDD *in utero* compared to vehicle-treated controls [209]. In addition, gestational exposure to TCDD at 0.40 and 1.0 µg/kg maternal body mass (single oral dose) causes a dose-related increase in plasma LH concentration after castration and administration of estradiol benzoate and progesterone in adulthood. Concurrently, the plasma LH concentrations were unaffected by progesterone in control males [210].

31.3.1.2 The AHR Regulates Development of Male Sexual Behaviors
Interestingly, *in utero* and lactational TCDD exposure also leads to feminized sexual behavior in male rats [209–211]. This result is not due to an effect on the sexual differentiation of the estrogen receptor system since ESR numbers and concentration are similar in the sexually dimorphic brain regions between TCDD-exposed and control rat brains [211]. Further, the volume of the sexually dimorphic brain regions is similar between TCDD-treated and control animals, indicating that the feminized sex behavior is not due to cell fate in the brain regions examined [211]. Interestingly, GAD 67 may be responsible for some of the expression of male sexual behaviors in the caudal medial preoptic nucleus (cMPN) of the hypothalamus, an area known to be important in these behaviors. Studies indicate that *in utero* exposure to TCDD (1 µg/kg maternal body mass; single oral dose) significantly reduces levels of GAD 67 mRNA in the cMPN of male rats compared to controls [5]. Since the AHR is colocalized with GAD 67 in GABA/Glu neurons, the activation of the AHR pathway could transcriptionally downregulate GAD 67 during fetal life leading to altered development of the cMPN and feminized sexual behavior in adulthood.

Collectively, these studies suggest involvement of the AHR in development of the brain regions responsible for puberty onset, male pattern gonadotropin release, and sexual behavior. More studies are needed to better understand the endogenous actions of the AHR in development and functioning of the male central nervous system. Additional studies could investigate the central nervous system in the male AHRKO, WT, and animals treated with various AHR ligands. For example, studies could compare GAD 67 levels or ESR levels and distribution in the sexual dimorphic brain regions of the AHRKO and WT mice. In addition, though the males are fertile, it has not been reported whether the AHRKO mice display normal male sexual behaviors.

Finally, it is unclear what role the AHR has in the development and functioning of the male anterior pituitary, though members of the AHR pathway have been detected in the pituitary as well as in the sexual dimorphic brain regions [198]. It would be important to examine pituitary function in AHRKO mice and in animals exposed to AHR ligands since it is an important player in neuroendocrine functions. Furthermore, as mentioned earlier, the pituitary is a promising target for future studies given that the Ahr promoter contains DNA binding sites for factors vital for normal pituitary development and germline mutations in the aryl hydrocarbon receptor interacting protein result in pituitary cancers [36, 37].

31.3.2 The AHR Regulates Male Sex Determination

Sex determination is partly dependent on the Y chromosome and activation of male determining factors. In the presence of the Y chromosome and induction of a cascade of male determining factors, the fetus develops as a male [73]. Therefore, in normal conditions, males have an X chromosome, which they inherit from their mother, and a Y chromosome, which they inherit from their father [212]. Epidemiological studies conducted in an accidentally exposed population to TCDD have found an association between paternal exposure to TCDD and decreased births of male offspring with respect to birth of female offspring [213, 214]. Additional studies conducted using rats found that TCDD exposure in utero and lactationally affects the development of male gonads in offspring (F1), leading to changes in the sex ratio of the subsequent generation (F2) without altering litter size [215, 216]. Specifically, the sex ratio (percentage of male pups) of F2 offspring was significantly reduced in the TCDD-exposed group compared with controls. Upon closer examination, the investigators found that the sex ratio of the offspring may be decreased at fertilization and not during the spermatozoa stage [217]. This is supported by studies showing that Y-bearing/X-bearing sperm ratio was not significantly decreased in the TCDD-exposed males, but the sex ratio of the two-cell embryos of the TCDD-exposed animals was significantly lower than that of the control group [217]. Interestingly, the master *sex-determining region Y* (*Sry*) gene can translocate from the Y chromosome resulting in males with a 46, XX karyotype [151]. Based on this observation, it would be interesting to evaluate whether AHR signaling controls the expression of male determining genes such as *Sry*, which determine the development of the indifferent gonad into a testis and are critical for spermatogenesis [218–220].

31.3.3 The AHR Regulates Testis Function

The testis is the male gonad and consists of an oval structure containing seminiferous tubules and intertubular interstitium. The epithelium inside the seminiferous tubules mainly contains germ cells and Sertoli cells, whereas the intertubular tissue mainly contains Leydig cells [221]. The testis has two major functions, one is to produce spermatozoa, which transmit the male genes to the embryo, and the other function is to produce hormones, which are required for the maintenance of reproductive functions in adulthood [3]. Several lines of evidence suggest that the AHR helps regulate testicular function. The first line of evidence comes from the observation that the AHR is present in the testis of distinct species. Specifically, AHR mRNA has been detected in whole human testis using quantitative PCR [199]. Further, AHR protein has been detected in germ cells, Sertoli cells, and Leydig cells in a number of species including human, rat, mouse, and guinea pig using techniques such as immunohistochemistry, AHR ligand binding assays, and Western blotting [200–205]. The second line of evidence for a role of AHR signaling in testicular function comes from the observation that prenatal and perinatal exposure to TCDD (1–50 μg/kg; single i.p. dose) decreases testicular weight in adulthood [222, 223]. Importantly, two studies using different AHRKO mouse lines have shown that AHR deletion reduces testicular weight in adulthood [205, 224]. Conversely, overexpression of rat AHR isoforms in transgenic mouse testes significantly increases their weight in adulthood [225]. The third line of evidence supporting a role of AHR signaling in testicular function comes from studies showing that AHR deletion and TCDD exposure affect testicular function. This evidence is detailed below according to specific functional roles of the testis.

31.3.3.1 The AHR Does Not Regulate Spermatogenesis
Spermatogenesis is a complex biological process that produces spermatozoa from spermatogonial stem cells. During spermatogenesis, spermatogonial cells first multiply by repeated mitotic divisions to produce spermatocytes. The spermatocytes then divide by meiosis to produce spermatids. The spermatids then differentiate into highly compacted spermatozoa [226]. Two studies using AHRKO mice have shown that the loss of a functional AHR in mice does not directly impair spermatogenesis [205, 224]. Specifically, these studies showed no differences in the number of spermatozoa and precursors in WT and AHRKO testicular histological sections [205], as well as no differences in testicular

daily sperm production between WT and AHRKO mice [224]. Collectively, these studies suggest that the AHR does not participate in the development of sperm.

31.3.3.2 The AHR Regulates Sertoli Cell Function Sertoli cells are columnar cells that are closely associated with germ cells in the seminiferous epithelium [221]. The main functions of Sertoli cells are to control release of mature spermatids into the tubular lumen, as well as to support the developing germ cells during spermatogenesis by secreting fluid, proteins, and growth factors [226]. In addition, Sertoli cells contribute to estrogen biosynthesis in the testis by producing aromatase and converting testosterone to E2, a hormone essential for directing spermatogenesis [227]. Whether the AHR has a role in the function of Sertoli cells has not been studied in AHRKO animal models. Some studies using TCDD, however, suggest that activation of the AHR signaling pathway may play a role in modulating intercellular interactions of Sertoli–Sertoli/germ cells and steroidogenesis. Specifically, TCDD exposure (0.3–25 μg/kg; single i.p. dose) in adulthood causes wider intercellular spaces between Sertoli–Sertoli cells and between Sertoli–germ cells compared to controls [228]. Further, 24 h of *in vitro* treatment with TCDD (0.2–2000 pg/mL) upregulates sertolin and downregulates testin mRNAs in isolated rat Sertoli cells, gene markers for cell–cell interactions in the rat testis [229–231]. The alterations in sertolin and testin are correlated with an upregulation of CYP1A1 mRNA [229], an indicator of the activation of the AHR signaling pathway [232]. Lai et al. also reported that *in vitro* TCDD exposure results in an induction of aromatase transcripts and increased E2 secretion in isolated Sertoli cells [229]. Collectively, these studies suggest that intercellular interactions of Sertoli–Sertoli/germ cells and E2 biosynthesis can be modulated by the activation of the AHR signaling pathway in Sertoli cells. Future studies using AHRKO models, however, are needed to directly test this hypothesis.

31.3.3.3 The AHR Regulates Androgen Production and Leydig Cell Maintenance Leydig cells are polygonal in shape and are the major cell type within the interstitium of the testis. These cells are often found adjacent to blood vessels and the seminiferous tubules. The main function of Leydig cells is to produce testosterone, which supports spermatogenesis [233]. Although no studies have directly focused on the role of the AHR in Leydig cell function, a few studies indicate that this receptor may play a role in the steroidogenic function of Leydig cells. Specifically, one study showed that AHR deletion in mice leads to decreased serum testosterone levels, which coincide with decreased levels of the steroidogenic enzymes that participate in the synthesis of testosterone such as hydroxy-delta-5-steroid dehydrogenase, 3beta- and steroid delta-isomerase 1 (HSD3B1) and steroidogenic acute regulatory protein (STAR) mRNAs [205, 234].

It is uncertain whether the AHR is needed for maintaining the number of Leydig cells in the testis. While one study demonstrates that AHR deletion does not impair the number of Leydig cells [205], studies using TCDD (12.5–50 μg/kg body weight; single i.p. dose) show that activation of the AHR may lead to reduced volume, decreased number, and reduced size of Leydig cells in the testis 4 weeks after TCDD treatment [201, 235]. Furthermore, one *in vitro* study showed that overexpression of the AHR in a mouse testicular Leydig cell line leads to suppression of Leydig cell growth [236]. Taken together, these studies suggest that the AHR is not absolutely required for the development of Leydig cells, but it may participate in the maintenance of these testicular steroidogenic cells.

31.3.4 The AHR Regulates Epididymal Function

Once spermatogenesis is completed in the testis, sperm transit through the epididymis and are stored in the last segment, the cauda epididymis, as well as in vas deferens until ejaculation [221]. Baba et al. found that AHRKO mice have reduced sperm numbers in the cauda epididymis compared to WT mice [205]. Since the AHR does not seem to modulate the production of sperm in the testis (see Section 31.3.3), it is possible that the reduction of sperm numbers in the cauda epididymis in AHRKO mice is due to defects in sperm transit rate in the epididymis. This hypothesis has not been studied in AHRKO animal models, but two studies using prenatal exposure to TCDD have addressed this possibility. The results, however, are not consistent. For example, sperm transit estimated by dividing daily sperm production per testis into the number of sperm in whole epididymis in male rats prenatally exposed with a single dose of TCDD (0.5–2.0 μg/kg) was decreased compared to control male rats [237]. In contrast, sperm transit monitored by [^{3}H]-thymidine in male rats prenatally exposed to a single dose of TCDD (1.0 μg/kg) was similar to control male rats [238]. It is possible that the differences observed in sperm transit rate are due to differences in the techniques utilized in both studies. These studies, however, raise the question as to whether the AHR regulates epididymal function by altering the transit of sperm or by increasing sperm phagocytosis, as suggested by Sommer et al. [238].

31.3.5 The AHR Regulates the Male Accessory Sex Glands

Male accessory sex glands mainly include the seminal vesicles and prostate. Their function is to provide the constituents of seminal plasma in the semen at ejaculation [3]. Seminal vesicles are two glands that open into the ejaculatory ducts. Their main function is to provide proteins, antioxidants, and other constituents to sperm at ejaculation [198]. The AHR protein has been identified in rat seminal vesicles by Western blotting and immunohistochemistry analyses [202]. Loss of functional AHR has been shown to impair the development of seminal vesicles since studies using AHRKO mice showed that AHR deletion results in regression of seminal vesicles in aged mice (24–52 weeks old) [205]. Development, maintenance, and functioning of seminal vesicles are influenced by androgens produced by testis [3]. Since Baba et al. observed reduced levels of testosterone in adult AHRKO mice compared to adult WT mice, and no difference in androgen receptor protein in WT and AHRKO seminal vesicles, they proposed that regression of seminal vesicles in aged AHRKO mice is due to impaired testosterone synthesis [205].

Another possible explanation for the regression of seminal vesicles in mice lacking a functional AHR is that the AHR may be needed for the development of seminal vesicles in fetal and/or prepubertal life. This possibility is supported by studies showing that *in utero* and lactational exposure to TCDD (1.0–5.0 μg/kg maternal body mass; single i.p. dose) alters development of seminal vesicles from treated pups as evidenced by decreased weight and decreased epithelium proliferation in the tissue on PNDs 32–35 [239–241]. This point of development corresponds to peak androgen levels within the seminal vesicles [239] and to a period of rapid proliferation and differentiation within the seminal vesicle epithelium [242].

The prostate is a gland that consists of three lobes (ventral, dorsolateral, and anterior) and is located below the base of the bladder completely encircling the proximal portion of the urethra. During fetal life, testicular androgens bind to androgen receptors in the urogenital mesenchyme and stimulate cell proliferation from prostate ductal progenitors (prostatic buds) to progressively form the prostate [198]. Both protein and mRNA of the AHR have been identified in the developing fetal prostate as well as in adult prostates in humans and rats using techniques such as Western blotting, immunohistochemistry, Northern blotting, and quantitative PCR [199, 202, 243, 244]. Several studies have suggested that the AHR is required for anterior and dorsolateral prostate development, and these studies have been extensively reviewed by Vezina et al. [245]. The first evidence for a role of the AHR signaling pathway in prostate development came from the observation that *in utero* and lactational exposure to TCDD (0.064–1.0 μg/kg body weight; single i.p. dose) retarded prostate growth in rats [246]. This observation was subsequently shown in other studies in rats and mice [240, 246–249]. Furthermore, Lin et al. elucidated that the AHR is mostly needed for prostate development in the fetal stage since pups treated with TCDD *in utero* (oral maternal dose; 5 μg/kg body weight) on GD 13 displayed inhibited dorsolateral prostate development from GD 16 [240]. In addition, studies suggest that AHR signaling participates in prostate development by antagonizing the actions of fibroblast growth factor 10 (FGF10) [250], a factor required for prostatic bud initiation in fetal mice and rats [251]. Taken together, these studies suggest that the AHR participates in fetal prostate development by modulating prostatic budding-related signaling pathways during fetal life.

31.4 THE AHR MAY REGULATE HUMAN REPRODUCTION

31.4.1 Single Nucleotide Genetic Polymorphisms in Members of the AHR Pathway and Their Effects on Human Reproduction

Throughout this chapter, reproductive outcomes in respect to activation of the AHR pathway by various ligands or the absence of the AHR were described extensively for animal models. However, outside of studies conducted in humans accidentally exposed to AHR ligands, only a few studies tested whether SNPs in the AHR pathway are associated with adverse reproductive outcomes in men and women. Some studies have tested whether SNPs in the AHR pathway are associated with adverse reproductive outcomes in men and women. Specifically, some studies have examined genetic polymorphisms in *AHR* gene (Arg554Lys; rs2066853), located at exon 10, in the *trans*-activating domain of the receptor protein that is responsible for the initiation of transcription from the target genes and reproductive functions in humans. This SNP was not significantly associated with endometriosis [252–254] or defective spermatogenesis [255, 256]. Similarly, the synonymous polymorphism in *ARNT* (Val189Val; rs2228099) at exon 7 was not predictive of male fertility potential [256] or endometriosis [252–254].

When a polymorphism at exon 5 of the *AHRR* gene (Pro185Ala; rs2292596) was evaluated in relation to endometriosis, conflicting results were reported. Tsuchiya et al. found that the polymorphism was associated with both susceptibility and severity of endometriosis [252]. The risk of endometriosis was approximately 2.5 times higher among Japanese women carrying any polymorphism of *AHRR* (Pro/Ala + Ala/Ala) compared to women who did not carry the polymorphism (Pro/Pro) [252]. The investigators suggested that these polymorphisms potentially facilitate proliferation of endometrial cells through the diminished downregulation of AHR-mediated signaling [252]. Nonetheless, these findings are not in agreement with those of Watanabe et al. [253] who found no association between the SNP and endometriosis or with those of Kim et al. [254] who found a marginal effect of the polymorphism in relation to endometriosis. Interestingly, Kim et al. [254] reported a significant increase in the risk of endometriosis when the polymorphism in *AHRR* was combined with a polymorphism in the *glutathione-S-transferase T1* (*GSTT1*) gene [254]. In Japanese men, the same SNP of the *AHRR* gene was shown to contribute significantly to a genetic predisposition to reduced male fertility and the presence of micropenis [255, 257, 258]. More specifically, the presence of the Pro185 allele was associated with reduced male fertility potential and abnormal genitalia development [255, 257, 258]. Interestingly, in a study of Estonian men, the Ala185 allele was more prevalent among patients with fertility problems than patients without fertility problems [256]. In addition, this SNP was negatively associated with sperm count [256].

The reasons this SNP is associated with male fertility are unknown, but the investigators suggest that the SNP is located in the heterodimerization domain of AHRR, and thus it may alter the function of the repressor, leading to a weaker negative feedback and exacerbating toxic effects of environmental pollutants [256, 257]. Interestingly, Tiido et al. reported a moderate positive association between serum levels of 2,2′,4,4′,5,5′-hexachlorobiphenyl (CB 153) and dichlorodiphenyl dichloroethene (p,p'-DDE) and the proportion of Y-bearing spermatozoa among Swedish fishermen [259].

These findings were further investigated to evaluate whether androgen receptor polymorphisms, which result in various CAG and GGN repeat lengths, as well as the AHR (Arg554Lys) and AHRR (Pro185Ala) variants play a role in modifying the effect of exposure to CB 153 and p,p'-DDE in regard to sperm Y:X ratio [259]. The investigators reported a trend toward a more pronounced increase in Y-sperm in men with short androgen receptor CAG repeats and the *AHRR* (Pro185Ala) Ala allele, indicative of possible gene–environment interactions in relation to impairment of male reproductive function in response to persistent organohalogen pollutant exposure [260].

Collectively, although there are limited data on the role of the AHR pathway in human reproductive function, some studies indicate that SNPs in the AHR pathway are associated with adverse reproductive outcomes in men and women. Future studies should be conducted to evaluate whether genetic polymorphisms in the AHR and its associated proteins are linked with adverse reproductive outcomes. Such studies will help elucidate the role of the AHR in human reproduction.

31.5 CONCLUSIONS AND SUMMARY

Exposure to various AHR ligands during specific developmental stages, deletion of the AHR in animal models, and SNPs in members of the AHR pathway result in deleterious effects in the reproductive system that may be due to disruption of the endogenous functioning of this receptor. Though endogenous functions for the AHR are still under investigation, the evidence presented in this chapter supports pleiotropic functions during development and functioning of the reproductive system. In summary, the experimental evidence supports involvement of the AHR in regulating female and male fertility by controlling the following reproductive processes: (1) development of neuroendocrine functions in the female AVPV, (2) regulating the ability of the hypothalamus and pituitary to coordinate ovulation, (3) regulation of ovarian function by controlling follicular endowment, the number and growth of antral follicles, steroid production, signaling and metabolism, and the ability to reach ovulation, (4) providing the optimum environment for fertilization, nourishing the embryo, and maintaining pregnancy, (5) development of the female external genitalia, (6) cyclicity and timing of reproductive senescence, (7) development of the male-specific hypothalamic nuclei responsible for gonadotropin release patterns and male sexual behaviors, (8) male sex determination, (9) testis function by regulating Sertoli cell interactions, androgen production, and Leydig cell maintenance, and (10) sperm transfer by regulating development and functioning of the male accessory sex glands. Though many mechanisms have been suggested for how the AHR functions in the reproductive system, it is still unknown whether the AHR has endogenous ligands and/or functions in the absence of ligands. Regardless of whether the AHR has endogenous ligands, understanding how the receptor functions in the reproductive system will help elucidate etiologies of infertility and diseases of the reproductive tract such as endometriosis and cancers of the pituitary and prostate gland. Ultimately, this information could lead to development of novel therapies for prevention or treatment of these disorders.

REFERENCES

1. Freeman, M. E. (2006). Neuroendocrine control of the ovarian cycle of the rat. In *Knobil and Neill's Physiology of Reproduction*, 3rd edition (Neill, J. D., et al., Eds). Elsevier, San Diego, CA, pp. 2327–2388.

2. Neill, J. D., Wassarman, P. M., Richards, J. S., de Kretser, D. M., Challis, J. R. G., Pfaff, D. W., and Plant, T. M. (2006). *Knobil and Neill's Physiology of Reproduction*, 3rd edition. Elsevier, San Diego, CA.

3. Johnson, M. H. and Everitt, B. J. (1995). *Essential Reproduction*. Blackwell Science, Cambridge, MA.

4. Pravettoni, A., Colciago, A., Negri-Cesi, P., Villa, S., and Celotti, F. (2005). Ontogenetic development, sexual differentiation, and effects of Aroclor 1254 exposure on expression of the arylhydrocarbon receptor and of the arylhydrocarbon receptor nuclear translocator in the rat hypothalamus. *Reproductive Toxicology*, 20, 521–530.

5. Hays, L. E., Carpenter, C. D., and Petersen, S. L. (2002). Evidence that GABAergic neurons in the preoptic area of the rat brain are targets of 2,3,7,8-tetrachlorodibenzo-*p*-dioxin during development. *Environmental Health Perspectives*, 110 (Suppl. 3), 369–376.

6. Petersen, S. L., Curran, M. A., Marconi, S. A., Carpenter, C. D., Lubbers, L. S., and McAbee, M. D. (2000). Distribution of mRNAs encoding the arylhydrocarbon receptor, arylhydro-

carbon receptor nuclear translocator, and arylhydrocarbon receptor nuclear translocator-2 in the rat brain and brainstem. *Journal of Comparative Neurology*, 427, 428–439.

7. Mukai, M., Lin, T. M., Peterson, R. E., Cooke, P. S., and Tischkau, S. A. (2008). Behavioral rhythmicity of mice lacking AhR and attenuation of light-induced phase shift by 2,3,7,8-tetrachlorodibenzo-p-dioxin. *Journal of Biological Rhythms*, 23, 200–210.

8. Güldner, F. H. (1982). Sexual dimorphisms of axo-spine synapses and postsynaptic density material in the suprachiasmatic nucleus of the rat. *Neuroscience Letters*, 28, 145–150.

9. Vida, B., Hrabovszky, E., Kalamatianos, T., Coen, C. W., Liposits, Z., and Kalló, I. (2008). Oestrogen receptor alpha and beta immunoreactive cells in the suprachiasmatic nucleus of mice: distribution, sex differences and regulation by gonadal hormones. *Journal of Neuroendocrinology*, 20, 1270–1277.

10. Horvath, T. L., Cela, V., and van der Beek, E. M. (1998). Gender-specific apposition between vasoactive intestinal peptide-containing axons and gonadotrophin-releasing hormone-producing neurons in the rat. *Brain Research*, 795, 277–281.

11. Kriegsfeld, L. J., Silver, R., Gore, A. C., and Crews, D. (2002). Vasoactive intestinal polypeptide contacts on gonadotropin-releasing hormone neurones increase following puberty in female rats. *Journal of Neuroendocrinology*, 14, 685–690.

12. Lauber, A. H., Romano, G. J., and Pfaff, D. W. (1991). Gene expression for estrogen and progesterone receptor mRNAs in rat brain and possible relations to sexually dimorphic functions. *Journal of Steroid Biochemistry and Molecular Biology*, 40, 53–62.

13. de la Iglesia, H. O. and Schwartz, W. J. (2006). Minireview. Timely ovulation: circadian regulation of the female hypothalamo-pituitary–gonadal axis. *Endocrinology*, 147, 1148–1153.

14. Barbacka-Surowiak, G., Surowiak, J., and Stokłosowa, S. (2003). The involvement of suprachiasmatic nuclei in the regulation of estrous cycles in rodents. *Reproductive Biology*, 3, 99–129.

15. Gibson, E. M., Williams, W. P., 3rd, and Kriegsfeld, L. J. (2009). Aging in the circadian system: considerations for health, disease prevention and longevity. *Experimental Gerontology*, 44, 51–56.

16. Downs, J. L. and Wise, P. M. (2009). The role of the brain in female reproductive aging. *Molecular and Cellular Endocrinology*, 299, 32–38.

17. Dolwick, K. M., Schmidt, J. V., Carver, L. A., Swanson, H. I., and Bradfield, C. A. (1993). Cloning and expression of a human Ah receptor cDNA. *Molecular Pharmacology*, 44, 911–917.

18. Karchner, S. I., Powell, W. H., and Hahn, M. E. (1999). Identification and functional characterization of two highly divergent aryl hydrocarbon receptors (AHR1 and AHR2) in the teleost fundulus heteroclitus. Evidence for a novel subfamily of ligand-binding basic helix–loop–helix–Per–ARNT–Sim (bHLH–PAS) factors. *Journal of Biological Chemistry*, 274, 33814–33824.

19. Böttner, M., Christoffel, J., Jarry, H., and Wuttke, W. (2006). Effects of long-term treatment with resveratrol and subcutaneous and oral estradiol administration on pituitary function in rats. *Journal of Endocrinology*, 189, 77–88.

20. Schlecht, C., Klammer, H., Jarry, H., and Wuttke, W. (2004). Effects of estradiol, benzophenone-2 and benzophenone-3 on the expression pattern of the estrogen receptors (ER) alpha and beta, the estrogen receptor-related receptor 1 (ERR1) and the aryl hydrocarbon receptor (AhR) in adult ovariectomized rats. *Toxicology*, 205, 123–130.

21. Baldridge, M. G. and Hutz, R. J. (2007). Autoradiographic localization of aromatic hydrocarbon receptor (AHR) in rhesus monkey ovary. *American Journal of Primatology*, 69, 681–691.

22. Khorram, O., Garthwaite, M., and Golos, T. (2002). Uterine and ovarian aryl hydrocarbon receptor (AHR) and aryl hydrocarbon receptor nuclear translocator (ARNT) mRNA expression in benign and malignant gynaecological conditions. *Molecular Human Reproduction*, 8, 75–80.

23. Nestler, D., Risch, M., Fischer, B., and Pocar, P. (2007). Regulation of aryl hydrocarbon receptor activity in porcine cumulus–oocyte complexes in physiological and toxicological conditions: the role of follicular fluid. *Reproduction*, 133, 887–897.

24. Bussmann, U. A. and Barañao, J. L. (2006). Regulation of aryl hydrocarbon receptor expression in rat granulosa cells. *Biology of Reproduction*, 75, 360–369.

25. Davis, B. J., McCurdy, E. A., Miller, B. D., Lucier, G. W., and Tritscher, A. M. (2000). Ovarian tumors in rats induced by chronic 2,3,7,8-tetrachlorodibenzo-p-dioxin treatment. *Cancer Research*, 60, 5414–5419.

26. Robles, R., Morita, Y., Mann, K. K., Perez, G. I., Yang, S., Matikainen, T., Sherr, D. H., and Tilly, J. L. (2000). The aryl hydrocarbon receptor, a basic helix–loop–helix transcription factor of the PAS gene family, is required for normal ovarian germ cell dynamics in the mouse. *Endocrinology*, 141, 450–453.

27. Hasan, A. and Fischer, B. (2003). Epithelial cells in the oviduct and vagina and steroid-synthesizing cells in the rabbit ovary express AhR and ARNT. *Anatomy and Embryology*, 207, 9–18.

28. Sakuma, Y. (2009). Gonadal steroid action and brain sex differentiation in the rat. *Journal of Neuroendocrinology*, 21, 410–414.

29. Petersen, S. L., Krishnan, S., and Hudgens, E. D. (2006). The aryl hydrocarbon receptor pathway and sexual differentiation of neuroendocrine functions. *Endocrinology*, 147 (6 Suppl.), s33–s42.

30. Abbott, B. D., Birnbaum, L. S., and Perdew, G. H. (1995). Developmental expression of two members of a new class of transcription factors. I. Expression of aryl hydrocarbon receptor in the C57BL/6N mouse embryo. *Developmental Dynamics*, 204, 133–143.

31. Abbott, B. D. and Probst, M. R. (1995). Developmental expression of two members of a new class of transcription factors. II. Expression of aryl hydrocarbon receptor nuclear

translocator in the C57BL/6N mouse embryo. *Developmental Dynamics*, 204, 144–155.

32. Thomsen, J. S., Kietz, S., Ström, A., and Gustafsson, J. A. (2004). HES-1, a novel target gene for the aryl hydrocarbon receptor. *Molecular Pharmacology*, 65, 165–171.

33. Gray, L. E., Jr. and Ostby, J. S. (1995). In utero 2,3,7,8-tetrachlorodibenzo-*p*-dioxin (TCDD) alters reproductive morphology and function in female rat offspring. *Toxicology and Applied Pharmacology*, 133, 285–294.

34. Gray, L. E., Wolf, C., Mann, P., and Ostby, J. S. (1997). *In utero* exposure to low doses of 2,3,7,8-tetrachlorodibenzo-*p*-dioxin alters reproductive development of female Long–Evans hooded rat offspring. *Toxicology and Applied Pharmacology*, 146, 237–244.

35. Kelberman, D., Rizzoti, K., Lovell-Badge, R., Robinson, I. C., and Dattani, M. T. (2009). Genetic regulation of pituitary gland development in human and mouse. *Endocrine Reviews*, 30, 790–829.

36. Harper, P. A., Riddick, D. S., and Okey, A. B. (2006). Regulating the regulator: factors that control levels and activity of the aryl hydrocarbon receptor. *Biochemical Pharmacology*, 72, 267–279.

37. Daly, A. F., Tichomirowa, M. A., and Beckers, A. (2009). The epidemiology and genetics of pituitary adenomas. *Best Practice & Research Clinical Endocrinology and Metabolism*, 23, 543–554.

38. Swain, A. (2006). Sex determination and differentiation. In *Knobil and Neill's Physiology of Reproduction*, 3rd edition (Neill, J. D., et al., Eds). Elsevier, San Diego, CA, pp. 245–260.

39. Edson, M. A., Nagaraja, A. K., and Matzuk, M. M. (2009). The mammalian ovary from genesis to revelation. *Endocrine Reviews*, 30, 624–712.

40. Chassot, A. A., Gregoire, E. P., Magliano, M., Lavery, R., and Chaboissier, M. C. (2008). Genetics of ovarian differentiation: Rspo1, a major player. *Sexual Development*, 2, 219–227.

41. Hirshfield, A. N. (1991). Development of follicles in the mammalian ovary. *International Review of Cytology*, 124, 43–101.

42. Tingen, C., Kim, A., and Woodruff, T. K. (2009). The primordial pool of follicles and nest breakdown in mammalian ovaries. *Molecular Human Reproduction*, 15, 795–803.

43. Pepling, M. E. (2006). From primordial germ cell to primordial follicle: mammalian female germ cell development. *Genesis*, 44, 622–632.

44. Pepling, M. E. and Spradling, A. C. (1998). Female mouse germ cells form synchronously dividing cysts. *Development*, 125, 3323–3328.

45. Hirshfield, A. N. and DeSanti, A. M. (1995). Patterns of ovarian cell proliferation in rats during the embryonic period and the first three weeks postpartum. *Biology of Reproduction*, 53, 1208–1221.

46. Guigon, C. J. and Magre, S. (2006). Contribution of germ cells to the differentiation and maturation of the ovary: insights from models of germ cell depletion. *Biology of Reproduction*, 74, 450–458.

47. Li, H. and Clagett-Dame, M. (2009). Vitamin A deficiency blocks the initiation of meiosis of germ cells in the developing rat ovary *in vivo*. *Biology of Reproduction*, 81, 996–1001.

48. Bowles, J., Knight, D., Smith, C., Wilhelm, D., Richman, J., Mamiya, S., Yashiro, K., Chawengsaksophak, K., Wilson, M. J., Rossant, J., Hamada, H., and Koopman, P. (2006). Retinoid signaling determines germ cell fate in mice. *Science*, 312, 596–600.

49. Bowles, J. and Koopman, P. (2007). Retinoic acid, meiosis and germ cell fate in mammals. *Development*, 134, 3401–3411.

50. Koubova, J., Menke, D. B., Zhou, Q., Capel, B., Griswold, M. D., and Page, D. C. (2006). Retinoic acid regulates sex-specific timing of meiotic initiation in mice. *Proceedings of the National Academy of Sciences of the United States of America*, 103, 2474–2479.

51. Gondos, B. (1978). Oogonia and oocytes in mammals. In *The Vertebrate Ovary* (Jones, R. E.,Ed.). Plenum Press, New York, pp. 83–120.

52. Ghafari, F., Gutierrez, C. G., and Hartshorne, G. M. (2007). Apoptosis in mouse fetal and neonatal oocytes during meiotic prophase one. *BMC Developmental Biology*, 7, 1–12.

53. McClellan, K. A., Gosden, R., and Taketo, T. (2003). Continuous loss of oocytes throughout meiotic prophase in the normal mouse ovary. *Developmental Biology*, 258, 334–348.

54. Pepling, M. E. and Spradling, A. C. (2001). Mouse ovarian germ cell cysts undergo programmed breakdown to form primordial follicles. *Developmental Biology*, 234, 339–351.

55. Greenfeld, C. R., Pepling, M. E., Babus, J. K., Furth, P. A., and Flaws, J. A. (2007). Bax regulates follicular endowment in mice. *Reproduction*, 133, 865–876.

56. Benedict, J. C., Lin, T. M., Loeffler, I. K., Peterson, R. E., and Flaws, J. A. (2000). Physiological role of the aryl hydrocarbon receptor in mouse ovary development. *Toxicological Sciences*, 56, 382–388.

57. Matikainen, T., Perez, G. I., Jurisicova, A., Pru, J. K., Schlezinger, J. J., Ryu, H. Y., Laine, J., Sakai, T., Korsmeyer, S. J., Casper, R. F., Sherr, D. H., and Tilly, J. L. (2001). Aromatic hydrocarbon receptor-driven Bax gene expression is required for premature ovarian failure caused by biohazardous environmental chemicals. *Nature Genetics*, 28, 355–360.

58. Matikainen, T., Moriyama, T., Morita, Y., Perez, G. I., Korsmeyer, S. J., Sherr, D. H., and Tilly, J. L. (2002). Ligand activation of the aromatic hydrocarbon receptor transcription factor drives Bax-dependent apoptosis in developing fetal ovarian germ cells. *Endocrinology*, 143, 615–620.

59. Changge, F., Caiqiao, Z., Huili, Q., Guoliang, X., and Yaoxing, C. (2001). Sexual difference in gonadal development of embryonic chickens after treatment of polychlorinated biphenyls. *Chinese Science Bulletin*, 46, 1900–1903.

60. Flaws, J. A., Sommer, R. J., Silbergeld, E. K., Peterson, R. E., and Hirshfield, A. N. (1997). *In utero* and lactational exposure to 2,3,7,8-tetrachlorodibenzo-*p*-dioxin (TCDD) induces genital dysmorphogenesis in the female rat. *Toxicology and Applied Pharmacology*, 147, 351–362.

61. Krysko, D. V., Diez-Fraile, A., Criel, G., Svistunov, A. A., Vandenabeele, P., and D'Herde, K. (2008). Life and death of

female gametes during oogenesis and folliculogenesis. *Apoptosis*, 13, 1065–1087.

62. Cohen, P. E., Pollack, S. E., and Pollard, J. W. (2006). Genetic analysis of chromosome pairing, recombination, and cell cycle control during first meiotic prophase in mammals. *Endocrine Reviews*, 27, 398–426.

63. Hunt, P. A. and Hassold, T. J. (2008). Human female meiosis: what makes a good egg go bad? *Trends in Genetics*, 24, 86–93.

64. McGee, E. A. and Hsueh, A. J. (2000). Initial and cyclic recruitment of ovarian follicles. *Endocrine Reviews*, 21, 200–214.

65. Pedersen, T. and Peters, H. (1968). Proposal for a classification of oocytes and follicles in the mouse ovary. *Journal of Reproduction and Fertility*, 17, 555–557.

66. Makabe, S., Naguro, T., and Stallone, T. (2006). Oocyte–follicle cell interactions during ovarian follicle development, as seen by high resolution scanning and transmission electron microscopy in humans. *Microscopy Research and Technique*, 69, 436–439.

67. Eppig, J. J. (2001). Oocyte control of ovarian follicular development and function in mammals. *Reproduction*, 122, 829–838.

68. Kidder, G. M. and Mhawi, A. A. (2002). Gap junctions and ovarian folliculogenesis. *Reproduction*, 123, 613–620.

69. Jamnongjit, M. and Hammes, S. R. (2006). Ovarian steroids: the good, the bad, and the signals that raise them. *Cell Cycle*, 5, 1178–1183.

70. Stouffer, R. L. (2006). Structure, function, and regulation of the corpus luteum. In *Knobil and Neill's Physiology of Reproduction* 3rd edition (Neill, J. D.,Ed.). Elsevier, San Diego, CA, pp. 475–526.

71. Champlin, A. K., Dorr, D. L., and Gates, A. H. (1973). Determining the stage of the estrous cycle in the mouse by the appearance of the vagina. *Biology of Reproduction*, 8, 491–494.

72. Gosden, R. and Spears, N. (1997). Programmed cell death in the reproductive system. *British Medical Bulletin*, 52, 644–661.

73. Matzuk, M. M. and Lamb, D. J. (2008). The biology of infertility: research advances and clinical challenges. *Nature Medicine*, 14, 1197–11213.

74. Hoyer, P. B. (2005). Damage to ovarian development and function. *Cell and Tissue Research*, 322, 99–106.

75. Baldridge, M. G., Stahl, R. L., Gerstenberger, S. L., Tripoli, V., and Hutz, R. J. (2003). Modulation of ovarian follicle maturation in Long–Evans rats exposed to polychlorinated biphenyls (PCBs) in-utero and lactationally. *Reproductive Toxicology*, 17, 567–573.

76. Barnett, K. R., Tomic, D., Gupta, R. K., Miller, K. P., Meachum, S., Paulose, T., and Flaws, J. A. (2007). The aryl hydrocarbon receptor affects mouse ovarian follicle growth via mechanisms involving estradiol regulation and responsiveness. *Biology of Reproduction*, 76, 1062–1070.

77. Barhoover, M. A., Hall, J. M., Greenlee, W. F., and Thomas, R. S. (2009). The AHR regulates cell cycle progression in human breast cancer cells via a functional interaction with CDK4. *Molecular Pharmacology*, 2, 195–201.

78. Bussmann, U. A., Bussmann, L. E., and Baranao, J. L. (2006). An aryl hydrocarbon receptor agonist amplifies the mitogenic actions of estradiol in granulosa cells: evidence of involvement of the cognate receptors. *Biology of Reproduction*, 74, 417–426.

79. Matsuda-Minehata, F., Inoue, N., Goto, Y., and Manabe, N. (2006). The regulation of ovarian granulosa cell death by pro- and anti-apoptotic molecules. *Journal of Reproduction and Development*, 52, 695–705.

80. Hussein, M. R. (2005). Apoptosis in the ovary: molecular mechanisms. *Human Reproduction Update*, 11, 162–178.

81. Benedict, J. C., Miller, K. P., Lin, T. M., Greenfeld, C., Babus, J. K., Peterson, R. E., and Flaws, J. A. (2003). Aryl hydrocarbon receptor regulates growth, but not atresia, of mouse preantral and antral follicles. *Biology of Reproduction*, 68, 1511–1517.

82. Heimler, I., Trewin, A. L., Chaffin, C. L., Rawlins, R. G., and Hutz, R. J. (1998). Modulation of ovarian follicle maturation and effects on apoptotic cell death in Holtzman rats exposed to 2,3,7,8-tetrachlorodibenzo-*p*-dioxin (TCDD) in utero and lactationally. *Reproductive Toxicology*, 12, 69–73.

83. Findlay, J. K., Britt, K., Kerr, J. B., O'Donnell, L., Jones, M. E., Drummond, A. E., and Simpson, E. R. (2001). The road to ovulation: the role of oestrogens. *Reproduction, Fertility and Development*, 13, 543–547.

84. Huhtaniemi, I. (2000). The Parkes lecture. Mutations of gonadotropin and gonadotropin receptor genes: what do they teach us about reproductive physiology? *Journal of Reproduction and Fertility*, 119, 173–186.

85. Drummond, A. E. (2006). The role of steroids in follicular growth. *Reproductive Biology and Endocrinology*, 4, 16.

86. Fitzpatrick, S. L., Carlone, D. L., Robker, R. L., and Richards, J. S. (1997). Expression of aromatase in the ovary: down-regulation of mRNA by the ovulatory luteinizing hormone surge. *Steroids*, 62, 197–206.

87. Baba, T., Mimura, J., Nakamura, N., Harada, N., Yamamoto, M., Morohashi, K., and Fujii-Kuriyama, Y. (2005). Intrinsic function of the aryl hydrocarbon (dioxin) receptor as a key factor in female reproduction. *Molecular and Cellular Biology*, 25, 10040–10051.

88. Barnett, K. R., Tomic, D., Gupta, R. K., Babus, J. K., Roby, K. F., Terranova, P. F., and Flaws, J. A. (2007). The aryl hydrocarbon receptor is required for normal gonadotropin responsiveness in the mouse ovary. *Toxicology and Applied Pharmacology*, 223, 66–72.

89. Chaffin, C. L., Peterson, R. E., and Hutz, R. J. (1996). In utero and lactational exposure of female Holtzman rats to 2,3,7,8-tetrachlorodibenzo-*p*-dioxin: modulation of the estrogen signal. *Biology of Reproduction*, 55, 62–67.

90. Myllymäki, S. A., Haavisto, T. E., Brokken, L. J., Viluksela, M., Toppari, J., and Paranko, J. (2005). In utero and lactational exposure to TCDD: steroidogenic outcomes differ in male and female rat pups. *Toxicological Sciences*, 88, 534–544.

91. Dasmahapatra, A. K., Wimpee, B. A., Trewin, A. L., Wimpee, C. F., Ghorai, J. K., and Hutz, R. J. (2000).

Demonstration of 2,3,7,8-tetrachlorodibenzo-*p*-dioxin attenuation of P450 steroidogenic enzyme mRNAs in rat granulosa cell *in vitro* by competitive reverse transcriptase-polymerase chain reaction assay. *Molecular and Cellular Endocrinology*, 164, 5–18.
92. Grochowalski, A., Chrzaszcz, R., Pieklo, R., and Gregoraszczuk, E. L. (2001). Estrogenic and antiestrogenic effect of *in vitro* treatment of follicular cells with 2,3,7,8-tetrachlorodibenzo-*p*-dioxin. *Chemosphere*, 43, 823–827.
93. Heimler, I., Rawlins, R. G., Owen, H., and Hutz, R. J. (1998). Dioxin perturbs, in a dose- and time-dependent fashion, steroid secretion, and induces apoptosis of human luteinized granulosa cells. *Endocrinology*, 139, 4373–4379.
94. Moran, F. M., Vandevoort, C. A., Overstreet, J. W., Lasley, B. L., and Conley, A. J. (2003). Molecular target of endocrine disruption in human luteinizing granulosa cells by 2,3,7,8-tetrachlorodibenzo-*p*-dioxin: inhibition of estradiol secretion due to decreased 17alpha-hydroxylase/17,20-lyase cytochrome P450 expression. *Endocrinology*, 144, 467–473.
95. Tian, Y., Ke, S., Thomas, T., Meeker, R. J., and Gallo, M. A. (1998). Regulation of estrogen receptor mRNA by 2,3,7,8-tetrachlorodibenzo-*p*-dioxin as measured by competitive RT-PCR. *Journal of Biochemical and Molecular Toxicology*, 12, 71–77.
96. Tian, Y., Ke, S., Thomas, T., Meeker, R. J., and Gallo, M. A. (1998). Transcriptional suppression of estrogen receptor gene expression by 2,3,7,8-tetrachlorodibenzo-*p*-dioxin (TCDD). *Journal of Steroid Biochemistry and Molecular Biology*, 67, 17–24.
97. Hirakawa, T., Minegishi, T., Abe, K., Kishi, H., Ibuki, Y., and Miyamoto, K. (2000). Effect of 2,3,7,8-tetrachlorodibenzo-*p*-dioxin on the expression of luteinizing hormone receptors during cell differentiation in cultured granulosa cells. *Archives of Biochemistry and Biophysics*, 15, 371–376.
98. Hirakawa, T., Minegishi, T., Abe, K., Kishi, H., Inoue, K., Ibuki, Y., and Miyamoto, K. (2008). Effect of 2,3,7,8-tetrachlorodibenzo-*p*-dioxin on the expression of follicle-stimulating hormone receptors during cell differentiation in cultured granulosa cells. *Endocrinology*, 141, 1470–1476.
99. Hayes, C. L., Spink, D. C., Spink, B. C., Cao, J. Q., Walker, N. J., and Sutter, T. R. (1996). 17Beta-estradiol hydroxylation catalyzed by human cytochrome P450 1B1. *Proceedings of the National Academy of Sciences of the United States of America*, 93, 9776–9781.
100. Spink, D. C., Eugster, H. P., Lincoln, D. W., 2nd, Schuetz, J. D., Schuetz, E. G., Johnson, J. A., Kaminsky, L. S., and Gierthy, J. F. (1992). 17Beta-estradiol hydroxylation catalyzed by human cytochrome P450 1A1: a comparison of the activities induced by 2,3,7,8-tetrachlorodibenzo-*p*-dioxin in MCF-7 cells with those from heterologous expression of the cDNA. *Archives of Biochemistry and Biophysics*, 293, 342–348.
101. Badawi, A. F., Cavalieri, E. L., and Rogan, E. G. (2001). Role of human cytochrome P450 1A1, 1A2, 1B1, and 3A4 in the 2-, 4-, and 16alpha-hydroxylation of 17beta-estradiol. *Metabolism*, 50, 1001–1003.
102. Dasmahapatra, A. K., Trewin, A. L., and Hutz, R. J. (2002). Estrous cycle-regulated expression of CYP1B1 mRNA in the rat ovary. *Comparative Biochemistry and Physiology, Part B*, 133, 127–134.
103. Bengtsson, M. and Rydstrom, J. (1983). Regulation of carcinogen metabolism in the rat ovary by the estrous cycle and gonadotropin. *Science*, 219, 1437–1438.
104. Dey, A. and Nebert, D. W. (1998). Markedly increased constitutive CYP1A1 mRNA levels in the fertilized ovum of the mouse. *Biochemical and Biophysical Research Communications*, 251, 657–661.
105. Thackaberry, E. A., Bedrick, E. J., Goens, M. B., Danielson, L., Lund, A. K., Gabaldon, D., Smith, S. M., and Walker, M. K. (2003). Insulin regulation in AhR-null mice: embryonic cardiac enlargement, neonatal macrosomia, and altered insulin regulation and response in pregnant and aging AhR-null females. *Toxicological Sciences*, 76, 407–417.
106. Kayampilly, P. P. and Menon, K. M. (2006). Dihydrotestosterone inhibits insulin-stimulated cyclin D2 messenger ribonucleic acid expression in rat ovarian granulosa cells by reducing the phosphorylation of insulin receptor substrate-1. *Endocrinology*, 147, 464–471.
107. Rice, S., Pellatt, L., Ramanathan, K., Whitehead, S. A., and Mason, H. D. (2009). Metformin inhibits aromatase via an extracellular signal-regulated kinase-mediated pathway. *Endocrinology*, 150, 4794–4801.
108. Garzo, V. G. and Dorrington, J. H. (1984). Aromatase activity in human granulosa cells during follicular development and the modulation by follicle-stimulating hormone and insulin. *American Journal of Obstetrics & Gynecology*, 148, 657–662.
109. Espey, L. L. and Richards, J. S. (2006). Ovulation. In *Knobil and Neill's Physiology of Reproduction*, 3rd edition (Neill, J. D., et al., Eds). Elsevier, San Diego, CA, pp. 425–474.
110. Levine, J. E. (1997). New concepts of the neuroendocrine regulation of gonadotropin surges in rats. *Biology of Reproduction*, 56, 293–302.
111. Mahesh, V. B. and Brann, D. W. (1998). Regulation of the preovulatory gonadotropin surge by endogenous steroids. *Steroids*, 63, 616–629.
112. Murdoch, W. J., Hansen, T. R., and McPherson, L. A. (1993). A review: role of eicosanoids in vertebrate ovulation. *Prostaglandins*, 46, 85–115.
113. Li, X. L., Johnson, D. C., and Rozman, K. K. (1995). Reproductive effects of 2,3,7,8-tetrachlorodibenzo-*p*-dioxin (TCDD) in female rats: ovulation, hormonal regulation, and possible mechanism(s). *Toxicology and Applied Pharmacology*, 133, 321–327.
114. Li, X. L., Johnson, D. C., and Rozman, K. K. (1995). Effects of 2,3,7,8-tetrachlorodibenzo-*p*-dioxin (TCDD) on estrous cyclicity and ovulation in female Sprague-Dawley rats. *Toxicology Letters*, 78, 219–222.
115. Gao, X., Son, D. S., Terranova, P. F., and Rozman, K. K. (1999). Toxic equivalency factors of polychlorinated dibenzo-*p*-dioxins in an ovulation model: validation of the toxic

equivalency concept for one aspect of endocrine disruption. *Toxicology and Applied Pharmacology*, 157, 107–116.

116. Son, D. S., Ushinohama, K., Gao, X., Taylor, C. C., Roby, K. F., Rozman, K. K., and Terranova, P. F. (1999). 2,3,7,8-Tetrachlorodibenzo-*p*-dioxin (TCDD) blocks ovulation by a direct action on the ovary without alteration of ovarian steroidogenesis: lack of a direct effect on ovarian granulosa and thecal–interstitial cell steroidogenesis *in vitro*. *Reproductive Toxicology*, 13, 521–530.

117. Gao, X., Terranova, P. F., and Rozman, K. K. (2000). Effects of polychlorinated dibenzofurans, biphenyls, and their mixture with dibenzo-*p*-dioxins on ovulation in the gonadotropin-primed immature rat: support for the toxic equivalency concept. *Toxicology and Applied Pharmacology*, 163, 115–124.

118. Petroff, B. K., Gao, X., Rozman, K. K., and Terranova, P. F. (2000). Interaction of estradiol and 2,3,7,8-tetrachlorodibenzo-*p*-dioxin (TCDD) in an ovulation model: evidence for systemic potentiation and local ovarian effects. *Reproductive Toxicology*, 14, 247–255.

119. Ushinohama, K., Son, D. S., Roby, K. F., Rozman, K. K., and Terranova, P. F. (2001). Impaired ovulation by 2,3,7,8-tetrachlorodibenzo-*p*-dioxin (TCDD) in immature rats treated with equine chorionic gonadotropin. *Reproductive Toxicology*, 15, 275–280.

120. Mizuyachi, K., Son, D. S., Rozman, K. K., and Terranova, P. F. (2002). Alteration in ovarian gene expression in response to 2,3,7,8-tetrachlorodibenzo-*p*-dioxin: reduction of cyclooxygenase-2 in the blockage of ovulation. *Reproductive Toxicology*, 16, 299–307.

121. Petroff, B. K., Roby, K. F., Gao, X., Son, D. S., Williams, S., Johnson, D., Rozman, K. K., and Terranova, P. F. (2001). A review of mechanisms controlling ovulation with implications for the anovulatory effects of polychlorinated dibenzo-*p*-dioxins in rodents. *Toxicology*, 158, 91–107.

122. Gao, X., Mizuyachi, K., Terranova, P. F., and Rozman, K. K. (2001). 2,3,7,8-Tetrachlorodibenzo-*p*-dioxin decreases responsiveness of the hypothalamus to estradiol as a feedback inducer of preovulatory gonadotropin secretion in the immature gonadotropin-primed rat. *Toxicology and Applied Pharmacology*, 170, 181–190.

123. Shearman, L. P., Zylka, M. J., Reppert, S. M., and Weaver, D. R. (1999). Expression of basic helix–loop–helix/PAS genes in the mouse suprachiasmatic nucleus. *Neuroscience*, 89, 387–397.

124. Sellix, M. T., Yoshikawa, T., and Menaker, M. (2010). A circadian egg timer gates ovulation. *Current Biology*, 20, R266–R267.

125. Li, X., Johnson, D. C., and Rozman, K. K. (1997). 2,3,7,8-Tetrachlorodibenzo-*p*-dioxin (TCDD) increases release of luteinizing hormone and follicle-stimulating hormone from the pituitary of immature female rats *in vivo* and *in vitro*. *Toxicology and Applied Pharmacology*, 142, 264–269.

126. Chaffin, C. L., Stouffer, R. L., and Duffy, D. M. (1999). Gonadotropin and steroid regulation of steroid receptor and aryl hydrocarbon receptor messenger ribonucleic acid in macaque granulosa cells during the periovulatory interval. *Endocrinology*, 140, 4753–4760.

127. McCartney, C. R., Eagleson, C. A., and Marshall, J. C. (2002). Regulation of gonadotropin secretion: implications for polycystic ovary syndrome. *Seminars in Reproductive Medicine*, 20, 317–326.

128. Messinis, I. E. (2006). Ovarian feedback, mechanism of action and possible clinical implications. *Human Reproduction Update*, 12, 557–571.

129. Rabii, J. and Ganong, W. F. (1976). Responses of plasma "estradiol" and plasma LH to ovariectomy, ovariectomy plus adrenalectomy, and estrogen injection at various ages. *Neuroendocrinology*, 20, 270–281.

130. Ojeda, S. R., Kalra, P. S., and McCann, S. M. (1975). Further studies on the maturation of the estrogen negative feedback on gonadotropin release in the female rat. *Neuroendocrinology*, 18, 242–255.

131. Gao, X., Petroff, B. K., Rozman, K. K., and Terranova, P. F. (2000). Gonadotropin-releasing hormone (GnRH) partially reverses the inhibitory effect of 2,3,7,8-tetrachlorodibenzo-*p*-dioxin on ovulation in the immature gonadotropin-treated rat. *Toxicology*, 147, 15–22.

132. Pocar, P., Augustin, R., and Fischer, B. (2004). Constitutive expression of CYP1A1 in bovine cumulus oocyte-complexes *in vitro*: mechanisms and biological implications. *Endocrinology*, 145, 1594–1601.

133. Pocar, P., Brevini, T. A. L., Antonini, S., and Gandolfi, F. (2006). Cellular and molecular mechanisms mediating the effect of polychlorinated biphenyls on oocyte *in vitro* maturation. *Reproductive Toxicology*, 22, 242–249.

134. Puga, A., Hoffer, A., Zhou, S., Bohm, J. M., Leikauf, G. D., and Shertzer, H. G. (1997). Sustained increase in intracellular free calcium and activation of cyclooxygenase-2 expression in mouse hepatoma cells treated with dioxin. *Biochemical Pharmacology*, 54, 1287–1296.

135. Vogel, C., Schuhmacher, U. S., Degen, G. H., Bolt, H. M., Pineau, T., and Abel, J. (1998). Modulation of prostaglandin H synthase-2 mRNA expression by 2,3,7,8-tetrachlorodibenzo-*p*-dioxin in mice. *Archives of Biochemistry and Biophysics*, 351, 265–271.

136. Chaffin, C. L. and Stouffer, R. L. (2002). Local role of progesterone in the ovary during the periovulatory interval. *Reviews in Endocrine & Metabolic Disorders*, 3, 65–72.

137. Stocco, C., Telleria, C., and Gibori, G. (2007). The molecular control of corpus luteum formation, function, and regression. *Endocrine Reviews*, 28, 117–149.

138. Stouffer, R. L., Martínez-Chequer, J. C., Molskness, T. A., Xu, F., and Hazzard, T. M. (2001). Regulation and action of angiogenic factors in the primate ovary. *Archives of Medical Research*, 32, 567–575.

139. Ishimura, R., Kawakami, T., Ohsako, S., and Tohyama, C. (2009). Dioxin-induced toxicity on vascular remodeling of the placenta. *Biochemical Pharmacology*, 77, 660–669.

140. Lyons, R. A., Saridogan, E., and Djahanbakhch, O. (2006). The reproductive significance of human Fallopian tube cilia. *Human Reproduction Update*, 12, 363–372.

141. Suarez, S. S. (2006). Gamete and zygote transport. In *Knobil and Neill's Physiology of Reproduction*, 3rd edition

(Neill, J. D., et al., Eds). Elsevier, San Diego, CA, pp. 113–145.

142. Elbi, C., Misteli, T., and Hager, G. L. (2002). Recruitment of dioxin receptor to active transcription sites. *Molecular Biology of the Cell*, 13, 2001–2015.

143. Hombach-Klonisch, S., Pocar, P., Kauffold, J., and Klonisch, T. (2006). Dioxin exerts anti-estrogenic actions in a novel dioxin-responsive telomerase-immortalized epithelial cell line of the porcine oviduct (TERT-OPEC). *Toxicological Sciences*, 90, 519–528.

144. Steinhauer, N., Boos, A., and Günzel-Apel, A. R. (2004). Morphological changes and proliferative activity in the oviductal epithelium during hormonally defined stages of the oestrous cycle in the bitch. *Reproduction in Domestic Animals*, 39, 110–119.

145. Shao, R., Weijdegård, B., Fernandez-Rodriguez, J., Egecioglu, E., Zhu, C., Andersson, N., Thurin-Kjellberg, A., Bergh, C., and Billig, H. (2007). Ciliated epithelial-specific and regional-specific expression and regulation of the estrogen receptor-beta2 in the Fallopian tubes of immature rats: a possible mechanism for estrogen-mediated transport process in vivo. *American Journal of Physiology: Endocrinology and Metabolism*, 293, E147–E158.

146. Amso, N. N., Crow, J., Lewin, J., and Shaw, R. W. (1994). A comparative morphological and ultrastructural study of endometrial gland and Fallopian tube epithelia at different stages of the menstrual cycle and the menopause. *Human Reproduction*, 9, 2234–2241.

147. Ohtake, F., Baba, A., Fujii-Kuriyama, Y., and Kato, S. (2008). Intrinsic AhR function underlies cross-talk of dioxins with sex hormone signalings. *Biochemical and Biophysical Research Communications*, 370, 541–546.

148. Pocar, P., Fischer, B., Klonisch, T., and Hombach-Klonisch, S. (2005). Molecular interactions of the aryl hydrocarbon receptor and its biological and toxicological relevance for reproduction. *Reproduction*, 129, 379–389.

149. Peters, J. M. and Wiley, L. M. (1995). Evidence that murine preimplantation embryos express aryl hydrocarbon receptor. *Toxicology and Applied Pharmacology*, 134, 214–221.

150. Hutt, K. J., Shi, Z., Albertini, D. F., and Petroff, B. K. (2008). The environmental toxicant 2,3,7,8-tetrachlorodibenzo-*p*-dioxin disrupts morphogenesis of the rat pre-implantation embryo. *BMC Developmental Biology*, 8, 1.

151. Hess, A. P., Nayak, N. R., and Giudice, L. C. (2006). Oviduct and endometrium: cyclic changes in the primate oviduct and endometrium. In *Knobil and Neill's Physiology of Reproduction*, 3rd edition (Neill, J. D., et al., Eds). Elsevier, San Diego, CA, pp. 337–381.

152. Buchanan, D. L., Sato, T., Peterson, R. E., and Cooke, P. S. (2000). Antiestrogenic effects of 2,3,7,8-tetrachlorodibenzo-*p*-dioxin in mouse uterus: critical role of the aryl hydrocarbon receptor in stromal tissue. *Toxicological Sciences*, 57, 302–311.

153. Küchenhoff, A., Seliger, G., Klonisch, T., Tscheudschilsuren, G., Kaltwasser, P., Seliger, E., Buchmann, J., and Fischer, B. (1999). Arylhydrocarbon receptor expression in the human endometrium. *Fertility and Sterility*, 71, 354–360.

154. Pitt, J. A., Feng, L., Abbott, B. D., Schmid, J., Batt, R. E., Costich, T. G., Koury, S. T., and Bofinger, D. P. (2001). Expression of AhR and ARNT mRNA in cultured human endometrial explants exposed to TCDD. *Toxicological Sciences*, 62, 289–298.

155. Zhao, D., Pritts, E. A., Chao, V. A., Savouret, J. F., and Taylor, R. N. (2002). Dioxin stimulates RANTES expression in an *in-vitro* model of endometriosis. *Molecular Human Reproduction*, 8, 849–854.

156. Bhattacharyya, K. K., Brake, P. B., Eltom, S. E., Otto, S. A., and Jefcoate, C. R. (1995). Identification of a rat adrenal cytochrome P450 active in polycyclic hydrocarbon metabolism as rat CYP1B1. Demonstration of a unique tissue-specific pattern of hormonal and aryl hydrocarbon receptor-linked regulation. *Journal of Biological Chemistry*, 270, 11595–11602.

157. Bofinger, D. P., Feng, L., Chi, L. H., Love, J., Stephen, F. D., Sutter, T. R., Osteen, K. G., Costich, T. G., Batt, R. E., Koury, S. T., and Olson, J. R. (2001). Effect of TCDD exposure on CYP1A1 and CYP1B1 expression in explant cultures of human endometrium. *Toxicological Sciences*, 62, 299–314.

158. Yang, J. H. (1999). Expression of dioxin-responsive genes in human endometrial cells in culture. *Biochemical and Biophysical Research Communications*, 257, 259–263.

159. Wang, H., Eriksson, H., and Sahlin, L. (2000). Estrogen receptors alpha and beta in the female reproductive tract of the rat during the estrous cycle. *Biology of Reproduction*, 63, 1331–1340.

160. Mueller, M. D., Vigne, J. L., Streich, M., Tee, M. K., Raio, L., Dreher, E., Bersinger, N. A., and Taylor, R. N. (2005). 2,3,7,8-Tetrachlorodibenzo-*p*-dioxin increases glycodelin gene and protein expression in human endometrium. *Journal of Clinical Endocrinology and Metabolism*, 90, 4809–4815.

161. Kretzschmar, G., Papke, A., Zierau, O., Möller, F. J., Medjakovic, S., Jungbauer, A., and Vollmer, G. (2010). Estradiol regulates aryl hydrocarbon receptor expression in the rat uterus. *Molecular and Cellular Endocrinology*, 2, 253–257.

162. Fernandez-Salguero, P. M., Ward, J. M., Sundberg, J. P., and Gonzalez, F. J. (1997). Lesions of aryl-hydrocarbon receptor-deficient mice. *Veterinary Pathology*, 34, 605–614.

163. Abbott, B. D., Schmid, J. E., Pitt, J. A., Buckalew, A. R., Wood, C. R., Held, G. A., and Diliberto, J. J. (1999). Adverse reproductive outcomes in the transgenic Ah receptor-deficient mouse. *Toxicology and Applied Pharmacology*, 155, 62–70.

164. Hasan, A. and Fischer, B. (2001). Hormonal control of arylhydrocarbon receptor (AhR) expression in the preimplantation rabbit uterus. *Anatomy and Embryology*, 204, 189–196.

165. Tscheudschilsuren, G., Hombach-Klonisch, S., Küchenhoff, A., Fischer, B., and Klonisch, T. (1999). Expression of the arylhydrocarbon receptor and the arylhydrocarbon receptor nuclear translocator during early gestation in the rabbit uterus. *Toxicology and Applied Pharmacology*, 160, 231–237.

166. Kitajima, M., Khan, K. N., Fujishita, A., Masuzaki, H., Koji, T., and Ishimaru, T. (2004). Expression of the arylhydrocarbon receptor in the peri-implantation period of the mouse uterus

and the impact of dioxin on mouse implantation. *Archives of Histology and Cytology*, 67, 465–474.

167. Li, B., Liu, H. Y., Dai, L. J., Lu, J. C., Yang, Z. M., and Huang, L. (2006). The early embryo loss caused by 2,3,7,8-tetrachlorodibenzo-*p*-dioxin may be related to the accumulation of this compound in the uterus. *Reproductive Toxicology*, 21, 301–306.

168. Detmar, J., Rennie, M. Y., Whiteley, K. J., Qu, D., Taniuchi, Y., Shang, X., Casper, R. F., Adamson, S. L., Sled, J. G., and Jurisicova, A. (2008). Fetal growth restriction triggered by polycyclic aromatic hydrocarbons is associated with altered placental vasculature and AhR-dependent changes in cell death. *American Journal of Physiology: Endocrinology and Metabolism*, 295, E519–E530.

169. Ietta, F., Wu, Y., Winter, J., Wang, J., Post, M., and Caniggia, I. (2006). Dynamic HIF1A regulation during human placental development. *Biology of Reproduction*, 75, 112–121.

170. Daucher, J. A., Clark, K. A., Stolz, D. B., Meyn, L. A., and Moalli, P. A. (2007). Adaptations of the rat vagina in pregnancy to accommodate delivery. *Obstetrics & Gynecology*, 109, 128–135.

171. Gray, L. E., Jr. and Ostby, J. S. (1995). *In utero* 2,3,7,8-tetrachlorodibenzo-*p*-dioxin (TCDD) alters reproductive morphology and function in female rat offspring. *Toxicology and Applied Pharmacology*, 133, 285–294.

172. Dienhart, M. K., Sommer, R. J., Peterson, R. E., Hirshfield, A. N., and Silbergeld, E. K. (2000). Gestational exposure to 2,3,7,8-tetrachlorodibenzo-*p*-dioxin induces developmental defects in the rat vagina. *Toxicological Sciences*, 56, 141–149.

173. Hurst, C. H., Abbott, B., Schmid, J. E., and Birnbaum, L. S. (2002). 2,3,7,8-Tetrachlorodibenzo-*p*-dioxin (TCDD) disrupts early morphogenetic events that form the lower reproductive tract in female rat fetuses. *Toxicological Sciences*, 65, 87–98.

174. Wolf, C. J., Ostby, J. S., and Gray, L. E., Jr., (1999). Gestational exposure to 2,3,7,8-tetrachlorodibenzo-*p*-dioxin (TCDD) severely alters reproductive function of female hamster offspring. *Toxicological Sciences*, 51, 259–264.

175. Jablonska, O., Shi, Z., Valdez, K. E., Ting, A. Y., and Petroff, B. K. (2010). Temporal and anatomical sensitivities to the aryl hydrocarbon receptor agonist 2,3,7,8-tetrachlorodibenzo-*p*-dioxin leading to premature acyclicity with age in rats. *International Journal of Andrology*, 33, 405–412.

176. Shirota, M., Mukai, M., Sakurada, Y., Doyama, A., Inoue, K., Haishima, A., Akahori, F., and Shirota, K. (2006). Effects of vertically transferred 3,3′,4,4′,5-pentachlorobiphenyl (PCB-126) on the reproductive development of female rats. *Journal of Reproduction and Development*, 52, 751–761.

177. Franczak, A., Nynca, A., Valdez, K. E., Mizinga, K. M., and Petroff, B. K. (2006). Effects of acute and chronic exposure to the aryl hydrocarbon receptor agonist 2,3,7,8-tetrachlorodibenzo-*p*-dioxin on the transition to reproductive senescence in female Sprague-Dawley rats. *Biology of Reproduction*, 74, 125–130.

178. Shi, Z., Valdez, K. E., Ting, A. Y., Franczak, A., Gum, S. L., and Petroff, B. K. (2007). Ovarian endocrine disruption underlies premature reproductive senescence following environmentally relevant chronic exposure to the aryl hydrocarbon receptor agonist 2,3,7,8-tetrachlorodibenzo-*p*-dioxin. *Biology of Reproduction*, 76, 198–202.

179. Warner, M., Eskenazi, B., Olive, D. L., Samuels, S., Quick-Miles, S., Vercellini, P., Gerthoux, P. M., Needham, L., Patterson, D. G., and Mocarelli, P. (2007). Serum dioxin concentrations and quality of ovarian function in women of Seveso. *Environmental Health Perspectives*, 115, 336–340.

180. Schmidt, J. V. and Bradfield, C. A. (1996). AH receptor signaling pathways. *Annual Review of Cell and Developmental Biology*, 12, 55–89.

181. Nebert, D. W., Brown, D. D., Towne, D. W., and Eisen, H. J. (1984). Association of fertility, fitness and longevity with the murine Ah locus among (C57BL/6N) (C3H/HeN) recombinant inbred lines. *Biology of Reproduction*, 30, 363–373.

182. Hombach-Klonisch, S., Pocar, P., Kietz, S., and Klonisch, T. (2005). Molecular actions of polyhalogenated arylhydrocarbons (PAHs) in female reproduction. *Current Medical Chemistry*, 12, 599–616.

183. Reichard, J. F., Dalton, T. P., Shertzer, H. G., and Puga, A. (2006). Induction of oxidative stress responses by dioxin and other ligands of the aryl hydrocarbon receptor. *Dose Response*, 3, 306–331.

184. McNulty, W. P. (1985). Toxicity and fetotoxicity of TCDD, TCDF and PCB isomers in rhesus macaques (*Macaca mulatta*). *Environmental Health Perspectives*, 60, 77–88.

185. Meredith, S. and Butcher, R. L. (1985). Role of decreased numbers of follicles on reproductive performance in young and aged rats. *Biology of Reproduction*, 32, 788–794.

186. Wise, P. M., Smith, M. J., Dubal, D. B., Wilson, M. E., Rau, S. W., Cashion, A. B., Böttner, M., and Rosewell, K. L. (2002). Neuroendocrine modulation and repercussions of female reproductive aging. *Recent Progress in Hormone Research*, 57, 235–256.

187. Wise, P. M. (1989). Aging of the female reproductive system: a neuroendocrine perspective. In *Neuroendocrine Perspectives* (Muller, E. E. and MacLeod, R. M.,Eds). Springer, New York, pp. 117–168.

188. Eskenazi, B., Warner, M., Marks, A. R., Samuels, S., Gerthoux, P. M., Vercellini, P., Olive, D. L., Needham, L., and Patterson, D., Jr., (2005). Serum dioxin concentrations and age at menopause. *Environmental Health Perspectives*, 113, 858–862.

189. Valdez, K. E., Shi, Z., Ting, A. Y., and Petroff, B. K. (2009). Effect of chronic exposure to the aryl hydrocarbon receptor agonist 2,3,7,8-tetrachlorodibenzo-*p*-dioxin in female rats on ovarian gene expression. *Reproductive Toxicology*, 28, 32–37.

190. Benedict, J. C., Lin, T. M., Loeffler, I. K., Peterson, R. E., and Flaws, J. A. (2000). Physiological role of the aryl hydrocarbon receptor in mouse ovary development. *Toxicological Sciences*, 56, 382–388.

191. O'Donnell, L., Meachem, S. J., Stanton, P. G., and McLachlan, R. I. (2006). Endocrine regulation of spermatogenesis. In *Knobil and Neill's Physiology of Reproduction*, 3rd

edition (Neill, J. D., et al., Eds). Elsevier, San Diego, CA, pp. 1017–1069.

192. Matsumoto, T., Shiina, H., Kawano, H., Sato, T., and Kato, S. (2008). Androgen receptor functions in male and female physiology. *Journal of Steroid Biochemistry and Molecular Biology*, 109, 236–241.

193. Collins, L. L., Lee, H. J., Chen, Y. T., Chang, M., Hsu, H. Y., Yeh, S., and Chang, C. (2003). The androgen receptor in spermatogenesis. *Cytogenetic and Genome Research*, 103, 299–301.

194. Walker, W. H. (2009). Molecular mechanisms of testosterone action in spermatogenesis. *Steroids*, 74, 602–607.

195. Nitta, H., Bunick, D., Hess, R. A., Janulis, L., Newton, S. C., Millette, C. F., Osawa, Y., Shizuta, Y., Toda, K., and Bahr, J. M. (1993). Germ cells of the mouse testis express P450 aromatase. *Endocrinology*, 132, 1396–1401.

196. Robaire, B., Hinton, B. T., and Orgibin-Crist, M. C. (2006). The epididymis. In *Knobil and Neill's Physiology of Reproduction*, 3rd edition (Neill, J. D., et al., Eds). Elsevier, San Diego, CA, pp. 1071–1148.

197. Risbridger, G. P. and Taylor, R. A. (2006). Physiology of the male accessory sex structures: the prostate gland, seminal vesicles, and bulbourethral glands. In *Knobil and Neill's Physiology of Reproduction*, 3rd edition (Neill, J. D., et al., Eds). Elsevier, San Diego, CA, pp. 1149–1172.

198. Huang, P., Rannug, A., Ahlbom, E., Hökansson, H., and Ceccatelli, S. (2000). Effect of 2,3,7,8-tetrachlorodibenzo-*p*-dioxin on the expression of cytochrome P450 1A1, the aryl hydrocarbon receptor, and the aryl hydrocarbon receptor nuclear translocator in rat brain and pituitary. *Toxicology and Applied Pharmacology*, 169, 159–167.

199. Yamamoto, J., Ihara, K., Nakayama, H., Hikino, S., Satoh, K., Kubo, N., Iida, T., Fujii, Y., and Hara, T. (2004). Characteristic expression of aryl hydrocarbon receptor repressor gene in human tissues: organ-specific distribution and variable induction patterns in mononuclear cells. *Life Sciences*, 74, 1039–1049.

200. Coutts, S. M., Fulton, N., and Anderson, R. A. (2007). Environmental toxicant-induced germ cell apoptosis in the human fetal testis. *Human Reproduction*, 22, 2912–2918.

201. Johnson, L., Dickerson, R., Safe, S. H., Nyberg, C. L., Lewis, R. P., and Welsh, T. H., Jr., (1992). Reduced Leydig cell volume and function in adult rats exposed to 2,3,7,8-tetrachlorodibenzo-*p*-dioxin without a significant effect on spermatogenesis. *Toxicology*, 76, 103–118.

202. Roman, B. L., Pollenz, R. S., and Peterson, R. E. (1998). Responsiveness of the adult male rat reproductive tract to 2,3,7,8-tetrachlorodibenzo-*p*-dioxin exposure: Ah receptor and ARNT expression, CYP1A1 induction, and Ah receptor down-regulation. *Toxicology and Applied Pharmacology*, 150, 228–239.

203. Schultz, R., Suominen, J., Värre, T., Hakovirta, H., Parvinen, M., Toppari, J., and Pelto-Huikko, M. (2003). Expression of aryl hydrocarbon receptor and aryl hydrocarbon receptor nuclear translocator messenger ribonucleic acids and proteins in rat and human testis. *Endocrinology*, 144, 767–776.

204. Gasiewicz, T. A. and Rucci, G. (1984). Cytosolic receptor for 2,3,7,8-tetrachlorodibenzo-*p*-dioxin. Evidence for a homologous nature among various mammalian species. *Molecular Pharmacology*, 26, 90–98.

205. Baba, T., Shima, Y., Owaki, A., Mimura, J., Oshima, M., Fujii-Kuriyama, Y., and Morohashi, K. I. (2008). Disruption of aryl hydrocarbon receptor (AhR) induces regression of the seminal vesicle in aged male mice. *Sexual Development*, 2, 1–11.

206. Fetissov, S. O., Huang, P., Zhang, Q., Mimura, J., Fujii-Kuriyama, Y., Rannug, A., Hökfelt, T., and Ceccatelli, S. (2004). Expression of hypothalamic neuropeptides after acute TCDD treatment and distribution of Ah receptor repressor. *Regulatory Peptides*, 119, 113–124.

207. Forger, N. G., Rosen, G. J., Waters, E. M., Jacob, D., Simerly, R. B., and de Vries, G. J. (2004). Deletion of Bax eliminates sex differences in the mouse forebrain. *Proceedings of the National Academy of Sciences of the United States of America*, 101, 13666–13671.

208. Ikeda, M., Mitsui, T., Setani, K., Tamura, M., Kakeyama, M., Sone, H., Tohyama, C., and Tomita, T. (2005). *In utero* and lactational exposure to 2,3,7,8-tetrachlorodibenzo-*p*-dioxin in rats disrupts brain sexual differentiation. *Toxicology and Applied Pharmacology*, 205, 98–105.

209. Bjerke, D. L. and Peterson, R. E. (1994). Reproductive toxicity of 2,3,7,8-tetrachlorodibenzo-*p*-dioxin in male rats: different effects of *in utero* versus lactational exposure. *Toxicology and Applied Pharmacology*, 127, 241–249.

210. Mably, T. A., Moore, R. W., Goy, R. W., and Peterson, R. E. (1992). *In utero* and lactational exposure of male rats to 2,3,7,8-tetrachlorodibenzo-*p*-dioxin. 2. Effects on sexual behavior and the regulation of luteinizing hormone secretion in adulthood. *Toxicology and Applied Pharmacology*, 114, 108–117.

211. Bjerke, D. L., Brown, T. J., MacLusky, N. J., Hochberg, R. B., and Peterson, R. E. (1994). Partial demasculinization and feminization of sex behavior in male rats by *in utero* and lactational exposure to 2,3,7,8-tetrachlorodibenzo-*p*-dioxin is not associated with alterations in estrogen receptor binding or volumes of sexually differentiated brain nuclei. *Toxicology and Applied Pharmacology*, 127, 258–267.

212. Wilhelm, D., Palmer, S., and Koopman, P. (2007). Sex determination and gonadal development in mammals. *Physiological Reviews*, 87, 1–28.

213. Mocarelli, P., Gerthoux, P. M., Ferrari, E., Patterson, D. G., Jr., Kieszak, S. M., Brambilla, P., Vincoli, N., Signorini, S., Tramacere, P., Carreri, V., Sampson, E. J., Turner, W. E., and Needham, L. L. (2000). Paternal concentrations of dioxin and sex ratio of offspring. *Lancet*, 355, 1858–1863.

214. Ryan, J. J., Amirova, Z., and Carrier, G. (2002). Sex ratios of children of Russian pesticide producers exposed to dioxin. *Environmental Health Perspectives*, 110, A699–A701.

215. Ikeda, M., Tamura, M., Yamashita, J., Suzuki, C., and Tomita, T. (2005). Repeated *in utero* and lactational 2,3,7,8-tetrachlorodibenzo-*p*-dioxin exposure affects male gonads in offspring, leading to sex ratio changes in F2 progeny. *Toxicology and Applied Pharmacology*, 206, 351–355.

216. Ishihara, K., Warita, K., Tanida, T., Sugawara, T., Kitagawa, H., and Hoshi, N. (2007). Does paternal exposure to 2,3,7,8-tetrachlorodibenzo-*p*-dioxin (TCDD) affect the sex ratio of offspring? *Journal of Veterinary Medical Science*, 69, 347–352.

217. Ishihara, K., Ohsako, S., Tasaka, K., Harayama, H., Miyake, M., Warita, K., Tanida, T., Mitsuhashi, T., Nanmori, T., Tabuchi, Y., Yokoyama, T., Kitagawa, H., and Hoshi, N. (2010). When does the sex ratio of offspring of the paternal 2,3,7,8-tetrachlorodibenzo-*p*-dioxin (TCDD) exposure decrease: in the spermatozoa stage or at fertilization? *Reproductive Toxicology*, 29, 68–73.

218. Harley, V. R. and Goodfellow, P. N. (1994). The biochemical role of SRY in sex determination. *Molecular Reproduction and Development*, 39, 184–193.

219. Li, Z., Haines, C. J., and Han, Y. (2008). "Micro-deletions" of the human Y chromosome and their relationship with male infertility. *Journal of Genetics and Genomics*, 35, 193–199.

220. Delbridge, M. L. and Graves, J. A. (1999). Mammalian Y chromosome evolution and the male-specific functions of Y chromosome-borne genes. *Reviews of Reproduction*, 4, 101–109.

221. Kerr, J. B., Loveland, K. L., O'Bryan, M. K., and Kretser, D. M. (2006). Cytology of the testis and intrinsic control mechanisms. In *Knobil and Neill's Physiology of Reproduction*, 3rd edition (Neill, J. D., et al., Eds). Elsevier, San Diego, CA, pp. 827–947.

222. Choi, J. S., Kim, I. W., Hwang, S. Y., Shin, B. J., and Kim, S. K. (2007). Effect of 2,3,7,8-tetrachlorodibenzo-*p*-dioxin on testicular spermatogenesis-related panels and serum sex hormone levels in rats. *British Journal of Urology International*, 101, 250–255.

223. Jin, M. H., Ko, H. K., Hong, C. H., and Han, S. W. (2008). *In utero* exposure to 2,3,7,8-tetrachlorodibenzo-*p*-dioxin affects the development of reproductive system in mouse. *Yonsei Medical Journal*, 49, 843–850.

224. Lin, T. M., Ko, K., Moore, R. W., Buchanan, D. L., Cooke, P. S., and Peterson, R. E. (2001). Role of the aryl hydrocarbon receptor in the development of control and 2,3,7,8-tetrachlorodibenzo-*p*-dioxin-exposed male mice. *Journal of Toxicology and Environmental Health, Part A*, 64, 327–342.

225. Pohjanvirta, R. (2009). Transgenic mouse lines expressing rat AH receptor variants: a new animal model for research on AH receptor function and dioxin toxicity mechanisms. *Toxicology and Applied Pharmacology*, 236, 166–182.

226. Hess, R. A. and Renato de Franca, L. (2008). Spermatogenesis and cycle of the seminiferous epithelium. *Advances in Experimental Medicine Biology*, 636, 1–15.

227. Carreau, S., Genissel, C., Bilinska, B., and Levallet, J. (1999). Sources of oestrogen in the testis and reproductive tract of the male. *International Journal of Andrology*, 22, 211–223.

228. Rune, G. M., de Souza, P., Krowke, R., Merker, H. J., and Neubert, D. (1991). Morphological and histochemical pattern of response in rat testes after administration of 2,3,7,8-tetrachlorodibenzo-*p*-dioxin (TCDD). *Histology and Histopathology*, 6, 459–467.

229. Lai, K. P., Wong, M. H., and Wong, C. K. (2005). Effects of TCDD in modulating the expression of Sertoli cell secretory products and markers for cell–cell interaction. *Toxicology*, 206, 111–123.

230. Cheng, C. Y., Grima, J., Stahler, M. S., Lockshin, R. A., and Bardin, C. W. (1989). Testins are structurally related Sertoli cell proteins whose secretion is tightly coupled to the presence of germ cells. *Journal of Biological Chemistry*, 264, 21386–21393.

231. Mruk, D. D. and Cheng, C. Y. (1999). Sertolin is a novel gene marker of cell–cell interactions in the rat testis. *Journal of Biological Chemistry*, 274, 27056–27068.

232. Whitlock, J. J. (1999). Induction of cytochrome P4501A1. *Annual Review of Pharmacology and Toxicology*, 39, 103–125.

233. Ge, R., Chen, G., and Hardy, M. P. (2008). The role of the Leydig cell in spermatogenic function. *Advances in Experimental Medicine and Biology*, 636, 255–269.

234. Stocco, D. M. and McPhaul, M. J. (2006). Physiology of testicular steroidogenesis. In *Knobil and Neill's Physiology of Reproduction*, 3rd edition (Neill, J. D., et al., Eds). Elsevier, San Diego, CA, pp. 977–1016.

235. Johnson, L., Wilker, C. E., Safe, S. H., Scott, B., Dean, D. D., and White, P. H. (1994). 2,3,7,8-Tetrachlorodibenzo-*p*-dioxin reduces the number, size, and organelle content of Leydig cells in adult rat testes. *Toxicology*, 89, 49–65.

236. Iseki, M., Ikuta, T., Kobayashi, T., and Kawajiri, K. (2005). Growth suppression of Leydig TM3 cells mediated by aryl hydrocarbon receptor. *Biochemical and Biophysical Research Communications*, 331, 902–908.

237. Wilker, C., Johnson, L., and Safe, S. (1996). Effects of developmental exposure to indole-3-carbinol or 2,3,7,8-tetrachlorodibenzo-*p*-dioxin on reproductive potential of male rat offspring. *Toxicology and Applied Pharmacology*, 141, 68–75.

238. Sommer, R. J., Ippolito, D. L., and Peterson, R. E. (1996). *In utero* and lactational exposure of the male Holtzman rat to 2,3,7,8-tetrachlorodibenzo-*p*-dioxin: decreased epididymal and ejaculated sperm numbers without alterations in sperm transit rate. *Toxicology and Applied Pharmacology*, 140, 146–153.

239. Roman, B. L., Sommer, R. J., Shinomiya, K., and Peterson, R. E. (1995). *In utero* and lactational exposure of the male rat to 2,3,7,8-tetrachlorodibenzo*p*-dioxin: impaired prostate growth and development without inhibited androgen production. *Toxicology and Applied Pharmacology*, 134, 241–250.

240. Lin, T. M., Ko, K., Moore, R. W., Simanainen, U., Oberley, T. D., and Peterson, R. E. (2002). Effects of aryl hydrocarbon receptor null mutation and *in utero* and lactational 2,3,7,8-tetrachlorodibenzo-*p*-dioxin exposure on prostate and seminal vesicle development in C57BL/6 mice. *Toxicological Sciences*, 68, 479–487.

241. Hamm, J. T., Sparrow, B. R., Wolf, D., and Birnbaum, L. S. (2000). *In utero* and lactational exposure to 2,3,7,8-tetrachlorodibenzo-*p*-dioxin alters postnatal development of seminal vesicle epithelium. *Toxicological Sciences*, 52, 424–430.

242. Fawell, S. E. and Higgins, S. J. (1986). Tissue distribution, developmental profile and hormonal regulation of androgen-responsive secretory proteins of rat seminal vesicles studied by immunocytochemistry. *Molecular and Cellular Endocrinology*, 48, 39–49.

243. Sommer, R. J., Sojka, K. M., Pollenz, R. S., Cooke, P. S., and Peterson, R. E. (1999). Ah receptor and ARNT protein and mRNA concentrations in rat prostate: effects of stage of development and 2,3,7,8-tetrachlorodibenzo-*p*-dioxin treatment. *Toxicology and Applied Pharmacology*, 155, 177–189.

244. Kashani, M., Steiner, G., Haitel, A., Schaufler, K., Thalhammer, T., Amann, G., Kramer, G., Marberger, M., and Schöller, A. (1998). Expression of the aryl hydrocarbon receptor (AhR) and the aryl hydrocarbon receptor nuclear translocator (ARNT) in fetal, benign hyperplastic, and malignant prostate. *Prostate*, 37, 98–108.

245. Vezina, C. M., Lin, T. M., and Peterson, R. E. (2009). AHR signaling in prostate growth, morphogenesis, and disease. *Biochemical Pharmacology*, 77, 566–576.

246. Mably, T. A., Moore, R. W., and Peterson, R. E. (1992). In utero and lactational exposure of male rats to 2,3,7,8-tetrachlorodibenzo-*p*-dioxin. 1. Effects on androgenic status. *Toxicology and Applied Pharmacology*, 114, 97–107.

247. Gray, L. E., Ostby, J. S., and Kelce, W. R. (1997). A dose–response analysis of the reproductive effects of a single gestational dose of 2,3,7,8-tetrachlorodibenzo-*p*-dioxin in male Long–Evans hooded rat offspring. *Toxicology and Applied Pharmacology*, 146, 11–20.

248. Simanainen, U., Adamsson, A., Tuomisto, J. T., Miettinen, H. M., Toppari, J., Tuomisto, J., and Viluksela, M. (2004). Adult 2,3,7,8-tetrachlorodibenzo-*p*-dioxin (TCDD) exposure and effects on male reproductive organs in three differentially TCDD-susceptible rat lines. *Toxicological Sciences*, 81, 401–407.

249. Theobald, H. M. and Peterson, R. E. (1997). In utero and lactational exposure to 2,3,7,8-tetrachlorodibenzo-rho-dioxin: effects on development of the male and female reproductive system of the mouse. *Toxicology and Applied Pharmacology*, 145, 124–135.

250. Vezina, C. M., Hardin, H. A., Moore, R. W., Allgeier, S. H., and Peterson, R. E. (2010). 2,3,7,8-Tetrachlorodibenzo-*p*-dioxin inhibits fibroblast growth factor 10-induced prostatic bud formation in mouse urogenital sinus. *Toxicological Sciences*, 113, 198–206.

251. Donjacour, A. A., Thomson, A. A., and Cunha, G. R. (2003). FGF-10 plays an essential role in the growth of the fetal prostate. *Developmental Biology*, 261, 39–54.

252. Tsuchiya, M., Katoh, T., Motoyama, H., Sasaki, H., Tsugane, S., and Ikenoue, T. (2005). Analysis of the AhR, ARNT, and AhRR gene polymorphisms: genetic contribution to endometriosis susceptibility and severity. *Fertility and Sterility*, 84, 454–458.

253. Watanabe, T., Imoto, I., Kosugi, Y., Fukuda, Y., Mimura, J., Fujii, Y., Isaka, K., Takayama, M., Sato, A., and Inazawa, J. (2001). Human arylhydrocarbon receptor repressor (AHRR) gene: genomic structure and analysis of polymorphism in endometriosis. *Journal of Human Genetics*, 46, 342–346.

254. Kim, S. H., Choi, Y. M., Lee, G. H., Hong, M. A., Lee, K. S., Lee, B. S., Kim, J. G., and Moon, S. Y. (2007). Association between susceptibility to advanced stage endometriosis and the genetic polymorphisms of aryl hydrocarbon receptor repressor and glutathione-S-transferase T1 genes. *Human Reproduction*, 22, 1866–1870.

255. Watanabe, M., Sueoka, K., Sasagawa, I., Nakabayashi, A., Yoshimura, Y., and Ogata, T. (2004). Association of male infertility with Pro185Ala polymorphism in the aryl hydrocarbon receptor repressor gene: implication for the susceptibility to dioxins. *Fertility and Sterility*, 82 (Suppl. 3), 1067–1071.

256. Merisalu, A., Punab, M., Altmäe, S., Haller, K., Tiido, T., Peters, M., and Salumets, A. (2007). The contribution of genetic variations of aryl hydrocarbon receptor pathway genes to male factor infertility. *Fertility and Sterility*, 88, 854–859.

257. Fujita, H., Kosaki, R., Yoshihashi, H., Ogata, T., Tomita, M., Hasegawa, T., Takahashi, T., Matsuo, N., and Kosaki, K. (2002). Characterization of the aryl hydrocarbon receptor repressor gene and association of its Pro185Ala polymorphism with micropenis. *Teratology*, 65, 10–18.

258. Soneda, S., Fukami, M., Fujimoto, M., Hasegawa, T., Koitabashi, Y., and Ogata, T. (2005). Association of micropenis with Pro185Ala polymorphism of the gene for aryl hydrocarbon receptor repressor involved in dioxin signaling. *Endocrine Journal*, 52, 83–88.

259. Tiido, T., Rignell-Hydbom, A., Jönsson, B., Giwercman, Y. L., Rylander, L., Hagmar, L., and Giwercman, A. (2005). Exposure to persistent organochlorine pollutants associates with human sperm Y:X chromosome ratio. *Human Reproduction*, 20, 1903–1909.

260. Tiido, T., Rignell-Hydbom, A., Jönsson, B. A., Rylander, L., Giwercman, A., and Giwercman, Y. L. (2007). Modifying effect of the AR gene trinucleotide repeats and SNPs in the AHR and AHRR genes on the association between persistent organohalogen pollutant exposure and human sperm Y:X ratio. *Molecular Human Reproduction*, 13, 223–229.

32

THE AHR IN THE CONTROL OF CELL CYCLE AND APOPTOSIS

CORNELIA DIETRICH

32.1 INTRODUCTION

The AHR was originally discovered due to its stimulation by a variety of planar aromatic hydrocarbons, such as benzo[a]pyrene (B[a]P), 2,3,7,8-tetrachlorodibenzo-p-dioxin (TCDD), and polychlorinated biphenyls (PCBs). It is generally accepted that the toxic responses of these environmental pollutants are the direct consequence of AHR activation. One of the most potent ligands known so far is TCDD that causes a plethora of toxic effects including chloracne, tumor promotion, immunosuppression, and teratogenesis. As presented in the first chapter, the AHR is a transcription factor that—after ligand binding—finally leads to induction of gene transcription of several phase I and phase II enzymes, for example, *cytochrome P450s* (*CYP1A1, CYP1A2, CYP1B1*). This canonical AHR-dependent pathway at least partially explains the carcinogenicity of polycyclic aromatic hydrocarbons that are metabolized to genotoxic compounds by these enzymes. However, it does not help to understand the molecular mechanisms of toxic effects induced by nongenotoxic AHR ligands, such as TCDD, which is not metabolized. No mechanistic link between CYP induction and TCDD-mediated hepatotoxicity or immunotoxicity could be established so far [1, 2]. In line with this observation, *in vivo* studies in two genetically different rat strains indicate that AHR-driven CYP1A1 induction and tumor promotion can be uncoupled from each other (reviewed in Ref. 3). Moreover, experiments in AHR-null mice revealed that the AHR—in the absence of exogenous ligands—is involved in several physiological processes. These observations point to AHR functions beyond xenobiotic metabolism (reviewed in Ref. 4; see also many chapters in this book).

Proper organ development and function is governed by a fine-tuned regulation of cell proliferation and apoptosis. Conversely, disturbance of cell cycle control and cell death may lead to toxic processes, such as tumor promotion, immunosuppression, and teratogenicity. A number of studies using different cell culture and animal models have clearly shown that the AHR regulates and deregulates cell proliferation and apoptosis. However, the observed effects on proliferation and apoptosis are enormously diverse: depending on the cell system studied, AHR activation may lead to either stimulation or inhibition of proliferation or apoptosis. Moreover, the physiological function of the AHR obviously differs from its toxicological role after exogenous ligand binding. This is also reflected by *in vivo* experiments. On the one hand, it was demonstrated that mice expressing a constitutively active AHR show an increase in the development of stomach tumors [5] as well as in liver tumors in a two-stage model of hepatocarcinogenesis [6]. On the other hand, a recent study revealed that the AHR may also possess tumor suppressor activities in liver [7]. In this chapter, I will summarize the current knowledge about the role of the AHR in cell proliferation and apoptosis *in vitro* and *in vivo* in mammalian systems. I will focus on studies using nongenotoxic AHR ligands. Data obtained with genotoxic AHR ligands, such as B[a]P, will be excluded since the perturbations of cell proliferation and survival here result from DNA damage and are therefore mechanistically different. I will further present studies addressing the physiological role of the AHR in the control of cell proliferation and apoptosis. Possible mechanisms will be discussed.

The AH Receptor in Biology and Toxicology, First Edition. Edited by Raimo Pohjanvirta.
© 2012 John Wiley & Sons, Inc. Published 2012 by John Wiley & Sons, Inc.

32.2 THE AHR AS A STIMULATOR OF PROLIFERATION

32.2.1 *In Vitro* Studies

Liver cells have been a main focus in AHR research as the prototypic nongenotoxic AHR ligand TCDD is one of the most potent tumor promoters ever tested in rodents, and one major target organ is liver [8]. Already more than 20 years ago, the group of K.-W. Bock revealed that hepatocytes isolated from rats that had been exposed to 3,4,3′,4′-tetrachlorobiphenyl (TCB), an AHR ligand, and cultured under serum-free conditions respond with higher growth rates after addition of serum or epidermal growth factor (EGF) [9]. A similar comitogenic effect was observed in rat hepatocytes cultured under serum-free conditions and cotreated with TCDD and EGF or insulin [10] or TCDD and ethinylestradiol (E2) [11, 12]. Interestingly, the comitogenic effect of TCDD was seen only at picomolar concentrations that are two orders of magnitude lower than those required for CYP1A1 induction. At such higher concentrations, even inhibition of DNA synthesis was detected [12]. No stimulatory TCDD effect could be observed under conditions allowing high rates of proliferation, that is, in the presence of serum [10] or when hepatocytes were isolated from juvenile rats that are more sensitive toward EGF than hepatocytes from adult animals [11]. These observations were confirmed in primary mouse hepatocytes [13]. Again, the comitogenic effect of TCDD and EGF was observed only under low basal cell proliferation, that is, at higher cell density, and at very low TCDD concentrations (0.03 pM). Higher concentrations (3 pM) of TCDD inhibited EGF-induced proliferation. The fact that a 100-fold higher concentration of TCDD was required for the comitogenic effect in hepatocytes isolated from a mouse strain expressing a low-affinity AHR suggests an involvement of the AHR in this response.

However, the proliferative effect of TCDD is highly cell type dependent. In contrast to primary hepatocytes in which TCDD exhibits a comitogenic response, but is ineffective when given alone, a proliferative effect of TCDD could be shown in the rat liver oval cell line WB-F344 at least under special culture conditions. In WB-F344 cells, TCDD and other nongenotoxic AHR ligands induce a release from contact inhibition [12, 14–19]. When confluent cultures of WB-F344 cells are exposed to TCDD, saturation density is increased twofold and multilayered foci can be detected [12, 14, 20]. Interestingly, TCDD-mediated loss of contact inhibition is potentiated by tumor necrosis factor-α when TCDD is given at picomolar concentrations [21]. That neither exponentially growing cells nor serum-deprived cultures respond to the growth stimulatory effect of TCDD indicates that TCDD does not exert a mitogenic effect per se, but specifically interferes with the signaling cascade of contact inhibition [20]. Although TCDD-dependent release from contact inhibition is not restricted to this single cell line, but rather appears to be a phenomenon observed in several cell lines, such as newborn human foreskin keratinocytes [22], oral squamous carcinoma cells [23], Madin–Darby canine kidney cells [19], and Madin–Darby bovine kidney cells [24], the effect is cell type specific since it is not observed in 5L hepatoma, Hepa, primary murine hepatocytes, or the human keratinocyte cell line HaCaT [19] (our own unpublished observations). Likewise, no effect of TCDD on cell proliferation could be detected in more than 20 cell lines neither when growing exponentially nor at confluence [25].

Hyperproliferation is considered to be involved in the teratogenic effects of TCDD in mice, that is, hydronephrosis and development of cleft palate. In embryonic ureteric as well as in palatal medial edge epithelial cells, TCDD exposure leads to increased proliferation [26–28] (see Chapter 17). Also, in lung alveolar carcinoma cells and immortalized breast epithelial cells, AHR overexpression or TCDD treatment, respectively, results in acceleration of proliferation [29–31].

32.2.2 *In Vivo* Studies

A variety of studies have been performed addressing AHR-dependent proliferation in mouse and rat liver, most of them using TCDD as model compound either alone or after application of a tumor initiator. Christian and Peterson revealed that in male (but not female) Sprague Dawley (SD) rats TCDD increases DNA synthesis in liver after 1/3 hepatectomy [32]. No effect could be detected without or after 2/3 hepatectomy—the latter reflecting high proliferative activity in the absence of TCDD—which is in line with the above-mentioned *in vitro* studies. In contrast, stimulation of hepatocyte proliferation was found in female (but not male) SD rats and male mice when measured 24 h after TCDD exposure in the absence of any further mitogenic stimulus [33]. Repeated TCDD treatment also leads to an increase in hepatocyte labeling index in female SD rats after 2 weeks although exclusively in the periportal, but not the centrilobular region where rather a decrease in proliferation occurs [34]. Using a two-stage model for hepatocarcinogenesis in female SD rats with diethylnitrosamine as initiator followed by repeated TCDD treatment for 30 weeks, a stimulation of normal hepatocyte proliferation could be observed although the labeling index in altered hepatic foci was not affected [35–38]. Common to all studies is the high variability of the proliferative response among the animals. Furthermore, under these conditions enhancement of proliferation seems to be restricted to later time points (30 and 60 weeks), since a decrease in proliferation was observed after 14 weeks [38]. No enhancement of proliferation could be observed in ovariectomized rats indicating dependence on ovarian hormones, probably E2, at least in SD rats [35], which is in line with the comitogenic effect of TCDD and E2 in primary hepatocytes as described above. However, in other rat strains, TCDD-induced proliferation is independent of ovarian hormones since studies in male Han/Wistar or male Fischer 344 rats show increased hepatic proliferation in response to initiation and subsequent treatment with the AHR ligands TCB or β-naphthoflavone, respectively [39, 40]. Additional support for a proliferative role of the AHR comes from the observation that pretreatment with TCDD enhances the mitogenic effect of the constitutive androstane receptor agonist TCPOBOP in intact mouse liver [41].

32.2.3 Mechanistic Aspects

The signaling pathways involved in AHR-dependent proliferation have not been fully elucidated so far. A large body of evidence coming from *in vivo* and *in vitro* studies suggests that the epidermal growth factor receptor (EGFR) pathway is activated in response to TCDD or other AHR ligands. Phosphorylation of the EGFR has been detected in rat liver after exposure to TCB [39]. Matsumura's group reported decreased EGF binding in response to TCDD in rat and guinea pig liver [42] and other tissues [43] indicating EGFR downregulation that is commonly interpreted as a result of EGFR activation [44]. Further evidence comes from the above-mentioned two-stage carcinogenesis models in rat liver where downregulation of the EGFR was observed [45]. Since TCDD is not an EGFR ligand itself, other factors must account for EGFR activation and several mechanisms have been described. One possible explanation is the observed phosphorylation and activation of the protein kinase c-Src, which is known to phosphorylate the EGFR. Levels of phosphorylated c-Src increase about twofold after TCDD treatment in the plasma membrane of rat, mouse, and guinea pig liver and several other organs [46]. No effect could be detected in a TCDD-resistant mouse strain. Indeed, translocation of c-Src to the plasma membrane and association of c-Src with the EGFR in response to AHR activation have been shown *in vitro* in WB-F344 cells and *in vivo* including mouse liver [47–50]. A mechanistic basis for these effects is provided by the observation of a crosstalk between the AHR/Hsp90/AIP/p23 complex and the c-Src/Hsp90/cdc37 complex [51]. While overall lethality of TCDD was reduced in c-Src knockout mice, hepatomegaly was not affected and to date a role of c-Src in TCDD-mediated proliferation and/or tumor promotion has not been finally proven [52]. Although TCDD leads to a translocation of c-Src to the plasma membrane in confluent WB-F344 cells [47], no increase in ERK phosphorylation could be detected downstream strongly arguing against a functional role of c-Src at least in TCDD-mediated loss of contact inhibition in WB-F344 cells [53].

Some lines of evidence indicate altered expression of the EGFR ligand transforming growth factor-α (TGF-α) in response to TCDD exposure. For instance, mRNA of TGF-α increases in mouse liver after administration of a single dose of TCDD [54, 55] as well as in HepG2 cells [56]. In contrast, Lin et al. failed to detect any accumulation of TGF-α in murine liver [57]. TCDD-mediated upregulation of TGF-α in human keratinocyte cultures occurs due to stabilization of mRNA [58, 59].

Another explanation for TCDD-induced activation of EGFR is given by upregulation of the EGFR ligands amphiregulin and epiregulin that has been detected in the developing ureter *in vivo* as well as in several cell lines including liver cells [31, 60–62]. Whether amphiregulin or epiregulin are increased after TCDD exposure in liver *in vivo* has to be determined. Transcriptional activation of epiregulin is probably mediated by direct binding of the AHR to a xenobiotic-responsive element (XRE, also called dioxin-responsive element (DRE)) in the promoter of the epiregulin gene [61]. Amphiregulin expression is very likely also dependent on the AHR, but downstream cAMP is increased and hence phosphorylation of the transcription factor CREB that then binds to a CRE consensus site in the amphiregulin promoter [62].

A consequence of EGFR activation should be binding of exchange factors and activation of the Ras–Raf–ERK pathway. Indeed, TCDD-induced association of Grb2/SOS and subsequent phosphorylation of Shc have been detected in primary rat hepatocytes [63]. TCDD activates $p21^{ras}$ in (male) rat and murine liver, in adipose tissue from guinea pig, and increases $p21^{ras}$ in membrane fractions in murine lung [64–66]. In gene expression studies in human hepatoma cells, upregulation of K-ras and SOS mRNA could be observed after TCDD exposure [67]. Increased expression of c-Raf was demonstrated especially in neoplastic liver nodules in female (but not male) SD rats initiated with DEN and promoted with PCBs [68]. Although XREs could be identified in the promoter of human and rat *c-Raf* and binding of the AHR to these XREs has been shown *in vitro* [69], others failed to detect upregulation of *c-Raf* expression and—even more importantly—c-Raf activation [70]. Downstream effectors of the Ras–Raf–ERK pathway are transcription factors of the AP-1 family including c-Jun, JunD, and c-Fos. Increased activity of AP-1 after TCDD exposure has been observed in liver and adipose tissue of guinea pigs as well as in endocervical epithelial cells derived from macaques [71–73]. Dere et al. identified increased TCDD-dependent JunD mRNA expression in murine liver [74]. Induction of c-Fos transcription was described in skin lesions from TCDD-exposed patients with chloracne [75]. In line, several *in vitro* studies revealed AHR-dependent expression of c-Jun [12, 76–78] and JunD [19, 79]. However, others failed to detect elevated AP-1 expression in TCDD-treated cells [80].

A link between AHR-dependent increased AP-1 activity and cell cycle progression was established in confluent WB-F344 cells (see above). We could show that TCDD exposure results in an increase in JunD (but not c-Jun) expression that in association with ATF-2 induces transcriptional activation of cyclin A [19]. Interestingly, no increase in the activity of any of the three MAPKs (ERK, p38, JNK) could be observed in WB-F344 cells in response to TCDD [53] (our own unpublished observations). Functional interference with AHR and ARNT revealed that induction of JunD, transcriptional activation of cyclin A, and release from contact inhibition in response to TCDD are absolutely dependent on the AHR, but very likely independent of ARNT. The functional significance of this novel pathway *in vivo* has to be determined. Hence, although some studies addressing selective components of the Ras–Raf–ERK–AP-1 pathway may suggest an involvement of this signaling cascade in response to TCDD exposure, others have failed to do so and it is important to note that stimulation of the entire pathway including ERK activation has not been shown in one single tissue or cell line so far.

32.3 THE AHR AS AN INHIBITOR OF PROLIFERATION

32.3.1 *In Vitro* Studies

In a large number of primary cells and cell lines including liver cells, activation of the AHR actually leads to inhibition of proliferation and cell cycle arrest. TCDD was shown to decrease saturation density in mouse teratoma cells [81] and to inhibit proliferation in primary hepatocytes [82] as well as in the rat hepatoma cell line 5L, but not in its AHR-deficient counterpart BP8 [83]. Restoration of AHR function in BP8 cells fully reconstitutes sensitivity to TCDD, which is reflected by a delay in G1–S progression [84]. These observations strongly suggest an involvement of the AHR in impairment of proliferation. Similar inhibitory effects on cell cycle after TCDD exposure have been described in B lymphocytes [85], murine neural precursor cells [86], human neuronal cells [87], insulin-treated breast cancer cells [88, 89], fetal thymocytes [90, 91], developing thymocytes [92], and cultures of intrathymic progenitor cells [93, 94]. Likewise, overexpression of a constitutively active AHR leads to G1 arrest in Jurkat cell [95]. It has been shown that arrest in cell cycle progression occurs at early stages of thymopoiesis [93] and appears to be an important pathway involved in TCDD-dependent thymic involution. Interestingly, TCDD also inhibits growth of some cancer cell lines such as human gastric [96], endometrial [97], prostate [98, 99], and pancreatic cancer cells [100].

32.3.2 *In Vivo* Studies

In contrast to the stimulatory effect of TCDD on hepatocyte proliferation after 1/3 hepatectomy, TCDD clearly inhibits hepatocyte cell cycle progression after 2/3 hepatectomy in rats and mice [101, 102]. This discrepancy can be explained by the higher basal proliferation obtained by 2/3 hepatectomy compared to 1/3 hepatectomy. Studies on liver regeneration using either AHR-null mice or mice expressing a constitutively active AHR to better understand the role of the AHR in liver regeneration have not been published to date. Inconsistent results have been obtained in the two-stage hepatocarcinogenesis model. While Walker et al. determined increased hepatocyte proliferation in initiated and promoted SD rats after 30 and 60 weeks, a decrease was detected after 14 weeks in the same study [38]. However, no significant change in hepatocyte labeling index was found in similar experimental settings by Buchmann et al. and Stinchcombe et al. [103, 104]. Although mechanistically not understood so far, the rate of basal proliferation, zonal region of hepatocytes, time point of analysis, concentration, and animal strain seem to determine the outcome of TCDD on proliferation in liver.

Further evidence for an inhibitory role of the AHR in proliferation comes from observations in fetal heart and mammary development. When pregnant mice are exposed to TCDD, fetal mice show impairment of cardiac growth and mammary development due to inhibition of cell cycle progression [105–107].

32.3.3 Mechanistic Aspects

Different mechanisms account for TCDD-induced cell cycle arrest. Groundbreaking studies in 5L hepatoma cells have revealed that repression of S-phase specific genes and transcriptional activation of cell cycle inhibitors are involved and that the AHR and ARNT are both required for cell cycle arrest [90, 108, 109].

A key player in TCDD-mediated cell cycle arrest seems to be the retinoblastoma protein (pRB). Two groups have independently shown that the AHR directly interacts with pRB, thereby mediating G1 arrest [110, 111]. Two pRB binding sites within the AHR could be identified, an LXCXE motif in the PAS domain and a second motif in the transactivation domain [110]. The AHR/pRB complex displaces the histone acetyl transferase p300 from E2F-dependent promoters, thereby inhibiting expression of S-phase specific genes such as *dihydrofolate reductase, cyclin E, Cdk2,* and *DNA polymerase α* [108, 109].

In addition, it has been shown convincingly in various cell lines and in murine liver that one consequence of TCDD treatment is induction of the cell cycle inhibitor $p27^{Kip1}$ [86, 87, 90, 109, 112]. Accumulation of $p27^{Kip1}$ leads to increased association with cyclin E/Cdk2, thereby inhibiting phosphorylation of pRB (reviewed in Ref. 113). A potential XRE was identified in the promoter of $p27^{Kip1}$, and reporter gene assays using 3-methylcholanthrene indeed suggest binding of the AHR to this XRE [114]. Probably, pRB also plays a coactivating role in AHR/ARNT-dependent transcriptional activation of $p27^{Kip1}$ [109].

Alternatively, TCDD or expression of a constitutively active AHR may lead to induction of the cell cycle inhibitor p21$^{Waf1/Cip1}$ [95, 100], particularly in cancer cells, as well as downregulation of cyclin D1 or cyclin E [98, 102]. The functional role of the AHR in upregulation of p21$^{Waf1/Cip1}$ has not been clarified so far. The mechanism underlying cyclin D1 or cyclin E decrease is also not understood. Considering that the AHR is part of the Cullin 4B ubiquitin ligase complex (see below and Chapter 10) and Cullin 4B is involved in degradation of cyclin E [115], it is tempting to speculate that activation of the AHR causes increased ubiquitination and proteolysis of cyclin E.

Prolonged treatment of MCF-7 cells results in secretion of TGF-β, a potent antimitogen [116]. Furthermore, crosstalks with other signal transduction pathways and transcription factors, such as NF-κB, have been described (see Chapter 8). The implication of these crosstalks for cell cycle regulation still awaits elucidation. Crosstalk with the estrogen receptor (ER) (see Chapters 9 and 10) and its impact on proliferation in breast cancer cells will be summarized in the next subsection.

32.3.4 AHR-Dependent Inhibition of Proliferation in Breast Cancer Cells

TCDD and a variety of AHR ligands inhibit E2-mediated proliferation in MCF-7 and other ER-proficient breast cancer cells [117–123]. Together with the observation that expression of a constitutively active AHR in MCF-7 cells blocks E2-dependent proliferation [124] and, vice versa, inhibiting AHR function attenuates the growth inhibitory effect of TCDD [125, 126] strongly argues for a pivotal role of the AHR in this response. Inhibition of proliferation is reflected by G1 arrest, which is mediated by inhibition of E2-induced cyclin D1 expression, Cdk4 and Cdk2 activity, and phosphorylation of pRB [127]. Also, early immediate genes, such as early growth response-α, and enzymes required for DNA synthesis, such as thymidylate synthase, are downregulated by TCDD [128]. Although inhibitory XREs have been described, for instance, in the promoter of c-fos that could account for reduced c-fos transcription and thereby attenuation of proliferation [129], the most convincing mechanism proposed is AHR-mediated degradation of ER through activation of proteasomes [130, 131]. A mechanistic explanation is given by the fact that the AHR itself possesses E3 ubiquitin ligase activity and is part of a novel Cullin 4B ubiquitin ligase complex [132] (see Chapter 10). The functional relevance of these observations for development or treatment of breast cancer is not clear to date. Kociba et al. revealed an inhibition of spontaneous mammary (and uterine) tumors in a 2-year carcinogenicity study in rats [133]. TCDD and other congeners also inhibit growth of chemically induced rat mammary tumors [134, 135]. However, in these studies tumor growth has only been observed for 3 weeks. In a MCF-7 xenograft model, it was demonstrated that TCDD transiently attenuates E2-dependent tumor growth, but then the tumor recovered. Likewise, TCDD initially represses expression of ERα, but chronic exposure of TCDD results in partial reexpression of the receptor [136]. Moreover, high levels of constitutively active AHR have been found in mammary tumors, and it has been suggested that the AHR drives tumor progression toward an invasive, metastatic phenotype (reviewed in Refs 137 and 138).

Very recently, it was reported that TCDD and MCDF inhibit proliferation of at least seven ER-negative breast cancer cell lines as well as tumor growth in athymic mice when one of the cell lines (MDA-MB-468) was injected into the mammary fat pad [139]. Downregulation of the AHR diminished the growth inhibitory action of the AHR ligands. Nothing is known about the underlying mechanisms.

32.4 THE AHR AS AN ENDOGENOUS REGULATOR OF CELL CYCLE

The pioneering work of Ma and Whitlock provided direct evidence for a physiological role of the AHR in cell cycle regulation in the absence of any exogenous ligand. AHR-defective mouse hepatoma cells exhibit a prolonged doubling time and a slower progression through G1 [140]. Several studies followed indicating that fibroblasts derived from AHR-null mice progress more slowly through G1 phase [141–143]. Importantly, reexpression of the AHR fully restores proliferation [143]. The key factor in this response appears to be TGF-β, which is a well-known antimitogen (and inducer of apoptosis) in epithelial cells including hepatocytes. Fibroblasts and hepatocytes derived from AHR-null mice show high secretion of TGF-β [141, 143, 144], which is in full accordance with *in vivo* observations in murine liver [144] (a detailed description of AHR/TGF-β interaction is given in Ref. 145). Downregulation of the AHR by siRNA in human hepatoma cells also results in a delayed G1–S progression [126]. A recent work by Barhoover et al. also points to a stimulatory function of the AHR in ER-negative and ER-positive breast cancer cell lines including MCF-7. They further provide a mechanistic link between the endogenous function of the AHR in cell cycle control and its toxicological role after TCDD treatment. They demonstrate that in exponentially growing cells, that is, in the presence of serum, the AHR binds to the cyclin D1/Cdk4 complex, thereby promoting cell cycle progression. Hence, the AHR may act as a scaffolding protein and may facilitate recruitment of pRB. Functional assays including downregulation of AHR expression have not been performed in this study. However, when TCDD is added, cyclin D1/Cdk4 association is disrupted leading to hypophosphorylation of pRB and G1 arrest [146].

Indirect evidence for a promotional role of the AHR in proliferation comes from studies in lung cancer cells. When

A549 cells are exposed to TGF-β, a decrease in proliferation is observed that correlates with downregulation of AHR as well as the proto-oncogenes c-Myc and cyclin A [147]. In contrast to the above-mentioned inhibitory role of the AHR in 5L rat hepatoma cells after TCDD treatment, the same cells show transient AHR activation and CYP1A1 induction when serum-starved cultures are stimulated with serum [148]. Interestingly, pharmacological inhibition of CYP1A1 results in cell cycle arrest and induction of $p27^{Kip1}$. The authors postulate that CYP1A1 negatively regulates an unknown endogenous AHR ligand, thereby preventing prolonged AHR activation that would then lead to cell cycle arrest. The fact that transient activation of the AHR can be detected *in vivo* after partial hepatectomy in the absence of any AHR ligand further supports the idea that the AHR physiologically is able to promote cell cycle progression [102]. Hence, another factor determining the cell's response concerning proliferation versus cell cycle arrest seems to be duration of AHR activation. Whether the AHR regulates cell cycle progression ligand independently or in response to a still-to-be-defined endogenous ligand or is maybe modulated by intracellular second messengers such as cAMP [149] is an open question.

32.5 THE AHR AS A STIMULATOR OF APOPTOSIS

32.5.1 *In Vitro* Studies

Since immunotoxicity including thymic involution and suppression of immune response is an important toxic response in rodents after TCDD exposure, thymocytes and T lymphocytes have been in the focus in research on TCDD-mediated apoptosis. Already more than two decades ago, McConkey et al. provided evidence for TCDD-mediated apoptosis when measuring DNA fragmentation in suspension of immature thymocytes derived from young rats [150]. However, this occurred at a high concentration of TCDD (100 nM) and detection of apoptosis *in vitro* at lower concentrations has evolved to be difficult [151]. Other studies suggest that TCDD provokes apoptosis in stimulated, but not resting T lymphocytes [152–154]. The relative resistance of resting compared to stimulated T lymphocytes is probably due to the lack of downregulation of the FLICE inhibitory protein (c-FLIP) [155]. A further explanation is provided by the fact that the genes coding for AHR and ARNT are silenced in quiescent, but not stimulated T lymphocytes [156]. In line with a possible AHR-dependent induction of apoptosis, overexpression of a constitutively active AHR in Jurkat cells that endogenously lack AHR expression leads not only to cell cycle arrest (see above), but also to induction of apoptosis [95]. In contrast to these observations, Lai et al. did not reveal any apoptosis in fetal thymus organ culture after TCDD treatment [91].

Very recently, it was reported that TCDD induces apoptosis in neural growth factor (NGF)-differentiated pheochromocytoma PC12 cells [157]. However, high concentrations of TCDD (EC_{50} about 200 nM) were required and the fact that only resveratrol, but not α-naphthoflavone partially inhibited apoptosis questions the role of the AHR.

32.5.2 *In Vivo* Studies

Detection of apoptosis of T lymphocytes *in vivo* is generally hampered by the fact that apoptotic cells are rapidly removed by phagocytes [158]. In line, several groups could not reveal any sign of apoptosis in the thymus of TCDD-exposed mice [159, 160]. However, when early time points were studied (8–12 h after TCDD application), Kamath et al. were able to detect TCDD-induced apoptosis by different methodological approaches [151]. TCDD also induces apoptosis in fetal precursor T lymphocytes after oral application in pregnant mice [161]. Moreover, in line with the *in vitro* studies, TCDD-induced apoptosis is predominantly detected in activated T lymphocytes, that is, after antigen challenge [162]. Indirect evidence for TCDD-mediated induction of apoptosis comes from studies in Fas- and FasL-deficient mice. Fas and FasL are well-known mediators of the extrinsic apoptotic pathway [163]. For instance, TCDD-mediated thymic atrophy is reduced in Fas-deficient mice, particularly at low doses of TCDD, and the decrease in activated T lymphocytes is attenuated in Fas- and FasL-deficient animals [153, 154, 160, 162, 164]. These observations suggest that apoptosis is involved in thymic atrophy and deletion of activated T lymphocytes. In contrast, the premature loss of antigen-specific T lymphocytes that is seen in TCDD-treated mice following antigen challenge was not affected by Fas deficiency, and the expression of Fas on T lymphocytes was actually decreased by TCDD in wild-type animals questioning Fas-mediated apoptosis in the TCDD response [165].

Evidence for TCDD-mediated apoptosis in other organs than thymus was recently provided by Yoshizawa et al. who revealed apoptosis in acinar cells in the pancreas in TCDD-exposed female rats [166].

32.5.3 Mechanistic Aspects

As stated above, in some, but not all studies TCDD-dependent thymus atrophy and impairment of T lymphocyte response are blocked in Fas- or FasL-deficient mice. Hence, TCDD possibly triggers the extrinsic, death receptor-mediated apoptotic pathway. In line with these observations, gene expression studies *in vitro* and *in vivo* revealed upregulation of several players involved in the extrinsic pathway, such as Fas, FasL, caspase 8, TRAF5, and others [55, 95, 167, 168], and an increase in soluble FasL has been found in serum of TCDD-treated mice [164]. It is hypothesized that TCDD leads to transcriptional activation of Fas in T lymphocytes whereas FasL expression is induced in thymic stromal cells. While induction of Fas is probably mediated by binding of the AHR to a DRE in the Fas promoter, transcriptional activation of FasL may be due to activation of NF-κB [167, 168].

High concentrations of TCDD may also lead to upregulation of players of the intrinsic mitochondrial pathway, such as Bad, Bax, and Bid [55, 168]. However, crosstalk of the extrinsic and intrinsic pathways is not required for TCDD-induced apoptosis as it is still seen in Bid-deficient mice [168]. This observation might also explain that bcl-2 transgenic mice (bcl-2 is antiapoptotic) are sensitive to TCDD-induced thymic involution [91, 169].

32.6 THE AHR AS AN INHIBITOR OF APOPTOSIS

32.6.1 *In Vitro* Studies

In contrast to the apoptotic function of the AHR as described above, activation of the AHR may also lead to inhibition of apoptosis in several cell lines, such as liver cells, MCF10A breast cancer cells, and some lymphoma cell lines. In rat primary hepatocytes, TCDD attenuates UVC light- and 2-acetylaminofluorene-induced apoptosis, but not when TGF-β, ochratoxin A, and cycloheximide are used as apoptotic stimuli [170–173]. The fact that TCDD-mediated inhibition of apoptosis can be prevented by an AHR antagonist argues for an involvement of the AHR [173]. However, other groups failed to detect TCDD-mediated inhibition of apoptosis in several rat, mouse, and human liver cell lines [70, 174, 175]. TCDD also attenuates apoptosis in the ER-negative breast cancer cell line MCF-10A in response to several apoptotic stimuli and in lymphoma cells after UV light very likely in an AHR-dependent manner [176–180].

32.6.2 *In Vivo* Studies

First evidence that TCDD inhibits apoptosis *in vivo* was provided by Stinchcombe et al. In a two-stage hepatocarcinogenicity model in the rat, they could demonstrate that TCDD strongly inhibits apoptosis in preneoplastic foci whereas proliferation is only marginally increased [104]. Apoptosis in normal hepatocytes was not altered. TCDD also attenuates apoptosis in liver of c-myc transgenic mice [181]. In a study by Pääjärvi et al., it was then demonstrated that TCDD pretreatment leads to inhibition of apoptosis in rat liver when measured 24 h after DEN exposure [182]. Similar results were obtained very recently with B[*a*]P as apoptotic stimulus in murine liver [183].

32.6.3 Mechanistic Aspects

The antiapoptotic effect of TCDD in liver is far from being understood. However, some evidence coming from *in vivo* and *in vitro* studies points to p53 as a potential player in the antiapoptotic function of TCDD. Pääjärvi et al. revealed that pretreatment of rats with TCDD attenuates the p53 liver response to DEN and reduces p53 protein and Ser15-phosphorylated p53. The decrease in phosphorylation was also detected in livers in wild-type mice, but not in AHR-null mice [182]. Immunohistological studies indicated that TCDD modulates Mdm2 protein levels, which was accompanied by increased phosphorylation at Ser166, thereby leading to enhanced p53 degradation. The authors did not investigate upstream signaling events. As described above, TCDD leads to EGFR activation in a number of independent studies. Since activation of growth factor signaling also leads to stimulation of survival pathways, that is, the PI3K/Akt pathway that among others results in phosphorylation of Mdm2 [184], a link between EGFR activation and inhibition of apoptosis might be possible.

Another explanation for TCDD-mediated decrease in p53 protein is given by the work of Reyes-Hernández et al. They identified the ubiquitin-conjugating enzyme Ube2l3 as a novel AHR target gene in murine liver, thereby promoting p53 ubiquitination and subsequent degradation via the proteasome pathway [183]. The authors conclude that TCDD-mediated decrease in p53 accounts for attenuation of apoptosis induced by B[*a*]P. However, TCDD was given before B[*a*]P application. To rule out that B[*a*]P metabolism was altered due to increased CYP1A1 induction—a protective rather than toxifying enzyme [2]—the amount of genotoxic damage should be determined. In HepG2 cells, TCDD also partially prevents etoposide-induced stabilization of p53 [185]. It was further revealed that TCDD induces expression of anterior gradient-2 (AGR2) by an AHR-dependent pathway that in turn silences p53 function. However, all these observations have been obtained at very early time points after the apoptotic stimulus, and nothing is known about desregulation of p53 function in TCDD-mediated liver tumor promotion. Moreover, functional studies demonstrating a causal role of p53 in TCDD-dependent apoptosis are still lacking.

It has further been revealed that TCDD increases protein levels of VDAC2 in 5L (but not in the AHR-deficient BP cell line), which is known to inhibit activation of the proapoptotic regulator Bak and the mitochondrial pathway [186]. However, the functional significance of this observation was not studied, and it is unclear if VDAC2 is upregulated by TCDD *in vivo*.

The antiapoptotic function of TCDD in MCF10A cells is mediated by enhanced production of TGF-α, thereby leading to EGFR activation [177]. Consequently, TCDD-mediated inhibition of apoptosis can be blocked by inhibitors of c-Src and ERK as well as by TGF-α antibodies [179]. Since the effect is blocked by an AHR inhibitor and, vice versa, overexpression of AHR in MCF10A cells leads to apoptotic resistance [30], a functional role of the AHR is very likely.

In several lymphoma cell lines, TCDD leads to increased expression of COX-2, which is AHR dependent. As a consequence, PGE2 is induced, thereby increasing levels of the antiapoptotic proteins Bcl-xl and Mcl-1 [180]. Interestingly, these changes also occur in lymphoma in mice induced by a prolonged TCDD treatment.

32.7 THE AHR AS AN ENDOGENOUS REGULATOR OF APOPTOSIS

Evidence for an endogenous role of the AHR in the regulation of apoptosis came from studies in AHR-null mice. These mice display decreased liver size, hepatic fibrosis, and increased rates of apoptosis [144, 187]. A reasonable explanation is provided by the observation of increased levels of TGF-β1 and TGF-β3 proteins that could be detected *in vivo* and in isolated hepatocytes or embryonal fibroblast derived from AHR-null mice [141, 144, 187]. TGF-β is expressed as a latent immature precursor that has to be cleaved by proteases to become active. One important factor in promoting activation of latent secreted TGF-β is latent TGF-β binding protein 1 (LTBP-1). Accordingly, upregulation of LTBP-1 mRNA and protein was detected in fibrotic regions in the livers of AHR-null mice [188]. Other possible mechanisms leading to activation of TGF-β, that is, altered expression of plasminogen activator inhibitor 2 or retinoids, are discussed by Schwarz et al. [70]. Furthermore, posttranscriptional stabilization of TGF-β1 mRNA has been described [143]. In line with a protective function of the AHR in apoptosis, AHR-defective Hepa 1c1c7 cells are more sensitive in response to apoptotic stimuli, such as UV, hydrogen peroxide, or serum deprivation. The impaired survival is explained by deregulation of the PI3K/Akt pathway and to a lesser extent by decreased EGFR activation [189]. When MEFs derived from AHR-null mice are transfected with a doxycycline-sensitive TET-OFF vector encoding for the AHR, and expression is inhibited by addition of doxycycline, a higher rate of oxidative stress and E2F-dependent apoptosis is observed [190]. In contrast, expression and activation of the AHR by TCDD protects the cells against etoposide-mediated apoptosis. The authors further revealed that the AHR directly binds to E2F1, thereby inhibiting expression of E2F1-regulated genes, such as the proapoptotic Apaf1 and p73. Consistent with an antiapoptotic role of the AHR, the same group revealed very recently that liver tumors are increased in number and size in AHR-null mice compared to wild-type mice after treatment with DEN [7]. These tumors display higher rates of oxidative stress, DNA double-strand breaks, apoptosis, and proliferation. Although mechanistically not understood, the authors propose an endogenous tumor suppressor function of the AHR. A hypothetical explanation is given by its role as an E3 ubiquitin ligase. For instance, the Cullin 4B complex is involved in degradation of Cdt1, which is crucial for regulating replication licensing [191]. Importantly, it is proteolyzed during G1 phase after DNA damage to inhibit replication. Conversely, overexpression of Cdt1 leads to aberrant replication and decreased genomic stability. It is tempting to speculate that Cdt1 degradation is regulated by the AHR and attenuated in AHR-null mice. Moreover, AHR ubiquitin ligase activity is involved in degradation of β-catenin in murine intestine and hence protection against colon carcinogenesis [192]. In AHR-null mice, β-catenin protein levels accumulate in the intestine, and the mice develop tumors in the cecum. Whether the AHR regulates β-catenin levels in liver has not been studied so far.

In contrast to these studies, AHR-dependent promotion of apoptosis has also been described. In Hepa 1c1c7 cells, addition of tumor necrosis factor-α (TNF-α) in combination with cycloheximide results in lysosomal disruption, hence leading to apoptosis that is triggered by the release of cathepsin D and then leading to Bid cleavage and activation of caspases 3/7 [193]. Interestingly, activation of caspase 8 is not involved. An AHR-deficient variant of this cell line is refractory to the treatment, but reexpression of the AHR in this mutant cell line restores the apoptotic signaling. In another study, it was revealed that reintroduction of the AHR in the AHR-negative BP8 hepatoma cell line as well as in primary hepatocytes derived from AHR-null mice enhances apoptosis in response to stimulation with FasL [194]. Treatment with FasL causes increased caspase activity, especially caspase 9, and release of mitochondrial cytochrome c. No alteration of the antiapoptotic Bcl-2 proteins could be detected. Importantly, AHR-null mice are protected against Fas-mediated liver apoptosis as it is partially prevented by downregulation of ARNT by siRNA.

The reasons for these discrepancies are not known so far and again demonstrate the diversity of the AHR response that depends on the cell type studied, treatment, and experimental conditions.

32.8 CONCLUSIONS

Besides its well-known role in xenobiotic metabolism, the AHR regulates cell cycle and apoptosis. Depending on the cell type studied, animal strain, experimental conditions, treatment regimen, time point of analysis, and duration of activation, the AHR may lead to either stimulation or inhibition of cell proliferation and cell death. Moreover, its endogenous role in the absence of any exogenous ligand seems to differ from its toxicological function. The reasons for the heterogenicity of the AHR response are not understood so far.

Hyperproliferation is among other factors involved in the teratogenic effects of TCDD, that is, development of cleft palate and hydronephrosis. Whether increased proliferation is required for TCDD-mediated liver tumor promotion at any stage is reasonable to assume, but still not known. The most convincing experimental observation to date is inhibition of apoptosis in preneoplastic lesions. Studies using a constitutively active AHR as well as the use of different animal strains expressing a high- or low-affinity AHR strongly argue for a pivotal role of the AHR in tumor promotion. However, studies in AHR-null mice, which would provide a final proof, are still lacking.

Although other factors such as skewed differentiation or desregulation of migration account for TCDD-mediated immunotoxicity [195], inhibition of cell proliferation in thymocytes and probably induction of apoptosis in T lymphocytes play crucial roles in AHR-dependent thymic involution and attenuation of T lymphocyte function.

The inhibitory effect of TCDD on proliferation, particularly in ER-positive breast cancer cells as well as in some animal models, has led to the development of selective AHR modulators, which exhibit high affinity for the AHR, but less toxicity than TCDD, as a potential new therapeutic strategy in breast cancer treatment [196]. However, the role of the AHR in breast cancer development is discussed controversially and there is some evidence for an increased risk of breast cancer in women exposed to TCDD [197, 198].

The endogenous role of the AHR is far from being understood. In the absence of any exogenous ligand, the AHR seems to be required for proper cell proliferation, for instance, during liver regeneration. However, studies in knockout mice to further evaluate the role of the AHR in liver regeneration have not been performed to date. Although the AHR may predispose for apoptosis in some cases, most of the studies point to an antiapoptotic endogenous function of the AHR. Moreover, the AHR seems to be required for the maintenance of genomic integrity by a yet unknown molecular mechanism. This might explain why AHR-null animals show an increased liver tumor development after genotoxic insult.

Hence, elucidating the molecular pathways in response to exogenous or endogenous stimulation of the AHR to better understand the pleiotropic and often contradictory effects of the AHR remains a continuing challenge.

ACKNOWLEDGMENTS

The critical reading of the manuscript by Beate Köberle, Peter Münzel, and Nancy Kerkvliet is gratefully acknowledged. My work was supported by the Deutsche Forschungsgemeinschaft and ECNIS (Environmental Cancer Risk, Nutrition and Individual Susceptibility), a network of excellence operating within the European Union 6th Framework Program. I warmly thank all my students for their contributions and, in particular, my postdoctoral fellow Dagmar Faust for her long-lasting excellent work.

REFERENCES

1. Nukaya, M., Moran, S., and Bradfield, C. A. (2009). The role of the dioxin-responsive element cluster between the Cyp1a1 and Cyp1a2 loci in aryl hydrocarbon receptor biology. *Proceedings of the National Academy of Sciences of the United States of America, 106*, 4923–4928.

2. Uno, S., Dalton, T. P., Sinclair, P. R., Gorman, N., Wang, B., Smith, A. G., Miller, M. L., Shertzer, H. G., and Nebert, D. W. (2004). Cyp1a1(−/−) male mice: protection against high-dose TCDD-induced lethality and wasting syndrome, and resistance to intrahepatocyte lipid accumulation and uroporphyria. *Toxicology and Applied Pharmacology, 196*, 410–421.

3. Tuomisto, J. (2005). Does mechanistic understanding help in risk assessment: the example of dioxin. *Toxicology and Applied Pharmacology, 207*, S2–S10.

4. Barouki, R., Coumoul, X., and Fernandez-Salguero, P. M. (2007). The aryl hydrocarbon receptor, more than a xenobiotic-interacting protein. *FEBS Letters, 581*, 3608–3615.

5. Andersson, P., McGuire, J., Rubio, C., Gradin, K., Whitelaw, M. L., Pettersson, S., Hanberg, A., and Poellinger, L. (2002). A constitutively active dioxin/aryl hydrocarbon receptor induces stomach tumors. *Proceedings of the National Academy of Sciences of the United States of America, 99*, 9990–9995.

6. Moennikes, O., Loeppen, S., Buchmann, A., Andersson, P., Ittrich, C., Poellinger, L., and Schwarz, M. (2004). A constitutively active dioxin/aryl hydrocarbon receptor promotes hepatocarcinogenesis in mice. *Cancer Research, 64*, 4707–4710.

7. Fan, Y., Boivin, G. P., Knudsen, E. S., Nebert, D. W., Xia, Y., and Puga, A. (2010). The aryl hydrocarbon receptor functions as a tumor suppressor of liver carcinogenesis. *Cancer Research, 70*, 212–220.

8. Pitot, H. C., Goldsworthy, T., Campbell, H. A., and Poland, A. (1980). Quantitative evaluation of the promotion by 2,3,7,8-tetrachlorodibenzo-p-dioxin of hepatocarcinogenesis from diethylnitrosamine. *Cancer Research, 40*, 3616–36200.

9. Wölfle, D., Münzel, P., Fischer, G., and Bock, K.-W. (1988). Altered growth control of rat hepatocytes after treatment with 3,4,3′,4′-tetrachlorobiphenyl in vivo and in vitro. *Carcinogenesis, 9*, 919–924.

10. Wölfle, D., Becker, E., and Schmutte, C. (1993). Growth stimulation of primary rat hepatocytes by 2,3,7,8-tetrachlorodibenzo-p-dioxin. *Cell Biology and Toxicology, 9*, 15–31.

11. Schrenk, D., Karger, A., Lipp, H.-P., and Bock, K.-W. (1992). 2,3,7,8-Tetrachlorodibenzo-p-dioxin and ethinylestradiol as co-mitogens in cultured rat hepatocytes. *Carcinogenesis, 13*, 453–456.

12. Münzel, P., Bock-Hennig, B., Schieback, S., Gschaidmeier, H., Beck-Gschaidmeier, S., and Bock, K.-W. (1996). Growth modulation of hepatocytes and rat liver epithelial cells (WB-F344) by 2,3,7,8-tetrachlorodibenzo-p-dioxin (TCDD). *Carcinogenesis, 17*, 197–202.

13. Schrenk, D., Schäfer, S., and Bock, K.-W. (1994). 2,3,7,8-Tetrachlorodibenzo-p-dioxin as growth modulator in mouse hepatocytes with high and low affinity AhR. *Carcinogenesis, 15*, 27–31.

14. Dietrich, C., Faust, D., Budt, S., Moskwa, M., Kunz, A., Bock, K. W., and Oesch, F. (2002). 2,3,7,8-Tetrachlorodibenzo-*p*-dioxin-dependent release from contact inhibition in WB-F344 cells: involvement of cyclin A. *Toxicology and Applied Pharmacology*, 183, 117–126.

15. Chramostová, K., Vondráček, J., Sindlerová, L., Vojtěšek, B., Kozubík, A., and Machala, M. (2004). Polycyclic aromatic hydrocarbons modulate cell proliferation in rat hepatic epithelial stem-like WB-F344 cells. *Toxicology and Applied Pharmacology*, 196, 136–148.

16. Vondráček, J., Chramostová, K., Plíšková, M., Bláha, L., Brack, W., Kozubík, A., and Machala, M. (2004). Induction of aryl hydrocarbon receptor-mediated and estrogen receptor-mediated activities, and modulation of proliferation by dinaphthofurans. *Environmental Toxicology and Chemistry*, 23, 2214–2220.

17. Vondráček, J., Machala, M., Bryja, V., Chramostová, K., Krcmář, P., Dietrich, C., Hampl, A., and Kozubík, A. (2005). Aryl hydrocarbon receptor-activating polychlorinated biphenyls and their hydroxylated metabolites induce cell proliferation in contact-inhibited rat liver epithelial cells. *Toxicological Sciences*, 83, 53–63.

18. Zatloukalová, J., Sviháľková-Sindlerová, L., Kozubík, A., Krcmář, P., Machala, M., and Vondráček, J. (2007). β-Naphthoflavone and 3′-methoxy-4′-nitroflavone exert ambiguous effects on Ah receptor-dependent cell proliferation and gene expression in rat liver 'stem-like' cells. *Biochemical Pharmacology*, 73, 1622–1634.

19. Weiss, C., Faust, D., Schreck, I., Ruff, A., Farwerck, T., Melenberg, A., Schneider, S., Oesch-Bartlomowicz, B., Zatloukalová, J., Vondráček, J., Oesch, F., and Dietrich, C. (2009). TCDD deregulates contact inhibition in rat liver oval cells via Ah receptor, JunD and cyclin A. *Oncogene*, 27, 2198–2207.

20. Dietrich, C., Faust, D., Moskwa, M., Kunz, A., Bock, K.-W., and Oesch, F. (2003). TCDD-dependent downregulation of γ-catenin in rat liver epithelial cells (WB-F344.). *International Journal of Cancer*, 103, 435–439.

21. Umannová, L., Zatloukalová, J., Machala, M., Krcmář, P., Májková, Z., Hennig, B., Kozubík, A., and Vondráček, J. (2007). Tumor necrosis factor-alpha modulates effects of aryl hydrocarbon receptor ligands on cell proliferation and expression of cytochrome P450 enzymes in rat liver "stem-like" cells. *Toxicological Sciences*, 99, 79–89.

22. Milstone, L. and Lavigne, J. (1984). 2,3,7,8-Tetrachlorodibenzo-*p*-dioxin induces hyperplasia in confluent cultures of human keratinocytes. *Journal of Investigative Dermatology*, 82, 532–534.

23. Hébert, C. D., Cao, Q. L., and Birnbaum, L. S. (1990). Inhibition of high-density growth arrest in human squamous carcinoma cells by 2,3,7,8-tetrachlorodibenzo-*p*-dioxin (TCDD). *Carcinogenesis*, 11, 1335–1342.

24. Fiorito, F., Pagnini, U., De Martino, L., Montagnaro, S., Ciarcia, R., Florio, S., Pacilio, M., Fucito, A., Rossi, A., Iovane, G., and Giordano, A. (2008). 2,3,7,8-Tetrachlorodibenzo-*p*-dioxin increases bovine herpesvirus type-1 (BHV-1) replication in Madin–Darby bovine kidney (MDBK) cells in vitro. *Journal of Cellular Biochemistry*, 103, 221–233.

25. Knutson, J. C. and Poland, A. (1980). 2,3,7,8-Tetrachlorodibenzo-*p*-dioxin: failure to demonstrate toxicity in twenty-three cultured cell types. *Toxicology and Applied Pharmacology*, 54, 377–383.

26. Abbott, B. D. and Birnbaum, L. S. (1990). Effects of TCDD on embryonic ureteric epithelial EGF receptor expression and cell proliferation. *Teratology*, 41, 71–84.

27. Bryant, P. L., Reid, L. M., Schmid, J. E., Buckalew, A. R., and Abbott, B. D. (2001). Effects of 2,3,7,8-tetrachlorodibenzo-*p*-dioxin (TCDD) on fetal mouse urinary tract epithelium in vitro. *Toxicology*, 162, 23–34.

28. Abbott, B. D. and Birnbaum, L. S. (1989). TCDD alters medial epithelial cell differentiation during palatogenesis. *Toxicology and Applied Pharmacology*, 99, 276–286.

29. Shimba, S., Komiyama, K., Moro, I., and Tezuka, M. (2002). Overexpression of the aryl hydrocarbon receptor (AhR) accelerates the cell proliferation of A549 cells. *Journal of Biochemistry*, 132, 795–802.

30. Wong, P. S., Li, W., Vogel, C. F., and Matsumura, F. (2009). Characterization of MCF mammary epithelial cells overexpressing the aryl hydrocarbon receptor. *BMC Cancer*, 9, 234.

31. Ahn, N.-S., Hu, H., Park, J.-S., Park, J.-S., Kim, J.-S., An, S., Kong, G., Aruoma, O. I., Lee, J.-S., and Kang, K.-S. (2005). Molecular mechanisms of the 2,3,7,8-tetrachlorodibenzo-*p*-dioxin-induced inverted U-shaped dose responsiveness in anchorage independent growth and proliferation of human breast epithelial cells with stem cell characteristics. *Mutation Research*, 579, 189–199.

32. Christian, B. J. and Peterson, R. E. (1983). Effects of 2,3,7,8-tetrachlorodibenzo-*p*-dioxin on [^{3}H]thymidine incorporation into rat liver deoxyribonucleic acid. *Toxicology*, 8, 133–146.

33. Büsser, M.-T. and Lutz, W. K. (1987). Stimulation of DNA synthesis in rat and mouse liver by various tumor promoters. *Carcinogenesis*, 8, 1433–1437.

34. Fox, T. R., Best, L. L., Goldsworthy, S. M., Mills, J. J., and Goldsworthy, T. L. (1993). Gene expression and cell proliferation in rat liver after 2,3,7,8-tetrachlorodibenzo-*p*-dioxin exposure. *Cancer Research*, 53(10 Suppl.) 2265–2271.

35. Lucier, G. W., Tritscher, A., Goldsworthy, T., Foley, J., Clark, G., and Goldstein, J. (1991). Ovarian hormones enhance 2,3,7,8-tetrachlorodibenzo-*p*-dioxin-mediated increases in cell proliferation and preneoplastic foci in a two-stage model for rat hepatocarcinogenesis. *Cancer Research*, 51, 1391–1397.

36. Maronpot, R. R., Foley, J. F., Takahashi, K., Goldsworthy, T., Clark, G., Tritscher, A., Portier, C., and Lucier, G. (1993). Dose response for TCDD promotion of hepatocarcinogenesis in rats initiated with DEN: histologic, biochemical, and cell proliferation endpoints. *Environmental Health Perspectives*, 101, 634–642.

37. Tritscher, A., Clark, G., Sewall, C., Sills, R., Maronpot, R., and Lucier, G. (1995). Persistence of TCDD-induced hepatic cell proliferation and growth of enzyme altered foci after chronic exposure followed by cessation of treatment in DEN initiated female rats. *Carcinogenesis*, 16, 2807–2811.

38. Walker, N., Miller, B., Kohn, M., Lucier, G., and Tritscher, A. (1998). Differences in kinetics of induction and reversibility of TCDD-induced changes in cell proliferation and Cyp1A1 expression in female Sprague-Dawley rat liver. *Carcinogenesis*, 19, 1427–1435.

39. Wölfle, D., Münzel, P., Fischer, G., and Bock, K.-W. (1988). Altered growth of rat hepatocytes after treatment with 3,4,3′,4′-tetrachlorobiphenyl in vivo and in vitro. *Carcinogenesis*, 9, 919–924.

40. Dewa, Y., Nishimura, J., Muguruma, M., Jin, M., Saegusa, Y., Okamura, T., Tasaki, M., Umemura, T., and Mitsumori, K. (2008). β-Naphthoflavone enhances oxidative stress responses and the induction of preneoplastic lesions in a diethylnitrosamine-initiated hepatocarcinogenesis model in partially hepatectomized rats. *Toxicology*, 244, 179–189.

41. Mitchell, K. A., Wilson, S. R., and Elferink, C. J. (2010). The activated aryl hydrocarbon receptor synergizes mitogen-induced murine liver hyperplasia. *Toxicology*, 276, 103–109.

42. Madhukar, B. V., Brewster, D. W., and Matsumura, F. (1984), Effects of in vivo-administered 2,3,7,8-tetrachlorodibenzo-*p*-dioxin on receptor binding of epidermal growth factor in the hepatic plasma membrane of rat, guinea pig, mouse and hamster. *Proceedings of the National Academy of Sciences of the United States of America*, 81, 7407–7411.

43. Enan, E., El-Sabeawy, F., Scott, M., Overstreet, J., and Lasley, B. (1998). Alterations in the growth factor signal transduction pathways and modulators of the cell cycle in endocervical cells from macaques exposed to TCDD. *Toxicology and Applied Pharmacology*, 151, 283–293.

44. Sorkin, A. (2001). Internalization of the epidermal growth factor receptor: role in signalling. *Biochemical Society Transactions*, 29(Pt 4), 480–484.

45. Sewall, C. H., Lucier, G., Tritscher, A., and Clark, G. (1993). TCDD-mediated changes in hepatic epidermal growth factor receptor may be a critical event in the hepatocarcinogenic action of TCDD. *Carcinogenesis*, 14, 1885–1893.

46. Bombick, D. W. and Matsumura, F. (1987). 2,3,7,8-Tetrachlorodibenzo-*p*-dioxin causes elevation of the levels of the protein tyrosine kinase pp60^{c-src}. *Journal of Biochemical Toxicology*, 2, 141–154.

47. Köhle, C., Gschaidmeier, H., Lauth, D., Topell, S., Zitzer, H., and Bock, K.-W. (1999). 2,3,7,8-Tetrachlorodibenzo-*p*-dioxin (TCDD)-mediated membrane translocation of c-Src protein kinase in liver WB-F344 cells. *Archives of Toxicology*, 73, 152–158.

48. Enan, E. and Matsumura, F. (1996). Identification of c-Src as the integral component of the cytosolic Ah receptor complex, transducing the signal of 2,3,7,8-tetrachlorodibenzo-*p*-dioxin (TCDD) through the protein phosphorylation pathway. *Biochemical Pharmacology*, 52, 1599–1612.

49. Blankenship, A. and Matsumura, F. (1997). 2,3,7,8-Tetrachlorodibenzo-*p*-dioxin-induced activation of a protein tyrosine kinase, pp60src, in murine hepatic cytosol using a cell-free system. *Molecular Pharmacology*, 52, 667–675.

50. Randi, A. S., Sanchez, M. S., Alvarez, L., Cardozo, J., Pontillo, C., and Kleiman de Pisarev, D. L. (2008). Hexachlorobenzene triggers AhR translocation to the nucleus, c-Src activation and EGFR transactivation in rat liver. *Toxicology Letters*, 177, 116–122.

51. Park, S., Dong, B., and Matsumura, F. (2007). Rapid activation of c-Src kinase by dioxin is mediated by the Cdc37–HSP90 complex as part of Ah receptor signaling in MCF10A cells. *Biochemistry*, 46, 899–908.

52. Matsumura, F., Enan, E., Dunlap, D. Y., Pinkerton, K. E., and Peake, J. (1997). Altered in vivo toxicity of 2,3,7,8-tetrachlorodibenzo-*p*-dioxin (TCDD) in c-SRC deficient mice. *Biochemical Pharmacology*, 53, 1397–1404.

53. Hölper, P., Faust, D., Oesch, F., and Dietrich, C. (2005). Evaluation of the role of c-Src and ERK in TCDD-dependent release from contact inhibition in WB-F344 cells. *Archives of Toxicology*, 79, 201–207.

54. Vogel, C. F., Zhao, Y., Wong, P., Young, N. F., and Matsumura, F. (2003). The use of c-src knockout mice for the identification of the main toxic signaling pathway of TCDD to induce wasting syndrome. *Journal of Biochemical and Molecular Toxicology*, 17, 305–315.

55. Zeytun, A., McKallip, R. J., Fisher, M., Camacho, I., Nagarkatti, M., and Nagarkatti, P. S. (2002). Analysis of 2,3,7,8-tetrachlorodibenzo-*p*-dioxin-induced gene expression profile in vivo using pathway-specific cDNA arrays. *Toxicology*, 178, 260.

56. Frueh, F. W., Hayashibara, K. C., Brown, P. O., and Whitlock, J. P., Jr., (2001). Use of cDNA microarrays to analyze dioxin-induced changes in human liver gene expression. *Toxicology Letters*, 122, 189–203.

57. Lin, F. H., Clark, G., Birnbaum, L. S., Lucier, G. W., and Goldstein, J. A. (1991). Influence of the Ah locus on the effects of 2,3,7,8-tetrachlorodibenzo-*p*-dioxin on the hepatic epidermal growth factor receptor. *Molecular Pharmacology*, 39, 307–313.

58. Choi, E. J., Toscano, D. G., Ryan, J. A., Riedel, N., and Toscano, W. A., Jr., (1991). Dioxin induces transforming growth factor-α in human keratinocytes. *Journal of Biological Chemistry*, 266, 9591–9597.

59. Gaido, K. W., Maness, S. C., Leonard, L. S., and Greenlee, W. F. (1992). 2,3,7,8-Tetrachlorodibenzo-*p*-dioxin-dependent regulation of transforming growth factors-alpha and -beta 2 expression in a human keratinocyte cell line involves both transcriptional and post-transcriptional control. *Journal of Biological Chemistry*, 267, 24591–24595.

60. Choi, S. S., Miller, M. A., and Harper, P. A. (2006). In utero exposure to 2,3,7, 8-tetrachlorodibenzo-*p*-dioxin induces amphiregulin gene expression in the developing mouse ureter. *Toxicological Sciences*, 94, 163–174.

61. Patel, R. D., Kim, D. J., Peters, J. M., and Perdew, G. H. (2006). The aryl hydrocarbon receptor directly regulates expression of the potent mitogen epiregulin. *Toxicological Sciences*, 89, 75–82.

62. Du, B., Altorki, N. K., Kopelovich, L., Subbaramaiah, K., and Dannenberg, A. J. (2005). Tobacco smoke stimulates the transcription of amphiregulin in human oral epithelial cells: evidence of a cyclic AMP-responsive element binding

protein-dependent mechanism. *Cancer Research*, 65, 5982–5988.
63. Park, R., Kim, D. H., Kim, M. S., So, H. S., Chung, H. T., Kwon, K. B., Ryu, D. G., and Kim, B. R. (1998). Association of Shc, Cbl, Grb2, and Sos following treatment with 2,3,7,8-tetrachlorodibenzo-*p*-dioxin in primary rat hepatocytes. *Biochemical and Biophysical Research Communication*, 253, 577–581.
64. Tullis, K., Olsen, H., Bombick, D. W., Matsumura, F., and Jankun, J. (1992). TCDD causes stimulation of c-ras expression in the hepatic plasma membranes *in vivo* and *in vitro*. *Journal of Biochemical Toxicology*, 7, 107–116.
65. Ramakrishna, G. and Anderson, L. M. (1998). Levels and membrane localization of the c-K-ras p21 protein in lungs of mice of different genetic strains and effects of 2,3,7,8-tetrachlorodibenzo-*p*-dioxin (TCDD) and Aroclor 1254. *Carcinogenesis*, 19, 463–470.
66. Enan, E. and Matsumura, F. (1994). Significance of TCDD-induced changes in protein phosphorylation in the adipocyte of male guinea pigs. *Journal of Biochemical Toxicology*, 9, 159–170.
67. Puga, A., Maier, A., and Medvedovic, M. (2000). The transcriptional signature of dioxin in human hepatoma HepG2 cells. *Biochemical Pharmacology*, 60, 1129–1142.
68. Jenke, H. S., Deml, E., and Oesterle, D. (1994). C-raf expression in early rat liver tumorigenesis after promotion with polychlorinated biphenyls or phenobarbital. *Xenobiotica*, 24, 569–580.
69. Borlak, J. and Jenke, H. S. (2008). Cross-talk between aryl hydrocarbon receptor and mitogen-activated protein kinase signaling pathway in liver cancer through c-raf transcriptional regulation. *Molecular Cancer Research*, 6, 1326–1336.
70. Schwarz, M., Buchmann, A., Stinchcombe, Kalkuhl, A., and Bock, K.-W. (2000). Ah receptor ligands and tumor promotion: survival of neoplastic cells. *Toxicology Letters*, 112–113 869–877.
71. Ashida, H., Nagy, S., and Matsumura, F. (2000). 2,3,7,8-Tetrachlorodibenzo-*p*-dioxin (TCDD)-induced changes in activities of nuclear protein kinases and phosphatases affecting DNA binding activity of c-Myc and AP-1 in the livers of guinea pigs. *Biochemical Pharmacology*, 59, 741–751.
72. Enan, E. and Matsumura, F. (1995). Regulation by 2,3,7,8-tetrachlorodibenzo-*p*-dioxin (TCDD) of the DNA binding activity of transcriptional factors via nuclear protein phosphorylation in guinea pig adipose tissue. *Biochemical Pharmacology*, 50, 1199–1206.
73. Enan, E., El-Sabeawy, F., Scott, M., Overstreet, J., and Lasley, B. (1998). Alterations in the growth factor signal transduction pathways and modulators of the cell cycle in endocervical cells from macaques exposed to TCDD. *Toxicology and Applied Pharmacology*, 151, 283–293.
74. Dere, E., Boverhof, D. R., Burgoon, L. D., and Zacharewski, T. R. (2006). In vivo–in vitro toxicogenomic comparison of TCDD-elicited gene expression in Hepa1c1c7 mouse hepatoma cells and C57BL/6 hepatic tissue. *BMC Genomics*, 7, 80.
75. Tang, N. J., Liu, J., Coenraads, P. J., Dong, L., Zhao, L. J., Ma, S. W., Chen, X., Zhang, C. M., Ma, X. M., Wei, W. G., Zhang, P., and Bai, Z. P. (2008). Expression of AhR, CYP1A1, GSTA1, c-fos and TGF-alpha in skin lesions from dioxin-exposed humans with chloracne. *Toxicology Letters*, 177, 182–187.
76. Puga, A., Nebert, D. W., and Carrier, F. (1992). Dioxin induces expression of c-fos and c-jun proto-oncogenes and a large increase in transcription factor AP-1. *DNA and Cell Biology*, 11, 269–281.
77. Puga, A., Barnes, S. J., Chang, C., Zhu, H., Nephew, K. P., Khan, S. A., and Shertzer, H. G. (2000). Activation of transcription factors activator protein-1 and nuclear factor-kappaB by 2,3,7,8-tetrachlorodibenzo-*p*-dioxin. *Biochemical Pharmacology*, 59, 997–1005.
78. Weiss, C., Faust, D., Dürk, H., Kolluri, S. K., Pelzer, A., Schneider, S., Dietrich, C., Oesch, F., and Göttlicher, M. (2005). TCDD induces c-jun expression via a novel Ah (dioxin) receptor-mediated p38-MAPK-dependent pathway. *Oncogene*, 24, 4975–4983.
79. Andrysik, Z., Vondrácek, J., Machala, M, Krcmár, P., Svihálková-Sindlerová, L., Kranz, A., Weiss, C., Faust, D., Kozubík, A., and Dietrich, C. (2007). The aryl hydrocarbon receptor-dependent deregulation of cell cycle control induced by polycyclic aromatic hydrocarbons in rat liver epithelial cells. *Mutation Research*, 615, 87–97.
80. Gohl, G., Lehmköster, T., Münzel, P., Schrenk, D. Viebahmn, R., and Bock, K.-W. (1996). TCDD-inducible plasminogen activator inhibitor type 2 (PAI-2) in human hepatocytes, HepG2 and monocytic U937 cells. *Carcinogenesis*, 17, 443–449.
81. Gierthy, J. F. and Crane, D. (1984). Reversible inhibition of *in vitro* epithelial cell proliferation by 2,3,7,8-tetrachlorodibenzo-*p*-dioxin. *Toxicology and Applied Pharmacology*, 74, 91–98.
82. Hushka, D. R. and Greenlee, W. F. (1995). 2,3,7,8-Tetrachlorodibenzo-*p*-dioxin inhibits DNA synthesis in rat primary hepatocytes. *Mutation Research*, 333, 89–99.
83. Göttlicher, M., Cikryt, P., and Wiebel, F. J. (1990). Inhibition of growth by 2,3,7,8-tetrachlorodibenzo-*p*-dioxin in 5L rat hepatoma cells is associated with the presence of Ah receptor. *Carcinogenesis*, 11, 2205–2210.
84. Weiss, C., Kolluri, S. K., Kiefer, F., and Göttlicher, M. (1996). Complementation of Ah receptor deficiency in hepatoma cells: negative feedback regulation and cell cycle control by the Ah receptor. *Experimental Cell Research*, 226, 154–163.
85. Morris, D. L., Karras, J. G., and Holsapple, M. P. (1993). Direct effects of 2,3,7,8-tetrachlorodibenzo-*p*-dioxin (TCDD) on responses to lipopolysaccharide (LPS) by isolated murine B-cells. *Immunopharmacology*, 26, 105–112.
86. Latchney, S. E., Lioy, D. T., Henry, E. C., Gasiewicz, T. S., Strathmann, F. G., Mayer-Proschel, M., and Opanashuk, L. A. (2011). Neural precursor cell proliferation is disrupted through activation of the aryl hydrocarbon receptor by 2,3,7,8-tetrachlorodibenzo-*p*-dioxin. *Stem Cells and Development*, 20, 313–326.

87. Jin, D. Q., Jung, J. W., Lee, Y. S., and Kim, J. A. (2004). 2,3,7,8-Tetrachlorodibenzo-*p*-dioxin inhibits cell proliferation through aryl hydrocarbon receptor-mediated G1-arrest in SK-N-SH human neuronal cells. *Neuroscience Letters, 363*, 69–72.

88. Liu, H., Biegel, L., Narasimhan, T. R., Rowlands, C., and Safe, S. (1992). Inhibition of insulin-like growth factor-I responses in MCF-7 cells by 2,3,7,8-tetrachlorodibenzo-*p*-dioxin and related compounds. *Molecular and Cellular Endocrinology, 87*, 19–28.

89. Liu, H. and Safe, S. (1996). Effects of 2,3,7,8-tetrachlorodibenzo-*p*-dioxin (TCDD) on insulin-induced responses in MCF-7 human breast cancer cells. *Toxicology and Applied Pharmacology, 138*, 242–250.

90. Kolluri, S. K., Weiss, C., Koff, A., and Göttlicher, M. (1999). p27^{Kip1} induction and inhibition of proliferation by the intracellular Ah receptor in developing thymus and hepatoma cells. *Genes & Development, 13*, 1742–1753.

91. Lai, Z. W., Fiore, N. C., Hahn, P. J., Gasiewicz, T. A., and Silverstone, A. E. (2000). Differential effects of diethylstilbestrol and 2,3,7,8-tetrachlorodibenzo-*p*-dioxin on thymocyte differentiation, proliferation, and apoptosis in bcl-2 transgenic mouse fetal thymus organ culture. *Toxicology and Applied Pharmacology, 168*, 15–24.

92. McMillan, B. J., McMillan, S. N., Glover, E., and Bradfield C. A. (2007). 2,3,7,8-Tetrachlorodibenzo-*p*-dioxin induces premature activation of the KLF2 regulon during thymocyte development. *Journal of Biological Chemistry, 282*, 12590–12597.

93. Laiosa, M. D., Wyman, A., Murante, F. G., Fiore, N. C., Staples, J. E., Gasiewicz, T. A., and Silverstone, A. E. (2003). Cell proliferation arrest within intrathymic lymphocyte progenitor cells causes thymic atrophy mediated by the aryl hydrocarbon receptor. *Journal of Immunology, 171*, 4582–4591.

94. Laiosa, M. D., Mills, J. H., Lai, Z. W., Singh, K. P., Middleton, F. A., Gasiewicz, T. A., and Silverstone, K. E., (2010). Identification of stage-specific gene modulation during early thymocyte development by whole-genome profiling analysis after aryl hydrocarbon receptor activation. *Molecular Pharmacology, 77*, 773–783.

95. Ito, T., Tsukumo, S., Suzuki, N., Motohashi, H., Yamamoto, M., Fujii-Kuriyama, Y., Mimura, J., Lin, T. M., Peterson, R. E., Tohyama, C., and Nohara, K. (2004). A constitutively active arylhydrocarbon receptor induces growth inhibition of Jurkat T cells through changes in the expression of genes related to apoptosis and cell cycle arrest. *Journal of Biological Chemistry, 279*, 25204–25210.

96. Peng, T. L., Chen, J., Mao, W., Liu, X., Tao, Y., Chen, L. Z., and Chen, M. H. (2009). Potential therapeutic significance of increased expression of aryl hydrocarbon receptor in human gastric cancer. *World Journal of Gastroenterology, 15*, 1719–1729.

97. Castro-Rivera, E., Wormke, M., and Safe, S. (1999). Estrogen and aryl hydrocarbon responsiveness of ECC-1 endometrial cancer cells. *Molecular and Cellular Endocrinology, 150*, 11–21.

98. Barnes-Ellerbe, S., Knudsen, K. E., and Puga, A. (2004). 2,3,7,8-Tetrachlorodibenzo-*p*-dioxin blocks androgen-dependent cell proliferation of LNCaP cells through modulation of pRB phosphorylation. *Molecular Pharmacology, 66*, 502–511.

99. Morrow, D., Quin, C., Smith, R., 3rd, and Safe, S. (2004). Aryl hydrocarbon receptor-mediated inhibition of LNCaP prostate cancer cell growth and hormone-induced transactivation. *Journal of Steroid Biochemistry and Molecular Biology, 88*, 27–36.

100. Koliopanos, A., Kleeff, J., Xiao, Y., Safe, S., Zimmermann, A., Büchler, M. W., and Friess, H. (2002). Increased arylhydrocarbon receptor expression offers a potential therapeutic target for pancreatic cancer. *Oncogene, 21*, 6059–6070.

101. Bauman, J. W., Goldsworthy, T. L., Dunn, C. S., and Fox, T. R. (1995). Inhibitory effects of 2,3,7,8-tetrachlorodibenzo-*p*-dioxin on rat hepatocyte proliferation induced by 2/3 partial hepatectomy. *Cell Proliferation, 28*, 437–451.

102. Mitchell, K. A., Lockhart, C. A., Huang, G., and Elferink, C. J. (2006). Sustained aryl hydrocarbon receptor activity attenuates liver regeneration. *Molecular Pharmacology, 70*, 163–170.

103. Buchmann, A., Stinchcombe, S., Körner, W., Hagenmaier, H., and Bock, K. W. (1994). Effects of 2,3,7,8-tetrachloro- and 1,2,3,4,6,7,8-heptachlorodibenzo-*p*-dioxin on the proliferation of preneoplastic liver cells in the rat. *Carcinogenesis, 15*, 1143–1150.

104. Stinchcombe, S., Buchmann, A., Bock, K. W., and Schwarz, M. (1995). Inhibition of apoptosis during 2,3,7,8-tetrachlorodibenzo-*p*-dioxin-mediated tumour promotion in rat liver. *Carcinogenesis, 16*, 1271–1275.

105. Thackaberry, E. A., Jiang, Z., Johnson, C. D., Ramos, K. S., and Walker, M. K. (2005). Toxicogenomic profile of 2,3,7,8-tetrachlorodibenzo-*p*-dioxin in the murine fetal heart: modulation of cell cycle and extracellular matrix genes. *Toxicological Sciences, 88*, 231–241.

106. Thackaberry, E. A., Nunez, B. A., Ivnitski-Steele, I. D., Friggins, M., and Walker, M. K. (2005). Effect of 2,3,7,8-tetrachlorodibenzo-*p*-dioxin on murine heart development: alteration in fetal and postnatal cardiac growth, and postnatal cardiac chronotropy. *Toxicological Sciences, 88*, 242–249.

107. Lew, B. J., Collins, L. L., O'Reilly, M. A., and Lawrence, B. P. (2009). Activation of the aryl hydrocarbon receptor during different critical windows in pregnancy alters mammary epithelial cell proliferation and differentiation. *Toxicological Sciences, 111*, 151–162.

108. Huang, G. and Elferink, C. E. (2005). Multiple mechanisms are involved in Ah receptor-mediated cell cycle arrest. *Molecular Pharmacology, 67*, 88–96.

109. Marlowe, J., Knudsen, E. S., Schwemberger, S., and Puga A. (2004). The aryl hydrocarbon receptor displaces p300 from E2F-dependent promoters and represses S phase-specific gene expression. *Journal of Biological Chemistry, 279*, 29013–29022.

110. Ge, N.-L. and Elferink, C. E. (1998). A direct interaction between the aryl hydrocarbon receptor and the retinoblastoma protein. *Journal of Biological Chemistry, 273*, 22708–22713.

111. Puga, A., Barnes, S. J., Dalton, T. P., Chang, C., Knudsen, E. S., and Maier, M. A. (2000). Aromatic hydrocarbon receptor interaction with the retinoblastoma protein potentiates repression of E2F-dependent transcription and cell cycle arrest. *Journal of Biological Chemistry, 275*, 2943–2950.

112. Rininger, J. A., Stoffregen, D. A., and Babish, J. G. (1997). Murine hepatic p53, RB, and CDK inhibitory protein expression following acute 2,3,7,8-tetrachlorodibenzo-*p*-dioxin (TCDD) exposure. *Chemosphere, 34*, 1557–1568.

113. Cobrinik, D. (2005). Pocket proteins and cell cycle control. *Oncogene, 24*, 2796–2809.

114. Pang, P. H., Lin, Y. H., Lee, Y. H., Hou, H. H., Hsu, S. P., and Juan, S. H. (2008). Molecular mechanisms of p21 and p27 induction by 3-methylcholanthrene, an aryl-hydrocarbon receptor agonist, involved in antiproliferation of human umbilical vascular endothelial cells. *Journal of Cellular Physiology, 215*, 161–171.

115. Higa, L. A., Yang, X., Zheng, J., Banks, D., Wu, M., Ghosh, P., Sun, H., and Zhang, H. (2006). Involvement of CUL4 ubiquitin E3 ligases in regulating CDK inhibitors Dacapo/p27^{Kip1} and cyclin E degradation. *Cell Cycle, 5*, 71–77.

116. Vogel, C. and Abel, J. (1995). Effect of 2,3,7,8-tetrachlorodibenzo-*p*-dioxin on growth factor expression in the human breast cancer cell line MCF-7. *Archives of Toxicology, 69*, 259–265.

117. Biegel, L. and Safe, S. (1990). Effects of 2,3,7,8-tetrachlorodibenzo-*p*-dioxin (TCDD) on cell growth and the secretion of the estrogen-induced 34-,52- and 160-kDa proteins in human breast cancer cells. *Journal of Steroid Biochemistry and Molecular Biology, 37*, 725–732.

118. Chaloupa, K., Krishnan, V., and Safe, S. (1992). Polynuclear aromatic hydrocarbon carcinogens as anti-estrogens in MCF-7 human breast cancer cells: role of the Ah receptor. *Carcinogenesis, 13*, 2233–2239.

119. Zacharewski, T., Harris, M., Biegel, L., Morrison, V., Merchant, M., and Safe, S. (1992). 6-Methyl-1,3,8-trichlorodibenzofuran (MCDF) as an antiestrogen in human and rodent cancer cell lines: evidence for the role of the Ah receptor. *Toxicology and Applied Pharmacology, 113*, 311–318.

120. Fernandez, P. and Safe, S. (1992). Growth inhibitory and antimitogenic activity of 2,3,7,8-tetrachlorodibenzo-*p*-dioxin (TCDD) in T47D human breast cancer cells. *Toxicology Letters, 61*, 185–197.

121. Liu, H., Wormke, M., Safe, S. H., and Bjeldanes, L. F. (1994). Indolo[3,2-b]carbazole: a dietary-derived factor that exhibits both antiestrogenic and estrogenic activity. *Journal of the National Cancer Institute, 86*, 1758–1765.

122. Chen, I., McDougal, A., Wang, F., and Safe, S. (1998). Aryl hydrocarbon receptor-mediated anti-estrogenic and antitumorigenic activity of diindolylmethane. *Carcinogenesis, 19*, 1631–1639.

123. Oenga, G. N., Spink, D. C., and Carpenter, D. O. (2004). TCDD and PCBs inhibit breast cancer cell proliferation *in vitro*. *Toxicology In Vitro, 18*, 811–819.

124. Köhle, C., Hassepass, I., Bock-Hennig, B. S., Bock, K.-W., Poellinger, L., and McGuire, J. (2002). Conditional expression of a constitutively active aryl hydrocarbon receptor in MCF-7 human breast cancer cells. *Archives of Biochemistry and Biophysics, 402*, 172–179.

125. Merchant, M., Krishnan, V., and Safe, S. (1993). Mechanism of action of alpha-naphthoflavone as an Ah receptor antagonist in MCF-7 human breast cancer cells. *Toxicology and Applied Pharmacology, 120*, 179–185.

126. Abdelrahim, M., Smith, R., 3rd, and Safe, S. (2003). Aryl hydrocarbon receptor gene silencing with small inhibitory RNA differentially modulates Ah-responsiveness in MCF-7 and HepG2 cancer cells. *Molecular Pharmacology, 63*, 1373–1381.

127. Wang, W., Smith, R., 3rd, and Safe, S. (1998). Aryl hydrocarbon receptor-mediated antiestrogenicity in MCF-7 cells: modulation of hormone-induced cell cycle enzymes. *Archives of Biochemistry and Biophysics, 356*, 239–248.

128. Chen, I., Hsieh, T., Thomas, T., and Safe, S. (2001). Identification of estrogen-induced genes downregulated by AhR agonists in MCF-7 breast cancer cells using suppression subtractive hybridization. *Gene, 262*, 207–214.

129. Duan, R., Porter, W., Samudio, I., Vyhlidal, C., Kladde, M., and Safe, S. (1999). Transcriptional activation of c-fos protooncogene by 17beta-estradiol: mechanism of aryl hydrocarbon receptor-mediated inhibition. *Molecular Endocrinology, 13*, 1511–1521.

130. Wormke, M., Stoner, M., Saville, B., and Safe, S. (2000). Crosstalk between estrogen receptor alpha and the aryl hydrocarbon receptor in breast cancer cells involves unidirectional activation of proteasomes. *FEBS Letters, 478*, 109–112.

131. Wormke, M., Stoner, M., Saville, B., Walker, K., Abdelrahim, M., Burghardt, R., and Safe, S. (2003). The aryl hydrocarbon receptor mediates degradation of estrogen receptor alpha through activation of proteasomes. *Molecular and Cellular Biology, 23*, 1843–1855.

132. Ohtake, F., Baba, A., Takada, I., Okada, M., Iwasaki, K., Miki, H., Takahashi, S., Kouzmenko, A., Nohara, K., Chiba, T., Fujii-Kuriyama, Y., and Kato, S. (2007). Dioxin receptor is a ligand-dependent E3 ubiquitin ligase. *Nature, 446*, 562–566.

133. Kociba, R. J., Keyes, D. G., Beyer, J. E., Carreon, R. M., Wade, C. E., Dittenber, D. A., Kalnins, R. P., Frauson, L. E., Park, C. N., Barnard, S. D., Hummel, R. A., and Humiston, C. G. (1978). Results of a two-year chronic toxicity and oncogenicity study of 2,3,7,8-tetrachlorodibenzo-*p*-dioxin in rats. *Toxicology and Applied Pharmacology, 46*, 279–303.

134. Holcomb, M. and Safe, S. (1994). Inhibition of 7,12-dimethylbenzanthracene-induced rat mammary tumor growth by 2,3,7,8-tetrachlorodibenzo-*p*-dioxin. *Cancer Letters, 82*, 43–47.

135. McDougal, A., Wilson, C., and Safe, S. (1997). Inhibition of 7,12-dimethyl-benz[a]anthracene-induced rat mammary tumor growth by aryl hydrocarbon receptor agonists. *Cancer Letters, 120*, 53–63.

136. Marquez-Bravo, L. G. and Gierthy, J. F. (2008). Differential expression of estrogen receptor alpha (ERalpha) protein in MCF-7 breast cancer cells chronically exposed to TCDD. *Journal of Cellular Biochemistry, 103*, 636–647.

137. Schlezinger, J. J., Liu, D., Farago, M., Seldin, D. C., Belguise, K., Sonenshein, G. E., and Sherr, D. H. (2006). A role for the aryl hydrocarbon receptor in mammary gland tumorigenesis. *Biological Chemistry*, 387, 1175–1187.
138. Dietrich, C. and Kaina, B. (2010). The aryl hydrocarbon receptor (AhR) in the regulation of cell–cell contact and tumor growth. *Carcinogenesis*, 31, 1319–1328.
139. Zhang, S., Lei, P., Liu, X., Li, X., Walker, K., Kotha, L., Rowlands, C., and Safe, S. (2009). The aryl hydrocarbon receptor as a target for estrogen receptor-negative breast cancer chemotherapy. *Endocrine-Related Cancer*, 16, 835–844.
140. Ma, Q. and Whitlock, J. P., Jr., (1996). The aromatic hydrocarbon receptor modulates Hepa1c1c7 cell cycle and differentiated state independently of dioxin. *Molecular and Cellular Biology*, 16, 2144–2150.
141. Elizondo, G., Fernandez-Salguero, P., Sheikh, M. S., Kim, G. Y., Fornace, A. J., Lee, K. S., and Gonzalez, F. J. (2000). Altered cell cycle control at the G(2)/M phases in aryl hydrocarbon receptor-null embryo fibroblast. *Molecular Pharmacology*, 57, 1056–1063.
142. Tohkin, M., Fukuhara, M., Elizondo, G., Tomita, S., and Gonzalez, F. J. (2000). Aryl hydrocarbon receptor is required for p300-mediated induction of DNA synthesis by adenovirus E1A. *Molecular Pharmacology*, 58, 845–851.
143. Chang, X., Fan, Y., Karyala, S., Schwemberger, S., Tomlinson, C. R., Sartor, M. A., and Puga, A. (2007). Ligand-independent regulation of transforming growth factor beta1 expression and cell cycle progression by the aryl hydrocarbon receptor. *Molecular and Cellular Biology*, 27, 6127–6139.
144. Zaher, H., Fernandez-Salguero, P. M., Letterio, J., Sheikh, M. S., Fornace, A. J., Jr., Roberts, A. B., and Gonzalez, F. J. (1998). The involvement of aryl hydrocarbon receptor in the activation of transforming growth factor-beta and apoptosis. *Molecular Pharmacology*, 54, 313–321.
145. Haarmann-Stemmann, T., Bothe, H., and Abel, J. (2009). Growth factors, cytokines and their receptors as downstream targets of aryl hydrocarbon receptor (AhR) signaling pathways. *Biochemical Pharmacology*, 77, 508–520.
146. Barhoover, M. A., Hall, J. M., Greenlee, W. F., and Thomas, R. S. (2010). Aryl hydrocarbon receptor regulates cell cycle progression in human breast cancer cells via a functional interaction with cyclin-dependent kinase 4. *Molecular Pharmacology*, 77, 195–201.
147. Döhr, O. and Abel, J. (1997). Transforming growth factor-beta1 coregulates mRNA expression of aryl hydrocarbon receptor and cell-cycle-regulating genes in human cancer cell lines. *Biochemical and Biophysical Research Communications*, 241, 86–91.
148. Levine-Fridman, A., Chen, L., and Elferink, C. (2004). Cytochrome P4501A1 promotes G1 phase cell cycle progression by controlling aryl hydrocarbon receptor activity. *Molecular Pharmacology*, 65, 461–4698.
149. Oesch-Bartlomowicz, B., Huelster, A., Wiss, O., Antoniou-Lipfert, P., Dietrich, C., Arand, M., Weiss, C., Bockamp, E., and Oesch, F. (2005). Aryl hydrocarbon receptor activation by cAMP vs. dioxin: divergent signaling pathways. *Proceedings of the National Academy of Sciences of the United States of America*, 102, 9218–9223.
150. McConkey, D. J., Hartzell, P., Duddy, S. K., Håkansson, H., and Orrenius, S. (1988). 2,3,7,8-Tetrachlorodibenzo-p-dioxin kills immature thymocytes by Ca^{2+}-mediated endonuclease activation. *Science*, 242, 256–259.
151. Kamath, A. B., Xu, H., Nagarkatti, P. S., and Nagarkatti, M. (1997). Evidence for the induction of apoptosis in thymocytes by 2,3,7,8-tetrachlorodibenzo-p-dioxin in vivo. *Toxicology and Applied Pharmacology*, 142, 367–377.
152. Pryputniewicz, S. J., Nagarkatti, M., and Nagarkatti, P. S. (1998). Differential induction of apoptosis in activated and resting T cells by 2,3,7,8-tetrachlorodibenzo-dioxin (TCDD) and its repercussion on T cell responsiveness. *Toxicology*, 129, 211–226.
153. Camacho, I. A., Hassuneh, M. R., Nagarkatti, M., and Nagarkatti, P. S. (2001). Enhanced activation-induced cell death as a mechanism of 2,3,7,8-tetrachlorodibenzo-dioxin (TCDD)-induced immunotoxicity in peripheral T cells. *Toxicology*, 165, 51–63.
154. Camacho, I. A., Nagarkatti, M., and Nagarkatti, P. S. (2002). 2,3,7,8-Tetrachlorodibenzo-p-dioxin (TCDD) induces Fas-dependent activation-induced cell death in superantigen-primed T cells. *Archives of Toxicology*, 76, 570–580.
155. Singh, N. P., Nagarkatti, M., and Nagarkatti, P. (2008). Primary peripheral T cells become susceptible to 2,3,7,8-tetrachlorodibenzo-p-dioxin-mediated apoptosis in vitro upon activation and in the presence of dendritic cells. *Molecular Pharmacology*, 73, 1722–1735.
156. Azkargorta, M., Fullaondo, A., Laresgoiti, U., Aloria, K., Infante, A., Arizmendi, J. M., and Zubiaga, A. M. (2010). Differential proteomics analysis reveals a role for E2F2 in the regulation of the AhR pathway in T lymphocytes. *Molecular & Cellular Proteomics*, 9, 2184–2194.
157. Sanchez-Martin, F. J., Fernandez-Salguero, P. M., and Merino, J. M. (2010). 2,3,7,8-Tetrachlorodibenzo-p-dioxin induces apoptosis in neural growth factor (NGF)-differentiated pheochromocytoma PC12 cells. *Neurotoxicology*, 31, 267–276.
158. Savill, J., Fadok, V., Henson, P., and Haslett, C. (1993). Phagocyte recognition of cells undergoing apoptosis. *Immunology Today*, 14, 131–136.
159. Silverstone, A. E., Frazier, D. E., Jr. Fiore, N. C., Soults, J. A., and Gasiewicz, T. A. (1994). Dexamethasone, beta-estradiol, and 2,3,7,8-tetrachlorodibenzo-p-dioxin elicit thymic atrophy through different cellular targets. *Toxicology and Applied Pharmacology*, 126, 248–259.
160. Dearstyne, E. A. and Kerkvliet, N. I. (2002). Mechanism of 2,3,7,8-tetrachlorodibenzo-p-dioxin (TCDD)-induced decrease in anti-CD3-activated $CD4^+$ T cells: the roles of apoptosis, Fas, and TNF. *Toxicology*, 170, 139–151.
161. Besteman, E. G., Zimmerman, K. L., and Holladay, S. D. (2005). Tetrachlorodibenzo-p-dioxin (TCDD) inhibits differentiation and increases apoptotic cell death of precursor T-cells in the fetal mouse thymus. *Journal of Immunotoxicology*, 2, 107–114.

162. Rhile, M. J., Nagarkatti, M., and Nagarkatti, P. S. (1996). Role of Fas apoptosis and MHC genes in 2,3,7,8-tetrachlorodibenzo-*p*-dioxin (TCDD)-induced immunotoxicity of T cells. *Toxicology*, 110, 153–167.

163. Krammer, P. H. (2000). CD95's deadly mission in the immune system. *Nature*, 407, 789–795.

164. Kamath, A. B., Camacho, I., Nagarkatti, P. S., and Nagarkatti, M. (1999). Role of Fas–Fas ligand interactions in 2,3,7,8-tetrachlorodibenzo-*p*-dioxin (TCDD)-induced immunotoxicity: increased resistance of thymocytes from Fas-deficient (lpr) and Fas ligand-defective (gld) mice to TCDD-induced toxicity. *Toxicology and Applied Pharmacology*, 160, 141–155.

165. Funatake, C. J., Dearstyne, E. A., Steppan, L. B., Shepherd, D. M., Spanjaard, E. S., Marshak-Rothstein, A., and Kerkvliet, N. (2004). Early consequences of 2,3,7,8-tetrachlorodibenzo-*p*-dioxin exposure on the activation and survival of antigen-specific T cells. *Toxicological Sciences*, 82, 129–142.

166. Yoshizawa, K., Marsh, T., Foley, J. F., Cai, B., Peddada, S., Walker, N. J., and Nyska, A. (2005). Mechanisms of exocrine pancreatic toxicity induced by oral treatment with 2,3,7,8-tetrachlorodibenzo-*p*-dioxin in female Harlan Sprague-Dawley Rats. *Toxicological Sciences*, 85, 594–606.

167. Fisher, M. T., Nagarkatti, M., and Nagarkatti, P. S. (2003). Combined screening of thymocytes using apoptosis-specific cDNA array and promoter analysis yields novel gene targets mediating TCDD-induced toxicity. *Toxicological Sciences*, 78, 116–124.

168. Camacho, I. A., Singh, N., Hegde, V. L., Nagarkatti, M., and Nagarkatti, P. S. (2005). Treatment of mice with 2,3,7,8-tetrachlorodibenzo-*p*-dioxin leads to aryl hydrocarbon receptor-dependent nuclear translocation of NF-kappaB and expression of Fas ligand in thymic stromal cells and consequent apoptosis in T cells. *Journal of Immunology*, 175, 90–103.

169. Staples, J. E., Fiore, N. C., Frazier, D. E., Jr. Gasiewicz, T. A., and Silverstone, A. E. (1998). Overexpression of the anti-apoptotic oncogene, bcl-2, in the thymus does not prevent thymic atrophy induced by estradiol or 2,3,7,8-tetrachlorodibenzo-*p*-dioxin. *Toxicology and Applied Pharmacology*, 151, 200–210.

170. Wörner, W. and Schrenk, D. (1996). Influence of liver tumor promoters on apoptosis in rat hepatocytes induced by 2-acetylaminofluorene, ultraviolet light, or transforming growth factor beta 1. *Cancer Research*, 56, 1272–1278.

171. Wörner, W. and Schrenk, D. (1998). 2,3,7,8-Tetrachlorodibenzo-*p*-dioxin suppresses apoptosis and leads to hyperphosphorylation of p53 in rat hepatocytes. *Environmental Toxicology and Pharmacology*, 6, 239–247.

172. Chopra, M., Dharmarajan, A. M., Meiss, G., and Schrenk, D. (2009). Inhibition of UV-C light-induced apoptosis in liver cells by 2,3,7,8-tetrachlorodibenzo-*p*-dioxin. *Toxicological Sciences*, 111, 49–63.

173. Chopra, M., Gährs, M., Haben, M., Michels, C., and Schrenk, D. (2010). Inhibition of apoptosis by 2,3,7,8-tetrachlorodibenzo-*p*-dioxin depends on protein biosynthesis. *Cell Biology and Toxicology*, 26, 391–401.

174. Christensen, J. G., Gonzales, A. J., Cattley, R. C., and Goldsworthy, T. L. (1998). Regulation of apoptosis in mouse hepatocytes and alteration of apoptosis by nongenotoxic carcinogens. *Cell Growth & Differentiation*, 9, 815–825.

175. Reiners, J. J., Jr. and Clift, R. E. (1999). Aryl hydrocarbon receptor regulation of ceramide-induced apoptosis in murine hepatoma 1c1c7 cells. A function independent of aryl hydrocarbon receptor nuclear translocator. *Journal of Biological Chemistry*, 274, 3505–2510.

176. Davis, J. W., II, Melendez, K., Salas, V. M., Lauer, F. T., and Burchiel, S. W. (2000). 2,3,7,8-Tetrachlorodibenzo-*p*-dioxin (TCDD) inhibits growth factor withdrawal-induced apoptosis in the human mammary epithelial cell line, MCF-10A. *Carcinogenesis*, 21, 881–886.

177. Davis, J. W., II, Lauer, F. T., Burdick, A. D., Hudson, L. G., and Burchiel, S. W. (2001). Prevention of apoptosis by 2,3,7,8-tetrachlorodibenzo-*p*-dioxin (TCDD) in the MCF-10A cell line: correlation with increased transforming growth factor α production. *Cancer Research*, 61, 3314–3320.

178. Davis, J. W., Jr. Burdick, A. D., Lauer, F. T., and Burchiel, S. W. (2003). The aryl hydrocarbon receptor antagonist, 3′-methoxy-4′-nitroflavone, attenuates 2,3,7,8-tetrachlorodibenzo-*p*-dioxin-dependent regulation of growth factor signaling and apoptosis in the MCF-10A cell line. *Toxicology and Applied Pharmacology*, 188, 42–49.

179. Park, S. and Matsumura, F. (2006). Characterization of anti-apoptotic action of TCDD as a defensive cellular stress response reaction against the cell damaging action of ultraviolet irradiation in an immortalized normal human mammary epithelial cell line, MCF10A. *Toxicology*, 217, 139–146.

180. Vogel, C. F. A., Li, W., Sciullo, E., Newman, J., Hammock, B., Reader, J. R., Tuscano, J., and Matsumura, F. (2007). Pathogenesis of aryl hydrocarbon receptor-mediated development of lymphoma is associated with increased cyclooxygenase-2 expression. *American Journal of Pathology*, 171, 1538–1548.

181. Schrenk, D., Müller, M., Merlino, G., and Thorgeirsson, S. S. (1997). Interactions of TCDD with signal transduction and neoplastic development in c-myc transgenic and TGF-alpha transgenic mice. *Archives of Toxicology Supplement*, 19, 367–375.

182. Pääjärvi, G., Viluksela, M., Pohjanvirta, R., Stenius, U., and Högberg, J. (2005). TCDD activates Mdm2 and attenuates the p53 response to DNA damaging agents. *Carcinogenesis*, 26, 201–208.

183. Reyes-Hernández, O. D., Mejía-García, A., Sánchez-Ocampo, E. M., Cabañas-Cortés, M. A., Ramírez, P., Chávez-González, L., Gonzalez, F. J., and Elizondo, G. (2010). Ube2l3 gene expression is modulated by activation of the aryl hydrocarbon receptor: implications for p53 ubiquitination. *Biochemical Pharmacology*, 80, 932–940.

184. Llovet, J. M. and Bruix, J. (2008). Molecular targeted therapies in hepatocellular carcinoma. *Hepatology*, 48, 1312–1327.

185. Ambolet-Camoit, A., Bui, L. C., Pierre, S., Chevallier, A., Marchand, A., Coumoul, X., Garlatti, M., Andreau, K., Barouki, R., and Aggerbeck, M. (2010). 2,3,7,8-Tetrachlorodibenzo-*p*-dioxin counteracts the p53 response to a

genotoxicant by upregulating expression of the metastasis marker agr2 in the hepatocarcinoma cell line HepG2. *Toxicological Sciences, 115*, 501–512.

186. Sarioglu, H., Brandner, S., Haberger, M., Jacobsen, C., Lichtmannegger, J., Wormke, M., and Andrae, U. (2008). Analysis of 2,3,7,8-tetrachlorodibenzo-*p*-dioxin-induced proteome changes in 5L rat hepatoma cells reveals novel targets of dioxin action including the mitochondrial apoptosis regulator VDAC2. *Molecular & Cellular Proteomics, 7*, 394–410.

187. Fernandez-Salguero, P., Pineau, T., Hilbert, D. M., McPhail, T., Lee, S. S., Kimura, S., Nebert, D. W., Rudikoff, S., Ward, J. M., and Gonzalez, F. J. (1995). Immune system impairment and hepatic fibrosis in mice lacking the dioxin-binding Ah receptor. *Science, 268*, 722–726.

188. Corchero, J., Martín-Partido, G., Dallas, S. L., and Fernández-Salguero, P. M. (2004). Liver portal fibrosis in dioxin receptor-null mice that overexpress the latent transforming growth factor-beta-binding protein-1. *International Journal of Experimental Pathology, 85*, 295–302.

189. Wu, R., Zhang, L., Hoagland, M. S., and Swanson, H. I. (2007). Lack of the aryl hydrocarbon receptor leads to impaired activation of AKT/protein kinase B and enhanced sensitivity to apoptosis induced via the intrinsic pathway. *Journal of Pharmacology and Experimental Therapeutics, 320*, 448–457.

190. Marlowe, J. L., Fan, Y., Chang, X., Peng, L., Knudsen, E. S., Xia, Y., and Puga, A. (2008). The aryl hydrocarbon receptor binds to E2F1 and inhibits E2F1-induced apoptosis. *Molecular Biology of the Cell, 19*, 3263–3271.

191. Tada, S. (2007). Cdt1 and geminin: role during cell cycle progression and DNA damage in higher eukaryotes. *Frontiers in Bioscience, 12*, 1629–1641.

192. Kawajiri, K., Kobayashi, Y., Ohtake, F., Ikuta, T., Matsushima, Y., Mimura, J., Pettersson, S., Pollenz, R. S., Sakaki, T., Hirokawa, T., Akiyama, T., Kurosumi, M., Poellinger, L., Kato, S., and Fujii-Kuriyama, Y. (2009). Aryl hydrocarbon receptor suppresses intestinal carcinogenesis in Apc$^{Min/+}$ mice with natural ligands. *Proceedings of the National Academy of Sciences of the United States of America, 106*, 13481–13486.

193. Caruso, J. A., Mathieu, P. A., Joiakim, A., Zhang, H., and Reiners, J. J. Jr. (2006). Aryl hydrocarbon receptor modulation of tumor necrosis factor-alpha-induced apoptosis and lysosomal disruption in a hepatoma model that is caspase-8-independent. *Journal of Biological Chemistry, 281*, 10954–10967.

194. Park, K. T., Mitchell, K. A., Huang, G., and Elferink, C. J. (2005). The aryl hydrocarbon receptor predisposes hepatocytes to Fas-mediated apoptosis. *Molecular Pharmacology, 67*, 612–622.

195. Majora, M., Frericks, M., Temchura, V., Reichmann, G., and Esser, C. (2005). Detection of a novel population of fetal thymocytes characterized by preferential emigration and a TCRgammadelta+ T cell fate after dioxin exposure. *International Immunopharmacology, 5*, 1659–1674.

196. Safe, S. and Wormke, M. (2003). Inhibitory aryl hydrocarbon receptor–estrogen receptor alpha cross-talk and mechanisms of action. *Chemical Research in Toxicology, 16*, 807–816.

197. Warner, M., Eskenazi, B., Mocarelli, P., Gerthoux, P. M., Samuels, S., Needham, L., Patterson, D., and Brambilla, P. (2002). Serum dioxin concentrations and breast cancer risk in the Seveso Women's Health Study. *Environmental Health Perspectives, 110*, 625–628.

198. Pesatori, A. C., Consonni, D., Rubagotti, M., Grillo, P., and Bertazzi, P. A. (2009). Cancer incidence in the population exposed to dioxin after the "Seveso accident": twenty years of follow-up. *Environmental Health, 8*, 39.

33

THE AHR REGULATES CELL ADHESION AND MIGRATION BY INTERACTING WITH ONCOGENE AND GROWTH FACTOR-DEPENDENT SIGNALING

Angel Carlos Roman, Jose M. Carvajal-Gonzalez, Sonia Mulero-Navarro, Aurea Gomez-Duran, Eva M. Rico-Leo, Jaime M. Merino, and Pedro M. Fernandez-Salguero

33.1 INTRODUCTION

The intracellular aryl hydrocarbon (dioxin) receptor (AHR) has distinctive functional and structural properties among the large family of basic helix–loop–helix (bHLH) proteins [1]. AHR is a member of the class VII of bHLH transcriptional regulators having a PAS (period (Per)–aryl hydrocarbon receptor nuclear translocator (ARNT)–single-minded (Sim)) domain to interact with and to become activated by high-affinity low molecular weight ligands [1, 2]. The consensus mechanism for AHR-dependent control of gene expression proposes that the receptor binds a specific ligand in the cytosol and then translocates to the nucleus where, after releasing a chaperone complex containing Hsp90, XAP2, and p23, it will heterodimerize with the partner bHLH protein ARNT [2–5]. Following interaction with a relatively characterized set of coactivators or corepressors, the AHR–ARNT heterodimer binds to xenobiotic-responsive elements (XREs; 5′-GCGTG-3′) located in the upstream regulatory sequence of target genes [3, 6]. Once transcriptional regulation has occurred (e.g., cytochrome P450 genes [6–9]), the AHR–ARNT complex is released, exported to the cytosol, and degraded by the proteasome to avoid permanent activation of the pathway [10–14]. In addition to such xenobiotic-devoted signaling, AHR also interacts with other pathways that regulate physiology and the cell response to nontoxicological stimuli, as will be analyzed in detail later.

The intense investigation focused on AHR over the past few years has led to the conclusion that this receptor has a dual role in the cell. On the one hand, toxicology-driven studies have revealed that this receptor is essential for the induction of genes responsible for the metabolic biotransformation of xenobiotics [15] and for the toxic and carcinogenic effects of such environmental contaminants as TCDD (2,3,7,8-tetrachlorodibenzo-p-dioxin) and benzo[a]pyrene [16, 17]. On the other hand, the early presence of AHR in metazoans and its high degree of conservation among species [18, 19], the fact that some invertebrate AHR heterologs modulate development but do not bind dioxins [20], and the altered phenotypes observed in liver [15, 21, 22], immune system [15, 23, 24], heart [25–27], and ovary [28] of mouse models lacking AHR ($Ahr^{-/-}$) strongly support the idea that AHR has relevant roles in cell physiology and tissue homeostasis. In fact, microarray and genome-wide experiments have shown that lack of AHR significantly alters gene expression in mouse liver [29] and in hepatoma [30] and immune cells [31]. Taken together, it can be argued that toxicological, physiological, and developmental signaling converge at some point to AHR through interconnected pathways, probably because xenobiotics act by exacerbating endogenous processes [2–4].

One intriguing feature of AHR is that some of its functions are greatly influenced by the phenotype of the target cell. For instance, AHR expression can either promote (in human breast tumor MCF-7 and human lung tumor A549 cells) or inhibit (mouse hepatoma Hepa-1 and human Jurkat T cells) cell proliferation *in vitro* and tumor development *in vivo* [3, 32]. Remarkably, the effects of AHR on cell adhesion and migration, although they are just beginning to emerge, also

---

The AH Receptor in Biology and Toxicology, First Edition. Edited by Raimo Pohjanvirta.
© 2012 John Wiley & Sons, Inc. Published 2012 by John Wiley & Sons, Inc.

seem to depend on the cell phenotype, as will be discussed below.

In this chapter, we will address the role of AHR in cell adhesion and migration both under physiological cell conditions (i.e., xenobiotic unrelated) and in the presence of exogenous ligands such as TCDD. We will use data from AHR-null mice to analyze how the phenotype of the target cell influences adhesion and migration in the absence of xenobiotics in fibroblastic, epithelial, and endothelial cells. Special attention will be given to the role of the stroma in migration and to the contribution of growth factors such as TGFβ (transforming growth factor β) and VEGF (vascular endothelial growth factor). The final goal will be to provide experimental support for the importance of AHR in normal cell adhesion and migration and to recommend this receptor as a novel potential target in human health-threatening diseases such as metastatic cancer. Some important issues on AHR, related to this chapter, but out of its main scope, have been discussed in previous work focused on cell proliferation and tumorigenesis [3, 33, 34], interaction with pRb and MAP kinase pathways [5, 32], regulation of downstream targets EGF, TNF-α [35], and TGFβ [35, 36], and crosstalk with signaling pathways that modulate cell adhesion and matrix remodeling [37].

33.2 AHR IN THE MODULATION OF CELL ADHESION

Morphology is an important characteristic of the cell that results from the balanced activity of different signaling pathways that coordinately regulate cytoskeleton dynamics through F-actin polymerization and the recycling of cell–cell and cell–substratum interactions [38]. The acquisition of migratory properties is accompanied by changes in cellular plasticity that often originates from a decrease in the strength of cell–cell and cell–substratum interactions. This is relevant not only during development and tissue repair, but also at early stages of the epithelial-to-mesenchymal transition (EMT) that presumably occurs in tumor metastasis [39–41]. Recent evidences from unrelated experimental systems strongly suggest that AHR has an active role in the control of cell adhesion, spreading, and migration under normal cell conditions and after treatment with xenobiotics. Cell adhesion and migration are very close processes that share many intermediate regulatory molecules and signaling pathways. We will first focus the discussion on the work analyzing the role of AHR in adhesion to later comment on the relatively more abundant studies dedicated to discover its function in migration.

Although the interest to understand the endogenous roles (e.g., xenobiotic unrelated) of AHR has greatly increased over the past few years, its implication in cell adhesion was already suggested in early reports. Suspension of adherent normal keratinocytes in culture readily increased the expression of the classical AHR target gene cytochrome P450 1A1 (CYP1A1) in the absence of xenobiotics, indicating that AHR activation was linked to the rupture of cell–cell and cell–substratum interactions [42]. A similar effect was also found in mouse hepatoma Hepa-1 cells in monolayer culture [43], further supporting that the transcriptional activity of AHR was modulated by the adhesion status of the cell. The transcriptional activation of AHR by inhibition of cell–cell contacts appears to be a general mechanism since it was also observed after suspension of several clonal lines of C3H10T1/2 fibroblasts [44]. Comparison of the kinetics of CYP induction by cell suspension versus TCDD treatment revealed that the former was more transient than the latter, probably as a consequence of a reduced transcriptional activity of AHR in the former condition [44]. Consistent with the importance of cell–cell adhesion in AHR activation, it has been found that cell density modulates the intracellular localization of this receptor. Low cell densities favored nuclear translocation of AHR in human keratinocyte HaCaT cells while high cell densities (i.e., confluent cultures) kept the transcriptionally inactive AHR in the cytosol [44, 45]. However, these studies could not demonstrate a causal role of AHR in cell adhesion. AHR activation by exogenous ligands such as TCDD and benzo[a]pyrene repressed the expression of the adhesion protein T-cadherin (T-Cad) in vascular smooth muscle cells, indicating not only that T-Cad is a novel AHR target gene but also that this receptor participates in maintaining cell–cell interactions [46].

With respect to the involvement of AHR in cell–substratum interactions, a previous study showed that rat mammary epithelial cells grown on Matrigel or collagen I extracellular matrices (ECM) increased AHR and ARNT expression and induced CYP1A1 transcription quickly after plating [47], pointing to the fact that early cell adhesion to the ECM was sufficient to activate AHR-dependent signaling. However, the contribution of AHR to cell–substratum interaction is probably more complex since AHR expression and/or activation can also modulate composition and remodeling of the ECM. Several separate studies suggest the interesting possibility that AHR could coordinate individual aspects of cell adhesion at the ECM level [37]. First, the inhibition of caudal fin regeneration in zebrafish by TCDD correlates with AHR activation and with impairment of ECM remodeling [48, 49]; second, the activities of proteases that modulate the ECM composition such as matrix metalloproteinase MMP-2 [50, 51] and cysteine proteases cathepsin B and cathepsin D are regulated in an AHR-dependent manner [52, 53]; third, AHR expression maintains basal levels of the ECM proteases PA/plasmin and elastase in fibroblasts, both being enzymes relevant to the activation of latent forms of growth factors (particularly TGFβ) that modulate cell migration [50]; and fourth, AHR-null mice have phenotypic alterations associated with defects in matrix remodeling in liver parenchyma [15, 21, 22, 54, 55], portal system [22, 55], and terminal end buds of the mammary gland [56].

We have recently described that AHR expression modulates cell morphology, adhesion, and cell–substratum interactions in both immortalized (T-FGM) and primary mouse embryo (MEF) fibroblasts [57, 58]. T-FGM and MEF cells lacking AHR had increased cell area and lower polarity as compared with the corresponding AHR-expressing cell lines, and AHR had a causal role in such morphological phenotype because it could be reverted by reexpressing the receptor. The increase in adhesion and spreading observed in $Ahr^{-/-}$ fibroblasts was maximal under substratum-restricted conditions and decreased when the cells were plated on more permissive supports such as fibronectin or collagen, further sustaining the relevance of the ECM composition in AHR-mediated adhesion. Remarkably, enhanced spreading of $Ahr^{-/-}$ fibroblasts involved a large increase in the number of focal adhesions (FAs) that were distributed along the cell periphery [57], and that coincided with a significant reduction in focal adhesion kinase (FAK) activity in these cells [58]. We anticipated that such properties of AHR-null fibroblasts could contribute to their lower migration rates since FAK-null fibroblasts, unexpectedly, had increased amounts of FAs and reduced migration potential [38, 59].

Altogether, enough experimental evidence is currently available to propose that AHR can have an active role in maintaining cell–cell interactions and ECM remodeling, likely by regulating the expression of proteins forming adhesion complexes and by activating proteases that remodel the ECM. Since these activities on cell adhesion do not fully depend on the presence of exogenous ligands, they are probably reflecting conserved endogenous functions of AHR.

33.3 AHR REGULATES CELL MIGRATION IN A CELL TYPE-SPECIFIC MANNER

It seems reasonable to assume that by modulating cell morphology and adhesion AHR could also regulate cell migration. A limited number of reports from the past few years have provided experimental support for that hypothesis, although many questions still remain unsolved. The available data suggest that similarly to its effects on cell proliferation [3], AHR influences cell migration depending on the phenotype of the target cell and on the presence of exogenous ligands. For example, the differential effect of AHR on epithelial versus mesenchymal cell migration offers interesting possibilities to examine its role in development and disease.

33.3.1 Control of Mesenchymal Cell Migration by AHR

Most of the studies exploring the implication of AHR in migration have been performed using epithelial, endothelial, and immune cells. We nevertheless sought to analyze how AHR could affect mesenchymal fibroblast migration based on the assumption that this cell phenotype is essential for organ development and for tumor cells to acquire motility and dissemination. As mentioned above, T-FGM and MEF fibroblasts from $Ahr^{-/-}$ mice had increased adhesion that was concomitant with reduced migration both in collagen matrices and during wound healing in culture [36, 58]. The existence of strong F-actin stress fibers in slowly migrating $Ahr^{-/-}$ fibroblasts suggested the involvement of small GTPases of the Rho/Rac family, in particular, Rac1 and RhoA [60, 61]. By combining different approaches including pull-down assays for activated Rac1 and RhoA, pharmacological inhibitors for those small GTPases, expression microarrays, and RNA interference (RNAi), we found that the lower migration ability of AHR-null cells was apparently due to an imbalance in the Rac1 and RhoA activities, which could explain the alterations in cytoskeleton structure and migration observed in those mutant cells. Detailed mechanistic analysis revealed that T-FGM and MEF fibroblasts lacking AHR had a marked decrease in the endogenous expression of the oncogene Vav3 [57], the canonical GTP exchange factor (GEF) to modulate Rac1 and RhoA activities [62–64]. Vav3 promoter cloning and activity assays, site-directed mutagenesis, and chromatin immunoprecipitation (ChIP) allowed us to demonstrate that Vav3 is a novel target gene whose constitutive expression is regulated by AHR [57, 65]. This finding is relevant because it identifies a tumor-related oncogene as an AHR target under normal cell conditions in fibroblast cells, and expands the current battery of endogenous AHR-regulated genes including $p21^{Cip1}$ [66], $p27^{Kip1}$ [67], c-jun and jun-D [68], T-Cad [46], c-Myc [69], and pRb [70]. We have thus defined a molecular mechanism that can help explain how AHR becomes integrated in the signaling pathway controlling cell adhesion and migration in fibroblasts cells (Fig. 33.1). In our model, AHR can act at two related but perhaps independent levels: one as a transcription factor in the nucleus and another as a nonnuclear protein interacting with FAK-related signaling molecules. Our data provide strong support for the former pathway in which AHR and the oncogene Vav3 have predominant roles. In such a scenario, AHR would keep constitutive Vav3 levels, which in turn would maintain the physiological balance between Rac1 and RhoA activities and the proper formation of membrane ruffles and stress fibers, thus allowing fibroblast cell migration. AHR could also interact with intermediate molecules that control focal adhesion dynamics such as FAK (whose activity is reduced in the absence of AHR expression). Interestingly, the FAK-regulated proteins phosphatidylinositol 3-kinase (PI3K), and its downstream targets PKB and ERK1/2, had decreased activity in $Ahr^{-/-}$ fibroblasts [58]. Since PI3K contributes to activate Rac1 and RhoA [71], it is possible that AHR-dependent FAK activity could cooperate with Vav3 in the control of cell migration. These novel

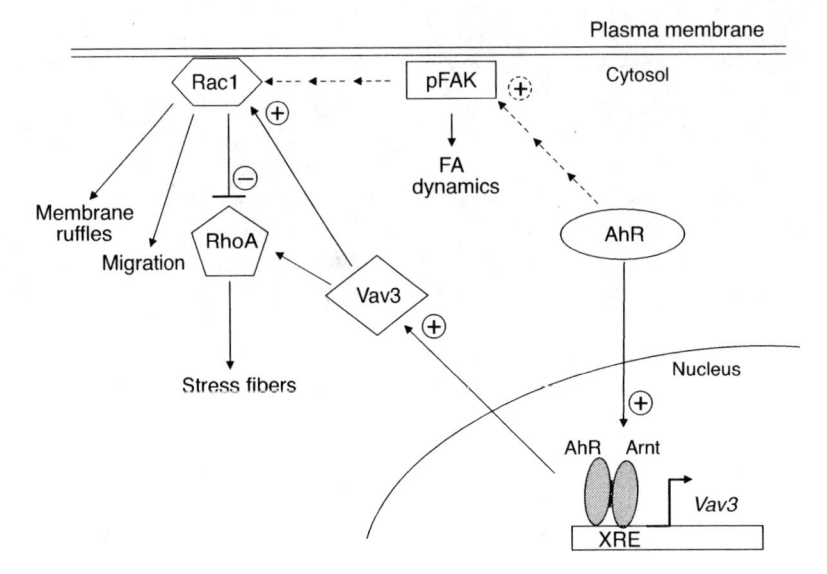

FIGURE 33.1 Proposed model for the AHR regulation of mesenchymal fibroblast cell migration. Endogenous AHR expression maintains the constitutive levels of its target gene *Vav3*, which is a critical GEF for the small GTPases Rac1 and RhoA. Basal activity of Rac1 is mainly responsible for the proper formation of membrane ruffles (lamellipodia) during cell migration. Low levels of RhoA (due to the inhibitory effect exerted by Rac1) limit the amount of F-actin stress fibers and favor cell migration. In addition, AHR could also modulate fibroblast migration by the FAK signaling pathway. By a yet unclear mechanism, AHR would contribute to FAK activation and to the subsequent signaling through PI3K and ERK1/2, which will converge to keep Rac1 activity. Broken arrows indicate unknown intermediate steps and unidentified molecules.

mechanisms underline unanticipated AHR functions in normal cell physiology, and their study will be of interest to address the role of AHR in migration of mesenchymal fibroblasts during organogenesis and in cells suffering EMT during tumor metastasis. Indeed, the carcinogen and AHR ligand benzo[*a*]pyrene induced vascular smooth muscle cell migration and invasion by increasing the expression of matrix metalloproteinases, again suggesting that AHR contributes to cell motility [72].

33.3.2 AHR Modulates Epithelial Cell Migration

Several studies have explored the implication of AHR in epithelial cell migration, most of them in the presence of xenobiotics. Treatment of human breast tumor MCF-7 cells with the high-affinity AHR ligand TCDD resulted in cell scattering due to loss of cell–cell interactions, to formation of lamellipodia, and to increased adhesion and migration [73]. Since expression of a constitutive form of AHR in MCF-7 cells mimicked the effects of TCDD, it appears that AHR has a causal effect in epithelial cell migration. Mechanistically, increased MCF-7 cell migration correlated with higher levels of Jun N-terminal kinase (JNK) activity and with reduced expression of the adherent junction protein of epithelial cells E-cadherin (E-Cad) [73]. Additional work indicated that these effects on JNK and E-Cad were mediated by and required the expression of the metastasis marker Nedd9/Hef1/Cas-L, which turned out to be a TCDD-inducible AHR target gene in that cell line [74, 75]. These reports therefore indicate that AHR activation by TCDD promotes tumor cell migration by induction of the AHR target gene and metastasis marker Nedd9/Hef1/Cas-L, which could serve as a link between AHR activation and tumor dissemination by environmental toxic pollutants. In this context, a possible association between ligand-activated AHR and hypoxia-induced HIF-1α (another class VII bHLH–PAS protein) in promoting cell migration of MCF-7 cells was investigated. Treatment with TCDD under hypoxic conditions activated the transcription factor nuclear factor of activated T cells (NFAT), which in turn induced the expression of autotoxin (ATX), a known promoter of tumor cell migration [76]. However, while AHR modulated MCF-7 migration, HIF-1α did not, suggesting that hypoxia could inhibit signaling downstream of ATX. Similar effects of TCDD were found in human gastric cancer cells, where AHR activation increased their invasiveness through c-Jun-dependent induction of matrix metalloproteinase 9 [77].

Fewer studies addressing the role of AHR in epithelial cell migration in the absence of xenobiotics are available. By increasing AHR activity in transgenic mice carrying a constitutively active form of the receptor targeted to keratinocytes, it was found that these animals developed severe skin lesions resembling atopic dermatitis [78], which could indicate that AHR overactivation impairs skin wound healing.

We have used AHR-null mice to analyze whether an alteration in the physiological levels of this receptor affects keratinocyte migration and skin healing *in vivo* [79]. By using primary keratinocyte cultures, skin explants, and *in vivo* wound healing experiments, we observed that, contrary to that seen in mesenchymal fibroblast cells, AHR-null epithelium had increased migration and faster wound healing. These effects were genuine since a preliminary macroscopic study performed in a different AHR-null line reported similar results [80]. An important observation made *in vivo* was that the mechanism controlling migration of $Ahr^{-/-}$ epithelial cells involved an active interaction between the epithelium and the underlying stroma. The stroma of $Ahr^{-/-}$ mice wounds had higher numbers of mesenchymal fibroblasts and increased deposition of collagen. These effects occurring at the stromal compartment were associated with increased TGFβ activity and with paracrine overactivation of the TGFβ pathway in the surrounding keratinocytes. Remarkably, inhibition of TGFβ activity by a neutralizing antibody or downmodulation of AHR by antisense oligonucleotides rescued the wild-type phenotype in $Ahr^{-/-}$ null mice, opening the possibility for AHR antisense technology to be used as a tool to improve skin repair and for the treatment of surgical or chronic wounds. The fact that AHR expression limits migration of epithelial cells while increases mesenchymal fibroblasts migration makes reasonable to hypothesize that this receptor could modulate migration during EMT by controlling the availability of active TGFβ, as will be discussed below. The importance of TGFβ activity in promoting wound healing gets further support from a study showing that TGFβ1 enhanced reepithelialization in β3-integrin-deficient mice [81].

33.3.3 Physiological AHR Expression Regulates Endothelial Cell Migration

One of the earliest characterized phenotypes in AHR-null mice is their altered hepatic vascularization including the existence of a portosystemic shunt in adult animals [22, 55]. This pathology is AHR dependent since treatment of AHR hypomorph mice with TCDD during gestation rescued a normal hepatic vascular system [82]. Furthermore, knocking out the expression of the AHR partner ARNT also induced vascular defects associated with a diminished response to glucose and to oxygen deprivation [83], indicating that the AHR–ARNT pathway has a relevant role in angiogenesis.

Activation of AHR by toxic exogenous ligands has been reported to inhibit angiogenesis in model cell systems. Thus, treatment of human umbilical vascular endothelial cells (HUVECs) with 3-methylcholanthrene (3-MC) inhibited vascular permeability, cell adhesion, and tube formation by a process involving decreased FAK activation concomitant to increased RhoA signaling [84, 85]. Altogether, these works imply that AHR activation by xenobiotics could adversely affect angiogenesis. Studies using *in vivo* models of ischemia-induced angiogenesis after femoral artery ligation in wild-type and AHR-null mice have revealed that absence of AHR expression improved the response to ischemia-induced angiogenesis, probably associated with an elevation in VEGF expression [86]. In agreement, treatment of ischemic wild-type mice with the ligand benzo[*a*]pyrene reduced angiogenesis whereas such effect was markedly attenuated in ischemic AHR-null animals [87]. Thus, under stress conditions or after treatment with xenobiotics, AHR signaling appears to inhibit angiogenesis.

However, little is known about the contribution of AHR to endothelial cell migration and angiogenesis under physiological conditions in the absence of xenobiotics. We have analyzed this question in primary mouse aortic endothelial cells (MAECs), aortic ring explants, and melanoma tumors induced in wild-type and AHR-null mice [88]. MAECs in culture, or endothelial cells from aortic rings lacking AHR expression, had lower ability to emigrate, a reduced potential to form tubes in Matrigel, and an impaired efficiency to branch. Importantly, AHR had a causal role in these phenomena since its downmodulation by RNA interference switched the wild-type to the mutant phenotype [88]. The mechanisms involved in the AHR-dependent regulation of angiogenesis are complex and require, at the least, the contribution of the endothelium and the stroma. Regarding the endothelium, $Ahr^{-/-}$ MAECs had a marked reduction in HIF-1α expression and in VEGF-A activity that blocked angiogenesis because restoring VEGF-A levels in $Ahr^{-/-}$ cells was enough to rescue the wild-type phenotype. Thus, endothelial cells expressing AHR produced higher levels of autocrine VEGF-A activity positively affecting angiogenesis, as already known in other systems [89–92]. With respect to the role of the stroma, a similar interaction to that seen for the epithelium was observed for the endothelium. The increase in TGFβ activity produced by AHR-null stromal fibroblasts significantly inhibited tube formation and branching of endothelial cells, while neutralization of TGFβ activity partially restored angiogenesis in $Ahr^{-/-}$ endothelium. Moreover, endothelial VEGF-A cooperated with stromal paracrine TGFβ because simultaneous manipulation of both growth factors was sufficient to switch the angiogenic response of wild-type endothelial cells to the AHR-null phenotype, and vice versa. Taken together, we propose that under normal cell conditions, the balance between positive autocrine endothelial VEGF-A and negative paracrine stromal TGFβ activities determines endothelial angiogenesis, and that the impaired angiogenesis observed in $Ahr^{-/-}$ endothelium could result from an additive effect of lower VEGF-A and higher TGFβ activities. Consistently, gene expression profiling of smooth muscle cells from aorta of AHR-null mice revealed significant reductions in the mRNAs of *Tgfβ3* and of additional molecules of TGFβ processing and signaling [93].

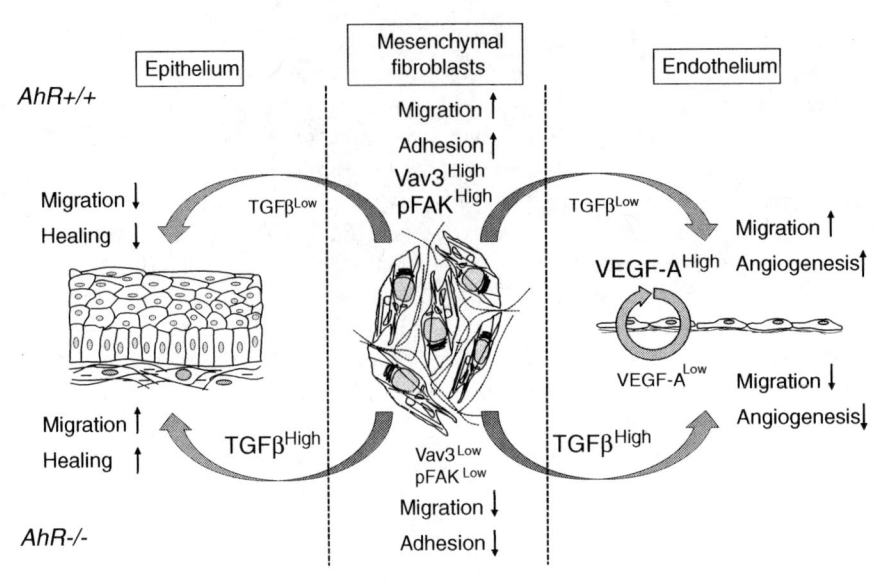

FIGURE 33.2 Proposed models for AHR-dependent migration in mesenchymal fibroblasts, and epithelial and endothelial cells. *Middle panel*: Mesenchymal fibroblasts have endogenous levels of Vav3/Rac1/RhoA signaling and pFAK activity that maintain basal adhesion and migration. $Ahr^{-/-}$ fibroblasts have both pathways inhibited, thus resulting in enhanced cell adhesion and lower migration. *Left panel*: Epithelial migration is modulated by paracrine TGFβ from stromal fibroblasts. $Ahr^{-/-}$ keratinocytes migrate faster for being exposed to higher levels of TGFβ activity. Under these conditions, epithelialization also occurs earlier in AHR-null wounds. *Right panel*: Migration and angiogenesis in endothelial cells (e.g., tube formation and branching) depend on the balance between the angiogenesis promoting effect of VEGF-A (produced by endothelial cells) and the antiangiogenesis activity of TGFβ (secreted by the stromal fibroblasts). AHR-null endothelium has reduced angiogenesis potential for being simultaneously affected by low levels of autocrine VEGF-A and high levels of paracrine TGFβ.

The functional interaction between endothelium and stromal fibroblasts closely resembles that of the epithelium discussed previously. Therefore, AHR expression in different cell types and tissue compartments can establish specific profiles of growth factor activity whose interaction would ultimately determine cell migration.

A summary of these results and a proposed mechanism are schematized in Fig. 33.2. Mesenchymal fibroblast cells expressing AHR have endogenously high levels of Vav3/Rac1/RhoA^{High} and FAK^{High} activities that maintain normal adhesion and migration. AHR-null fibroblasts, in contrast, have both pathways inhibited (Vav3/Rac1/RhoA^{Low} and FAK^{Low}), with increased adhesion and spreading and decreased migration. In addition, AHR expression in stromal fibroblasts modulates epithelial and endothelial cell migration by paracrine-acting TGFβ. Migration of keratinocytes can be positively affected by paracrine TGFβ secreted by fibroblasts at the underlying mesenchyme. AHR-null keratinocytes are thus exposed to increased levels of active TGFβ, which will increase their migration and favor their efficiency to repair skin wounds *in vivo*. Endothelial cell migration is determined by the balanced effect of, at the least, two growth factors. Migration will be modulated by the equilibrium between the angiogenesis promoting activity of autocrine VEGF-A (secreted by endothelial cells) and the antiangiogenic effect of paracrine TGFβ (secreted by mesenchymal fibroblasts). We propose that AHR-null endothelial cells will be simultaneously exposed to low VEGF-A activity and to high levels of TGFβ, collectively resulting in a low angiogenic potential, decreased migration, and reduced capability for tube formation and branching.

33.3.4 Effect of AHR Expression in the Migration of Additional Cell Types

AHR expression by exogenous ligands also affects migration of immune and neuronal cells. Experiments performed in murine thymic organ explants treated with AHR ligands revealed that emigration was more relevant in CD4$^-$CD8$^-$ negative cells expressing the CD44v7 and Cd44v10 isoforms [94]. Similar findings were obtained *in vivo* since AHR activation promoted emigration of CD4$^-$CD8$^-$ cells to the periphery and their accumulation in the spleen [95]. Neuronal cells are also targets for AHR-dependent control of cell migration. *Caenorhabditis elegans* lacking aryl hydrocarbon receptor-1 (*ahr-1*) expression exhibited aberrant cell migration and defective axon branching and neuronal processes [96, 97], indicating that physiological AHR expression is required for normal nervous system development in that organism. Therefore, AHR appears to participate in the control of cell adhesion and migration in many different cell types, making this novel function of AHR of potentially wide interest in development, toxicology, and pathology.

33.4 AHR AS A REGULATOR OF CELL MIGRATION DURING DEVELOPMENT AND PATHOLOGY

The implication of AHR in cell migration will gain considerable biological interest if eventually linked to development and/or pathology. Our knowledge about the molecular mechanisms through which AHR presumably participates in the control of cell adhesion and migration *in vivo* is still very incomplete; however, we will summarize here some situations involving cell migration where AHR has relevant roles.

33.4.1 Role of AHR in Cell Migration During Palate Closure

Closure of the palate is a required step in development that should be completed before birth. Failure to close results in a common human birth defect called cleft palate that is also found in rodents. It is believed that many genetic traits, environmental factors, and intracellular signaling pathways converge to the control of palate development [98, 99]. Interestingly enough, two accepted molecular regulators of palate fusion in rodents and perhaps in humans are AHR [100] and TGFβ3 [101]. The causal role of these proteins in palate formation was demonstrated by analyzing mouse models lacking their expression. *In utero* AHR activation by TCDD treatment of pregnant female mice resulted in a high incidence of cleft palate in newborn pups that was almost completely abolished when the experiments were done in $Ahr^{-/-}$ female mice [102]. Similarly, *Tgfβ3*−/− newborn mice consistently presented, under physiological conditions, a high incidence of a phenotypically similar cleft palate [103, 104]. Palate fusion involves interaction between epithelial and mesenchymal cells and requires cell migration along the palatal shelves [99, 105]. In addition, secondary palate formation is considered to require, at least partially, a process of epithelial-to-mesenchymal transition driven by TGFβ [41, 106]. As a result, since endogenous AHR expression reduces epithelial motility but maintains mesenchymal fibroblast migration, and because AHR regulates TGFβ activity in fibroblasts, it is possible that the palate establishes a physiological profile of AHR expression and/or activation during EMT that controls proper closure of the organ, and that overactivation of AHR signaling by teratogenic compounds disrupts EMT, thus blocking epithelial cell migration and mesenchyme formation.

33.4.2 AHR in Cell Adhesion and Migration During Tumor Development

The influence of AHR expression in tumor development has been mostly analyzed in cell lines and mouse models, whereas our knowledge about its role in human tumors is yet significantly scarcer. Different studies have addressed how AHR expression modulates growth and migration of isolated tumor cell lines (see above and Refs 3, 65, 75, and 107). In animal models of cancer, endogenous AHR has a critical role in mediating xenobiotic-induced tumors [17], while constitutively active AHR promotes carcinogen-induced [108] and spontaneous tumors in mice [109]. The analysis of human tumor samples has revealed that increased AHR expression seems to correlate with development of upper gastrointestinal tract tumors [110, 111] and with increased malignancy of the prostate [112], lung [113], urothelial [114], and pancreas [115]. However, to date, little is known about the precise molecular mechanisms through which AHR modulates tumor promotion and progression. Our hypothesis is that AHR expression is relevant for cancer at two major compartments: the tumor cell itself and the surrounding stroma. In support of the first, we have found that tumorigenic transformed fibroblasts lacking AHR expression had an impaired ability to induce xenograft tumors in nude mice and a lower migration potential, suggesting that AHR expression in these transformed cells is relevant for tumor growth [58]. Regarding the second, AHR expressing murine B16F10 melanoma cells had a significantly reduced competence to induce tumors in $Ahr^{-/-}$ mice, probably because the AHR-null stroma was less efficient in supporting angiogenesis [88]. Although AHR expression seems to coordinate functional interactions between epithelium, stroma, and endothelium through the secretion of the growth factors VEGF-A and TGFβ, additional and more detailed studies are needed to analyze whether AHR expression or activity varies during tumor invasiveness and if this receptor has causal roles in those processes.

One promising target for study is the EMT that allows tumor cells to invade neighboring tissues. AHR appears to be related to key EMT regulators of the zinc finger family of proteins Slug (Snai2) and Snail (Snai1). Previous work has shown that, in fact, Slug is a transcriptional target of AHR in human HaCaT keratinocytes, and that both proteins colocalize at the nucleus of actively migrating cells in wound healing experiments in culture [116]. Further, Slug [117] and Snail (Roman et al., unpublished observations) interact with AHR under basal cell conditions (e.g., in the absence of xenobiotics), again suggesting that these proteins could cooperate to regulate gene expression during cell migration. Indeed, this appears a very likely possibility because by genome-wide analysis we have identified and characterized a novel short interspersed DNA element (SINE) of the B1 family of mouse retrotransposons (B1-X35S) having conserved binding sites

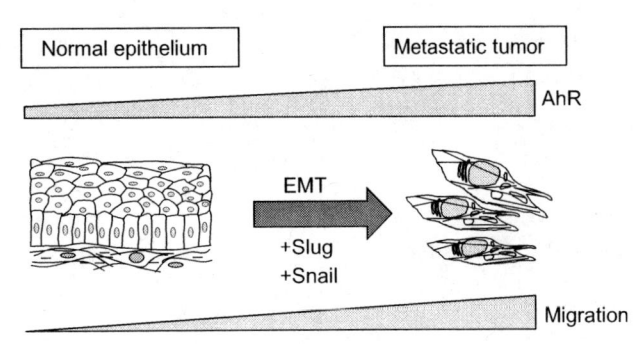

FIGURE 33.3 Proposed mechanism for the implication of AHR in epithelial-to-mesenchymal transition during tumor metastasis. The observed increase in AHR expression during progression of several human tumors would favor its interaction with major EMT regulators Slug and Snail. The ability of AHR to bind Slug and Snail would eventually favor the regulation of target genes and signaling pathways controlling migration and metastasis. Such functional interaction would be potentially more relevant for genes having a B1-X35S retrotransposon in their promoters because its presence would allow their coordinated regulation by AHR–Slug or AHR–Snail transcriptional complexes.

for AHR and Slug/Snail that is able to regulate gene expression *in vitro* and *in vivo* [117]. The fact that B1-X35S is present in over 14,000 instances in the mouse genome, many of them at gene promoters, suggests that coordinated binding of AHR and Slug/Snail to B1-X35S-containing genes can be functionally relevant. Altogether, we propose a model to integrate AHR in the acquisition of a migratory phenotype during EMT and tumor metastasis (Fig. 33.3). Considering the data available on AHR expression in human tumor samples, we suggest that the increase in AHR levels during tumor progression would collaborate with Slug/Snail in promoting EMT and the acquisition of a metastatic phenotype. Relevant questions need to be answered regarding how the increase in AHR, Slug, and Snail expression is coordinated in such process, which target genes are involved other than E-Cad or γ-catenin, how these proteins interact with adhesion and migration signaling pathways, and what is the relevance of this mechanism in particular tumor types and specific patients. Future studies will surely help answer these issues.

33.5 CONCLUSIONS

The recent discoveries made in the AHR field have greatly expanded the relevance of this transcription factor into normal cell physiology far beyond its fundamental toxicological function. It is now widely accepted that AHR most probably plays important roles in organ development and in homeostasis of the hepatic, vascular, immune, and reproductive systems, among others. Main cellular functions such as proliferation, apoptosis, and differentiation also appear to require AHR activity under endogenous, xenobiotic-unrelated conditions. Recently, the control of cell adhesion and migration has emerged as a novel AHR function whose study will unveil unexpected interactions with different signaling pathways. It is remarkable that certain endogenous functions of AHR are cell type specific, and this has been precisely demonstrated for cell proliferation, adhesion, and migration. The fact that under normal cell conditions, AHR reduces epithelial cell migration but maintains the motility of mesenchymal fibroblastic cells, opens the intriguing possibility that changes in the expression of this receptor could be relevant to modulate the epithelial-to-mesenchymal transition that takes place in development and cancer. Indeed, AHR expression seems to increase during progression of certain human tumors, and this observation will be important to address whether or not this receptor is a driver of tumor metastasis. Endothelial cell migration is also AHR dependent and defective receptor expression severely impairs angiogenesis *in vivo*, in agreement with the phenotypes found in AHR-null mice. The functional interaction between AHR and growth factors relevant not only for cell adhesion and migration but also for cell proliferation represents an additional field of interest in AHR research. New molecular mechanisms are interconnected and perhaps direct these many activities of AHR. Among them, the establishment of nonnuclear functions with intermediate signaling molecules involved in cell adhesion and migration appears interesting. In summary, AHR is reemerging as a genuine physiological regulator with potential implications in an increasingly intricate set of biological activities that modulate cell function in health and disease.

ACKNOWLEDGMENTS

This work was supported by grants to P. M. F-S. from the Spanish Ministry of Science and Innovation (MICINN) (SAF2005-00130 and SAF2008-00462), the Junta de Extremadura (GRU09001), and the Red Temática de Investigación Cooperativa en Cáncer (RTICC) (RD06/0020/1016, Fondo de Investigaciones Sanitarias (FIS), Carlos III Institute, Spanish Ministry of Health). All Spanish funding is cosponsored by the European Union FEDER program. J. M. C.-G and A. C. R. were supported by the MICINN.

REFERENCES

1. Massari, M. E. and Murre, C. (2000). Helix–loop–helix proteins: regulators of transcription in eucaryotic organisms. *Molecular and Cellular Biology*, 20, 429–440.
2. Furness, S. G., Lees, M. J., and Whitelaw, M. L. (2007). The dioxin (aryl hydrocarbon) receptor as a model for adaptive responses of bHLH/PAS transcription factors. *FEBS Letters*, 581, 3616–3625.

3. Barouki, R., Coumoul, X., and Fernandez-Salguero, P. M. (2007). The aryl hydrocarbon receptor, more than a xenobiotic-interacting protein. *FEBS Letters*, *581*, 3608–3615.
4. Bock, K. W. and Kohle, C. (2006). Ah receptor: dioxin-mediated toxic responses as hints to deregulated physiologic functions. *Biochemical Pharmacology*, *72*, 393–404.
5. Puga, A., Tomlinson, C. R., and Xia, Y. (2005). Ah receptor signals cross-talk with multiple developmental pathways. *Biochemical Pharmacology*, *69*, 199–207.
6. Nebert, D. W. and Dalton, T. P. (2006). The role of cytochrome P450 enzymes in endogenous signalling pathways and environmental carcinogenesis. *Nature Reviews Cancer*, *6*, 947–960.
7. Hankinson, O. (1995). The aryl hydrocarbon receptor complex. *Annual Review of Pharmacology and Toxicology*, *35*, 307–340.
8. Hankinson, O. (2005). Role of coactivators in transcriptional activation by the aryl hydrocarbon receptor. *Archives of Biochemistry and Biophysics*, *433*, 379–386.
9. Nebert, D. W., Dalton, T. P., Okey, A. B., and Gonzalez, F. J. (2004). Role of aryl hydrocarbon receptor-mediated induction of the CYP1 enzymes in environmental toxicity and cancer. *Journal of Biological Chemistry*, *279*, 23847–23850.
10. Davarinos, N. A. and Pollenz, R. S. (1999). Aryl hydrocarbon receptor imported into the nucleus following ligand binding is rapidly degraded via the cytoplasmic proteasome following nuclear export. *Journal of Biological Chemistry*, *274*, 28708–28715.
11. Ma, Q. and Baldwin, K. T. (2000). 2,3,7,8-Tetrachlorodibenzo-p-dioxin-induced degradation of aryl hydrocarbon receptor (AhR) by the ubiquitin-proteasome pathway. Role of the transcription activation and DNA binding of AhR. *Journal of Biological Chemistry*, *275*, 8432–8438.
12. Pollenz, R. S. (2002). The mechanism of AH receptor protein down-regulation (degradation) and its impact on AH receptor-mediated gene regulation. *Chemico-Biological Interactions*, *141*, 41–61.
13. Santiago-Josefat, B. and Fernandez-Salguero, P. M. (2003). Proteasome inhibition induces nuclear translocation of the dioxin receptor through an Sp1 and protein kinase C-dependent pathway. *Journal of Molecular Biology*, *333*, 249–260.
14. Santiago-Josefat, B., Pozo-Guisado, E., Mulero-Navarro, S., and Fernandez-Salguero, P. M. (2001). Proteasome inhibition induces nuclear translocation and transcriptional activation of the dioxin receptor in mouse embryo primary fibroblasts in the absence of xenobiotics. *Molecular and Cellular Biology*, *21*, 1700–1709.
15. Fernandez-Salguero, P., Pineau, T., Hilbert, D. M., McPhail, T., Lee, S. S., Kimura, S., Nebert, D. W., Rudikoff, S., Ward, J. M., and Gonzalez, F. J. (1995). Immune system impairment and hepatic fibrosis in mice lacking the dioxin-binding Ah receptor. *Science*, *268*, 722–726.
16. Fernandez-Salguero, P. M., Hilbert, D. M., Rudikoff, S., Ward, J. M., and Gonzalez, F. J. (1996). Aryl-hydrocarbon receptor-deficient mice are resistant to 2,3,7,8-tetrachlorodibenzo-p-dioxin-induced toxicity. *Toxicology and Applied Pharmacology*, *140*, 173–179.
17. Shimizu, Y., Nakatsuru, Y., Ichinose, M., Takahashi, Y., Kume, H., Mimura, J., Fujii-Kuriyama, Y., and Ishikawa, T. (2000). Benzo[a]pyrene carcinogenicity is lost in mice lacking the aryl hydrocarbon receptor. *Proceedings of the National Academy of Sciences of the United States of America*, *97*, 779–782.
18. Hahn, M. E. (1998). The aryl hydrocarbon receptor: a comparative perspective. *Comparative Biochemistry and Physiology, Part C*, *121*, 23–53.
19. Hahn, M. E. (2002). Aryl hydrocarbon receptors: diversity and evolution. *Chemico-Biological Interactions*, *141*, 131–160.
20. Crews, S. T. and Brenman, J. E. (2006). Spineless provides a little backbone for dendritic morphogenesis. *Genes & Development*, *20*, 2773–2778.
21. Corchero, J., Martin-Partido, G., Dallas, S. L., and Fernandez-Salguero, P. M. (2004). Liver portal fibrosis in dioxin receptor-null mice that overexpress the latent transforming growth factor-beta-binding protein-1. *International Journal of Experimental Pathology*, *85*, 295–302.
22. Lahvis, G. P., Lindell, S. L., Thomas, R. S., McCuskey, R. S., Murphy, C., Glover, E., Bentz, M., Southard, J., and Bradfield, C. A. (2000). Portosystemic shunting and persistent fetal vascular structures in aryl hydrocarbon receptor-deficient mice. *Proceedings of the National Academy of Sciences of the United States of America*, *97*, 10442–10447.
23. Hogaboam, J. P., Moore, A. J., and Lawrence, B. P. (2008). The aryl hydrocarbon receptor affects distinct tissue compartments during ontogeny of the immune system. *Toxicological Sciences*, *102*, 160–170.
24. Laiosa, M. D., Wyman, A., Murante, F. G., Fiore, N. C., Staples, J. E., Gasiewicz, T. A., and Silverstone, A. E. (2003). Cell proliferation arrest within intrathymic lymphocyte progenitor cells causes thymic atrophy mediated by the aryl hydrocarbon receptor. *Journal of Immunology*, *171*, 4582–4591.
25. Fernandez-Salguero, P. M., Ward, J. M., Sundberg, J. P., and Gonzalez, F. J. (1997). Lesions of aryl-hydrocarbon receptor-deficient mice. *Veterinary Pathology*, *34*, 605–614.
26. Lund, A. K., Goens, M. B., Nunez, B. A., and Walker, M. K. (2006). Characterizing the role of endothelin-1 in the progression of cardiac hypertrophy in aryl hydrocarbon receptor (AhR) null mice. *Toxicology and Applied Pharmacology*, *212*, 127–135.
27. Vasquez, A., Atallah-Yunes, N., Smith, F. C., You, X., Chase, S. E., Silverstone, A. E., and Vikstrom, K. L. (2003). A role for the aryl hydrocarbon receptor in cardiac physiology and function as demonstrated by AhR knockout mice. *Cardiovascular Toxicology*, *3*, 153–163.
28. Benedict, J. C., Lin, T. M., Loeffler, I. K., Peterson, R. E., and Flaws, J. A. (2000). Physiological role of the aryl hydrocarbon receptor in mouse ovary development. *Toxicological Sciences*, *56*, 382–388.
29. Tijet, N., Boutros, P. C., Moffat, I. D., Okey, A. B., Tuomisto, J., and Pohjanvirta, R. (2006). Aryl hydrocarbon receptor

regulates distinct dioxin-dependent and dioxin-independent gene batteries. *Molecular Pharmacology, 69,* 140–153.

30. Sartor, M. A., Schnekenburger, M., Marlowe, J. L., Reichard, J. F., Wang, Y., Fan, Y., Ma, C., Karyala, S., Halbleib, D., Liu, X., Medvedovic, M., and Puga, A. (2009). Genomewide analysis of aryl hydrocarbon receptor binding targets reveals an extensive array of gene clusters that control morphogenetic and developmental programs. *Environmental Health Perspectives, 117,* 1139–1146.

31. Frericks, M., Burgoon, L. D., Zacharewski, T. R., and Esser, C. (2008). Promoter analysis of TCDD-inducible genes in a thymic epithelial cell line indicates the potential for cell-specific transcription factor crosstalk in the AhR response. *Toxicology and Applied Pharmacology, 232,* 268–279.

32. Puga, A., Ma, C., and Marlowe, J. L. (2009). The aryl hydrocarbon receptor cross-talks with multiple signal transduction pathways. *Biochemical Pharmacology, 77,* 713–722.

33. Elferink, C. J. (2003). Aryl hydrocarbon receptor-mediated cell cycle control. *Progress in Cell Cycle Research, 5,* 261–267.

34. Marlowe, J. L. and Puga, A. (2005). Aryl hydrocarbon receptor, cell cycle regulation, toxicity, and tumorigenesis. *Journal of Cellular Biochemistry, 96,* 1174–1184.

35. Haarmann-Stemmann, T., Bothe, H., and Abel, J. (2009). Growth factors, cytokines and their receptors as downstream targets of arylhydrocarbon receptor (AhR) signaling pathways. *Biochemical Pharmacology, 77,* 508–520.

36. Gomez-Duran, A., Carvajal-Gonzalez, J. M., Mulero-Navarro, S., Santiago-Josefat, B., Puga, A., and Fernandez-Salguero, P. M. (2009). Fitting a xenobiotic receptor into cell homeostasis: how the dioxin receptor interacts with TGFbeta signaling. *Biochemical Pharmacology, 77,* 700–712.

37. Kung, T., Murphy, K. A., and White, L. A. (2009). The aryl hydrocarbon receptor (AhR) pathway as a regulatory pathway for cell adhesion and matrix metabolism. *Biochemical Pharmacology, 77,* 536–546.

38. Mitra, S. K., Hanson, D. A., and Schlaepfer, D. D. (2005). Focal adhesion kinase: in command and control of cell motility. *Nature Reviews Molecular Cell Biology, 6,* 56–68.

39. Lee, J. M., Dedhar, S., Kalluri, R., and Thompson, E. W. (2006). The epithelial–mesenchymal transition: new insights in signaling, development, and disease. *Journal of Cell Biology, 172,* 973–981.

40. Li, S., Guan, J. L., and Chien, S. (2005). Biochemistry and biomechanics of cell motility. *Annual Review of Biomedical Engineering, 7,* 105–150.

41. Thiery, J. P., Acloque, H., Huang, R. Y., and Nieto, M. A. (2009). Epithelial–mesenchymal transitions in development and disease. *Cell, 139,* 871–890.

42. Sadek, C. M. and Allen-Hoffmann, B. L. (1994). Cytochrome P450IA1 is rapidly induced in normal human keratinocytes in the absence of xenobiotics. *Journal of Biological Chemistry, 269,* 16067–16074.

43. Sadek, C. M. and Allen-Hoffmann, B. L. (1994). Suspension-mediated induction of Hepa 1c1c7 Cyp1a-1 expression is dependent on the Ah receptor signal transduction pathway. *Journal of Biological Chemistry, 269,* 31505–31509.

44. Cho, Y. C., Zheng, W., and Jefcoate, C. R. (2004). Disruption of cell–cell contact maximally but transiently activates AhR-mediated transcription in 10T1/2 fibroblasts. *Toxicology and Applied Pharmacology, 199,* 220–238.

45. Ikuta, T., Kobayashi, Y., and Kawajiri, K. (2004). Cell density regulates intracellular localization of aryl hydrocarbon receptor. *Journal of Biological Chemistry, 279,* 19209–19216.

46. Niermann, T., Schmutz, S., Erne, P., and Resink, T. (2003). Aryl hydrocarbon receptor ligands repress T-cadherin expression in vascular smooth muscle cells. *Biochemical and Biophysical Research Communications, 300,* 943–949.

47. Larsen, M. C., Brake, P. B., Pollenz, R. S., and Jefcoate, C. R. (2004). Linked expression of Ah receptor, ARNT, CYP1A1, and CYP1B1 in rat mammary epithelia, *in vitro,* is each substantially elevated by specific extracellular matrix interactions that precede branching morphogenesis. *Toxicological Sciences, 82,* 46–61.

48. Andreasen, E. A., Mathew, L. K., Lohr, C. V., Hasson, R., and Tanguay, R. L. (2007). Aryl hydrocarbon receptor activation impairs extracellular matrix remodeling during zebra fish fin regeneration. *Toxicological Sciences, 95,* 215–226.

49. Zodrow, J. M. and Tanguay, R. L. (2003). 2,3,7,8-Tetrachlorodibenzo-p-dioxin inhibits zebrafish caudal fin regeneration. *Toxicological Sciences, 76,* 151–161.

50. Gomez-Duran, A., Mulero-Navarro, S., Chang, X., and Fernandez-Salguero, P. M. (2006). LTBP-1 blockade in dioxin receptor-null mouse embryo fibroblasts decreases TGF-beta activity: role of extracellular proteases plasmin and elastase. *Journal of Cellular Biochemistry, 97,* 380–392.

51. Santiago-Josefat, B., Mulero-Navarro, S., Dallas, S. L., and Fernandez-Salguero, P. M. (2004). Overexpression of latent transforming growth factor-β binding protein 1 (LTBP-1) in dioxin receptor-null mouse embryo fibroblasts. *Journal of Cell Science, 117,* 849–859.

52. Kohle, C., Hassepass, I., Bock-Hennig, B. S., Walter Bock, K., Poellinger, L., and McGuire, J. (2002). Conditional expression of a constitutively active aryl hydrocarbon receptor in MCF-7 human breast cancer cells. *Archives of Biochemistry and Biophysics, 402,* 172–179.

53. Wang, F., Samudio, I., and Safe, S. (2001). Transcriptional activation of cathepsin D gene expression by 17β-estradiol: mechanism of aryl hydrocarbon receptor-mediated inhibition. *Molecular and Cellular Endocrinology, 172,* 91–103.

54. Peterson, T. C., Hodgson, P., Fernandez-Salguero, P., Neumeister, M., and Gonzalez, F. J. (2000). Hepatic fibrosis and cytochrome P450: experimental models of fibrosis compared to AHR knockout mice. *Hepatology Research, 17,* 112–125.

55. Lahvis, G. P., Pyzalski, R. W., Glover, E., Pitot, H. C., McElwee, M. K., and Bradfield, C. A. (2005). The aryl hydrocarbon receptor is required for developmental closure of the ductus venosus in the neonatal mouse. *Molecular Pharmacology, 67,* 714–720.

56. Hushka, L. J., Williams, J. S., and Greenlee, W. F. (1998). Characterization of 2,3,7,8-tetrachlorodibenzofuran-dependent suppression and AH receptor pathway gene expression in the developing mouse mammary gland. *Toxicology and Applied Pharmacology*, 152, 200–210.

57. Carvajal-Gonzalez, J. M., Mulero-Navarro, S., Roman, A. C., Sauzeau, V., Merino, J. M., Bustelo, X. R., and Fernandez-Salguero, P. M. (2009). The dioxin receptor regulates the constitutive expression of the vav3 proto-oncogene and modulates cell shape and adhesion. *Molecular Biology of the Cell*, 20, 1715–1727.

58. Mulero-Navarro, S., Pozo-Guisado, E., Perez-Mancera, P. A., Alvarez-Barrientos, A., Catalina-Fernandez, I., Hernandez-Nieto, E., Saenz-Santamaria, J., Martinez, N., Rojas, J. M., Sanchez-Garcia, I., and Fernandez-Salguero, P. M. (2005). Immortalized mouse mammary fibroblasts lacking dioxin receptor have impaired tumorigenicity in a subcutaneous mouse xenograft model. *Journal of Biological Chemistry*, 280, 28731–28741.

59. Ilic, D., Furuta, Y., Kanazawa, S., Takeda, N., Sobue, K., Nakatsuji, N., Nomura, S., Fujimoto, J., Okada, M., and Yamamoto, T. (1995). Reduced cell motility and enhanced focal adhesion contact formation in cells from FAK-deficient mice. *Nature*, 377, 539–544.

60. Bustelo, X. R., Sauzeau, V., and Berenjeno, I. M. (2007). GTP-binding proteins of the Rho/Rac family: regulation, effectors and functions *in vivo*. *Bioessays*, 29, 356–370.

61. Fukata, M., Nakagawa, M., and Kaibuchi, K. (2003). Roles of Rho-family GTPases in cell polarisation and directional migration. *Current Opinion in Cell Biology*, 15, 590–597.

62. Hornstein, I., Alcover, A., and Katzav, S. (2004). Vav proteins, masters of the world of cytoskeleton organization. *Cellular Signalling*, 16, 1–11.

63. Movilla, N. and Bustelo, X. R. (1999). Biological and regulatory properties of Vav-3, a new member of the Vav family of oncoproteins. *Molecular and Cellular Biology*, 19, 7870–7885.

64. Couceiro, J. R., Martin-Bermudo, M. D., and Bustelo, X. R. (2005). Phylogenetic conservation of the regulatory and functional properties of the Vav oncoprotein family. *Experimental Cell Research*, 308, 364–380.

65. Fernandez-Salguero, P. M. (2010). A remarkable new target gene for the dioxin receptor: the Vav3 proto-oncogene links AhR to adhesion and migration. *Cell Adhesion & Migration*, 4, 172–175.

66. Barnes-Ellerbe, S., Knudsen, K. E., and Puga, A. (2004). 2,3,7,8-Tetrachlorodibenzo-p-dioxin blocks androgen-dependent cell proliferation of LNCaP cells through modulation of pRB phosphorylation. *Molecular Pharmacology*, 66, 502–511.

67. Kolluri, S. K., Weiss, C., Koff, A., and Gottlicher, M. (1999). p27[Kip1] induction and inhibition of proliferation by the intracellular Ah receptor in developing thymus and hepatoma cells. *Genes & Development*, 13, 1742–1753.

68. Hoffer, A., Chang, C. Y., and Puga, A. (1996). Dioxin induces transcription of fos and jun genes by Ah receptor-dependent and -independent pathways. *Toxicology and Applied Pharmacology*, 141, 238–247.

69. Yang, X., Liu, D., Murray, T. J., Mitchell, G. C., Hesterman, E. V., Karchner, S. I., Merson, R. R., Hahn, M. E., and Sherr, D. H. (2005). The aryl hydrocarbon receptor constitutively represses c-myc transcription in human mammary tumor cells. *Oncogene*, 24, 7869–7881.

70. Marlowe, J. L., Knudsen, E. S., Schwemberger, S., and Puga, A. (2004). The aryl hydrocarbon receptor displaces p300 from E2F-dependent promoters and represses S phase-specific gene expression. *Journal of Biological Chemistry*, 279, 29013–29022.

71. Sasaki, A. T., Chun, C., Takeda, K., and Firtel, R. A. (2004). Localized Ras signaling at the leading edge regulates PI3K, cell polarity, and directional cell movement. *Journal of Cell Biology*, 167, 505–518.

72. Meng, D., Lv, D. D., Zhuang, X., Sun, H., Fan, L., Shi, X. L., and Fang, J. (2009). Benzo[*a*]pyrene induces expression of matrix metalloproteinases and cell migration and invasion of vascular smooth muscle cells. *Toxicology Letters*, 184, 44–49.

73. Diry, M., Tomkiewicz, C., Koehle, C., Coumoul, X., Bock, K. W., Barouki, R., and Transy, C. (2006). Activation of the dioxin/aryl hydrocarbon receptor (AhR) modulates cell plasticity through a JNK-dependent mechanism. *Oncogene*, 25, 5570–5574.

74. Barouki, R. and Coumoul, X. (2009). Cell migration and metastasis markers as targets of environmental pollutants and the aryl hydrocarbon receptor. *Cell Adhesion & Migration*, 4, 72–76.

75. Bui, L. C., Tomkiewicz, C., Chevallier, A., Pierre, S., Bats, A. S., Mota, S., Raingeaud, J., Pierre, J., Diry, M., Transy, C., Garlatti, M., Barouki, R., and Coumoul, X. (2009). Nedd9/Hef1/Cas-L mediates the effects of environmental pollutants on cell migration and plasticity. *Oncogene*, 28, 3642–3651.

76. Seifert, A., Rau, S., Kullertz, G., Fischer, B., and Santos, A. N. (2009). TCDD induces cell migration via NFATc1/ATX-signaling in MCF-7 cells. *Toxicology Letters*, 184, 26–32.

77. Peng, T. L., Chen, J., Mao, W., Song, X., and Chen, M. H. (2009). Aryl hydrocarbon receptor pathway activation enhances gastric cancer cell invasiveness likely through a c-Jun-dependent induction of matrix metalloproteinase-9. *BMC Cell Biology*, 10, 27.

78. Tauchi, M., Hida, A., Negishi, T., Katsuoka, F., Noda, S., Mimura, J., Hosoya, T., Yanaka, A., Aburatani, H., Fujii-Kuriyama, Y., Motohashi, H., and Yamamoto, M. (2005). Constitutive expression of aryl hydrocarbon receptor in keratinocytes causes inflammatory skin lesions. *Molecular and Cellular Biology*, 25, 9360–9368.

79. Carvajal-Gonzalez, J. M., Roman, A. C., Cerezo-Guisado, M. I., Rico-Leo, E. M., Martin-Partido, G., and Fernandez-Salguero, P. M. (2009). Loss of dioxin-receptor expression accelerates wound healing *in vivo* by a mechanism involving TGFbeta. *Journal of Cell Science*, 122, 1823–1833.

80. Ikuta, T., Namiki, T., Fujii-Kuriyama, Y., and Kawajiri, K. (2009). AhR protein trafficking and function in the skin. *Biochemical Pharmacology*, 77, 588–596.

81. Reynolds, L. E., Conti, F. J., Lucas, M., Grose, R., Robinson, S., Stone, M., Saunders, G., Dickson, C., Hynes, R. O., Lacy-Hulbert, A., and Hodivala-Dilke, K. (2005). Accelerated re-epithelialization in beta3-integrin-deficient mice is associated with enhanced TGF-beta1 signaling. *Nature Medicine*, 11, 167–174.

82. Walisser, J. A., Bunger, M. K., Glover, E., and Bradfield, C. A. (2004). Gestational exposure of Ahr and Arnt hypomorphs to dioxin rescues vascular development. *Proceedings of the National Academy of Sciences of the United States of America*, 101, 16677–16682.

83. Maltepe, E., Schmidt, J. V., Baunoch, D., Bradfield, C. A., and Simon, M. C. (1997). Abnormal angiogenesis and responses to glucose and oxygen deprivation in mice lacking the protein ARNT. *Nature*, 386, 403–407.

84. Chang, C. C., Tsai, S. Y., Lin, H., Li, H. F., Lee, Y. H., Chou, Y., Jen, C. Y., and Juan, S. H. (2009). Aryl-hydrocarbon receptor-dependent alteration of FAK/RhoA in the inhibition of HUVEC motility by 3-methylcholanthrene. *Cellular and Molecular Life Sciences*, 66, 3193–3205.

85. Juan, S. H., Lee, J. L., Ho, P. Y., Lee, Y. H., and Lee, W. S. (2006). Antiproliferative and antiangiogenic effects of 3-methylcholanthrene, an aryl-hydrocarbon receptor agonist, in human umbilical vascular endothelial cells. *European Journal of Pharmacology*, 530, 1–8.

86. Ichihara, S., Yamada, Y., Ichihara, G., Nakajima, T., Li, P., Kondo, T., Gonzalez, F. J., and Murohara, T. (2007). A role for the aryl hydrocarbon receptor in regulation of ischemia-induced angiogenesis. *Arteriosclerosis, Thrombosis, and Vascular Biology*, 27, 1297–1304.

87. Ichihara, S., Yamada, Y., Gonzalez, F. J., Nakajima, T., Murohara, T., and Ichihara, G. (2009). Inhibition of ischemia-induced angiogenesis by benzo[*a*]pyrene in a manner dependent on the aryl hydrocarbon receptor. *Biochemical and Biophysical Research Communications*, 381, 44–49.

88. Roman, A. C., Carvajal-Gonzalez, J. M., Rico-Leo, E. M., and Fernandez-Salguero, P. M. (2009). Dioxin receptor deficiency impairs angiogenesis by a mechanism involving VEGF-A depletion in the endothelium and transforming growth factor-beta overexpression in the stroma. *Journal of Biological Chemistry*, 284, 25135–25148.

89. Carmeliet, P., Ferreira, V., Breier, G., Pollefeyt, S., Kieckens, L., Gertsenstein, M., Fahrig, M., Vandenhoeck, A., Harpal, K., Eberhardt, C., Declercq, C., Pawling, J., Moons, L., Collen, D., Risau, W., and Nagy, A. (1996). Abnormal blood vessel development and lethality in embryos lacking a single VEGF allele. *Nature*, 380, 435–439.

90. Ferrara, N., Carver-Moore, K., Chen, H., Dowd, M., Lu, L., O'Shea, K. S., Powell-Braxton, L., Hillan, K. J., and Moore, M. W. (1996). Heterozygous embryonic lethality induced by targeted inactivation of the VEGF gene. *Nature*, 380, 439–442.

91. Lu, P. and Werb, Z. (2008). Patterning mechanisms of branched organs. *Science*, 322, 1506–1509.

92. Ruhrberg, C., Gerhardt, H., Golding, M., Watson, R., Ioannidou, S., Fujisawa, H., Betsholtz, C., and Shima, D. T. (2002). Spatially restricted patterning cues provided by heparin-binding VEGF-A control blood vessel branching morphogenesis. *Genes & Development*, 16, 2684–2698.

93. Guo, J., Sartor, M., Karyala, S., Medvedovic, M., Kann, S., Puga, A., Ryan, P., and Tomlinson, C. R. (2004). Expression of genes in the TGF-beta signaling pathway is significantly deregulated in smooth muscle cells from aorta of aryl hydrocarbon receptor knockout mice. *Toxicology and Applied Pharmacology*, 194, 79–89.

94. Esser, C., Temchura, V., Majora, M., Hundeiker, C., Schwarzler, C., and Gunthert, U. (2004). Signaling via the AHR leads to enhanced usage of CD44v10 by murine fetal thymic emigrants: possible role for CD44 in emigration. *International Immunopharmacology*, 4, 805–818.

95. Temchura, V. V., Frericks, M., Nacken, W., and Esser, C. (2005). Role of the aryl hydrocarbon receptor in thymocyte emigration *in vivo*. *European Journal of Immunology*, 35, 2738–2747.

96. Qin, H. and Powell-Coffman, J. A. (2004). The *Caenorhabditis elegans* aryl hydrocarbon receptor, AHR-1, regulates neuronal development. *Developmental Biology*, 270, 64–75.

97. Qin, H., Zhai, Z., and Powell-Coffman, J. A. (2006). The *Caenorhabditis elegans* AHR-1 transcription complex controls expression of soluble guanylate cyclase genes in the URX neurons and regulates aggregation behavior. *Developmental Biology*, 298, 606–615.

98. Murray, J. C. (2002). Gene/environment causes of cleft lip and/or palate. *Clinical Genetics*, 61, 248–256.

99. Murray, J. C. and Schutte, B. C. (2004). Cleft palate: players, pathways, and pursuits. *Journal of Clinical Investigation*, 113, 1676–1678.

100. Mimura, J., Yamashita, K., Nakamura, K., Morita, M., Takagi, T. N., Nakao, K., Ema, M., Sogawa, K., Yasuda, M., Katsuki, M., and Fujii-Kuriyama, Y. (1997). Loss of teratogenic response to 2,3,7,8-tetrachlorodibenzo-p-dioxin (TCDD) in mice lacking the Ah (dioxin) receptor. *Genes to Cells*, 2, 645–654.

101. Jezewski, P. A., Vieira, A. R., Nishimura, C., Ludwig, B., Johnson, M., O'Brien, S. E., Daack-Hirsch, S., Schultz, R. E., Weber, A., Nepomucena, B., Romitti, P. A., Christensen, K., Orioli, I. M., Castilla, E. E., Machida, J., Natsume, N., and Murray, J. C. (2003). Complete sequencing shows a role for MSX1 in non-syndromic cleft lip and palate. *Journal of Medical Genetics*, 40, 399–407.

102. Peters, J. M., Narotsky, M. G., Elizondo, G., Fernandez-Salguero, P. M., Gonzalez, F. J., and Abbott, B. D. (1999). Amelioration of TCDD-induced teratogenesis in aryl hydrocarbon receptor (AhR)-null mice. *Toxicological Sciences*, 47, 86–92.

103. Kaartinen, V., Voncken, J. W., Shuler, C., Warburton, D., Bu, D., Heisterkamp, N., and Groffen, J. (1995). Abnormal lung development and cleft palate in mice lacking TGF-beta 3 indicates defects of epithelial–mesenchymal interaction. *Nature Genetics*, 11, 415–421.

104. Proetzel, G., Pawlowski, S. A., Wiles, M. V., Yin, M., Boivin, G. P., Howles, P. N., Ding, J., Ferguson, M. W., and Doetschman, T.

(1995). Transforming growth factor-beta 3 is required for secondary palate fusion. *Nature Genetics*, *11*, 409–414.

105. Rice, R., Spencer-Dene, B., Connor, E. C., Gritli-Linde, A., McMahon, A. P., Dickson, C., Thesleff, I., and Rice, D. P. (2004). Disruption of Fgf10/Fgfr2b-coordinated epithelial–mesenchymal interactions causes cleft palate. *Journal of Clinical Investigation*, *113*, 1692–1700.

106. Martinez-Alvarez, C., Blanco, M. J., Perez, R., Rabadan, M. A., Aparicio, M., Resel, E., Martinez, T., and Nieto, M. A. (2004). Snail family members and cell survival in physiological and pathological cleft palates. *Developmental Biology*, *265*, 207–218.

107. Dietrich, C. and Kaina, B. (2010). The aryl hydrocarbon receptor (AhR) in the regulation of cell–cell contact and tumor growth. *Carcinogenesis*, *31*, 1319–1328.

108. Moennikes, O., Loeppen, S., Buchmann, A., Andersson, P., Ittrich, C., Poellinger, L., and Schwarz, M. (2004). A constitutively active dioxin/aryl hydrocarbon receptor promotes hepatocarcinogenesis in mice. *Cancer Research*, *64*, 4707–4710.

109. Andersson, P., Rubio, C., Poellinger, L., and Hanberg, A. (2005). Gastric hamartomatous tumours in a transgenic mouse model expressing an activated dioxin/Ah receptor. *Anticancer Research*, *25*, 903–911.

110. Peng, T. L., Chen, J., Mao, W., Liu, X., Tao, Y., Chen, L. Z., and Chen, M. H. (2009). Potential therapeutic significance of increased expression of aryl hydrocarbon receptor in human gastric cancer. *World Journal of Gastroenterology*, *15*, 1719–1729.

111. Roth, M. J., Wei, W. Q., Baer, J., Abnet, C. C., Wang, G. Q., Sternberg, L. R., Warner, A. C., Johnson, L. L., Lu, N., Giffen, C. A., Dawsey, S. M., Qiao, Y. L., and Cherry, J. (2009). Aryl hydrocarbon receptor expression is associated with a family history of upper gastrointestinal tract cancer in a high-risk population exposed to aromatic hydrocarbons. *Cancer Epidemiology, Biomarkers and Prevention*, *18*, 2391–2396.

112. Gluschnaider, U., Hidas, G., Cojocaru, G., Yutkin, V., Ben-Neriah, Y., and Pikarsky, E. (2010). β-TrCP inhibition reduces prostate cancer cell growth via upregulation of the aryl hydrocarbon receptor. *PLoS One*, *5*, e9060.

113. Chang, J. T., Chang, H., Chen, P. H., Lin, S. L., and Lin, P. (2007). Requirement of aryl hydrocarbon receptor overexpression for CYP1B1 up-regulation and cell growth in human lung adenocarcinomas. *Clinical Cancer Research*, *13*, 38–45.

114. Ishida, M., Mikami, S., Kikuchi, E., Kosaka, T., Miyajima, A., Nakagawa, K., Mukai, M., Okada, Y., and Oya, M. (2010). Activation of the aryl hydrocarbon receptor pathway enhances cancer cell invasion by upregulating the MMP expression and is associated with poor prognosis in upper urinary tract urothelial cancer. *Carcinogenesis*, *31*, 287–295.

115. Koliopanos, A., Kleeff, J., Xiao, Y., Safe, S., Zimmermann, A., Buchler, M. W., and Friess, H. (2002). Increased arylhydrocarbon receptor expression offers a potential therapeutic target for pancreatic cancer. *Oncogene*, *21*, 6059–6070.

116. Ikuta, T. and Kawajiri, K. (2006). Zinc finger transcription factor Slug is a novel target gene of aryl hydrocarbon receptor. *Experimental Cell Research*, *312*, 3585–3594.

117. Roman, A. C., Benitez, D. A., Carvajal-Gonzalez, J. M., and Fernandez-Salguero, P. M. (2008). Genome-wide B1 retrotransposon binds the transcription factors dioxin receptor and Slug and regulates gene expression *in vivo*. *Proceedings of the National Academy of Sciences of the United States of America*, *105*, 1632–1637.

34

THE PHYSIOLOGICAL ROLE OF AHR IN THE MOUSE IMMUNE SYSTEM

CHARLOTTE ESSER

34.1 INTRODUCTION

Responding to environmental signals in a relevant fashion is of paramount importance to all organisms. As described in many chapters of this book, AHR is part of the adaptive response to chemicals by controlling xenobiotic metabolizing enzymes, and thus clearing low molecular weight chemicals from the body. Several lines of evidence have suggested early that the AHR has importance beyond such xenobiotic metabolism. First, AHR expression is not restricted to the liver or other organs with strong metabolic functions, albeit it is not ubiquitous either. Interestingly, AHR expression is strong in immunological barrier organs, that is, the lung, skin, and, as we could recently show, the gut. Similarly, expression is high in the thymus, the site of T-cell generation. The tissue distribution of transcripts or proteins is considered physiologically and toxicologically relevant. Therefore, the abundance of AHR in tissues and cells of the immune system was thought-provoking. A second line of evidence came from early *in vitro* and *in vivo* immunotoxicological studies with dioxins and other AHR binding substances, which strongly and persistently activate AHR. They showed that the immune system is a very sensitive target in all animal species analyzed so far. Immunotoxic AHR effects are the topic of Chapter 19. Finally, and maybe most importantly, mouse models for systemic or cell-specific AHR deficiency, AHR hypomorphs, AHR transgenics, AHR reporter mice, or natural low-affinity mutants have been studied in recent years, and the results led to a growing understanding that the AHR is a critical regulator of cell differentiation programs. In this chapter, I will discuss the state of the art regarding AHR and immunity.

34.2 A BRIEF OUTLINE OF MAJOR PLAYERS IN THE IMMUNE SYSTEM

Dealing with pathogenic organisms and the toxins they produce is a vital function for both vertebrate and invertebrate organisms. The immune system has evolved into a complex organ suited to this task. Pathogens can enter body via all epithelia (nose, gut, lung, skin, etc.), via wounds into tissues, and directly into the bloodstream. It is therefore no surprise that cells of the immune system are dispersed throughout the body, abundant at major pathogen entry sites, and can move to sites of infections as well. The immune system is made up of various cell types, which have highly diverse functions and interact with each other at short or long distance. Communication is via cell–cell contact using receptor–ligand surface structures, or over considerable distances via lymphokines and chemokines. Lymph nodes situated throughout the body along the lymph vessels provide relevant spatial structures as "meeting points" for direct communication of immune cells.

There are two main arms of the immune system, the so-called innate immunity (which is evolutionary older) and the adaptive immune system (which evolved only in vertebrates). Macrophages, neutrophils, and dendritic cells (DCs) belong to the former, and T cells and B cells to the latter. Cells of the innate immune system have the so-called "pathogen recognition receptors" for structures found exclusively on or are produced by bacteria, fungi, or viruses, such as lipopolysaccharides, flagellins, unmethylated CpG, or dsRNA. Upon recognition of such structures, innate immune cells immediately fight the infection by, for example, phagocytosing the bacteria and oxidative burst. Moreover, some innate immune

---

The AH Receptor in Biology and Toxicology, First Edition. Edited by Raimo Pohjanvirta.
© 2012 John Wiley & Sons, Inc. Published 2012 by John Wiley & Sons, Inc.

cells have the additional capacity to digest pathogen proteins and display them as small peptide pieces on their surface. This instructs and directs antigen-specific T cells that danger is at hand, and thus starts the adaptive immune response. Specialists for this "antigen presentation" are the dendritic cells, which incidentally are rich in AHR expression (see below). T cells need the interaction with such antigen-presenting cells to mature into effector cells. In turn, B cells need help from T cells to function. Only a subset of B cells, the so-called CD5 B cells, produces antibodies without the help of T cells.

All cells of the immune system are generated throughout life from the common hematopoietic stem cell. They have individual life spans ranging from a few days to many years. Continuous hematopoiesis, migration of cells, and mature cell homeostasis are distinct features of the immune system. Immunotoxic substances could interfere with any of these processes.

Immune cells follow their intrinsic programs, and/or adapt to external cues, relayed into the cells by surface receptors coupled to signal transduction pathways. Thus, transcription factors are pivotal in shaping the immune response, as all immune cells pass at some point through the executive steps of up- or downregulation of genes. Major pathways in immune cells are G protein-coupled receptors, the MAP kinases, NF-κB, or the Janus kinase (JAK)–STAT pathways. Another is direct activation by ligand of latent transcription factors (such as glucocorticoid receptors). Table 34.1 lists major cellular players.

As discussed elsewhere in this book, AHR activation by 2,3,7,8-tetrachlorodibenzo-p-dioxin (TCDD) and structurally related halogenated hydrocarbons is immunotoxic; that is, they dose-dependently impair functions in most of these cells.[1]

This might result from direct or indirect effects. Due to the complex and manifold interactions, any AHR-dependent impairment in one cell type may ultimately lead to dysfunction of other cell types as well. In any case, it is important to keep in mind that the immune system is a continuously dynamic spatial and temporal organ, and misbalancing its regulatory network can be devastating.

TABLE 34.1 Major Cellular Players in the Immune System

Cell Name	Subsets	Function
Innate immune cells		
Macrophage		Phagocytosis
Granulocytes	Neutrophils	Phagocytosis
	Basophils	
	Eosinophils	
Dendritic cells	Myeloid DCs	Antigen recognition, transport into lymph nodes, and presentation to T cells
	Langerhans cells	
	Plasmacytoid DCs	
	$CD103^+$ gut DCs	
	$CD103^-$ gut DCs	
Natural killer cells		Killing of virus-infected cells
Adaptive immune cells		
αβ TCR T cells	CD4 helper T cells	Cytokine production
	T_h1	IFN-γ (and others)
	T_h2	IL-4 (and others)
	T_h17	IL-17, IL-22
	Regulatory T cells	Downmodulation of immune response
	Inducible	Differentiate after antigen contact
	Natural	Generated in the thymus
	CD8 killer T cells	Cytotoxicity
	NKT	Cytotoxicity
γδ TCR	Invariant TCR (skin)	Epithelial integrity, other functions unclear
	Invariant TCR (gut)	Epithelial integrity, unclear
	Variant TCR	Unclear
B cells	CD5 B cells	T-independent antibody production against major bacterial antigens (carbohydrates)
	Conventional B cells	Humoral immune response

34.3 THE IMMUNE PHENOTYPE OF AHR-NULL MICE

One way to elucidate the role of a gene *in vivo* is to delete it and analyze the effects. In the 1990s, three groups independently generated AHR-deficient mouse mutants, by deleting either exon 1 or exon 2 [1–3]. Recently, I reviewed the differences and common defects of these AHR mutant mice and other AHR mutant strains [4]. As expected, all AHR-null mice failed to induce xenobiotic metabolizing enzymes upon TCDD exposure; TCDD lethality and TCDD-mediated toxicity were abrogated. All mice were made with embryonic

---

[1] (Immuno)toxicity is of course a matter of dose and of toxicokinetics. For xenobiotic AHR ligands, their varying affinities to AHR have been correlated with toxicity and carcinogenicity, and the so-called "toxic equivalency factor" is used to calculate overall acute toxicity in mixtures of AHR ligands (e.g., dioxins, polychlorinated biphenyls, and furans). TCDD is assigned the value 1. It may be physiologically more important, though, how quickly an AHR ligand is degraded in the cell. As TCDD is degraded at an extremely slow rate, AHR signalling is sustained, which some authors referred to as "overactivation." The prototypes of persistent xenobiotic and transient endogenous ligands are TCDD and FICZ, respectively.

stem cells from the 129 mouse strain, which were transferred into C57BL/6 blastocysts; thus, many early studies were done with mice in genetically mixed C57BL/6 x129 backgrounds. In general, studies done after 2000 (B6.129-Ahr^{tmBra}), 2003 (B6.129-Ahr^{tmGonz}), and 2004 (Ahr^{tmYfk}) use mice that were backcrossed often enough to become congenic to C57BL/6.

Common features of these AHR-deficient mouse strains are reduced fecundity and a somewhat shorter life span. They also exhibit skin and especially liver inflammatory pathology without exposure to exogenous chemicals [2, 4, 5]. Several chapters in this book deal with AHR-dependent physiological alterations on the cardiac, metabolic, vascular, and reproductive systems.

AHR-deficient mice kept in clean or specific pathogen-free laboratory conditions did not have a lethal or even obviously devastating immunological phenotype. Problems, if they became obvious, appeared only with aging. This was unexpected considering the high sensitivity of the immune system after AHR activation with environmental chemicals. One of the AHR-deficient strains had decreased lymphocyte numbers in the spleen and lymphocyte infiltration of lung, intestine, and urinary tract. However, these immunological alterations were not detected in the second widely used mouse strain [6]. These discrepancies have not been resolved, but might be related to different animal facilities or the mixed genetic background (see above).

Adult thymus cellularity and overall pattern of lymphocyte subpopulations (determined by surface staining and FACS analysis) were not significantly affected. Induction of cellular and humoral immune responses with model antigens was successful in AHR-deficient mice [7]. The mice produced normal immunoglobulin titers and class switching [7, 8]. Likewise, AHR-deficient CD4 and CD8 T cells responded normally to alloantigen *in vivo* as reflected by downregulation of CD62L and upregulation of CD69, CD25, CD28, GITR, and CTLA-4 [9, 10]. Alloantigen-induced T-cell proliferation was also normal except for a small increase in cycling of AHR-deficient CD8 T cells [10]. However, contact hypersensitivity could not be successfully induced in AHR-deficient mice [11]. Contact hypersensitivity models competent T cell-mediated allergic response to antigens entering via the skin, orchestrated by appropriate antigen presentation by Langerhans cells (LCs), and cytokine availability for T-cell responses. AHR-deficient Langerhans cells, which are impaired in maturation, may fail to initiate the response properly. Concomitantly, there may be a failure of T cells to produce IL-2, whose gene is a direct target of AHR [11, 12].

34.4 INFECTIOUS MODELS SHOW THE IMPORTANCE OF AHR FOR IMMUNITY

In contrast to models of humoral or cellular immune responses described above, infectious diseases reflect more realistically a possible role for AHR in immunity. Several infections have been tested, including intracellular parasites, bacteria, and viruses (see Table 34.2). Although, in general, AHR-deficient mice were more susceptible to infections, results depended on the pathogen.

Many AHR-deficient mice suffer from rectal prolapse when they get older [17]. The group of Frank Gonzalez showed that these mice harbor *Helicobacter hepaticus* in their gut crypts, an opportunistic infection indicating immunodeficiency. Rectal prolapse could be related to this infection [17, 19]. It is not clear whether the causes are specific defects in the immune system, or some other damages resulting from AHR deficiency. Interestingly, we found that a specific subset of immune cells found exclusively in the gut, γδTCR/CD8αα T cells, is reduced in AHR-deficient mice (S. Chmill, Dissertation, Düsseldorf, Germany, 2010). These cells are the main producers of GMCSF and important for maintaining the epithelial barrier of the gut [20]. They appear to be replaced with normal cytotoxic T cells.

Survival rate in a lethal *Streptococcus pneumoniae* infection model was slightly enhanced in AHR-null mice, albeit less than after AHR with TCDD [16]. Viral infections and

TABLE 34.2 Outcome of Infectious Disease in AHR-Deficient Mice

Name of Pathogen	Type	Outcome in $Ahr^{-/-}$ Mice
Influenza	Respiratory tract virus	Normal susceptibility [13]
T. gondii	Protozoa; targets neural tissue in mice	Higher susceptibility [14]
L. monocytogenes	Gram-positive bacterium; intracellular in macrophages	Higher susceptibility, better resistance to reinfection [15]
S. pneumoniae	Gram-positive bacterium; colonizes body surfaces and epithelia; causes inflammatory infections	Slightly higher survival rate [16]
H. hepaticus	Gram-negative bacterium; colonizes the gut; associated with colitis	Higher susceptibility [17]
Bacterial lipopolysaccharide treatment	(LPS is part of wall structure of Gram-negative bacteria)	Higher sensitivity to lethal septic shock [18]

AHR have been studied for influenza. While activation of AHR with TCDD was immunosuppressive and decreased survival [21, 22], AHR deficiency as such did not influence neutrophil frequency or infection-driven IFN-γ secretion in the lungs, which were comparable to wild-type mice [13]. Interestingly, the viral studies suggested a role of AHR in lung epithelial barrier changes as well, albeit this has not been studied in any detail so far [23].

AHR-deficient mice infected with *Listeria monocytogenes*, an intracellular bacterium, were more susceptible to infection but developed enhanced resistance to reinfection [15]. Concomitant to infection, and possibly caused by high bacterial burdens rather than some AHR-driven gene transcription, the anti-inflammatory cytokine IL-10 and IL-12 levels increased upon infection. Cytokine producing *Listeria*-specific T-cell numbers after the infection equalled or surpassed in both AHR-null mice and wild-type mice. Moreover, macrophages retained their ability to ingest *Listeria* or inhibit parasite growth [15].

More recently, it was shown that *Toxoplasma gondii*-infected AHR-deficient mice succumbed to the parasite significantly faster than congenic wild-type mice and displayed greater liver damage. Again, these data showed that the AHR contributes to an optimal immune response. AHR-deficient mice had higher serum levels of tumor necrosis factor (TNF)-α or nitric oxide (NO). While shifts in cytokine production might be very important in pathogenicity, the mechanisms responsible for enhanced susceptibility are not entirely clear, and obviously more research needs to be done.

Although not strictly an infectious model, it was shown that AHR-deficient mice are much more sensitive to LPS lethal shock treatment, a mimic of bacterial sepsis. Underlying mechanism appears to be a higher production of the inflammatory cytokine IL-6, which is normally inhibited by an interaction of AHR with signal transducer and activator of transcription 1 (Stat1) and NF-κB [18].

Besides systemic immunosuppression, AHR activation by TCDD causes changes in the differentiation of immune cells, in particular, thymus atrophy. This might reflect a distinctive role of AHR in hematopoiesis and immune cell homeostasis. Results obtained with AHR-deficient mice suggest, however, that this is not a major function of AHR. Rather, as detailed below, AHR function is important for ongoing immune responses, integrating, for example, environmental signals.

34.5 THE ROLE OF LIGANDS IN SHAPING AHR IMMUNE RESPONSES

Although it is possible that AHR might function in the absence of a ligand, overwhelming evidence suggests that ligands are necessary [24]. AHR has not been crystallized, and little information is available regarding conformational changes induced by different ligands. Ligand-protected protease digestion studies indicated that only one binding pocket exists in AHR; moreover, we found that different ligands (TCDD, 3,3′,4,4′-PCB, indolo[3,2-b]carbazole, and β-naphthoflavone) protected AHR protein differentially against protease digestion [25] (own unpublished observations). It should be stressed that the issue of "endogenous versus exogenous ligand" has not been resolved. As described in Chapter 4, AHR ligands are numerous. They can be classified into exogenous (natural or anthropogenic) and endogenous (e.g., generated by the body, by bacteria in the body, or by UV as an exogenous trigger). Notwithstanding that many are found at effective concentrations in the body [26, 27], it remains a puzzle which of these are indeed relevant in the physiological situation. Note that for all endogenous ligands known so far tagging them as "made for the purpose" proved difficult or was not attempted. Physiological use of exogenous ligands (as in vitamins: needed but not generated in the body) is similarly enigmatic. Most likely, the answer will be that several different ligands are used and shape the function of AHR differentially. Thus, while an endogenous ligand might be important for homeostatic differentiation of immune cells, other ligands might sensitize immune responses via AHR to the actual environmental situation.

Two particularly interesting ligands, tryptophan dimers and tryptophan metabolic breakdown products, highlight this point. Tryptophan dimers were identified already in the 1980s as potent AHR ligands [28]. They are generated by UV and visible light in solutions *in vitro* and in skin *in vivo* [29, 30]. Intensive studies to characterize the tryptophan photoproducts identified them as 6-formylindolo[3,2-b]carbazole (FICZ) and 6,12-diformylindolo[3,2-b]carbazole (dFICZ). FICZ is a planar molecule and has a very high affinity for AHR, but is quickly and efficiently degraded in cells by AHR-induced CYP1A1, CYP1A2, and CYP1B1, giving it low intracellular steady-state levels. The discovery of a chemical messenger molecule with high AHR affinity explains cutaneous and extracutaneous induction of CYP1A1 after dermal exposure to UV light, and might be pivotal for rapid elimination of a topical carcinogenic substance. Recently, AHR was shown to be part of skin pigmentation and needed for melanocyte proliferation in response to UV irradiation, and conceivably, FICZ is the triggering ligand [31]. Catabolic breakdown of tryptophan by the immunosuppressive enzyme indoleamine 2,3-dioxygenase (IDO) could also involve AHR. AHR-deficient dendritic cells cannot upregulate IDO anymore, and gene expression profiling of Langerhans cells from AHR-deficient versus wild-type mice showed a virtual shutdown of IDO transcription by AHR deficiency [11]. IDO catalyzes the rate-limiting first step of tryptophan degradation to kynurenine, and kynurenine is a potential AHR ligand [27, 32, 33]. The physiological consequences of this are currently not known, but it was suggested that IDO and tryptophan catabolites might be involved in the formation of regulatory T cells and tolerance [34, 35]. Certainly, this line of evidence will attract more attention in the future.

34.6 IMMUNOLOGICALLY RELEVANT TRANSCRIPTION FACTOR ACTIVITY OF AHR

AHR is a latent cytosolic transcription factor. The first signal transduction pathway discovered acts via dimerization with ARNT (also known as HIF-1β) in the nucleus and binding to AHREs. Target genes include, but are not limited to, genes of the xenobiotic metabolism, often dubbed the "AHR gene battery." PCR, Northern blot, and chip array studies done with TCDD in various cell types showed that hundreds of genes can be targets of the AHR. Not all of these genes have AHREs[2] in their promoters. Noteworthy, AHR activation can also suppress gene expression, a phenomenon that needs further biochemical–genetic analysis.

Evidence suggested for a long time that AHR also interferes with other signaling pathways, for example, influences the activities of MAP kinase, NF-κB, and Nrf2 pathways The complexity of this is discussed in Chapters 7–10. Briefly, AHR can bind to the estrogen receptor, or to subunits of NF-κB, and induce transcription of genes with ERE of NF-κB binding sites in their promoters. On the other hand, other signaling can substantially modify the AHR–ARNT response; indeed, this is part of the basis of the high cell-specific action of AHR. With regard to the immune system, the interactions between NF-κB, Stat1, or Stat5 are especially intriguing. NF-κB is a transcription factor involved in many immune functions and used by many immune cells. NF-κB signaling is triggered by, for example, inflammatory cytokines IL-1β, IL-6, and TNF-α and by lipopolysaccharide. At the same time, these cytokines suppress constitutive and AHR-induced CYP1A1 expression, suggesting a link between inflammation, oxidative stress, and CYP expression [37]. Conversely, AHR ligands can repress or transactivate NF-κB responses [37, 38].

AHR-mediated gene expression is highly cell and context specific. In immune cells, many TCDD-inducible genes have been identified by now using various methods. Genes responsive to TCDD include, for instance, cytokine genes (*IL-1β, IL-2, TNF-α*), apoptosis genes (*FasL, TRAIL*), differentiation markers (*Notch-1, CD44, CD69*), and costimulatory signals (*CD40, CD80*), depending on the cell type and cell differentiation stage. Many cytokine genes have AHREs in their promoters and AHR-deficient mice verified involvement of AHR in cytokine constitutive/inducible expression. IL-5 production increased in AHR-deficient mice in two models of allergy [39, 40]. However, IL-5 was not increased, but reduced, after nonallergic immunization of AHR-deficient mice by subcutaneous injection of OVA in Freund's adjuvant. T-cell proliferation and IL-4 production were unaffected. In these null mutant mice, stimulation of spleen cells with either ConA or restimulation with OVA triggered higher IFN-γ and IL-12 protein production in spleen cells [8]. In agreement with this, we found increased transcription of IFN-γ, but not IL-4 in purified, naive CD4 T cells from AHR-deficient mice [41]. No data exist on IFN-γ production by AHR-deficient CD8$^+$ cells, NKT cells, or other possible sources for IFN-γ. The data point to a role of AHR in balancing T_h1 versus T_h2 cytokines, for example, by keeping IFN-γ and IL-12 at low levels in normal mice [8]. To what extent AHR contributes to lineage decisions of naïve T cells—directing them to become regulatory or inflammatory T cells—is the focus of much current research.

34.7 EXPRESSION LEVELS OF AHR IN IMMUNE CELLS

Liver and lung are known for their high AHR expression. However, certain hematopoietic stem cells, some dendritic cells, subsets of thymocytes, and T cells display equal or even higher levels of AHR expression than the liver [42–44]. Immune functions and immune cells can be targeted directly or indirectly by AHR activity. Early studies looking at AHR expression in entire lymphoid organs gave the impression that AHR is ubiquitous in the immune system. More recently, analysis in defined immune cell subsets gave a much more differentiated picture; Table 34.3 compiles the current knowledge in this respect. Unfortunately, data on AHR protein levels are not available for many immune cell subpopulations, limiting the interpretation of causative AHR ligand effects. Microarrays and studies using cell sorting combined with RT-PCR and Western blotting have identified Lin$^-$Sca$^+$ and Sca$^-$ progenitor cells in bone marrow (BM), double negative (CD4$^-$CD8$^-$) DN4 cells in thymus, CD4$^+$T$_h$17 cells, BM-derived dendritic cells, and Langerhans cells as subpopulations with high AHR levels [11, 41, 45, 53]. It is noteworthy that dendritic cells—as far as analyzed—express high AHR levels and high levels of AHR repressor at the same time. A careful and wide-ranging reanalysis of highly pure immune cell subsets for AHR expression would be helpful in filling up the gaps and untangling the cell-specific actions of AHR signaling. This is particularly important for the question whether AHR directly generates regulatory T cells, a claim currently under debate.

---

[2] The target sequence for AHR in promoters is 5′-T/GNGCGTGCC/GAN-3′ (AHR/ARNT binding core sequence underlined). It is called "dioxin-responsive element" (DRE), xenobiotic-responsive element" (XRE), or AHR-responsive element (AHRE) in the literature. For a position weight matrix, see Ref. 36.

TABLE 34.3 Expression of AHR in Immune Cells

Cell Subset (Source)[a]	RNA	Protein	References
T cells			
Splenic CD4+ *in vitro* differentiated			
T_h0 cells	–[b]	+/–	[39, 45–47]
T_h1 cells	+/–	–	
T_h2 cells	–	+/–	
T_h17 cells	+ + +	+ + +	
Inducible T_{reg} cells	+/–	+/–	
Human differentiated T_h17	+ + +	n.d.	[45]
CD4, CD8 splenic T cells, *ex vivo*	+	n.d.	
FoxP3+ regulatory T cells, *ex vivo*	+	+	[41, 48]
Dendritic epidermal T cells (a skin-specific subset of γδ T cells)	+ +	+ +	C. Esser, unpublished observations
CD8αα–γδ T cells (gut-specific subset)	+ +	n.d.	[49]
CD4+CCR6+ γδ T cells (a subset of IL-17 secreting γδ T cells isolated from lymph nodes after immunization with heat-killed *Mycobacterium*)	+ +	n.d.	[50]
CD4+CCR6–	–	n.d.	
Dendritic cells			
Langerhans cells (skin)	+ +	+ +	[11]
Human ex vivo LCs/in vitro from CD34+ progenitors generated LCs/monocytes/granulocytes/ CD34+ progenitors		+ +/+ +/+/–/ –/very weak	[51]
In vitro BM-derived CD11c+		+ +	[11]
CD11c+ cells from Flt3 ligand-treated mouse, sorted (spleen)		+	[52]
CD11c+ DCs (spleen)	+ +	n.d.	[41]
CD103+ DCs (mesenteric lymph node)	+	n.d.	[49]
B cells			
Splenic B cells		+/–	[52]
LPS- or IL-4-treated splenic B cells		+ +	[52]
Others			
Lin–Sca1– bone marrow progenitors	+ +		[53]
Keratinocytes	+ + +	+ + +	[11, 30]
Melanocytes	+ +	+ +	[31]
Intestinal epithelial cells	+ +	n.d.	[49]
Human MonoMac macrophage cell line		+	[54]
Human U937 monocytic cells		+	[38]
Human macrophages		+	[55]

[a] All cells are from mouse; human cells are in italics.
[b] Published RNA or protein levels offer relative levels, for example, in relation to housekeeping genes. To make data comparable, such a relative abundance is given as + + (high abundance in respect to the other measurements in the same paper), + (medium), +/– (little, but detectable), and – (not detectable). In our study, AHR levels measured by PCR or as protein correlated well. n.d. = not done.

34.8 AHR AND T-CELL SUBSETS: A BALANCING EXERCISE

Both dendritic cells and T cells are needed for an efficient immune response (Fig. 34.1). As evident from Table 34.3, subsets of both cell types are rich in AHR, and thus potential targets of AHR-activating ligands. Current evidence shows that AHR is relevant for efficient immune responses and has a special role in tilting the balance within effector T-cell subsets and of effector versus regulatory T cells. Indeed, some researchers have looked at the potential of AHR ligands to manipulate immune responses. In particular, two compounds with AHR binding capacity were studied. VAF347, a water-soluble ligand of AHR, was found to suppress allergies and could suppress graft rejection in mouse transplantation experiments [56]. The evidence presented so far suggests that shifts in T helper subset frequencies as well as effects on dendritic cells are relevant to the outcome [54]. Another ligand, a compound named M50367, has been tested in models of allergy [57]. The authors suggest that the compound-suppressed allergy is due to suppressed T_h2 differentiation from naïve T cells, presumably via inhibition of the

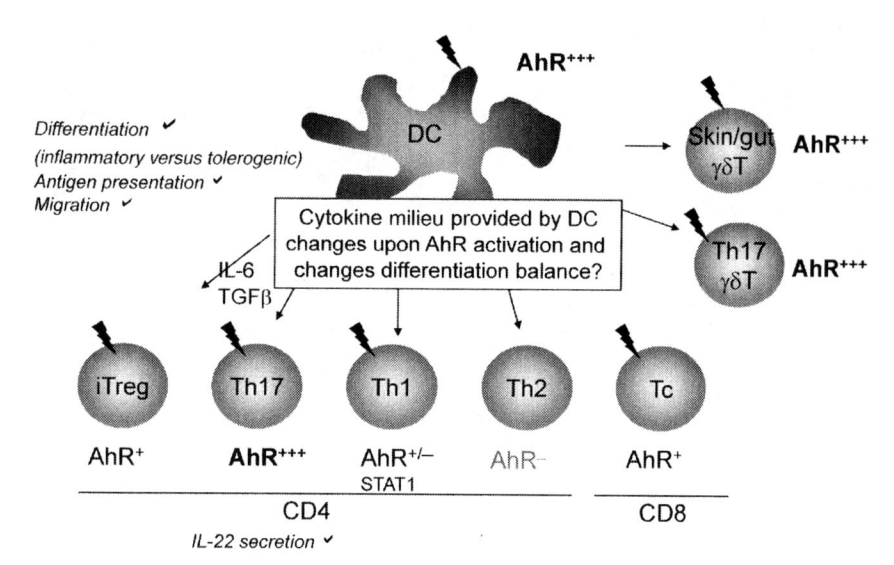

FIGURE 34.1 AHR involvement in DC–T-cell interaction. DCs can be affected by AHR-activating substances, resulting in impaired interaction with T cells. As a result, T-cell differentiation might differ. This could be a toxic event or a pathogen-driven event. T cells can become direct targets of AHR ligands as well, in particular those with high AHR expression. Arrows: effects of AHR ligands experimentally shown. For details and references, see the text.

T_h2-driving transcription factor GATA3 in naïve T cells. While their findings are intriguing, it must be pointed out that neither naïve T cells nor T_h1 express AHR at high levels, and that possibly dose-dependent cell subset-specific toxicity has not been rigorously addressed.

Generation of T cells with regulatory functions by strong and persistent AHR activation with TCDD has been described as well [9, 58, 59]. Possibly, FoxP3, the marker transcription factor for T_{reg}, is induced by AHR via the DREs in its promoter [48]. The frequency of FoxP3$^+$ regulatory is the same in wild-type mice as in AHR-deficient mice, indicating that AHR is not necessary for homeostasis of this cell subset [45]. The potential induction of T_{reg} by TCDD has been used experimentally to assess manipulation of autoimmune diseases. TCDD suppressed the induction of experimental autoimmune encephalitis in association with an expanded population of FoxP3$^+$ T_{reg} cells [48]. In another study, chronic treatment of NOD mice with TCDD potently suppressed the development of autoimmune type 1 diabetes and expanded regulatory T cells [60]. In experimental autoimmune uveoretinitis (an inflammation of the eye), activation of AHR by TCDD markedly suppressed autoimmune uveoretinitis through mechanisms that expand regulatory T cells and interfere with the activation of T_h1 and T_h17 cells [59]. While these results may be exciting, ligands are decisive in the outcome. Thus, using FICZ (a very labile high-affinity AHR ligand) rather than TCDD (a persistent high-affinity ligand) could completely turn tables on regulatory versus inflammatory response [45, 48]. Understanding ligand-specific interference with immune responses will be pivotal in understanding T effector cell misbalance, and thus for any therapeutic approach [62]. Especially one important aspect, which needs further evaluation, is the possibility of different susceptibilities against TCDD-induced cell death, and of parallel TCDD systemic effects that might be relevant. Also, ligand metabolites could be differentially cell toxic [63]. Moreover, it is self-evident that attempts to manipulate the immune system with compounds that bind to a transcription factor present also in many nonimmune cells, and whose systemic triggering can be highly toxic[3] must be viewed with the utmost caution. Obviously, TCDD cannot be considered a pharmacological compound. Research on nontoxic, "selective AHR modulators" is ongoing, and of high interest also in cancer research [64].

A well-defined role for the AHR in the immune system exists for T_h17 cells, a T helper cell subset discovered only a few years ago. *In vitro*-generated T_h17 cells express high levels of AHR [45] and AHR activation results in an expansion of T_h17 cells as well as increased cytokine production. T_h17 cells are linked to autoimmune diseases, albeit their true role is in fighting bacteria. In T_h17 cells, activation of AHR is a prerequisite for the production of the cytokine IL-22. IL-22 is linked to proinflammatory processes such as dermal inflammation, psoriasis, inflammatory bowel disease, and Crohn's disease. In addition, IL-22 induces the production of antimicrobial proteins that are needed for the defense against pathogens in the skin and gut [65, 66]. IL-22 is found strongly upregulated in skin lesions from psoriasis patients [67].

---

[3] The LD_{50} dose for TCDD in C57BL/6 mice is approximately 110 μg/kg body weight. Most studies regarding the immune system used single dose schemes of 10 μg/kg body weight. While mice did not die from this dose and had no overt signs of acute or long-term toxicity, possible subtle effects on organs beyond the IS usually were often not assessed in parallel. I do not use the term "toxicity" with a connotation of "acute" in my text.

T_h17 cells from AHR-deficient mice are unable to produce IL-22, and paracrine cytokines secreted under inflammatory conditions (e.g., IL-23 produced by dendritic cells) cannot overcome the effect [45]. FICZ increases the proportion of T_h17 cells and their production of IL-22 by activating the AHR [45], thus demonstrating a role for AHR in mounting an elevated response against infections. Activating the AHR with FICZ in a model of experimental autoimmune encephalitis (EAE) exacerbated the disease, and mice with a genetic variation of AHR that lowers its affinity for TCDD were reported to be resistant to EAE induction [48] and highly susceptible to infection with the yeast *Candida albicans* [68], both important features of T_h17 responses. This link between a transcription factor that responds to a wide range of environmental pollutants and the T_h17 program is intriguing. Since environmental factors clearly contribute to autoimmune disease manifestation, such as multiple sclerosis, lupus, or rheumatoid arthritis, further studies of AHR involvement should provide important clues to understand disease parameters such as relapses or induction.

34.9 AHR AND DENDRITIC CELLS: THE LINK TO INNATE IMMUNITY

Dendritic cells phagocytose antigen, present it to T cells, provide costimulatory signals, and secrete cytokines. Many subsets of dendritic cells are characterized by surface markers and—if known—by function. All subsets analyzed so far have strikingly high AHR levels, and they react to AHR activation by TCDD, benzo[*a*]pyrene, or FICZ. Immunotoxic outcomes observed were impaired maturation, functional differentiation, upregulation of costimulatory signals, and cytokine secretion [69–73]. The tolerogenic effect of the low molecular weight compound VAF347 not only is mediated via regulatory T cells, but also affects dendritic cells [56]; that is, transfer of splenic $CD11c^+$ dendritic cells from VAF347-tolerized hosts resulted in donor-specific graft acceptance in a transplantation model. The effect appears to be due to an inhibition of proinflammatory cytokine IL-6 secretion by the dendritic cells, which is AHR dependent [54]. In dendritic cells of the skin, the so-called Langerhans cells, we made an unexpected finding. Langerhans cells express high levels of AHR and need it for maturation and proper function, as evident from AHR-deficient mice. Congruent with the finding, AHR-deficient mice mounted significantly weaker contact hypersensitivity reactions, which need fully functioning Langerhans cells. Curiously, Langerhans cells expressed very high levels of the repressor of AHR, AHRR. Treatment with TCDD did not induce any genes of the AHR battery; indeed, the transcriptome remains largely inert to TCDD exposure. We have interpreted this finding to reflect a risk strategy of Langerhans cells against unwanted immune activation by potential skin allergens. The fact that AHR is needed nonetheless for proper maturation of cells is intriguing and suggests that AHR in these cells uses "noncanonical" pathways (see above), possibly the NF-κB pathway. AHR-deficient Langerhans cells have many genes differentially regulated from their wild-type counterparts. Particularly interesting was the blockade of IDO expression in AHR-deficient cells. As described above, IDO is a tolerogenic enzyme and catalyzes the first degradation step of tryptophan, generating kynurenine (an AHR ligand). Tryptophan is easily dimerized into FICZ by UV irradiation and light, and conceivably, IDO could help in preventing unhealthy FICZ levels in the skin. However, by now high AHRR levels have been found in other AHR expressing dendritic cells as well. This is still very much unclear, but an interesting aspect will be the interaction of AHR with the NF-κB pathway. Why this? It is known that activated AHR can bind to NF-κB subunits and activate NF-κB-responsive genes [74]. The signal pathway of NF-κB starts on the cell surface with Toll-like receptors (TLRs), that is, those receptors that sense pathogen structure such as LPS or CpG. This could reveal important links of AHR to pathogens. At least for a subset of IL-17 producing T cells[4] ($\gamma\delta TCR^+ CCR6^+$ cells), such a link was recently shown. These cells express Toll-like receptors 1 and 2 (in contrast to $\alpha\beta TCR^+$ T_h17 cells), expand in response to bacterial antigen, produce IL-23, and recruit neutrophils, thereby orchestrating an effective response [50]. Dendritic cells express Toll-like receptors as well, and it will be very interesting to dissect the AHR–NF-κB signaling as an adaptive response to TLR-recognized pathogens.

34.10 CONCLUSIONS

The AHR has several natural functions: it controls xenobiotic metabolism, it controls differentiation and lineage decisions, and it links immune responses to environmental cues. Toxic AHR overactivation by environmental pollutants remains of concern for public health and calls for action. However, new insights into AHR biology point to chances for pharmacological manipulation [75, 76]. AHR seems particularly relevant for the differentiation and balance of T-cell subsets in ongoing immune responses, and for the decision of the immune system to tolerate or fight antigens. Comparisons of wild-type and AHR-deficient mice indicate that AHR is

---

[4] The T-cell receptor (TCR) is a dimeric protein. Two TCR types are known, made from either an α and β chain or a γ and δ chain. A given T cell expresses either the αβ or the γδ dimer, never both. The genes for these four proteins are generated by a genetic process unique to the immune system, the so-called gene rearrangement. During this process, the final gene for each protein chain is pieced together from hundreds of variant gene segments to give an α, β, γ, or δ chain with a highly variable N-terminus that binds to antigen. Skin and gut residing γδTCR are special in that they have very little variation because they use only one or two of these variant gene segments.

indeed needed for constitutive transcription of many genes in the immune system [41]. The reasons are unclear but likely due to a networking of AHR-regulated genes with multiple other signaling pathways, such as the NF-κB, hypoxia, or the estrogenic pathway [77]. Albeit the AHR can in principle act in the absence of any endogenous ligand, it is likely that a relevant constitutive activation of the AHR signaling pathway exists, for example, by endogenous or food-derived ligands [42, 78, 79]. The extent and control of this is completely unexplored.

34.11 OUTLOOK

The AHR links the immune response to environmental factors and may help control the risk of developing adverse immune reactions. Further research will have to focus on the role of individual ligands in shaping these responses, elucidating the environmental risks for autoimmunity, allergy, and cancer, and how the AHR pathway can be exploited pharmacologically. Eventually, also the role of the AHR in the aging immune system with its lifelong experience of environmental insults—including AHR ligands in environment and food—will be of high interest. Currently, the bulk of our knowledge regarding AHR is won in mice and rats. Human epidemiological studies to verify links between AHR, immune dysfunctions, and AHR ligands will be needed. Finally, it will be very important to address functional consequences of "nontoxic" ligands, in particular, dietary components.

ACKNOWLEDGMENTS

The work of the author has been funded mainly by the Deutsche Forschungsgemeinschaft and the German Bundesministerium für Umwelt. I would particularly like to acknowledge the contributions by my Ph.D. and Diploma students whose scientific enthusiasm over the years was always gratifying.

REFERENCES

1. Fernandez-Salguero, P., Pineau, T., Hilbert, D. M., McPhail, T., Lee, S. S., Kimura, S., Nebert, D. W., Rudikoff, S., Ward, J. M., and Gonzalez, F. J. (1995). Immune system impairment and hepatic fibrosis in mice lacking the dioxin-binding Ah receptor. *Science*, 268, 722–726.
2. Mimura, J., Yamashita, K., Nakamura, K., Morita, M., Takagi, T. N., Nakao, K., Ema, M., Sogawa, K., Yasuda, M., Katsuki, M., and Fujii-Kuriyama, Y. (1997). Loss of teratogenic response to 2,3,7,8-tetrachlorodibenzo-p-dioxin (TCDD) in mice lacking the Ah (dioxin) receptor. *Genes to Cells*, 2, 645–654.
3. Schmidt, J. V., Su, G. H., Reddy, J. K., Simon, M. C., and Bradfield, C. A. (1996). Characterization of a murine Ahr null allele: involvement of the Ah receptor in hepatic growth and development. *Proceedings of the National Academy of Sciences of the United States of America*, 93, 6731–6736.
4. Esser, C. (2009). The immune system of Ahr null mutant mouse strains: not a simple mirror of xenobiotic receptor over-activation. *Biochemical Pharmacology*, 77, 597–607.
5. Lahvis, G. P. and Bradfield, C. A. (1998). Ahr null alleles: distinctive or different? *Biochemical Pharmacology*, 56, 781–787.
6. Schmidt, J. V., Su, G. H., Reddy, J. K., Simon, M. C., and Bradfield, C. A. (1996). Characterization of a murine Ahr null allele: involvement of the Ah receptor in hepatic growth and development. *Proceedings of the National Academy of Sciences of the United States of America*, 93, 6731–6736.
7. Vorderstrasse, B. A., Steppan, L. B., Silverstone, A. E., and Kerkvliet, N. I. (2001). Aryl hydrocarbon receptor-deficient mice generate normal immune responses to model antigens and are resistant to TCDD-induced immune suppression. *Toxicology and Applied Pharmacology*, 171, 157–164.
8. Rodriguez-Sosa, M., Elizondo, G., Lopez-Duran, R. M., Rivera, I., Gonzalez, F. J., and Vega, L. (2005). Over-production of IFN-gamma and IL-12 in AhR-null mice. *FEBS Letters*, 579, 6403–6410.
9. Funatake, C. J., Marshall, N. B., Steppan, L. B., Mourich, D. V., and Kerkvliet, N. I. (2005). Cutting edge: activation of the aryl hydrocarbon receptor by 2,3,7,8-tetrachlorodibenzo-p-dioxin generates a population of $CD4^+$ $CD25^+$ cells with characteristics of regulatory T cells. *Journal of Immunology*, 175, 4184–4188.
10. Funatake, C. J., Marshall, N. B., and Kerkvliet, N. I. (2008). 2,3,7,8-Tetrachlorodibenzo-p-dioxin alters the differentiation of alloreactive $CD8^+$ T cells toward a regulatory T cell phenotype by a mechanism that is dependent on aryl hydrocarbon receptor in $CD4^+$ T cells. *Journal of Immunotoxicology*, 5, 81–91.
11. Jux, B., Kadow, S., and Esser, C. (2009). Langerhans cell maturation and contact hypersensitivity are impaired in aryl hydrocarbon receptor-null mice. *Journal of Immunology*, 182, 6709–6717.
12. Jeon, M. S. and Esser, C. (2000). The murine IL-2 promoter contains distal regulatory elements responsive to the Ah receptor, a member of the evolutionarily conserved bHLH–PAS transcription factor family. *Journal of Immunology*, 165, 6975–6983.
13. Neff-LaFord, H. D., Vorderstrasse, B. A., and Lawrence, B. P. (2003). Fewer CTL, not enhanced NK cells, are sufficient for viral clearance from the lungs of immunocompromised mice. *Cellular Immunology*, 226, 54–64.
14. Sanchez, Y., Rosado, J. D., Vega, L., Elizondo, G., Estrada-Muniz, E., Saavedra, R., Juarez, I., and Rodriguez-Sosa, M. (2010). The unexpected role for the aryl hydrocarbon receptor on susceptibility to experimental toxoplasmosis. *Journal of Biomedicine and Biotechnology*, 2010, 505694.
15. Shi, L. Z., Faith, N. G., Nakayama, Y., Suresh, M., Steinberg, H., and Czuprynski, C. J. (2007). The aryl hydrocarbon receptor

is required for optimal resistance to *Listeria monocytogenes* infection in mice. *Journal of Immunology*, 179, 6952–6962.

16. Vorderstrasse, B. A. and Lawrence, B. P. (2006). Protection against lethal challenge with *Streptococcus pneumoniae* is conferred by aryl hydrocarbon receptor activation but is not associated with an enhanced inflammatory response. *Infection and Immunity*, 74, 5679–5686.

17. Fernandez-Salguero, P. M., Ward, J. M., Sundberg, J. P., and Gonzalez, F. J. (1997). Lesions of aryl-hydrocarbon receptor-deficient mice. *Veterinary Pathology*, 34, 605–614.

18. Kimura, A., Naka, T., Nakahama, T., Chinen, I., Masuda, K., Nohara, K., Fujii-Kuriyama, Y., and Kishimoto, T. (2009). Aryl hydrocarbon receptor in combination with Stat1 regulates LPS-induced inflammatory responses. *Journal of Experimental Medicine*, 206, 2027–2035.

19. Ward, J. M., Anver, M. R., Haines, D. C., Melhorn, J. M., Gorelick, P., Yan, L., and Fox, J. G. (1996). Inflammatory large bowel disease in immunodeficient mice naturally infected with *Helicobacter hepaticus*. *Laboratory Animal Science*, 46, 15–20.

20. Gennari, R., Alexander, J. W., Gianotti, L., Eaves-Pyles, T., and Hartmann, S. (1994). Granulocyte macrophage colony-stimulating factor improves survival in two models of gut-derived sepsis by improving gut barrier function and modulating bacterial clearance. *Annals of Surgery*, 220, 68–76.

21. Lawrence, B. P. and Vorderstrasse, B. A. (2004). Activation of the aryl hydrocarbon receptor diminishes the memory response to homotypic influenza virus infection but does not impair host resistance. *Toxicological Sciences*, 79, 304–314.

22. Jin, G. B., Moore, A. J., Head, J. L., Neumiller, J. J., and Lawrence, B. P. (2010). Aryl hydrocarbon receptor activation reduces dendritic cell function during influenza virus infection. *Toxicological Sciences*, 116, 514–522.

23. Head, J. L. and Lawrence, B. P. (2009). The aryl hydrocarbon receptor is a modulator of anti-viral immunity. *Biochemical Pharmacology*, 77, 642–653.

24. Bunger, M. K., Moran, S. M., Glover, E., Thomae, T. L., Lahvis, G. P., Lin, B. C., and Bradfield, C. A. (2003). Resistance to 2,3,7,8-tetrachlorodibenzo-p-dioxin toxicity and abnormal liver development in mice carrying a mutation in the nuclear localization sequence of the aryl hydrocarbon receptor. *Journal of Biological Chemistry*, 278, 17767–17774.

25. Kronenberg, S., Esser, C., and Carlberg, C. (2000). An aryl hydrocarbon receptor conformation acts as the functional core of nuclear dioxin signaling. *Nucleic Acids Research*, 28, 2286–2291.

26. Nguyen, L. P. and Bradfield, C. A. (2008). The search for endogenous activators of the aryl hydrocarbon receptor. *Chemical Research in Toxicology*, 21, 102–116.

27. Denison, M. S. and Nagy, S. R. (2003). Activation of the aryl hydrocarbon receptor by structurally diverse exogenous and endogenous chemicals. *Annual Review of Pharmacology and Toxicology*, 43, 309–334.

28. Rannug, A., Rannug, U., Rosenkranz, H. S., Winqvist, L., Westerholm, R., Agurell, E., and Grafstrom, A. K. (1987). Certain photooxidized derivatives of tryptophan bind with very high affinity to the Ah receptor and are likely to be endogenous signal substances. *Journal of Biological Chemistry*, 262, 15422–15427.

29. Oberg, M., Bergander, L., Håkansson, H., Rannug, U., and Rannug, A. (2005). Identification of the tryptophan photoproduct 6-formylindolo[3,2-b]carbazole, in cell culture medium, as a factor that controls the background aryl hydrocarbon receptor activity. *Toxicological Sciences*, 85, 935–943.

30. Fritsche, E., Schafer, C., Calles, C., Bernsmann, T., Bernshausen, T., Wurm, M., Hubenthal, U., Cline, J. E., Hajimiragha, H., Schroeder, P., Klotz, L. O., Rannug, A., Furst, P., Hanenberg, H., Abel, J., and Krutmann, J. (2007). Lightening up the UV response by identification of the arylhydrocarbon receptor as a cytoplasmatic target for ultraviolet B radiation. *Proceedings of the National Academy of Sciences of the United States of America*, 104, 8851–8856.

31. Jux, B., Kadow, S., Luecke, S., Rannug, A., Krutmann, J., and Esser, C. (2011). The aryl hydrocarbon receptor mediates UVB radiation-induced skin tanning. *Journal of Investigation Dermatology*. 131, 203–210.

32. Dinatale, B. C. and Perdew, G. H. (2010). Protein function analysis: rapid, cell-based siRNA-mediated ablation of endogenous expression with simultaneous ectopic replacement. *Cytotechnology*, 62, 95–100.

33. Rieber, N. and Belohradsky, B. H. (2010). AHR activation by tryptophan: pathogenic hallmark of Th17-mediated inflammation in eosinophilic fasciitis, eosinophilia-myalgia-syndrome and toxic oil syndrome? *Immunology Letters*, 128, 154–155.

34. Mellor, A. L. and Munn, D. H. (2004). IDO expression by dendritic cells: tolerance and tryptophan catabolism. *Nature Reviews Immunology*, 4, 762–774.

35. Frumento, G., Rotondo, R., Tonetti, M., Damonte, G., Benatti, U., and Ferrara, G. B. (2002). Tryptophan-derived catabolites are responsible for inhibition of T and natural killer cell proliferation induced by indoleamine 2,3-dioxygenase. *Journal of Experimental Medicine*, 196, 459–468.

36. Sun, Y. V., Boverhof, D. R., Burgoon, L. D., Fielden, M. R., and Zacharewski, T. R. (2004). Comparative analysis of dioxin response elements in human, mouse and rat genomic sequences. *Nucleic Acids Research*, 32, 4512–2453.

37. Zordoky, B. N. and El-Kadi, A. O. (2009). Role of NF-kappaB in the regulation of cytochrome p450 enzymes. *Current Drug Metabolism*, 10, 164–178.

38. Vogel, C. F., Sciullo, E., Li, W., Wong, P., Lazennec, G., and Matsumura, F. (2007). RelB, a new partner of aryl hydrocarbon receptor-mediated transcription. *Molecular Endocrinology*, 21, 2941–2955.

39. Negishi, T., Kato, Y., Ooneda, O., Mimura, J., Takada, T., Mochizuki, H., Yamamoto, M., Fujii-Kuriyama, Y., and Furusako, S. (2005). Effects of aryl hydrocarbon receptor signaling on the modulation of TH1/TH2 balance. *Journal of Immunology*, 175, 7348–7356.

40. Lawrence, B. P., Denison, M. S., Novak, H., Vorderstrasse, B. A., Harrer, N., Neruda, W., Reichel, C., and Woisetschlager, M. (2008). Activation of the aryl hydrocarbon receptor is essential for mediating the anti-inflammatory effects of a novel low molecular weight compound. *Blood*, 112, 1158–1165.

41. Frericks, M., Temchura, V. V., Majora, M., Stutte, S., and Esser, C. (2006). Transcriptional signatures of immune cells in aryl hydrocarbon receptor (AHR)-proficient and AHR-deficient mice. *Biological Chemistry*, 387, 1219–1226.
42. Frericks, M., Meissner, M., and Esser, C. (2007). Microarray analysis of the AHR system: tissue-specific flexibility in signal and target genes. *Toxicology and Applied Pharmacology*, 220, 320–332.
43. Veldhoen, M., Hirota, K., Christensen, J., O'Garra, A., and Stockinger, B. (2009). Natural agonists for aryl hydrocarbon receptor in culture medium are essential for optimal differentiation of Th17 T cells. *Journal of Experimental Medicine*, 206, 43–49.
44. Hirabayashi, Y. and Inoue, T. (2009). Aryl hydrocarbon receptor biology and xenobiotic responses in hematopoietic progenitor cells. *Biochemical Pharmacology*, 77, 521–535.
45. Veldhoen, M., Hirota, K., Westendorf, A. M., Buer, J., Dumoutier, L., Renauld, J. C., and Stockinger, B. (2008). The aryl hydrocarbon receptor links TH17-cell-mediated autoimmunity to environmental toxins. *Nature*, 453, 106–109.
46. Alam, M. S., Maekawa, Y., Kitamura, A., Tanigaki, K., Yoshimoto, T., Kishihara, K., and Yasutomo, K. (2010). Notch signaling drives IL-22 secretion in CD4$^+$ T cells by stimulating the aryl hydrocarbon receptor. *Proceedings of the National Academy of Sciences of the United States of America*, 107, 5943–5948.
47. Kimura, A., Naka, T., Nohara, K., Fujii-Kuriyama, Y., and Kishimoto, T. (2008). Aryl hydrocarbon receptor regulates Stat1 activation and participates in the development of Th17 cells. *Proceedings of the National Academy of Sciences of the United States of America*, 105, 9721–9726.
48. Quintana, F. J., Basso, A. S., Iglesias, A. H., Korn, T., Farez, M. F., Bettelli, E., Caccamo, M., Oukka, M., and Weiner, H. L. (2008). Control of T_{reg} and T_H17 cell differentiation by the aryl hydrocarbon receptor. *Nature*, 453, 65–71.
49. Chmill, S., Kadow, S., Winter, M., Weighardt, H., and Esser, C. (2010). 2,3,7,8-Tetrachlorodibenzo-p-dioxin impairs stable establishment of oral tolerance in mice. *Toxicological Sciences*, 118, 98–107.
50. Martin, B., Hirota, K., Cua, D. J., Stockinger, B., and Veldhoen, M. (2009). Interleukin-17-producing gammadelta T cells selectively expand in response to pathogen products and environmental signals. *Immunity*, 31, 321–330.
51. Platzer, B., Richter, S., Kneidinger, D., Waltenberger, D., Woisetschlager, M., and Strobl, H. (2009). Aryl hydrocarbon receptor activation inhibits *in vitro* differentiation of human monocytes and Langerhans dendritic cells. *Journal of Immunology*, 183, 66–74.
52. Tanaka, G., Kanaji, S., Hirano, A., Arima, K., Shinagawa, A., Goda, C., Yasunaga, S., Ikizawa, K., Yanagihara, Y., Kubo, M., Kuriyama-Fujii, Y., Sugita, Y., Inokuchi, A., and Izuhara, K. (2005). Induction and activation of the aryl hydrocarbon receptor by IL-4 in B cells. *International Immunology*, 17, 797–805.
53. Singh, K. P., Casado, F. L., Opanashuk, L. A., and Gasiewicz, T. A. (2009). The aryl hydrocarbon receptor has a normal function in the regulation of hematopoietic and other stem/progenitor cell populations. *Biochemical Pharmacology*, 77, 577–587.
54. Lawrence, B. P., Denison, M. S., Novak, H., Vorderstrasse, B. A., Harrer, N., Neruda, W., Reichel, C., and Woisetschlager, M. (2008). Activation of the aryl hydrocarbon receptor is essential for mediating the anti-inflammatory effects of a novel low-molecular-weight compound. *Blood*, 112, 1158–1165.
55. Podechard, N., Lecureur, V., Le, F. E., Guenon, I., Sparfel, L., Gilot, D., Gordon, J. R., Lagente, V., and Fardel, O. (2008). Interleukin-8 induction by the environmental contaminant benzo(*a*)pyrene is aryl hydrocarbon receptor-dependent and leads to lung inflammation. *Toxicology Letters*, 177, 130–137.
56. Hauben, E., Gregori, S., Draghici, E., Migliavacca, B., Olivieri, S., Woisetschlager, M., and Roncarolo, M. G. (2008). Activation of the aryl hydrocarbon receptor promotes allograft-specific tolerance through direct and dendritic cell-mediated effects on regulatory T cells. *Blood*, 112, 1214–1222.
57. Negishi, T., Kato, Y., Ooneda, O., Mimura, J., Takada, T., Mochizuki, H., Yamamoto, M., Fujii-Kuriyama, Y., and Furusako, S. (2005). Effects of aryl hydrocarbon receptor signaling on the modulation of TH1/TH2 balance. *Journal of Immunology*, 175, 7348–7356.
58. Marshall, N. B., Vorachek, W. R., Steppan, L. B., Mourich, D. V., and Kerkvliet, N. I. (2008). Functional characterization and gene expression analysis of CD4$^+$ CD25$^+$ regulatory T cells generated in mice treated with 2,3,7,8-tetrachlorodibenzo-p-dioxin. *Journal of Immunology*, 181, 2382–2391.
59. Marshall, N. B. and Kerkvliet, N. I. (2010). Dioxin and immune regulation: emerging role of aryl hydrocarbon receptor in the generation of regulatory T cells. *Annals of the New York Academy of Sciences*, 1183, 25–37.
60. Kerkvliet, N. I., Steppan, L. B., Vorachek, W., Oda, S., Farrer, D., Wong, C. P., Pham, D., and Mourich, D. V. (2009). Activation of aryl hydrocarbon receptor by TCDD prevents diabetes in NOD mice and increases Foxp3 T cells in pancreatic lymph nodes. *Immunotherapy*, 1, 539–547.
61. Zhang, L., Ma, J., Takeuchi, M., Usui, Y., Hattori, T., Okunuki, Y., Yamakawa, N., Kezuka, T., Kuroda, M., and Goto, H. (2010). Suppression of experimental autoimmune uveoretinitis by inducing differentiation of regulatory T cells via activation of aryl hydrocarbon receptor. *Investigative Ophthalmology & Visual Science*, 51, 2109–2117.
62. Stevens, E. A. and Bradfield, C. A. (2008). Immunology: T cells hang in the balance. *Nature*, 453, 46–47.
63. Singh, N. P., Nagarkatti, M., and Nagarkatti, P. (2008). Primary peripheral T cells become susceptible to 2,3,7,8-tetrachlorodibenzo-p-dioxin-mediated apoptosis in vitro upon activation and in the presence of dendritic cells. *Molecular Pharmacology*, 73, 1722–1735.
64. Safe, S. and McDougal, A. (2002). Mechanism of action and development of selective aryl hydrocarbon receptor modulators for treatment of hormone-dependent cancers (review). *International Journal of Oncology*, 20, 1123–1128.

65. Aujla, S. J., Chan, Y. R., Zheng, M., Fei, M., Askew, D. J., Pociask, D. A., Reinhart, T. A., McAllister, F., Edeal, J., Gaus, K., Husain, S., Kreindler, J. L., Dubin, P. J., Pilewski, J. M., Myerburg, M. M., Mason, C. A., Iwakura, Y., and Kolls, J. K. (2008). IL-22 mediates mucosal host defense against Gram-negative bacterial pneumonia. *Nature Medicine*, *14*, 275–281.

66. Wolk, K., Witte, E., Wallace, E., Docke, W. D., Kunz, S., Asadullah, K., Volk, H. D., Sterry, W., and Sabat, R. (2006). IL-22 regulates the expression of genes responsible for antimicrobial defense, cellular differentiation, and mobility in keratinocytes: a potential role in psoriasis. *European Journal of Immunology*, *36*, 1309–1323.

67. Wolk, K., Kunz, S., Witte, E., Friedrich, M., Asadullah, K., and Sabat, R. (2004). IL-22 increases the innate immunity of tissues. *Immunity*, *21*, 241–254.

68. Ashman, R. B., Fulurija, A., and Papadimitriou, J. M. (1996). Strain-dependent differences in host response to *Candida albicans* infection in mice are related to organ susceptibility and infectious load. *Infection and Immunity*, *64*, 1866–1869.

69. Laupeze, B., Amiot, L., Sparfel, L., Le Ferrec, E., Fauchet, R., and Fardel, O. (2002). Polycyclic aromatic hydrocarbons affect functional differentiation and maturation of human monocyte-derived dendritic cells. *Journal of Immunology*, *168*, 2652–2658.

70. Vorderstrasse, B. A., Dearstyne, E. A., and Kerkvliet, N. I. (2003). Influence of 2,3,7,8-tetrachlorodibenzo-p-dioxin on the antigen-presenting activity of dendritic cells. *Toxicological Sciences*, *72*, 103–112.

71. Ruby, C. E., Leid, M., and Kerkvliet, N. I. (2002). 2,3,7,8-Tetrachlorodibenzo-p-dioxin suppresses tumor necrosis factor-alpha and anti-CD40-induced activation of NF-kappaB/Rel in dendritic cells: p50 homodimer activation is not affected. *Molecular Pharmacology*, *62*, 722–728.

72. Lee, J. A., Hwang, J. A., Sung, H. N., Jeon, C. H., Gill, B. C., Youn, H. J., and Park, J. H. (2007). 2,3,7,8-Tetrachlorodibenzo-p-dioxin modulates functional differentiation of mouse bone marrow-derived dendritic cells. Downregulation of RelB by 2,3,7,8-tetrachlorodibenzo-p-dioxin. *Toxicology Letters*, *173*, 31–40.

73. Bankoti, J., Burnett, A., Navarro, S., Miller, A. K., Rase, B., and Shepherd, D. M. (2010). Effects of TCDD on the fate of naive dendritic cells. *Toxicological Sciences*, *115*, 422–434.

74. Vogel, C. F., Sciullo, E., Li, W., Wong, P., Lazennec, G., and Matsumura, F. (2007). RelB, a new partner of aryl hydrocarbon receptor-mediated transcription. *Molecular Endocrinology*, *21*, 2941–2955.

75. Selgrade, M. K. (2007). Immunotoxicity: the risk is real. *Toxicological Sciences*, *100*, 328–332.

76. Hauben, E., Gregori, S., Draghici, E., Migliavacca, B., Olivieri, S., Woisetschlager, M., and Roncarolo, M. G. (2008). Activation of the aryl hydrocarbon receptor promotes allograft specific tolerance through direct- and DC-mediated effects on regulatory T cells. *Blood*, *112*, 1214–1222.

77. Frericks, M., Burgoon, L. D., Zacharewski, T. R., and Esser, C. (2008). Promoter analysis of TCDD-inducible genes in a thymic epithelial cell line indicates the potential for cell-specific transcription factor crosstalk in the AhR response. *Toxicology and Applied Pharmacology*, *232*, 268–279.

78. McMillan, B. J. and Bradfield, C. A. (2007). The aryl hydrocarbon receptor sans xenobiotics: endogenous function in genetic model systems. *Molecular Pharmacology*, *72*, 487–498.

79. Ito, S., Chen, C., Satoh, J., Yim, S., and Gonzalez, F. J. (2007). Dietary phytochemicals regulate whole-body CYP1A1 expression through an arylhydrocarbon receptor nuclear translocator-dependent system in gut. *Journal of Clinical Investigation*, *117*, 1940–1950.

35

AHR AND THE CIRCADIAN CLOCK

SHELLEY A. TISCHKAU

35.1 INTRODUCTION

Members of the protein family designated by amino acid sequences labeled Per–Arnt–Sim (PAS) domains are critical elements of homeostatic regulatory networks that mediate responsiveness to environmental change [1]. PAS domains are multifunctional structural motifs that allow protein–protein interactions among family members, typically forming heterodimeric transcription factors to affect the transcription of target genes. The PAS domains of certain family members will also bind small molecules, providing a mechanism to discriminate changes in the environment, such as oxygen saturation or the presence of toxins. Prototypical PAS domain-dependent pathways include the circadian clock network and metabolic regulation of the xenobiotic response through the aryl hydrocarbon receptor (AHR). Although these two pathways regulate diverse and seemingly unrelated biological functions, their dominance by PAS domain proteins evokes the possibility that these pathways are interconnected. PAS domain proteins are notoriously promiscuous, forming many different combinations of partnerships, which creates competition for binding partners and leads to commodious potential for crosstalk and interdependent regulation of downstream events [2–5].

Molecular regulation of biological timing on a daily basis requires the PAS domain-dependent activity of circadian locomotor output cycles kaput (CLOCK) or its homolog neuronal PAS-containing protein 2 (NPAS2) with its partner brain and muscle ARNT-like protein 1 (BMAL1), which together drive transcription of the PAS domain-containing *Period* (*Per*) genes. Production of PER proteins provides a mechanism of feedback to inhibit further activity of CLOCK/BMAL1, thereby defining the circadian cycle. Disruption or genetic deletion of these PAS proteins severely cripples the clockworks, leading to aberrant expression of core clock genes and to behavioral arrhythmicity [6]. More importantly, clock disruption has been associated with pathologies that include sleep disturbances and mental health issues, cancer, heart disease, and metabolic disorders, including diabetes [7–11]. Dioxin exposure results in a strikingly similar pathological profile. The considerable accordance in pathologies associated with circadian clock disruption and dioxin toxicity raises the intriguing possibility that the two pathways are interrelated.

The PAS domain proteins AHR and ARNT are well documented as central regulators of the biological response to exposure to xenobiotics, such as polycyclic aromatic hydrocarbons. Activation of AHR by these small molecules leads to PAS domain-dependent interactions of AHR with ARNT and activation of target genes, including the cytochrome P450 enzymes that metabolize the xenobiotics into metabolites and eventually allow their degradation by phase II detoxifying enzymes. Interestingly, several aspects of AHR/ARNT biology are impacted by the circadian clockworks. AHR and ARNT are expressed with diurnal variation consistent with regulation by the clock. There is significant sequence homology between ARNT and BMAL1, suggesting coregulation or perhaps even overlapping function. Finally, naturally occurring AHR ligands are produced by light-induced metabolism of tryptophan, suggesting a role for AHR in physiological modulation of circadian rhythms. This chapter, therefore, explores the significant, albeit underinvestigated interrelationships between the circadian clock and the AHR.

35.2 THE bHLH–PAS Domain Family

AHR is a member of the bHLH–PAS domain family of transcriptional regulators that typically mediate biological

---

The AH Receptor in Biology and Toxicology, First Edition. Edited by Raimo Pohjanvirta.
© 2012 John Wiley & Sons, Inc. Published 2012 by John Wiley & Sons, Inc.

responses to environmental change. Basic helix–loop–helix (bHLH) describes the structural motif of a DNA binding domain, constructed from two alpha helices with an intervening loop sequence. The larger of the alpha helices binds DNA. bHLH transcription factors typically bind to prototypic E-box enhancer sequences (CANNTG) in the promoters of genes. The bHLH–PAS proteins are a subgroup of the larger bHLH family. Similar to other bHLH family members, bHLH–PAS transcription factors bind E-box elements in the promoters of target genes to activate transcription in response to a stimulus. The PAS domain family of proteins is named for the first three proteins identified, the protein product of the *Drosophila* circadian clock gene, Period (PER), a component of the dioxin signaling pathway, human aryl hydrocarbon receptor nuclear transporter (ARNT), and a gene that regulates neuronal development in flies, Simple-minded (SIM). The PAS domain was originally defined based upon sequence homology in an ~275-amino acid region of these three proteins. Further analysis of this region revealed that two ~70-amino acid repeats, delineated as PAS A and PAS B, respectively, mediate protein function [1]. Sequence alignment has determined that PAS repeats occur in thousands of proteins throughout all kingdoms of life [12]. Typically, PAS domain-containing proteins act as environmental and/or physiological sensors to exact changes through protein–protein interactions and ultimately through alteration of gene expression. PAS A and PAS B repeats, however, have distinctly different roles in implementing PAS domain protein function.

Functional analysis supports the concept that PAS A repeats negotiate homotypic interactions between two PAS domain proteins. The importance of PAS A for dimerization is exemplified by deletion experiments examining AHR/ARNT interactions. Transcriptional activation downstream of dioxin signaling progresses after AHR binds ARNT, forming a heterodimeric transcription factor that effects gene regulation through *cis*-elements, known as dioxin response elements (DREs), located in the promoters of target genes [13–15]. After deletion of the PAS A domain from AHR, binding to ARNT, and ultimately transcriptional activation of target genes, is significantly attenuated [14, 16, 17]. Alternatively, AHR retains the ability to bind ARNT after deletion of the PAS B domain. Furthermore, the AHR repressor protein (AHRR), which inhibits the activity of AHR/ARNT, binds ARNT, despite lacking the PAS B domain [18, 19]. Thus, although PAS A is necessary and sufficient for protein–protein interactions among family members, whether PAS A determines the specificity of the interaction remains unknown.

Although PAS B can also promote homotypic interactions between PAS domain proteins, as observed in the interactions between HIF1A and ARNT [20, 21], PAS B also mediates interactions with other cellular proteins and small molecules. For example, the PAS B domain of AHR binds with heat shock protein 90 (hsp90) to hold unliganded AHR stably in the cytoplasm [22]. Ligand binding of AHR also occurs at least partially within the PAS B domain [1]. Heme or carbon monoxide binding to the PAS B domain of NPAS2 alters NPAS2/BMAL1 complexes leading to their inactivation [23, 24]. PAS B is required for the oxygen-sensing activity of HIF1a [24]. Thus, the PAS B domain may confer specificity of interaction with other family members, as well as function, including targeting to particular promoters.

Dioxin signaling through the AHR is a prototypical example of how bHLH–PAS domain proteins act as sensors to allow organisms to mount a response to environmental change. AHR binds to a variety of small molecules, including those of physiological origin, but most notably AHR mediates the biological response to xenobiotics produced as byproducts of industrial processes. Ligand binding, primarily to the PAS B repeat, reveals a nuclear localization signal. After phosphorylation by protein kinase C, AHR translocates into the nucleus and is released from hsp90, p23, and Ara9 [14, 16, 25]. The PAS A repeat of nuclear AHR binds ARNT, forming a heterodimeric transcription factor that binds specifically to DREs, recruits a number of coactivators, including the general transcriptional machinery associated with RNA polymerase II, and activates target genes [13, 26, 27]. Target genes include *CYP1A1*, *CYP1A2*, and *CYP1B1* phase 1 metabolic enzymes, the PAS domain protein, AHRR, and elements of the circadian clock. Thus, PAS domain proteins are important regulators of xenobiotic metabolism. PAS domains mediate interactions between (1) AHR and its ligands, (2) AHR and ARNT, and (3) ARNT and AHRR. Furthermore, recent evidence indicates that at least one core component of the circadian clock gene network, the *Per1* gene, is regulated by interactions of AHR with another clock component and PAS domain protein, BMAL1, suggesting a functional link between the xenobiotic response pathway and the circadian clock [28–30]. This chapter explores the functional connections between these PAS domain-dominated regulatory networks.

35.3 MOLECULAR CONTROL OF CIRCADIAN RHYTHMS

Perhaps the most consistent environmental factor present throughout the course of evolution of complex organisms is the 24 h pattern of light and darkness caused by the rotation of the Earth upon its axis, which is defined as a day. The importance of this environmental background is underscored by the inclusion of a genetic program that measures time on a near 24 h (circadian) scale in organisms ranging from cyanobacteria to mammals, including humans. Thus, organisms call upon this molecular program to enable them to sustain circadian rhythms in the absence of environmental cues. This molecular circadian "clock" drives daily cycles of behavior, physiology, endocrinology, and metabolism to provide organisms with adaptive mechanisms beneficial for survival. Molecular circadian clock mechanisms are present in nearly all cells. In mammals, however, individual cellular clocks are synchronized to each other and to solar time (light–dark cycles) by a "master clock" in the suprachiasmatic nuclei (SCN) of the anterior hypothalamus [31, 32]. As its name suggests, the SCN lie nestled in the optic chiasm and receive direct input from a specific subset of photosensitive retinal ganglion cells via the retinohypothalamic tract [33]. The SCN detects levels of illumination in the environment and then transmits this time-of-day information to coordinate clocks throughout the body, thereby synchronizing processes in the body with each other and the environment. Thus, peripheral clocks maintain normal homeostatic 24 h rhythms in vigilance, temperature, autonomic function, hormone secretion, metabolism, and cell division through direction of the SCN [31, 32]. The importance of homeostatic regulation of systemic circadian processes to maintenance of health is only beginning to be understood. Clock disruption brought on by shift work in human populations predisposes the affected to numerous health risks, including increased risk of breast and prostate cancers, increased risk of cardiovascular disease, and increased incidence of type 2 diabetes [7–11]. This chapter explores the intersection of the AHR and its signaling pathways with this molecular clock.

Cellular circadian rhythms are generated by self-sustaining, interlocking transcription–translation feedback loops regulated by another group of PAS domain-containing proteins, known as clock genes (Fig. 35.1) [33]. The PAS A domains of *Clock* (or the clock paralog *Npas2* [34]) and *Bmal1* proteins form a heterodimeric transcription factor that recruits numerous coactivators and drives the expression of a large number of target genes that contain E-box *cis*-regulatory enhancer sequences [35]. Within the core clock, the positive elements, CLOCK and BMAL1 proteins, are central to rhythmicity and act by driving transcription of the repressor genes *Period* (*Per1*, *Per2*, *Per3*) and *Cryptochrome* (*Cry1*, *Cry2*) during the day. PER/CRY proteins accumulate after a 6 h delay, and ultimately feed back to repress their own transcription by acting on the CLOCK/BMAL1 complex. Posttranslational modifications of PER and CRY proteins are important in regulating their feedback functions. Casein kinases 1 delta and 1 epsilon phosphorylate PER and CRY progressively throughout the early night, ultimately targeting these proteins for polyubiquitination and degradation by the 26S proteosome [36–40]. Degradation of PER and CRY proteins during the night [41] releases transcriptional repression and allows the cycle to begin anew. This primary cycle takes approximately 24 h to complete, thus defining the circadian period.

The circadian clock is self-regulating and stable, yet adaptive to environmental change. An accessory loop that involves the orphan nuclear receptor gene *Rev-erbα* and the gene *Rora* (retinoic acid receptor-related orphan receptor alpha) stabilizes the core loop through coordinated regulation of *Bmal1* [42–44]. CLOCK/BMAL heterodimers activate *Rev-erbα*, which subsequently feeds back to repress *Bmal1* transcription. *Rora*, which is necessary for *Bmal1* transcription and normal locomotor rhythms, competes with

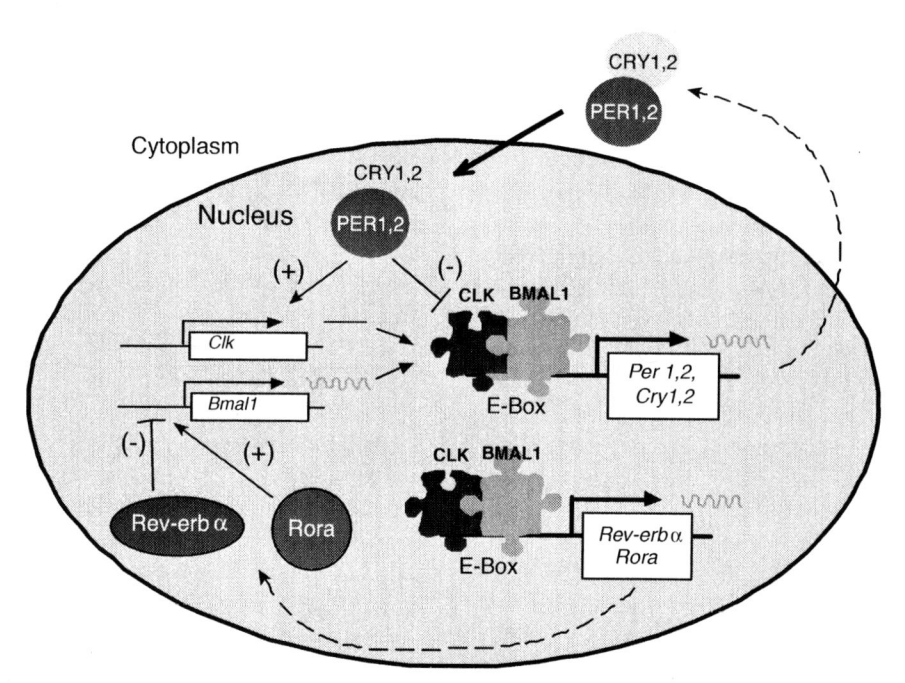

FIGURE 35.1 Schematic of the molecular circadian clockworks.

Rev-erbα for the same promoter element and acts at this site to promote *Bmal1* transcription [45, 46]. Genetic deletion of either *Rev-erbα* or *Rora* disturbs the rhythmic expression of *Bmal1* and ultimately destabilizes the primary clockworks, as evidenced by shortened periodicity and highly variable circadian output [44–46].

It is thus readily apparent that, similar to the mechanisms underlying xenobiotic metabolism, the circadian clock regulatory network relies heavily on PAS domain proteins to effectuate its vital homeostatic function. The primary loop of the clock is driven by PAS domain-dependent interactions of CLOCK (or NPAS2) with BMAL1, whereas negative feedback relies on actions of the PAS protein Per. The remainder of this chapter explores crosstalk between these two pathways.

35.4 CLOCK-CONTROLLED REGULATION OF AHR SIGNALING

Approximately 10% of all mRNA transcripts oscillate in a circadian manner in the primary clock in the SCN, as well as in most peripheral tissues, but less than 50 of these oscillating mRNAs are common among all tissues [47–49]. Although the literature is sparse regarding studies that examine circadian rhythmicity in AHR signaling, a few studies do indicate that certain components of the AHR pathway, including AHR, ARNT, and CYP1A1, are among the transcripts and proteins that vary across a 24 h period [29, 50–52]. When animals are maintained on a light/dark cycle, AHR transcripts vary across the day in both the SCN and peripheral tissues, with peak levels observed at the time of lights off (zeitgeber time 12, or ZT 12, where ZT 0 is defined as the time of lights on in the animal colony) in SCN and 4 h earlier in liver, at ZT 8 [29]. Daily oscillations persist when mice are placed in constant darkness, although the time of peak shifted; peak levels were observed at the time of lights on (circadian time 0, or CT 0, defined with regard to the previous light/dark cycle in this case) in the liver and at CT 8 in the SCN [29]. These data are consistent with regulation of AHR by the molecular circadian clockworks in both SCN and liver. The amplitude of the AHR transcript oscillation was maintained under constant conditions in the liver, but was dampened somewhat in the SCN. Interestingly, this same amplitude suppression was observed for the BMAL1 transcript in liver, suggesting that genes may be coregulated [29].

The AHR target gene, *Cyp1a1*, was also expressed with circadian variation, albeit out of phase with *AHR* (4–8 h delay), which implies that activation of the AHR signaling pathway is under control of the circadian clock [29]. Diurnal expression of AHR protein levels is consistent with the possibility that there are circadian changes in the availability of AHR for activation [50, 51]. Furthermore, similar to AHR, the amplitude of CYP1A1 expression in the liver was preserved in constant darkness and the time of peak levels for both AHR and CYP1A1 was advanced. The effects on time of peak may reflect a shortening of the circadian period under constant conditions. It is also intriguing to speculate that light may stimulate an endogenous AHR ligand to activate AHR signaling and thus cause attenuation of the amplitude of the circadian rhythm of AHR signaling under conditions where light is present [30] (see below).

Changes in AHR and ARNT over the course of the day imply that physiological as well as pharmacological and toxicological sequelae associated with AHR signaling are also regulated by the circadian clock. Circadian changes in endogenous CYP1A1 transcripts support the concept that activation of AHR is rhythmic under physiological conditions [29]. Induction of the AHR target genes *Cyp1a1* and *Cyp1b1* in mouse mammary glands and liver in response to treatment with the prototypical AHR ligand, TCDD, is also temporally regulated [53, 54]. *Cyp1a1* and *Cyp1b1* transcript levels were low in untreated animals, although a small diurnal variation was observed with an increase in basal expression during the day. Furthermore, both genes were significantly induced by TCDD treatment regardless of the time of day when TCDD was administered. The response of both genes to TCDD was, however, substantially greater when TCDD was given during the night [53, 54]. The most convincing evidence that the diurnal changes in activity of the AHR signaling pathway are controlled by the circadian clock derives from studies in transgenic mice with genetic alteration of the *Per* genes. Diurnal variation in induction of *Cyp1a1* in response to TCDD was abolished in mice bearing targeted disruption of *Per1* and/or *Per2* [54]. Whether these effects are specifically due to interactions between Per1/2 and AHR signaling components, or are more generally attributable to the disrupted clock function in these mice, remains an open question.

35.5 EFFECTS OF AHR ON ENDOGENOUS RHYTHMICITY

Despite multifarious research efforts, the physiological function of AHR remains largely unknown. Structural similarity to established circadian clock elements, particularly BMAL1, has led to speculation that AHR and/or ARNT may be involved in the regulation of circadian rhythmicity. The bHLH and PAS domains of BMAL1 are most closely related to ARNT [55]. Genomic organization, including the intron/exon splice pattern and a unique conservation of five exons comprising the PAS domain, fosters the perception that BMAL1 and AHR are common derivatives of the same ancestral gene [56]. Genetic deletion of BMAL1 provokes a more severe disruption of circadian rhythmicity than any other single clock gene mutation [57]; BMAL1-null mice display an immediate and complete loss of rhythmicity upon placement in constant darkness, underscored by altered expression of core clock genes. The formidable sequence homology between ARNT and BMAL1 and the conservation of genomic organization between AHR and BMAL1 led to the hypothesis that AHR is a constituent of the network of genes that regulate circadian rhythms.

A potential role for AHR in endogenous regulation of rhythmicity has been explored by investigating the overt circadian phenotype in mice bearing a genetic deletion of the AHR [29]. Initial studies indicated that AHR-null mice have relatively normal behavioral circadian rhythms. $AHR^{-/-}$ animals readily entrained to a light/dark cycle reset their clock in response to both phase delaying and phase advancing "jet lag"-type stimuli and maintained a free-running period that was, on average, not significantly different from their wild-type littermates [29]. Further analysis of the behavior of these animals has revealed some deficits in circadian behavior (S. Tischkau, unpublished observations). $AHR^{-/-}$ mice experience an increase in daytime activity and less well-defined activity onset compared to wild-type mice. Increased interanimal variability, which is commonly observed after deletion of other genes associated with clock stability, has also been observed [45, 46]. Finally, $AHR^{-/-}$ mice display an altered pattern of activity at night. As is common for the C57BL/6 strain, $AHR^{-/-}$ mice have a period of lesser activity or inactivity during the night, which is often referred to as a "siesta" or "nap." The timing of this "nap" is altered in the AHR-null mice. $AHR^{-/-}$ mice display a nadir in nocturnal activity that occurs approximately 1 h earlier than in their wild-type counterparts (S. Tischkau, unpublished observations). Interestingly, this is similar to a previous study where B6.D2NAahrd/j mice, which are genetically altered in the region that includes the AHR locus on chromosome 12, also displayed a delayed nocturnal "nap" time [58]. Collectively, these data indicate a role for AHR in behavioral rhythms that is supportive in nature, not unlike the functions of other orphan nuclear receptors such as rev-erbα or Rora [45, 46].

Activation of AHR is an alternative strategy to explore its role in regulation of endogenous circadian rhythmicity. TCDD exposure reportedly alters circadian rhythms, feeding behavior, locomotor activity, hormone levels, and gene expression [28, 59–65]. AHR and ARNT are expressed within specific nuclei of the hypothalamus known to control feeding, hormone secretion, and circadian rhythms and TCDD treatment alters hypothalamic gene expression, including an induction of the known AHR target gene, CYP1A1 [66–70].

With respect to feeding behavior, it is well established that high doses of TCDD cause hypophagia leading to wasting syndrome and eventually to death [71–74]. More relevant, however, to the topic of this chapter are the effects of TCDD on food intake patterns [63]. After exposure to a large dose of TCDD (1000 μg/kg), food intake patterns in TCDD-resistant H/W rats (see Chapter 12) were altered such that animals consumed approximately 50% of their daily feed during the first 5 h of the morning. Control animals, in contrast, consumed only 6% during this same time, displaying the expected diurnal rhythmicity of food intake by eating the majority of their food during the night, when they are naturally active. Although this study did not explore changes in locomotor activity, which, if altered, might cause the observed changes in feeding, these results definitively demonstrate that TCDD exposure attenuates the normal diurnal rhythm in feeding behavior; treated animals seemed to be unable to identify the appropriate time to eat. The mechanism for this change remains to be determined.

Plasma levels of numerous hormones, including prolactin, corticosterone, thyroid hormones, and melatonin, are controlled by the circadian clock. Corticosterone, which is usually elevated mid-morning in rodents, is rendered arrhythmic after TCDD exposure [61]. Similarly, the prolactin rhythm is eliminated and thyroid hormone levels are significantly attenuated. Finally, the melatonin peak is also altered by TCDD treatment; TCDD suppresses the nocturnal melatonin peak in both TCDD-sensitive L-E and TCDD-resistant H/W rats [64, 75]. Thus, a number of different hormones are similarly affected by TCDD, and the common denominator is that rhythmicity of each of these factors is controlled by the circadian clock in the SCN. Although it is tempting to speculate that alteration of hormone patterns by TCDD occurs subsequent to direct interactions of TCDD with the clockworks in the SCN, further mechanistic studies are required.

A few studies have begun to explore the direct effects of AHR activation on the master clock in the SCN. High doses of TCDD reportedly cause splitting of the rhythm and/or behavioral arrhythmicity in mice [76]. Deer mice treated with a low dose (100 ng/kg body weight) of TCDD experienced altered circadian locomotor activity rhythms. TCDD caused a phase advance of the activity rhythm; treated mice became active at an earlier time than would be predicted from their endogenous behavioral pattern prior to exposure [65]. Changes in behavioral rhythms were associated with a concomitant divergence in the expression pattern of PER1 protein in the SCN of these mice. Rhythms of clock gene transcripts, specifically *Per1* and *Bmal1*, are also perturbed in the SCN of TCDD-exposed mice [29]. Thus, evidence is mounting that AHR activation leads to alteration of the primary circadian clock, ultimately causing phase alterations and even arrhythmicity in key circadian-regulated behaviors.

Perhaps more important than the consequences of AHR activation on the master clock function of the SCN are the effects of AHR on circadian rhythmicity in peripheral tissues. Among myriad pathological manifestations associated with dioxin exposure are cancer; immune deficiencies; endocrine disruption, including reproductive abnormalities and thyroid disorders; diabetes and metabolic syndrome; liver damage; skin rashes; insomnia; and fatigue (reviewed in Ref. 77). Similarly, circadian clock disruption is associated with sleep disturbances, metabolic disorders, cardiovascular disease, and cancer [78–81]. Given that AHR/ARNT and many of the core circadian clock elements are PAS proteins capable of promiscuity of interaction, it seems reasonable to pose the hypothesis that clock disruption might be a common feature that underlies disease processes consequent to dioxin exposure. The literature on dysregulation of circadian rhythms in peripheral tissues after AHR activation is limited. However, one study examined the effects of TCDD treatment on proliferation and differentiation of hematopoietic progenitor cells, a process that is directly under the control of the circadian clock [59]. AHR activation subsequent to TCDD administration significantly altered the phase, period, and amplitude of circadian rhythms in a number of antigenically and functionally defined hematopoietic progenitor cell classes in bone marrow. TCDD-induced changes were associated with dampening of the rhythms of the clock genes *Per1* and *Per2* within these same cell populations, which may imply a direct crosstalk between AHR signaling and the core circadian clock [59].

A recent microarray analysis indicated that TCDD treatment alters the expression of genes associated with cholesterol and fatty acid synthesis, glucose metabolism, and circadian rhythm in the liver [82]. Many of the same genes affected by TCDD treatment are controlled physiologically by the circadian clock. The gene that encodes 3-hydroxy-3-methylglutaryl-coenzyme A reductase (HMGCR), the rate-limiting enzyme in the synthesis of cholesterol, and the sterol element binding protein (SREBP), which is important in the regulation of sterol synthesis, are both expressed with circadian periodicity under physiological conditions and are downregulated by TCDD treatment [82–86]. Glucokinase gene expression and activity in the liver is also regulated by the circadian clock [82, 87]; TCDD treatment reduced glucokinase expression, which contributes to an elevation in blood glucose, which may explain the increased incidence of type 2 diabetes in human populations exposed to high levels of dioxins [77]. A causal relationship between TCDD-induced downregulation of genes and enzymes associated with critical metabolic processes and the circadian clock has only begun to be explored. TCDD does alter expression of core clock genes [28, 29, 82]. The Per1 rhythm in the liver is altered by low-dose TCDD treatment; time of peak expression is significantly delayed in livers of TCDD-treated animals [28]. Perhaps more importantly, however, TCDD significantly attenuates the amplitude of the Per1 rhythm [28], making a compelling case that constitutive activation of AHR by TCDD severely weakens the fidelity of the clockworks in the liver. Although research in this area remains in its infancy, the limited amount of data available suggests that interactions of AHR signaling with the circadian clockworks may provide an alternative mechanism to explain the pathological effects on metabolism that are associated with exposure to toxic or even subtoxic levels of dioxins.

35.6 CROSSTALK BETWEEN AHR SIGNALING AND LIGHT SIGNALING IN THE CIRCADIAN CLOCK

An essential function of the primary circadian clock in the SCN is the ability to adjust to changes in the timing of the onset of light and darkness. Abrupt shifts in the light/dark cycle (such as the one occurring when humans fly across time zones) are perceived by the SCN, which is responsible ultimately for readjusting the rhythms of the entire body to the new light/dark cycle. A strong argument can be made that perception of the phase of the environmental light/dark cycle is the most important function of the SCN. When light occurs at night, the SCN recognizes the presence of illumination as an error signal, and subsequently sets in motion a series of events to rectify the error. Light in the early night is perceived as an extension of daytime; thus, the SCN clock will respond by delaying the rhythm the next day. Alternatively, light in the late night is interpreted by the SCN as an early beginning to the forthcoming day. The clock responds to late night light by moving ahead, or phase advancing, to the ensuing day. The mechanism by which the clock resets itself in response to changes in environmental light has been the subject of intense investigation in the field of biological rhythms [88–93]. Essentially, phase resetting of the SCN clock in response to light proceeds through activation of NMDA-type glutamate receptors on retinorecipient SCN neurons, leading to an influx of calcium, activation of mitogen-activated protein kinases and CREB, and ultimately to induction of the clock gene *Per1* [88, 94–103]. Although glutamate is essential as the primary neurotransmitter that mediates the effect of light on the SCN, numerous other signals can interact with the glutamatergic input to modulate the quality of the light signal. For example, pituitary adenylate cyclase activating peptide (PACAP) interacts with glutamate to increase the magnitude of phase delays and to attenuate the intensity of phase advances [104]. Serotonin, or simply injection of the serotonin precursor, tryptophan, also acts to temper the magnitude of light- and glutamate-induced responses in the SCN [105–107].

A role for interaction of AHR signaling with light/glutamate responses in the SCN is predicted from the recent discovery that photoproducts of tryptophan have high affinity for the AHR and may, in fact, fit the criteria as the elusive physiological ligand of this orphan receptor [108–110]. Ultraviolet light metabolizes tryptophan, thereby producing a number of compounds with affinity for AHR, including the high-affinity ligand 6-formylindolo[3,2-*b*]carbazole (FICZ).

But does AHR activation with FICZ or other ligands affect light signaling to the SCN? Application of FICZ directly to SCN brain slices *in vitro* had no effects on the rhythmic firing of SCN neurons. However, similar to the effects of serotonin, FICZ blocks the glutamate-induced phase resetting of the SCN clock [30]. In this same investigation, FICZ did affect the expression of clock genes, including *Per1* in SCN neurons. Systemic injection of FICZ did not, however, affect light-induced phase resetting of locomotor activity rhythms. In contrast, light-induced phase resetting was substantially attenuated in animals exposed to TCDD. The discrepancy between these experiments may be technical in nature. It is not clear that FICZ could have reached the SCN in concentrations sufficient to cause a response, owing to either the dose used or the presence of the blood–brain barrier. Thus, further studies are required to substantiate the results. Nevertheless, the fact that phase resetting effects of light were attenuated after TCDD exposure *in vivo* or after application of FICZ directly onto the SCN *in vitro* provides compelling evidence for crosstalk between AHR signaling and the circadian clock.

35.7 MOLECULAR INTERACTIONS OF AHR AND CIRCADIAN CLOCK COMPONENTS

The molecular mechanisms that govern interactions between AHR signaling and the circadian clockworks remain largely unknown. Protein–protein interactions regulated by PAS domain-containing proteins, however, dominate both pathways [1]. Regulation of both circadian rhythmicity and AHR signaling requires transcriptional regulation executed by heterodimeric transcription factors formed via interactions of their PAS domains. PAS domain proteins are notoriously promiscuous. Heterodimers can form between many alternative family members, providing ample opportunity for diverse functionality and crosstalk. Not surprisingly given the indiscriminate nature of PAS domain interactions, it seems that the two pathways are intertwined in a relatively complex manner. Activation of AHR leads to alteration of the clockworks [28–30, 59], and reciprocally, the clock modulates the activity of AHR [53, 54, 111].

Rhythmic expression of *Per1* is controlled under physiological conditions through regulation of the CLOCK/BMAL1 heterodimeric transcription factor. CLOCK/BMAL1 drives transcription of *Per1*, and the PER1 protein then feeds back to inhibit the activity of CLOCK/BMAL1 in a typical negative feedback fashion [1]. Upon activation, AHR enters the nucleus where it can form a heterodimer with BMAL1. The AHR/BMAL1 heterodimer acts as a transcriptional repressor by blocking the activity of CLOCK/BMAL1 directly on the *Per1* promoter [28]. Inhibition of CLOCK/BMAL1 activity impedes the production of PER1 (Fig. 35.2). Difficulty in producing PER1 will dampen the robustness of the rhythm and slow the progression of the clock. PER1 rhythms in liver are attenuated and phase delayed after TCDD exposure [28], which is compelling evidence that AHR slows and weakens circadian rhythmicity.

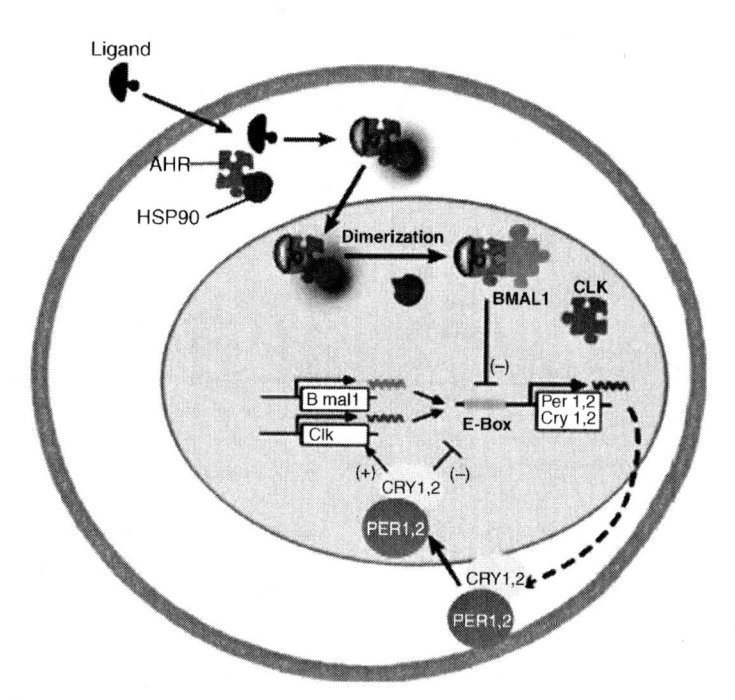

FIGURE 35.2 Working model for AHR interactions with the circadian clockworks.

Conversely, disruption of the clockworks, specifically in the presence of aberrant PER1 expression, precipitates changes in AHR signaling. Activity of the AHR target genes *Cyp1a1* and *Cyp1b1* is significantly enhanced after *Per1* knockdown using siRNA, or in animals bearing mutated *Per1* genes [53, 54, 111]. Although the molecular events underlying the heightened AHR activity remain unknown, it is tempting to speculate that PER1 may physically interact with the AHR/ARNT heterodimer to function as a transcriptional repressor, similar to the effects of PER1 on CLOCK/BMAL1. Collectively, these studies definitely demonstrate reciprocal interactions between AHR signaling components and the core circadian clockworks.

35.8 CONCLUSIONS

An ability to sense and adapt to changes in the external environment is critical to the success of any species. Perhaps the most persistent environmental factor under which life has evolved on this planet is the rotation of the Earth on its own axis creating a day that is consistently 24 h in length. The value of the propensity to measure time on a daily scale is reflected in the incorporation of circadian rhythmicity into the genetic architecture of nearly all organisms. This internal clock not only generates and maintains rhythms in the absence of external stimuli, but also responds to specific cues that allow synchronization with the solar cycle. Competency in mounting a robust defense in the face of exposure to environmental toxins is similarly adaptive and critical to an organism's survival. The bHLH–PAS domain family of proteins is central to regulation of each of these critical biological processes. Until recently, however, the possibility that the circadian clock regulatory pathway might intersect with AHR signaling elements has received little attention. Recent studies that demonstrate the importance of circadian clock fidelity in health and well-being highlight the considerable congruence of pathologies associated with either clock disruption or dioxin toxicity. This chapter has summarized compelling evidence that these two pathways intersect in a complex and reciprocal manner. Components of the AHR signaling pathway are expressed in a diurnal pattern. Moreover, activation of AHR pathway is influenced by the circadian clock; sensitivity of the pathway to activation by high-affinity agonists varies across the day. Conversely, both activation and downregulation of the AHR pathway alter the normal rhythmicity. In both cases, the clock gene, *Per1*, appears to be a central regulator. Further studies are clearly necessary to reveal the molecular interactions associated with the convergence of these two integral signaling pathways. A better understanding of the crosstalk between these pathways will provide insight into the physiological role of the AHR, as well as provide novel insight into the pathological events associated with dioxin exposure.

REFERENCES

1. McIntosh, B. E., Hogenesch, J. B., and Bradfield, C. A. (2010). Mammalian Per–Arnt–Sim proteins in environmental adaptation. *Annual Review of Physiology*, 72, 625–645.
2. Chilov, D., Hofer, T., Bauer, C., Wenger, R. H., and Gassmann, M. (2001). Hypoxia affects expression of circadian genes PER1 and CLOCK in mouse brain. *FASEB Journal*, 15, 2613–2622.
3. Hogenesch, J. B., Gu, Y. Z., Jain, S., and Bradfield, C. A. (1998). The basic-helix–loop–helix–PAS orphan MOP3 forms transcriptionally active complexes with circadian and hypoxia factors. *Proceedings of the National Academy of Sciences of the United States of America*, 95, 5474–5479.
4. Moffett, P., Reece, M., and Pelletier, J. (1997). The murine Sim-2 gene product inhibits transcription by active repression and functional interference. *Molecular and Cellular Biology*, 17, 4933–4947.
5. Woods, S. L. and Whitelaw, M. L. (2002). Differential activities of murine single minded 1 (SIM1) and SIM2 on a hypoxic response element. Cross-talk between basic helix–loop–helix/per–Arnt–Sim homology transcription factors. *Journal of Biological Chemistry*, 277, 10236–10243.
6. Reppert, S. M. and Weaver, D. R. (2001). Molecular analysis of circadian rhythms. *Annual Review of Physiology*, 63, 647–676.
7. Davis, S., Mirick, D. K., and Stevens, R. G. (2001). Night shift work, light at night, and risk of breast cancer. *Journal of the National Cancer Institute*, 93, 1557–1562.
8. Copertaro, A., Bracci, M., Barbaresi, M., and Santarelli, L. (2008). Assessment of cardiovascular risk in shift healthcare workers. *European Journal of Cardiovascular Preventative Rehabilitation*, 15, 224–229.
9. Chalernvanichakorn, T., Sithisarankul, P., and Hiransuthikul, N. (2008). Shift work and type 2 diabetic patients' health. *Journal of the Medical Association of Thailand*, 91, 1093–1096.
10. Biggi, N., Consonni, D., Galluzzo, V., Sogliani, M., and Costa, G. (2008). Metabolic syndrome in permanent night workers. *Chronobiology International*, 25, 443–454.
11. Conlon, M., Lightfoot, N., and Kreiger, N. (2007). Rotating shift work and risk of prostate cancer. *Epidemiology*, 18, 182–183.
12. Hefti, M. H., Francoijs, K. J., de Vries, S. C., Dixon, R., and Vervoort, J. (2004). The PAS fold. A redefinition of the PAS domain based upon structural prediction. *European Journal of Biochemistry*, 271, 1198–1208.
13. Bacsi, S. G., Reisz-Porszasz, S., and Hankinson, O. (1995). Orientation of the heterodimeric aryl hydrocarbon (dioxin) receptor complex on its asymmetric DNA recognition sequence. *Molecular Pharmacology*, 47, 432–438.
14. Fukunaga, B. N., Probst, M. R., Reisz-Porszasz, S., and Hankinson, O. (1995). Identification of functional domains of the aryl hydrocarbon receptor. *Journal of Biological Chemistry*, 270, 29270–29278.
15. Hankinson, O. (1995). The aryl hydrocarbon receptor complex. *Annual Reviews of Pharmacology and Toxicology*, 35, 307–340.

16. Dolwick, K. M., Swanson, H. I., and Bradfield, C. A. (1993). *In vitro* analysis of Ah receptor domains involved in ligand-activated DNA recognition. *Proceedings of the National Academy of Sciences of the United States of America, 90,* 8566–8570.

17. Reisz-Porszasz, S., Probst, M. R., Fukunaga, B. N., and Hankinson, O. (1994). Identification of functional domains of the aryl hydrocarbon receptor nuclear translocator protein (ARNT). *Molecular & Cellular Biology, 14,* 6075–6086.

18. Mimura, J., Ema, M., Sogawa, K., and Fujii-Kuriyama, Y. (1999). Identification of a novel mechanism of regulation of Ah (dioxin) receptor function. *Genes & Development, 13,* 20–25.

19. Evans, B. R., Karchner, S. I., Allan, L. L., Pollenz, R. S., Tanguay, R. L., Jenny, M. J., Sherr, D. H., and Hahn, M. E. (2008). Repression of aryl hydrocarbon receptor (AHR) signaling by AHR repressor: role of DNA binding and competition for AHR nuclear translocator. *Molecular Pharmacology, 73,* 387–398.

20. Erbel, P. J., Card, P. B., Karakuzu, O., Bruick, R. K., and Gardner, K. H. (2003). Structural basis for PAS domain heterodimerization in the basic helix–loop–helix–PAS transcription factor hypoxia-inducible factor. *Proceedings of the National Academy of Sciences of the United States of America, 100,* 15504–15509.

21. Card, P. B., Erbel, P. J., and Gardner, K. H. (2005). Structural basis of ARNT PAS-B dimerization: use of a common beta-sheet interface for hetero- and homodimerization. *Journal of Molecular Biology, 353,* 664–677.

22. Petrulis, J. R. and Perdew, G. H. (2002). The role of chaperone proteins in the aryl hydrocarbon receptor core complex. *Chemical and Biological Interactions, 141,* 25–40.

23. Dioum, E. M., Rutter, J., Tuckerman, J. R., Gonzalez, G., Gilles-Gonzalez, M. A., and McKnight, S. L. (2002). NPAS2: a gas-responsive transcription factor. *Science, 298,* 2385–2387.

24. Yang, J., Zhang, L., Erbel, P. J., Gardner, K. H., Ding, K., Garcia, J. A., and Bruick, R. K. (2005). Functions of the Per/ARNT/Sim domains of the hypoxia-inducible factor. *Journal of Biological Chemistry, 280,* 36047–36054.

25. Ikuta, T., Kobayashi, Y., and Kawajiri, K. (2004). Phosphorylation of nuclear localization signal inhibits the ligand-dependent nuclear import of aryl hydrocarbon receptor. *Biochemical and Biophysical Research Communications, 317,* 545–550.

26. Beischlag, T. V., Wang, S., Rose, D. W., Torchia, J., Reisz-Porszasz, S., Muhammad, K., Nelson, W. E., Probst, M. R., Rosenfeld, M. G., and Hankinson, O. (2002). Recruitment of the NCoA/SRC-1/p160 family of transcriptional coactivators by the aryl hydrocarbon receptor/aryl hydrocarbon receptor nuclear translocator complex. *Molecular and Cellular Biology, 22,* 4319–4333.

27. Hankinson, O. (2005). Role of coactivators in transcriptional activation by the aryl hydrocarbon receptor. *Archives of Biochemistry and Biophysics, 433,* 379–386.

28. Xu, C. X., Krager, S. L., Liao, D. F., and Tischkau, S. A. (2010). Disruption of CLOCK–BMAL1 transcriptional activity is responsible for aryl hydrocarbon receptor-mediated regulation of Period1 gene. *Toxicological Sciences, 115,* 98–108.

29. Mukai, M., Lin, T. M., Peterson, R. E., Cooke, P. S., and Tischkau, S. A. (2008). Behavioral rhythmicity of mice lacking AhR and attenuation of light-induced phase shift by 2,3,7,8-tetrachlorodibenzo-p-dioxin. *Journal of Biological Rhythms, 23,* 200–210.

30. Mukai, M. and Tischkau, S. A. (2007). Effects of tryptophan photoproducts in the circadian timing system: searching for a physiological role for aryl hydrocarbon receptor. *Toxicological Sciences, 95,* 172–181.

31. Buijs, R. M., Scheer, F. A., Kreier, F., Yi, C., Bos, N., Goncharuk, V. D., and Kalsbeek, A. (2006). Organization of circadian functions: interaction with the body. *Progress in Brain Research, 153,* 341–360.

32. Kalsbeek, A., Kreier, F., Fliers, E., Sauerwein, H. P., Romijn, J. A., and Buijs, R. M. (2007). Minireview. Circadian control of metabolism by the suprachiasmatic nuclei. *Endocrinology, 148,* 5635–5639.

33. Takahashi, J. S., Hong, H. K., Ko, C. H., and McDearmon, E. L. (2008). The genetics of mammalian circadian order and disorder: implications for physiology and disease. *Nature Reviews Genetics, 9,* 764–775.

34. DeBruyne, J. P., Weaver, D. R., and Reppert, S. M. (2007). CLOCK and NPAS2 have overlapping roles in the suprachiasmatic circadian clock. *Nature Neuroscience, 10,* 543–545.

35. Gachon, F., Nagoshi, E., Brown, S. A., Ripperger, J., and Schibler, U. (2004). The mammalian circadian timing system: from gene expression to physiology. *Chromosoma, 113,* 103–112.

36. Eide, E. J., Kang, H., Crapo, S., Gallego, M., and Virshup, D. M. (2005). Casein kinase I in the mammalian circadian clock. *Methods in Enzymology, 393,* 408–418.

37. Gallego, M. and Virshup, D. M. (2007). Post-translational modifications regulate the ticking of the circadian clock. *Nature Reviews Molecular and Cellular Biology, 8,* 139–148.

38. Lee, C., Etchegaray, J. P., Cagampang, F. R., Loudon, A. S., and Reppert, S. M. (2001). Posttranslational mechanisms regulate the mammalian circadian clock. *Cell, 107,* 855–867.

39. Lee, C., Weaver, D. R., and Reppert, S. M. (2004). Direct association between mouse PERIOD and CKIepsilon is critical for a functioning circadian clock. *Molecular and Cellular Biology, 24,* 584–594.

40. Meng, Q. J., Logunova, L., Maywood, E. S., Gallego, M., Lebiecki, J., Brown, T. M., Sladek, M., Semikhodskii, A. S., Glossop, N. R., Piggins, H. D., Chesham, J. E., Bechtold, D. A., Yoo, S. H., Takahashi, J. S., Virshup, D. M., Boot-Handford, R. P., Hastings, M. H., and Loudon, A. S. (2008). Setting clock speed in mammals: the CK1 epsilon tau mutation in mice accelerates circadian pacemakers by selectively destabilizing PERIOD proteins. *Neuron, 58,* 78–88.

41. Eide, E. J., Woolf, M. F., Kang, H., Woolf, P., Hurst, W., Camacho, F., Vielhaber, E. L., Giovanni, A., and Virshup, D. M. (2005). Control of mammalian circadian rhythm by

CKIepsilon-regulated proteasome-mediated PER2 degradation. *Molecular and Cellular Biology*, 25, 2795–2807.

42. Triqueneaux, G., Thenot, S., Kakizawa, T., Antoch, M. P., Safi, R., Takahashi, J. S., Delaunay, F., and Laudet, V. (2004). The orphan receptor Rev-erbalpha gene is a target of the circadian clock pacemaker. *Journal of Molecular Endocrinology*, 33, 585–608.

43. Guillaumond, F., Dardente, H., Giguere, V., and Cermakian, N. (2005). Differential control of Bmal1 circadian transcription by REV-ERB and ROR nuclear receptors. *Journal of Biological Rhythms*, 20, 391–403.

44. Preitner, N., Damiola, F., Lopez-Molina, L., Zakany, J., Duboule, D., Albrecht, U., and Schibler, U. (2002). The orphan nuclear receptor REV-ERBalpha controls circadian transcription within the positive limb of the mammalian circadian oscillator. *Cell*, 110, 251–260.

45. Akashi, M. and Takumi, T. (2005). The orphan nuclear receptor RORalpha regulates circadian transcription of the mammalian core-clock Bmal1. *Nature Structural and Molecular Biology*, 12, 441–448.

46. Sato, T. K., Panda, S., Miraglia, L. J., Reyes, T. M., Rudic, R. D., McNamara, P., Naik, K. A., FitzGerald, G. A., Kay, S. A., and Hogenesch, J. B. (2004). A functional genomics strategy reveals Rora as a component of the mammalian circadian clock. *Neuron*, 43, 527–537.

47. Ueda, H. R., Chen, W., Adachi, A., Wakamatsu, H., Hayashi, S., Takasugi, T., Nagano, M., Nakahama, K., Suzuki, Y., Sugano, S., Iino, M., Shigeyoshi, Y., and Hashimoto, S. (2002). A transcription factor response element for gene expression during circadian night. *Nature*, 418, 534–539.

48. Panda, S., Antoch, M. P., Miller, B. H., Su, A. I., Schook, A. B., Straume, M., Schultz, P. G., Kay, S. A., Takahashi, J. S., and Hogenesch, J. B. (2002). Coordinated transcription of key pathways in the mouse by the circadian clock. *Cell*, 109, 307–320.

49. Lowrey, P. L. and Takahashi, J. S. (2004). Mammalian circadian biology: elucidating genome-wide levels of temporal organization. *Annual Review of Genomics and Human Genetics*, 5, 407–441.

50. Shimba, S. and Watabe, Y. (2009). Crosstalk between the AHR signaling pathway and circadian rhythm. *Biochemical Pharmacology*, 77, 560–565.

51. Richardson, V., Santostefano, M., and Birnbaum, L. S. (1998). Daily cycle of bHLH–PAS proteins, Ah receptor and Arnt, in multiple tissues of female Sprague-Dawley rats. *Biochemical and Biophysical Research Communications*, 252, 225–231.

52. Huang, P., Ceccatelli, S., and Rannug, A. (2002). A study on diurnal mRNA expression of CYP1A1, AHR, ARNT and PER2 in rat pituitary and liver. *Environmental Toxicology and Pharmacology*, 11, 119–126.

53. Qu, X., Metz, R. P., Porter, W. W., Cassone, V. M., and Earnest, D. J. (2007). Disruption of clock gene expression alters responses of the aryl hydrocarbon receptor signaling pathway in the mouse mammary gland. *Molecular Pharmacology*, 72, 1349–1358.

54. Qu, X., Metz, R. P., Porter, W. W., Neuendorff, N., Earnest, B. J., and Earnest, D. J. (2010). The clock genes period 1 and period 2 mediate diurnal rhythms in dioxin-induced Cyp1A1 expression in the mouse mammary gland and liver. *Toxicology Letters*, 196, 28–32.

55. Takahata, S., Sogawa, K., Kobayashi, A., Ema, M., Mimura, J., Ozaki, N., and Fujii-Kuriyama, Y. (1998). Transcriptionally active heterodimer formation of an Arnt-like PAS protein, Arnt3, with HIF-1a, HLF, and clock. *Biochemical and Biophysical Research Communications*, 248, 789–794.

56. Yu, W., Ikeda, M., Abe, H., Honma, S., Ebisawa, T., Yamauchi, T., Honma, K., and Nomura, M. (1999). Characterization of three splice variants and genomic organization of the mouse BMAL1 gene. *Biochemical and Biophysical Research Communications*, 260, 760–767.

57. Bunger, M. K., Wilsbacher, L. D., Moran, S. M., Clendenin, C., Radcliffe, L. A., Hogenesch, J. B., Simon, M. C., Takahashi, J. S., and Bradfield, C. A. (2000). Mop3 is an essential component of the master circadian pacemaker in mammals. *Cell*, 103, 1009–1017.

58. Hofstetter, J. R., Svihla-Jones, D. A., and Mayeda, A. R. (2007). A QTL on mouse chromosome 12 for the genetic variance in free-running circadian period between inbred strains of mice. *Journal of Circadian Rhythms*, 5, 7.

59. Garrett, R. W. and Gasiewicz, T. A. (2006). The aryl hydrocarbon receptor agonist 2,3,7,8-tetrachlorodibenzo-*p*-dioxin alters the circadian rhythms, quiescence, and expression of clock genes in murine hematopoietic stem and progenitor cells. *Molecular Pharmacology*, 69, 2076–2083.

60. Yellon, S. M., Singh, D., Garrett, T. M., Fagoaga, O. R., and Nehlsen-Cannarella, S. L. (2000). Reproductive, neuroendocrine, and immune consequences of acute exposure to 2,3,7,8-tetrachlorodibenzo-p-dioxin in the Siberian hamster. *Biology of Reproduction*, 63, 538–543.

61. Jones, M. K., Weisenburger, W. P., Sipes, I. G., and Russell, D. H. (1987). Circadian alterations in prolactin, corticosterone, and thyroid hormone levels and down-regulation of prolactin receptor activity by 2,3,7,8-tetrachlorodibenzo-p-dioxin. *Toxicology and Applied Pharmacology*, 87, 337–350.

62. Pohjanvirta, R., Laitinen, J., Vakkuri, O., Linden, J., Kokkola, T., Unkila, M., and Tuomisto, J. (1996). Mechanism by which 2,3,7,8-tetrachlorodibenzo-p-dioxin (TCDD) reduces circulating melatonin levels in the rat. *Toxicology*, 107, 85–97.

63. Pohjanvirta, R. and Tuomisto, J. (1990). 2,3,7,8-Tetrachlorodibenzo-p-dioxin enhances responsiveness to post-ingestive satiety signals. *Toxicology*, 63, 285–299.

64. Pohjanvirta, R., Tuomisto, J., Linden, J., and Laitinen, J. (1989). TCDD reduces serum melatonin levels in Long–Evans rats. *Pharmacology and Toxicology*, 65, 239–240.

65. Miller, J., Settachan, D., Frame, L., and Dickerson, R. (1999). 2,3,7,8-Tetrachlorodobenzo-p-dioxin phase advance the deer mouse (Peromyscus maniculatus) circadian rhythm by altering expression of clock proteins. *Organohalogen Compounds*, 42, 23–28.

66. Korkalainen, M., Linden, J., Tuomisto, J., and Pohjanvirta, R. (2005). Effect of TCDD on mRNA expression of genes

encoding bHLH/PAS proteins in rat hypothalamus. *Toxicology*, *208*, 1–11.

67. Linden, J., Korkalainen, M., Lensu, S., Tuomisto, J., and Pohjanvirta, R. (2005). Effects of 2,3,7,8-tetrachlorodibenzo-p-dioxin (TCDD) and leptin on hypothalamic mRNA expression of factors participating in food intake regulation in a TCDD-sensitive and a TCDD-resistant rat strain. *Journal of Biochemistry and Molecular Toxicology*, *19*, 139–148.

68. Tuomisto, J., Pohjanvirta, R., MacDonald, E., and Tuomisto, L. (1990). Changes in rat brain monoamines, monoamine metabolites and histamine after a single administration of 2,3,7,8-tetrachlorodibenzo-p-dioxin (TCDD). *Pharmacology and Toxicology*, *67*, 260–265.

69. Petersen, S. L., Curran, M. A., Marconi, S. A., Carpenter, C. D., Lubbers, L. S., and McAbee, M. D. (2000). Distribution of mRNAs encoding the arylhydrocarbon receptor, arylhydrocarbon receptor nuclear translocator, and arylhydrocarbon receptor nuclear translocator-2 in the rat brain and brainstem. *Journal of Comparative Neurology*, *427*, 428–439.

70. Huang, P., Rannug, A., Ahlbom, E., Hakansson, H., and Ceccatelli, S. (2000). Effect of 2,3,7,8-tetrachlorodibenzo-p-dioxin on the expression of cytochrome P450 1A1, the aryl hydrocarbon receptor, and the aryl hydrocarbon receptor nuclear translocator in rat brain and pituitary. *Toxicology and Applied Pharmacology*, *169*, 159–167.

71. Christian, B. J., Inhorn, S. L., and Peterson, R. E. (1986). Relationship of the wasting syndrome to lethality in rats treated with 2,3,7,8-tetrachlorodibenzo-p-dioxin. *Toxicology and Applied Pharmacology*, *82*, 239–255.

72. Pohjanvirta, R., Unkila, M., and Tuomisto, J. (1994). TCDD-induced hypophagia is not explained by nausea. *Pharmacology Biochemistry and Behavior*, *47*, 273–282.

73. Seefeld, M. D., Corbett, S. W., Keesey, R. E., and Peterson, R. E. (1984). Characterization of the wasting syndrome in rats treated with 2,3,7,8-tetrachlorodibenzo-p-dioxin. *Toxicology and Applied Pharmacology*, *73*, 311–322.

74. Tuomisto, J. T., Pohjanvirta, R., Unkila, M., and Tuomisto, J. (1995). 2,3,7,8-Tetrachlorodibenzo-p-dioxin-induced anorexia and wasting syndrome in rats: aggravation after ventromedial hypothalamic lesion. *European Journal of Pharmacology*, *293*, 309–317.

75. Linden, J., Pohjanvirta, R., Rahko, T., and Tuomisto, J. (1991). TCDD decreases rapidly and persistently serum melatonin concentration without morphologically affecting the pineal gland in TCDD-resistant Han/Wistar rats. *Pharmacology and Toxicology*, *69*, 427–432.

76. Frame, L., Li, W., Miller, J., and Dickerson, R. (2004). A proposed role for the arylhydrocarbon receptor (AhR) in non-photic feedback to the master circadian clock. *Toxicologist*, *78*, 954.

77. Schecter, A., Birnbaum, L., Ryan, J. J., and Constable, J. D. (2006). Dioxins: an overview. *Environmental Research*, *101*, 419–428.

78. Gery, S. and Koeffler, H. P. (2010). Circadian rhythms and cancer. *Cell Cycle*, *9*, 1097–1103.

79. Silva, C. M., Sato, S., and Margolis, R. N. (2010). No time to lose: workshop on circadian rhythms and metabolic disease. *Genes & Development*, *24*, 1456–1464.

80. Anea, C. B., Zhang, M., Stepp, D. W., Simkins, G. B., Reed, G., Fulton, D. J., and Rudic, R. D. (2009). Vascular disease in mice with a dysfunctional circadian clock. *Circulation*, *119*, 1510–1517.

81. Viswambharan, H., Carvas, J. M., Antic, V., Marecic, A., Jud, C., Zaugg, C. E., Ming, X. F., Montani, J. P., Albrecht, U., and Yang, Z. (2007). Mutation of the circadian clock gene Per2 alters vascular endothelial function. *Circulation*, *115*, 2188–2195.

82. Sato, S., Shirakawa, H., Tomita, S., Ohsaki, Y., Haketa, K., Tooi, O., Santo, N., Tohkin, M., Furukawa, Y., Gonzalez, F. J., and Komai, M. (2008). Low-dose dioxins alter gene expression related to cholesterol biosynthesis, lipogenesis, and glucose metabolism through the aryl hydrocarbon receptor-mediated pathway in mouse liver. *Toxicology and Applied Pharmacology*, *229*, 10–19.

83. Acimovic, J., Fink, M., Pompon, D., Bjorkhem, I., Hirayama, J., Sassone-Corsi, P., Golicnik, M., and Rozman, D. (2008). CREM modulates the circadian expression of CYP51, HMGCR and cholesterogenesis in the liver. *Biochemical and Biophysical Research Communications*, *376*, 206–210.

84. Le Martelot, G., Claudel, T., Gatfield, D., Schaad, O., Kornmann, B., Sasso, G. L., Moschetta, A., and Schibler, U. (2009). REV-ERBalpha participates in circadian SREBP signaling and bile acid homeostasis. *PLoS Biology*, *7*, e1000181.

85. Brewer, M., Lange, D., Baler, R., and Anzulovich, A. (2005). SREBP-1 as a transcriptional integrator of circadian and nutritional cues in the liver. *Journal of Biological Rhythms*, *20*, 195–205.

86. Matsumoto, E., Ishihara, A., Tamai, S., Nemoto, A., Iwase, K., Hiwasa, T., Shibata, S., and Takiguchi, M. (2010). Time-of-day and nutrients in feeding govern daily expression rhythms of the gene for sterol regulatory element-binding protein (SREBP)-1 in the mouse liver. *Journal of Biological Chemistry*, *285*, 33028–33036.

87. Oishi, K., Amagai, N., Shirai, H., Kadota, K., Ohkura, N., and Ishida, N. (2005). Genome-wide expression analysis reveals 100 adrenal gland-dependent circadian genes in the mouse liver. *DNA Research*, *12*, 191–202.

88. Gillette, M. U. and Tischkau, S. A. (1999). Suprachiasmatic nucleus: the brain's circadian clock. *Recent Progress in Hormone Research*, *54*, 33-58; discussion 58–39.

89. Antle, M. C., Smith, V. M., Sterniczuk, R., Yamakawa, G. R., and Rakai, B. D. (2009). Physiological responses of the circadian clock to acute light exposure at night. *Reviews of Endocrine and Metabolic Disorders*, *10*, 279–291.

90. Golombek, D. A., Agostino, P. V., Plano, S. A., and Ferreyra, G. A. (2004). Signaling in the mammalian circadian clock: the NO/cGMP pathway. *Neurochemistry International*, *45*, 929–936.

91. Hirota, T. and Fukada, Y. (2004). Resetting mechanism of central and peripheral circadian clocks in mammals. *Zoological Sciences*, *21*, 359–368.

92. Yan, L. (2009). Expression of clock genes in the suprachiasmatic nucleus: effect of environmental lighting conditions. *Reviews of Endocrine and Metabolic Disorders*, 10, 301–310.
93. Golombek, D. A. and Rosenstein, R. E. (2010). Physiology of circadian entrainment. *Physiological Reviews*, 90, 1063–1102.
94. Cao, R., Lee, B., Cho, H. Y., Saklayen, S., and Obrietan, K. (2008). Photic regulation of the mTOR signaling pathway in the suprachiasmatic circadian clock. *Molecular and Cellular Neuroscience*, 38, 312–324.
95. Pizzio, G. A., Hainich, E. C., Ferreyra, G. A., Coso, O. A., and Golombek, D. A. (2003). Circadian and photic regulation of ERK, JNK and p38 in the hamster SCN. *Neuroreport*, 14, 1417–1419.
96. Dziema, H., Oatis, B., Butcher, G. Q., Yates, R., Hoyt, K. R., and Obrietan, K. (2003). The ERK/MAP kinase pathway couples light to immediate-early gene expression in the suprachiasmatic nucleus. *European Journal of Neuroscience*, 17, 1617–1627.
97. Butcher, G. Q., Doner, J., Dziema, H., Collamore, M., Burgoon, P. W., and Obrietan, K. (2002). The p42/44 mitogen-activated protein kinase pathway couples photic input to circadian clock entrainment. *Journal of Biological Chemistry*, 277, 29519–29525.
98. Akashi, M. and Nishida, E. (2000). Involvement of the MAP kinase cascade in resetting of the mammalian circadian clock. *Genes & Development*, 14, 645–649.
99. Obrietan, K., Impey, S., and Storm, D. R. (1998). Light and circadian rhythmicity regulate MAP kinase activation in the suprachiasmatic nuclei. *Nature Neuroscience*, 1, 693–700.
100. Tischkau, S. A., Mitchell, J. W., Tyan, S. H., Buchanan, G. F., and Gillette, M. U. (2003). Ca^{2+}/cAMP response element-binding protein (CREB)-dependent activation of Per1 is required for light-induced signaling in the suprachiasmatic nucleus circadian clock. *Journal of Biological Chemistry*, 278, 718–723.
101. Ding, J. M., Faiman, L. E., Hurst, W. J., Kuriashkina, L. R., and Gillette, M. U. (1997). Resetting the biological clock: mediation of nocturnal CREB phosphorylation via light, glutamate, and nitric oxide. *Journal of Neuroscience*, 17, 667–675.
102. Naruse, Y., Oh-hashi, K., Iijima, N., Naruse, M., Yoshioka, H., and Tanaka, M. (2004). Circadian and light-induced transcription of clock gene Per1 depends on histone acetylation and deacetylation. *Molecular and Cellular Biology*, 24, 6278–6287.
103. Yan, L. and Silver, R. (2002). Differential induction and localization of mPer1 and mPer2 during advancing and delaying phase shifts. *European Journal of Neuroscience*, 16, 1531–1540.
104. Chen, D., Buchanan, G. F., Ding, J. M., Hannibal, J., and Gillette, M. U. (1999). Pituitary adenylyl cyclase-activating peptide: a pivotal modulator of glutamatergic regulation of the suprachiasmatic circadian clock. *Proceedings of the National Academy of Sciences of the United States of America*, 96, 13468–13473.
105. Glass, J. D., Selim, M., Srkalovic, G., and Rea, M. A. (1995). Tryptophan loading modulates light-induced responses in the mammalian circadian system. *Journal of Biological Rhythms*, 10, 80–90.
106. Quintero, J. E. and McMahon, D. G. (1999). Serotonin modulates glutamate responses in isolated suprachiasmatic nucleus neurons. *Journal of Neurophysiology*, 82, 533–539.
107. Sterniczuk, R., Stepkowski, A., Jones, M., and Antle, M. C. (2008). Enhancement of photic shifts with the 5-HT1A mixed agonist/antagonist NAN-190: intra-suprachiasmatic nucleus pathway. *Neuroscience*, 153, 571–580.
108. Bergander, L., Wincent, E., Rannug, A., Foroozesh, M., Alworth, W., and Rannug, U. (2004). Metabolic fate of the Ah receptor ligand 6-formylindolo[3,2-b]carbazole. *Chemical and Biological Interactions*, 149, 151–164.
109. Heath-Pagliuso, S., Rogers, W. J., Tullis, K., Seidel, S. D., Cenijn, P. H., Brouwer, A., and Denison, M. S. (1998). Activation of the Ah receptor by tryptophan and tryptophan metabolites. *Biochemistry*, 37, 11508–11515.
110. Helferich, W. G. and Denison, M. S. (1991). Ultraviolet photoproducts of tryptophan can act as dioxin agonists. *Molecular Pharmacology*, 40, 674–678.
111. Qu, X., Metz, R. P., Porter, W. W., Cassone, V. M., and Earnest, D. J. (2009). Disruption of period gene expression alters the inductive effects of dioxin on the AhR signaling pathway in the mouse liver. *Toxicology and Applied Pharmacology*, 234, 370–377.

INDEX

(t = table, f = figure; reference to a whole chapter is indicated by "c" followed by chapter number in bold)

A549 human lung carcinoma cell line   105, 114, 345, 472, 485
Abbott, Barbara   257
Abel, Josef   11, 109
2-Acetylaminofluorene   249, 359, 473
N-Acetylcysteine   353-4
ACHN human renal carcinoma cell line   105
Aconitase   230
Actinomycin D   350–4
Activating transcription factor 3 (see ATF3 transcription factor)
Activator protein-1 (see AP-1 transcription factor)
Adaptive immune system/response   277–8, 499–500, t34.1
Adrenal-4 binding protein/steroidogenic factor-1 (A4BP/SF-1)   135
Agent Orange   277, 309–11, 313, 432
Ah locus   4–5, 7, 12, 37, 183
AHA-1   406, 408
AHR (see Aryl hydrocarbon receptor)
AHR1   185, 299, 390–3, 395
AHR-1   162–3, 390, 393, 406–7, 409, 490
   AHR-1-deficient mutants   406–407
AHR2   185, 292, t20.1, 299–302, 390–393, 395
AHR3   391
AHR-deficient mice   11, 38, 53, 113, 129, 134, 165, 183, 250, 252, 268, 289, 293, 299, 379, 489–491
   Cardiovascular abnormalities   113, 359, 414–5, 425–431
   Circadian rhythms   515
   Colon tumors   150
   Development of reproductive organs and functions   358, 437, 440–6, 448–51, 453–455
   General characteristics   14, 110
   Immune phenotype   360–1, 500–503, t34.2, 505–506
   Liver phenotype   413–5, f29.1, f29.2
AHR degradation promoting factor   39
AHR-modified mouse models (see also AHR-deficient mice)   113, 183–4, 188, 207, 217–8, 221–2, 259–61, 289, c**26**, 415–6, 428–31
AHR nuclear translocator (ARNT)   38, 49, 82, 95–6, 258, t17.1, t17.2, 287–9, 344, 362, 419, 439, 447, 452, 514
   Coactivator role   40, 130
   Discovery   6
   Orthologs and paralogs   300, 406–8
   Partner for HIF-1α   40, 110, f8.1, 417
   Structure   38, f2.1
   SUMOylation   104
AHR/RelB dimer   86–87, 208, 361
AHR repressor (AHRR)   11, 84–85, c**7**, 111–2, f8.2, 164, 291, 347–8,
   Action mechanisms   103–4, f7.6
   Expression in cells and tissues   102–103, 105, 452, 506
   Nucleocytoplasmic shuttling   103
   Orthologs and paralogs   101–102, 300, 390–2, 395
   Physiological functions   105–107, 300, 395, 443, 455–6
   Polymorphisms   107, 455–6
   Structure   101–102, f7.1
AHR-response element (AHRE)   5–6, t2.1, 37–8, t2.2, 71–2, 81–4, t5.1, 300, 334, 471
   AHRE-II   6, 83
AIP (see XAP2)
AKT protein kinase   246, 350, 356, 473–4
Ameloblasts   286–7, 290
δ-Amino levulinic acid (ALA) synthetase   4
Aldehyde dehydrogenase 3A1 (ALDH3A1)   36, t2.1, 165, 186

The AH Receptor in Biology and Toxicology, First Edition. Edited by Raimo Pohjanvirta.
© 2012 John Wiley & Sons, Inc. Published 2012 by John Wiley & Sons, Inc.

INDEX

Aldolase A 111
Alkaline phosphatase (ALP) t20.1, 289, 291, 381
Aminoflavone 114, 344
cAMP 55, t3.1, f3.4, 469
cAMP-dependent protein kinase (PKA) 201–8, f13.3, f13.4
cAMP-response element-binding protein (CREB) 165, 205, 469, 516
CREB-binding protein (CBP) 71, t6.1, 96–8, f7.6, 111, 132, 143, 165, 344–5, 352, 417, 449
Amphiregulin 469
Andersson, Patrik 373
Androgen receptor (AR) 56, 85, 104, 127, 130–131, 147–150, f10-5, 455–6
Angiogenesis 113–115, 165, 416–8, 446, 489–90, f33.2
Angiotensin converting enzyme (ACE) 424–5, 427–9, f30.6
Ankyrin-repeat family A protein 2 (ANKRA2) 104–5, f7.6
Anterior gradient-2 249–50
Antioxidant-response element (ARE) t2.1, 40, 109, f8.3, 116, 118, 346, 349
Anwar-Mohamed, Anwar 343
AP-1 transcription factor 114–5, 128, f9.3, 130–1, 133, 199–200, 202, 207, 246, 279, 348, 352, 354, f25.1, 469–70,
Apaf1 gene 249–50
APC gene 150
Apigenin 114, 334
Apoptosis 41, 54–6, 113, 163, 201, 218, 232, 246–51, 345, 415–6, 440, 442, 467, 472–4, 503
ARA9 (*see XAP2*)
Arachidonic acid 13, 134, 201, 350, 360, 432
ARNT (*see AHR nuclear translocator*)
ARNT-deficient mice 112–113, 184
Aromatase (*see CYP19*)
Arsenic (As^{3+}) 348–50
Aryl hydrocarbon hydroxylase (AHH) 4–5, f1.1, 7, 36–7, 181
Aryl hydrocarbon receptor (AHR; *see also titles of individual Chapters*)
  Alternative signaling pathways 87, 109, 147–151, f10-5, f10–6, c13, 344, 361, 419, 470, 506
  Apparent molecular mass 82
  Basic-helix-loop-helix-Per/Arnt/Sim (bHLH-PAS) motifs 37, 63, 82, 512
  Canonical signaling pathway 36–39, f2–2, f3–1,183–4
  Cell cycle regulation f1.1, 12–3, 41, 86, 106, 164–5, 246, 250, 300, 415, 442, 468–72 5–16, 10–21, 31–16
  Circadian rhythms in expression 514
  Constitutively active 113, 116, 131, 147–50, 163–4, 252, 280, 287, t20.1, 300, 359, c26, 393, 407, 419, 467, 470–2, 488, 491
  Cytosolic complex 49, f3.1
  Degradation 38–39, 101, 146–7, f10–3,
  Expression in tissues and cells throughout development 258–60, t17.1, 287–9, 358–9, 413, 424–5, 438–9, 447, 452, 503, t34.3
  Expression in tumors 491–2
  Half-life 38, 49
  Human polymorphisms 455
  Induction of drug-metabolizing enzymes 35–39, 96–98, 514
  Inhibitory domain 37
  Interactions
    Clock genes 514–8, f35.2
    HIF-1α 111–113, 348, 417
    MAPK pathway 261–3, 345–6, 348
    NF-κB 85, f13.4, 207–8, 347–8, 503, 506
    NRF2 84, 116–8, f8-4
    Nuclear receptors 85, 129–35, 147–50, f10.4, f10.5, 346–7
    PKC 345
    RB/E2F1 86
    SP1 86
    TGFβ 261–2
    Tyrosine kinases 346
  Ligands 13–14, 40, 64–66, t4–1, f4–1, 318–9, t23.1, t23.2, 322–3, 331–6, 502–3
  Ligand-binding domain 66–70, 184–5, 393–4
  Ligand-binding pocket 63–4, 67–70, f4-2, f4-3, 344
  Nuclear localization and export signals 38, 48, 52, 63, 66, 345
  Nucleocytoplasmic shuttling 48, 52
  Phylogenetic aspects 299–300, c27, 405–6
  Physiological functions 14, t2–2, 113, 163, 197, 251–2, 394–5, 416–9, f29.3, 439–56, c32, 486–91, 501–2, 504–6, f34.1
  Polymorphic variants 393, 455–6
  Regulation of expression 1–18, 7–2, 8–18, 9–12, 10–9, 11–27
  Selective modulators 70, 72, 98, 332–5, f24.2, f24.3, f24.4
  Structure f2–1, f3–2
  Superinduction 39
  Target genes 8, t2–1, t2–2, t5–1, 83–84, 301, 473, 487, 501, 503
  Transactivation domain 51, 95–6, 147, f12–2,184–9, 289, 470
ATF-3 transcription factor 236
ATL (adult T cell leukemia) cell line 357
Autoimmunity 127, 208, 277, 281, 335, 361, 505–506
Autoradiography 416
5-Aza-2′-deoxycytidine 167
Azo dye N-demethylase 4, 36
Azo dyes 3, 36, 116

B16F10 mouse melanoma cell line 491
B-cell activating factor of the tumor necrosis factor family (BAFF) 86, 361
Barr body 159
BAX (*see BCL2-associated X protein*)
BCL-1 B cell line (AHR-deficient) 279, 348
BCL2-associated X protein (BAX) t2.2, 72, 232, 236, 249–50, 440, 452, 473
Bell, David 267
Benzo[*a*]pyrene 5, 35–40, 54, 68, 130, 167–9, t20.1, 293, 335–6, 345, 348–350, t25.1, 353–4, t25.2, 357–60, 417, 473, 486, 488–9, 506
Benzo[*a*]pyrene hydroxylase (*see Aryl hydrocarbon hydroxylase*)
BeWo human choriocarcinoma cell line 246
bHLH/PAS proteins 82, 109, 348, 511–2
Bifunctional enzyme inducers 116, 346
Bilirubin t2.1, 40, t4.1, t12.1, t12.2, 336, 349–50
Birnbaum, Linda 307
Bisphenol A (*see Xenoestrogens*)
B lymphocytes 113, 279, 374, 378, 470, 499–500, t34.3
B-lymphocyte chemoattractant (BLC) 86
B lymphocyte-induced maturation protein-1 (Blimp-1) 279–80

BMAL1 (*see Clock genes*)
Bonati, Laura   63
Bone formation   285
Bone sialoprotein   291
Bono1   290
Boutros, Paul   181, 217, 220–2, t14.1, 224
Boverhof, Darrell   219–22, t14–1, 224
Bovine aortic endothelial cells (BAECs)   431
BP8 rat hepatoma cells (AHR-deficient)   184, 470, 474
Bradfield, Christian   6, 9, 12–4, 217, 223
Brahma-related gene 1 (BRG-1)   t6.1, 96–7, 111, 143, 167, 344
Breast cancer resistance protein (BCRP)   111
Breast cancer susceptibility gene 1 (BRCA1)   84, 95–6, 111
Brunnberg, Sara   127, 373
Buthionine-(S,R)-sulfoximine   353–5
Butylated hydroxyanisole (BHA)   116–7, 230–1
Butylated hydroxytoluene   116
Butyrate   105, 166

C2C12 mouse myoblast cell line   236, 250
C3H10T1/2 mouse embryonic fibroblast cell line   133, 167, 468
C57BL/6 mouse strain   4, 36, 114, 181, 183, 185, 188–9, 208, 219–21, t14.1, t14.2, 229–231, 234–5, 237, 246, 257–8, 260–2, 287, 290–1, 359–60, 417, 501, 505, 515
Caco-2 human colorectal adenocarcinoma cell line   134
Cadherins   486, 488
Cadmium (Cd^{2+})   353, t25.2, 361
Caenorhabditis elegans   51, 162–3, 390–1, 405–9, 490
Calcium (Ca^{2+})   198–9, f13.2, 201–5, 209–10, 234–5, 237–8, 292, 361, 423–4, 516
Cap-N-Collar family of basic region leucine-zipper (CNC-bZIP) proteins   84, 115
Carbaryl   t4.1, 71
Carbonic anhydrase IX (CA9)   111, 114
Carlos Roman, Angel   485
Carlson, Erik   221–2, t14–2
Carvajal-Gonzalez, Jose   485
Casein kinase I   513
Casein kinase II (CK2)   206–7
Caspases   54, 232, 250, 361, 472, 474
Catalase   231, 235–6
β-Catenin   11, 114–5, 143, 150, f10–6, 302, 440, 474
Cathepsins   130, 381, f26.2, 474, 486
CBP/p300   t6–1, 96–8, f7.6, 111, 132, 143, 160, 165, 167–8, 344–5, 352, 354, 417, 449, 470
CC-chemokine ligand 1 (CCL1)   86
CCR7 chemokine receptor   279
CD proteins   11, t2.2, 113, 278–81, 361, 378, 490, t34.1, 501, 503, t34.3
Cell adhesion   113, 291, c33
C/EBP   134, 205–7, 209–10
Cell cycle kinase proteins (cyclins; cyclin-dependent kinases)   86, 106, 442–3, 470–2
Cell migration   113, 164, f13.3, 258, 261–2, 279, 406–7, 439, 475, c33, 500
Cell stress response   197–201, 206–7, 210–11, 250, 343, 405, 416
Cerebellar granule cells   237
CH12.LX murine lymphoma cell line   279, 348

CH223191   t4.1, f4.1, 69
Chadalapaka, Gayathri   331
Chaperones (*see HSP90, XAP2 and P23*)
Chicken ovalbumin upstream promoter transcription factor 1 (*see COUP-TFI*)
Chloracne (*see 2,3,7,8-Tetrachlorodibenzo-p-dioxin, Skin toxicity*)
Chopra, Martin   245
Chromatin immunoprecipitation (ChIP)   94, 96–8, 106, 146–7, 149, 151, 160, 168, 224, 335, 361, 487
Chromium (Cr^{6+})   353–5, t25.2
Chrysin   114, 334
Chrysoeriol   117
Circadian locomotor output cycles kaput (*see CLOCK*)
CK2 (*see Casein kinase II*)
CL3 human lung adenocarcinoma cells   349, t25.1
Cleft palate (*see 2,3,7,8-Tetrachlorodibenzo-p-dioxin, Developmental toxicity*)
Clock genes   6, 37, 66, 511–8, f35.1
Cobalt chloride ($CoCl_2$; *see Iron-chelating agents*)
Coiled Coil Coactivator (CoCoA)   95–6
Coimmunoprecipitation (*see Immunoprecipitation*)
Colony-stimulating factor-1 (CSF-1)   201–2
Confocal microscopy   292
Connexins (*see Intercellular communication*)
Conney, Allan   3–4
Constitutive androstane receptor (CAR)   134–5, 346
Copper (Cu^{2+})   354, t25.2, 360–1
Coregulatory proteins (*see also individual coactivators and corepressors*)   9, 39–40, 70–1, 85, c6, f7.6, 111, 127–8, f9.3, 130–1, 144, 147–8, 165–8, 189, 332, 344–5, 380
COS-1 monkey kidney cell line   49, 52
COS-7 monkey fibroblast or fibroblast-like cell line   104, 300
COUP-TFI   85, 135
COX2 (*see Cyclooxygenase-2*)
"CpG islands"   97, 157, 159–61, 163, 166, 499, 506
CRABPII gene   133
Cre-lox technology   183–4, 416, 430
CREB (*see cAMP-response element-binding protein*)
CREB binding protein (CBP; *see CBP/p300*)
C-terminus of Hsp70-interacting protein (CHIP)   146
Cullin-RING ubiquitin ligases (CRLs)   145–6, 150, 152
Cullin 4B ubiquitin-ligase complex   130, 471
Curcumin   117–8, 231, f24.4, 335
CV-1 monkey kidney cell line (AHR-deficient)   345
CXCR4 chemokine receptor   114
1-[2-Cyano-3,12-dioxooleana-1,9(11)-dien-28-oyl] imidazole   116
Cyclin-dependent kinases (*see Cell cycle kinase proteins*)
Cyclins (*see Cell cycle kinase proteins*)
Cycloheximide   39, 111, 132, 350–2, 354, 473–4
Cyclooxygenase-2 (COX2)   109, 184, f13.2, 201–7, t13.1, 209–10, t20.1, 291, 301, 334, 424, 444, 446, 474
CYP1A1   5–6, 8, 10, 35–9, t2.1, f2.2, t5.1, 83–7, 94–98, 116, 183, 232, 301, 347, 358, 360–2, 425, 430, 432, 443, 518
CYP1A2   8, 35–6, t2.1, 83–4, 112, 183, 232, 259, 347–8, 358, 360–2
CYP1B1   36, t2.1, 84, 97–8, 116, 166–7, 183, 232, 301, 345, 347, 358–60, 362, 425, 430, 443, 518

CYP1C1   102, 301
CYP1C2   301
CYP2A5   t2.1, 36, 165
CYP2B10   116
CYP2C11   361
CYP2C12   361
CYP2C19   359
CYP2E1   361
CYP2J2   111
CYP2S1   36, t2.1, 111, 165, 343
CYP3A2   361
CYP3A4   111
CYP3A5   359
CYP3A7   359
CYP17A1   451
CYP19 (aromatase)   130, 135, 147, 163, 201–2, f13.3, 206, 358, 441–3, 454
Cytochrome P450c17   (*see CYP17A1*)
Cytochrome P450 side chain cleavage (P450scc)   131, 442–3
Cytokines (*see also individual substances*)   85, 110–1, 163–4, t12.1, 183, 204, 209, 252, 278–81, 348, 360–1, 416, 424, 433, t34.1, 501–3, 505–6
Cytosolic phospholipase A2 (cPLA2)   200–3, f13.2, t13.1, 205, 209–10
Cytotoxic T lymphocytes (CTL)   185, 278–80, 501

Damaged DNA-binding protein 2 (DDB2)   146, 149, 152
DBA/2 mouse strain   18, 37, 181, 183, 185, 229–30, 233–4, 237, 246, 261, 279, 359
DeGroot, Danica   63
Dendritic cells   13, t12.1, 208, 499–500, 502–4, t34.3, 506
Denison, Michael   63
Dentin sialophosphoprotein   290
Desferrioxamine (*see Iron-chelating agents*)
Dexamethasone   132, 211, 347
Dibenzoylmethane   335
Diethylnitrosamine (DEN)   163–4, 245, 249, 251–2, 379, 468–9, 473–4
Diethylstilbestrol (*see Xenoestrogens*)
Dietrich, Cornelia   467
3,3'-Diindolylmethane   t4.1, 70, 98, 167, 334
7,12-Dimethylbenz[a]anthracene (DMBA)   4, 36, 72, 117–8, t25.2, 358, 440, 443
Dimethyl sulfoxide (DMSO)   102, f4.4
Dioxin (*see 2,3,7,8-Tetrachlorodibenzo-p-dioxin*)
Dioxins (*see also 2,3,7,8-Tetrachlorodibenzo-p-dioxin and Polychlorinated biphenyls*)   f1.1, 5, 7, t4.1, 71, 109, 144, 152, 168, 189, 235, 286, 317–24
  Additivity   321
  Estrogenic and antiestrogenic effects   147
  Exposure   307–8
  Relative potencies   186, 245, t23.1
  Sources   307–8
  Tolerable daily intake   373
Dioxin response element (DRE; *see AHR-response element*)
DNA footprinting   81
DNA methyltransferase 1 (DNMT1)   97, 159, 167–8
DNA single-strand breaks   230–1, 233–7

Drosophila melanogaster   6, 37, 51, 66, 159–62, 390, 405, 407–9, 512
Drug-metabolizing enzymes   49, t2–1, 109, 184, 188, 197, 346, 357
Drug transporters (*see Transporter proteins*)
Ductus venosus (*see AHR-deficient mice, vascular abnormalities*)

E box   82, 512–3
E2F transcription factors   11, 86, 106, 150, 250, 470, 474
E3 ubiquitin ligase   11, 87, 104, 110, 114–6, 144–7, f10.4, 149–51, f10.7, 223, 302, 471, 474
Early growth response genes (EGR-1)   114, 471
EGTA-AM (intracellular calcium-chelator)   202
Eisen, Howard   12
Electrophoretic mobility shift assay (*see Gel shift assay*)
El-Kadi, Ayman   343
Ellagic acid   230–1, 236–7
Embryonic stem (ES) cells   (*see Stem cells*)
Endothelin-1 (ET-1)   164, 359, 423–5, 427–8, 433
Endotoxin (*see Lipopolysaccharide*)
Endrin   230, 234–5
Enzyme-linked immunosorbent assay   83
Epidermal growth factor (EGF)   134, 200, 203–4, 209, 249, 251, t17.1, 260–3, 290, 468–9, 486
Epidermal growth factor receptor (EGFR)   109, 114, t12.1, 203–4, 233, 249, 258, 260–3, t17.2, 287, 290, 469, 473
Epigallocatechin-3-gallate (EGCG)   50, t4.1, 114, f24.4, 335
Epigenetic inheritance   157–61
Epiregulin   86, 261, 263, 469
Epoxide hydrolase   116, 357
Epstein-Barr virus nuclear antigen-3 (EBNA-3)   53, t3.1, 84
ERK1/2   f1.2, 134, 199–201, f13.2, 233, 249–51, 261–2, 345, 348, 350, 352, 354, 431, 469–70, 473, 487, f33.1
ERR (*see Estrogen-related receptor*)
ERR-deficient mice   135
Erythropoietin (EPO)   40, 110–1, f8.1
Esser, Charlotte   499
17β-Estradiol (E2)   40, 85, 96, 106, 127–30, f9.3, 135, 144–5, f10.2, 147–50, 163, 165, 246, 250–1, 312–3, 332, 346–7, 358, 380, f31.1, 439, 441–6, 448, 450–2, f31.3, 454, 468, 471
  Reactive metabolites   251, 443
Estrogen receptors (ERα [=ESR1], ERβ [=ESR2])   11, 85–7, 95–6, t6.1, 106, 127, f9.3, 130, 135, 144, 147–50, f10.4, f10.5, 165, 332, 346–7, 430, 439, 442–3, 447–8, 452, 471
  Membrane receptors   128, f9.2, 198
Estrogen receptor associated protein 140 (ERAP 140) co-activator   167
Estrogen-related receptor (ERR)   135
Estrogen-response element (ERE)   85, f9.2, 147–8, 503
ET cortical thymic epithelial cell line   114
Ethanol   118
Etoposide   249–50, 473–4
Euchromatin   157–9
Extracellular matrix (ECM)   164–5, 258, 261, 301–2, 486–7
Extracellular signal-regulated kinase (*see ERK*)

FasL/Fas signaling   279, 472–4, 503

F4/80 macrophage marker 207, 278
F-box proteins 145–6, 149–50
Fenton, Suzanne 307
Fernandez-Salguero, Pedro 485
Fibroblast growth factor 165, 262, 455
FK506 49, f3.3, 250
Flaveny, Colin 222, t14–2
Flavonoids (*see also individual substances*) 13, 69, 109, 114, 332, 334, 336, 394, 432
Flaws, Jodi 437
Fluorescence-activated cell sorting (FACS) 37, 378, 501
Fluorescence recovery and photobleaching (FRAP) analysis 97
Fluorescence resonance energy transfer (FRET) 95
Follicle-stimulating hormone (FSH) 312, 358, 437, f31.1, 439, 441–5, 450–2, f31.3
Follicle-stimulating hormone receptor (FSHR) 358, 441–3, 445
6-Formylindolo [3,2-b]-carbazole (FICZ) 13, f4.1, f24.5, 336, 500, 502, 505–6, 516–7
Forskolin 55, 206, 208
c-*Fos* gene 115, 129–31, 147, 246, 252, 292, 346, 348, 469, 471
Four and a half LIM domain 2 (FHL2) 131
FoxQ1b transcription factor t20.1, 292, 301
Foxp3 transcription factor 13, 208, 280–1, 361, t34.3, 505
Fraccalvieri, Domenico 63
Fujii-Kuriyama, Yoshiaki 5–6, 9, 11, 101
1-furan-2-yl-3-pyridin-2-yl-propenone (FPP-3) 118

Gamma-amino butyric acid (GABA) 68, 162, 406–7, 439
    GABA/Glutamate neurons 439, 452
Gap junctions (*see Intercellular communication*)
Gasiewicz, Thomas 3, 5, 8, 13
GC box 86, 102, f7.3, 166
Geldanamycin 48, 50, 70
Gel shift assay 37, 81, 116, 166, 206, 208, 291, 300, 354, 408
Genistein 334
Glucocorticoids t3.1, 54, 98, 132, 211
Glucocorticoid receptor 47–49, t3.1, 53–6, 85, 127, 132, 150, 346–7, f25.1, 500
Glucocorticoid response element (GRE) 132, 347
GLUT4 glucose transporter 207
Glutamic acid decarboxylase (GAD 67) 439, 452–3
Glutathione 109, 230–2, 235–6, 238, 349–50, 352–4
Glutathione peroxidase 229–31, 235–7
Glutathione reductase 231, 236–7
Glutathione S-transferase (GST; *see also individual GST subtypes*) 8, 35, 84, 94, 109, 165, 231, 343, 455
Glycodelin 448
GM0637 human fibroblast cell line 353, t25.2
GM4429 human fibroblast cell line 353, t25.2
Gomez-Duran, Aurea 485
Gonadotropin-releasing hormone (GnRH) 437, f31.1, 439, 444–5, 450, f31.3, 452
GPR30 transmembrane estrogen receptor 129
Graft-versus-host (GVH) reaction 279–81, 361
Granzyme B 164, 280–1
Greenlee, William 5, 8
GRIP1-associated coactivator 63 (GAC63) 95–6, t6.1

Growth factors (*see also individual substances*) 11, 54, 85, 110, t12.1, 183, 204, 210–11, 220, 259, 261, 285, 440, 454, 486, 489–92
GST1 117
GST A1 35–6, t2.1, 39, 116, 165, 343, 346
GST A2 116
GST M1 117
GST M2 117
GST M3 117
GST M6 117
GST P2 117
GST Pi 117
GST T1 455
GST T2 117
GTPases 487, f33.1
Guanylate cyclase A receptor 129
Gustafsson, Jan-Åke 5, 127

H1L6.1c3 mouse hepatoma cell line 432
H7 (inhibitor of PKC) 204
H89 (inhibitor of PKA) 201, f13.3, 205–6, 208
Haarmann-Stemmann, Thomas 109
HaCaT human keratinocyte cell line 468, 486, 491
Hahn, Mark 7, 389
Hairy and enhancer of split homolog-1 (HES-1) 439
Halogenated aromatic hydrocarbons (HAHs) [*see also Polychlorinated biphenyls (PCBs), Polychlorinated dibenzo-p-dioxins (PCDDs) and Polychlorinated dibenzofurans (PCDFs)*] 35, 63, t4.1, 229, 249, 257, 336, 343, 393, 405, 417, 432
Hanberg, Annika 373
Hankinson, Oliver 5–6, 12, 93
Han/Wistar (*Kuopio*) [H/W] rat strain 181, 185–9, t12.2, f12.3, f18.2, f18.3, t18.2, 221, 223, 231, 245, 249, t18.1, 273, t20.1, 289, 293, 319, 380, 469, 515
Hassoun, Ezdihar 229
HC11 mouse mammary cells 96, 130
He, Guochun 63
HeLa human cervical adenocarcinoma cell line 97, 105, 150, 350
Heme 4, 66, 183, 232, 336, 349, 351, 353, 356, 409, 424, 512
Heme oxygenase-1 (HO-1) 110, 116, 346, 349, 352–4, 356
Henry, Ellen 3
Hepa-1 (*see Mouse hepatoma cell lines*)
Hepatitis B virus X protein t3.1
Hepatitis B virus X protein-associated protein 2 (*see XAP2*)
HepG2 human hepatocellular carcinoma cells 68, 97, 105, 112, f8.2, 147, 167, 203, t13.1, 249, 262, 334–5, 348–50, t25.1, 352–4, t25.2, 356, 361, 416, 469, 473
Hep3B human hepatoma cells 111
Herbimycin A 344
Hernández-Ochoa, Isabel 437
Heterochromatin 157–60
Hexamethonium 425, 429–30
HIF (*see Hypoxia-inducible factor*)
HIF-1α-deficient mice 112
HIF-2α-deficient mice 113
Histone proteins 93, 157, 344

Histone acetyltransferase (HAT)   71, 93, t6.1, 97, 143, 160, 165, 167
"Histone code"   93, 158, 160–1
Histone deacetylases (HDACs)   93, 97, 104–5, f7.6, 133–4, 159, 165–8, 207, 344, 354
Histone demethylases   93, 161
Histone methyltransferases   93, 130, 158–60
Homeobox transcription factors   168, 439
HomoloGene database of homologs   219
Homology modeling   66
HSP70   146, 151
HSP90   6, f1.2, 11, 37–9, f2.1, 47–52, f3.1, f3.2, t3.1, f3.3, 54–7, 63, 66, 68, 70, 81, f7.6, 111, 114, 132, 146, 184, 186, 252, 259, 343–5, 347, 350, 374, 405–6, 417, 469, 512
HT-1197 human bladder carcinoma cell line   105
Huh-7 human hepatoma cell line   249, 334, 350, t25.1
Human vascular endothelial cells (HUVECs)   113, 431–2, 489
Hyaloid artery   113, 415, 425
Hydronephrosis (see 2,3,7,8-Tetrachlorodibenzo-p-dioxin, Developmental toxicity)
8-Hydroxydeoxyguanosine (8-OHDG)   234
5-Hydroxytryptamine (5-HT)   237
Hypothalamus-pituitary-genital (ovary/testis) axis   437–8, f31.1, 444–6, 450–2, f31.3
Hypoxia   6, 40, 110–3, 115, 118, 343, 348, 356, 416–9, 423–4 426–2, 433, 449, 488
Hypoxia-inducible factor (HIF)   6, 38, 40, 66–7, 84, 95–6, 110–4, f8.1, 118, 346, 348, 356, f25.1, 359–60, 406, 417, 419, 427, 429, 449, 488–489, 512
Hypoxia response element (HRE)   110–1, 348, 417

Ichihara, Sahoko   413
IGF binding protein 1   111
IL-1   85, 111, t12.1, 203, t13.1, 209, 278, 360–1, 503
IL-2   t2.2, 111, 281, 501, 503
IL-4   34–11
IL-5   503
IL-6   t2.2, 85, 114, t12.1, 279, 360–1, 502–3, 506
IL-8   86, 204, 206–8, 279, 361
IL-10   279–81, 502
IL-12   278–80, 502–3
IL-21   281
IL-22   t2.2, t34.1, 505–6
IL-23   506
IL-27   281
Immunoglobulins   312, 374, 501
Immunohistochemistry   287, 374, f26.2, 438, 453, 455
Immunophilins   49, 52, f3.3, 55
Immunoprecipitation (see also Chromatin immunoprecipitation)   51–2, 94–5, 104, 347
Importins   52, f7.6
Imprinted genes   157, 161, 166
Indirubin   40, t4.1, f4.1, 69–70, f4.4, 185, 189, 319, 335
Indole-3-acetic acid (IAA)   152–3
Indole-3-carbinol   40, 148, 150, 323, 334
Indoleamine 2,3-dioxygenase (IDO)   13, 208, 279, 502–3, 506
Indolo[3,2-b]carbazole (ICZ)   13, 251, 344, 502
Indomethacin N-octylamide   209

Induced pluripotent stem cells (iPS)   158
Induction equivalency (IEQ)   322
Induction of drug-metabolizing enzymes (see Aryl hydrocarbon receptor)
Inflammation   13, 40, 55, 111, 115, 134, 164–5, 184, 197–9, 202–3, 205–11, 218, 238, 280, 360–1, 416, 444, 503, 505
Innate immune system   211, 278, 499, t34.1, 506
INS-1E insulin-secreting beta cell line   231
In situ hybridization   209, 287, 290, 292, 374, 438
Insulin   110, 182, t12.1, 200, 207, 221, 231, 238, 311, 325, 443–4, 468, 470
Insulin-like growth factor (IGF)   110–1, 129, 162, 166, f13.1
Intercellular communication   251–2, 441
Interferon (IFN)   111, 278, 361, 418, t34.1, 502–3
Interferon γ response factor 3 (IFR3)   86
Interleukin (see IL)
Iron   110, 230, 232–4, 236, 349
Iron-chelating agents   110–1, 233

JAK/STAT pathway   500
JAR human trophoblast-like cell line   232, 234
Jun NH2-terminal kinase (JNK)   134, 199, 246, 261, 345, 348, 352, 354, 431. 470, 488
c-Jun transcription factor   115, 131, 236, 246, 252, 292, 345, 348, 431, 469, 487–8
Jurkat human T cell lymphoblast-like cell line   470–1, 485
Jutooru, Indira   331

K562 chronic myeloid leukemia cell line   357
Karchner, Sibel   389
Karman, Bethany   437
Kato, Shigeaki   143
Kawajiri, Kaname   101
KC (see IL-8)
Kelch-like ECH-associated protein 1 (KEAP1)   115–6, f8.3, 146, 152, 346
  KEAP1-deficient mice   116
Kende, Andrew   5–6
Kerkvliet, Nancy   277
Keyhole limpet hemocyanin (KLH)   277–8
Klaassen, Curtis   11
KLF4 transcription factor   158
Ko, Chia-I   157
Kociba, Richard   14
Korkalainen, Merja   181, 285
5L rat hepatoma cell line   468, 470, 472–3

LacZ mice   237, 258
Lead (Pb^{2+})   352–3, t25.2, 362
Leptomycin B   147, 249
Leydig cells   t12.1, f31.3, 452–454, 456
Lindane   230
Lipid peroxidation   t12.2, 229–38
Lipofuscin   237
Lipopolysaccharide (LPS)   85, 111, 250, 279, 348, 361, 499, t34.2, 502–3, t34.3, 506
Lipoxin A4   40, 64
LNCaP human prostate adenocarcinoma cell line   131–2

Long-Evans (*Turku/AB*) [L-E] rat strain   181, 186, t12.2, f12.3, 188–9, 245, 249, 273, t20.1, 289, 293, 380, 515
Low-density lipoprotein (LDL)   13, 64, 163–4, 359, 430, 432–3
LS-180 human colon carcinoma cell line   105
Luteinizing hormone (LH)   312, 358, 437, f31.1, 439, 441–6, f31.3, 452
Luteinizing hormone receptor (LHCGR)   441, 443, 445
LXXLL motif   95

Ma, Qiang   9, 35, 471
Mably, Thomas   267–73, f18.1, f18.2
Macrophages   103, f7.4, 114, t12.1, 202–3, f13.3, t13.1, 207–8, 232–3, 278, 361, 378, 499, t34.1, t34.2, 502, t34.3
Madin-Darby bovine kidney cell line   468
Madin-Darby canine kidney cell line   468
MAFP (inhibitor of cPLA2)   f13.2, 201–2
Malondialdehyde (*see Lipid peroxidation*)
Matrix metalloproteinases   133, 164–5, 290, 486, 488
Matsumura, Fumio   11, 197
MC3T3-E1 mouse calvarial clonal preosteoblastic cell line   287, t20.1
MCF-7 human breast cancer cell line   t2.2, 96–8, 106, 117–8, 130–1, 133, 147, 164–5, 167, 250–1, 332, 334–5, 345–6, 374, 380, 471, 485, 488
MCF-10A human mammary epithelial cell line   200–2, t13–1, f13.3, 206, 246, 249–50, 473
MCF-10AT1 cell line   250
MCP-1 (*see Monocyte chemo-attractant protein-1*)
MDA-MB-468 breast cancer cell line   246, 471
MDM2 ubiquitin ligase   249–50, 473
Mediator of RNA polymerase II transcription subunit 1 (Med1)   t6–1, 96–7
Membrane receptors of steroids   128
Membrane fluidity   231, 234–5
2-Mercapto-5-methoxybenzimidazole   70
Mercury (Hg^{2+})   350–2, t25.2, 361
Merino, Jaime   485
Metallothionein   361, 416
Methoxychlor   162
3′-Methoxy-4′-nitroflavone   68–9, 250, 291, 332, 344
Methylation of genes   157–9
3-Methylcholanthrene (3-MC)   3–5, 36–7, f4.1, 70–2, 105, 113, 116–7, 131, 133–4, 147, 149–50, 169, 181, 230, 233, 302, 346, 348, 350, 353, t25.1, t25.2, 360, 379–80, 417, 470, 489
   Estrogenic activity   130
   Irregular induction potency   334
3′-Methyl-4-dimethylaminoazobenzene   3, 36
6-Methyl-1,3,8-trichlorodibenzofuran (MCDF)   14, t4.1, 72, 332, f24.2, 471
MG132 proteasome inhibitor   38–9, 149
MHC Class II molecules   278–9
Microarray analysis   98, 148, 189, 217–221, 252, 279, 290, 292, 416, 418, 485, 487, 503, 516
Microsomes   4, 117, 182–3, t12.1, 229–30, 232–4, 236, 238, 350
Miettinen, Hanna   285
Mitochondria   115, 131, t12.1, 231–6, 238, 250, 349, 473–4
Mitogen-activated protein kinase (MAPK) (*see also Jun NH2-terminal kinase and ERK1/2*)   11, 109–10, 113–5, f8.3, 118, 236, 246, 261, 345–6, 348, 352, 354, 356, f25.1, 470, 503

MMDD1 mouse kidney macula densa cell line   t13–1, 209–10
Moffat, Ivy   181
Molybdate   48
Monocyte chemo-attractant protein-1 (MCP-1)   204, 207, 209, 278
Monofunctional enzyme inducers   131
Mouse embryonic fibroblast cells (MEFs)   105, 116, 134, 167, 208–9, 250, 345, 474, 487
Mouse hepatoma cell lines   6, 37, 68, 71, 93, 96–8, 167, 210, 234, 246, 334, 346, 349–50, t25.1, 352–3, t25.2, 471, 485–6
Mucins   375–6
Mulero-Navarro, Sonia   485
Multidrug resistance-associated proteins (MRPs)   t2.1, 36, 40, 116
Multiple sclerosis   127, 277, 281, 335, 362, 506
Murray, Iain   47
Muscle ARNT-like protein 1 (*see BMAL1*)
c-Myc transcription factor   40, t2.2, 158, 165, 246, 292, 472–3, 487

NADPH oxidase (Nox)   233, 424, 427
NAD(P)H:quinone oxidoreductase 1 (NQO1)   35–6, t2.1, 39–40, 84, 109, f8.3, 116–8, f8.4, 165, 186, 220, 222, 224, 301, 343, 346
α–Naphtoflavone   113, 133–4, 147, 250–1, 259, t20.1, 292, 332, 344, 472
β-Naphthoflavone   51, 63, 70, 116, 135, 147, 149–50, 167, 302, 347–8, 350, t25.1, t25.2, 360, 393, 405–7, 431–2, 442, 447, 469, 502
Natural killer (NK) cells   t12.1, 278, t34.1
NCoA (*see Nuclear receptor coactivator*)
Nebert, Daniel   4, 5, 8, 12
NEC14 human embryonal testis carcinoma cell line   105
Nedd9/Hef1/Cas-L metastasis marker   488
Neuroblastoma cells   234
Neutrophils   203, 278, 499, t34.1, 502, 506
NF-κB   f1.2, 11, 40, 85–7, 102–3, f7.3, 109, 111–2, 115, 150, 165, 168, 206–8, 250, 345–7, 349, t25.1, 352–4, t25.2, 356, f25.1, 360–1, 473, 500, 502–3, 506–7
NF-E2-related factor-2 (*see Nrf2*)
Nitric oxide (NO)   111, 279, 424–5, 429–430, 502
Nitric oxide synthase (NOS)   110–1, 278, 424–5, 429
Nitroarginine   11, 430
NMR spectroscopy   66–7
Noradrenaline   (*see Norepinephrine*)
Norepinephrine   237, 423–5
Nrf2   f1.2, 11, 40, 83–4, t5.1, 109–110, 115–8, f8.3, f8.4, 346, 349, f25.1, 432
   Nrf2-deficient mice   116–7, 346
Nuclear factor-1 (NF-1)   346
Nuclear factor erythroid 2-related factor-2 (*see Nrf2*)
Nuclear factor –κB (*see NF-κB*)
Nuclear receptor corepressor (NCoR)   128, 132
Nuclear receptor coactivator (NCoA)   66, t6–1, 95–7, 131, 167
Nuclear receptors (*see also individual receptors*)   53, 95, 127, 132–3, f9–1, 143, 151, 165, 198, 331, 346, 442, 513, 515

OCT3/4 transcription factor   158–9
Ochratoxin A   473
Odontoblasts   t12.1, 286–7, 289–90

Ohtake, Fumiaki 143
Okey, Allan 3, 5, 9, 181, 220
OMC-3 human ovarian carcinoma cell line 105
Omeprazole 40, f4.1, 68, 70, 135, 335, 344
Oocytes t2.2, 158, 161, 358, f31.1, 438, 440–1, 443–6
Ossification types 285
Osteoblasts t12.1, 285, 287–9, t20.1, 291–3, 381
Osteocalcin t20.1, 289, 291
Osteoclasts 164, t12.1, 285, 287–9, t20.1, 292–3, 380–1, f26.2
Osteopontin 164, 252, t20.1, 291, 377–8, 427
Ovalbumin 85, 135, 277
Ovarian folliculogenesis 437, f31.1, 441
Ovarian steroid metabolism 442
Oxidative stress (*see also Reactive oxygen species*) f1.2, 115–6, 118, 152, t12.1, 185, 199, 207, 229–38, 246, 346, 348–50, 352–3, f25.1, 358, 361, 427, 474, 503
8-Oxo-deoxyguanosine (8-oxo-DG) 234, 246

P21$^{Waf1/Cip1}$ cyclin-dependent kinase inhibitor 290, 432, 471, 487
P23 chaperone 6, 49, f3.2, 52, t3.1, 56, f7.6, 146, 345, 469
P27^{Kip1} cell cycle inhibitor t2.2, 470, 472, 487
P53 tumor suppressor 168, 221, 232, 235–7, 249–252, 292, 473
 P53-deficient mice 235, 237
P300 coactivator (*see CBP/p300*)
P300/CBP-associated factor (PCAF) 97
PAI-2 (*see Plasminogen activator inhibitor-2*)
Paired box 5 isoform a (Pax5a) 279
Pam3CSK4 103, f7.4
Pandini, Allessandro 63
Paraoxonase 1 (Pon-1) 72, 109, 334
Paraquat 234
PC-3 human prostate cancer cell line 131
PC12 rat pheochromocytoma-derived cell line 472
PCNA 106, 375–6, 442
Pendulin 49
Perdew, Gary 6, 8, 11, 13, 47
PERIOD (PER) (*see Clock genes*)
Peripheral quantitative computed tomography (pQCT) 286, 293
Peroxisome proliferator-activated receptors (PPARα, PPARγ, and PPARδ) 53–4, t3.1, 133–5, 147
Persistent organic pollutants (POPs) 109, 322, 325
Peterson, Richard 14
Phases I and II of xenobiotic metabolism (*see also individual enzymes*) 84, 109, f8.3, 116, 118, 134, t12.2, 343, 346
Phenobarbital 68, 135
Phorbol diesters (*see TPA and PMA*)
Phorbol 12-myristate 13-acetate (*see PMA*)
Phosphatidylinositol-3-kinase (PI3K) 110, 114–5, f8.3, 356, 473–4, 487, f33.1
Phosphodiesterases 51, t3.1, 55, f3.4
Photobiomodulation 231
PIAS proteins 104
Picrocam (inhibitor of COX2) f13.3, 206
Piroxicam (inhibitor of COX2) 202
PKAs (*see cAMP-dependent protein kinases*)
cPLA2 (*see Cytosolic phospholipase A2*)

Plasminogen activator inhibitor-2 (PAI-2) 133, f13.1, 361, 416, 474
PMA 352, 354
Poellinger, Lorenz 5–6, 373
Pohjanvirta, Raimo 9, 181, 220–1
Poland, Alan 3–7
Pollenz, Richard 5, 9
Polybrominated diphenylethers (PBDEs) 322, 336
Polychlorinated biphenyls (PCBs) 7, t4.1, 168, t16.1, 307, 502
 Antagonism of TCDD 68, 336
 Binding to the AHR 66, 68–9
 Biochemical effects 234–8, 249, 251, 360, 469
 Health effects 245–6, 286, 313, 325, 357, 377, 440–1, 446,
 Resistant populations in fish 394
 Structures f23.2, f24.1
 TEF values 317–8, t23.1, t23.2, 320–3
Polychlorinated dibenzo-*p*-dioxins (PCDDs; *see Dioxins*)
Polychlorinated dibenzofurans (PCDFs; *see Dioxins*)
Polycomb repressive complex (PRC) 159–60
Polycyclic aromatic hydrocarbons (PAHs) [*see also individual PAH compounds*] 3–5, 7, 35–6, t2.1, t2.2, 63–4, t4.1, 68–9, 72, 113–4, 116–7, 181, 336, 348, t25.1, 352–3, t25.2, 357–8, 360, 362, 393, 430, 432, 449, 467
Porphyria cutanea tarda (*see 2,3,7,8-Tetrachlorodibenzo-p-dioxin, Toxicity, Skin*)
Powell-Coffman, Jo Anne 405
PP-2 (inhibitor of Src tyrosine kinase) 200–1, f13.3
Pregnane X receptor (PXR) 66, 69, 134–5
Prazosin 425, 429–30
Primaquine 70
Progesterone 134, 312, 417, f31.1, 441, 444, 446–7, 449, 452
Progesterone receptor 48–9, 56, 128–9, 134, 444, 446–7
Programmed cell death (*see Apoptosis*)
Proliferating cell nuclear antigen (*see PCNA*)
Prolyl hydroxylase domain containing protein 2 (PHD2) f8.1, 114
Prostaglandin E2 (PGE2) 206, 209, 360, 474
Prostaglandin G2 (PGG2) 40, f4.1, 64
Prostate specific antigen (PSA) 131
Proteasomal degradation 10, 38–9, 101, 104, 110, f8.1, 114–5, f9.3, 130, 143–7, f10.1, f10.3, 149–51, f10.6, 352, 471, 473
Protein arginine methyltransferase-1 (PRMT1) 168
Protein kinase C (PKC) f1.2, 115, f8.3, 118, 199, 201, 203–5, 237–8, 251, 345–6, 512
pS2 gene t2.2, 130
Puga, Alvaro 9, 157

Qin, Hongtao 405
Quercetin 69, 72, 114, 334

Ras oncogenes (and proteins) 246, 469–70
RAW264.7 mouse leukemic monocyte macrophage cell line t20.1, 293
RDH9 gene 133
Reactive oxygen species (ROS) 109, 115, f8.3, f8.4, 229–38, 246, 251, 349, 351, 353–4, 361, 424, 432
Real-time RT-PCR assays f7.4, 105, f8.2, 166, f12.3, 218, 232, 290, 453, 455, t34.3

Receptor-associated coactivator 3 (RAC3) 132
Receptor-interacting protein 140 (RIP140) t6–1, 130, 344–6
Regulatory T cells (*see Treg cells*)
REH acute lymphoblastic leukemia cell line 357
RelA (*see NF-κB*)
RelB (*see NF-κB*)
RelBAHRE 208
Renin-angiotensin-system (RAS) 423–5, 428, 433
Reporter gene assays 68, 94, 96, 101, 104, 111–2, 116, 187, t20.1, 292, 322, 333, 347, 349, 406, 470
Resveratrol t4.1, 231, 259, t20.1, 291, f24.4, 335, 472
RET51 tyrosine receptor kinase 54, f3.4
Retinoblastoma protein (pRB) 11, 86, 95, 432, 470–1, 486–7
Retinoid acid receptors (RAR and RXR) 132–4, 147, 150, 290, 347, f25.1
Retinoic acid response element (RARE) 132–3, 347
Retinoids 132–3, 163, 218, 335, 415, 474
Rhodopsin 408–9
Rico-Leo, Eva 485
RNA polymerase II (pol II) 70, 82, 93, 96–7, f10.1, 157, 160, 166–8, 335, 350, 512
Rolipram (phosphodiesterase inhibitor) 55
R-spondin1 302
RUNX2 t20.1, 290–1, 381

Safe, Stephen 6–8, 14, 331
SB203580 (MAPK inhibitor) t4.1, 352, 431
SB600125 (JNK inhibitor) 352
SCC-4 human tongue squamous cancer cell line 133
Schrenk, Dieter 245
Scott, Jason 423
Selective AHR modulators (*see Aryl hydrocarbon receptor*)
Selenium 230
Serial analysis of gene expression (SAGE) 416
Serotonin (*see 5-Hydroxytryptamine*)
Sertoli cells t12.1, f31.3, 452–4, 456
Seveso accident 115, 129, 252, 286, 308–13, 321, 325
Sexually dimorphic hypothalamic regions 438–9, 451–3
SGA 360 (a novel AHR ligand) 72
Shear stress 163, 359, 424, 430–3
Sheep red blood cells (SRBC) 277
shRNA experiments 94
Silencing mediator for retinoid and thyroid hormone receptors (SMRT) 128, 130, 133, 167, 344, 346
Silkworth, Jay 221
Simonich, Michael 299
Sinclair, Peter 8
siRNA experiments 54, 96–7, 105–6, 114, 132, 202, 208, 249–250, 349, 432, 471, 474, 518
Site-directed mutagenesis 4–10, 4–11, 8–18
Slug zinc-finger protein 491–2, f33.3
Smad transcription factors 11, 165, 236, 262, 281
Snail zinc-finger protein 49–2
Smith, Andrew 8
Sodium butyrate 105
SOX2 transcription factor 158
SOX9b transcription factor 102, 292, 301–2

SP1 transcription factor 84, 86, 102, 128, f9.2, f9.3, 130, 163, 166
Spare receptors 188
Species differences 48, 52, f4–3, 132, 181, 185–6, c14, 229, 249, 257, 268, 299, 393–4
   Binding to, and activation of, the AHR 68–9
   TEFs 319
Spineless 162, 407–9
c-Src tyrosine kinase 11, 109, 114, 134, 187, 246, 249–50, 200–4, f13.3, t13.1, 208–9, 252, 469, 473
   cSrc-deficient mice and cells 208–9, 211
SRC (*see Nuclear receptor coactivator*)
Stable repression of gene expression 11–8
STAT proteins 5–15
Staurosporine 16–13
Stem cells 13–4, t1.1, 158–61, 164, 251, 285, 288, 291–2, 419, 503
Steroidogenic acute regulatory protein (StAR) 131, 454
Steroid receptor coactivator (*see Nuclear receptor coactivator*)
Stohs, Sidney 229
Strain differences, mouse (*see DBA/2 mouse strain*)
Strain differences, rat (*see Han/Wistar (Kuopio) [H/W] rat strain*)
Sucrose density gradient assay 47, 49, 183, 185
Sulforaphane 118
SUMOylation of proteins 104, f7.5, f7.6
Superoxide anion (*see Reactive oxygen species*)
Superoxide dismutase 231, 236–7
Suppressive subtraction hybridization 377
Suppressor of cytokine signaling 2 (SOCS2) 279
Suprachiasmatic nucleus 438, 513
Survivin (BRIC5) t3.1, 54, f3.4
Sutter, Thomas 8
Swanson, Hollie 81
Swedenborg, Elin 127
Sympathetic nervous system (SNS) 423–5, 428–30
3T3-L1 preadipocyte cell line 133–4, t13–1, 204, 206–7
293T human embryonic kidney 293 cell line 345

T lymphocytes t2.2, 113, 164, t12.1, 185, 208, 278–81, 335, 357, 360–1, 378, 488, 499–506, t34.1, t34.3, f34.1
   Th17 T helper cells t2.2, 150, 164, 208, 281, t34.1, 503–6, t34.3, f34.1
   Treg cells 13, 150, 164, 208, 279–81, 336, 361, t34.1, t34.3, 503–6, f34.1
T-47D human breast cancer cell line 71, 96, 131, 169, 334–5, 349, t25.1
Tango 407–9
Tanguay, Robert 299
Tartrate-resistant acid phosphatase (TRACP) 292
TATA Binding Protein (TBP) 96–7, 167
T-cadherin (*see Cadherins*)
TCDD (*see 2,3,7,8-Tetrachlorodibenzo-p-dioxin*)
TCDD equivalent (TEQ) 245, 317, 321–5, 336–7
TCDD-inducible poly(ADP-ribose)polymerase (TiPARP) t2.2, t5.1, 84
Tea melanin 231
TERT gene 147
Testosterone 128, 130–1, 182, t12.1, 236, 312, 358, 443, f31.3, 452, 454–5

2,3,7,8-Tetrachlorodibenzo-*p*-dioxin (TCDD)
    Docking in AHR ligand-binding pocket f4–2, 68–70
    Half-life in humans 323
    Induction of drug-metabolizing enzymes (*see Aryl hydrocarbon receptor*)
    Inhibition of tumor growth 129, 357, 471
    Lowest adverse effect level (LOAEL) 268, t18–1
    Release from contact inhibition 468–70
    Structure f23.1, f24.1
    Toxicity 12, t2.2, 164, 181–3, t12–1
        Adipose tissue 133
        Adrenal gland 238
        Birds 299
        Bone 286–7, 291–293
        Cancer 16, 235–6, c**16**, 309–11, 356–7
        Developmental 165, 183–6, 209–10, 237, c**17**, c**18**, 290–3, 300–1, 312–3, 359, 379, 439, 443, 447, 449, 452, 470, 491
        Endocrine (*see also* 2,3,7,8-Tetrachlorodibenzo-*p*-dioxin, *Reproductive organs and Pancreas*) t2.2, 182, t12–1, 238, 311–2, 325, 362, 443, 445, 449–52, 515
        Fish 102, 292, 299–301
        Heart 113, 164, 237–238, 312, 359–60
        Immune system (*see also* 2,3,7,8-Tetrachlorodibenzo-*p*-dioxin, *Thymus*) t2.2, 164, c**19**, 312, 360–1, 501–2
        Kidney 132, 209, 238, 379
        Liver 164, 183, 217, 245–6, t16.1, 468–9
        Nervous system 236–7, 361–2
        Pancreas 238, 472
        Placenta 113–4, t12.2, 230, 237, 418–9
        Porphyria 4, 8, t12.1, 183, 185, 230, 232–4, 332
        Reproductive organs 236, 270, 312–3, 446, 453–5
        Skin 4, 162, 164, t12.1, 246, t16.1, 311
        Stomach t12.1, 376
        Teeth 286, 289–91
        Thymus t2.2, t12.1, t12.2, 188, 204, 230–1, 238, t16.1, 378, 472
        Transgenerational 9, 300–1, 313
        Vasculature 113–115, 163, 312, 417, 432
        Wasting syndrome 133, 181–2, 186, t12.2, 204, 208, 211, 231, 277, 332, 515
12-*O*-Tetradecanoylphorbol-13-acetate (*see TPA*)
4b,5,9b,10-Tetrahydroindeno[1,2-b]indole 231
TGF-α t12.1, 200, 204, 208, 246, 249–50, t17.1, 260–3, t17.2, 290, 469, 473
TGF-β 11, 113, 133, 164–5, t12.1, 236, 249–50, 252, t17.1, 260–3, t17.2, 279–81, 290, 415, 471–4, 486, 489–91, f33.2
Thalidomide 152
Thrombospondin-1 419, 432
Thyroid hormone receptors 53–5, t3.1, 95, 134
Thyroid Hormone Receptor/ Retinoblastoma Interacting Protein 230 (TRIP230) 95–6, t6.1, 111
TiPARP [*see TCDD-inducible poly(ADP-ribose)polymerase*]
Tischkau, Shelley 511
TNF-α 85, 111, 134, 168, t12.1, 203, t13.1, 209, 233, 238, 278–9, 347, 360–1, 468, 474, 486, 502–3
Tooth development 285, 287, 290

Toxic equivalence factor (TEF) 317–24, t23.1, t23.2
TPA 103, 199–200, 204, 345, 348
TRAMP (transgenic adenocarcinoma of the mouse prostate) mice 14
    AHR-deficient TRAMP mice 163, 252, 417
Transcription factor NF-E2 p45-regulated factor (*see Nrf2*)
Transcriptional intermediary factor 2 (TIF2) 128, 132
Transforming growth factors (*see TGF-α and TGF-β*)
Transglutaminase 133, 415
Transporter proteins 6, t2.1, 36, 40, 116, 209, 258
Trans-3,4′,5-trihydroxystilbene (*see Resveratrol*)
TRAP/DRIP/ARC/Mediator complex 167
2,4,5-Trichlorophenol 308
2,4,5-Trichlorophenoxyacetic acid (2,4,5-T) 308
Trichostatin A 104, 133, 166
6,2′,4,-Trimethoxyflavone 69
Trypan blue perfusion 416
Tryptophan metabolites and derivatives (*see also 6-Formylindolo [3,2-b]-carbazole*) 13, 40, 64, 109, 334, 336, 502–3, 506, 511, 516
Tumor necrosis factor-α (*see TNF-α*)
Tuomisto, Jouko 317
Type I collagen 291, 381

U0126 (ERK inhibitor) 352
U937 human macrophages 114, t13–1, 202, 208
Ubiquitination 38–9, 52, 93, 101, 110, 131, 144–7, f10.2, f10.3, 149–50, f10.6, 158, 160, 471, 473, 513
UDP-glucuronosyltransferase (UGT; see also individual UGT subtypes) 35, 84, 109, f8.3, 116, f8.4, 134
UGT1A1 36, t2.1, 116–7, 165, 186
UGT1A3 116
UGT1A4 116
UGT1A5 117
UGT1A6 8, 36, t2.1, 116–7,165, 343, 346, 379
UGT1A7 116
UGT1A8 116
UGT1A9 36, t2.1, 116–7
UGT1A10 116
UGT2B34 117
UGT2B35 117
UGT2B36 117
UMR-106 rat osteosarcoma cell line t20.1, 291
UV light 9, 109, 143, 146, 249–51, 345, 473–4, 502, 506

V79 Chinese hamster fibroblast cell line 251
Vanadium (V^{5+}) 354, 356
Vascular endothelial growth factor (VEGF) f8.1, 110, 112–4, f8.2, 147, 165, 206, 346, 348, 356, 359, 417–8, 427, 429, 449, 489–91, f33.2
    VEGF-deficient mice 112
Vav3 oncogen 487, f33.1, f33.2, 490
VEGF receptor Flt1 (VEGFR-1) 110, 113, 418
Viluksela, Matti 285
Vinclozolin 162
Vitamin A (retinol) 132–3, t12.1, 183, t12.2, 230–1, 347, 415
Vitamin C (ascorbic acid) 230
Vitamin E 230–1, 236–7

Walker, Mary   30–1
WAY-169916 (a selective estrogen receptor and AHR modulator)   13, 72
WB-F344 rat hepatic epithelial progenitor cell line   204, 251, 468–70
Wejheden, Carolina   373
Western blot analysis   94, 111, 187, 438, 453, 455, 503
White, Sally   307
Whitelaw, Murray   9
Whitlock, James Jr.   5, 81, 471
Whole-genome duplications   390, 392–4
WNT signaling (*see also β-Catenin*)   150, 165, 302, 440
WY (PPARα agonist)   134

Xanthine oxidase/dehydrogenase   230

XAP2 (AIP, ARA9)   6, 14, 32, 49–56, f3.2, f3.3, 66, 146, 345, 406 469
   Association with pituitary adenoma   54–5, 439
Xenobiotic response element (XRE; *see AHR-response element*)
Xenoestrogens   128, 162
Xenopus spp. (frogs)   f7.2, 185, t27.1, 392
X-ray microtomography   289
XRE-II (*see AHRE-II*)

Yeast two-hybrid assay   49, 53, t3.1, 71, 95, 104, 408
Yu-Cheng accident   286, 324

Zacharewski, Timothy   9, 219
Ziv-Gal, Ayelet   437
ZR75 human breast carcinoma cell line   112